Stefan Odenbach (Ed.)
Magnetic Hybrid-Materials

Also of interest

Multiferroics
Fundamentals and Applications
Andres Cano, Dennis Meier, Morgan Trassin (Eds.), 2021
ISBN 978-3-11-058097-6, e-ISBN (PDF) 978-3-11-058213-0,
e-ISBN (EPUB) 978-3-11-058104-1

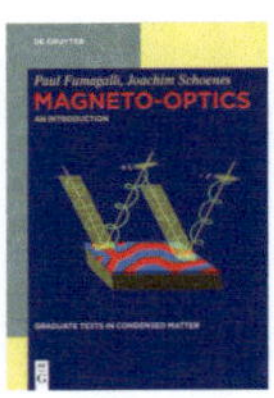

Magneto-Optics
An introduction
Paul Fumagalli, Joachim Schoenes, 2021
ISBN 978-3-11-063522-5, e-ISBN (PDF) 978-3-11-063530-0,
e-ISBN (EPUB) 978-3-11-063545-4

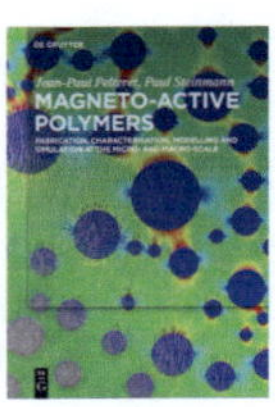

Magneto-Active Polymers
Fabrication, characterisation, modelling and simulation at the micro- and macro-scale
Jean-Paul Pelteret, Paul Steinmann, 2019
ISBN 978-3-11-041951-1, e-ISBN (PDF) 978-3-11-041857-6,
e-ISBN (EPUB) 978-3-11-041862-0

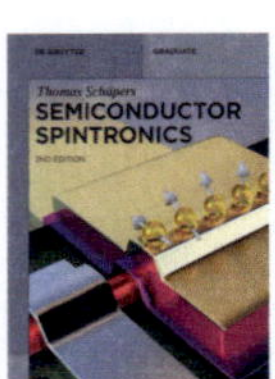

Semiconductor Spintronics
Thomas Schäpers, 2021
ISBN 978-3-11-063887-5, e-ISBN (PDF) 978-3-11-063900-1,
e-ISBN (EPUB) 978-3-11-063932-2

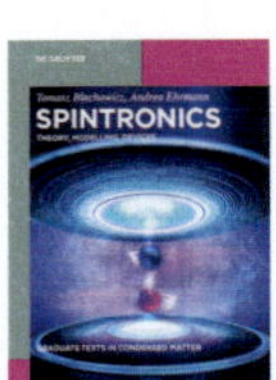

Spintronics
Theory, Modelling, Devices
Tomasz Blachowicz, Andrea Ehrmann, 2019
ISBN 978-3-11-049062-6, e-ISBN (PDF) 978-3-11-049063-3,
e-ISBN (EPUB) 978-3-11-049069-5

Magnetic Hybrid-Materials

Multi-scale Modelling, Synthesis, and Applications

Edited by
Stefan Odenbach

DE GRUYTER

Editor
Prof. Dr. rer. nat. habil. Stefan Odenbach
TU Dresden
Magnetofluiddynamik, Mess-
und Automatisierungstechnik
George-Bähr-Str. 3
01069 Dresden
Germany
stefan.odenbach@tu-dresden.de

ISBN 978-3-11-056877-6
e-ISBN (PDF) 978-3-11-056963-6
e-ISBN (EPUB) 978-3-11-056883-7

Library of Congress Cataloging-in-Publication Data
A CIP catalog record for this book has been applied for at the Library of Congress.

Bibliographic information published by the Deutsche Nationalbibliothek
The Deutsche Nationalbibliothek lists this publication in the Deutsche Nationalbibliografie; detailed bibliographic data are available on the Internet at http://dnb.dnb.de.

Cover image: GarryKillian/iStock/Getty Images Plus
Typesetting: TNQ Technologies Pvt. Ltd.
Printing and binding: CPI books GmbH, Leck

www.degruyter.com

Preface

Multifunctional hybrid materials represent a forward-looking class of materials in the context of modern materials research, which allow a wide range of possible applications due to the wide range of their material properties. In particular, stimuli-responsive materials, which react to changes in the environment by changing their properties, can drive the development of new technological innovations and thus stimulate significant improvements right into everyday life. Chemists, physicists and material scientists are involved in the current rapid development of new material properties and contribute to an innovative interdisciplinary field of research.

The use of magnetic fields as an external stimulus to control material properties is of considerable technical interest, as magnetic fields are technically easy to generate and well controlled. Magnetically controlled materials such as suspensions of magnetic nano- or microparticles - ferrofluids and magneto-rheological fluids - show the ability to strongly change the material behaviour at reasonable technical effort. However, in the case of the aforementioned fluids, the matrix in which the particles are located, i.e. the carrier fluid, represents only a thermal bath that changes the typical time constants of the material but does not offer any specific interaction between the particles and the matrix. In contrast, magneto-rheological elastomers, in which magnetic particles are embedded in an elastic matrix, represent a first step towards magnetic hybrid materials with controllable particle-matrix interaction.

In such materials consisting of a particulate magnetic component in a complex matrix, the mutual influence of particles and matrix provides an additional set of parameters in the material behaviour. With this, novel material properties can be generated via magnetically controlled changes. Knowledge of the interaction between the particles and the surrounding matrix is an important building block for understanding the material behaviour itself and thus the basis for the targeted development of such materials for novel applications in actuators and sensors. At the same time, understanding the interaction between functionalised particles and the matrix is also the basis for an expanded understanding of the behaviour of magnetic nanoparticles in biomedical applications. The interaction of the particles in contact with cells and biological tissue is of crucial importance both for the tissue uptake of the particles and for their biodistribution as well as for the relaxation behaviour of the particles, such as used for magnetically assisted imaging.

Within the framework of a priority programme of the German Research Foundation, five key questions were posed to gain a deeper understanding of the material behaviour of magnetorheological elastomers: First, it had to be clarified how (1) the material behaviour of a magnetically controllable hybrid material can be influenced via the particle-matrix interaction and how corresponding materials can be synthesised. Understanding the material behaviour required (2) a cross-scale consistent

https://doi.org/10.1515/9783110569636-201

description that explains the magnetic controllability of the material properties on a microscopic basis. This material modelling was also necessary to generate material laws for the application based on a detailed understanding of the material. Closely linked to the description concepts was (3) the experimental investigation of material behaviour in the magnetic field and thus the question of what changes in material properties can be produced by varying their internal structure in the magnetic field. Building on the understanding of the magnetic hybrid materials, it was then possible to clarify both (4) what possibilities they offer in novel actuator and sensory applications and (5) how the effectiveness of the biomedical use of magnetic nanoparticles can be improved by controlling the interaction between functionalised particles and tissue.

The present publication summarises in 29 contributions, the main results of the research within the priority programme and provides a fundamental approach to a cross-scale description of magnetic hybrid materials.

Stefan Odenbach

Contents

List of contributing authors

Prof. Christoph Alexiou
Department of Otorhinolaryngology
Head and Neck Surgery
Section of Experimental Oncology and Nanomedicine (SEON)
Else Kröner-Fresenius-Stiftung-Professorship
Universitätsklinikum Erlangen
Erlangen
Germany
christoph.alexiou@uk-erlangen.de

Ingo Appel
Institute of Catalysis Research and Technology
Karlsruhe Institute of Technology
Postfach 3640
76021 Karlsruhe
Germany

Abdolhamid Attaran
Institute of Solid Mechanics
TU Dresden
Dresden
Germany
hamid.attaran@gmail.com

Anton V. Bakhtiiarov
State Scientific Research Institute for Chemical Technologies of Organoelement Compounds
105118 Moscow
Russia

Lester Barnsley
Forschungszentrum Jülich GmbH
Jülich Centre for Neutron Science at MLZ
Lichtenbergstr. 1
85748 Garching
Germany

Tatiana I. Becker
Technical Mechanics Group
Faculty of Mechanical Engineering
Technische Universität Ilmenau
Ilmenau
Germany
tatiana.becker@tuilmenau.de

Silke Behrens
Institute of Catalysis Research and Technology
Karlsruhe Institute of Technology
Postfach 3640
76021 Karlsruhe
Germany
silke.behrens@kit.edu

Philip Biehl
Friedrich-Schiller-University Jena
Jena
Germany
and Jena Center for Soft Matter (JCSM)
Friedrich-Schiller-University Jena
Jena
Germany
philip.biehl@uni-jena.de

Kerstin Birster
Experimentalphysik, Universität des Saarlandes
Campus D2 2
D-66123 Saarbrücken
Germany

Valter Böhm
Faculty of Mechanical Engineering
Ostbayerische Technische Hochschule Regensburg,
Regensburg
Germany
valter.boehm@oth-regensburg.de

Dmitry Borin
Institute of Mechatronic Engineering
Technische Universität
Dresden
Dresden
01062 Germany
dmitry.borin@tu-dresden.de

Dmitry Yu. Borin
Institute of Mechatronic Engineering
Technische Universität Dresden
01062 Dresden
Germany

https://doi.org/10.1515/9783110569636-202

Helmut R. Brand
Department of Physics
University of Bayreuth
Bayreuth,
Germany
brand@uni-bayreuth.de

Jörg Brummund
Chair of Computational and Experimental Solid Mechanics
Institute of Soild Mechanics
TU Dresden
01062 Dresden
Germany

Jhohan Chavez
Department of Mechanical Engineering
Technical Mechanics Group
TU Ilmenau
Ilmenau
Germany
jhohan-harvey.chavez-vega@tu-ilmenau.de

Eike Clasen
Institute of Physical Chemistry
Hamburg University
Grindelallee 117
20146 Hamburg
Germany

Joachim H. Clement
Jena Center for Soft Matter (JCSM)
Friedrich-Schiller-University Jena
Jena
Germany
and Klinik für Innere Medizin II
Abteilung Hämatologie und Internistische Onkologie
Universitätsklinikum Jena
Jena, Germany
Joachim.Clement@med.uni-jena.de

Annelies Coene
Department of Electromechanical Systems and Metal Engineering, Ghent University,
9052 Ghent
Belgium;
and Cancer Research Institute Ghent (CRIG)
9000 Ghent
Belgium

Johanna Demut
Klinik für Innere Medizin II
Abteilung Hämatologie und Internistische Onkologie
Universitätsklinikum Jena
Jena
Germany
Johanna.Demut@med.uni-jena.de

Andreas Diermeier
Mechatronics Research Unit
OTH-Regensburg
Regensburg
Germany

Sebastian Draack
Institute for Electrical Measurement Science and Fundamental Electrical Engineering
Technical University Braunschweig
Braunschweig, Germany

Silvio Dutz
Institute of Biomedical Engineering and Informatics (BMTI)
Technische Universität Ilmenau
Ilmenau, Germany
and Department of Nano Biophotonics
Leibniz
Institute of Photonic Technology (IPHT)
Jena
Germany
silvio.dutz@tu-ilmenau.de

Dietmar Eberbeck
Physikalisch-Technische Bundesanstalt
Abbestraße 2-12,
10587 Berlin
Germany

Alexey Eremin
Otto von Guericke University
Institute of Physics
Magdeburg, Germany
alexey.eremin@ovgu.de

Elena Felgenhauer
Institute of Physical Chemistry
Hamburg University
Grindelallee 117
20146 Hamburg
Germany

Dagmar Fischer
Jena Center for Soft Matter (JCSM)
Friedrich-Schiller-University Jena
Jena
Germany
and Institute of Pharmacy, Pharmaceutical Technology und Biopharmacy
Friedrich-Schiller-University Jena
Jena, Germany
dagmar.fischer@uni-jena.de

Dr. Ralf P. Friedrich
Department of Otorhinolaryngology
Head and Neck Surgery
Section of Experimental Oncology and Nanomedicine (SEON)
Else Kröner-Fresenius-Stiftung-Professorship
Universitätsklinikum Erlangen
Erlangen
Germany
ralf.friedrich@ukerlangen

Lisa Fruhner
Jülich Centre for Neutron Science (JCNS-1) and Institute for Complex Systems (ICS-1), Forschungszentrum Jülich GmbH
Leo-Brandt-Straße, 52425 Jülich
Germany
and Institute of Physical Chemistry
RWTH Aachen University
Landoltweg 2
52056 Aachen, Germany

Simon Gast
Technische Universität Ilmenau
Technische Mechanik
Ilmenau
Germany

Philipp Gebhart
Institute of Solid Mechanics
TU Dresden
Dresden
Germany
philipp.gebhart@tu-dresden.de

Christine Gräfe
Klinik für Innere Medizin II
Abteilung Hämatologie und Internistische Onkologie
Universitätsklinikum Jena
Jena
Germany
christine.graefe@med.uni-jena.de

Dr. Christine Gräfe
Department of Internal Medicine II
Hematology and Medical Oncology
Jena University Hospital
Jena
Germany,
christine.graefe@med.uni-jena.de

Lennart Gresing
Department of Internal Medicine II
Hematology and Medical Oncology
Jena University Hospital
Jena
Germany,
elena.mueller@med.uni-jena.de

Larissa Großmann
Institute of Physical Chemistry
Hamburg University
Grindelallee 117
20146 Hamburg
Germany

Martin Hähsler
Institute of Catalysis Research and Technology
Karlsruhe Institute of Technology
Postfach 3640
76021 Karlsruhe
Germany

Birgit Hankiewicz
Institute of Physical Chemistry
Hamburg University
Grindelallee 117
20146 Hamburg
Germany
birgit.hankiewicz@chemie.uni-hamburg.de

Thomas Heinze
Institute of Organic Chemistry and Macromolecular Chemistry
Center of Excellence for Polysaccharide Research
Friedrich Schiller University of Jena
Humboldtstraße 10
D-07743 Jena
Germany
thomas.heinze@uni-jena.de

Melissa Hess
Chemistry Department
Universität zu Köln
Köln
Germany

Stephan Hinrichs
Institute of Physical Chemistry
Hamburg University
Grindelallee 117
20146 Hamburg
Germany

Christian Holm
Institute for Computational Physics
University of Stuttgart
Stuttgart
Germany

Christina Janko
Department of Otorhinolaryngology
Head and Neck Surgery
Section of Experimental Oncology and Nanomedicine (SEON)
Else Kröner-Fresenius-Stiftung-Professorship
Universitätsklinikum
Erlangen
Germany,
Christina.Janko@uk-erlangen.de

Sandra Jungmann
Institute of Pharmacy, Pharmaceutical Technology und Biopharmacy
Friedrich-Schiller-University Jena
Jena
Germany
sandra.jungmann@uni-jena.de

Karl A. Kalina
Chair of Computational and Experimental Solid Mechanics
Institute of Soild Mechanics
TU Dresden
01062 Dresden
Germany

Sofia S. Kantorovich
Institute of Natural Sciences and Mathematics
Ural Federal University
Ekaterinburg
Russia
Faculty of Physics
Computational and Soft Matter Physics
University of Vienna,
Vienna
Austria

Markus Kästner
Chair of Computational and Experimental Solid Mechanics
Institute of Soild Mechanics
TU Dresden
01062 Dresden
Germany
markus.kaestner@tudresden.de

Sabine H. L. Klapp
Institut für Theoretische Physik
Technische Universität Berlin
Hardenbergstraße 36
10623 Berlin
Germany

Karin Koch
Chemistry Department
Universität zu Köln
Köln
Germany

Patrick Kreissl
Institute for Computational Physics
University of Stuttgart
Stuttgart
Germany

Margarita Kruteva
Jülich Centre for Neutron Science (JCNS-1)
and Institute for Complex Systems (ICS-1),
Forschungszentrum Jülich GmbH
Leo-Brandt-Straße
52425 Jülich, Germany

Janna Kuchinka
Institute of Organic Chemistry and
Macromolecular Chemistry
Center of Excellence for Polysaccharide Research
Friedrich Schiller University of Jena
Humboldtstraße 10
D-07743 Jena
Germany

Joachim Landers
Faculty of Physics and Center for Nanointegration
Duisburg-
Essen (CENIDE)
University of Duisburg-Essen
Duisburg
Germany

Jonathan Leliaert
Department of Solid State Sciences
Ghent University
9000 Ghent, Belgium
Philine Jauch
Leibniz Institute of Surface Engineering (IOM)
Permoserstrasse 15
04318 Leipzig
Germany;
and Division of Surface Physics
Department of Physics and Earth Sciences,
University of Leipzig
Linnéstrasse 5
04103 Leipzig
Germany

Maik Liebl
Physikalisch-Technische Bundesanstalt
Abbestraße 2-12
10587 Berlin
Germany
maik.liebl@ptb.de

Julia M. Linke
TU Dresden
Faculty of Mechanical Engineering
George-Bähr-Str. 3, 01062 Dresden,
Germany

Hartmut Löwen
Theoretische Physik II: Weiche Materie
Heinrich-Heine-Universität Düsseldorf
Universitätsstr. 1
40225 Düsseldorf
Germany

Niklas Lucht
Institute of Physical Chemistry
Hamburg University
Grindelallee 117
20146 Hamburg
Germany

Frank Ludwig
Institut für Elektrische Messtechnik und
Grundlagen der Elektrotechnik
TU Braunschweig
D-38106 Braunschweig
Germany
f.ludwig@tu-bs.de

Moritz von der Lühe
Institute of Organic Chemistry and
Macromolecular Chemistry (IOMC)
Friedrich-Schiller-University Jena
Jena
Germany
and Jena Center for Soft Matter (JCSM)
Friedrich-Schiller-University Jena
Jena
Germany
moritz.von-derluehe@uni-jena.de

Stefan Lyer
Department of Otorhinolaryngology
Head and Neck Surgery
Section of Experimental Oncology and
Nanomedicine (SEON)
Else Kröner-Fresenius-Stiftung-Professorship
Universitätsklinikum
Erlangen
Germany,
Stefan.Lyer@uk-erlangen.de

Kai Mayer
Department of Biomaterials
Universität Bayreuth
Prof.–Rüdiger-Bormann-Straße 1
95447 Bayreuth
Germany

Stefan G. Mayr
Leibniz Institute of Surface Engineering (IOM)
Permoserstrasse 15
04318 Leipzig
Germany;
and Division of Surface Physics
Department of Physics and Earth Sciences,
University of Leipzig
Linnéstrasse 5
04103 Leipzig
Germany

Andreas M. Menzel
Institut für Physik
Otto-von-Guericke-Universität
Magdeburg
Universitätsplatz 2
39106 Magdeburg
Germany;
Theoretische Physik II: Weiche
Materie
Heinrich-Heine-Universität Düsseldorf
Universitätsstr. 1
40225 Düsseldorf
Germany
a.menzel@ovgu.de

Philipp Metsch
Chair of Computational and Experimental Solid
Mechanics
Institute of Soild Mechanics
TU Dresden
01062 Dresden
Germany

Gareth J. Monkman
Mechatronics Research Unit
OTH-Regensburg
Regensburg,
Germany
gareth.monkman@oth-regensburg.de

Elena K. Müller
Department of Internal Medicine II
Hematology and Medical Oncology
Jena University Hospital
Jena
Germany,
elena.mueller@med.uni-jena.de

Robert Müller
Leibniz-Institute of Photonic Technology (IPHT)
P.O.B. 100239
D-07702 Jena
Germany

Anton Musikhin
Ural Federal University
Lenina Ave 51
620083, Ekaterinburg
Russia
Dmitry Chirikov,
Ural Federal University
Lenina Ave 51
620083, Ekaterinburg
Russia

Hajnalka Nádasi
Otto von Guericke University
Institute of Physics
Magdeburg, Germany

Stefan Odenbach
TU Dresden
Faculty of Mechanical Engineering
George-Bähr-Str. 3
01062 Dresden
Germany
stefan.odenbach@tu-dresden.de

Catharina Pelz
Institute of Physical Chemistry
Hamburg University
Grindelallee 117
20146 Hamburg
Germany

Stavros D. Peroukidis
Department of Chemical Engineering
University of Patras
26504 Patras
Greece
and Hellenic Open University
26222, Patras
Greece

Harald Pleiner
Max Planck Institute for Polymer Research
Mainz
Germany
pleiner@mpipmainz.mpg.de

Nina Prem
Mechatronics Research Unit
OTH-Regensburg
Regensburg
Germany

Alexander Raßloff
Chair of Computational and Experimental Solid Mechanics
Institute of Soild Mechanics
TU Dresden
01062 Dresden
Germany

Patricia Radon
Physikalisch-Technische Bundesanstalt
Berlin
Germany
patricia.radon@ptb.de

Yuriy L. Raikher
Laboratory of Physics and Mechanics of Soft Matter
Institute of Continuous Media Mechanics
Russian Academy of Sciences (Ural Branch)
Perm, Russia
raikher@icmm.ru

Hilke Remmer
Institut für Elektrische Messtechnik und Grundlagen der Elektrotechnik
TU Braunschweig
D-38106 Braunschweig
Germany

Soma Salamon
Faculty of Physics and Center for Nanointegration Duisburg-
Essen (CENIDE)
University of Duisburg-Essen
Duisburg
Germany

Felix H. Schacher
Friedrich-Schiller-University Jena
Jena
Germany
and Jena Center for Soft Matter (JCSM)
Friedrich-Schiller-University Jena
Jena
Germany
felix.schacher@uni-jena.de

Thomas Scheibel
Department of Biomaterials
Universität Bayreuth
Prof.–Rüdiger-Bormann-Straße 1
95447 Bayreuth
Germany
Bayreuth Center for Colloids and Interfaces (BZKG), Universität Bayreuth
Universitätsstraße 30
95440 Bayreuth
Germany; Bayreuth
Center for Molecular Biosciences (BZMB)
Universität Bayreuth
Universitätsstraße 30
95440 Bayreuth,
Germany
Bayreuth Center for Material Science (BayMAT)
Universität Bayreuth
Universitätsstraße 30,
95440 Bayreuth, Germany
and Bavarian Polymer Institute (BPI)
Universität Bayreuth
Universitätsstraße 30
95440 Bayreuth
Germany
Thomas.Scheibel@bm.uni-bayreuth.de

Meinhard Schilling
Institute for Electrical Measurement Science and Fundamental Electrical Engineering
Technical University Braunschweig
Braunschweig, Germany

Mia Schliephake
TU Dresden
Faculty of Mechanical Engineering
George-Bähr-Str. 3
01062 Dresden
Germany
mia.schliephake@tu-dresden.de

Annette M. Schmidt
Chemistry Department
Universität zu Köln
Köln
Germany
annette.schmidt@uni-koeln.de

Christoph Schopphoven
Experimentalphysik, Universität des Saarlandes
Campus D2 2
D-66123 Saarbrücken
Germany

Malte Schümann
Faculty of Mechanical Engineering
TU Dresden
Dresden
Germany
Malte.Schuemann@tu-dresden.de

Rouven Schweitzer
Experimentalphysik, Universität des Saarlandes
Campus D2 2
D-66123 Saarbrücken
Germany

Max Schwenk
Institute of Physical Chemistry
Hamburg University
Grindelallee 117
20146 Hamburg
Germany

Julian Seifert
Chemistry Department
Universität zu Köln
Köln
Germany

Gaurav P. Shrivastav
Institute für Theoretical Physics
Technische Universität Wien
Wiedner Hauptstr. 8-10
1040 Vienna
Austria

Nima H. Siboni
Institut für Theoretische Physik
Technische Universität Berlin,
Hardenbergstraße 36
10623 Berlin, Germany
hamidisiboni@tu-berlin.de

Dirk Sindersberger
Mechatronics Research Unit
OTH-Regensburg
Regensburg
Germany

Ralf Stannarius
Otto von Guericke University
Institute of Physics
Magdeburg, Germany
ralf.stannarius@ovgu.de

Gennady Stepanov
State Research Center GNIICHTEOS
105118, Moscow
Russia

Andreas Tschöpe
Experimentalphysik
Universität des Saarlandes
Campus E2 6
D-66123 Saarbrücken
Germany
a.tschoepe@nano.uni-saarland.de

Gennady V. Stepanov
State Scientific Research Institute for Chemical
Technologies of Organoelement Compounds
105118 Moscow
Russia
gstepanov@mail.ru

Oleg V. Stolbov
Laboratory of Physics and Mechanics of Soft
Matter
Institute of Continuous Media Mechanics
Russian Academy of Sciences (Ural Branch)
Perm, Russia
Pedro A. Sánchez
Institute of Natural Sciences and Mathematics
Ural Federal University
Ekaterinburg, Russia
and Wolfgang Pauli Institute
University of Vienna
Vienna
Austria

Pavel A. Storozhenko
State Scientific Research Institute for Chemical
Technologies of Organoelement Compounds
105118 Moscow
Russia

Stephen Strassburg
Department of Biomaterials
Universität Bayreuth
Prof.–Rüdiger-Bormann-Straße 1
95447 Bayreuth
Germany

Tamara Szecsey
Mechatronics Research Unit
OTH-Regensburg
Regensburg
Germany

Harald Unterweger
Department of Otorhinolaryngology
Head and Neck Surgery
Section of Experimental Oncology and
Nanomedicine (SEON)
Else Kröner-Fresenius-Stiftung-Professorship
Universitätsklinikum
Erlangen
Germany
Harald.Unterweger@uk-erlangen.de

Thilo Viereck
Institute for Electrical Measurement Science and
Fundamental
Electrical Engineering
Technical University Braunschweig
Braunschweig
Germany
t.viereck@tu-bs.de

Thomas Wallmersperger
Institute of Solid Mechanics
TU Dresden
Dresden
Germany
thomas.wallmersperger@tu-dresden.de

Paul Warncke
Institute of Pharmacy, Pharmaceutical
Technology und Biopharmacy
Friedrich-Schiller-University Jena
Jena
Germany
paul.warncke@uni-jena.de

Samira Webers
Faculty of Physics and Center for Nanointegration
Duisburg-
Essen (CENIDE)
University of Duisburg-Essen
Duisburg
Germany

Rudolf Weeber
Institute for Computational Physics
University of Stuttgart,
Stuttgart
Germany
rudolf.weeber@icp.uni-stuttgart.de

Andreas Weidner
Institute of Biomedical Engineering and
Informatics (BMTI)
Technische Universität Ilmenau
Ilmenau
Germany
Andreas.Weidner@tu-ilmenau.de

Heiko Wende
Faculty of Physics and Center for Nanointegration
Duisburg-
Essen (CENIDE)
University of Duisburg-Essen
Duisburg
Germany
heiko.wende@uni-due.de

Frank Wiekhorst
Physikalisch-Technische Bundesanstalt
Abbestraße 2-12,
10587 Berlin
Germany

Maximilian Wollner
Chair of Computational and Experimental Solid
Mechanics
Institute of Soild Mechanics
TU Dresden
01062 Dresden
Germany

Igor Zeidis
Technische Universität Ilmenau
Technische Mechanik
Ilmenau
Germany

Klaus Zimmermann
Technical Mechanics Group
Faculty of Mechanical Engineering
Technische Universität Ilmenau
Ilmenau
Germany
klaus.zimmermann@tu-ilmenau.de

Andrey Zubarev
Ural Federal University
Lenina Ave 51
620083, Ekaterinburg
Russia
and M. N. Mikheev Institute of Metal Physics of the Ural Branch of the Russian Academy of Sciences Ekaterinburg
Russia
A.J.Zubarev@urfu.ru

Julian Seifert, Karin Koch, Melissa Hess and Annette M. Schmidt*

1 Magneto-mechanical coupling of single domain particles in soft matter systems

Abstract: Combining inorganic magnetic particles with complex soft matrices such as liquid crystals, biological fluids, gels, or elastomers, allows access to a plethora of magnetoactive effects that are useful for sensing and actuation perspectives, allowing inter alia to explore and manipulate material properties on the nanoscale. The article provides a comprehensive summary of recent advancement on employing magnetic nanoparticles either as tracers for dynamic processes, or as nanoscopic actuating units. By variation of the particle characteristics in terms of size, shape, surface functionality, and magnetic behavior, the interaction between the probe or actuator particles and their environment can be systematically tailored in wide ranges, giving insight into the relevant structure–property relationships.

Keywords: complex magnetic fluids, ferrogels, magnetic particle nanorheology, magnetic tracers, magneto-mechanical coupling, particle-matrix interaction

1.1 Introduction

The design of hybrid materials that can be manipulated by magnetic fields is possible by incorporating inorganic magnetic nanoobjects into complex soft matrices such as gels [1], elastomers [2], liquid crystals [3], or biological fluids [4] in a predetermined way. A proper setup allows the application of external magnetic fields to either analyze or alter the material properties of such systems on a nanoscopic scale. The resulting hybrid structures have promising responsive properties that can be categorized into applications of interest either for actuation or sensing.

The fascinating prospective arising from magnetic nanoparticles dispersed in liquids (ferrofluids) [5] and the manipulation of their structure [6] and dynamics [7] by magnetic fields have led to a broad field of applications. On the other side of the material spectrum, magnetoactive elastomers attract increasing attention due to the theoretical prediction and experimental observation of even more sophisticated effects such as magnetoresponsive shape changes or mechanical properties, as often observed for elastomers [8–10].

***Corresponding author: Annette M. Schmidt**, Chemistry Department, Universität zu Köln, Köln, Nordrhein-Westfalen, Germany, E-mail: annette.schmidt@uni-koeln.de. https://orcid.org/0000-0002-7284-7847

Julian Seifert, Karin Koch and Melissa Hess, Chemistry Department, Universität zu Köln, Köln, Nordrhein-Westfalen, Germany

This article has previously been published in the journal Physical Sciences Reviews. Please cite as: J. Seifert, K. Koch, M. Hess and A. M. Schmidt "Magneto-mechanical coupling of single domain particles in soft matter systems" *Physical Sciences Reviews* [Online] 2020, 5. DOI: 10.1515/psr-2019-0092 | https://doi.org/10.1515/9783110569636-001

The magnetic probe particles are further useful nanoscopic magnetic tracers, allowing the investigation of the nanoparticle interaction with their surrounding matrix by using dynamic magnetic methods on small sample volumes in a nondestructive way [11]. Of particular interest is the option to investigate the mechanical properties of soft matter at the length scale of the probe particle, giving access to local information in microstructured samples that are hardly accessible by classical methods [12, 13].

Research of this kind strongly benefits from new developments in the controlled synthesis of prospective tracer particles with improved uniformity, stability, controlled surface characteristics, and novel coupling mechanisms [10, 14–16]. In addition, the development and availability of new and better resolved methods for the investigation of structure and dynamics in such soft matter hybrid systems facilitates the detection and establishment of novel coupling measuring modes [17]. Recent studies demonstrate that by variation of the particle characteristics in terms of size [18], shape [19], surface functionality [20], and magnetic behavior [21], the interaction between the probe or actuator particles and their environment can be tailored in wide limits.

1.2 Particle dynamics in fluids: Magnetic particle nanorheology

The flow properties of complex fluids, as generally investigated by rheology, are relevant for the processing and application properties of all kinds of products in daily life, such as cosmetics and food. A broad spectrum of rheological methods is available for their assessment, spanning a large range of time and size regimes as well as measuring geometries.

The employment of microscopic approaches for the investigation of complex fluids is often related to the application of small tips or particles as probes. While passive methods, like particle tracking microscopy [22], diffusing wave spectroscopy [23], or fluorescence correlation spectroscopy [24], have been shown to yield good results for many materials, active methods where the probes are exposed to active displacement, generally benefit from a broader frequency range and higher possible torques enabling the analysis of even elastic samples [25]. Already in 1950 Crick et al. showed that the microrheological approach has significant benefits for the analysis of biological samples. As low sample volume is needed, the mechanical stress subjected to the sample is low, and a wide frequency range is accessible [26]. The original idea of this technique is to obtain macrorheological results on small samples in a nondestructive way. While this is fulfilled for many relevant samples, an analogous approach using probe particles in a size more similar to the relevant length scales of the matrix material opens even the pathway to the investigation of size-dependent properties, as relevant for internal dynamics in systems such as nanocomposites and cells [27, 28].

1.2.1 Methodology

The strong size-dependent properties that are measured in nanorheological experiments on microstructured samples make this method a promising candidate for the development of a method capable of determining local rheological properties on a nanoscale. The use of magnetically blocked, nanoscopic particles as tracers under the influence of an oscillating magnetic field is a particularly promising approach. The response of the particles in terms of their magnetic susceptibility is recorded, and in analogy to classical macrorheology, the phase shift of the response signal delivers information on the complex susceptibility and thus on the frequency-dependent excitation of the particle rotation.

The magnetic nanoparticles (MP) that are used as probes in this approach ideally need to fulfill a number of requirements in order to facilitate a simple and meaningful data analysis. In order to result in a direct correlation of the magnetic susceptibility data and the hydrodynamic characteristics of the probes, it is important to make sure that they are well-dispersed in the sample material on a single-particle level in a nonagglomerated way, and that no significant adsorption of matrix components on the particle surface occurs. Further, it is beneficial if the size distribution is narrow, and the basic magnetic properties, such as the magnetic moment, of each individual particle are uniform. Finally, the magnetic anisotropy is required to be high enough to ensure that the particles are magnetically blocked and accordingly relax predominantly by Brownian rotation [29]. Within these limits, there are still a variety of tracer particles of different size, shape, and chemical composition possible, and meanwhile a range of different tracers are successfully employed for magnetic particle nanorheology (MNP) and related methods [18, 19, 30].

Among the particle systems suitable for application as tracers in MNP, magnetically blocked, single domain cobalt ferrite ($CoFe_2O_4$) nanoparticles, as shown in Figure 1.1 are particularly versatile. The inherent magnetocrystalline anisotropy of $CoFe_2O_4$ leads to strong pinning of the magnetization vector in the crystal lattice, thus providing a well-defined crystalline structure. This results in a high effective anisotropy constant and high volume-based saturation magnetization. Further, the particles can be stabilized in organic as well as aqueous media, and in order to tailor the hydrodynamic size of the probe particle, they can be decorated with a silica shell of variable thickness [18].

A suitable experimental setup is based on AC susceptometry. In a typical experiment, the sample is placed between a pair of excitation coils, while two pairs of detection coils are employed, one measuring the sample and the other measuring a blank. By amplification of the subtracted signals using a lock-in amplifier, the complex susceptibility is accessible. During the measurement a sinusoidal AC magnetic field with small amplitude is applied, which can be varied in frequency. The alternating field activates the magnetic nanoparticles in the sample to a frequency-dependent

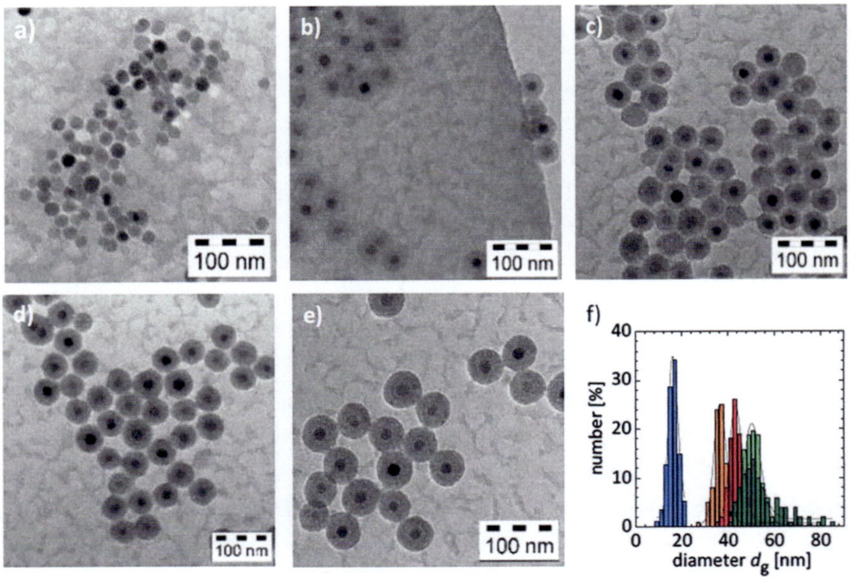

Figure 1.1: TEM images of tracer particles of the same magnetic core size and varying SiO_2 shell thickness of a) 12 nm; b) 25 nm; c) 29 nm; d) 37 nm; and e) 42 nm. In f) the size distribution of the geometrical diameter d_g with corresponding log normal function is shown for the different probe particles (a) (blue); b) (orange); c) (red); d) (light green); e) (green)). Reproduced by permission of the PCCP Owner Societies from [18].

relaxation behavior. This rotational relaxation process of the particles occurs against the effective sample viscosity that might be influenced by the probe size and might further depend on frequency.

The simplest model to describe the frequency-dependent response function of the particle is given by the Debye model where the real part χ' and the imaginary part χ'' of the susceptibility are expressed as functions of the excitation frequency ω and the relaxation time τ of the particle [31, 32],

$$\chi'(\omega) = \frac{\chi_0}{1 + (\omega\tau)^2} \tag{1.1}$$

$$\chi''(\omega) = \frac{\chi_0 \omega\tau}{1 + (\omega\tau)^2} \tag{1.2}$$

$$\chi_0(\omega) = \frac{\mu_0 n m^2}{3k_B T} \tag{1.3}$$

Here, n denotes the number density of particles and m is the particle's magnetic moment. For thermally blocked MP, the Brownian relaxation is dominant, and hence the corresponding relaxation time is accessible from a fit of the experimental data according to the Debye model. By application of the well-known Stokes-Einstein equation the obtained relaxation time τ_B is related to the viscosity η of the medium, the hydrodynamic radius of the particle r_h, and the rotational diffusion coefficient D_r.

$$D_r = \frac{k_B T}{8\pi\eta r_h^3} = \frac{1}{2\tau_B} \tag{1.4}$$

This approach describes the behavior of monodisperse MP in Newtonian fluids. Once a real, polydisperse, particle system is analyzed, a broadening of the relaxation is observed, and the quality of the data description decreases significantly. In order to compensate for this, extended versions of the Debye model have been developed allowing a sufficiently good description of the measured data by introducing a size distribution term [17, 32, 33].

$$\chi* = \chi_0 \int_0^\infty \frac{1}{1 - i\omega\tau_B(r_h)} f(r_h) dr_h \tag{1.5}$$

$$f(r_h) = \frac{1}{\sqrt{2\pi}\sigma r_h} e^{\frac{\ln\left(r_h/r_{h,\mathrm{eff}}\right)^2}{2\sigma^2}} \tag{1.6}$$

Here, $\chi*$ is the complex susceptibility, $f(r_h)$ is the lognormal size distribution of the particle system and τ_B the Brownian relaxation time. While this approach works well for Newtonian fluids, as soon as more complex, non-Newtonian samples are analyzed, the extended Debye model fails to describe the data, therefore strongly limiting the applicability of this otherwise extremely versatile and attractive technique. Especially for the analysis of biological samples, this technique is promising as low sample amounts of 200 µL are characterized concerning their rheological data on the nanoscale in a non-destructive way employing small amounts of MP as tracers.

1.2.2 Application to complex fluids

The rheological properties of polymer-based fluids and soft, viscoelastic solids, e.g., gels or melts, typically show non-Newtonian behavior depending on the volume fraction and the average chain length of the polymer component. Concerning the nanorheological analysis of such systems, characteristic length scales of the polymer component become relevant in relation to the probe, such as mesh size in gels, the radius of gyration for polymer coils, or the tube diameter a and the correlation length ξ in polymer solutions. Rubinstein et al. developed a theory differentiating three categories with respect to the relation between probe size and characteristic length scales in the sample [34]. Here, particles with a hydrodynamic size smaller than the correlation length ξ are classified as small particles and their translational diffusion is determined by the solvent viscosity. The tracers are classified as large particles if their diameter is larger than the tube diameter a of the polymer chain. For this case, a (quasi)continuous behavior for long time scales is expected. In the intermediate size regime (ξ < MP < a), an effective viscosity is observed, which is related to the properties of the polymer strand on the probe scale [34]. The dependence of the observed viscosity on the probe size and characteristic length scales in the polymer solution leads to frequency-dependent rheological properties, as instead of a single relaxation process, as observed in Newtonian fluids, different relaxation processes, including relevant processes within the matrix itself, contribute to the signal. This offers the possibility to extract frequency-dependent rheological properties by modification of the data analysis approach.

Hence, the concept of MPN is established. Here, the German-DiMarzio-Bishop (GDB) model is employed and modified in order to correlate the measured frequency-dependent susceptibility to the complex modulus G^* of the polymer solution. A frequency-dependent friction coefficient is introduced leading to the following equations for the loss modulus $G''(\omega)$ and the storage modulus $G'(\omega)$ [17].

$$G''(\omega) = \frac{\chi''_N}{K\left(\chi'^2_N + \chi''^2_N\right)} \tag{1.7}$$

$$G'(\omega) = \frac{\dfrac{\chi'_N}{\chi'^2_N + \chi''^2_N} - 1}{K} \tag{1.8}$$

with the constant K

$$K = \frac{4\pi r_h^3}{k_B T} \tag{1.9}$$

and the abbreviations

$$\chi'_N = \frac{\chi' - \chi_\infty}{\chi_0 - \chi_\infty} \tag{1.10}$$

$$\chi''_N = \frac{\chi''}{\chi_0 - \chi_\infty} \tag{1.11}$$

With this in hand, a powerful method is provided that allows the determination of frequency-dependent rheological properties of soft matter systems on the length scale of the tracer particles employed.

1.2.2.1 MPN of Polymer Solutions

By employing magnetically blocked nanoparticles as nanoscopic, quasi-dipolar magnetic probes, the complex dynamic behavior of soft matter at the particle scale is investigated in fluid media of increasing complexity. This approach allows the extraction of the complex, frequency-dependent rheological information from AC-susceptibility measurements of tracer particles and polymer solutions. In Figure 1.2, exemplary AC-susceptibility curves of well-defined cobalt ferrite particles dispersed in aqueous solutions of ethylene glycol, triethylene glycol, and its polymeric analogues as analyzed by MPN are shown [17].

Depending on the volume fraction of polymer, a shift of the maximum of the imaginary part of the susceptibility to lower frequencies is found, that is attributed to an increase of the viscosity of the dispersion medium in Eq. (1.4). For small chain length of the ethylene glycol derivatives (Figure 1.2a, b), no change of the shape of the relaxation peaks is observed and the spectra may be described by an extended Debye model, considering an increased viscosity. For larger chain length (Figure 1.2c, d), the shift is significantly more pronounced, but also the shape of the relaxation peak changes. In chapter 1.2.2, this effect is attributed to the presence of several simultaneously traced relaxation processes in the sample that are dependent on the relative size between probe and polymer length scale. Employing the analysis according to the GDB model, frequency depending moduli can be obtained.

In Figure 1.3a, this analysis is performed for PEG–water mixtures with high molar mass and compared to macrorheological analysis [35]. This allows extending the accessible frequency range of rheological data by several orders of magnitude [17, 18, 35]. The non-Newtonian behavior in polymer solutions leads to a complex frequency-

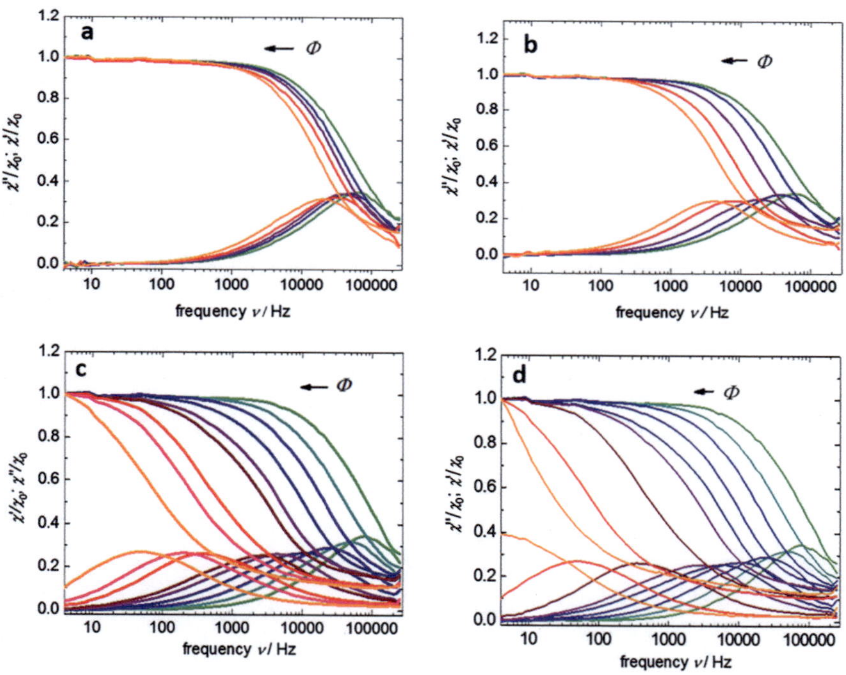

Figure 1.2: ACS spectra for aqueous ethylene glycol (a); triethylene glycol (b); polyethylene glycol ($M = 1000$ g mol^{-1}) (c); and polyethylene glycol ($M = 8000$ g mol^{-1}) (d) solutions with $CoFe_2O_4$ nanoparticles as tracers. Reprinted by permission from Springer Nature: Springer, Colloid and Polymer Science, from reference [17], Copyright (2014).

dependent flow behavior for macrorheological experiments. The crossover point where G' and G'' intersect is typically related to the longest relaxation time of the polymer chain, e.g., by reptation. It's dependence on the polymer fraction gives access to the determination of the respective polymer fraction critical for entanglement formation (Figure 1.3b) [35]. A second relaxation time typically related to polymer segment mobility becomes visible at higher frequency values, as is discussed in more detail in **section 1.2.2.3.**

1.2.2.2 Size effects

The rheological properties of non-Newtonian fluids strongly depend on the size and volume fraction of dissolved macromolecules, as discussed in section 1.2.2. For the nanorheological analysis, the system gets even more complex, as the probe is only sensitive to certain parts of the polymer depending on the probe's size. In order to get a full picture of a sample system, it is therefore interesting to probe the sample on different time and length scales. For this, the combination of macro-, micro- and nanorheological methods is a promising tool. This is done in a comparative study of three different PEG–water mixtures with the same macroscopic zero-shear viscosity, but different chain length and volume fraction of the dissolved PEG macromolecules. Here, cobalt ferrite (d = 16.4 nm) as well as anisotropic Ni nanorods (l = 233 nm, d = 23.9 nm) are employed as tracers, thereby probing significantly different length scales. In addition to classical macrorheology and MPN, oscillating field optical transmission (OFOT) and DC-overlayed AC-susceptometry are used. The rotational dynamics of these probes with different sizes and shapes are correlated to the macrorheological analysis. Here, a strong dependence of the rheological properties of the samples depending on the length scales of the polymer compared to the size of the nanoparticle is observed. If the probe is smaller than the correlation length of the polymer, the apparent viscosity is found to be lower than the value determined by

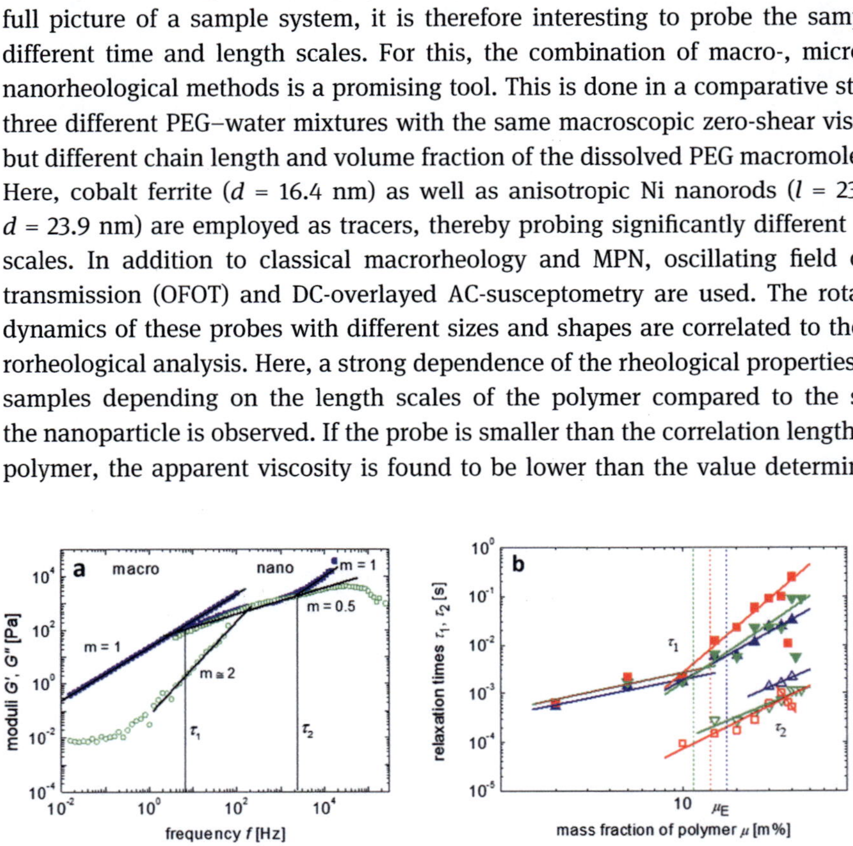

Figure 1.3: a) Comparison of the frequency-dependent macroscopic (circles) and nanoscopic (squares) loss (G'', filled) and storage moduli (G', unfilled) for an aqueous solution of PEG 35k (35 m%), b) relaxation times τ_1 (filled) and τ_2 (unfilled) versus polymer mass fraction μ_{PEG} for PEG8k (blue), PEG20k (green) and PEG35k (red).

macrorheology and frequency-dependent with decreasing values for increasing frequency. If the probe on the other hand is large compared to tube diameter an average equivalent to the macroscopic measurement is obtained indicated by no frequency dependence and good agreement with the macroscopically determined zero-shear viscosity, see Figure 1.4 [12].

In Figure 1.5, macro- and nanorheologically obtained loss moduli are presented. If the chain length of the polymer in the complex fluid is low (Figure 1.5a), generally good agreement of macro- and nanorheological properties is found, indicating nearly Newtonian behavior. However, for polymers with a molar mass larger than the entanglement molar mass (Figure 1.5b), strong deviations between macro- and nanorheological results are found, which is attributed to the particle size being similar to the polymer correlation length in the solution.

In order to obtain a complete picture of the polymer in solution, the sample can be probed with probes of different size. A systematic study in this context uses a size-tuneable silica shell around the probe particle to perform this experiment [18]. The systematic evaluation is shown in Figure 1.5c. If the particle is large compared to the size of the polymer chain, hence larger than the tube diameter a, a mean viscosity comparable to the macro-rheological result is obtained, as this curve and slope basically

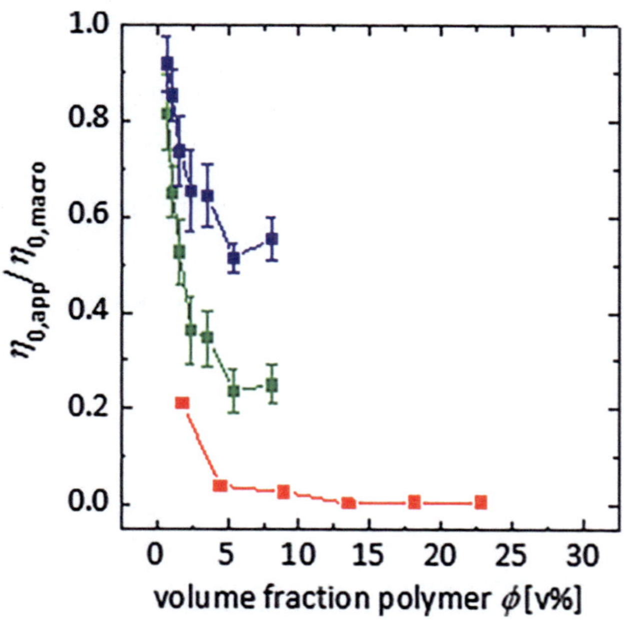

Figure 1.4: Apparent zero shear viscosity $\eta_{0,\,app,}\,(\omega)$ of aqueous PEG300k solutions normalized by the macroscopic viscosity $\eta_{0,\,macro}\,(\omega)$ in dependence on the polymer volume fraction ϕ for rod-like tracer particles with a length of 550 nm (blue) and 140 nm (green) and for spherical tracers with a diameter of 13.3 nm (red).

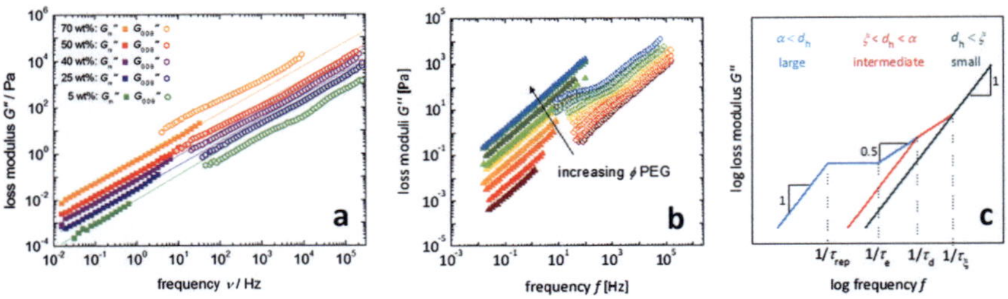

Figure 1.5: Frequency-dependent loss modulus G'' as obtained from macrorheology (solid symbols) and calculated according to the modified DGB model (open symbols) with 13 nm tracer particles; a) for PEG ($M = 1.0$ kg mol^{-1}) (adapted with permission from Springer Nature: Springer, Colloid and Polymer Science, from reference [17], Copyright (2014)); and b) for PEG ($M = 35$ kg mol^{-1}); c) scaling behavior of the loss modulus with frequency as expected from theoretical prediction, (b and c adapted by permission of the PCCP Owner Societies from [18]).

describe a Newtonian flow behavior. The corresponding relaxation time extractable from the nanorheological experiment therefore is given by the whole polymer chain. If the particle is significantly smaller than the correlation length of the polymer, on average the solvent in between the polymer segments is analyzed by the probe and again a scaling factor of 1.0 as for the Newtonian fluid is obtained. In the intermediate regime confinement of the probe particle by the polymer chain is expected leading to a subdiffusive behavior indicated by a scaling of 0.5 in Figure 1.5c. With a small number of independent experiments employing probes of different sizes this approach allows the determination of four different relaxation times, namely the relaxation of the whole polymer τ_{rep}, the relaxation time of an entangled polymer strand τ_e, the relaxation time of a polymer segment in the size range of the probe particle τ_d, and the relaxation time of the polymer connected to the correlation length τ_2 [18]. The possibility to determine and distinguish multiple relaxation times inherent to soft matter systems is of importance for future investigations, as will be under more addressed in chapter 1.2.2.3.

1.2.2.3 Determination of relaxation times

The crossover point between the storage and loss modulus in a rheological experiment contains valuable information, as it indices a drastic change in flow behavior that is commonly related to a relaxation time of the material. A MPN experiment allows the determination of the frequency-dependent storage and loss modulus and therefore can also be used to extract the relevant relaxation times. As the material is probed on the nanoscale, with this approach also local relaxations are accessible. A sample system that is especially suitable for the analysis of relaxation times are dynamic physical gels, as these systems offer a variety of relaxation processes and an interesting flow behavior [36, 37]. One way to achieve this interesting architecture is to induce more specific dynamic interactions in the polymer matrix by introducing terpyridine-based end groups to star-shaped PEG molecules. These end groups show affinity to divalent transition metal cations in the form of 2:1 complexes, as shown in Figure 1.6c). Therefore, upon dissolving in water in the presence of metal ions, complexation of the latter is observed leading to the formation of dynamic supramolecular networks with intrinsic self-healing properties. The rheological properties of these dynamic networks can be probed on the nanoscale employing MPN, and afterwards related to macroscopically determined storage and loss moduli, see Figure 1.6. Here, the macroscopic results show the typical flow behavior of viscoelastic liquids as described by the Maxwell model with a relaxation time $\tau_{l,\mathrm{macro}}$ found at the crossover point of storage and loss modulus. This relaxation time contains contributions from relaxation of whole star-shaped PEG molecules and the dissociation time of metal-ligand complexes leading to a chain extension, thereby actively slowing the molecular relaxation. For the nanorheological analysis, distinct differences are found in the shape of curves determined at different stoichiometric ratios $r_{\mathrm{M/P}}$, which is defined as the molar ratio of the divalent metal ions to the total terpyridine functionalities. For this system bidentate complexes are expected, therefore a homogeneous network could be formed at a stoichiometric ratio of 2.0. For lower values broader meshes are expected [38].

Depending on the stoichiometric ratio $r_{\mathrm{M/P}}$, in the dynamic networks fundamentally different flow behavior is found by MPN. For low values of $r_{\mathrm{M/P}}$ the loss modulus is higher than the storage modulus indicating a pseudo-Newtonian-like flow behavior. This is also strengthened by the finding of only one relaxation time $\tau_{\mathrm{l,nano}}$. In the case of higher values of the $r_{\mathrm{M/P}}$ a rubbery plateau is found in the nanorheological investigation. Here, two relaxation times are accessible. In addition to $\tau_{\mathrm{l,nano}}$, which is associated to the breaking of the metal ligand bond, also a fast relaxation $\tau_{f,\mathrm{nano}}$ is found and attributed to network-strand fluctuations. These data are used to on the one hand extract the mean mesh sizes of the dynamic network from the plateau storage modulus according to rubber elasticity theory and on the other hand to determine the

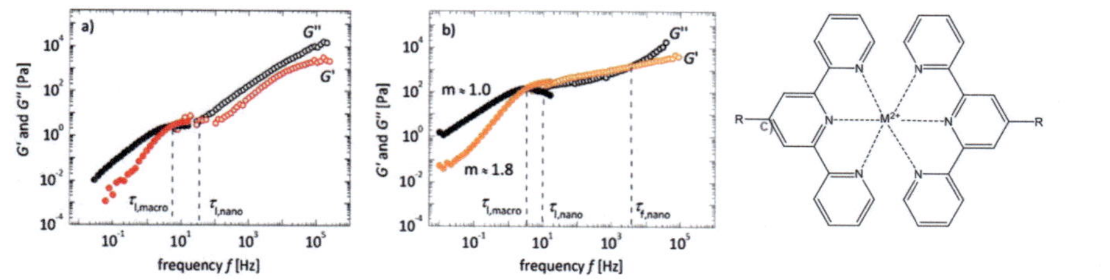

Figure 1.6: Comparison of macroscopic (full symbols) and nanoscopic (empty symbols) storage G' (colored) and loss moduli G'' (black) in dependence on frequency for Zn^{2+} as counter ion to terpyridine functional tetrafunctional PEG stars with the stoichiometric ratios a) $r_{\mathrm{M/P}} = 0.65$ and b) $r_{\mathrm{M/P}} = 1.09$, reproduced by permission from The Royal Society of Chemistry from [38]. c) Molecular structure of the formed complex. M^{2+} denotes the employed metal cation (Mn^{2+}, Zn^{2+}, or Co^{2+}) and R is the rest of the tetrafunctional ligand.

respective relaxation times for all analyzed samples, depending on the stoichiometric ratio and the type of the metal ion, as shown in Figure 1.7 [38, 39].

1.2.2.4 Phase transitions

The determination of phase transitions of materials is of significant importance for daily life due to the properties of materials changing drastically upon the transition, thereby leading to potentially dangerous situations as material is out of the parameter range for application. The transition of the rheological properties of supramolecular networks as shown in **chapter 1.2.2.3** from fluid-like to solid-like can be described as a sol-gel transition depending on the amount of metal ions in the sample. This shows the principle applicability of MPN to extract phase transition behavior.

In order to expand this concept, we will consider different phase transitions in this chapter. One example of a phase transition that should be detectable by MPN is the sol–gel transition, where the reaction mixture starts in a fluid phase and proceeds to a viscoelastic solid. Connected to this transition is a drastic change of the macroscopically measurable rheological properties that translates to the nanoscopically observed properties depending on the mesh size and the characteristic length scales in the gel [12, 18, 38].

A possible example for such systems are poly(acrylamide) (PAAm) hydrogels that are commonly used model systems. By analyzing samples with different crosslinking density as well as polymer volume fraction, the mechanical properties as well as the nanoscale structure of the gel samples are varied. For these well-defined model hydrogels rubber elasticity theory is successfully used to calculate the mean mesh size of the polymer networks from the plateau storage modulus obtained from macro-rheological analysis [38, 39]. Analyzing the network formation kinetics, an approach frequently used for macroscopic gels and elastomers is the cure-curve. Here, a liquid

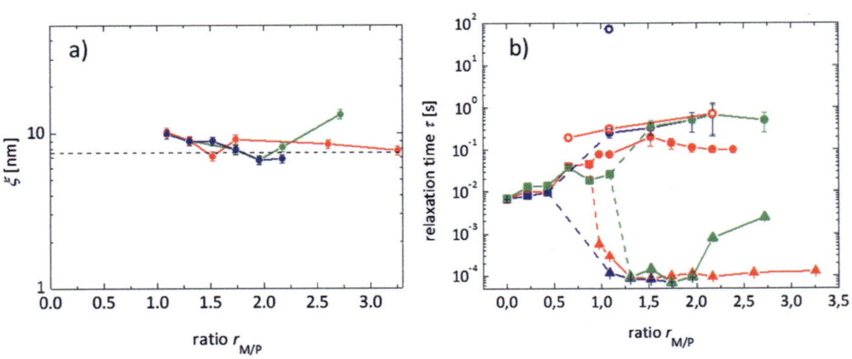

Figure 1.7: a) Mesh size ξ and b) relaxation times for different systems consisting of metal ion and tetrafunctional PEG (Mn^{2+} (green), Zn^{2+} (red), Co^{2+} (blue)) depending on the stoichiometric ratio, reproduced by permission from The Royal Society of Chemistry from [38].

reaction mixture is placed in a rheometer and the sample is analyzed at a constant frequency and strain in fixed time intervals. The obtained curve shows a transition from a liquid-like sample to a solid (visco-)elastic sample depending on the time. Using MPN, comparable to a macrorheologically determined cure–curve, the reaction kinetics can be determined by time-dependent measurements at a constant frequency yielding the dependence of the reaction rate on the initiator concentration.

Using the MPN approach as described in section 1.2.2, the frequency-dependent storage and loss modulus of PAAm hydrogels can be determined. In Figure 1.8 the modulus at the crossing point is shown depending on the macroscopically determined mean mesh size of the hydrogel. For mesh sizes smaller than the average particle size a strong increase of the modulus is found upon decreasing the mesh size, as the particle is trapped inside the meshes of the network. Upon reaching a mesh size that is equal to the particle diameter, a sharp drop of the modulus and a local minimum is found for the course of the cross modulus indicating that the size relation of probe to characteristic length scale in the sample is of great importance for the analysis and that the structure is in fact probed on a nanoscale.

Gelation of samples can also be induced by a change in temperature. A prominent example for this is gelatin. The sol–gel phase transition is triggered by a change in the temperature of the medium leading to a change in the solvent condition of the

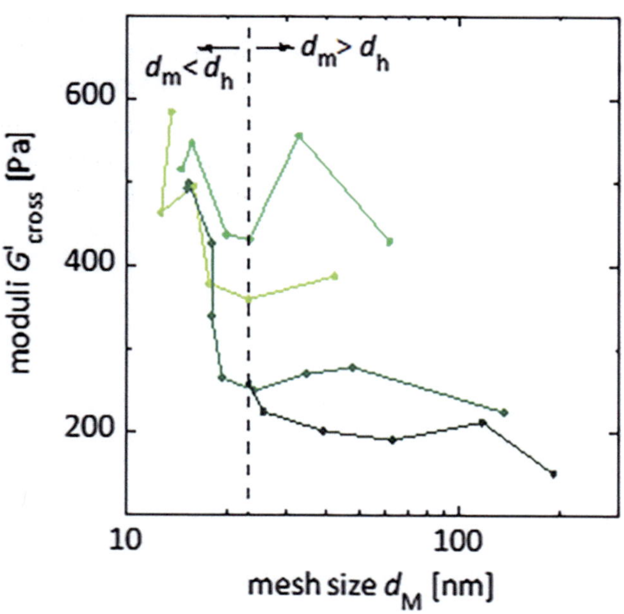

Figure 1.8: Modulus at the crossover point for PAAm hydrogels with different polymer volume fraction (light green: $\phi = 0.07$, green: $\phi = 0.06$, olive: $\phi = 0.05$, dark green: $\phi = 0.04$) in correlation with the experimental mesh size of the polymer network.

dissolved macromolecules, as opposed to the gelation of polyacrylamide as described above, where the gelation is initiated by a chemical reaction forming the macromolecules *in situ* from respective monomers. For the analysis of the progressing gelation of gelatin the time-dependence is shown to be observable by a shift of the imaginary part of the magnetic susceptibility to lower frequencies [40, 41].

The most common phase transitions are temperature triggered, such as melting and glass transitions. Polymer solutions here are a rich and versatile sample system to study, as they show a number of thermal transitions that can oftentimes be tuned and shifted by the chain length and volume fraction of the dissolved polymer. Analyzing such systems with magnetic probes as tracers by zero-field-cooled-field-cooled (ZFC-FC) magnetization measurements clear features are observed that are related to the thermal transitions as found in differential scanning calorimetry (DSC) studies. In this way, it is shown that by doping materials with a tracer amount of MP it is possible to determine the temperature-dependent phase behavior of polymer solutions in a simple nondestructive way [42].

1.3 MP as tracers and actuators in nanostructured solids

The combination of MP and elastic, polymeric matrices leads to magnetosensitive hybrid materials, which are due to their unique properties discussed extensively for biomedical applications as well as in soft robotics [8, 43–45]. On example that is especially popular is the combination of MP and gels to so called ferrogels. The gel, here, is a polymeric or macromolecular network that may be formed by either permanent covalent crosslinks or dynamic physical bonds. In this swollen network structure MP are embedded [46]. By variation of the particle surface functionalization the interactions between the elastic matrix and the particles can be tuned in type and strength, ranging from weak van der Waals interaction over Coulomb and coordinative bonds to strong covalent bonds [14, 47–49]. Analyzing the static and dynamical properties of ferrogels, the obtained relations strongly depend on the type and strength of the mutual interactions between MP and the matrix. Here, *Kickelbick* established a concept for the classification of hybrid materials that has been adapted for ferrogels in a review of our group in cooperation with AG Holm [45, 50]. *Kickelbick* classifies hybrid materials in two categories based on the strength of their intermaterial interactions. In Class I materials with weak physical interactions like hydrogen bonding and van der Waals or weak electrostatic interactions are found. Class II materials are composites with permanent chemical interactions or strong physical interactions.

1.3.1 Conventional ferrogels

The most common way to build up a gel network is to introduce crosslinks into a system consisting of long polymeric chains. These chains may either be directly introduced to the system and crosslinked by a proper agent or be generated *in situ* by polymerization of monomers in presence of a crosslinker. The interaction of MP dispersed in these matrices are predominately determined by van der Waals interactions and typically strongly depend on the mesh size of the gel network relative to the particle size.

One important issue for the macroscopic material's properties is the particle mobility in such structures on the particle scale. Oftentimes in elastic matrices especially for larger particles no information on the particle mobility is obtained by nanorheological investigations, as the relaxation processes are on time and length scales that are not accessible. Hence in a novel approach, Mössbauer spectroscopy has been used to gain information on the particle mobility of a sample system of spindle type hematite particles dispersed in crosslinked PAAm hydrogels with varying crosslinking degree and therefore also mesh size. Here, it is found that while all samples show no particle mobility in the dynamic magnetic field of a nanorheological investigation, a distinct trend of the line broadening in Mössbauer spectroscopy shows mobility for all particles on small time and length scales that is also depending on the mesh size of the polymer network [51].

For smaller particles, the mesh size change of the hydrogel matrix can have a significantly stronger effect on the particle mobility, as the particle size is in the range of the mesh size. As mentioned in section 1.2.1, MP exhibit two main mechanisms of relaxation after excitation by an external magnetic field, the Brown and the Néel mechanism. The Brownian relaxation is connected to a reorientation of the particle magnetic moment by rotation of the whole particle, while during the Néel mechanism the magnetic moment rotates inside the particle against the effective magnetocrystalline anisotropy. Hence, both mechanisms are fundamentally different, and also the heat dissipated by each respective mechanism is different and depends severely on the matrix and the properties of the magnetic particle. One way to separate these mechanisms and determine the individual contribution to the overall heating is presented by analysis of mesh size tunable PAAm hydrogels as a tissue model. Here, by varying the ratio of Aam to BIS at a constant polymer volume fraction the mesh size of prepared PAAm hydrogels is tuned systematically. Performing a macrorheological analysis, the obtained plateau storage modulus is correlated to the mean mesh size of the hydrogel network. By synthesizing the polymer network in the presence of magnetic iron oxide nanoparticles a homogeneous distribution of the MP in the resulting meshes is achieved. Analysis of magnetic heating experiments yields a distinct decrease of the heating efficiency with the mean mesh size, as presented in Figure 1.9. This drop in heating efficiency as represented by the specific loss power SLP is attributed to a gradual immobilization of the MP in meshes of the hydrogel and a blocking of the Brownian particle relaxation mechanism. Hereby, it is shown that the Brownian relaxation mechanism contributes to approximately 35 % of the overall the particle heating power of this particle system at the given field conditions [52].

One major focus of the research interest regarding ferrogels is based on anisotropic shape changes for possible applications as actuators. For this purpose, microgels are promising candidates due to their low switching times with respect to stimuli induced changes of the swelling degree. Magnetoresponsive microgels are obtained by immobilizing MP in the polymer network. Oftentimes poly(*N*-isopropylacrylamide) is employed, which is known for its thermoresponsive properties manifested in the volume phase transition temperature at a physiologically relevant temperature of 32 °C, therefore generating a multiresponsive system with temperature as well as magnetic field as stimuli. Loading of these microgel beads with MP leads to magnetoresponsive materials, where an anisotropic volume shrinkage is found upon application of a magnetic field [53]. The microgel bead is elongated parallel to the magnetic field and contracts perpendicular the field, which is in line with theoretical calculations for a comparable system [54].

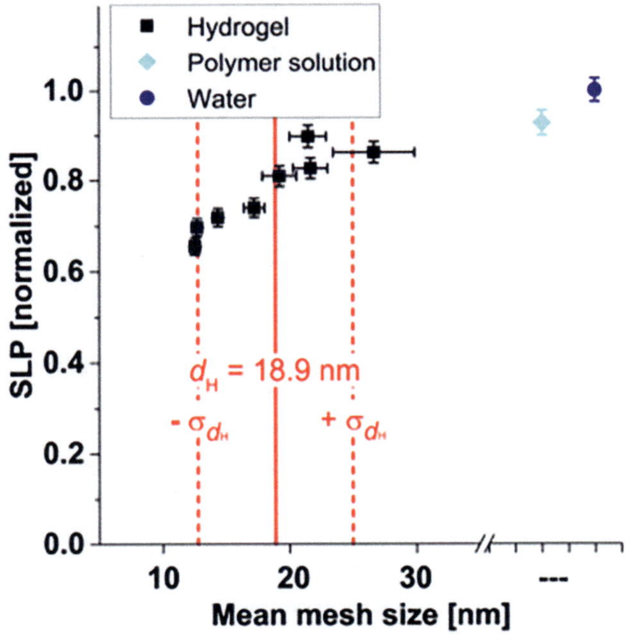

Figure 1.9: Normalized SLP values of MNP with a hydrodynamic diameter d_h of (18.9 ± 6.1) nm, as dispersed in water, in polymer solution, and gradually immobilized in hydrogels with decreasing mean mesh size. Adapted from reference [52] with permission from Elsevier.

1.3.2 Particle-crosslinked ferrogels

The architecture with the strongest interactions between MP and polymer chains is represented by particle-crosslinked ferrohydrogels. This type of stimuli responsive material gives rise to improved material properties due to a strong magneto-mechanical coupling and a novel response mechanism to an external magnetic field [55–60]. A simple effect of the covalent attachment is that in contact with swelling media no particle loss is observed due to diffusion, and the materials may therefore be used over many cycles even in contact with biological tissue [61–63]. If the network is solely crosslinked by the magnetic nanoparticles, strongly particle content depending properties are obtained [14, 49]. As particle component in addition to classic spherical particles also anisotropic particles can be used as shown in Figure 1.10, where the structure and homogeneity is confirmed by cryo-TEM investigations. Here, the particle-crosslinked ferrohydrogel is based on spindle-type hematite particles as sole cross-linkers of a PAAm network [64].

Analyzing the properties of particle-crosslinked ferrohydrogels, a strong decrease of the swelling degree upon increasing the volume fraction of particulate crosslinkers is found. The data obtained for the swelling degree are further used to quantify the crosslinker functionality of the particles and the average molar mass of polymer segments between the particles, as shown in Figure 1.11. By comparison with macro-rheological measurements and extraction of the same properties, we get to a more sophisticated image of the architecture of the solely particle-crosslinked ferrohydrogels, as shown in Figure 1.10b, where the network is built up by only a small fraction

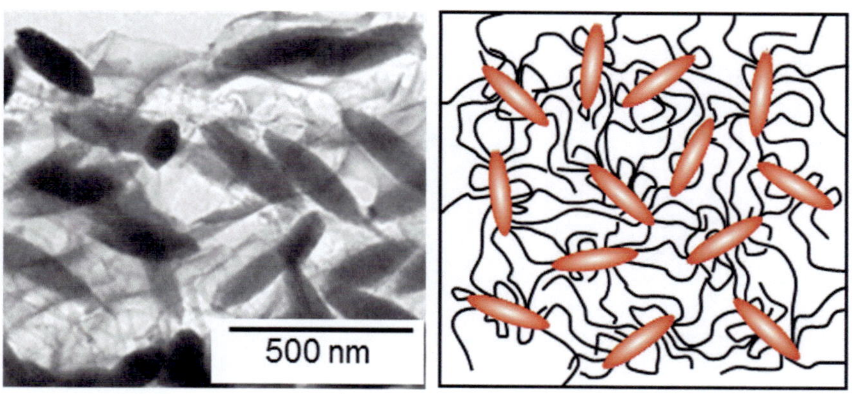

Figure 1.10: a) Cryo-TEM image of a particle-crosslinked ferrohydrogel, crosslinked with spindle-type hematite particles. b) Schematic representation of the network architecture of the particle-crosslinked ferrohydrogel. Adapted with permission from [64]. Copyright 2014 American Chemical Society.

of elastically active particle-linking polymer chains and a large number of loops and dangling ends [64].

When thermally blocked MP dispersed in a fluid medium are subjected to a homogeneous magnetic field, an orientation of the dispersed particles is observed. This anisotropic orientation can be immobilized by crosslinking the particles into the polymer backbone while applying a homogeneous magnetic field, leading to an angle-dependent magnetization behavior [65]. In this way, when employing spindle-type hematite particles as crosslinkers, a novel class of particle-crosslinked ferrohydrogels is prepared showing a permanent magnetic anisotropy as well as a geometric anisotropy since the magnetic properties of the spindle-type hematite particles are coupled to their geometric anisotropy. Due to the anisotropic magnetic properties, in magnetization measurements, a dependence of the magnetization curve on the angle between synthesis field and probe field is found, that is characterized in detail and presented in Figure 1.12. In these measurements two main components next to residual particle mobility can be found. In hard anisotropic ferrohydrogels, a strong dependence of the remanence magnetization on the respective angle is found and can be attributed to an immobilization of the particles in the elastic matrix with a preferred orientation of the magnetic moment, which is in agreement with a theoretical description by the Stoner–Wolfarth model. In soft anisotropic ferrohydrogels on the other hand a plastic deformation effect is found, leading to a significant deviation from the proposed Stoner–Wolfarth model and an attribution of the resulting magnetization behavior to an adaptive preferred orientation of the magnetization axis [66].

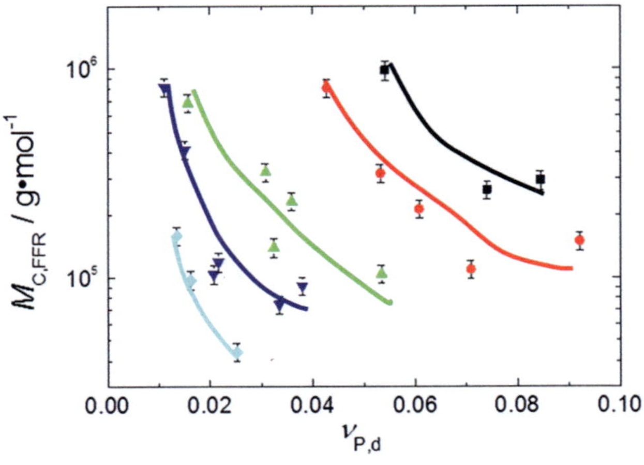

Figure 1.11: Average network segment length $M_{c,FFR}$ based on the Frenkel–Flory–Rehner model as a function of the particle volume fraction reprinted with permission from [64]. Copyright 2014 American Chemical Society.

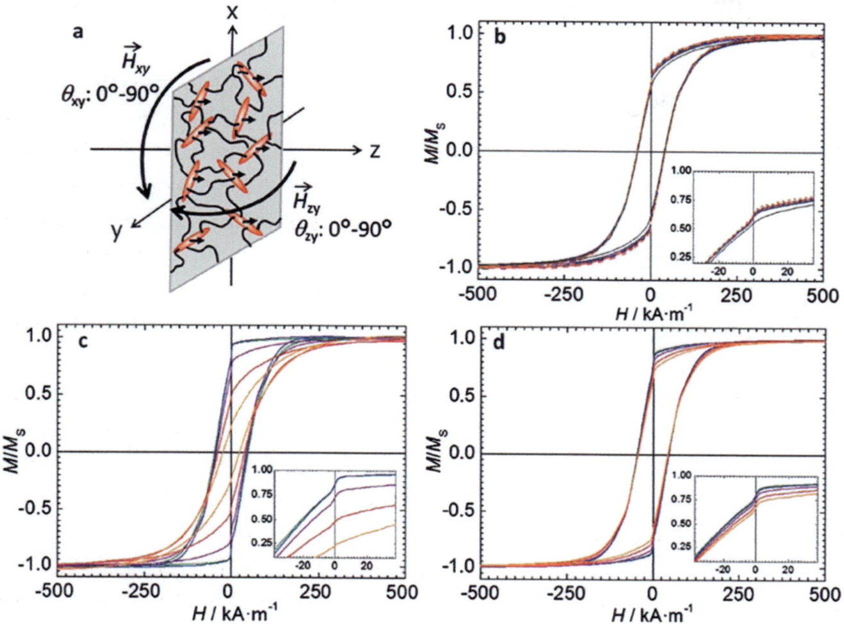

Figure 1.12: a) Schematic illustration of the setup geometry and axes annotations. Magnetization curves of b) the isotropic hard gel in comparison with the particles measured as powder (grey); c) the anisotropic hard gel; d) the anisotropic soft gel; solid line: *zy*-plane, dashed line: *xy*-plane; θ_{xy} or θ_{zy}: 0° (green), 22.5° (blue), 45° (purple), 67.5° (red), 90° (orange); reproduced by permission of the PCCP Owner Societies from [66].

1.3.3 Nanostructuring by self-assembly

Structuring materials on a nanoscale opens the possibility for various applications ranging from surface modifications to filler networks in polymeric composite materials. Here, self-assembly has proven to be a very effective, yet elegant way to generate ordered structures on a nanoscale.

The self-assembly of 12-hydroxyoctadecanoic acid (12-HOA) in organic media leads to the formation of network structures that are formed from fibrillous strands of 12-HOA and therefore exhibit an intrinsically anisotropic shape of the meshes due to the highly anisotropic fibers building them. The dispersion of a commercial ferrofluid inside the pregel allows the preparation of organo-ferrogels. By application of a magnetic field during the synthesis, an anisotropic distribution of the MP can be conserved in the gelation process, leading to anisotropic ferrogels. In comparsion to the isotropic case, a significant difference of the birefringence is found with higher values of the saturation birefringence observed for the anisotropic samples [67].

Another interesting aspect is the self-organization of magnetic nanoparticles. Nanostructured magnetic materials are of interest for applications for data storage or actuation, if the anisotropy can be coupled to the mechanical properties of a suitable matrix, as described above. Spindle-type hematite nanoparticles, here, are a scientifically interesting model system as they show a magnetic as well as geometric anisotropy, a tunable aspect ratio, a comparably low polydispersity, and present an unusual magnetic orientation in external magnetic fields with their long axis perpendicular to the field vector [68–70]. The well described orientation mechanism for these particles in a homogeneous magnetic field leads to an orientation, where the long particle axis is distributed in a plane perpendicular to magnetic field vector. A novel orientation mechanism for this particle system could be established by application of a rotating magnetic field. Here, a truly uniaxial orientation could be achieved, as shown depicted on TEM images in Figure 1.13. The long particle axis is perpendicular to the applied field vector for all cases, but the rotating nature of the applied field therefore induces a uniaxial orientation of the particle. Analyzing the samples with small angle X-ray scattering (SAXS), the successful orientation of the particles is shown for larger ensembles by reproduction of the observed scattering patterns with simulations [71].

1.4 Summary

The incorporation of magnetic nanoparticles into various structures is a versatile approach to find access to the rich world of magnetically responsive materials. Key to understanding and optimizing the composite systems is a detailed knowledge of the underlying particle–matrix interactions in such hybrid materials. These interactions are based on the surface chemistry of the particles and their coupling to the surrounding matrix, but also fundamentally depend on the relative size of the particles to characteristic sizes in the matrix, such as mesh size or correlation length. The latter dependence allows using MP as probes in order to extract frequency-dependent rheological properties from measurements of the AC-susceptibility in a simple experiment using the approach of MPN. Here, low concentrations of MP and small sample volumes are needed, making the non-destructive experiments an ideal candidate for the analysis of local micro- and nanostructures in polymer solutions with complex rheological behavior and biological fluids. In combination with elastic matrices, the mutual interaction of particle and matrix can be used to gain more insight in the coupling behavior, or information about the properties of the elastic and magnetic component. If the particles are trapped in the elastic matrix, magnetic fields can even be used to induce actuation by an anisotropic deformation of the elastic matrix. Direct anisotropic distribution of the magnetic particles inside elastic matrices is shown to generate different behavior compared to isotropic samples as well as angle-dependent properties. While all these investigations demonstrate the

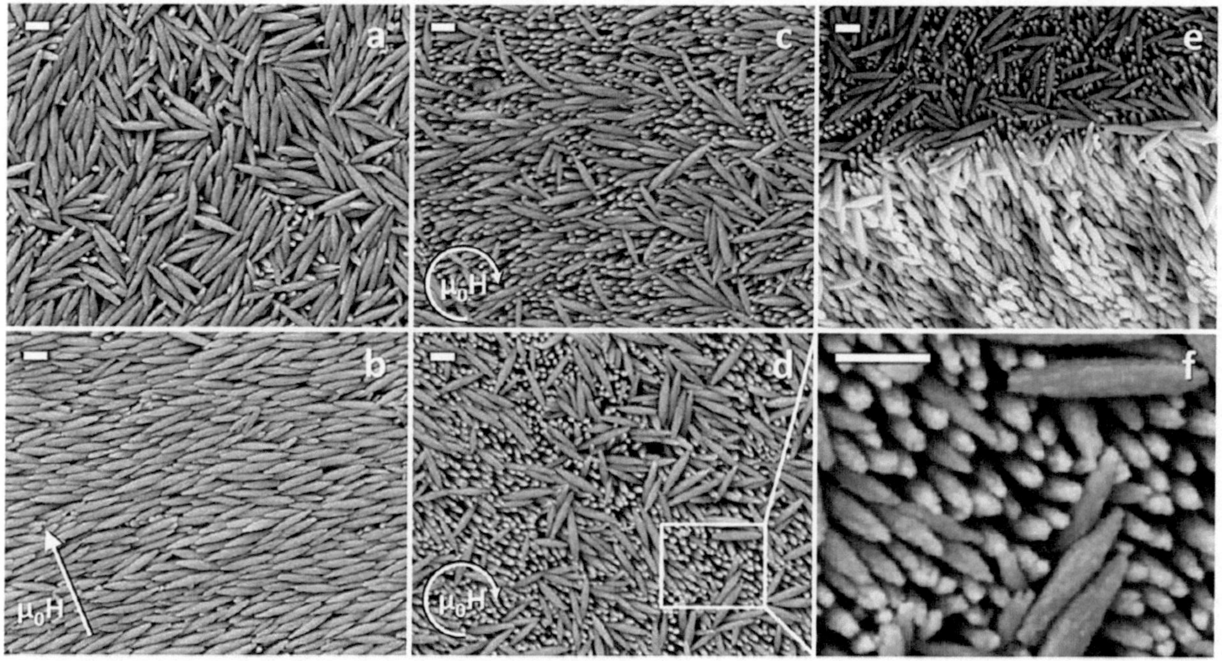

Figure 1.13: SEM images of hematite nanospindles assembled in a) absence of magnetic field; b) static field of 38 mT in the indicated direction; c) and d) rotating magnetic field of 200 Hz and 20 mT, along with e) side view of a breaking edge of the sample assembled in a rotating field and f) detailed view. Scale bars represent 200 nm. Reproduced by permission of The Royal Society of Chemistry from [71].

utility of magnetically responsive hybrid materials, many open questions still remain on the coupling of MP and complex matrices and on the applicability of these responsive materials.

Acknowledgments: Financial support is acknowledged from DFG-SPP 1681 "Feldgesteuerte Partikel-Matrix-Wechselwirkungen" (SCHM1747/10).
Author contribution: All the authors have accepted responsibility for the entire content of this submitted manuscript and approved submission.
Research funding: They study was financially supported from DFG-SPP 1681 "Feldgesteuerte Partikel-Matrix-Wechselwirkungen" (SCHM1747/10).
Conflict of interest statement: The authors declare no conflicts of interest regarding this article.

References

1. Schopphoven C, Birster K, Schweitzer R, Lux C, Huang S, Kästner M, et al. Elastic deformations in semi-dilute Ni nanorod/hydrogel composites. Arch Appl Mech 2019;89:119–32.
2. Sánchez PA, Gundermann T, Dobroserdova A, Kantorovich SS, Odenbach S. Importance of matrix inelastic deformations in the initial response of magnetic elastomers. Soft Matter 2018;14: 2170–83.
3. Appel I, Nádasi H, Reitz C, Sebastián N, Hahn H, Eremin A, et al. Doping of nematic cyanobiphenyl liquid crystals with mesogen-hybridized magnetic nanoparticles. Phys Chem Phys 2017;19: 12127–35.
4. Gräfe C, Weidner A, Lühe MVD, Bergemann C, Schacher FH, Clement JH, et al. Intentional formation of a protein corona on nanoparticles: serum concentration affects protein corona mass, surface charge, and nanoparticle-cell interaction. Int J Biochem Cell Biol 2016;75:196–202.
5. Odenbach S. Ferrofluids. In: Buschow KHJ, editor. Handbook of magnetic materials. Amsterdam: Elsevier; 2006, vol 16:1–534 pp.
6. Rosensweig RE. Ferrohydrodynamics. Mineola: Courier Corporation; 2013.
7. Odenbach S, Störk H. Shear dependence of field-induced contributions to the viscosity of magnetic fluids at low shear rates. J Magn Mater 1998;183:188–94.
8. Szabo D, Szeghy G, Zrínyi M. Shape transition of magnetic field sensitive polymer gels. Macromolecules 1998;31:6541–8.
9. Winger J, Schümann M, Kupka A, Odenbach S. Influence of the particle size on the magnetorheological effect of magnetorheological elastomers. J Magn Mater 2019;481:176–82.
10. Menzel AM. Tuned, driven, and active soft matter. Phys Rep 2015;554:1–45.
11. Gratz M, Tschöpe A. Optical transmission versus ac magnetization measurements for monitoring colloidal Ni nanorod rotational dynamics. J Phys D Appl Phys 2017;50:015001.
12. Hess M, Gratz M, Remmer H, Webers S, Landers J, Borin D, et al. Scale-dependent particle diffusivity and apparent viscosity in polymer solutions as probed by dynamic magnetic nanorheology. Soft Matter 2020;16:7562–75.
13. Schrittwieser S, Ludwig F, Dieckhoff J, Tschoepe A, Guenther A, Richter M, et al. Direct protein detection in the sample solution by monitoring rotational dynamics of nickel nanorods. Small 2014;10:407–11.

14. Messing R, Frickel N, Belkoura L, Strey R, Rahn H, Odenbach S, et al. Cobalt Ferrite nanoparticles as multifunctional cross-linkers in PAAm ferrohydrogels. Macromolecules 2011;44:2990–9.
15. Becker TI, Zimmermann K, Borin DY, Stepanov GV, Storozhenko PA. Dynamic response of a sensor element made of magnetic hybrid elastomer with controllable properties. J Magn Mater 2018;449: 77–82.
16. Monkman GJ, Sindersberger D, Diermeier A, Prem N. The magnetoactive electret. Smart Mater Struct 2017;26:075010.
17. Roeben E, Roeder L, Teusch S, Effertz M, Deiters UK, Schmidt AM. Magnetic particle nanorheology. Colloid Polym Sci 2014;292:2013–23.
18. Hess M, Roeben E, Rochels P, Zylla M, Webers S, Wende H, et al. Size effects on rotational particle diffusion in complex fluids as probed by magnetic particle nanorheology. Phys Chem Phys 2019; 21:26525–39.
19. Remmer H, Gratz M, Tschope A, Ludwig F. Magnetic field dependence of Ni nanorod Brownian relaxation. IEEE Trans Magn 2017;53:2015–18.
20. Frickel N, Messing R, Gelbrich T, Schmidt AM. Functional silanes as surface modifying primers for the preparation of highly stable and well-defined magnetic polymer hybrids. Langmuir 2010;26: 2839–46.
21. Behrens S, Appel I. Magnetic nanocomposites. Curr Opin Biotechnol 2016;39:89–96.
22. Wirtz D. Particle-tracking microrheology of living cells: principles and applications. Annu Rev Biophys 2009;38:301–26.
23. Pine DJ, Weitz DA, Chaikin PM, Herbolzheimer E. Diffusing wave spectroscopy. Phys Rev Lett 1988; 60:1134–7.
24. Rathgeber S, Beauvisage H-J, Chevreau H, Willenbacher N, Oelschlaeger C. Microrheology with fluorescence correlation spectroscopy. Langmuir 2009;25:6368–76.
25. Waigh TA. Microrheology of complex fluids. Rep Prog Phys 2005;68:685–742.
26. Crick FHC, Hughes AFW. The physical properties of cytoplasm. Exp Cell Res 1950;1:37–80.
27. Maldonado-Camargo L, Yang C, Rinaldi C. Scale-dependent rotational diffusion of nanoparticles in polymer solutions. Nanoscale 2017;9:12039–50.
28. Gratz M, Tschöpe A. Size effects in the oscillatory rotation dynamics of Ni nanorods in poly(ethylene oxide) solutions. Macromolecules 2019;52:6600–12.
29. Dieckhoff J, Eberbeck D, Schilling M, Ludwig F. Magnetic-field dependence of Brownian and Néel relaxation times. J Appl Phys 2016;119:043903.
30. Santiago-Quiñonez DI, Rinaldi C. Enhanced rheological properties of dilute suspensions of magnetic nanoparticles in a concentrated amphiphilic surfactant solution. Soft Matter 2012;8: 5327–33.
31. Debye P. Polar molecules. New York: Chemical Catalog Company; 1929.
32. Ludwig F, Guillaume A, Schilling M, Frickel N, Schmidt AM. Determination of core and hydrodynamic size distributions of $CoFe_2O_4$ nanoparticle suspensions using ac susceptibility measurements. J Appl Phys 2010;108:033918.
33. Feyen M, Heim E, Ludwig F, Schmidt AM. Magnetic nanorotors with tailored field-induced dynamics. Chem Mater 2008;20:2942–8.
34. Rubinstein M, Colby RH. Polymer physics. New York: OUP Oxford; 2003.
35. Hermes M, Roeben E, Kibkalo L, Schmidt AM. Magnetic particle nanorheology in complex fluids. Ann Trans Nord Rheol Soc 2017;25:97–105.
36. Monkenbusch M, Krutyeva M, Pyckhout-Hintzen W, Antonius W, Hövelmann CH, Allgaier J, et al. Molecular view on supramolecular chain and association dynamics. Phys Rev Lett 2016;117: 147802.

37. Gold BJ, Hövelmann CH, Lühmann N, Pyckhout-Hintzen W, Wischnewski A, Richter D. The microscopic origin of the rheology in supramolecular entangled polymer networks. J Rheol 2017; 61:1211–26.
38. Hess M, Roeben E, Habicht A, Seiffert S, Schmidt AM. Local dynamics in supramolecular polymer networks probed by magnetic particle nanorheology. Soft Matter 2019;15:842–50.
39. Treloar LRG. The Physics of rubber elasticity. Oxford: Oxford University Press; 2005.
40. Remmer H, Roeben E, Schmidt AM, Schilling M, Ludwig F. Dynamics of magnetic nanoparticles in viscoelastic media. J Magn Mater 2017;427:331–5.
41. Remmer H, Dieckhoff J, Tschöpe A, Roeben E, Schmidt AM, Ludwig F. Dynamics of $CoFe_2O_4$ single-core nanoparticles in viscoelastic media. Phys. Procedia 2015;75:1150–7.
42. Webers S, Hess M, Landers J, Schmidt AM, Wende H. Effect of phase transitions in polymer solutions on the magnetic response of embedded nanoparticles. ACS Appl Polym Mater 2020;2: 2676.
43. Zubarev AY, Musikhin AY, Iskakova LY, Bulytcheva SV. Shear modulus of isotropic ferrogels. J Magn Mater 2019;477:136–41.
44. Zhao X, Kim J, Cezar CA, Huebsch N, Lee K, Bouhadir K, et al. Active scaffolds for on-demand drug and cell delivery. Proc Natl Acad Sci Unit States Am 2011;108:67–72.
45. Weeber R, Hermes M, Schmidt AM, Holm C. Polymer architecture of magnetic gels: a review. J Phys Condens Matter 2018;30:063002.
46. Zrínyi M, Barsi L, Büki A. Ferrogel: a new magneto-controlled elastic medium. Polym Gels Netw 1997;5:415–27.
47. Van Berkum S, Dee JT, Philipse AP, Erné BH. Frequency-dependent magnetic susceptibility of magnetite and cobalt ferrite nanoparticles embedded in PAA hydrogel. Int J Mol Sci 2013;14: 10162–77.
48. Roeder L, Bender P, Tschöpe A, Birringer R, Schmidt AM. Shear modulus determination in model hydrogels by means of elongated magnetic nanoprobes. J Polym Sci, Part B: Polym Phys 2012;50: 1772–81.
49. Frickel N, Messing R, Schmidt AM. Magneto-mechanical coupling in $CoFe_2O_4$-linked PAAm ferrohydrogels. J Mater Chem 2011;21:8466.
50. Kickelbick G. Hybrid materials. Weinheim: Wiley; 2006.
51. Landers J, Roeder L, Salamon S, Schmidt AM, Wende H. Particle-matrix interaction in cross-linked PAAm-hydrogels analyzed by mössbauer spectroscopy. J Phys Chem C 2015;119:20642–8.
52. Engelmann UM, Seifert J, Mues B, Roitsch S, Ménager C, Schmidt AM, et al. Heating efficiency of magnetic nanoparticles decreases with gradual immobilization in hydrogels. J Magn Mater 2019; 471:486–94.
53. Backes S, Witt MU, Roeben E, Kuhrts L, Aleed S, Schmidt AM, et al. Loading of PNIPAM based microgels with $CoFe_2O_4$ nanoparticles and their magnetic response in bulk and at surfaces. J Phys Chem B 2015;119:12129–37.
54. Fischer L, Menzel AM. Magnetostriction in magnetic gels and elastomers as a function of the internal structure and particle distribution. J Chem Phys 2019;151:114906.
55. Annunziata MA, Menzel AM, Löwen H. Hardening transition in a one-dimensional model for ferrogels. J Chem Phys 2013;138:204906.
56. Ilg P. Stimuli-responsive hydrogels cross-linked by magnetic nanoparticles. Soft Matter 2013;9: 3465–8.
57. Weeber R, Kantorovich S, Holm C. Ferrogels cross-linked by magnetic nanoparticles-deformation mechanisms in two and three dimensions studied by means of computer simulations. J Magn Mater 2015;383:262–6.
58. Weeber R, Kantorovich S, Holm C. Ferrogels cross-linked by magnetic particles: field-driven deformation and elasticity studied using computer simulations. J Chem Phys 2015;143:154901.

59. Weeber R, Kantorovich S, Holm C. Deformation mechanisms in 2D magnetic gels studied by computer simulations. Soft Matter 2012;8:9923–32.
60. Weeber R, Kreissl P, Holm C. Studying the field-controlled change of shape and elasticity of magnetic gels using particle-based simulations. Arch Appl Mech 2019;89:3–16.
61. Fuhrer R, Athanassiou EK, Luechinger NA, Stark WJ. Crosslinking metal nanoparticles into the polymer backbone of hydrogels enables preparation of soft, magnetic field-driven actuators with muscle-like flexibility. Small 2009;5:383–8.
62. Bonini M, Lenz S, Giorgi R, Baglioni P. Nanomagnetic sponges for the cleaning of works of art. Langmuir 2007;23:8681–5.
63. Barbucci R, Pasqui D, Giani G, De Cagna M, Fini M, Giardino R, et al. A novel strategy for engineering hydrogels with ferromagnetic nanoparticles as crosslinkers of the polymer chains. Potential applications as a targeted drug delivery system. Soft Matter 2011;7:5558.
64. Roeder L, Reckenthäler M, Belkoura L, Roitsch S, Strey R, Schmidt AM. Covalent ferrohydrogels based on elongated particulate cross-linkers. Macromolecules 2014;47:7200–7.
65. Bender P, Günther A, Tschöpe A, Birringer R. Synthesis and characterization of uniaxial ferrogels with Ni nanorods as magnetic phase. J Magn Mater 2011;323:2055–63.
66. Roeder L, Bender P, Kundt M, Tschöpe A, Schmidt AM. Magnetic and geometric anisotropy in particle-crosslinked ferrohydrogels. Phys Chem Phys 2015;17:1290–8.
67. Nádasi H, Corradi Á, Stannarius R, Koch K, Schmidt AM, Aya S, et al. The role of structural anisotropy in the magnetooptical response of an organoferrogel with mobile magnetic nanoparticles. Soft Matter 2019;15:3788–95.
68. Ozaki M, Kratohvil S, Matijević E. Formation of monodispersed spindle-type hematite particles. J Colloid Interface Sci 1984;106:146–51.
69. Nack A, Seifert J, Passow C, Wagner J. Hindered nematic alignment of hematite spindles in poly(N-isopropylacrylamide) hydrogels: a small-angle X-ray scattering and rheology study. J Appl Crystallogr 2018;51:87–96.
70. Reufer M, Dietsch H, Gasser U, Grobety B, Hirt AM, Malik VK, et al. Magnetic properties of silica coated spindle-type hematite particles. J Phys Condens Matter 2011;23:065102.
71. Hoffelner D, Kundt M, Schmidt AM, Kentzinger E, Bender P, Disch S. Directing the orientational alignment of anisotropic magnetic nanoparticles using dynamic magnetic fields. Faraday Discuss 2015;181:449–61.

Gennady V. Stepanov*, Dmitry Yu. Borin, Anton V. Bakhtiiarov and Pavel A. Storozhenko

2 Hybrid magnetic elastomers prepared on the basis of a SIEL-grade resin and their magnetic and rheological properties

Abstract: Hybrid magnetic elastomers (HMEs) belong to a novel type of magneto-controllable elastic materials capable of demonstrating extensive variations of their parameters under the influence of magnetic fields. Like all cognate materials, HMEs are based on deformable polymer filled with a mixed or modified powder. The complex of properties possessed by the composite is a reflection of interactions occurring between the polymer matrix and the particles also participating in interactions among themselves. For example, introduction of magnetically hard components into the formula results in the origination of a number of significantly different behavioral features entirely unknown to magnetorheological composites of the classic type. Optical observation of samples based on magnetically hard filler gave the opportunity to establish that initial magnetization imparts magnetic moments to initially unmagnetized grains, as a result of which chain-like structures continue to be a feature of the material even after external field removal. In addition, applying a reverse field causes them to turn into the polymer as they rearrange into new ring-like structures. Exploration of the relationship between the rheological properties and magnetic field conducted on a rheometer using vibrational mechanical analysis showed an increase of the relative elastic modulus by more than two orders of magnitude or by 3.8 MPa, whereas the loss factor exhibited steady growth with the field up to a value of 0.7 being significantly higher than that demonstrated by elastomers with no magnetically hard particles. At the same time, measuring the electroconductivity of elastomers filled with a nickel-electroplated carbonyl iron powder made it possible to observe that such composites demonstrated an increase of variation of the resistivity of the composite influenced by magnetic field in comparison to elastomers containing untreated iron particles. The studies conducted indicate that this material exhibits both magnetorheological and

***Corresponding author: Gennady V. Stepanov,** State Scientific Research Institute for Chemical Technologies of Organoelement Compounds, 105118 Moscow, Russia, E-mail: gstepanov@mail.ru. https://orcid.org/0000-0003-0053-1883
Dmitry Yu. Borin, Institute of Mechatronic Engineering, Technische Universität Dresden, 01062 Dresden, Germany. https://orcid.org/0000-0003-3842-1487
Anton V. Bakhtiiarov and Pavel A. Storozhenko, State Scientific Research Institute for Chemical Technologies of Organoelement Compounds, 105118 Moscow, Russia. https://orcid.org/0000-0002-2258-5850 (A.V. Bakhtiiarov)

This article has previously been published in the journal Physical Sciences Reviews. Please cite as: G. V. Stepanov, D. Y. Borin, A. V. Bakhtiiarov and P. A. Storozhenko "Hybrid magnetic elastomers prepared on the basis of a SIEL-grade resin and their magnetic and rheological properties" *Physical Sciences Reviews* [Online] 2020, 5. DOI: 10.1515/psr-2020-0008 | https://doi.org/10.1515/9783110569636-002

magnetoresistive effect and does indeed have the potential for use in various types of devices.

Keywords: hybrid, magnetoactive, magnetoresistivity, magnetorheological, ring-like structures

2.1 Introduction

2.1.1 Background

This article is predominantly dedicated to a selective consideration of the magnetic and elastic features of hybrid magnetic composites. At the same time, in light of the fact that to a significant degree the "smart" status of the material is determined by the quality of the polymer to be flexible under stress of physical force, it is of certain interest to discuss its capability to conduct electric current depending on external magnetic field, also known as magnetoresistivity. Unlike conventional electroconductive substances, the family of composites being studied exhibits the capability of varying resistivity within a range of several orders of magnitude when influenced by moderate magnetic fields. Affected by a field, the particles tend to change their positions with respect to the neighbors to form structures more or less ordered into chains capable of conducting electric current. Owing to the fact that the intensity of this process depends on the possibility of every particle to make sufficient moves, this brings up the question of whether the polymer matrix can afford such deformations, which directly refers to its elastic properties.

As has become apparent from the experimental results obtained over the last year, these materials possess a significantly more extended set of specific features in comparison to classic magnetorheological elastomers (MREs). In view of the difficulties of studying the aforementioned properties in hybrid elastomers containing magnetically hard and soft components simultaneously, it seemed wise to investigate the magnetoresistivity phenomenon and magnetic/elastic features in pure form separately, which assumed preparation of samples solely filled with particles of either type.

A material based on magnetically hard filler is expected to exhibit good damping properties without the necessity to supply the overall unit with a bulky system of permanent magnets or electromagnets. Despite the fact that such a device may not be controlled externally, which will imminently result in a lower effectiveness, its simplicity will be a compensation leading to extensive application. As was expected initially, substances of this kind would possess a remnant magnetization and thus an internal magnetic field resulting in enhanced damping performance. In addition, it was established that under the influence of external magnetic fields, they exhibited an asymmetry of their elastic features determined by the direction of initial magnetizing [1–4]. All in all, these results found confirmation in Ref. [5].

Another distinctive quality of the new materials is that their loss-factor magnitudes were found to be noticeably higher in comparison to those of other MREs [6]. The new elastomer was assumed to show a strong magnetostriction, positive or negative depending on the directions of the initial magnetizing and external magnetic field vectors, which, however, had no experimental validation. As was established in later investigations, the phenomenon of rotation of magnetically hard particles embedded in the polymer as the external magnetic field changes direction was the cause. Owing to the fact that the samples were prepared on the basis of sufficiently soft silicone matrices, the elastic properties of the polymer made it possible for the rotation to occur in weak magnetic fields. This explains the fact that the samples practically always demonstrated codirectionality between the remanence vector and the external field independently of its direction. The interpretation of the rotation mechanism is brought in Refs. [7–10].

Studying the magnetorheological properties of hybrid magnetic elastomer (HME) on an oscillating rheometer in the plane-plane system has been found to be challenging. The difficulties are connected to the fact that the sample suffers significant deformation when subjected to magnetizing. The edges of a cylinder-shaped specimen wind strongly and roll up turning the cylinder into an ellipsoid. Work with such materials requires a special philosophy of measurement.

Soft matrices came useful in magnetoresistivity phenomenon studies. On the basis of the comparative method of investigation, the polymer was filled with untreated carbonyl iron particles and those subjected to electroplating with nickel. In light of the fact that nickel is more electroconductive and corrosion-proof than pure iron, samples with the nickel-electroplated magnetic filler were expected to show more extensive field-induced resistivity variations as compared to those containing no nickel. Measured at moderate magnetic fields using alternating current, the samples with nickel-plated particles did demonstrate the effect. However, as was found out later, the intensity of the phenomenon was not as strong as assumed initially. A possible reason for this is that galvanic treatment causes some agglomeration among the particles resulting in the formation of elongated grains. Embedded in polymer, such formations may experience difficulties moving, in light of which the elastic features of the matrix remain crucial.

2.1.2 Elastomeric materials

Initial attempts to prepare polymer-based magnetic composite materials were made using various elastic substances. Among them were natural rubber [11, 12] and aqueous gels prepared on the basis of polyvinyl alcohol and glutaraldehyde [13–16]. In his research, T. Mitsumata used N-isopropylacrylamide and carrageenan [17–20]. Meanwhile, interest expressed toward natural and synthetic rubbers has been significantly more intense. In particular, M. Lokander has produced magnetorheological elastomers on the basis of natural and nitrile rubbers containing acrylonitrile and polybutadiene-rubber additives [21–23]. The most frequent use, at the same time, has been made of polyurethane and silicone resins. The advantages of two-can urethane compositions originate from the possibility to vary product elasticity in quite wide ranges and their higher durability in comparison to other elastomers having close elastic properties. Thus, 15 papers published by T. Mitsumata et al. have been dedicated to polyurethane elastomers as the basic material in research [14, 24]. Composites with polyurethane matrices have also been the object of study in the work conducted by A. Boczkowska [25–27].

Despite that, silicone elastomer remains a material most extensively employed in scientific investigations. In total, 70 to 80% of studies dedicated to magnetorheological and magnetoactive elastomers have been done using silicones. Like urethanes, their elastic properties may easily be preset in ranges even more extensive than those of the former. The second advantage is the simplicity of control of polymerization rate performed by temperature variation. Synthesis of a good sample of magnetic elastomer requires a degassing procedure following mixture preparation and followed by thermal treatment. This technology offers an easy possibility to obtain products with the desired mechanical features giving silicone resins advantage over urethanes. Indeed, in comparison to silicone, the rate of polymerization of urethane elastomer depends on the micro-amounts of catalyst introduced into the mixture being prepared. After that, there is only a limited time for the conduction of polymerization of the overall composition including inconvenient and time-consuming vacuuming, which requires certain knowledge and skills. At the same time, silicone is a commercial product and control of its elasticity may be performed even by a researcher representing an area other than polymer chemistry. Thus, according to the literature, polyaddition silicones are the polymers most frequently used by scientists. Unlike cold-cure compositions, their polymerization may quickly be initiated by simple heating. Their employment allows the researcher to spend as much time as necessary on the preparation of a sample with the desired formula and then quickly turn the semiproduct into polymer by heating. Bulky specimens may be easily prepared by cooking in the microwave oven [7]. Silicone elastomers of this type have been developed at Dow Corning (US Patent 3697473 (1973), US Patent 4322320, US Patent 4340709 (1983)) and Russian StateResearch Institute for Chemical Technologies of Organoelement Compounds (SIEL-grade resin, SU Patent 564315 (1975), SU Patent 639267 (1978), SU Patent 1086787 (1983)).

Most frequently, our samples have been fabricated on the basis of a SIEL-grade resin polymerizing according to the polyaddition mechanism; the semifabricate is a product of Russian State Research Institute for Chemical Technologies of Organoelement Compounds.

2.2 Synthesis and measurements

2.2.1 Synthesis of magnetic elastomers based on SIEL-grade silicone resin

Owing to streamlined synthesis, a tailored elastomer is formed. The polyaddition reaction, whose path is shown schematically further, features one of the basic polymerization mechanisms (Figure 2.1.1):

As may be seen from the diagram, no secondary output is formed as a result of the polymerization, which allows to utilize such compositions not only as filmy or coating ones, but also as potting compounds to be used in enclosed volumes of every shape. Mold-cured silicone-based materials may have a range of physical states from gel-like or elastomeric to glass-like. The advantage of these elastomers is in the wide operation temperature interval extending from −90 to +350 °C. Depending on the task specifics, silicone-based polyaddition compounds may solidify at moderate temperatures from room temperature to 160 °C and over various time periods ranging from one day to 30 min; these materials are capable of polymerizing with the formation of layers of different thicknesses. Owing to their good capability to cling to such surfaces as metal, glass, or ceramics, these elastomers demonstrate high adhesion to NdFeB or Fe particles coated with hydride-containing silicones.

The surface-modifying process is based on mixing a modifying agent and magnetic powders in toluene, during which the modifying agent precipitates on the surfaces of the grains forming a coating. Among other compounds, oligo-(methyl (or ethyl) hydrogen)siloxanes $[RSiHO]_n$ ($R = CH_3$, C_2H_5; $n = 10–15$) are considered to be most effective. In a mixture with silicone oil, these silicones may be used for the creation of more bulky coatings intended to provide regularly spaced distribution of particles inside the silicone resin.

The hydrophobization mechanism of magnetic powders is based on the chemical interaction of their surface hydroxyl-groups, forming as a result of reacting with adsorbed water, with Si–H-bonds of the modifying agent leading to the addition of silicone groups to the surface of a particle and formation of water molecules (Figure 2.1.2):

According to this principle, all hydroxyl groups become substituted with siloxane groups, thus imparting wettability toward silicones to the surface of the magnetic particle.

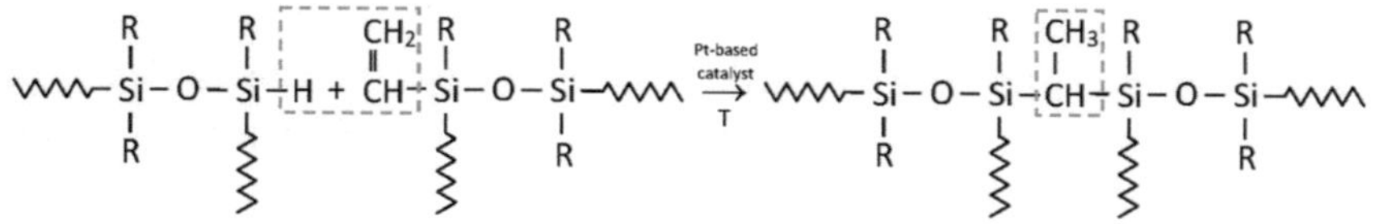

Figure 2.1.1: Polyaddition reaction; R being CH_3, C_2H_5, $CH_2CH_2CF_3$, or C_6H_5.

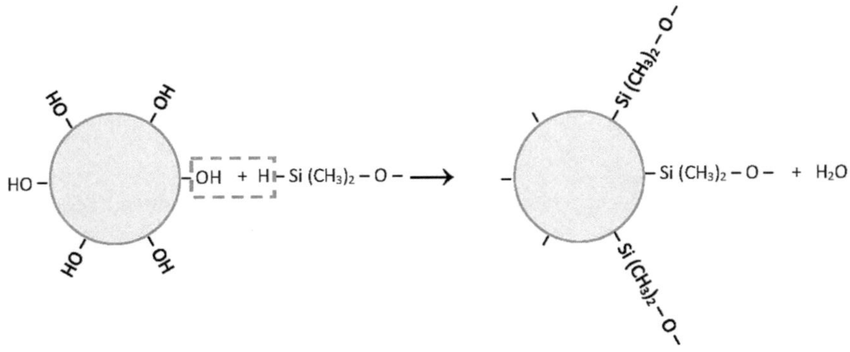

Figure 2.1.2: Interaction of hydroxyl groups with Si–H-bonds.

A study of thermal aging processes occurring in the polymers has revealed their heat stability retained over a long period of time, during which such basic parameters of elastomeric compositions as durability and relative strain continue to meet an adequate standard.

In view of the aforementioned properties, silicone elastomer is considered a promising material to be used as elastic matrix in magnetic elastomers fabrication.

2.2.2 Magnetic properties of HME

The key feature among the magnetic properties of a composite containing both magnetically hard and soft components is the newly discovered magnetization curve demonstrating an overall resemblance with but having a more complicated shape than that of a pure high-coercivity powder. Whereas an enhanced magnetic susceptibility peak, whose occurrence is determined by the presence of the magnetically soft component, most frequently being carbonyl iron, and another peak originating due to the availability of the magnetically hard component are observed in the areas of zero-field and filling-material demagnetizing force, respectively, increase of the magnetization loop amplitude or fabrication of samples based on softer matrices first results in the dying down of the "coercivity" peak followed by its shifting into the zero-field vicinity. At this, a composite filled with magnetically hard particles demonstrates the behavior rather characteristic of magnetically soft substances, owing to which most investigations directed at the detection and identification of processes taking place inside this magnetic elastomer were carried out on specimens prepared with no magnetically soft component. Therefore, all observations were limited to the phenomena demonstrated by magnetically hard particles.

In previous studies, attention was paid to the behavioral features of coercivity in HME, which took very low magnitudes [7–9] and might even be shifted into the area of fields of opposite polarity [10]. These effects are determined by the rotation of magnetically hard particles embedded in the polymer matrix. Magnetized, they start, at one moment, exhibiting the tendency to restore codirectionality with the external field after it changes its direction to opposite, rather that suffer interior polarity reversal themselves. A description of this effect is brought in our previous publications [7–10].

At the same time, the remnant magnetization demonstrated by the samples also tends to slide from high magnitudes, characteristic of magnetically hard filler, down to zero in intensity. It was supposed that whereas the cause of the phenomenon of zero magnetic flux on the surface of a magnetized specimen, which assumably had to demonstrate quite to the contrary, was not apparent initially, the explanation of this effect is based on the specifics of structuring of the magnetically hard filler inside the polymer. When the external magnetic field weakens to 0, the magnetized particles show the tendency to form ring-like structures to confine the flux lines. Depending on the concentration of magnetic particles and their vicinities, the rings may contain various numbers of members starting from three (Figure 2.2.1). It should be noted that this assembling into circular formations is energetically favorable. The possibility of their existence had been considered for magnetically soft particles dispersed in liquid by S. Kantorovich in her theoretical work [28, 29] and by D. Borin and A. Zubarev in Ref. [30]. In a special experiment these assumptions were confirmed.

Typical magnetization curves recorded for two samples prepared on the basis of matrices with different rigidities (Figure 2.2.2) are presented further. Containing magnetically hard filler of the same type at equal concentrations of 25 vol%, the specimens were obtained with shear moduli of 140 and 210 kPa, as was determined using a Thermo Scientific Haake MARS III rheometer. The essential behavioral dissimilarities they demonstrate are obvious: as may be noticed, whereas the rigid elastomer produces a wide hysteresis loop, that of the soft specimen is significantly narrower. In addition, the ascending curve pierces the zero-field area very close to the origin. It may, in particular, be seen from Figure 2.2.2 (b), which depicts the central section of the magnetization hysteresis loop recorded for a soft HME sample ($G' = 140$ kPa) containing plate-like NdFeB particles.

Meanwhile, a more profound comprehension of the mechanisms of the processes occurring inside the composite under the influence of different magnetic fields may be

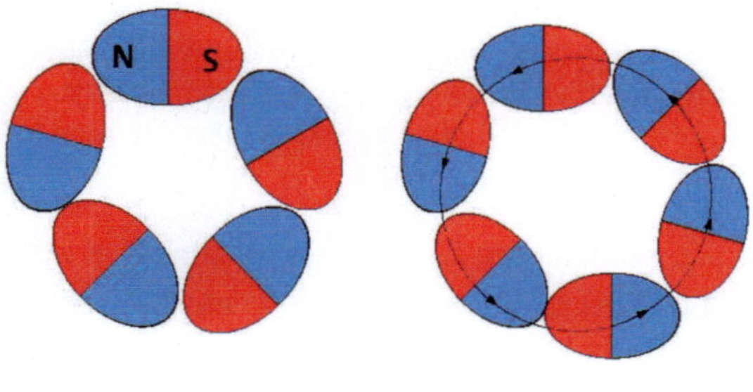

Figure 2.2.1: Ring-like structures formation phenomenon.

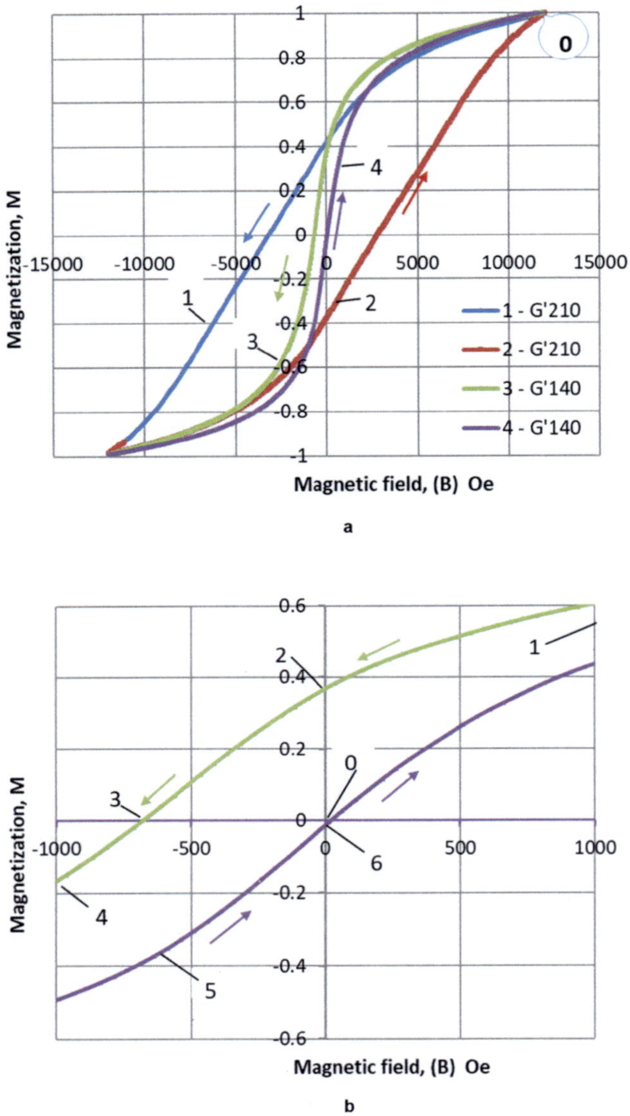

Figure 2.2.2: a) Magnetization hysteresis loops recorded for two magnetic elastomers filled with NdFeB particles and based on a rigid (G′ = 210 kPa, lines 1 and 2) and soft (G′ = 140 kPa; lines 3 and 4) matrix; b) central section of the magnetization hysteresis loop corresponding to the most significant structural rearrangement occurring inside a soft HME sample. The path of magnetization variation with magnetic field is shown by the arrows. The initial magnetization state corresponds to point 0. The numbered points correspond to the structures shown in Figure 2.2.3.

obtained by means of a qualitative experiment. Figure 2.2.3 (a–f) present photographic images of different arrangements of particles contained in a monolayer of the soft magnetic elastomer placed in different external conditions.

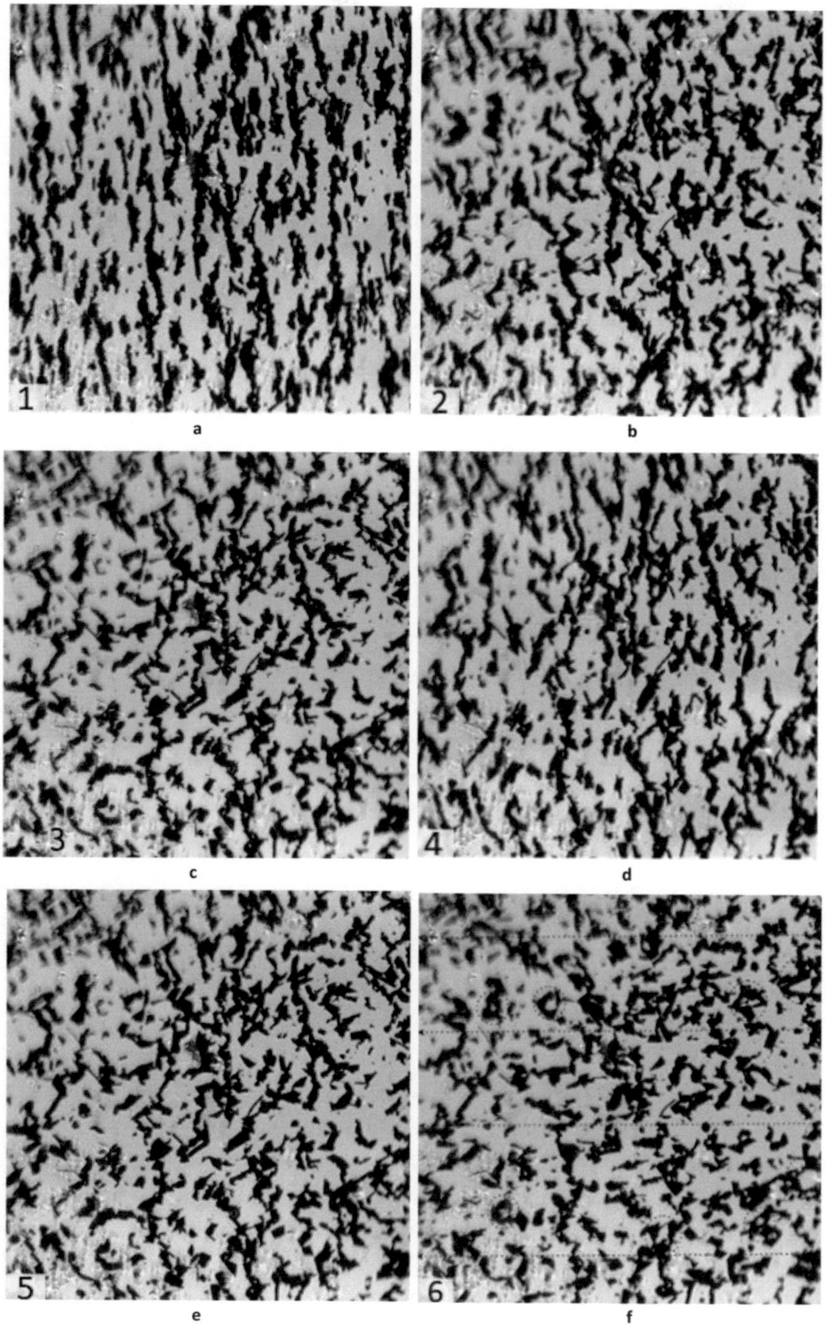

Figure 2.2.3: Structures formed by magnetized YMM-Q-grade NdFeB-alloy particles: a) in a magnetic field of 9000 Oe (point 1); b) at zero field reduced down to from 9000 Oe (point 2); c) in a reverse field of 800 Oe (point 3); d) in a reverse field of 1550 Oe (point 4); e) in a reverse field of 800 Oe reduced down to from a reverse field of 1550 Oe (point 5); f) at zero field reduced down to from a nonzero reverse field (point 6). Ring-like structures and the preferred orientation direction of the particles at zero field are denoted by dashed lines.

2.2.2.1 Plate-like particles

The experiments, the results of which are presented further, were carried out with a monolayer specimen based on a soft matrix filled with plate-like YMM-Q-grade NdFeB-alloy particles with a coercivity of 250 mT. Before being studied in magnetic fields of different intensities and polarities, a nonmagnetic sample was subjected to magnetizing at 1.7 T, by which an internal structure was created. The images presented in Figure 2.2.3 (a–f) show the specifics of structural changes determined by the exterior influence complemented by the elastic forces of the polymer. The ordinal numbers of the images correspond to the numbered points on the magnetization curves presented in Figure 2.2.2.

Exhibiting a pronounced orientation in a field codirectional with initial magnetization (a), the chains partly lose ordering at zero field as a result of the tendency of the elastic forces of the polymer to destroy them (b). As the reverse magnetic field becomes stronger, the chains formed previously continue to break. Showing little change in weak reverse magnetic fields (c), the particles start turning over in growing field to restore codirectionality with its force lines becoming building material for a structure forming anew. Finally, exhaust of the potential for turning is observed, indicative of that all the particles are now engaged. On further increasing the reverse magnetic field, the chains demonstrate some elongation and straightening along the force lines (d). On the other hand, decreasing the reverse magnetic field causes destruction of the chains by the elastic forces of the polymer matrix (e). As the reverse field weakens to 0, ring-like structures are formed, whereas the particles themselves and the previously built chains demonstrate a tendency to gain an orientation other than being colinear with the vector of the previously applied field (f). Thus, remnant magnetization is the factor preventing the destruction of the chains, which results in the formation of intermediate arrays. The polymer matrix remains strained inside owing to the deformations caused by the interacting magnetic particles it contains. The most illustrative video depicting structural rearrangements inside an initially magnetized sample being placed in a reverse magnetic field is available at Ref. [31].

2.2.3 Measuring technique of the rheological properties of HME

The rheological properties of NdFeB-based HME samples were studied on a Thermo Scientific HAAKE MARS III rheometer supplied with a magnetic measuring cell (Figure 2.3.1 (a)) consisting of a coil and a yoke. The cell is capable of generating magnetic fields ranging from 0 to 500 mT and was calibrated by means of measuring the field inside it in the area where the specimen was supposed to be placed later, as a function of the electric current feeding the coil. It should be noted, however, that estimated at a level of 95% in the space immediately adjacent to the upper face of the stem of the yoke, the homogeneity of the magnetic field decreases with distance along the vertical axis dropping to 90% at 2 mm away from the surface. Therefore, whereas the greater heights of the specimen can provide a more pronounced rheological response, one always has to remember that such tests are limited by the necessity to strike a happy medium between the size of the sample and field uniformity.

As follows from Figure 2.3.1 (b), the specimen in the initial (I) state before being subjected to the magnetizing procedure bears against the rotor and base-plate surfaces

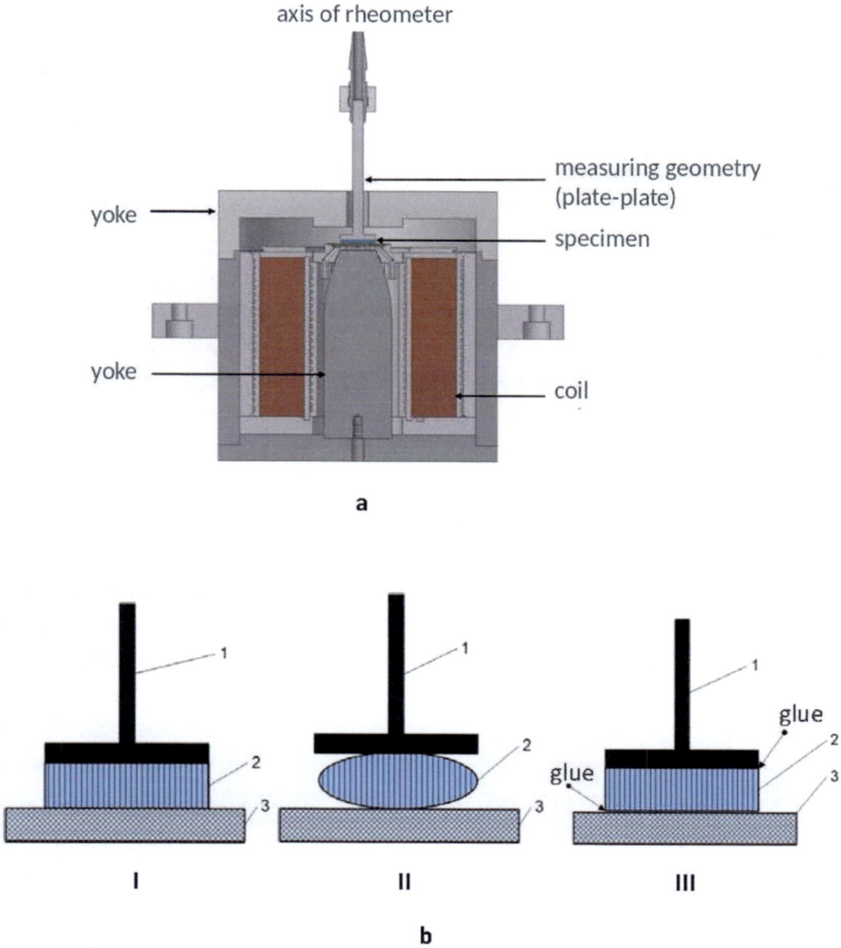

Figure 2.3.1: a) Magnetic measuring cell, a cross-sectional view (the device was designed by the Chair of Magnetofluiddynamics, Measuring and Automation Technology, TU Dresden[1]); b) HME sample in the rheometer: (I) before being magnetized; (II) in a magnetized state; and (III) glued to the surfaces of the rotor and base plate and magnetized. 1, 2, and 3 are the rotor, sample, and base, respectively.

1 The authors thank Prof. Stefan Odenbach for providing laboratory facilities of the Chair of Magnetofluiddynamics, Measuring and Automation Technology at TU Dresden.

quite fully. However, after getting magnetized in the vertical direction (II), the elastomer shows a significant remanence, which results in strong attraction forces between the opposite poles of the sample causing its distortion into an ellipsoid. After that, even the application of strong compression forces along the vertical axis cannot restore the fullness of contact between the sample and the upper and lower surfaces. It is important to note that the standard technique of measurement using rheometers assumes sample compression by means of two planes from above and from below. At this, the rheological features of the material change significantly making all attempts to compare samples from different research groups meaningless, because of unacceptably high variations of compression from sample to sample and its relationship with the elasticity of a concrete specimen.

Gluing the sample to the surfaces of the rotor and base plate (III) makes it possible to minimize this irregularity and by that leave the sample undistorted, as a result of which measurement begins practically at zero compression. Caused by zero specimen sliding against the surfaces, the experimental indeterminacy is thus brought to a negligible level. In view of this, all the rheological measurements carried out within the frames of this project were performed on cylinder-shaped samples being 14 and 4 mm in diameter and height, respectively, and glued in according to the aforementioned method.

2.2.4 Electric conductivity improvement by means of electroplating iron powders with nickel

Electroplating of carbonyl iron particles was carried out according to the technology described in Ref. [32]. The procedure was conducted in a plastic container in an ethylene-glycol-based electrolyte composed by dissolving commercial nickel chloride in the solvent at concentrations ranging from 60 to 300 g/L. Despite the fact that galvanic operations are most frequently performed in aqueous solutions, selection of the glycol over water was determined by the capability of this solvent to wet even lyophobic surfaces. In addition, ethylene-glycol-based electrolytes allow high electric currents and remain efficient at temperatures close to boiling.

With the anode featuring a graphite bar immersed in the liquid and the cathode being a thin stainless-steel rod dipped into the bulk of powder, the process was started at ambient temperature. The powder being electroplated was held in contact with the cathode while simultaneously shuffled by a magnetic field created by a rotating constant magnet. During the process, as a result of the application of electric currents as strong as 1–2 A, the temperature of the electrolyte quickly increased to 60–80 °C. After such treatment for 1 h on average, the powders were decanted and washed with distilled water followed by rinsing with aqueous ammonia, then water again, and finally with isopropyl alcohol to be dried out at 60–70 °C in the oven.

Preliminary examination of the product recovered was performed by observation for the presence of hard agglomerates and impurities, after which their quality was evaluated by measuring their conducting properties. The measurements were done by placing small amounts of powder into a Teflon die between the face planes of two cylindrical-shaped rods being the upper and lower contacts, after which the readings of the ohmmeter (R) were recorded. At this, the assemblage was loaded with a 1.5 kg weight installed on top of the upper contact, which, at a 5 mm inner diameter of the die, corresponded to a pressure of 7.5 kgf/cm^2 applied axially. Aimed at being 3 mm in height, the amount of sample was controlled with a Vernier caliper. The necessity to know the geometrical parameters was dictated by the advantage of determining the more fundamental characteristics: the resistivities (ρ) of the products were calculated according to the formula:

$$\rho = \mathrm{R}\frac{\pi d^2}{4l} \tag{2.1}$$

with R, d, and l being the resistance of the sample, inner diameter of the die, and height of the sample, respectively. Samples of magnetic elastomer filled with electroplated powders were fabricated according to the technology described earlier. However, unlike samples produced for rheological tests, these specimens were polymerized in a polyurethane mold with a certain thickness (h) and diameter (D) of the cutout, pressed between two copper plates being the contacts.

Besides purely chemical investigations, the goals of this study assumed developing a better understanding of the relationship between the chemical formula of the magnetic filler and physical parameters of the overall composite. For this, the physical-measurement part of research was carried out on the basis of comparison of samples containing the products of electroplating with those solely filled with untreated carbonyl iron particles. All prepared isotropic, the specimens were subjected to the determination of their resistances (R) and capacitances (C) at different magnetic fields and various frequencies assigned by means of an E20-7 LCR-meter in the parallel-mode setting. The data recorded in the field and frequency ranges 0–420 mT and 1–50 kHz, respectively, were processed into a series of frequency-reliant dependences of resistivity (ρ) and permittivity (ε) on external magnetic field (Eqs. (2.1) and (2.2), respectively). Ratios of resistivity and permittivity values determined at the strongest field and at zero magnetic field were used as measures of the effectiveness of the elastomers (Eqs. (2.3) and (2.4), respectively):

$$\varepsilon = \frac{4Ch}{\pi D^2 \varepsilon_0} \tag{2.2}$$

$$\mathrm{n}_\rho = \frac{\rho_{\mathrm{B}\approx 0}}{\rho_{\min}} \tag{2.3}$$

$$\mathrm{n}_\varepsilon = \frac{\varepsilon_{\max}}{\varepsilon_{\mathrm{B}\approx 0}} \tag{2.4}$$

with D = 18.0 mm and h = 1.5 mm, C and ε_0 being the capacitance of the sample and electric permittivity of free space equal to 8.854•10^{-12} F/m, respectively.

2.3 Discussion

2.3.1 Structuring mechanism

As is evident from the observations presented earlier, the overall mechanism of structuring is determined by a few factors. On the one hand, external magnetic field, opposing the interior resistance of the material, tends to order magnetic particles along its vector. On the other hand, interactions occurring inside the composite feature a combination of two competing phenomena being the elastic resistance of the polymer and dipole interactions among magnetic particles. In addition, whereas the response of pure polymer is solely based on physical movements in the matrix and remains indifferent toward field influence, interplay among the particles exhibits a strong field dependence. At the same time, the vector of the strongest magnetic field ever applied to the material introduces the overriding factor. Indeed, a freshly prepared HME sample features unstrained polymerized resin filled with nonmagnetic powder and imposition of a magnetic field imparts magnetic moments to the particles. Interactions among the newly created magnetic dipoles immediately result in their multiple moves straining the polymer. Enhanced by the primary external field, magnetic interactions remain sufficiently intensive to hold the particles assembled into straight chains. Meanwhile, after turning the field off, magnetic forces pulling the particles together decrease in intensity, thus subjecting the chains to the influence of elastic forces tending to diminish interior tension and return the grains into their initial positions. However, as has been shown earlier, removal of the field is followed by only a partial destruction of the pattern. Owing to the strong remanence possessed by high-coercivity materials, magnetized particles continue to participate in dipole interactions remaining connected to each other. In addition, despite the reduction of their degree of order, the chains retain the direction, in which the primary structure was formed.

Abrupt switching to a magnetic field of opposite polarity creates a situation when its vector is contradirectional to the magnetic moments of the particles. In order to restore codirectionality, they might exhibit two different types of behavior depending on the elastic properties of the polymer: while tightly embedded in a rigid matrix, the grains demonstrate the classic mechanism of internal rearrangement, first changing their state to nonmagnetic, then to opposite polarity; soft matrices allow the particles to turn inside. Increasing the oppositely directed magnetic field does not lead to the formation of as well-ordered chains, however. Such an asymmetry may be explained by noticeably stronger elastic forces affecting the order of the internal chain-like structure being formed by turned-over particles as compared to when structuring requires no rotation and therefore doing additional work against the restoring forces of the polymer.

The most interesting case of structuring is observed when the reversal magnetic field is gradually reduced to 0. As the field-induced magnetic forces holing the particles gathered in chains weaken, their tendency to turn back becomes more pronounced. As a result, at moderate fields, the chains are subjected to significant distortions making the impression that the system tends to rearrange the buildups in the way perpendicular to the influence vector. At the same time, magnetic interactions complemented by elastic forces are directed at diminishing the overall system energy. Driven by their remnant magnetization, particles assemble into rings confining circular magnetic field. It should be noticed that the ascending branch of the magnetization hysteresis loop lies close to the zero point, which corresponds to the low coercivity and remnant magnetization magnitudes related to it. In addition to that, it is possible to conclude that the asymmetry of the magnetization loop originates from the initial magnetizing-field memory phenomenon based on the existence of the easy magnetization direction possessed by every particle.

2.3.2 Rheological properties of HME

Determination of the rheological properties of HME was one of the purposes of the research. Meanwhile, owing to the complexity of the structuring mechanism, using the direct approach in their study was inefficient. As has been mentioned earlier, simplification of data interpretation suggested that all the experimental specimens be fabricated with magnetically hard particles only. At the same time, it was of interest to investigate the influence of particle size on the rheological behavior of the material. In order to do that, a commercial magnetic NdFeB-alloy powder with grains of sizes ranging up to 250 μm was sieved into three fractions. Thus, three elastomer samples containing particles smaller than 40 μm, those falling in the range 40–80 μm, and grains bigger than 80 μm, were fabricated. The fourth sample filled with unfractionated polydisperse powder was synthesized for comparative purposes.

The relationship between shear modulus and deformation exhibited by the sample based on the smallest particles influenced by various magnetic fields is shown in Figure 2.3.2. As is possible to notice, the curves are of a standard shape and resemble those demonstrated by MRE containing magnetically soft particles. At deformations not exceeding 0.01%, the shear modulus attains saturation and no longer exhibits the tendency to grow. Increasing deformation from 0.01 to 10% results in lower shear modulus magnitudes. At the same time, at stronger magnetic fields, the dependence of the shear modulus on deformation is pronounced more noticeably. Whereas at zero field, the shear modulus decreases by a factor of 3.3, which corresponds to a decrease from 25.4 kPa observed at a 0.01% deformation to 7.5 kPa observed at a 10% deformation, at 490 mT the modulus changes by a factor of 35 decreasing from 3849 to 108 kPa. A similar relationship between the elasticity of filled elastomers and deformation, also known as the Payne effect, had been observed in samples filled with magnetically soft particles during testing on a rheometer in in-plane shear mode [33] and in stretch mode [34].

Comparison of the rheological behaviors demonstrated by the sample containing the finest fraction of particles and one filled with medium-size (40–80 µm) particles makes it possible to see that the latter exhibits a similar relationship with deformation.

The lines lie very close to each other. On the one hand, it makes sense that the curve corresponding to the finest-filler sample runs slightly above that of its counterpart. On the other hand, the other sample might demonstrate a stronger elasticity increase when influenced by a magnetic field. However, the fact that this does not happen gives grounds for the conclusion that a high magnetorheological effect may indeed be observed with small particles only. At small deformations occurring in magnetic field, the elastic modulus increases by a factor of 150 or by in increment of 3.8 MPa, whereas at a 10% deformation the modulus increases by a factor of 15, which corresponds to an increment of 120 kPa.

A comparative diagram for the loss factors demonstrated by the two samples as functions of deformation is brought in Figure 2.3.4.

Dependences of loss factor on deformation, grain size, and external magnetic field are tangly. At zero field, deformation increasing from 0.01 to 10% is followed by a small increase of the loss factor by 0.1. At the same time, magnetic field causes more significant incremental changes. It is interesting to note that its maximum value is observed at deformations ranging from 0.1 to 1%. In addition, samples filled with bigger (40–80 µm) particles show even higher loss-factor values.

The sample containing the biggest (≥80 µm) particles shows results similar to those produced by the other two samples brought in Figure 2.3.2 and 2.3.3. However, the initial elastic modulus of this specimen equals 190 kPa, which is one order of magnitude greater in comparison with the other specimens. As a result, the sample demonstrates a

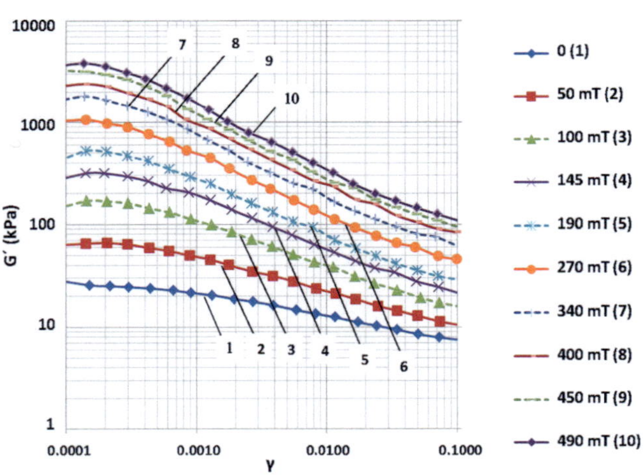

Figure 2.3.2: HME sample containing the smallest (≤40 µm) particles. Shear modulus (G′) as a function of deformation at different magnetic fields.

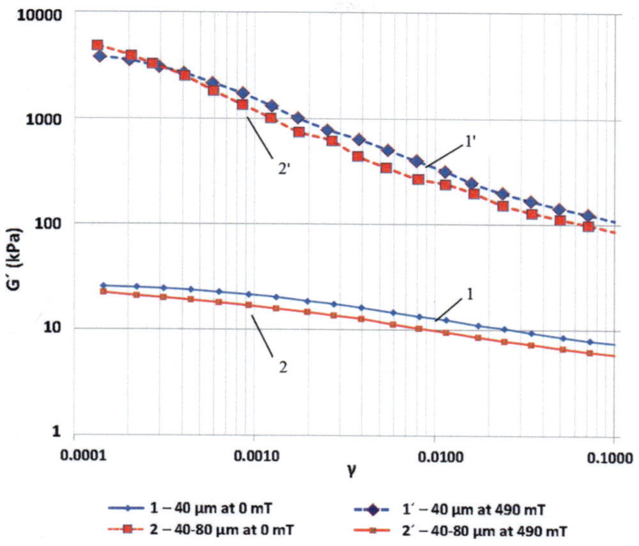

Figure 2.3.3: HME samples containing the smallest (≤40 μm) and medium-size (40–80 μm) particles. Comparison of their shear moduli (G′) as functions of deformation at zero magnetic field and 490 mT.

very close absolute MR effect equal to 3.1 MPa, whereas at a deformation of 0.01% its relative magnitude turns out to be only 16 (Figure 2.3.5).

Finally, determination of the rheological parameters of the specimen containing unfractionated polydisperse particles was carried out for a better understanding of the tendencies demonstrated by the three samples considered previously and the dependences of their properties on filler dispersity. The proximity of its elastic modulus

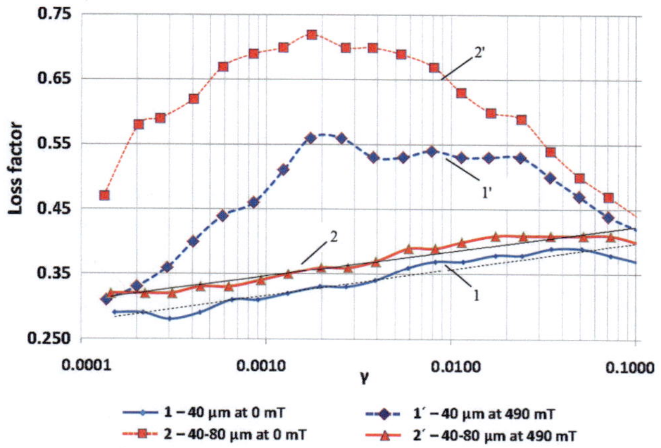

Figure 2.3.4: HME samples containing the smallest (≤40 μm) and medium-size (40–80 μm) particles. Comparison of their loss factors as functions of deformation at zero magnetic field and 490 mT.

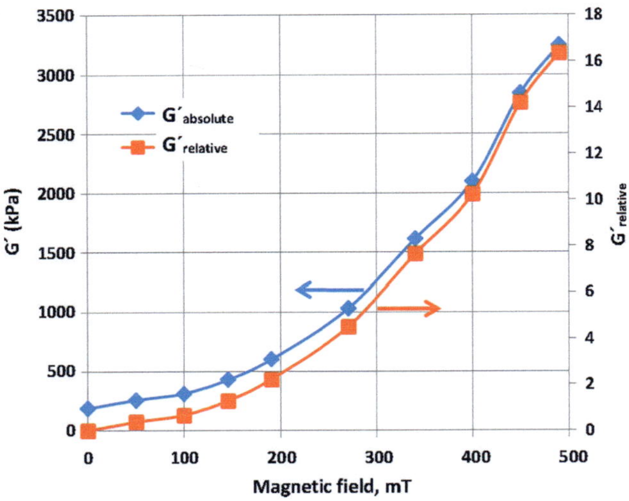

Figure 2.3.5: HME sample containing the largest (≥80 μm) particles. Absolute and relative MR effect in relationship with external magnetic field.

being 28 kPa to those of the three samples gives such a possibility. A comparative diagram for the sample filled with unfractionated particles and that containing the finest fraction is presented in Figure 2.3.6.

Meanwhile, comparison of the sample containing the smallest particles with those based on less-disperse powders unveiled very similar behavioral features in spite of the fact that the former shows a stronger magnetization and thus is expected to undergo a stronger structuring in magnetic fields. At the same time, the cumulative influence of such factors as high concentration of the filler, big size of its particles, and their plate-like shapes may evidently be the cause of steric difficulties experienced by the grains attempting to line up along the magnetic field force lines.

It is necessary to note that whereas iron powders become saturated at 500 mT, NdFeB-alloy particles require a 1–2 T field to attain saturation. As may be noticed from the following drawing, the magnetization and MR effect do not level out at 500 mT and continue to grow, which is also seen well on the magnetization curves (Figure 2.3.7). A detailed description of these magnetic features is given in Ref. [16]. Most strongly the MR effect appears at small deformations. For instance, G′ increases by a factor of 120 at a deformation of 0.01% and by a factor of 30 at a deformation of 1% (Figure 2.3.8).

As may be seen from Figure 2.3.7, owing to the fact that the magnetization and remnant magnetization magnitudes demonstrate monotonous growth over the entire interval of fields ranging up to 1500 mT, the rheological studies carried out at fields not exceeding 500 mT cannot unveil the complete potential of the composite material.

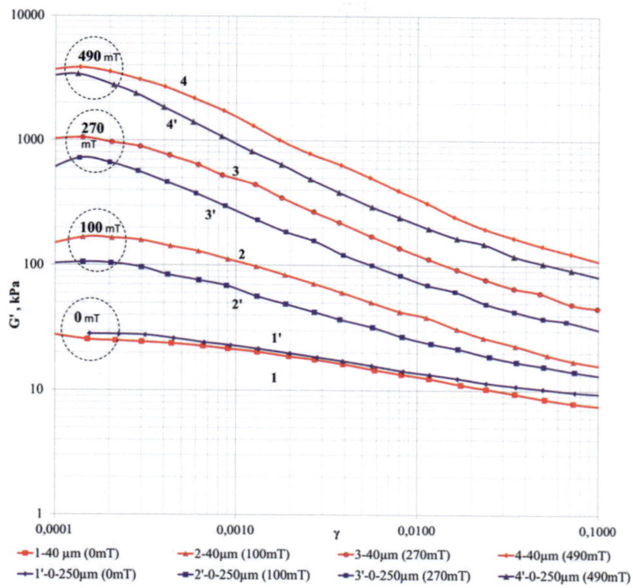

Figure 2.3.6: HME samples containing the smallest (≤40 μm) particles and unfractionated filler (0–250 μm). Comparison of their shear moduli (G′) as functions of deformation at four magnitudes of external magnetic field.

The loss factor is an important index for MRE. Its magnitude is proportional to the capability of the material to dissipate energy and absorb vibrations. Comparison of specimens fabricated on the basis of particles with different sizes makes it possible to conclude that this parameter grows with external magnetic field strength and grain size.

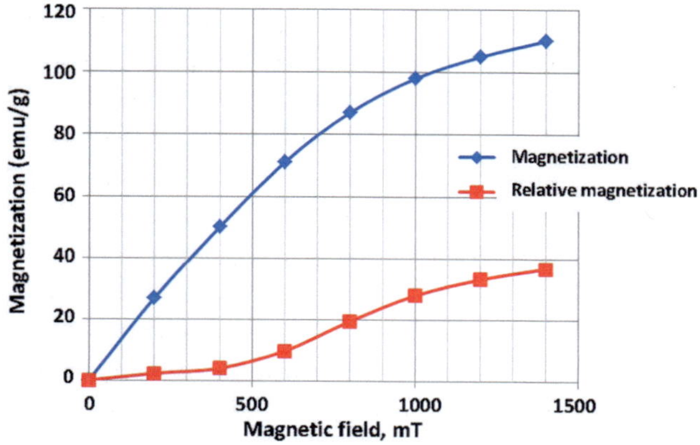

Figure 2.3.7: Magnetization and remnant magnetization as functions of external magnetic field.

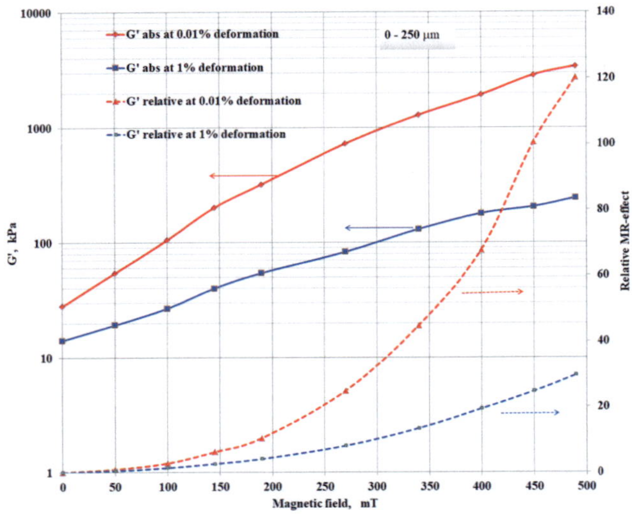

Figure 2.3.8: HME sample with polydisperse filler (0–250 μm). Absolute and relative MR effect as a function of magnetic field at different deformations.

2.3.3 Electrical properties of fillers and magnetic elastomers on their basis

Accumulation of experimental results has offered a possibility to draw primary conclusions as well as do a step forward. On the one hand, carbonyl iron powders taken from different sources and having quite high resistivities become significantly more conductive after treatment. On the other hand, whereas ethylene-glycol-based electroplating has proven itself to be productive, we have not yet been able to make it fully controllable. All uncertainties evidently come from secondary chemical processes requiring additional study. So far, it has been noticed that the resistivities of the recovered products show a degree of variation; however, the set of factors having a direct effect of the merit of nickel-plated powders remain unknown. While nickel chloride concentration and high electric current most probably have a positive influence on the product quality, there is no clear understanding of such factors as the presence of Fe^{3+} cations in the electrolyte and its pH or freshly preparedness. Meanwhile, the range, within which most of the results have fallen, extends from 6.5 to 131.0 mΩ·m being orders of magnitude narrower as compared to the spread of values demonstrated by the resistivities of untreated carbonyl iron powders. For example, whereas a commercial carbonyl iron powder containing spherical grains being 3–8 μm in size exhibited a resistivity of $6.50{\cdot}10^3$ Ω·m, the product of its three-step grinding demonstrated a magnitude of 1.31 Ω·m. As is indicated by our experimental observations, the conductive properties of the initial material have a positive effect on the degree of conductivity improvement. In the example brought just earlier, this correlation may be explained by the fact that the grinding results in peeling an oxidized layer off the grains being treated. Thus, demonstrating a better conductivity before electroplating, such powders more stably and vigorously participate in the target electrochemical process. As has been established by means of the X-ray excited optical luminescence method, in most cases concentrations of metal nickel deposited on the iron particles fall within the limits of an interval extending from 1 to 4 wt.%. Powders with better initial conductive properties correspond to higher nickel contents in the products, however.

The products of electroplating feature fluffy powders with color tones slightly darker than that of untreated carbonyl iron and contain no visible agglomerates. At the same time, although the overall method does prevent massive particle agglomeration manifesting itself as the formation of small stones, remnant magnetization turns up to be the factor responsible for the aggregation of particles into elongated gatherings (Figure 2.4.1).

As may be seen from the photographic images, the electroplated powder consists of agglomerates 50–100 μm long and 20–30 μm wide, which is noticeably more massive in comparison to the initial 3–8 μm spherical particles.

A powder containing 1 wt.% of metal nickel and exhibiting a resistivity of 45.5 mΩ·m obtained from a carbonyl iron powder with an initial resistivity of 25–27 Ω·m was used for the synthesis of experimental magnetic elastomer samples to be tested for magnetic field-sensing properties. Their physical parameters were measured against similar specimens fabricated on the basis of the initial carbonyl iron. All the samples were prepared isotropic and contained 80 wt.% of filler.

Presented in Figure 2.4.2, the graphs demonstrate frequency-reliant dependences of resistivity (ρ) and permittivity (ε) on external magnetic field. Whereas such features as the hysteresis phenomenon, the positive and negative frequency-dependent axial displacements of the ρ- and ε-loops, respectively, and their geometrical orientations in the physical parameter – magnetic field coordinate systems remain similar, dissimilarities deserve a more attentive examination. First of all, the hysteresis loops recorded for the samples containing electroplated particles demonstrate higher widths resulting from a more reluctant response to the decreasing of external field. Not observed in samples filled with untreated iron powders, this retardation effect also manifests itself as a shifting of the extremum value from the corner of the loop along the reverse line to a lower magnetic field.

Prepared on the basis of soft polymer, the samples must allow easy particles displacements accompanied by their structuring in magnetic fields. Meanwhile, the reluctance observed in samples based on electrochemically treated powder containing large elongated particles most probably reflects the difficulties they experience moving inside the matrix resulting in an interval of fields, within which both the resistivity and permittivity continue to retain the variation tendency they showed while the external

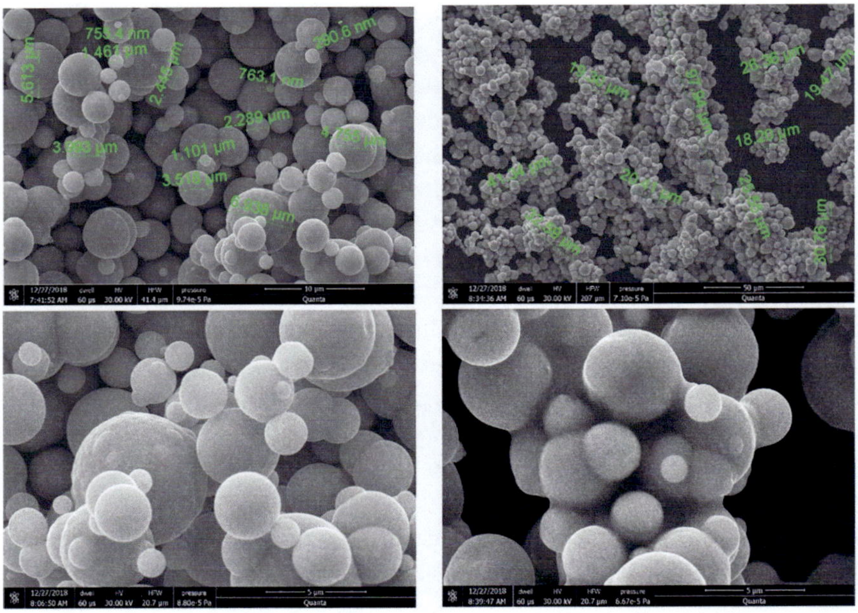

a b

Figure 2.4.1: Microscopic images of (a) untreated carbonyl iron particles (scale bar 10 μm above and 5 μm below) and (b) nickel-electroplated carbonyl iron particles (scale bar 50 μm above and 5 μm below).

field was changing toward its maximum value. In the case of samples filled with spherical particles of untreated carbonyl iron, the corner of the hysteresis loop corresponding to the highest magnetic field is coincident with the extrema of the physical parameters being measured.

Substituting the extremal values and those the samples exhibit at zero field upon completion of a full hysteresis loop, into the expressions for n_ρ and n_ε brought in Part 2.4, we have established experimentally that introduction of nickel-electroplated particles indeed results in a higher performance of the magnetic elastomer. As may be seen from Tables 2.1 and 2.2 listing the magnitudes corresponding to the exceptional points of the frequency-reliant dependences and efficiency indices computed on their basis, the nickel-containing composite stably demonstrates a more extensive variation of its parameters, thus indicating a better magnetic field sensitivity.

It should be noted, however, that the capability to conduct electric current is determined not only by the low resistivity of the filler, but also by the possibility for its particles to line up in chains. Indeed, the magnetoresistivity phenomenon is based on internal structural changes similar to those described in Part 3.1. At the

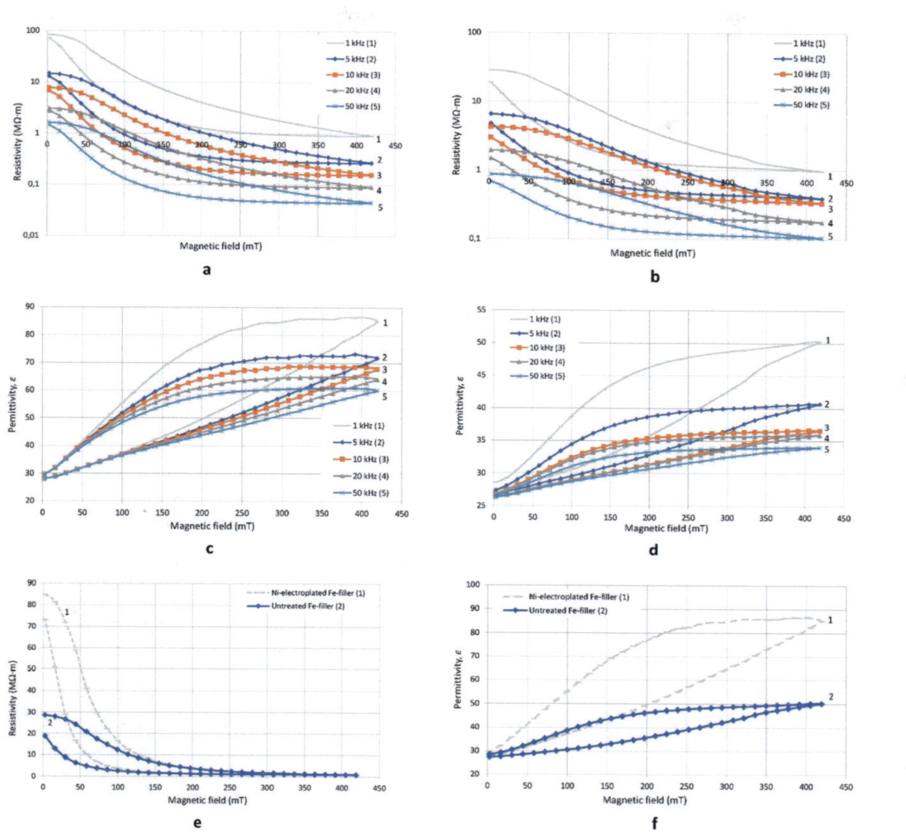

Figure 2.4.2: Physical parameters of samples of elastomer filled with magnetically soft particles as functions of external magnetic field: resistivity (ρ) of elastomer filled with nickel-electroplated (a) and untreated (b) Fe particles; permittivity (ε) of elastomer filled with nickel-electroplated (c) and Fe-untreated (d) particles. Comparative diagrams showing the resistivity (e) and permittivity (f) variations demonstrated by samples with different fillers under the influence of external magnetic field, measurements done at 1 kHz; the resistivity values are plotted along a linear scale.

same time, as is suggested by the closeness of the values of parameters observed at zero field in the initial ($B_1 \approx 0$) and final ($B_2 \approx 0$) points of the cycle, these materials containing no magnetically hard components are unlikely to develop an interior preferred orientation on primary magnetizing. Therefore, it may be assumed that the magnetic field sensitivity demonstrated by the sample filled with nickel-coated grains might probably be pronounced more strongly either with a more elastic matrix or if the electroplated particles had retained their spherical shapes, thus experiencing less intensive steric hindrances when driven by the magnetic field.

Table 2.1: Resistivities at the starting point of the hysteresis loop ($B_1 \approx 0$), at the highest magnetic field ($B = 420$ mT), at a point of minimum (ρ_{min}), and upon completion of a full cycle ($B_2 \approx 0$) recorded for a sample filled with nickel-electroplated particles against one containing untreated carbonyl iron particles; resistivity-based performance index magnitudes determined for both specimens in combination with the relative resistivity-based performance characteristic of the former sample.

Frequency (kHz)	Sample	ρ ($B_1 \approx 0$) (Ω·m)	ρ ($B = 420$ mT) (Ω·m)	ρ_{min} (Ω·m)	ρ ($B_2 \approx 0$) (Ω·m)	n_ρ	n_ρ (Fe/Ni)/n_ρ (Fe)
1	Ni/Fe	$8.48{\cdot}10^7$	$9.21{\cdot}10^5$	$8.99{\cdot}10^5$	$7.31{\cdot}10^7$	81.3	4.26
	Fe	$2.86{\cdot}10^7$	$9.85{\cdot}10^5$	$9.85{\cdot}10^5$	$1.88{\cdot}10^7$	19.1	
5	Ni/Fe	$1.48{\cdot}10^7$	$2.73{\cdot}10^5$	$2.68{\cdot}10^5$	$1.30{\cdot}10^7$	48.5	4.01
	Fe	$6.53{\cdot}10^6$	$3.94{\cdot}10^5$	$3.94{\cdot}10^5$	$4.76{\cdot}10^6$	12.1	
10	Ni/Fe	$7.75{\cdot}10^6$	$1.58{\cdot}10^5$	$1.55{\cdot}10^5$	$6.85{\cdot}10^6$	44.2	4.97
	Fe	$4.27{\cdot}10^6$	$3.34{\cdot}10^5$	$3.34{\cdot}10^5$	$2.96{\cdot}10^6$	8.9	
20	Ni/Fe	$3.07{\cdot}10^6$	$9.13{\cdot}10^4$	$8.89{\cdot}10^4$	$2.78{\cdot}10^6$	31.3	3.77
	Fe	$1.95{\cdot}10^6$	$1.79{\cdot}10^5$	$1.79{\cdot}10^5$	$1.49{\cdot}10^6$	8.3	
50	Ni/Fe	$1.62{\cdot}10^6$	$4.55{\cdot}10^4$	$4.44{\cdot}10^4$	$1.46{\cdot}10^6$	32.9	5.06
	Fe	8.70·105	1.05·105	1.05·105	6.86·105	6.5	

Table 2.2: Permittivities at the starting point of the hysteresis loop ($B_1 \approx 0$), at the highest magnetic field ($B = 420$ mT), at a point of maximum (ε_{max}), and upon completion of a full cycle ($B_2 \approx 0$) recorded for a sample filled with nickel-electroplated particles against one containing untreated carbonyl iron particles; permittivity-based index performance magnitudes determined for both specimens in combination with the relative permittivity-based performance characteristic of the former sample.

Frequency (kHz)	Sample	ε ($B_1 \approx 0$)	ε ($B = 420$ mT)	ε_{max}	ε ($B_2 \approx 0$)	n_ε	n_ε (Fe/Ni)/n_ε (Fe)
1	Ni/Fe	28.0	85.2	86.5	29.7	2.91	1.65
	Fe	27.4	50.3	50.3	28.6	1.76	
5	Ni/Fe	28.0	71.9	73.2	29.6	2.47	1.66
	Fe	26.8	40.7	40.7	27.4	1.49	
10	Ni/Fe	28.1	67.9	68.6	29.6	2.32	1.69
	Fe	26.4	36.7	36.7	26.8	1.37	
20	Ni/Fe	28.1	64.2	65.0	29.6	2.20	1.64
	Fe	26.4	35.9	35.9	26.8	1.34	
50	Ni/Fe	28.0	60.1	60.9	29.5	2.06	1.61
	Fe	26.3	34.0	34.0	26.6	1.28	

2.4 Conclusions

Study of the behavior exhibited by composite materials containing magnetic filler is essentially based on research on systems consisting of strongly interacting elements. Their mutual influence is complex and results in the origination of effects having a certain subordination manifesting itself as an asymmetry of physical properties and memory effects.

Analysis of the experimental observations has led us to the understanding that introduction of high-coercivity components into the formula is determinative of the mechanism of interactions among magnetized particles contained in the elastomer. The detailed consideration of the formation of various patterns brought in Parts 2.2 and 3.1 helps develop a vision of how the interior processes affect the behavior of HME observed during testing. In particular, such phenomena as low coercivity and weak remnant magnetization exhibited by the sample accompanied by its overall polarity reversal at moderate reverse magnetic fields take place owing to the capability of the magnetized particles embedded in its soft matrix to turn. At this, conditioned by a certain relation among the demagnetizing force of the pure filler, its remanence, and polymer elasticity, the rotation effect may turn out to be quite intensive resulting in a significant asymmetry of the magnetization hysteresis loop showing as shifting its ascending branch to the left of the origin and thus demonstrating abnormal signs of coercivity and magnetization at moderate reverse magnetic fields [10]. In addition, the closeness of the magnetic moment demonstrated by HME on the ascending magnetization curve is reflective of rearrangements among the magnetized particles leading to the formation of ring-like assemblages as the reverse magnetic field is decreasing in intensity approaching zero. At the same time, the specifics of the structuring processes remain geometrically predetermined by the direction of initial magnetization manifesting itself as a disproportional response to external magnetic vector variations.

Determined by these interior structuring processes, the rheological properties exhibited by HME also have certain specifics. For instance, unlike "classic" elastomers filled with magnetically soft particles, samples of hybrid magnetic composite demonstrate a tendency of the loss factor to increase whereas the field, the material is exposed to, becomes stronger. Showing an elastic modulus several times that of MRE with magnetically soft filler, HME, nevertheless, demonstrates a similar tendency of its viscosity and elastic modulus to grow with field.

This resemblance of the material of interest with elastomers containing magnetically soft particles gives a chance to consider the influence, the shapes of particles have on its overall performance, from a different angle. Owing to the fact that measurements of the magnetoresistivity effect assuming successive testings of the same sample over the same interval of magnetic fields are impossible, the sameness of field-induced stiffening comes useful making it possible to substitute a sample containing high-coercivity particles with one filled with magnetically soft powder. As has been mentioned in Part 3.3, electroplating results in some agglomeration leading to the formation of elongated assemblages. Although the nickel-coated particles occur orders of magnitude more electroconductive in comparison to their untreated carbonyl iron counterparts, alone, this property remains a useful but insufficient factor for making the overall composite better conducting. At the same time, the experimental results suggest that the softness of the polymer matrix and the capability of the particles of forming chain-like structures under the influence of magnetic fields are the required supplementary conditions. Whereas the performance of a sample filled with electroplated particles, which have retained their initial spherical morphologies, has not yet been observed, using experimentally obtained powders has indeed boosted magnetic field sensitivity. The resistivity- and permittivity-based performance indices demonstrated by a sample prepared with nickel-coated particles exceed those shown by one containing untreated carbonyl iron by a factor of 3.77–5.06 and 1.61–1.69, respectively. Meanwhile, the wider hysteresis loops resulting from a more sterically hindered, and thus reluctant, motion of the electroplated particles are probably indicative of a somewhat more pronounced tendency of the material to retain interior structures.

Keeping in mind that this research has been inspired by the purpose to develop a novel material with extraordinary features, collection of scientific facts about elastomeric materials containing magnetically hard particles is definitely a step forward. Study of the magnetoresistivity phenomenon is a contribution to the knowledge about sensors.

Author contribution: All the authors have accepted responsibility for the entire content of this submitted manuscript and approved submission.
Research funding: The reported study was funded by RFBR according to the research project 19-53-12039 and by Deutsche Forschungsgemeinschaft (DFG) under Grant BO 3343/2-1 within PAK 907.
Conflict of interest statement: The authors declare no conflicts of interest regarding this article.

References

1. Stepanov GV, Chertovich AV, Kramarenko EY. Magnetorheological and deformation properties of magnetically controlled elastomers with hard magnetic filler. J Magn Magn Mater 2012;324: 3448–51.
2. Linke JM, Borin DY, Odenbach S. First-order reversal curve analysis of magnetoactive elastomers. RSC Adv 2016;6:100407–16.
3. Borin DY, Stepanov GV, Odenbach S. Tuning of the tensile modulus of the magnetorheological elastomer with magnetic hard powder. J Phys Conf 2013;412:012040.
4. Borin DY, Stepanov GV. Oscillation measurements on magnetoactive elastomers with complex composition. J Optoelectron Adv Mater 2013;15:249–53.
5. Kramarenko EY, Chertovich AV, Stepanov GV, Semisalova AS, Makarova LA, Perov NS, et al. Magnetic and viscoelastic response of elastomers with hard magnetic filler. Smart Mater Struct 2015;24:035002.
6. Wen Q, Wang Yu, Gong X. The magnetic field dependent dynamic properties of magnetorheological elastomers based on hard magnetic particles. Smart Mater Struct 2017;26:075012.
7. Stepanov GV, Borin DY, Kramarenko EY, Bogdanov VV, Semerenko DA, Storozhenko PA. Magnetoactive elastomer based on magnetically hard filler: synthesis and study of viscoelastic and damping properties. Polym Sci Ser A 2014;56:603–13.

8. Stepanov GV, Borin DY, Storozhenko PA. Rotation of magnetic particles inside the polymer matrix of magnetoactive elastomers with a hard magnetic filler. JMMM 2017;431:138–40.
9. Stepanov GV, Borin DY, Bakhtiiarov AV. Magnetic properties of hybrid elastomers with magnetically hard fillers: rotation of particles. Smart Mater Struct 2017;26:035060.
10. Stepanov GV, Borin DY, Bakhtiiarov AV, Storozhenko PA. Negative coercivity of magnetic elastomers filled with magnetically hard particles. JMMM 2019;498:166125.
11. Jolly MR, Carlson JD, Munoz BC. The magnetoviscoelastic response of elastomer composites consisting of ferrous particles embedded in a polymer matrix. J Intell Mater Syst Struct 1996;7: 613–22.
12. Jolly MR, Carlson JD, Munoz BC. A model of the behaviour of magnetorheological materials. Smart Mater. Structure 1996;5:607–14.
13. Zrínyi M, Barsi L, Büki A. Deformation of ferrogels induced by nonuniform magnetic fields. J Chem Phys 1996;104:8750–4.
14. Barsi L, Büki A, Szabó D, Zrinyi M. Gels with magnetic properties. Prog Colloid Polym Sci 1996;102: 57–63.
15. Zrínyi M, Barsi L, Szabó D, Kilian H-G. Direct observation of abrupt shape transition in ferrogels induced by nonuniform magnetic field. J Chem Phys 1997;106:5685.
16. Zrínyi M, Barsi L, Büki A. Ferrogel: a new magneto-controlled elastic medium. Polym Gels Netw 1997;5:415–27.
17. Mitsumata T, Ikeda K, Gong J, Osada Y, Szabŕo D, Zrinyi M. Magnetism and compressive modulus of magnetic fluid containing gels. J Appl Phys 1999;85:8452–5.
18. Mitsumata T, Abe N. Magnetic-field sensitive gels with wide modulation of dynamic modulus. Chem Lett 2009;38:922–3.
19. Mitsumata T, Kosugi Y, Ouchi Sh. Effect of particles alignment on giant reduction in dynamic modulus of hydrogels containing needle-shaped magnetic particles. Prog Colloid Polym Sci 2009; 136:163–70.
20. Ouchi Sh, Mitsumata T. Magnetorheological effect of magnetic gels containing Fe_2O_3. Trans Mater Res Soc of Japan 2009;34:459–60.
21. Lokander M, Reitberger T, Stenberg B. Oxidation of natural rubber-based magnetorheological elastomers. Polym Degrad Stabil 2004;86:467–71.
22. Lokander M, Stenberg B. Performance of isotropic magnetorheological rubber materials. Polym Test 2003;22:245–51.
23. Lokander M, Stenberg B. Improving the magnetorheological effect in isotropic magnetorheological rubber materials. Polym Test 2003;22:677–80.
24. Endo H, Kato Sh, Watanebe M, Kikuchi T, Mitsumata T. Magnetically tunable vibration transmissibility for polyurethane magnetic elastomers. Polymers 2018;10:104.
25. Boczkowska A, Awietjan SF. Smart composites of urethane elastomers with carbonyl iron. J Mater Sci 2009;44:4104–11.
26. Boczkowska A, Awietjan SF. Magneto-rheological urethane elastomers activated by a magnetic field. Polimery 2009;54:26–30.
27. Boczkowska A, Awietjan SF. Urethane magnetorheological elastomers – manufacturing, microstructure and properties. Solid State Phenom 2009;154:107–12.
28. Kantorovich SS, Ivanov AO, Rovigatti L, Tavares JM, Sciortino F. Nonmonotonic magnetic susceptibility of dipolar hard-spheres at low temperature and density. Phys Rev Lett 2013;110: 148306.
29. Prokopieva TA, Danilov VA, Kantorovich SS, Holm C. Ground state structures in ferrofluid monolayers. Phys Rev E 2009;80:031404.
30. Borin DY, Odenbach S, Zubarev A. Non-ergodic tube structures in magnetic gels and suspensions. Soft Matter 2018;14:8537–44.

31. https://magnetolab.ru/osn_napr_elast_en.html.
32. Bakhtiiarov AV, Stepanov GV, Storozhenko PA. Patent RU 2684295, 2018.
33. Sorokin VV, Ecker E, Stepanov GV, Shamonin M, Monkman GJ, Kramarenko EY, et al. Experimental study of magnetic field enhanced Payne effect in magnetorheological elastomers. Soft Matter 2014;10:8765–76.
34. Stepanov GV, Abramchuk SS, Grishin DA, Nikitin LV, Kramarenko EY, Khokhlov AR. Effect of a homogeneous magnetic field on the viscoelastic behavior of magnetic elastomers. Polymer 2007; 48:488–95.

Martin Hähsler, Ingo Appel and Silke Behrens*

3 Magnetic hybrid materials in liquid crystals

Abstract: The integration of nanoparticles with magnetic, ferroelectric or semiconducting properties into liquid crystals (LCs) has attracted great interest both for fundamental investigations and for technological applications. Here, an overview of hybrid materials based on magnetic nanoparticles (MNPs) and thermotropic LCs is given. After a general introduction to thermotropic LCs and LC-MNP hybrid materials, various preparation methods established by us are presented. The synthesis of shape-(an)isotropic MNPs, their functionalization by tailored (pro)mesogenic ligands with linear or dendritic structures and their integration into LC hosts are discussed. The characterization of the MNPs, (pro)mesogenic ligands and resulting MNP-LC hybrid materials is described to show the influence of MNP functionalization on the MNP-LC interactions including aspects such as colloidal stability and structuring in the LC host. Overall, we show that the physical properties of the hybrid material are significantly influenced not only by the MNPs (i.e., their size, shape and composition) but also by their surface properties (i.e., the structure of the (pro)mesogenic ligands).

Keywords: inorganic/organic hybrid materials, magnetic nanoparticles, magneto-optical properties, nematic liquid crystals, (pro)mesogenic ligands

3.1 Introduction

The integration of nanoparticles with magnetic, ferroelectric or semiconducting properties in liquid crystals (LCs) has attracted a lot of interest for both fundamental investigation and technological application [1]. Long-distance orientational interaction in LCs leads to a strong influence of the dispersed particles on the mesogenic properties of the LC and *vice versa*. In general, two major directions are pursued (1) the LC host directs the organization of the particles into ordered arrays with synergistic collective behaviors or (2) the particle dopants modulate and improve the properties of the LC. Dopants of magnetic nanoparticles (MNPs) can effectively modify the electro- and/or magneto-optical responses and other physical characteristics of LCs. These

***Corresponding author: Silke Behrens,** Institute of Catalysis Research and Technology, Karlsruhe Institute of Technology, Postfach 3640, 76021 Karlsruhe, Germany; and Institute of Inorganic Chemistry, Ruprecht-Karls University Heidelberg, Im Neuenheimer Feld 270, 69120 Heidelberg, Germany, E-mail: silke.behrens@kit.edu, https://orcid.org/0000-0003-4328-9564
Martin Hähsler and Ingo Appel, Institute of Catalysis Research and Technology, Karlsruhe Institute of Technology, Postfach 3640, 76021 Karlsruhe, Germany; and Institute of Inorganic Chemistry, Ruprecht-Karls University Heidelberg, Im Neuenheimer Feld 270, 69120 Heidelberg, Germany

This article has previously been published in the journal Physical Sciences Reviews. Please cite as: M. Hähsler, I. Appel and S. Behrens "Magnetic hybrid materials in liquid crystals" *Physical Sciences Reviews* [Online] 2020, 5. DOI: 10.1515/psr-2019-0090 | https://doi.org/10.1515/9783110569636-003

hybrid materials reveal a great potential to improve current liquid crystal display (LCD) technologies, for example, through new or modified switching modes, lower operating voltages and larger contrast-ratios [2]. Recently, hybrid materials of LCs and MNPs with magnetically controllable and erasable characteristics (in particular magneto-chromic properties) were developed which could be interesting for application as magnetic paper [3]. Magnetic fields could be directly visualized by combining the magneto-optical and electro-optical response of ferromagnetic LCs [4]. Polymeric LCs have also been doped with MNPs which allows for their mechanical deformation or heating by the action of external (magnetic) stimuli [5]. Transparent magnets with flexible and optically homogeneous properties, for example, were achieved by linking polymeric, side-chain LCs with a siloxane backbone to MNPs [6].

LCs thermodynamically range between the highly ordered crystalline and disordered isotropic liquid state and are therefore often referred to as mesophase (Figure 3.1a). This mesophase combines anisotropic properties of crystals (such as optical birefringence) with flow properties of ordinary liquids. Anisotropic properties arise from ordering of constituent entities, while fluid properties are provided by their concomitant mobility in LC state. In general, molecules, macromolecules, supramolecular aggregates or nanoparticles may act as constituent entities in LCs. Here, we focus on LCs formed by low-molecular weight molecules. Depending on the mesophase, LCs are classified into two categories, namely thermotropic and lyotropic LCs. Lyotropic LCs can be found in solutions of amphiphilic molecules forming micellar aggregates, where concentration largely determines the type of LC phase formed. In thermotropic LCs, LC ordering is a function of temperature (Figure 3.1a). Thermotropic LCs are formed by single organic molecules as constituent entities (i. e. mesogens) or mixtures thereof,

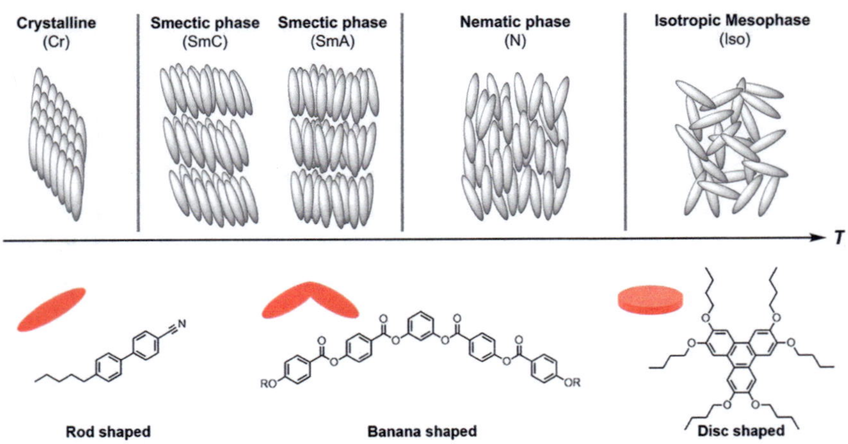

Figure 3.1: (a) Organization in the crystalline, liquid crystal (LC) and liquid state with common LC phases formed by calamitic mesogens. (b) Selected structures of mesogens forming LCs.

which are further distinguished according to the geometric shape into calamitic (rod-like) , discotic (disc-like), sanidic (brick-like or lath-like) or bent-core (Figure 3.1b) [7]. LCs are symmetry-broken, ordered fluids in which one or more angular and/or positional degrees of freedom are frozen in. While mesogens in nematic mesophases show long-range, orientational ordering along a preferred direction (director $\boldsymbol{n}$), they additionally reveal positional ordering in smectic phases. Various types of thermotropic mesophases occur at different temperatures; some common types formed by calamitic mesogens are displayed in Figure 3.1b [7].

Already in the 1970s, Brochard and de Gennes proposed the embedding of MNPs into LC hosts to increase the magnetic response or sensitivity, respectively, of LCs to magnetic fields [8]. Meanwhile, a number of lyotropic and thermotropic MNP-LC hybrids with interesting magneto-optical properties have been realized experimentally [9]. In this context, ferronematics represent stable colloidal suspensions of MNPs in nematic LCs. Mertelj et al. have reported on ferromagnetic nematics by the stabilization of ferromagnetic Sc-doped barium hexaferrite platelets in the nematic LC 4-pentyl-4′-cyanobiphenyl (5CB) [10]. The interplay of nematic-mediated repulsive interaction and magnetic attraction forces lead to stabilization of the MNPs and their ferromagnetic ordering in the LC [11]. Multiferroic properties were achieved due to the coupling of the nematic director $\boldsymbol{n}$ and the magnetization $\boldsymbol{M}$ of the platelets [12]. The electro-optical effect of the ferromagnetic LC, which did not differ from the pure LC, was accompanied by an opposite magneto-electric and an (indirect) magneto-optical effect. The birefringence of LCs consisting of shape-anisotropic organic pigments was controlled by MNPs forming chain-like structures in the magnetic field [13]. As shown by molecular dynamic simulations, the size and number of chain-like MNP clusters determined the alignment of these pigments [14]. Recently, Stannarius, Schmidt and Eremin et al. also demonstrated the magneto-optical response of isotropic and anisotropic fibrillous organogels by mobile MNPs [15]. Orientational ordering is known not only to affect the electro-optical but also the rheological properties of colloidal dispersions. Nematics, for example, reveal a complex flow behavior which depends on the flow direction and type and involves different viscosities (so-called Leslie coefficients). In this case, magneto-viscous effects which have been described for ferrofluids could also open up novel possibilities for MNP dispersions in LC hosts [16]. Just recently, an interesting magneto-viscous effect was observed by Odenbach et al. for doping of isotropic, micellar potassium laurate/water systems with MNPs [17]. The ternary mixture (potassium laurate/water/decanol) also forms a lyotropic mesophase which could give rise to interesting magneto-rheological properties [18]. In an LC mixture of magnetic and nonmagnetic nanorods, a nonmonotonic dependence of the shear stress on the strength of an external magnetic field was observed by molecular dynamics simulations, which is in contrast to the monotonic behavior in conventional ferrofluids [19]. Recently, translational and rotational motions of ferromagnetic liquid droplets were precisely actuated by an external magnetic field [20]. The droplets could be reconfigured into different shapes while preserving the magnetic properties of solid ferromagnets with classic north–south dipole interactions. Here, the droplets were built from isotropic phases and MNP-surfactants, but this could also inspire future studies on active matter and programmable liquid constructs based on MNP-LC hybrids.

However, the embedding of particles in LC matrices is by no means trivial. As shown by us and others, aggregation of MNPs up to the complete phase separation is highly challenging. The stability of MNP dispersions is typically low, and formation of aggregates often occurs even in colloids with low MNP content. In some cases, the stability of the colloidal LC dispersion seemed to be hardly longer than the measurement time of the experiments. Visible to the eye, macroscopic MNP precipitates may rapidly form below the isotropic–nematic transition leading to a spontaneous, complete phase separation into colorless LC phase and brown MNP precipitate. MNP dispersions, however, can also appear macroscopically homogeneous below the isotropic–nematic transition, whereby no precipitate is formed, while examination with the optical microscope reveals the formation of (micro)aggregates [10, 21]. Even microaggregate formation may affect the magneto–optical behavior of MNP dispersions in LCs, as demonstrated by polarizing microscopy (POM) in the magnetic field [22]. Local response of the director $\boldsymbol{n}$ near the MNP microaggregates orienting in the magnetic field led to a first nonthreshold region of the phase retardation, while the collective response of the LC with the homogeneously dispersed MNPs occurred at larger magnetic fields (≥95 mT). MNP aggregates may be removed, for example, applying an external magnetic field gradient ($Nd_2Fe_{14}B$ laboratory magnet), which leads to colloidally stable MNP dispersions where no aggregates are detected by optical microscopy [21]. However, this may not exclude the formation of small MNP clusters well dispersed in the LC and not visible under the optical microscope [23]. In particular high MNP fractions may also affect the LC viscosity and thereby the fluidity of the sample, and even gelatinous nonfluid materials have been reported for high MNP fractions [22]. Preparation procedures made also use of rapid thermal quench (i. e., by rapid cooling from the isotropic to the nematic phase) to keep the particles, which are in general better dispersible in the isotropic liquid phase, from aggregating in the nematic phase [24].

In general, a stabilization of MNPs in LCs requires not only the compatibility of the particles with the structure and dimensions of the LC host as well as specific surface properties, but also a balanced interplay of LC-mediated and magnetic forces that might occur. The coating of the MNPs with tailored (pro)mesogenic ligands is very promising, particularly for stabilizing MNPs in thermotropic LCs. For particles of semiconductors or metals (in particular small Au nanoparticles [25] and nanorods up to 50 nm [26]), the use of (pro)mesogenic ligands has been successfully demonstrated. Desymmetrization by replacing spherical Au particles by nanorods as well as the coating with chiral ligands, induced a much tighter helical distortion and an amplification of chirality in the LC host [26]. The functionalization of ZnO nanoparticles, spindle-shaped TiO_2 and α-Fe_2O_3 particles with (pro)mesogenic ligands is also described [27]. Surprisingly, only few examples describe the use of (pro)mesogenic ligands for stabilization of MNPs in LCs. As compared to oleic acid-stabilized nanorods, nanorods coated with 4-*n*-octyloxybiphenyl-4-carboxylic acid revealed an enhanced stability in 5CB [91]. (Pro)mesogenic ligands with dendritic structure are particularly interesting and were first exploited for the functionalization of MNPs by Dermortière et al. [28]. Recently, stable dispersions of CdSe@ZnS quantum dots and MNPs were also achieved in 5CB by using dendritic ligands [22, 25c, 29].

In the following, we give an overview on the synthesis of MNPs of various elemental compositions and defined spherical or anisotropic shape as well as their integration in low-molecular weight LCs. We report on the synthesis of various (pro) mesogenic ligands and the successive surface engineering of the MNPs for preparing stable colloidal dispersions in nematic LCs.

3.2 Discussion

3.2.1 Size and shape-controlled synthesis of magnetic nanoparticles (MNPs)

In general, the magnetic properties of MNPs depend on their size, shape and elemental composition. The size, shape, topology and the magnetic properties of the MNPs are very important because they determine the interactions between the particles inserted into a medium that is partially ordered such as a nematic LC. Various synthetic methods are available to prepare uniform MNPs with a defined size, shape and various elemental compositions, including co-precipitation [21], thermal decomposition [30], microemulsion [31] and hydrothermal/solvothermal [32] synthesis. Exemplary MNPs of various sizes, shapes and elemental compositions which were obtained here are depicted in Figure 3.2 [18].

Bulk $CoFe_2O_4$ displays an inverse spinel structure, a high chemical and physical stability together with an unusually high magnetocrystalline anisotropy (~2×10^5 J/m^3). Electrostatically stabilized $CoFe_2O_4$ MNPs, for example, were obtained *via* coprecipitation of the Co^{2+} and Fe^{3+} salts [21, 33]. A NaOH solution was injected into the vigorously stirred, acidic solution of the Co^{2+} and Fe^{3+} precursors to yield ultrasmall uniform $CoFe_2O_4$ MNPs with a particle size of 2.5 (±0.6) nm. As these MNPs were electrostatically stabilized, the (pro)mesogenic ligands could be directly bound to the MNP surface, and no ligand exchange was required. The functionalization of MNPs is discussed in detail below. Thermal decomposition of $Co(acac)_2$ and $Fe(acac)_3$ yielded 7.4 (±0.9)-nm-sized $Co_{0.6}Fe_{2.4}O_4$ MNPs stabilized by oleyl amine/oleic acid by modifying a procedure originally described by Sun et al. (Figure 3.2a) [30a]. The shape of the MNPs is another important aspect, and particles with shape anisotropy are of particular interest. While size (and composition) determines superparamagnetic or magnetically blocked

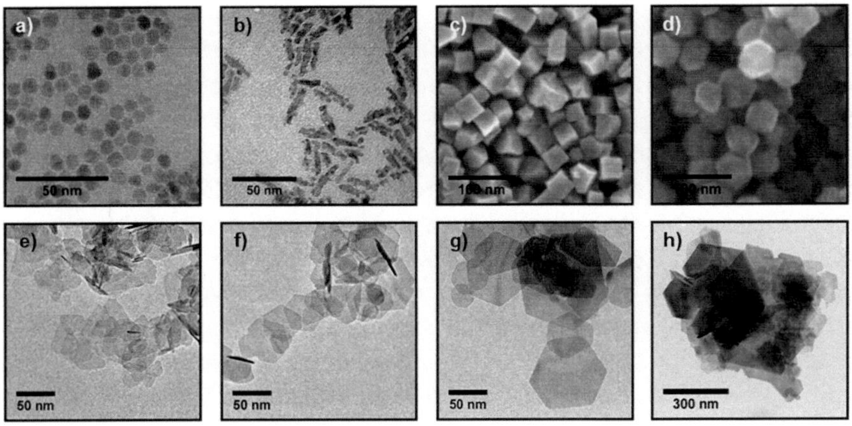

Figure 3.2: Transmission electron microscopy micrographs of (a) spherical $Co_{0.6}Fe_{2.4}O_4$ MNPs (7.4 (±0.9) nm) and (b) Fe_3O_4 nanorods (dimensions 27.0 (±9.1) × 5.5 (±1.0) nm). (c)–(d) scanning electron microscopy images of (c) nanocubes (edge length 27.0 (±3.0) nm) and (d) cuboctahedra with a particle size of 37.5 (±5.7) nm. (e)–(h) TEM micrographs of scandium-doped barium hexaferrite nanodiscs. The higher the reaction temperature was, the bigger the as-formed nanodiscs were: (e) 210 °C, (f) 260 °C, (g) 310 °C, and (h) 340 °C (reproduced with permission of the American Chemical Society and Elsevier from refs. [36, 40, 41]).

properties due to size-dependent spin exchange effects, shape is related to switching properties of MNPs due to shape-induced, magnetic anisotropy [34]. If the shape of the MNPs is anisotropic, they may further adopt a certain orientation with respect to the nematic order. Platelet-shaped MNPs, for example, resulted in LC hybrids where an equilibrated interplay between attractive magnetic interactions and LC-mediated repulsive forces induced MNP alignment and ferromagnetic ordering [10]. Effects of nanoparticle shape on transition temperature, order parameter and mobility in LC-based dispersions were further elucidated by Monte Carlo computer simulations [35]. In case of plasmonic Au particles, a remarkable amplification of chirality was recently also achieved for nematic LCs by desymmetrization of spherical nanoparticles to nanorods [26]. The synthesis of monodispersive nanorods (<100 nm) is challenging. In order to break the structural symmetry of, for example, Fe_3O_4, polymers or soft, micellar templates have to be used to induce anisotropic nanorod growth. Fe_3O_4 nanorods (27.0 (±9.1) × 5.5 (±1.0) nm), for example, have been synthesized *via* chemical transformation of nanorod seeds (Figure 3.2b) [36]. The nanorod seeds already displayed the desired anisotropic structure but they were composed of a ferri/ferro (oxide-)hydroxide phase and their magnetic properties were rather poor. $Fe_3O_4/CoFe_2O_4$ nanorods were further obtained by a simple seed-mediated synthesis. Nanorod seeds were exploited here as a platform for both chemical phase change and growth of $CoFe_2O_4$ by thermal codecomposition of cobalt(II) and iron(III) acetylacetonate precursors [36]. Fe_3O_4 nanocubes (Figure 3.2c) were obtained by thermal decomposition of $Fe(acac)_3$ in benzyl ether in the presence of (1,1′-biphenyl)-4-carboxylic acid at 290 °C

Table 3.1: Summary of composition, particle sizes and magnetic properties of the magnetic nanoparticles (MNPs) shown in Figure 3.2.

MNPs	Figure no.	Size (nm)	M_S (A m^2 kg^{-1})	H_C (298 K) (mT)
$Co_{0.6}Fe_{2.4}O_4$	Figure 3.2a	7.4 (±0.9)	37.2	0.0
Fe_3O_4	Figure 3.2b	27 (±9.1)	14.4	0.0
Fe_3O_4	Figure 3.2c	28 (±3.0)	62.9	0.0
Fe_3O_4	Figure 3.2d	38 (±5.7)	63.4	0.0
Sc-doped $BaFe_{12}O_{19}$ (Ba/Sc = 1:1.0)	Figure 3.2e	37 (±15)	35.5	43
Sc-doped $BaFe_{12}O_{19}$(Ba/Sc = 1:1.1)	Figure 3.2f	52 (±18)	38.5	50
Sc-doped $BaFe_{12}O_{19}$(Ba/Sc = 1:0.8)	Figure 3.2g	59 (±20)	39.2	57
Sc-doped $BaFe_{12}O_{19}$(Ba/Sc = 1:0.9)	Figure 3.2h	168 (±193)	n.d.	n.d.

[37]. The use of 4′-hexyl-(1,1′-biphenyl)-4-carboxylic acid instead yielded Fe_3O_4 cuboctahedra (Figure 3.2d).

Bulk hexaferrites such as $BaFe_{12}O_{19}$ are very attractive due to their low cost, hard magnetic properties and stability in air [38]. They display a hexagonal crystal structure with closed packed layers of oxygen ions. Trivalent metal cations (Fe^{3+}) are located in interstitial sites, while the heavy ions (e. g., Ba^{2+}) enter substitutionally the oxygen layers [39]. Particles of scandium-doped barium hexaferrite with disc-like shape were obtained by hydrothermal synthesis. The lateral dimensions of the nanodiscs were dependent on the reaction temperature and increased from 7 (±4) nm (160 °C) over 70 (±38) nm (240 °C) to 168 (±193) nm (340 °C) (Figure 3.2e–h) while their height was approximately 5 nm [40]. Table 3.1 summarizes the magnetic properties (saturation magnetization (M_S) and coercitivity (H_c)), sizes and composition of MNPs shown in Figure 3.2.

3.2.2 LC matrices

LCs are composed of anisotropic building units (i.e. mesogens) which are spontaneously oriented along a common direction (along the so-called director ***n***). In the simplest case of a nematic LC, these mesogens show only orientational ordering along the director ***n*** but no positional order (Figure 3.1a). LCs are usually further distinguished into two categories, thermotropic and lyotropic LCs. The influence of different particle parameters on the stability of lyotropic hybrid systems has been systematically investigated and is summarized in Ref. [18]. Thermotropic LCs, where the ordering is a function of the temperature, are typically further distinguished according to the shape of their mesogens, e.g. calamitic (rod-like), discotic (disc-like), sanidic (brick-like or lath-like) or bent-core mesogens (Figure 3.1b). The majority of LCs is formed by calamitic mesogens which are typically composed of a ridged core (e. g. a biphenyl group) and

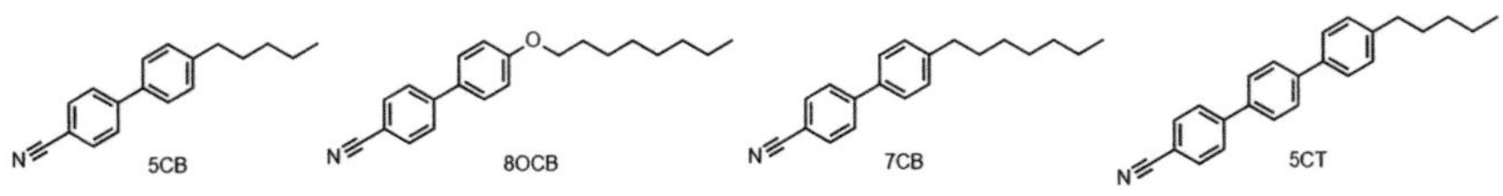

Figure 3.3: Structure of calamitic mesogens forming nematic LCs: 4-cyano-4′-pentylbiphenyl (5CB), 4-cyano-4′-octylbiphenyl (8OCB), 4-cyano-4′-n-octyloxybiphenyl (8OCB) and 8 wt% 4-cyano-4″-pentyl-p-terphenyl (5CT), respectively. E7 is a mixture of different mesogens, i. e. 5CB (51 wt%), 7CB (25 wt%), 8OCB (16 wt%) and 5CT (8 wt%).

flexible end groups (e. g. alkyl or alkoxy chains) (Figure 3.3). A nitrile residue attached to the biphenyl results in a permanent dipole moment enabling directional alignment in the external electric field. 5CB (Figure 3.3), for example, is a calamitic mesogen forming a nematic LC at ambient temperature (T_{CN} = 18 °C; T_{NI} = 35 °C). While the single 5CB molecule has a length of 1.9 nm, dimers with a length of approx. 2.5 nm are formed both in the isotropic and nematic phase by partial overlapping of the biphenyl moieties. Also, mixing of different mesogens may further enhance LC properties. For example, E7 contains mainly 5CB (51 wt%), but also 4-cyano-4′-heptylbiphenyl (7CB) (25 wt%), 4-cyano-4′-n-octyloxybiphenyl (8OCB) (16 wt%) and 4-cyano-4″-pentyl-p-terphenyl (5CT) (8 wt%) (Figure 3.3) [42]. It likewise forms a nematic phase at ambient temperature but its clearing temperature is shifted to higher temperatures (T_{NI} = 61 °C).

LCs can be (re)oriented by electric and/or magnetic fields due to the anisotropy of their electric permittivity (ε_a) or diamagnetic susceptibility (χ_a), respectively [43]. The dielectric anisotropy is in the order of unity (e. g. $\varepsilon_{a,5CB} = 11$) and the required voltages are in the order of a few volts [44]. Due to their very low anisotropy of the diamagnetic susceptibility ($\chi_a \sim 10^{-6}$–10^{-7}), however, LCs are less sensitive to the magnetic field and their realignment may require a large magnetic field strength being in the order of 1 T [45]. Doping of LCs with MNPs enhances their response to the applied magnetic field, an idea which dates back to Brochard and de Gennes [8].

3.2.3 Integration of MNPs in LC matrices

Entropic alignment of mesogens is the origin of the isotropic–nematic transition in an LC. If particles are immersed in a nematic LC, deformations and topological defects arise in the LC in response to the foreign inclusions. One important feature of LC colloidal dispersions is that the elastic distortions of the director and/or the disturbance of the local order parameter in the vicinity of the MNPs in the host LC crystal lead to LC-mediated interactions between the particles immersed in it, while such interactions are absent in usual colloidal dispersions with isotropic host fluids [45]. Therefore, the tendency of MNPs to agglomerate is much stronger in the anisotropic LC phase than in the corresponding isotropic phase. 5CB in the nematic mesophase is shown in Figure 3.3a. Oleic acid-functionalized Fe_3O_4 MNPs (particle size 10 nm), for

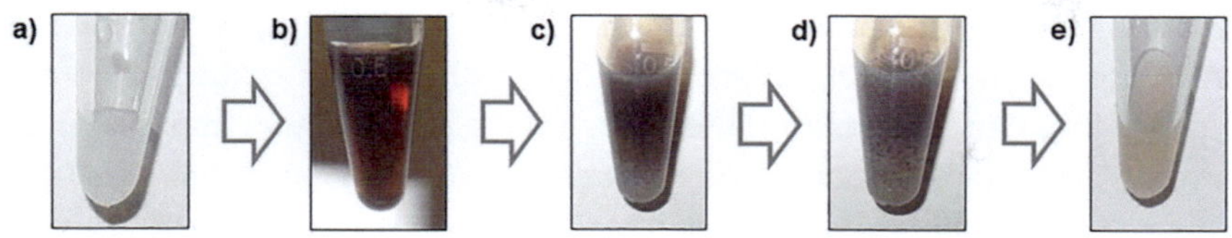

Figure 3.4: Spontaneous agglomeration of oleic acid-stabilized Fe_3O_4 magnetic nanoparticles (MNPs) (particle size 10 nm) in 5CB during phase transition from the isotropic to the nematic LC phase: (a) 5CB reference; (b) colloidal MNPs in isotropic 5CB host (40 °C; MNP concentration 1 wt%); (c, d) phase transition to the nematic LC phase (<35 °C) where MNP aggregates form; (e) after magnetic separation of the agglomerates, the MNP content in the nematic 5CB host is very low (0.003 wt%), so that the physical properties are hardly affected [33].

example, initially form a homogeneous, colloidally stable dispersion in the isotropic phase of 5CB (40 °C) (Figure 3.3b) [33]. Under these conditions, they form a ferrofluid which is also stable in the magnetic field. If the temperature is decreased below the clearing temperature of 5CB (<35 °C), the nematic mesophase forms and simultaneously, spontaneous agglomeration of the MNPs occurs (Figure 3.4c–d). After magnetic separation of the MNP agglomerates, only very low concentrations of MNPs remain in the 5CB LC host (as indicated by the pale light brownish color in Figure 3.4e) which behaves comparable to undoped 5CB and shows no significant effect in the magnetic field. This is also observed for $CoFe_2O_4$ MNPs stabilized by conventional organic acids (such as caproic acid, lauric acid, myristic acid, palmitic acid or oleic acid) even if their particle size is only 2.5–3 nm [33]. Thus, surfactants typically employed for preparing conventional ferrofluids in isotropic hosts are not well-suited to colloidally stabilize MNPs in higher concentrations in anisotropic LC hosts such as 5CB. Indeed, ferronemates described in the literature are usually characterized by limited colloidal stability and very low concentrations of MNPs [91, 46].

The stability of MNP dispersions is typically low and formation of aggregates often occurs even in colloids with low MNP content. In general, the size, shape, topology as well as magnetic and surface properties of the MNPs influence the interactions between MNPs inserted into a medium that is partially ordered, such as a nematic LC [47]. Coagulation of the immersed MNPs is caused by LC-mediated, van der Waals and/or magnetic dipole–dipole interactions which may lead to subsequent phase separation by gravitational forces or magnetic field gradients. The origin of this instability depends on the particle size [24, 25c, 48]: For relatively large particles ($r_{part} > 100$ nm), the director $\boldsymbol{n}$ of the LC is distorted around the particles which gives rise to topological defects and strong attractive orientational elastic interactions between the particles, encouraging MNP aggregation. The elastic distortions of the host LC arise from the anchoring of the mesogens on the particle surface. The force of these interactions has a strong topological signature [47]. For small particles ($r_{part} \ll 100$ nm), distortion of the LC director is not favored energetically resulting in a macroscopically uniform alignment and thus, orientational elastic forces do not contribute significantly to the aggregation process. In this case, interactions due to disturbance of the local order parameter S of the LC in the vicinity of the nanoparticles might play an important role [25b]. However, since the behavior of MNPs immersed in an LC is more complex and not only influenced by particle size alone, the transition between these two size regimes may be fluent. In order to prevent aggregation of the MNPs and phase separation, the surface of the MNPs must therefore be considered at the molecular level and the particle-matrix interaction must be specifically tailored by use of suitable ligands.

3.2.3.1 Synthesis of long-chain, (pro)mesogenic ligands

General concepts for stabilization of MNPs in conventional isotropic hosts can't be simply transferred to LC hosts. In ferrofluids, for example, the surface of the MNP surface is typically coated by conventional surfactants (such as oleic acid) to increase the (electro)steric repulsion radius and thus kinetically stabilize the MNPs in the isotropic host. In the case of nematic LC hosts, however, the interaction of the surfactants with the MNP surface and the LC host needs to be specifically tailored in such a way that the disturbance of the LC order in the proximity of the particles is minimized as much as possible. This may be achieved by use of (pro)mesogenic ligands. The role of these (pro)mesogenic ligands is not only steric repulsion by a large exclusion volume, but also to smooth out the disturbance of the local LC director at the MNP-LC interface [25c]. (Pro)mesogenic ligands are typically composed of three structural parts: (1) a functional group that allows for binding to the respective MNPs which is linked by (2) a flexible spacer to (3) the (pro)mesogenic structural unit.

A series of long-chain, (pro)mesogenic ligands (**1**–**8**) was synthesized in which the (pro)mesogenic structural unit was either cyanobiphenyl or octylbiphenyl. This (pro) mesogenic unit was linked *via* an alkyl chain of different lengths ($-(CH_2)_n-$ with $n = 6, 7, 14, 15, 16, 17$, or 25) to a functional group for nanoparticle binding (a carboxylate, phosphate or amino group) (Figure 3.5) [21]. Ligands **1**–**3** were obtained with a yield of

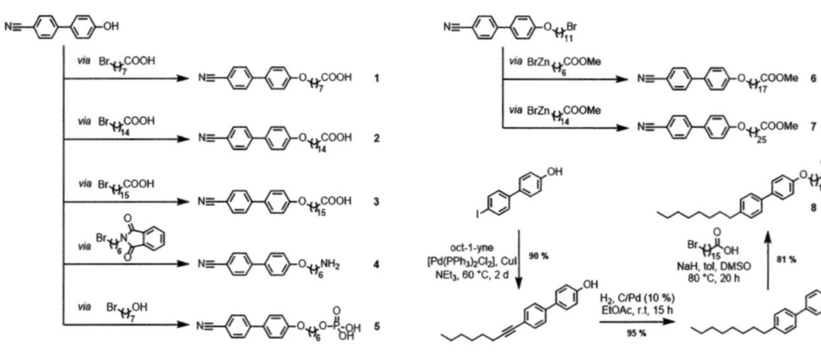

Figure 3.5: Synthesis of the long-chain, aliphatic, (pro)mesogenic ligands **1–8** [21, 49].

76–79% *via* etherification. A Gabriel reaction with an imide, followed by deprotection, gave the amine-functionalized ligand **4**. The organophosphate ligand **5** was received in a yield of 47%. For $n = 17$ and 25, a modified Negishi coupling was employed to yield the ligands **6** and **7**. The corresponding carbonic acids, however, were insoluble in common organic solvents and thus, only **6** has been deprotected and used without further characterization for the coating of MNPs [49]. Ligand **8** was synthesized in a three-step procedure using (1) a Sonogashira cross-coupling, (2) followed by a Pd-catalyzed reduction of the triple bond with hydrogen and (3) an etherification [49].

3.2.3.2 Synthesis of dendritic, (pro)mesogenic ligands

The tendency of MNPs to aggregate in LC hosts can be considerably reduced by use of long-chain (pro)mesogenic ligands enabling the preparation of homogeneous, colloidal dispersions of very small MNP in 5CB (see below). If the MNPs are immersed in the nematic LC, however, the mutual molecular alignment not only disturbs the local order of the LCs in the vicinity of the MNPs, but also the originally isotropic, spherical ligand shell at the MNP surface (Figure 3.5a). The (pro)mesogenic ligands get aligned parallel to the local director field leading to an increased ligand density at the MNP poles and to their equatorial depletion. The distortion of the ligand shell from spherical to tactoidal further encourages MNP agglomeration (Figure 3.6b) [25b]. In general, however, the process of agglomeration in these systems seems to be more complex.

(Pro)mesogenic ligands with dendritic structure reduce the distortion of the ligand shell and prevent equatorial ligand depletion, thus minimizing the tendency for MNP agglomeration [22, 25c]. However, the preparation of these (pro)mesogenic dendritic ligands typically requires multistep synthetic procedures which yield only small amounts of the target ligands [50]. In this context, we developed a convergent synthetic procedure for different ligands with mesogenic structural units and dendritic structure

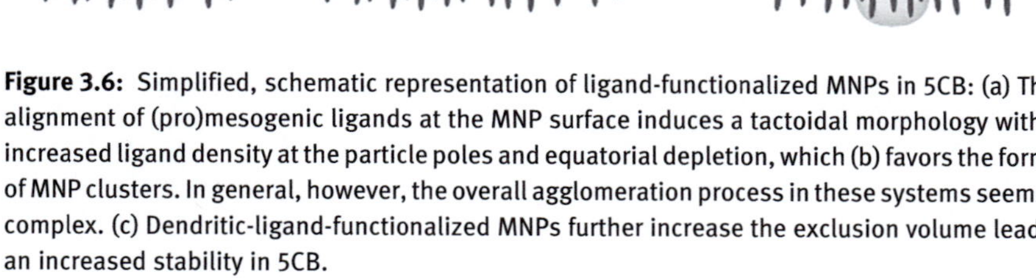

Figure 3.6: Simplified, schematic representation of ligand-functionalized MNPs in 5CB: (a) The alignment of (pro)mesogenic ligands at the MNP surface induces a tactoidal morphology with increased ligand density at the particle poles and equatorial depletion, which (b) favors the formation of MNP clusters. In general, however, the overall agglomeration process in these systems seems to be complex. (c) Dendritic-ligand-functionalized MNPs further increase the exclusion volume leading to an increased stability in 5CB.

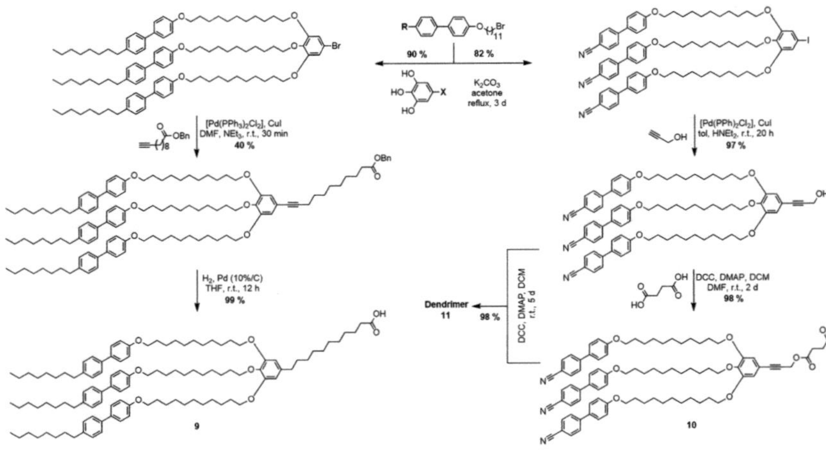

Figure 3.7: Synthetic procedure for different dendritic ligands with mesogenic structural units. In general, this strategy may be exploited for further tailoring the structure of dendritic ligands (e.g. in terms of spacer length, MNP binding group and (pro)mesogenic structural unit) or to obtain also dendrimers as shown for **11** [49].

(Figure 3.7) [49]. A three-step sequence of etherification, Sonogashira cross-coupling and esterification or deprotection the (pro)mesogenic dendritic ligands in overall good yields. Starting from literature known 5-iodobenzene-1,2,3-triol, for example, yielded the dendritic ligand **10** with an overall yield of 77 %. Since this procedure is highly versatile, it could be further extended to dendritic ligands with different spacer lengths and MNP binding groups, e.g. **11** (Figure 3.7). In preliminary experiments, electrostatically stabilized MNPs with larger diameter were coated by these dendritic ligands and stabilized in the nematic phase of 5CB. These experiments are very promising with regard to the future stabilization of larger (anisotropic) MNPs.

In order to get insights into the temperature-dependent, LC behavior of the dendritic ligands **10** and dendrimer **11**, we investigated both compounds *via* POM. For **10**, a mesophase was formed but it was stable only in a small temperature range ($\Delta T = 0.2$ °C): The transition from the isotropic to the smectic A-phase took place at $T_{\mathrm{SmA\ I}} = 88.2$ °C (Figure 3.8a), while the crystalline phase was formed at $T_{\mathrm{Cr\ SmA}} = 88.0$ °C. The characteristic, fan-shaped textures of the smectic A-phase was also observed for dendrimer **11**, but the textures were much more pronounced (Figure 3.8b) [51]. In case of dendrimer **11**, the transition from the isotropic to the smectic A-phase occurred at $T_{\mathrm{SmA\ I}} = 104.9$ °C and at $T_{\mathrm{SmC\ SmA}} = 98$ °C a smectic C-phase was formed (Figure 3.8c). In addition to the fan-shaped textures, star-like defects were observed [51]. The smectic C-phase was maintained down to $T_{\mathrm{Cr\ SmC}} = 86$ °C.

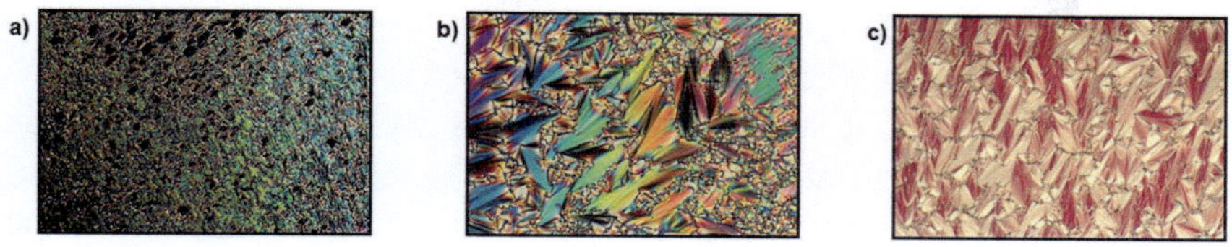

Figure 3.8: Polarizing microscopy (POM) images (under crossed polarizers) of dendritic ligand **10** and dendrimer **11**: The samples (a) **10** and (b) dendrimer **11** were placed between glass slides and show the characteristic, fan-shaped texture of a smectic A-phase. (c) The POM image of dendrimer **11** in an LC test cell (10 µm; planar cell rubbing) shows the characteristic texture of the smectic C-phase (Characteristic phase transition temperatures: **10**: $T_{SmA\ I}$ = 88.2 °C, $T_{Cr\ SmA}$ = 88.0 °C; dendrimer **11**: $T_{Cr\ SmC}$ = 86 °C, $T_{SmC\ SmA}$ = 98 °C, $T_{SmA\ I}$ = 104.9 °C) [41].

3.2.3.3 Coating of magnetic nanoparticles with (pro)mesogenic ligands

The absence of strongly coordinating ligands in case of electrostatically stabilized MNPs facilitates post-synthetic surface engineering of the MNPs. Electrostatically stabilized MNPs (e. g. Fe_3O_4, $CoFe_2O_4$, $CuFe_2O_4$, $NiFe_2O_4$, $ZnFe_2O_4$) with very small particle sizes

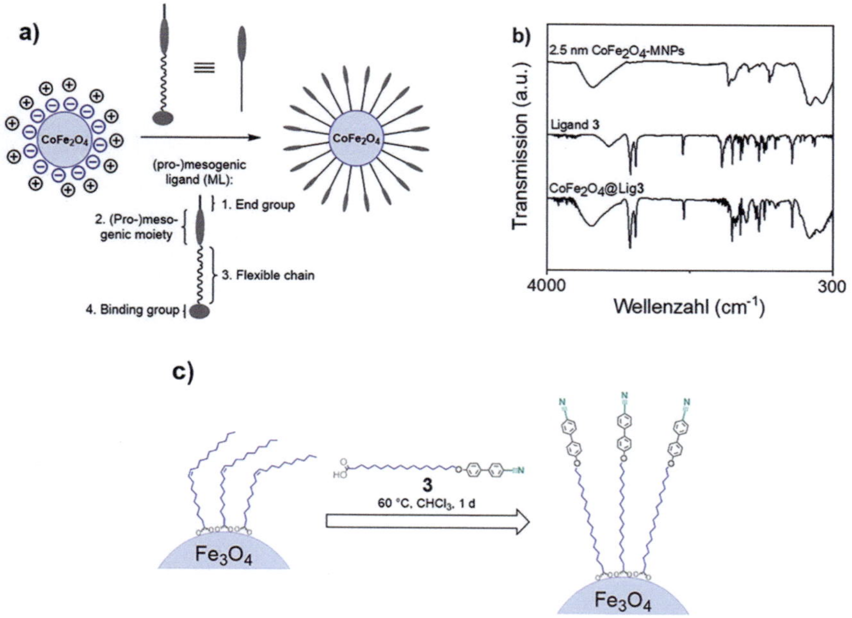

Figure 3.9: Functionalization of $CoFe_2O_4$ MNPs with conventional aliphatic and (pro)mesogenic ligands: (a) Schematic representation of the approach followed to generate functionalized MNPs. (b) Comparison of IR spectra confirmed a successful functionalization with (pro)mesogenic ligand **3** of $CoFe_2O_4$ MNPs. (c) Ligand exchange-based approach was exploited to functionalize different MNPs, e. g. 5.3 nm Fe_3O_4@oleate MNPs to yield 5.3 nm Fe_3O_4@Lig**3**-MNPs. The exchange reaction and the purity of the as-synthesized MNPs was verified *via* IR spectroscopy.

were functionalized and purified modularly with the (pro)mesogenic ligands using a process developed by us (Figure 3.9a) [21, 33]. DMF was used in the coating procedure as a solvent for the electrostatically stabilized MNPs and the (pro)mesogenic ligands. The MNPs coated with ligands were then thoroughly purified by repeated magnetic separation. IR spectra demonstrate the successful functionalization of the MNPs and the attachment of the ligands to the particle surface (Figure 3.9b). The amount of ligand loading on the MNP surface was determined by thermogravimetry (TGA) . The saturation magnetization of the particles was small due to the small particle size and the large surface-to-volume ratio. Spin canting at the MNP surface, crystal defects or formation of a nonmagnetic dead layer usually lead to lower saturation magnetizations in MNPs compared to bulk material. M_{1T} (298 K) of the 2.5 nm size, ligand coated MNPs was in the range of 2.4–5.5 A m^2 kg^{-1} depending on the type of ligand coating (in some cases, no saturation was reached for the applied magnetic field (1 T)).

Thermal decomposition of metal precursors, e.g., at high reaction temperatures in the presence of strongly coordinating ligands yields MNPs of high crystallinity and often, enhanced magnetic properties. In general, these ligands control nucleation and growth processes in the synthetic procedure and they stabilize the as-formed MNPs. Any change to this reaction parameter will critically influence the quality of the final MNPs. Therefore, in case of MNPs where the synthesis required the presence of strongly coordinating ligands, another strategy had to be pursued to functionalize the MNPs by (pro)mesogenic ligands. In this case, the (pro)mesogenic ligands were introduced *via* post-synthetic ligand exchange (Figure 3.9c). This ligand exchange reaction critically depends on various reaction parameters (e. g. binding group, solubility and size of MNPs) and needs to be optimized for every MNP-(pro)mesogenic ligand system. Chloroform at 60 °C, for example, turned out to be a suitable medium to functionalize 5.3 nm, oleic acid-stabilized Fe_3O_4 MNPs with the (pro)mesogenic ligand **3** (corresponding IR spectrum is not shown here). Purification of the MNP functionalized with (pro)mesogenic ligands is important and none of the conventional ligand used for previous MNP synthesis and also no free (pro)mesogenic ligands should remain in the sample.

3.2.3.4 Influence of the structure of the (pro)mesogenic ligands on the interaction of the MNPs with the LC host

The MNPs which were coated with (pro)mesogenic ligands were then integrated in the LC host. After ultrasonic treatment, 2.5 nm sized $CoFe_2O_4$ MNPs coated with (pro) mesogenic ligands, for example, formed a colloidal dispersion in the isotropic phase of 5CB (Figure 3.10b). During the transition to the nematic 5CB phase, some macroscopic agglomerates were formed, which could be easily separated with a laboratory magnet (Figure 3.10c,d). A brown-colored, colloidal MNP dispersion was obtained in the nematic 5CB phase, which was stable in a magnetic field gradient (laboratory magnet) and also showed long-term colloidal stability in the LC host (>5 months) (Figure 3.10e). No macroscopic agglomerates were observed under crossed polarizers. For $CoFe_2O_4$ MNPs coated with **3** (M_{1T} (298 K) = 5.5 A m^2 kg^{-1}), for example, a colloidally stable $CoFe_2O_4$-5CB hybrid material with a concentration of 0.12 wt% could be synthesized. More details on magnetic properties and magnetization curves of $CoFe_2O_4$ MNPs coated with **3** are reported in refs. [21, 33]. While the undoped 5CB exhibited the usual diamagnetic behavior with identical magnetization curves in the nematic and isotropic phases, the magnetic $CoFe_2O_4$-5CB hybrid (0.085 wt%) behaved as a superparamagnet at low magnetic field (displaying no hysteresis) with a sharp knee in the magnetization curve [23]. Due to interaction between mesogens and MNPs *via* ligands, the change of the mesogen orientation also reorients the MNPs, a behavior observed as a sharp knee in the magnetization curve. No significant increase of the clearing temperature T_{N-I} (0.5 °C *via* differential scanning calorimetry (DSC) was observed after doping 5CB with MNPs. The structure of the (pro)mesogenic ligands was highly important and influenced the MNP content of the final colloidal MNP dispersion in the LC host. We could show, for example, that carboxylate binding groups were suitable for binding of (pro)mesogenic ligands to the surface of e. g. $CoFe_2O_4$ MNPs. Ligand loadings of up to 50 wt% were achieved for $CoFe_2O_4$ MNPs using ligand **3** [33]. If the size of the **3**-coated $CoFe_2O_4$ MNPs was slightly increased from 2.5 nm over 3.0 –4.4 nm, the MNP concentration decreased from 0.12 wt% over 0.08 to 0.04 wt%, respectively. It is important to note here, that only MNPs were considered that remained colloidally stable and dispersed in the nematic phase of the 5CB host after exposure to a magnetic field gradient. As shown in Figure 3.10f, the length of the flexible linker ($-(CH_2)_n-$) also influenced the effective

Figure 3.10: Integration of 2.5 nm size $CoFe_2O_4$ MNPs coated with (pro)mesogenic ligand **3** in 5CB: (a) 5CB, (b) colloidal MNP dispersion in the isotropic 5CB phase, (c) after transition to the nematic 5CB phase, (d) magnetic separation of macroscopic agglomerate and (e) stable colloidal MNP dispersion (0.12 wt%) in the nematic phase of 5CB. The corresponding POM images (under crossed polarizers) of the thin film samples in a LC test cell (25 µm, parallel cell rubbing) are shown below (Reproduced from Ref. [20] with permission from the Physical Chemistry Chemical Physics Owner Societies). (f) The concentration of MNPs (mass fraction of cobalt ferrite as determined by inductively coupled plasma optical emission spectroscopy (ICP-OES)) in their stable colloidal dispersion in the nematic 5CB phase depends on the chain lengths of the (pro)mesogenic ligand [41].

particle concentration in the stable colloidal MNP dispersion after magnetic separation (n = 7, 14, 15, 17 for **1**, **2**, **3** and **6**, respectively).

For $CoFe_2O_4$ MNPs, the particle content of the 5CB LC increased with the chain length of the linker (Figure 3.10f). However, it seemed to reach a maximum for n = 15 and further increase of the alkyl chain length to n = 17 did not allow for stabilizing higher MNP concentrations in the 5CB LC. As the length of the alkyl chain increases, the flexibility of the (pro)mesogenic units of the ligands on the MNP surface is improved. This does not only increase the steric MNP repulsion by a larger exclusion volume but also seemed to effectively smooth out the disturbance of the LC in the vicinity of the MNPs. However, the longer the ligands get, the less they resemble also the structure of 5CB which counteracts the stabilization by the bigger exclusion volume. Moreover, the structure of the mesogenic unit influences the MNP stability in the nematic phase of 5CB. MNPs functionalized by (pro)mesogenic ligands with either octylbiphenyl (**8**) or cyanobiphenyl group (**3**) and otherwise similar structure showed a significantly different behavior when immersed in 5CB (Figure 3.11). No MNPs remained in the 5CB LC

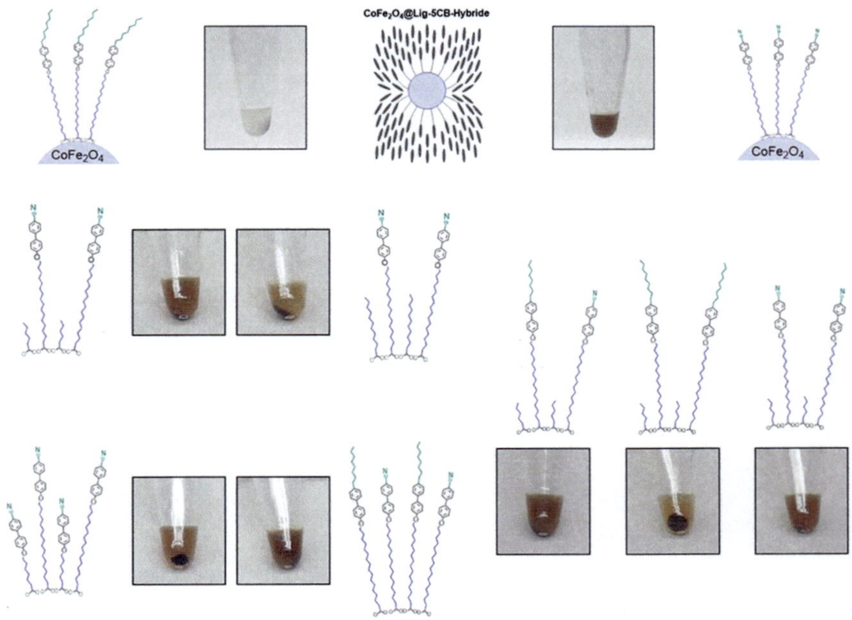

Figure 3.11: Influence of the end group on the stability: schematic representation and photograph, respectively, of the hybrids. The comparison of (pro)mesogenic ligands with different group on the biphenyl moiety shows that ligands with nitrile groups stabilizes the MNPs more efficiently in 5CB. Here, also the influence of coligands (caproic acid, lauric acid) and mixed (pro)mesogenic ligands (ML1/ML3; ML3/ML8) on the colloidal stability of the MNPs in 5CB is shown. This illustrates the importance of MNP-matrix interactions in these systems and the need for a highly specific surface engineering of the MNPs [41].

after magnetic separation of the **8**-coated MNPs, while the brown color of the sample indicates the successful stabilization of the MNPs coated by **3**. Interestingly, if the MNPs were coated with a mixture of both ligands **3** and **8** (Figure 3.11), even higher MNP concentrations were achieved.

The additional use of smaller co-ligands was reported to allow for increasing and optimizing the contact area between the LC mesogens and the mesogenic structural units on the surface of the MNPs [25a, 25c]. They also increase the angle-dependent mobility of the (pro)mesogenic ligand, enabling their optimized alignment with respect to the 5CB host. Both the molar ratio of the (pro)mesogenic ligand/coligand and their relative lengths play a role here. In a systematic series of experiments, the influence of different coligands (caproic acid, lauric acid) and molar ratios (1:1; 1:2) on the stability of MNPs in 5CB was investigated (Figure 3.11). In these experiments, however, it turned out that the correlation between the functionalized surface and the stability in 5CB is complex.

3.2.4 Characterization of MNPs in LC hosts

The magneto-optical properties of the MNP-5CB hybrids were investigated in electric ($\boldsymbol{n} \perp \boldsymbol{E}$), magnetic ($\boldsymbol{n} \perp \boldsymbol{B}$) and combined fields ($\boldsymbol{n} \parallel \boldsymbol{B} \perp \boldsymbol{E}$ and $\boldsymbol{n} \perp \boldsymbol{B} \parallel \boldsymbol{E}$) in collaboration with R. Stannarius and A. Eremin [21]. The Fréedericksz transition in the electric and/or magnetic field was investigated by capacitance and optical measurements of the 2.5 nm size $CoFe_2O_4$ MNPs coated by linear, (pro)mesogenic ligands (**1**, **3**, **4**, **5**) in commercial LC test cells. For MNPs with different linear, (pro)mesogenic ligands (**1**, **3**, **4**, **5**) and different concentrations in 5CB (0.03–0.12 wt%), the electrical Fréedericksz transition was not significantly affected by the embedding of the MNPs in the LC. Also the birefringence of the MNP-5CB hybrids was the same as for pure 5CB. A decrease of the threshold for the magnetic Fréedericksz transition, however, was observed in the external magnetic field. For MNPs with the same particle size (2.5 nm) and different ligands (**1**, **3**, **4**, **5**), the magnetic Fréedericksz threshold B_c linearly decreased with increasing MNP

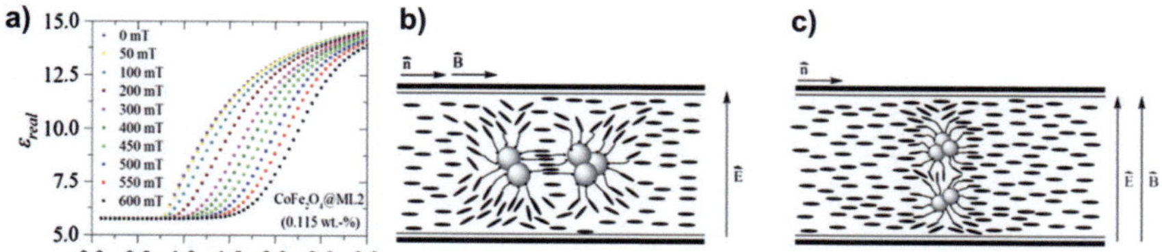

Figure 3.12: Electro- and magneto-optical effects of $CoFe_2O_4$ MNPs coated with a linear, (pro) mesogenic ligand (**3**): (a) Dependence of the electric threshold on the magnetic induction $\boldsymbol{B}$ for crossed fields ($\boldsymbol{n} \parallel \boldsymbol{B} \perp \boldsymbol{E}$). Cartoon illustrating the potential behavior of anisotropic MNP aggregates in (b) crossed ($\boldsymbol{n} \parallel \boldsymbol{B} \perp \boldsymbol{E}$) and (c) parallel ($\boldsymbol{n} \perp \boldsymbol{B} \parallel \boldsymbol{E}$) fields (Reproduced from Ref. [21] with permission from the PCCP Owner Societies).

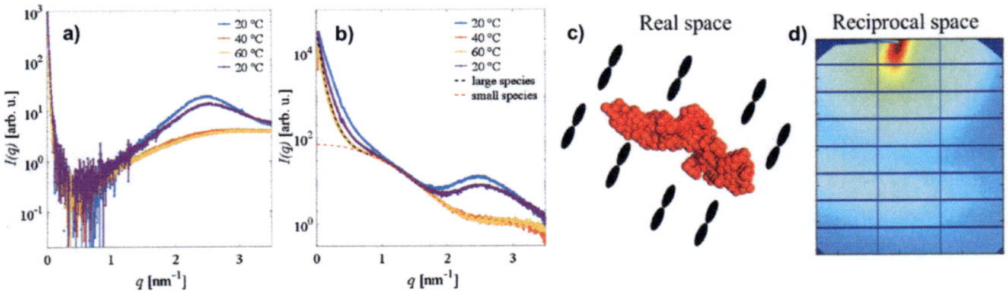

Figure 3.13: Temperature-dependent SAXS intensity of (a) the 5CB reference and (b) 3 nm sized, **3**-coated $CoFe_2O_4$ MNPs in 5CB. (c) *Ab initio* model showing aggregate orientation with respect to the 5CB dimers. (d) 2D SAXS pattern at $T = 20$ °C (reproduced with permission of the Royal Society of Chemistry from ref. [23]).

concentration in 5CB. This indicated that the magnetic properties of the MNPs and their interaction with the 5CB host were responsible for the observed shifts. The electrical Fréederick transition was also investigated under a bias magnetic field either parallel ($\boldsymbol{n} \perp \boldsymbol{B} \parallel \boldsymbol{E}$) or perpendicular ($\boldsymbol{n} \parallel \boldsymbol{B} \perp \boldsymbol{E}$) to the applied electric field. In this case, the alignment of the director $\boldsymbol{n}$ of the LC host lead either to an increase ($\boldsymbol{n} \parallel \boldsymbol{B} \perp \boldsymbol{E}$) or decrease ($\boldsymbol{n} \perp \boldsymbol{B} \parallel \boldsymbol{E}$) of the electric threshold field. If the magnetic field was perpendicular to the electric field ($\boldsymbol{n} \parallel \boldsymbol{B} \perp \boldsymbol{E}$), the MNPs seemed to counteract the influence of the external magnetic field on the alignment of the 5CB mesogens (Figure 3.12a). Based on the Burylov-Raikher theory, the formation of anisotropically-shaped MNP clusters in the 5CB was suggested to explain this counterintuitive behavior (Figure 3.12b,c). Due to the small size of the MNPs, formation of the anisotropic aggregates should not be induced by magnetic dipole interaction but rather by interaction of the MNPs with the surrounding LC host.

Based on these results, the interaction of the LC with the MNPs was further investigated by small-angle X-ray scattering (SAXS) and by a superconducting quantum interference device magnetometer (SQUID) [23] In SAXS measurements of 3 nm sized, **3**-coated $CoFe_2O_4$ MNPs in 5CB, a strong signal was observed for MNPs at smaller scattering vectors $\boldsymbol{q}$ besides the scattering contribution of 5CB (Figure 3.13). Interestingly, this signal was anisotropic and pointed to the direction of the LC nematic director $\boldsymbol{n}$. If the LC was heated above the clearing point T_{NI} (40 °C), where it forms the isotropic phase, the SAXS patterns became also isotropic. If the sample was cooled below T_{NI}, the SAXS pattern became anisotropic again, but to a lesser extent. Refinement of the SAXS data yielded a mean size of the MNPs of 3.8 nm, which was in good agreement with the results of the analysis of the MNPs by TEM and dynamic light scattering. However, the characteristic shape of the pair-distance distribution function and the anisotropic scattering pattern in the nematic 5CB phase also strongly pointed to the formation of small, anisotropically-structured MNP aggregates (with a long axis of approximately 100 nm). Overall, the 5CB LC was well oriented with the long axis of the MNP clusters oriented perpendicular to the 5CB long axis (Figure 3.13). Magnetic measurements of the MNPs in 5CB revealed the typical behavior of a superparamagnetic system up to a field strength of approximately 557 kA m^{-1}

where the magnetic order seemed to change. As the magnetic field strength was further increased, only the diamagnetic contribution from 5CB was observed.

3.3 Conclusion and outlook

Hybridization of MNPs and LCs can effectively modify the electro- and magneto-optic responses and other physical characteristics of the LCs. The MNP-matrix interactions in these systems, however, are highly complex and depend on various parameters related to the size, shape, topology, surface structure and magnetic properties of the MNPs on one side and the structure and dimensions of the LC mesogens on the other side. Elastic distortions of the host LC crystal can mediate attractive interactions between the MNPs, while such interactions are absent in usual colloidal dispersions with isotropic ferrofluids. Therefore, the tendency of MNPs to agglomerate is much stronger in the anisotropic LC phase than in the corresponding isotropic phase. Surface engineering of MNPs using tailored (pro)mesogenic ligands can tackle this challenge. In this context, a library of different (pro)mesogenic ligands has been established where the structural parts (i. e., MNP binding group, spacer length of alkyl chain and mesogenic unit) were systematically addressed and varied. The structure of the (pro)mesogenic ligands critically influences the MNP–matrix interactions and the stability of the colloidal MNP dispersion in the LC host. The increase in spacer length to up to 15 C atoms increased the overall MNP stability in 5CB, while its further increase decreased stability. Moreover, the structure of the mesogenic unit was highly important and a nitrile group was shown to be beneficial for MNP stabilization. Small MNPs were functionalized by these (pro)mesogenic ligands and stabilized in the nematic phase of 5CB. The doping of 5CB LC with these MNPs leads to a decrease of the magnetic Fréedericksz transition threshold. As shown by SAXS measurements, the MNPs formed anisotropic structures in 5CB which lead to an interesting behavior particularly in crossed electric and magnetic fields. However, with these ligands the stabilization in 5CB was restricted to small MNP sizes. If the particle size was further increased, colloidal dispersions of lower MNP content were achieved. In this context, sterically more demanding, (pro)mesogenic ligands with dendritic structures might be highly interesting. This may enable the functionalization of larger (anisotropic) MNPs and their successive stabilization in 5CB, where interesting effects on the magneto-optical properties may be expected.

The contactless nature of magnetic manipulation and versatility of magnetic, magneto-optical and magneto-electro-optical effects are extremely attractive for various applications (e. g. in LCDs or sensors). In this context, further enhancement of the sensitivity to the magnetic field and decrease of the magnetic Fréedericksz threshold may be achieved by long-term colloidal stabilization of larger MNPs with improved magnetic properties (e. g. higher M_S). In addition, the stabilization of MNPs with anisotropic shape (such as nanorods) may further enhance the switching properties. Here, also the modulation of the LC structure (e. g. in terms of chirality) is certainly another very interesting aspect. Moreover, LC-mediated assembly of nanoparticles has recently attracted a lot of interest. It may not only lead to LCs with ferromagnetic properties but also to novel and complex, nanoparticle-based superstructures. The parameter space is large here which opens up many possibilities for future studies on active matter and programmable liquid constructs.

Author contribution: All the authors have accepted responsibility for the entire content of this submitted manuscript and approved submission.
Research funding: This work was supported by Deutsche Forschungsgemeinschaft (BE 2243/2-1, BE 2243/2-2, BE 2243/2-3).
Conflict of interest statement: The authors declare no conflicts of interest regarding this article.

References

1. a) Shen Y, Dierking I. Perspectives in liquid-crystal-aided nanotechnology and nanoscience. Appl Sci 2019;9:2512–59. b) Chernyshuk SB, Lev BI, Yokoyama H. Paranematic interaction between nanoparticles of ordinary shape. Phys Rev E 2005;71:062701-1–4.
2. Qi H, Hegmann T. Impact of nanoscale particles and carbon nanotubes on current and future generations of liquid crystal displays. J Mater Chem 2008;18:3288–94.
3. Hu W, Zhao H, Shan L, Song L, Cao H, Yang Z, et al Magnetite nanoparticles/chiral nematic liquid crystal composites with magnetically addressable and magnetically erasable characteristics. Liq Cryst 2010;37:563–9.
4. Rupnik PM, Lisjak D, Čopič M, Mertelj A. Ferromagnetic liquid crystals for magnetic field visualisation. Liq Cryst 2015;42:1684–8.
5. a) Zadoina L, Lonetti B, Soulantica K, Mingotaud AF, Respaud M, Chaudret B, et al Liquid crystalline magnetic materials. J Mater Chem 2009;19:8075–8. b) Riou O, Lonetti B, Davidson P, Tan RP, Cormary B, Mingotaud A-F, et al Liquid crystalline polymer–co nanorod hybrids: structural analysis and response to a magnetic field. J Phys Chem B 2014;118:3218–25. c) Riou O, Lonetti B, Tan RP, Harmel J, Soulantica K, Davidson P, et al Room-temperature, strain-tunable orientation of magnetization in a hybrid ferromagnetic co nanorod-liquid crystalline elastomer nanocomposite. Angew Chem Int Ed 2015;54:10811–15. d) Kaiser A, Winkler M, Krause S, Finkelmann H, Schmidt AM. Magnetoactive liquid crystal elastomer nanocomposites. J Mater Chem 2009;19:538–43. e) Garcia-Márquez A, Demortière A, Heinrich B, Guillon D, Bégin-Colin S, Donnio B. Iron oxide nanoparticle-containing main-chain liquid crystalline elastomer: towards soft magnetoactive networks. J Mater Chem 2011;21:8994–6.
6. Song HM, Kim JC, Hong JH, Lee YB, Choi J, Lee JI, et al Magnetic and transparent composites by linking liquid crystals to ferrite nanoparticles through covalent networks. Adv Funct Mater 2007;17:2070–6.
7. a) Fleischmann E-K, Zentel R. Liquid-crystalline ordering as a concept in materials science: from semiconductors to stimuli-responsive devices. Angew Chem Int Ed 2013;52:8810–27. b) Tschierske C. Development of structural complexity by liquid-crystal self-assembly. Angew Chem Int Ed 2013; 52:8828–78. c) Bisoyi HK, Li Q. Light-driven liquid crystalline materials: from photo-induced phase transitions and property modulations to applications. Chem Rev 2016;116:15089–166.
8. Brochard F, de Gennes PG. Swelling equilibrium and light spectroscopy in swollen polymeric networks at theta conditions. J Phys (France) 1970;31:691–708.

9. a) Chen S-H, Amer NM. Observation of macroscopic collective behavior and new texture in magnetically doped liquid crystals. Phys Rev Lett 1983;51:2298–301. b) Bacri JC, Neto AMF. Dynamics of lyotropic ferronematic liquid crystals submitted to magnetic fields. Phys Rev E 1994; 50:3860–4. c) Ponsinet V, Fabre P, Veyssié M. Transition of a ferrosmectic in a very weak magnetic field. Europhys Lett 1995;30:277–82. d) Berejnov V, Cabuil V, Perzynski R, Raikher Y. Lyotropic system potassium laurate/1-decanol/water as a carrier medium for a ferronematic liquid crystal: phase diagram study. J Phys Chem B 1998;102:7132–8. e) Berejnov V, Raikher Y, Cabuil V, Bacri JC, Perzynski R. Synthesis of stable lyotropic ferronematics with high magnetic content. J Colloid Interface Sci 1998;199:215–7. f) Berejnov V, Bacri JC, Cabuil V, Perzynski R, Raikher Y. Lyotropic ferronematics: magnetic orientational transition in the discotic phase. Europhys Lett 1998;41: 507–12. g) Ramos L, Fabre P, Fruchter L. Magnetic field induced instabilities of a doped lyotropic hexagonal phase. Eur Phys J B 1999;8:67–72. h) Cruz CD, Sandre O, Cabuil V. Phase behavior of nanoparticles in a thermotropic liquid crystal. J Phys Chem B 2005;109:14292–9. i) Buluy O, Nepijko S, Reshetnyak V, Ouskova E, Zadorozhnii V, Leonhardt A, et al Magnetic sensitivity of a dispersion of aggregated ferromagnetic carbon nanotubes in liquid crystals. Soft Matter 2011;7: 644–9. j) Vallooran JJ, Bolisetty S, Mezzenga R. Macroscopic alignment of lyotropic liquid crystals using magnetic nanoparticles. Adv Mater 2011;23:3932–7. k) Kopčanský P, Tomašovičová N, Koneracká M, Timko M, Závišová V, Džarová A, et al Phase transitions in liquid crystal doped with magnetic particles of different shapes. Int J Thermophys 2011;32:807–17. l) Podoliak N, Buchnev O, Bavykin DV, Kulak AN, Kaczmarek M, Sluckin TJ. Magnetite nanorod thermotropic liquid crystal colloids: synthesis, optics and theory. J Colloid Interface Sci 2012;386:158–66. m) Garbovskiy Y, Baptist JR, Thompson J, Hunter T, Lim JH, Min SG, et al Increasing the switching speed of liquid crystal devices with magnetic nanorods. Appl Phys Lett 2012;101:181109-1–5.
10. Mertelj A, Lisjak D, Drofenik M, Čopič M. Ferromagnetism in suspensions of magnetic platelets in liquid crystal. Nature 2013;504:237–41.
11. Shuai M, Klittnick A, Shen Y, Smith GP, Tuchband MR, Zhu C, et al Spontaneous liquid crystal and ferromagnetic ordering of colloidal magnetic nanoplates. Nat Commun 2016;7:10394.
12. Mertelj A, Osterman N, Lisjak D, Čopič M. Magneto-optic and converse magnetoelectric effects in a ferromagnetic liquid crystal. Soft Matter 2014;10:9065–72.
13. a) May K, Eremin A, Stannarius R, Peroukidis SD, Klapp SHL, Klein S. Colloidal suspensions of rodlike nanocrystals and magnetic spheres under an external magnetic stimulus: experiment and molecular dynamics simulation. Langmuir 2016;32:5085–93. b) May K, Eremin A, Stannarius R, Szabó B, Börzsönyi T, Appel I, et al Exceptionally large magneto-optical response in dispersions of plate-like nanocrystallites and magnetic nanoparticles. J Magn Mater 2017;431:79–83.
14. Peroukidis SD, Klapp SHL. Orientational order and translational dynamics of magnetic particle assemblies in liquid crystals. Soft Matter 2016;12:6841–50.
15. Nádasi H, Corradi Á, Stannarius R, Koch K, Schmidt AM, Aya S, et al The role of structural anisotropy in the magnetooptical response of an organoferrogel with mobile magnetic nanoparticles. Soft Matter 2019;15:3788–95.
16. a) Odenbach S, Thurm S. Magnetoviscous effects in ferrofluids. In: Odenbach S, editor. Ferrofluids: magnetically controllable fluids and their applications. Berlin, Heidelberg: Springer Berlin Heidelberg; 2002:185–201 pp.b) Potisk T, Pleiner H, Svenšek D, Brand HR. Magneto-optic dynamics in a ferromagnetic nematic liquid crystal. Phys Rev E 2018;97:042705-1–13. c) Potisk T, Svenšek D, Pleiner H, Brand HR. Continuum model of magnetic field induced viscoelasticity in magnetorheological fluids. J Chem Phys 2019;150:174901-1–12.
17. Arantes FR, Odenbach S. The magnetoviscous effect of micellar solutions doped with water based ferrofluids. J Magn Mater 2015;390:91–5.
18. Appel I, Behrens S. Influence of the particle parameters on the stability of magnetic dopants in a ferrolyotropic suspension. J Magn Mater 2017;431:49–53.

19. Siboni NH, Shrivastav GP, Klapp SHL. Non-monotonic response of a sheared magnetic liquid crystal to a continuously increasing external field. J. Chem. Phys 2020;152:024505-1-12.
20. Liu X, Kent N, Ceballos A, Streubel R, Jiang Y, Chai Y, et al Reconfigurable ferromagnetic liquid droplets. Science 2019;365:264–7.
21. Appel I, Nadasi H, Reitz C, Sebastian N, Hahn H, Eremin A, et al Doping of nematic cyanobiphenyl liquid crystals with mesogen-hybridized magnetic nanoparticles. Phys Chem Phys 2017;19:12127–35.
22. Prodanov MF, Buluy OG, Popova EV, Gamzaeva SA, Reznikov YO, Vashchenko VV. Magnetic actuation of a thermodynamically stable colloid of ferromagnetic nanoparticles in a liquid crystal. Soft Matter 2016;12:6601–9.
23. Gdovinova V, Schroer MA, Tomasovicova N, Appel I, Behrens S, Majorosova J, et al Structuralization of magnetic nanoparticles in 5CB liquid crystals. Soft Matter 2017;13:7890–6.
24. Stamatoiu O, Mirzaei J, Feng X, Hegmann T. Nanoparticles in liquid crystals and liquid crystalline nanoparticles. In: Tschierske C, editor. Liquid crystals: materials design and self-assembly. Berlin, Heidelberg: Springer Berlin Heidelberg; 2012:331–93 pp.
25. a) Qi H, Kinkead B, Marx VM, Zhang HR, Hegmann T. Miscibility and alignment effects of mixed monolayer cyanobiphenyl liquid-crystal-capped gold nanoparticles in nematic cyanobiphenyl liquid crystal hosts. ChemPhysChem 2009;10:1211–8. b) Draper M, Saez IM, Cowling SJ, Gai P, Heinrich B, Donnio B, et al Self-assembly and shape morphology of liquid crystalline gold metamaterials. Adv Funct Mater 2011;21:1260–78. c) Prodanov MF, Pogorelova NV, Kryshtal AP, Klymchenko AS, Mely Y, Semynozhenko VP, et al Thermodynamically stable dispersions of quantum dots in a nematic liquid crystal. Langmuir 2013;29:9301–9. d) Mirzaei J, Urbanski M, Kitzerow H-S, Hegmann T. Synthesis of liquid crystal silane-functionalized gold nanoparticles and their effects on the optical and electro-optic properties of a structurally related nematic liquid crystal. ChemPhysChem 2014;15:1381–94. e) Feng X, Sosa-Vargas L, Umadevi S, Mori T, Shimizu Y, Hegmann T. Discotic liquid crystal-functionalized gold nanorods: 2- and 3D self-assembly and macroscopic alignment as well as increased charge carrier mobility in hexagonal columnar liquid crystal hosts affected by molecular packing and π-π interactions. Adv Funct Mater 2015;25: 1180–92. f) Nealon GL, Greget R, Dominguez C, Nagy ZT, Guillon D, Gallani JL, et al Liquid-crystalline nanoparticles: hybrid design and mesophase structures. Beilstein J Org Chem 2012;8: 349–70. g) Khatua S, Manna P, Chang W-S, Tcherniak A, Friedlander E, Zubarev ER, et al Plasmonic nanoparticles–liquid crystal composites. J Phys Chem C 2010;114:7251–7.
26. Nemati A, Shadpour S, Querciagrossa L, Li L, Mori T, Gao M, et al Chirality amplification by desymmetrization of chiral ligand-capped nanoparticles to nanorods quantified in soft condensed matter. Nat Commun 2018;9:3908-1–13.
27. a) Kanie K, Sugimoto T. Organic–inorganic hybrid liquid crystals: hybridization of calamitic liquid-crystalline amines with monodispersed anisotropic TiO_2 nanoparticles. J Am Chem Soc 2003;125: 10518–9. b) Kanie K, Muramatsu A. Organic–inorganic hybrid liquid crystals: thermotropic mesophases formed by hybridization of liquid-crystalline phosphates and monodispersed α-Fe_2O_3 particles. J Am Chem Soc 2005;127:11578–9. c) Saliba S, Coppel Y, Davidson P, Mingotaud C, Chaudret B, Kahn ML, et al Liquid crystal based on hybrid zinc oxide nanoparticles. J Mater Chem 2011;21:6821–3.
28. Demortière A, Buathong S, Pichon PB, Panissod P, Guillon D, Bégin-Colin S, et al Nematic-like organization of magnetic mesogen-hybridized nanoparticles. Small 2010;6:1341–6.
29. Popova EV, Gamzaeva SA, Krivoshey AI, Kryshtal AP, Fedoryako AP, Prodanov MF, et al Dielectric properties of magnetic nanoparticles' suspension in a ferroelectric liquid crystal. Liq Cryst 2015; 42:334–43.
30. a) Sun S, Zeng H, Robinson DB, Raoux S, Rice PM, Wang SX, et al Monodisperse MFe2O4 (M = Fe, Co, Mn) nanoparticles. J Am Chem Soc 2004;126:273–9. b) Behrens S, Essig S. A facile procedure for magnetic fluids using room temperature ionic liquids. J Mater Chem 2012;22:3811–6. c) Essig S,

Behrens S. Ionic liquids as size- and shape-regulating solvents for the synthesis of cobalt nanoparticles. Chem Ing Tech 2015;87:1741–7. d) Gorschinski A, Khelashvili G, Schild D, Habicht W, Brand R, Ghafari M, et al A simple aminoalkyl siloxane-mediated route to functional magnetic metal nanoparticles and magnetic nanocomposites. J Mater Chem 2009;19:8829–38.
31. Liu C, Zou B, Rondinone AJ, Zhang ZJ. Reverse micelle synthesis and characterization of superparamagnetic $MnFe_2O_4$ spinel ferrite nanocrystallites. J Phys Chem B 2000;104:1141–5.
32. Wang X, Zhuang J, Peng Q, Li Y. A general strategy for nanocrystal synthesis. Nature 2005;437: 121–4.
33. Appel I [Ph.D. thesis]. Magnetische Hybridmaterialien auf Basis komplexer, flüssigkristalliner Systeme. Ruprecht-Karls University Heidelberg: Heidelberg; 2017.
34. a) Zhou Z, Zhu X, Wu D, Chen Q, Huang D, Sun C, et al Anisotropic shaped iron oxide nanostructures: controlled synthesis and proton relaxation shortening effects. Chem Mater 2015; 27:3505–15. b) Leslie-Pelecky DL, Rieke RD. Magnetic properties of nanostructured materials. Chem Mater 1996;8:1770–83.
35. Orlandi S, Benini E, Miglioli I, Evans DR, Reshetnyak V, Zannoni C. Doping liquid crystals with nanoparticles. A computer simulation of the effects of nanoparticle shape. Phys Chem Phys 2016; 18:2428–41.
36. Hähsler M, Landers J, Nowack T, Salamon S, Zimmermann M, Heißler S, et al Magnetic properties and Mössbauer spectroscopy of Fe_3O_4/$CoFe_2O_4$ nanorods. Inorg Chem 2020;59:3677–85.
37. Kim D, Lee N, Park M, Kim BH, An K, Hyeon T. Synthesis of uniform ferrimagnetic magnetite nanocubes. J Am Chem Soc 2009;131:454–5.
38. a) Ovtar S, Lisjak D, Drofenik M. Barium hexaferrite suspensions for electrophoretic deposition. J Colloid Interface Sci 2009;337:456–63. b) Lisjak D, Jenuš P, Mertelj A. Influence of the morphology of ferrite nanoparticles on the directed assembly into magnetically anisotropic herarchical structures. Langmuir 2014;30:6588–95.
39. Pullar RC. Hexagonal ferrites: a review of the synthesis, properties and applications of hexaferrite ceramics. Prog Mater Sci 2012;57:1191–334.
40. Hähsler M, Zimmermann M, Heißler S, Behrens S. Sc-doped barium hexaferrite nanodiscs: tuning morphology and magnetic properties. J Magn Mater 2020;500:166349-3–5.
41. Hähsler M [Ph.D. thesis]. Responsive magnetische Hybridmaterialienauf Basis funktionalisierter form(an)isotroper Nanopartikel und Flüssigkristalle. Ruprecht-Karls University Heidelberg: Heidelberg; 2020.
42. Derouiche Y, Dubois F, Douali R, Legrand C, Maschke U. Some properties of nematic liquid crystal E7/acrylic polymer networks. Mol Cryst Liq Cryst 2011;541:201/[439]–210/[448].
43. Gdovinová V, Schroer MA, Tomašovičová N, Appel I, Behrens S, Majorošová J, et al Structuralization of magnetic nanoparticles in 5CB liquid crystals. Soft Matter 2017;13:7890–6.
44. Gray GW, Harrison KJ, Nash JA. New family of nematic liquid crystals for displays. Electron Lett 1973;9:130–1.
45. Fukuda J-I, Stark H, Yoneya M, Yokoyama H. Interaction between two spherical particles in a nematic liquid crystal. Phys Rev E 2004;69:041706-1–10.
46. Kopčanský P, Potočová I, Koneracká M, Timko M, Jansen AGM, Jadzyn J, et al The anchoring of nematic molecules on magnetic particles in some types of ferronematics. J Magn Mater 2005;289: 101–4.
47. Muševič I. Nematic liquid-crystal colloids. Materials 2018;11:24.
48. a) Reznikov Y. Ferroelectric colloids in liquid crystals. In: Li Q, editor. Liquid crystals beyond displays: chemistry, physics, and applications. John Wiley & Sons: Hoboken, New Jersey; 2012: 403–26 pp.b) Voloschenko D, Pishnyak OP, Shiyanovskii SV, Lavrentovich OD. Effect of director distortions on morphologies of phase separation in liquid crystals. Phys Rev E 2002;65:060701.

49. Hähsler M, Behrens S. Dendritic ligands for magnetic suspensions in liquid crystals. Eur J Org Chem 2019;2019:7820–30.
50. Prodanov MF, Vashchenko OV, Vashchenko VV. A synthetic strategy toward branched oligomesogenic phosphonic acids: comparison of alternative pathways. Tetrahedron Lett 2014;55: 275–8.
51. Dierking I. Textures of liquid crystals. Wiley VCH; WILEY-VCH Verlag GmbH & Co. KGaA, Weinheim; 2003.

Niklas Lucht, Stephan Hinrichs, Larissa Großmann, Catharina Pelz, Elena Felgenhauer, Eike Clasen, Max Schwenk and Birgit Hankiewicz*

4 Synthesis of magnetic ferrogels: a tool-box approach for finely tuned magnetic- and temperature-dependent properties

Abstract: Multi responsive hydrogels have many potential applications in the field of medicine as well as technical fields and are of great interest in fundamental research. Here we present the synthesis and characterization of tailored magnetic hydrogels – micro- as well as macrogels – which consist of iron oxide and cobalt ferrite, varying in phase and morphology, embedded in a thermoresponsive polymer. We introduce new ways to synthesize magnetic particles and revisit some common strategies when dealing with particle synthesis. Subsequently we discuss the details of the thermoresponsive matrix and how we can influence and manipulate the thermoresponsive properties, i.e. the lower critical solution temperature. Ultimately, we present the particle-hydrogel composite and show two exemplary applications for particle matrix interactions, i.e. heat transfer and reorientation of the particles in a magnetic field.

Keywords: cobalt ferrite; hyperthermia; iron oxides; magnetite; SPIONs; thermoresponsive polymer.

4.1 Introduction

Ferrogels combine the advantages of a gel- or polymer matrix with the magnetic properties of a ferrofluid [1–4]. As matrix materials hydrogels can be utilized to incorporate water. Many of those hydrogels can respond to external stimuli like temperature, pH or light. Those kinds of hydrogels are known as responsive hydrogels [4–7]. Their characteristics have a variety of potential applications in medicine and technics [2, 6–9].

An even more versatile class of materials are multiresponsive hydrogels. When the hydrogels are combined with magnetic particles, the particles can heat up the polymer matrix when exposed to an alternating current magnetic field, thus the magnetic properties of the particles influence the thermoresponsive properties of the hydrogel.

***Corresponding author: Birgit Hankiewicz,** Institute of Physical Chemistry, Hamburg University, Grindelallee 117, 20146 Hamburg, Germany, E-mail: birgit.hankiewicz@chemie.uni-hamburg.de
Niklas Lucht, Stephan Hinrichs, Larissa Großmann, Catharina Pelz, Elena Felgenhauer, Eike Clasen and Max Schwenk, Institute of Physical Chemistry, Hamburg University, Grindelallee 117, 20146 Hamburg, Germany

This article has previously been published in the journal Physical Sciences Reviews. Please cite as: N. Lucht, S. Hinrichs, L. Großmann, C. Pelz, E. Felgenhauer, E. Clasen, M. Schwenk and B. Hankiewicz "Synthesis of magnetic ferrogels: a tool-box approach for finely tuned magnetic- and temperature-dependent properties" *Physical Sciences Reviews* [Online] 2021, 6. DOI: 10.1515/psr-2019-0120 | https://doi.org/10.1515/9783110569636-004

Those materials can be used in heat therapy for cancer treatment [10] or in controlled drug release applications [11] as well as for shape-memory actuators [3].

When an alternating magnetic field is applied, the particles can rotate either by spin-flip processes within the particles, known as Néel relaxation, or by rotation of the whole particle, known as Brownian relaxation. Which process dominates depends on the particle size and the magnetic anisotropy. The magnetic relaxation mechanism can be tuned to affect the possible heating efficiency via induction [12–14]. Hard magnetic materials like cobalt ferrite are already dominated by Brownian relaxation above 5 nm, whereas soft magnetic particles like magnetite require larger particle sizes [14]. The shape of the particles plays an important role as well.

Anisotropic particles are more responsive to external magnetic fields compared to spherical particles due to their higher coercivity force [15, 16]. Cubic particles are suitable particles for magnetic resonance therapy [17] or as heat mediator in general purpose [18, 19].

Finally, an important parameter is the magnetic moment. For most applications, a high magnetic moment is necessary. Hence, there is a demand for magnetic particles with a high saturation magnetization. Cobalt, iron and other transition-metal composites have an inherently high magnetic moment. However, they suffer from their high sensitivity to oxidation and the synthesis is often challenging [20], therefore iron oxides like magnetite and maghemite are primarily used despite their lower magnetic moment. Depending on the application, the right choice of elements used for magnetic particles is crucial, apart from their shape and size.

In the Section 4.2, we will sum up the synthesis procedures for magnetic nanoparticles with the focus on pure and mixed iron oxides in water. In addition, we will highlight the preparation of anisotropic particles with a low magnetic moment like goethite and hematite which can be oriented in very low magnetic fields due to their inherent anisotropy. In Section 4.3 we will summarize the possibilities to embed the nanoparticles within a thermoresponsive matrix. In Section 4.4 we will focus on a possible application for magnetic hydrogels: clinical hyperthermia. We will present a proof of concept for hyperthermia induced into a poly(*N*-isopropylacrylamide) (pNIPAM) matrix by cobalt ferrite nanoparticles. Section 4.5 of this report will cover the details about the experimental methods.

4.2 Preparation of magnetic particles

One of the main challenges during the synthesis is the stabilization right after particle formation to prevent agglomeration as early as possible. Once particles are agglomerated it is hard to separate them again. The stronger the magnetic dipole-dipole interactions are the higher the tendency to form agglomerates. Especially for biomedical as well as technological applications particles with high magnetization are necessary which must be still stable in dispersion. Particles with high magnetic moments like pure metals, e.g., iron and cobalt, suffer from oxidization and are sensitive to handle in general. Therefore, most common magnetic nanoparticles are the respective oxides; magnetite most prominently. Despite being an oxide magnetite is vulnerable to further oxidization into a less magnetic iron oxide phase.

Another important factor to consider is that most biomedical applications require water-based systems. There is a plethora of aqueous synthesis routes to prepare aqueous iron oxides which can be mainly divided into two categories [20]. One is the direct bottom–up synthesis of nanoparticles using iron salt precursors the other is thermal decomposition of organic iron complexes. The latter needs a very high-boiling organic solvent which makes a phase transfer as post modification step obligatory [21]. The main advantage of this method is that very narrow size distributions can be achieved and superparamagnetic iron oxide nanoparticles (SPIONs) below 20 nm can be produced [22]. Apart from the additional phase-transfer step required, the toxicity of the organic solvent is another drawback. The common direct aqueous approach is based on a coprecipitation method. The main advantage of this method is that the synthesis is performed in directly water and the particles can be stabilized without phase transfer [23].

Using an autoclave as the reaction vessel, the synthesis can take place at temperatures above the boiling point of water, so-called hydrothermal synthesis. Hydrothermal synthesis using an aqueous solution of simple metal salts offers a very facile, fast and low-cost procedure [24]. Additionally there is a tendency for higher crystallinity using a hydrothermal approach compared to other methods [25].

In the next section we will present the outcome of our research on how to synthesize anisotropic magnetite particles in water. We will discuss the possibility to use a reduction step to prepare magnetite and discuss cubic magnetite particles in more details. In addition, we show the potential of cobalt ferrite as a model system for Brownian relaxation.

4.2.1 Anisotropic iron oxide particles

A general rule of thumb for magnetic particles is that with increasing particle size the tendency to agglomerate grows. One approach to circumvent this problem is the synthesis of a low magnetic moment iron oxide species which is stabilized in solution, e.g., by synthesis of a silica-shell around the particle, to prevent agglomeration. In a second step a phase transformation to a stronger magnetic iron phase like magnetite can be performed. Hereafter we present promising precursors with the focus on anisotropic particles and delineate approaches to transform the particle phase.

4.2.1.1 Goethite, hematite and akaganeite

Literature describes many different ways to synthesize anisotropic goethite, hematite and akaganeite. Goethite is the most stable iron oxide phase. Goethite results via aging of iron(III)-salts at high pH values [26]. Depending on the temperature and concentration different shapes can be synthesized, i.e., long cylinders can be prepared keeping an iron(III)-solution for nine days at 70 °C [27]. The resulting cylinders are of about 500 nm in length and 10 nm in width (Figure 4.1A). Most of the known goethite syntheses lead to polydisperse systems (Figure 4.1B). By transferring the aging process into a hydrothermal reactor, we can gain some more control over the shape of the goethite rods [28, 29]. Surface active substances can be used to tune the particle shape [30, 31]. In short, akaganeite (described further down) is synthesized by hydrolysis and used as a precursor to age in a hydrothermal reactor. The pH is adjusted to 12 and poly vinyl pyrrolidone (PVP) is added, the amount depending on the desired particle size. The hydrothermal process takes part at 160 °C under 10 bar nitrogen pressure. The reaction is stirred for 1 h and then heated during another hour, the temperature is kept at 160 °C for 2 h before cooling to ambient temperatures in an ice bath. The final product, i.e., length and width of the cylinders, depends on the stirring speed, the amount of PVP and the exact timing of PVP addition, as well as on the precursor quality. Further details will be discussed elsewhere [32].

Hematite can be prepared at weakly alkaline conditions [26]. The typical crystalline structure of hematite is cubical, therefore it is the easiest to prepare cubes [33]. To demonstrate this, an iron-(III)-chloride solution has been aged for eight days. Depending

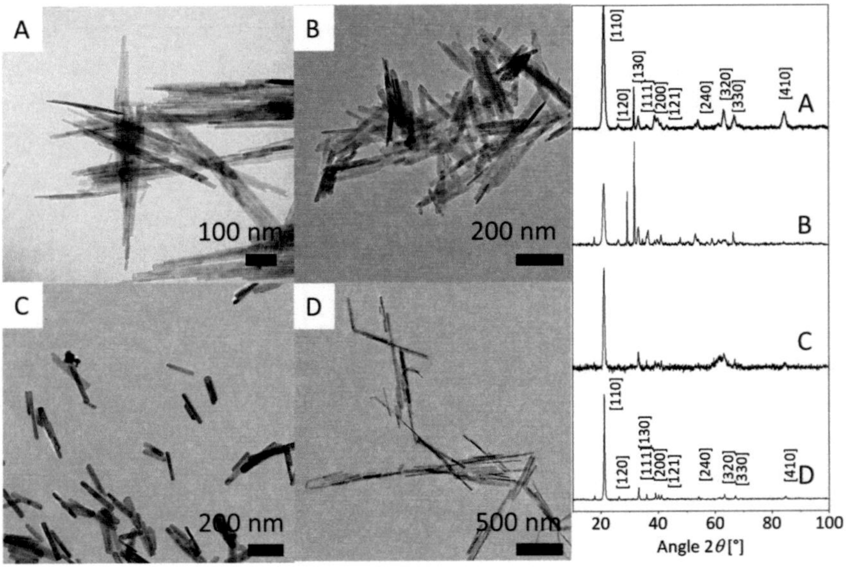

Figure 4.1: TE micrographs (left) and XRD (right) of anisotropic goethite nanoparticles, (A) aged at high pH at 70 °C for nine days, (B) aged at high pH over several days at room temperature, (C) aged in a hydrothermal reactor at pH 12 over 4 h at 160 °C (D) aged in a hydrothermal reactor at pH 12 over 4 h at 160 °C with PVP. The peaks of the XRD pattern can be indexed by the goethite reference indicated in black.

on the temperature and the initial aging solution different particle sizes can be prepared [34]. This yields cubic particles between 30 nm and 1 μm (Figure 4.2A). XRD confirms phase pure hematite. Compared to goethite, monodisperse particles are easier to prepare.

The most common approach for the synthesis of anisotropic hematite is to use phosphate ions. Phosphate ions have a strong coordination to the c-axis of the hexagonal crystal system, therefore the particle shape results in spindles (Figure 4.2B) [33]. The aspect ratio of hematite spindles can be varied via the phosphate ion concentration [35, 36].

Surface active substances, typically cetyltrimethylammonium bromide (CTAB) or polyethylene glycol (PEG), play a huge role in particle shape during their formation [37, 38]. As an example a 0.2 M iron(III)-solution with 0.02 M CTAB at 120 °C creates cylinders with a length of about 480 nm and width of about 60–80 nm (Figure 4.2C). Analogous the addition of PEG also yields cylindrical structures. The aspect ratio of the particles is directly influenced by the molecular weight of the PEG. In a hydrothermal reactor, cylindrical particles between 500 nm and 1 μm length and a width of about 20 nm can be synthesized within 24 h at 150 °C (Figure 4.3A). Here a mixture of a magnetite phase and a hematite phase is produced (Figure 4.3A).

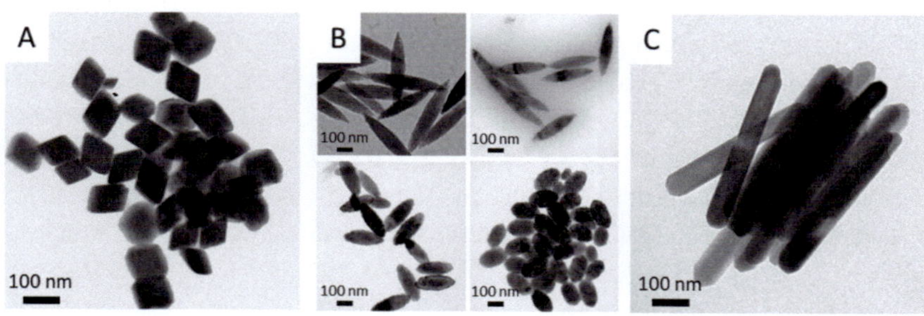

Figure 4.2: TE micrographs of hematite particles, (A) aged at lower pH for eight days at 100 °C, (B) prepared with different amounts of hydrogen phosphate at 100 °C (C) with CTAB at 120 °C.

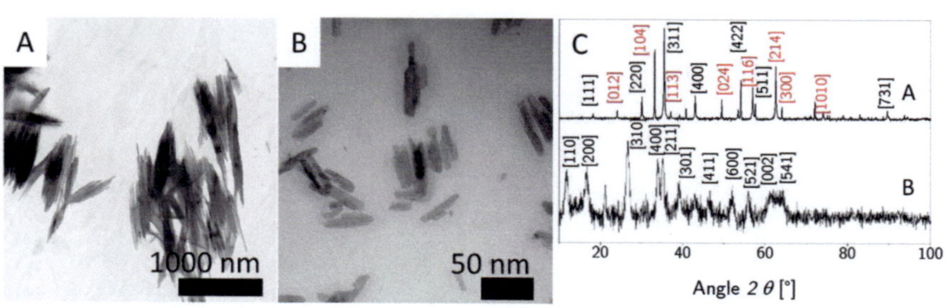

Figure 4.3: TE micrographs of (A) magnetite with hematite impurities in the presence of PEG and (B) β-FeOOH and (C) XRD for magnetite with hematite (A) and β-FeOOH (B) impurities. The peaks are indexed in black for magnetite and in red for hematite according the references.

Akaganeite is mostly an intermediate phase at low temperatures during the synthesis to goethite or hematite (Figure 4.3B), it consists of cylinders with a length of about 20–50 nm and width of about 5 nm [39].

Further synthetic approaches for magnetic anisotropic particles like cubes [34], wires [40], propellers [41] and plates are known from literature [42, 43]. However to the best of our knowledge not many hydrogels are known with embedded anisotropic particles [44, 45].

4.2.1.2 Reduction to magnetite—keeping the shape?

To increase the magnetic moment, iron(III)-oxides can be reduced to magnetite which has a higher magnetic moment compared to goethite and hematite. Magnetite can be oxidized to maghemite which is a more stable iron oxide phase, with a weak oxidizing agent in a second step. This has already been done for hematite spindles [46, 47]. However, the reduction of hematite takes place at high temperatures (400 °C) under hydrogen atmosphere. The oxidization can then take place using nitric acid [48].

The disadvantage is that the particles need to be redispersed in water after the reduction. To achieve this, the particles are typically covered by a silica shell [49, 50]. Another possibility is the transfer of the process to hydrothermal conditions [51]. Crassou et al. [52] reported a method to stabilize the particles with a thin silica layer and to subsequently incorporate them into a hydrogel. The thickness of the shell can be tuned via the amount of tetraethoxysilane (TEOS) (Figure 4.4A–C). This procedure can be adapted to goethite and akaganeite by following Graf et al. [49]. As a reduction agent sodium borohydride, hydrazine and sodium phosphite were already reportedly used for aqueous systems [40, 53–57]. The redox potential (see Table 4.1) shows that iron ions

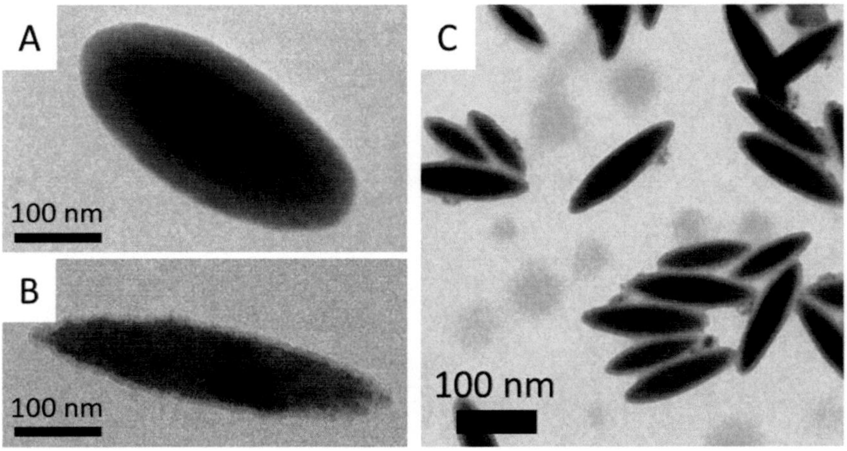

Figure 4.4: TE micrographs of the silica covered hematite spindles with different thicknesses of the silica shell from about 5 to 40 nm (A–C).

Table 4.1: Standard redox potential of the reducing agents and of iron.

Substance	Reaction	Standard potential E0/V vs. NHE 25 °C
Fe^{3+}	$Fe^{3+} + e^- \rightarrow Fe^{2+}$	0.77 [58]
Fe^{2+}	$Fe^{2+} + 2e^- \rightarrow Fe$	−0.44 [57]
$NaBH_4$	$BH_4^- + 4\,OH^- \rightarrow BO_2^- + 2\,H_2O + 2H_2 + 4e^-$	−1.73 [57]
N_2H_4	$N_2H_4 + 4\,OH^- \rightarrow N_2 + 4H_2O + 4e^-$	−1.16 [57]

can even be reduced to their elementary state. Nevertheless oxygen from the air often results in a reoxidation which needs to be taken into account for the synthesis conditions [57]. Therefore all reductions need to be performed under nitrogen atmosphere.

With sodium borohydride the particles could be reduced to magnetite, however the resulting shape strongly depends on the rate of sodium borohydride addition.

A fast addition rate (40 mL/h) leads to spherical particles (Figure 4.5A) and a slow addition rate (5 mL/h) can partially keep the particle shape (Figure 4.5B). A slow addition rate does not lead to complete reduction of the product although the same total amount of sodium borohydride is used. Not only the addition rate but also the pH, the ratio of Fe^{3+} to the reduction agent, pressure and temperature take an important role in the reductive synthesis of magnetite.

Although it proved to be difficult to keep the shape in a reproducible way, there are several synthetics possibilities to use the anisotropic particles from Section 4.2.1.1 made of goethite, hematite or akaganeite to produce magnetite particles via reduction with anisotropic shape. Exemplarily we will show it on cubic particles in the following section.

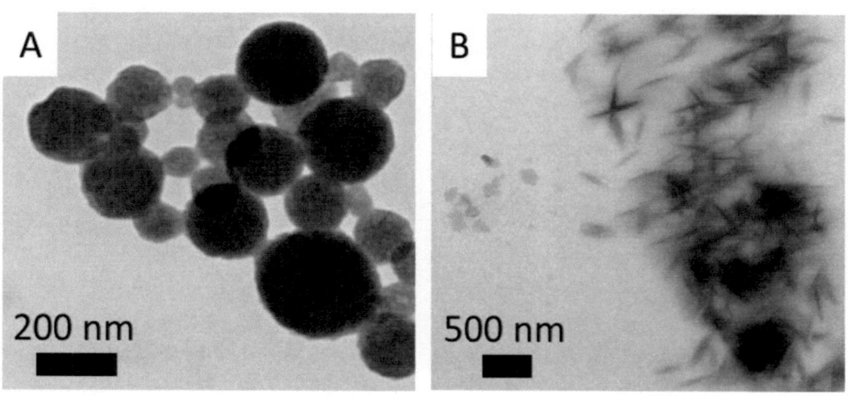

Figure 4.5: TE micrographs after reduction of iron(III)-oxide with sodium borohydride with fast (A) and slow addition rate (B) of the reduction agent.

4.2.1.3 Cubic magnetite

One reduction approach was deemed especially promising to produce cubic magnetite particles [59]. In the first step a precursor consisting of akaganeite is synthesized by treating an iron(III)-solution with NaH_2PO_4 at 80 °C [34]. The resulting precursor particles consist of small cylinders (20–55 nm length) (similar to Figure 4.3B). In the second step of the synthesis the precursor is reduced to magnetite by hydrazine. In order to optimize this reaction, we tested a whole array of parameters, employed in various research articles [38, 60–62]. We worked at temperatures between 80 and 160 °C, varied the ratio of iron educt to reducing agent, the concentration of the akaganeite precursor, the variety and concentration of surfactants as well as the duration of the whole reaction. We observed the best results when using a 0.2 M precursor solution with hydrazine in a ratio of 12:1 iron:hydrazine. The reaction was heated to 160 °C and held for 24 h while stirring. As structure forming agent the leftover NaH_2PO_4 from the precursor particles is used. This reaction produced weakly anisotropic, but phase pure magnetite cubes with an edge length of about 25 nm (Figure 4.6A and B). We studied the synthesis during different time frames and observed that during the reduction after 5–10 h big agglomerates form in the reaction medium (Figure 4.7A and B). These dissolve over the course of the reaction to form spherical particles after around 20 h (Figure 4.7C).

Afterward these particles convert to a cubic shape consisting of pure magnetite (Figure 4.7D–F). X-ray diffraction revealed that during the early stages of the reaction a goethite rich phase is formed which is subsequently reduced to magnetite (Figure 4.6B).

This reaction was scaled from a 25 mL reactor up to a 200 mL reactor and even to 500 mL. The up-scaled process yields magnetite particles of the same dimensions and with the same phase purity as the 25 mL version. We even observed a decrease in polydispersity, which is caused by the increased pressure during the reaction, caused by the increased amount of hydrazine paired with the rapid dissolution of NaOH. The product contained cubic/rhombic magnetite in the dimensions of 25 nm edge length (Figure 4.6A) with a saturation magnetization of about 70 Am^2/kg (Figure 4.6C) which

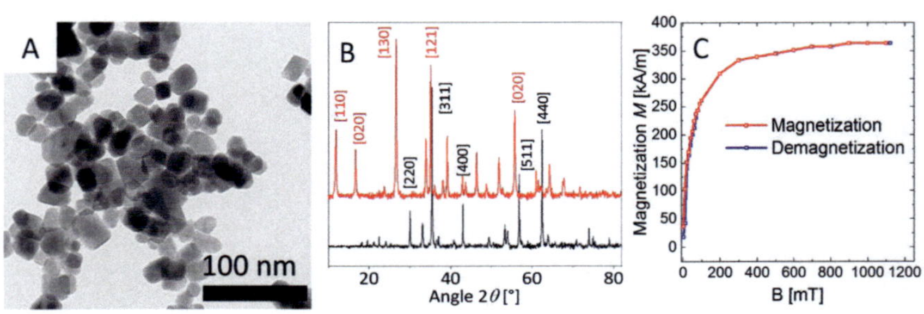

Figure 4.6: Cubic magnetite after reduction from an akaganeite precursor (A) TE micrograph (B) and XRD before (red) and after the reduction (black), peaks are indexed accordingly with akaganeite in red and magnetite in black, as well as static magnetization curve (C) of the particles in A.

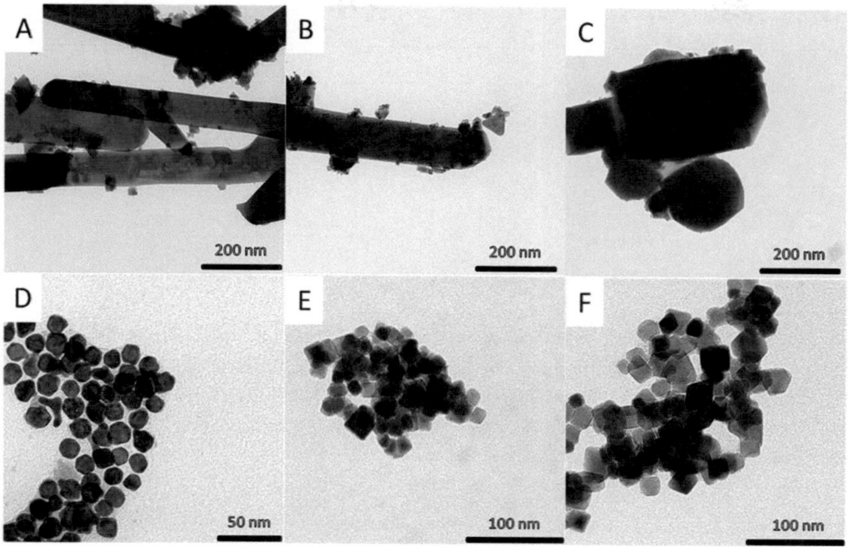

Figure 4.7: Reaction time-dependent TE micrographs study of the akaganeite reduction with 5 h (A), 10 h (B), 15 h (C), 20 h (D), 48 h (E) and 72 h (F) reaction time. All measurements were taken from single, isolated reactions.

is about 75% of the bulk magnetization. This up-scaled synthesis allowed for approx. 0.5 g of magnetite per reaction.

4.2.1.4 Multicore particles

Our hydrothermal treatment allows for an easy and universal conversion of akaganeite to magnetite by reduction and in some cases prior oxidation of an iron salt. We succeeded to conduct a quantitative conversion from akaganeite to magnetite without the creation of a harmful by-product. Further experiments suggest that this reaction pathway is indeed universally applicable to convert any akaganeite impurities to magnetite.

The material of interest in this case are multicore particles (Figure 4.8C) [63, 64]. Those multicore particles are precipitated with a slow addition of sodium hydrogen carbonate (about 450 µl/min). However, using such slow addition rate leads to some impurities of various iron oxide phases. Therefore, we treated the sample by adding hydrazine in a 12:1 ratio iron:hydrazine and put it into a Teflon lined stainless steel hydrothermal reactor. The reactor was heated to 160 °C over the course of 2 h while stirring at 500 rpm.

These conditions were kept for 24 h before the reactor was cooled down to room temperature and the black precipitate was transferred into a beaker. The precipitate was highly magnetic and was washed three times by magnetic decantation until the pH of the supernatant was neutral.

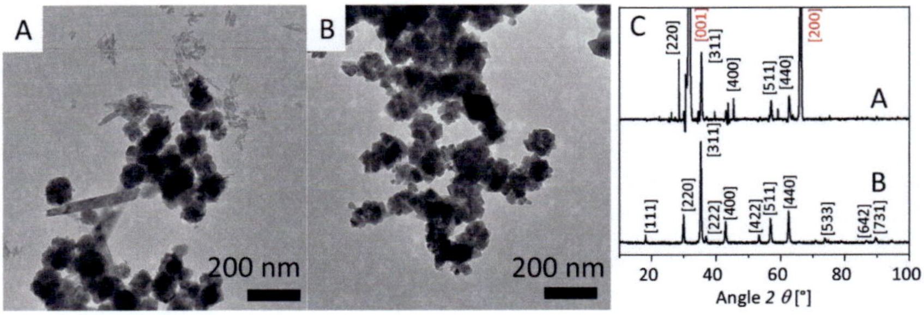

Figure 4.8: Multicore particles before hydrothermal treatment (A) and after (B). TE micrograph with different phases visible prior to hydrothermal treatment. After the treatment only one species is present (C). (C) XRD of the multicore particles before (A) and after (B) the hydrothermal treatment. The peaks are indexed in black for magnetite and in red for akaganeite impurities. They further show the phase purity after hydrothermal treatment.

The product was characterized by TEM and XRD directly after the synthesis (Figure 4.8). The resulting multicore particles consist of cubic and spherical particles alike, all of them pure magnetite as indicated by XRD (Figure 4.8C). This result indicates that this method is capable of transforming iron oxide phases to magnetite without disrupting the multicore cohesion. This can be used to improve the magnetic properties of a wide variety of mixed phase iron oxide samples that would be otherwise unusable.

4.2.2 Cobalt ferrite particles

Cobalt ferrite is a ferrimagnetic mixed ferrite consisting of a stoichiometric composition of one cobalt and two iron ions. While cobalt ferrite has the same crystalline structure as magnetite and is ferrimagnetic as well the size difference between the cobalt and the iron atoms results in a very high anisotropy energy resulting in cobalt ferrite being hard-magnetic with a very high coercive force in bulk. In nanomaterials this manifests in a dominating relaxation mechanism in dependence of particle size. While magnetite particles can grow well above 20 nm before being Brownian relaxation dominated, cobalt ferrite can already be Brownian dominated at particle sizes around 10 nm [14]. This makes cobalt ferrite a very interesting system to compare relaxation mechanisms of particles and a good material for hyperthermia (as discussed in Section 4.4). Cobalt ferrite also has a synthetic advantage, due to being a mixed ferrite it is far more unlikely to have phase impurities. When synthesizing magnetite, a typical problem is the partial oxidation to maghemite as an impurity during the precipitation process. Another possible, unwanted side effect is the full oxidation of magnetite to the less magnetic hematite when storing the particles.

Our cobalt ferrite is synthesized in a straightforward coprecipitation method derived from Nappini et al. [65]. 7.6 g cobalt(II)-nitrate and 17.3 g iron(III)-chloride are dissolved in 100 mL Millipore water. Two milliliter concentrated nitric acid is added and the solution is heated to the boiling point. Four-hundred milliliter of 1 M NaOH solution are prepared and heated to the boiling point. The boiling salt solution is then rapidly transferred into the NaOH solution under vigorous stirring. The temperature is held for another 90 min before cooling down the solution to room temperature. The particles are separated by magnetic decantation, washed twice with water and finally suspended in 40 mL 2 M HNO_3. The suspension is then heated to the boiling point and a boiling mixture of 56 mL iron(III)-chloride and 28 mL cobalt(II)-nitrate is added rapidly under vigorous stirring. After 30 min, the solution is repeatedly washed with water through magnetic decantation until a neutral pH of the supernatant is reached. The particles are then dispersed in 25 mL 0.25 M aqueous tetramethylammonium hydroxide solution and slightly stirred overnight. The mass percentage of the product is determined through gravimetry. For the final stabilization step, the particle solution is added to a 100 mM solution of citric acid with a target weight percentage of about 0.25 wt% and are stirred for at least 2 h. After another magnetic separation step, the particles are dispersed in the same volume of 20 mM trisodium citrate solution and are stirred for another hour. Lastly, the volume was reduced by about 75% through evaporation and the concentrated solution dialyzed against Millipore water for one week. The weight percentage of the final product is once again determined through gravimetry.

The cobalt ferrite particles mainly used in this project have a varying median size between 15 and 20 nm, the exemplified TE micrograph in Figure 4.9A shows particles with 18 ± 4 nm. The particles are quite polydisperse and have a rough surface which is fabricated during the synthesis to increase stability in solution. The particles are stable for months in water.

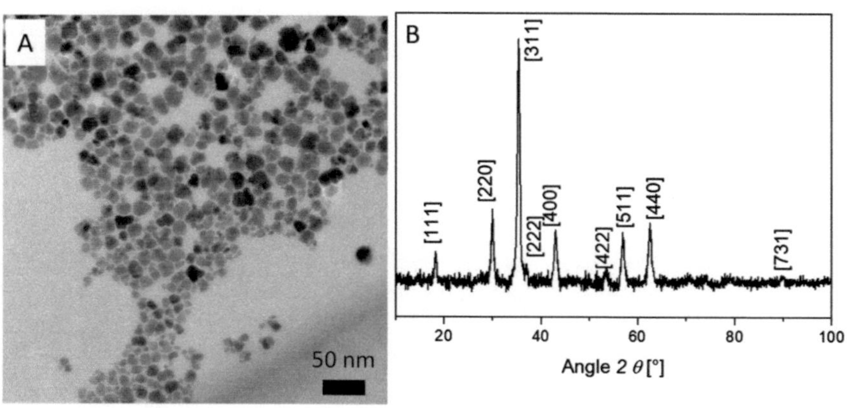

Figure 4.9: (A) TE micrograph and (B) XRD of the prepared phase-pure cobalt ferrite particles. The black indices were set by a cobalt ferrite reference.

The nanoparticles are phase pure as shown in Figure 4.9B. Utilizing the Scherrer-equation a crystallite size 15 ± 3 nm is derived which is a bit smaller compared to the TEM result due to the surface roughening which leads to amorphous domains at the outer layer of the particles. The particles have high saturation magnetization of about 52 Am^2/kg [66]. The saturation moment of cobalt ferrite is very comparable to magnetite of the same particle size. This fact makes it very interesting to compare particle species as magnetite below 20 nm is typically still dominated by Néel relaxation while cobalt ferrite is already dominated by Brownian relaxation which has been shown by Draack et al. [67]. The comparatively small particle size needed for Brownian dominated relaxation is also very interesting for the application in hyperthermia. A detailed study on the dynamics of cobalt ferrite in viscoelastic media are done within the priority program SPP 1681 from the groups of Tschoepe et al. [68]. Similar mobility investigation are done for magnetite in a dextran matrix—as biocompatible composites—within the priority program [69]. There are several publications raising qualms about the biocompatibility of nanoparticles in general as it was shown that nanoparticles can easily penetrate most bodily membranes given their small size [70]. Cobalt ferrite especially raises suspicion due to its cobalt content. Elemental cobalt is a very toxic material. Cobalt ions can exchange calcium ions in the body blocking intracellular Ca^{2+} binding proteins, moreover, cobalt can also induce hypoxia. To alleviate those suspicions initial toxicological assessments of our particle system were conducted with Jurkat cells. In the first 24 h there was no apparent cytotoxicity up to concentrations of 400 μg/mL. After 48 h cobalt ferrite expresses significant toxicity above 100 μg/mL which can be reduced by adding a silica shell to the particles [66].

4.3 Magnetic hydrogels

One application of multi responsive hydrogels is a controlled drug release for medical purposes. Carbon, magnetic nanoparticles or gold nanoparticles are often used to manipulate the hydrogel matrix as switches for the coil-to-globule transition. The triggers for the aforementioned switch would be radio frequency, an alternating magnetic field or light. Recently Satarkar et al. [71] highlighted the developments in those remote controlled biomaterials.

In general hydrogel systems can be distinguished by their size into micro- and macrogels (Figure 4.10). Microgels with a single particle as core are often called core-shell-microgels (Figure 4.10A) [59]. Another variant are systems with a distribution of nanoparticles over the whole microgel particle (Figure 4.10B) [72]. Macrogels can be either cross-linked microgel particles or bulk-hydrogels [71]. The synthesis process for the macrogel decides if the particles are mobile [69, 73, 74] or fixated in the network [71, 74–76]. A comparative study of both macrogels was done by Messing et al. [74] of cobalt ferrite within ferrohydrogels.

Figure 4.10: (A) Nanoparticles (black) covered with a hydrogel (blue). (B) Nanoparticles embedded in microgel microparticles. (C) Macrogel with embedded nanoparticles, red depicts a crosslinker. The scale of the particles increases from left to right. Left shows a few nanometers in total size and C up to the millimeter range.

We will discuss magnetic core–shell microgels (Figures 4.10A and 4.11A) and macrogels which are cross-linked microgel particles (Figure 4.11B and C). The microgel particles are able to diffuse and to rotate freely. In case of the macrogels we distinguish two types of behavior: fixed nanoparticles on network knots and free particles in the network. In this former case the core–shell microgels (Figure 4.11A) will be cross-linked additionally to a macrogel (Figure 4.11B).

In this case the particles are hindered in rotation as well as in diffusion. The embedded nanoparticles are typically covered with an additional thin silica layer. Another type of macrogel network does not completely hinder the particles in their mobility (Figure 4.11C). In this case a thermoresponsive cross-linked microgel is synthesized independently of nanoparticles which are added in a second, separate step. While the diffusion of the particles is hindered by the gel network, the particles—if small enough—can still freely rotate [73].

pNIPAM is a common and widely researched thermoresponsive polymer that has the inherent advantage that the lower critical solution temperature (LCST) lies close the human body temperature with 34 °C. The LCST describes a temperature above which two

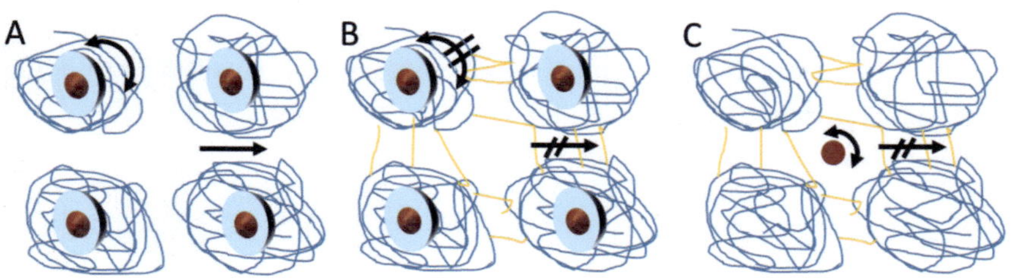

Figure 4.11: Silica coated nanoparticles (brown in light blue spheres) with random coils of a thermoresponsive polymer. The microgel particles can freely rotate and translate inside the solvent (A). When the particles are cross-linked into a macrogel (yellow lines depict the crosslinker) the translation and rotation are hindered (B). A microgel and particles can be synthesized separately and mixed prior to the crosslinking step leading to immobile particles that can still rotate (C).

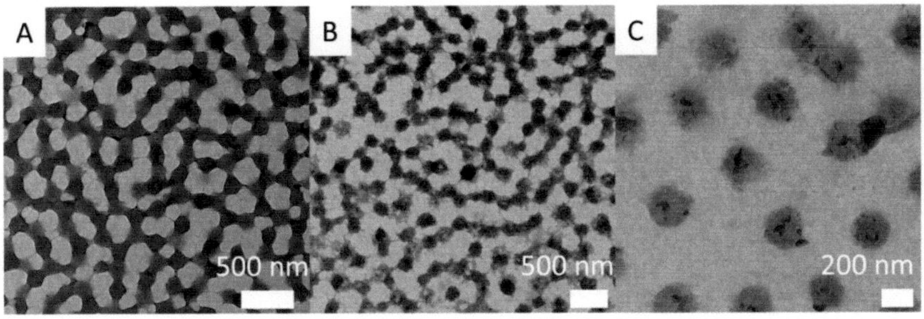

Figure 4.12: TE micrographs of macrogel particles with varying crosslinker densities, (A) 0.1%, (B) 0.05%, (C) 0.025 % total mass of crosslinker.

substances (gel and solvent) are no longer miscible and a polymer undergoes a coil-to-globule transition, to minimize its surface area. Usually an additional crosslinking agent is used to improve the reversibility of the coil-to-globule transition. The most common crosslinking agent for acrylamides is *N,N′*-methylene-bis-acrylamide (BIS). The amount as well as the distribution of the crosslinker within the microgel affects the swelling behavior. Varga et al. [77], showed that a homogenous distribution leads to a larger swelling ratio than a heterogeneous distribution at low temperature. The influence of the crosslinker on the LCST is quite low.

4.3.1 Tuning of the LCST

The LCST of pNIPAM cannot be tuned in a wide range. As an example of a more versatile matrix material copolymers like poly[*N*-acryloyl piperidine-*random*-*N*-acryloylpyrrolidine] allow the tuning of the LCST in a wide range between 3 °C and 45 °C by varying the ratio of the monomers before the polymerization [78]. Another promising material class is poly[oligo(ethylene oxide)] which is biocompatible. The LCST of these polymers can be manipulated by adjusting the side chain length of the oligo ethylene oxide [79].

The right choice of material and a good synthesis control have the potential to create highly functional and tight fit materials. For medical applications an LCST in the range between 37 and 40 °C would be ideal, for other application the ideal LCST could be different.

4.3.2 Core-shell hydrogel as matrix

A possible route to synthesize core–shell microgels is described by Karg [80, 81]. In a first approach silica nanoparticles are used as cores. The silica core is functionalized with 3-(trimethoxysilyl)propyl methacrylate (TPM) and then covered with a pNIPAM shell. The general swelling behavior and the LCST do not change significantly when altering the crosslinker density. This has the neat side effect that the hydrodynamic volume of the particles can be tuned independently of the thermoresponsiveness by varying the crosslinking density, i.e., the ratio of BIS and NIPAM as well as the distribution of the BIS inside the gel which has also direct influence on the temperature-dependent gel size [82].

4.3.3 Macrogel matrix

4.3.3.1 Core–shell macrogels

Starting from a core–shell microgel, the transition to a macrogel is rather easy. After the particles were covered by silica and while pNIPAM was polymerized onto the TPM functionalization, the NIPAM molecules were densely crosslinked with BIS to form a tight microgel sphere. The outside of this sphere was then functionalized by allylamine (AA) to provide amine functional groups which are necessary for the crosslinking into a macrogel. These amine functions are then crosslinked by glutaraldehyde (GA) at temperatures above the LCST to form a viscous macrogel [83].

4.3.3.2 Macrogels

Another approach toward particle loaded macrogels was taken by Nack et al. [73], here the microgel is made of pure pNIPAM crosslinked directly by glutaraldehyde. This synthesis is done above the LCST, so pNIPAM particles are in their shrunken state and form dense spheres that are subsequently crosslinked by GA. GA forms a bond between two NIPAM Amino groups but this bond exists in a rather unstable equilibrium, which means that the crosslinks can be easily rearranged [84]. The degree of crosslinking, i.e., the density of the polymer network, can be regulated by the GA content (Figure 4.12). Since this reaction is also done above the LCST, sodium dodecyl sulfate (SDS) is deployed to keep the polymerizing NIPAM chains in solution [85]. The size of the forming polymer globuli during the synthesis is determined by the concentration of SDS given. We used this to tune the size of the microgel spheres inside the macrogel (Figure 4.13).

4.3.4 Magnetic particles fixed within the network—magnetic core–shell macrogels

To synthesize core–shell microgels with magnetic nanoparticles as cores, the nanoparticles are first covered by a silica shell [49, 50]. Then the particles are treated as described above. A first demonstration of magnetic particles fixed in a thermoresponsive

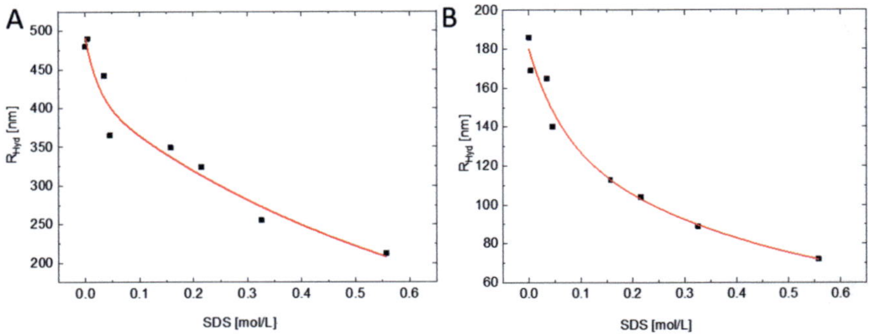

Figure 4.13: Hydrodynamic radii of pNIPAM spheres in dependence of the SDS concentration during synthesis above (A) and below (B) the LCST. The red lines are a guide to the eye roughly extrapolated as an exponential decay.

polymer was done with hematite cubes [59]. First the hematite cubes are synthesized like described by Ozaki [86] (Figure 4.2A in Section 4.2.1.1) and then covered by a thin silica layer functionalized with TPM. To grow the pNIPAM chains, a radical polymerization was performed. In addition to NIPAM the crosslinker BIS was used and the surface was covered with AA [80, 81]. As described above, the amino groups of the AA are used to crosslink the microgel to a macrogel using glutaraldehyde. This macrogel still has the same thermoresponsive behavior as the microgel. At low temperature, the microgel is swollen and has a radius of about 172 nm. Above the LCST at about 33 °C, the microgel particles collapse and the size shrinks to 55 nm.

This transition is also visible with the naked eye. Above the LCST, the microgel turns turbid or milky (Figure 4.14A). This is also directly visible in the macrogel (Figure 4.14B). The gelation process can be investigated by rheology *in-situ* [59]. The higher the temperature the faster the gelation process works. The same procedure is also possible for other magnetic cores [59].

4.3.5 Magnetic particles mobile within the network

An additional property of form-anisotropic particles is that they align in a magnetic field, i.e., the hematite spindles orient perpendicular to their length axis within the field

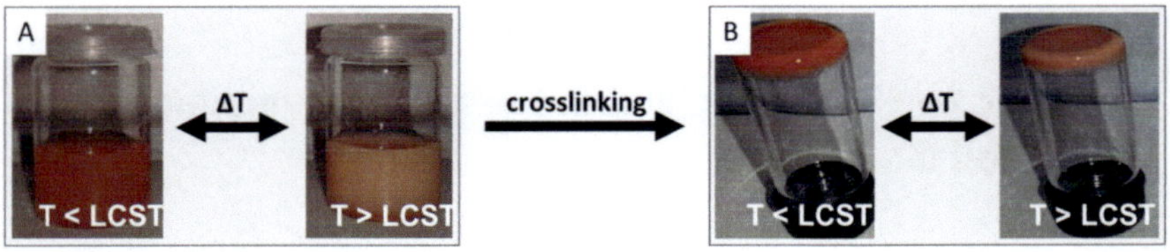

Figure 4.14: Photograph of liquid microgel particles below and above the LCST (A) and after crosslinking to a solid macrogel (B).

[35, 36, 87]. If they were put into a macrogel, the hematite particles are hindered in their orientational behavior [73]. Investigation on the mobility of hematite spindles were done from groups of Wende and Schmidt [76, 88] within the priority program 1681. This can be similarly achieved using goethite cylinders which should orient parallel in very low magnetic fields (<35 mT) and perpendicular at higher field strengths [43, 89]. Goethite particles were synthesized according to Section 4.2.1.1 (Figure 4.1C). According to TEM, the particle length was about 210 ± 61 nm and the particle width was about 27 ± 9 nm, resulting in an aspect ratio of about 8. At a field strength of about 550 mT, the goethite particles align, which is visible as an anisotropic pattern in the small angle X-ray scattering data (Figure 4.15A). Compared to an isotropic pattern (indicated by a circle in the data) the minimum of the intensity shifts to higher q-values perpendicular to the field and to lower q-values parallel to the field. By averaging over a defined wave vector range, i.e., between 0.0223 and 0.0266 nm^{-1}, the intensity peaks at 135° and −45° (Figure 4.15B), which indicates that the particles align perpendicular to the field direction.

The goethite particles were embedded in a pNIPAM matrix. These macrogels were synthesized as described in Section 4.3.3.2 similar to Nack et al. [73]. Within this macrogel—even at a high field strength of about 550 mT—there is no visible alignment at room temperature (Figure 4.17A). The averaged normalized intensity at a defined wave vector range between 0.0223 and 0.0266 nm^{-1} results in a parallel line. Increasing the temperature to 30 °C, close to the LCST, makes an alignment of the particles possible (Figure 4.16B). An anisotropic pattern is directly visible in the SAXS data which leads to a peak at −45° and 135 °C (Figure 4.15B). A further increase of the temperature does not lead to further alignment (Figures 4.15B and 4.16C). Returning to room temperature while under influence of the field shows that the particles stay aligned even after turning off the

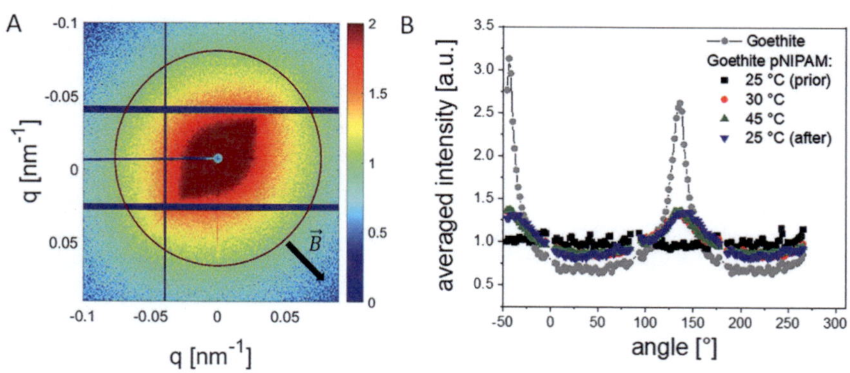

Figure 4.15: (A) U-SAXS of goethite particles in water in a field of about 550 mT. The field direction is indicated with a black arrow. The ring is a guideline for the eye. (B) Averaged intensity at a defined wave vector range, i.e., between 0.0223 and 0.0266 nm^{-1} versus angle for goethite particles in water and goethite particles in a pNIPAM matrix at 25 °C prior to the first heating cycle, at 30 °C, 45 °C and at 25 °C after the heating cycle.

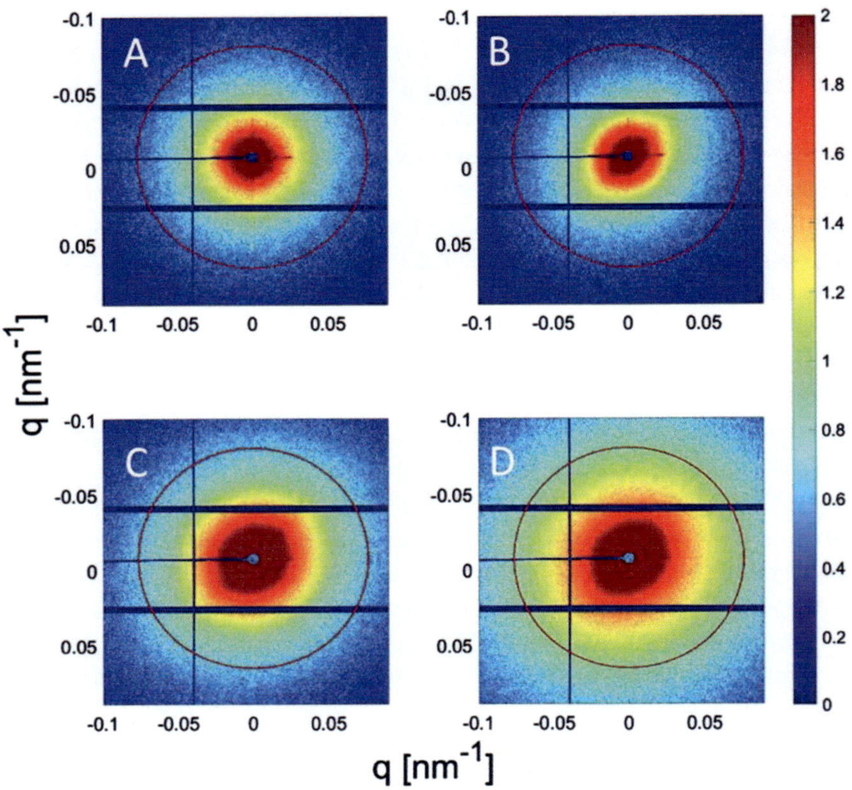

Figure 4.16: SAXS images of a pNIPAM macrogel with embedded goethite particles in a permanent magnetic field of about 550 mT at (A) RT prior to the heating cycle, (B) 30 °C, (C) 45 °C and (D) again at RT after the heating cycle.

field (Figures 4.15B and 4.16D). A detailed investigation on how this alignment affects the properties of the pNIPAM gel and the particle dynamics is running.

4.3.6 Outlook: particles within network cavities

When particles are embedded into a gel as described in Section 4.3.4, they are fixated in the matrix, i.e., the particles cannot move isolated from the matrix. The described particles can basically be split into three compartments: the magnetic nanoparticle, the silica shell and the hydrogel shell. The magnetic nanoparticle itself and the hydrogel have a distinct function in the composite material while the silica shell is used to create a stable composite material which becomes obsolete in the finished material. One approach to tune the composite can be the removal the silica shell [90–93], which should increase mobility of the magnetic particle and therefore open the opportunity to utilize applications that require mobile particles, e.g., for hyperthermia.

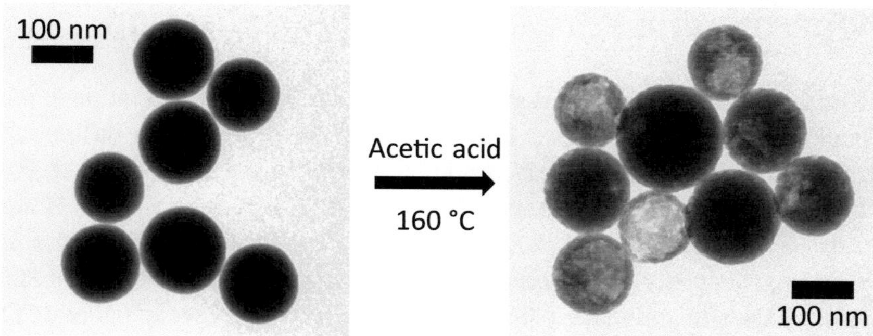

Figure 4.17: TE micrograph of the evolution of hollow silica spheres through etching with acetic acid before and after.

Silica shells are formed by a condensation reaction of SiOH. The most common reagent TEOS carries four such alcohol groups. Apart from condensation reactions these groups can also associate by hydrogen bond formation, which is less stable but significantly faster and the resulting bond-length is shorter. During the rapid formation of the silica particles the condensation typically is partial in the center. The degree of condensation typically increases closer to the surface of the particles because the increased curvature forces the Si to form the longer condensed Si–O–Si-bonds. We attempted several approaches to dissolve/etch the silica shell [90–93]. The most commonly reported approach is using sodium hydroxide as etching agent which has the inherent disadvantage that it also catalyzes the condensation reaction [91]. This renders the whole process very difficult to control. Another seemingly more controlled process was the use of acetic acid [90]. The described approach is very straightforward: Bring a solution of silica or core–shell particles to pH 3.9 and heat it to 160 °C for 24 h in an autoclave. The first results can be seen in Figure 4.17 and show hollow silica spheres (B) in comparison to the particles before treatment (A). This approach lacks reproducibility and the results are very inconsistent leading us to abandoning this approach.

A combination of a very fine-tuned sodium hydroxide etching with a subsequent dialysis step yielded promising results at etching away about 95% of the silica shell. This approach does not produce hollow spheres but removes the silica shell in its entirety. This process has the disadvantage that it has a direct influence on the swelling behavior of the gel material and—while being very reproducible—is therefore problematic to transfer into a practical application. A newer approach was recently discovered using ammonium fluoride to etch the silica shell [92]. Ammonium fluoride works as efficient as hydrofluoric acid while being far less dangerous to handle. First experiments showed a very clean etching of the silica shell while leaving the gel's swelling behavior unaffected. This approach will be further pursued in the project to create mobile particles in the center of a gel bead.

4.4 Hyperthermia—the step toward application

Hyperthermia is a medical symptom describing an elevation of the temperature in the human body as a reaction to certain pathogens. Medical hyperthermia utilizes an external heating source to induce that overheating. The main disadvantage of the 'classic' approach to medical hyperthermia is that it requires methods that are either highly invasive in case of microwaves or need direct contact to the body in case of heating elements. Magnetic hyperthermia is an approach to realize specific and targeted heating while being minimally invasive and noncontact. Magnetic radiation of reasonable field strength is harmless to the human body and easily penetrates tissue without causing damage. To induce heat with magnetic radiation magnetic material must be brought to a target area. In our project we were not concerned with targeting an area but rather with the proof of concept of noncontact heating utilizing magnetic nanoparticles in an external ac field. For this purpose, we build a custom setup to measure heat induction in nanoparticle solutions. The core of the setup is a series resonant circuit which includes a 2 Ω power resistor and is powered by an audio amplifier. The temperature is measured with a fiber optic thermometer directly in solution [66].

In general, there are two approaches how to evaluate acquired data from hyperthermia. The main approach to assess heating efficiency of magnetic material is a calorimetric approach that mainly takes the medium around the material into account while normalizing the data to the mass m of particles used, the heat capacity of the medium c and the slope of the temperature vs. time curve $\Delta T/\Delta t$:

$$\mathrm{SLP} = \frac{c}{m}\frac{\Delta T}{\Delta t} \tag{4.1}$$

The intrinsic approach takes the actual magnetic characteristics of the material and the excitation field into account. The relation of specific loss power (SLP) and intrinsic loss power (ILP) can be described using the frequency f and the strength H of the excitation field:

$$\mathrm{ILP} = \frac{\mathrm{SLP}}{f * H^2} \tag{4.2}$$

The SLP can be directly accessed via temperature versus time measurements of a nanoparticle solution using the specific heat capacity of the solvent and/or matrix material. For comparability the data is usually normalized to the mass of particles used for a measurement.

Figure 4.18 shows heating versus time curves in dependence on the external magnetic field strength. The linear region in the first approximate 180 s can be utilized to calculate the SLP, which is shown in reliance to the respective field strength. The particles show a high heating power even at low frequent fields. This can be further tuned by optimizing the excitation frequency or increasing the viscosity of the solution.

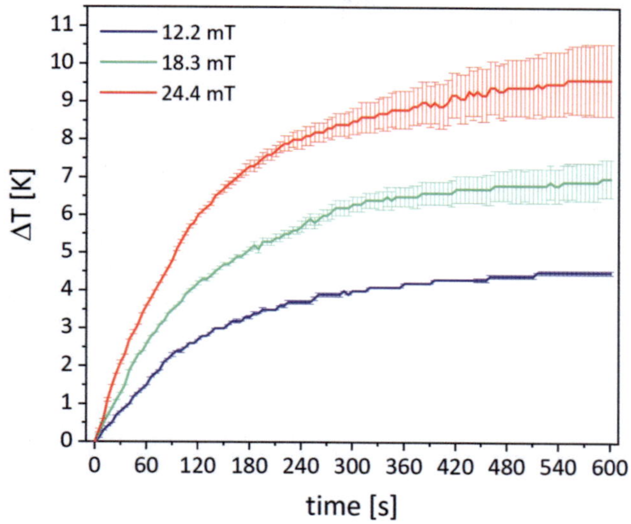

Figure 4.18: Temperature versus time curves comparing the heating ability of cobalt ferrite in pure water at different field strengths. With higher field strength the heating efficiency increases but also the error from coil heating.

pNIPAM can be synthesized as beads in aqueous solution and mixed with water dispersible particles. Figure 4.19 shows the hydrodynamic radius in dependence on the temperature for pure pNIPAM beads in water. A steep phase transition can be seen at roughly 30 °C. As a proof of principle these beads were mixed with cobalt ferrite particles and measured in our setup.

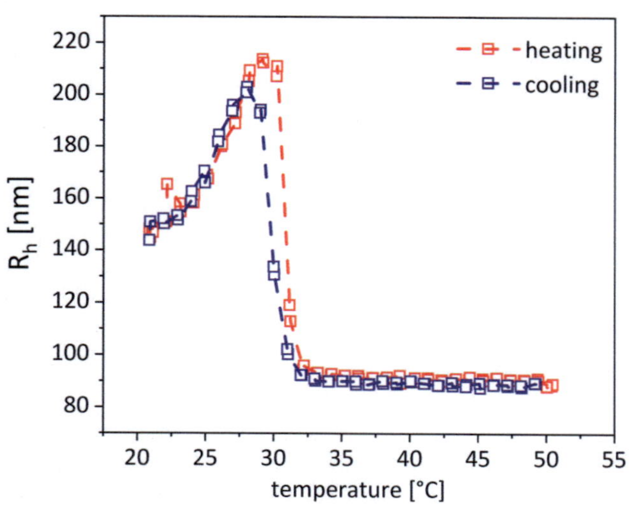

Figure 4.19: Temperature-dependent hydrodynamic radii of pNIPAM gel beads in water. A steep decrease of the hydrodynamic radius at around 30 °C can be seen indicating the phase transition.

Figure 4.20 exemplarily shows the heating versus time curves of two composite mixtures in comparison to a pure particle sample starting from ambient temperature. Evidently the composite mixture reaches roughly the double temperature increase while having about 60% of the particle content. This efficiency increase of about 300% can be explained by the increased Brownian relaxation time of the particles due to the higher viscosity of the solution. This increased relaxation time shifts the ideal excitation frequency of the particles to lower frequencies and therefore we get increased efficiency at the given frequency.

4.5 Methods

4.5.1 TEM

A drop of diluted particle suspension was deposited on a carbon coated copper grid. The solvent was removed by a filter paper. TE micrographs were recorded on a FEI Tecnai G2 spirit TWIN at an accelerating voltage of 120 kV.

4.5.2 VSM

The static magnetic characterization was conducted directly in solution using the commercially available EZ9 VSM system by Microsense in a range of ±2.5 T.

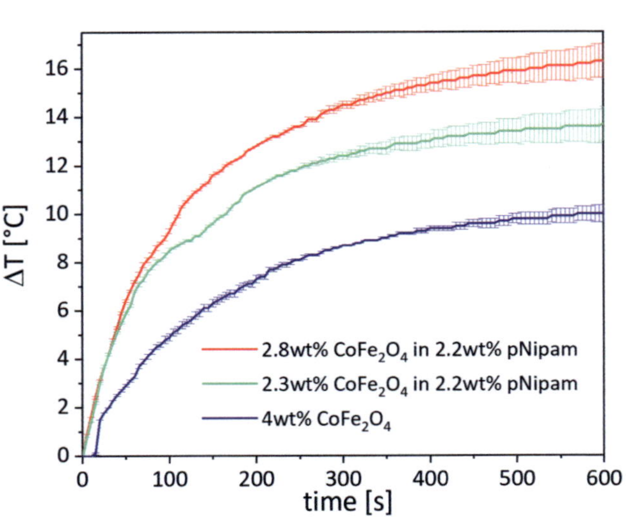

Figure 4.20: Temperature versus time curves comparing the heating ability of cobalt ferrite in pure water with cobalt ferrite mixed in a solution of pNIPAM gel beads with error bars.

4.5.3 XRD

The sample was washed to neutral pH. Then about 100 μL of the particle dispersion were deposited on a silicon waver and dried in air. The X-ray diffraction was done on a Philips X`Pert PRO MPD. The radiation was Cu-K_α with a wavelength of 1.54 Å. For the evaluation of the crystal phases present, we used the software Highscore X'pert PRO by PanAnalytical with the PDF Nr. 00-034-1266 for akaganeite, PDF Nr. 00-022-1086 for cobalt ferrite, PDF Nr. 00-024-0072 for hematite, PDF Nr. 01-081-0462 for goethite and PDF Nr. 01-086-1344 for magnetite. All samples were background corrected to compensate for X-ray fluorescence at the given X-ray wavelength.

4.5.4 DLS

The dynamic light scattering (DLS) measurements were performed with an ALV/CGS-3 Compact Goniometer-System using an ALV/LSE-5004 Multiple Tau Digital Correlator (V.1.7.9.) in combination with a COBOLT SAMBA 50 laser (Nd:YAG, 532 nm, 400 mW) and the ALV Digital Correlator Software 3.0. The measuring angle was set to 90° for all measurements and every individual measurement was conducted for 60 s. The sample vials consisted of quartz glass and were placed into a measurement cell filled with toluene. The temperature-dependent viscosity and refractive index of the solvents were automatically corrected according to tabulated values. The toluene bath and thus the samples were tempered by a JULABO F25 thermostat working with a mixture of water and ethylene glycol and delivering a temperature accuracy of 0.01 °C.

4.5.5 Hyperthermia

Hyperthermia was measured with a custom-built coil setup. The frequency was set using the STEMlab web interface of the red pitaya amplified by a t.amp TSA 2200 audio amplifier from Thomann powering a series resonant circuit including an additional 2 Ω power resistor. The factual field strength was calculated directly from the measured current driving the excitation coil. The current was measured with a Sensitec 3025 sensor. The signal was processed within STEMlab of the red pitaya. The temperature was measured with a fiberoptic gallium arsenide measurement device from Optocon. The sample holder is a 3D-printed ABS piece to fit 5 mL rotilab glass vials from Carl Roth and is thermally insulated by a 2 mm water-cooled polystyrene hose.

4.5.6 U-SAXS

The U-SAXS measurements were performed at the Coherence Beamline P10 at the German Synchrotron Facility (DESY) in Hamburg. The energy was set to 7.049 keV, which lies under the absorption edge of iron. The distance from detector to sample was about 21.2 m. An Eiger 4M detector with 2070 × 2167 pixels was used with a pixel size of about 75 × 75 μm^2. The samples were filled in Quartz capillaries prior to the measurement and sealed with hot glue. For the low magnetic field strengths, a coil setup was utilized generating a DC magnetic field up to 100 mT. For stronger magnetic field strengths permanent magnets were used. The field strength was measured by a hall effect magnetometer.

4.6 Summary and outlook

We succeeded in the synthesis of a variety of iron oxide phases in different sizes and shapes with unique magnetic properties. The iron oxides are usually very stable in their respective oxidation phases and maintain their respective properties for months. All particles can be additionally coated with a silica shell which further prevents agglomeration and offers the possibility for further functionalization via the very flexible silicon hydroxide function on the outer shell. We additionally showed the combination of our particle systems with pNIPAM as a proof-of-principle of a composite material for how the thermoresponsive matrix can be influenced by a magnetic field. A variety of different analysis methods could be applied to the particles.

Magnetite generally has very good magnetic properties but proved to be very sensitive to handle. We could show with various scattering techniques that impurities of maghemite or magnetically far inferior hematite were a very common byproduct of the synthesis. We succeeded to establish a hydrothermal route that enables us to synthesize phase pure magnetite nanoparticles not only from pre-prepared akageneite precursors but also from impure samples as a means for purification. These akaganeite precursors are very simple and controlled in their synthetic route and can also be used to create antiferromagnetic goethite particles which have unique magnetic properties as well. Cobalt ferrite as a hard-magnetic counterpart to magnetite can also be synthesized straightforward and was stabilized for months in watery dispersion.

All mentioned particles can be reliably coated with a silica shell that is tunable in its size. This shell offers the possibility of a wide range of functionalization through silane chemistry. The silica shell itself seems to offer quite sophisticated properties regarding biocompatibility and opsonization behavior in biological fluids like serum or blood. We could show that our cobalt ferrite nanoparticles express a seemingly low toxicity under 48 h which could be further reduced through the silica shell. Preliminary tests showed that proteins in fetal calf serum coordinate well around silica coated cobalt ferrite particles which has potential in a future application to mask particles from an answer of the bodily immune system.

An array of interactions between our different particle systems and the thermoresponsive matrix pNIPAM could be observed and reported in this work. pNIPAM has a significant influence on the orientational behavior which can be switched by a temperature change. We could also induce enough heat to achieve the phase transition via an alternating current magnetic field into a gel without direct contact.

The proofs of principle presented in this work will be further pursued in the future. The cobalt ferrite particles will be tuned in their stability in biological fluids as serum and blood. Hemocompatibility tests as well as toxicity assays with different cell lines are planned as the next step. Different sizes of cobalt ferrite particles and a comparative toxicological assay dependent on the particle size and magnetic properties. Parallel to biocompatibility in general, cobalt ferrite and magnetite nanoparticles will be further examined on their hyperthermia behavior in gel matrices with a special focus on the difference between particles that are just mixed into a gel and particles that are covalently bound into the network. Especially temperature-dependent dynamic magnetic analysis is expected to deliver very interesting insights to particle matrix interactions. Further X-ray scattering experiments on goethite nanorods are planned to get an even better insight into the orientational behavior and the change of that orientational behavior inside different gel matrices.

Acknowledgment: The authors would also like to thank various work groups inside the SPP1681 for the fruitful cooperation that lead to different publications during the time of the project. The authors would like to acknowledge Andreas Weidner and the group of Sylvio Dutz from Ilmenau for the cooperation in the synthesis of multicore particles. The authors would like to thank Sebastian Draack and the groups of Thilo Viereck and Frank Ludwig for fruitful discussions and mutual publication of the research. The authors would also like to thank Ralf P. Friedrich and the group of Christoph Alexiou for much insight into biocompatibility of nanoparticles and work toward a publication in cooperation with Sebastian Draack. The authors would like to thank Almut Barck and Margarethe Fritz for measuring XRD. The authors acknowledge Fabian Westermeier and Michael Sprung for their help during beamtime at DESY P10.
Author contributions: All the authors have accepted responsibility for the entire content of this submitted manuscript and approved submission.
Research funding: The Deutsche Forschungsgemeinschaft is kindly acknowledged for funding our project under the SPP 1681 "Field Controlled Particle-Matrix Interactions" with the Project FI 1235/2-1 and FI 1235/2-2. Stephan Hinrichs would like to acknowledge the doctoral scholarship of the University of Hamburg for funding the first funding period of his work.
Conflict of interest statement: The authors declare no conflicts of interest regarding this article.

References

1. Lattermann G, Krekhova M. Thermorevers Ferrogels Macromol Rapid Commun 2006;27:1373–9.
2. Messing R, Schmidt AM. Perspectives for the mechanical manipulation of hybrid hydrogels. Polym Chem 2011;2. https://doi.org/10.1039/c0py00129e.

3. Thevenot J, Oliveira H, Sandre O, Lecommandoux S. Magnetic responsive polymer composite materials. Chem Soc Rev 2013;42:7099–116.
4. Menzel AM. Tuned, driven, and active soft matter. Phys Rep 2015;554:1–45.
5. Gil ES, Hudson SM. Stimuli-responsive polymers and their bioconjugates. Prog Polym Sci 2004;29: 1173–222.
6. Smith AE, Xu X, McCormick CL. Stimuli-responsive amphiphilic (co)polymers via RAFT polymerization. Prog Polym Sci 2010;35:45–93.
7. de Las Heras Alarcon C, Pennadam S, Alexander C. Stimuli responsive polymers for biomedical applications. Chem Soc Rev 2005;34:276–85.
8. Calvert P. Hydrogels for soft machines. Adv Mater 2009;21:743–56.
9. Li Y, Huang G, Zhang X, Li B, Chen Y, Lu T, et al. Magnetic hydrogels and their potential biomedical applications. Adv Funct Mater 2013;23:660–72.
10. Jordan A, Scholz R, Maier-Hauff K, Johannsen J, Wust P, Nadobny J, et al. Presentation of a new magnetic field therapy system for the treatment of human solid tumors with magnetic fluid hyperthermia. J Magn Magn Mater 2001;225. https://doi.org/10.1016/s0304-8853(00)01239-7.
11. Yadavalli T, Ramasamy S, Chandrasekaran G, Michael I, Therese HA, Chennakesavulu R. Dual responsive PNIPAM–chitosan targeted magnetic nanopolymers for targeted drug delivery. J Magn Magn Mater 2015;380:315–20.
12. Büscher K, Helm CA, Gross C, Glöckl G, Romanus E, Weitschies W. Nanoparticle composition of a ferrofluid and its effects on the magnetic properties. Langmuir 2004;20. https://doi.org/10.1021/la030261x.
13. Hergt R, Dutz S, Roder M. Effects of size distribution on hysteresis losses of magnetic nanoparticles for hyperthermia. J Phys Condens Matter 2008;20:385214.
14. Rosensweig RE. Heating magnetic fluid with alternating magnetic field. J Magn Magn Mater 2002; 252:370–4.
15. Ozaki M, Egami T, Sugiyama N, Matijević E. Agglomeration in colloidal hematite dispersions due to weak magnetic interactions: II. The effects of particle size and shape. J Colloid Interface Sci 1988; 126:212–9.
16. Ozaki M, Suzuki H, Takahashi K, Matijević E. Reversible ordered agglomeration of hematite particles due to weak magnetic interactions. J Colloid Interface Sci 1986;113:76–80.
17. Lee N, Kim H, Choi SH, Park M, Kim D, Kim H-C, et al. Magnetosome-like ferrimagnetic iron oxide nanocubes for highly sensitive MRI of single cells and transplanted pancreatic islets. Proc Natl Acad Sci USA 2011;108:2662–7.
18. Materia ME, Guardia P, Sathya A, Pernia Leal M, Marotta R, Di Corato R, et al. Mesoscale assemblies of iron oxide nanocubes as heat mediators and image contrast agents. Langmuir 2015;31:808–16.
19. Wang C, Peng S, Lacroix L-M, Sun S, et al. Synthesis of high magnetic moment CoFe nanoparticles via interfacial diffusion in core/shell structured Co/Fe nanoparticles. Nano Res 2010;2:380–5.
20. Santoyo Salazar J, Perez L, de Abril O, Truong Phuoc L, Ihiawakrim D, Vazquez M, et al. Magnetic iron oxide nanoparticles in 10–40 nm range: composition in terms of magnetite/maghemite ratio and effect on the magnetic properties. Chem Mater 2011;23:1379–86.
21. Hufschmid R, Arami H, Ferguson RM, Gonzales M, Teeman E, Brush LN, et al. Synthesis of phase-pure and monodisperse iron oxide nanoparticles by thermal decomposition. Nanoscale 2015;7: 11142–54.
22. Palanisamy S, Wang YM. Superparamagnetic iron oxide nanoparticulate system: synthesis, targeting, drug delivery and therapy in cancer. Dalton Trans 2019;48:9490–515.
23. Massart R, Dubois E, Cabuil V, Hasmonay E. Preparation and properties of monodisperse magnetic fluids. J Magn Magn Mater 1995;149:1–5.
24. Kaman O, Kulickova J, Herynek V, Koktan J, Maryško M, Dědourková T, et al. Preparation of Mn-Zn ferrite nanoparticles and their silica-coated clusters: magnetic properties and transverse relaxivity. J Magn Magn Mater 2017;427:251–7.

25. Hayashi H, Hakuta Y. Hydrothermal synthesis of metal oxide nanoparticles in supercritical water. Materials 2010;3:3794–817.
26. Cornell RM, Schwertmann U. Adsorption of ions and molecules. In: The iron oxides. Weinheim: Wiley-VCH Verlag GmbH & Co. KGaA; 2003:253–96 pp.
27. Thies-Weesie DME, de Hoog JP, Mendiola MHH, Petukhov AV, Vroege GJ. Synthesis of goethite as a model colloid for mineral liquid crystals. Chem Mater 2007;19:5538–46.
28. Răditoiu V, Diamandescu L, Cosmin Corobea M, Răditoiu A, Popescu-Pogrion N, Nicolae CA. A facile hydrothermal route for the synthesis of α-FeOOH with controlled morphology. J Cryst Growth 2012; 348:40–6.
29. Zamiri R, Ahangar HA, Zakaria A, Zamiri G, Bahari HR, Drummen GPC. Hydrothermal synthesis of goethite (α-FeOOH) nanorods in the presence of ethylenediamine:thiourea. J Nano Res 2014;16: 1–10.
30. Frens G. Controlled nucleation for the regulation of the particle size in monodisperse gold suspensions. Nat Phys Sci (Lond) 1973;241:20–2.
31. Leff DV, Ohara PC, Heath JR, Gelbart WM. Thermodynamic control of gold nanocrystal size: experiment and theory. J Phys Chem 1995;99:7036–41.
32. Hinrichs S, Hankiewicz B. Goethite nanorods: synthesis and investigation of the size effect on their orientation within a magnetic field by SAXS. Nanomaterials 2020;10:2526.
33. Sugimoto S, Wang Y, Itoh H, Muramatsu A. Systematic control of size, shape and internal structure of monodisperse α-Fe_2O_3 particles. Colloids Surf, A Physicochem Eng Asp 1998;265–79.
34. Malik V, Grobety B, Trappe V, Dietsch H, Schurtenberger P. A closer look at the synthesis and formation mechanism of hematite nanocubes. Colloids Surf A-Physicochem Eng Asp 2014;445: 21–9.
35. Ozaki M, Kratohvil S, Matijevic E. Formation of monodispersed spindle-type hematite particles. J Colloid Interface Sci 1984;102:146–51.
36. Märkert C, Fischer B, Wagner J. Small-angle scattering from spindle-shaped colloidal hematite particles in external magnetic fields. J Appl Crystallogr 2011;44:441–7.
37. Pu ZF, Cao MH, Jing Y, Hu C. Controlled synthesis and growth mechanism of hematite nanorhombohedra, nanorods and nanocubes. Nanotechnology 2006;17:799–804.
38. Harraz FA. Polyethylene glycol-assisted hydrothermal growth of magnetite nanowires: synthesis and magnetic properties. Phys E Low-dimens Syst Nanostruct 2008;40:3131–6.
39. Sugimoto T, Muramatsu A. Formation mechanism of monodispersed α-Fe_2O_3 particles in dilute $FeCl_3$ solutions. J Colloid Interface Sci 1996;184:626–38.
40. Wiogo H, Lim M, Munroe P, Amal R. Understanding the formation of iron oxide nanoparticles with acicular structure from iron(III) chloride and hydrazine monohydrate. Cryst Growth Des 2011;11: 1689–96.
41. Yang W-H, Lee C-F, Tang HY, Shieh D-B, Yeh C-S. Iron oxide nanopropellers prepared by a low-temperature solution approach. J Phys Chem B 2006;110:14087–91.
42. Mertelj A, Lisjak D, Drofenik M, Čopič M. Ferromagnetism in suspensions of magnetic platelets in liquid crystal. Nature 2013;504:237–41.
43. Lemaire BJ, Davidson P, Panine P, Jolivet JP. Magnetic-field-induced nematic-columnar phase transition in aqueous suspensions of goethite (α-FeOOH) nanorods. Phys Rev Lett 2004;93: 267801.
44. Sánchez-Ferrer A, Mezzenga R, Dietsch H. Orientational behavior of ellipsoidal silica-coated hematite nanoparticles integrated within an elastomeric matrix and its mechanical reinforcement. Macromol Chem Phys 2011;212:627–34.
45. Haberl JM, Sanchez-Ferrer A, Mihut AM, Dietsch H, Hirt AM, Mezzenga R. Liquid-crystalline elastomer-nanoparticle hybrids with reversible switch of magnetic memory. Adv Mater 2013;25: 1787–91.

46. Ozaki M, Matijević E. Preparation and magnetic properties of monodispersed spindle-type γ-Fe_2O_3 particles. J Colloid Interface Sci 1985;107:199–203.
47. Ngo AT, Pileni MP. Cigar-shaped ferrite nanocrystals: orientation of the easy magnetic axes. J Appl Phys 2002;92:8–4649.
48. Bee A, Massart R, Neveu S. Synthesis of very fine maghemite particles. J Magn Magn Mater 1995; 149:6–9.
49. Graf C, Vossen DLJ, Imhof A, van Blaaderen A. A general method to coat colloidal particles with silica. Langmuir 2003;19:6693–700.
50. Zou J, Peng YG, Tang YY. A facile bi-phase synthesis of Fe_3O_4@SiO_2 core-shell nanoparticles with tunable film thicknesses. RSC Adv 2014;4:9693–700.
51. Zhang L, Chen L, Qan Q-H. Preparation of uniform magnetic microspheres through hydrothermal reduction of iron hydroxide nanoparticles embedded in a polymeric matrixes. Chem Mater 2008; 20:3345–53.
52. Crassous JJ, Mihut AM, Dietsch H, Pravaz O, Ackermann-Hirschi L, Hirt AM, et al. Advanced multiresponsive comploids: from design to possible applications. Nanoscale 2014;6:8726–35.
53. Huang KC, Ehrman SH. Synthesis of iron nanoparticles via chemical reduction with palladium ion seeds. Langmuir 2007;23:1419–26.
54. Rozenson I, Heller-Kallai L. Reduction and oxidation of Fe^{3+} in dioctahedral smectites-1: reduction with hydrazine and dithionite. Clay Clay Miner 1976;24:271–82.
55. Anik T, Touhami ME, Himm K, Schireen S, Belkhmima RA, Abouchane M, et al. Influence of pH solution on electroless copper plating using sodium hypophosphite as reducing agent. Int J Electrochem Sci 2012;7:2009–18.
56. Klačanová K, Fodran P, Šimon P, Rapta P, Boča R, Jorik V, et al. Formation of Fe(0)-nanoparticles via reduction of Fe(II) compounds by amino acids and their subsequent oxidation to iron oxides. J Chem 2013;2013:1–10.
57. Iida H, Nakanishi T, Takada H, Osaka T. Preparation of magnetic iron-oxide nanoparticles by successive reduction-oxidation in reverse micelles: effects of reducing agent and atmosphere. Electrochim Acta 2006;52:292–6.
58. Hayes PC, Algie SH. Process principles in minerals and materials production. Cincinnati: Hayes Publishing; 1993: 660 ff p.
59. Hinrichs S, Nun N, Fischer B. Synthesis and characterization of anisotropic magnetic hydrogels. J Magn Magn Mater 2017;431:237–40.
60. Peng Z, Wu M, Xiong Y, Wang J, Chen Q. Synthesis of magnetite nanorods through reduction of β-FeOOH. Chem Lett 2005;34:636–7.
61. Fan R, Chen XH, Gui Z, Liu L, Chen ZY. A new simple hydrothermal preparation of nanocrystalline magnetite Fe_3O_4. Mater Res Bull 2001;36:497–502.
62. Nemati Z, Salili SM, Alonso J, Ataie A, Das R, Phan MH, et al. Superparamagnetic iron oxide nanodiscs for hyperthermia therapy: does size matter? J Alloys Compd 2017;714:709–14.
63. Dutz S, Clement JH, Eberbeck D, Gelbrich T, Hergt R, Müller R, et al. Ferrofluids of magnetic multicore nanoparticles for biomedical applications. J Magn Magn Mater 2009;321:1501–4.
64. Grafe C, von der Luhe M, Weidner A, Globig P, Clement JH, Dutz S, et al. Protein corona formation and its constitutional changes on magnetic nanoparticles in serum featuring a polydehydroalanine coating: effects of charge and incubation conditions. Nanotechnology 2019; 30:1–15.
65. Nappini S, Magnano E, Bondino F, Píš I., Barla A., Fantechi E, et al. Surface charge and coating of $CoFe_2O_4$ nanoparticles: evidence of preserved magnetic and electronic properties. J Phys Chem C 2015;119:25529–41.
66. Lucht N, Friedrich RP, Draack S, Alexiou C, Viereck T, Ludwig F, et al. Biophysical characterization of (Silica-coated) cobalt ferrite nanoparticles for hyperthermia treatment. Nanomaterials 2019;9: 1–13.

67. Draack S, Lucht N, Remmer H, Martens M., Fischer B., Schilling M, et al. Multiparametric magnetic particle spectroscopy of $CoFe_2O_4$ nanoparticles in viscous media. J Phys Chem C 2019;123: 6787–801.
68. Remmer H, Dieckhoff J, Tschöpe A, Roeben E, Schmidt AM, Ludwig F. Dynamics of $CoFe_2O_4$ single-core nanoparticles in viscoelastic media. Physics Procedia 2015;75:1150–7.
69. Müller R, Zhou M, Liebert T, Landers J, Salamon S, Webers S, et al. Mobility investigations of magnetic nanoparticles in biocomposites. Mater Chem Phys 2017;193:364–70.
70. Bossi E, Zanella D, Gornati R, Bernardini G. Cobalt oxide nanoparticles can enter inside the cells by crossing plasma membranes. Sci Rep 2016;6:1–9.
71. Satarkar NS, Biswal D, Hilt JZ. Hydrogel nanocomposites: a review of applications as remote controlled biomaterials. Soft Matter 2010;6:2364–71.
72. Witt MU, Hinrichs S, Moller N, Backes S, Fischer B, von Klitzing R. Distribution of $CoFe_2O_4$ nanoparticles inside PNIPAM-based microgels of different cross-linker distributions. J Phys Chem B 2019;123: 2405–13.
73. Nack A, Seifert J, Passow C, Wagner J. Hindered nematic alignment of hematite spindles in poly(N-isopropylacrylamide) hydrogels: a small-angle X-ray scattering and rheology study. J Appl Crystallogr 2018;51:87–96.
74. Messing R, Frickel N, Belkoura L, Strey R, Rahn H, Odenbach S. Cobalt ferrite nanoparticles as multifunctional cross-linkers in PAAm ferrohydrogels. Macromolecules 2011;44:2990–9.
75. Hu Z, Huang G. A new route to crystalline hydrogels, guided by a phase diagram. Angew Chem Int Ed Engl 2003;42:4799–802.
76. Roeder L, Reckenthäler M, Belkoura L, Roitsch S, Strey R, Schmidt AM. Covalent ferrohydrogels based on elongated particulate cross-linkers. Macromolecules 2014;47:7200–7.
77. Acciaro R, Gilanyi T, Varga I. Preparation of monodisperse poly(N-isopropylacrylamide) microgel particles with homogenous cross-link density distribution. Langmuir 2011;27:7917–25.
78. Lucht N, Eggers S, Abetz V. Cononsolvency in the 'drunken' state: the thermoresponsiveness of a new acrylamide copolymer in water-alcohol mixtures. Polym Chem 2017;8:1196–205.
79. Bozorg M, Hankiewicz B, Abetz V. Soft Matter 2020;16:1066–81.
80. Karg M, Pastoriza-Santos I, Liz-Marzan LM, Hellweg T. A versatile approach for the preparation of thermosensitive PNIPAM core-shell microgels with nanoparticle cores. Chem Phys Chem 2006;7: 2298–301.
81. Karg M, Wellert S, Prevost S, Schweins R, Dewhurst C, Liz-Marzán LM, et al. Well defined hybrid PNIPAM core-shell microgels: size variation of the silica nanoparticle core. Colloid Polym Sci 2011; 289:699–709.
82. Nun N, Hinrichs S, Schroer MA, Sheyfer D, Grübel G, Fischer B. Tuning the size of thermoresponsive poly(N-isopropyl acrylamide) grafted silica microgels. Gels 2017;3:3–13.
83. Hu Z, Huang G. A new route to crystalline hydrogels, guided by a phase diagram. Angew Chem Int Ed Engl 2003;42:4799–802.
84. Migneault I, Dartiguenave C, Bertrand MJ, Waldron KC. Glutaraldehyde: behavior in aqueous solution, reaction with proteins, and application to enzyme crosslinking. Biotechniques 2004;37: 798–802.
85. Wu X, Pelton RH, Hamielec AE, Woods DR, McPhee W. The kinetics of poly(N-isopropylacrylamide) microgel latex formation. Colloid Polym Sci 1994;272:467–77.
86. Ozaki M. Preparation and properties of well-defined magnetic particles. MRS Bull 1989;14:35–40.
87. Wagner J, Markert C, Fischer B, Müller L. Direction dependent diffusion of aligned magnetic rods by means of x-ray photon correlation spectroscopy. Phys Rev Lett 2013;110:048301-1–048301-5.
88. Landers J, Roeder L, Salamon S, Jamet JP, Panine P, Dozov I, et al. Particle–matrix interaction in cross-linked PAAm-hydrogels analyzed by Mössbauer spectroscopy. J Phys Chem C 2015;119: 20642–8.

89. Lemaire BJ, Davidson P, Ferré J, Jamet JP, Panine P, Dozov I, et al. Outstanding magnetic properties of nematic suspensions of goethite (alpha-FeOOH) nanorods. Phys Rev Lett 2002;88:12–125507.
90. Ding H, Zhang Y, Xu S, Li G. A wrinkle to sub-100 nm yolk/shell Fe_3O_4@SiO_2 nanoparticles. Nano Research 2016;9:3632–43.
91. Meng Q, Xiang S, Zhang K, Wang M, Bu X., Xue P, et al. A facile two-step etching method to fabricate porous hollow silica particles. J Colloid Interface Sci 2012;384:22–8.
92. Kalem S. Synthesis of ammonium silicon fluoride cryptocrystals on silicon by dry etching. Appl Surf Sci 2004;236:336–41.
93. Liu Z, Chen L, Ye X, Yu H, Li J, Zeng F. Selective basic etching of bifunctional core–shell composite particles for the fabrication of organic functionalized hollow mesoporous silica nanospheres. New J Chem 2016;40:825–31.

Robert Müller, Janna Kuchinka and Thomas Heinze*

5 Studies about the design of magnetic bionanocomposite

Abstract: Magnetic nanocomposites are a class of smart materials that have attracted recent interest as drug delivery systems or as medical implants. A new approach toward the biocompatible nanocomposites suitable for remote melting is presented. It is shown that magnetite nanoparticles (MNPs) can be embedded into a matrix of biocompatible thermoplastic dextran esters. For that purpose, fatty acid esters of dextran with adjustable melting points in the range of 30–140 °C were synthesized. Esterification of the polysaccharide by activation of the acid as iminium chlorides guaranteed mild reaction conditions leading to high-quality products as confirmed by Fourier-transform infrared (FTIR) and nuclear magnetic resonance (NMR) spectroscopy as well as by gel permeation chromatography (GPC). A method for the preparation of magnetically responsive bionanocomposites (BNCs) was developed consisting of combined dissolution/suspension of the dextran ester and hydrophobized MNPs in an organic solvent followed by homogenization with ultrasonication, casting of the solution, drying and melting of the composite for a defined shaping. This process leads to a uniform distribution of MNPs in BNC as revealed by scanning electron microscope (SEM). Samples of different geometries were exposed to high-frequency alternating magnetic field (AMF). It could be shown that defined remote melting of such biocompatible nanocomposites is possible for the first time. This may lead to a new class of magnetic remote-control systems, which are suitable for controlled release applications or self-healing materials. BNCs containing biocompatible dextran fatty acid ester melting close to human body temperature were prepared and loaded with Rhodamine B (RhB) or green fluorescent protein (GFP) as model drugs to evaluate their potential use as drug delivery system. The release of the model drugs from the magnetic BNC investigated under the influence of a high-frequency AMF (20 kA/m at 400 kHz) showed that on-demand release is realized by applying the external AMF. The BNC possessed a long-term stability (28 d) of the incorporated iron oxide particles after incubation in artificial body fluids. Temperature-dependent mobility investigations of MNP in the molten BNC were carried out by optical microscopy, magnetometry,

***Corresponding author: Thomas Heinze,** Institute of Organic Chemistry and Macromolecular Chemistry, Center of Excellence for Polysaccharide Research, Friedrich Schiller University of Jena, Humboldtstraße 10, D-07743 Jena, Germany, E-mail: thomas.heinze@uni-jena.de
Robert Müller, Leibniz-Institute of Photonic Technology (IPHT), P.O.B. 100239, D-07702 Jena, Germany
Janna Kuchinka, Institute of Organic Chemistry and Macromolecular Chemistry, Center of Excellence for Polysaccharide Research, Friedrich Schiller University of Jena, Humboldtstraße 10, D-07743 Jena, Germany

This article has previously been published in the journal Physical Sciences Reviews. Please cite as: R. Müller, J. Kuchinka and T. Heinze "Studies about the design of magnetic bionanocomposite" *Physical Sciences Reviews* [Online] 2020, 5. DOI: 10.1515/psr-2019-0122 | https://doi.org/10.1515/9783110569636-005

alternating current (AC) susceptibility, and Mössbauer spectroscopy measurements. Optical microscopy shows a movement of agglomerates and texturing in the micrometer scale, whereas AC susceptometry and Mössbauer spectroscopy investigations reveal that the particles perform diffusive Brownian motion in the liquid polymer melt as separated particles rather than as large agglomerates. Furthermore, a texturing of MNP in the polymer matrix by a static magnetic field gradient was investigated. First results on the preparation of cross-linkable dextran esters are shown. Cross-linking after irradiation of the BNC prevents melting that can be used to influence texturing procedures.

Keywords: drug release, magnetic nanocomposite, thermoplastic dextran

5.1 Introduction

Magnetic bionanocomposites (BNCs) can open promising opportunities for the design of novel tools for diagnostics and therapies. Today, magnetic BNCs are intensively studied in the field of hyperthermia, [1–3] microfluidic devices, [4, 5], and controlled drug release [6, 7]. Magnetically responsive systems for drug delivery belong to exogenous stimuli-responsive materials, which can react on alternating current (AC) or oscillating magnetic field through two heating mechanisms: (a) hysteresis loss for ferromagnetic particles and/or (b) Néel and Brownian relaxation for superparamagnetic particles and on direct current (DC) magnetic field through targeting [8, 9].

A new approach toward magnetically responsive biomaterials is the use of thermoplastic biopolymers containing magnetic nanoparticles (MNPs) . In 2011, an elegant method could be developed for the synthesis of pure dextran fatty acid esters by activation of the carboxylic acids with iminium chloride, which is formed from *N,N*-dimethylformamide (DMF) and oxalyl chloride [10]. The bio-based dextran esters obtained are hydrophobic and possess-adjustable melting temperature in the range from room temperature to about 140 °C depending on the detailed structure. The melting temperature could be tailored by the type of substituent introduced, the degree of substitution (DS), and the molecular weight (M_w) of the polymer.

In the frame of own studies, polysaccharide esters were investigated regarding the preparation of magnetic BNC and the properties of the resulting biomaterials. In this chapter, we are going to discuss the synthesis of proper starting thermoplastic polymers dextran esters, the fabrication of the BNC. Moreover, thermoplastic and cross-linkable dextran esters were prepared for the first time and were doped with MNP in order to get barrier forming materials by irradiation with UV light. Studies about the magnetic and thermal behavior were in the center of interest. Moreover, the MNP movement in the molten matrix under a static magnetic field gradient and release properties using model drugs will be highlighted for this new type of magnetic BNC. To get a deeper insight in the new BNC regarding their heating abilities and magnetic relaxation process, AC susceptometry (ACS) studies were included. ACS is sensitive to the reorientation of magnetic moments in small magnetic fields, realized by Brownian motion or Néel relaxation [11, 12]. Mössbauer spectroscopy was applied as well that represents another versatile technique able to quantify parameters of Néel relaxation. In case of Brownian particle motion, a broadening of absorption lines occurs that allows to determine hydrodynamic radii or the viscosity of the surrounding liquid, respectively [13, 14].

5.2 Materials and methods

5.2.1 Materials

Oxalyl chloride, *N,N*-dimethylacetamide (DMAc), and DMF were obtained from Arcos Organics, Geel, Belgium. $FeCl_3 \cdot 6H_2O$ and $FeCl_2 \cdot 4H_2O$, tetrahydrofuran (THF), and oleic acid were obtained from Merck, Darmstadt, Germany. Palmitic acid (PA), myristic acid (MA), and lauric acid (LA) were purchased from Carl Roth GmbH, Karlsruhe, Germany. *N,N′*-carbonyldiimidazole (CDI) was obtained from abcr GmbH, Karlsruhe, Germany. Dimethyl sulfoxide (DMSO) was purchased from Fisher Scientific, Loughborough (U.K). Dextran from *Leuconostoc mesenteroides* ssp. (M_w 6,000, 15,000, and 60,000 g/mol from Sigma Aldrich, Steinheim, Germany) was treated in vacuum at 100 °C for 24 h prior to use. LiCl was supplied by Sigma Aldrich, Steinheim, Germany, and was treated in vacuum for 48 h at 100 °C. 2-[(4-methyl-2-oxo-2*H*-chromen-7-yl)oxy]acetic acid (MUB) was prepared according to [15]. All other reagents were used without further purification.

5.2.2 Synthesis of dextran ester

5.2.2.1 Dextran fatty acid ester (2a-l): Dextran fatty acid esters were prepared by activation of fatty acids (LA, MA, and PA) as iminium chlorides following a method described by Liebert et al. [10]. General procedure: Dextran samples of different molar mass (M_w 6,000, 15,000, and 60,000 g/mol) were dissolved in DMAc/LiCl at 100 °C for 1 h under stirring. The fatty acid iminium chloride was prepared by conversion of DMF with oxalyl chloride and subsequent reaction with LA, MA, or PA at – 20 °C for 30 min. Subsequently, the solution of the carboxylic acid iminium chloride was added to the dissolved biopolymer to form the corresponding dextran ester (16 h, 60 °C). After precipitation in isopropanol, the product was dried in vacuum at room temperature for 2 days.

Peracetylation of the dextran esters was carried out to determine the DS by means of ^1H NMR spectroscopy according to literature [16]. Details for the esterification and peracetylation of dextran esters are described in Ref. [17].

5.2.2.2 Dextran-2-[(4-methyl-2-oxo-2*H*-chromen-7-yl)oxy]acetic acid ester (3a,b): Cross-linkable dextran ester was prepared by *in situ* activation of 2-[(4-methyl-2-oxo-2*H*-chromen-7-yl)oxy]acetic acid (MUB) with CDI. To a solution of dextran (M_w 6,000 g/mol, 3.0 g, 18.5 mmol) in DMSO (90 mL), MUB (4.33 g, 18.5 mmol) and CDI (3.0 g, 18.5 mmol) were added and allowed to react at 80 °C for 20 h under stirring. The product was precipitated in 1000 mL ethanol and washed three times. Subsequently, the product was dried under vacuum for 2 days (room temperature) yielding a white solid.

5.2.2.3 Dextran–MUB fatty acid ester (4a,b): The synthesis of the dextran–MUB fatty acid ester (**4a,b**) was carried out according to the same procedure as described for dextran esters **2a-l** using dextran esters **3a,b** as starting materials.

5.2.3 Preparation of magnetic nanoparticles

The preparation of MNPs was performed at the Leibniz-Institute of Photonic Technologies (IPHT) and by the group of S. Dutz [18] following a modification of a basic precipitation method described by S. Dutz [19]. Briefly, a solution of 1 M $NaHCO_3$ was slowly added to $FeCl_2/FeCl_3$ solution (Fe^{3+}/Fe^{2+}: 4.2 M/3.3 M) with a rate of 2 mL/min under permanent stirring up to pH 7–7.5, leading to the formation of a brownish precipitate. After that, the solution with the precipitate was boiled for 10 min to form an almost black precipitate. After cooling to room temperature, the suspension was washed with distilled water and dried. Details can be found in [20].

The method described generates bigger MNP than superparamagnetic iron oxide nanoparticles (SPIONs, mean diameter about 15 nm above the superparamagnetic limit) that have larger specific heating power at "high" magnetic field amplitude (Example: 20 kA/m) [21]. The MNPs tend to agglomerate due to the high surface energy and magnetic interaction.

The obtained MNPs were then coated with oleic acid in order to make their surface hydrophobic for a better compatibility within the hydrophobic dextran ester. Oleic acid and concentrated KOH solution were mixed with aqueous suspension of the particles to promote the surfactant in the form of oleate to adhere to the particle surface. The oleate was converted into oleic acid by adding concentrated HNO_3. The mixing steps were done in a ball mill (pulverisette 5, Fritsch, Germany) for 2 h and 150 rpm in order to destroy large agglomerates. The product was washed several times with acetone and hot water alternately. After sedimentation in acetone, supernatant liquid was removed and the substance was dried. Details are described in [22].

5.2.4 Spectroscopic-, chromatographic- and thermal measurements

Fourier-transform infrared (FTIR) spectra were recorded on a Nicolet Protégé 460 spectrometer with 64 scans and a resolution of 4 cm^{-1} (KBr technique). KBr tablets were dried at 100 °C for 1 h to remove moisture prior to the measurement. Nuclear magnetic resonance (NMR) spectra were acquired on a Bruker AMX 250 spectrometer at room temperature with 16 scans for ^1H NMR, 10,240 scans for ^{13}C NMR measurements, and $^1H/^{13}C$ Heteronuclear Single Quantum Coherence (HSQC) spectra (40 mg sample/mL in $CDCl_3$). The DS values of dextran esters **2a-l** were calculated from ^1H NMR spectra of peracetylated samples (according to Ref. [16]), $DS_{fatty\ acid} = 3 - (7 \times I_{H,\ acetyl}) / (3 \times I_{H,\ AGU})$. The DS values of dextran esters (**3a,b** and **4a,b**) were calculated from elemental analysis (EA), performed with a VARIO EL III CHNS analyzer (Elementar Analysensysteme GmbH Hanau, Germany). The molecular weight was obtained by gel permeation chromatography (GPC, Shimadzu, Japan, calibrated with polystyrene, 370–128,000 g/mol). The samples were dissolved in chloroform/isopropanol/triethylamine (94/2/4) and measured with a flow rate of 1 mL/min at 40 °C. The thermal properties of 10 mg samples were measured by differential scanning calorimetry (DSC, Netzsch DSC 204 F1 Phoenix) in an aluminum pan under nitrogen environment. The heating rate was 10 K/min and cooling rate was 20 K/min. The sample was first cooled to –50 °C and heated up to 200 °C. The heating/cooling cycle was carried out twice. The morphology and distribution of MNP in the polymer matrix were investigated by field emission scanning electron microscopy

(SEM, sputtered with graphite,15 kV, FE-SEM JSM 6300F, JEOL, BSE contrast). One sample with 2.2 wt% MNP was split into two parts by cryogenic break under liquid nitrogen to observe the BNC morphology in the cross-sectional area. UV–Vis spectroscopic investigations were carried out on Lambda 25 UV–Vis Spectrometer (Perkin Elmer, Singapore).

5.2.5 Magnetic measurements

Static magnetic measurements were carried out with a vibrating sample magnetometer (VSM) Micro-Mag^{TM} 3900 (Princeton Measurements Corp., USA). Magnetization curves *M*(H) were measured on the unmodified magnetic particles (to get the magnetic phase after preparation), on oleic acid coated particles, and on the BNC for determination of the particle content.

The magnetic alternating current susceptibility (from ACS measurements) of 1 wt% MNP in dextran myristic ester–based biopolymer was measured using a Quantum Design MPMS-5S in the frequency range of 10^2–10^3 Hz at temperatures of 20–120 °C with an applied AC magnetic field amplitude of 4 Oe. Additionally, temperature-dependent magnetization curves of both samples were recorded at 5–390 K with an applied magnetic field of 10 mT following the common zero-field-cooled/field-cooled (ZFC/FC) protocol.

Mössbauer spectra of dextran myristate and palmitate ester–based BNC with 1 wt% of MNP were measured in transmission geometry at temperatures of 20–110 °C using a custom-built setup with a 40 mCi ^{57}Co(Rh) in constant acceleration mode. A BNC sample with thickness of 9 mm was used, corresponding to about 9 mg/cm^2 of nanoparticles. Mössbauer spectra at 4.3 K were measured using a He cryostat containing a superconducting magnet in split-coil geometry providing a uniform magnetic field of up to 5 T. ACS and Mössbauer investigations have been done by research group of Dr. Wende, [23] University of Duisburg-Essen. For details, see their contribution in this book and 35.

5.2.6 Alternating magnetic field studies

BNCs with different sample geometries, type of polymer, and MNP content were fabricated as described in Section 5.3.3, in order to study their influence on heating ability by AMF. Disk-shaped BNCs (2 wt% MNP) were made with thickness of 1.4 mm and radius of 8 mm by casting into cylindrical mold. Films of the magnetic BNCs with different MNP concentrations were coated on object glasses by lab applicator into about 30 mm × 20 mm square with thickness of 600 ± 20, 50 ± 3, and 5 ± 1 µm (Table 5.3). BNC coatings on object glass and disk-shaped samples were placed in the middle of a coil (3 turns and 5.5 cm diameter, water cooled) in air at room temperature and subjected to an AMF amplitude of 20 kA/m and frequency 400 kHz for 5 min. The deviation of the field amplitude in the relevant area inside the coil was measured as smaller than 10 %. The surface temperature of thin film and disk-shaped sample in 1 s interval was monitored by infrared (IR) thermography using a thermal camera (NEC Avio infrared Technology) 20 s before the start of magnetic induction for 5 min and analyzed by InfReC Analyzer NS9500 Standard (frame rate: 1 fps). The results were averaged over three points on the IR thermography. The spatial resolution of the camera combined with the arrangement of the experiment (distance between camera and sample) is about 0.3 mm.

The specific absorption rate (SAR) of the BNC (2 wt% MNP) was determined by measuring the initial heating rate with a fiberoptical sensor (OPTOcon, Dresden, Germany) on a bulk-shaped sample of 0.10 g BNC (granules < 1 mm) in 0.90 g gelatin gel, using the equation SAR = $c \times m_F/m_c \times \Delta T/\Delta t$ with c as the specific sample's heat capacity (value of water), m_c the mass of BNC, m_F the fluid mass, and $\Delta T/\Delta t$ the maximum value of the linear slope of the heating curve after subjecting the sample into an AMF (20 kA/m, 400 KHz) inside of a polystyrene isolation.

5.3 Results and discussion

5.3.1 Synthesis and characterization of dextran esters

Esterification of polysaccharides with carboxylic acids is one of the most versatile transformations to create biopolymers with valuable properties [24]. Efficient esterification techniques using special methods and *in situ* procedures for activation of carboxylic acids are a very useful alternative to the conventional reagents (acid chloride or acid anhydride) for the preparation of pure polysaccharide esters [16, 24]. For this purpose, various activating reagents are available, e.g., sulfonic acid chlorides, dialkylcarbodiimide, 1,1′-carbonyldiimadazole (CDI), or activation via iminium chlorides [10, 24].

Studies about esterification of dextran with fatty acids have shown that iminium chlorides of carboxylic acids are in particular efficient [10]. The iminium chlorides of carboxylic acids are formed by the conversion of DMF with chlorinating agents (e.g., phosphoryl chloride, phosphorus trichloride, or oxalyl chloride) and subsequently reaction with the carboxylic acid forming only gaseous by-products. It is a very mild and efficient method for esterification of polysaccharides resulting in pure products [10], which enables biological and medical applications. The polysaccharide dextran, which finds widespread use in medical and pharmaceutical applications, was used for the preparation of thermoplastic bio-based material.

5.3.1.1 Long-chain fatty acid esters of dextran

Esterification of dextran with long-chain fatty acids is a path to generate thermoplastic products [24]. The synthesis of fatty acid esters was performed by activation of the acids via iminium chlorides using oxalyl chloride as chlorinating agent (Figure 5.1).

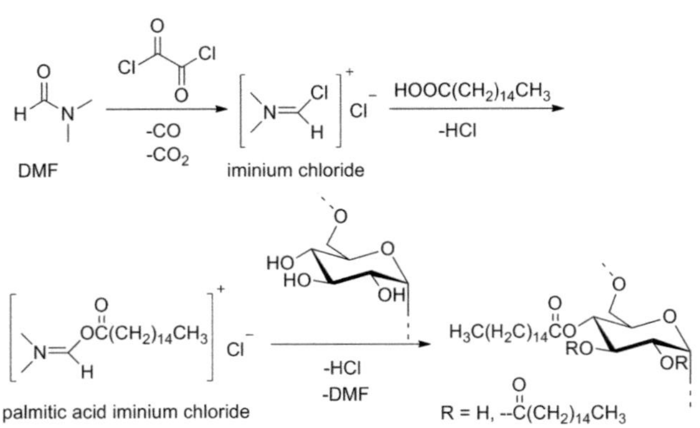

Figure 5.1: Reaction schema of the conversion of dextran with palmitic acid activated as iminium chloride.

Table 5.1: Reaction conditions for and results of the esterification of dextran (M_w 6,000 g/mol) with different fatty acids activated as iminium chlorides. (AGU: anhydroglucose unit).

No	Carboxylic acid (C-number)	Molar ratio of AGU:acid	Degree of substitution	Range of melting temperature (°C)
2a	Lauric (12)	1:2	1.06	60–70
2b	Lauric (12)	1:3	1.47	90–98
2c	Lauric (12)	1:5	2.51	30–40
2d	Myristic (14)	1:2	0.98	70–75
2e	Myristic (14)	1:3	1.50	96–105
2f	Myristic (14)	1:5	2.56	40–45
2g	Palmitic (16)	1:2	1.42	100–105
2h	Palmitic (16)	1:3	0.92	130–140
2i	Palmitic (16)	1:5	2.69	42–55
2k[a]	Palmitic (16)	1:2.5	1.59	88–95
2l[b]	Palmitic (16)	1:1.8	1.25	100–105[c]

[a]Prepared with dextran M_w of 15,000 g/mol. [b]Prepared with dextran M_w of 60,000 g/mol. [c]Incomplete melt.

The melting range of the dextran esters can be adjusted by the DS, the type of substituent introduced, and the molecular weight of the polysaccharide. The DS value can be easily controlled by the amount of reagent used. A summary of reaction conditions and results is given in Table 5.1.

Dextran esters **2a-l** with melting areas in the range of 30 and 140 °C were obtained. Compared to the acids introduced and the M_w of the dextran, the DS value has a significant influence on the melting temperature (Table 5.1). An increasing DS mostly leads to a lower melting temperature. However, the melting depends on the supramolecular structure of the polymer, i.e., from the interactions of the polymer chains with each other. Thus, the DS and the distribution of the ester moieties may determine the temperature range of the melting, which must not give steady decrease of the temperature with increasing DS. This behavior was also found for the samples with an average DS of about 1.5 that possesses a higher melting temperature compared to samples of lower DS. The structure and the purity of the dextran esters were confirmed with FTIR and NMR spectroscopy.

5.3.1.2 Cross-linkable dextran ester

In order to generate biomaterial with different magnetic microsegments, dextran was functionalized with a photochemically cross-linkable moiety. Thus, substructures that may have a barrier effect can be created through irradiation with UV light. To produce cross-linkable dextran ester, 2-[(4-methyl-2-oxo-2*H*-chromen-7-yl)oxy]acetic acid (MUB) was chosen as a representative from the well-known coumarin family [25]. Dextran–MUB esters (**3a,b**) were prepared by *in situ* activation of MUB with CDI and conversion with the biopolymer (Figure 5.2). The dextran–MUB esters (**3a,b**) were

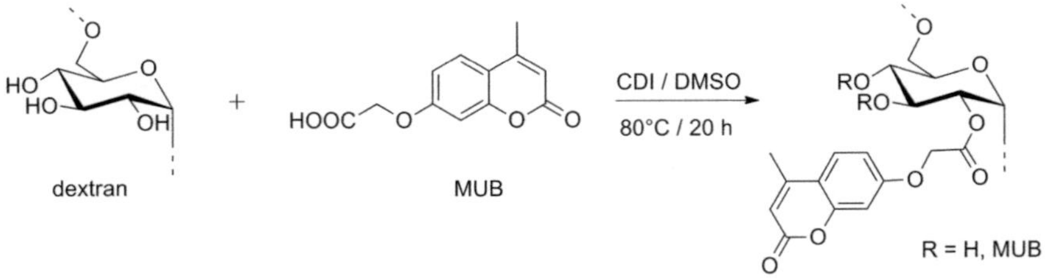

Figure 5.2: Reaction schema for the preparation of dextran-2-[(4-methyl-2-oxo-2*H*-chromen-7-yl)oxy] acetic acid ester via *in situ* activation of the carboxylic acid with CDI.

further esterified with palmitic acid to get thermoplastic biopolymer derivatives (see Section 5.3.1.1). The DS value can be easily adjusted by the molar ratio both of AGU to MUB and modified AGU to palmitic acid, respectively. A summary of reaction conditions and results is given in Table 5.2.

Cross-linkable dextran esters **3a** and **3b** possess DS values of 0.19 and 0.62. Thermoplastic and cross-linkable dextran esters were obtained with a range of melting temperature from 58 to 65 °C (**4a**) and from 75 to 80 °C (**4b**). It can be concluded that the temperature of the melting area increased with increasing DS of MUB and decreased with increasing DS of palmitate. The structure and the purity of the different dextran esters (**3a,b** and **4a,b**) were evaluated by means of ^{1}H and ^{13}C NMR spectroscopy.

5.3.2 Photo-cross-linking

Light-induced cross-linking opens up a possibility for the generation of polymeric networks. It is especially attractive because photo-cross-linking is an ecological and nondestructive process [26]. Moreover, a high level of control over the network formation and cross-linking density can be achieved. Photo-cross-linkable polysaccharides can be generated through functionalization with various photosensitive groups such as cinnamate, anthracene, or coumarin derivatives [27]. These groups are

Table 5.2: Reaction conditions for and results of the esterification of dextran (M_w 6,000 g/mol) with MUB and of dextran–MUB ester (**3a,b**) with palmitic acid iminium chloride (**4a,b**).

No	Carboxylic acid	Molar ratio (AGU:acid)	Degree of substitution		Range of melting temperature (°C)
			MUB	Palmitate	
3a	MUB	1:0.4	0.19	–	–
3b	MUB	1:1.0	0.62	–	–
4a	Palmitic	1:5.0	0.19	1.92	58–65
4b	Palmitic	1:5.0	0.62	1.66	75–80

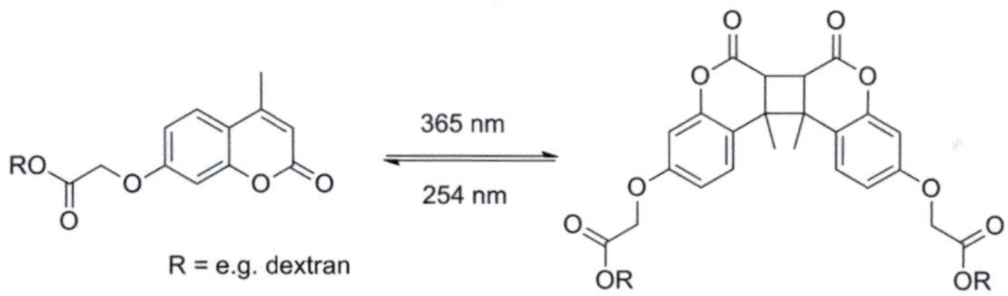

Figure 5.3: Schematic presentation of the [2π + 2π] cycloaddition of 2-[(4-methyl-2-oxo-2H-chromen-7-yl)oxy]acetic acid derivative.

able to form covalent bonds under controlled UV-light irradiation via photodimerization, which may be reversible depending on the wavelength of the light. Coumarin derivatives, e.g., undergo a [2π + 2π] cycloaddition forming a cyclobutane during irradiation at wavelength above 300 nm, whereas cleavage occurs at 254 nm (Figure 5.3)[27].

The photochemical response of dextran esters **4a** and **4b** was studied in solution by means of UV–Vis spectroscopy under irradiation with light of a wavelength of 365 nm (Figure 5.4). Upon irradiation, the absorption at 316 nm decreases indicating the formation of the MUB dimer via [2π + 2π] cycloaddition. After 180 min, a degree of dimerization of 50 % for **4a** and around 70 % for **4b** was achieved, concluding that the material with a higher DS of cross-linkable groups shows better cross-linking behavior. Furthermore, the absorbance increases with increasing DS of MUB because of the relative higher content of coumarin moieties.

Photo-cross-linking experiments with films of dextran esters **4a** and **4b**, prepared with a lab applicator, were carried out. The films were irradiated for 3 h with light of a wavelength λ = 365 nm to investigate the cross-linking ability in the solid state. The films obtained show changes of properties, including solubility and melting behavior. Irradiated films of dextran ester **4b** do not melt anymore, whereas films prepared from **4a** are still partially meltable. It can be concluded that a DS of MUB of 0.19 is not sufficient to obtain a complete cross-linking of the material.

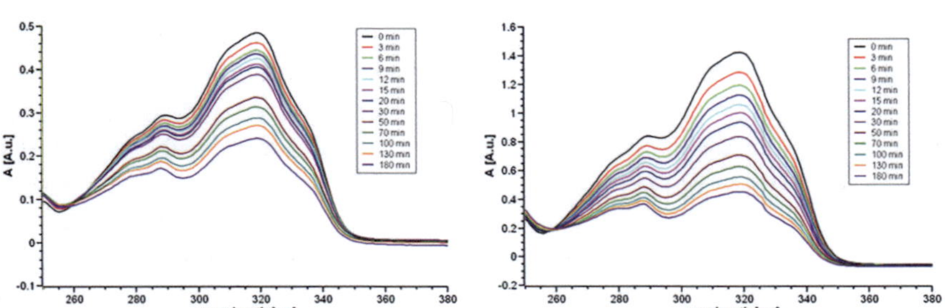

Figure 5.4: UV–Vis spectra of the photodimerization of **4a** (left) and **4b** (right) dissolved in $CHCl_3$ after different irradiation times.

5.3.3 Preparation and analysis of magnetic bionanocomposite (BNC)

The dextran ester and the MNPs were analyzed by different methods including FTIR spectroscopy, DSC, thermal gravimetric analysis (TGA), viscosimetry, dilatometry, hot table microscope (for polymers), and dynamic light scattering (DLS) and vibrating sample magnetometry (for MNP).

Modification of iron oxide nanoparticle with carboxylic acid results from the interactions between the carboxylic/carboxylate moieties and the surface metal ion [28]. Uncoated MNP shows a coercivity H_C of 3.1 kA/m and a saturation magnetization M_S of 74.8 Am^2/kg, oleic acid–coated MNP a decreased H_C of 2.5 kA/m and M_S of 68.8 Am^2/kg. From M_S measurements the content of oleic acid was calculated. It is in the range of 8 wt%. Dextran myristate (**2f**) and dextran palmitate (**2i**), showing melting points in the region suitable for applications in the biological field, were chosen for fabrication of magnetic hybrid material and were applied for the following experiments.

In the frame of these studies, three different approaches were investigated for fabrication of magnetic BNCs, namely mixing of the polymer and the MNP, followed by melt extrusion, dissolution/suspension of polymer, and film formation on glass by spin coating [17]. However, only dissolution/suspension combined with solution casting and subsequent coating with lab applicator on glass after melting was used to prepare BNC for the studies, bulk samples and layers, respectively, for all types of dextran ester. In detail, magnetic BNCs with MNP concentrations between 0.05 and 2.2 wt% were obtained by following steps: dissolution of the dextran ester (2.0 g including MNP) and suspending hydrophobized MNPs in the organic solvent tetrahydrofuran (THF) followed by homogenization with ultrasonication (Elma Transsonic 460/H) for 20 min and evaporation of the solvent by drying of thin layers under air flow. The material was collected in the form of granulates, dried in vacuum to remove residual THF, and shaped in the molten state. The BNCs were coated on object glasses into $20 \times 30\ mm^2$ square with thickness of 5, 50, and 600 µm or casted to disks of up to 1.4 mm thickness. The used preparation method gives a uniform dispersion of MNP in the polymer matrix. The parameters of the different samples can be seen in Table 5.3.

The spatial distribution and morphology of MNP in the polymer were studied by SEM and optical microscope (Figure 5.5). In a sample containing 2.2 wt% of MNP, the MNP and small agglomerates in range between 100 nm and 1 µm are uniformly distributed in the polymer matrix (**2i**, Figure 5.5). The SEM picture shows that the particles are also well distributed. The evenly distributed particles could also be seen in the cross-sectional direction of a sliced sample (Figure 5.5, right). Less aggregates are formed in samples with a smaller MNP content [17]. The MNPs are interlocked in the polymer matrix.

Rheological measurements on dextran palmitate in a magnetic field of 100 kA/m have been performed by D. Borinx [30]. Control of the gradient-free sample temperature

Table 5.3: Summary of composition and geometry of magnetic BNC fabricated by solution casting and their maximal heating response by IR thermography on alternating magnetic field (AMF) heating.

No	Composition	Sample geometry	Thickness (μm)	Maximal ΔT (°C)[a]	Heating rate (mK/s)[b]
C1	**2i** + 2 wt% MNP	Coating film	600 ± 20	25	114
C2	**2i** + 2 wt% MNP	Coating film	50 ± 3	7	32
C3	**2i** + 2 wt% MNP	Coating film	5 ± 1	1	5
C4	**2i** + 1 wt% MNP	Coating film	50 ± 3	4	14
C5	**2i** + 0.3 wt% MNP	Coating film	50 ± 3	1	4
C6	**2i** + 0.05 wt% MNP	Coating film	50 ± 3	0	0
C7	**2i** (control sample)	Coating film	50 ± 3	0	0
C8	**2i** + 2 wt% MNP	Disk sample (r = 8 mm)	1400 ± 50	38	136
C9	**2i** + 2 wt% MNP	Disk sample (r = 8 mm)	1400 ± 50	33	138
C10	0.1 g **2i** + 2 wt% MNP in 0.9 g gelatin	Granules (< 1 μm)	-	30	88

[a]Subtracted by control sample. [b]Heating rate (maximum temperature divided by the time, measuring frequency: 1 Hz).

was ensured with a Peltier hood. After the tempering procedure, an examined sample has been pre-sheared at a shear rate of 100 1/s at least during 120 s and the external magnetic field was applied using the same shearing conditions. The measurements show only a weak effect of the magnetic field of about 0.8 % viscosity increase at 60 °C and 2.8 % increase at 100 °C, respectively, that is a further indication of the weak magnetic interaction between the particles. The magnetic effect is reversible.

A further indication of the spatially averaged, predominant particle interaction depending on the particle distance can be obtained from the Henkel plot [31] that can

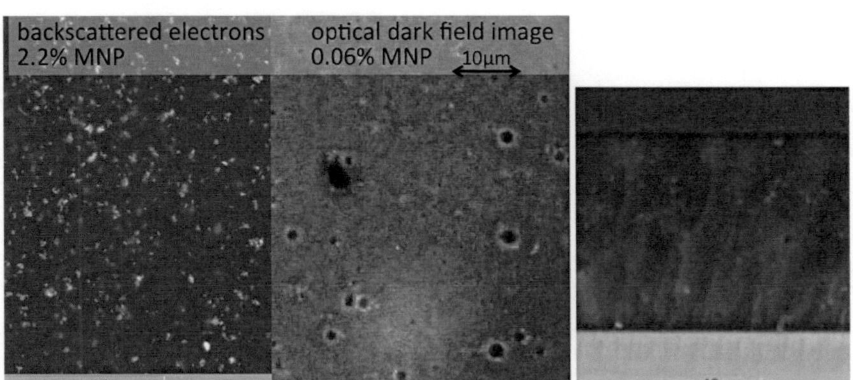

Figure 5.5: SEM (left) and optical image (center) of MNP in dextran ester 2i and SEM image (right) of cross section of the 2.2 w.t% MNP layer (fracture surface). © [2015] IEEE. Reprinted, from [45].

be derived from the initial and the demagnetization remanence curves. Both remanence curves are related via the relation of Wohlfarth [32] in the absence of interaction (straight line in Figure 5.7). In Figure 5.7, only weak particle interaction in samples with dispersed particles is shown, almost independent from the MNP concentration. In order to show the effect of a particle agglomeration in the Henkel plot, the polymer matrix of a composite sample with 2.2 wt% of MNP was dissolved and the sample was measured in liquid state, i.e., after agglomeration.

Magnetic measurements by VSM were carried out to compare magnetic particle–particle interaction that depends on the mean particle distance. Increasing hysteresis parameters (H_c, remanence ratio M_r/M_s) with decreasing particle content were found that suggests a decreasing magnetic interaction, i.e., a better separation of particles on microscopic scale (Figure 5.6) [29]. The particle content was calculated from magnetization values.

DSC measurements were carried out to evaluate if the embedment of particles modifies the thermal behavior of the dextran esters. The first heating curve of dextran palmitic ester and BNC obtained with this sample (1 and 2 wt% of MNP) is shown in

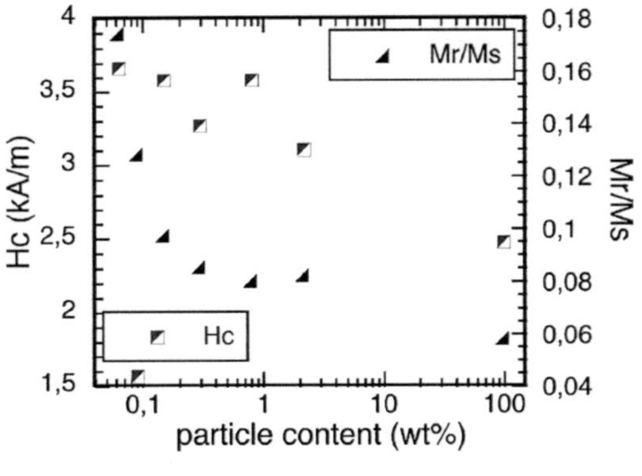

Figure 5.6: Hysteresis parameters vs. MNP concentration. © [2015] IEEE. Reprinted, from [45].

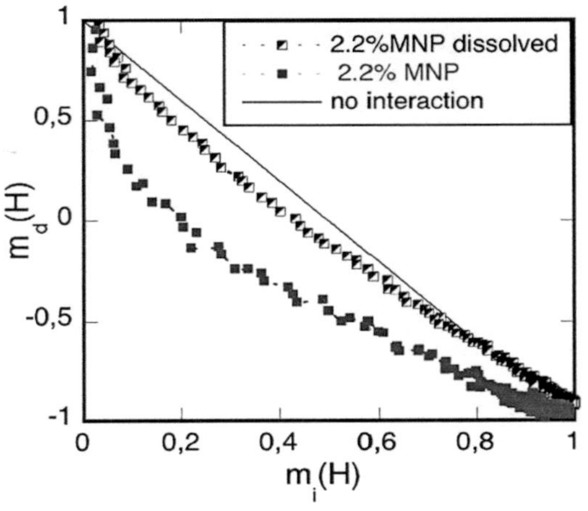

Figure 5.7: Henkel plot (magnetic interaction) of different MNP arrangement. © [2015] IEEE. Reprinted, from [45].

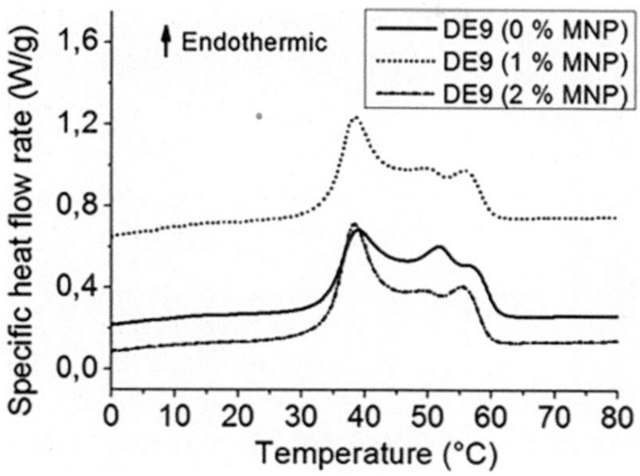

Figure 5.8: Thermal behavior of dextran palmitic ester and BNC characterized by DSC, solid line, 2i; point, 2i with 1 w.t% MNP; dash-point, 2i with 2 w.t% MNP. Reprinted from 17 Copyright (2015) American Chemical Society.

Figure 5.8. An endothermic peak is found at 39 °C for dextran palmitic ester and two additional peaks at 50 °C and 56 °C that is the same temperature range observed optically as melting region. Detailed assignment of these peaks is not yet possible. Nevertheless, the DSC measurements of the BNC of **2i** show the same signals with comparable intensities, which means the presence of the MNP does not affect the thermal behavior of dextran ester.

5.3.4 Alternating magnetic field heating experiments

The internal melting behavior of the BNC depends on type of polymer, MNP content, and geometry of the sample. Heating experiments in an AMF based on magnetic losses of the MNP in the composite material have been done to find temperature limits using certain geometrical arrangements of the samples in comparison to melting temperatures of the dextran ester and the specific absorption rate of the BNC. Such experiments are precondition for further thermal-driven drug release experiments.

The heating response of magnetic BNC on glass with different geometries (thickness 5, 50, and 600 µm) and MNP content to AMF was studied in detail (Table 5.3) [17]. BNC films on glass were placed in the middle of a coil at room temperature and exposed to an AMF of 20 kA/m and frequency of 400 kHz for 5 min. The surface temperature of the thin films was monitored by IR thermography using a thermal camera 20 s before and 5 min after the magnetic induction. Details on these measurements are given in [17]. A typically uniform heating response of the sample (**C2**) on application of AMF is generated due to the good dispersion of MNP in the BNC layer (Figure 5.9, left). AMF for longer times increases the surface temperature, correspondingly. The heating response is strongly dependent on the thickness of the layer [17] because of heat dissipation through conduction to the glass substrate and convection to air depending on the ratio of the surface area to volume (Figure 5.9, right). The maximal increase in temperature after 5 min and the corresponding heating rates measured on the coating films without

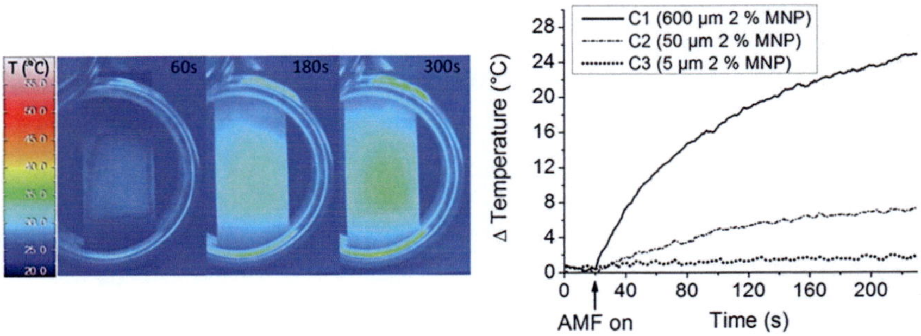

Figure 5.9: IR thermography of a film prepared from sample **C2** after 60, 180, and 300 s in AMF (left). The graph (right) shows the surface temperature (subtracted by control sample **C7**) of the BNC films **C1–C3** subjected to continuous AMF. Reprinted from 17 Copyright (2015) American Chemical Society.

MNP are listed in Table 5.3. When the MNP loading is below 1 wt% (**C5, C6**), there is no significant heating response for samples with thickness of 50 µm.

Disk-like shaped samples with thickness of 1.4 mm and radius of 8 mm were prepared by casting into a cylindrical mold to diminish heat dissipation compared to the films on glass and were placed on a polystyrene holder in the middle of coil. After 2–3 min, such disk-type samples were heated above the melting temperature (**C8**: 50 °C, **C9**: 45 °C) of the dextran ester, i.e., these samples have a much better heating rate than the films (Table 5.3). After switching off the AMF, the BNCs were quickly cooled in air under the melting temperature and solidified within 2 min (Figure 5.10, right) that confirms that the material is in principle very well suitable for remote melting, e.g., in controlled release applications.

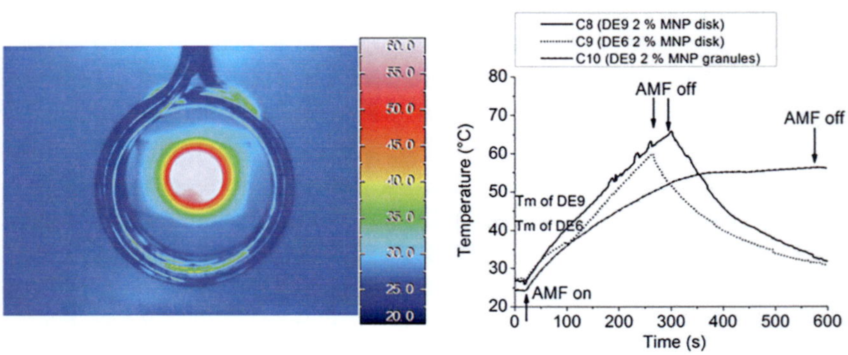

Figure 5.10: IR thermography of sample **C8** after 240 s in an AMF (left); surface temperature of BNC disk of samples **C8** and **C9** in comparison to sample **C10** subjected to continuous AMF (right). Reprinted from 17 Copyright (2015) American Chemical Society.

Further heating experiments were performed in an arrangement more comparable to a medical application, like drug release, on small bulk-shaped samples in water or in gelatin as biological model system. Measuring the temperature of the surrounding water at known mass ration sample/water, the SAR of the sample material can be calculated from the initial heating rate. Here the temperature was measured with a fiber-optical sensor (OPTOcon, Dresden, Germany).

The SAR of the BNC **C10** (10 wt% in gelatin, 2.2 wt% MNP) was determined as 6.3 W/g (Figure 5.10, right), which was subjected to AMF (20 kA/m, 400 kHz). A specific heat capacity of 1.5 kJ/kg·K was assumed. A plateau in the heating curve at 55 °C was found and lasted for about 80 s. The temperature went to 54 °C until the field was turned off. The heating rate is smaller compared to disk samples, because the heat was absorbed by water. Nevertheless, even in this setup melting of the BNC was readily achieved.

5.3.5 Texturing experiments

In a molten polymer matrix (dextran palmitate **2i** with different content of MNPs, by AMF or external heating), the MNP can be moved inside the matrix by a magnetic field gradient. Our experiments revealed such a movement by application of an external static magnetic field (NdFeB magnet, field gradient about 15–20 T/m). In samples with a low concentration of particles (< 0.2 %), the movement of single particles in the matrix is microscopically observable. Particle velocities of up to 0.25 mm/s were observed at sample temperatures of about 100 °C leading to a viscosity of about 1 Pa s. This movement can be influenced by a superposition with convection or viscosity deviations by a thermal gradient in the case of a nonuniform heating source. In samples with a higher MNP concentration (> 0.5 %), we observed a "magnetic texturing" of the BNC under a static field, which alters the SAR after switching off the DC field (Figure 5.11).

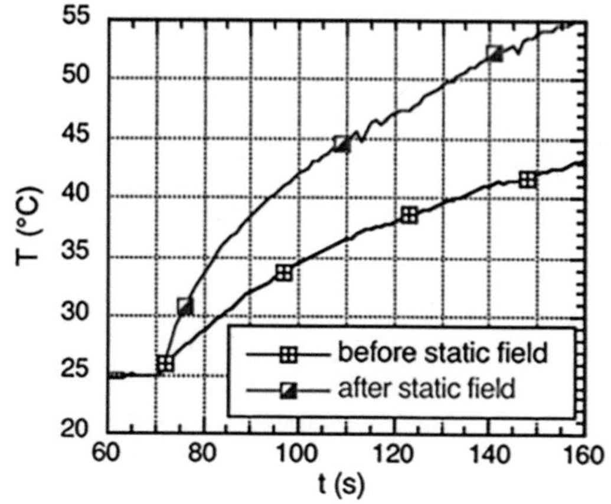

Figure 5.11: Maximum temperature of the BNC in an AMF before and after texturing in a static magnetic field. Texturing axis was perpendicular to the BNC layer = parallel to AMF. © [2015] IEEE. Reprinted, from [45].

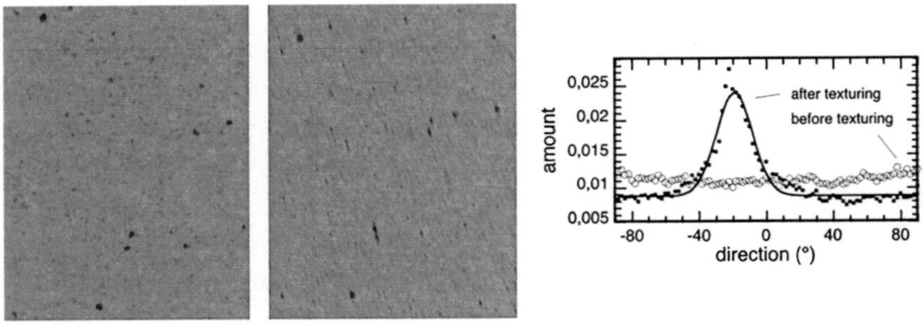

Figure 5.12: Proof of particle texturing in a static magnetic field under external heating at 80 °C/5 min: Optical image before (left) and after (right) texturing and corresponding pixel analysis (0° = y-direction). © [2015] IEEE. Reprinted, from [45].

The structural reason for that (orientation of MNP and agglomerates or increase of the local MNP concentration) is not investigated yet.

In addition to optical micrographs, a "magnetic texturing" can be seen by pixel analysis (in-plane direction of pixel structures, [33] for details, see [34]) in Figure 5.12. The pixel analysis reveals a distribution of the area of "similar pixels" (not of particles). The texturing axis corresponds roughly with the direction of the field gradient. Before texturing, the image reveals no preferred direction of pixel structures, i.e., there is no particle texturing caused by formation of the BNC films. The dependence of the quasi-static hysteresis losses (area of the VSM magnetization loop) on the direction concerning the texture axis was investigated on a textured sample of about 10 mm^3 at a maximum field strength (20 kA/m) similar to the AMF amplitude (Figure 5.13). The losses measured perpendicular to the texturing axis are higher by a factor of 1.7 compared to these measured parallel to the texturing axis (losses per magnetization cycle: $2.0 \cdot 10^{-8}$ J vs. $1.2 \cdot 10^{-8}$ J). Further minor loop parameters are given in [35].

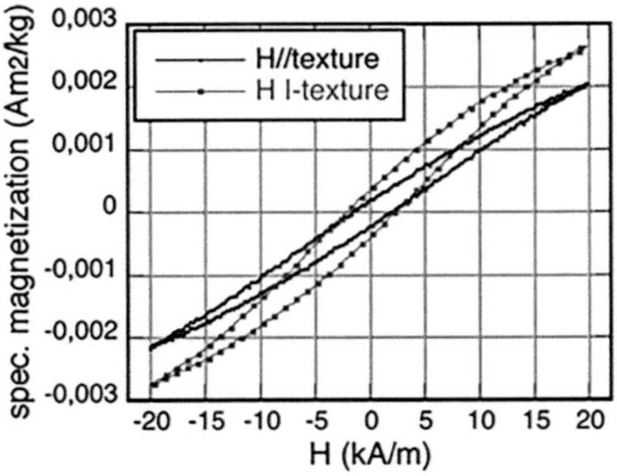

Figure 5.13: Minor magnetization curves measured parallel and perpendicular to the texturing axis. © [2015] IEEE. Reprinted, from [45].

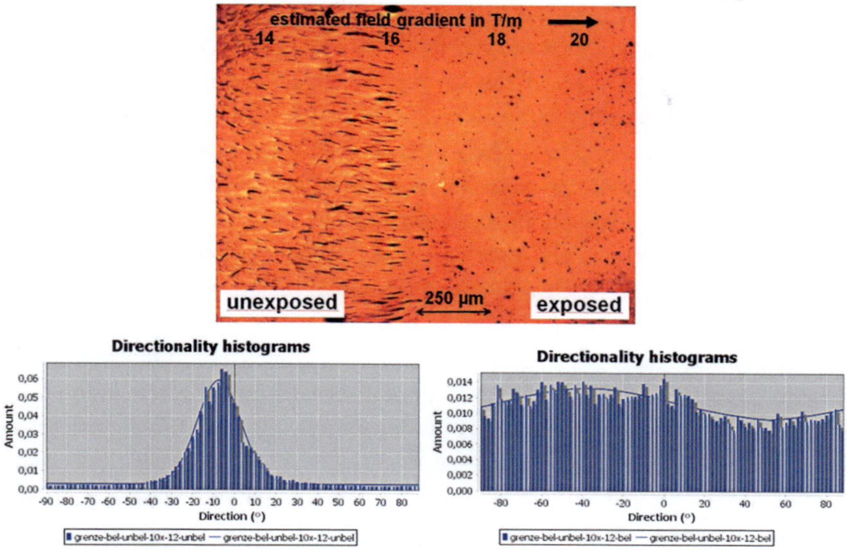

Figure 5.14: Particle texturing in a static magnetic field under external heating at 110 °C/5 min: optical image without (left side) and with previous irradiation at 365 nm (right side) after texturing and corresponding pixel analysis (0° = y-direction).

Furthermore, texturing experiments of particles in cross-linkable biopolymer (dextran ester **4b** with 2 wt% MNP, film thickness 250 µm, external heating (110 °C/ 5 min)) by a static magnetic field were performed. The BNC was coated on object glasses and a defined sector was irradiated with UV light. A movement of MNP could be only observed in the area of unexposed BNC (Figure 5.14). Pixel analysis (in-plane direction of pixel structures) [33, 34] of the corresponding areas of the optical image and magnetic measurements confirm the texturing.

5.3.6 Magnetometry and AC susceptometry experiments

The general magnetic relaxation behavior of dextran myristate **2f** and palmitate **2i** ester with MNP was investigated in [35] by recording ZFC/FC magnetization curves between 5 and 390 K using an applied magnetic field of 10 mT. Since the irreversibility temperature at which both magnetization curves coincide is not reached, one can conclude that in both samples there is at least a minor fraction of larger multicore particles showing sufficiently slow Brownian relaxation to be considered as magnetically blocked on the magnetometry time scale.

The average hydrodynamic diameter of the MNP (or agglomerates) moving in the liquid polymer melt can be determined by measurements of the magnetic AC susceptibility [36]. The imaginary component χ'' of the magnetic susceptibility is given by

$\chi'' = \chi'\omega\tau/(1+(\omega\tau)^2)$, $\tau = \pi\eta V_h/k_B T$ (ω: angular frequency, τ: relaxation time, η: dynamic viscosity, $k_B T$: thermal energy, χ': static susceptibility). Effects of the particle diameter distribution, resulting in a shift of the average Brownian relaxation time and a broadened peak in χ'', were considered as described in [35]. The influence of the oleic acid coating with about 1–2 nm in thickness can be neglected compared to the total particle size. The thickness of the oleic acid coating of ferrite particles coated by a similar procedure was found by SANS investigations [37].

ACS shows measurements of the imaginary part χ'' of the magnetic susceptibility of 1 wt% MNPs in the dextran myristate–based sample at different temperatures. For temperatures below the melting region, χ'' is found to be relatively constant that is presumably caused by slow Néel relaxation processes with a broad distribution of relaxation times and is in agreement with the absence of a distinct Néel relaxation peak in ZFC magnetization curves. Whereas far above the melting region, a broad peak at the low frequency regime can be seen, moving toward higher frequencies with increasing temperature [35].

Assuming a polymer melt viscosity of about 0.5 Pa s at 393 K, an average particle diameter of approx. 64 nm and a width of the log normal size distribution of 36 nm could be calculated [35]. Taking into account the size of the multicore particles of about 30–80 nm found by SEM, this indicates that Brownian motion in the liquid polymer melt is dominated by the movement of separate particles rather than large agglomerates. However, high relaxation times of large agglomerates in a high viscous melt might be outside of the accessible measurement range using ACS.

5.3.7 Mössbauer spectroscopy

Mössbauer investigations on magnetic BNC were carried out by the group of Dr. H. Wende (University of Duisburg-Essen) [23]. In the present paper only a short summary of the results is given. Details can be found in [35].

Mössbauer spectra of the dextran myristate **2f** BNC were measured at different temperatures with and without an applied magnetic field of 5 T, respectively. Figure 5.15 shows an example of a measurement (temperature: 4.3 K, magnetic field: 0T). Further spectra are available in [35]. The spectra display a magnetically ordered sextet structure reproducible by three sextet subspectra, [38] Fe^{3+} and Fe^{2+} on octahedral sites and Fe^{3+} on tetrahedral sites, a mixture of magnetite and maghemite within the particles, and an insignificant fraction of paramagnetic material. Furthermore, the relative line intensities indicate a low degree of spin canting that may suggest a well-ordered magnetic structure of the studied iron-oxide particles. The influence of Néel relaxation taking into account the lognormal particle diameter distribution [39] could be seen by measurements at different temperatures [35]. Details of reproducing the spectra can be found in [35]. Considering limitations in the time constant of Mössbauer spectroscopy and of magnetic exchange and interaction effects, only a rough approximation of the

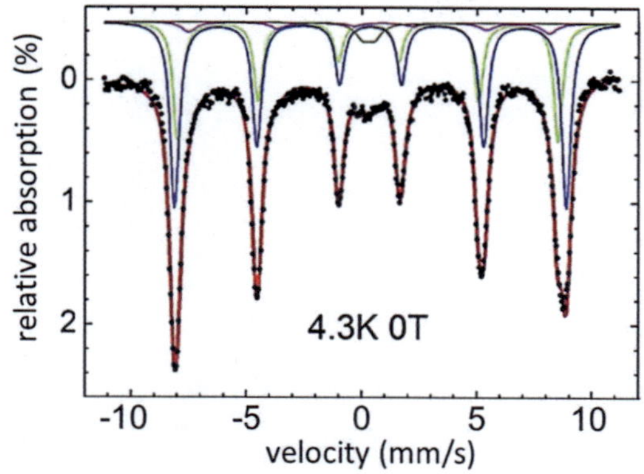

Figure 5.15: Mössbauer spectrum of dextran myristate–based sample measured at 4.3 K without an applied magnetic field along γ-ray propagation direction, showing a typical sextet structure and the deconvolution of the spectrum.

particles' anisotropy was possible. An effective anisotropy constant of about 18 kJm^{-3} was estimated assuming an average size of about 15 nm of the crystallites.

Effects of particle motion can also be observed by Mössbauer spectroscopy, measuring the broadening of the absorption lines caused by the diffusive translational motion of the particles within the polymer melts above the melting temperature. The line broadening is independent of the Néel relaxation process. Details on the analysis of the data are given in [35]. A first considerable line broadening of dextran myristate and palmitate BNC can be observed in the melting region of the samples. A slow, exponential increase in line width caused by the decrease in polymer melt viscosity at higher temperatures indicates an enhanced particle mobility as it was illustrated by AC susceptometry data.

5.3.8 Cell experiments/biocompatibility studies

For a possible biomedical application of the dextran ester BNC cell experiments for biocompatibility studies are necessary and were carried out by the group of J. Clement (University Hospital, Jena) [40]. Human brain microvascular endothelial cells (HBMECs) representing the human blood–brain barrier were used for testing biocompatibility of the dextran ester and their coating *in vitro*. The cells were seeded on the dextran ester–coated coverslips within 24 well plates and cultured for 48 h. For microscopic investigation of cell viability, adherent cells were washed with phosphate buffered saline, fixed in neutral buffered formalin, and subsequently, the cell membrane was permeabilized in 0.1 % Triton-X 100 (Sigma-Aldrich Chemie, Steinheim, Germany). For visualization and microscopic characterization, both F-actin (a protein that forms microfilaments in the cytoskeleton) and nuclei of the cells were stained simultaneously. Fluorescence was analyzed using the confocal laser scanning microscope LSM 510 META (Carl Zeiss Microscopy GmbH, Jena, Germany). Details are given in [17].

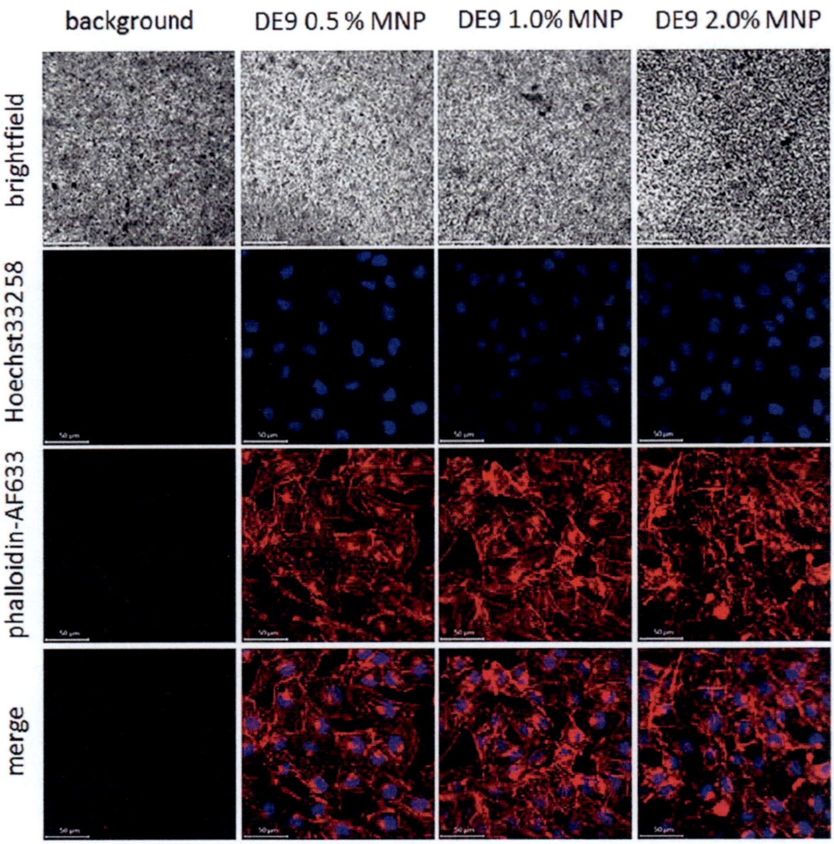

Figure 5.16: HBMC were seeded on dextran ester **2i** containing MNP concentrations of 0.5, 1.0, and 2.0 % immobilized on glass coverslips. Upon fixation and permeabilization, nuclei and F-actin were stained with Hoechst33258 (blue) and Alexa Fluor phalloidin 633 (red), respectively. Stained samples were analyzed by confocal laser scanning microscopy with 400-fold magnification. Scale bars indicate 50 µm. Reprinted from 17 Copyright (2015) American Chemical Society.

In the experiment, HBMECs were covered by increasing concentrations of MNP embedded within dextran ester **2i**. The cells attach to the dextran ester surface in a well-distributed manner and show a regular cellular growth compared to control cells on glass coverslips without any composite coating (Figure 5.16). The staining of F-actin indicates that the presence of dextran ester with particle concentrations between 0.5 and 2.0 % does not influence the cell morphology and cell–cell contacts do not appear disrupted. Additionally, a recovery of a numbers of cells from the dextran-coated surface was observed. That means these results confirm a biocompatible nature of this BNC containing up to 2.0 % magnetic particles.

Furthermore, investigations on the biodegradation of the BNC in artificial body fluids (simulated body fluid, SBF; pH 7.4) and artificial lysosomal fluid (ALF, pH 4.5) that may influence the applicability of BNC in medicine were carried out by D. Fischer [41]. SBF simulates a neutral body environment (pH 7.4) that can be found in the blood stream, the extracellular matrix, or in the cytoplasm. ALF was used to simulate the degradation of potentially released iron oxide particles from the BNC and to assess the shielding effect of the dextran fatty acid ester in acidic environments. The iron release from the incorporated MNPs was used as a measure for the stability of the material. The measurements are described in [42]. Only 0.31 % (dextran palmitate) or 0.19 % (dextran myristate) of the total iron content was released during storage in SBF indicating the stability of the MNP in the BNC, whereas in ALF higher cumulative amount of 5.4 % (dextran palmitate) or 2.17 % (dextran myristate) was mobilized after 28 d at 37 °C, which is in accordance with pH-dependent results of Gutiérrez et al. [43]. It might be that the hydrophobicity of the dextran esters prevents degradation by limiting the penetration of the aqueous simulation media into deeper regions of the BNC and therefore only superficial areas degraded. This indicates a sufficient stability of the BNC that could be used as remote-controlled drug delivery systems or for the induction of hyperthermia [44] over a long period. The dextran fatty acid esters are able to protect the incorporated MNP from degradation even at acidic conditions (ALF). There were no significant differences regarding the stability of the BNC between the two investigated polymers dextran myristic ester and dextran palmitic ester.

5.4 Conclusion

It is shown that thermoplastic magnetic BNC can be prepared from dextran fatty acid ester and MNPs with melting temperatures slightly above human body temperature and beyond (>100 °C). The heating response of BNC to the application of high-frequency AMF depends on MNP content and geometry of the sample. With an optimal content of MNP between 1 and 2 wt% and a thickness of at least 50 μm, heating above the melting temperature is possible. In the molten state of the BNC, a texturing of MNP is possible by means of a static magnetic field that improves the heating ability of the material. AC susceptometry reveals that the MNPs perform diffusive Brownian motion in the liquid polymer melt as separated particles rather than large agglomerates. Mössbauer spectroscopy confirms this result in respect to a clear increase of particle mobility at temperatures of 50–110 °C and further indicates a well-ordered magnetic structure within the particles. Furthermore, meltable dextran ester can be functionalized with a photochemically cross-linkable moiety. Irradiation of cross-linkable BNC with UV light leads to changed properties, including solubility and melting behavior.

Acknowledgment: The financial support of the German Science Foundation DFG, priority program SPP 1681, contracts HE2054/14-1, HE2054/14-2, HE2054/22-1, and MU2382/4-1, MU2382/4-2, MU2382/5-1 is gratefully acknowledged.The authors are very thankful to former coworkers in the project: M. Zhou, Dr. N. Kuhl, and A. Dellith.The

authors thank the colleagues of the priority program SPP 1681 for useful discussions and especially Dr. D. Borin/Prof. Dr. S. Odenbach (Technical University of Dresden) for rheological measurements, Prof. Dr. S. Dutz and coworkers (Technical University of Ilmenau) for the preparation of the magnetic nanoparticles, Dr. J. Clement (University Hospital, Jena) for cell experiments for biocompatibility studies, Prof. Dr. D. Fischer and coworkers (Friedrich-Schiller University of Jena, no SPP member) for investigations on the biodegradation of the bionanocomposites in artificial body fluids, and Dr. J. Landers/Prof. Dr. H. Wende and coworkers (University of Duisburg-Essen) for ACS and Mössbauer investigations.
Author contribution: All the authors have accepted responsibility for the entire content of this submitted manuscript and approved submission.
Research funding: Deutsche Forschungsgemeinschaft Award Numbers :HE2054/14-1HE2054/14-2HE2054/22-1MU2382/4-1MU2382/4-2MU2382/5-1.
Conflict of interest statement: The authors declare no conflicts of interest regarding this article.

References

1. Gupta AK, Gupta M. Synthesis and surface engineering of iron oxide nanoparticles for biomedical applications. Biomaterials 2005;26:3995–4021.
2. Sadhukha T, Wiedmann TS, Panyam J. Inhalable magnetic nanoparticles for targeted hyperthermia in lung cancer therapy. Biomaterials 2013;34:5163–71.
3. Sadhukha T, Wiedmann TS, Panyam J. Enhancing therapeutic efficacy through designed aggregation of nanoparticles. Biomaterials 2014;35:7860–9.
4. Satarkar NS, Zhang W, Eitel RE, Hilt JZ. Magnetic hydrogel nanocomposites as remote controlled microfluidic valves. Lab Chip 2009;9:1773–9.
5. Wilson SA, Jourdain RP, Zhang Q, Dorey RA, Bowen CR, Willander M, et al. New materials for micro-scale sensors and actuators: an engineering review. Mater Sci Eng R Rep 2007;56:1–129.
6. Satarkar NS, Biswal D, Hilt JZ. Hydrogel nanocomposites: a review of applications as remote controlled biomaterials. Soft Matter 2010;6:2364–71.
7. Satarkar NS, Hilt JZ. Magnetic hydrogel nanocomposites for remote controlled pulsatile drug release. J Contr Release 2008;130:246–51.
8. Mura S, Nicolas J, Couvreur P. Stimuli-responsive nanocarriers for drug delivery. Nat Mater 2013; 12:991–1003.
9. Qiu Y, Park K. Environment-sensitive hydrogels for drug delivery. Adv Drug Deliv Rev 2001;53:321–39.
10. Liebert T, Wotschadlo J, Laudeley P, Heinze T. Meltable dextran esters as biocompatible and functional coating materials. Biomacromolecules 2011;12:3107–13.
11. Chung S-H, Hoffmann A, Guslienko K, Bader SD, Liu C, Kay B, et al. Biological sensing with magnetic nanoparticles using brownian relaxation. J Appl Phys 2005;97:10R101.
12. Zeisberger M, Dutz S, Müller R, Hergt R, Matoussevitch N, Bönnemann H. Metallic cobalt nanoparticles for heating applications. J Magn Magn Mater 2007;311:224–7.
13. Keller H, Kündig W. Mössbauer studies of brownian motion. Solid State Commun 1975;16:253–6.
14. Landers J, Salamon S, Remmer H, Ludwig F, Wende H. Simultaneous study of brownian and Néel relaxation phenomena in ferrofluids by mössbauer spectroscopy. Nano Lett 2016;16:1150–5.

15. Chimichi S, Boccalini M, Cosimelli B. A new convenient route to 2-oxoethoxycoumarins: key intermediates in the synthesis of natural products. Tetrahedron 2002;58:4851–8.
16. Liebert TF, Heinze T. Tailored cellulose esters: synthesis and structure determination. Biomacromolecules 2005;6:333–40.
17. Zhou M, Liebert T, Müller R, Dellith A, Gräfe C, Clement JH, et al. Magnetic biocomposites for remote melting. Biomacromolecules 2015;16:2308–15.
18. Preparation of magnetic nanoparticles was carried out by Dr. S. Dutz. Institute of Biomedical Engineering and Informatics. Germany: Technical University of Ilmenau. 98693.
19. Dutz S, Andrä W, Hergt R, Müller R, Oestreich C, Schmidt C, et al. Influence of dextran coating on the magnetic behaviour of iron oxide nanoparticles. J Magn Magn Mater 2007;311:51–4.
20. Müller R, Dutz S, Habisreuther T, Zeisberger M. Investigations on magnetic particles prepared by cyclic growth. J Magn Magn Mater 2011;323:1223–7.
21. Müller R, Dutz S, Neeb A, Cato ACB, Zeisberger M. Magnetic heating effect of nanoparticles with different sizes and size distributions. J Magn Magn Mater 2013;328:80–5.
22. Müller R, Hiergeist R, Steinmetz H, Ayoub N, Fujisaki M, Schüppel W. Barium hexaferrite ferrofluids–preparation and physical properties. J Magn Magn Mater 1999;201:34–7.
23. Mössbauer investigations were carried out by the group of Dr. H. Wende. Faculty of Physics and Center for Nanointegration Duisburg-Essen. Germany: University of Duisburg-Essen. 47057.
24. Heinze T, Liebert T, Koschella A. Esterification of Polysaccharides. Berlin, Heidelberg, Germany: Springer Science & Business Media; 2006.
25. Wondraczek H, Heinze T. Efficient synthesis and characterization of new photoactive dextran esters showing nanosphere formation. Macromol Biosci 2008;8:606–14.
26. Defize T, Thomassin J-M, Ottevaere H, Malherbe C, Eppe G, Jellali R, et al. Photo-cross-linkable coumarin-based poly (ε-Caprolactone) for light-controlled design and reconfiguration of shape-memory polymer networks. Macromolecules 2018;52:444–56.
27. Trenor SR, Shultz AR, Love BJ, Long TE. Coumarins in polymers: from light harvesting to photo-cross-linkable tissue scaffolds. Chem Rev 2004;104:3059–78.
28. Soler MAG, Alcantara GB, Soares FQ, Viali WR, Sartoratto PPC, Fernandez JRL, et al. Study of molecular surface coating on the stability of maghemite nanoparticles. Surf Sci 2007;601: 3921–5.
29. Dutz S, Hergt R. The role of interactions in systems of single domain ferrimagnetic iron oxide nanoparticles. J Nano Electronic Phy; 2012;4:20101–7.
30. Rheological measurements were carried out by Dr. D. Borin. Institute of Fluid Mechanics, Germany. Technical University of Dresden. 01062.
31. Henkel O. Remanenzverhalten und wechselwirkungen in hartmagnetischen teilchenkollektiven. Phys Status Solidi 1964;7:919–29.
32. Wohlfarth EP. Relations between different modes of acquisition of the remanent magnetization of ferromagnetic particles. J Appl Phys 1958;29:595–6.
33. Available at: https://imagej.net/Fiji.
34. Available at: https://imagej.net/Directionality.
35. Müller R, Zhou M, Liebert T, Landers J, Salamon S, Webers S, et al. Mobility investigations of magnetic nanoparticles in biocomposites. Mater Chem Phys 2017;193:364–70.
36. Krishnan KM. Fundamentals and Applications of Magnetic Materials. Oxford, UK: Oxford University Press; 2016.
37. Hoell A, Muller R, Heinemann A, Wiedenmann A. Structure investigations of barium hexaferrite ferrofluids and their precursors by sans with polarized neutrons. Magnetohydrodynamics 2003; 39:111–18.

38. Darbandi M, Stromberg F, Landers J, Reckers N, Sanyal B, Keune W, et al. Nanoscale size effect on surface spin canting in iron oxide nanoparticles synthesized by the microemulsion method. J Phys Appl Phys 2012;45:195001.
39. Landers J, Stromberg F, Darbandi M, Schöppner C, Keune W, Wende H. Correlation of superparamagnetic relaxation with magnetic dipole interaction in capped iron-oxide nanoparticles. J Phys Condens Matter 2014;27: 026002.
40. Cell experiments for biocompatibility studies were carried out by Dr. J. Clement, Department Hematology/Oncology, University Hospital of Jena. Germany: Friedrich-Schiller University of Jena. 07747.
41. Investigations on the biodegradation of the bionanocomposites in artificial body fluids were carried out by Prof. Dr. D. Fischer, Department of Pharmaceutical Technology and Biopharmacy. Germany: Friedrich-Schiller University of Jena. 07743Germany.
42. Heinze T, Müller R, Zhou M, Rabel M, Warncke P, Fischer D. Studies on the controlled release of drugs from magnetic nanobiocomposites. IJFAC 2019;4:1–8.
43. Gutiérrez L, Romero S, da Silva GB, Costo R, Vargas MD, Ronconi CM, et al. Degradation of magnetic nanoparticles mimicking lysosomal conditions followed by AC susceptibility. Biomed Eng/ Biomedizinische Technik 2015;60:417–25.
44. Hilger I. In vivo applications of magnetic nanoparticle hyperthermia. Int J Hyperther 2013;29: 828–34.
45. Müller R, Zhou M, Liebert T, Dellith A, Dutz S, Borin D, et al Performance analysis of carrier-less modulation schemes for wireless nanosensor networks. In: IEEE Proceedings of the 15th International Conference on Nanotechnology (IEEE-NANO) Rome, Italy: IEEE-NANO); 2015. p. 1–4. https://doi.org/10.1109/NANO.2015.7388731.

Silvio Dutz*, Andreas Weidner, Moritz von der Lühe, Christine Gräfe, Philip Biehl, Johanna Demut, Paul Warncke, Sandra Jungmann, Dagmar Fischer, Felix H. Schacher and Joachim H. Clement

6 Hybrid nanomaterials of biomolecule corona coated magnetic nanoparticles and their interaction with biological systems

Abstract: Magnetic nanoparticles (MNPs) are interesting for various applications in medicine. If administered to a biological system like the human body, a so-called biomolecule corona is formed on the surface of the particles, which highly determines the biological fate of the particles. To elucidate whether a preconditioning of the MNPs by incubation with biomolecules influences biocompatibility and bioavailability, the formation of such a corona was investigated in more detail. For this, the influence of particle characteristics, e.g., surface charge, as well as various incubation parameters on the resulting corona was investigated. It was found that the biomolecule corona is

---

***Corresponding author: Silvio Dutz,** Institute of Biomedical Engineering and Informatics (BMTI), Technische Universität Ilmenau, Ilmenau, Germany; and Department of Nano Biophotonics, Leibniz Institute of Photonic Technology (IPHT), Jena, Germany, e-mail: silvio.dutz@tu-ilmenau.de, https://orcid.org/0000-0002-7258-0943

Andreas Weidner, Institute of Biomedical Engineering and Informatics (BMTI), Technische Universität Ilmenau, Ilmenau, Germany, e-mail: Andreas.Weidner@tu-ilmenau.de

Moritz von der Lühe, Philip Biehl and Felix H. Schacher, Institute of Organic Chemistry and Macromolecular Chemistry (IOMC), Friedrich-Schiller-University Jena, Jena, Germany; and Jena Center for Soft Matter (JCSM), Friedrich-Schiller-University Jena, Jena, Germany, e-mail: moritz.von-der-luehe@uni-jena.de (M. von der Lühe), philip.biehl@uni-jena.de (P. Biehl), felix.schacher@uni-jena.de (F.H. Schacher)

Christine Gräfe and Johanna Demut, Klinik für Innere Medizin II, Abteilung Hämatologie und Internistische Onkologie, Universitätsklinikum Jena, Jena, Germany, e-mail: christine.graefe@med.uni-jena.de (C. Gräfe), Johanna.Demut@med.uni-jena.de (J. Demut)

Paul Warncke and Sandra Jungmann, Institute of Pharmacy, Pharmaceutical Technology und Biopharmacy, Friedrich-Schiller-University Jena, Jena, Germany, e-mail: paul.warncke@uni-jena.de (P. Warncke), sandra.jungmann@uni-jena.de (S. Jungmann)

Dagmar Fischer, Jena Center for Soft Matter (JCSM), Friedrich-Schiller-University Jena, Jena, Germany; and Institute of Pharmacy, Pharmaceutical Technology und Biopharmacy, Friedrich-Schiller-University Jena, Jena, Germany, e-mail: dagmar.fischer@uni-jena.de

Joachim H. Clement, Jena Center for Soft Matter (JCSM), Friedrich-Schiller-University Jena, Jena, Germany; and Klinik für Innere Medizin II, Abteilung Hämatologie und Internistische Onkologie, Universitätsklinikum Jena, Jena, Germany, e-mail: Joachim.Clement@med.uni-jena.de. https://orcid.org/0000-0002-6601-2456

This article has previously been published in the journal Physical Sciences Reviews. Please cite as: S. Dutz, A. Weidner, M. von der Lühe, C. Gräfe, P. Biehl, J. Demut, P. Warncke, S. Jungmann, D. Fischer, F. H. Schacher and J. H. Clement "Hybrid nanomaterials of biomolecule corona coated magnetic nanoparticles and their interaction with biological systems" *Physical Sciences Reviews* [Online] 2020, 5. DOI: 10.1515/psr-2019-0110 | https://doi.org/10.1515/9783110569636-006

formed immediately after bringing together the particles with the biomolecule source. By variation of the biomolecule content of the incubation medium, the size of the corona can be modulated. Regarding the interaction of the nanoparticles with cells, it was shown that the presence of a biomolecule corona reduces the interaction and that a more pronounced biomolecule corona leads to a reduced uptake of the magnetic nanohybrids into the cells. Cell viability tests confirmed biocompatibility of the biomolecule-coated particles. A more pronounced corona promotes a higher cell viability. By using a shell-less hen's egg model, no or reduced adverse effects of all biomolecule-coated MNP for this *in vivo* test were found. Resulting from these investigations, we were able to demonstrate that our newly developed nanohybrids significantly reduce *in vivo* toxicity compared to uncoated MNPs.

Keywords: biomolecule corona; hybrid materials; magnetic nanoparticles; protein corona.

6.1 Introduction

Magnetic nanoparticles (MNPs) and their biocompatible suspensions (ferrofluids) are very promising materials for biomedical applications [1–4]. Due to their high surface-to-volume ratio and small size, MNPs exhibit properties differing from that of the bulk material, like superparamagnetism or stability against sedimentation. They possess magnetic properties that enable magnetic manipulation or heating by external magnetic fields, so that they cannot only be used in therapy, but also in diagnostics and as tracers. A wide range of applications like hyperthermia [5–7], drug targeting [2, 8, 9] or as contrast and tracer agent for medical imaging [10] are under development.

Therefore, a lot of hybrid MNP systems with specific cores and coatings have been developed and tested for different applications. Due to their high biocompatibility, especially iron oxide MNP is of great interest as a core material [11–13]. The so-called magnetic multicore nanoparticles (MCNPs) show superferrimagnetism, which means ferrimagnetic behavior in presence of an external magnetic field and superparamagnetic behavior in its absence. Such particles consist of clusters made of small single domain primary cores. Due to statistical orientation of easy axis of the primary cores within the clusters, the resulting magnetization without an external field is relatively low in comparison to single core particles in the size of the clusters. This means, these large particles show a very weak remnant magnetization like superparamagnetic particles und thus only a very low tendency to form agglomerates. This makes such particles relatively stable against sedimentation which is a good basis for medical applications. On the other hand, when exposed to an external magnetic field, these particles behave like bigger ferrimagnetic particles with a pronounced coercivity due to exchange interactions between primary particles.

The MCNPs are very promising for medical application [14], especially for magnetic manipulation due to their large particle volume and for hyperthermia because of their high heating performance [15–17]. Superparamagnetic iron oxide nanoparticles (SPIONs) are of particular interest for application as tracer and contrast agents for magnetic particle imaging (MPI) and magnetic resonance imaging (MRI), respectively [2, 13, 18]. Typically, MNP of all different core types are covered by diverse coatings like inorganic shells (e.g., silica) or organic ones (e.g., dextrans or polymers) to prevent agglomeration, resulting in sedimentation stable fluids of the MNPs for medical application. Additionally, the coating serves as an anchor layer for functionalization with a pharmaceutical agent or dyes.

One of the main obstacles for the medical use of these promising materials is their behavior when applied to biological liquids and biological systems, like the human body. For decades chemists, biologists, and engineers have developed methods to stabilize the MNP in aqueous suspensions and test biocompatibility *in vitro* and *in vivo*. After application of MNP into a biological system, it was found that the MNPs are immediately covered by biomolecules available within the body. A so-called "biomolecule corona" [19–21] is formed that changes their behavior drastically and gives them a new biological identity [22]. This additional biomolecule coating, consisting mainly of a protein corona and lipids, directly influences the properties of the used MNP. The biomolecule corona alters surface chemistry and thus their stability against agglomeration and sedimentation [23–25], as well as their interaction with biological systems *in vitro* and *in vivo* [19, 26, 27]. To enable a safe and reliable administration of MNP to biological systems for medical reasons, the knowledge of the influencing factors on structure and amount of an evolving biomolecule corona and the possibilities to control and to include the corona formation as a positive event is of major importance.

A deeper understanding of the complex interactions of MNP with biomolecules and biological systems is still in progress [19, 28–31] and first promising results could be made based on the concept of a "hard" and "soft" corona [32]. While the "hard" corona describes the fact that several biomolecules are strongly adsorbed to the surface of the MNP and only slow changes of the composition can be observed, the "soft" corona changes dynamically within time and depending on the environment [30, 33, 34]. Due to its temporal stability, the "hard" corona can be investigated with established measurements regarding its size, shape, hydrodynamic size, surface charge, as well as methods to determine their composition and integrity [32, 35–38]. Due to their magnetic properties, the MNP enable the use of dynamic magnetic measurements, what gives additional information on particle–matrix interactions and can help to understand the dynamic process of corona formation and protein cross-linking in more detail [39–41].

In the recent research on corona formation, it was found that the corona formation and the composition and amount of the final "hard" corona is depending on key parameters like particle size, surface charge, temperature, or incubation time [20, 42–48], what gives the possibility to shape, control and modify the corona and thus the particle–matrix interactions to avoid harmful effects like unwanted agglomeration or sedimentation and supporting beneficial effects like a longer blood half-life and lower toxicity [49–54].

That is why the main objectives of our work on this topic have been to develop and to show ways for reproducible preparation of sedimentation stable and biocompatible hybrid core–shell MNP (magnetic nanohybrid) suspensions for medical application.

Main approach to achieve these objectives was the pre-incubation of these MNP with different coatings in a biological fluid, serving as a source for the formation of the biomolecule corona under controlled conditions, to investigate the evolving biomolecule corona and to study the interactions of the magnetic nanohybrids with biological systems *in vitro* and *in vivo*. For this, we investigated and optimized MNP systems with different cores (MCNP, SPION) and different coatings featuring different surface charges and charge patterns (positive, negative, neutral, zwitterionic). We tested the influence of these parameters together with the parameters for the preincubation procedure on the resulting structure and amount of biomolecules bound to the surface of the MNP. Additionally, we investigated the interactions of these magnetic nanohybrids with biological systems like cells in established toxicity assays, as well as in quasi *in vivo* assays using the shell-less hen's egg test on the chick area vasculosa (HET-CAV). Furthermore, we developed strategies to sterilize and store the magnetic nanohybrids for a longer term, which is crucial towards their application in medicine. In this paper, we present a review of our main findings on these questions, obtained from our work in the Priority Program SPP1681 of the German Research Foundation Deutsche Forschungsgemeinschaft over the past six years [50, 55–62].

6.2 Methods

6.2.1 Preparation of nanohybrids

The magnetic nanohybrids used in this study mainly consist of a magnetic core of magnetic iron oxide (predominantly maghemite) in a multicore structure. This core is coated with a shell, which serves as an anchor layer for a further functionalization of these core/shell particles with the biomolecules. For the anchor layers, two different groups of material are used (see Figure 6.1). First, several types of biopolymers (dextrans with different functional groups) and second, polyelectrolytes and polyampholytes based on polydehydroalanine (PDha) [60, 63, 64]. On the surface of the anchor layer, biomolecules from a natural source (fetal calf serum – FCS) are coupled by an adhesive process.

6.2.1.1 Magnetic cores: The magnetic iron oxide nanoparticles used in this paper were prepared similar to the well-known wet chemical precipitation methods [65] but using another alkaline medium [66, 67]. For this, a $NaHCO_3$ solution was directly added to a $FeCl_2/FeCl_3$ solution and a brownish precipitate of nonmagnetic iron carbonates occurred. After the addition of distilled water, the particles were boiled for 5 min at 100 °C. In this way, magnetic iron oxides were formed under the release of CO_2 and the color

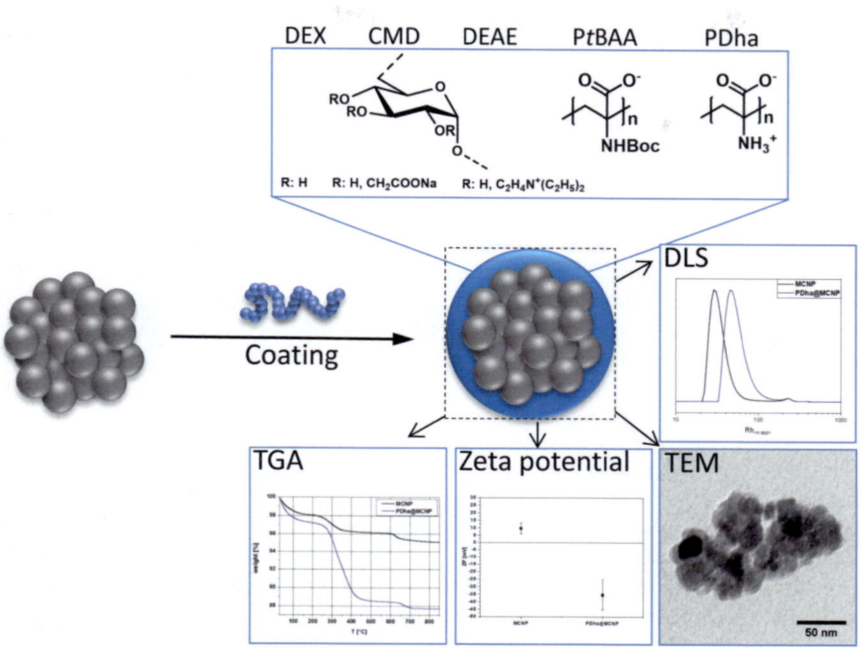

Figure 6.1: Schematic representation of the different core/shell magnetic nanoparticles (MNPs) and the technologies for their characterization.

of the solution turned black. Afterward, the obtained MNP suspension was washed by magnetic separation with distilled water using a high-performance permanent magnet to remove excess educts. The addition rate of the alkaline medium was varied to enable preparation of monodomain single core particles or to prepare clusters of about 50 nm consisting of several monodomain single cores, showing a superferrimagnetic behavior due to their multicore structure [67–69]. The particles size was controlled by adjusting the temperature and addition rate of the alkaline medium during particle preparation.

6.2.1.2 Anchor layer

6.2.1.2.1 Dextrans: For the investigation of the influence of particle surface charge on the formation of the biomolecule corona, MNP were coated with different materials (dextran – DEX, carboxymethyldextran – CMD, diethylaminoethyldextran – DEAE; all purchased from Sigma-Aldrich, Schnelldorf, Germany). These materials have a neutral dextran backbone but different substitution patterns and thus enable a variation of surface charge (DEX – neutral, CMD – negative, DEAE – positive). For coating the MNP with dextran and its derivatives, the nanoparticles were dispersed by ultrasonic treatment (Sonopuls GM200, Bandelin electronic, Berlin, Germany) for 1 min and sonicated for 1 min in an ultrasonic bath (S100H, Elmasonic, Germany). HCl was added to adjust the pH value at 2–3 and the suspension was tempered at 45 °C in a water bath and stirred. At the same time, the coating agents were dissolved in distilled water in a mass ratio (coating/core) of 1:1. The so prepared coating solution was steadily added to the nanoparticle suspension and stirred for 1 h at 45 °C. Afterward the suspension was treated with ultrasound as described above, washed magnetically twice with distilled water to remove excess coating material, and the desired concentration was adjusted by adding distilled water.

6.2.1.2.2 Polyelectrolytes and polyampholytes: To study the influence of net charge and charge density, polyampholytic and polyzwitterionic coatings were investigated with regard to the corona formation [55]. Here, the magnetic cores were coated with polyanionic or polyzwitterionic polymeric shells based on PDha. For this, polyanionic poly(*tert*-butoxycarbonylamino acrylic acid) (P*t*BAA) and polyzwitterionic PDha were used as obtained by deprotection of poly(*tert*-butoxycarbonylamino methacrylate) (P*t*BAMA) as described earlier [64]. The corresponding polyelectrolyte was dissolved in 0.01 M NaOH and titrated to pH = 5 using 0.01 M HCl. The MNP were treated with ultrasound for 10 min and magnetically separated afterward. Subsequently, the respective MNPs were dispersed in the polyelectrolyte solution at constant weight ratios (MNP:polyelectrolyte = 1:8) and the mixture was ultrasonicated for 1 h, magnetically separated and washed with aqua bidest five times. We further investigated the dependence of shell thickness and stability on different parameters such as solution pH, polymer/MNP ratio, and reaction time or temperature [60].

6.2.1.3 Biomolecule coating: For producing a protein/biomolecule corona around MNP, the particles have to be incubated in a natural protein source, which leads to an accumulation of proteins on the surface of the MNP. For our studies, FCS was used as natural protein source. FCS was obtained from Biochrom GmbH (Berlin), tested for mycoplasma and viruses, tested for endotoxins, and heat-inactivated at 56 °C for 1 h prior to use.

FCS incubation of MNP was performed by water bath heating resulting in a homogeneous temperature distribution throughout the sample. For water bath incubation, FCS was tempered at defined temperatures in a water bath. A suspension volume containing 15 mg of prepared MNP was added to 2 mL of tempered FCS and kept at the same temperature for a defined time. Incubation time starts with the addition of the suspension. During the time of incubation ultrasonic treatment at the given temperature was carried out (S100H, Elmasonic, Germany) to redisperse possible agglomerates. After incubation, the suspensions were taken out of the water bath and put on a magnet for magnetic separation, excess FCS was withdrawn and distilled water was added. The washed incubated nanoparticle suspensions were kept at 4 °C for short-term storage or at −80 °C for long time storage [57].

Besides the here described general procedure to prepare a biomolecule/protein corona coating on the surface of MNP, the used incubation parameters were varied in a broad range to study the influence of the incubation procedure on the evolving corona, as described in Section 6.2.2.

6.2.1.4 Size-dependent fractionation: To study the influence of particle size distribution on the stability of the resulting nanohybrid fluids against agglomeration and sedimentation and their interaction with biological systems, size-dependent fractionation experiments by using centrifugation were performed.

In the first approach, the core/shell MNP with the anchor layer only were centrifuged (1–2.5 min at 1000–5000×*g*) to remove agglomerates and obtain narrowly size distributed particle ensembles. Afterward, the obtained samples with narrowed size distribution were biomolecule coated by FCS incubation as described above.

In a second approach, the size-dependent fractionation was performed after the corona formation. For this, the unfractionated core/shell MNPs were incubated in FCS at 37 °C for 10 min to obtain FCS precoated MNP. To remove excess FCS and loosely bound biomolecules to obtain MNPs with a so-called "hard corona", incubated core/shell MNPs have been washed magnetically four times with distilled water and were redispersed by vortexing and ultrasonication. Afterward, MNPs were centrifuged (1–3 min at 1000–5000×*g*) and the supernatants, containing narrowly size distributed MNPs, were used for further investigation.

6.2.2 Control of corona formation

For the investigation, if the total amount of resulting biomolecule corona on the surface of the MNP can be modified by the incubation medium, FBS (Biochrom, Germany) diluted in cell culture medium (RPMI 1640 + GlutaMAX™, Life Technologies, USA) was used for the incubation of commercially polyethylenimine (PEI)-coated MNP (fluidMAG-PEI/B and fluorescence marker-labeled nanoscreenMAG/G-PEI/750/O, chemicell, Berlin, Germany). Additionally, we used these cell toxic MNP with a PEI coating, to investigate a putative protecting effect of a biomolecule corona with regard to cytotoxicity. For the incubation, a water bath heating at 37 °C for 10 min was used. To modulate the amount of proteins bound to MNP the composition of the incubation medium was varied. Starting from pure FCS, the amount of FCS in RPMI 1640 was reduced continuously down to an FCS concentration of 0% (v/v). The resulting biomolecule corona was characterized regarding its protein composition and total biomolecule amount on particles surface.

For the investigation of the influence of particle surface charge on the formation of the biomolecule corona, coatings with a neutral dextran backbone but different substitution patterns (DEX, CMD, DEAE) resulting in different surface charge (DEX – neutral, CMD – negative, DEAE – positive) were incubated in FCS and the resulting biomolecule corona was characterized.

During the preparation of the biomolecule corona, the process parameters incubation temperature and time were varied to study their influence on the resulting biomolecule corona. For this, the incubation was performed in a water bath as described above at defined incubation temperatures (25, 37, 50, 70 °C). To study the influence of incubation time, the suspensions were taken out of the water bath (at different temperatures) at defined incubation time points (1, 5, 10, 20 min) and put on a magnet for magnetic separation. Excess FCS was withdrawn and distilled water was added to terminate the incubation process.

To investigate, if the biomolecule corona formation can be prevented or at least reduced, corona formation on polyzwitterion-coated MNP was performed for the above described parameters and the resulting corona was compared to that one obtained on negatively charged (similar charge as the biomolecules) MNP. For this, polyzwitterionic (PDha) and anionic (P*t*BAA)-coated MNP were used for incubation and the resulting biomolecule corona was characterized regarding its protein composition and total biomolecule amount on the particle surface.

6.2.3 Sterilization/preservation

For a potential application of the biomolecule-coated MNPs in animals or humans, the particles have to be free from biological pathogens or any biological contamination. For particles smaller than 220 nm, filtration can be used for sterilization. Unfortunately, this method is not suitable for sterilization of herein used larger biomolecule-coated particles (>220 nm) because the membrane cut off for sterile filtration is 220 nm.

Furthermore, a method for preservation of biomolecule-coated MNP is needed, which guarantees storage for several months without an alteration of the integrity of the biomolecule coating. In previous experiments, we found a complete decay of the protein coating after a few days for storing in the fridge (4 °C).

Up to now, standard methods for sterilization and preservation of proteins, which are the main proportion of the biomolecule corona, are UV sterilization and lyophilization, respectively. To our knowledge, such procedures were never tested for sterilization and preservation of biomolecule-coated MNP. Therefore, the aim of our study was to establish procedures, which allow a reliable sterilization and preservation of larger biomolecule-coated MNP without any damaging effect on the integrity of the biomolecule corona. For this, biomolecule-coated DEAE-MNP were prepared as described above, treated by different procedures for sterilization and preservation (UV sterilization, autoclaving, freezing, and lyophilization), and characterized afterward regarding possible damaging effects on the integrity of the biomolecule coating as well as other negative effects like agglomeration etc. More details about the used protocols for the treatments are described in Dutz et al. [57].

6.2.3.1 UV sterilization: To sterilize the biomolecule-coated MNPs, the magnetic nanohybrids were exposed to a UV-C radiation with a wavelength in the range from 200 to 280 nm. About 254 nm is the optimum wavelength to damage nucleic acids and thus the DNA of organisms [46], which kills the potentially existing germs within the sample. The exposure time was adjusted for a first sample set to be 150 min, representing the minimum necessary time for the sterilization as confirmed in a previous study [57], and for a second sample set to be 240 min, representing a treatment time for sterilization including a safety margin.

6.2.3.2 Autoclaving: A standard protocol for autoclaving [70] was used to sterilize the biomolecule-coated MNP. The biomolecule coated MNP suspension was placed into a Varioklav 25 T (Thermo Fisher, USA) and exposed to a temperature of 121 °C for 20 min at a pressure of 3 bar. At this increased temperature, the DNA of pathogens in the sample is damaged due to protein denaturation and leads to a die-off of the pathogens.

6.2.3.3 Freezing: The reproduction of bacteria is slowed at low temperatures and reaches a standstill at temperatures below 4 °C [71]. The samples were frozen at −15 °C (freezing) and −80 °C (deep-freezing) to check the impact of freezing-time on the structure of the nanohybrids. For this, at each time point six samples of 1 mL of the frozen protein-coated MNP suspension were stored at −15 °C and −80 °C for up to 6 weeks. At distinct time points (1 day, 2 days, 1 week, 2, 4, 6 weeks) after freezing, the samples were defrosted and characterized immediately.

6.2.3.4 Lyophilization: Lyophilization is a well-established method for preservation of aqueous protein-based pharmaceuticals [72]. During the lyophilization procedure, all the water from the sample is removed, which leads to a complete cessation of the vital function of the pathogens as well as to a complete standstill of their reproduction. Biomolecule-coated MNP samples used in this study were lyophilized, and the dry samples were stored at 4 °C. At distinct time points (1, 3, and 6 weeks after lyophilization), the dry powder samples were redispersed and characterized regarding the occurrence of larger agglomerates within the suspensions and if the lyophilization procedure influences the integrity of the biomolecule structure. To determine the influence of additives on the redispersion of the dry powder samples, a second set of samples was mixed with tetramethylammonium hydroxide (TMAH) or polyethylene glycol (PEG), which are known as additives to enhance the suspensions stability against agglomeration and sedimentation of ferrofluids, and treated as described above.

6.2.4 Characterization of the samples

The magnetic characterization of all types of particles was done by vibrating sample magnetometry (VSM, Micromag 3900, Princeton Measurement Corporation, Princeton, USA) at room temperature. From liquid samples and dry powders, the saturation magnetization (Ms), the coercivity (Hc), and the relative remanence (Mr/Ms) were determined. Usually, drying of ferrofluids to powders leads to an alteration of the magnetic properties of the single cores due to increasing magnetic interactions between the single cores, when the distance between the cores is reduced [73]. Therefore, the obtained values for Hc and Mr/Ms are interpreted qualitatively only. Since Ms is mainly unaffected from magnetic interactions, this value can be used for quantitative analysis. Taking into account the measured Ms of the plain magnetic cores, the particle concentration within the fluidic samples was calculated. From the Ms of the dry samples, the proportion of nonmagnetic material on the surface of the MNP, which can be attributed to the coating material and/or the biomolecule corona, was estimated.

The inner structure of the magnetic MCNP was investigated by means of X-ray diffraction (XRD, Panalytical X'pert Pro, Malvern Panalytical, Almelo, The Netherlands). The results of the XRD investigations gave information about magnetic phase composition [74] and the mean sizes of the primary cores, which form the multicore particle. The size of the primary cores was calculated from measurements of the XRD line width by using the Scherrer formula. The physical cluster size and shape as well as the agglomeration behavior of all different particles was derived from transmission electron microscopy (TEM, Tecnai™ G^2 20, FEI company, Hillsboro, USA) images. Hydrodynamic diameters (dh) and its polydispersity index, as well as surface charge (zeta potential) of the core/shell MNP and the hybrid nanoparticles were determined by using dynamic light scattering (DLS, Zetasizer nano ZS, Malvern Instruments, Malvern, UK). Before the measurement, samples were diluted in the ratio 1:30 with distilled water and treated in an ultrasonic bath. For size measurements, the z-average of the intensity weighted normalization was used. To measure the zeta potential, the medium viscosity and dielectric constant were taken from water at 25 °C with 0.8872 cP and 78.5, respectively. All measurements were performed in three consecutive runs and obtained values were averaged.

For the determination of the mass of biomolecules bound to the particles surface, thermogravimetric analysis (TGA, STA409, Netzsch, Selb, Germany) was used. Samples were freeze-dried to obtain fine dry powders for TGA experiments. These samples were heated from room temperature up to 330 °C and the corresponding mass loss was continuously determined. The obtained curves for biomolecule-coated MNP were normalized to curves for plain MNP and core/shell MNP.

The determination of the composition and the amount of the proteins of the biomolecule corona on the surface of the nanohybrids was performed by means of sodium dodecyl sulfate polyacrylamide gel electrophoresis (SDS-PAGE). For this, 2 × Laemmli sample buffer (Bio-Rad, Munich, Germany) supplemented with 2-mercaptoethanol (final conc. 355 mM) was added to the samples in the first step and heated up to 95 °C to destroy both secondary and tertiary structure of proteins. Afterward, the denatured proteins were separated by molecular weight with PAGE on a 4–12% Bis-Tris gel (Bio-Rad, Munich, Germany). After gel electrophoresis, the proteins were visualized by highly sensitive silver staining (SilverXpress Silver Staining Kit, Invitrogen, Heidelberg, Germany). The gel images were processed by ImageJ (National Institutes of Health, Bethesda, USA) [75]. As references, a molecular weight standard protein collection Kaleidoscope marker (Bio-Rad, Munich, Germany) and untreated FBS were used.

To investigate time dependent processes of corona formation, magnetorelaxometry (MRX) was performed to investigate the Brownian relaxation behavior of the MNP as function of corona thickness. For this, a setup, which utilizes fluxgate sensors (FG-MRX) for measuring the magnetization decay, was used [76].

6.2.5 Interaction with biological systems

6.2.5.1 Cell viability studies: Cell viability studies were performed utilizing human brain microvascular endothelial cells (HBMECs). HBMECs were cultivated at 37 °C and 5% CO_2 in RPMI 1640 + GlutaMAX™ I (Invitrogen, Germany) supplied with 10% (v/v) FBS (Biochrom, Germany), 100 U/mL penicillin, and 0.1 mg/mL streptomycin (Life Technologies, USA). Afterward, the cells were seeded into black-walled 96-well plates (μ-Clear, F-bottom, Greiner Bio-One, Frickenhausen, Germany) in triplicate and then incubated with particle concentrations between 5 and 100 μg/cm^2 (corresponding to 19 - 378 μg/mL), water (negative control [NC]), or PEI-coated PEI-M particles (positive control, micromod Partikeltechnologie GmbH, Rostock, Germany). The "PrestoBlue" cell viability Reagent (Thermo Fisher Scientific, Waltham, USA) was added 3 and 24 h after the seeding and the incubation was continued for further 30 min at 37 °C. After magnetic separation of the MNP to the outer walls of the wells and an excitation of the cells with light of 550 nm (20 nm bandwidth), a fluorescence signal at 600 nm (40 nm bandwidth) was detected by using the CLARIOstar microplate reader (BMG LABTECH GmbH, Orthenberg, Germany). By measuring cell free wells with added particles, the particle-associated auto fluorescence effect was measured and used for correction of cell measurements. Similarly, quenching effects were considered by fluorescence measurements immediately before and after addition of particles to the cells in the wells. From these investigations, the cell survival rate (which is an indicator for cell toxicity of the particles) was derived for different particle concentrations for the above mentioned time points of 3 and 24 h after start of particle-cell incubation.

6.2.5.2 Flow cytometry: Furthermore, the interaction of biomolecule-coated MNP and living HBMECs was investigated by means of flow cytometry based on the cellular loading with fluorochrome-labeled nanoparticles (λ_{abs} = 476 nm, λ_{em} = 490 nm). First, incubated cells were washed and harvested using accutase (PAA Laboratories, Austria). Second, after a centrifugation step (5 min, 200 rcf, 4 °C) the cell pellet was resuspended in 10% (w/v) formalin (Sigma-Aldrich, Germany) and fixed at room temperature for 15 min. This washing procedure was performed in triplicate. After that, the cells were analyzed with the FACSCalibur cytometer (Becton Dickinson, USA) with 2500–10,000 cell counts per sample depending on recoverable cells and cytotoxicity of respective particles. By application of the Becton Dickinson's software "CellQuest Pro", the geometric mean value for fluorescence from cells only was determined and correlated to the condition of the cells in dependence of the amount of incubated particles.

6.2.5.3 Confocal laser scanning microscopy: More detailed investigations on cellular particle interaction and uptake were performed by using confocal laser scanning microscopy. For this, HBMECs were seeded on glass cover slips (12 mm diameter, Menzel, Germany) placed within a 24-well plate. The HBMECs were incubated with fluorescence-labeled MNPs as described above. After the incubation process, the cells were washed three times with phosphate buffered saline (Invitrogen, Germany) and fixed in 10% (v/v) neutral buffered formalin (Sigma-Aldrich, Germany) for 15 min.

This was followed by a permeabilization of the cell membrane in 0.1% (v/v) Triton X 100 (Sigma-Aldrich, Germany) for 15 min and cytoskeletal actin and nuclei were stained simultaneously with Alexa Fluor® 647 Phalloidin and Hoechst33258 (both Invitrogen, Germany) for 60 min at 37 °C. The fluorescence from the stained structures was analyzed by using a confocal laser scanning microscope LSM 510 META (Carl Zeiss Microscopy GmbH, Germany) with 400-fold magnification. For this, the samples were scanned with a slice thickness of 1 µm layer by layer, resulting in *z*-stacked images of 10–15 slices. As a result, the spatial distribution of the MNP within the cell layer is obtained. A quantification of internalized MNPs was performed by overlapping signals derived from the cytoskeletal actin and the MNP channels by using MATLAB (MathWorks, USA) to analyze the single image stacks.

6.2.5.4 Shell-less hen's egg model: Core/shell MNP and biomolecule-coated MNP were tested for their biocompatibility after systemic injections using a shell-less hen's egg model (see Figure 6.2) according to Schlenk et al. [77]. Briefly, fertilized chicken eggs were obtained from a local supplier and incubated for 72 h. The contents of the fertilized chicken eggs were transferred into petri dishes (Greiner Bio-One GmbH, Frickenhausen, Germany) containing Ringer's solution pH 7.0 to obtain the vascularized CAV at stages 14–17 according to Hamburger and Hamilton [78]. After further 24 h incubation, MNP samples were injected (2 µL) into the vitelline vein using a microinjector (Sutter Instrument, Novato, USA) connected to a borosilicate glass capillary. Directly before injections, selected MNP and biomolecule-coated MNP dispersions were diluted with sterile water to the final test concentrations of 160, 500, or 1000 µg/mL. Different fractionated and nonfractionated samples were tested. Branched PEI (25 kDa bPEI, 25 mg/mL dissolved in sterile water, kindly provided by BASF, Ludwigshafen, Germany) was

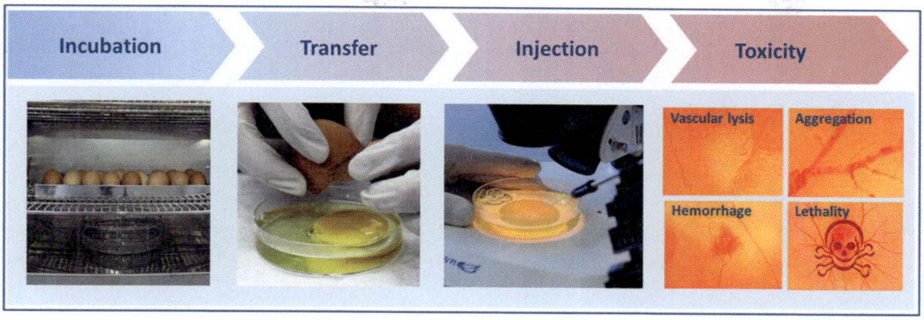

Figure 6.2: Schematic representation of the single work stages during the investigation of *in vivo/ex ovo* toxicity by means of the shell-less hen's egg model.

injected as positive control. According to the recommendations of the Interagency Coordinating Committee on the Validation of Alternative Method (ICCVAM) for the HET-CAM test [79], 0.9% NaCl and sterile water were selected as NC and solvent control, respectively. Eggs were inspected for toxic (hemorrhage, vascular lysis, aggregation) or lethal (stop of the heartbeat) reactions by macroscopic and microscopic observations at different time points (0, 1, 2, 4, 8 and 24 h) after injection. Historical lab control values were used to confirm the validity of the test procedure. Each sample was tested in two independent experiments with a total number of at least nine eggs.

6.3 Results and discussion

6.3.1 Properties of magnetic cores/shell particles

Magnetic nanoparticle cores were prepared from a wet chemical precipitation process. By variation of reaction temperature and velocity, single core and multicore structures were obtained.

Figure 6.3A shows a typical TEM image of an ensemble of single core MNP. For this type of MNP, the core diameter was determined by means of XRD and TEM and shows a typical value of about 10 nm. The analysis of the 440-peak in the diffractogram confirmed a spinel structure of the lattice with a dominating maghemite (γ-Fe_2O_3) phase and small proportion of magnetite (Fe_3O_4). The freeze-dried sample shows in VSM measurements at room temperature a saturation magnetization of about 68 Am^2/g, a coercivity of less than 0.2 kA/m, and a relative remanence of about 0.005. The obtained values are in the typical range for maghemite nanoparticles of this size and confirm a superparamagnetic behavior. A size derived from the magnetic properties by applying the Chantrell method [80] revealed a value of 9.6 nm, which is in good accordance to size from XRD and TEM, which is a further evidence of superparamagnetic behavior.

For the multicore MNPs, a cluster size of about 50 nm was measured by TEM (see Figure 6.3B) and the primary core size of the single cores forming the cluster was 11 nm,

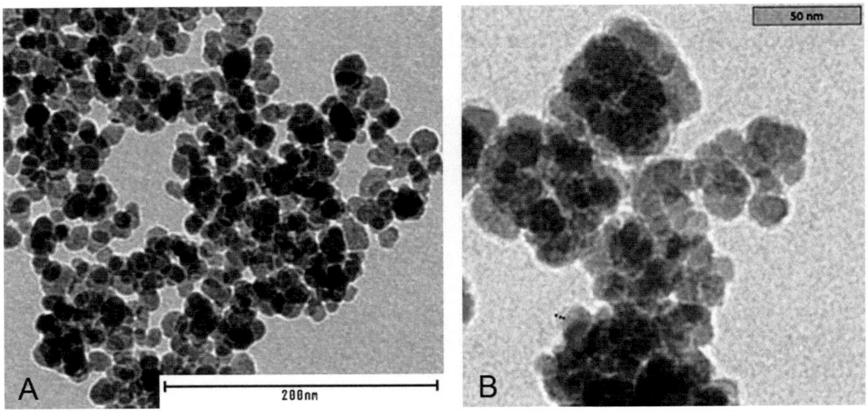

Figure 6.3: Typical transmission electron microscopy (TEM) image of the prepared magnetic nanoparticles; (A) pristine single core magnetic nanoparticles (MNPs) and (B) multicore MNP coated with polydehydroalanine (PDha).

measured by means of XRD. From these findings, it was confirmed by TEM that the multicore structure exists already for the pristine particles and is not an effect of the coating process. Analysis of the 440-peak in XRD revealed that the magnetic phase composition of the MCNP is slightly shifted towards magnetite, compared to the phase of the single cores. Dry powders of the plain MCNP show a Ms of about 74 Am^2/kg, a coercivity of about 2.6 kA/m, and a relative remanence of about 0.06. These measurements confirm a ferrimagnetic behavior of the MCNP in the presence of a magnetic field.

After coating the magnetic cores (single cores as well as multicores) with different dextrans (Dex, CMD, DEAE) and dispersing the core/shell particles in water, sedimentation stable ferrofluids were obtained. Depending on the coating, the zeta potential of the MNP (+35 mV) was shifted to +60 mV for DEAE, +2 mV for Dex, and −33 mV for CMD. For all different types of coating, a variance of the zeta potential in the range of ±5 mV was observed, which results from the accuracy of the measurement technology.

For the coating of the MNP with PDha and P*t*BAA, it could be shown using a combination of TEM, DLS, and TGA that well-defined polymeric shells of a few nanometer thickness could be formed [58]. The net surface charge was determined for both coatings to be about −40 mV. The surface charge pattern is polyanionic for P*t*BAA and polyzwitterionic for PDha-coated MNP. In addition, it turned out that MNP@PDha core–shell hybrid nanoparticles exhibit similar charge characteristics as the shell material (PDha) itself, allowing to invert the surface charge of the hybrid particles [61]. This could later on be used to demonstrate reversible adsorption/desorption of either polyelectrolytes, small charged molecules as model for the removal of pollutants from aqueous media, or even BSA as model protein [81].

6.3.2 Formation of the biomolecule corona

For producing a protein/biomolecule corona around MNP, the particles were incubated in a natural protein source (FCS), which leads to an accumulation of proteins on the surface of the MNP. In this section, we discuss the general formation process of a biomolecule corona around MNP. The influence of the surface of the different core/shell MNP and the incubation parameters on the evolving corona are discussed in Section 6.3.3.

Figure 6.4 shows the zeta potential of pristine uncoated single core MNP, as well as for MNP with DEAE, DEX, and CMD coating prior and after incubation in FCS for 10 min at 25 °C. It is clearly to see that the serum incubation changes the zeta potential of the samples significantly. For all samples, the surface charge turns to negative (−32 to −41 mV) after serum incubation. Because proteins and their amino acids, as well as other biomolecules like lipids have a negative charge at pH of 7, these results confirm the formation of a biomolecule corona on the surface of all tested particles. Furthermore, it can be recognized that, a weaker negative/stronger positive surface charge leads to stronger values of the negative surface charge after incubation. This might be explained by a binding of more negative proteins to positive surfaces. This effect is discussed in more detail in Section 6.3.3.

To obtain a general information about the amount of biomolecules bound to the surface of the MNP, freeze-dried powders of pristine MNP incubated in FCS were investigated by means of VSM and TGA (see Figure 6.5). From the VSM measurements (see Figure 6.5A), it can be seen that the Ms of the samples decreases for incubated particles. This effect can be explained by a proportion of nonmagnetic material within the incubated sample, which might be attributed to (nonmagnetic) biomolecules and

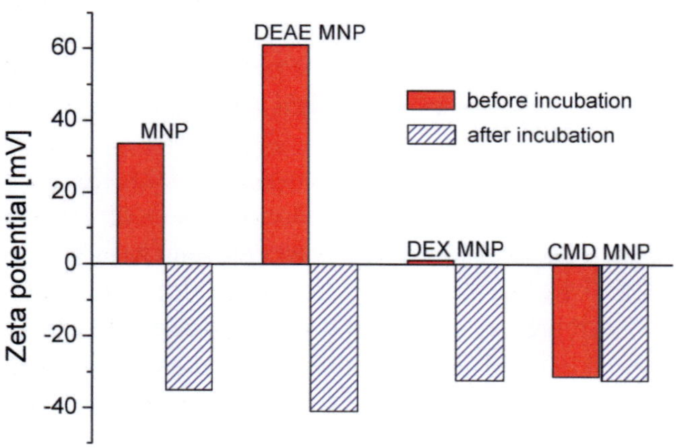

Figure 6.4: Zeta potential of magnetic nanoparticles (MNPs) and for MNPs coated with diethylaminoethyldextran (DEAE)-dextran, dextran, and CM-dextran before (red columns) and after serum incubation (blue hatched columns) confirm formation of a biomolecule corona around magnetic nanoparticles during serum incubation [62].

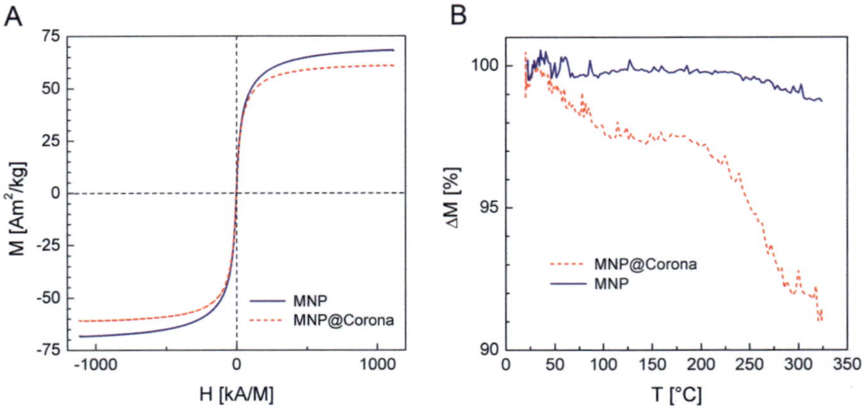

Figure 6.5: (A) Hysteresis curves and (B) TGA investigations of dried pristine, as well as biomolecule-coated magnetic nanoparticles (MNPs) confirm a proportion of biomolecules of about 10% by mass for the magnetic nanohybrids [62].

confirms the formation of a biomolecule corona on the surface of the MNP. Taking into account a Ms of 68.2 Am^2/kg for the pristine MNP, this decreased to 60.8 Am^2/kg for the incubated particles, the proportion of biomolecules on the particles was estimated to be 10% by mass.

This value was confirmed by TGA. For this, the same sample (before and after incubation) was heated up to 330 °C and the mass loss was recorded. For the pristine MNP, a mass loss of about 1% was found, which is attributed to the evaporation of adsorbed water and crystal water. Investigation of incubated MNPs (see Figure 6.5B) revealed a mass loss of about 9% in two steps. First, the water evaporates and then the biomolecules on the surface decompose. Comparison of both curves leads to a mean proportion of biomolecules on the nanohybrids of about 8% by mass. A possible explanation of the lower mass loss in TGA compared to VSM measurements is the fact that heating up to 330 °C might lead to an incomplete decomposition of the biomolecules.

In a further study, the temporal effects of the biomolecule corona formation were investigated. Because the binding of biomolecules to the surface of the MNP leads to significant changes in relaxation time, the kinetics of corona formation were investigated by means of Flugate-MRX [76]. For this, FCS was added to a core/shell MNP ferrofluid at room temperature and MRX measurement started immediately after the mixing process. Core/shell MNP with DEAE coating was used since these particles seem to have a strong affinity to bind biomolecules (see Figure 6.4). The core size was adjusted to be multicore with a cluster size of 50 nm, since such particles show a blocked magnetism at room temperature, which is a prerequisite to perform MRX measurements.

In the relaxation curves for the investigations on the short-term scale up to 5 min after mixing (see Figure 6.6A), it can be seen that already for the first measurement 15 s after mixing, a significant change in the relaxation curve compared to the relaxation curve of the nonincubated core/shell MNP occurs. An increase of the relaxation time was observed, which can be attributed to an increase of the hydrodynamic diameter, caused by the formation of a biomolecule corona on the surface of the MNP or an agglomeration of the particles. No significant deviations from this relaxation behavior to the curves for the following 5 min after mixing were found. This means that the size of the corona remains constant within the investigated time range of 5 min. This was also confirmed by quantitative SDS-PAGE, where a relatively constant amount of bound biomolecules during the first minutes of incubation was observed (data not shown).

On a longer time scale (up to 4 days after mixing), a further decrease of relaxation time was observed (see Figure 6.6B). The most probable explanation for this behavior might be a cross-linking of the surface proteins, leading to the formation of agglomerates as described before [82].

From these experiments, it can be seen that the formation of the biomolecule corona occurs within seconds, and no significant changes takes place within the following minutes. This observed temporal stability is true for the size of the biomolecule corona only, since from these experiments no information about the change of the corona composition as function of time can be derived. The proportion of biomolecules on the magnetic nanohybrids is in the order of 10% by mass.

6.3.3 Corona formation and manipulation

For the investigation, if the total amount of resulting biomolecule corona on the surface of the MNP can be controlled by the incubation medium, FCS diluted in cell culture medium (RPMI) was used for the incubation of the core/shell MNP. To modulate the

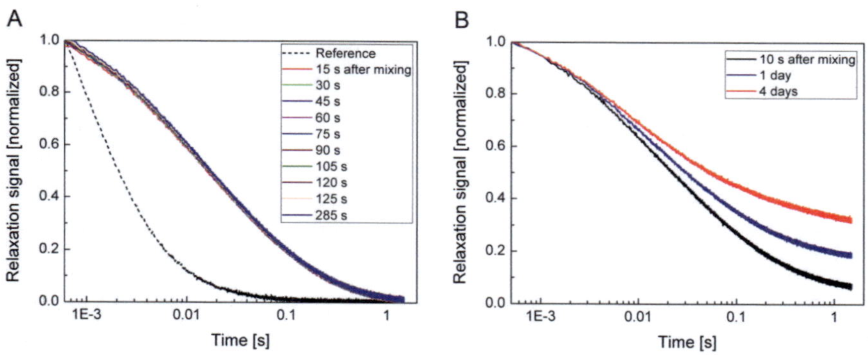

Figure 6.6: Temporal evolution of the magnetorelaxometry (MRX) signal after adding 140 µL fetal calf serum (FCS) to 10 µL diethylaminoethyldextran (DEAE)-coated magnetic nanoparticles (MNPs) suspension (A) for the first 285 s after mixing and (B) for incubation times up to 4 days, curves are normalized to be "1" for time point 1 ms. For comparison, MRX signal measured on reference sample (10 µL DEAE-coated MNPs suspension diluted with 140 µL DI H_2O) is shown in (A) [62].

amount of proteins bound to MNP, the composition of the incubation medium was varied from pure FCS down to an FCS concentration of 0%.

For an increasing concentration of FCS in the incubation medium, an alteration of the surface charge of the incubated particles was confirmed. Starting with a zeta potential of +48.4 ± 7.8 mV (three independent experiments) for the nonincubated particles, the zeta potential decreased down to zero and turned to negative values for higher FCS concentrations and pure FCS (see Figure 6.7). From this behavior, it can be concluded that a higher FCS amount in the incubation medium leads to the formation of a thicker biomolecule corona. As discussed in Section 6.3.2, the biomolecules are negatively charged at pH7 and thus, an accumulation of biomolecules on the surface of the MNP leads to changes in the net surface charge depending on bound biomolecule amount. Corresponding to the measured values, a saturation of zeta potential was observed for FCS concentrations above 75 vol-%. A possible explanation for this behavior might be that above a certain FCS concentration, the resulting corona leads to a complete covering of the surface of the particles, and a further increase of FCS concentration does not cause a further growth of the biomolecule corona.

To obtain more detailed information on the amount of biomolecules on the surface and additionally, if the composition of the incubation medium influences the composition of the resulting protein corona too, samples were studied by means of SDS-PAGE. From visual investigation of the gels (see Figure 6.8A), it becomes obvious that a higher amount of FCS in the incubation medium leads to a more pronounced biomolecule corona. ImageJ analysis of the optical density of the protein bands (see Figure 6.8B) reveals again a saturation effect of bound biomolecules on the particles for

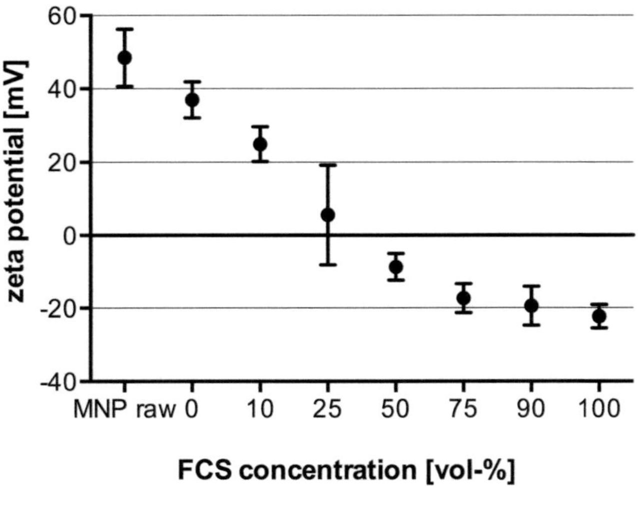

Figure 6.7: Zeta potential of polyethylenimine-magnetic nanoparticles (PEI-MNPs) incubated in RPMI 1640 cell medium containing different amount of fetal bovine serum (FCS) as indicated in a study by Grafe et al. [50].

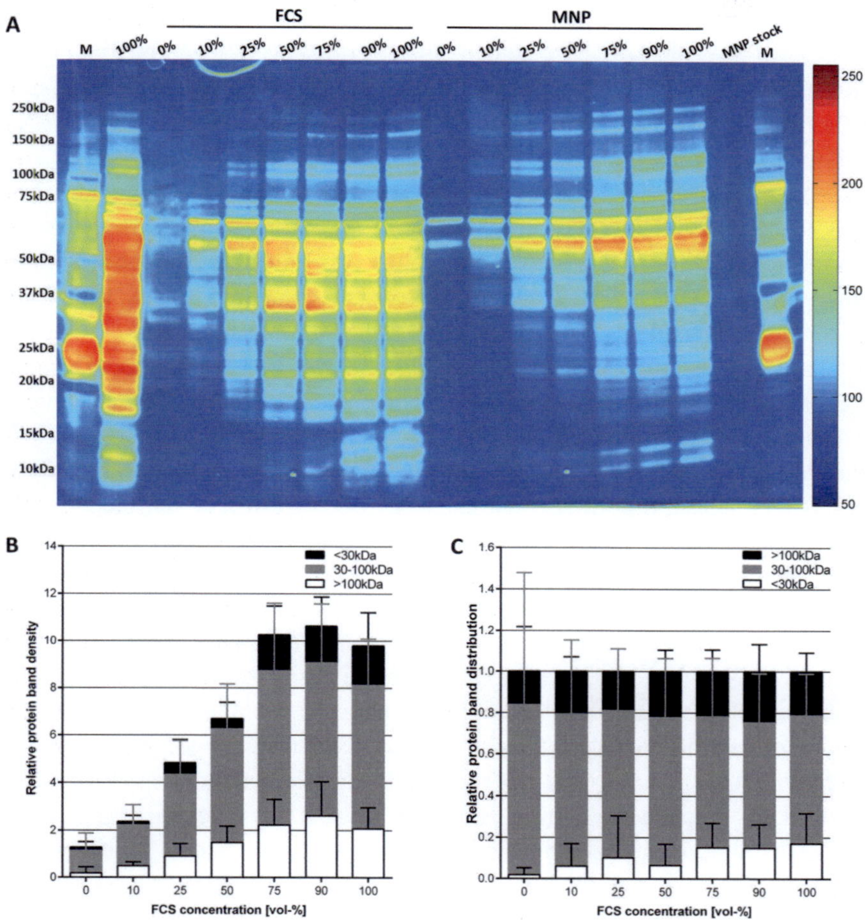

Figure 6.8: Concentration-dependent effect of fetal calf serum (FCS) on biomolecule corona formation on magnetic nanoparticles (MNPs). MNPs incubated with indicated concentrations of FCS (% [v/v]) were subjected to sodium dodecyl sulfate polyacrylamide gel electrophoresis (SDS-PAGE). Corona-derived proteins were visualized via highly sensitive silver staining. Lanes 10–16 comprise protein bands originated from MNPs, lanes 3–9 show protein bands of respective FCS solution prior to MNP incubation. (A) For better visualization, images are presented in alternative color mapping with low densities shown in blue and high densities in red. (B) Additionally optical density measurements of protein bands of low (<30 kDa), medium (30–100 kDa), and high (>100 kDa) molecular weight are shown relative to 0% (v/v) FCS-treated lanes and (C) their distribution within the respective lanes (C). M is a protein standard marker [50].

FCS concentrations of 75 vol-% and above, which is in good agreement with the zeta potential measurements.

In general, the protein profiles on the surface of the particles show a similar molecular weight distribution compared to the protein bands for initial FCS used for incubation. A quantitative analysis of the relative distribution of three specified protein size classes within each lane by means of ImageJ revealed that the medium-sized proteins (30–100 kDa) are dominating (see Figure 6.8C). When the MNP were incubated in media with higher FCS concentrations (75–100% FCS), the fraction of low-molecular-weight corona proteins (<30 kDa) increased compared to media with FCS concentrations between 0 and 50% at the expense of the medium-sized proteins (30–100 kDa). No influence of the composition of the incubation medium on the amount of large-sized proteins within the biomolecule corona was found and the ratio of high-molecular-weight proteins (>100 kDa) remains constant compared to the total corona protein content independent of the applied FCS concentrations.

Resulting from these investigations it can be confirmed that the amount of formed biomolecule corona can be tuned by the content of FCS within the incubation medium. Above a certain concentration of FCS within the incubation medium (here 75 vol-%), the resulting corona leads to a complete coverage of the particle surface and a further increase of FCS concentration does not cause a further growth of the biomolecule corona. Furthermore, it is possible to alter the resulting composition of the protein corona by variation of the FCS concentration of the incubation medium.

To investigate the influence of incubation temperature and incubation time, both parameters were varied during FCS incubation of pristine MNP as well as P*t*BAA- and PDha-coated MNP. Figure 6.9 shows the results of SDS-PAGE and TGA investigation of the incubated anionic P*t*BAA MNP. For all temperatures, the total amount of bound proteins increases with escalating incubation time. Similar results were obtained for pristine MNP [58]. When comparing different incubation temperatures for analogous incubation times, a higher amount of proteins is observed for higher temperatures. Protein size distribution analysis showed that for temperatures up to 50 °C mostly low-molecular-weight proteins are bound. At the same time, the amount of medium and high-molecular-weight proteins decreases. Reverse effects were observed for 70 °C, where the fraction of low-molecular-weight proteins decreases with increasing incubation time, and the fraction of medium- and high-molecular-weight proteins increases at the same time. For the incubated pristine MNP very similar results were observed.

TGA investigations for MNP incubated at 25 °C and 37 °C showed no remarkable increase in the amount of bound proteins for different incubation times. For temperatures of 50 and 70 °C, an increasing incubation time leads to a higher amount of absorbed proteins. Very similar tendencies were found for incubated pristine MNP.

A different behavior was found for incubated PDha-coated MNP. From SDS-PAGE, it was found that, for all temperatures except 50 °C a higher protein amount on the surface results for longer incubation times. This effect is very strong at 70 °C. For 50 °C, no influence of the incubation time on the amount of adsorbed proteins was found [58].

To investigate, if the formation of a biomolecule corona can be prevented or at least reduced, FCS incubation of polyzwitterion-coated (PDha) MNP was performed for selected incubation temperatures and durations and the resulting corona was compared to that one obtained for same incubation parameters on negatively charged (P*t*BAA) and uncoated MNP.

From SDS-PAGE, it was found that the highest amount of biomolecules was adsorbed to uncoated (cationic) MNP, followed by (anionic) P*t*BAA-coated MNP (see Figure 6.10A, B). For the (polyzwitterion) PDha, a significantly reduced protein amount on the surface of the MNP was determined. For all studied samples the majority of the corona consists of medium molar weight proteins (see Figure 6.10C). For PDha instead the proportion of medium molar weight proteins increases with increasing temperatures and incubation times. Also for PDha, the smallest proportion of low-molecular-weight proteins was observed. These results were confirmed by TGA (see Figure 6.10D). Again, the highest amount of biomolecules was found for uncoated MNP (7–12%), followed by P*t*BAA (5–9%), and significantly lower for PDha (0.5–2.5%).

Concluding from these results it can be stated that a clear influence of incubation time and temperature on absorbed protein amount during incubation exists. Usually, longer incubation time and/or higher incubation temperature results in a larger amount of proteins on the surface of the MNP. Additionally, also an influence of surface charge or charge patterns on the protein amount exists. Whilst positive and negative coatings result in very similar results regarding bound proteins on the surface, for

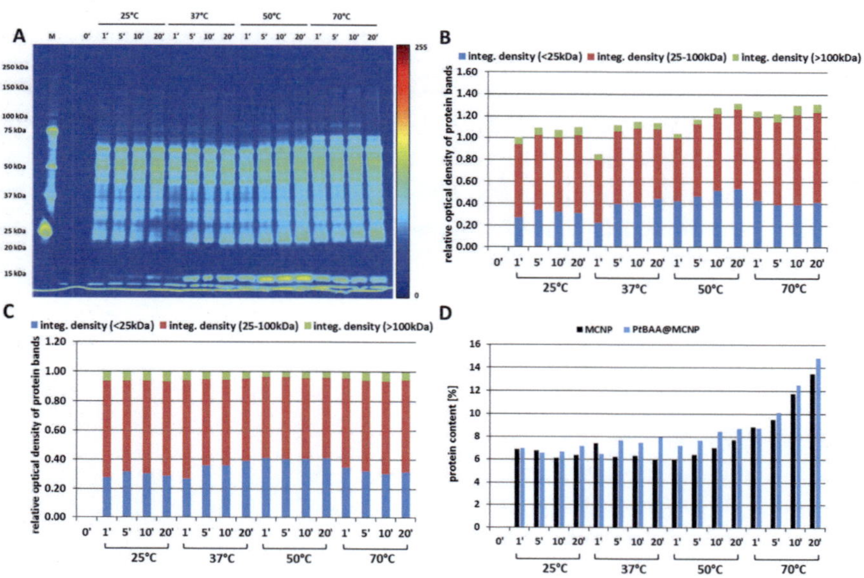

Figure 6.9: Determination of protein content and protein composition of P*t*BAA-multicore nanoparticles (MCNPs) incubated at different temperatures and times: (A) pseudo color image of protein bands upon sodium dodecyl sulfate polyacrylamide gel electrophoresis (SDS-PAGE) and silver staining, (B–C) relative optical densities of protein bands abstracted from (A), and (D) protein contents calculated from TGA measurements, where the black bars correspond to the pristine MCNPs and the blue bars refer to P*t*BAA-MCNPs [58].

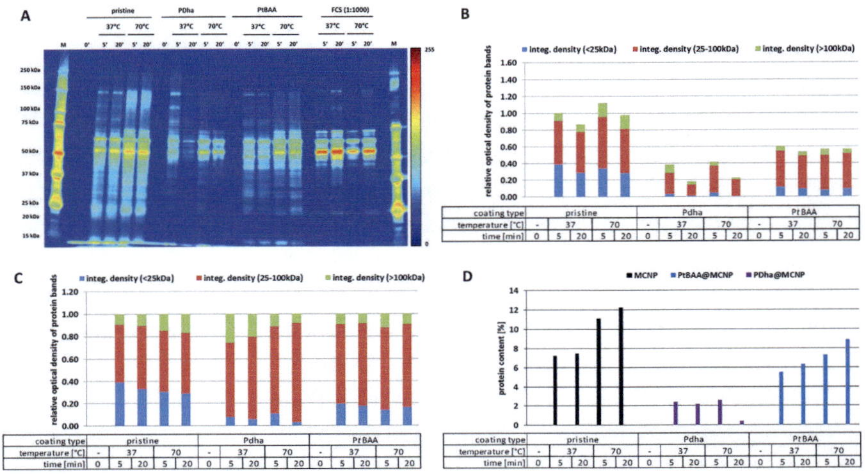

Figure 6.10: Comparison of protein content and protein composition of pristine multicore nanoparticles (MCNPs), polydehydroalanine (PDha)-MCNPs and P*t*BAA-MCNPs incubated at selected temperatures and for different incubation times: (A) pseudo color image of protein bands upon sodium dodecyl sulfate polyacrylamide gel electrophoresis (SDS-PAGE) and silver staining, (B–C) relative optical densities of protein bands abstracted from (A), and (D) protein contents calculated from TGA measurements [58].

zwitterionic coatings a significantly lower amount of protein was adsorbed to the surface. This might be interesting for the preparation of stealth coatings. As well as for the amount of protein also an influence of the discussed parameters on the composition of the corona was confirmed. But this correlation is too complex to allow description by a simple model and needs support by additional composition analysis to support our findings.

6.3.4 Sterilization/preservation

6.3.4.1 UV sterilization

After performing the UV sterilization procedure, samples with 150 and 240 min exposure time showed no considerable changes in the mean hydrodynamic particles size (see Figure 6.11A), particle size distribution, as well as zeta potential due to exposure to UV radiation. Samples before and after UV sterilization show a zeta potential of about −30 mV, which confirms the presence of a biomolecule corona and that there is no influence of UV exposure on the biomolecule amount on the surface of the magnetic nanohybrids.

SDS-PAGE revealed no evidence for a UV radiation caused change of the protein integrity. The pattern of the pseudo color greyscale image (see Figure 6.11B) of the original sample (UV-0) looks very similar to the samples exposed to UV radiation for 150 min (UV-150) and for 240 min (UV-240). ImageJ analysis of the PAA gel (see Figure 6.11C) shows that in the frame of the measurement accuracy the corona proteins for the different samples have an identic size distribution.

From these measurements we conclude that exposure to UV radiation for up to 240 min causes no relevant changes in the protein amount and integrity und thus, UV sterilization is suitable for the application to biomolecule-coated MNP.

6.3.4.2 Autoclaving

After autoclaving of the biomolecule-coated MNP, a significant decrease of the hydrodynamic particles size and particle size distribution was observed. A higher density of the surface proteins results due to coagulation of the proteins during the autoclaving at elevated temperatures up to 121 °C could be a probable explanation. This leads to a

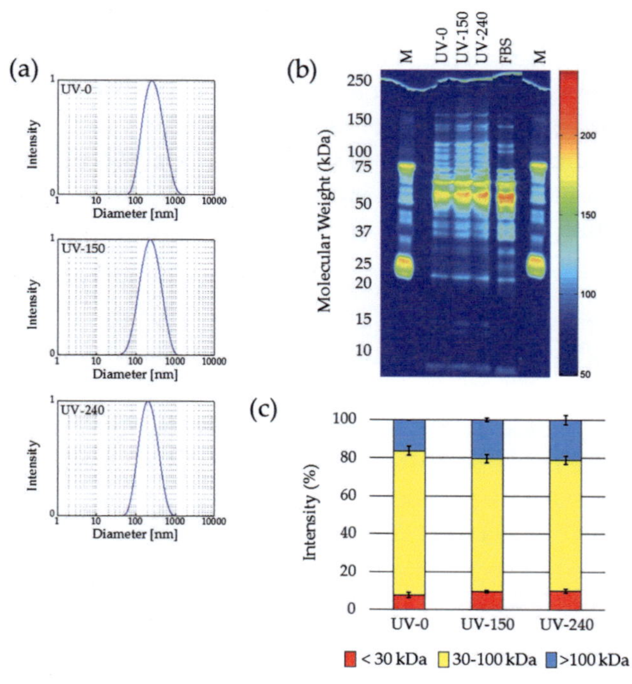

Figure 6.11: Ultraviolet (UV) sterilization of biomolecule-coated nanoparticles does not affect the protein corona. (A) DLS measurements show that the hydrodynamic diameter of the protein-coated nanoparticles decreases slightly with increasing treatment time. (B) Protein size distribution is not altered as demonstrated by gel electrophoresis (silver staining and false color representation). (C) Variations of the protein size ranges over the storage period of 6 weeks [57].

lower volume of the biomolecule coating on the core/shell MNP and thus a smaller size is measured. SDS-PAGE investigations revealed strong damages in the protein structure. In the PAA gel, the protein bands are no longer separable but appear as a smear, especially for the high-molecular-weight proteins. All of the typical peaks are vanished after autoclaving, which is attributed to a strong coagulation, denaturation, and degradation of the proteins [57].

From the obtained results, we conclude that autoclaving is not suitable to sterilize biomolecule-coated MNP.

6.3.4.3 Freezing

The visual investigation of thawed samples showed large agglomerates and significant changes in the rheology of the fluids for storing times of two weeks or longer at −15 and −80 °C. This was confirmed by DLS where agglomerates up to 1000 nm were found for samples stored for 6 weeks [57]. Despite the zeta potential provided a constant value of about −30 mV for all samples, it became obvious from the size measurements that during the storage at −15 and −80 °C a significant change of the integrity of the biomolecule coating takes place. This was confirmed by SDS-PAGE. For longer storage times, a decrease of the high-molecular weight proteins was observed combined with an increased proportion of the low-molecular weight proteins. This behavior can be interpreted as a degradation of the larger proteins to smaller fragments.

In summary, freezing and storing biomolecule-coated MNP at −15 and −80 °C is not suitable for long-term storage of such magnetic nanohybrids. Short-term storage of up to one week seems to be possible.

6.3.4.4 Lyophilization

The biomolecule-coated MNP samples were freeze-dried in the condition as prepared (plain) and after addition of PEG or TMAH.

PEG was used because it is suggested as cryoprotectant in combination with sugars, and the inherent sugars in serum were expected to be sufficient for this approach.

Similar behavior in particle size and agglomeration was found in each single sample series (plain, PEG, TMAH) for all three different re-dispersion time points (see Figure 6.12A). After redispersion the PEG samples show a similar size and size distribution like the original sample before drying. The plain samples show a slightly higher mean hydrodynamic diameter caused by a weak proportion of not redispersed agglomerates. For TMAH sample, a large proportion of agglomerates remains after redispersion, and the fluid is not stable against sedimentation. The zeta potential showed a constant value of about −30 mV for the original and all Plain and PEG samples and the zeta potential of all TMAH samples was in the range of −42 to −44 mV. From these observations, a change of the biomolecule integrity due to the addition of TMAH can be expected.

No remarkable changes in the integrity of the biomolecule coating can be observed in SDS-PAGE for plain and PEG sample for a storage time of one week (see Figure 6.12B). TMAH samples show an alteration of the protein integrity already after one week storage. After 3 weeks storage, the sample plain and PEG exhibit an increase in the low-molecular-weight fraction, but the protein distribution remains unchanged. After 6 weeks storage, a very slight degradation of the larger proteins occurs and the fraction of small protein fragments increases (see Figure 6.12C).

In summary, the lyophilization of biomolecule-coated MNP after the addition of PEG might be a suitable method for the preservation of such particles for some weeks. The redispersed particles show similar size and stability like the original sample. For storage times longer than 1 week, a partial degradation of the corona proteins can be

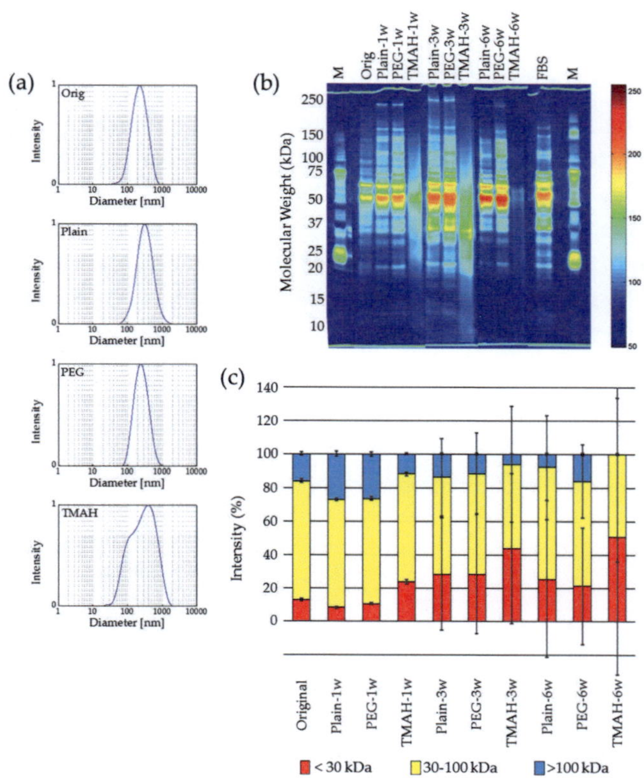

Figure 6.12: Lyophilization of biomolecule-coated nanoparticles affects the integrity of the protein corona during long-term storage. (A) After storage of the dry powders for 6 weeks the DLS measurements show that the hydrodynamic diameter of resuspended plain and polyethylene glycol (PEG) samples is very similar to that of the original sample, but for tetramethylammonium hydroxide (TMAH) an agglomeration results; (B) Proteins are degraded during lyophilization, especially in the presence of TMAH, resulting in smaller sizes after several weeks of storage demonstrated by gel electrophoresis (silver staining and false color representation); (C) Variations of the protein size ranges over the storage period (based on (B)) [57].

observed. The importance of this partial degradation for the storage of such particles is not clear until now.

6.3.5 Interaction with biological systems

6.3.5.1 Cell viability studies

To validate the biocompatibility of all used core/shell particles prior and after formation of a biomolecule corona on their surface, *in vitro* viability studies by using HBME cells were performed. No cytotoxic effects were confirmed for pristine MNP, PDha, and P*t*BAA-coated nanoparticles (polymer coatings), before and after FCS incubation, for single core as well as for multicore particles (see Figure 6.13). Similar results were found for the dextran-coated (Dex, CMD, DEAE) core/shell MNP before and after incubation in FCS.

Additionally, from the data in Figure 6.13 it can be seen that the formation of a biomolecule corona might lead to an even enhanced cell viability. Interestingly, a slight reduction in cell viability was observed for the pristine multicore MNP before FCS incubation. This effect still remains for the pristine single core MNP even after the formation of the biomolecule corona. A possible explanation for this observation might be a faster dissociation or degradation on the single core MNP with a higher surface-to-volume ratio compared to the multicore MNP.

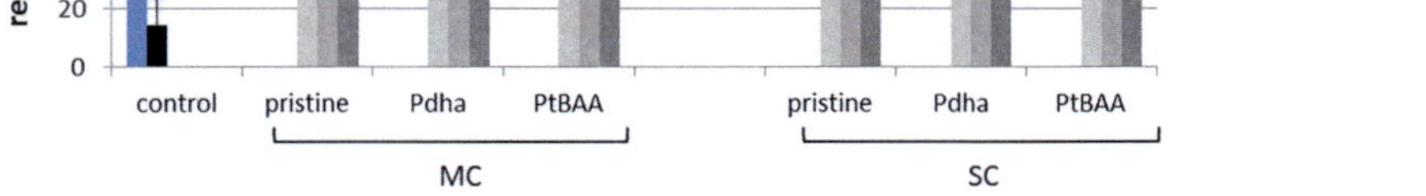

Figure 6.13: Cell viability changes in response to differently coated single core (SC) and multicore (MC) magnetic nanoparticles (MNPs) incubated at different temperatures and times. Human brain microvascular endothelial cells (HBMECs) seeded into 96-well plates in triplicate were incubated with 100 μg/cm^2 (equal to 378 μg/mL) of indicated particles for 24 h. Positive controls and NCs were analogously treated with aqua bidest or cationic polyethylenimine-coated PEI-M, respectively. Detected fluorescence upon incubation with PrestoBlue reagent is normalized to positive controls [58].

6.3.5.2 Flow cytometry

To analyze the MNP loading to cells in dependence of a present biomolecule corona, flow cytometry was used. For this characterization method, fluorescence-labeled MNP are needed and therefore commercially available PEI-coated MNP coupled with a fluorescence dye were used as core/shell particles and incubated in FBS.

From Figure 6.14A, it becomes obvious that the intensity of MNP-associated fluorescence in and on the cells is reduced for cells incubated with MNP bearing a biomolecule corona compared to bare core/shell MNP. This can be interpreted as a reduced cellular loading for biomolecule corona–coated MNP. This means a reduced interaction of biomolecule-coated MNP with cells of about 60 and 47% compared to bare core/shell particles for incubation times of 5 and 30 min, respectively (see Figure 6.14B). This result might be explained by an alteration of the surface potential. Due to the incubation with FCS the surface potential turns from positive to negative and the negative charge of cell membranes might hinder interactions with negatively charged nanoparticles.

6.3.5.3 Confocal laser scanning microscopy

No particle aggregation could be observed for the serum incubated PEI-coated nanoparticles.

For a more detailed investigation of this, HBMECs were incubated with MNP bearing biomolecule coronas of gradually increasing extent (by changing the FCS content of the incubation medium) and investigated by means of confocal laser scanning microscopy. Data analysis of the image stacks revealed that, the presence of a biomolecule corona reduces the particles adhesion to the cells and that the adhesion decreases with an increasing extent of the corona [50]. MATLAB analysis of the obtained stacks showed that MNP-derived signals are detectable at cells incubated with MNP of increasing biomolecule corona for 5 and 30 min (see Figure 6.15). It can be seen

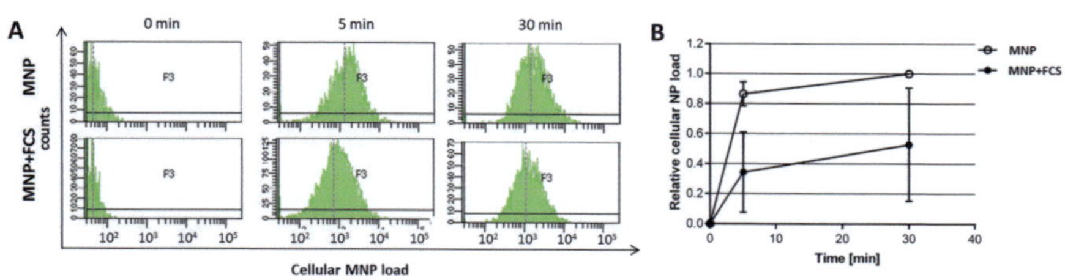

Figure 6.14: Interactions of polyethylenimine (PEI)-magnetic nanoparticles (MNPs) with human brain microvascular endothelial cells (HBMECs) as function of a present biomolecule corona. Cells were incubated for 5 or 30 min with particles without (MNPs) or with (MNP + FCS) a biomolecule corona. It can be seen that for both time points the presence of a biomolecule corona leads to a reduction of the MNP load to the cells [50].

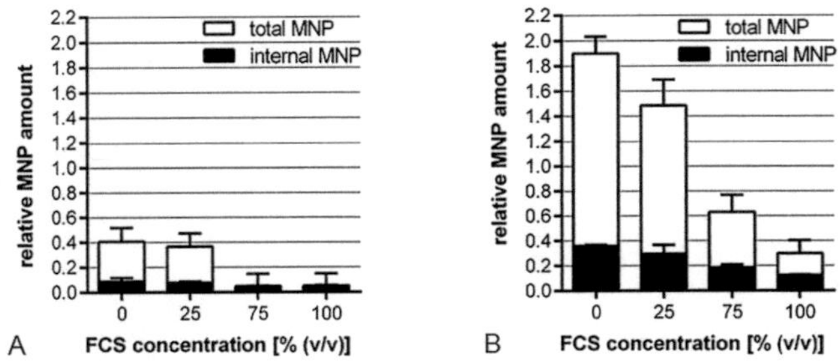

Figure 6.15: Adhesion and uptake of magnetic nanoparticles (MNPs) by human brain microvascular endothelial cells (HBMECs) depends on particles biomolecule corona. For (A) or (B) 30 min, HBMEC were incubated with 25 µg/cm^2 fluorescence-labeled MNPs and imaged by means of laser scanning microscopy. Image stacks were analyzed by MATLAB where the amount of internalized MNPs was calculated from overlapping signals derived from the cytoskeletal actin and the MNP channels. Shown are means ± standard deviation of 10–15 slices [50].

that a more pronounced biomolecule corona leads to a lower uptake of the magnetic nanohybrids into the cells for incubation times up to 30 min. Thus, we could demonstrate that the cellular uptake and adhesion processes can be controlled by the extent of a biomolecule coating on MNP.

6.3.5.4 Shell-less hen's egg model

Furthermore, core/shell MNP and biomolecule-coated MNP were tested for their biocompatibility after systemic injection using a shell-less hen's egg model [77]. It was tested, if the prepared magnetic nanohybrids cause any adverse effects when administered to a living biological system. Another question was, if a possible irritation of the biological system can be masked with a biomolecule corona and if agglomeration of particles within the biological system can be avoided by a biomolecule coating and/or changing the particle size distribution by removal of large particles from the ensemble by means of size-dependent fractionation.

For unfractionated samples, the concentration of injected particles was adjusted to be 160 μg_{MNP}/mL. Figure 6.16 shows the probability of the occurrence of vascular lysis, hemorrhage, aggregation, and lethality for different time points (up to 24 h) after injection. The validity of the assay was ensured by several controls. 0.9% sodium chloride was used as NC as recommended control by the ICCVAM [79] and sterile water as solvent control to set up the baseline for irritative effects. Sterile water showed no adverse effects.

It can be seen that the FCS-incubated CMD and PDha core/shell particles showed no aggregation and only sporadic hemorrhagic and lethal events (0–1/9 eggs) during this test comparable to the control values. The positive control, highly cationic charged PEI polymer caused a strong aggregation effect and finally a high lethality. A possible explanation for this might be the high attraction between the cationic surface of the

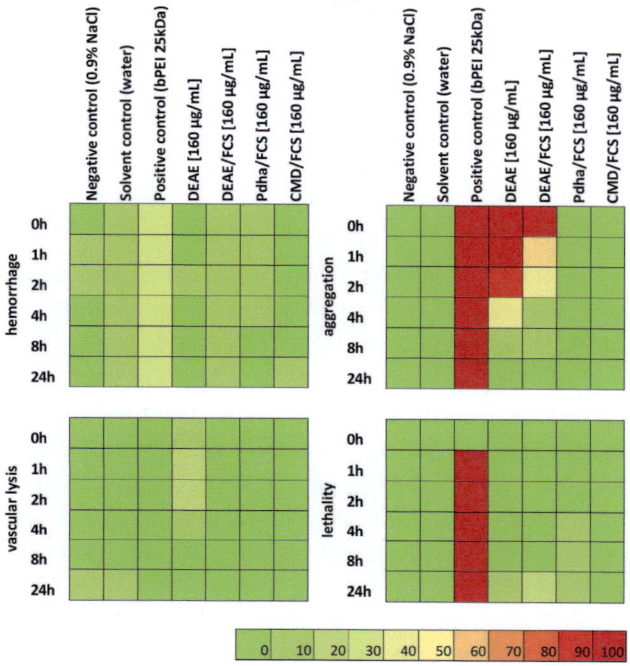

Figure 6.16: Percentual probability for the occurrence of adverse effects (aggregation, hemorrhage, vascular lysis, and lethality) for different particles and time points within the shell-less hen's egg model serving as a test to evaluate the *in vivo* toxicity of the magnetic nanohybrids.

polymer and anionic charged proteins present in the blood, leading to the formation of polymer/protein aggregates. For DEAE-MNP, which shows a similar surface potential like PEI, also a tendency for the occurrence of aggregation was observed. After coating these particles with biomolecules, a reduction of aggregation and a faster reversibility of the effects were obtained. Probably, the biomolecule coating leads to a less pronounced aggregate formation, because aggregates are disintegrated by shear forces of the flowing blood within the first hour after administration.

When using fractionated DEAE-MNP (d_h = 124 nm) instead of the previously used unfractionated ones (d_h = 160 nm) at a concentration of 500 µg/mL, there is only a low tendency (<20%) to form agglomerates for DEAE-MNP prior and after formation of the biomolecule corona. This might be explained by the removal of the larger agglomerates by centrifugation that block the blood flow in the vessels after injection (and a reduced agglomeration potential of smaller FCS-coated MNP). For this concentration, also no adverse effects were found for fractionated incubated CMD-MNP (n = 5, data not shown). In preliminary experiments also for DEX-MNP and PEI-MNP comparable effects were observed even up to 1 mg/mL (data not shown).

Resulting from these investigations, we were able to show that the *ex ovo* toxicity of our newly developed nanohybrids under the tested conditions could be highly improved by the FCS coating, as well as by the fractionation process.

6.4 Conclusion

The formation of a biomolecule corona is an inevitable event when nanoparticles enter biological fluids. Our investigations reveal that within seconds after bringing MNPs in contact with a protein source (FCS) a biomolecule corona is formed on the surface of the particles. This effect can be of major importance for the use of MNPs in medical applications where such particles have to be administered to the vascular system. There, the blood serves as a protein source and the biomolecule (protein) corona around the particles determines the biological fate of the particles within the body. Possibly, a precoating of the core/shell MNP with biomolecules might lead to an altered interaction of the preconditioned particles with biological systems, which can help to improve the biocompatibility of such magnetic nanohybrids.

As measured by means of MRX, DLS, and TGA, the biomolecule corona formation leads to an increase of the particle size. Due to cross-linking of the proteins on the surface of the particles, a stronger agglomeration of the biomolecule-coated MNP occurs for longer incubation times in the range from hours to days. The amount of biomolecules adsorbed to the surface was determined to be around 10% by mass of the complete magnetic nanohybrid and its surface charge is in the range from –30 to –40 mV.

Different coatings (Dex, CMD, DEAE, PDha, and P*t*BAA) were used for the preparation of core/shell MNP to study the influence of surface charge on the formation of the biomolecule corona and its composition. Furthermore, the influence of incubation time and temperature on the formation was studied. Slight variations in the zeta potential of the obtained biomolecule-coated MNP are first hints for differences in amount and composition of the biomolecule corona at different incubation times and temperatures and different coating materials.

Since the formation of the biomolecule corona is a very complex process, which is influenced by a lot of different factors, it is not possible to derive a general model for the corona formation for all used coatings. For the most of the here tested core/shell MNP, our TGA results show that at lower temperatures (25 and 37 °C) the amount of biomolecules does not increase with increasing incubation time whereas this is the case at elevated temperatures (50 and 70 °C). In that regard, polyzwitterionic PDha is a highly interesting material for coating of MNP since a distinctly lower amount of biomolecules was adsorbed during the FCS incubation compared to all other coating materials. The amount of biomolecules adsorbed to the surface of the core/shell MNPs can be tuned by the composition of the incubation medium. A higher amount of biomolecules in the incubation medium leads to an increasing volume of the biomolecule corona and thus a larger thickness of the biomolecule coating.

A semi-quantitative analysis of the gels obtained from SDS-PAGE confirmed the results from TGA and provided additional information about the composition of the proteins contributing to the formed biomolecule corona. Although for all coating materials and temperatures an influence on the ratios between low-molecular-weight proteins (<25 kDa), medium-molecular-weight proteins (25–100 kDa), and high-molecular-weight (>100 kDa) proteins was found, no general trends can be derived from the obtained data so far.

Before application of such particles to biological systems it has to be assured that the particle preparations are free from pathogens and also procedures have to be established, which enable the long-term storage of the particles.

For the sterilization by means of UV radiation, we found no damaging effects for biomolecule-coated MNP. Contrary, the treatment of biomolecule-coated particles by means of autoclaving leads to significant changes in the integrity of the proteins and thus this method is not suitable for sterilization. Freezing at −15 and −80 °C allows storage for up to 1 week. After that, a remarkable degradation of the larger proteins in combination with an agglomeration of the particles takes place. During lyophilization, the addition of PEG to the samples revealed well dispersed fluids after resuspension of the powders. For storage times longer than one week, a slight degradation of the proteins was observed. This storage period might be extended by using optimized storing conditions (e.g., lower temperature, oxygen exclusion, reduced humidity). Furthermore, it has to be investigated, which consequences this slight degradation for the application of such particles to biological systems means. Therefore, future studies are needed to elucidate the treatment-dependent changes on the biomolecule corona, as well as on the single protein level with regard to cytotoxicity and immunogenicity. In conclusion, we state that the UV sterilization followed by lyophilization under addition of PEG is the most promising route for the preparation of sterilized long-term durable biomolecule-coated MNP.

In cell viability tests, no cytotoxicity was found for all tested biomolecule-coated MNP. A higher amount of biomolecules adsorbed to the surface lead to an increased cell viability. By flow cytometry, it was shown that the presence of a biomolecule corona reduces the interaction of nanoparticles with cells during short-term incubation in the range from 5 to 30 min. Laser scanning microscopy revealed that a more pronounced biomolecule corona leads to a retarded uptake of the magnetic nanohybrids into the cells. Thus, we could demonstrate that for short-term incubation the cellular uptake and adhesion processes can be controlled by the extent of a biomolecule coating on MNP. Nonetheless, possible effects of long-term incubation have to be investigated. By using a shell-less hen's egg model, no adverse effects of all biomolecule-coated MNP for this *in vivo/ex ovo* test were found. Resulting from these investigations, we were able to show that the *in vivo* biocompatibility of our newly developed nanohybrids under the tested conditions could be highly improved by the biomolecule coating.

In summary, we developed and described a novel type of core/shell nanoparticles with improved biocompatibility for medical application. These magnetic nanohybrids bear a second coating of biomolecules, which enables the control of the interaction of these particles with biological systems. The biomolecule corona may mask adverse effects on biological systems of cytotoxic core/shell MNP. In ongoing studies, the biocompatibility has to be tested for mammals and the consequences of variations in the composition of the biomolecule corona for the application have to be investigated in more detail.

Acknowledgments: This work was supported by the Deutsche Forschungsgemeinschaft (DFG) in the framework of SPP 1681 (FKZ: CL202/3-1, CL202/3-2, CL202/3-3, DU1293/4-1, DU1293/4-2, DU1293/7-3, and SCHA1640/7-1). The authors thank all partners within the SPP 1681 for the fruitful collaboration and especially Prof. Stefan Odenbach (TU Dresden) for the excellent coordination of the DFG Priority Programme.
Figures 6.7, 6.8, 6.14, and 6.15 are reprinted from "Gräfe, C.; Weidner, A.; von der Lühe, M.; Bergemann, C.; Schacher, F.H.; Clement, J.H.; Dutz, S. Intentional formation of a protein corona on nanoparticles: Serum concentration affects protein corona mass, surface charge, and nanoparticle-cell interaction. Int. J. Biochem. Cell Biol. 75: 196-202 (2016)", with permission from Elsevier.
Figures 6.9, 6.10, and 6.13 are reprinted from "Gräfe, C.; von der Lühe, M.; Weidner, A.; Globig, P.; Clement, J.H.; Dutz, S.; Schacher, F.H. Protein corona formation and its constitutional changes on magnetic nanoparticles in serum featuring a polydehydroalanine coating: effects of charge and incubation conditions. Nanotechnology 30: 265707 (2019)", © IOP Publishing. Reproduced with permission. All rights reserved.
Author contribution: All the authors have accepted responsibility for the entire content of this submitted manuscript and approved submission.
Research funding: Deutsche Forschungsgemeinschaft (DFG) in the framework of SPP 1681 (FKZ: CL202/3-1, CL202/3-2, CL202/3-3, DU1293/4-1, DU1293/4-2, DU1293/7-3, and SCHA1640/7-1).
Conflict of interest statement: The authors declare no conflicts of interest regarding this article.

References

1. Cardoso VF, Francesko A, Ribeiro C, Banobre-Lopez M, Martins P, Lanceros-Mendez S. Advances in magnetic nanoparticles for biomedical applications. Adv Healthc Mater 2018;7. https://doi.org/10.1002/adhm.201700845.
2. Krishnan KM. Biomedical nanomagnetics: a spin through possibilities in imaging, diagnostics, and therapy. IEEE Trans Magn 2010;46:2523–58.

3. Pankhurst QA, Connolly J, Jones SK, Dobson J. Applications of magnetic nanoparticles in biomedicine. J Phys D Appl Phys 2003;36:R167–81.
4. Wu K, Su DQ, Liu JM, Saha R, Wang JP. Magnetic nanoparticles in nanomedicine: a review of recent advances. Nanotechnology 2019;30. https://doi.org/10.1088/1361-6528/ab4241.
5. Dutz S, Hergt R. Magnetic particle hyperthermia-a promising tumour therapy?. Nanotechnology 2014;25. https://doi.org/10.1088/0957-4484/25/45/452001.
6. Gilchrist RK, Medal R, Shorey WD, Hanselman RC, Parrott JC, Taylor CB. Selective inductive heating of lymph nodes. Ann Surg 1957;146:596–606.
7. Perigo EA, Hemery G, Sandre O, Ortega D, Garaio E, Plazaola F, et al. *Fundamentals and advances in magnetic hyperthermia. Appl Phys Rev 2015;2. https://doi.org/10.1063/1.4935688.*
8. Estelrich J, Escribano E, Queralt J, Busquets MA. Iron oxide nanoparticles for magnetically-guided and magnetically-responsive drug delivery. Int J Mol Sci 2015;16:8070–101.
9. Lubbe AS, Bergemann C, Riess H, Schriever F, Reichardt P, Possinger K, et al. *Clinical experiences with magnetic drag targeting: a phase I study with 4′-epidoxorubicin in 14 patients with advanced solid tumors. Cancer Res 1996;56:4686–93.*
10. Gleich B, Weizenecker R. Tomographic imaging using the nonlinear response of magnetic particles. Nature 2005;435:1214–7.
11. Dadfar SM, Roemhild K, Drude NI, von Stillfried S, Knuchel R, Kiessling F, et al. *Iron oxide nanoparticles: diagnostic, therapeutic and theranostic applications. Adv Drug Deliv Rev 2019;138: 302–25.*
12. Wu W, Jiang CZ, Roy VAL. Designed synthesis and surface engineering strategies of magnetic iron oxide nanoparticles for biomedical applications. Nanoscale 2016;8:19421–74.
13. Wu W, Wu ZH, Yu T, Jiang CZ, Kim WS. Recent progress on magnetic iron oxide nanoparticles: synthesis, surface functional strategies and biomedical applications. Sci Technol Adv Mater 2015; 16. https://doi.org/10.1088/1468-6996/16/2/023501.
14. Dutz S. Are magnetic multicore nanoparticles promising candidates for biomedical applications?. IEEE Trans Magn 2016;52. https://doi.org/10.1109/tmag.2016.2570745.
15. Dutz S, Kettering M, Hilger I, Muller R, Zeisberger M. Magnetic multicore nanoparticles for hyperthermia-influence of particle immobilization in tumour tissue on magnetic properties. Nanotechnology 2011;22:265102.
16. Kostopoulou A, Brintakis K, Fragogeorgi E, Anthousi A, Manna L, Begin-Colin S, et al. *Iron oxide colloidal nanoclusters as theranostic vehicles and their interactions at the cellular level. Nanomaterials 2018;8. https://doi.org/10.3390/nano8050315.*
17. Lartigue L, Hugounenq P, Alloyeau D, Clarke SP, Levy M, Bacri J-C, et al. *Cooperative organization in iron oxide multi-core nanoparticles potentiates their efficiency as heating mediators and MRI contrast agents. ACS Nano 2012;6:10935–49.*
18. Kratz H, Taupitz M, de Schellenberger AA, Kosch O, Eberbeck D, Wagner S, et al. *Novel magnetic multicore nanoparticles designed for MPI and other biomedical applications: from synthesis to first in vivo studies. PloS One 2018;13. https://doi.org/10.1371/journal.pone.0190214.*
19. Del Pino P, Pelaz B, Zhang Q, Maffre P, Nienhaus GU, Parak WJ. Protein corona formation around nanoparticles – from the past to the future. Mater Horiz 2014;1:301–13.
20. Tenzer S, Docter D, Rosfa S, Wlodarski A, Kuharev J, Rekik A, et al. *Nanoparticle size is a critical physicochemical determinant of the human blood plasma corona: a comprehensive quantitative proteomic analysis. ACS Nano 2011;5:7155–67.*
21. Vroman L. Effect of adsorbed proteins on the wettability of hydrophilic and hydrophobic solids. Nature 1962;196:476–7.
22. Nel AE, Maedler L, Velegol D, Xia T, Hoek EMV, Somasundaran P, et al. *Understanding biophysicochemical interactions at the nano-bio interface. Nat Mater 2009;8:543–57.*

23. Dai Q, Bertleff-Zieschang N, Braunger JA, Bjornmalm M, Cortez-Jugo C, Caruso F. Particle targeting in complex biological media. Adv Healthc Mater 2018;7. https://doi.org/10.1002/adhm.201700575.
24. Eberbeck D, Kettering M, Bergemann C, Zirpel P, Hilger I, Trahms L. Quantification of the aggregation of magnetic nanoparticles with different polymeric coatings in cell culture medium. J Phys D Appl Phys 2010;43. https://doi.org/10.1088/0022-3727/43/40/405002.
25. Illes E, Szekeres M, Toth IY, Farkas K, Foldesi I, Szabo A, et al. *PEGylation of superparamagnetic iron oxide nanoparticles with self-organizing polyacrylate-PEG brushes for contrast enhancement in MRI diagnosis. Nanomaterials 2018;8. https://doi.org/10.3390/nano8100776.*
26. Jain P, Pawar RS, Pandey RS, Madan J, Pawar S, Lakshmi PK, et al. *In-vitro in-vivo correlation (IVIVC) in nanomedicine: is protein corona the missing link?. Biotechnol Adv 2017;35: 889–904.*
27. Mahmoudi M, Lynch I, Ejtehadi MR, Monopoli MP, Bombelli FB, LaurentS. et al. *Protein-nanoparticle interactions: opportunities and challenges. Chem Rev 2011;111:5610–37.*
28. Boraschi D, Italiani P, Palomba R, Decuzzi P, Duschl A, Fadeel B, et al. *Nanoparticles and innate immunity: new perspectives on host defence. Semin Immunol 2017;34:33–51.*
29. Lynch I, Dawson KA. Protein-nanoparticle interactions. Nano Today 2008;3:40–7.
30. Pearson RM, Juettner VV, Hong S. Biomolecular corona on nanoparticles: a survey of recent literature and its implications in targeted drug delivery. Front Chem 2014;2:108–108.
31. Winzen S, Schoettler S, Baier G, Rosenauer C, Mailaender V, Landfester K,et al. *Complementary analysis of the hard and soft protein corona: sample preparation critically effects corona composition. Nanoscale 2015;7:2992–3001.*
32. Cedervall T, Lynch I, Lindman S, Berggard T, Thulin E, Nilsson H, et al. *Understanding the nanoparticle-protein corona using methods to quantify exchange rates and affinities of proteins for nanoparticles. Proc Natl Acad Sci U S A 2007;104:2050–5.*
33. Pederzoli F, Tosi G, Vandelli MA, Belletti D, Forni F, Ruozi BProtein corona and nanoparticles: how can we investigate on? Wiley Interdiscip Rev Nanomed Nanobiotechnol 2017;9. https://doi.org/10.1002/wnan.1467.
34. Weber C, Morsbach S, Landfester K. Possibilities and limitations of different separation techniques for the analysis of the protein corona. Angew Chem Int Ed 2019;58:12787–94.
35. Brancolini G, Bellucci L, Maschio MC, Di Felice R, Corni S. The interaction of peptides and proteins with nanostructures surfaces: a challenge for nanoscience. Curr Opin Colloid Interface Sci 2019; 41:86–94.
36. Langevin D, Lozano O, Salvati A, Kestens V, Monopoli M, Raspaud E, et al. *Inter-laboratory comparison of nanoparticle size measurements using dynamic light scattering and differential centrifugal sedimentation. Nanoimpact 2018;10:97–107.*
37. Nel AE, Madler L, Velegol D, Xia T, Hoek EMV, Somasundaran P, et al. *Understanding biophysicochemical interactions at the nano-bio interface. Nat Mater 2009;8:543–57.*
38. Roecker C, Poetzl M, Zhang F, Parak WJ, Nienhaus GU. A quantitative fluorescence study of protein monolayer formation on colloidal nanoparticles. Nat Nanotechnol 2009;4:577–80.
39. Bohorquez AC, Unni M, Belsare S, Chiu-Lam A, Rice L, Pampo C, et al. *Stability and mobility of magnetic nanoparticles in biological environments determined from dynamic magnetic susceptibility measurements. Bioconjug. Chem. 2018;29:2793–805.*
40. Eberbeck D, Wiekhorst F, Steinhoff U, Trahms L. Quantification of biomolecule agglutination by magnetorelaxometry. Appl Phys Lett 2009;95. https://doi.org/10.1063/1.3267054.
41. Soukup D, Moise S, Cespedes E, Dobson J, Telling ND. In situ measurement of magnetization relaxation of internalized nanoparticles in live cells. ACS Nano 2015;9:231–40.
42. Bekdemir A, Liao SY, Stellacci F. On the effect of ligand shell heterogeneity on nanoparticle/protein binding thermodynamics. Colloids Surf B Biointerfaces 2019;174:367–73.

43. Casals E, Pfaller T, Duschl A, Oostingh GJ, Puntes V. Time evolution of the nanoparticle protein corona. ACS Nano 2010;4:3623–32.
44. Huhn D, Kantner K, Geidel C, Brandholt S, De Cock I, Soenen SJH, et al. *Polymer-coated nanoparticles interacting with proteins and cells: focusing on the sign of the net charge. ACS Nano 2013;7:3253–63.*
45. Kumar S, Yadav I, Aswal VK, Kohlbrecher J. Structure and interaction of nanoparticle-protein complexes. Langmuir 2018;34:5679–95.
46. Lundqvist M, Stigler J, Elia G, Lynch I, Cedervall T, Dawson KA. Nanoparticle size and surface properties determine the protein corona with possible implications for biological impacts. Proc Natl Acad Sci U S A 2008;105:14265–70.
47. Mahmoudi M, Abdelmonem AM, Behzadi S, Clement JH, Dutz S, Ejtehadi MR, et al. *Temperature: the "ignored" factor at the NanoBio interface. ACS Nano 2013;7:6555–62.*
48. Partikel K, Korte R, Mulac D, Humpf HU, Langer K. Serum type and concentration both affect the protein-corona composition of PLGA nanoparticles. Beilstein J Nanotechnol 2019;10: 1002–15.
49. Bolanos K, Kogan MJ, Araya E. Capping gold nanoparticles with albumin to improve their biomedical properties. Int J Nanomed 2019;14:6387–406.
50. Grafe C, Weidner A, von der Luhe M, Bergemann C, Schacher FH, Clement JH, et al. *Intentional formation of a protein corona on nanoparticles: serum concentration affects protein corona mass, surface charge, and nanoparticle-cell interaction. Int J Biochem Cell Biol 2016;75:196–202.*
51. Gulati NM, Stewart PL, Steinmetz NF. Bioinspired shielding strategies for nanoparticle drug delivery applications. Mol Pharm 2018;15:2900–9.
52. Mariam J, Sivakami S, Dongre PM. Albumin corona on nanoparticles - a strategic approach in drug delivery. Drug Deliv 2016;23:2668–76.
53. Monteiro-Riviere NA, Samberg ME, Oldenburg SJ, Riviere JE. Protein binding modulates the cellular uptake of silver nanoparticles into human cells: implications for in vitro to in vivo extrapolations?. Toxicol Lett 2013;220:286–93.
54. Schottler S, Becker G, Winzen S, Steinbach T, Mohr K, Landfester K, et al. *Protein adsorption is required for stealth effect of poly(ethylene glycol)- and poly(phosphoester)-coated nanocarriers. Nat Nanotechnol 2016;11:372–7.*
55. Biehl P, von der Luhe M, Dutz S, Schacher FH. Synthesis, characterization, and applications of magnetic nanoparticles featuring polyzwitterionic coatings. Polymers 2018;10. https://doi.org/10.3390/polym10010091.
56. Billing M, Grafe C, Saal A, Biehl P, Clement JH, Dutz S, et al. *Zwitterionic iron oxide (-Fe2O3) nanoparticles based on P(2VP-grad-AA) copolymers. Macromol Rapid Commun 2017;38. https://doi.org/10.1002/marc.201600637.*
57. Dutz S, Wojahn S, Grafe C, Weidner A, Clement JH. Influence of sterilization and preservation procedures on the integrity of serum protein-coated magnetic nanoparticles. Nanomaterials 2017; 7. https://doi.org/10.3390/nano7120453.
58. Grafe C, von der Luhe M, Weidner A, Globig P, Clement JH, Dutz S, et al. *Protein corona formation and its constitutional changes on magnetic nanoparticles in serum featuring a polydehydroalanine coating: effects of charge and incubation conditions. Nanotechnology 2019; 30. https://doi.org/10.1088/1361-6528/ab0ed0.*
59. Muller EK, Grafe C, Wiekhorst F, Bergemann C, Weidner A, Dutz S, et al. *Magnetic nanoparticles interact and pass an in vitro co-culture blood-placenta barrier model. Nanomaterials 2018;8. https://doi.org/10.3390/nano8020108.*
60. von der Luhe M, Gunther U, Weidner A, Grafe C, Clement JH, Dutz S, et al. *SPION@polydehydroalanine hybrid particles. Rsc Adv 2015;5:31920–9.*

61. von der Luhe M, Weidner A, Dutz S, Schacher FH. Reversible electrostatic adsorption of polyelectrolytes and bovine serum albumin onto polyzwitterion-coated magnetic multicore nanoparticles: implications for sensing and drug delivery. ACS Appl Nano Mater 2018;1:232–44.
62. Weidner A, Gräfe C, Von der Luhe M, Remmer H, Clement JH, Eberbeck D, et al. *Preparation of core-shell hybrid materials by producing a protein corona around magnetic nanoparticles. Nanoscale Res Lett 2015;10:992.*
63. Billing M, Festag G, Bellstedt P, SchacherAmphiphilic FH. And double hydrophilic block copolymers containing a polydehydroalanine block. Polym Chem 2017;8:936–45.
64. Guenther U, Sigolaeva LV, Pergushov DV, Schacher FH. Polyelectrolytes with tunable charge based on polydehydroalanine: synthesis and solution properties. Macromol Chem Phys 2013;214: 2202–12.
65. Khalafalla SE, Reimers GW. Preparation of dilution-stable aqueous magnetic fluids. IEEE Trans Magn 1980;16:178–83.
66. Dutz S, Andrae W, Hergt R, Mueller R, Oestreich C, Schmidt C, et al. *Influence of dextran coating on the magnetic behaviour of iron oxide nanoparticles. J Magn Magn Mater 2007;311:51–4.*
67. Dutz S, Clement JH, Eberbeck D, Gelbrich T, Hergt R, Mueller R, et al. *Ferrofluids of magnetic multicore nanoparticles for biomedical applications. J Magn Magn Mater 2009;321:1501–4.*
68. Hergt R, Dutz S, Mueller R, Zeisberger M. Magnetic particle hyperthermia: nanoparticle magnetism and materials development for cancer therapy. J Phys Condens Matter 2006;18:S2919–34.
69. Suzdalev IP, Maksimov YV, Buravtsev VN, Imshennik VK, Kazakevich AG, Novichikhin SV. The formation and properties of a system of iron oxide nanoclusters. Colloid J 2000;62:224–33.
70. Bodenschatz W. Handbuch für den Desinfektor in Ausbildung und Praxis. München/Germany: Elsevier; 1993.
71. Kurzhals H. Kühlen und Gefrieren von Lebensmitteln. Hamburg/Germany: BEHRS Verlag; 2007.
72. Wang W. Lyophilization and development of solid protein pharmaceuticals. Int J Pharm 2000;203: 1–60.
73. Dutz S, Hergt R. The role of interactions in systems of single domain ferrimagnetic iron oxide nanoparticles. J. Nano- Electron. Phys. 2012;4:20101–7.
74. Dutz S, Hergt R, Murbe J, Topfer J, Muller R, Zeisberger M, et al. *Magnetic nanoparticles for biomedical heating applications. Z Phys Chem 2006;220:145–51.*
75. Schneider CA, Rasband WS, Eliceiri KW. NIH Image to ImageJ: 25 years of image analysis. Nat Methods 2012;9:671–5.
76. Ludwig F, Mauselein S, Heim E, Schilling M. Magnetorelaxometry of magnetic nanoparticles in magnetically unshielded environment utilizing a differential fluxgate arrangement. Rev Sci Instrum 2005;76. https://doi.org/10.1063/1.2069776.
77. Schlenk F, Werner S, Rabel M, Jacobs F, Bergemann C, Clement JH, et al. *Comprehensive analysis of the in vitro and ex ovo hemocompatibility of surface engineered iron oxide nanoparticles for biomedical applications. Arch Toxicol 2017;91:3271–86.*
78. Hamburger V, Hamilton HL. A series of normal stages in the development of the chick embryo. J Morphol 1951;88:49&#x2013.
79. NIH. In vitro cytotoxicity test methods for estimating starting doses for acute oral systemic toxicity testing ICCVAM test method evaluation report 2006. NIH Publication No. 07-4519.
80. Chantrell RW, Popplewell J, Charles SW. Measurements of particle-size distribution parameters in ferrofluids. IEEE Trans Magn 1978;14:975–7.
81. Biehl P, von der Luhe M, Schacher FH. Reversible adsorption of methylene blue as cationic model cargo onto polyzwitterionic magnetic nanoparticles. Macromol Rapid Commun 2018;39. https://doi.org/10.1002/marc.201800017.
82. Fasold H, Klappenberger J, Meyer C, Remold H. Bifunctional reagents for the crosslinking of proteins. Angew Chem Int Ed 1971;10:795.

Stephen Strassburg, Kai Mayer and Thomas Scheibel*

7 Functionalization of biopolymer fibers with magnetic nanoparticles

Abstract: Hybrid fibers consisting of biopolymers and inorganic nanoparticles are receiving increasing attention due to their unique properties. Commonly, the nanoparticles are chosen for their intrinsic properties such as magnetic, thermal, or electrical conductivity. The biopolymer component of the hybrid fiber is chosen for its mechanical properties and ability to act as a scaffold or matrix for the nanoparticles. While there are many fiber-forming synthetic polymers, there has been a recent interest in replacing these systems with biopolymers due to their sustainability, biocompatibility, nontoxicity, and biodegradability. Fibers made from biopolymers have one additional benefit over synthetic polymers as they make good scaffolds for embedding nanoparticles without the need of any additional bonding agents. In particular, naturally occurring biopolymers such as proteins exhibit a myriad of interactions with nanoparticles, including ionic, H-bonding, covalent, Van der Waals, and electrostatic interactions. The diverse range of interactions between magnetic nanoparticles and biopolymers makes resulting hybrid fibers of particular interest as magnetic-responsive materials. Magnetically responsive hybrid biopolymer fibers have many features, including enhanced thermal stabilities, strong mechanical toughness, and perhaps most interestingly multifunctionality, allowing for a wide range of applications. These applications range from biosensing, filtration, UV shielding, antimicrobial, and medical applications, to name a few. Here, we review established hybrid fibers consisting of biopolymers and nanoparticles with a primary focus on biopolymers doped with magnetic nanoparticles and their various putative applications.

Keywords: electrospinning, magnetic nanoparticle, microfluidics, polysaccharide, silk, wet spinning

***Corresponding author: Thomas Scheibel,** Department of Biomaterials, Universität Bayreuth, Prof.–Rüdiger-Bormann-Straße 1, 95447 Bayreuth, Germany; Bayreuth Center for Colloids and Interfaces (BZKG), Universität Bayreuth, Universitätsstraße 30, 95440 Bayreuth, Germany; Bayreuth Center for Molecular Biosciences (BZMB), Universität Bayreuth, Universitätsstraße 30, 95440 Bayreuth, Germany; Bayreuth Center for Material Science (BayMAT), Universität Bayreuth, Universitätsstraße 30, 95440 Bayreuth, Germany; and Bavarian Polymer Institute (BPI), Universität Bayreuth, Universitätsstraße 30, 95440 Bayreuth, Germany, E-mail: Thomas.Scheibel@bm.uni-bayreuth.de. https://orcid.org/0000-0002-0457-2423

Stephen Strassburg and Kai Mayer, Department of Biomaterials, Universität Bayreuth, Prof.–Rüdiger-Bormann-Straße 1, 95447 Bayreuth, Germany

This article has previously been published in the journal Physical Sciences Reviews. Please cite as: S. Strassburg, K. Mayer and T. Scheibel "Functionalization of biopolymer fibers with magnetic nanoparticles" *Physical Sciences Reviews* [Online] 2020, 5. DOI: 10.1515/psr-2019-0118 | https://doi.org/10.1515/9783110569636-007

7.1 Introduction

Biopolymers are a family of macromolecules containing natural or recombinant proteins, polysaccharides, and synthetic biodegradable polymers (Figure 7.1). Characteristics of biopolymers may include biocompatibility, biodegradation, and renewability, which make these materials excellent candidates to be used in different fields of research and applications [1–5]. Despite this, synthetic polymers are simply easier and cheaper to mass produce and thus have been subsequently more commonly used as fiber materials [6]. Recently that trend appears to be changing as advances in protein and polysaccharide extraction as well as better methods of producing recombinant proteins have significantly dropped the cost of these biopolymers [7–10]. As environmental concerns grow and more renewable materials are necessary, there appears to be a growing shift from synthetic polymeric materials to renewable biopolymers [9].

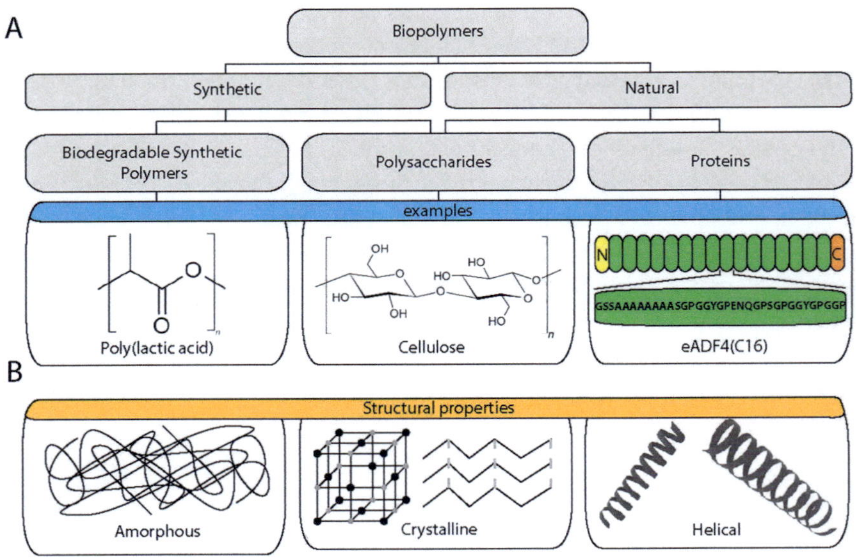

Figure 7.1: (A) Overview of different biopolymers including synthetic (biodegradable) and naturally derived polymers. Biopolymers include polymers derived from natural materials or synthetic polymers that are biodegradable. Natural biopolymers include polysaccharides and proteins. An example of a biodegradable synthetic polymer, (poly(lactic acid)), a polysaccharide, (cellulose), and the protein eADF4(C16), a recombinant major ampullate silk derived from *Araneus diadematus* is shown on the bottom of (A). (B) Structural properties found in biopolymers. Synthetic polymers and polysaccharides often have multiple structural properties (i.e., semicrystallinity: amorphous and crystalline domains) as do proteins (helical and β-sheet conformations).

7.2 Biopolymer-based materials

Biopolymers are highly adaptable and offer more functionality than traditional synthetic polymers. For example, many proteins have evolved over billions of years to carry out a myriad of diverse tasks such as catalysis, molecular recognition, and the storage of energy or information. Chances are, if there is a material needed for a specific application, nature has already designed a protein or polysaccharide for the job. For example, spider silk fibers outperform most of their synthetic polymer counterparts in toughness [5, 11, 12]. Proteins such as hemoglobin and ferritin are excellent binders of metal ions such as iron. Ferritin has even been used as a scaffold for the synthesis of gold nanoparticles [13]. Chitin and cellulose are often used to make thermally stable films [1]. In addition, more recent studies have used cellulose for optically clear glasses and papers. [14, 15]. These are but a few examples and do not even begin to scratch the surface of the myriad of functionalities biopolymers can have. But that does not mean nature is perfect, since natural materials have evolved to fulfill multiple criteria. In fact, many of these systems can be improved for specific applications simply by adding synthetic materials, such as inorganic nanoparticles. Using biopolymers as scaffolds offers a diversity that cannot be matched by traditional synthetic polymer systems; and the addition of nanoparticles into biopolymer fibers allows a tunability of the material that would not otherwise be realized. These hybrid composites allow for antimicrobial, thermally, electrically, magnetically, and/or light-responsive materials with unmatched renewability, biocompatibility, and mechanical properties.

Despite the diversity of biopolymers there are still drawbacks, the main one being their processability. The difficulty with processability is different for proteins compared to polysaccharides. Typically, proteins are difficult to obtain in high yields due to recovery methods being costly and time-consuming, while polysaccharides tend to be insoluble, often requiring harsh organic solvents to fully dissolve. More recently, advances in protein recovery and unique variants of solvents (such as ionic liquids and deep eutectic solvents) for solubilizing polysaccharides have led to biopolymers becoming an increasingly more appealing option compared to traditional crude-oil-based polymers [8, 9, 16].

7.2.1 Polysaccharides for material's applications

The two most commonly studied and also the most abundant polysaccharides in nature are cellulose and chitin [17]. Chitin is derived from many crustaceans, insects, mollusks, and fish and is most commonly treated with a deacetylation process to form chitosan [2]. Conversely, cellulose typically is derived from plants, such as wood from trees, cotton, algae, and even can be secreted by bacteria, making it the most abundant biopolymer on Earth [17]. Both cellulose and chitosan can be processed into fibers, films, and gels with good mechanical properties, and it should come as no surprise that they are common materials for use in bionanocomposites [1, 2, 15].

7.2.2 Proteins for material's applications

Many natural proteins derived from animal or plant organisms are used for producing fibers. Protein-based composite materials can be made from a wide range of differing proteins. Globular proteins such as bovine serum albumin, lysozyme, and cytochrome *c* are all used in conjunction with nanoparticles for drug delivery applications [18–20]. Conversely, many fibrous and structural proteins such as collagen, silk, and keratin are common materials for fibers, films, and gels used in material applications [1, 21–23]. The most abundant of these proteins examined in literature is collagen. Despite this abundance of literature, there are only a few examples of collagen nanoparticle composite materials examined and most of these studied systems use a collagen–Ag nanoparticle composite for antimicrobial applications [24, 25].

Natural silk fibers are produced by insects, arachnids, and even some crustaceans [26]. Silk is a well-known natural fiber used in textile industry and other applications. Specifically, silkworm silk has become a multibillion-dollar textile business, a long way from the initial silk textile beginnings several thousand years ago in China [27].

Most of the silk produced today comes from domesticated *Bombyx mori* silkworms [28]. While not as strong as dragline spider silk (16×10^4 J kg^{-1}), it is more abundant and easily usable for textiles [29]. There is a large portion of literature working with silkworm silk and nanoparticles. Much like collagen and other protein-based materials, much of the biopolymer–nanoparticle composite literature focuses on antimicrobial textiles and nonwoven meshes [30].

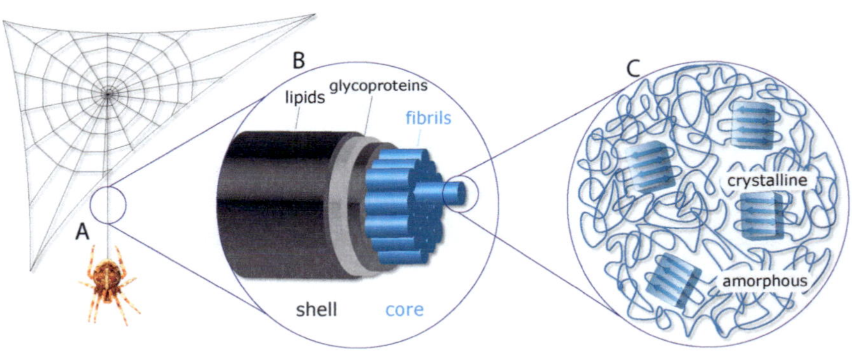

Figure 7.2: Schematic of a spider silk web from (A) *Araneus diadematus*, (B) the macroscopic structure of dragline spider silk, and (C) the microscopic molecular structure of the MAS protein dragline proteins. Crystalline regions are predominately β-sheets while the amorphous regions consist of α-helices and random coils. Originally published by Heidebrecht et al. [38] and slightly adapted with permissions.

A very interesting and highly complex proteinaceous material from nature is spider silk [34]. Spiders produce a variety of elaborate proteins among which major ampullate silks (MAS) are the basis for a fiber with outstanding mechanical properties combining a high tensile strength and moderate elasticity [22]. MAS fibers are used by spiders as a scaffold upon which to attach other silks during the creation of a web and as a lifeline when it is necessary to escape from predators. The combination of strength and stretchiness is mainly derived from the domains of crystalline β-sheets and flexible helices/amorphous regions within the polypeptide chain (Figure 7.2), imparting a toughness that is greater than that of bone, synthetic rubber, Kevlar, and high tensile steel or other man-made fibers [11, 12, 35–37].

7.3 Magnetic nanoparticles

Nanoparticles are small particles ranging between 1 and 100 nm in size and can form a variety of shapes such as, spherical, cubic, conical, rod-like, and many others [39, 40]. They can be made from organic, metallic, or ceramic molecules, and often their nanoscale properties differ from their bulk properties (even between materials of the same composition) [40]. Nanoparticles are of particular scientific relevance because of the diverse range of interactions they can experience. These interactions include: ionic, H-bonding, covalent, Van der Waals, and electrostatic interactions, which allow for a highly robust material that can be used for many applications [41].

Magnetic nanoparticles are a subset of nanoparticles that exhibit a magnetic response when exposed to a magnetic or electric field. Magnetically responsive materials may appear a niche topic when compared to nanoparticle research in anti-microbial, thermally, or electrically responsive biocomposites, but their application

Table 7.1: Examples of magnetic nanoparticles, their shape, and application.

Examples	Size	Shape	Application
Fe_3O_4 [50, 51]	12 nm–µm	Sphere, spindle, cube, rods octahedral,	Drug delivery, medical applications, thermal coatings
$BaFe_{12}O_{19}$ [52]	<100 nm	Sphere	Medical applications
$SrFe_{12}O_{19}$ [53]	30–70 nm	Sphere, spindle, cube, rods	High-frequency devices, permanent magnets
$ZnFe_2O_4$ [54, 55]	<100 nm	Sphere	Medical applications
$CoFe_2O_4$ [50]	8.5 nm	Sphere	Drug delivery
$InCu_2Fe_2O_4$ [56]	28–37 nm	Sphere, cube	Electronic applications
$CuFe_2O_4$ [56]	28–37 nm	Sphere, cube	Electronic applications
Ni/NiO [54, 57]	13 nm	Sphere	Medical applications
$MnFe_2O_4$ [50, 54]	<100 nm	Sphere, cube	Medical applications

introduces a diverse field of responsive materials where morphology and topology can be controlled via magnetic fields [42]. Furthermore, tunable morphology invites a diverse range of applications from membranes, to filtration, and even for biosensor and electronic applications [21, 45–49].

Typically, magnetic nanoparticles used for biocomposite materials are a variant of iron oxide [58–62]. However, nanoparticles of cobalt [47, 63, 64], platinum [65], manganese–zinc [54], and yttrium–iron [66] can also be used as magnetic materials. The myriad of differing magnetic nanoparticles not only offers diverse chemical interactions but also comes in a wide range of sizes and shapes. Table 7.1 shows the diversity of magnetic nanoparticle composition, size, shape, and examples of biocomposite material applications found in literature.

7.4 Nanoparticle-doped biopolymer fibers

7.4.1 Processing techniques

There are three main processing techniques for biopolymer–nanocomposite based fibers, wet spinning, electrospinning and microfluidics. Each technique produces fibrous materials but the size of the materials will vary depending on the spinning method. Choosing the correct technique will largely depend on the biopolymer–nanoparticle composite that is being examined as well as the final application for the material.

7.4.1.1 Wet spinning

The most commonly used method of producing biopolymer fibers with diameters in the micrometer or larger range is wet spinning [67]. Wet spinning is conducted by extruding an aqueous or organic biopolymer solution into a coagulation bath. Upon contact with the coagulation bath, the biopolymer precipitates into a continuous fiber while the solution solvent diffuses to the bulk of the coagulation bath (Figure 7.3A) [67]. The fiber can then be collected or post-treated/post-stretched in a continuous or batch process. The diameters of fibers produced by wet spinning can be controlled by the concentration of the spinning dope, needle diameter, and posttreatment [67]. Typical solvents used for coagulation baths of biopolymers are polar organic solvents such as methanol, ethanol, acetone, propanol, or isopropanol [68], as well as aqueous solutions with varying pHs, ionic strengths, and salt compositions [69].

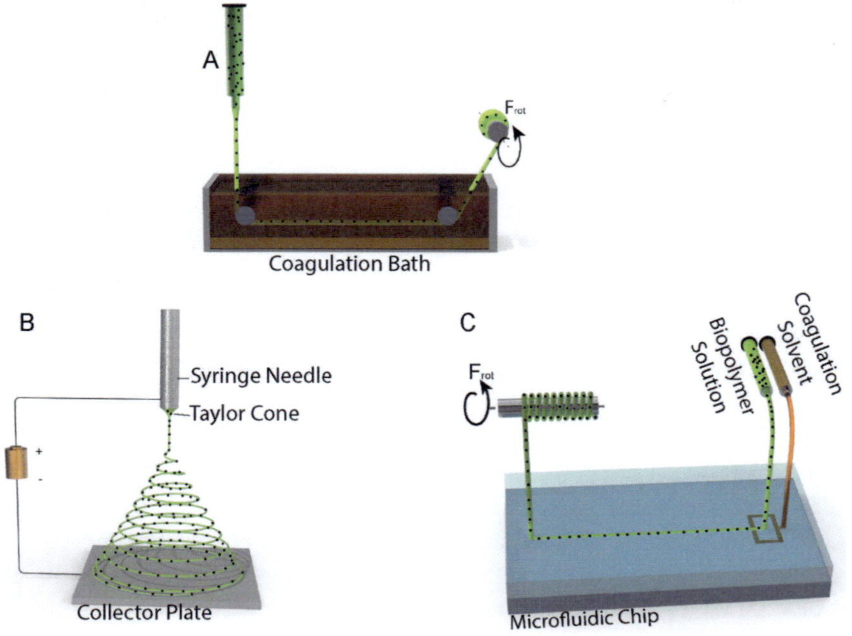

Figure 7.3: Schematics of (A) wet spinning, (B) electrospinning, and (C) microfluidics. The green color represents the biopolymer, the black dots represent nanoparticles, and the orange/brownish color represents coagulation bath or solvent. Wet spinning produces larger fiber diameters than both electrospinning and microfluidics with electrospinning producing the smallest fibers. Wet spinning and microfluidics use a coagulation bath that precipitates the dissolved biopolymer while the solvent in format mismatch! electrospinning is dried during the stretching and whipping process on the way to the collector plate.

7.4.1.2 Electrospinning

Electrospinning is an electrostatically driven process allowing for the production of fibers with diameters ranging from micrometers down to a few nanometers [70]. The basic setup comprises a needle, which is charged (±0–30 kV) and through which a biopolymer solution is extruded. [71] The high electric field induces electrostatic repulsion at the surface of an extruded droplet and leads to the formation of a Taylor cone through coulombic forces (Figure 7.3B). When the electrostatic forces overcome the solvent surface tension and the chain entanglement in the biopolymer solution is sufficiently high, a droplet is formed into a continuous jet [4]. At a certain distance from the Taylor cone, the jet experiences whipping instabilities and is stretched and dries on its path to the oppositely charged collector plate [72]. Fibers are "randomly" deposited on the collector plate as a nonwoven mesh. The diameters of fibers produced by electrospinning can be well controlled, typically in the micro- or nanometer range, by adaption of the spinning dope properties (biopolymer concentration, viscosity, conductivity, solution volatility, and surface tension of the solvent) and process parameters (voltage, distance to deposition plate, flow rate, temperature, humidity) [4]. Fibers produced via electrospinning tend to have smaller diameters than those produced through wet spinning [73].

7.4.1.3 Microfluidics

Microfluidics, while not as well studied as wet spinning or electrospinning, is an increasingly common method of fiber formation explored in literature [74]. Similar to wet spinning, microfluidic spinning allows for decreasing the scale of the input and coagulation solvent feeds (typically in the submilliliter range). The smaller size of the inlet streams leads to precise control of the fluid dynamics as well as the introduction of capillary forces that are often ignored in traditional wet spinning. Precise control offers advantages including small volumes of solvents, small size of the spinning set up, narrower fiber diameters than wet spinning and low energy consumption but with the trade-off of lower fiber yield [75]. There are many types of microfluidic setups including both batch and continuous flow processes [74, 76]. An example of one such batch setup is shown in Figure 7.3C.

7.4.2 Examples of nanoparticle-doped fibers

There is a growing literature on nanoparticle fiber doping using noble metals. Using biopolymers as scaffolds for the growth or adhesion of noble metal nanoparticles offers many advantages over traditional synthetic polymers. One unique consideration for interactions with metal nanoparticles is the structure of the protein or polysaccharide. Proteins in particular, with their diverse amino acid sequences, can adopt different secondary structures, such as random coil, β-sheet, or α-helix, which have been shown to have a major effect on the capture and release of noble metal nanoparticles [77]. This consideration leads to a finely tunable system but also adds an additional layer of complexity to the material production.

Work with *B. mori* silk fibers by Calamak et al. and *Pholcus phalangioides* spider silk by Singh et al. are two excellent examples of the effect protein structure has on biocomposite materials [43, 77]. In the first case, Calamak et al. examined the release profiles of Ag nanoparticles when embedded in *B. mori* silk scaffolds consisting of either a random coil or β-sheet conformation. The results showed a clear distinction in the release profiles, with the crystalline β-sheet structure showing a cumulative lower release rate [77]. Similarly, Singh et al. used spider silk, harvested from webs of *P. phalangioides* spiders, as a scaffold for growth of gold nanoparticles, which could later be used for vapor detection via changes in conductivity. Not only did the gold nanoparticles grown in this manner showed a rapid response time in conductivity on exposure to methanol, the nanoparticles showed excellent cycling efficiency indicating a strong binding to the silk fiber, likely a result of the β-sheet conformation of the spider silk protein [43]. Similarly, Tang et al. saw similar strong adhesion of Ag nanoparticles to commercial silk fabrics that could withstand many washing cycles [78]. Thus it appears for applications where nanoparticle–protein adhesion is important, β-sheet conformation is desired; while for applications where nanoparticle release is important, random coil or helix conformations are more suitable.

A subset of nanoparticle hybrid fibers, using magnetic nanoparticles, are of growing interest because of their magnetically responsive behavior allowing their use in many applications such as in medical devices [1], as antimicrobial surfaces [79], or to exploit their advantageous thermal and electrical properties [48, 82–86]. Nevertheless, the design and modification of these responsive materials are dependent upon the biopolymer and nanoparticle interactions. Often that requires choosing the correct biopolymer and nanoparticle combination and there are several to choose from [15, 87]. In some cases, modification of the biopolymer can be involved, which can be a nontrivial process. Even after an adequate system is chosen, a correct solvent and spinning method may be difficult to find. Still, the rewards for finding a system that successfully fits all of the criteria necessary to obtain biocomposite hybrid materials are vast, and there is an increasing amount of literature discussing magnetic nanoparticle–biopolymer interactions and processing techniques [11, 42, 44, 47, 58, 60, 64, 66, 88, 89].

One such reward involves a lesser known characteristic of metallic nanoparticles, their ability to absorb light, specifically, they have remarkable UV shielding. This is well known in the sunscreen industry where nanoparticles are readily used as additives [90]. Using similar nanoparticles as in sun screen, such as TiO_2 and CeO_2, several studies have shown enhanced UV shielding could be applied to silk textile fibers [31, 32]. In addition, not only did the TiO_2 and CeO_2 nanoparticles show remarkable adhesion to the silk fabrics tested, they also provided antibacterial capabilities and good thermal stabilities. Of particular interest, Li et al. reported a strong correlation in the decrease of transmitted UV light with increased TiO_2 content. Equally interesting, upon exposure to UV illumination, a photolytic self-cleaning was observed for the TiO_2-loaded silk fabrics [31].

The plethora of magnetic nanoparticle characteristics, ranging from thermal stability to electrical and magnetic responsiveness to UV shielding and antimicrobial properties, make magnetic nanoparticle biocomposites a rich field for exploring many different applications. These applications of magnetic nanoparticle biocomposites include medical devices, membranes, biosensors, and even bioseparations and filtration. Several interesting applications for nanoparticle–biopolymer composite systems are examined in section 7.6.

7.5 Biopolymer–nanoparticle composite properties

7.5.1 Morphology

Homogeneity of the fiber, density and distribution of nanoparticles, uniformity of fiber diameters, surface quality, and spinning artifacts are but a few of the important morphological features that are necessary to examine. There are different techniques for examination of these properties including: scanning electron microscopy (SEM), energy-dispersive X-ray spectroscopy (EDS), transmission electron microscopy (TEM), and atomic force microscopy (AFM), to name a few of the more commonly used techniques.

For evaluating surface morphology and interfaces of hybrid fibers, SEM and AFM are the most popular techniques [91, 92]. Briefly, SEM uses back-scattered electrons to detect differences in electron densities [91]. The combination of high magnification, large depth of focus, great resolution, ease of sample observation, and high contrast between metal nanoparticles and biopolymers makes SEM a ubiquitous characterization technique in the field of biopolymer–nanoparticle fibers. Most, if not all, of the different applications examined in section 7.6 use SEM to characterize the presence of nanoparticles and examine the interface between biopolymer and nanoparticle. While a qualitative technique, the addition of an EDS detector allows for the probing of elemental information that is not apparent in the SEM image [91]. EDS exploits the unique interactions between elemental atomic structure and their emission of X-rays. Since each element has a unique atomic structure, the emitted X-rays likewise have a unique energy spectra allowing for precise determination of elemental makeup of the surface [91]. EDS is particularly useful for distinguishing between nanoparticles (higher energy emissions) and biopolymers (lower energy emissions) and can help to elucidate nanoparticle density and distribution [91].

The less used, but more powerful technique, AFM, allows the user to examine many aspects of the surface of the hybrid material that could not otherwise be noticed in SEM [92]. AFM uses a force sensitive cantilever to tap or drag on the surface of the fiber creating a 2D surface map that is sensitive to the surface roughness, stiffness, and phase (delayed force response). As an example, recent work by Carapeto et al. used AFM to examine films of 1,6-hexanediamine-*graft*-cellulose doped with Ag nanoparticles [93]. They examined the surface of the hybrid film using height, amplitude, and phase images of the material. There they found no large aggregates and a random distribution of the Ag nanoparticles, which could have also been observed in SEM. However, what makes this work particularly interesting is by combining information on the three surface images, and estimating the shape of the nanoparticles to be a sphere, the authors were capable of quantifying the volume of each individual nanoparticle as well as the total volume fraction of the nanoparticles at the surface of the material [93]. Using a quartz crystal microbalance as a comparison, the quantitative estimations were only ~7% different [93].

Perhaps the most powerful technique, but also the most difficult to work with is TEM. Unlike SEM, TEM requires electrons passing through the material (thus samples are required to be quite thin: 1–100 nm in thickness). Electrons passing through the sample experience a slight loss of energy, which can arise from differences in thickness, density, atomic element, crystal structure, and/or orientation [94]. TEM can be used for many diverse characterization methods. Uses of TEM include determining particle size or examining the cross-sectional area of fibers [77, 79], probing fibril formation [4, 95, 96], and examining electron diffraction patterns, to name a few. The main advantage of using TEM is the ability to probe the bulk of the material that the aforementioned techniques cannot replicate.

7.5.2 Thermal properties

Thermal gravimetric analysis (TGA) and dynamic scanning calorimetry (DSC) are the two most common methods of thermal analysis [97]. TGA measures the change (loss) of mass of a sample with increasing temperature. DSC examines the change in heat capacities of a sample compared with that of a control (air) at varying temperatures.

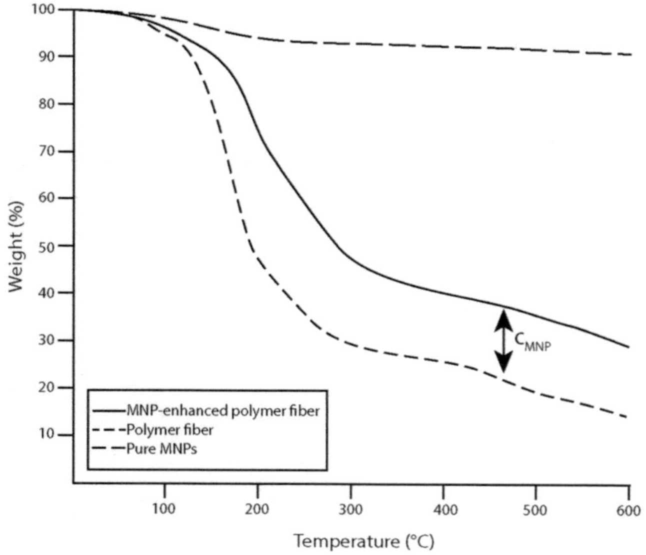

Figure 7.4: TGA schematic curves of a pure biopolymer (short dashed line), pure magnetic nanoparticle (long dashed line), and magnetic nanoparticle-doped biopolymer (solid line). The biopolymer doped with magnetic nanoparticles exhibits enhanced thermal properties compared to that of the pure biopolymer. The difference of mass loss at very high temperature (shown by double arrow) is directly proportional to the concentration of nanoparticles in the doped fiber.

Biopolymer fibers with and without magnetic nanoparticles demonstrate a different behavior under varying thermal conditions. Unlike mechanical properties, increasing the concentration of magnetic nanoparticles has often been reported in literature as increasing the thermal stability of the material [32, 33, 77, 78, 98, 99]. A schematic of a TGA plot including pure nanoparticles, pure biopolymers, and nanoparticle–biocomposite hybrid fibers is shown in Figure 7.4. The pure biopolymer fibers lose mass at a much lower temperature than pure nanoparticles. When a composite material of nanoparticles and biopolymers is produced, the composite material loses mass at a temperature and rate somewhere in the middle that of pure biopolymer and pure nanoparticles [100].

DSC can characterize the material glass transition temperature, crystallinity, crystallization/melting temperatures, and transitional enthalpies. Magnetic nanoparticles can influence many of these transitions. As mentioned earlier, Calamak et al. found that increasing the crystallinity (β-sheet) of silk fibers lowers the total cumulative Ag release compared to amorphous (random coil) fibers [77]. The authors also found that the thermal degradation temperature increased slightly with the addition of Ag nanoparticles and was higher for the crystalline structures as opposed to the amorphous fibers. Only the amorphous fibers showed a glass transition (190 °C) [77]. These results show that understanding the different thermal transitions in DSC is an important factor in designing nanoparticle–biopolymer composite materials.

7.5.3 Mechanical properties

Mechanical properties describe the physical properties of a material under the application of a force (i.e. resistance to shear or strain). Examples of mechanical properties include the modulus of elasticity, tensile strength, elongation at break, hardness, fatigue limit, and toughness. Biopolymers are known to have a wide range of mechanical properties. For example, natural dragline spider silk has extraordinary mechanical properties due to a combination of strength and extensibility [22, 101]. Using biopolymers, such as dragline spider silk, for their strong mechanical properties is an appealing prospect for many applications, but there is no guarantee that after the addition of magnetic nanoparticles the extraordinary mechanical properties would remain.

Work by Mayes et al. examined how the mechanical properties from dragline spider silk from *Nephila edulis* behaved after the addition of magnetite nanoparticles [11]. Magnetite nanoparticles were added via dipping fibers in colloidal suspension of $FeCl_3$ for 30 min followed by air-drying overnight. The authors found that the dragline silk retained its flexibility and its toughness even after the addition of magnetic nanoparticles. Hysteresis cycles were performed at ambient temperature and humidity using a custom-built stress–strain gauge. The fact that there was no discernible difference between unmodified spider silk (no nanoparticles added) and magnetite-doped silk fibers even after repeated loading and unloading of the fibers was a promising sign that nanoparticles situated on a fiber's surface did not adversely affect the mechanical properties of the biopolymer fibers [11]. However, this is not always the case. Work by Li et al. examined cellulose fibers soaked in an $AgNO_3$ aqueous solution at 80 °C for 24 h. The cellulose–AG nanocomposite fibers showed a decrease in stiffness and elasticity with increasing wt% of Ag NP [99]. Another consideration when doping biopolymers with magnetic nanoparticles is the minimum amount of magnetic ingredients in the composite fiber to obtain a magnetic effect. Zhou et al. showed that the strength of a wet-spun single-walled carbon nanotube–permalloy nanoparticle–poly(vinyl alcohol) (PVA) fiber exhibited a failure strength of around 0.7 GPa when at saturation load of magnetic nanoparticle. This failure strength was decreased significantly from ~25 GPa for the pure single-walled carbon nanotube–PVA mixture without the permalloy MNP [102]. Though the work by Zhou did not use a biopolymer, it highlights an important trade-off between mechanical strength and magnetic properties that must be considered when designing magnetically responsive fibers.

7.5.4 Magnetic properties

Magnetic properties of hybrid fibers can be tested using different physical methods, though all experimental setups feature an external magnetic field, which elicits a reaction of the magnetic material. The simplest test of the magnetic behavior is the manual detection of the influence of an external magnet on the sample. More advanced techniques such as a vibrating sample magnetometer (VSM) and superconducting quantum interference device (SQUID) are far more sensitive and capable of detecting minute changes in magnetic fields [103–105].

It is rare in literature that more advanced setups are used for magnetic responsiveness characterization. One such example is work by Munaweera et al., which used yttrium iron garnet nanoparticles embedded in cellulose acetate nonwovens (more information further) [66]. Before embedding into the nonwoven material, the yttrium iron garnet nanoparticle powder was repeatedly cycled using the SQUID to saturation magnetization at room temperature. The authors found that no magnetic hysteresis was observed, but the particle size had a strong effect on the saturation magnetization [66].

7.6 Applications

7.6.1 Antimicrobial

The most abundant use of biopolymer–nanocomposite materials is for antimicrobial applications. Silver nanoparticles are the most commonly used ones due to the natural

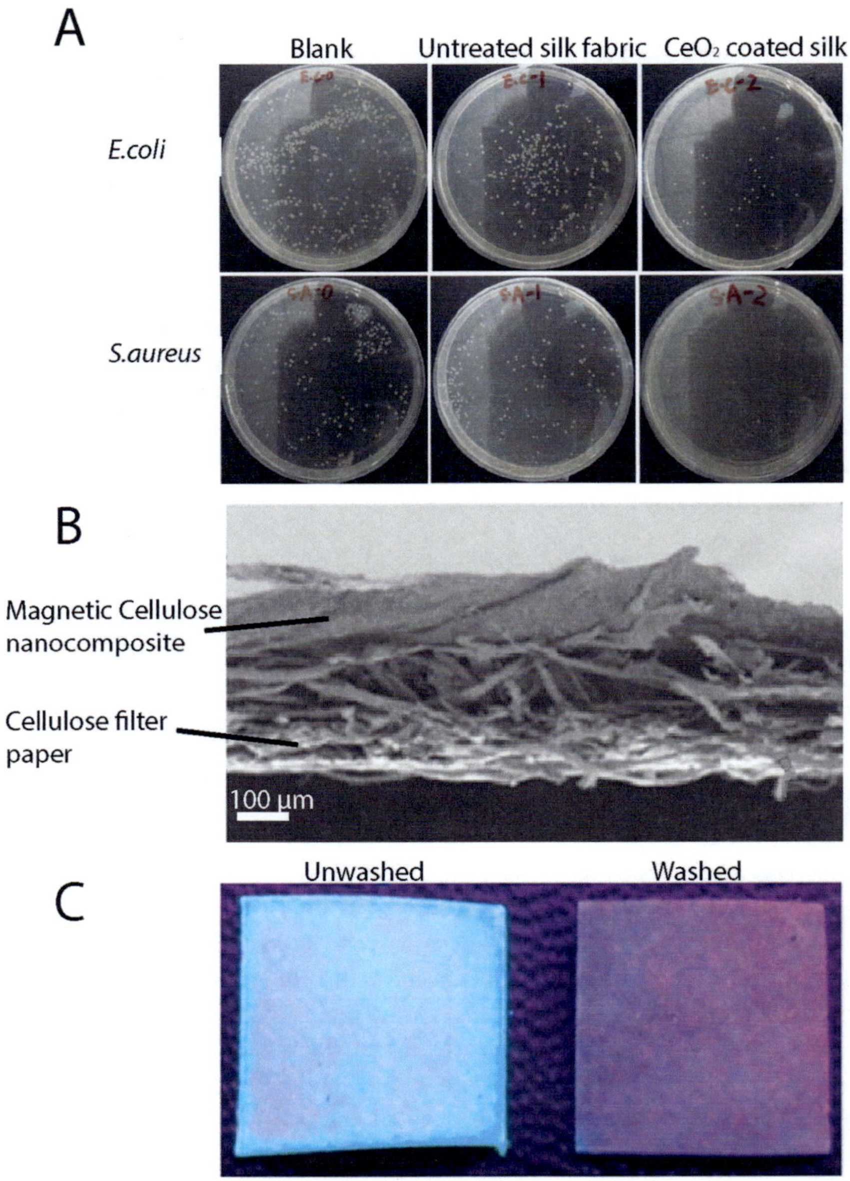

Figure 7.5: (A) Antimicrobial effects of uncoated and CeO_2 nanoparticle-loaded silk fabrics on *E. coli* and *S. aureus* cultured in LB medium under constant shaking at 37 °C overnight. The media was then exposed to 2 mm × 2 mm squares of untreated and CeO_2-loaded silk fabrics for 2 h. The contents were plated onto agar plates and cultured for another 24 h. The CeO_2-treated solutions indicated a significant decrease in bacterial colony survival compared to the control (blank) and the untreated silk fabrics. Adapted with permissions from Lu et al. [32]. (B) SEM image of the yttrium iron garnet nanoparticle–loaded cellulose nonwoven mesh attached to cellulose filter paper. The electrospun nonwoven retains its magnetic responsiveness even after attachment to the cellulose filter paper. (C) Images of unwashed and washed filter papers from (B). The unwashed filter paper shows a strong fluorescence when exposed to UV irradiation at 245 nm indicating that the FITC-BSA has been successfully bound to the filter paper. Upon washing with deionized water, the fluorescence disappears indicating that the FITC-BSA has been removed. (B) and (C) were reproduced with permissions from Munaweera et al. [66].

antimicrobial characteristics of the silver ions, but the use of the biopolymer can vary. Examples include: silk (spider or silk worm) [33, 78, 80], cotton [106], cellulose [81, 99, 107], chitosan [108, 109], and even biocompatible synthetic fibers [110, 111].

Silver nanoparticles are generally either embedded into the biopolymer fiber or grown on the fiber itself regardless of the biopolymer in use. Differing approaches can be used for the embedding of particles. Lu et al. successfully developed an efficient green method for attaching Ag nanoparticles to polydopamine-functionalized silk that resulted in a uniform distribution [112]. The strong absorption to the polydopamine-functionalized silk was induced by using a reducing agent. This process is not unique to silk either. A similar method was developed by Chen et al. They demonstrated that using cotton fibers grafted with glycidyl-methacrylate–iminodiacetic acid could strongly adsorb and bind Ag nanoparticles [106]. For both of the aforementioned cases, the nanoparticles were attached via metal coordination bonds, but Ag nanoparticles can also be attached covalently [31]; a process that is especially useful for protein–Au nanocomposites because of the high affinity for Au nanoparticles to bind to cysteine amino acid residues in many proteins [113].

Alternative to embedding, nanoparticles can be grown onto the fiber. Both Lu et al. and Chang et al. successfully grew Ag nanoparticles on silk fibers via UV-or γ-irradiation, respectively [33, 80]. For either of the aforementioned cases, the silk fibers were soaked in $AgNO_3$ solution and irradiated for at least 1 h, resulting in uniformly distributed Ag nanoparticles on the fiber surface. The process is highly tunable allowing for the density and size of the nanoparticles to be controlled by the initial concentration of $AgNO_3$ in the starting solution and irradiation time. Furthermore, these nanoparticles were strongly adhered to the surface and showed strong stability even after repeated washing cycles [33, 80]. In addition to their work with Ag nanoparticles, Lu et al. also showed similar antimicrobial results using CeO_2 nanoparticles immobilized on the surface of silk fabrics [32]. These fabrics showed strong antimicrobial behavior to *Escherichia coli*, *Staphylococcus aureus*, *Pseudomonas aeruginosa*, and *B. subtilis* (Figure 7.5A). Perhaps equally interesting, these fabrics also exhibited excellent UV-shielding properties, paving the way for multi-functional biopolymer–nanocomposite fibers.

Regardless of how nanoparticles are attached, their main purpose is to release ions to prevent bacterial growth on the fiber and/or destroy bacteria. The required strength of the adhesion of nanoparticles will vary depending on the application, but this is also what makes the biopolymer–nanocomposite fiber so robust. Take, for example, the work by Xu et al. with Ag–poly(lactic acid) nanocomposite fibers [110]. Their work showed a steady release of Ag nanoparticles over time indicating weak adhesion to the poly(lactic acid) fiber. The fibers did show very strong antimicrobial characteristics against *S. aureus* and *E. coli* (as high as 94–98% prevention of bacterial growth) for up to 20 days when even the poly(lactic acid) fiber degraded [110]. A short-term biodegradable nanofiber composite would be particularly useful in wound dressings or as a skin/surface drug delivery system. In contrast, examining work by He et al., development of Ag nanoparticles synthesized in porous cellulose was found to be especially stable over a much longer time period [107]. In addition, work by He et al. showed a remarkable antimicrobial effect on to *E. coli*, *S. aureus*, *Penicillium glaucum*, and *Saccharomyces cerevisiae*. This long-term stability is useful for applications involving implants or biological scaffolds. Simply by tuning the strength of adhesion between biopolymer and nanoparticle, a plethora of different applications can be found.

7.6.2 Magnetically responsive materials

Cellulose is a common polysaccharide used for magnetic-responsive materials. Notably, Biliuta et al. examined cellulose fibers doped with Fe_3O_4 nanoparticles (both uncoated and coated with oleic acid) using 2,2,6,6-(tetramethlpiperidin-1-yl)oxyl (TEMPO) as an oxidation method for attaching the nanoparticles to the fibers [88]. The authors found that oxidation of the fibers increased the activity and attachment of the magnetic nanoparticles even in the absence of the oleic acid coating, which prior to this discovery was often deemed necessary for attaching nanoparticles to cellulose fibers. This methodology indicates a robust method of attachment of magnetic nanoparticles without the need for stabilizers such as oleic acid. While the attachment density of noncoated Fe_3O_4 nanoparticles was lower than those of the coated particles, the uncoated particles showed a much higher magnetization value [88], thus paving the way forward for new methods of nanoparticle attachment in the absence of a stabilizer.

As another example of cellulose-based hybrid materials, cellulose acetate fibers were doped with yttrium iron garnet nanoparticles [66]. Munaweera et al. electrospun novel magnetic nonwoven mats using cellulose acetate and yttrium iron garnet nanoparticles. Nonwovens produced in this manner showed a magnetic response and could easily be picked up by an external magnet. Interestingly, when electrospun onto cellulose filter paper (Figure 7.5B), the mats retained their magnetic responsiveness and were capable of filtering fluorescently labeled bovine serum albumin protein from aqueous carbonate solutions (Figure 7.5C). These filter papers were found to separate from 0.0039 to 0.0069 μmol of protein per 1 cm^2 of filter paper. Furthermore, bovine serum albumin could then be recovered by washing the magnetic filter paper with deionized water [66]. While the amount of protein recovered may be small, this example highlights the more promising applications for magnetically responsive biopolymer meshes.

In a similar study using chitosan, the reverse effect was elucidated. Electrospun cross-linked chitosan nanofibers were added with Fe_3O_4 for targeted protein release. Bovine serum albumin was again used as a model protein, this time for drug delivery, and the release rate was found to be controllable by varying the cross-link density (by increasing the molarity of the cross-linker) and magnetic stimulation [60]. What is interesting is that both of these examples show opposite effects by changing the magnetic nanoparticle chemistry, thus indicating a potential application in magnetically assisted bioseparations and/or drug delivery simply by choosing a specific nanoparticle for the necessary application.

For protein hybrid materials, there appears to be only one recent example of magnetic nanoparticles used as composite materials in conjunction with collagen fibers. Therein, cross-linked collagen fibers were embedded with oleic-acid-coated iron oxide nanoparticles. These embedded fibers were found to be only weakly magnetic but remained stable in organic solutions of heptane. Perhaps most interesting, the collagen–iron oxide nanocomposite fiber was capable of rapid, efficient absorption of used motor oil from water [58], thus making these materials a potentially promising new absorption material for applications in environmental protection and cleanup industries. Aside from collagen, hemolysin protein derived from *P. aeruginosa* was used as a glue to connect $Co_2Fe_2O_4$ nanoparticles. These materials could later be lyophilized from solution and formed a fiber-like structure that aligned when exposed to a magnetic field [47, 64].

Silk-based proteins from silkworms and spiders also show scant literature (though more than most other proteins) [11, 42, 44, 89]. A recent publication by Zhou et al. reported that Fe_3O_4 nanoparticles could be embedded in silk fabrics through a swelling-fixing method [44]. The authors reported that these fabrics showed a decrease in crystallinity after nanoparticle embedding but their thermal and mechanical stability remained unaffected. Furthermore, the composite fabrics showed good magnetic responsiveness but the responsiveness decreased by ~15% after several washing cycles [44]. Despite the decrease in magnetic responsiveness after washing, the authors believe this was a good first step in determining a facile method of functionalizing silk textiles with magnetic nanoparticles.

A more commonly examined protein–nanoparticle biocomposite fiber found in literature is spider silk. Recombinant spider silk makes an ideal material for use with hybrid magnetically responsive nanoparticles, but only recently has a recombinant spider silk fiber been produced with a toughness matching that of naturally spun spider silk [22, 101]. It should come then as no surprise that after this breakthrough, the literature of spider silk–nanoparticle biocomposites has begun to grow. One of the more recent examples includes work by Liu et al. where they report a facile, environmentally friendly method to prepare Ag-Fe_3O_4 silk fiber nanocomposites with high antimicrobial activities against both *E. coli* and *S. aureus*. These materials also appear to be responsive at relatively low magnetic strengths (actuated with a small household magnet) [42]. What makes this work of particular interest is the ability to “regenerate” or recycle the released antimicrobial Ag ions. The authors suggest that due to the synergistic nature of the Ag-Fe_3O_4 nanoparticles, ions of Ag released upon contact with bacteria are later reabsorbed by the remaining Fe_3O_4 nanoparticles allowing for a regenerative material that is also magnetically responsive [42].

Similar work with the goal of environmentally friendly composite materials is ongoing within our group. A recent, 2018 paper by Herold et al. examined Au nanoparticle coupling with recombinant spider silk proteins for use in sustainable hydrogen production [114]. The authors demonstrated that spider silk with Au or TiO_2 binding moieties could be processed into films. These films could then be used for light-induced hydrolysis of water to produce H_2 gas [114]. Not only are these films responsive to external stimuli (in this case light) but also use materials that have been shown in literature to be antimicrobial (a topic touched upon previously). These results mark a seminal first step in the production of biopolymer–nanoparticle composite materials and have spawned several new projects examining different methods of producing biopolymer–nanoparticle composites.

Many of the other examinations of silk–magnetic nanoparticle biocomposites in literature still use natural spider silk fibers, typically obtained from the webs of spiders in nature or harvested from spiders themselves. Perhaps one of the earliest example of spider silk–magnetic biocomposites is work with MAS from *Nephila edulis* and superparamagnetic magnetite (Fe_3O_4) nanoparticles. The inaugural work indicated that spider silk fibers could be easily coated with magnetic (and nonmagnetic) nanoparticles via dip coating the fibers in a colloidal solution of nanoparticles. Furthermore, these silk fibers retained their mechanical toughness and elasticity even after addition of magnetite particles to the fiber scaffold. However, it was also noted that the coating of Fe_3O_4 nanoparticles was susceptible to delamination upon mechanical flexing, which may explain why there is no adverse effect to the mechanical properties of the dragline spider silk fiber [11]. While techniques and methodology for nanoparticle attachment have improved over the past 20 years, this remains a seminal starting point to magnetic nanoparticle biocomposites.

More advanced methods of attaching magnetic nanoparticles to natural spider silk are demonstrated in the work by Singh et al., which used spider silk webs from *Crossopriza lyoni* and Fe_3O_4 nanoparticles [89]. Their work is unique in that it uses ionic liquids and deep eutectic solvents for the dispersion of spider silk and magnetic nanoparticles. The dispersions of silk and magnetic nanoparticles indicated nanoscale structural distribution and rapid attachment of the Fe_3O_4 nanoparticles. Regenerated fibers from these dispersions did not appear to inhibit the growth of mammalian cells *in vitro* and appear to show antimicrobial activity against *E. coli*, *Pseudomonas stutzeri*, and *Bacillus licheniformis* [89].

Similar, ongoing work by Grill et al. successfully electrospun meshes of recombinant spider silk protein (derived from *Araneus diadematus*, see also Figure 7.1A) in a hexafluoroisopropanol, HFIP, solution (Figure 7.6A) and successfully attached Au-coated magnetic nanoparticles to the fiber mesh via dip coating (Figure 7.6B). A high affinity between the recombinant spider silk and the Au magnetic nanoparticles was observed in the fiber meshes. Unfortunately, the low solubility of the Au nanoparticles in the HFIP dope solution prevented electrospinning from a single dope solution. In an effort to circumvent this low solubility, other functionalized magnetic nanoparticles with various shapes and sizes were explored. These included but are not limited to: hematite iron oxide spindles synthesized by the Wagner group (University of Rostock), biophosphate-coated

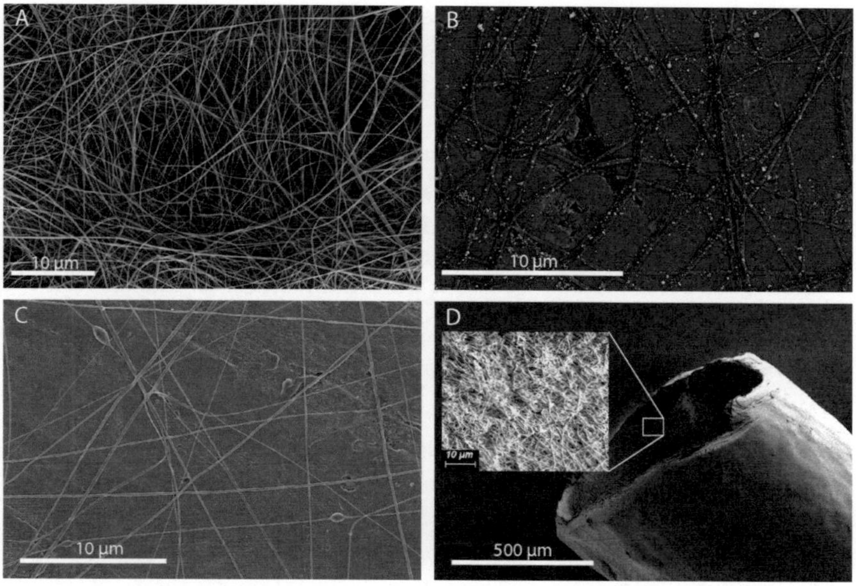

Figure 7.6: SEM images of electrospun fiber meshes of recombinant spider silk (derived from *Araneus diadematus*) (A) without Au-coated magnetic nanoparticles and (B) with Au-coated magnetic nanoparticles following dip coating. The bright clusters located on the fiber mesh indicate high affinity between the recombinant spider silk and the gold-coated magnetic nanoparticles. Because the Au nanoparticles were insoluble in the HFIP solutions, attempts at improving solubility were tested with other coated nanoparticles. The SEM images of electrospun fiber meshes of recombinant spider silk (C) with hematite iron oxide spindles indicate a somewhat successful single pot production of nanoparticle-doped spider silk meshes. Building upon these results a recombinant spider silk mesh was applied to a chitosan film. The resulting film indicated no loss of rollability after the mesh was applied (D). The inset of (D) indicates that the electrospun spider silk mesh is present on the inner layer of the chitosan roll, but it is not possible at this magnification to determine the presence of nanoparticles.

magnetite spheres synthesized by the Dutz group (University of Ilmenau), and gelatin-coated nickel magnetic rods synthesized by the Tschöpe group (University of Saarland). Each nanoparticle was electrospun into a magnetic mesh to varying degrees of success. The hematite iron oxide spindles did cluster on the spider silk mesh but only sparsely and as large aggregates (Figure 7.6C). This is likely due to the low solubility of the hematite iron oxide in the HFIP solutions. The biophosphate-coated magnetite and gelatin-coated nickel magnetic nanoparticles did show higher solubility in the dope solution but did not adhere to fiber meshes. While not completely solving the issues of solubility, a magnetic mesh from a single dope solution did allow for applications with previous materials developed by Aigner et al. [115]. Using "rollable" chitosan films previously described in literature, spider silk–magnetic nanoparticle meshes were successfully electrospun onto the surface of chitosan films (Figure 7.6D) though it is unclear from the SEM images if the nanoparticles remain adhered after rolling. The thickness of the mesh varied depending on the spinning time, which ranged from 30 s to 7 min. Surprisingly, the mesh thickness did not have an effect on the rolling ability of the chitosan films, thus providing a potential posttreatment method for producing magnetically responsive materials without sacrificing the initial material elasticity.

It is important to note that the aforementioned examples are the first few of a rapidly growing field. Whether it be for use as therapeutic, structural, or many of the other applications, spider silk–nanoparticle biocomposites are quickly becoming an interesting case study for magnetic-responsive materials.

7.7 Summary and future directions

In addition to antimicrobial properties, nanoparticles offer many other benefits in combination with biopolymer fibers, most notably their thermal absorption characteristics. In fact, many of the mentioned examples not only showed enhanced antimicrobial or magnetic properties of the biopolymer–nanocomposite fibers, but also the thermal stability was significantly increased in the presence of metal nanoparticles [32, 33, 77, 78, 98, 99]. In addition to high thermal stabilities, many nanoparticles have a wide range of dielectrics. This makes them appealing materials for use in organic electronics. Recently, work by Deshmukh et al. has examined SiO_2 nanoparticle reinforced poly(vinyl alcohol) and poly(vinyl pyrrolidone) blends for use in flexible organic electronics [49], adding to the growing list of uses for biopolymer–nanoparticle composites.

Due to the many advantageous properties of biopolymer nanocomposites, it should come as no surprise that addition of nanoparticles brings with it multiple benefits. These benefits allow for a variety of applications into many different fields, some of which yield surprising results. A niche but growing application for nanoparticles is their using in catalytic systems. As an example, Ouyang et al. created a hollow polyethersulfone fiber with Au nanoparticles embedded inside the hollow fiber for use in catalysis reactions [116]. Though polyethersulfone is not a biopolymer, similar work has been examined using biopolymers and nanoparticles. In addition to work on antimicrobial fibers, both He et al. and Xu et al. examined attached metallic nanoparticles as catalytic reaction sites [107, 110]. The embedded Au nanoparticles could then act as a catalytic reduction site for 4-nitrophenol when soaked in a solution of $NaBH_4$. More recently, Marks et al. and Kamal et al. have developed systems of cellulose fibers embedded with metal nanoparticles for the removal of contaminants in water [98, 117]. For the former case, Marks et al. used the weakly magnetic Pd nanoparticles embedded in hollow cellulose fibers, which later were added into a hydrogel membrane as channels for water treatment applications [117]. The latter case, Kamal et al. successfully embedded stabilized ferromagnetic Co nanoparticles in bacterial cellulose, which showed a very high affinity for test pollutants, 2,6-dinitrophenol, and methylene blue dye. These nanoparticles embedded in bacterial cellulose fibers continued to be very stable after five catalytic cycles, showing no significant difference in performance after each cycle [98].

Responsive materials, purification, filtration, antimicrobial properties, and bioreactors are but a few of the growing applications for magnetically doped biopolymer composites The applications for magnetically responsive biocomposite materials will only grow as we learn more about the interaction between biopolymers and nanoparticles. Given the fact that many nanoparticles (and biopolymers) have multiple desired material properties, it is likely that literature will start to see more multipurpose materials. With an increased emphasis on biocompatibility, recyclability, and "green" environmentally friendly materials, we expect to see a continued shift away from synthetic polymer materials. As we have described in this review article, while still early in development, magnetically responsive biocomposite materials can be used for many applications, and the future looks very promising for their continued study.

Acknowledgments: Funding was generously provided by the German Research Foundation (DFG) within the priority program SPP 1681. The authors would like to thank Carolin Grill for her help with the nanoparticle information comprised in Table 7.1.
Author contribution: All the authors have accepted responsibility for the entire content of this submitted manuscript and approved submission.
Research funding: This work was supported by Deutsche Forschungsgemeinschaft (SPP 1681).
Conflict of interest: The authors declare no conflict of interest.

References

1. Bealer EJ, Kavetsky K, Dutko S, Lofland S, Hu X. Protein and polysaccharide-based magnetic composite materials for medical applications. Int J Mol Sci 2019;21:186.
2. Ravi Kumar MNV. A review of chitin and chitosan applications. React Funct Polym 2000;46:1–27.
3. Perotto G, Cittadini M, Tao H, Kim S, Yang M, Kaplan DL, et al. Fabrication of tunable, high-refractive-index titanate-silk nanocomposites on the micro-and nanoscale. Adv Mater 2015;27:6728–32.
4. Humenik M, Lang G, Scheibel T. Silk nanofibril self-assembly versus electrospinning. Wiley Interdiscip Rev Nanomed Nanobiotechnol 2018;10:e1509.
5. Huang W, Ling S, Li C, Omenetto FG, Kaplan DL. Silkworm silk-based materials and devices generated using bio-nanotechnology. Chem Soc Rev 2018;47:6486–504.
6. Sadasivuni KK, Saha P, Adhikari J, Deshmukh K, Ahamed MB, Cabibihan J-J. Recent advances in mechanical properties of biopolymer composites: a review. Polym Compos 2020;41:32–59.
7. Li N, Wang Y, Xu K, Huang Y, Wen Q, Ding X. Development of green betaine-based deep eutectic solvent aqueous two-phase system for the extraction of protein. Talanta 2016;152:23–32.
8. Xu K, Wang Y, Huang Y, Li N, Wen Q. A green deep eutectic solvent-based aqueous two-phase system for protein extracting. Anal Chim Acta 2015;864:9–20.

9. Mazzola PG, Lopes AM, Hasmann FA, Jozala AF, Penna TCV, Magalhaes PO, et al. Liquid–liquid extraction of biomolecules: an overview and update of the main techniques. J Chem Technol Biotechnol 2008;83:143–57.
10. Shamshina JL, Barber PS, Gurau G, Griggs CS, Rogers RD. Pulping of Crustacean waste using ionic liquids: to extract or not to extract. ACS Sustain Chem Eng 2016;4:6072–81.
11. Mayes EL, Vollrath F, Mann S. Fabrication of magnetic spider silk and other silk-fiber composites using inorganic nanoparticles. Adv Mater 1998;10:801–5.
12. Junghans F, Morawietz M, Conrad U, Scheibel T, Heilmann A, Spohn U. Preparation and mechanical properties of layers made of recombinant spider silk proteins and silk from silk worm. Appl Phys A Mater Sci Process 2006;82:253–60.
13. Keyes JD, Hilton RJ, Farrer J, Watt RK. Ferritin as a photocatalyst and scaffold for gold nanoparticle synthesis. J Nanopart Res 2011;13:2563–75.
14. Lagerwall JPF, Schütz C, Salajkova M, Noh JH, Park JH, Scalia G, et al. Cellulose nanocrystal-based materials: from liquid crystal self-assembly and glass formation to multifunctional thin films. NPG Asia Mater 2014;6:e80.
15. Kalia S, Dufresne A, Cherian BM, Kaith BS, Avérous L, Njuguna J, et al. Cellulose-based bio- and nanocomposites: a review. Int J Polym Sci 2011;2011:837875.
16. Wang D, Yang X-H, Tang RC, Yao F. Extraction of keratin from rabbit hair by a deep eutectic solvent and its characterization. Polymers 2018;10:993.
17. Klemm D, Heublein B, Fink HP, Bohn A. Cellulose: fascinating biopolymer and sustainable raw material. Angew Chem Int Ed Engl 2005;44:3358–93.
18. Cai H, Yao P. In situ preparation of gold nanoparticle-loaded lysozyme–dextran nanogels and applications for cell imaging and drug delivery. Nanoscale 2013;5:2892–900.
19. Yu Z, Yu M, Zhang Z, Hong G, Xiong Q. Bovine serum albumin nanoparticles as controlled release carrier for local drug delivery to the inner ear. Nanoscale Res Lett 2014;9:343.
20. Steiert E, Ewald J, Wagner A, Hellmich UA, Frey H, Wich PR. pH-Responsive protein nanoparticles via conjugation of degradable PEG to the surface of cytochrome *c*. Polym Chem 2020;11:551–9.
21. Lang G, Jokisch S, Scheibel T. Air filter devices including nonwoven meshes of electrospun recombinant spider silk proteins. J Vis Exp 2013;75:e50492.
22. Anton AM, Heidebrecht A, Mahmood N, Beiner M, Scheibel T, Kremer F. Foundation of the outstanding toughness in biomimetic and natural spider silk. Biomacromolecules 2017;18: 3954–62.
23. Gupta P, Nayak KK. Compatibility study of alginate/keratin blend for biopolymer development. J Appl Biomater Funct Mater 2015;13:E332–9.
24. Guo J, Wang X, Miao P, Liao X, Zhang W, Shi B. One-step seeding growth of controllable Ag@Ni core–shell nanoparticles on skin collagen fiber with introduction of plant tannin and their application in high-performance microwave absorption. J Mater Chem 2012;22:11933–42.
25. He L, Gao S-Y, Wu H, Liao X-P, He Q, Shi B. Antibacterial activity of silver nanoparticles stabilized on tannin-grafted collagen fiber. Mater Sci Eng C 2012;32:1050–6.
26. Sutherland TD, Young JH, Weisman S, Hayashi CY, Merritt DJ. Insect silk: one name, many materials. Annu Rev Entomol 2010;55:171–88.
27. Peters MA. The ancient silk road and the birth of merchant capitalism. Educ Philos Theor 2019:1–7. https://doi.org/10.1080/00131857.2019.1691481.
28. Scheibel T, Zahn H, Krasowski A. Silk. In: Ullmann's encyclopedia of industrial chemisty. Weinheim: Wiley-VCH; 2016:1–15 pp.
29. Shao ZZ, Vollrath F. Materials: surprising strength of silkworm silk. Nature 2002;418:741.
30. Kundu J, Chung Y-I, Kim YH, Tae G, Kundu SC. Silk fibroin nanoparticles for cellular uptake and control release. Int J Pharm 2010;388:242–50.

31. Li G, Liu H, Zhao H, Gao Y, Wang J, Jiang H, et al. Chemical assembly of TiO_2 and TiO_2@Ag nanoparticles on silk fiber to produce multifunctional fabrics. J Colloid Interface Sci 2011;358: 307–15.
32. Lu Z, Mao C, Meng M, Liu S, Tian Y, Yu L, et al. Fabrication of CeO_2 nanoparticle-modified silk for UV protection and antibacterial applications. J Colloid Interface Sci 2014;435:8–14.
33. Lu Z, Meng M, Jiang Y, Xie J. UV-assisted in situ synthesis of silver nanoparticles on silk fibers for antibacterial applications. Colloid Surface Physicochem Eng Aspect 2014;447:1–7.
34. Smith A, Scheibel T. CHAPTER 12 Hierarchical protein assemblies as a basis for materials. Materials design inspired by nature: function through inner architecture. Cambridge, UK: The Royal Society of Chemistry; 2013:256–81 pp.
35. Lintz ES, Scheibel TR. Dragline, egg stalk and byssus: a comparison of outstanding protein fibers and their potential for developing new materials. Adv Funct Mater 2013;23:4467–82.
36. Heim M, Keerl D, Scheibel T. Spider silk: from soluble protein to extraordinary fiber. Angew Chem Int Ed 2009;48:3584–96.
37. Doblhofer E, Heidebrecht A, Scheibel T. To spin or not to spin: spider silk fibers and more. Appl Microbiol Biotechnol 2015;99:9361–80.
38. Heidebrecht A, Scheibel T. Chapter four – recombinant production of spider silk proteins. In: Sariaslani S, Gadd GM, editors. Advances in applied microbiology. Oxford, UK: Academic Press; 2013:115–53 pp.
39. Sau TK, Rogach AL. Complex-shaped metal nanoparticles. Singapore: Wiley-VCH Verlag GmbH & Co. KGaA; 2012.
40. Kulkarni N, Muddapur U. Biosynthesis of metal nanoparticles: a review. J Nanotechnol 2014; 2014:1–8.
41. Pan Y, Du X, Zhao F, Xu B. Magnetic nanoparticles for the manipulation of proteins and cells. Chem Soc Rev 2012;41:2912–42.
42. Liu X, Yin G, Yi Z, Duan T. Silk fiber as the support and reductant for the facile synthesis of Ag-$Fe_{(3)}O_{(4)}$ nanocomposites and its antibacterial properties. Materials 2016;9:501.
43. Singh A, Hede S, Sastry M. Spider silk as an active scaffold in the assembly of gold nanoparticles and application of the gold–silk bioconjugate in vapor sensing. Small 2007;3:466–73.
44. Zhou J, Zhao Z, Hu R, Fan T, Liu Y, Lu M. Magnetic silk fabrics through swelling-fixing method with Fe_3O_4 nanoparticles. Surf Coating Technol 2018;342:23–8.
45. Jokisch S, Neuenfeldt M, Scheibel T. Silk-based fine dust filters for air filtration. Adv Sustain Syst 2017;1:1700079.
46. Jokisch S, Scheibel T. Spider silk foam coating of fabric. Pure Appl Chem 2017;89:1769–76.
47. Pham TA, Schreiber A, Sturm Nee Rosseeva EV, Schiller S, Colfen H. Hemolysin coregulated protein 1 as a molecular gluing unit for the assembly of nanoparticle hybrid structures. Beilstein J Nanotechnol 2016;7:351–63.
48. Alshehri AH, Jakubowska M, Młożniak A, Horaczek M, Rudka D, Free C, et al. Enhanced electrical conductivity of silver nanoparticles for high frequency electronic applications. ACS Appl Mater Interfaces 2012;4:7007–10.
49. Deshmukh K, Ahamed MB, Sadasivuni KK, Ponnamma D, AlMaadeed MA-A, Deshmukh RR, et al. Fumed SiO_2 nanoparticle reinforced biopolymer blend nanocomposites with high dielectric constant and low dielectric loss for flexible organic electronics. J Appl Polym Sci 2017;134:44427.
50. Stojak K, Srikanth H, Mukherjee P, Phan M-H, Thanh NTK. Size- and shape-variant magnetic metal and metal oxide nanoparticles: synthesis and properties. In: Sau TK, Rogach AL, editors. Complex-shaped metal nanoparticles: bottom-up syntheses and application. Weinheim, DE: Wiley-VCH; 2012.
51. Issa B, Obaidat IM. Magnetic nanoparticles as MRI contrast agents. In: Manchev L, editor. Magnetic resonance imaging. London, UK: IntechOpen Limited; 2019.

52. Xu P, Han X, Wang M. Synthesis and magnetic properties of $BaFe_{12}O_{19}$ hexaferrite nanoparticles by a reverse microemulsion technique. J Phys Chem C 2007;111:5866–70.
53. Das A, Roychowdhury A, Pati SP, Bandyopadhyay S, Das D. Structural, magnetic and hyperfine properties of single-phase $SrFe_{12}O_{19}$ nanoparticles prepared by a sol–gel route. Phys Scripta 2015; 90:025802.
54. Gupta P, Asmatulu R, Claus R, Wilkes G. Superparamagnetic flexible substrates based on submicron electrospun Estane® fibers containing MnZnFe.Ni nanoparticles. J Appl Polym Sci 2006; 100:4935–42.
55. Vinosha PA, Mely LA, Jeronsia JE, Krishnan S, Das SJ. Synthesis and properties of spinel $ZnFe_2O_4$ nanoparticles by facile co-precipitation route. Optik 2017;134:99–108.
56. Manikandan V, Vanitha A, Ranjith Kumar E, Chandrasekaran J. Effect of in substitution on structural, dielectric and magnetic properties of $CuFe_2O_4$ nanoparticles. J Magn Magn Mater 2017; 432:477–83.
57. Jeon YT, Moon JY, Lee GH, Park J, Chang Y. Comparison of the magnetic properties of metastable hexagonal close-packed Ni nanoparticles with those of the stable face-centered cubic Ni nanoparticles. J Phys Chem B 2006;110:1187–91.
58. Alliraja C, Rao JR, Thanikaivelan P. Magnetic collagen fibers stabilized using functional iron oxide nanoparticles in non-aqueous medium. RSC Adv 2015;5:20939–44.
59. Liu F, Ni Q-Q, Murakami Y. Preparation of magnetic polyvinyl alcohol composite nanofibers with homogenously dispersed nanoparticles and high water resistance. Textil Res J 2012;83:510–8.
60. Nicknejad ET, Ghoreishi SM, Habibi N. Electrospinning of cross-linked magnetic chitosan nanofibers for protein release. AAPS PharmSciTech 2015;16:1480–6.
61. Perera AS, Zhang S, Homer-Vanniasinkam S, Coppens MO, Edirisinghe M. Polymer-magnetic composite fibers for remote-controlled drug release. ACS Appl Mater Interfaces 2018;10:15524–31.
62. Wang J-T, Li L-L, Feng L, Li J-F, Jiang L-H, Shen Q. Directly obtaining pristine magnetic silk fibers from silkworm. Int J Biol Macromol 2014;63:205–9.
63. Kriha O, Becker M, Lehmann M, Kriha D, Krieglstein J, Yosef M, et al. Connection of hippocampal neurons by magnetically controlled movement of short electrospun polymer fibers—a route to magnetic micromanipulators. Adv Mater 2007;19:2483–5.
64. Pham TA, Schreiber A, Schiller SM, Cölfen H. Toroidal protein adaptor assembles ferrimagnetic nanoparticle fibers with constructive magnetic coupling. Adv Funct Mater 2017;27:1604532.
65. Song T, Zhang Y, Zhou T, Lim CT, Ramakrishna S, Liu B. Encapsulation of self-assembled FePt magnetic nanoparticles in PCL nanofibers by coaxial electrospinning. Chem Phys Lett 2005; 415:317–22.
66. Munaweera I, Aliev A, Balkus KJ Jr. Electrospun cellulose acetate-garnet nanocomposite magnetic fibers for bioseparations. ACS Appl Mater Interfaces 2014;6:244–51.
67. Chawla KK. Fibrous reinforcements for composites: overview. In: Buschow KHJ, Cahn KHJ, Flemings KHJ, Mahajan KHJ, Ilschner KHJ, Kramer KHJ, editors. Encyclopedia of materials: science and technology. Oxford: Elsevier; 2001:3160–7 pp.
68. Chen X, Cai H, Ling S, Shao Z, Huang Y. Conformation transition of Bombyx mori silk protein monitored by time-dependent fourier transform infrared (FT-IR) spectroscopy: effect of organic solvent. Appl Spectrosc 2012;66:696–9.
69. Rammensee S, Slotta U, Scheibel T, Bausch AR. Assembly mechanism of recombinant spider silk proteins. Proc Natl Acad Sci USA 2008;105:6590–5.
70. Sundaray B, Subramanian V, Natarajan TS. Electrospinning of continuous aligned polymer fibers. Appl Phys Lett 2004;84:1222–4.
71. Garg K, Bowlin GL. Electrospinning jets and nanofibrous structures. Biomicrofluidics 2011;5: 13403.
72. Reneker DH, Yarin AL. Electrospinning jets and polymer nanofibers. Polymer 2008;49:2387–425.

73. Greiner A, Wendorff JH. Elektrospinnen: eine faszinierende methode zur präparation ultradünner fasern. Angew Chem 2007;119:5770–805.
74. Zhang J, Yan S, Yuan D, Alici G, Nguyen N-T, Warkiani ME, et al. Fundamentals and applications of inertial microfluidics: a review. Lab Chip 2016;16:10–34.
75. Haynl C, Hofmann E, Pawar K, Forster S, Scheibel T. Microfluidics-produced collagen fibers show extraordinary mechanical properties. Nano Lett 2016;16:5917–22.
76. Suh YK, Kang S. A review on mixing in microfluidics. Micromachines 2010;1:82–111.
77. Calamak S, Aksoy EA, Ertas N, Erdogdu C, Sagıroglu M, Ulubayram K. Ag/silk fibroin nanofibers: effect of fibroin morphology on Ag+ release and antibacterial activity. Eur Polym J 2015;67:99–112.
78. Tang B, Sun L, Kaur J, Yu Y, Wang X. In-situ synthesis of gold nanoparticles for multifunctionalization of silk fabrics. Dyes Pigments 2014;103:183–90.
79. Abdelwahab NA, Shukry N. Synthesis, characterization and antimicrobial properties of grafted sugarcane bagasse/silver nanocomposites. Carbohydr Polym 2015;115:276–84.
80. Chang S, Kang B, Dai Y, Chen D. Synthesis of antimicrobial silver nanoparticles on silk fibers via γ-radiation. J Appl Polym Sci 2009;112:2511–5.
81. Son WK, Youk JH, Lee TS, Park WH. Preparation of antimicrobial ultrafine cellulose acetate fibers with silver nanoparticles. Macromol Rapid Commun 2004;25:1632–7.
82. Ahmad S, Sultan A, Mohammad F. Electrically conductive polyaniline/silk fibroin composite for ammonia and acetaldehyde sensing. Polym Polym Compos 2018;26:177–87.
83. Huang XY, Wang S, Zhu M, Yang K, Jiang P, Bando YB, et al. Thermally conductive, electrically insulating and melt-processable polystyrene/boron nitride nanocomposites prepared by in situ reversible addition fragmentation chain transfer polymerization. Nanotechnology 2015;26:015705.
84. Ji YM, Li YZ, Chen GQ, Xing TL. Fire-resistant and highly electrically conductive silk fabrics fabricated with reduced graphene oxide via dry-coating. Mater Des 2017;133:528–35.
85. Jung R, Kim H-S, Kim Y, Kwon S-M, Lee HS, Jin H-J. Electrically conductive transparent papers using multiwalled carbon nanotubes. J Polym Sci B Polym Phys 2008;46:1235–42.
86. Varesano A, Antognozzi B, Tonin C. Electrically conducting-adhesive coating on polyamide fabrics. Synth Met 2010;160:1683–7.
87. Rojas J, Bedoy M, Ciro Y. *Current trends in the production of cellulose nanoparticles and nanocomposites for biomedical applications*. London: IntechOpen Limited; 2005.
88. Biliuta G, Coseri S. Magnetic cellulosic materials based on TEMPO-oxidized viscose fibers. Cellulose 2016;23:3407–15.
89. Singh N, Mondal D, Sharma M, Devkar RV, Dubey S, Prasad K. Sustainable processing and synthesis of nontoxic and antibacterial magnetic nanocomposite from spider silk in neoteric solvents. ACS Sustain Chem Eng 2015;3:2575–81.
90. Deng Y, Ediriwickrema A, Yang F, Lewis J, Girardi M, Saltzman WM. A sunblock based on bioadhesive nanoparticles. Nat Mater 2015;14:1278–85.
91. Goldstein JI, Newbury DE, Michael JR, Ritchie NWM, Scott JHJ, Joy DC. Scanning electron microscopy and X-ray microanalysis, 4th ed. New York, NY, U.S.A.: Springer; 2018.
92. Nasrollahzadeh M, Atarod M, Sajjadi SM, Issaabadi Z. Chapter 6 – plant-mediated green synthesis of nanostructures: mechanisms, characterization, and applications. In: Nasrollahzadeh M, Mohammad Sajadi S, Sajjadi M, Issaabadi Z, Atarod M, editors. Interface science and technology. Oxford, UK: Elsevier; 2019:199–322 pp.
93. Carapeto PA, Ferraria MA, Botelho do Rego MA. Silver nanoparticles on cellulose surfaces: quantitative measurements. Nanomaterials 2019;9:780.
94. Levin BD, Padgett E, Chen C-C, Scott MC, Xu R, Theis W, et al. Nanomaterial datasets to advance tomography in scanning transmission electron microscopy. Sci Data 2016;3:160041.

95. Suhre MH, Hess S, Golser AV, Scheibel T. Influence of divalent copper, manganese and zinc ions on fibril nucleation and elongation of the amyloid-like yeast prion determinant Sup35p-NM. J Inorg Biochem 2009;103:1711–20.
96. Humenik M, Mohrand M, Scheibel T. Self-Assembly of spider silk-fusion proteins comprising enzymatic and fluorescence activity. Bioconjugate Chem 2018;29:898–904.
97. Wunderlich B. Thermal analysis of polymeric materials. In: Jürgen Buschow KH, Cahn RW, Flemings MC, Ilschner B, Kramer EJ, Mahajan S, et al., editors. Encyclopedia of materials: science and technology. Oxford: Elsevier; 2001:9134–41 pp.
98. Kamal T, Ahmad I, Khan SB, Asiri AM. Bacterial cellulose as support for biopolymer stabilized catalytic cobalt nanoparticles. Int J Biol Macromol 2019;135:1162–70.
99. Li R, He M, Li T, Zhang L. Preparation and properties of cellulose/silver nanocomposite fibers. Carbohydr Polym 2015;115:269–75.
100. Stone R, Hipp S, Barden J, Brown PJ, Mefford OT. Highly scalable nanoparticle-polymer composite fiber via wet spinning. J Appl Polym Sci 2013;130:1975–80.
101. Heidebrecht A, Eisoldt L, Diehl J, Schmidt A, Geffers M, Lang G, et al. Biomimetic fibers made of recombinant spidroins with the same toughness as natural spider silk. Adv Mater 2015;27:2189–94.
102. Zhou G, Wang Y-Q, Byun J-H, Yi J-W, Yoon S-S, Cha H-J, et al. High strenght single walled carbon nanotube/permalloy nanoparticle/poly(vinyl alcohol) multifunctional nanocompsite fiber. ACS Nano 2015;9:11414–21.
103. Chwala A, Smit JP, Stolz R, Zakosarenko V, Schmelz M, Fritzsch L, et al. Low temperature SQUID magnetometer systems for geophysical exploration with transient electromagnetics. Supercond Sci Technol 2011;24:125006.
104. Colclough MS, Gough CE, Keene M, Muirhead CM, Thomas N, Abell JS, et al. Radio-frequency SQUID operation using a ceramic high-temperature superconductor. Nature 1987;328:47–9.
105. Lee LP, Char K, Colclough MS, Zaharchuk G. Monolithic 77 K dc SQUID magnetometer. Appl Phys Lett 1991;59:3051–3.
106. Chen C-Y, Chiang C-L. Preparation of cotton fibers with antibacterial silver nanoparticles. Mater Lett 2008;62:3607–9.
107. He J, Kunitake T, Nakao A. Facile in situ synthesis of noble metal nanoparticles in porous cellulose fibers. Chem Mater 2003;15:4401–6.
108. Lee SJ, Heo DN, Moon JH, Ko WK, Lee JB, Bae MS, et al. Electrospun chitosan nanofibers with controlled levels of silver nanoparticles. Preparation, characterization and antibacterial activity. Carbohydr Polym 2014;111:530–7.
109. Zhao Y, Zhou Y, Wu X, Wang L, Xu L, Wei S. A facile method for electrospinning of Ag nanoparticles/poly (vinyl alcohol)/carboxymethyl-chitosan nanofibers. Appl Surf Sci 2012;258: 8867–73.
110. Xu X, Yang Q, Wang Y, Yu H, Chen X, Jing X. Biodegradable electrospun poly(l-lactide) fibers containing antibacterial silver nanoparticles. Eur Polym J 2006;42:2081–7.
111. Celebioglu A, Aytac Z, Umu OCO, Dana A, Tekinay T, Uyar T. One-step synthesis of size-tunable Ag nanoparticles incorporated in electrospun PVA/cyclodextrin nanofibers. Carbohydr Polym 2014; 99:808–16.
112. Lu Z, Xiao J, Wang Y, Meng M. In situ synthesis of silver nanoparticles uniformly distributed on polydopamine-coated silk fibers for antibacterial application. J Colloid Interface Sci 2015;452:8–14.
113. Dolati A, Imanieh I, Salehi F, Farahani M. The effect of cysteine on electrodeposition of gold nanoparticle. Mater Sci Eng B 2011;176:1307–12.
114. Herold HM, Aigner TB, Grill CE, Krüger S, Taubert A, Scheibel T. SpiderMAEn: recombinant spider silk-based hybrid materials for advanced energy technology. Bioinspired, Biomimetic Nanobiomaterials 2019;8:99–108.

115. Aigner TB, Haynl C, Salehi S, O'Connor A, Scheibel T. Nerve guidance conduit design based on self-rolling tubes. Mater Today Bio 2020;5:100042.
116. Ouyang L, Dotzauer DM, Hogg SR, Macanás J, Lahitte J-F, Bruening ML. Catalytic hollow fiber membranes prepared using layer-by-layer adsorption of polyelectrolytes and metal nanoparticles. Catal Today 2010;156:100–6.
117. Marks R, Seaman J, Perez-Calleja P, Kim J, Nerenberg R, Doudrick K. Catalytic hydrogel membrane reactor for treatment of aqueous contaminants. Environ Sci Technol 2019;53:6492–500.

Alexey Eremin*, Hajnalka Nádasi and Ralf Stannarius*

8 Multifunctionality by dispersion of magnetic nanoparticles in anisotropic matrices

Abstract: Interactions between magnetic nanoparticles and an anisotropic environment give rise to a variety of new magneto-optical, rheological and mechanical phenomena. This opens new avenues for developing novel multifunctional materials. In the course of this project, we investigated three types of anisotropic systems: dispersions of shape-anisotropic nanocrystals, magnetically doped molecular and colloidal liquid crystals, and organoferrogels. They were investigated by means of magneto-optical observations and by a magneto-mechanical torsion pendulum method.

Keywords: magnetic gels, magnetic suspensions, liquid crystals

8.1 Aims and concepts

Magnetic fluids are composed of nanometre- to micrometre-sized solid particles in a liquid carrier. A combination of the magnetic properties of the dispersed particles with the fluidity of the matrix enables the development of broadly applicable magneto-responsive materials. In these fluids, interactions with external magnetic fields occur with the individual magnetic particles (MPs). Thus, the interaction energies are not large with respect to thermal energies, in contrast to magnetic solids of the same material as the MPs. In the absence of external magnetic fields, ferrofluids lack a remanent magnetization, they behave like superparamagnets. Confining the magnetic liquid phase in a viscoelastic environment by gelation brings about mechanical and diffusive constraints resulting in materials with distinct sensitivity to mechanical stress and/or magnetic field.

Liquid-crystalline phases are characterized by spontaneous collective orientational order of the molecules in certain temperature or concentration ranges. However, the common liquid crystal (LC) mesogens are diamagnetic, and their interactions with magnetic fields in the mesophases are governed by the anisotropy of the diamagnetic susceptibility of the order of 10^{-6}. The collective arrangement of the mesogens allows us to switch their preferential direction, the director $\vec{n}$, even in weak magnetic fields of

***Corresponding authors: Alexey Eremin,** Otto von Guericke University, Institute of Physics, Magdeburg, Germany, E-mail: alexey.eremin@ovgu.de; and **Ralf Stannarius,** Otto von Guericke University, Institute of Physics, Magdeburg, Germany, E-mail: ralf.stannarius@ovgu.de
Hajnalka Nádasi, Otto von Guericke University, Institute of Physics, Magdeburg, Germany

This article has previously been published in the journal Physical Sciences Reviews. Please cite as: A. Eremin, H. Nádasi and R. Stannarius "Multifunctionality by dispersion of magnetic nanoparticles in anisotropic matrices" *Physical Sciences Reviews* [Online] 2020, 5. DOI: 10.1515/psr-2019-0111 | https://doi.org/10.1515/9783110569636-008

a few hundreds mT. This allows to employ magnetic fields for magneto-optical or magnetomechanical switching even without MNPs.

It is a promising but also challenging concept to combine these two areas of soft-matter physics, magnetic fluids and LCs, to create new classes of multifunctional materials that combine a strong magnetic response with collective magneto-optical and magnetomechanical effects. A promising strategy is the dispersion of functionalized MNPs in anisotropic environment such as suspensions of shape-anisotropic (anisometric) microparticles, lyotropic nematic matrices or even thermotropic LCs. We describe how interactions of MNPs with complex anisotropic matrices drive self-assembly and affects the physical properties of the materials. Primary focus is the influence of the matrix structure on the magneto-optical and viscoelastic properties of the composites. An external magnetic field triggers self-organization of the MNPs, which affects the anisotropic environment (matrix). The feedback of the matrix determines the structure formation dynamics of the magnetic subphase, as well as optical and mechanical properties. We examine three different matrix systems: non-MPs (pigment crystallites) dispersed in an isotropic liquid, organic LCs, and organogels with filamentary internal structure. A special aspect is the transition from isotropic to anisotropic matrices.

The following section discusses mixtures of magnetic nanoparticles with shape-anisotropic (rod-like and plate-like) nanocrystallites in suspension. These suspensions may form lyotropic phases at larger particle concentrations, but even at low concentrations the steric interactions with codispersed MNPs enhance the macroscopic response to magnetic fields considerably. The idea is to exploit these interactions to command the larger nonmagnetic constituents via magnetic torques of external fields on MNPs and agglomerations of the latter. In Section 8.3, we briefly describe attempts to suspend surface-functionalized MNPs in thermotropic nematic LCs. Here, the mesogens are thermodynamically stable LC mesophases. The surface-functionalized MNPs interact with the director field to transfer magnetic alignment to the nematic host. In Section 8.4, the complexity is increased by physical cross-linking of the host phase. This leads to self-organized arrangements of the MNPs in the gel matrix and to the dependence of mechanical and optical sample parameters on the magnetic history of the material.

In the final section, we introduce a mechanical characterization technique that can be employed to all of these systems. It can be used to characterize magnetomechanical properties of the fluids, in particular the transfer of torques from the magnetic field to the carrier fluid and sample containers.

8.2 Colloidal suspensions of anisometric particles

Colloidal suspensions of anisometric nanoparticles particles are scientifically attractive because of their ability to form microstructured and nanostructured phases [1–6].

Phases with broken rotational (nematics) and translational (smectics, columnar) symmetries have been discovered in various colloidal materials including clay [7–9], goethite [10, 11], TiO_2 rods [12], viruses [13–16], or even microparticles. Due to a coupling between the orientational degrees of freedom and external fields, these dispersions show a complex rheological response and a distinctive behaviour in electric fields [4, 17].

The crucial role of entropy-driven self-assembly in the colloidal systems is particularly pronounced in binary mixtures of rod-shaped or platelet-shaped nonmagnetic nanoparticles (NPs) with MNPs [18]. These mixtures exhibit a sterically-induced orientation transfer (Onsager-Lekkerkerker effect) [18, 19]. A field-induced alignment of the MNP subsystem is transferred to the nonmagnetic components. This does not only work with shape-anisotropic MNPs. A uniform external magnetic field can cause the formation of small, stable clusters of spherical MNPs which, in turn, may command the orientation of the nonmagnetic constituents.

One example of the sterical orientation transfer occurs in dispersions of rod-shaped pigment nanocrystallites doped with MNPs. Pigment particles investigated in our study (C.I. Pigment Red) have an average length of 230 ± 70 nm and a diameter of 46 ± 20 nm [20]. The particles form a stable dispersion in dodecane with the commercially available dispersant Solsperse 11200 (Lubrizol, Brussels, Belgium). The magnetic dopant is a commercially available ferrofluid (APG 935, Ferrotec), containing magnetite NPs with an average diameter of about 10 nm suspended in hydrocarbons. In our study, we investigated dispersions with various concentrations of MNPs, as well as pigment nanorods (NRs).

8.2.1 Magneto-optical behaviour

The orientational order of the pigment NRs determines the birefringence of the dispersions, which was measured using the optical modulation technique [21]. Diluted dispersions of the ferrofluid exhibit only weak birefringence saturating above 200 mT. Dispersions of pigment NRs show even weaker magneto-optical response compared to the magnetic fluid. The alignment of the pigment NRs was demonstrated by measurements of the linear dichroism in strong magnetic fields [22]. The particles align perpendicular to the field direction in fields of about 5 T.

By contrast, mixtures of the pigment NRs and MNPs exhibit much higher birefringence in comparison to that of the pigment-only suspensions, where the birefringence is nearly zero, and also in comparison to that of pure MPs. Figure 8.1 shows the birefringence of the NR/MNP mixtures with pigment particle concentration $c_r = 5$ vol% and varying concentration of MNPs (c_s). The curves exhibit a saturating at about 200–300 mT, which is comparable to that of the ferrofluid. Nevertheless, the birefringence is strongly enhanced by the presence of the NRs. The maximum birefringence Δn_{max} attained at 640 mT is shown in Figure 8.2.

At a constant c_s and low NR concentrations, the birefringence increases linearly, while for high c_r, a different behaviour is observed (Figure 8.2). This dramatic increment of birefringence can be explained by an onset of an orientationally ordered phase of the pigment suspensions at high concentrations. At the same time, the maximum birefringence at a constant c_r shows a linear dependence on c_s (see Figure 8.2) in the investigated concentration range.

8.2.2 Sterical orientation transfer

The sterical orientation transfer can be described by considering the free energy expansion for a binary mixture of nonmagnetic colloidal rods and magnetic rods. In a mixture of N_1 magnetic and N_2 nonmagnetic rods ($N = N_1 + N_2$) with corresponding lengths and diameters L_1, D_1 and L_2, D_2, respectively, the free energy expansion can be given by [23, 24]:

$$\frac{\Delta F}{Nk_BT} = \frac{\mu^0}{k_BT} - 1 + \ln c + (1-x)\ln(1-x) + x\ln x + (1-x)\sigma_1 + x\sigma_2 + c \left[b_{11}(1-x)^2\rho_{11} + 2b_{12}x(1-x)\rho_{12} + b_{22}x^2\rho_{22} \right] - ax \int \cos\beta f_2(\Omega)\, d\Omega \tag{8.1}$$

where μ^0 is the chemical potential, c is the net number density of all particles, x is the fraction the magnetic rods of the total number of rods, $(1–x)$ is the fraction of nonmagnetic rods, $a = (\mu H)/(k_bT)$, $b_{ij} = \frac{\pi}{8}(D_i + D_j)L_iL_j$, and β is the angle between **H** and

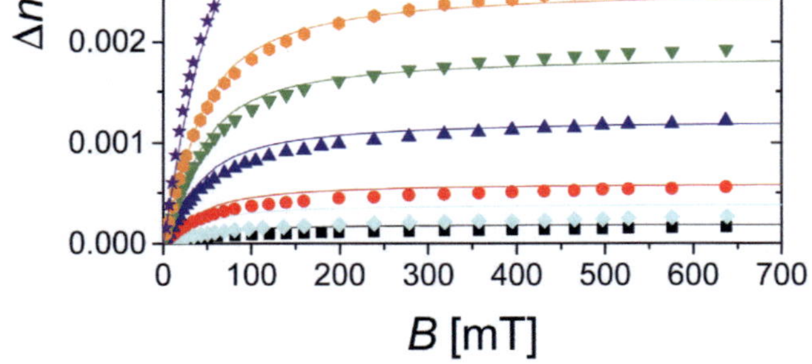

Figure 8.1: Birefringence of mixtures with pigment particle concentration c_r = 5 vol% and varying MPs concentration (c_s). (Adopted from a study by May et al. [21]).

$\boldsymbol{\mu}$ (i.e. the magnetic moment of the magnetized rods). The integration is over the solid angle Ω describing the orientation of the particles.

$$\sigma_i = \int f_i(\Omega) \ln 4\pi f_j(\Omega)\, d\Omega, i = 1,2 \tag{8.2}$$

and

$$\rho_{ij} = \iint |\sin\gamma| f_i(\Omega_i) f_j(\Omega_j)\, d\Omega\, d\Omega', i,j = 1,2 \tag{8.3}$$

where γ is the angle between two vectors defined by the angles Ω and Ω′, $f_1(\Omega)$ and $f_2(\Omega)$ are normalized distribution functions of the two particle types.

For the calculations, it is assumed that both particle types are monodispersed and the chains of MPs are treated as magnetic rods with the lengths of two spherical particles and the widths of one. Thus, the MPs are approximated as short cylinders 28 × 14 nm in size and with the magnetic moments of two spherical MPs. The calculated distribution function determines the orientational order parameters S_i (i = s,r) of both particle types:

$$S_i = \int \frac{1}{2} f_i (3\cos^2\theta - 1)\, d\Omega, \tag{8.4}$$

with θ being the angle between the long axis of a rod and the director. The resulting birefringence is

$$\Delta n = \Delta n_{0,\mathrm{r}}\, c_\mathrm{r}\, S_\mathrm{r} + \Delta n_{0,\mathrm{s}}\, c_\mathrm{s}\, S_\mathrm{m}. \tag{8.5}$$

where $\Delta n_{0,\mathrm{s}}$ and $\Delta n_{0,\mathrm{r}}$ are specific birefringences of the magnetic and rod-shaped particles, respectively. As expected, the measured birefringence and calculated order parameters for constant concentrations are approximately proportional to each other. The birefringence curves $\Delta n(B)$ could be fitted well with Eq. (8.5) containing numerically estimated order parameters S_r and S_s and the specific birefringence $\Delta n_{0,\mathrm{r}} = 0.7$. The numerical results fit the experimental data well; in particular, the behaviour of the maximum birefringence at constant pigment particle concentration shows good agreement. In contrast, the maximum birefringence at high c_r deviates from the numerical results. This behaviour is a consequence of polydispersity, which leads to a decreased transition concentration to the nematic phase [19].

8.2.3 Molecular dynamics simulation

The molecular dynamics (MD) simulations were performed by Stavros D. Peroukidis and Sabine H. L. Klapp at the TU Berlin [21]. The model fluid consisted of a binary mixture of N_r uniaxial rods and N_s magnetic spheres. The rods were represented as prolate, nonmagnetic ellipsoids with a length l and the width σ_0 (l/σ_0=3) that interact via the Gay-Berne (GB) potential, using a standard parametrization [21]. The magnetic spheres are modelled via dipolar soft spheres (DSSs), with an embedded central dipole

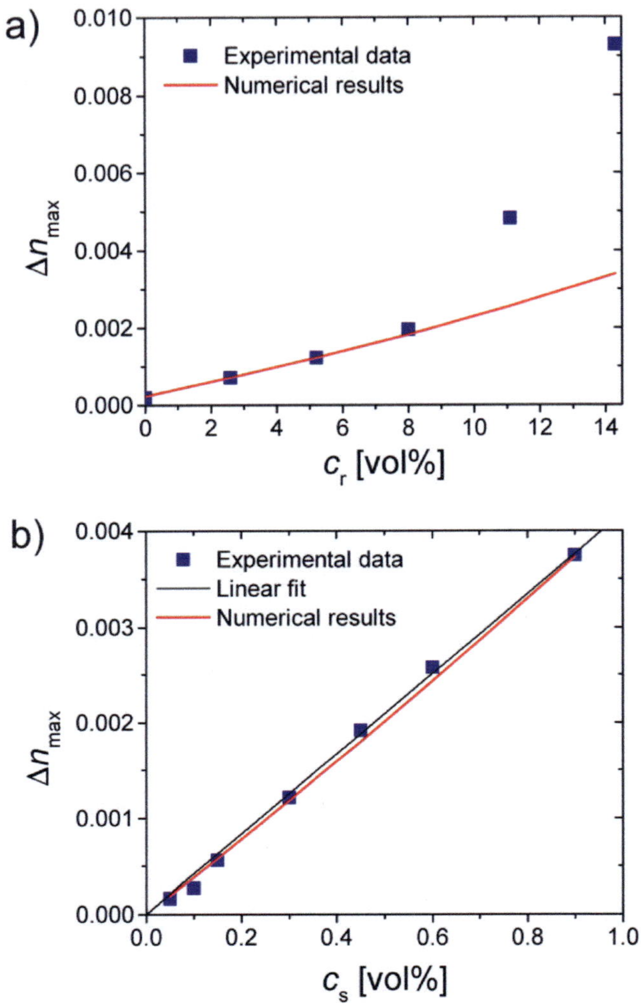

Figure 8.2: Saturation birefringence: (a) at a constant magnetic particle concentration of $c_s = 0.3$ vol % and different pigment particle concentrations c_r, and (b) at constant pigment particle concentration $c_r = 5$ vol% and different MNP concentrations c_s. (Adopted from a study by May et al. [21]).

moment $\boldsymbol{\mu}$. They interact via a soft repulsive potential and a dipole-dipole interaction [25, 26]. The DSS particles have a diameter σ_S which was set to 1/4 the width of the rods, i.e. $\sigma_S^* = \sigma_S/\sigma_O = 0.25$, which is consistent with the estimated relative sizes of the experimental species.

In the simulation, GB/DSS mixtures were studied in a uniform magnetic field H at various field strengths $H^* = \mu H/k_B T$, where μ is the dipole moment of a particle. The reduced density is defined as $\rho^* = N\sigma_O^3/V$ and V is the volume of the system with $(N = N_s + N_r)$, fraction of particles $x_a = N_a/N$ (where a = r,s for rods and spheres, respectively). Five states were considered: $[(\rho^*,x_S) = (0.350,0.20)]$, $[(\rho^*,x_S) = (0.560,0.50)]$, $[(\rho^*,x_S) = (0.778,0.64)]$, $[(\rho^*,x_S) = (1.00,0.72)]$, and $[(\rho^*,x_S) = (1.40,0.80)]$,, that correspond to the isotropic phase of a field-free GB-DSS mixture. For these states the concentration of rods was kept the same ($c_r = 44\,\%$) whereas the concentration of DSS, c_S, was varied (it increases by increasing the concentration x_S).

Figure 8.3 shows the calculated order parameter $S^{(r)}$ and the magnetization $\langle M \rangle$ as functions of the magnetic field. Both magnetization and order parameter exhibit saturating behaviour. The saturation order parameter $S_{saturation}$ has a linear dependence on the concentration of the magnetic spheres as shown in Figure 8.4. Since the birefringence is proportional to $S^{(r)}$, the result in Figure 8.4 demonstrates a qualitative agreement between the theory and the experiment. The simulations also show that the optical response of the binary mixtures is expected to decrease if the magnetic interactions between the spheres are reduced. Therefore, not only the concentrations of the MPs but also the interparticle interactions determine the response of the binary systems to an external magnetic field.

8.3 Dispersions of magnetic NPs in a nematic LC

Dispersions of MNPs in a nematic LC are called ferronematics (FN). We investigated the coupling between the nematic director and the magnetic order of $CoFe_2O_4$-based functionalized MNPs, which were synthesized in the group of Silke Behrens (KIT) [27]. The particle size was around 2.5 nm. In the MNP-doped system, a significant decrease of the magnetic Fréedericksz threshold was found. This suggested the existence of the coupling between the nematic director and the magnetization of small clusters of the MNPs in a particular manner. The MNP clusters represent inclusions in the director field that distort the surrounding director field. The behaviour of the LC dispersions was described using the Raikher-Burylov (RB) model for some of these systems and estimated the magnetization-director coupling strength [27].

At a combined application of electric and magnetic fields to a nematic LCM in a planar sandwich cell, one can increase the electric Fréedericksz threshold field by choosing the direction of the magnetic field perpendicular to the electric field, in the cell plane. In absence of MNP doping, the diamagnetic torque on the director stabilizes the ground state. The addition of MNPs reverses this effect, the MNP aggregates align with the magnetic field and destabilize the director ground state. The details of these experiments are found in the contribution by S. Behrens et al. in this volume.

8.4 Mobile magnetic NPs in fibrillous gels

Ferrogels are composite materials consisting of MPs embedded in a viscoelastic matrix [28]. Their magnetoelastic properties enable applications in the fields of transducers,

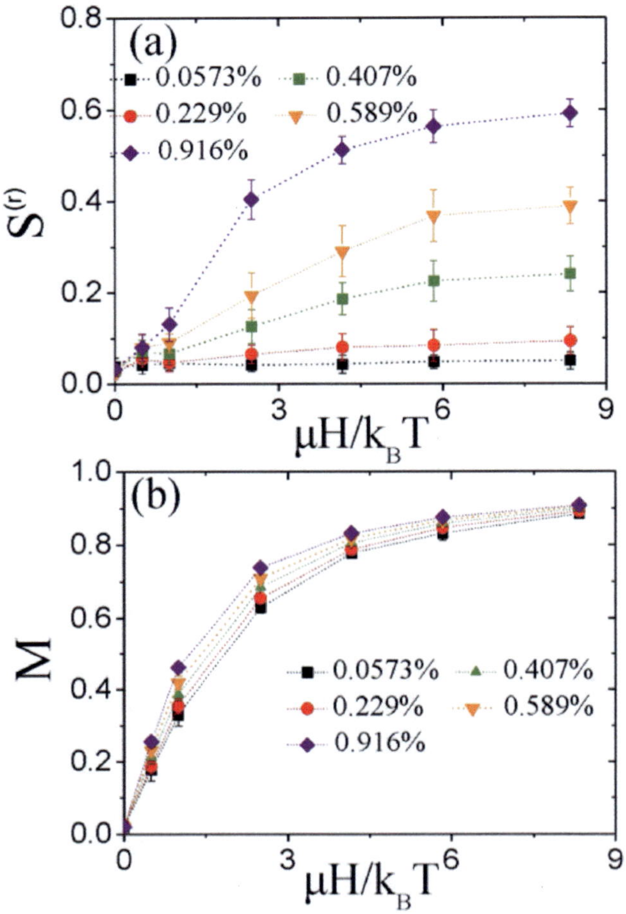

Figure 8.3: (a) Nematic order parameter of the rod species $S^{(r)}$ and (b) magnetization $\langle M \rangle$ of the MPs as a function of the external magnetic field strength for a GB-DSS mixture with reduced magnetic dipole moment $\mu^* = 2.4$ of the MNPs. The results are taken for constant concentration of rods $c_r = 44$ % by varying the concentration of magnetic spheres c_S. (Adopted from a study by May et al. [21]).

sensors and actuators [29]. Depending on the viscoelastic properties of the matrix, the magnetic properties of the particles and the strength of the coupling between them, their response to external stimuli like, e.g. magnetic fields or mechanical stress can be tuned in a wide range [30, 31]. Especially strong magnetomechanical response is expected when the MPs are micronsized and connected to the viscoelastic matrix. In this case the viscoelastic matrix directly experiences the torque of the Brownian rotation exhibited by the MPs when they align to the applied magnetic field [28]. On the other hand, a strong magneto-optical response can be expected from ferrogels where the MPs are not connected to the matrix and their alignment and chain formation along the

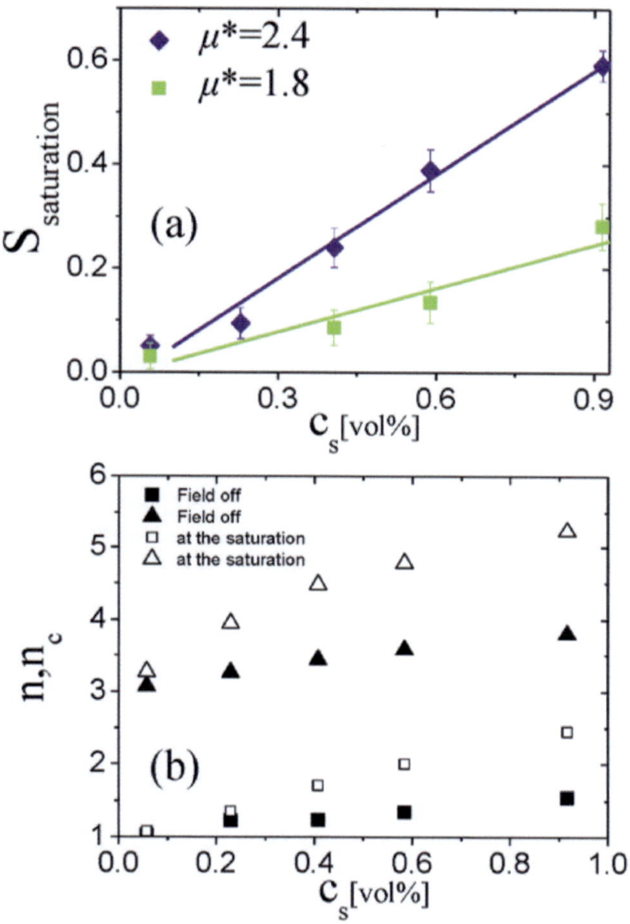

Figure 8.4: (a) Saturation value of the nematic order parameter of rods $S^{(r)}$ as a function of c_s for constant $c_r = 44\,\%$, and various values of dipolar moment μ^*. (b) Mean size (number of particles) of all clusters $\langle n \rangle$ -open squares- (irrespective of their type) and mean size of chainlike clusters $\langle n_c \rangle$ -open triangles- at the saturation value of $S^{(r)}$. The corresponding values of $\langle n \rangle$ and $\langle n_c \rangle$ when the field is off are shown by solid squares and triangles, respectively. (Adopted from a study by May et al. [21]).

magnetic field is only determined by the void accessibility in the confined volume of cavities in the mesh [32–36].

The ferrogels prepared in our laboratory belong to the latter category [37]. They consist of commercially available APG2135 ferrofluid diluted with n-dodecane and gelled by 12-hydroxyoctadecanoic acid (12-HOA) gelator. APG2135 consists of single-domain superparamagnetic magnetite particles surrounded by a surfactant layer to avoid aggregation and suspended in a synthetic hydrocarbon carrier. The average core diameter of the particles is 10 nm. The saturation magnetization is 17885 A m^{-1} and the volume fraction of the particles is 3.9 vol%. The particle size distribution can be described by a lognormal distribution with an average hydrodynamic diameter of $D_0 = 14.5$ nm and a standard deviation $\sigma = 1.2$ [38, 39]. Diluting APG2135 with the nonpolar solvent n-dodecane results in stable suspensions. The initial birefringence of those ferrofluids is insignificant and indicates the presence of some aggregates in the ferrofluid. The origin of birefringence in magnetic fluids can be explained in the following way. In a magnetic field, the MNPs are equivalents of magnetic dipoles aligning along the field in head-to-tail fashion. In the electric field of linearly polarized light they become oscillating dipoles. Depending on the orientation of the chains with respect to the direction of the polarization of the light, the oscillating dipole interaction between the particles is asymmetric, resulting in a strong optical anisotropy. Thus, the birefringence of the ferrofluid increases with increasing magnetic field [40–48]. The maximum birefringence at a given field strength is only dependent on the concentration of the MPs, as can be seen in Figure 8.7a.

Gelling the ferrofluids brings about several constraints into the system. Since the gelator network divides the volume into cavities with certain amounts of MNPs, there is a limited amount of available particles to form chains. Additionally, due to adsorption along the cavity walls MNPs become immobilized depleting the MNP supply. As a result there will be a smaller number of MPs available in the vicinity of a given particle to form chains than in a ferrofluid.

The mobilities of single particles will be heterogenous and spatially dependent because the alignment of chains along the magnetic field is restricted by the restrained void accessibility. The translational degree of freedom of the particles is constrained by the cavity walls leading to imperfect alignment of the dipoles in the confined space. The torque exhibited by the translations and rotations of the MNPs and their aggregates leads to a deformation of the gel network. These irreversible (inelastic) deformations happen when the ferrogel is exposed to a magnetic field for the first time [49] (Figure 8.5).

The gelator network is formed by 12-hydroxyoctadecanoic acid (12-HOA) molecules. This chiral low-molecular-weight gelator [50–53] crystallises in organic solvents and due to the anisotropy of strong intermolecular interactions between the individual gelator molecules gives rise to a helical self-assembly in one dimension. The anisotropy of the interfacial free energy of the resulting strands leads to bundling and the evolution of fibrillar structures. These aggregates intertwine and develop into a three-dimensional gelator network. Since this self-assembled fibrillar network (SAFiN) is based on intermolecular interactions, i.e. H-bond, London dispersion forces, their formation is thermoreversible. Depending on the gelator concentration the gel-sol transition occurs around 65–80 °C. The SAFiN-mesh structure depends on the cooling rate during the preparation process. The slower the cooling rates the longer the annealing time of the fibre growth resulting in a small number of permanent nodes originating from crystalline mismatch branching. If the high temperature sol is quenched to room temperature, the branching grade intensifies, a SAFiN containing high numbers of nodes evolves [54–65].

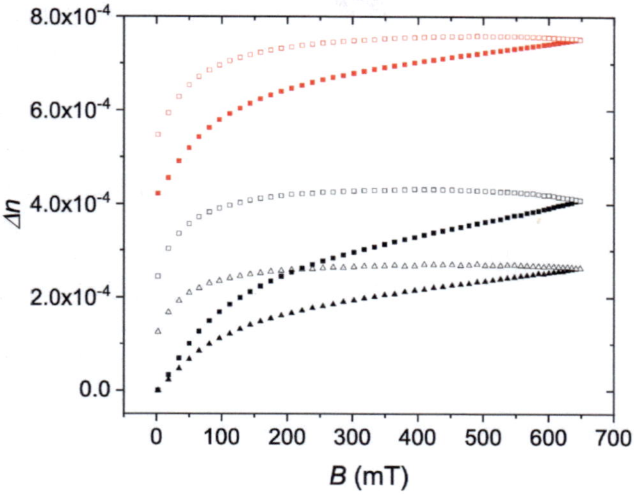

Figure 8.5: Magnetically induced birefringence $\Delta n(B)$ in an isotropic (black squares – 1st run, black triangles – 2nd run) and an anisotropic (red triangles) gel with 10 wt% MNP and 12.5 wt% gelator concentration at room temperature. Filled symbols correspond to increasing field and empty symbols to decreasing field. (Adopted from a study by Nádasi et al. [37]).

The gel structure was characterized by atomic force microscopy and scanning electron microscopy (SEM). A mosaic of spherulitic domains with a radial distribution of fibres can be seen in the low-magnification SEM images (Figure 8.6a), whose appearance is consistent with the textures observed by polarizing optical microscopy. High-magnification images (Figure 8.6b) reveal an interwoven structure of thin 32 ± 6 nm fibres in a gel with 10 wt% gelator. At this length-scale, the fibres appear disordered, forming a mesh with an average size of 60 ± 25 nm. This mesh size is significantly larger than the particle diameter (10 nm). Therefore, one mesh cell may accommodate short chains with 4–8 MNPs. The solvent-gelator compatibility of APG2135 with 12-HOA is comparable to that of a nonpolar solvent, since the ferrofluid can be considered as a nonpolar liquid with superparamagnetic properties. However, exactly the response of the sol in a magnetic field enables additional fine-tuning of the properties of the ferrogel.

To prepare the isotropic ferrogels, the sample was quenched to room temperature (cooling rate > 50 K min^{-1}). In the case of anisotropic ferrogels, the heated sample was cooled down at a controlled cooling rate (1 K min^{-1}) in a magnetic field of 650 mT. The samples were prepared in 100-μm thick rectangular capillaries. Bars of ferrogels easily bend upon bringing them in a magnetic field gradient. This demonstrates that despite the MNPs are dispersed in the liquid subphase, there is coupling between magnetic and elastic degrees of freedom.

This coupling is confirmed by our magneto-optical measurements. In Figure 8.5, the magnetic field dependence of the birefringence of an isotropic gel measured by optical modulation technique is shown. The two distinctively characteristic features of those curves are the hysteresis originating from the difference between the birefringence measured in increasing and decreasing magnetic field and the lowering of the birefringence values in the second measurement.

In the first measurement, the isotropic gel is exposed to a magnetic field for the first time. As described above, the viscoelastic matrix hinders the perfect alignment of the embedded MPs along the magnetic field. The competition between the aligning magnetic dipoles and the viscoelastic deformation of the matrix leads to some irreversible deformations in favour of better alignment of the magnetic dipoles. In the meantime, this interaction facilitates the adsorption of the particles along the cavity walls, resulting in a fewer number of free MPs contributing to the chain formation in the second run. Therefore, the birefringence of the isotropic gel in the second run is lower. The hysteresis is the consequence of the viscoelastic deformations of the gel network. The relaxation of the network to the initial state is a long-term process. In fact, after a full measurement cycle (0 mT → 650 mT → 0 mT), it takes at least 10 h for the gel to relax in absence of the magnetic field to a constant birefringence close to the initial one. The hysteresis of the first run is always bigger than that of the second run corroborating the idea of the matrix inelastic deformations.

Anisotropic gels prepared by a slow cooling from the sol state exhibit a striped birefringent texture in polarizing microscopy, where the stripes are aligned along the field direction applied during the preparation.

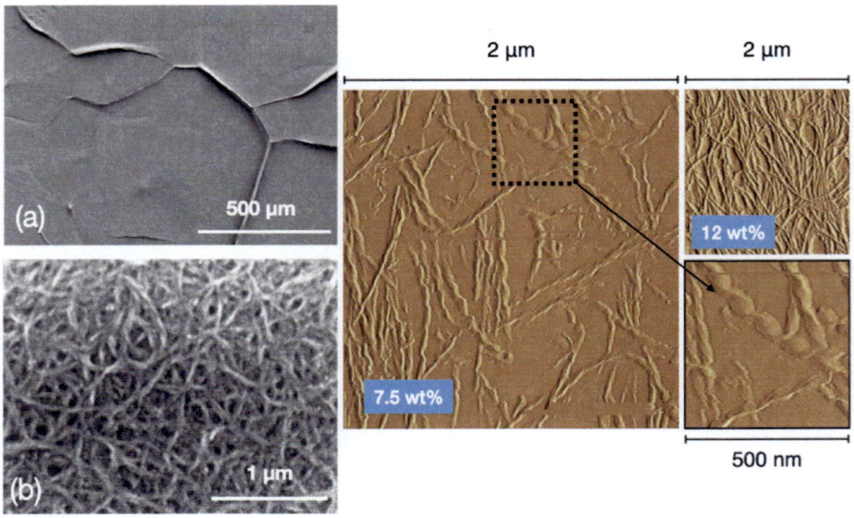

Figure 8.6: Scanning electron microscopy images of an isotropic gel with 7 wt% of the MNPs and 10 wt% of the gelator at (a) low magnification and (b) high magnification. (U = 1.5) kV. (c) Atomic force microscopy images of the gels with 7.5 wt% and 12.5 wt% of the gelator (4.5 wt% of MNPs). The inset shows a magnified image of a helical nanofilament. (Adopted from a study by Nádasi et al. [37]).

The high birefringence of the anisotropic gels in Figure 8.7d can be explained by the formation of MNP chains in the initial high temperature sol state. As in the ferrofluid, the higher the applied field the bigger is the birefringence of the sol, and no hysteresis can be observed (Figure 8.7c). Since the magnetic field is applied throughout the whole preparation process, the chains do not fall apart and they give rise to a direction dependent evolution of the gel network. Additionally, the gradual slow cooling increases the annealing time of the gel fibre growth resulting in less branching. Altogether, the mesh network is expected to be looser, especially in the direction of the applied preparation field. If no magnetic field is present, the MNP chains fall apart, yet the direction dependent growth of the fibres produces an optical anisotropy, i.e. birefringence. That is why the initial birefringence of the anisotropic gels is always higher than of the isotropic ones (Figure 8.7d). Since the MNP chain formation in the gel is facilitated in the external field direction, i.e. the optical axis is aligned to the magnetic field, the following phenomena are observed:

– the $\Delta n(B)$ curves coincide in repeated measurements,

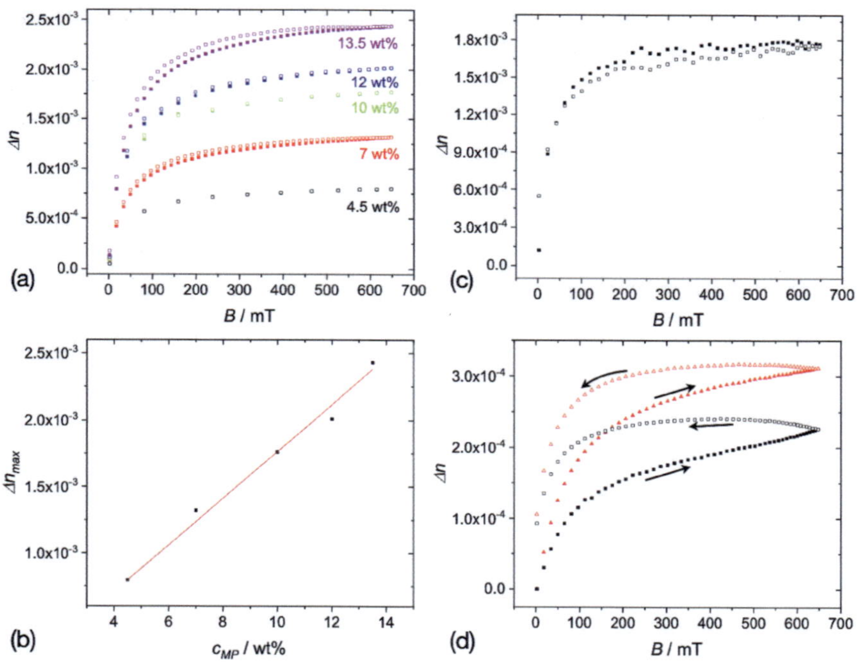

Figure 8.7: (a) Birefringence $\Delta n(B)$ of the ferrofluid APG2135 diluted with n-dodecane. (b) Saturation birefringence Δn_{max} as a function of the concentration of the MNPs c_{MP} in the ferrofluids. (c) Birefringence $\Delta n(B)$ of a dispersion of 13.5 wt% MNPs and 10 wt% 12-HOA at $T = 80\ ^{\circ}C$ (sol). No significant hysteresis behaviour can be found for a dwell time of 30 s. (d) Magnetically-induced birefringence $\Delta n(B)$ in an isotropic (black squares) and an anisotropic (red triangles) gel with 4.5 wt% MNP and 7.5 wt% gelator concentration at room temperature. Filled symbols correspond to increasing field and empty symbols to decreasing field. (Adopted from a study by Nádasi et al. [37]).

- the birefringence is higher in the anisotropic gel with the same c_G and c_{MP} for any applied field,
- the hysteresis of the anisotropic gels is always smaller than for the isotropic gels with the same c_G and c_{MP}.

The birefringence decreases, changing its sign if the measuring field is perpendicular to the anisotropy axis. Varying the concentration of the MNPs (c_{MP}) and the gelator (c_G) enables fine-tuning of the optical properties of the organoferrogel (Figures 8.8 and 8.9a, b).

An increasing gelator concentration provides negative feedback on the magneto-optical response: the saturating birefringence decreases with increasing gelator concentration (Figure 8.9c, d). However, the dependence $\Delta n(c_G)$ remains linear in the investigated concentration range (Figure 8.11b, d).

There is a clear difference between the isotropic and the anisotropic gels if one considers the slopes $s(c_G) = d\Delta n(c_{MP})/dc_{MP}$. The slope s for low concentration of the gelator is nearly independent on the concentration. This can be understood by assuming that at low gelator concentrations, the mesh is loose in the anisotropy direction. The gel mesh has less confinement effect on the growth of the magnetic aggregates. For higher gelator concentration (12.5 wt%), the gel network restricts the growth and the slope $s(c_G)$ becomes reduced. The hysteresis $h(c_{MP})$ increases with increasing MNP concentration c_{MP} in both isotropic and anisotropic gels. However, the isotropic gels have a systematically lower birefringence and larger magneto-optical hysteresis (Figure 8.10).

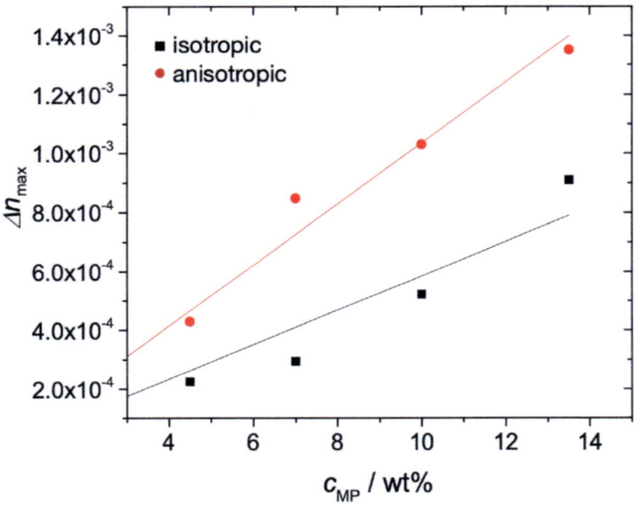

Figure 8.8: Magnetic particle concentration dependence of the saturation birefringence $\Delta n_{max}(c_{MP})$ for isotropic and anisotropic gels with 10 wt% gelator. The straight lines are linear fits under the constraint $\Delta n_{max}(0) = 0$. (Adopted from a study by Nádasi et al. [37]).

Figure 8.7d shows the field dependence of the birefringence $\Delta n(B)$ for a gel with a gelator concentration of 10 wt%. The field dependence of the birefringence $\Delta n_+(B)$ and $\Delta n_-(B)$ on increasing and decreasing magnetic field B, respectively, differ considerably. Only $\Delta n_-(B)$ can be satisfactory fitted with the Eq. (8.8) in gels with the gelator concentration below 10 wt%. To estimate the hysteresis strength, we introduce a dimensionless parameter h:

$$h = \frac{\int_0^{B_{\max}} \Delta n_-(B)\mathrm{dB} - \int_0^{B_{\max}} \Delta n_+(B)\mathrm{dB}}{\int_0^{B_{\max}} \Delta n_-(B)\mathrm{dB}} \tag{8.6}$$

where $B_{\max}\rightarrow\infty$. In our experiment, $B_{\max}$ was limited to 650 mT.

It is important to mention that these gels show no magnetic hysteresis. The magnetic relaxation in our system is governed by the Néel relaxation mechanism with corresponding relaxation rates in the range of 100 MHz. Thus, the magnetization rapidly adjusts to the applied magnetic field on a very short (compared to our experiment) time scale. The birefringence response Δn has two important contributions: the birefringence Δn_{MP} attributed to the aligned magnetic chains and the birefringence Δn_G

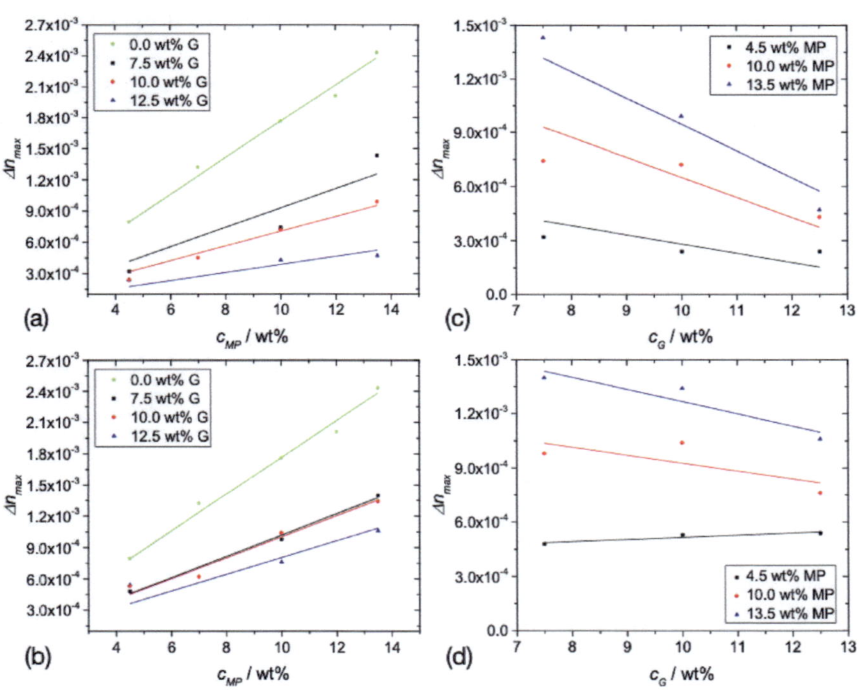

Figure 8.9: Dependence of the saturation birefringence $\Delta n_{\max}$ of the ferrogel on the concentration of the MPs c_{MP} in the isotropic (a) and the anisotropic (b) gels. Dependence of the saturation birefringence $\Delta n_{\max}$ of the ferrogel on the concentration of the gelator c_G in the isotropic (c) and the anisotropic (d) gels. (Adopted from a study by Nádasi et al. [37]).

caused by a deformation of the gel matrix. In a dilute limit, the relaxation times for Δn_{MP} can be estimated as the time an MNP requires to diffuse over a distance of particle radius R, $\tau = 6\pi\eta R^3/kT$, where η is the viscosity of the matrix, T is the temperature. For dodecane, $\tau \approx 0.2\,\mu s$. In the gel state, the situation is much more complex. The viscosity increases by several orders of magnitude and the environment of the particles becomes strongly heterogeneous. Particles interact with the gel network. Already at $c_G = 3$ wt%, τ reaches the order of magnitude of a second. A slow increase of the birefringence in a magnetic field can also be attributed to slow deformations of the gel network and the adhesion of MNPs to the gel fibres. This also results in a residual birefringence which relaxes on the time scale of days. The dynamical properties of the mobile NPs in the gel matrix will be discussed in an upcoming paper.

The behaviour of the saturating birefringence Δn_{max} and the hysteresis parameter h show the effect of the local anisotropy of the gel, which is a surprising finding because the sizes of the particles are smaller than the mesh size of the gel. Larger Δn_{max} and smaller h suggests that MNPs have more space to rearrange and form larger aggregates along the anisotropy axis. Birefringence in MNP agglomerates results from the mutual polarizability of adjacent MNPs [42, 66]. The association of MNPs into chains was described in the early seventies by P.G. de Gennes and P.C. Jordan using hard-sphere models [40, 41, 67]. The formation of MNP dimers reduces the entropy of the system and shifts the equilibrium from single particles to dimers [67, 68].

The effective dielectric tensor $\hat{\varepsilon}_{eff}$ of the composite fluid can be decomposed into the isotropic part $\hat{\varepsilon}_{host}$ of the host fluid and the anisotropic part of the inclusions $\hat{\varepsilon}_{inc}$:

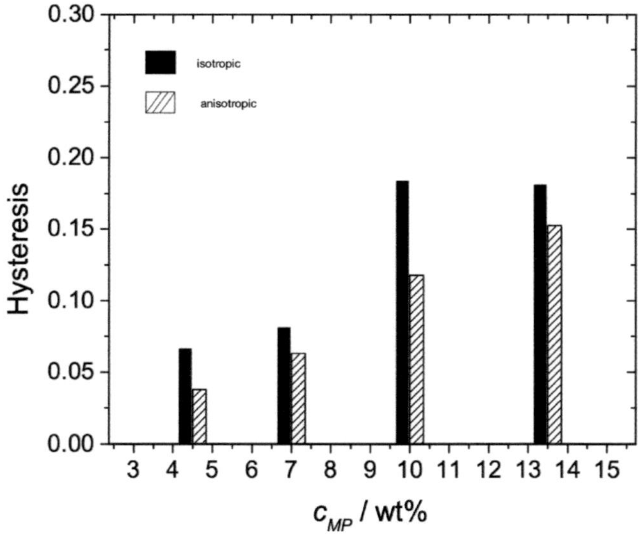

Figure 8.10: Dependence of the hysteresis parameter h on the concentration of the magnetic particles c_{MP}. (Adopted from a study by Nádasi et al. [37]).

$\widehat{\varepsilon}_{\rm eff} = \widehat{\varepsilon}_{\rm host} + \widehat{\varepsilon}_{\rm inc}$. The dielectric polarizability α of the anisometric MNP aggregates is anisotropic. In case of magnetic chains, the length of the chain and the intrinsic optical properties of MNPs determine the polarizability anisotropy $\Delta\alpha$. In the thermal equilibrium, the alignment of the chains is subject to thermal fluctuations. The components of the dielectric permittivity of the chains are given by the average

$$\varepsilon_{\rm inc_j} = \frac{1}{\varepsilon_0} \sum_{k=1}^{\infty} \varphi(k) \langle \alpha_j \rangle_{\theta,\phi} \tag{8.7}$$

where the summation runs over all chains of the length k with the volume fraction $\varphi(k)$, respectively. The averaging is done over all possible orientations θ, ϕ.

The field-dependence of the birefringence induced by the dimerization in a magnetic fluid was studied by several authors [42, 44, 66, 69–72]. Assuming a Boltzmann orientational distribution of the particle pairs, the birefringence attributed to the dimers can be found from Janssen's dependence [66, 72]:

$$\Delta n(x) = \varphi_2 A\left(1 - 3\left(\frac{\coth(x)}{x} - \frac{1}{x^2}\right)\right) \tag{8.8}$$

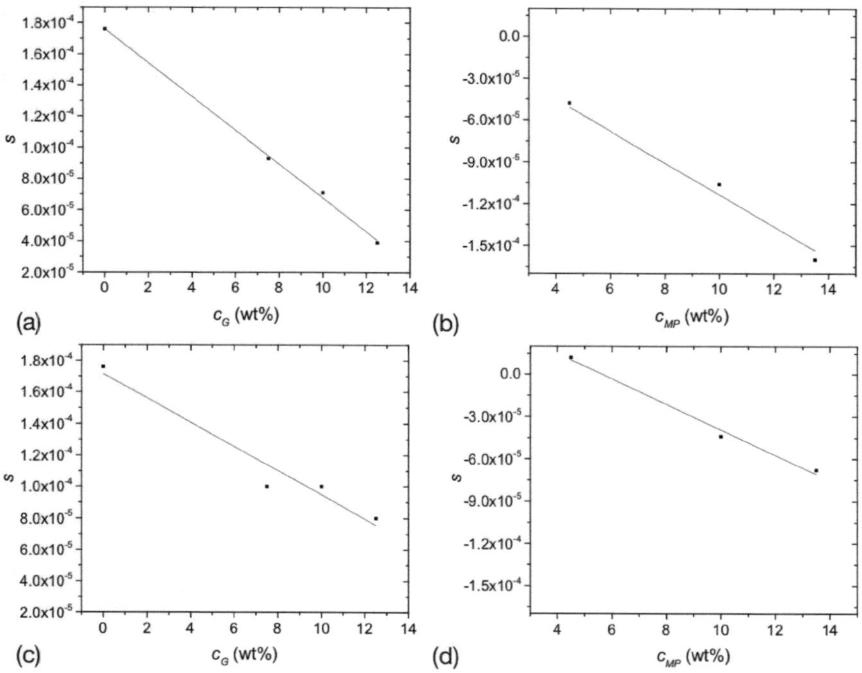

Figure 8.11: (a) and (c) show the slopes s of linear fits of $\Delta n_{\rm max}(c_{\rm MP})$ as functions of the gelator concentration in the isotropic and anisotropic gels, respectively. Figures (b) and (d) show the slopes s of the linear fits of $\Delta n_{\rm max}(c_{\rm G})$ as functions of the MNP concentration of the isotropic and anisotropic gels, respectively (Adopted from a study by Nádasi et al. [37])

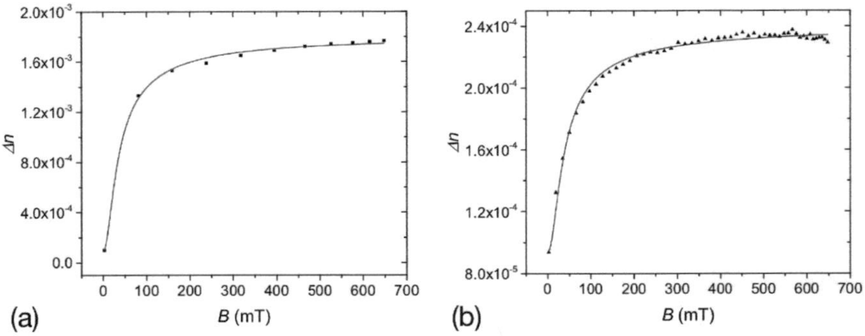

Figure 8.12: Magnetically induced birefringence $\Delta n(B)$ as a function of the field strength B in (a) ferrofluid with $c_{MP} = 10.4$ wt% and (b) ferrogel with $c_{MP} = 4.5$ wt% of MNPs and $c_G = 12.5$ wt%. The solid red lines are fits using Eq. (8.8).

where $x = \mu_{aggr}B/kT\text{b}$, and μ_{aggr} is the magnetic moment of the chain aggregate, T is the temperature and k_b is the Boltzmann constant. The factor A is determined by the polarizabilities of the particles.

Equation (8.8) describes well the experimental observations in pure ferrofluids and in the sol state (Figure 8.12). From the fit of $\Delta n(B)$, we deduce the magnetic moment of the chain-aggregate $\mu_{aggr} = 4.28 \cdot 10^{-19} \text{Am}^2$. This corresponds roughly to two MNPs. The volume fraction of magnetic dimers is determined by the magnetic field and the temperature. In the gel state, however, this description fails. Satisfactory fits can be found mostly for Δn with decreasing field B for low concentrations of MNP and the gelator. The reason for that is the nonequilibrium behaviour of $\Delta n(B)$. The deformation of the gel network occurs on much slower time scales than the rearrangement of the MNPs.

8.5 Magnetic suspensions in rotating fields

8.5.1 Rotational effect

Among the phenomena that attracted particular attention of researchers of magnetic liquids, the magnetic torque transfer in rotating or oscillating magnetic fields remains incompletely understood. Macroscopic samples of magnetic fluids have a tendency to develop vortex flow when exposed to rotating magnetic fields. For this phenomenon, the term rotational effect [73–79] has been coined. It turns out that this effect depends on sample geometries, free surfaces of the fluid, the composition (polydispersity) of the magnetic fluid, as well as frequencies and magnitudes of the magnetic field in a complex way. This effect can not only be used to transfer torque to the container or immersed objects [78], it can be exploited to study negative viscosity effects [80, 81], to rectify thermal motion [82, 83] or to characterize structural sample properties [84, 85].

Numerous studies in literature were devoted to magnetic fluids exposed to external rotating or oscillating magnetic fields, see, e.g. references in a study by Storozhenko et al. [84]. We focus here on two detailed aspects of this effect, on the study of diluted ferrofluids in order to reveal the role of viscosity on the efficiency of the torque transfer [84] and an analysis of the frequency and field strength dependence on the composition and microscopic parameters of the magnetic fluid.

8.5.2 Torsion pendulum setup

When a magnetic fluid is exposed to a magnetic field that exerts a permanent torque on the MNPs, this torque is transferred to the carrier fluid to create a vortex flow, and further to the container by shear forces of the flowing liquid at the container walls. If the container is freely suspended, the rotation is continuously accelerated [82], the measurement of the torque from the angular velocity profile of the container is difficult. Therefore, a torsion pendulum is a reasonable alternative, it measures the distortion of the suspending wire in equilibrium of the magnetic field–induced torque with the restoring elastic torque of the wire. Figure 8.13 sketches the setup [83, 84].

The magnetic field is generated by two sinusoidal currents, with 90° phase shift, in two pairs of coaxial coils, and monitored with Hall probes. The samples are enclosed by spherical glass containers of approximately 1 mL volume. The sensitivity of the setup

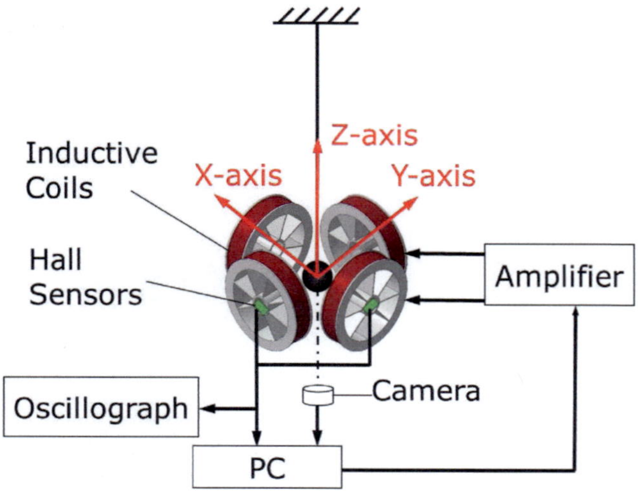

Figure 8.13: Torque balance setup for the measurement of the rotational effect in ferrofluids. The magnetic field is generated with two pairs of coaxial coils [83, 84]. The torque is measured from the twist of the suspending thread. The sample container is partly immersed in an oil bath to dampen the pendulum oscillations in order to reach the equilibrium deflection faster, and to attenuate influences of external noise. Figure adapted from a study by Storozhenko [84].

allows us to measure torque densities of the order of 10^{-4} N/m^2 (torques of 10^{-10} Nm on the samples), angular frequencies were typically of the order of 50 s^{-1} to 5000 s^{-1}, in some experiments up to 30,000 s^{-1}, field strengths up to a few mT were applied. The sample was suspended by a thin glass fibre, with very low restoring torque, so that the eigenfrequency of the pendulum was below 100 mHz.

8.5.3 Diluted ferrofluids and viscosity effects

The transfer of the magnetic torque onto the suspending fluid occurs in two steps: in the first step, the external magnetic field interacts with the magnetization of the MNPs. In a second step, this torque is transferred to the carrier fluid by viscous forces. The torque excerted by a field $\vec{B}$ on the magnetic moment $\vec{m}$ of an individual NP is $\vec{T} = \vec{m} \times \vec{B}$. This torque is proportional to sinα, with α being the angle between $\vec{m}$ and $\vec{B}$. In the at moderate rotation rates $\omega = \dot{\varphi}$ of $\vec{B} = B(\cos\varphi, \sin\varphi, 0)$, the magnetization follows the field after a short transient with the same angular velocity but a constant phase lag α. Two different mechanisms compete in the reorientation of $\vec{m}$. The particles themselves may rotate, with $\vec{m}$ being fixed relative to the MNP. This process is usually referred to as Brownian relaxation. It depends upon the viscosity of the carrier fluid that dampens the rotation of particles relative to the matrix. The higher this viscosity, the larger is the phase lag α and the larger is the magnetic torque at given B, $|\vec{m}|$, and ω. The second process is the reorientation of the magnetization $\vec{m}$ relative to the particle orientation, referred to as Néel relaxation. When the particles are small, this is the prevailing mechanism. Here, α does not depend upon the rotation state of the MNP, and is thus

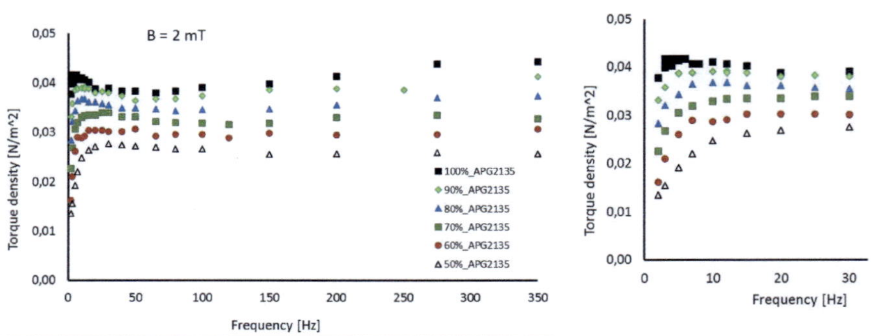

Figure 8.14: Torque density in samples with different concentrations of the original ferrofluid. The right graph is an expansion of the low-frequency region of the graph at the left. Figure adapted from a study by Storozhenko et al. [84].

independent of viscosity, even though the transfer of the magnetic torque from the MNPs to the carrier fluid still proceeds by viscous shear forces.

In order to explore the influence of viscosity of the magnetic fluid on the torque transfer, a commercial ferrofluid (APG2135, *Ferrotec Corp. (Japan)*) with a viscosity of 1.5 Pas was diluted with different concentrations of durasyn (viscosity 46 mPas). If the internal structure of the magnetic fluid were independent of the applied magnetic field, the torque would be proportional to the MP concentrations, and it would continuously increase with increasing rotation rate of the magnetic field. This is the consequence of an increasing phase lag between the sample magnetization and $\vec{B}$ with increasing ω. This condition is in fact fulfilled only for well-diluted samples. Experimental results are seen in Figure 8.14.

The decrease of the torque density is smaller than expected from the decrease of MNP concentration in the range of up to 50 % dilution. On the other hand, the change of viscosity by roughly one order of magnitude apparently has only little effect on the graphs. A certain saturation of the torque density is found at comparable rotation rates. If the torque was primarily generated from Brownian relaxation, one would expect that it would substantially decrease in the diluted samples.

A characteristic difference between the high-concentration and diluted systems is a slight elevation of the torque characteristics at rotation frequencies below 10 Hz. This was attributed to the formation of small particle aggregates at higher fields, which would increase the magnetic response. At larger rotation rates, these agglomerates decompose since they cannot follow the field direction. Then, the torque drops in the range between 10 and 50 Hz for the highly concentrated suspensions.

The torque transfer is also influenced by the rheological properties of the carrier fluid. This was demonstrated by addition of nonmagnetic platelet-shaped pigment particles [86]. When the isotropic solvent (n-dodecane) is replaced by anisometric crystallites, the character of the torque characteristics changes substantially owing to the non-Newtonian character of the concentrated pigment particle suspensions

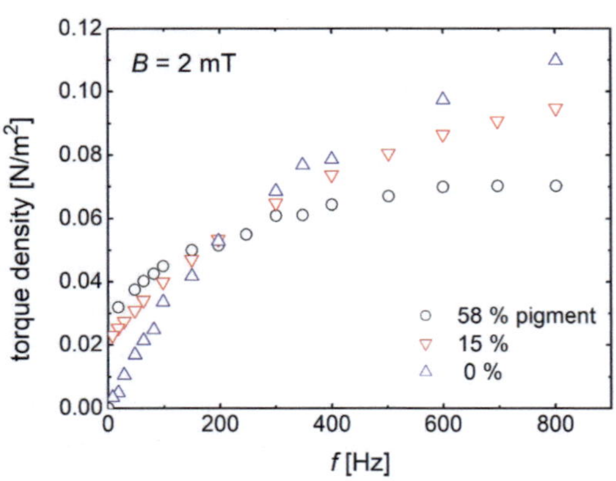

Figure 8.15: Torques on ternary mixtures of ferrofluid, dodecane and Permanent Rubine pigment particles. The ferrofluid concentration is 40%, the concentrations of pigment particles are given in the legend, the rest is dodecane. Printed with kind permission of A. Storoshenko [84].

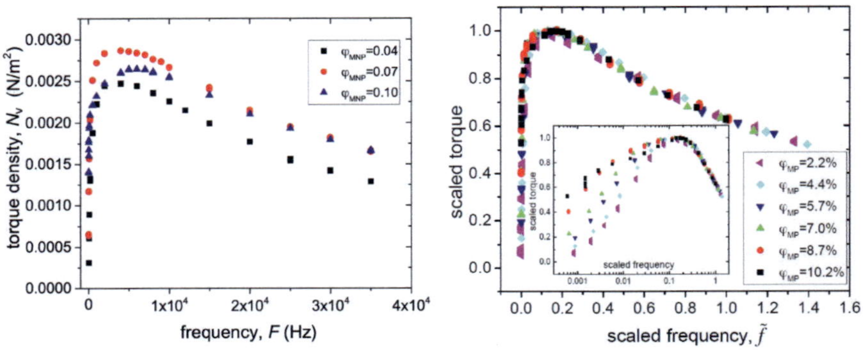

Figure 8.16: Left: Frequency dependence of the magnetic torque density in dispersions with various volume fractions φ_{MNP} of magnetic particles. Right: Scaled torque dependencies $N_v/N_{max}(\tilde{f})$ for dispersions with different volume fractions of magnetic particles. Note the extended frequency range as compared to Figure 8.14. Reprinted from a study by Usadel et al. [85].

Figure 8.15. The saturation of the torque shifts to lower frequencies. This reflects the interactions of the flow generated by the MNPs and the anisometric platelets.

8.5.4 Néel and Brownian relaxation in rotating fields

So far, we have disregarded the nature of the interactions between the MNP magnetizations and the external field. A closer analysis shows that in fact, the Néel relaxation time τ is dominant for most of the magnetic NPs contained in commercial ferrofluids. The quantitative analysis shows that the crossover from predominant Néel relaxation of small MNPs to Brownian relaxation of large MNPs in polydisperse samples occurs near 10 nm particle diameter. Smaller MNPs represent by far the majority of particles in typical ferrofluids, so that most of the particles are characterized by an internal relaxation of the magnetization. On the other hand, the magnetization of individual particles grows with the third power of their diameter, so that relatively few large particles still contribute noticeably to the magnetic torque. These processes were analysed recently in a combined theoretical and experimental study [85]. Magnetic torques were measured with the torque balance technique over an extended frequency range up to 35 kHz.

The original ferrofluid synthesized in the Scientific Research Laboratory of Applied Ferrohydrodynamics (ISPU, Ivanovo) by chemical condensation consists of single-domain spherical nanoparticles of magnetite Fe_3O_4 stabilized by oleic acid and dispersed in kerosene. It was stepwise diluted with dodecane (viscosity: 1.34 mPas) so that the concentration of MNPs ranged from 10.2 to 2.2%, while the viscosity dropped with increasing dilution from about 30 mPas down to 2 mPas. Figure 8.16 shows the experimental findings. At higher frequencies (beyond the torque maximum) all graphs can be scaled to a master curve. At low frequencies, there are some systematic deviations that most probably stem from aggregation of single MNPs in the higher concentration range.

Measurements of the magnetic torque of diluted dispersions of MNPs in rotating magnetic fields and the comparison to numerical simulations have demonstrated the crucial importance of the distribution of particle sizes. From a theoretical point of view, it was shown that the rigid dipole model (RDM), which fixes the magnetization in the particle, cannot explain the experimental observations. It predicts that the torque and other dynamical features depend only on the product of the viscosity and rotation rate of the external field. In order to understand the experimental results, it is necessary to take both Brownian and Néel relaxation into account [85]. The theoretical results show that in a rotating magnetic field a transition takes place from a state with the magnetic moment locked to the anisotropy axis of the MNPs to a state with precessional spin dynamics. This transition depends on the anisotropy energy. The dependence of the anisotropy energy on the magnetic volume of the MNPs causes an essential dependence on particle sizes. This must be taken into account in all experiments performed with diluted ferrofluids of a broad distribution of particle sizes.

For MNPs at room temperature with magnetic parameters typical for iron oxides (such as magnetite), the magnetic moment can be considered as being locked if the magnetic radius is larger than about 10^{-8} m [85]. Such particles can be described analytically within the RDM. For smaller particles, both Brownian and Néel dynamics are relevant, and a numerical analysis is required. The faster Néel relaxation decreases the relaxation time, and consequently the phase lag of the magnetization and the transferred torque considerably.

These conclusions apply to most of the commercial ferrofluids widely used in applications, which are commonly characterized by broad distributions of particle sizes. The theoretical analysis shows that much larger torques could be achieved with particles with larger anisotropy constant. Such particles would be similarly stable, e. g. with respect to aggregation, but the critical radius above which Brownian relaxation dominates would become much smaller, so that more particles contribute to a large torque. An experimental confirmation of this theoretical prediction is a desirable goal of future research.

8.6 Summary

Experimental and MD simulation of cosuspensions of non-magnetic anisotropic microcrystallites and magnetic nanoparticles have demonstrated an efficient orientation transfer from the magnetic subphase to the non-magnetic one, even with shape-isotropic MNPs. In relatively low magnetic fields (up to 700 mT), suspensions of the non-magnetic pigment particles alone show very low magnetically induced birefringence. The magneto-optical response is drastically enhanced by addition of a small volume fraction of MNPs. The form anisotropy of the non-MPs plays a crucial role in this effect. The gain of translational entropy resulting from the alignment of the elongated non-MPs with anisometric aggregates of the MNP dopants drives the reorientation of the nonmagnetic subphase. This alignment is particularly well pronounced in platelet-shaped nanocrystallites of Pigment Red 176, which shows an exceptionally large magneto-optical response [87].

Addition of MNPs to thermotropic liquid-crystalline phases also generates magneto-optical effects or alters the magneto-optical response of the doped LCs [27]. However, the measured effects are substantially weaker than with the magnetically doped suspensions. The stability of the samples is a substantial problem, the nematic host prefers to expel the dopants or to collect it in defect structures. The interactions of the MNPs with the LC matrix depend crucially on the MNP surface chemistry. The doped LC materials did not show any qualitatively new features, but quantitative effects are observed, and they can be explained satisfactorily.

In fibrillous gels, the motion of included MNPs is restricted by the gelator network with a mesh size comparable to the MNPs. We investigated the structure and the magneto-optical response of isotropic and anisotropic fibrillous organoferrogels with mobile MNPs. The presence of the gel network restricts the magneto-optical response of the ferrogel. Even though the ferrogel exhibits no magnetic hysteresis, an optical hysteresis has been found. This suggests that the optical response is primarily determined by the dynamics of self-assembly of the embedded MNPs into shape-anisotropic agglomerates. The optical anisotropy of the system can be fine-tuned by varying the concentrations of the gelator and the MNPs. The optical response in structurally anisotropic gels is orientation-dependent, revealing an intricate interplay between the confining mesh and the MNPs [37].

The torque measurement technique introduced in Section 8.5 can determine magnetic torques with nanonewton accuracy. It was demonstrated to provide useful results both for ferrofluids and for suspensions of nonmagnetic pigment crystallites and magnetic nanoparticles. Torques generated in conventional ferrofluids composed of spherical particles are quite small because Néel relaxation that dominates the dynamics of small nanoparticles is much faster than usual rotation rates of the magnetic field. The transferred torque depends upon the phase lag of the sample magnetization respective to the rotating external magnetic field, which is extremely small in typical spherical MNPs [85]. A strategy to achieve substantially larger torques would be the use of shape-anisotropic MNPs, in which the magnetization is pinned to the particle and the relaxation is governed by Brownian motion of the MNPs. The phase lag is substantially larger in that case even at slow rotation rates. Potential applications of this method are manifold, including the study of rectified thermal motions of MNPs in an oscillating field, using a ratchet effect [83, 84].

Acknowledgment: The German Science Foundation is acknowledged for funding within projects STA 425/36-(1-3) in the priority program SPP1681, and the Council for grants of the President of the Russian Federation is acknowledged for Grant No. 075-15-2019-743. The authors acknowledge important contributions by K. Usadel to theoretical modeling, and A. Storoshenko and Á. Corradi to the experiments. Valuable collaborations with the groups of A. Schmidt, S. Klapp, and S. Behrens within SPP 1681 are gratefully acknowledged.
Author contribution: All the authors have accepted responsibility for the entire content of this submitted manuscript and approved submission.
Research funding: German Science Foundation STA 425/36-(1-3) in the priority program SPP1681, Council for Grants of the President of the Russian Federation Grant No. 075-15-2019-743.
Conflict of interest statement: The authors declare no conflicts of interest regarding this article.

References

1. Onsager L. Anisotropic solutions of colloids. Phys Rev 1942;62:558.
2. Onsager L. The effects of shape on the interaction of colloidal particles. Ann N Y Acad Sci 1949;51: 627–59.
3. Vroege GJ, Lekkerkerker HNW. Phase transitions in lyotropic colloidal and polymer liquid crystals. Rep Prog Phys 1992;55:1241–309.
4. Davidson P, Gabriel J-CP. Mineral liquid crystals. Curr Opin Colloid Interf 2005;9:7–7.
5. Gompper G, Schick M. Soft matter: volume 3: colloidal order: entropic and surface forces. Germany: Wiley-VCH; 2007.
6. Grelet E, Lettinga MP, Bier M, van Roij R, van der Schoot P. Dynamical and structural insights into the smectic phase of rod-like particles. J Phys Condens Matter 2008;20:494213.
7. van der Beek D, Petukhov AV, Oversteegen SM, Vroege GJ, Lekkerkerker HNW. Evidence of the hexagonal columnar liquid-crystal phase of hard colloidal platelets by high-resolution SAXS. Eur Phys J E 2005;16:253–8.
8. van der Beek D, Reich H, van der Schoot P, Dijkstra M, Schilling T, Vink R, et al. Isotropic-nematic interface and wetting in suspensions of colloidal platelets. Phys Rev Lett 2006;97:087801.
9. van der Beek D, Petukhov AV, Davidson P, Ferré J, Jamet JP, Wensink HH, et al. Magnetic-field-induced orientational order in the isotropic phase of hard colloidal platelets. Phys Rev E 2006;73: 041402.
10. Lemaire BJ, Davidson P, Ferré J, Jamet J-P, Petermann D, Panine P, et al. The complex phase behaviour of suspensions of goethite (alpha-FeOOH) nanorods in a magnetic field. Faraday Discuss 2005;128:271–83.
11. Lemaire BJ, Davidson P, Ferré J, Jamet JP, Panine P, Dozov I, et al. Outstanding magnetic properties of nematic suspensions of goethite (alpha-FeOOH) nanorods. Phys Rev Lett 2002;88:125507.
12. Meuer S, Oberle P, Theato P, Tremel W, Zentel R. Liquid crystalline phases from polymer-functionalized TiO2 nanorods. Adv Mater 2007;19:2073.
13. Fraden S, Hurd AJ, Meyer RB, Cahoon M, Caspar DLD. Magnetic-field-induced alignment and instabilities in ordered collloids of tobacco mosaic virus. J Phys Colloq 1985;46:85–113.

14. Dogic Z, Fraden S. Smectic phase in a colloidal suspension of semiflexible virus particles. Phys Rev Lett 1997;78:2417–20.
15. Dogic Z, Fraden S. Cholesteric phase in virus suspensions. Langmuir 2000;16:7820–4.
16. Purdy KR, Dogic Z, Fraden S, Rühm A, Lurio L, Mochrie SGJ. Measuring the nematic order of suspensions of colloidal fd virus by X-ray diffraction and optical birefringence. Phys Rev E 2003;67: 627.
17. Fraden S, Hurd A, Meyer RB. Electric-field-induced association of colloidal particles. Phys Rev Lett 1989;63:2373–6.
18. Kredentser S, Buluy O, Davidson P, Dozov I, Malynych S, Reshetnyak V, et al. Strong orientational coupling in two-component suspensions of rod-like nanoparticles. Soft Matter 2013;9:5061.
19. Lekkerkerker HNW, Coulonvan der Haegen PR, Deblieck R. On the isotropic-liquid crystal phase separation in a solution of rodlike particles of different lengths. J Chem Phys 1984; 80:3427.
20. Schmidt MU, Hofmann DW, Buchsbaum MC, Metz HJ. Crystal structures of pigment red 170 and derivatives, as determined by X-ray powder diffraction. Angew Chem Int Ed 2006;45:1313–17.
21. May K, Eremin A, Stannarius R, Peroukidis SD, Klapp SH, Klein S. Colloidal suspensions of rodlike nanocrystals and magnetic spheres under an external magnetic stimulus: experiment and molecular dynamics simulation. Langmuir 2016;32:5085–93.
22. Eremin A, Geng Y, Stannarius R, Ostapenko T, Challa PK, Gleeson JT, et al. Peculiarities of the magneto-optical response in dispersions of anisometric pigment nano-particles. RSC Adv 2016;6: 80666–9.
23. Herzfeld J, Berger AE. A highly convergent algorithm for computing the orientation distribution functions of rodlike particles. Macromolecules 1984;17:1718–23.
24. Slyusarenko K, Reshetnyak V, Reznikov Y. Magnetic field control of the ordering of two-component suspension of hard rods. Philos Trans A Math Phys Eng Sci 2013;371:20120250–20120250.
25. Peroukidis SD, Lichtner K, Klapp SHL. Tunable structures of mixtures of magnetic particles in liquid-crystalline matrices. Soft Matter 2015;11:5999–6008.
26. Peroukidis SD, Klapp SHL. Spontaneous ordering of magnetic particles in liquid crystals: from chains to biaxial lamellae. Phys Rev E 2015;92:010501.
27. Appel I, Nádasi H, Reitz C, Sebastián N, Hahn H, Eremin A, et al. Doping of nematic cyanobiphenyl liquid crystals with mesogen-hybridized magnetic nanoparticles. Phys Chem Chem Phys 2017; 504:1–9.
28. Zrínyi M, Barsi L, Büki A. Ferrogel: a new magneto-controlled elastic medium. Polym Gels Netw 1997;5:415–27.
29. Gollwitzer C, Turanov A, Krekhova M, Lattermann G, Rehberg I, Richter R. Measuring the deformation of a ferrogel sphere in a homogeneous magnetic field. J Chem Phys 2008;128:164709.
30. Varga Z, Filipcsei G, Zrínyi M. Smart composites with controlled anisotropy. Polymer 2005;46: 7779–87.
31. Jestin J, Cousin F, Dubois I, Ménager C, Schweins R, Oberdisse J, et al. Anisotropic reinforcement of nanocomposites tuned by magnetic orientation of the filler network. Adv Mater 2008;20:2533–40.
32. Krekhova M, Lattermann G. Thermoreversible organoferrogels: morphological, thermal and magnetic characterisation. J Mater Chem 2008;18:2842–8.
33. Krekhova M, Lattermann G, Schmalz H, Lang T, Richter R. Towards softer thermo-reversible magnetogels. Phys Procedia 2010;9:224–8.
34. Gollwitzer C, Krekhova M, Lattermann G, Rehberg I, Richter R. Surface instabilities and magnetic soft matter. Soft Matter 2009;5:2093–8.
35. Galicia JA, Cousin F, Dubois E, Sandre O, Cabuil V, Perzynski R. Static and dynamic structural probing of swollen polyacrylamide ferrogels. Soft Matter 2009;5:1–11.

36. Hasmonay E, Bee A, Bacri JC, Perzynski R. pH effect on an ionic ferrofluid: evidence of a thixotropic magnetic phase. J Phys Chem B 1999;103:6421–8.
37. Nadasi H, Corradi Á, Stannarius R, Koch K, Schmidt AM, Aya S, et al. The role of structural anisotropy in the magnetooptical response of an organoferrogel with mobile magnetic nanoparticles. Soft Matter 2019;15:3788–95.
38. Embs J, Müller HW, Krill CE, Meyer F, Natter H, Müller B, et al. Characterization of the grain size in ferromagnetic colloids: comparing torsional-pendulum measurements with standard complementary methods. Z Phys Chem 2006:153–71.
39. Embs JP, Huke B, Leschhorn A, Lücke M. Equilibrium and nonequilibrium behaviour of ferrofluids – experiments and theory. Z Phys Chem 2008;222:527–86.
40. de Gennes PG, Pincus PA. Pair correlations in a ferromagnetic colloid. Phys kondens Materie 1970; 11:189–98.
41. Jordan PC. Association phenomena in a ferromagnetic colloid. Mol Phys 1973;25:961–73.
42. Xu M, Ridler PJ. Linear dichroism and birefringence effects in magnetic fluids. J Appl Phys 1997;82: 326.
43. Ivanov AO, Kantorovich SS, Mendelev VS, Pyanzina ES. Ferrofluid aggregation in chains under the influence of a magnetic field. J Magn Magn Mater 2006;300:e206–9.
44. Ivanov AO, Kantorovich SS. Chain aggregate structure and magnetic birefringence in polydisperse ferrofluids. Phys Rev E 2004;70:021401.
45. Kantorovich SS. Chain aggregate structure in polydisperse ferrofluids: different applications. J Magn Magn Mater 2005;289:203–6.
46. Butter K, Bomans PH, Frederik PM, Vroege GJ, Philipse AP. Direct observation of dipolar chains in ferrofluids in zero field using cryogenic electron microscopy. J Phys Condens Matter 2003;15: S1451–70.
47. Odenbach S. Ferrofluids – magnetically controlled suspensions. Colloid Surface Physicochem Eng Aspect 2003;217:171–8.
48. Zubarev AY, Iskakova LY. Theory of structural transformations in ferrofluids: chains and "gas-liquid" phase transitions. Phys Rev E 2002;65:061406.
49. Sánchez PA, Gundermann T, Dobroserdova A, Kantorovich S, Odenbach S. Importance of matrix inelastic deformations in the initial response of magnetic elastomers. Soft Matter 2018;14:1–14.
50. Escuder B, Miravet JF, editors. Functional molecular gels. Soft Matter Series. Cambridge: Royal Society of Chemistry; 2013.
51. Guenet J-M. Organogels thermodynamics, structure, solvent role, and properties. Switzerland: Springer; 2016.
52. Terech P, Weiss RG. Low molecular mass gelators of organic liquids and the properties of their gels. Chem Rev 1997;97:3133–60.
53. van Esch JH, Feringa BL. New functional materials based on self-assembling organogels: from serendipity towards design. Angew Chem Int Ed 2000;39:2263–6.
54. Tachibana T, Mori T, Hori K. New type of twisted mesophase in jellies and solid films of chiral 12-hydroxyoctadecanoic acid. Nature 1979;278:578–9.
55. Co ED, Marangoni AG. Organogels: an alternative edible oil-structuring method. J Am Oil Chem Soc 2012;89:749–80.
56. Rogers MA, Wright AJ, Marangoni AG. Engineering the oil binding capacity and crystallinity of self-assembled fibrillar networks of 12-hydroxystearic acid in edible oils. Soft Matter 2008;4:1483–9.
57. Gao J. Solvent induced modifications to fiber nanostructure and morphology for 12HSA molecular gels [PhD thesis]. New Jersey: The State University of Rutgers; 2014.
58. Sakurai T, Masuda Y, Sato H, Yamagishi A, Kawaji H, et al. A comparative study on chiral and racemic 12-hydroxyoctadecanoic acids in the solutions and aggregation states: does the racemic form really form a gel?. Bull Chem Soc Jpn 2010;83:145–50.

59. Lan Y, Rogers MA. 12-Hydroxystearic acid SAFiNs in aliphatic diols – a molecular oddity. CrystEngComm 2015;17:8031–8.
60. Rogers MA, Marangoni AG. Kinetics of 12-hydroxyoctadecanoic acid SAFiN crystallization rationalized using hansen solubility parameters. Langmuir 2016;32:12833–41.
61. Terech P. 12-D-Hydroxyoctadecanoic acid organogels : a small angle neutron scattering study. J Phys II 1992;2:2181–95.
62. Lam R, Quaroni L, Pedersen T, Rogers MA. A molecular insight into the nature of crystallographic mismatches in self-assembled fibrillar networks under non-isothermal crystallization conditions. Soft Matter 2010;6:404–8.
63. Grahame DAS, Olauson C, Lam RS, Pedersen HT, Borondics F, Abraham S, et al. Influence of chirality on the modes of self-assembly of 12-hydroxystearic acid in molecular gels of mineral oil. Soft Matter 2011;7:7359–65.
64. Sato H, Hori K, Sakurai T, Yamagishi A. Long distance chiral transfer in a gel: experimental and ab initio analyses of vibrational circular dichroism spectra of R- and S-12-hydroxyoctadecanoic acid gels. Chem Phys Lett 2008;467:140–3.
65. Laupheimer M, Preisig N, Stubenrauch C. The molecular organogel n-decane/ 12-hydroxyoctadecanoic acid: sol–gel transition, rheology, and microstructure. Colloid Surface Physicochem Eng Aspect 2015;469:315–25.
66. Szczytko J, Vaupotic N, Osipov MA, Madrak K, Gorecka E. Effect of dimerization on the field-induced birefringence in ferrofluids. Phys Rev E 2013;87:062322.
67. Jordan PC. Field dependent chain formation by ferromagnetic colloids. Mol Phys 1979;38:769–80.
68. Kantorovich S, Ivanov AO, Rovigatti L, Tavares JM, Sciortino F. Nonmonotonic magnetic susceptibility of dipolar hard-spheres at low temperature and density. Phys Rev Lett 2013;110: 148306.
69. Osipov MA, Gorkunov MV. Effect of nanoparticle chain formation on dielectric anisotropy of nematic composites. Phys Rev E 2015;92:26.
70. Raşa M. Magnetic properties and magneto-birefringence of magnetic fluids. Eur. Phys. J. E 2000;2: 265–75.
71. Raşa M. Improved formulas for magneto-optical effects in ferrofluids. J Magn Magn Mater 1999; 201:170–3.
72. Janssen JJM, Perenboom JAAJ. Magneto-optical phenomena in magnetic fluids: the influence of orientation of anisotropic scatterers. J Magn Magn Mater 1989;81:14–24.
73. Pshenichnikov AF, Lebedev AV, Shliomis M. On the rotational effect in nonuniform magnetic fluids. Magnetohydrodynamics 2000;36:275–81.
74. Tsebers AO. Interfacial stresses in the hydrodynamics of liquids with internal rotations. Magnetohydrodynamics 1975;11:79–82.
75. Tsebers AO. Torques and hydrodynamics of a magnetic fluid in homogeneous rotating magnetic fields. Magnetohydrodynamics 1978;14:398–401.
76. Lebedev AV, Pshenichnikov AF. Motion of a magnetic fluid in a rotating magnetic field. Magnetohydrodynamics 1991;27:4–8.
77. Lebedev AV, Pschenichnikov AF. Rotational effect: the influence of free or solid moving boundaries. J Magn Magn Mater 1993;122:227–30.
78. Pshenichnikov AF, Lebedev AV. Action of a rotating magnetic field on a dielectric cylinder immersed in a magnetic fluid. J Appl Mech Tech Phys 1996;37:305–10.
79. Rosensweig RE, Popplewell J, Johnston RJ. Magnetic fluid motion in rotating field. J Magn Magn Mater 1990;85:171–80.
80. Bacri JC, Perzynski R, Shliomis MI, Burde GI. “Negative-Viscosity” effect in a magnetic fluid. Phys Rev Lett 1995;75:2128–31.

81. Sánchez JH, Rinaldi C. Magnetoviscosity of dilute magnetic fluids in oscillating and rotating magnetic fields. Phys Fluids 2010;22:043304–9.
82. Engel A, Müller H-W, Reimann P, Jung A. Ferrofluids as thermal ratchets. Phys Rev Lett 2003;91: 060602.
83. John T, Stannarius R. Experimental investigation of a brownian ratchet effect in ferrofluids. Phys Rev E 2009;80:050104.
84. Storozhenko AM, Stannarius R, Tantsyura AO, Shabanova IA. Measurement of the torque on diluted ferrofluid samples in rotating magnetic fields. J Magn Magn Mater 2017;431:66.
85. Usadel KD, Storozhenko A, Arefyev I, Nádasi H, Trittel T, Stannarius R, et al. Frequency-dependent conversion of the torque of a rotating magnetic field on a ferrofluid confined in a spherical cavity. Soft Matter 2019;15:9018.
86. Storozhenko A, Stannarius R, Eremin A, Aref'ev I. Torque on suspensions of anisotropic pigment particles and magnetic nanoparticles in a rotating magnetic field. Dresden: Poster at the 16th Ferrofluid Workshop; 2016.
87. May K, Eremin A, Stannarius R, Szabó B, Börzsönyi T, Appel I, et al. Exceptionally large magneto-optical response in dispersions of plate-like nanocrystallites and magnetic nanoparticles. J Magn Magn Mater 2017;431:79–83.

Maik Liebl*, Dietmar Eberbeck, Annelies Coene, Jonathan Leliaert, Philine Jauch, Margarita Kruteva, Lisa Fruhner, Lester Barnsley, Stefan G. Mayr and Frank Wiekhorst

9 Magnetic measurement methods to probe nanoparticle–matrix interactions

Abstract: Magnetic nanoparticles (MNPs) are key elements in several biomedical applications, e.g., in cancer therapy. Here, the MNPs are remotely manipulated by magnetic fields from outside the body to deliver drugs or generate heat in tumor tissue. The efficiency and success of these approaches strongly depend on the spatial distribution and quantity of MNPs inside a body and interactions of the particles with the biological matrix. These include dynamic processes of the MNPs in the organism such as binding kinetics, cellular uptake, passage through cell barriers, heat induction and flow. While magnetic measurement methods have been applied so far to resolve the location and quantity of MNPs for therapy monitoring, these methods can be advanced to additionally access these particle–matrix interactions. By this, the MNPs can further be utilized as probes for the physical properties of their molecular environment. In this review, we first investigate the impact of nanoparticle–matrix interactions on magnetic measurements in selected experiments. With these results, we then advanced the imaging modalities magnetorelaxometry imaging and magnetic microsphere tracking to spatially resolve particle–matrix interactions.

Keywords: AC susceptibility; magnetic nanoparticles; magnetic particle spectroscopy; magnetorelaxometry; multicolor imaging; particle–matrix interactions.

***Corresponding author: Maik Liebl,** Physikalisch-Technische Bundesanstalt, Abbestraße 2-12, 10587 Berlin, Germany, E-mail: maik.liebl@ptb.de

Dietmar Eberbeck and Frank Wiekhorst, Physikalisch-Technische Bundesanstalt, Abbestraße 2-12, 10587 Berlin, Germany

Annelies Coene, Department of Electromechanical, Systems and Metal Engineering, Ghent University, 9052 Ghent, Belgium; and Cancer Research Institute Ghent (CRIG), 9000 Ghent, Belgium

Jonathan Leliaert, Department of Solid State Sciences, Ghent University, 9000 Ghent, Belgium

Philine Jauch and Stefan G. Mayr, Leibniz Institute of Surface Engineering (IOM), Permoserstrasse 15, 04318 Leipzig, Germany; and Division of Surface Physics, Department of Physics and Earth Sciences, University of Leipzig, Linnéstrasse 5, 04103 Leipzig, Germany

Margarita Kruteva, Jülich Centre for Neutron Science (JCNS-1) and Institute for Complex Systems (ICS-1), Forschungszentrum Jülich GmbH, Leo-Brandt-Straße, 52425 Jülich, Germany

Lisa Fruhner, Jülich Centre for Neutron Science (JCNS-1) and Institute for Complex Systems (ICS-1), Forschungszentrum Jülich GmbH, Leo-Brandt-Straße, 52425 Jülich, Germany; and Institute of Physical Chemistry, RWTH Aachen University, Landoltweg 2, 52056 Aachen, Germany

Lester Barnsley, Forschungszentrum Jülich GmbH, Jülich Centre for Neutron Science at MLZ, Lichtenbergstr. 1, 85748 Garching, Germany

This article has previously been published in the journal Physical Sciences Reviews. Please cite as: M. Liebl, D. Eberbeck, A. Coene, J. Leliaert, P. Jauch, M. Kruteva, L. Fruhner, L. Barnsley, S. G. Mayr and F. Wiekhorst "Magnetic measurement methods to probe nanoparticle–matrix interactions" *Physical Sciences Reviews* [Online] 2021, 6. DOI: 10.1515/psr-2019-0112 | https://doi.org/10.1515/9783110569636-009

9.1 Magnetic measurement methods to support biomedical applications of magnetic nanoparticles

Magnetic nanoparticles (MNPs) used in biomedical applications typically consist of a magnetic iron oxide (e.g., magnetite, maghemite) core of 4–30 nm in diameter surrounded by an organic shell. Their unique magnetic properties and the small size of the MNPs allow them to function at a cellular level, making them attractive candidates for cell labeling, imaging, tracking and as carriers. MNPs serve as actuators and local probes in novel therapeutic and diagnostic applications (for review see e.g., [1–4]). For instance, in magnetic hyperthermia an alternating magnetic field is applied toward the MNPs for local heating of tumor tissue. In magnetic drug targeting, the MNPs serve as drug carriers that are remotely guided toward a tumor by magnetic field gradients to accumulate the drug on the tumor side. Other examples are targeted accumulation of genetic material (magnetofection) and contrast agents for magnetic resonance imaging and magnetic particle imaging (MPI).

In all these applications, the MNPs get into contact and thus interact with a biological environment or the surrounding physiological matrix such as tissue, cells, or the bloodstream. The safety and success of these applications vitally depends on a quantitative and spatially resolved knowledge of the interactions attributed to the biological environment. This can be accomplished by probing the surrounding matrix via magnetic measurement techniques capable of tracking changes in the magnetic behavior of MNPs. Examples are the investigation of cellular MNP uptake in target cells [5–7], MNP triggered cell growth [8, 9], the long-term fate and metabolism of MNPs [10], MNP-matrix interactions affecting hyperthermia application [11, 12], the interaction of MNPs with blood molecules [13, 14], biological matrices [15] and the passage of MNPs through cell barriers [16–19].

One possibility to analyze particle–matrix interactions (PMIs) is by probing the dynamics of the MNP magnetic moments. The ensemble magnetization responds with a time delay to a change in an external magnetic field. For immobilized MNPs, this delay is determined by the Néel relaxation time [20] necessary for the magnetic moments to overcome the energy barriers resulting from magneto-crystalline structure and/or shape anisotropy. If the MNPs are embedded in a matrix but are still free to mechanically rotate as a whole, Brownian relaxation [21] is possible. The Brownian relaxation time depends on the viscosity of the matrix and the hydrodynamic MNP size distribution. If both relaxation mechanisms are present, the faster mechanism dominates the resulting effective MNP relaxation.

In the following, we describe some magnetic measurements probing the rotational MNP motion to investigate PMIs for biomedical applications. In Section 9.2, we introduce the magnetic measurement techniques used to characterize MNP–matrix systems.

These are applied in Section 9.3 on well-defined MNP-matrix systems to investigate the effect of PMIs in the respective MNP response. In Sections 9.4 and 9.5, we incorporate these results into the development of novel imaging modalities for a quantitative and spatially resolved description of PMIs.

9.2 Magnetic measurement and imaging techniques

9.2.1 Magnetorelaxometry

In magnetorelaxometry (MRX) [22], the MNP response to a fast change in an applied magnetic field is detected by a sensitive magnetic field sensor, e.g., a Superconducting Quantum Interference Device (SQUID) [22] or a fluxgate [23]. MRX measurements consist of two consecutive phases. In a first magnetizing phase, a static magnetic field $\mathbf{H}_v$ with a typical amplitude of 1 mT/μ_0 is applied toward the MNP sample at position $\mathbf{r}_v$ in order to partially align the individual particle moments along the field direction resulting in a net magnetic moment $\mathbf{m}_0 = \chi_m \mathbf{H}_v X_{\mathrm{MNP}}$. This moment depends on the mass susceptibilityχ_m (m^3/kg) of the MNPs, the strength of the applied magnetic field $\mathbf{H}_v$ and the MNP mass X_{MNP} within the sample. In the subsequent measurement phase initiated by switching off the applied magnetic field, the decay of this net magnetic moment over time is detected. This decay is determined by Néel [20] and Brownian relaxation [21] processes and can be described as $\mathbf{m}(t) = \mathbf{m}_0 \cdot \kappa(t)$ where the monotonically decreasing relaxation function $\kappa(t)$ displays values between one and zero. Both relaxation mechanisms are characterized by a specific time constant:

$$\tau_B = \frac{3\eta V_{\mathrm{h}}}{k_{\mathrm{B}}T}, \tau_N = \tau_0 \exp\left(\frac{KV_{\mathrm{c}}}{k_B T}\right) \tag{9.1}$$

The Brownian relaxation time τ_B is determined by the hydrodynamic volume V_{h} of the nanoparticle, the local viscosity η of the particle surroundings, and its thermal agitation given by temperature T and Boltzmann constant k_{B}. The Néel relaxation time is defined by a time constant τ_0 with values in literature between 10^{-8} and 10^{-12} s [24], the anisotropy constant K and the volume of its magnetic core V_{c}. If both relaxation mechanisms occur, the particle relaxation is determined by an effective relaxation time τ_{eff}

$$\tau_{\mathrm{eff}}(V_c, V_h) = \frac{\tau_N(V_{\mathrm{c}})\tau_B(V_{\mathrm{h}})}{\tau_N(V_{\mathrm{c}}) + \tau_B(V_{\mathrm{h}})} \tag{9.2}$$

in which the fastest relaxation mechanism dominates the effective relaxation time. In practice, MNP systems exhibit a broad size distribution $P(V_{\mathrm{c}}, V_{\mathrm{h}})$. Hence, the relaxation function $\kappa(t)$ detected in MRX experiments is composed of effective relaxation times given by both the core and hydrodynamic MNP size distribution.

In the MRX measurement phase, the decaying net magnetic moment $\mathbf{m}(t)$ of a MNP source at position $\mathbf{r}_v$ gives rise to the MRX signal $B_s(t)$ in sensor s. The measured flux density of a point-like MNP source, $B_s(t)$ detected by sensor s at position $\mathbf{r}_s$ with a sensitive axis $\mathbf{n}_s = [n_x, n_y, n_z]$ (normal vector of the sensor) is given by [25]:

$$B_s(t) = \frac{\mu_0}{4\pi}\left(\frac{3\left(\mathbf{n}_s^{\mathrm{T}}(\mathbf{r}_s - \mathbf{r}_v)\right)(\mathbf{r}_s - \mathbf{r}_v)^{\mathrm{T}}}{\|(\mathbf{r}_s - \mathbf{r}_v)\|^5} - \frac{\mathbf{n}_s^{\mathrm{T}}}{\|(\mathbf{r}_s - \mathbf{r}_v)\|^3}\right)\mathbf{H}_v \chi_m \kappa(t) X_{MNP} \tag{9.3}$$

where μ_0 is the vacuum permeability and $\mathbf{H}_v$ is the magnetic field on position $\mathbf{r}_v$. The obtained MRX curve can be parametrized by the relaxation amplitude ΔB ($=B(t_1) - B(t_2)$ with $t_1 < t_2$) and the relaxation time $t_{1/e}$, i.e. the period after that $\kappa(t)$ has dropped by 36.7% of its value at $B(t_1)$. The relaxation amplitude is directly proportional to the MNP amount of the sample allowing for MNP quantification. The relaxation time parameter $t_{1/e}$ depends on the Brownian and Néel relaxation processes which will be employed to probe interactions between the nanoparticles and the physiological samples environment.

9.2.2 Magnetorelaxometry imaging

Magnetorelaxometry imaging (MRXI) is a sensitive and specific imaging modality for 3D quantification of MNP distributions. In MRXI, the MRX response of a MNP sample is measured on different locations by a sensor array [26, 27], where the spatial encoding is improved by applying a series of spatially constrained magnetic fields using a number of K excitation coils surrounding the measurement volume [25, 28, 29]. By combining all individual MRX measurements, the quantitative 3D reconstruction of the MNP distribution can be obtained as a solution of an inverse problem.

In classical MRX imaging, the presence of only one type of MNPs in the measurement volume is assumed, corresponding to one relaxation function $\kappa(t)$. Note that the only time dependency in the model of Equation (9.3) relies in $\kappa(t)$. It becomes static when only the difference in amplitude $\kappa(t_1, t_2) = \kappa(t_1) - \kappa(t_2)$ with $t_1 < t_2$ is of interest. Thus, the classical MRXI forward model calculates only the difference in amplitude between two time points, t_1 and t_2, of the MRX signal $Bs(t_1, t_2)$. In practice, the relaxation product $\chi_m\ \kappa(t_1, t_2)$ is obtained from an MRX measurement of a reference sample containing a known MNP amount.

Equation (9.3) can be simplified by including previous MNP material and geometry parameters of the MRXI setup in the sensitivity coefficient Lsv linking the particle mass in voxel v to the measurement in sensor s

$$B_s(t_1, t_2) = L_{sv} X_{\mathrm{MNP},v} \tag{9.4}$$

Equation (9.4) can be extended from just one sensor and one voxel to S sensors and V voxels:

$$\mathbf{B} = [B_1, \ldots, B_S]^{\mathrm{T}} = \sum_{v=1}^{V} L_{s,v} X_{MNP,v} = \mathbf{L}\mathbf{X}_{\mathrm{MNP}} \tag{9.5}$$

with the sensitivity matrix $\mathbf{L}$ having dimensions ($S \times V$) and $\mathbf{X}_{\mathrm{MNP}}$ being a vector containing the MNP masses for each voxel v $[X_{\mathrm{MNP},1}, \ldots, X_{\mathrm{MNP},V}]^{\mathrm{T}}$. In practice, the MRX measurements (Equation 9.5) are performed for K spatially varying magnetic fields generated by the excitation coils. This results in the following forward model

$$\mathbf{B}_{\mathrm{sim}} = [\mathbf{B}_1, \ldots, \mathbf{B}_K]^{\mathrm{T}} = [\mathbf{L}_1, \ldots, \mathbf{L}_K]^{\mathrm{T}} \mathbf{X}_{\mathrm{MNP}} \tag{9.6}$$

with $\mathbf{B}_k$ containing the S relaxation measurements for the k-th magnetic field pattern and $\mathbf{L}_k$ the corresponding sensitivity matrix. In our MRXI setup 30 excitation coils are applied sequentially to magnetize the MNP distribution with 30 spatially distinct magnetic fields. Thus, we obtain $K = 30 \times S$ single MRX measurements and a sensitivity matrix $\mathbf{L}$ of dimension ($KS \times V$). The MNP distribution consisting of V MNP masses $\mathbf{X}_{\mathrm{MNP}}$ is then recovered by solving

$$\boldsymbol{X}^{*}_{\mathrm{MNP}} = \underset{\mathbf{X}_{\mathrm{MNP}}}{\operatorname{argmin}} \| \mathbf{B}_{\mathrm{sim}} - \mathbf{B}_{\mathrm{meas}} \| \tag{9.7}$$

where the difference between the modeled measurements $\mathbf{B}_{\mathrm{sim}}$, and the actual MRX measurements $\mathbf{B}_{\mathrm{meas}}$, are minimized by searching for the most probable MNP distribution $\mathbf{X}_{\mathrm{MNP}}$ to cause $\mathbf{B}_{\mathrm{meas}}$. Equation (9.7) has been successfully solved in previous work using a non-negative least squares (NNLS) [30] and by truncated singular value decomposition (tSVD) [26, 29]. In NNLS, Equation (9.7) is iteratively solved with the constraints that all elements of the solution (i.e. the MNP amounts) are positive [31]

$$\mathbf{w} = \mathbf{L}^{\mathrm{T}} (\mathbf{B}_{\mathrm{sim}} - \mathbf{B}_{\mathrm{meas}}) \tag{9.8}$$

In this equation, $\mathbf{w}$ is the dual vector and should finally only contain elements smaller than or equal to zero, so that the solution $\mathbf{X}^{*}_{\mathrm{MNP}}$ only has positive elements.

9.2.3 MNP response to alternating magnetic fields

In contrast to the static magnetic field applied in MRX magnetization, PMIs can also be characterized by the MNP response to alternating magnetic fields. For moderate field amplitudes (0.5–1 mT/μ_0) this is referred to as AC susceptibility (ACS). ACS uses an induction coil system to measure the linear dynamic magnetic response of a sample to an alternating magnetic field within the frequency range 10 Hz–0.5 MHz. At these moderate magnetic field amplitudes, ACS data is well described by the Debye model [32], where

$$\chi'(\omega) = \frac{\chi_0}{1 + (\omega\tau_{\text{eff}})^2} \tag{9.9}$$

and

$$\chi''(\omega) = \chi_0 \frac{\omega\tau_{\text{eff}}}{1 + (\omega\tau_{\text{eff}})^2} \tag{9.10}$$

are the real $\chi'(\omega)$ and imaginary parts $\chi''(\omega)$ of the complex susceptibility $\chi(\omega)$. The equilibrium susceptibility χ_0 is given by

$$\chi_0 = \frac{\mu_0 n_p \mu_p^2}{3k_B T} \tag{9.11}$$

with μ_p the moment of a single particle and n_p the number of particles. The phase φ of the complex susceptibility is defined by $\varphi = \arctan(\chi''(\omega)/\chi'(\omega))$. Note that the effective relaxation time τ_{eff} is determined as the maximum in the imaginary part $\omega\tau_{\text{eff}} = 1$. Similar to MRXI, additional spatial encoding in ACS by multiple sensors or excitation coils allows for the development of MNP imaging approaches [33–36].

Increasing the amplitude of the AC magnetic field allows to obtain additional signal contributions of the MNPs by their nonlinear magnetic susceptibility. This is employed in Magnetic Particle Spectroscopy (MPS). Here, the MNP response to AC magnetic fields of some tens of mT/μ_0 (here $B = 25$ mT) at fixed frequency in the kHz range (here $f_0 = 25$ kHz) is detected by an induction coil system. The MPS signal does not only contain signal components at the excitation frequency f_0, but also MNP-specific higher harmonics (i.e. odd multiples of f_0). After obtaining the MPS spectrum from a Fourier transformation, two characteristic parameters can be extracted. These are the amplitude of the third harmonic M_3 and the harmonic ratio M_5/M_3, i.e. amplitude ratio between fifth and third harmonic. M_3 is directly proportional to the MNP amount of the sample. M_5/M_3 depends on the dynamic magnetic behavior of the MNPs and is often used as indicator for PMIs [12]. In magnetic particle imaging (MPI) [37], this concept is combined with a spatial encoding by magnetic gradient fields for the 3D quantitative imaging of MNP distributions.

9.3 Interaction of MNPs with matrices

Nanoparticle matrix systems generally consist of MNPs embedded in a nonmagnetic often soft material environment. Prominent technical applications are MNP-polymer systems where mechanical properties of the matrix like the elastic moduli shall be controlled by external magnetic fields (e.g., [38–40]). In a biological MNP-matrix system the MNP can be located in the cellular system (inside the cell, at the cell surface or in the intercellular space) and may be targeted to transport a drug by external magnetic fields. Since the magnetic response of the MNPs is sensitive to the particle surrounding, this offers a way to analyze PMIs. To reduce the complexity of biological systems, one often resorts to model systems like MNPs embedded in hydrogel (also called ferrogel). Here, the viscous hydrogel emulates the biological matrix with well-defined properties. Microscopically, the hydrogel is not a homogeneous matrix, but a polymer scaffold filled with an aqueous, solvating fluid.

The PMIs in hydrogels are mainly caused by (i) the MNP binding with the polymer scaffold and (ii) the drag of MNP movements determined by the (local) viscosity of the solvating fluid and the inner structure of the hydrogel, e.g., the pore size and its connectivity. On the other hand, the coating and the hydrodynamic size of the MNPs are important factors to study PMIs. For instance, the presence of functional groups like carboxylate groups (COO^-) in the coating may lead to a preferential binding of MNPs to positively charged domains of the hydrogel scaffold. The drag of the MNP motion is caused by the fluid resistance, which is mainly determined by the hydrodynamic size of the MNPs but also by reversible (weak) bonding caused by macromolecular entanglement between coating molecules and matrix molecules. The binding functionality can be measured using a magnetic assay [41], but in the following we focus on the physical structure of the coating, i.e. its thickness and density.

As magnetic measurement techniques we apply MRX, ACS and MPS (Section 9.2) and analyze PMIs by detecting changes in the Brownian and Néel relaxation of the MNPs. We used MNP systems with different core material and coating. For convenience they are labeled X_i*Y, where the first letter X indicates the core material, with X = M for magnetite and X = C for cobalt ferrite, followed by a consecutive sample number *i*. The term Y after the delimiting * denotes the key molecular part of the coating. Here, Y=COO^- indicates carboxylate groups, Y=CONH-PEG denotes amino-PEG (polyethylene-glycol) coupled to a carboxylate group and, Y=S stands for silica (SiO_2).

9.3.1 Probing the melting of a matrix by Brownian MNP relaxation

First, we investigated the melting behavior of ferrogels in gelatin. Here, we used the Brownian relaxation of the MNPs to probe the local viscosity of the matrix during its melting. To this end we used MPS and MRX as sensitive magnetic measurement techniques to detect variations in the MNP relaxation caused by matrix property changes during the melting (e.g., disintegration, reduced viscosity).

Figure 9.1 displays the MRX relaxation time and the MPS amplitude of the third harmonic normalized to iron mass M_3 during the melting process of gelatin, i.e. as a function of temperature. As expected, significant parameter changes are observed for temperature close to the gelatin melting point of about 308 K. This is attributed to the onset of Brownian rotation of the MNPs becoming possible above the melting point. Interestingly, there are small parameter changes already before the macroscopic melting point (Figure 9.1). These are interpreted as local melting [42], where the MNPs sense a maceration of the matrix without being disintegrated. This is further confirmed by a drop in the MRX relaxation amplitude (not shown in the graph).

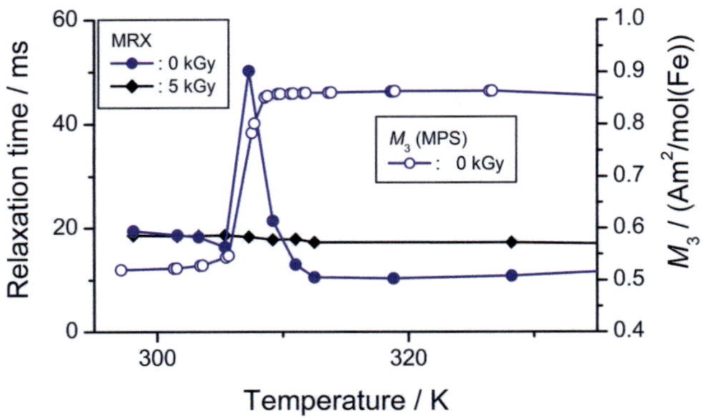

Figure 9.1: MPS third harmonic M_3 (open blue circles) and MRX relaxation time (closed blue symbols) during melting of gelatin with embedded MNPs. Black curve: Relaxation time after hardening of the ferrogel by electron irradiation with 5 kGy dose protecting the gelatin from melting up to at least 70 °C. Adapted with permission from Wisotzki et al. [42]. Copyright 2017 Royal Society of Chemistry.

While the M_3 clearly remains increased after crossing the melting point, the MRX relaxation time reaches a maximum at the melting point. This can be explained by the presence of Brownian and Néel relaxation at the same time: The MNP system exhibits a broad distribution of Néel relaxation times that dominate the magnetic behavior up to the melting point. A fraction of MNPs exhibits Néel relaxation times above the upper limit of the MRX measurement time window what explains the small initial value of the effective relaxation time. In the melting process, particles of those fraction now enter the MRX measurement time window by the Brownian relaxation mechanism, which leads to the maximum in the relaxation time. With increasing temperature, the MNPs sense a strong change from high to low viscosity resulting in a decrease of their viscosity-dependent Brownian relaxation times after the maximum. A more detailed and quantitative interpretation can be found in Ref. [42]. The black squares in Figure 9.1 are the relaxation times obtained from a ferrogel that was hardened by electron beam of strength 5 kGy. While no macroscopic melting was observed, small variations in the relaxation times during the temperate increase are again attributed to changes in the local viscosity.

Thus, we demonstrated that MRX and MPS are sensitive to both, the local and macroscopic melting behavior of ferrogels.

9.3.2 Probing the spatial arrangement of MNPs within a matrix by Néel relaxation

The Néel relaxation is defined as the thermal activated jump of the MNP magnetic moment over the barrier of its anisotropy energy, i.e. the internal flip of the magnetization vector. If MNPs aggregate, then dipole–dipole interactions between MNPs might

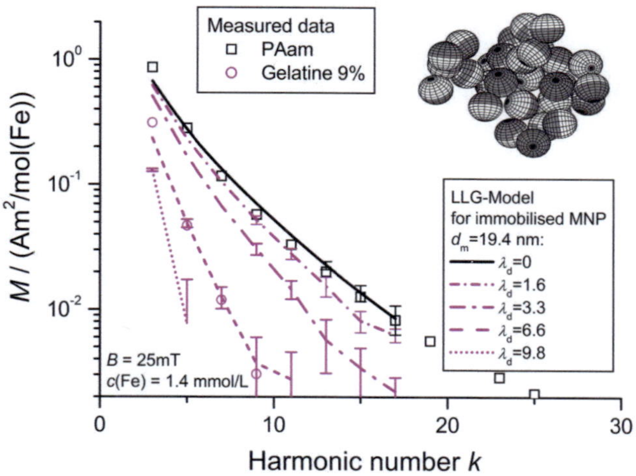

Figure 9.2: MPS-data of ferro-hydrogels on the base of polyacrylamide gel (PAam) and gelatin (symbols) together with the results of the LLG-model for clusters of 30 randomly arranged immobilized MNP of one size (d_m = 19.4 nm) taking into account the indicated interaction energies.

alter these internal energy barriers. This affects the magnetic susceptibility detected by ACS and MPS. Hence, changes in ACS or MPS signals provides information about the aggregation state of embedded MNP probes. To investigate these changes, we immobilized magnetite based MNP with a coating containing carboxylate groups on its surface [42] in two different matrices, polyacrylamide (PAam) and gelatin. The MNPs were added prior to the polymerization process to ensure their homogeneous distribution within the sample. After polymerization, the samples were measured by MPS. The observed spectra are depicted in Figure 9.2. The spectrum of the gelatin-based sample decays much faster and the third harmonic M_3 is only 31% of that of the PAam based sample.

Using the parameters magnetic moment distribution and minimal possible particle distance, we modeled the MPS spectra of immobile MNPs solving the Landau–Lifshitz–Gilbert (LLG) equation considering the dipole–dipole interactions between the MNPs (in collaboration with P. Ilg, University of Reading). We modeled the magnetization response of close packed clusters comprising 30 randomly arranged MNPs of one size, all having a mean volume diameter of 19.4 nm. A clear dependence of the modeled MPS spectra on the interaction parameter $\lambda_d = \mu_0 \mu^2/(4\pi r^3 k_B T)$ is found (Figure 9.2). The interaction parameter λ_d describes the dipole–dipole interaction between two magnetic moments μ at a distance r. While the model of noninteracting MNPs $\lambda_d = 0$ describes the data of the PAam sample, an interaction parameter $\lambda_d = 6.6$ nm is necessary for a description of the gelatin-based sample. This interaction parameter corresponds to the interaction energy of closely packed MNPs. Hence, we conclude that the MNPs are aggregated within the gelatin matrix.

We hypothesize that the observed aggregation is caused by negatively charged carboxylate groups of the MNPs interacting with the charged collagen molecules of the gelatin. While gelatin A is predominantly positively charged (pI = 8.5–9) PAam does not contain charged functional groups. Note, due to hydrolysis there may be a molecular content of carboxylate groups of about 2–6%.

9.3.3 Requirements on MNPs to probe particle–matrix interactions

9.3.3.1 Suppressing Néel relaxation to sensitively monitor the MNP rotation inside a matrix

The physical state of the matrix (e.g., soft, hard, charged) can be characterized by measuring the Brownian motion of embedded MNPs. Methods like ACS, MRX, and MPS are very sensitive to detect rotational MNP movement while a translational motion requires magnetic gradient fields to drag the MNPs through the matrix. To accurately measure the rotational motion of the MNP as a whole, the Néel relaxation process (i.e. the rotation of the magnetic moment within the MNP) should be suppressed by a sufficiently high anisotropy energy barrier, i.e. $\tau_N >> \tau_B$. Cobalt ferrite is an appropriate core material for such MNPs because its anisotropy constant is about 10 times higher than that of magnetite. Furthermore, translational motion requires a high saturation magnetization of the whole core–shell MNP and the overall size of the MNP should match the structure size (pore size) under study. Finally, the analysis of dipolar interactions (Section 9.3.2) requires MNPs with high magnetic moments and a thin coating layer.

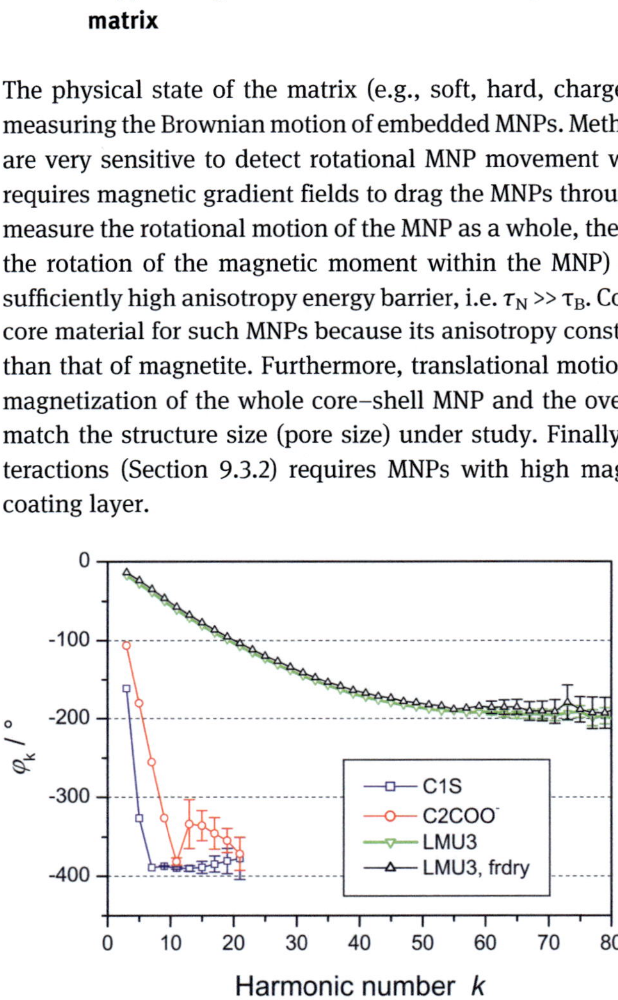

Figure 9.3: MPS phase data of the cobalt ferrite systems C1*S and C2*COO⁻ in comparison to magnetite based magnetosome particles LMU3 [43] dispersed in water and after freeze drying within a mannitol matrix.

The prevalence of the dominating relaxation mechanism (Brownian or Néel) can be assessed by the phase φ of the MPS data. For MNP systems with dominating Néel relaxation it is often observed that the MPS phase smoothly decays with the number of harmonics k asymptotically approaching to −180° (Figure 9.3). Such a characteristic phase curve is shown for mobile and immobilized magnetite based magnetosome particles LMU3 [43] in Figure 9.3. On the other hand, the Néel relaxation of cobalt ferrite MNPs with a core diameter larger than about 12 nm is strongly suppressed. Therefore, the Brownian relaxation dominates. The MPS phase curve of these MNPs decline nearly linearly and much faster with phases far below −180° for higher harmonics. The phase saturates at harmonic numbers, since here the signal of Néel relaxation of small MNPs prevails that of Brownian relaxation.

9.3.3.2 Opsonization as consequence of MNP-biomolecule interactions

The coating of MNPs influences their functionality and clearance in biomedical applications. The properties of the coating (thickness, density) might further be influenced by the medium in which the MNPs are dispersed. A prominent example is the opsonization of MNPs by proteins that may occur in blood and determines the interaction of the MNPs with the organism. Here, we combined the information gained by the magnetic measurement method ACS and Small Angle Neutron Scattering (SANS) to probe the opsonization of MNPs by the thickness and density of their opsonization layer. The thickness of the MNP coating can be estimated from the difference between hydrodynamic diameter d_h and core diameter d_c. MRX and ACS are promising integral measurement methods for the estimation of the distribution of d_h [44, 45]. The complex susceptibility $\chi(\omega)$ as measured by ACS is well suited to estimate d_h. We modeled $\chi(\omega)$ of noninteracting MNP with a lognormal core diameter distribution, $f(d_c)$, assuming each MNP surrounded by a shell of thickness δ_s

$$\chi(\omega) = \frac{1}{\overline{d_c^3}} \int \frac{f(d_c) d_c^3 L(H, M_s, d_c, T) \chi_0}{1 - i\omega\tau_{\text{eff}}} d_c \tag{9.12}$$

Here, M_s denotes the saturation magnetization and L the Langevin function. The hydrodynamic volume is given by $V_h = \pi(d_c + \delta_s)^3/6$. This becomes of particular interest when measuring in a (biological) environment. We estimated the thickness of the MNP's shell, δ_s, fitting the model of Equation (9.12) to $\chi(\omega)$.

Small Angle Neutron Scattering (SANS) with contrast variation provides a very sensitive access to the thickness of the MNP coating [46–48]. The scattering intensity of the sample scales with the square of the difference of scattering length densities of the particle core, ρ_p, the particle's shell, ρ_s, and that of the whole sample, ρ_0. Here, we varied ρ_0 by replacing H_2O by heavy water D_2O. In H_2O the contrast is mainly determined by the nanoparticle core scattering and in D_2O the details of shell structure are seen (see, e.g., [49]). The SANS measurement were performed on the instrument KWS-1

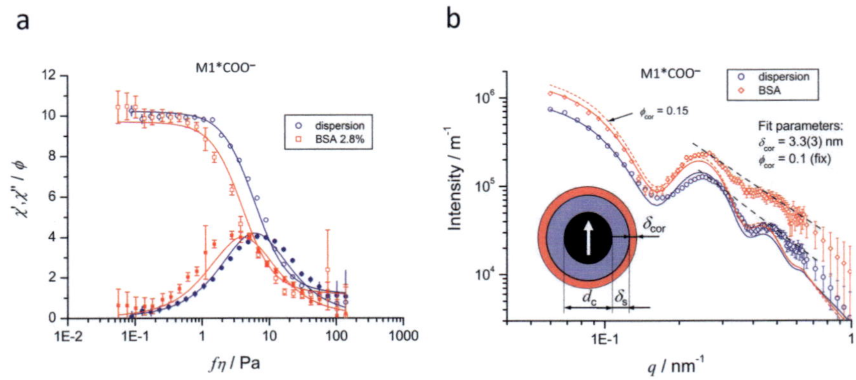

Figure 9.4: ACS-data (left) and SANS-data (right) of the commercially available nearly monodisperse MNP system SHP-30, denoted as M1*COO$^-$, before (blue symbols) and after opsonization by BSA (red symbols). The chosen for ACS delineation already eliminates the effect of the sample viscosity η on the data. The lines represent the best fit of a model of lognormally distributed magnetite cores with a coating and a corona layer of one thickness (inset). The short dash line represents the best fit model with corona with an adjusted density volume fraction of the corona, ϕ_{cor}, illustrating the sensitivity of SANS with respect to ϕ_{cor}. The SANS data of the sample with the opsonized MNP (in BSA) suffers from a quite large systematic uncertainty because of the high intensity of the background scattering of the BSA-solution. A solvation layer thickness of 1.5 nm was assumed in the fit. ϕ and η denote the volume fraction of core material, here magnetite, and the viscosity of the sample, respectively.

[50, 51] at the MLZ in Garching, Germany. The incident neutron wavelength λ was 7 Å ($\Delta\lambda/\lambda = 10\%$). Data was obtained from two different detector and collimation distances: the detector distances of 1.5 and 8 m and a collimation distance of 8 m leading to a total Q range of 0.05–1.5 nm^{-1}. The samples were measured in quartz cells with beam path of 1 mm for H_2O and 2 mm for D_2O. The measurements were done at room temperature. The size of the sample aperture was set as 6 × 6 mm. The data presented here were converted to an absolute intensity unit of m^{-1} taking into account the sample thickness, transmission, the scattering from a standard sample and the background from electronic noise, the solvent and the quartz cell. Data reduction has been done using the QtiKWS software [52].

Fitting a model [53, 54] describing the SANS data by noncorrelated core–shell particles (Figure 9.4) with the corresponding scattering length densities, we could estimate the thickness δ_S and density ϕ_S of the polymeric shell of the MNP system M1*COO$^-$ to $\delta_S = 2.9\,(1)$nm and $\phi_S = 1.0\,(2)$. Note, that only particles with a very narrow size distribution make the analysis robust (Figure 9.4b).

The particles M1*COO$^-$ were dispersed in a 4.3% bovine serum albumin (BSA) solution to probe their opsonization, i.e. a corona formation of BSA around the MNPs increasing their hydrodynamic diameter. For the estimation of the thickness of the second layer, i.e. the opsonization layer, we combined SANS and ACS data using a common model where both submodels share the same fit parameters of the particle

Table 9.1: Mean volume core diameter, d_{cv}, thickness of the coating shell, δ_S, and the thickness of the corona, δ_{cor}, obtained by fitting analysis of ACS, M(H), and additionally for M1*COO$^-$, SANS data.

Quantity	M1*COO$^-$	M2*COO$^-$
d_{cv}/nm	30.3 ± 0.2	30.0 ± 0.2
δ_S/nm	2.9 ± 0.1	2.8 ± 0.4
δ_{cor}/nm	3.3 ± 0.3	4.2 ± 0.2

structure (Figure 9.4). This was reasoned because of the relatively weak signal change in SANS by the opsonization and a relatively large uncertainty in SANS due to the large background signal. The fit yields an opsonization layer thickness of $\delta_{cor} = 3.3(3)$nm (Table 9.1). This matches well with the shortest dimension of an BSA molecule the shape of which is assumed similar to an oblate ellipsoidal shape with axes of $(3.4 \times 8.4 \times 8.4)$ nm^3 [54].

The relatively small signal increase measured following the opsonization (Figure 9.4b) reflects the low density (or volume fraction) of the corona formed by opsonization of about $\phi_{cor} = 0.1$. Note that ϕ_{cor} was adjusted manually because of fitting instabilities caused by the relatively large systematic uncertainties as mentioned above. Also, the applied core-shell-opsonization layer model (Figure 9.4b inset) might be an oversimplification of the real corona structure. For instance, the corona might have an inhomogeneous structure as suggested in [55]. The resulting lower fractal dimensionality would explain the slightly slower decay of $I(q)$ of opsonized MNPs within the Porod regime, as marked by dashed line. It was shown that ACS reliably allows to estimate the thickness of an opsonization layer while SANS additionally yields the density of this layer. By combined evaluation of ACS and SANS data, we could enhance the confidence of the results. In particular, with given thickness of the solvation layer, ACS sensitively displays the increase of the hydrodynamic diameter of the MNPs by the opsonization process.

9.3.4 Translational motion of MNPs with different coatings in matrices

To gain information about the migration of MNPs within a biological matrix, we investigate the translational motion of MNPs driven by an external magnetic field gradient through a collagen matrix. To achieve a large magnetic force, we utilized MNPs with a large magnetic moment of about 4 aAm^2. The MNPs were similar to those used in Ref. [42]. As mentioned above, the capability of the outer MNP shell to bind to the matrix determines the MNP's functionality. Therefore, we used two MNP systems with identical cores but different coatings. The carboxymethyl dextran shell of the first

Table 9.2: Mean volume core diameter, d_{cv}, thickness of the coating shell, δ_s, and the thickness of the corona, δ_{cor}, obtained by fitting analysis of ACS and M(H) data.

Quantity	M2*CONH-PEG
d_{cv}/nm	35.6 ± 0.2
δ_s/nm	20.5 ± 0.7
δ_{cor}/nm	0.5 ± 0.7

system, M2*COO⁻, presents carboxylate groups leading to a negative zeta potential at neutral pH. The second system, M2*CONH-PEG (Table 9.2), was prepared by coupling amino-PEG to M2*COO⁻. Theoretically, no charged functional groups are present in M2*CONH-PEG. This leads to the hypothesis that M2*CONH-PEG can be moved more effective by a field gradient through a hydrogel than M2*COO⁻. To verify this, we filled a tubular sample holder with collagen gel and loaded it with MNP dispersions. A homogeneous magnetic field gradient of 8 T/m was applied for 48 h. Afterward, the sample holder with the collagen was dissected into nine segments and the MNP content of each collagen segments was quantified by MPS. The results show that the concentration of M2*CONH-PEG within the second segment is 2.4 times higher than that of the

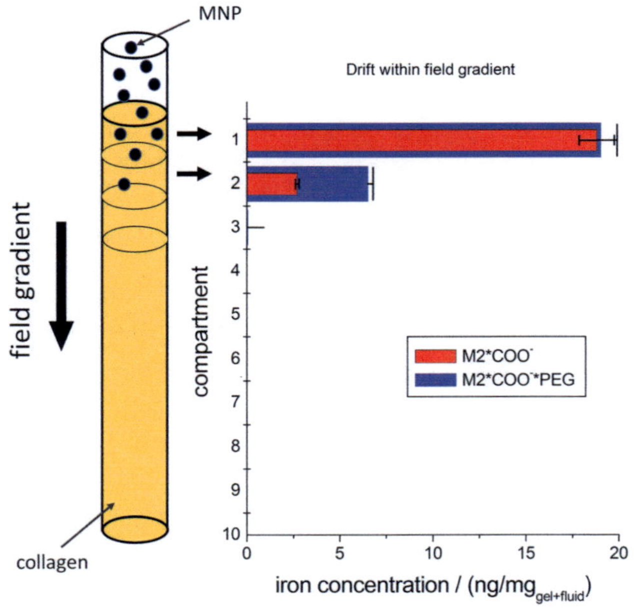

Figure 9.5: Left: Experimental set-up for the drag of MNP through a collagen gel by a magnet field gradient. Right: Concentration of MNP (iron) within each collagen segment (of nine) quantified by MPS.

corresponding M2*COO⁻ sample (Figure 9.5). Hence, M2*CONH-PEG has a substantial higher mobility in collagen than M2*COO⁻.

Then, we related the observed different mobility in a collagen gel to the opsonization by BSA, justified by the molecular similarity of collagen and BSA. We checked the ability of the MNP to bind the proteins BSA by opsonization measurements using ACS as it was done in Section 9.3.3.2, for M1*COO⁻. The fit of Equation (9.12) to the ACS data of samples prior to BSA incubation (Figure 9.6) reveals that the coating of M2*CONH-PEG is about 18 nm thicker than in M2*COO⁻ (Table 9.1). Obviously, this is the PEG-layer thickness. After BSA incubation the hydrodynamic size of M2*COO⁻ increases by 4.2 nm while it does not change for M2*CONH-PEG (Tables 9.1 and 9.2). M2*CONH-PEG was also not opsonized, even by fetal calf serum (FCS) albumin which contains, in contrast to BSA, a rich variety of proteins and other molecules. This observation supports the commonly accepted thesis of stealth properties of PEG decorated particles [56].

The apparent absence of any opsonization of M2*CONH-PEG correlates well with the observed higher mobility through the collagen gel as shown above. Furthermore, we expected that M2*CONH-PEG would have passed through the all gel segments of the tubular sample holder after 48 h since the large mean pore size of about 100–200 nm of the collagen gel, inferred from reference [57] where a PAam gel of nearly the same composition was investigated by TEM. Hence, it seems to be likely that the M2*CONH-PEG mobility is reduced by binding to the collagen matrix.

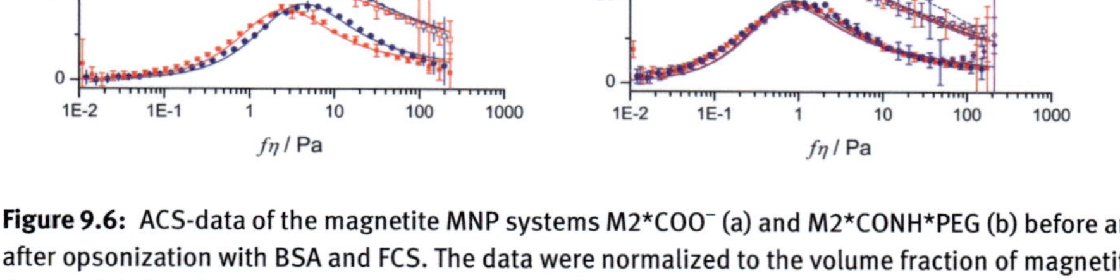

Figure 9.6: ACS-data of the magnetite MNP systems M2*COO⁻ (a) and M2*CONH*PEG (b) before and after opsonization with BSA and FCS. The data were normalized to the volume fraction of magnetite, ϕ. A model of noninteracting core–shell MNP was fitted simultaneously to M(H) and ACS-data (lines). The dashed line on the right axes in (b) represents the curve of M2*COO⁻ before opsonization, with adapted amplitude for better comparison.

9.4 Multicolor magnetorelaxometry imaging for quantitative visualization of particle–matrix interactions

9.4.1 Classical MRXI and multicolor magnetorelaxometry imaging

Classical MRXI focuses on the quantitative reconstruction of a spatial MNP distribution. Because the relaxation curve is very sensitive to the direct environment of MNPs, MRXI has been extended to also gain information about interactions between MNPs and their surrounding matrix [58, 59]. By this approach, called multicolor MRXI, the spatial distribution of MNPs interacting with different molecular environments (e.g., liquid, blood, cells, tissue matrix) or MNPs with distinct properties (e.g., size distributions) can be reconstructed and separated. For convenience, we further summarize such differences as phases. As an example, Figure 9.7 shows classical MRX imaging and the multicolor MRXI approach on two distinct MNP distributions, with the respective MRXI reconstructions in the insets.

Our MRXI setup consists of an MNP distribution phantom, a set of excitation coils on printed circuit boards mounted on top (not shown in Figure 9.7 for clarity) and bottom of the phantom and a 304 SQUID sensor system [60] for MNP relaxation detection. The

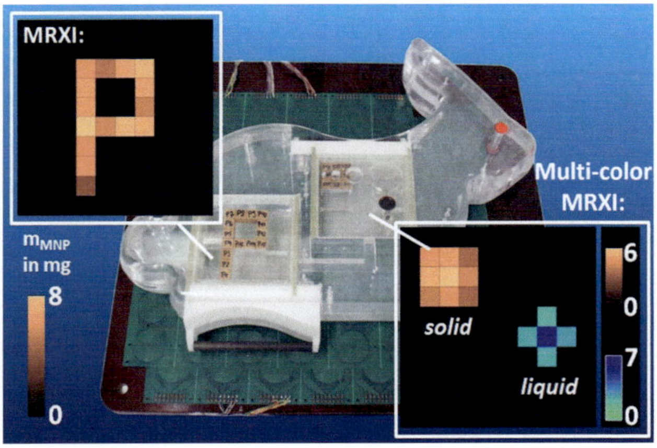

Figure 9.7: MRX imaging using multiple measurements on a rabbit phantom. The MNP phantom models defined MNP distributions with physical and physiological parameters of magnetic drug targeting application in rabbits. Left support: MNP test distribution in the tumor region of the rabbit phantom (brown cubes containing 6 mg of MNP each) and reconstructed MNP distribution after MRX imaging. Right support: multicolor MRXI with MNP support containing 1 mL MNP suspension (12 mg MNP) and 9 MNP loaded gypsum cubes (brown cubes containing 6 mg of MNP each) and MRXI reconstructions achieved by multicolor MRXI for MNPs in liquid phase (blue-green voxels) and MNPs solid phase (brown voxels).

phantom has a shape and size comparable to a rabbit and was developed to analyze the performance of MRXI in a magnetic drug targeting study done by Alexiou et al. [61, 62]. The phantom contains two supports, to model the regions-of-interest (ROIs) in the tumor region (left) and the liver region (right) of the rabbit. Each support allows the flexible arrangement of MNP distributions within a total volume of 9.6 cm × 9.6 cm × 6 cm. The circuit boards provide 15 excitation coils above and below each support. The coils are applied sequentially to generate the spatially constrained magnetic fields of the MRXI sequence.

For classical MRXI (which assumes all MNPs in the same phase), we formed in the ROI tumor an MNP distribution resembling the letter "P" out of 15 gypsum cubes (each loaded with 3.7 mg/cm^3 MNPs). Here, all MNPs are in a solid phase, i.e. fixed in a gypsum matrix, so that only Néel relaxation contributes. The MRXI reconstruction reveals a high correlation to the nominal MNP distribution (Pearson's correlation coefficient >95%) and a mass deviation of the phantom's total MNP amount below 10% [58].

In the multicolor MRXI experiment, we used an MNP distribution containing two phases (MNPs in a solid matrix and MNPs suspended in a liquid) arranged in the ROI liver (right support and inset of Figure 9.7). The MNP distribution in the solid phase is formed of nine gypsum cubes (each loaded with 3.7 mg/cm^3 MNPs) resembling a square and placed beside a 1 mL MNP suspension in a liquid phase (12 mg/cm^3 of MNPs) in a circular container of 2 cm diameter. For the chosen MNP relaxation type we mainly have Néel relaxation in the region of MNPs in the solid phase and Brownian relaxation in the region of MNPs in the liquid phase. This allows for a quantitative separation of MNPs in both phases in one MRXI measurement as visualized by different colormaps in Figure 9.7 (the brown color corresponds to the MNPs in solid phase and the blue color to the MNPs in the liquid phase). Applying multicolor MRXI, it was possible to simultaneously and quantitatively reconstruct and separate MNPs in both phases forming our MNP distribution with a total mass deviation below 10% [58]. In the following, we detail our multicolor MRXI approach and extend the phantoms from this two-phase system to multiphase systems [59].

9.4.2 Extending MRXI to quantitatively imaging particle–matrix interactions

While classical MRXI employs only the amplitude $\kappa(t_1, t_2)$ for image reconstruction, we make use of the complete relaxation curve shape $\kappa(t)$ in our multicolor MRXI approach [59]. This additional temporal information allows to extract portions of distinct MNP relaxation curves due to different phases out of a single MRX measurement. In our approach, the MRXI reconstruction is split up into two subproblems as schematically depicted in Figure 9.8. The first subproblem finds the relative contributions of N distinct MNP relaxation curves $\chi_n \kappa_n(t)$ (N distinct phases) to the measured signal in the sensors.

These are obtained from reference measurements of samples in a specific phase under well-known and controlled conditions. It is assumed that similar phases occur simultaneously in the sample under investigation. The second subproblem finds the associated MNP distribution to the previously determined relative contribution in the sensors. These N distributions are finally merged in a color-coded image in which the color represents the specific phase and the intensity reflects the amount of MNPs. In the reference measurement well-known MNP samples with controlled phase are positioned at location $\mathbf{r}_v$ and the MRX signal is normalized to applied magnetic field $\mathbf{H}_v$ and MNP mass $X_{\mathrm{ref},n}$ to obtain $\chi_n\kappa_n(t)$ for use in the first subproblem:

$$\chi_n\,\kappa_n(t) = \frac{B_{\mathrm{ref}}(t)}{\frac{\mu_0}{4\pi}\left(\frac{3\left(\mathbf{n}_s^{\mathrm{T}}(\mathbf{r}_s-\mathbf{r}_v)\right)(\mathbf{r}_s-\mathbf{r}_v)^{\mathrm{T}})\right)}{\|(\mathbf{r}_s-\mathbf{r}_v)\|^5} - \frac{\mathbf{n}_s^{\mathrm{T}}}{\|(\mathbf{r}_s-\mathbf{r}_v)\|^3}\right)}\mathbf{H}_v\,X_{\mathrm{ref},n}\,. \tag{9.13}$$

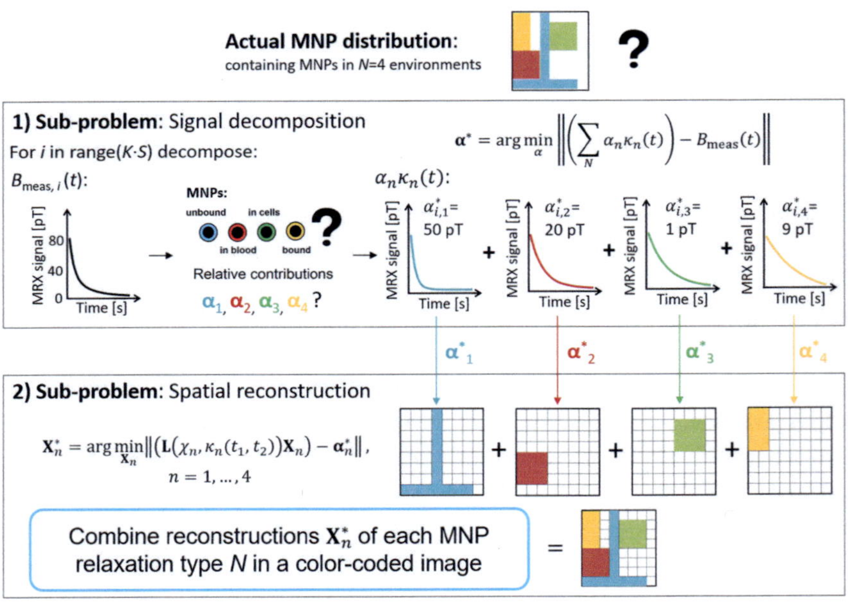

Figure 9.8: Principle of multicolor MRXI. An actual MNP distribution (top) contains $N = 4$ compartments where the MNPs are either suspended (blue), dispersed in blood (red), in cells (green) or bound to a fixed matrix (yellow). In a first subproblem, each measured relaxation signal B_{meas}, $i(t)$ of the MRXI sequence is decomposed into portions of four reference MNP relaxations $\alpha_n\kappa_n(t)$, recorded on MNP samples of known MNP mass and in the respective environment. By this, a coefficient matrix $\boldsymbol{\alpha}^*$ of dimension (KS × N) is obtained, containing the contributions of each MNP relaxation type row-wise. In the second subproblem, an inverse problem $\mathbf{X}^*_n$= argmin‖($\mathbf{L}(\chi_n, \kappa_n(t_1, t_2))\mathbf{X}_n) - \boldsymbol{\alpha}^*_n\|$ is solved independently for each MNP relaxation type and the result $\mathbf{X}^*_n$ assigned to a different color. Finally, an overlay of all $\mathbf{X}^*_n$ is created to generate the multicolor MRX image.

The measured relaxation in the experiment $B_{meas}(t)$ is then modeled as a linear combination of the N reference relaxations $\alpha_n \kappa_n(t)$. Consequently, the first subproblem can be solved by estimating the coefficients $\boldsymbol{\alpha}^*$ in [T]

$$\boldsymbol{\alpha}^* = \underset{\alpha}{\operatorname{argmin}} \left\| \left(\sum_N \alpha_n \kappa_n(t) \right) - B_{meas}(t) \right\| \tag{9.14}$$

to split-up $B_{meas}(t)$ into the contributions of the different phases. This problem is solved using NNLS (Equation 9.7) assuming the absolute contributions to be positive values, which is a correct approximation after signal filtering. Thus, $\boldsymbol{\alpha}^*$ (dimension $1 \times N$) contains the relative contributions of the N phases in a single MRX measurement. Applying Equation (9.14) to the complete MRXI data set, i.e. K MRX measurements of S sensors, yields the final coefficient matrix $\boldsymbol{\alpha}^*$ of dimension $(KS \times N)$.

The solution of this first subproblem in the example in Figure 9.8 separates the signal into $N = 4$ phases originating from MNPs in suspension, in blood, in cells and immobilized in a tissue matrix, respectively. Therefore, we first prepare MNP reference samples of known MNP mass X_{ref} that emulate best the respective phase. Two reference MNP samples are simply suspended MNPs and immobilized MNPs, e.g., freeze-dried in a sugar matrix. The other two samples are prepared by diluting the MNP suspension in whole blood and the cell type of interest. These (small) samples are then consecutively measured by MRX to parameterize our data model (Equation 9.13) with $\chi_n \kappa_n(t)$. The four rows of the coefficient matrix $\boldsymbol{\alpha}^*$ obtained by Equation (9.14) contain the signal contributions of each phase for all MRX measurements of the MRXI sequence. Note that the signal separation can be advanced by the choice of appropriate time windows to evaluate each $\kappa_n(t)$ based on the reference properties (e.g., using only time frames when the signal-to-noise ratio of $\kappa_n(t)$ is above a certain limit) [59].

With this coefficient matrix, we enter the second subproblem which is the imaging problem. Here, Equation (9.7) is solved independently for each phase:

$$\mathbf{X}_n^* = \underset{\mathbf{X}_n}{\operatorname{argmin}} \left\| \left(\mathbf{L}(\chi_n, \kappa_n(t_1, t_2)) \mathbf{X}_n \right) - \boldsymbol{\alpha}_n^* \right\|, \quad n = 1, \ldots, N \tag{9.15}$$

Since the time-dependence $\kappa(t)$ was already employed in the first subproblem for signal separation, it is sufficient to use the amplitude difference of $\kappa_n(t_1, t_2)$ in the imaging problem. As a result, we obtain independent 3D MRXI reconstructions $\mathbf{X}_n^*$ for each phase n. The overlay of all N fractional MNP distributions $\mathbf{X}_n^*$ is the reconstructed MNP distribution obtained by multicolor MRXI. In our example in Figure 9.8, the second subproblem is solved independently for each row of $\boldsymbol{\alpha}^* = [\boldsymbol{\alpha}_1^*, \boldsymbol{\alpha}_2^*, \boldsymbol{\alpha}_3^*, \boldsymbol{\alpha}_4^*]$ to obtain four individual quantitative MNP reconstructions that belong to each phase of the MNP distribution. The final multicolor MRXI image is then the overlay of the four MNP distributions encoded by color to show the phase.

In Figure 9.9, we experimentally demonstrate multicolor MRXI on a MNP distribution measured in the ROI tumor of the rabbit-sized phantom [59]. The MNP distribution was assembled of MNPs in $N = 4$ phases as listed in Table 9.3, while Figure 9.9 shows a photograph of the MNP support (a), the nominal MNP distribution (b), the resulting four fractional MRXI reconstructions (c) and their overlay (d). Note that except for the freeze-dried sample of F200 in which the particles were embedded in a solid phase, all MNPs were suspended in a liquid in our experiment. Hence, we mainly varied the MNP size distributions to generate distinct phases to emulate different PMIs. These systems allow us to investigate multicolor MRXI under well-defined experimental conditions. The MNP systems applied are fluid MAG-D (chemicell GmbH, Germany) and nanoMAG-D (micromod Partikeltechnologie GmbH, Germany).

The MNPs in the four phases are only placed one voxel (1.2 cm) from each other. In this case, the complexity of both multicolor MRXI subproblems increases compared to the previous MNP distribution (Figure 9.7) in which only MNPs in two spatially well-

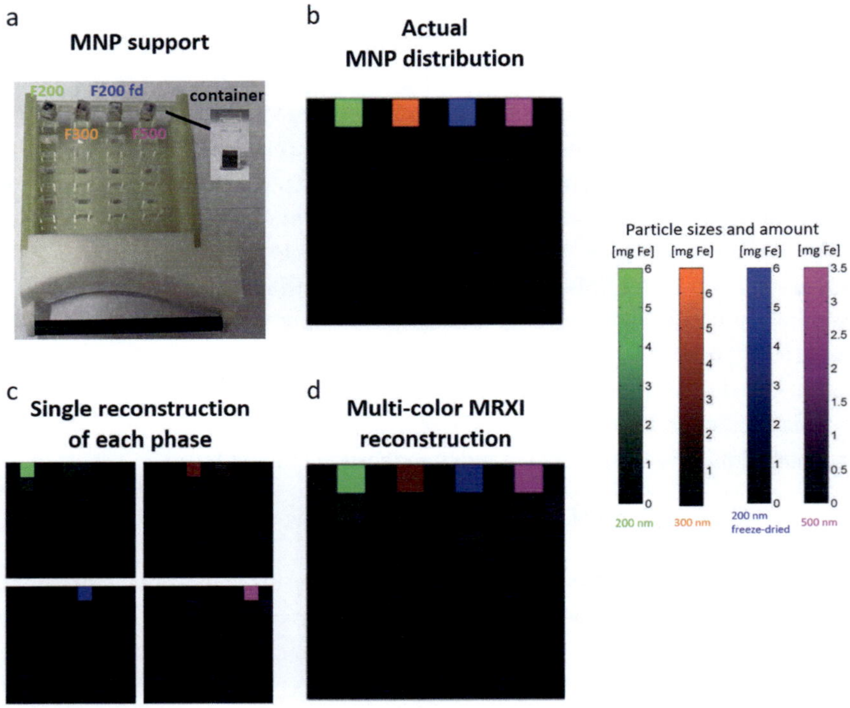

Figure 9.9: Multicolor MRX imaging of the MNP distribution assembled in the MNP support ROI tumor in the rabbit-sized phantom. (a) Photograph of the MNP support showing the spatial arrangement of MNPs in four different phases within their respective sample container ($V = 0.5$ mL). (b) The nominal MNP distribution for the four MNPs phases with 200 nm MNPs (F200: green), 300 nm MNPs (F300: orange), 200 nm freeze-dried MNPs (F200 fd: blue) and 500 nm MNPs (F500: violet). Each sample fills a single voxel in the reconstruction grid of 10×10 voxel with a spatial resolution of 1.2 cm, respectively. The four single-color reconstructions $\mathbf{X}_n^*$ are shown in (c) and their overlay is shown in (d). A high degree of similarity to the nominal MNP distribution is visible. Adapted with permission from Coene et al. [57]. Copyright 2017 IOP Publishing.

Table 9.3: MNP properties under study.

Name	Type	Size (nm)	Amount (mg Fe)
F200	FluidMAG-D	200	6
F200 freeze-dried	FluidMAG-D	200	6
F300	FluidMAG-D	300	6.7
N500	NanoMAG-D	500	3.5

separated phases are present. The multicolor MRXI reconstructions, however, show excellent correspondence to the nominal MNP distribution. MNPs in the four phases are spatially well separated in the image that shows a correlation above 87% for all MNP distributions$\mathbf{X}_n^*$ and even 96% for F200. Even in imaging this four-phase MNP system, multicolor MRXI accomplishes a quantitative reconstruction with a total mass deviation below 30%.

This clearly demonstrates the ability of multicolor MRXI to quantitatively image PMIs. MNP distributions containing MNPs in up to four different phases were quantitatively resolved with a sensitivity in the milligram/cubic centimeter range.

9.5 Flow induced particle–matrix interactions

9.5.1 Microscale visualization of particle–matrix interactions in flow

In this section, we investigate interactions between particles and flowing media. By magnetic monitoring of these interactions, we aim to employ the particles as local flow probes. On the one hand, this is important for biomedical MNP applications, where particle interactions with the circulatory system influence the efficiency of biomedical applications. On the other hand, this might enable the detection of perturbations in the hemodynamics, as, e.g., attributed to pathologic narrowing of vessels (atherosclerosis).

Inside the circulatory system, MNPs experience Stokes friction and shear forces generated by the blood flow. These forces result in translational and rotational MNP motion. In medical imaging (e.g., [63, 64]), translational MNP motion is visualized by mapping the transit of a contrast agent MNP bolus moving with the blood stream. However, the dispersion of MNPs in blood hardens blood flow quantification based on this motion. At the same time, rotational MNP motion that might give additional information on the hemodynamics is not visible.

Here, we detail a novel method called magnetic microsphere tracking (MMT) [65]. It might overcome these limitations by magnetic monitoring of translational and rotational particle motion. A single permanent magnetic microsphere (MM) with a diameter of a few micrometer locally probes the flow while passing through a vessel. The 304 SQUID sensor system [60] detects these movements and allows for high-resolution (millisecond/millimeter) 3D tracking of the translational and rotational motion of the sphere's magnetic moment in a field-of-view (FOV) of 15 × 20 cm. By this, we successfully localize and evaluate diameter constrictions in an arteria phantom. The size of the applied MM (d = 34.6 μm) is close to that of red blood cells (6–8 μm) and demonstrates the potential for biomedical MMT application for blood flow quantification.

The concept of MMT is sketched in Figure 9.10. Here, a MM with permanent magnetic moment moves inside a vessel due to the hydrodynamic forces applied by the flow profile. The diameter of the vessel decreases along the x-coordinate due to increasing grades of stenosis. Thus, the flow vectors of the quadratic flow profile increase along x. A shear force acts on the MM due to variations of the flow vectors over the cross-section of the sphere. This force induces a rotational motion to the MM while drag forces translate it. Hence, the angle δ between MM magnetic moment and x-axis oscillates along x at a certain frequency determined by the shear force. Thus, the rotational frequency of these oscillations increases with narrowing vessel diameter due to increasing shear forces. At the same time, the motions of the MM magnetic moment are remotely detected by a SQUID sensor at location $\mathbf{r}_s$ above the vessel. The amplitude of the SQUID signal depends on the actual position of the MM $\mathbf{r}_v(t) = [r_x(t), r_y(t), r_z(t)]$ relative to the SQUID. The distance dependency

$$B_s(t) = \frac{\mu_0}{4\pi}\left(\frac{3(\mathbf{n}^{\mathrm{T}}(\mathbf{r}_s - \mathbf{r}_v(t))(\mathbf{r}_s - \mathbf{r}_v(t))^{\mathrm{T}})}{\|(\mathbf{r}_s - \mathbf{r}_v(t))\|^5} - \frac{\mathbf{n}_s^{\mathrm{T}}}{\|(\mathbf{r}_s - \mathbf{r}_v(t))\|^3}\right)\mathbf{m}(t) \tag{9.16}$$

is employed to localize the actual position $\mathbf{r}_v(t)$ and magnetic moment $\mathbf{m}(t) = [m_x(t), m_y(t), m_z(t)]$ of the MM by the SQUID array iteratively for each discrete measured sample $t = [T_1, \ldots, T_{\mathrm{end}}]$. This can be accomplished by solving the inverse problem

$$\Omega(\mathbf{r}_v(t),\ \mathbf{m}(t)) = \underset{\Omega}{\operatorname{argmin}} \|\ [B_1(t), \ldots,\ B_S(t)]^{\mathrm{T}} - \mathbf{B}_{\mathrm{meas}}(t)\|,\ \ t = T_1, \ldots, T_{\mathrm{end}}. \tag{9.17}$$

using a Levenberg–Marquart optimizer [66, 67] to iteratively minimize the functional$\Omega(\mathbf{r}_v(t),\ \mathbf{m}(t))$ by parameter variation $\mathbf{r}_v(t)$ and $\mathbf{m}(t)$. The inverse problem was simplified by the constant MM magnetic moment of $|\boldsymbol{m}| = 10\ \mathrm{nAm}^2$ that allows to reduce the model parameters to three MM coordinates in 3D space and the two angles α and β defining the orientation of its magnetic moment, i.e. five degrees of freedom. This dipole localization was applied for each measured sample recorded during the MM passage through a flow phantom with a sampling frequency of 750 Hz, i.e. a temporal resolution of round 1.3 ms. We calculated the translational MM movement as first derivate of $\mathbf{r}_v'(t)$ over time and estimated the MM rotational frequency by the oscillations of the angles α and β. For further details see [65].

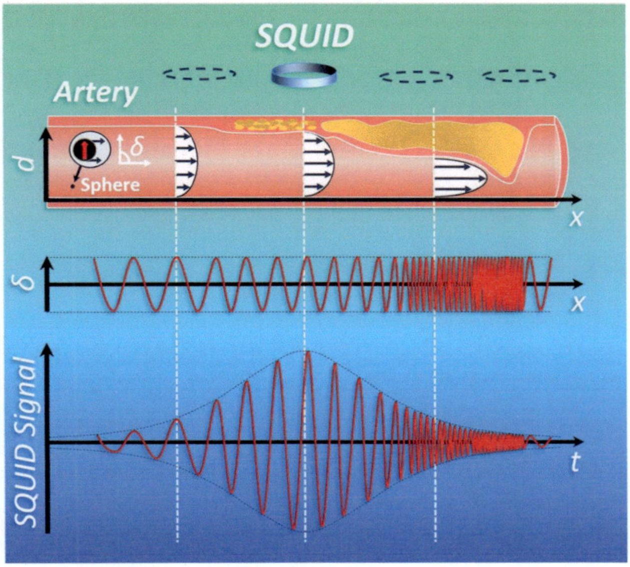

Figure 9.10: Principle of magnetic microsphere tracking: a microsphere with permanent magnetic moment passes an artery with increasing diameter constrictions (grades of stenosis) and thus, increasing velocity vectors of the quadratic flow profile. Due to the shear forces, the angle delta between magnetic moment and x-axis will oscillate along x with increasing frequency. This oscillation is then detected by a SQUID and the actual sphere position tracked by a sensor array. Reproduced from Ref. [63] under the Creative Commons Attribution 4.0 International License.

9.5.2 Tracking a single magnetic microsphere to monitor the fluid dynamics

Here, we apply MMT to visualize PMIs in the flow phantom depicted in Figure 9.11a [65]. It consists of a straight tube section with a constriction region of 1 cm length. Here, variable diameter reductions dr of the original tube diameter (d_{tube} = 1.5 mm) of up to 50% can be adjusted. An additional curved tube section provides a secondary flow (Dean effect [68]) resulting in lateral migrations of the MM whether toward the tube center or wall. Note that these migrations would either lead to an increase in velocity and decrease in rotational frequency when the MM migrates toward the centerline or vice-versa. A constriction, however, results in a temporary increase in both, MM velocity and rotational frequency. The MM was separated from commercial Magnequench® powder (MQP™-S-11-9) using a cell selector (ALS®) and magnetized at a flux density of 5 T. The flow media was pumped by a syringe pump (Landgraf®, LA800) at a constant flow rate of 30 mL/min (Re = 212).

The localized x- and y-positions obtained by MMT are depicted in Figure 9.11b. The path of the MM is encoded by color to show the rotational MM frequency f_{rot}. For an

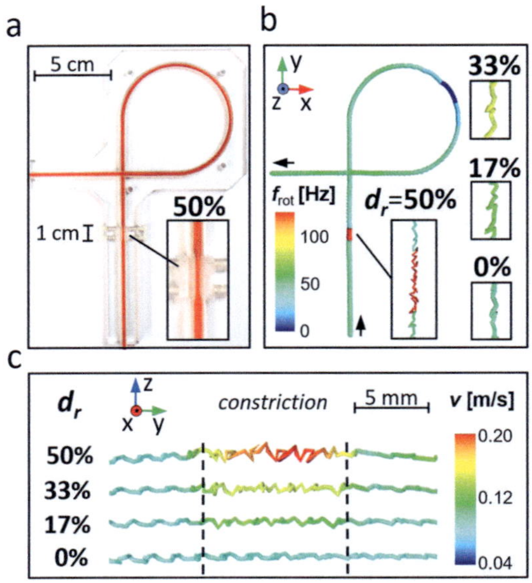

Figure 9.11: Magnetic Microsphere Tracking in a flow phantom. (A) The flow phantom comprising a straight and a curved tube section. In the straight section variable constrictions can be applied over a length of 1 cm using four nylon screws. (B) Localized *x*, *y* – position of the MM color coded to show the rotational MM frequency. The insets show an enlargement of the constriction region. A systematically increasing rotational MM frequency with increasing diameter reduction d_r is visible. (C) Enlargement of the localized *y*, *z* – positions of the MM in the constriction region for different diameter reductions. The MM trace is color coded to show the MM velocity. A clear acceleration is seen when the MM enters the constriction and it decelerates when it leaves the constriction. Reproduced from Ref. [63] under the Creative Commons Attribution 4.0 International License.

adjusted diameter constriction of $d_r = 50\%$, a clear increase of the rotational frequency is visible when the MM passes the constriction region and it remains constant in the absence of a constriction. The enlargements in Figure 9.11b show MM path and rotational frequency through the constriction region for variable diameter constrictions. A decrease of the rotational frequency is observed with increasing cross-section of the tube.

As expected, a decrease of the rotational frequency is also visible in the curved tube section when the MM migrates toward the tube center due to the applied secondary flow.

Figure 9.11c shows an enlarged view of the localized MM positions (y–z plane) during the passage through the constriction region for different diameter reductions. The color of the MM path is encoded by color to show the local velocity v. An abrupt acceleration is visible when the MM enters the constriction, and it decelerates again when it has left the constriction. Inside the constriction region, the MM velocity increase systematically with decreasing tube diameter, while it remains unchanged in the absence of a constriction. Hence, the degree of constriction has a clear impact on the MMT localization parameters frequency and velocity.

Hence, MMT allows to visualize microscale PMIs in flow by tracking a single microsphere (size range of a red blood cell) with a temporal resolution of 1.3 ms and a spatial resolution below 1 mm. From the measurements we were able to quantify the flow conditions experienced by the microsphere. This enables localization and quantification of constricted volumes and distinction from lateral MM migrations. Additionally, separating influences of Stokes friction and shear flow makes MMT an ideal tool for rheology investigations and might help to understand complex flow dynamics. By its properties (quantitative, high resolution, no radiation, no tissue contributions, low marker dose [ng]) MMT holds great potential for biomedical blood flow quantification (more details in [65]).

9.5.3 Sensing of particle–matrix interactions in flow on the nanoscale

In the following, we investigate flow-induced PMIs in the nanoscale using MRX [69]. To accomplish this, we combine our MRX device comprising of a SQUID detector and a magnetizing coil with a flow phantom as sketched in Figure 9.12a. For flow experiments, 3 mL of a homogeneous MNP solution was injected into a 5 m long PVC tube with inner diameter of 0.8 mm. Inside the magnetizing coil of the MRX device, i.e. the sensing volume, the tube was arranged to form a coil with five turns. The field vectors of the applied magnetic field were oriented perpendicular to the flow direction. The MNP

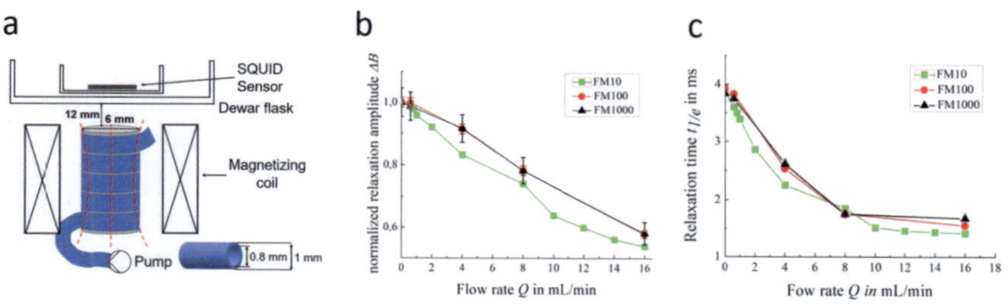

Figure 9.12: MRX measurements in flow. (a) Scheme of the experimental setup for the MRX measurements of MNPs under flow conditions. Driven by a syringe pump, the MNPs are flowing through the tube inside the MRX magnetizing coil of the MRX measurement device. To increase the sample volume inside the magnetizing coil, the tube is coiled up on a nonmagnetic core with five turns. The magnetizing coil is placed under the bottom of the SQUID sensor for detection of the magnetic relaxation. On the right, relaxation amplitudes ΔB (b) and relaxation times $t_{1/e}$ (c) as function of the flow rate Q are depicted. These were obtained for dilutions of the original MNP suspension by a ratio of 1:10 (FM10), 1:100 (FM100), 1:1000 (FM1000) with BSA. Adapted with permission from Slabu et al. [67]. Copyright 2017 IOP Publishing.

suspension was pumped through the tubing system by a syringe pump (LA-120, New Era Pump Systems, USA) at a constant flow rate. MRX measurements were performed with constant flow rates adjusted between 0.6 up to 16 mL/min. We used homogeneous fluidMAG/12-HS/130 (chemicell GmbH, Germany) suspensions of iron concentration $c = (2.8, 0.28, 0.028\ \mathrm{mg/mL})$, respectively. In the MRX magnetizing phase, an external magnetic field of 2 mT was applied to the MNP suspension for magnetizing time of 1 s. In the measurement phase, the resulting MRX signal was recorded by the SQUID sensor for a measurement time of 0.5 s at a sampling frequency of 100 kHz.

The characteristic MRX parameters relaxation amplitude ΔB and relaxation time $t_{1/e}$ are depicted in Figure 9.12b and c as a function of the flow rate Q, respectively. Independent on MNP concentration, a clear decrease in both parameters with increasing flow rate is visible. Hence, shear-stress induced rotational MNP motion fastens Brownian relaxation and decreases the relaxation amplitude. Interestingly, while ΔB decreases proportional to the applied flow rate, the decrease of $t_{1/e}$ nearly saturates for flow rates above 8 mL/min. This effect is reproducibly visible in all three MNP concentrations. This saturation behavior indicates a change in the tube flow profile. Slight concentration dependent variations, i.e. faster decreasing relaxation amplitudes and relaxation times for the higher concentrated FM10 are visible as well. This might be due to MNP provoked shear thinning and structure formation as, e.g., nanoparticle chain formations [69].

To summarize, we sensitively detect flow-induced changes in the Brownian relaxation of MNPs by MRX measurements. By analyzing the characteristic MRX parameters relaxation amplitude and relaxation time, we apply the MNPs as local probes to quantify flow and determine perturbations in the flow profile.

9.6 Summary and conclusion

We reviewed magnetic measurement methods that sensitively detect particle–matrix interactions and thereby, access the physical properties of the molecular MNP environment. For selected MNP-matrix systems, we demonstrated that the combination of different noninvasive and sensitive magnetic measurement techniques for specific MNP detection is a powerful approach to provide insights into dynamic processes of the MNPs as, e.g., binding kinetics, cellular uptake, passage through cell barriers and heat induction.

Furthermore, we developed functional imaging approaches that allow for spatially resolved investigation of particle–matrix interactions. Here, we utilized the impact of particle–matrix interactions on the MNP relaxation to advance MRX imaging with a multicolor approach and to resolve the location, quantity and molecular environment of MNPs. This technique might become a valuable tool to visualize and quantify particle–matrix interactions in preclinical research and propelling the development of future particle mediated therapies.

As MNPs in nearly all biomedical applications are subject to body fluids and their prevailing fluid dynamics, we investigated flow-induced particle interactions. These can be detected by sensitive magnetic measurement methods to quantify the fluid dynamics acting on particles in the micro- and nanoscale. Magnetic microsphere tracking allows for the spatially resolved quantification of the fluid dynamics based on the particle velocity and rotational frequency with excellent temporal and spatial resolution. This novel method has great potential to become a valuable tool for rheology or even biomedical blood flow quantification and to pave the way for a deeper understanding of flow-induced particle dynamics inside the body.

Though countless MNP systems with sizes in the nanometer and micrometer range have been developed and applied for biomedical applications in recent years, there still is strong demand for further MNP research. To facilitate this, an accompanying development and improvement of established measurement techniques and analysis methods for MNPs and their interaction with biological environment is a mandatory prerequisite.

Author contributions: All the authors have accepted responsibility for the entire content of this submitted manuscript and approved submission.
Research funding: The authors gratefully acknowledge financial support by the DFG priority program SPP1681 (grant nos. WI 4230/1-3, MA 2432/6-2, KR 3929/2-1/2). A. C. and J. L. are supported by the Research Foundation – Flanders (FWO) through a postdoctoral fellowship. AC and JL were guest scientists at PTB with the support of PTB and the FWO (V409915N, AC and V410015N, JL).
Conflict of interest statement: The authors declare no conflicts of interest regarding this article.

References

1. Gupta AK, Gupta M. Synthesis and surface engineering of iron oxide nanoparticles for biomedical applications. Biomaterials 2005;26:3995–4021.
2. Hergt R, Dutz S. Magnetic particle hyperthermia-biophysical limitations of a visionary tumour therapy. J Magn Magn Mater 2007;311:187–92.
3. Pankhurst QA, Connolly J, Jones SK, Dobson J. Applications of magnetic nanoparticles in biomedicine. J Phys D Appl Phys 2003;36:R167.
4. Dutz S, Hergt R. Magnetic particle hyperthermia – a promising tumour therapy? Nanotechnology 2014;25:452001.
5. Poller JM, Zaloga J, Schreiber E, Unterweger H, Janko C, Radon P, et al. Selection of potential iron oxide nanoparticles for breast cancer treatment based on in vitro cytotoxicity and cellular uptake. Int J Nanomed 2017;12:3207.
6. Lugert S, Unterweger H, Mühlberger M, Janko C, Draack S, Ludwig F, et al. Cellular effects of paclitaxel-loaded iron oxide nanoparticles on breast cancer using different 2D and 3D cell culture models. Int J Nanomed 2019;14:161.

7. Slabu I, Roeth AA, Engelmann UM, Wiekhorst F, Buhl EM, Neumann UP, et al. Modeling of magnetoliposome uptake in human pancreatic tumor cells in vitro. Nanotechnology 2019;30: 184004.
8. Ottersbach A, Mykhaylyk O, Heidsieck A, Eberbeck D, Rieck S, Zimmermann K, et al. Improved heart repair upon myocardial infarction: combination of magnetic nanoparticles and tailored magnets strongly increases engraftment of myocytes. Biomaterials 2018;155:176–90.
9. Vosen S, Rieck S, Heidsieck A, Mykhaylyk O, Zimmermann K, Bloch W, et al. Vascular repair by circumferential cell therapy using magnetic nanoparticles and tailored magnets. ACS Nano 2016; 10:369–76.
10. Poller WC, Pieber M, Boehm-Sturm P, Ramberger E, Karampelas V, Möller K, et al. Very small superparamagnetic iron oxide nanoparticles: long-term fate and metabolic processing in atherosclerotic mice. Nanomed Nanotechnol Biol Med 2018;14:2575–86.
11. Engelmann UM, Roeth AA, Eberbeck D, Buhl EM, Neumann UP, Schmitz-Rode T, et al. Combining bulk temperature and nanoheating enables advanced magnetic fluid hyperthermia efficacy on pancreatic tumor cells. Sci Rep 2018;8:13210.
12. Wells J, Löwa N, Paysen H, Steinhoff U, Wiekhorst F. Probing particle–matrix interactions during magnetic particle spectroscopy. J Magn Magn Mater 2019;475:421–8.
13. Weidner A, Gräfe C, von der Lühe M, Remmer H, Clement JH, Eberbeck D, et al. Preparation of core–shell hybrid materials by producing a protein corona around magnetic nanoparticles. Nanoscale Res Lett 2015;10:282.
14. Dutz S, Weidner A, von der Lühe M, Gräfe C, Biehl P, Demut J, et al. Hybrid nanomaterials of biomolecule corona coated magnetic nanoparticles and their interaction with biological systems. Phys Sci Rev 2020. https://doi.org/10.1515/psr-2019-0110. In press.
15. Deuflhard M, Eberbeck D, Hietschold P, Wilharm N, Mühlberger M, Friedrich RP, et al. Magnetically responsive composites: electron beam assisted magnetic nanoparticle arrest in gelatin hydrogels for bioactuation. Phys Chem Chem Phys 2019;21:14654–62.
16. Gräfe C, Müller EK, Gresing L, Weidner A, Radon P, Friedrich RP, et al. Magnetic hybrid materials interact with biological matrices. Phys Sci Rev 2020. https://doi.org/10.1515/psr-2019-0114. In press.
17. Gräfe C, Slabu I, Wiekhorst F, Bergemann C, Von Eggeling F, Hochhaus A, et al. Magnetic particle spectroscopy allows precise quantification of nanoparticles after passage through human brain microvascular endothelial cells. Phys Med Biol 2016;61:3986.
18. Heid S, Unterweger H, Tietze R, Friedrich RP, Weigel B, Cicha I, et al. Synthesis and characterization of tissue plasminogen activator—functionalized superparamagnetic iron oxide nanoparticles for targeted fibrin clot dissolution. Int J Mol Sci 2017;18:1837.
19. Müller EK, Gräfe C, Wiekhorst F, Bergemann C, Weidner A, Dutz S, et al. Magnetic nanoparticles interact and pass an in vitro co-culture blood-placenta barrier model. Nanomaterials 2018;8:108.
20. Nèel L. Théorie du traînage magnétique des ferromagnétiques en grains fins avec application aux terres cuites. Ann Geophys 1949;5:99–136.
21. Einstein A. Investigations on the theory of the Brownian movement. New York: Courier Corporation; 1956.
22. Kötitz R, Fannin PC, Trahms L. Time domain study of Brownian and Néel relaxation in ferrofluids. J Magn Magn Mater 1995;149:42–6.
23. Ludwig F, Heim E, Menzel D, Schilling M. Investigation of superparamagnetic Fe_3O_4 nanoparticles by fluxgate magnetorelaxometry for use in magnetic relaxation immunoassays. J Appl Phys 2006; 99:08P106.
24. Bessais L, Ben Jaffel L, Dormann JL. Relaxation time of fine magnetic particles in uniaxial symmetry. Phys Rev B 1992;45:7805.

25. Liebl M, Steinhoff U, Wiekhorst F, Haueisen J, Trahms L. Quantitative imaging of magnetic nanoparticles by magnetorelaxometry with multiple excitation coils. Phys Med Biol 2014;59: 6607–20.
26. Baumgarten D, Liehr M, Wiekhorst F, Steinhoff U, Münster P, Miethe P, et al. Magnetic nanoparticle imaging by means of minimum norm estimates from remanence measurements. Med Biol Eng Comput 2008;46:1177.
27. Adolphi NL, Butler KS, Lovato DM, Tessier TE, Trujillo JE, Hathaway HJ, et al. Imaging of Her2-targeted magnetic nanoparticles for breast cancer detection: comparison of SQUID-detected magnetic relaxometry and MRI. Contrast Media Mol Imaging 2012;7:308–19.
28. Steinhoff U, Wiekhorst F, Baumgarten D, Haueisen J, Trahms L. Bildgebung magnetischer nanopartikel basierend auf magnetrelaxometrie mit sequentieller aktivierung inhomogener anregungsfelder. Biomed Tech 2010;55:22–5.
29. Coene A, Crevecoeur G, Dupre L. Adaptive control of excitation coil arrays for targeted magnetic nanoparticle reconstruction using magnetorelaxometry. IEEE Trans Magn 2012;48:2842–5.
30. Liebl M, Steinhoff U, Wiekhorst F, Coene A, Haueisen J, Trahms L. Quantitative reconstruction of a magnetic nanoparticle distribution using a non-negativity constraint. Biomed Tech 2013;58(1): 000010151520134261.
31. Lawson CL, Hanson RJ. Linear least squares with linear inequality constraints solving least squares problems. Englewood Cliffs, NJ: Prentice-Hall; 1974:158–73 pp.
32. Debye PJW. Polar molecules. New York: Chemical Catalog Company, Incorporated; 1929.
33. Ficko BW, Nadar PM, Hoopes PJ, Diamond SG. Development of a magnetic nanoparticle susceptibility magnitude imaging array. Phys Med Biol 2014;59:1047.
34. Ficko BW, Nadar PM, Diamond SG. Spectroscopic AC susceptibility imaging (sASI) of magnetic nanoparticles. J Magn Magn Mater 2015;375:164–76.
35. Baffa O, Corá LA, Américo MF, Fonseca PR, Oliveira RB, Miranda JR. Magnetic images of pharmaceutical dosage forms in the human gastrointestinal tract. In: Annual international conference of the IEEE engineering in medicine and biology – proceedings; 2005.
36. Soares G, Próspero A, Calabresi M, Rodrigues D, Simoes L, Quini C, et al. Multichannel AC biosusceptometry system to map biodistribution and assess the pharmacokinetic profile of magnetic nanoparticles by imaging. IEEE Trans NanoBioscience 2019;18:456–62.
37. Gleich B, Weizenecker J. Tomographic imaging using the nonlinear response of magnetic particles. Nature 2005;435:1214.
38. Pessot G, Cremer P, Borin DY, Odenbach S, Löwen H, Menzel AM. Structural control of elastic moduli in ferrogels and the importance of non-affine deformations. J Chem Phys 2014;141:015005.
39. Deng H-X, Gong X-l, Wang L-H. Development of an adaptive tuned vibration absorber with magnetorheological elastomer. Smart Mater Struct 2006;15:N111.
40. Varga Z, Fehér J, Filipcsei G, Zrínyi M. Smart nanocomposite polymer gels. In: Macromolecular symposia; 2003.
41. Eberbeck D, Bergemann C, Wiekhorst F, Steinhoff U, Trahms L. Quantification of specific bindings of biomolecules by magnetorelaxometry. J Nanobiotechnol 2008;6:1–12.
42. Wisotzki EI, Eberbeck D, Kratz H, Mayr SG. Magnetic response of gelatin ferrogels across the sol–gel transition: the influence of high energy crosslinking on thermal stability. Soft Matter 2016;12: 3908–18.
43. Kraupner A, Eberbeck D, Heinke D, Uebe R, Schüler D, Briel A. Bacterial magnetosomes - nature's powerful contribution to MPI tracer research. Nanoscale 2017;9:5788–93.
44. Eberbeck D, Wiekhorst F, Steinhoff U, Trahms L. Aggregation behaviour of magnetic nanoparticle suspensions investigated by magnetorelaxometry. J Phys Condens Matter 2006;18:S2829.

45. Astalan AP, Ahrentorp F, Johansson C, Larsson K, Krozer A. Biomolecular reactions studied using changes in Brownian rotation dynamics of magnetic particles. Biosens Bioelectron 2004;19: 945–51.
46. Kammel M, Hoell A, Wiedenmann A. Structure of magnetite ferrofluids investigated by SANS with polarized neutrons. Scripta Mater 2001;44:2341–5.
47. Avdeev MV, Aksenov VL. Small-angle neutron scattering in structure research of magnetic fluids. Phys Usp 2011;53:971–93.
48. Nagornyi AV, Bulavin LA, Petrenko VI, Avdeev MV, Aksenov VL. Sensitivity of small-angle neutron scattering method at determining the structural parameters in magnetic fluids with low magnetite concentrations. Ukr J Phys 2013;58:735–41.
49. Narayanan T, Wacklin H, Konovalov O, Lund R. Recent applications of synchrotron radiation and neutrons in the study of soft matter. Crystallogr Rev 2017;23:160–226.
50. Frielinghaus H, Feoktystov A, Berts I, Mangiapia G. KWS-1: small-angle scattering diffractometer. J Large-Scale Res Facil JLSRF 2015;1:28.
51. Feoktystov AV, Frielinghaus H, Di Z, Jaksch S, Pipich V, Appavou MS, et al. KWS-1 high-resolution small-angle neutron scattering instrument at JCNS: current state. J Appl Crystallogr 2015.
52. Pipich V. QtiKWS: user-friendly program for reduction, visualization, analysis and fit of SA(N)S data. Software Toolbox for SANS Data; 2012. http://www.qtikws.de [Accessed 30 Mar 2021].
53. Szczerba W, Costo R, Veintemillas-Verdaguer S, Del Puerto Morales M, Thünemann AF. SAXS analysis of single- and multi-core iron oxide magnetic nanoparticles. J Appl Crystallogr 2017;50: 481–8.
54. Zhang F, Skoda MW, Jacobs RM, Martin RA, Martin CM, Schreiber F. Protein interactions studied by SAXS: effect of ionic strength and protein concentration for BSA in aqueous solutions. J Phys Chem B 2007;111:251–9.
55. Wïniewska M, Szewczuk-Karpisz K, Sternik D. Adsorption and thermal properties of the bovine serum albumin-silicon dioxide system. J Therm Anal Calorim 2015;120:1355–64.
56. Nosrati H, Salehiabar M, Fridoni M, Abdollahifar MA, Kheiri Manjili H, Davaran S, et al. New insight about biocompatibility and biodegradability of iron oxide magnetic nanoparticles: stereological and in vivo MRI monitor. Sci Rep 2019;9:1–10.
57. Muri HI, Hoang L, Hjelme DR. Mapping nanoparticles in hydrogels: a comparison of preparation methods for electron microscopy. Switzerland: Applied Sciences; 2018.
58. Liebl M, Wiekhorst F, Eberbeck D, Radon P, Gutkelch D, Baumgarten D, et al. Magnetorelaxometry procedures for quantitative imaging and characterization of magnetic nanoparticles in biomedical applications. Biomed Tech 2015;60:427–43.
59. Coene A, Leliaert J, Liebl M, Löwa N, Steinhoff U, Crevecoeur G, et al. Multi-color magnetic nanoparticle imaging using magnetorelaxometry. Phys Med Biol 2017;62:3139–57.
60. Schnabel A, Burghoff M, Hartwig S, Petsche F, Steinhoff U, Drung D, et al. A sensor configuration for a 304 SQUID vector magnetometer. Neurol Clin Neurophysiol Annu NCN 2004;2004:70.
61. Alexiou C, Arnold W, Klein RJ, Parak FG, Hulin P, Bergemann C, et al. Locoregional cancer treatment with magnetic drug targeting. Canc Res 2000;60:6641–8.
62. Wiekhorst F, Liebl M, Steinhoff U, Trahms L, Lyer S, Dürr S, et al. Magnetorelaxometry for in-vivo quantification of magnetic nanoparticle distributions after magnetic drug targeting in a rabbit carcinoma model. Springer Proc Phys 2012;140:301–5.
63. Nicol ED, Padley SPG. Non-invasive cardiac imaging: current and emerging roles for multi-detector row computed tomography. Part 2. Br J Cardiol 2007;14:237–41.
64. Gordon Y, Partovi S, Müller-Eschner M, Amarteifio E, Bäuerle T, Weber M-A, et al. Dynamic contrast-enhanced magnetic resonance imaging: fundamentals and application to the evaluation of the peripheral perfusion. Cardiovasc Diagn Ther 2014;4:147–64.

65. Liebl M, Gleich B, Eberbeck D, Radon P, Rahmer J, Trahms L, et al. Noninvasive monitoring of blood flow using a single magnetic microsphere. Sci Rep 2019;9:1–8.
66. Levenberg K. A method for the solution of certain non-linear problems in least squares. Q Appl Math 1944;2:164–8.
67. Marquardt DW. An algorithm for least-squares estimation of nonlinear parameters. J Soc Ind Appl Math 1963;11:431–41.
68. Dean WR. XVI. Note on the motion of fluid in a curved pipe. Lond Edinb Dublin Philos Mag J Sci 1927; 4:208–23.
69. Slabu I, Liebl M, Wiekhorst F, Eberbeck D. Magnetic relaxation of magnetic nanoparticles under the influence of shear flow. J Phys D: Appl Phys 2019;52:1–9.

Mia Schliephake*, Julia M. Linke and Stefan Odenbach*

10 Magnetic characterization of magnetoactive elastomers containing magnetic hard particles using first-order reversal curve analysis

Abstract: The use of new types of intelligent materials is becoming increasingly widespread. These include magnetoactive elastomers with hard magnetic filling components, which offer the unique chance to adapt active and passive material properties. In this context, this paper presents an overview of the experimental results on the study of the magnetic properties of elastic composites with a magnetic hard component. First-order reversal curves, which are recorded with a vibrating sample magnetometer, are used as method to characterize the magnetic material behavior. The influence of various parameters on the process of magnetization of composites is considered, including the stiffness of the polydimethylsiloxane-based matrix polymer, the particle ratio and the particle size as well as the so-called training effect.

Keywords: first-order reversal curve, interparticle interaction, magnetization behavior, smart materials

10.1 Introduction

Magnetic hybrid materials are a kind of smart materials consisting of a matrix ranging from Newtonian liquids to polymerized elastomers and a magnetic filler, i.e., magnetic particles which can be micron-as well as nanosized, and which can be made of magnetically soft or hard materials [1]. A prominent example of a material within this class of magnetic hybrid composites are magnetic particles embedded in a polymeric matrix. Important for the design of applications using magnetic hybrid materials is a strong change in the respective properties induced by a magnetic field. This necessity requires a high-particle volume fraction and a low elastic modulus of the matrix. Magnetic hybrid materials basied on such soft matrices are usually called soft magnetic elastomers as well as magnetoactive elastomers (MAEs). Caused by external stimuli, such as external magnetic fields, the physical properties of MAEs can be controlled and adapted to the specific application [2].

***Corresponding authors: Stefan Odenbach** and **Mia Schliephake,** TU Dresden, Faculty of Mechanical Engineering, George-Bähr-Str. 3, 01062 Dresden, Germany, E-mail: stefan.odenbach@tu-dresden.de (S. Odenbach), mia.schliephake@tu-dresden.de (M. Schliephake)
Julia M. Linke, TU Dresden, Faculty of Mechanical Engineering, George-Bähr-Str. 3, 01062 Dresden, Germany

This article has previously been published in the journal Physical Sciences Reviews. Please cite as: M. Schliephake, J. M. Linke and S. Odenbach "Magnetic characterization of magnetoactive elastomers containing magnetic hard particles using first-order reversal curve analysis" *Physical Sciences Reviews* [Online] 2021, 6. DOI: 10.1515/psr-2019-0124 | https://doi.org/10.1515/9783110569636-010

Conventionally, MAEs use magnetically soft particles providing a magnetic control due to external fields. This is usually called a magnetoactive material [3]. Alternatively, magnetically hard particles can be used as a filler, which lead to passive magneto-rheological properties of the composite [4]. It has also been shown that mixtures of magnetically hard and soft particles lead to new properties of the corresponding MAEs, which are not only a simple superposition of the passive and active effects but also which would be found in elastomers containing hard or soft particles only [5–7]. The overall properties of such mixed MAEs rely obviously on the complex interaction of the magnetically hard and soft particles. In such composites the coupling of the inner structure to the magnetic-field-induced changes of the properties is even more complicated than in MAEs containing one species of magnetic particles only, since the complex inner magnetic fields structure has to be evaluated and understood. For instance, when a remanent magnetization is induced in the hard magnetic component, magnetization and structuring of the soft magnetic components can as well take place even without an actively applied external field [8].

As it has been presented in [8], measurements of the first order reversal curves (FORCs) can be used as an effective method to study the magnetic interactions between the magnetically hard and soft particles inside the MAEs. The FORC method was firstly introduced by Roberts and Pike [9, 10] and provides a quantitative evaluation of the coercive field distribution and the local interaction field. The current work is an overview of the results obtained from the FORC measurements performed on the MAEs with mixed magnetically hard and soft components, as well as on the specimens containing magnetically hard particles only. In this context, the influence of various parameters, such as matrix stiffness particle ratio and training effect on the magnetization behavior of the hybrid material is presented. Furthermore, to understand the complex correlation between magnetization behavior and internal particle structure of MAEs, it is necessary to obtain a detailed view to the internal microstructures at different points in time of the FORC analysis. This overview also briefly addresses the microstructure analysis of the magnetized MAEs by X-ray computed microtomography, which provides important information about the relative structure and movements of the particles in relation to each other through a three-dimensional visualization of the particles. It assists in the interpretation of FORC analysis.

10.2 Samples and methods

10.2.1 Sample preparation

The MAE samples considered in this paper are based on a polydimethylsiloxane (PDMS) matrix. Usually a two-component composition with an organometallic cross-linking agent (Pt catalysts) is used. Tuning of the matrix stiffness is achieved by different mixing ratios of the individual polymer components and by dilution of the composition with a silicone oil. Details on a manufacturing processes are given elsewhere [5]. In addition particle-matrix-systems with an inelastic epoxy resin matrix are used as reference samples.

Magnetically hard NdFeB-alloy microparticles or combinations of hard magnetic NdFeB-alloy microparticles with soft magnetic carbonyl iron microparticles are used as filling components. The NdFeB-alloy particles have either a spherical (MQP-S-11 from Magnequench) or a flake-like shape (MQA-38-14 from Magnequench), with a size distribution of the spherical particles between 5 and 50 μm and a length distribution of the flake-like particles between 10 and 100 μm [11]. Before adding to the noncrosslinked matrix, the NdFeB particles are modified with a mixture of volatile alcohol and silicone oil to improve adhesion to the polymer chains [5]. The mean diameter of the precoated by the manufacturer spherical carbonyl iron particles (BASF, grade CC)) is ~5 μm.

The particles are added before the crosslinking reaction and distributed homogeneously during polymerization using mechanical stirring. The total particle content as well as the ratio of hard magnetic to soft magnetic components is varied for different samples as given in [8, 11].

10.2.2 FORC method

Magnetic characterization of MAEs was performed using vibrating sample magnetometers (VSM) of 7407 series from Lake Shore. All measurements have been conducted at room temperature. The magnetization behavior of the MAEs containing hard magnetic components used in this work was characterized using the FORC method.

The FORC method divides the measuring process (see Figure 10.1) into the following steps: (1) possibly training of the MAE sample by multiple repetition of the full magnetization cycle, (2) magnetization of the sample in a maximum positive field, (3) reduction of the magnetic field until the reverse field H_r is reached, (4) measurement of a FORC curve, (5) n-fold repetition of the steps (2) – (4) with regular intervals between H_r to $H_r = -H_{\max}$. Set of the reversal magnetization curves obtained in this way is exemplary demonstrated in Figure 10.1. The resulting set of functions $M\,(H_a, H_r)$ provides the magnetization of the examined MAE at any point depending on the respective reverse field H_r and the respectively applied magnetic field H_a. The FORC distribution is defined by the mixed derivation of the functional group according to H_r and H_a

$$\mu(H_r, H_a) = -\frac{1}{2}\frac{\partial^2 M(H_r, H_a)}{\partial H_r\,\partial H_a} \tag{10.1}$$

and the FORC diagrams are in turn the result of a coordinate transformation by 45°,

$$H_c = \frac{H_a - H_r}{2} \tag{10.2}$$

$$H_u = \frac{H_a + H_r}{2} \tag{10.3}$$

so that the horizontal axis now corresponds to the coercive field H_c and the vertical axis to the interaction field H_u. The FORCinel analysis software with a smoothing factor of 10 was used for the data treatment and calculations of FORC diagrams. In the measurements according to [8], 699 FORC curves were recorded based on the original algorithm of the MicroMag™ software of the Princeton Measurements Corporation. Further measurements include 250–350 curves based on an algorithm provided by the Lake Shore Software. The measuring time per point was between 1 and 2 s in all experiments.

10.2.3 Microstructure analysis

Computed X-ray microtomography (μCT) is a nondestructive method for analyzing the internal microstructure of MAEs [12]. Paired with modern algorithms of digital image processing, the high resolution μCT makes it possible to differentiate individual particles and to track their movements under the influence of magnetic fields as demonstrated in Figure 10.2 [13–15]. The microstructural analysis performed using μCT is attended to assists in the interpretation of FORC diagrams.

10.3 Results

10.3.1 Training effects

The so-called training effect of magnetoactive elastomers with magnetically hard, flake-shaped NdFeB alloy particles has been investigated in [11]. During the magnetization of the samples it was observed that the first magnetization curve of a virgin

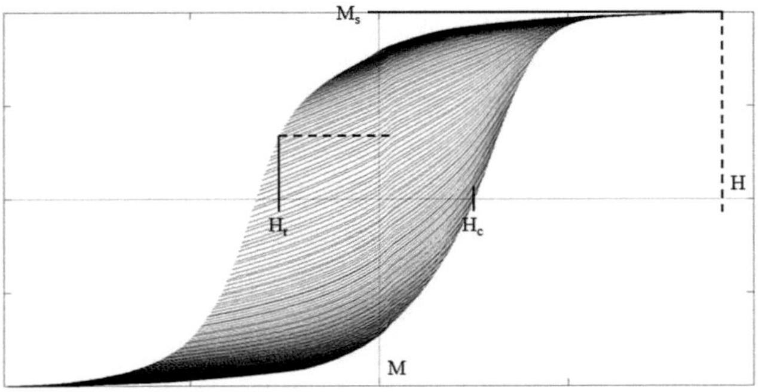

Figure 10.1: Function set of the reversal curves of an MAE with hard magnetic NdFeB particles and hard elastic matrix represented as magnetization M over the applied magnetic field H (H_a). A reversal or FORC curve starts at H_r and runs to H_{max} (or H_{Sat}). The saturation magnetization M_S is reached when more than 95% of the magnetic particles are magnetized. H_c (coercive field) indicates the particles own magnetic field which remains after external magnetization.

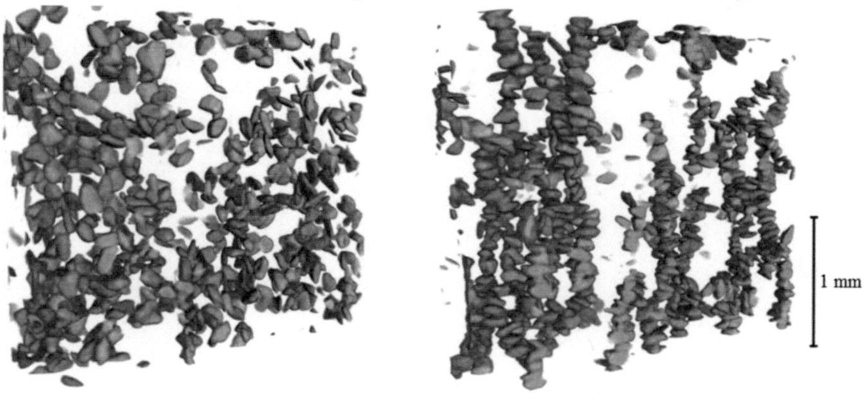

Figure 10.2: On the left side the µCT scan of a sample MAE with flake-like NdFeB-alloy particles before magnetization is shown. The particles are evenly distributed. The right side, in contrast, shows the scan of the same sample after it was magnetized at 2 T. Now the particles have shifted within the soft elastic matrix and chain-like structures have formed [14, 15].

sample differs significantly from the following curves. By stepwise magnetization of the sample with µCT scans in between, it was possible to follow the particle movements and the alignment of their main axes relative to the magnetic field direction, as it is show in Figure 10.3. From the angular distribution of the particles, which was evaluated from the µTC scans at the different magnetization stages, it was possible to conclude that the particle-matrix coupling in the first magnetization loop is strong and only a part of the particles rotate. Only after further magnetization loops and the saturation of the sample the particles were able to rotate reversibly and align themselves along the magnetic field. As a consequence of the µCT scans with the corresponding angular distributions and the simulation model introduced in, a first, indirect impression of the irreversible matrix deformation behavior was obtained.

In [17, 18] the phenomena of training of MAE with hard magnetic spherical NdFeB-alloy particles were also analyzed experimentally and theoretically. According to this, complex magnetization processes of the individual grains take place in the multigrain particles, which interact with each other. The training effect is based on the fact that the magnetic moments within the individual grains change their orientation along the anisotropy axis during the magnetization loops and switch moments in some of the grains. The more often the magnetization loop is passed through, the more stable the system becomes since grains turn by the same angle under the influence of the magnetic field, and the same groups of grains begin to switch their magnetic moment. The magnetization of microparticles thus depends on the strength of the applied magnetic field as well as on the number of grains and does not correspond to the saturation magnetization of NdFeB-alloy powder, which is about 7.3 T [17, 18].

Figures 10.4–10.6 demonstrate influence of the training effect on the FORC distribution obtained for the MAE with hard magnetic spherical NdFeB-alloy particles. The

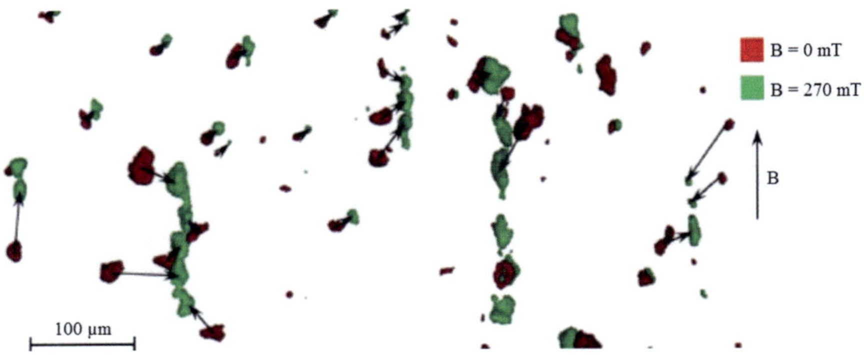

Figure 10.3: Combined µCT scans of a MAE with flake-shaped NdFeB-alloy particles, with and without an applied magnetic field, showing the movements and rotations of the particles by single particle tracking [15, 16].

sample has a soft elastic PDMS matrix (E ~50 kPa) and filled with 10 vol% of particles having a size of 5–20 µm. In Figure 10.4 the FORC distribution of a sample which was not premagnetized before the FORC measurement is presented, while Figure 10.5 shows the results from a sample which was initially magnetized (from $H = 0$ to H_{max}) and then passed through a magnetization loop. In contrast in Figure 10.6 a sample passing through six magnetization loops was used for FORC measurement. From the FORC diagrams it can be seen that the peak of the coercivity field in the untrained sample is relatively large due to the still strong particle–matrix interactions. As the magnetization state of the system becomes increasingly stable, analogous to [17], the magnetization becomes larger and the peak of the coercivity field narrower. The particles can rotate reversibly in the meanwhile stable particle–matrix-system and show comparable movement patterns when a magnetic field is applied again (e.g., during further FORC measurements).

10.3.2 Matrix influence

The elasticity of the polymer matrix has a significant influence on the magnetization behavior of magnetoactive elastomers [19]. The more soft the matrix is, the more the particles can move within it under the influence of the applied magnetic field. Various particle–matrix MAEs containing magnetic hard particles were considered using FORC analysis. Corresponding results are shown in Figures 10.7 and 10.8.

Figure 10.7 shows that the central peak moves to the right along the coercive field axis with increasing modulus of elasticity. This drastic shift can be explained by the increasing restriction of particle movement. The stiffer the matrix, the less the particles can move and rotate. After all, in completely inelastic matrices, only inner-particle domain processes can take place in response to an external magnetic field. The interaction between the different particles type is also restricted with increasing matrix

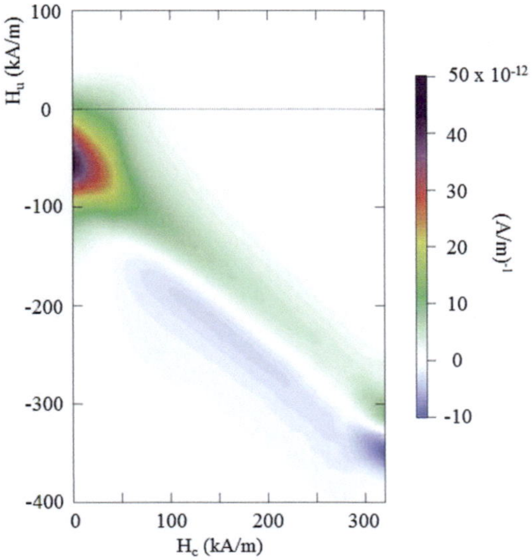

Figure 10.4: FORC diagram of an untrained sample of MAE with 10 vol% NdFeB particles (<20 μm) in soft elastic matrix (E ~50 kPa).

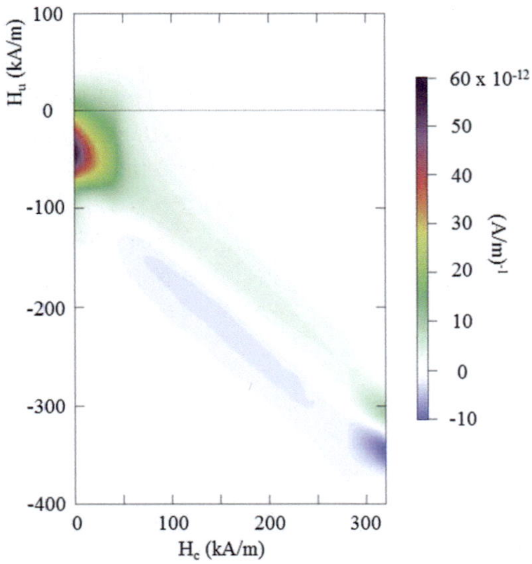

Figure 10.5: FORC diagram of MAE with 10 vol% NdFeB particles (<20 μm) in soft elastic matrix (E ~50 kPa), sample was premagnetized one time.

stiffness. The narrow ridge along the interaction axis represents the coupling between reversible magnetization of the soft magnetic iron particles and the irreversible magnetization state of the entire system. The diagonal feature in the FORC diagrams, however, indicates magnetostatic coupling and the movement of the iron particles in the matrix as discussed in [8].

FORC measurement performed on specimens solely based on magnetic hard powder confirmed the significant influence of matrix elasticity on the movement behavior of the particles under the influence of magnetic fields (Figure 10.8). The stiffer

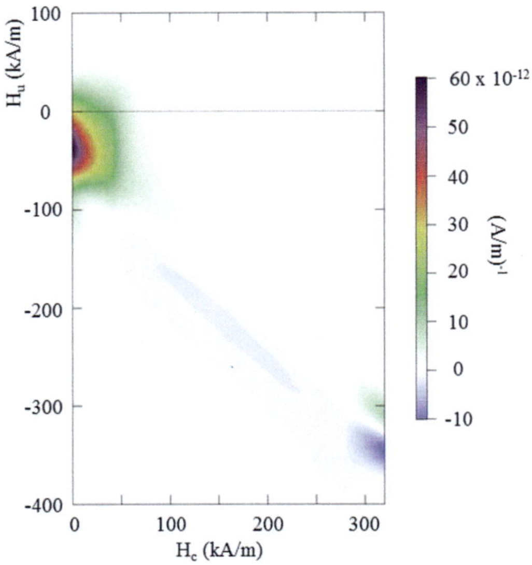

Figure 10.6: FORC diagram of MAE with 10 vol% NdFeB particles (<20 μm) in soft elastic matrix (E ~50 kPa), sample was premagnetized six times.

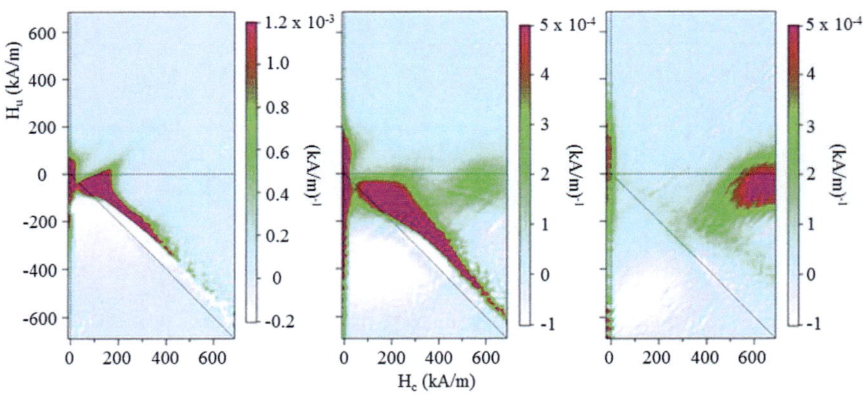

Figure 10.7: FORC diagrams of MAE with mixed magnetic filler components at a total particle content of 28 vol% and a mixing ratio of NdFeB particles to iron particles of 55:45. The elasticity of the polymer matrix increases from left to right: 50 kPa (left), 120 kPa (middle) and 440 kPa (right) [8].

the matrix system is, the less the particles can rotate and move. Inner-particle domain processes increase, the magnetization of the system decreases. In the hard epoxy matrix the particles are firmly anchored which means that the particles can no more rotate, and the magnetization of the system is based entirely on the inner-particle processes of the multidomain particles, whose individual regions align themselves along the magnetic field lines or, according to [17], switch over at sufficiently high fields. In addition, a weak diagonal ridge is visible in the FORC diagrams obtained for the samples without magnetic soft powder. According to [20], this could be due to soft

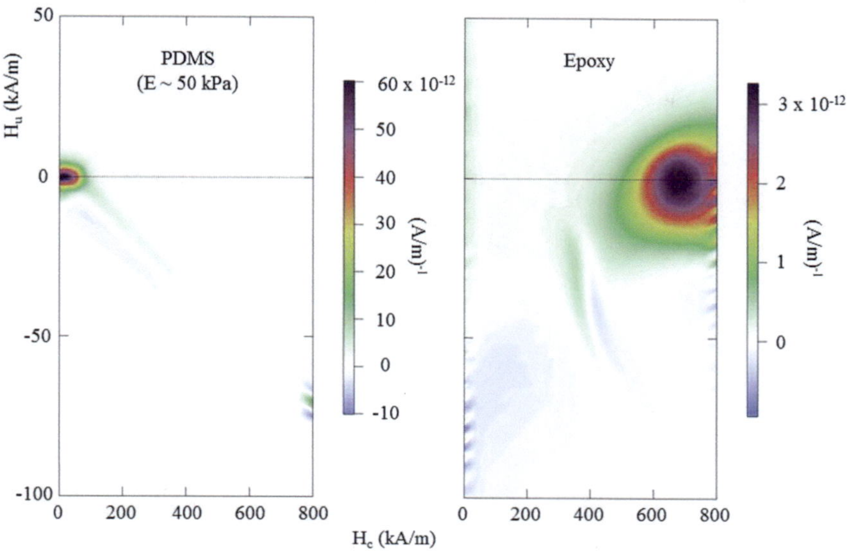

Figure 10.8: FORC diagrams of MAE with exclusively hard magnetic filling components at a particle content of 5 vol%. The samples are trained before passing FORC measurement.

magnetic fractions in the magnetically hard NdFeB particles, but this has not been proven, so the origin of this FORC feature is still to be investigated.

10.3.3 Particles influence

10.3.3.1 Particle ration

The influence of the ratio of the mixed magnetic filling components on the magnetization behavior of the MAE was investigated in [8] to evaluate whether there are interactions between the different particle types that have a possible influence on the material properties. For this purpose, MAEs with a constant total particle content of 28 vol% were used, but the ratio of magnetically hard NdFeB-alloy particles to magnetically soft carbonyl iron particles was varied. Figure 10.9 shows that as the iron particle content increases, the intensity of the irreversible features decreases while the reversible portions increase. The asymmetric degree along the H_u axis results from the coupling of the reversible magnetization behavior of the iron particles and the irreversible magnetization state of the whole system. In contrast, the areas of irreversible magnetization that occur at interaction fields of $H_u < -50$ kA/m are caused by the magnetostatic coupling and the movement of the particles within the matrix. In the case of higher contents of magnetically soft carbonyl iron particles, there are furthermore effects with positively and negatively applied fields. In the negative range, these are probably caused by a magnetization reverse of iron particles in the immediate

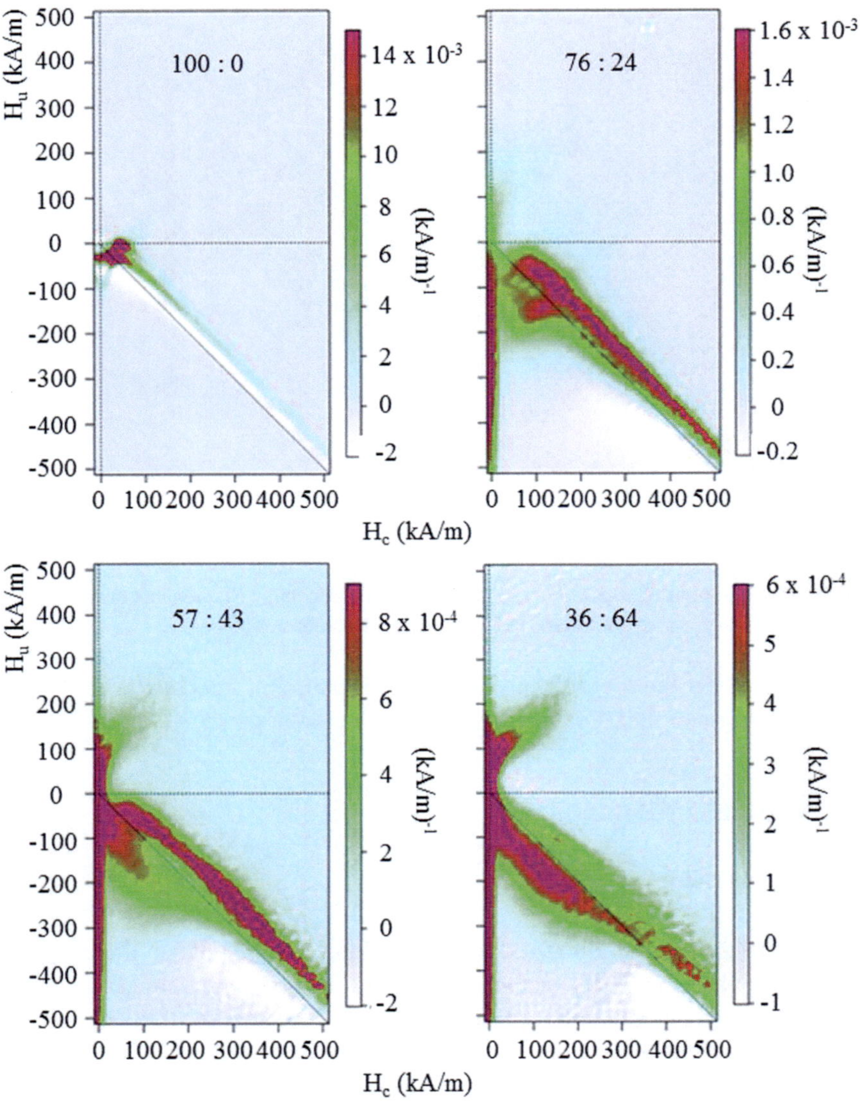

Figure 10.9: FORC diagrams of MAE with mixed magnetic filling components at a total particle content of 28 vol%, a constant matrix stiffness and a varied mixing ratio of NdFeB-alloy particles to iron particles. For example, the FORC diagram at the top on the left side has a ratio of 100% NdFeB particles to 0% Fe particles. The ratios of the other three diagrams are obtained in analogy (76:24, 57:43, 36:64) [8].

vicinity of positively magnetized NdFeB-alloy particles. The magnetization of these soft magnetic particles can only be reversed at this point by a negative field that compensates the positive remanent field of the nearby NdFeB-alloy particles. The effect in

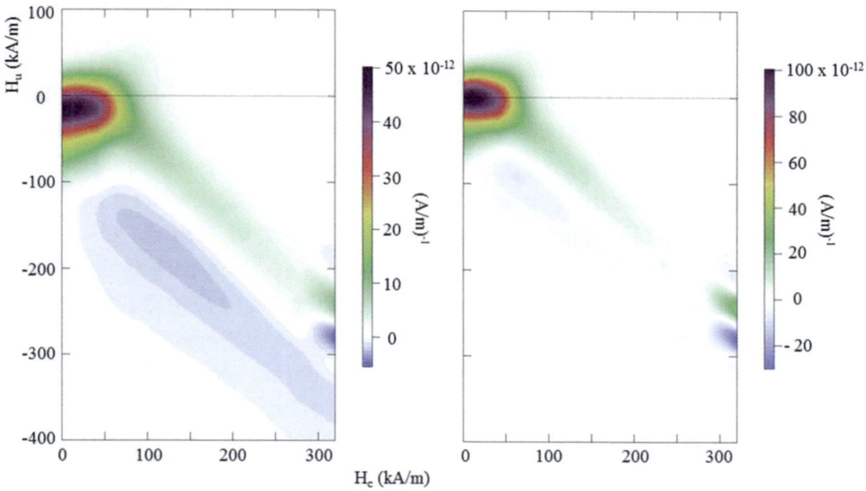

Figure 10.10: FORC diagrams of MAE with 10 vol% hard magnetic NdFeB particles in a polymer matrix with an elasticity of *E* ~50 kPa and varying particle size distribution: 20–39 μm (left) and 63–79 μm (right). Both samples were trained in advance.

the positive region probably comes from the movement and restructuring of the iron particles in the matrix.

10.3.3.2 Particle size

According to the manufacturer's specifications the diameter of the spherical NdFeB-alloy particles used by Magnequench is between 5 and 100 μm. To analyze the influence of particle size on the magnetic behavior of a composite, the particles were fractionated into different size ranges by the dry sieve method. Therefore the HAVER EML 200 Premium Remote Test Sieve Shaker with different nominal sizes of test sieves was used to fractionize the particles. FORC diagrams obtained for two samples with soft-elastic PDMS matrix (*E* ~50 kPa) but different particle size classes are compared in Figure 10.10. The larger particles show a much higher magnetization, which may be explained by the correspondingly higher number of domains within the multigrain particles. Furthermore, the feature corresponding to the coercive field is larger in the FORC diagram obtained for the sample with the smaller particles. This can be explained by a more tightly structured particlepolymer network, which, in analogy to Figure 10.8, is accompanied by an increase in the coercive field. In relation to the larger NdFeB alloy particles, the smaller particle fraction has higher total surface area, which potentially allows more rubbers between coated particles and polymer chains as discussed in [21]. However, the exact reasons for the different behavior is not clear and still need to be investigated.

10.4 Conclusion and outlook

Magnetoactive elastomers with magnetically hard particles or mixed magnetic filling components show a complex magnetization behavior that is influenced by several factors. This paper provides an overview of the influencing parameters for particlematrix systems with primary spherical NdFeB-alloy particles. The magnetic characterization is performed using the FORC method. While the training effect has already been demonstrated in the magnetization curves of such systems with magnetically hard NdFeB-alloy particles [11, 17, 18], its influence on the FORC distributions is discussed in this study. The FORC diagrams also show that the magnetization behavior depends on the training state of the sample. While the particle-matrix-coupling is at first much more intense in the untrained state, the particles are able to move better with each magnetization cycle until they can finally rotate reversibly. These observations indirectly indicate irreversible damage to the polymer structure as a result of training a MAE sample.

The polymer matrix also has a significant influence on the movement behavior of the NdFeB-alloy particles and therefore on their magnetization characteristics. Especially in soft elastic polymers, reversible rotation and restructuring of the particles under the influence of an applied magnetic field is possible. In this case, the magnetization of the system results from the orientation of the particles in the field as well as from interparticle-interactions. In contrast, in stiff matrix-systems, such as in an inelastic epoxy matrix, the particles are firmly anchored in the matrix structure and the relatively small magnetization of the material compared to the previous case results exclusively from particle-internal domain effects. The coercive field also increases drastically with stiffer matrices.

Furthermore, NdFeB-alloy particles of different sizes in the micrometer range show different magnetic characteristics. As expected, composites with smaller particles have lower magnetization and slightly higher coercivity. This can be reasoned with a more distinct particlematrix-bond. The influence of the particle size is of particular importance with regard to future technical applications, since a constantly reproducible magnetization behavior of the particlematrix-system is required for series products.

As a consequence of the knowledge obtained to date on the magnetization behavior of magnetically active elastomers with hard magnetic NdFeB-alloy particles, a detailed examination of the relationship between magnetization properties and particle movements is required. This can be realized using a coupling of μCT-Scans to obtain the three-dimensional representation of the particle structures with the results of stepwise FORC measurements to obtain respective macroscopic response of the material. The results of such investigations are of particular importance for the development and verification of theoretical models, which in turn must be used for a wide range of technical applications, e.g., in sensor technology [22–24].

Author contributions: All the authors have accepted responsibility for the entire content of this submitted manuscript and approved submission.
Research funding: Financial support by Deutsche Forschungsgesellschaft (DFG) under Grant SO 18/24-1 within SPP 1681 and DFG-RFBR PAK 907 is gratefully acknowledged.
Conflict of interest statement: The authors declare no conflicts of interest regarding this article.

References

1. Odenbach S. Microstructure and rheology of magnetic hybrid materials. Arch Appl Mech 2016;86: 269–79.
2. Odenbach S. Magnetic hybrid materials. Arch Appl Mech 2019;89:1–2.
3. Stepanov GV, Borin DY, Raikher YL, Melenev PV, Perov NS. Motion of ferroparticles inside the polymeric matrix in magnetoactive elastomers. J Phys Condens Matter 2008;20:204121.
4. Borin DY, Stepanov GV, Odenbach S. Tuning the tensile modulus of magnetorheological elastomers with magnetically hard powder. J Phys Conf 2013;412:012040.
5. Borin DY, Stepanov GV, Dohmen E. Hybrid magnetoactive elastomer with a soft matrix and mixed powder. Arch Appl Mech 2019;89:105–17.
6. Borin DY, Stepanov GV, Dohmen E. On anisotropic mechanical properties of heterogeneous magnetic polymeric composites. Phil Trans Math Phys Eng Sci 2019;377:20180212.
7. Borin DY, Stepanov GV. Magneto-mechanical properties of elastic hybrid composites. Phys Sci Rev 2020, (submitted to this issue).
8. Linke JM, Borin DY, Odenbach S. First-order reversal curve analysis of magneto-active elastomers. RSC Adv 2016;6:100407–16.
9. Roberts AP, Pike CR, Verosub KL. First-order reversal curve diagrams: a new tool for characterizing the magnetic properties of natural samples. J Geophys Res 2000;105:28461–75.
10. Roberts AP, Heslop D, Zhao X, Pike CR. Understanding fine magnetic particle systems through use of first-order reversal curve diagrams. Rev Geophys 2014;52:557–602.
11. Sánchez PA, Gundermann T, Dobrosedova A, Kantorovich S, Odenbach S. Importance of matrix inelastic deformations in the initial response of magnetic elastomers. Soft Matter 2018;14: 2170–83.
12. Borin DY, Günther D, Hintze C, Heinrich G, Odenbach S. The level of cross-linking and the structure of anisotropic magnetorheological elastomers. J Magn Magn Mater 2012;324:3452–4.
13. Schümann M, Odenbach S. The microstructure of magnetorheological materials characterized by means of computed X-ray microtomography. Phys Sci Rev 2020, (submitted to this issue).
14. Schümann M, Borin DY, Huang S, Auernhammer GK, Müller R, Odenbach S. A characterization of the magnetically induced movement of NdFeB-particles in magnetorheological elastomers. Smart Mater Struct 2017;26:095018.
15. Schümann M, Gundermann T, Odenbach S. Microscopic investigation of the reasons for field dependent changes of the properties of magnetic hybrid materials using X-ray micro tomography. Arch Appl Mech 2019;89:77–89.
16. Gundermann T, Odenbach S. Investigation of the motion of particles in magnetorheological elastomers by X-μCT. Smart Mater Struct 2014;23:105013.
17. Vaganov MV, Borin DY, Odenbach S, Raikher YL. Training effect in magnetoactive elastomers due to undermagnetization of magnetically hard filler. Physica B Phys Condens Matter 2020;578:411866.

18. Vaganov MV, Borin DY, Odenbach S, Raikher YL. Mesomagnetomechanics of hybrid elastomer composites: magnetization of elastically trapped particles. J Magn Magn Mater 2019;499:166249.
19. Krautz M, Werner D, Schrödner M, Funk A, Jantz A, Popp J, et al. Hysteretic behavior of soft magnetic elastomer composites. J Magn Magn Mater 2017;426:60–3.
20. Chen P-A, Yang C-Y, Chang S-J, Lee M-H, Tang N-K, Yen S-C, et al. Soft and hard natures of Nd2Fe14B permanent magnet explored by first-order-reversal-curves. J Magn Magn Mater 2014;370:45–53.
21. Fu S-Y, Feng X-Q, Lauke B, Mai Y-W. Effects of particle size, particle/matrix interface adhesion and particle loading on mechanical properties of particulate–polymer composites. Compos B Eng 2008;39:933–61.
22. Becker TI, Zimmermann K, Borin DY, Stepanov GV, Storozhenko PA. Dynamic response of a sensor element made of magnetic hybrid elastomer with controllable properties. J Magn Magn Mater 2018;449:77–82.
23. Chavez J, Böhm V, Becker TI, Gast S, Zeidis I, Zimmermann K. Actuators based on a controlled particle-matrix interaction in magnetic hybrid materials for applications in locomotion and manipulation systems. Phys Sci Rev 2020, (submitted to this issue).
24. Becker TI, Raikher YL, Stolbov OV, Böhm V, Zimmermann K. Magnetoactive elastomers for magnetically tunable vibrating sensor systems. Phys Sci Rev 2020, (submitted to this issue).

Joachim Landers, Soma Salamon, Samira Webers and Heiko Wende*

11 Microscopic understanding of particle-matrix interaction in magnetic hybrid materials by element-specific spectroscopy

Abstract: Mössbauer spectroscopy is a well-known technique to study complex magnetic structures, due to its sensitivity to electronic and magnetic interactions of the probed nucleus with its electronic surrounding. It has also been applied to the emerging fields of magnetic hybrid materials as well as to ferrofluids, as information on the magnetic alignment and the velocity of the probed nucleus, i.e. of the particle it is embedded in, can be inferred from the spectra in addition to the above-mentioned quantities. Considering the wide range of preparation methods and sample properties, including fluids, particle powders, sintered pellets, polymer matrices and viscoelastic hydrogels, a considerable advantage of Mössbauer spectroscopy is the usage of γ-photons. This allows measurements on opaque samples, for which optical experiments are usually not feasible, also making the technique relatively independent of specific sample geometries or preparation. Using iron oxide nanoparticles in glycerol solution as an exemplary material here, the variety of system parameters simultaneously accessible via Mössbauer spectroscopy can be demonstrated: Spectra recorded for particles of different sizes provided information on the particles' Brownian dynamics, including the effect of the shell thickness on their hydrodynamic diameter, the presence (or absence) and ballpark frequency of Néel superspin relaxation as well as the particles' average magnetic orientation in external magnetic fields. For single-core particles, this resulted in the observation of standard Langevin-type alignment behavior. Mössbauer spectra additionally provide information on the absolute degree of spin alignment, also allowing the determination of the degree of surface spin canting, which limits the maximum magnetization of ferrofluid samples. Analyzing the alignment behavior of agglomerated particles for comparison, we found a completely different trend, in which spin alignment was further hindered by the competition of easy magnetic directions. More complex particle dynamics are observed when going from ferrofluids to hybrid materials, where the particle mobility and alignability depends not only on the particles' shape and material, but also on the matrices' inner structure and the acting force-transfer mechanism between particles and the surrounding medium. In ferrohydrogels for example, particle mobility in terms of

***Corresponding author: Heiko Wende**, Faculty of Physics and Center for Nanointegration Duisburg-Essen (CENIDE), University of Duisburg-Essen, Duisburg, Germany, E-mail: heiko.wende@uni-due.de
Joachim Landers, Soma Salamon and Samira Webers, Faculty of Physics and Center for Nanointegration Duisburg-Essen (CENIDE), University of Duisburg-Essen, Duisburg, Germany

This article has previously been published in the journal Physical Sciences Reviews. Please cite as: J. Landers, S. Salamon, S. Webers and H. Wende "Microscopic understanding of particle-matrix interaction in magnetic hybrid materials by element-specific spectroscopy" *Physical Sciences Reviews* [Online] 2021, 6. DOI: 10.1515/psr-2019-0116 | https://doi.org/10.1515/9783110569636-011

Mössbauer spectroscopy was probed for different crosslinker concentrations, resulting in widely different mesh-sizes of the polymer network and different degrees of freedom. While a decrease in particle dynamics is clearly visible in Mössbauer spectroscopy upon rising crosslinker density, complementary AC-susceptometry experiments indicated no Brownian motion on the expected timescales. This apparent contradiction could, however, be explained by the different timescales of the experiments, probing either the relatively free Brownian motion on ultrashort timescales or the more bound state preventing extensive particle motion by interaction with the trapping mesh walls in the millisecond regime. However, it should also be considered that the effect of the surroundings on particle rotation in AC-susceptometry may also differ from the variation in translational motion, probed by Mössbauer spectroscopy. Being sensitive mainly to translational motion also results in a wide range of particles to be accessible for studies via Mössbauer spectroscopy, including larger agglomerates embedded in polymers, intended for remote-controlled heating. Despite the agglomerates' wide distribution in effective diameters, information on particle motion was found to be in good agreement with AC-susceptometry experiments at ultralow frequencies in and above the polymer melting region, while additionally giving insight into Néel relaxation of the individual nanoparticles and their magnetic structure.

Keywords: Mössbauer spectroscopy, nanoparticle dynamics, nanoviscosity, relaxation phenomena

11.1 Introduction

Mössbauer spectroscopy is an element-specific and non-destructive absorption technique, based on the effect of resonant nuclear absorption of γ-rays, discovered by Rudolf Mößbauer in 1958 [1]. Due to its ability to characterize a material's "fingerprint" hyperfine structure, it was established as a key technique in the fields of materials science, magnetism and molecular studies and even led to significant discoveries in geology, astrophysics and archeology [2–5].

In addition to one of its primary usages in the analysis of static magnetic properties, including phase composition, iron site occupation [6] and magnetic alignability, Mössbauer spectroscopy was found to exhibit several properties that make it a versatile technique for the study of magnetic hybrid materials (see also chapter 11.2):

- While there are several methods to access nanoparticle dynamics on the macroscopic scale, Mössbauer spectroscopy provides information on very local properties in combination with very short timescales. Thereby, one can yield valuable insight into the field of nanoviscosity, i.e. how the nanostructure of the surrounding medium affects the (sub-) diffusive behavior of magnetic nanoparticles (MNPs) on different time- and length scales.

- As it is sensitive to the dynamics of individual atomic magnetic moments instead of a macroscopic net magnetization, Mössbauer spectroscopy is able to clearly distinguish between different magnetic relaxation mechanisms. As will be illustrated in more detail below, this can be utilized to characterize Néel-type relaxation and Brownian motion separately, studying either the characteristic deformation of the spectrum affected by the frequency of internal reorientation of the particles' superspin or the symmetric broadening of the absorption lines evoked by the Doppler effect corresponding to diffusive nanoparticle Brownian motion.
- In combination with standard magnetometry, optical experiments belong to the most frequently used approaches for the characterization of MNPs in ferrofluids. In the case of higher nanoparticle concentrations or for the study of MNPs in solid matrices, this approach may not be feasible. Mössbauer spectroscopy utilizes γ-radiation in the keV-range (14.4 keV for the most common isotope ^{57}Fe), yielding appropriate penetration depths of ca. 5–15 mm for common matrix materials (water-based ferrofluids, hydrogels, polymer matrices), the latter being dependent on the density and atomic number of the used elements, being similar for the above-mentioned examples. Thereby, samples of sufficient iron concentration can be studied in detail relatively independent of their thickness and type of preparation (powder, fluid, solid), without being hindered by opaque matrix materials.
- Although less often applied in the preparation of magnetic hybrid materials, paramagnetic, superparamagnetic or antiferromagnetic (nano-) particles may resist the most commonly used magnetometry experiments for sample characterization. Mössbauer spectroscopy allows the analysis of such materials as well, due to its sensitivity to the interactions between the active nuclei und their electronic surroundings.

Based on these favorable properties, Mössbauer spectroscopy was applied to ferrofluids relatively soon after its invention, early studies being performed for example by Bhonchev et al. [7, 8], supported by a theoretical description on the influence of different types of diffusive motion on the spectral structure by Singwi and Sjölander [9]. Beside iron-based nanoparticles in fluids, as ferrofluids were a less intensively studied field back then, Mössbauer studies on diffusive processes included the atomic diffusion of ^{57}Fe atoms in crystal lattices [10, 11], molecular dynamics [12, 13] and even diffusive processes in nature [14]. In recent years, it has also been applied to magnetic nanoparticles in hydrogels, gels, polymer compounds, porous media and even liquid crystal systems [15–17].

To provide an overview over possible ways to apply Mössbauer spectroscopy, what has to be considered in the evaluation of experimental spectra and which quantities can be extracted from it, several examples will be given in chapter 11.3–11.5 including temperature-dependent as well as in-field experiments on ferrofluids, hydrogels and magnetic polymer compounds. The required theoretical background regarding the basics of Mössbauer spectroscopy and the different apprehended spectral features are presented in chapter 11.2.

11.2 Theoretical background

Mössbauer spectroscopy is a non-destructive measurement method based on the nuclear resonant emission and absorption of γ-photons. For atoms incorporated into crystalline solid bodies, there is a certain probability (given by the Lamb-Mössbauer factor, also called Debye-Waller factor) for such a photon to be emitted recoil-free, without the change in emission energy associated with the generation of phonon lattice vibrations, as well as for the same photon to be absorbed again by another ^{57}Fe nucleus, referred to as resonant emission and absorption [2].

In the most commonly used standard transmission setup shown schematically in Figure 11.1(a), the radiation source, often consisting of paramagnetic ^{57}Co atoms embedded in a metal foil (e.g. Rhodium), is mounted on an electromechanical Mössbauer drive, which oscillates the source back and forth, in order to modulate the emitted γ-photon energies via the optical Doppler effect. The sample is placed behind the radiation source, with the distance being dictated by the avoidance of too small spacings that would lead to angular dependent effects (so-called "cosine smearing"), or excessively large spacings, which would lead to a diminished number of photons arriving at the sample. Once having passed through and interacted with the sample, the remaining γ-photons are recorded by a detector, e.g., a proportional counter. Other types of Mössbauer spectroscopy include synchrotron setups, measurements in backscattering geometry or the detection of conversion electrons, each with its different advantages, as, e.g., enhanced surface sensitivity [18].

These 14.4 keV γ-photons are emitted after the decay of ^{57}Co source atoms to ^{57}Fe, when relaxing from the excited $I = 3/2$ ^{57}Fe nuclear state to the ground state. According to Heisenberg's uncertainty principle, the first excited nuclear state's relatively long half-life of 98 ns corresponds to a narrow natural linewidth of the emitted γ-radiation of less than 5 neV, being the origin of the method's extreme energy resolution. This enables the investigation of miniscule variations in absorption energy of the studied ^{57}Fe atoms, caused by the so-called hyperfine interactions between ^{57}Fe nuclei and the surrounding electrons. The effects most frequently discussed in the study of static magnetic properties are the isomer shift δ, i.e. the shift of the entire spectrum relative to a reference material, the nuclear Zeeman splitting quantified by the hyperfine magnetic field B_{hf}, causing the characteristic sextet line structure in case of magnetically ordered samples, and the quadrupole splitting ΔE_Q (or nuclear quadrupole level shift 2ε), see Figure 11.1.

When studying a magnetically ordered sample in an external magnetic field, there are two primary effects giving insight into the magnetic behavior of the material: The change in nuclear Zeeman splitting caused by the superposition of the internal

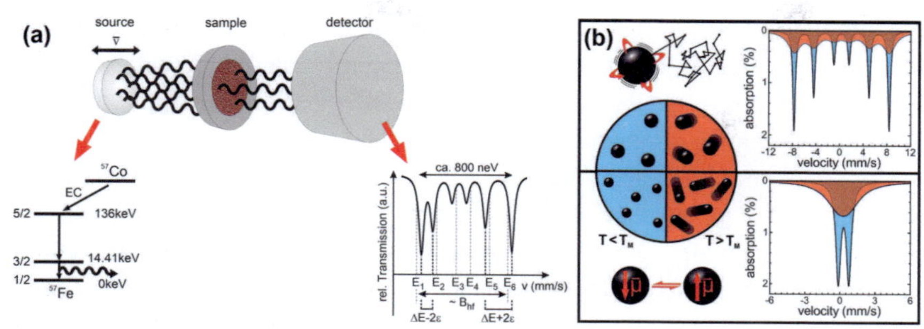

Figure 11.1: (a) Schematic illustration of nuclear energy-levels following the decay of ^{57}Co source material generating highly monochromatic γ-radiation and contributions to hyperfine structure, and (b) the effect of different relaxation mechanisms on the Mössbauer spectrum for small particles (Néel relaxation, bottom) and large particles (top) in rigid (left) or liquid matrices (right, broadened lines), adapted with permission from [19], Copyright (2016) American Chemical Society.

"hyperfine field" B_{hf} and the external magnetic field B_{appl}, and the variation in relative line intensities, in accordance to absorption probabilities of dipole radiation dependent on the angle between spin and the direction of the incident γ-photon, given by Fermi's golden rule. The first one can be determined directly as vectorial sum of B_{hf} and B_{appl}, leading, e.g. to a reduction of sextet splitting in ferromagnets, as B_{hf} is antiparallel to the iron atoms' magnetic moment, while the latter is often characterized by the so-called A_{23} line ratio as given in Eq. (11.2.1) dependent on the average spin canting angle θ between spin and γ-photon direction [20, 21]. This is also illustrated in detail for in-field Mössbauer spectra in chapter 11.3.2.

$$A_{23} = \frac{I_2}{I_3} = \frac{4\sin^2(\theta)}{1+\cos^2(\theta)} \tag{11.2.1}$$

The parameter A_{23} is therefore a representation of the average magnetic alignment of the system, dependent on the applied field amplitude and can thus be understood as comparable to a single data point in a standard field-dependent magnetometry experiment. However, A_{23} exhibits a fundamental difference, as it provides an absolute measure of magnetic orientation, a distinction that is most obvious, e.g., in the study of magnetic nanoparticles, often reported to have a reduced saturation magnetization M_S due to frustrated surface spins or reduced magnetic moments in the top atomic layers [22]. This expression "saturation magnetization" has to be considered with caution, as nanoparticle samples seemingly being saturated may exhibit values of A_{23} far from 0, a clear indication that in such systems the "reduction" in M_S can primarily be assigned to spin frustration instead of reduced magnetic moments, whereby the term "high-field magnetization" may be more appropriate instead.

In terms of magnetic hybrid materials, the time constants of Mössbauer spectroscopy are also of interest, i.e. the Larmor precession frequency (e.g. $\omega_{Larmor} \approx 200$ MHz $\rightarrow \tau_{Larmor} \approx 5$ ns for $B_{hf} \sim 50$ T) and the half-life of 98 ns, making the technique highly sensitive to magnetization dynamics on the nanosecond timescale. Thereby it is a valuable extension of the more commonly used approaches on soft magnetic hybrid systems via rheology or susceptometry experiments etc., as these usually work on the range of seconds to microseconds.

The thermally excited relaxation of a nanoparticle's magnetic moment (Néel superspin relaxation), present in case the thermal energy begins to overcome the magnetic anisotropy energy barrier of the particle, will lead to a reduction in sextet splitting when approaching the timescale of τ_{Larmor}, on which we observe the time-averaged nuclear Zeeman splitting [23]. Combined with lognormal distributions of particle core diameter and anisotropy energy characteristic for the result of nanoparticle growth processes, this will create easily recognizable asymmetric lineshapes, as different Néel relaxation times will result in almost unchanged values of B_{hf} for larger particles and strongly reduced ones for smaller particles, finally reaching the completely superparamagnetic doublet state [24, 25]. By using an adequate relaxation model as proposed e.g. by Blume et al. [26] or more complex many-state relaxation models [27, 28], one can reproduce the variation in lineshape and the degree of sextet collapse into the doublet state to extract the magnetic anisotropy energy of the particles, including effects of interparticle interaction decelerating Néel relaxation [29, 30].

Brownian particle motion, on the other hand, has a completely different effect on the spectral structure. As it introduces an additional relative instantaneous velocity between the absorbing ^{57}Fe atom and the ^{57}Co source, the diffusive motion also corresponds to a Doppler shift of the absorption energy, or, as the measured spectrum is averaged over a long time and a huge ensemble of particles moving statistically in all directions with the same probability, it leads to a Doppler broadening $\Delta\Gamma$ of the absorption line. The latter is proportional to the velocity of an absorbing atom, mainly generated by translational diffusion of the particle the ^{57}Fe atom is embedded in, as described in Eq. (11.2.2) [9]. Here, k denominates the wave vector corresponding to the γ-photon energy of 14.4 keV, D_T the translational diffusion coefficient, η the carrier fluid's dynamic viscosity, T the temperature and R_H the particle's hydrodynamic radius. Minor contributions to atomic motion are also caused by nanoparticle rotation, which is, however, seldomly studied in detail [31] and often neglected in first approximation calculations.

$$\Delta\Gamma = 2\hbar k^2 D_T \propto \frac{T}{R_H \eta} \tag{11.2.2}$$

With the half-life of only 98 ns, corresponding to the natural linewidth of ca. 5 neV or a half-width of 0.085 mm/s in the Mössbauer spectrum, a line broadening induced by Brownian particle motion may be detected being as small as 0.01 mm/s, yielding diffusion coefficients in the range of 10^{-16}–10^{-17} m^2/s, while linewidths up to several 100 mm/s have also been studied in case of sufficient spectral area (intensity). Thereby being sensitive to particle motion on the nanosecond timescale (corresponding to equally small length scales the particle is covering in these few nanoseconds), [32]

Mössbauer spectroscopy is, although applicable to Fe-bearing particles in liquids, even more promising for the study of magnetic hybrid materials. For these, short-time particle-matrix interactions may define the materials' behavior in terms of particle mobility and magnetic alignment. This statement is further reinforced by the fact that the probed motion is the component of particle diffusion along the γ-ray propagation direction, as atomic velocities in other directions do not contribute to the detected linear Doppler effect. Thereby even measurements of direction-resolved diffusion coefficients are in principle possible [33, 34], especially interesting for the study of either strongly elongated nanoparticles or the dynamics of particles in anisotropic nanostructures [35, 36]. In addition to the broadening of the absorption lines, Brownian motion also affects the probability for resonant absorption, usually described in terms of the Debye-Waller- or Lamb-Mössbauer factor f. Similar to lattice vibrations or molecular dynamics, accelerating diffusive motion will result in a decrease in resonant absorption for the benefit of quasi-elastic contributions. [37] However, due to the different timescales involved in the above-mentioned processes and further differences – does the atom perform free motion, bound diffusion or purely vibrational dynamics? – the dependence of resonant absorption on the covered mean square displacement of the ^{57}Fe nucleus will strongly depend on the specific mechanism and material at hand [13, 38, 39].

11.3 Accessing nanoparticle dynamics

11.3.1 Simultaneous determination of Néel and Brownian dynamics in ferrofluids

To obtain a general overview on particle magnetization dynamics studied via Mössbauer spectroscopy, we will start by looking at monodisperse nanoparticles in a Newtonian fluid, minimizing the number of further parameters apart from the beforementioned theory. Commercial ferrofluidic samples were used (Ocean NanoTech SHP-series), which we denote with regard to the used nanoparticles as S(mall), M(edium) and L(arge), as the contained iron-oxide nanoparticles were spherical in shape, with core diameters of 6.0, 14.8 and 26.4 nm. The particles were coated with an amphiphilic polymer and oleic acid and dispersed in 70 vol% glycerol solution. For Mössbauer experiments ca. 500 µl of each ferrofluid were used, to ensure sufficient spectral intensity. Due to the nanoparticle concentration of ca. 1.5 wt%, the absorption signal was relatively small compared to a common Mössbauer experiment on powder samples. Ca. 50 µl were used for complementary magnetometry and AC-susceptometry experiments.

As the nanoparticles used in sample S–L cover a wide range of core diameters (Figure 11.2(a)), going from predominant Néel superspin relaxation in the smallest particles to blocked Néel relaxation in larger ones, leaving Brownian motion as only mechanism

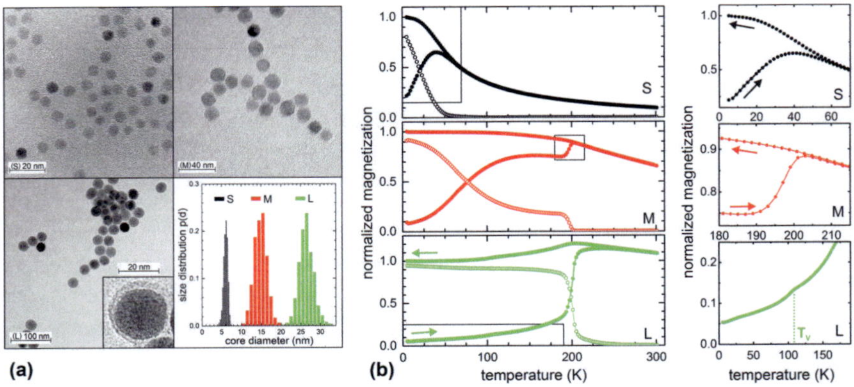

Figure 11.2: (a) TEM images and size distribution analysis, and (b) ZFC-FC magnetometry of ferrofluid samples S–L. The images verify the regular spherical shape as well as particle monodispersity. Temperature-dependent magnetometry demonstrates different magnetization dynamics, dominated by either the Néel (S) or the Brownian mechanism (L), or the combination of both (M). Focus regions of interest (right-hand side) are shown in magnification, i.e. Néel blocking temperatures (S), the onset of Brownian motion (M) and the Verwey transition at T_V (L), proving the presence of magnetite (particle core) material. Adapted with permission from [19], Copyright (2016) American Chemical Society.

for magnetic reorientation, temperature-dependent magnetometry already provides valuable reference information on magnetic dynamics. Here, the standard ZFC-FC methodology was applied, where the temperature-dependent magnetization is measured after cooling down the sample either without (zero-field-cooled) or with a minor external magnetic field (field-cooled), where a change in the splitting between both curves represents the onset of magnetic realignment processes, whereby identical curves M_{ZFC} and M_{FC} indicate the completely superparamagnetic state [40, 41]. The results of these experiments shown in Figure 11.2(b) display Néel-type blocking temperatures of ca. 20–30 K for sample S and ca. 80–100 K for sample M. Also, a relatively sharp increase in magnetization is visible at ca. 200 K, marking the onset of Brownian motion, allowing the reorientation of all previously blocked (large particle) magnetic moments in field direction on the timescale of the magnetometry experiment [42]. Similar behavior is seen at ca. 200 K for sample L, although here almost all magnetic moments are blocked at lower temperatures due to the larger particles' higher anisotropy energy E_A.

Although the observation of a superparamagnetic state depends on the time-scale of the experimental technique at hand, which is of course very different in magnetometry (seconds to hours) and Mössbauer spectroscopy (nanoseconds), this pre-characterization provides valuable insight and goes hand in hand with the observed general structure of the corresponding Mössbauer spectra visible in Figure 11.3. As has been found in a similar manner in studies on Néel relaxation in powder samples of small nanoparticles in sample S, we find a state of fast Néel relaxation (presumably the relaxation frequency is comparable to the Larmor frequency), where the sextet

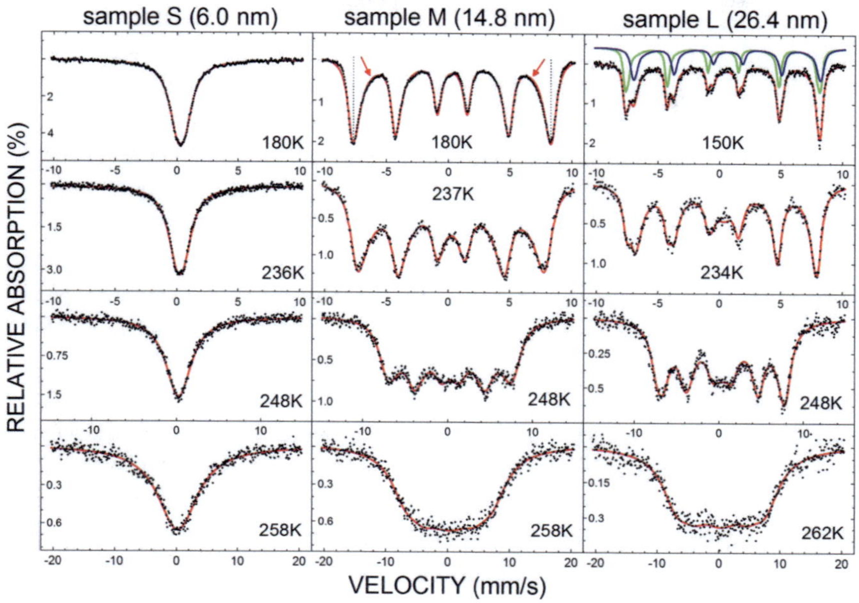

Figure 11.3: Mössbauer spectra of ferrofluid sample S–L recorded upon rising temperature at 234–262 K. Reference (static) Mössbauer spectra measured at 150–180 K are shown on top, which were used to extract trends in hyperfine parameters for the modeling of the spectra affected by Brownian line broadening. Reprinted with permission from [19], Copyright (2016) American Chemical Society.

structure has already collapsed, while the doublet structure of the "completely" superparamagnetic state is still obscured. The largest particles in sample L on the other hand display almost no sign of spin dynamics, showing more or less symmetric absorption lines, which allows us to assign subspectra positions corresponding to ^{57}Fe on the lattice sites of tetrahedral (A-site, green) and octahedral geometry (B-site, blue). Particles of intermediate core diameter (sample M) exhibit absorption lines with a characteristic asymmetric deformation, from which the distribution of anisotropy energies can be extracted using an appropriate theoretical model [28]. Comparing common magnetic characterization techniques, at this point the major difference between Mössbauer spectroscopy on the one hand and magnetometry and susceptometry on the other hand becomes evident: ZFC-FC magnetometry experiments display *at which temperature* a change in particle dynamics takes place, allowing only indirect deduction on the relevant parameters, and AC-susceptometry provides the *total* relaxation frequency of present relaxation mechanisms. Mössbauer spectroscopy on the other hand, while less direct in data evaluation, allows the simultaneous extraction of Néel as well as Brownian parameters (anisotropy energy E_A, diffusion coefficient D_T), as these processes affect different features of the Mössbauer spectrum [19, 43].

Now focusing on particle diffusion in the three ferrofluids, we extract the Brownian line broadening from the spectra. Regarding the spectra at rising temperatures, beginning at ca. 230–240 K, we see that the broadening of the lines becomes evident. In order to determine the line broadening and thereby nanoparticle diffusion coefficients, we have to reproduce the spectral structure with an appropriate fitting model and subtract the static linewidth Γ_{stat} from the total linewidth Γ. Therefore, the information on Brownian dynamics cannot be determined from the spectra without considering the influence of Néel relaxation on the lineshape and recording e.g. a reference spectrum at sufficiently low temperature, where the solid carrier medium does not allow Brownian motion and cancels the additional Doppler broadening of the absorption lines (providing Γ_{stat}). Alternatively, if the carrier fluid does not crystallize but instead exhibits a glass transition, as in the case presented here using glycerol solution, Γ_{stat} could also be extrapolated from spectra at sufficiently low temperatures (ultrahigh viscosities), where Doppler broadening is present but far smaller than the static linewidth [44, 45].

The particle diffusion coefficient can now be estimated by usage of Eq. (11.2.2) from the temperature-dependent line broadening shown in Figure 11.4(a). The same goes for the dynamic viscosity of the carrier medium in Figure 11.4(b) under the consideration that hydrodynamic diameters are known. The three particle core diameters present in sample S, M and L yield identical trends in $\eta\ (T)$ assuming an effective coating thickness of ca. 5 nm, the latter being in good agreement with coating thicknesses reported by the manufacturer [19]. Since the glycerol solution used as carrier medium is a simple Newtonian fluid, unlike in experiments discussed later on, where effects of nanoviscosity have to be considered, the agreement between the dynamic viscosity $\eta\ (T)$ obtained via Mössbauer spectroscopy with that determined in macroscopic rheological experiments from literature proves the applicability of the presented approach for the study of particle diffusion [46].

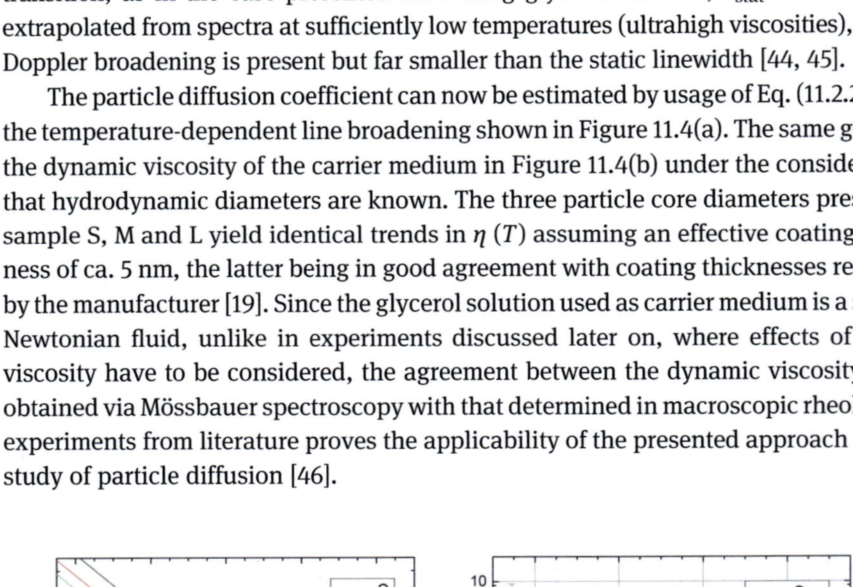

Figure 11.4: (a) Doppler line broadening of sample S–L extracted from Mössbauer spectra, with factors 1.58 and 1.36 originating from the different values of R_H in the ferrofluids, and (b) derived values of the carrier fluid's dynamic viscosity. Adapted with permission from [19], Copyright (2016) American Chemical Society.

11.3.2 Dynamics of nanoparticle agglomerates in polymer melts

Brownian particle dynamics are also studied for agglomerates in polymer melts, representing a system with similarities, as well as differences to the spherical particles in glycerol solutions discussed above. This gives us the opportunity to compare particle dynamics in Mössbauer spectroscopy on different scales of effective particle diameter and dynamic viscosity. The biopolymers used here are based on meltable dextran ester, designed to allow on-demand drug release through internal heating via the embedded nanoparticles when exposed to an AC-magnetic field [47, 48]. While the particles are of ca. 15 nm in diameter, it has to be considered that in the as-studied biopolymer composite material, a fraction of particles forms larger 30–80 nm multicore agglomerates that have a larger specific heating power than smaller superparamagnetic nanoparticles [49].

Therefore, despite the relatively high hydrodynamic particle diameter, Néel relaxation of the individual particle superspins has to be considered in the modeling of the experimental spectra as shown in Figure 11.5(c, d), recorded at temperatures between room temperature – below the polymer melting region – and temperatures of ca. 120 °C. Here, one observes similar relaxation features as for the above-mentioned intermediate sized particles in sample M. One should keep in mind that, when discussing effective anisotropy barriers, magnetic dipolar interaction between neighboring particles in agglomerates discussed here may be a considerable contribution, which can usually be neglected in ferrofluids of low MNP concentration [50].

In terms of Brownian motion, the rather high viscosity of the polymer melts of ca. 30–1 Pas in the temperature region of 55–100 °C in combination with the high effective diameters leads to far lower line broadening as compared to the ferrofluidic samples. At this point, size distribution effects have to be introduced for the first time, affecting the agglomerate (effective hydrodynamic) radius R_H much more than the relatively narrow core diameter distribution. As the main component of the Doppler line broadening is caused by translational Brownian motion, this line broadening can be considered as proportional to R_H^{-1}, with the signal intensity of agglomerates being directly dependent on the number of included ^{57}Fe-atoms, whereby the number distribution of hydrodynamic diameters has to be weighted with the particle volume when being applied to reproduce spectral features. To some extent, the distribution $P(D_H)$ can be extracted from the measurement by adding a distribution of linewidths to the spectrum, while utilizing fixed hyperfine parameters that were extrapolated from low temperature measurements. In agreement with the line broadening from spectra shown in Figure 11.5(c, d), an average effective agglomerate diameter of ca. 64 nm was found in AC-susceptometry measurements [51].

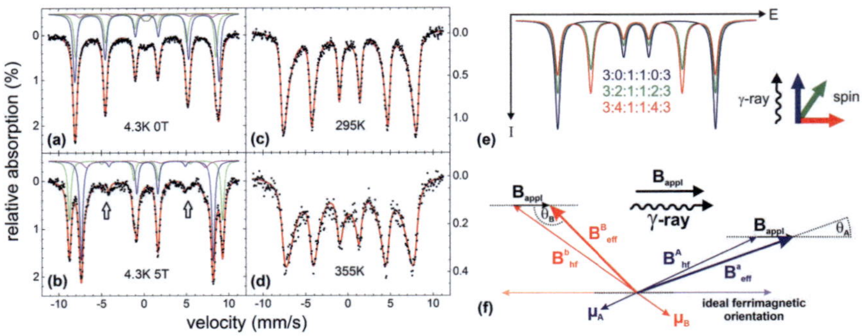

Figure 11.5: Mössbauer spectra of iron-oxide nanoparticles in biopolymers at 4.3 K with (b) and without (a) an external magnetic field of 5 T, at room temperature (c) and in the molten polymer state at 355 K (d), reprinted from [51], Copyright (2016), with permission from Elsevier. Arrows in (b) mark lines 2 and 5, whose relative intensity is determined by the average angle θ between spin and γ-ray propagation direction, as illustrated in (e), (f) effective magnetic fields in ferrimagnetic materials, adapted with permission from [20]. For symmetry reasons, usually the canting angles are given relative to ideal ferrimagnetic orientation in spinels, yielding values of 0–90° [52].

Additional information on the composite material was obtained from Mössbauer experiments performed at cryogenic temperatures of 4.3 K with (b) and without (a) an external magnetic field of 5 T along the γ-ray propagation direction (Figure 11.5). At such low temperatures there is no additional line broadening either by Néel relaxation or by Brownian diffusion, wherefore fine structures of the spectrum can be studied more easily, allowing the resolution of individual contributions of B-site Fe^{3+}(blue), B-site Fe^{2+} (violet) and A-site Fe^{3+} (green) due to their minor differences in isomer shift, hyperfine magnetic field B_{hf} and quadrupole splitting [29]. From the relative amount of Fe^{2+} ions (ca. 9%), the magnetite fraction in the nanoparticles can be estimated to roughly 25–30%, indicating a core of original magnetite material remaining from the as-synthesized state, surrounded by an oxidized maghemite passivation layer, slowing down further oxidation [53, 54]. A minute doublet in Figure 11.5(a) could indicate a minor paramagnetic contribution in the particles; the existence of a superparamagnetic doublet at such low temperatures is unlikely due to the minimum thermal energy at 4.3 K.

By applying an external magnetic field, we can also determine the total degree of particle magnetic alignment for each component and even the sublattices individually by comparing the relative sextet line intensities and using Eq. (11.2.1). By this method, we can extract canting angles of ca. 15° for the A-site and ca. 21° on average relative to ideal ferrimagnetic alignment for Fe^{2+} and Fe^{3+} on octahedral lattice positions from the in-field spectrum. In Figure 11.5(b) one can clearly see the shift of the tetrahedral A- and octahedral B-site subspectrum into different directions via the superposition of the external field $B_{appl} = 5$ T and the hyperfine magnetic fields $B_{hf,(4.3\,K)} \approx 50–52$ T, which are antiparallel for A- and B-site due to the material's ferrimagnetic structure, as showcased in Figure 11.5(f).

It is important to keep in mind that, although the spectrum shown in Figure 11.5(a) recorded at 4.3 K is helpful for the general characterization of the nanoparticle component of composite materials, it should not be used as a reference for the determination of line broadening in spectra above room temperature, deviating in hyperfine parameters as well as Néel relaxation due to temperature effects. For that purpose, instead a spectrum should be utilized, which was measured at only slightly lower temperatures, where the spectral structure is almost identical, while Brownian motion is still slow enough to create negligible line broadening, as visible e.g. through comparison of spectra (c) and (d) of Figure 11.5.

11.4 Magnetic alignment and dynamics: comparison of individual MNPs and NP-agglomerates

In the first section of the previous chapter we discussed the simultaneous analysis of individual magnetic relaxation processes (Néel relaxation and Brownian particle motion), while in the following section the method to study magnetic particle alignment via Mössbauer spectroscopy was introduced. Although the latter is more commonly applied in the study of static magnetic properties in solids, the change in the nanoparticles' magnetic orientation as a response to an external stimulus (magnetic or electric fields, matrix deformation, temperature change, etc.) can also be probed in fluid or viscoelastic hybrid materials. This is demonstrated in the following, by probing the superspin orientation in ferrofluidic samples in combination with the particles' Brownian dynamics as well as Néel relaxation. We will see that by the combination of the field-dependence of these quantities, we can also infer properties such as surface spin canting and the presence and size of nanoparticle agglomerates in the fluids.

Samples of the original ferrofluids S and M as introduced in chapter 11.3.1 were used for this in-field Mössbauer study, in combination with a newly prepared sample L, with a core diameter of 22.0 nm similar to the original sample [55]. To obtain a thorough and precise measurement of the magnetic orientation, spectra as shown in Figure 11.6 were recorded in external magnetic fields parallel as well as perpendicular to the γ-ray propagation direction, by using either an electromagnetic coil setup (maximum field ca. 1 T) or a set of permanent NdFeB ring magnets, providing homogeneous fields without blocking the γ-pathway (see Figure 11.6, schematic setup). Upon rising temperature, the exponential increase in linewidth leads to some superposition of the sextet of absorption lines. Still, up to a maximum linewidth of ca. 2–3 mm/s the variation in relative line intensities induced by the external magnetic field is visible even by eye, while at linewidths larger than 5–10 mm/s, the estimation of the parameter A_{23} is less precise even using professional modeling software. In parallel field geometry a decrease, and in perpendicular geometry an increase in A_{23} is visible, corresponding to the same trend of increasing magnetic alignment probed from two different directions, leading to lower, respectively higher angles between probing γ-ray and field (and preferred nanoparticle magnetic moment) direction. In these spectra, no distinct splitting between A- and B-site spinel subspectra is observable as it was in an external magnetic field of 5 T as shown in Figure 11.5. Although sufficient to align the magnetic nanoparticle moments in our ferrofluidic samples, the field amplitude of up to ca. 0.2 T is too small to lead to a clearly visible variation in relative line positions of the subspectra. Sample M exhibits some beginning Néel relaxation, leading to the characteristic asymmetric lineshapes, as was visible in Figure 11.3, apart from which it displays similar changes of spectral structure as in-field experiments on sample L do in Figure 11.6.

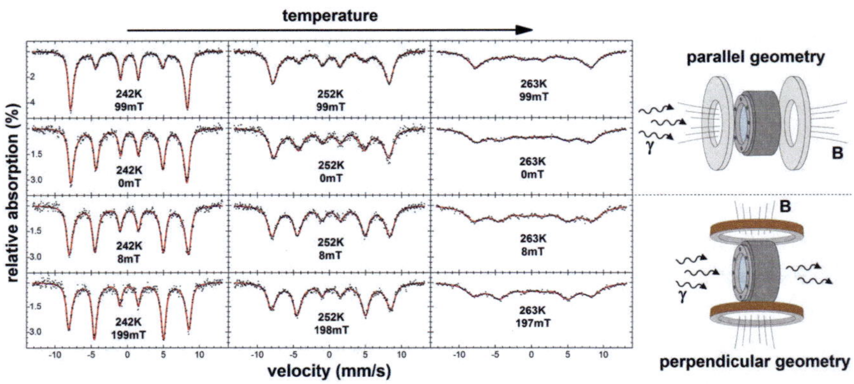

Figure 11.6: Mössbauer spectra of ferrofluid L recorded at 242–263 K in external magnetic fields of up to 199 mT applied either perpendicular (bottom) or parallel (top) to the γ-ray propagation direction, as illustrated in the adjacent schematic setups. Reprinted with permission from [55], Copyright (2019) American Chemical Society.

Values of A_{23} are displayed in Figure 11.7(a) against the rising magnetic field amplitude for intermediate sized particles in sample M. As discussed above, here the two clear branches represent the increasing degree in nanoparticle in-field alignment probed from two different directions, where $A_{23} = 4$ would match a spin alignment perpendicular to the γ-ray propagation and $A_{23} = 0$ coaxial (antiparallel) spin and γ-direction. Both branches start at $A_{23} = 2$, representing a 3D-random distribution of spin directions as expected of an ensemble of single-domain magnetic nanoparticles in a ferrofluid in the absence of external fields. Although we approach the maximum and minimum A_{23}-values mentioned above, the trend in A_{23} (H) clearly does not converge completely in the covered range of field amplitudes. Seemingly both branches A_{23} (H) converge towards lower degrees of ordering, caused by a minor degree of nanoparticle surface spin canting. We assume the degree in surface spin frustration to change moderately for different magnetic fields, whereby it would produce an almost constant background in Figure 11.7(a), defining the maximum attainable spin alignment before going to much higher field amplitudes.

Based on these considerations, A_{23} (H) is theoretically reproduced using standard Langevin theory to describe the distribution of superspin orientations in an external magnetic field at a given temperature. To consider spin frustration at the particle surface, the Langevin model function is weighted with $(1–x)$, with x being the fraction of canted surface spins, added to $2{\cdot}x$, as the frustrated spin fraction is assumed to be random in orientation (with $A_{23} = 2$). This model yields excellent agreement to experimental data, describing the first two regions of A_{23} (H) as shown in the insets of Figure 11.7(a): The low-field range I, where the random distribution of superspins yields $A_{23} \approx 2$, and the intermediate-field range II, where the superspins are primarily aligned, with a remaining degree of spin disorder from frustrated surface layers. In these two regions, the model yields a net magnetic moment of $6.6{\cdot}10^{-19}$ Am^2 for the alignable nanoparticle core moment in sample M, and a fraction of frustrated surface spins of ca. 13%. Combining these two parameters, the saturation magnetic moment of the particles is in the range of $7.6{\cdot}10^{-19}$ Am^2, in good agreement with estimations based on the maghemite bulk saturation magnetization and the particles' average volume. This total magnetic moment corresponds to a state of complete spin alignment, which is however attainable only for much higher fields (range III), starting also the (partial) alignment of surface moments, and is thus not feasible in our experiment. It is worth noticing that although values of A_{23} were extracted from spectra measured in a broad temperature interval with values of $\Delta\Gamma$ ranging from ca. 0.2–5 mm/s, no significant deviations in the extracted line ratios are observable, proving that this approach can be reproduced for samples of different particle mobility.

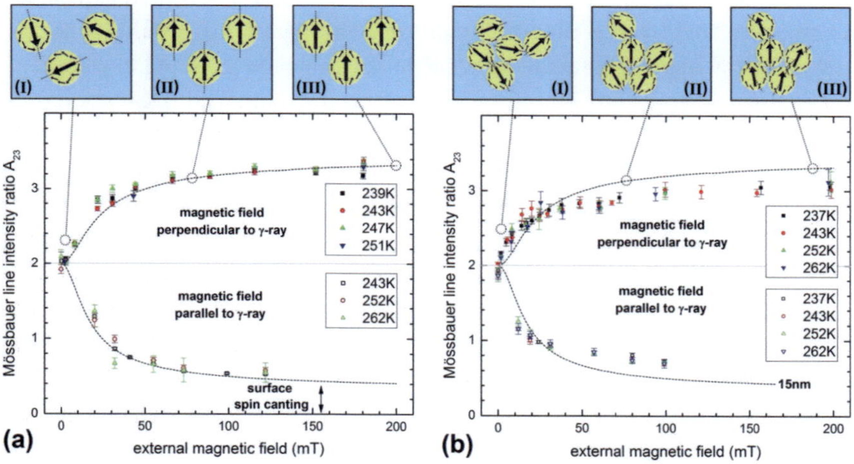

Figure 11.7: A_{23} line intensity ratio versus external magnetic field in perpendicular (top) and parallel field geometry (bottom) for ca. 15 nm particles in sample M (a) and larger 22 nm particles in sample L (b), evaluated at several temperatures at 237–262 K. Dashed lines represent the modeled trend in A_{23} (H) for sample M corresponding to the modified Langevin approach describing field- and temperature-dependent magnetic alignment. Adapted with permission from [55], Copyright (2019) American Chemical Society.

One could notice that the experiment effectively provided the particles' net magnetic moment via the Langevin parameter, which is also available by standard field-dependent magnetometry. As long as this parameter is the only quantity required, this is a correct statement. However, if a more thorough sample characterization is needed, this approach provides, as illustrated above, not only a total measure of spin alignment, but also the ability to characterize the particles' static magnetic structure, magnetization dynamics and orientation simultaneously from one spectrum or at least one series of experiments.

While results obtained on sample M were found to be in good agreement to theoretical models designed to describe spin and superspin alignment of isolated particles in ferrofluids exposed to magnetic fields, a different behavior is observed for larger particles in sample L, shown in Figure 11.7(b). In the low-field region the degree of particle alignment rises much faster as compared to sample M (dashed line), while above ca. 30 mT sample L shows either a slower convergence to a saturation state or a lower final degree of spin orientation. As experimental data of A_{23} (H) for both samples display a crossing at ca. 30 mT, we can deduce an additional effect to be present, as a variation in particle core diameter and net magnetic moment will affect the Langevin parameter but will never create a crossing. Additional information was taken from spectra of sample L (Figure 11.6), leading to an average hydrodynamic diameter R_H of ca. 68 nm estimated from the absorption line broadening. Since the original sample L, as discussed in Section 11.3 consisted of individual monodisperse nanoparticles, this indicates nanoparticle agglomeration in the second batch of sample L.

Based on the assumption of small particle agglomerates, their net magnetic moment was extracted by modeling only the low-field region of A_{23} (H) and M (H), where the relatively high agglomerate net magnetic moment is expected to yield high Langevin parameters and spin alignment. At the same time, the field is low enough to ensure not to increase the net magnetic moment by aligning the individual particle superspins out of their preferred direction defined by magnetic anisotropy. Comparing agglomerate magnetic moments from this approach and estimated magnetic moments of individual 22 nm particles, and also taking into account the agglomerates' saturation moment extracted from the high-field region in M (H), we estimate an average number of 8–22 particles per agglomerate. This translates to an effective hydrodynamic diameter of 64–90 nm, matching estimations made based on the line broadening (68 nm) and from AC-susceptometry (77 nm).

With the comparison of sample M and L leading to exciting insights regarding alignment behavior and agglomeration, it seems reasonable to perform a similar attempt for sample S. Here, the small anisotropy energy of the particles with only 6.0 nm core diameter results in a completely superparamagnetic doublet spectrum, from which the line broadening, i.e. particle dynamics, but no longer the line intensity ratio, i.e. spin orientation, can be inferred. Still, a rough estimation is possible when going to higher fields of >300 mT, where the external field deforms the anisotropy

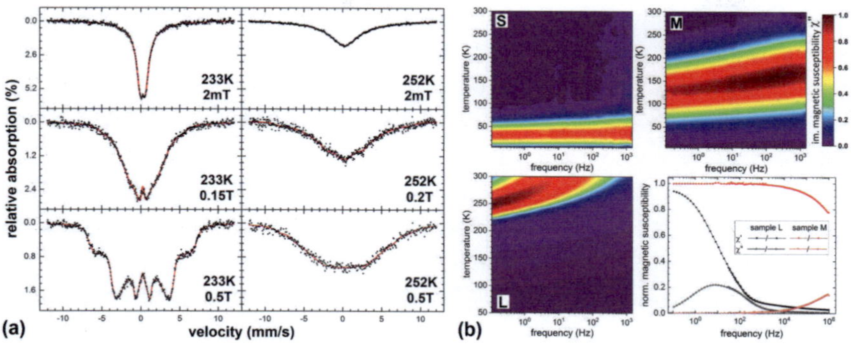

Figure 11.8: (a) Mössbauer spectra of 6.0 nm particles in sample S recorded in perpendicular field geometry up to 0.5 T and temperatures of 233–252 K, (b) colorplots of the imaginary part of magnetic susceptibility χ'' measured for sample S–L at 5–300 K and 0.1–1500 Hz, also measured on a wider frequency scale for samples M and L at room temperature. Adapted with permission from [55], Copyright (2019) American Chemical Society.

energy landscape sufficiently to suppress fast superspin relaxation, whereby the line ratio can again be extracted, reaching $A_{23(0.5\ T)} \approx 3.4$, corresponding to a low degree of orientation, matching the low Langevin parameter and high degree of surface spin canting expected for such small nanoparticles at moderate magnetic fields.

The observation of fast and beginning Néel relaxation in sample S and M, respectively, and blocked superspin relaxation in ferrofluid L was verified via detailed AC-susceptometry experiments as seen in Figure 11.8(b). Although for each individual sample only the dominant relaxation mechanism could be studied, values of the effective magnetic anisotropy in case of Néel relaxation in sample M from Mössbauer spectroscopy many-state relaxation modeling could be verified by this method, as well as the macroscopic dynamic viscosity of the carrier medium in sample L, where Brownian diffusion is the dominant process and results based on line broadening and particle rotation frequencies yielded matching viscosities.

11.5 Influence of matrix nanostructure on Brownian MNP motion

In discussions of magnetic hybrid materials, the interaction between the magnetic nanoparticles and the nanostructure of the surrounding matrix is essential and has not been covered in the previous chapters explicitly. To illustrate effects of the matrix properties on nanoparticle orientation and possible modes of motion, we study electrostatically stabilized spindle-shaped hematite nanospindles of ca. 387 × 85 nm. With an aspect ratio of ca. 4.5 the nanoparticles are expected to show distinctively anisotropic diffusion in fluid media, while also aligning perpendicular to external magnetic fields due to their peculiar magnetic structure [56]. Here, they are dispersed in a precursor solution when polymerization starts, whereby the manifesting polymer network forms meshes around the nanoparticles, limiting their mobility to a certain degree. The degree of constriction in the PAAm-hydrogel depends on the network crosslinking density defined by the concentration c_{MBA} of N,N′-Methylene bis(acrylamide) (MBA) crosslinker in the range of $1{\cdot}10^{-6}$–$2{\cdot}10^{-3}$ relative to the number of AAM monomers.

Dependent on the crosslinking, the hydrogels are widely tunable in their viscoelastic properties, with the "hardest" sample with highest c_{MBA} having a defined form and showing almost pure elastic response behavior, while the intermediate sample is softer and less resilient (Figure 11.9(b–d)). The "softest" hydrogel of minimum c_{MBA} is viscoelastic and easy to deform, with the ensemble of our hydrogel samples thereby covering a very broad range of macroscopic (visco-)elastic response behavior. While it stands to reason that hydrogels of lower crosslinking density and wider meshes will lead to less constraint of particle motion, the more important question is the following: What is the correlation between the probed mode of particle motion as well as the timescale of the experiment on the one hand, and the observed diffusive behavior on the other?

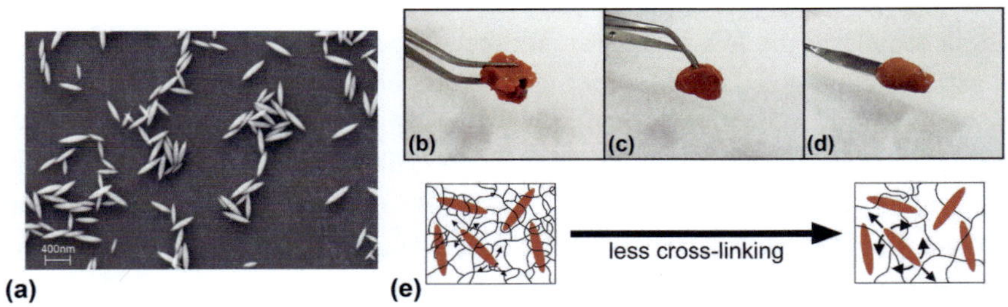

Figure 11.9: (a) SEM image of hematite nanospindles, and (b) hard, (c) medium and (d) soft hydrogel samples prepared using different crosslinking concentrations and (e) schematic illustration of increasingly free modes of particle motion. Adapted from [57] and with permission from [58] Copyright (2015) American Chemical Society.

Regarding this criterion, Mössbauer spectroscopy is a very local and relatively "fast" technique, meaning that even dynamic processes on the nanosecond timescale will produce considerable changes in spectral structure, while even slightly higher mean square displacements in atomic vibration/motion will yield far lower spectral intensity, as described by the Debye-Waller factor in Section 11.2. Both effects illustrate that via Mössbauer spectroscopy one will observe effects of particle-matrix interaction (if present) on very short time- and length scales. We should keep this in mind when studying the temperature-dependent line broadening of the three ferrohydrogels across their melting region. All samples have in common that instead of a sharply defined melting temperature we have a narrow region, in which the line broadening quickly increases before reaching the maximum value corresponding to the "liquid" hydrogel state, with the steepest point of the curve being at almost identical temperatures T_{melt} for all samples. Linewidths in the liquid region above T_{melt} are in the range of 12–30 mm/s and increase for lower crosslinker densities. Although this was to be expected, the relatively low variation in linewidth is somewhat surprising, considering the significant differences in macroscopic viscoelastic behavior. But is there a measure or intuitive physical understanding, based on the observed decrease in line broadening, to which extent the particle motion is constricted? There are at least two states we can put these hydrogel results in relation to:

On the one hand a perfectly rigid surrounding matrix, allowing no additional atomic motion as compared to an atom in the crystal lattice of the immobile nanoparticle, which will yield the static linewidth Γ_{stat}, and on the other hand the state of motion undisturbed by the polymer network, in which the particles would exhibit free Brownian motion in water. For the latter, a simple estimation using Eq. (11.2.2) in case of spherical particles or a more precise estimation based on the direction-dependent diffusion coefficients of spindle-shaped nanoparticles yield a dynamic line broadening of ca. 400 mm/s, based on the viscosity of water $\eta_{RT} \approx 1$ mPas [59]. Compared to this wide spectrum of possible states of motion, values of Γ of the same magnitude seem to indicate similar degrees of constriction of translational nanoparticle diffusion, at least when studied on the short Mössbauer timescale.

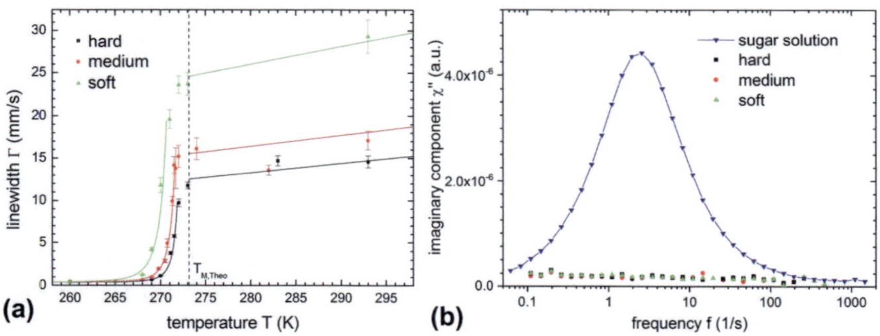

Figure 11.10: (a) Mössbauer absorption linewidth of hydrogels of different crosslinking density across the hydrogel melting region and (b) imaginary component of magnetic susceptibility of hydrogels compared to a 60 wt% sucrose reference solution, normalized to the concentration of included hematite nanospindles, adapted with permission from [58], Copyright (2015) American Chemical Society.

While in previous sections, AC-susceptometry provided reference values for individual samples, either in terms of Néel relaxation frequency or Brownian rotation frequency, here it is more essential to have it as a complementary technique, not only because of the much longer timescales that the particle dynamics are probed on. Of similar importance is the fact that via the combination of Mössbauer spectroscopy and AC-susceptometry, one has two techniques probing either translational or rotational motion of the particles, which may experience widely different degrees of constriction, especially for the here studied anisotropic hematite nanospindles.

For that purpose, the imaginary part of magnetic susceptibility of a 60 wt% sucrose reference solution is shown in Figure 11.10(b). As expected, we find a distinct narrow peak at ca. 2–3 Hz, which would correspond to an effective particle radius in case of spherical particles of ca. 70–80 nm, using literature values of ca. $\eta_{300K} \approx 40$ mPas [60]. If particle rotation would not be hindered in the studied hydrogels, matching the free rotation in water, accordingly we would expect the peak to shift to higher frequencies by a factor of 40. Instead, no χ'' – signal is visible in the entire covered frequency range aside from a minor background almost independent of frequency. Reference experiments performed at different temperatures including the frozen hydrogel state (not shown) indicate that it can be assigned completely to Néel-type relaxation, presumably in the easy magnetic plane of the hematite nanospindles, in which magnetic reorientation can be induced even by very moderate thermal excitation due to its ultralow magnetic anisotropy [61, 62].

We would explain this apparent contradiction of our two experimental approaches regarding particle mobility either by the type of motion that is probed, or by the methods' timescales: Dependent on the timescale of network formation and crosslinking as compared to the particle diffusion coefficient, the hydrogel meshes will more or less reproduce the shape of the hematite nanospindles, with a minimum average distance between particle surface and polymer strings dependent on the crosslinker density. This very simple model does not consider friction induced by continuous direct contact between particles and the polymer network, which would equally hinder all types of motion. Nevertheless, it implies that while translational as well as rotational motion about the short particle axes are only possible on very short timescales before getting in contact with the constricting surrounding, particle rotation about the long axis should still be possible. This could indicate that it is less the probed mode of motion and more the timescale of the experiments which allows – or prevents – the detection of particle motion. As linewidths from Mössbauer spectroscopy, which are far lower than that of free particle motion, rather seem to indicate a constricted type of translational motion with slight variations in the degree of confinement, it seems reasonable that only on the very short timescales probed in Mössbauer spectroscopy the particles in average cover a distance short enough to not inevitably lead to contact with the polymer network. To verify this assumption, similar experiments with spherical particles could be helpful, where the hydrogels would presumably form isotropic meshes without preferred direction of motion, as well as a study of the mobility of particles in a nanostructured surrounding, whose length scale is independent of the used nanoparticles [17, 63].

11.6 Conclusion

Summarizing insights made for different magnetic hybrid materials, we find the main advantage of Mössbauer spectroscopy to be the ability to access a plethora of critical system parameters simultaneously as it is shown in Figure 11.11, without the need to compare results recorded under different experimental conditions, although it should be mentioned that each parameter can be obtained individually at least with similar precision via a specialized complementary technique. The independence from sample transparency is also very helpful in combination with similar γ-ray absorption in water (ferrofluids), gels and many types of hydrocarbon-based systems (ferrogels and ferropolymers), covering most of the matrix materials currently used in the preparation of magnetic hybrid systems.

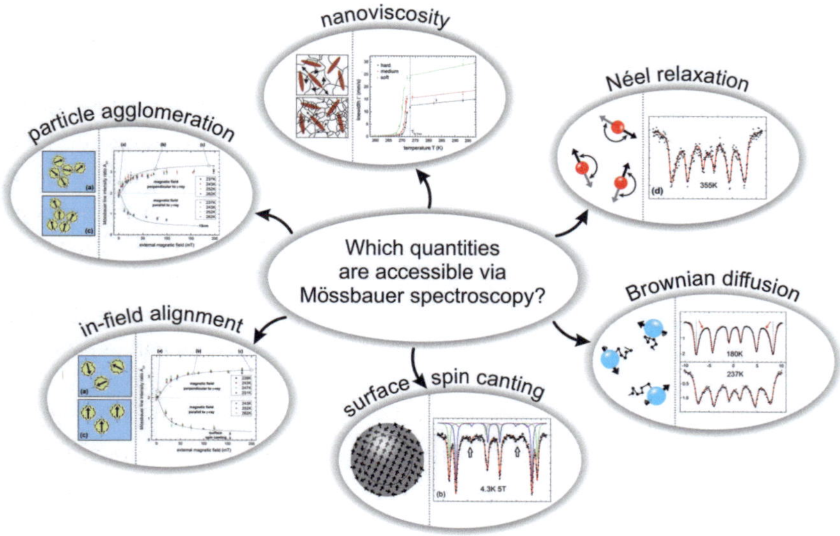

Figure 11.11: Quantities of interest in the study of magnetic soft hybrids via Mössbauer spectroscopy, adapted with permission from [19, 55, 58], Copyright (2015–2019) American Chemical Society, from [51], Copyright (2016), with permission from Elsevier and [57].

Apart from the experiments shown above, the special traits of Mössbauer spectroscopy could match specific problems perfectly, supporting future investigations of new hybrid materials. These include for example composites of magnetic nanoparticles and liquid-crystals as well as liquid-crystal elastomers, where the ability to probe particle diffusion and alignment at the same time may provide a better understanding of the interplay of magnetic particle dynamics with different self-ordered states of liquid crystal molecules. Connected to this, also the possibility of direction-resolved measurements of particle diffusion coefficients could be beneficial for studies on either anisotropic nanoparticles or particle diffusion in anisotropic surroundings. Similar to the field-dependent analysis of particle alignment presented in Section 11.4, Mössbauer experiments could be applied to analyze how nanoparticle chain formation behavior is affected by the nanostructure of surrounding matrix material, as chain formation will affect the degree of magnetic alignment as well as the particle mobility. As it is common to apply external magnetic fields in Mössbauer spectroscopy, one could also think of researching the properties of actively heated MNPs in AC-fields, to support the development of better materials for magnetic hyperthermia. Further information regarding the correlation of magnetic and spatial particle orientation could be extracted, e.g., for α-Fe_2O_3 nanospindles as discussed in Section 11.5, utilizing effects of the electric field gradient in hematite.

Author contribution: All the authors have accepted responsibility for the entire content of this submitted manuscript and approved submission.
Research funding: This research was supported by the Deutsche Forschungsgemeinschaft (SPP 1681, project WE 2623/7).
Conflict of interest statement: The authors declare no conflicts of interest regarding this article.

References

1. Mössbauer RL. Kernresonanzfluoreszenz von Gammastrahlung in Ir^{191}. Z Phys 1958;151:124–43.
2. Barb D. Grundlagen und Anwendungen der Mössbauerspektroskopie. Berlin: Akademie-Verlag; 1980.
3. Klingelhöfer G, DeGrave E, Morris RV, Van Alboom A, de Resende VG, De Souza PA Jr., et al. Mössbauer spectroscopy on Mars: goethite in the Columbia Hills at Gusev crater. Hyperfine Interact 2005;166:549–54.
4. Schünemann V, Winkler H. Structure and dynamics of biomolecules studied by Mössbauer spectroscopy. Rep Prog Phys 2000;63:263–353.
5. Wagner FE, Kyek A. Mössbauer spectroscopy in archaeology: introduction and experimental considerations. Hyperfine Interact 2004;154:5–33.
6. Concas G, Spano G, Cannas C, Musinu A, Peddis D, Piccaluga G. Inversion degree and saturation magnetization of different nanocrystalline cobalt ferrites. J Magn Magn Mater 2009;321:1893–7.
7. Bonchev T, Aidemirski P, Mandzhukov I, Nedyalkova N, Skorchev B, Strigachev A. A study of Brownian motion by means of the Mössbauer effect. Sov Phys JETP 1966;23:42–6.
8. Keller H, Kündig W. Mössbauer studies of Brownian motion. Solid State Commun 1975;16:253–6.
9. Singwi KS, Sjölander A. Resonance absorption of nuclear gamma rays and the dynamics of atomic motions. Phys Rev 1960;120:1093–102.
10. Anand HR, Mullen JG. Diffusion broadening of the Mössbauer line in Wüstite. Phys Rev B 1973;8: 3112–16.
11. Schwalbach P, Laubach S, Hartick M, Kankeleit E, Keck B, Menningen M, et al. Diffusion and isomer shift of interstitial iron in silicon observed via in-beam Mössbauer spectroscopy. Phys Rev Lett 1990;64:1274–7.
12. Knapp EW, Fischer SF, Parak F. The influence of protein dynamics on Mössbauer spectra. J Chem Phys 1983;78:4701–11.
13. Parak F, Achterhold K, Schmidt M, Prusakov V, Croci S. Protein dynamics on different timescales. J Non-Cryst Solids 2006;352:4371–8.
14. Bonchev T, Vassilev I, Sapundzhiev T, Evtimov M. Possibility of investigating movement in a group of ants by the Mössbauer effect. Nature 1968;217:96–8.

15. Bhide VG, Kandpal MC. Mössbauer study of Brownian motion of $FeSnO_3$ particles in liquid-crystal N-(p-ethoxybenzylidene)-p′-butylaniline (EBBA). Phys Rev B 1979;20:85–90.
16. Nienhaus GU, Plachinda AS, Fischer M, Khromov VI, Parak F, Suzdalev IP, et al. Temperature dependence of the dynamics of ultrafine particles in a polymeric network. Hyperfine Interact 1990; 56:1471–6.
17. Plachinda AS, Sedov VE, Khromov VI, Suzdalev IP, Goldanskii VI, Nienhaus GU, et al. Mössbauer studies of bound diffusion in a model polymer system. Phys Rev B 1992;45:7716–23.
18. Gütlich P, Bill E, Trautwein AX. Mössbauer spectroscopy and transition metal chemistry. Heidelberg: Springer; 2011.
19. Landers J, Salamon S, Remmer H, Ludwig F, Wende H. Simultaneous study of Brownian and Néel relaxation phenomena in ferrofluids by Mössbauer spectroscopy. Nano Lett 2016;16:1150–5.
20. Ammar S, Jouini N, Fiévet F, Beji Z, Smiri L, Moliné P, et al. Magnetic properties of zinc ferrite nanoparticles synthesized by hydrolysis in a polyol medium. J Phys Condens Matter 2006;18: 9055–69.
21. Greneche J-M. Structural and magnetic properties of nanostructured oxides investigated by ^{57}Fe Mössbauer spectometry. Hyperfine Interact 2003;148/149:79–89.
22. Kodama RH, Berkowitz A, McNiff EJ Jr., Foner S. Surface spin disorder in $NiFe_2O_4$ nanoparticles. Phys Rev Lett 1996;77:394–7.
23. Mørup S, Topsøe H. Mössbauer studies of thermal excitations in magnetically ordered microcrystals. Appl Phys 1976;11:63–6.
24. Dormann JL, D'Orazio F, Lucari F, Tronc E, Prene P, Jolivet JP, et al. Thermal variation of the relaxation time of the magnetic moment of γ-Fe_2O_3 nanoparticles with interparticle interactions of various strengths. Phys Rev B 1996;53:14291–7.
25. Madsen MB, Mørup S, Costa TV, Knudsen JM, Olsen M. Superparamagnetic component in the orgueil meteorite and Mössbauer spectroscopy studies in applied magnetic fields. Nature 1986; 321:501–3.
26. Blume M, Tjon JA. Mössbauer spectra in a fluctuating environment. Phys Rev 1968;165:446–56.
27. Chuev MA. Mössbauer spectra of single-domain particles in a weak magnetic field. J Phys Condens Matter 2008;20:505201.
28. Jones DH, Srivastava KK. Many-state relaxation model for the Mössbauer spectra of superparamagnets. Phys Rev B 1986;34:7542–8.
29. Landers J, Stromberg F, Darbandi M, Schöppner C, Keune W, Wende H. Correlation of superparamagnetic relaxation with magnetic dipole interaction in capped iron-oxide nanoparticles. J Phys Condens Matter 2015;27:026002.
30. van Lierop J, Ryan DH. Mössbauer spectra of single-domain fine particle systems described using a multiple-level relaxation model for superparamagnets. Phys Rev B 2001;63:064406.
31. Heilmann I, Olsen B, Jensen JH. Non-Lorentzian diffusion-broadened Mössbauer lines. J Phys C Solid State Phys 1974;7:4355–60.
32. Elliott JA, Hall HE, Bunbury DS. Study of liquid diffusion by Mössbauer absorption and Rayleigh scattering. Proc Phys Soc 1966;89:595–612.
33. Dzyublik AY. Mössbauer effect in ellipsoidal Brownian particles. Sov Phys JETP 1974;40:763–5.
34. Steinmetz KH, Vogl G, Petry W, Schroeder K. Diffusion of iron in copper studied by Mössbauer spectroscopy on single crystals. Phys Rev B 1986;34:107–16.
35. Nack A, Seifert J, Passow C, Wagner J. Hindered nematic alignment of hematite spindles in poly(N-isopropylacrylamide) hydrogels: a small-angle X-ray scattering and rheology study. J Appl Crystallogr 2018;51:87–96.
36. Roeder L, Bender P, Kundt M, Tschöpe A, Schmidt AM. Magnetic and geometric anisotropy in particle-crosslinked ferrohydrogels. Phys Chem Chem Phys 2015;17:1290–8.

37. Brand RA, Hippert F, Frick B. Iron dynamics in Al-Cu-Fe quasicrystals and approximants: Mössbauer and neutron experiments. J Phys Condens Matter 2009;21:045405.
38. Keller H, Debrunner PG. Evidence for conformational and diffusional mean square displacements in frozen aqueous solution of oxymyoglobin. Phys Rev Lett 1980;45:68–71.
39. Żukrowski J, Błachowski A, Komędera K, Ruebenbauer K, Gabka G, Bujak P, et al. Dynamics of ternary Cu-Fe-S_2 nanoparticles stabilized by organic ligands. J Phys Chem C 2017;121:6977–85.
40. Hansen MF, Morup S. Estimation of blocking temperatures from ZFC/FC curves. J Magn Magn Mater 1999;203:214–16.
41. Respaud M, Broto JM, Rakoto H, Fert AR, Thomas L, Barbara B, et al. Surface effects on the magnetic properties of ultrafine cobalt particles. Phys Rev B 1998;57:2925–35.
42. Webers S, Hess M, Landers J, Schmidt AM, Wende H. Effect of phase transitions in polymer solutions on the magnetic response of embedded nanoparticles. ACS Appl Polym Mater 2020;2: 2676–85.
43. Cherepanov VM, Gabbasov RR, Yurenya AY, Nikitin AA, Abakomov MA, Polikarpov MA, et al. Study of the Brownian broadening in the Mössbauer spectra of magnetic nanoparticles in colloids with different viscosities. Crystallogr Rep 2020;65:398–403.
44. Schröter K, Donth E. Viscosity and shear response at the dynamic glass transition of glycerol. J Chem Phys 2000;113:9101–9.
45. González JAT, Longinotti MP, Corti HR. The viscosity of glycerol-water mixtures including the supercooled region. J Chem Eng Data 2011;56:1397–406.
46. Segur JB, Oberstar HE. Viscosity of glycerol and its aqueous solutions. Ind Eng Chem 1951;43: 2117–20.
47. Müller R, Zhou M, Dellith A, Liebert T, Heinze T. Meltable magnetic biocomposites for controlled release. J Magn Magn Mater 2017;431:289–93.
48. Zhou M, Liebert T, Müller R, Dellith A, Gräfe C, Clement JH, et al. Magnetic biocomposites for remote melting. Biomacromolecules 2015;16:2308–15.
49. Müller R, Dutz S, Neeb A, Cato AC, Zeisberger M. Magnetic heating effect of nanoparticles with different sizes and size distributions. J Magn Magn Mater 2013;328:80–5.
50. Hansen MF, Koch CB, Mørup S. Magnetic dynamics of weakly and strongly interacting hematite nanoparticles. Phys Rev B 2000;62:1124–35.
51. Müller R, Zhou M, Liebert T, Landers J, Salamon S, Webers S, et al. Mobility investigations of magnetic nanoparticles in biocomposites. Mater Chem Phys 2017;193:364–70.
52. Roca AG, Niznansky D, Poltierova-Vejpravova J, Bittova B, González-Fernández MA, Serna CJ, et al. Magnetite nanoparticles with no surface spin canting. J Appl Phys 2009;105:114309.
53. Gallagher KJ, Feitknecht W, Mannweiler U. Mechanism of oxidation of magnetite to γ-Fe_2O_3. Nature 1968;217:1118–21.
54. Sidhu PS, Gilkes RJ, Posner AM. Mechanism of the low temperature oxidation of synthetic magnetites. J Inorg Nucl Chem 1977;39:1953–8.
55. Landers J, Salamon S, Remmer H, Ludwig F, Wende H. In-field orientation and dynamics of ferrofluids studied via Mössbauer spectroscopy. ACS Appl Mater Interfaces 2018;11:3160–8.
56. Bødker F, Hansen MF, Koch CB, Lefmann K, Mørup S. Magnetic properties of hematite nanoparticles. Phys Rev B 2000;61:6826–38.
57. Landers J. Study of magnetic relaxation dynamics in soft matter nanoparticle composite systems [Dissertation]. Hrsg.: U. o. Duisburg-Essen; 2016.
58. Landers J, Roeder L, Salamon S, Schmidt AM, Wende H. Particle-matrix interaction in cross-linked PAAm-hydrogels analyzed by Mössbauer spectroscopy. J Phys Chem C 2015;119:20642–8.
59. Perrin F. Mouvement Brownien d'un ellipsoide (II). Rotation libre et dépolarisation des fluorescences. Translation et diffusion de molécules ellipsoidales. J Phys Radium 1936;7:1–11.

60. Rampp M, Buttersack C, Lüdemann H-D. Self-diffusion of sucrose in molasses. Ind Eng Chem Res 2000;39:4400–7.
61. Besser PJ, Morrish AH, Searle CW. Magnetocrystalline anisotropy of pure and doped hematite. Phys Rev 1967;153:632–42.
62. Flanders PJ, Remeika JP. Magnetic properties of hematite single crystals. Philos Mag A 1965;11: 1271–88.
63. Hess M, Roeben E, Rochels P, Zylla M, Webers S, Wende H, et al. Size effects on rotational particle diffusion in complex fluids as probed by magnetic particle nanorheology. Phys Chem Chem Phys 2020;16:7562–75.

Sebastian Draack, Meinhard Schilling and Thilo Viereck*

12 Magnetic particle imaging of particle dynamics in complex matrix systems

Abstract: Magnetic particle imaging (MPI) is a young imaging modality for biomedical applications. It uses magnetic nanoparticles as a tracer material to produce three-dimensional images of the spatial tracer distribution in the field-of-view. Since the tracer magnetization dynamics are tied to the hydrodynamic mobility via the Brownian relaxation mechanism, MPI is also capable of mapping the local environment during the imaging process. Since the influence of viscosity or temperature on the harmonic spectrum is very complicated, we used magnetic particle spectroscopy (MPS) as an integral measurement technique to investigate the relationships. We studied MPS spectra as function of both viscosity and temperature on model particle systems. With multispectral MPS, we also developed an empirical tool for treating more complex scenarios via a calibration approach. We demonstrate that MPS/MPI are powerful methods for studying particle-matrix interactions in complex media.

Keywords: magnetic nanoparticles, magnetic particle imaging, magnetic particle spectroscopy, magnetic relaxation dynamics, particle-matrix interactions, viscosity, temperature

PACS: 87.85.Pq, 07.55.Jg, 29.30.Aj, 32.10.Dk, 33.15.Kr, 36.40.Cg, 61.46.Df, 65.80.n, 81.07.Bc, 82.35.Np, 87.64.K-

12.1 Introduction

Magnetic particle imaging (MPI) is a comparably young imaging modality, invented by B. Gleich and J. Weizenecker in 2005 [1]. It enables real-time 3D volume imaging of the spatial distribution of a magnetic nanoparticle (MNP) tracers [2–4]. MPI promises to deliver quantitative imaging, high specificity and exceptional tracer-to-background contrast. The magnetic nanoparticles can be used as nanoscale probes to provide information about their environmental surroundings. Stimulated by an applied magnetic field, the particles' magnetization reorients toward the external field direction. Since the dynamic reorientation process depends on various parameters,

***Corresponding author: Thilo Viereck,** Institute for Electrical Measurement Science and Fundamental Electrical Engineering, Technical University Braunschweig, Braunschweig, Germany,
E-mail: t.viereck@tu-bs.de. https://orcid.org/0000-0001-6814-6266
Sebastian Draack and Meinhard Schilling, Institute for Electrical Measurement Science and Fundamental Electrical Engineering, Technical University Braunschweig, Braunschweig, Germany

This article has previously been published in the journal Physical Sciences Reviews. Please cite as: S. Draack, M. Schilling and T. Viereck "Magnetic particle imaging of particle dynamics in complex matrix systems" *Physical Sciences Reviews* [Online] 2021, 6. DOI: 10.1515/psr-2019-0123 | https://doi.org/10.1515/9783110569636-012

e.g., temperature or viscosity, experimental results can provide detailed insights into particle-matrix interactions. With the introduction of multispectral MPI imaging, quantification of tracer concentrations is preserved when the dynamic particle behavior is changed and even enables visualization of particle-matrix interactions such as the particle binding state or heating due to hysteresis losses and surface friction.

The time evolution of the energy minimization interplay between the magnetic moment and the applied magnetic field on the MNPs, called relaxation, can occur via two different processes: An internal reorientation of the particle's magnetic moment without any interaction with the particles' surrounding is called Néel relaxation. In contrast, the relaxation via Brownian rotation is influenced by the shear force acting on the hydrodynamic surface of the particles. Thus, the Brownian process is affected by rheological properties of the solvent or matrix, in which the particle is embedded.

Magnetic particle spectroscopy (MPS) [5] is a powerful tool to investigate the dynamic relaxation of magnetic nanomaterials and has been explored as a tool to quantify viscosity, binding state and temperature [6–16]. It utilizes the nonlinear magnetization curve of magnetic nanoparticles by periodically forcing the magnetization of the sample into the saturation regime. The harmonic response to a sinusoidal magnetic field includes very sensitive information, since small changes of the matrix properties lead to significant alteration of the higher harmonic spectrum generated by the particles. Due to the close relationship between MPS and MPI, the investigation of particle-matrix interactions in MPS provides valuable insights into the dynamic particle behavior in MPI. Compared to direct measurements in MPI, MPS measurements are very straightforward and are performed with a much better signal-to-noise ratio due to an improved filling factor of the differential detection coil. MPS has very short measurement times (compared to other dynamic magnetic characterization methods, e.g., ac susceptometry [ACS] or magnetorelaxometry [MRX]) and thus enables multiparametric investigations by varying the magnetic field amplitude, excitation frequency, sample temperature or matrix properties in a series of sample measurements.

Measured spectra can be compared to results from effective field, Fokker-Planck or stochastic (coupled Langevin and Landau-Lifshitz-Gilbert equation) simulations covering the dynamic nonlinear magnetization process of the nanoparticles. Thus, compared to ACS, which can be described via the linear frequency-dependent Debye model, the modeling of the magnetization response in MPS / MPI is much more difficult. For that reason, we introduced multispectral MPS as a calibration-based tool for treating different magnetization spectra empirically.

The review article at hand is focused on MPS as the primary tool for investigating particle-matrix interactions (Section 12.2) and their impact on the harmonic magnetization spectrum of magnetic nanoparticles. However, we will also conclude with some exemplary MPI measurements on a more complex gelatin matrix (Section 12.3).

12.2 Magnetic particle spectroscopy (MPS)

In magnetic particle spectroscopy [5], a homogeneous pure sinusoidal magnetic excitation field is generated with a drive-field coil. Samples are prepared in a small vial and placed in the center of the drive-field generator. A pick-up coil is used to detect the change of the sample's magnetization via Faraday's law of induction. To suppress the fundamental feed-through of the excitation field, the pick-up coil is realized in a differential design. The induced voltage is then measured via a digital acquisition card, which further serves as a synchronized signal source for the power amplifier driving the excitation coil. The sinusoidal excitation field forces the magnetization of the sample periodically into the saturation regime. In a first approximation, the magnetization curve of magnetic nanoparticles is given by the Langevin function [17]:

$$M(H) = M_s \cdot \left(\coth\left(\frac{m\mu_0 H}{k_B T} \right) - \frac{k_B T}{m\mu_0 H} \right) \tag{12.1}$$

Note that the Langevin function is actually only applicable to static scenarios, since dynamics are neglected for the equilibrium case. However, the magnetic response of the particles can be described as the orthogonal projection of the applied field along the magnetization curve. Thus, the magnetization over time contains the fundamental oscillation of the excitation but also higher harmonics as a result of the saturated regime of the magnetization curve. The change of the magnetic flux induces a voltage $U_{\text{ind}} = -d\Phi/dt$ in the pick-up coil, which is typically transformed to the frequency domain via Fourier transform or digital lock-in method and plotted as magnitude (and phase) spectrum. The signal generation in MPS is graphically illustrated in Figure 12.1.

The institute's custom-built MPS setup (Figure 12.2) enables measurements with magnetic field strengths of up to $\mu_0\hat{H} \leq 30$ mT at selectable discrete frequencies of up to $f_0 \leq 25$ kHz. Quasi-continuous measurements are possible in the lower-frequency range of up to $f_0 \leq 1.0$ kHz. Series-resonant circuits with capacitors for discrete frequency values are required at higher frequencies. The drive-field coil encloses a Shapal™ Hi-M Soft ceramic rod, which is used as a coil bobbin for the detection coil. Shapal™ offers a high thermal conductivity and a high electrical insulation. Thus, the sample-housing rod can be temperature-controlled while eddy current-based power losses are inhibited. For measurements, the sample filled into a Thermo Scientific™ Microtiter™ well or a similar glass vial is placed in the center of the ceramic rod. Here, the excitation field is maximally homogeneous. The magnetic response is then acquired with a differential pick-up coil providing an attenuation of 69 dB at the fundamental feedthrough. A Peltier-element-based temperature controller enables temperature-dependent experiments with sample temperatures in the range $-20 \leq T \leq 120$ °C. An additional outer-mounted Helmholtz coil pair provides an option to superimpose a dc field with up to 30 mT.

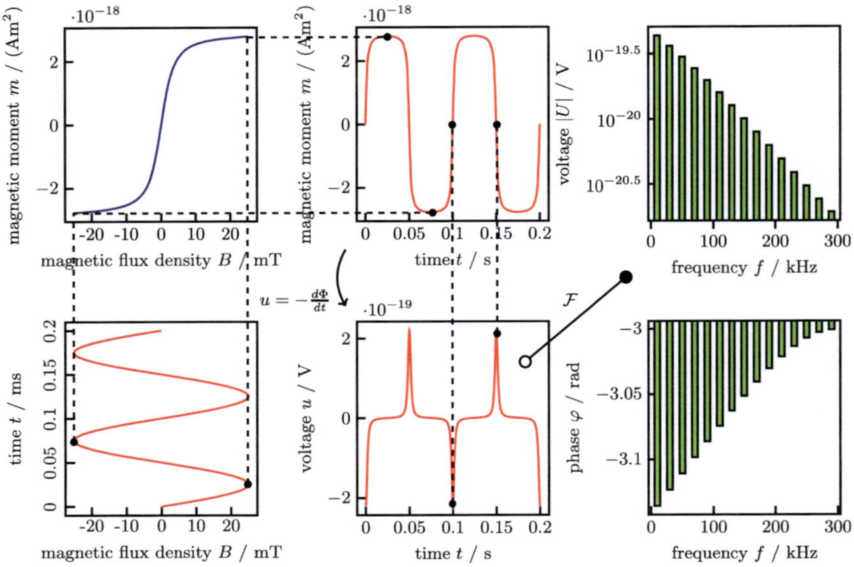

Figure 12.1: MPS signal generation. A sinusoidal drive-field excites the particle's magnetization via their specific magnetization curve (blue). The magnetization response is detected with a differential pick-up coil measuring the induced voltage, which is proportional to the phase-inverted time derivative. Typically, the harmonic response is evaluated in frequency domain consisting of magnitude and phase spectra (green).

(a) MPS CAD drawing

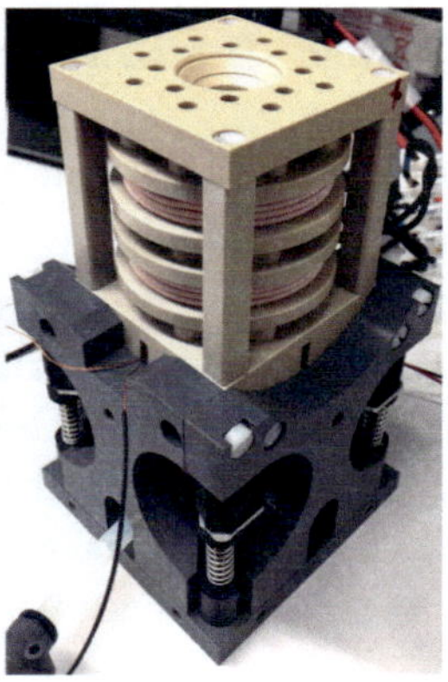

(b) MPS photography

Figure 12.2: MPS hardware setup: (a) CAD design and (b) final hardware of our MPS system.

12.2.1 Materials/particle systems

Typical tracer nanoparticle systems optimized for magnetic resonance imaging (MRI) or magnetic particle imaging (MPI) are made of Fe_3O_4 and have a multicore structure

[18–22]. FeraSpin™ XL is a size-selected agent derived from FeraSpin™ R produced by nanoPET Pharma GmbH (Berlin, Germany), which is characterized by mean hydrodynamic particle sizes of 50 nm ≤ d_h ≤60 nm. The hydrodynamic mean diameter of perimag® from micromod Partikeltechnologie GmbH (Rostock, Germany) is given with $d_h \approx 130$ nm. Both commercially available particle systems show a comparable broad size distribution. Therefore, the probability is high that a small amount of the particles have optimal magnetic properties for generating a broad harmonic response as required in MPI. However, most of the signals are the result of Néel-dominated relaxation. Thus, the influence of particle-matrix interactions on the harmonic response is small. Further, signal generation is hard to understand in detail and almost impossible to predict or model. To investigate the underlying physics of particle-matrix interactions, Brownian-dominated model systems are required. SHP-25 from Ocean NanoTech (San Diego, USA) are single-core nanoparticles with a narrow size distribution that can serve as a model system. Unfortunately, these particles show significant deviations between different production batches. Further, SHP-25 shows both Brownian and Néel contributions [23]. Although iron oxide materials are preferred for bio-compatibility reasons in medical diagnostics or therapy, tailored $CoFe_2O_4$ nanoparticles can be used to study the physical magnetic behavior in experimental applications [24]. Such particles were manufactured by Niklas Lucht from the working group of Dr. Birgit Hankiewicz within the priority program SPP1681 of the German Research Foundation DFG in a similar way as described in the study by Nappini et al. [25]. These particles were used in MPS viscosity experiments, which enable comparisons to experimental results of other particle systems. The main particle properties of the tracer materials used for experiments are summarized in Table 12.1.

Depending on the particle properties, the particles relax via two possible mechanisms: the Brownian rotation or the Néel relaxation, characterized by their specific relaxation times.

The Néel zero-field relaxation time τ_{N0} and Brownian zero-field relaxation time τ_{B0} can approximately be measured in magnetorelaxometry (MRX) experiments. In MRX, the net magnetic moment is aligned with a static externally applied magnetic field. After a certain time all magnetic moments are aligned and the field is switched off

Table 12.1: Particle properties.

	FeraSpin™ XL	perimag®	SHP-25	Lucht
Material	Fe_3O_4	Fe_3O_4	Fe_3O_4	$CoFe_2O_4$
Type	Multicore	Multicore	Single-core	Single-core
d_c	–	–	25 nm	15 nm
Shell	Dextran	Dextran	Carboxylic acid	Silica
d_h	50–60 nm	130 nm	35 nm	38 nm

again. Now, the effective net magnetic moment relaxes due to the thermal energy. The relaxation process follows an exponential function $m(t) \propto \exp(-t\tau^{-1})$ with a characteristic relaxation time τ. When switching off the applied field, the single magnetic moments of the particles reorient stochastically in all spatial directions such that the net magnetic moment of the particle ensemble, which is the sum of all single magnetic moments, vanishes to zero. When relaxing via the Néel mechanism, the relaxation time $\tau = \tau_{N0}$ depends on the effective anisotropy energy density K, the particle's core volume $V_c = (1/6)\pi d_c^3$ with core diameter d_c and thermal energy $E_T = k_B T$ including the Boltzmann constant k_B and the temperature T as well as the saturation magnetization M_s, the material-specific Gilbert damping factor $\alpha \approx 0.1$ and the gyromagnetic ratio $\gamma = 1.76 \times 10^{11}$ rad s^{-1}T^{-1} of an electron as expressed in (12.2).

$$\tau_{N0} = \underbrace{\frac{\sqrt{\pi}}{2\sqrt{\frac{KV_c}{k_B T}}} \overbrace{\frac{M_s(1+\alpha^2)}{2K\gamma\alpha}}^{\tau_0}}_{\tau_0^*} \exp\left(\frac{KV_c}{k_B T}\right) \tag{12.2}$$

Please note that all material-specific parameters have to be taken into account. However, for approximation purposes, prefactors of the exponential function are sometimes summarized as τ_0^* or $\tau_0 \approx 1$ ns. For Néel relaxation, there is no effect from particle-matrix interactions. On the other hand, when relaxing via Brownian rotation, the zero-field relaxation time $\tau = \tau_{B0}$ further depends on the surface friction expressed via the surroundings' viscosity η acting on the hydrodynamic volume $V_h = (1/6)\pi d_h^3$ with the hydrodynamic diameter d_h as described in (12.3).

$$\tau_{B0} = \frac{3\eta V_h}{k_B T} \tag{12.3}$$

In liquid solvents, the particles can typically relax via both mechanisms—depending on the measurement parameters. In this case, the mechanism with shorter relaxation time dominates the effective relaxation time τ_{eff} (cf. (4)), which can be understood as parallel arrangement of both relaxation times.

$$\tau_{\text{eff}}^{-1} = \tau_{B0}^{-1} + \tau_{N0}^{-1} \Rightarrow \tau_{\text{eff}} = \frac{\tau_{B0} \cdot \tau_{N0}}{\tau_{B0} + \tau_{N0}} \tag{12.4}$$

Relaxation times must always be considered carefully in dynamic excitation scenarios such as during periodical excitation in MPS or MPI measurements. Additionally, it should be noted that in modalities where the magnetic field pulls the net magnetic moment, zero-field relaxation times must be replaced by field-dependent ones. The field dependence of the relaxation times were investigated in the literature studies [26–29]. The field-dependent Néel relaxation time in (12.5) and the Brownian field-dependent relaxation time in (12.6) with $\xi = (m\mu_0 H)/(k_B T)$ being the ratio of magnetic and thermal energy significantly depend on the amplitude of the applied magnetic field strength H.

$$\tau_{NH} = \tau_0 \cdot \exp\left(\frac{KV_c}{k_B T} \cdot \left(1 - \frac{\mu_0 H M_s}{2K}\right)^2\right) \tag{12.5}$$

$$\tau_{BH} = \frac{\tau_{B0}}{\sqrt{1 + 0.126 \cdot \xi^{1.72}}} \tag{12.6}$$

Note that τ_{NH} is only valid for static fields and with the particle's easy axis aligned with the external field direction. In contrast to ACS, which uses small magnetic field strengths so that field dependence can be neglected, MPS and MPI require large amplitudes of the applied magnetic field to reach the saturation regime of the particles and to generate higher harmonics. Here, field dependence must explicitly be taken into account. Note that the dominating relaxation mechanism can be determined from temperature-dependent MPS measurements [23].

12.2.2 Temperature dependence

Magnetic particle spectroscopy (MPS) is well suited for the characterization of a particle system by means of multiparametric evaluation. Especially the variation of the excitation frequency and the sample temperature allows one to derive properties of the matrix from the acquired data. While at high excitation frequencies (>10 kHz), the Néel relaxation dominates (for MPS/MPI-typical particle types), Brownian rotation of the particles at low frequencies ($\ll$10kHz) provides access to the local interaction of the particles with their surrounding matrix. In this context, the variation of temperature is interesting for both Newtonian ferrofluids, where the temperature dependence of the dynamic viscosity η can be described by the Arrhenius–Andrade equation [30], as well as for gel systems.

With these gels, a temperature change in the sol-gel transition area leads to a significant change in the viscous or viscoelastic properties. Therefore, parametric measurements on such systems form the basis for modeling the MPS spectra as a function of matrix interaction. In order to investigate the temperature influences during the spectral characterization of such hybrid systems, a temperature-controlled MPS system was designed and realized [31]. With the help of the new design, the sample temperature can be varied in a temperature range from −20° to 120 °C during the measurements.

First temperature-dependent measurements were performed on FeraSpin™ XL (nanoPET Pharma GmbH, Berlin, Germany) [31] and SHP-25 (Ocean NanoTech, San Diego, California) at an excitation frequency of f_0 = 5 kHz and an excitation amplitude of $\mu_0 H$ = 25 mT, as exemplary shown in Figure 12.3.

In the quasi-static case, which can be explained by the Langevin function, an increase in temperature leads to a decrease in the MPS signal amplitudes at higher harmonics. The opposite behavior in Figure 12.3 for the freeze-dried state must therefore be attributed to a significant shortening of the relaxation times. This contradiction and the complex temperature dependence of the harmonics of the suspension (crossing of harmonic curves due to Brownian- to Néel-dominated transition from lower harmonics to higher harmonics) emphasizes the importance of particle magnetization dynamics in the description of MPS and MPI signals.

For the analysis of the magnetization behavior of the magnetic nanoparticles, dynamic magnetization curves were reconstructed (Figure 12.4), which allow an observation outside the frequency space of the MPS spectra and enable or simplify the comparison with other measurement methods. Another powerful evaluation method is the short-time fast Fourier transform (SFFT) analysis of harmonics during variation of the temperature. Figure 12.5 shows the temporally resolved course of the first 21 harmonics of a diluted SHP-25 sample ($10\%_{vol}$) during the phase transition of the suspension from the subcooled to the frozen state. This approach allows one to investigate the same sample in a mobile and in an immobile state. Since the new MPS can be used to investigate time-resolved temperature-dependent processes in the millisecond range, the system provides an innovative tool for investigating biologically [9, 32–34] and technically [35] relevant scenarios.

12.2.3 Viscosity dependence

Generally, typical Fe_3O_4 nanoparticle systems optimized for MPI show both Brownian and Néel relaxation whereas the Néel relaxation even dominates the signal generation. However, only the particle contribution via Brownian rotation provides information about the particles' surrounding. To investigate the impact of Brownian contributions on the harmonic response, Brownian-dominated particles are required as a model system. $CoFe_2O_4$ nanoparticles can serve as such a Brownian model system and help us revealing the underlying physics. The high anisotropy energy density of $CoFe_2O_4$

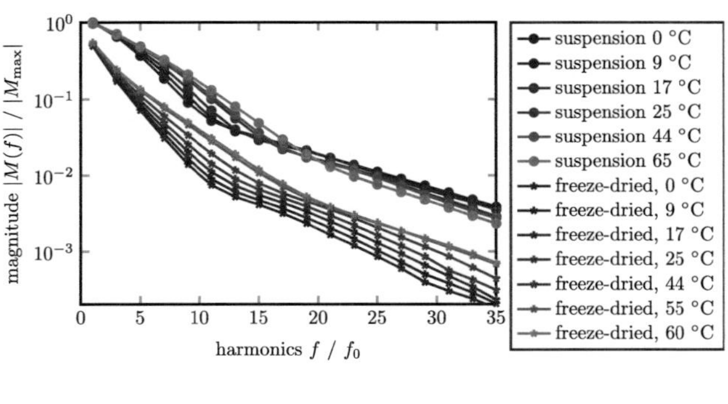

Figure 12.3: Temperature-dependent MPS spectra: The two samples (two sets of curves) are Ocean NanoTech SHP-25 in suspension and in freeze-dried immobilized form. The red color coding corresponds to the warm sample. The blue one marks the cooled—but not yet frozen—sample.

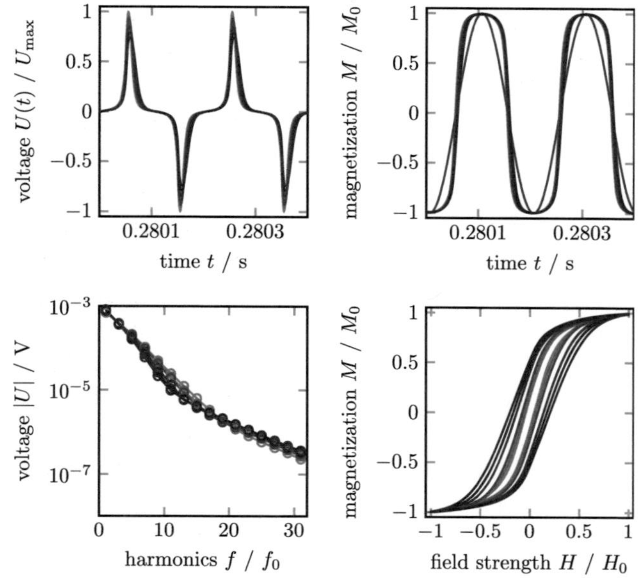

Figure 12.4: Temperature-dependent MPS measurement data of the Ocean NanoTech SHP-25 suspension as spectra of the harmonics (bottom left), as time derivative of the magnetization (top left), as reconstructed time-dependent magnetization curve (top right) and as reconstructed dynamic M(H) curve (bottom right). Color coding according to Figure 12.3.

blocks the Néel relaxation at measurement conditions ($T \approx 298$ K) almost completely for the given core size, which means that the particles are thermally blocked. Thus, the net magnetic moment of the particles only aligns via Brownian relaxation when applying an external magnetic field. The coupling of Brownian relaxation to dynamic viscosity of the particles' environment enables the investigation of the particle mobility influence. The impact of Brownian relaxation on the harmonic response was studied in the study by Draack et al. [36] in detail using a tailored $CoFe_2O_4$ particle system. For that, a logarithmically equidistant viscosity series was prepared. Viscosities of the samples were adjusted by water-glycerol mixtures as solvents and are listed in Table 12.2. For measurements, the samples were filled into glass vials with the amount of 150 mg (approx. 150 µL) each, as shown in Figure 12.6.

Table 12.2: Viscosities η/mPas of the logarithmically equidistant series of $CoFe_2O_4$ nanoparticle samples M_i to investigate the particle mobility.

M_1	M_2	M_3	M_4	M_5	M_6	M_7	M_8	M_9	M_{10}
989.43	686.41	475.05	329.62	228.43	158.21	109.76	76.07	52.73	36.07
M_{11}	M_{12}	M_{13}	M_{14}	M_{15}	M_{16}	M_{17}	M_{18}	M_{19}	M_{20}
25.31	17.55	12.17	8.43	5.85	4.05	2.81	1.95	1.35	0.94

Due to a trade-off between sensitivity of the inductive sensor and relaxation time of the MNP, further investigations focus on data acquired at $f_0 = 1.0$ kHz and $\mu_0\hat{H} = 25$ mT, as exemplary shown in Figure 12.7. Harmonic magnitude spectra (Figure 12.7a) and reconstructed dynamic $M(H)$ curves (Figure 12.7b) show continuous changes with increasing viscosity. Additionally, the residual error R (blank measurement subtracted by another blank measurement) is depicted to be able to distinguish between magnetic signals and noise. Samples with small viscosities show strong magnitudes for the fundamental and higher harmonics, whereas magnitudes of samples with high viscosities drop significantly toward higher harmonics. Reconstructed dynamic $M(H)$ curves show distinct quasi-saturated curves for small viscosities with small hystereses (phase shift due to Brownian relaxation) and elliptical shapes with pronounced hysteresis for high viscosities (particles cannot follow the field due to high friction force on the MNPs' hydrodynamic surface).

Furthermore, each sample of the viscosity series was measured with MPS at different excitation frequencies between 500 Hz $\le f_0 \le$ 25.0 kHz [18, 37] and magnetic excitation field amplitudes 5mT $\le \mu_0\hat{H} \le$ 25 mT.

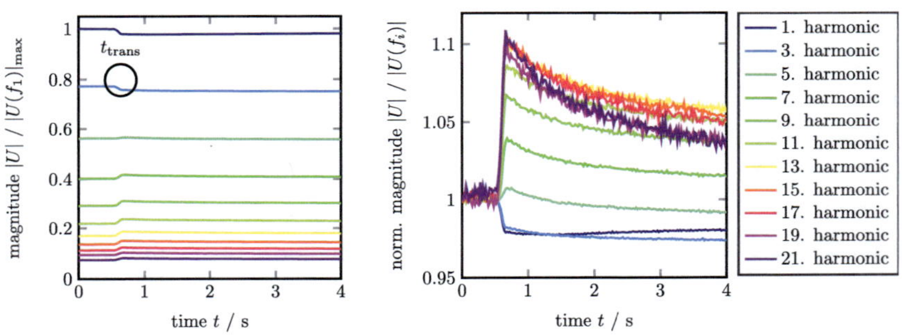

(a) Phase transition (b) Normalized phase transition

Figure 12.5: Representation of the time-resolved temperature-dependent change of the higher harmonics of a diluted Ocean NanoTech SHP-25 sample during the phase transition from subcooled suspension to the frozen state. Figure (b) shows the phase transition of each harmonic normalized to its respective starting value to emphasize the jump of the curve at the transition point marked in (a).

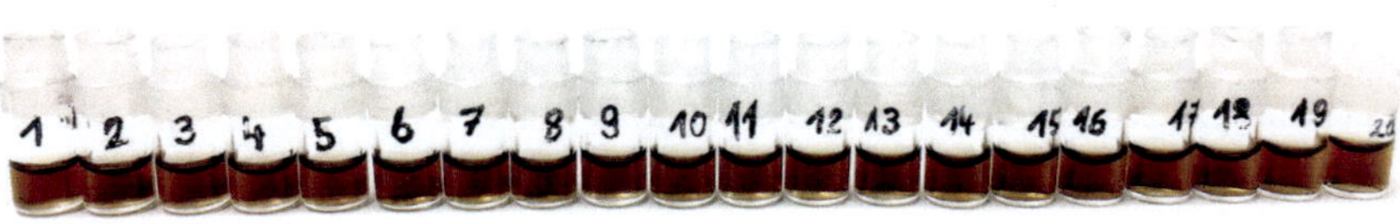

Figure 12.6: Photograph of the $CoFe_2O_4$ sample series providing 20 logarithmically equidistant viscosity values.

Coming from an ACS perspective, the following representation of data is common: real and imaginary parts of the fundamental frequency are plotted as a function of $\omega = 2\pi f_0$, i.e., the parameter along a curve varies in $\omega\tau$ in the range of $\omega_{min}\tau < \omega\tau < \omega_{max}\tau$. Since the relaxation time τ is constant and independent of excitation frequency, we can instead fix the excitation frequency and change $\tau \propto \eta$ along the curve to obtain the same shape of curve. One such plot, where real and imaginary parts are plotted against viscosity η, is now generated for each harmonic frequency of the MPS data. Experimental results of the viscosity series with 20 different viscosities is depicted in Figure 12.8. Multiparametric investigations by varying field amplitude (Figure 12.8a) and frequency (Figure 12.8b) and, e.g., fitting data to Fokker-Planck simulations enable detailed insights into matrix interactions at comparable high field strengths. However, such experimental studies require a comparably high effort of sample preparation and prior knowledge about physical particle properties by incorporating further characterization methods like ACS [36].It should be noted that the amount of change of real and imaginary parts is a measure of the sensitivity for viscosity changes. It is obvious that the shift of the whole harmonic spectrum by varying the field amplitude and excitation frequency controls the sensitive viscosity range covered by the MPS data. The fundamental (and low harmonic frequencies) is most sensitive at high viscosities, whereas the higher harmonics are more sensitive in the lower viscosity range. Unfortunately, higher harmonics are typically limited by signal-to-noise ratio. Still, evaluation of the higher harmonics (not only harmonic ratios of fifth and third harmonic magnitudes as used by other groups [6, 8, 32]) provides a very sensitive tool with respect to particle-matrix interactions.

12.2.4 Multi-spectral MPS

In the previous two chapters, the connection between temperature or viscosity of the sample and the harmonics spectra measured in MPS was explored. Obviously, there is no simple and direct relation apparent from the data. This is especially true, if the particle system relaxes via both the Néel relaxation mechanism as well as Brownian rotation. Most particle systems typically used for MPS/MPI contain superimposed signals of both relaxation processes. Such a tracer can only be described by complex physical models, such as Fokker–Planck equations [29, 38] or a stochastic differential equation set, where the Néel relaxation is described by the Landau–Lifshitz–Gilbert equation [39] and Brownian rotation is formulated via a forced damped oscillatory motion [17]. Since these models are very difficult and time consuming to solve and in most cases the required set of particle parameters (e.g., effective core diameter, anisotropy constant, core and hydrodynamic size distributions, etc.) is not accurately known, an alternative approach is required to cope with the complexity of these particles systems. Multispectral MPS is such a method that follows an empirical approach, where the particle system is described

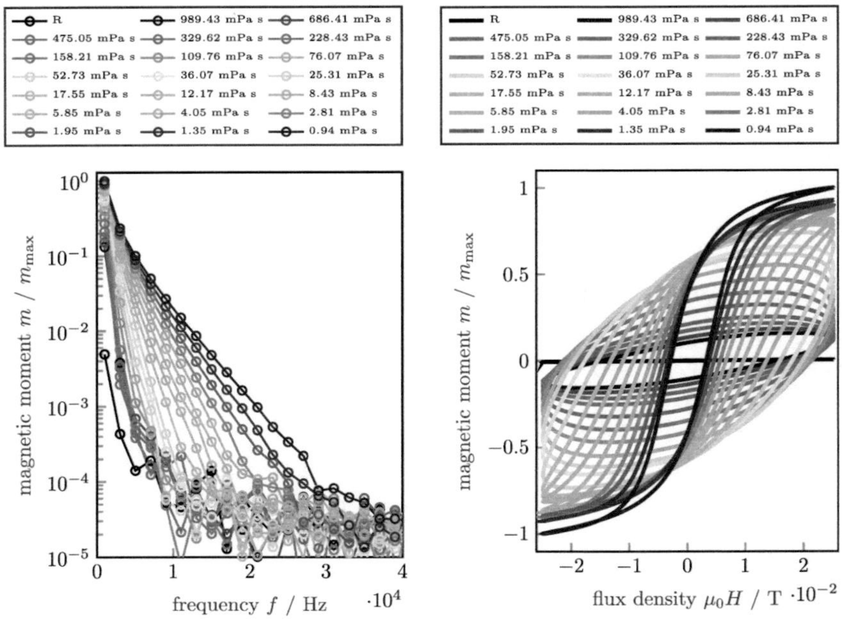

(a) Harmonic magnitude spectra **(b)** Dynamic M(H) curves

Figure 12.7: Harmonic spectra (a) and dynamic magnetization curves (b) of the $CoFe_2O_4$ sample series measured at $f_0 = 1.0$ kHz and $\mu_0\hat{H} = 25$ mT.

via a set of calibration measurements. Multispectral MPS draws a parallel to multispectral MPI reconstruction using the same underlying mathematical approach. Multispectral reconstruction solves a linear system of equations $Ax = b$ (Figure 12.9). Here, A is the reference (or system) matrix containing m frequency components of n reference samples, x is the concentration vector and b is the measurement vector (i.e., the harmonic spectrum). The frequency components (real and imaginary parts) are extracted from MPS raw data via the digital lock-in method using cross-correlation. Having a reference matrix A, intermediate values between the reference samples can be estimated by linear superposition, expressing the linear system of equations as inverse problem $x = A^{-1}b$. There are several approaches for solving the inverse problem, e.g., Tikhonov regularization [40] or iterative algorithms like the Karczmarz method [41, 42]. Here, we use the truncated SVD (tSVD) [43, 44], where the matrix A is approximated by $A = U\Sigma V^{\star} \approx U\Sigma_k V^{\star} = A_k$ by means of singular value decomposition (SVD), with the reduced rectangular diagonal matrix Σ_k (truncated to the k largest eigenvalues). Since the exact inverse A^{-1} is indeterminable due to the fact that A is nonsquare and noise-afflicted, the pseudo-inverse A^{+} is used in its place. It is constructed from the SVD via $A^+ = V\sum_k^{-1}U^*$. The reconstruction result x is then calculated as:

$$x = A^+b = V\sum_k^{-1}U^*b \tag{12.7}$$

U^* and V^* denote the Hermitian transpose of U and V, respectively. The concept was introduced in the study by Viereck et al. [45] to analyze particle mixtures and is applied to viscosity or temperature here. The entries of the system matrix A and measurement vector b are complex-valued, i.e., they are the complex-valued frequency components from the MPS harmonics spectra. While in most cases, the vector x is considered real-valued, i.e., a concentration (and other observable physical quantities) is real-valued, we will find it necessary to generalize x as a complex-valued vector when considering MPS data on viscous media in the upcoming section.

The estimation of a quantity y requires calibration of at least two different references (with different y values). Measurements of intermediate states $\hat{y}$ can be understood as linear superposition of the supplied references $y_{\text{ref},i-1} < \hat{y} < y_{\text{ref},i}$ (we solve a linear set of equations in (12.7)). Multispectral reconstruction therefore intrinsically assumes a linear relationship between supporting reference points. Nonlinear relationships are covered by introducing mapping functions $f(x) \mapsto y$ from the reconstruction values x to the actual physical quantity y. Between two references, the linear mapping functions can be written in two different ways as follows:

$$\hat{y}_a = y_{\text{ref},i-1} + x \cdot \left(y_{\text{ref},i} - y_{\text{ref},i-1}\right) \tag{12.8a}$$

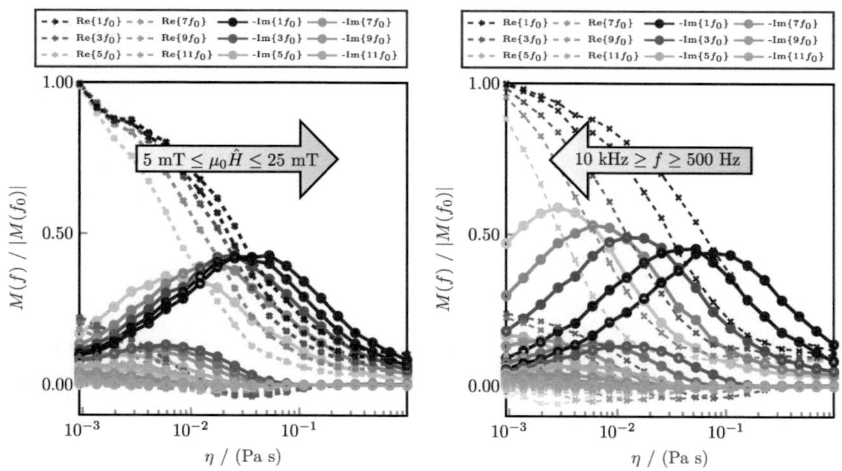

(a) Field-dependent measurements (b) Frequency-dependent measurements

Figure 12.8: Real and imaginary parts of odd harmonics ($1f_0$ to $11f_0$) as a function of sample viscosity η. Field-dependent measurements are shown in (a), frequency-dependent data is depicted in (b).

$$\widehat{y}_b = y_{\text{ref},i} - \left[(1-x)\cdot\left(y_{\text{ref},i} - y_{\text{ref},i-1}\right)\right] \tag{12.8b}$$

In Sections 12.2.4.1 and 12.2.4.2, the mapping functions for MPS viscosity measurements and temperature estimation are analyzed, respectively.

12.2.4.1 Viscosity mapping

The continuous change of the harmonic response suggests that intermediate values between reference points can be modeled as a superposition of the references. Figure 12.10 shows complex-valued tSVD reconstruction results of the sample viscosity series for two reference samples. For the reconstruction using two references, the most viscous sample (M_1, dark red) and the most liquid sample (M_{20}, dark blue) were used as references. Apparently, the reconstruction results are real-valued at the two reference points. The behavior in between the reference points is more complicated. Real and imaginary parts of both references show distinct asymmetries, overshoots and especially a value ambiguity. Clearly, overshoots are more pronounced for the more viscous reference sample since the information content is reduced due to the faster decay of higher harmonics in comparison with the pure water sample. Nevertheless, a similar behavior is observed for the liquid reference sample as can be seen from the detailed excerpt shown in Figure 12.10b. This observation implies that a complex-valued reconstruction is required for mapping since real-valued reconstruction would be ambiguous.

Adding an additional (third) reference point (M_{11}, light blue) leads to reconstruction results depicted in Figure 12.11. Again, reconstruction yields real-valued results at the reference points and overshoots in between. Evidently, a mapping function is required, which maps the reconstruction estimate x into the dynamic viscosity domain η. The complex relationship between reconstruction result $\widehat{x}$ and dynamic viscosity η requires a nonlinear mapping function. A viscosity mapping function using two references (as suggested in (12.8)) is given by

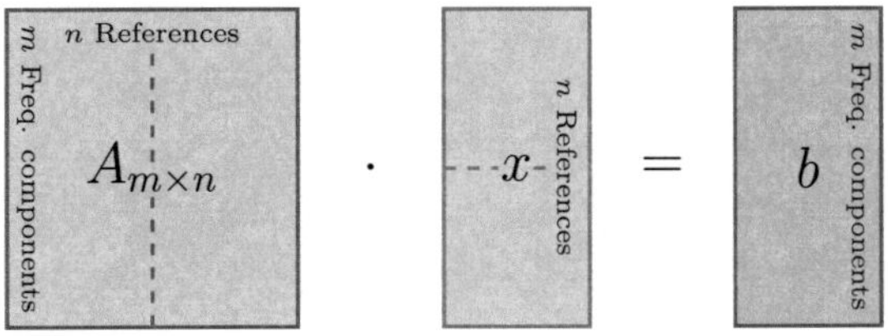

Figure 12.9: Dimensions of vectors/matrices of the linear system set of equations for the reconstruction problem. A is the system matrix, x the concentration vector and b the measurement vector or observed spectrum. The dotted lines symbolize a system with 2 references. In that case, A and x are a two-part compound matrix/vector, respectively.

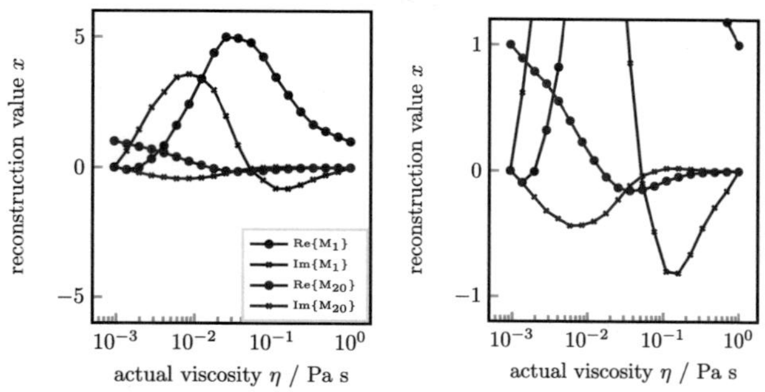

(a) Full value range (b) Detailed excerpt at small values

Figure 12.10: Spectral decomposition results of the $CoFe_2O_4$ viscosity series for two reference samples M_1 and M_{20}. Figure (a) shows the full value range of the reconstruction value. A detailed excerpt of small values is depicted in (b).

$$\widehat{\eta} = \eta_{\text{ref},i} - \left[(1 - \widehat{x}) \cdot \left(\eta_{\text{ref},i} - \eta_{\text{ref},i-1} \right) \right] \tag{12.9}$$

The reconstructed estimate $\widehat{\eta}$ of the viscosity of a sample between the two reference samples depends on the viscosity values of the first reference $\eta_{\text{ref},i-1}$, the viscosity of the second reference $\eta_{\text{ref},i}$ and the reconstructed value x. Here, $\widehat{x}$ is either the real part $\text{Re}\{x\}$, the imaginary part $\text{Im}\{x\}$ or the magnitude $|x|$ of the reconstruction value x. Figure 12.12 shows estimated viscosity results for two (Figure 12.12a) and three (Figure 12.12b) references using the proposed mapping function.

As can be seen from Figure 12.12a, the real-valued estimation $\text{Re}\{x\}$ is ambiguous for higher viscosities. A mapping function using the magnitude $|x|$ gives a unique assignment. At low viscosities, the mapping is very sensitive because the mapping function has a rapid increase of the estimated viscosity versus actual viscosity. On the other hand, the function has poor estimation resolution at the high viscosity limit. To overcome this problem, additional reference points can be consulted. Figure 12.12b shows

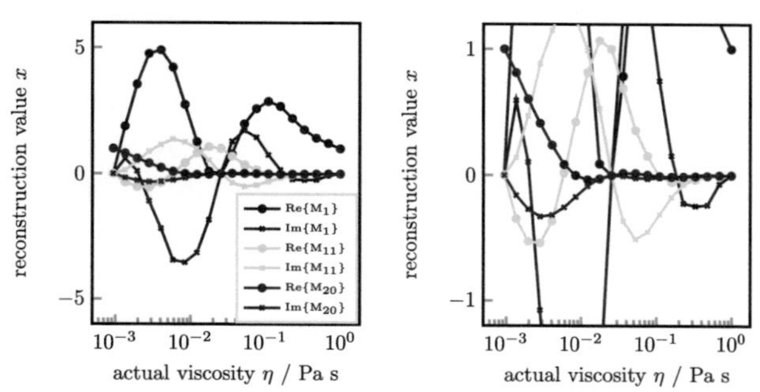

(a) Full value range (b) Detailed excerpt at small values

Figure 12.11: Spectral decomposition results of the $CoFe_2O_4$ viscosity series for three reference samples M_1, M_{11} and M_{20}. Figure (a) shows the full value range of the reconstruction value. A detailed excerpt of small values is depicted in (b).

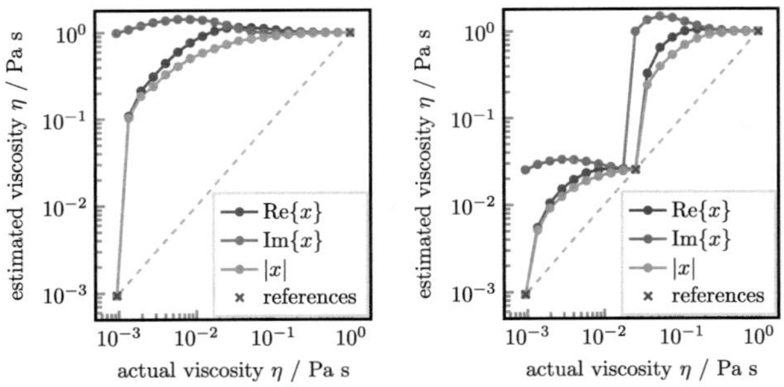

(a) Two viscosity references (M_1 and M_{20})

(b) Three viscosity references (M_1, M_{11} and M_{20})

Figure 12.12: Viscosity mapping functions for two (a) and three (b) $CoFe_2O_4$ reference samples. The dashed line represents a linear mapping function between the references to illustrate the difference from the reconstruction estimate.

the mapping function for three reference samples. The real-valued estimation is still ambivalent, although it is not as obvious as for two references due to scaling. Exact (direct) viscosity estimation is achieved at the reference points only, which is due to the modeling of the mapping function. While the direct evaluation gives inaccurate (too large) estimated viscosities, the curves obtained here can be used as calibration curves for correction. With that we can be significantly more accurate than with a "conventional" linear mapping approach. The viscosity estimation is most sensitive for viscosity values slightly larger than each reference point at which it further becomes most accurate when incorporating the nonlinear mapping relationship.

While the mapping functions in Figure 12.12 are obtained on a Brownian-only model system with $CoFe_2O_4$ particles, the same approach can be used for MPS/MPI-relevant tracers, such as FeraSpin™ XL. Figure 12.13 shows the viscosity mapping function for FeraSpin™ XL using real-valued reconstruction values x.

Due to the strong contribution of the Néel relaxation, which does not change with viscosity, here a real-valued mapping is sufficient to create a unique function over the whole viscosity range (two orders of magnitude, starting from water). The shape of the mapping function bears strong resemblance to the mapping function of the $CoFe_2O_4$ sample. However, Figure 12.13 shows that the mapping approach is readily applicable to iron oxide particle systems used in biomedical applications.

12.2.4.2 Temperature mapping

Similar to the mapping function for viscosity in the previous section, we can establish a mapping function for temperature estimation $\widehat{T}$:

$$\widehat{T} = T_{\text{ref},i} - \left[(1-x) \cdot \left(T_{\text{ref},i} - T_{\text{ref},i-1} \right) \right] \tag{12.10}$$

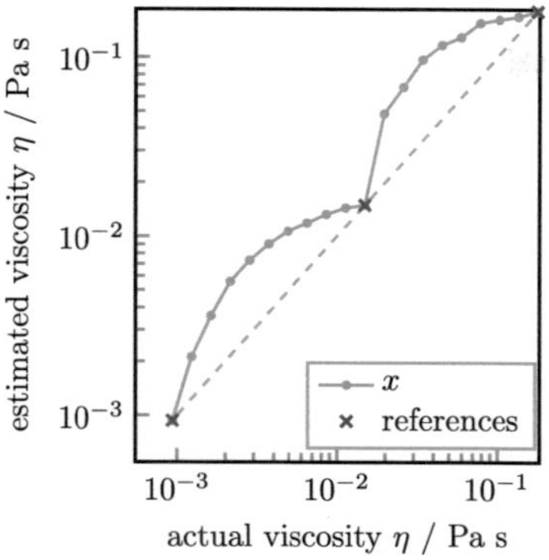

Figure 12.13: Viscosity mapping function (with three references) for FeraSpin™ XL. The dashed line represents a linear mapping function between all three references to illustrate the difference from the reconstruction estimate.

Figure 12.14 depicts the mapping function for SHP-25 particles. The temperature estimation via tSVD of the harmonic spectrum in MPS happens to be much more linear. No oscillating behavior is observed in the decomposition, so that Figure 12.14 could be constructed from a real-valued reconstruction x.

A strong correlation between reconstructed values and a linear superposition (denoted by the dashed line in Figure 12.14) is observed for this particle system over a wide temperature range between –10° and 50 °C (with a maximum temperature deviation of approximately 10 K). For biomedical applications, e.g., cell experiments or animal studies, a much smaller range of typically 35–45 °C is needed. Arguably, multispectral MPS (or MPI) could be used directly for temperature estimation. However, with a calibrated mapping function, with references adjusted for the application's temperature range, the estimation error for temperature can be reduced. As a general rule, the absolute estimation error can be reduced by inserting additional reference points. However, the signal-to-noise margin required for spectral decomposition poses an upper limit for the maximum number of reference points, e.g., using five reference points translates into at least five harmonics available in the measurement above the noise floor.

In conclusion, the multispectral MPS approach provides a very valuable tool for empirically treating real-world particle systems in order to use them for parameter estimation of both viscosity and temperature. Our investigations show that complex-valued reconstruction is required to estimate the particle mobility (for Brownian-dominated particle systems) due to ambiguousness of real-valued reconstruction. Since spectral decomposition in MPS is closely related to reconstruction in MPI, the findings can be transferred to mobility MPI as a quantitative medical imaging modality including information about the particles' binding state or temperature.

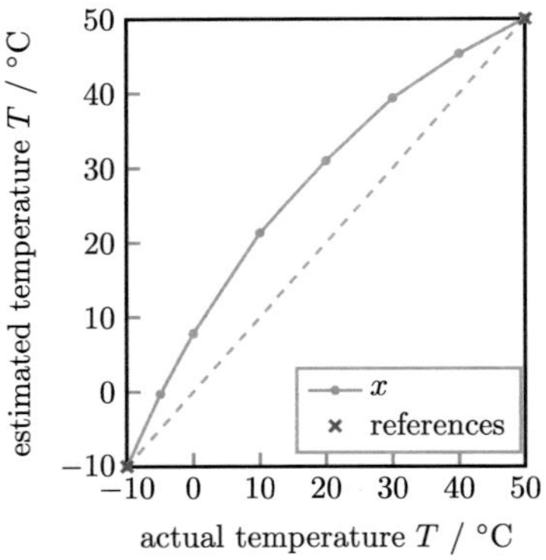

Figure 12.14: Temperature mapping function (with two references) for SHP-25. The dashed line represents a linear mapping function between both references to illustrate the difference from the reconstruction estimate.

12.3 Magnetic particle imaging (MPI)

Magnetic particle spectroscopy (MPS) was the primary tool for investigating particle-matrix interactions throughout the priority programme SPP 1681. However, our definitive goal at the end was to realize such experiments, where we obtain a physical quantity for temperature or viscosity inside an MPI instrument acquiring volume images of an object under investigation. This chapter gives a recap on our MPI system and measurements performed with it; gelatin hydrogels being the representative for a complex matrix system investigated in MPI concludes the project.

12.3.1 Dual-frequency MPI system

Using our 2.5D field-free point (FFP) MPI scanner, the spatially resolved iron concentration of magnetic nanoparticles as well as the rheological mobility of the particles can be visualized. This method is called mobility MPI (mMPI) [46–52]. Our MPI system works with two alternative excitation frequencies ($f_l = 10$ kHz and $f_h = 25$ kHz), whereby an explicit gain in contrast can be recorded in mobility imaging compared to commercial MPI systems, which only provide a single excitation frequency (typically 25 kHz) [50]. Figure 12.15 shows the pivotal coil assembly of the MPI system built at our institute with a bore of ≤35 mm at its center.

The coil assembly contains transmit and receive coils (equivalent to the MPS system described in Section 12.2) as well as the NdFeB selection field generator producing the spatial encoding field for MPI imaging.

12.3.2 Viscosity dependence

For initial experiments, the MPI scanner was operated with a standard two-dimensional Lissajous trajectory at 25 kHz and an imaging gradient of 3 Tm^{-1} in the isotropic imaging plane. The results show that it is possible to separate mobile (suspension) and completely immobilized particles (freeze-dried), which were acquired simultaneously in the imaging field of the scanner (Figure 12.16a), i.e., spatial mapping of particle mobility in MPI is possible. In order to achieve that, a multispectral reconstruction scheme [46, 48–50, 53] is employed. The reconstruction is based on the Kaczmarz method, which is considered the default algorithm in MPI [54–56]. Similar to the multispectral decomposition and reconstruction method in MPS (→ chapter 12.2), it relies on two or more calibration datasets being available. Multispectral reconstruction returns one image per provided reference. In Figure 12.16, we use two references at opposite ends of the mobility spectrum. For the final image, both reconstructed images are combined into a single false color image, where the yellow color denotes the contribution from high viscosity (freeze-dried or 100% glycerol) and green color corresponds to the low viscosity contribution (H_2O suspension).

Figure 12.16b shows the particle mobility contrast of FeraSpin™ XL in glycerol-water mixtures with $80\%_{vol}$ (left) and $20\%_{vol}$ (right) glycerol content, respectively. The samples were prepared with a particle content of $10\%_{vol}$, where the rest was filled with water-glycerol mixtures of variable ratios. In Figure 12.16b, the spatial resolution and ability to distinguish different viscosity levels can be evaluated. Also, since the references used for reconstruction of Figure 12.16b, i.e., $0\%_{vol}$ and $100\%_{vol}$ glycerol, were not identical to the viscosity of the sample points, artifacts from the reconstruction (ghost images at the sides) are observed. Similar to Figure 12.16a, the spatial resolution depends on the particle mobility. For (partially) immobilized samples, the spatial resolution is

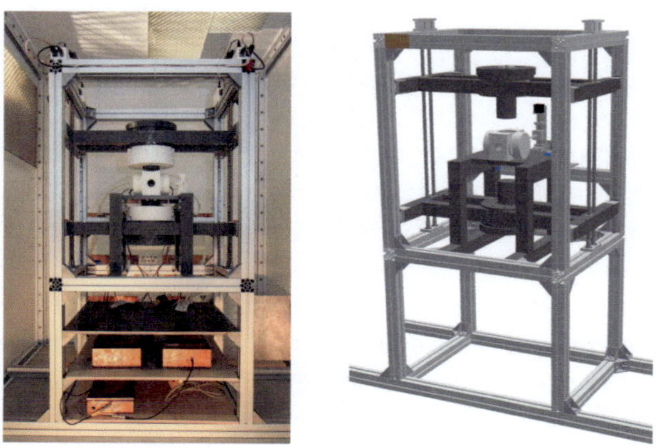

(a) Photo of coil assembly (b) CAD drawing of coil assembly

Figure 12.15: Overview of MPI system hardware.

deteriorated, which translates into the wider spot size of the sample with higher viscosity. It is still an open question, why the reconstruction (which intrinsically performs a deconvolution) is not able to equalize the different contributions in terms of apparent size (but also with respect to its estimated iron content). But it is logical to conclude that the very different signal-to-noise ratios of the two samples, i.e., smaller signal magnitude and faster decay toward higher harmonics, are responsible and not (yet) incorporated explicitly in the reconstruction process.

It should be noted that viscosity information can also be extracted by other means as discussed in the literature studies [47, 51, 57], mostly by evaluating the relaxation effects directly in the time-domain signal.

Since calibration in standard (Lissajous) 2D imaging is very time consuming, we switched to an alternative imaging scheme to be able to perform experiments on complex matrices more easily and time-efficient. In the following, we use Cartesian MPI [58], where the system employs a unidirectional excitation along the x-axis only, while the orthogonal y-direction is scanned consecutively (either by moving the sample or shifting the FFP by means of an offset field). The advantages of this method are the higher signal-to-noise ratio, which translates to a better image resolution along the trajectory axis in x-direction. Also, for evaluating the one-dimensional MPI signal, model-based or time-domain reconstruction methods can be applied.

The signal-to-noise ratio can be increased by selecting the excitation frequency of the system nearest to the expected particle response frequency. We therefore chose the lower available image frequency of 10 kHz to measure the slow relaxation of particles in the gelatin matrix. The samples were built with a visco-elastic gelatin matrix, which constitutes an application-relevant material. The matrix was prepared from $50\%_{vol}$ gelatin and $50\%_{vol}$ of commercial perimag® particles, resulting in an iron concentration of 12.5 mg/mL.

The phantom (shown in Figure 12.17) consists of three bars, where the center bar is filled with MNP in the gelatin matrix, whereas the outer bars (left and right) are loaded with H_2O-diluted perimag® only for reference. The gelatin matrix forms progressively over several hours, enabling MPI to observe the gelation process and its kinetics over time [59]. The gelation process is studied by Draack et al. [60], including variations in gelatin concentration, and observing reconstruction artifacts.

Figure 12.18 shows the MPI data measured on the phantom. Both images are recorded without any noticeable artifacts. In order to reconstruct the images, we use two different calibrations datasets as references. The image on the left was obtained on diluted perimag® (H_2O reference); the right image on perimag® in gelatin (gelatin reference). The two-dimensional images in the top row are displayed as surface plots to better visualize the "raw" values from the reconstructed images. As can be seen, the magnitudes depend on whether a sample point matches the calibration. For the water-filled outer bars, a higher value is observed for the H_2O reference, whereas the gelatin bar in the middle reveals a better match with the gelatin reference. X-/Y-dimensions are given in pixels on an isotropic grid (1×1 mm). We suggest that the mapping functions obtained from MPS (→ Section 12.2.4) can either be used directly in MPI or at least the

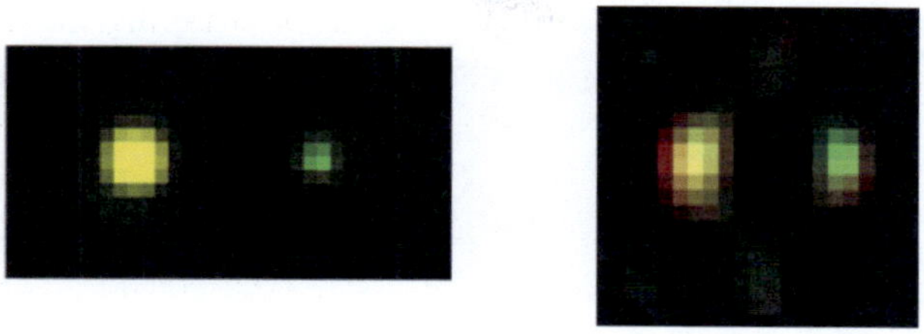

(a) Suspension vs. freeze-dried (b) Different viscosities

Figure 12.16: Multispectral MPI measurement results at 25 kHz depicting different mobility states of FeraSpin™ XL being separable. In (a) the particles are in H_2O suspension (left, yellow) versus freeze-dried (right, green). Figure (b) shows the tracer prepared in viscous glycerol-water mixtures with $80\%_{vol}$ (left) and $20\%_{vol}$ (right) glycerol.

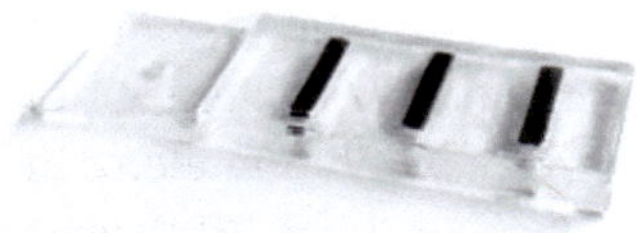

Figure 12.17: MPI sample geometry with perimag® particles: outer bars filled with H_2O suspension, center bar with tracer in gelatin matrix.

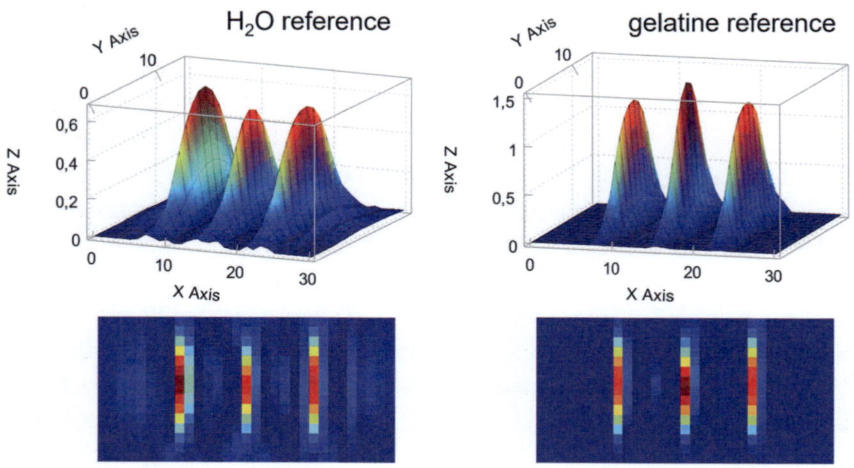

Figure 12.18: Viscosity in Cartesian MPI: Surface plots (upper row) and top view (lower row) of reconstructed perimag® dash samples. Outer (left and right) dashes are filled with particles in a water suspension and center dash with particles embedded in a gelatin-water matrix. Reconstruction was performed with the water reference (left column) and with the gelatin reference (right column). *X*-/*Y*-dimensions are given in pixels. *Z* constitutes the ‘raw’ reconstruction value.

same mapping approach should be applicable. However, it is still under investigation whether viscosity or temperature can be reliably quantified in MPI imaging.

12.4 Conclusion

Magnetic particle spectroscopy (MPS) is a very powerful tool for investigating magnetic nanoparticles. The method is simple, fast and especially very sensitive for small changes in dynamic magnetization behavior, even at low particle concentrations. We were able to establish a connection between the viscosity of the medium around a Brownian-dominated particle system and the harmonic spectrum observed in MPS. For particles relaxing via both the Néel relaxation and the Brownian rotation, which is typically the case for most MPS/MPI tracer systems, we also investigated multispectral analysis and reconstruction methods as an empirical method. These methods use a calibration-based approach to deal with the details of the nonlinear magnetization response of application-relevant particles. Therefore, MPS data were utilized to establish a foundation with respect to the underlying physical processes and for exploring mathematical magnetization models. Many aspects of the investigations performed in MPS translate into the imaging regime in MPI. We showed, that MPI is well capable of mapping the local particle environment. However, quantification of particle concentration and of viscosity or temperature remains challenging. Still, it can be concluded that the spectral characterization methods, the multiparametric evaluation in MPS as a function of excitation frequency and temperature, as well as mobility imaging in MPI, represent innovative tools for the investigation of particle-matrix interactions.

Acknowledgment: Financial support by the German Research Foundation DFG via priority program SPP1681 under grant numbers SCHI 383/2-1 and VI 892/1-1 and "Niedersächsisches Vorab" through "Quantum- and Nano-Metrology (QUANOMET)" initiative within the project NP-2 are gratefully acknowledged.
Author contribution: All the authors have accepted responsibility for the entire content of this submitted manuscript and approved submission.
Research funding: This article was supported by German Research Foundation DFG under grant numbers SCHI 383/2-1 and VI 892/1-1 and "Niedersächsisches Vorab" through "Quantum- and Nano-Metrology (QUANOMET)" initiative within the project NP-2.
Conflict of interest statement: The authors declare no conflicts of interest regarding this article.

References

1. Gleich B, Weizenecker J. Tomographic imaging using the nonlinear response of magnetic particles. Nature 2005;435:1214–17.

2. Weizenecker J, Gleich B, Rahmer J, Dahnke H, Borgert J. Three-dimensional real-time in vivo magnetic particle imaging. Phys Med Biol 2009;54:L1–10.
3. Knopp T, Buzug TM. Magnetic particle imaging: an introduction to imaging principles and scanner instrumentation. Heidelberg: Springer; 2012.
4. Borgert J, Schmidt JD, Schmale I, Bontus C, Gleich B, David B, et al. Perspectives on clinical magnetic particle imaging. Biomedizinische Technik/Biomed Eng 2013;58:551–6.
5. Biederer S, Knopp T, Sattel TF, Lüdtke-Buzug K, Gleich B, Weizenecker J, et al. Magnetization response spectroscopy of superparamagnetic nanoparticles for magnetic particle imaging. J Phys Appl Phys 2009;42:205007.
6. Rauwerdink AM, Giustini AJ, Weaver JB. Simultaneous quantification of multiple magnetic nanoparticles. Nanotechnology 2010;21:455101.
7. Rauwerdink AM, Hansen EW, Weaver JB. Nanoparticle temperature estimation in combined ac and dc magnetic fields. Phys Med Biol 2009;54:L51–5.
8. Rauwerdink AM, Weaver JB. Viscous effects on nanoparticle magnetization harmonics. J Magn Magn Mater 2010;322:609–13.
9. Rauwerdink AM, Weaver JB. Measurement of molecular binding using the Brownian motion of magnetic nanoparticle probes. Appl Phys Lett 2010;96:033702.
10. Weaver JB, Harding M, Rauwerdink AM, Hansen EW. The effect of viscosity on the phase of the nanoparticle magnetization induced by a harmonic applied field. In: Molthen RC, Weaver JB, editors. Medical imaging 2010: biomedical applications in molecular, structural, and functional imaging. San Diego: SPIE; 2010. https://doi.org/10.1117/12.845576.
11. Weaver JB, Kuehlert E. Measurement of magnetic nanoparticle relaxation time. Med Phys 2012;39: 2765–70.
12. Weaver JB, Rauwerdink AM. Chemical binding affinity estimation using MSB. In: Weaver JB, Molthen RC, editors. Medical imaging 2011: biomedical applications in molecular, structural, and functional imaging. Lake Buena Vista, Orlando: SPIE; 2011. https://doi.org/10.1117/12.878788.
13. Weaver JB, Rauwerdink AM, Hansen EW. Magnetic nanoparticle temperature estimation. Med Phys 2009;36:1822–9.
14. Enpuku K, Tsujita Y, Nakamura K, Sasayama T, Yoshida T. Biosensing utilizing magnetic markers and superconducting quantum interference devices. Supercond Sci Technol 2017;30:053002.
15. Du Z, Sun Y, Higashi O, Noguchi Y, Enpuku K, Draack S, et al. Effect of core size distribution on magnetic nanoparticle harmonics for thermometry. Jpn J Appl Phys 2019;59:010904.
16. Zhong J, Schilling M, Ludwig F. Magnetic nanoparticle thermometry independent of Brownian relaxation. J Phys Appl Phys 2017;51:015001.
17. Coffey WT, Kalmykov YP. Thermal fluctuations of magnetic nanoparticles: fifty years after brown. J Appl Phys 2012;112:121301.
18. Kuhlmann C, Khandhar AP, Ferguson RM, Kemp S, Wawrzik T, Schilling M, et al. Drive-field frequency dependent MPI performance of single-core magnetite nanoparticle tracers. IEEE Trans Magn 2015;51:1–4.
19. Ludwig F, Eberbeck D, Löwa N, Steinhoff U, Wawrzik T, Schilling M, et al. Characterization of magnetic nanoparticle systems with respect to their magnetic particle imaging performance. Biomed Tech/Biomed Eng 2013;58:535–45.
20. Ludwig F, Kuhlmann C, Wawrzik T, Dieckhoff J, Lak A, Kandhar AP, et al. Dynamic magnetic properties of optimized magnetic nanoparticles for magnetic particle imaging. IEEE Trans Magn 2014;50:1–4.
21. Ludwig F, Remmer H, Kuhlmann C, Wawrzik T, Arami H, Ferguson RM, et al. Self-consistent magnetic properties of magnetite tracers optimized for magnetic particle imaging measured by ac susceptometry, magnetorelaxometry and magnetic particle spectroscopy. J Magn Magn Mater 2014;360:169–73.

22. Ludwig F, Wawrzik T, Yoshida T, Gehrke N, Briel A, Eberbeck D, et al. Optimization of magnetic nanoparticles for magnetic particle imaging. IEEE Trans Magn 2012;48:3780–3.
23. Draack S, Viereck T, Nording F, Janssen K-J, Schilling M, Ludwig F. Determination of dominating relaxation mechanisms from temperature-dependent magnetic particle spectroscopy measurements. J Magn Magn Mater 2019;474:570–3.
24. Lucht N, Friedrich RP, Draack S, Alexiou C, Viereck T, Ludwig F, et al. Biophysical characterization of (silica-coated) cobalt ferrite nanoparticles for hyperthermia treatment. Nanomat: Appl Mag Nanomater 2019;9:1713.
25. Nappini S, Magnano E, Bondino F, Píš I, Barla A, Fantechi E, et al. Surface charge and coating of CoFeO nanoparticles: evidence of preserved magnetic and electronic properties. J Phys Chem C 2015;119:25529–41.
26. Yoshida T, Enpuku K. Simulation and quantitative clarification of AC susceptibility of magnetic fluid in nonlinear Brownian relaxation region. Jpn J Appl Phys 2009;48:127002.
27. Dieckhoff J, Eberbeck D, Schilling M, Ludwig F. Magnetic-field dependence of Brownian and Néel relaxation times. J Appl Phys 2016;119:043903.
28. Coffey WT, Cregg PJ, Kalmykov YUP. On the theory of Debye and Néel relaxation of single domain ferromagnetic particles. Advances in chemical physics. John Wiley & Sons, Inc.; 2007:263–464 p. https://doi.org/10.1002/9780470141410.ch5.
29. Deissler RJ, Wu Y, Martens MA. Dependence of Brownian and Néel relaxation times on magnetic field strength. Med Phys 2013;41:012301.
30. Gutsalyuk VM, Guly IS, Mel'nichenko YB, Klepko VV, Vasil'ev GI, Avdeev NN. Mutual diffusion in aqueous gel solutions. Polym Int 1994;33:359–65.
31. Draack S, Viereck T, Kuhlmann C, Schilling M, Ludwig F. Temperature-dependent MPS measurements. Int J Mag Part Imag 2017;3. doi:https://doi.org/10.18416/ijmpi.2017.1703018.
32. Löwa N, Seidel M, Radon P, Wiekhorst F. Magnetic nanoparticles in different biological environments analyzed by magnetic particle spectroscopy. J Magn Magn Mater 2017;427:133–8.
33. Engelmann UM, Buhl EM, Draack S, Viereck T, Ludwig F, Schmitz-Rode T, et al. Magnetic relaxation of agglomerated and immobilized iron oxide nanoparticles for hyperthermia and imaging applications. IEEE Mag Lett 2018;9:1–5.
34. Poller WC, Löwa N, Wiekhorst F, Taupitz M, Wagner S, Möller K, et al. Magnetic particle spectroscopy reveals dynamic changes in the magnetic behavior of very small superparamagnetic iron oxide nanoparticles during cellular uptake and enables determination of cell-labeling efficacy. J Biomed Nanotechnol 2016;12:337–46.
35. Müssig S, Granath T, Schembri T, Fidler F, Haddad D, Hiller K-H, et al. Anisotropic magnetic supraparticles with a magnetic particle spectroscopy fingerprint as indicators for cold-chain breach. ACS Appl Nano Mater 2019;2:4698–702.
36. Draack S, Lucht N, Remmer H, Martens M, Fischer B, Schilling M, et al. Multiparametric magnetic particle spectroscopy of CoFeO nanoparticles in viscous media. J Phys Chem C 2019;123:6787–801.
37. Wawrzik T, Kuhlmann C, Remmer H, Gehrke N, Briel A, Schilling M, et al. Effect of Brownian relaxation in frequency-dependent magnetic particle spectroscopy measurements. In: 2013 International Workshop on Magnetic Particle Imaging (IWMPI). Berkeley, CA, USA: IEEE; 2013. https://doi.org/10.1109/iwmpi.2013.6528371.
38. Weizenecker J. "The Fokker–Planck equation for coupled Brown–Néel-rotation. Phys Med Biol 2018;63:035004.
39. Gilbert T. Classics in magnetics a phenomenological theory of damping in ferromagnetic materials. IEEE Trans Magn 2004;40:3443–9.
40. Goncharsky A, Stepanov VV, Tikhonov AN, Yagola AG. Numerical methods for the solution of ill-posed problems. Netherlands: Springer; 1995, 325. https://doi.org/10.1007/978-94-015-8480-7.

41. Golub GH, von Matt U. Tikhonov regularization for large scale problems. Self; 1997. Available from: http://citeseerx.ist.psu.edu/viewdoc/summary?doi=10.1.1.51.409.
42. Parks PC. S. Kaczmarz (1895–1939). Int J Contr 1993;57:1263–7.
43. Hansen PC. The truncated SVD as a method for regularization. BIT Numer Math 1987;27:534–53.
44. Fierro RD, Golub GH, Hansen PC, O'Leary DP. Regularization by truncated total least squares. SIAM J Sci Comput 1997;18:1223–41.
45. Viereck T, Draack S, Schilling M, Ludwig F. Multi-spectral magnetic particle spectroscopy for the investigation of particle mixtures. J Magn Magn Mater 2019;475:647–51.
46. Haegele J, Vaalma S, Panagiotopoulos N, Barkhausen J, Vogt FM, Borgert J, et al. Multi-color magnetic particle imaging for cardiovascular interventions. Phys Med Biol 2016;61:N415–26.
47. Hensley D, Goodwill P, Croft L, Conolly S. Preliminary experimental x-space color MPI. In: 2015 5th International Workshop on Magnetic Particle Imaging (IWMPI). Istanbul, Turkey: IEEE; 2015. https://doi.org/10.1109/iwmpi.2015.7106993.
48. Rahmer J, Halkola A, Gleich B, Schmale I, Borgert J. First experimental evidence of the feasibility of multi-color magnetic particle imaging. Phys Med Biol 2015;60:1775–91.
49. Stehning C, Gleich B, Rahmer J. Simultaneous magnetic particle imaging (MPI) and temperature mapping using multi-color MPI. Int J Mag Part Imag 2016;2:1612001.
50. Viereck T, Kuhlmann C, Draack S, Schilling M, Ludwig F. Dual-frequency magnetic particle imaging of the Brownian particle contribution. J Magn Magn Mater 2017;427:156–61.
51. Wawrzik T, Kuhlmann C, Ludwig F, Schilling M. Estimating particle mobility in MPI. In 2013 International Workshop on Magnetic Particle Imaging (IWMPI). IEEE; 2013. https://doi.org/10.1109/IWMPI.2013.6528372.
52. Wawrzik T, Ludwig F, Schilling M. Magnetic particle imaging: exploring particle mobility. Springer proceedings in physics. Berlin, Heidelberg: Springer; 2012, vol 140. 21–5 p.
53. Möddel M, Meins C, Dieckhoff J, Knopp T. Viscosity quantification using multi-contrast magnetic particle imaging. New J Phys 2018;20:083001.
54. Knopp T, Rahmer J, Sattel TF, Biederer S, Weizenecker J, Gleich B, et al. Weighted iterative reconstruction for magnetic particle imaging. Phys Med Biol 2010;55:1577–89.
55. Kluth T, Jin B. Enhanced reconstruction in magnetic particle imaging by whitening and randomized SVD approximation. Phys Med Biol 2019;64:125026.
56. Weber A, Knopp T. Reconstruction of the magnetic particle imaging system matrix using symmetries and compressed sensing. Adv Math Phys 2015;2015:1–9.
57. Utkur M, Muslu Y, Saritas EU. Relaxation-based viscosity mapping for magnetic particle imaging. Phys Med Biol 2017;62:3422–39.
58. Werner F, Gdaniec N, Knopp T. First experimental comparison between the Cartesian and the Lissajous trajectory for magnetic particle imaging. Phys Med Biol 2017;62:3407–21.
59. Remmer H, Roeben E, Schmidt AM, Schilling M, Ludwig F. Dynamics of magnetic nanoparticles in viscoelastic media. J Magn Magn Mater 2017;427:331–5.
60. Draack S, Schilling M, Ludwig F, Viereck T. Dynamic gelation process observed in Cartesian magnetic particle imaging. J Magn Magn Mater 2020;522:167478.

Frank Ludwig* and Hilke Remmer

13 Rotational dynamics of magnetic nanoparticles in different matrix systems

Abstract: Dynamic magnetic measurements on magnetic nanoparticle (MNP) samples have been widely used for the determination of structural MNP parameters as well as for the realization of bioassays. On the other hand, proposed that the MNPs are thermally blocked, i.e., that the dynamics are dominated by the Brownian rotation, and knowing the distribution of their hydrodynamic size, information on the matrix properties can be obtained. In contrast to conventional rheology, the local environment of the MNPs is sensed on the nanoscale so that important information on the embedding of MNPs in the matrix and thus the particle-matrix interaction is obtained. Depending on the characteristic length scales of the matrix and the size of the MNPs, rheological parameters, such as viscosity and shear modulus, derived from nanorheological measurements can differ from the values obtained from conventional rheology. To measure the MNP dynamics, different experimental techniques can be applied. In this contribution, the focus lies on ac susceptometry and fluxgate magnetorelaxometry. The analysis of the complex ACS spectra is generally carried out within a modified Debye model. Different approaches for the estimation of rheological parameters from the complex ACS spectra will be presented. Two model systems will exemplarily be discussed in detail. As a Newtonian matrix system, water-glycerol mixtures were studied. It is demonstrated that the dynamic viscosity can accurately be estimated from ACS measurements on well thermally blocked single-core as well as on multicore MNP systems, which include Brownian and Néel dynamics. As a viscoelastic matrix system, aqueous gelatin solutions were studied. Gelatin is known to be a Voigt-Kelvin model system, in which elastic and viscous forces are parallel. In particular, we studied the gelation dynamics by repetitive measurements of the complex ACS spectrum. Different approaches to derive viscosity and shear modulus are applied and compared. In order to identify magnetoviscous effects in dynamic magnetic measurements, the magnetic field dependence of the Brownian relaxation time has to be eliminated. ACS measurements on various sufficiently strongly diluted aqueous MNP suspensions were performed in dependence of ac field amplitude and superimposed dc field strength and compared to theory. Excellent agreement was found.

Keywords: Brownian relaxation, Debye model, dynamic magnetic measurements, magnetic nanoparticles, Newtonian fluids, viscoelastic matrix

---

***Corresponding author: Frank Ludwig,** Institut für Elektrische Messtechnik und Grundlagen der Elektrotechnik, TU Braunschweig, D-38106 Braunschweig, Germany, E-mail: f.ludwig@tu-bs.de
Hilke Remmer, Institut für Elektrische Messtechnik und Grundlagen der Elektrotechnik, TU Braunschweig, D-38106 Braunschweig, Germany

This article has previously been published in the journal Physical Sciences Reviews. Please cite as: F. Ludwig and H. Remmer "Rotational dynamics of magnetic nanoparticles in different matrix systems" *Physical Sciences Reviews* [Online] 2021, 6. DOI: 10.1515/psr-2019-0115 | https://doi.org/10.1515/9783110569636-013

13.1 Introduction

The dynamics of magnetic nanoparticles (MNPs) have been widely studied theoretically [1–4] as well as experimentally [5–7]. It is characterized by two distinct mechanisms—the Brownian and the Néel one. In the former, the whole particle including shell rotates, and the Brownian relaxation time is given by

$$\tau_B = \frac{3\eta V_h}{k_B T} \tag{13.1}$$

with hydrodynamic volume V_h, dynamic viscosity of the medium η, temperature T, and Boltzmann constant k_B. In the latter one, the magnetic moment flips thermally activated between easy axes, and the Néel relaxation time—for uniaxial anisotropy—is given by [2]

$$\tau_N = \frac{\sqrt{\pi}\tau_{N0}}{2(KV_c/k_B T)^{1/2}} \exp\left(\frac{KV_c}{k_B T}\right). \tag{13.2}$$

Here V_c is the (magnetic) core volume, K is the effective anisotropy constant (or often denoted as effective anisotropy energy density) and t_{NO} is a constant of the order of 10^{-9} s [5]. For simplicity, Equation (13.2) is often simplified to

$$\tau_N = \tau_0 \ \exp\left(\frac{KV_c}{k_B T}\right). \tag{13.3}$$

If both processes can take place—for example when MNPs are suspended in a liquid—dynamics in the small field limit is characterized by an effective relaxation time [8]

$$\tau_{eff} = \frac{\tau_B \tau_N}{\tau_B + \tau_N}, \tag{13.4}$$

i.e., the faster of the two mechanisms dominates. Figure 13.1 depicts the relaxation times for typical parameters of magnetite (Fe_3O_4) MNPs in dependence of hydrodynamic diameter (spherical shapes are assumed, i.e., $V_h = (\pi/6)d_h^3$) for different values of the dynamic viscosity.

Dynamic magnetic measurements combined with Equations (13.1)–(13.4) have been widely used for the characterization of MNPs [6, 9–11]. While the Néel process contains information on the core size (to be more precise on the anisotropy energy KV_c), the Brownian relaxation times allows one to determine the hydrodynamic size of the MNPs, provided that the viscosity η is known.

If the Brownian relaxation time is smaller than the Néel one, i.e., MNPs is thermally blocked, dynamic magnetic measurements have also been proposed for the realization of homogeneous bioassays [12–19]. Here the basic idea is that the hydrodynamic size and consequently the Brownian relaxation time increase upon specific binding of analytes to the functionalized surface of the MNPs.

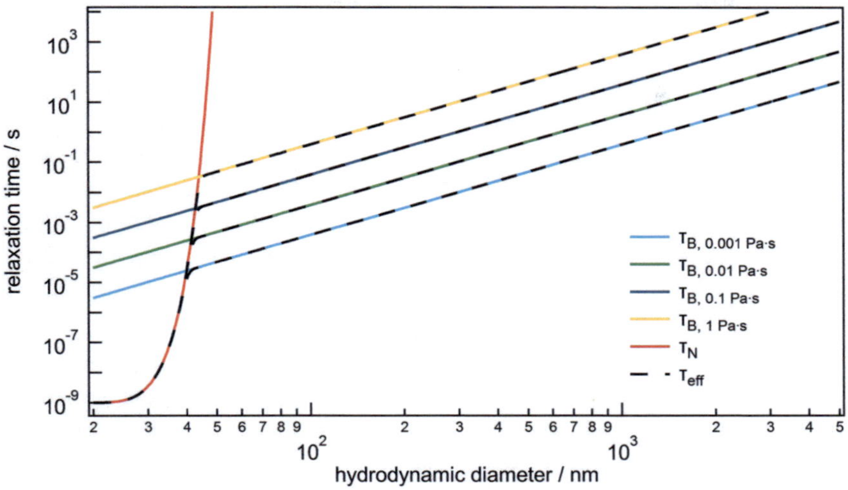

Figure 13.1: Relaxation time versus hydrodynamic diameter calculated for four different viscosities. Parameters: $K = 10^4$ J/m^3, $T = 296$ K, shell thickness $d_{sh} = 10$ nm, $\tau_0 = 1$ ns.

On the other hand, if the MNPs are thermally blocked (the magnetic moment is "pinned" to the easy axes) and if their hydrodynamic size is known, dynamic magnetic measurements can be applied for the determination of the (local) dynamic viscosity. The use of MNPs as probes for measuring the local viscosity was first reported by Bacri et al. [20]. Later on, similar studies, utilizing different magnetic markers and different experimental techniques, were presented by a number of groups extending the work to non-Newtonian media [21–25]. In contrast to conventional rheology, applying MNPs as probes, the local environment is sensed on the nanoscale, yielding to the term "nanorheology". Nanorheological investigations can also be applied to understand the embedding of MNPs in the matrix and thus the particle-matrix interaction, which is central topic of the SPP1681 and this book. Depending on the characteristic length scales of the matrix and the size of the MNPs, rheological parameters, such as viscosity and shear modulus, derived from nanorheological measurements, can differ from the values obtained from standard rheology.

In this contribution, the basics of nanorheology based on MNPs as local probes are summarized. In particular, experimental techniques will be described and compared, and models for the estimation of (nano)rheological parameters from the experimental data will be discussed. The application of nanorheology will be described for two model systems: First, the dynamics of MNPs in water-glycerol mixtures, known to be Newtonian media, will be presented. Then, the gelation kinetics of aqueous gelatin solutions and the analysis of experimental data with the Voigt-Kelvin model will be presented.

Regarding the control or material parameters by external magnetic fields, several groups reported on magnetoviscous effects applying conventional rheology [26]. Here one has to distinguish between the rotational viscosity, which is the hindrance of the rotation of single particles and which was theoretically described by Shliomis [27], and the increase of viscosity with rising magnetic field strength caused by dipolar interactions and thus structure formation [28]. While the former mechanism does—for geometric reasons—not take place in a dynamic magnetic measurement with stationary sample, the latter can principally occur. In order to identify magnetoviscous effects in nanorheology, one must, however, keep in mind that the Brownian relaxation time itself depends on the amplitude of the applied magnetic field. Therefore, the field dependence of the Brownian relaxation time is studied applying ac susceptometry and results are compared with theoretical models. It will be shown that excellent agreement with a theoretical model, based on solving the Fokker-Planck equation, is observed for Brownian relaxation.

13.2 Dynamic magnetic measurement techniques

In order to measure the dynamics of MNPs, several magnetic measurement techniques have been developed and established. Measurements can be carried out in the time or in the frequency domain. In time domain, one generally applies a magnetic field pulse, and measures the decay of the magnetic signal form the sample after switching off the field. This technique, also known as magnetorelaxometry (MRX), has been realized with various magnetic sensors. The pioneering MRX work at the PTB was realized using highly sensitive Nb SQUIDs as sensors [6, 29, 30]. In our work, we employ a custom-built MRX setup based on fluxgate magnetometers. Although not reaching the resolution of the mentioned SQUID MRX setup, the differential fluxgate setup provides a number of practical advantages [7, 11].

In the frequency domain, most commonly ac susceptometry (ACS) is used. A sinusoidal magnetic field with amplitude in the range of a few 100 μT is applied and a frequency sweep (at ideally constant field amplitude) is performed. The magnetic signal from the MNP sample is generally detected by a gradiometric detection coil. Many of the custom-built and the commercial ACS systems from Rise-Acreo [31] are designed for isothermal measurement around room temperature. Temperature-dependent measurements of the complex (ac) susceptibility ($\chi(T)$) are mainly performed utilizing commercial systems, such as the Magnetic Property Measurement System MPMS from Quantum Design [32]. The custom-built ACS systems applied in this work cover frequency ranges from 10 Hz to 10 kHz (NF system with field amplitude of 567 μT) and 200 Hz–1 MHz (HF system with field amplitude of 90 μT). Another system operating in the frequency domain is the fluxgate-based setup, which was originally realized for measurements of the response of the sample's magnetic moment to a rotating magnetic field [33]. The rotating magnetic field is generated by two Helmholtz coils. Frequency can be varied between 2 and 9 kHz. Using just one of the two Helmholtz coils, the system can also be operated as an ac susceptometer. In contrast to the ACS systems mentioned above, this system allows one to apply ac magnetic fields with amplitudes up to 9 mT. Alternatively, a small ac field can be superimposed by static magnetic fields having magnitudes of up to 9 mT with directions parallel or perpendicular to the ac field. Thus, ACS measurements can also be carried out in the nonlinear regime, which allows one to study, e.g., the magnetic field dependence of relaxation times. In addition, this system allows a variation of sample temperature from room temperature to about 80 °C.

13.3 Models for analysis of dynamic magnetic measurements

13.3.1 AC susceptibility of MNPs in Newtonian fluids

The complex (ac) susceptibility is generally described by the Debye model and it is given by

$$\chi(\omega) = \frac{\chi_0}{1 + i\omega\tau_{eff}}. \tag{13.5}$$

Here ω is the angular frequency $2\pi f$, and χ_0 is the static susceptibility given by

$$\chi_0 = \frac{\mu_0 n m^2}{3k_B T} \tag{13.6}$$

with vacuum permeability μ_0, number density of magnetic nanoparticles n, and magnetic moment of a single MNP m. Splitting the complex susceptibility into real and imaginary parts, the following equations are obtained:

$$\chi'(\omega) = \frac{\chi_0}{1 + (\omega\tau_{eff})^2} \tag{13.7}$$

$$\chi''(\omega) = \chi_0 \frac{\omega\tau_{eff}}{1 + (\omega\tau_{eff})^2} \tag{13.8}$$

Figure 13.2 shows real (χ') and imaginary part (χ'') versus the normalized frequency $\omega\tau_{\text{eff}}$. Most pronounced is the maximum in the imaginary part, which lies at $\omega\tau_{\text{eff}} = 1$. Consequently, from knowing the frequency of the maximum, the effective relaxation can be determined. In practice, there is a distribution of relaxation times $f(\tau_{\text{eff}})$, so that Equation (13.5) modifies to

$$\chi(\omega) = \chi_0 \int_0^\infty \frac{f(\tau_{eff})}{1 + i\omega\tau_{eff}} d\tau_{eff} + \chi_\infty. \tag{13.9}$$

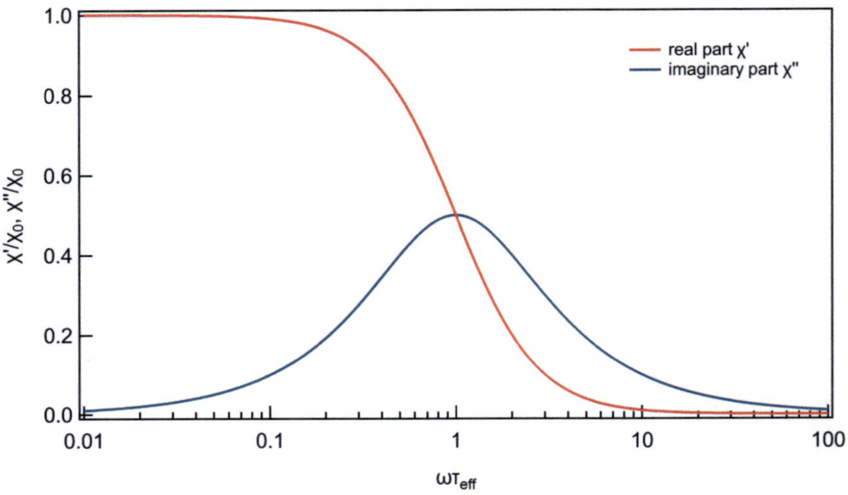

Figure 13.2: Real and imaginary parts of the ac susceptibility versus $\omega\tau_{\text{eff}}$ within the Debye model.

Here χ_∞ denotes the finite value of the real part at high frequencies, which is caused by either small superparamagnetic MNPs with relaxation times well below 1 µs and/or by intra-potential-well contributions [34–36].

Provided that the MNPs are thermally blocked, so that the effective relaxation time is given by the Brownian one, and assuming that the hydrodynamic size of the MNPs has some distribution with probability density function *f(V_h)*, Equation (13.9) modifies to

$$\chi(\omega) = \chi_0^{**} \int_0^\infty \frac{f(d_h)}{1 + i\omega\tau_B} dd_h + \chi_\infty \tag{13.10}$$

with

$$\chi_0^{**} = \frac{\mu_0 n M_s^2}{3k_B T}\left(\frac{\pi}{6}\bar{d}_c^3\right)^2.$$

Here M_s is the saturation magnetization and $\bar{d}_c$ the mean core diameter (assuming spherical cores). If the MNPs are not fully blocked, i.e., the dynamics of a certain portion is dominated by the Néel mechanism, the following, more complex model has to be applied [37]:

$$\chi(\omega) = \chi_0^{*} \int_0^\infty \int_0^\infty \frac{d_c^6 f(V_h, V_c)}{1 + i\omega\tau_{eff}} dd_c dd_h + \chi_\infty \tag{13.11}$$

with

$$\chi_0^{*} = \frac{\mu_0 n M_s^2}{3k_B T}\left(\frac{\pi}{6}\right)^2.$$

Here $f(d_c, d_h)$ is a bivariate probability density function, which accounts for correlations between core and hydrodynamic size (for single-core MNP with very thin shell, the correlation factor between core and hydrodynamic size $\rho = 1$). For fitting ACS spectra measured on MNPs, generally a bimodal lognormal distribution is assumed:

$$f(d_h, d_c) = \frac{1}{2\pi d_c d_h \sigma_c \sigma_h \sqrt{1-\rho^2}} \exp\left[-\frac{1}{2(1-\rho^2)}\left(\frac{\ln^2\left(\frac{d_c}{\mu_c}\right)}{\sigma_c^2} + \frac{\ln^2\left(\frac{d_h}{\mu h}\right)}{\sigma_h^2} - 2\rho \frac{\ln\left(\frac{d_c}{\mu_c}\right)}{\sigma_c}\frac{\ln\left(\frac{d_h}{\mu_h}\right)}{\sigma_h}\right)\right]$$

These and various other generalizations of the Debye model for the determination of MNP structure parameters are described in detail in the study by Ludwig et al. [37].

13.3.2 AC susceptibility of MNPs in non-Newtonian media

Assuming again that the MNPs are thermally blocked so that only Brownian relaxation takes place, the Debye model can be modified to non-Newtonian media. Analogously to the model by DiMarzio and Bishop for the dielectric case [38], the spectrum of the complex susceptibility can be modified by replacing the dynamic viscosity η by a complex value $\eta^* = \eta' + i\eta''$. With the complex shear modulus $G^* = G' + iG'' = i\omega\eta^*$, Equation (13.5) reads

$$\chi(\omega) = \frac{\chi_0}{1 + AG^*} + \chi_\infty . \tag{13.12}$$

Here $A = \pi d_h^3/(2k_B T)$ is a geometry factor, as introduced in the study by Roeben et al. [21]. Ignoring a distribution of hydrodynamic size, Equation (13.12)—normalized to χ_0, corrected for χ_∞ and splitted into real and imaginary parts—can easily been used to determine both storage G' and loss modulus G'' for each frequency [21]. The situation is more complicated if a distribution *f(A)* must be considered. In the studies by Gratz and Tschöpe [39] and Sriviriyakula et al. [23], the distribution *f(A)* is determined on a sample with known viscosity (generally using a Newtonian fluid) and assumed to remain unchanged when analyzing susceptibility spectra measured for non-Newtonian systems. A similar approach was proposed by Roeben et al. [21]. The proposed procedures turned out to work at least for non-Newtonian media with weak elastic contributions, such as PEG solutions [24].

The situation changes, if a certain viscoelastic model can be assumed. The simplest viscoelastic models are the Maxwell model with a viscous and an elastic term in series and the Voigt-Kelvin model with viscous and elastic terms in parallel. Basic rules for the arbitrary combination of viscous and elastic forces (torques) provide the following expressions for the complex shear modulus for the two cases:

$$G^* = \frac{i\omega\eta G}{G + i\omega\eta} \quad \text{for the Maxwell model}$$

$$G^* = \eta + i\omega G \quad \text{for the Voigt-Kelvin model}$$

Here η is the dynamic viscosity and $G = G'$ is the storage part of the complex shear modulus. For example, for MNPs with a distribution of hydrodynamic diameters $f(d_h)$

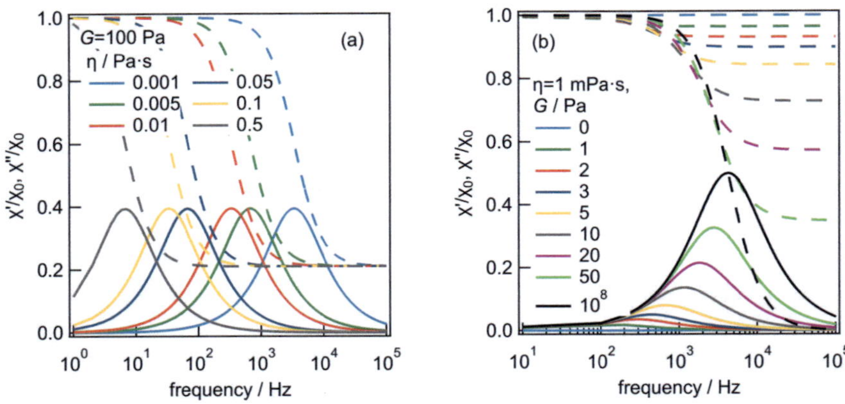

Figure 13.3: Simulated spectra for MNPs in a Maxwell-type medium. (a) Variation of viscosity and (b) variation of shear modulus (d_h = 46 nm, T = 296 K).

embedded in a Voigt-Kelvin-like medium, the normalized complex susceptibility is given by

$$\chi_n(\omega) = \frac{\chi(\omega) - \chi_\infty}{\chi_0^{**}} = \int_0^\infty \frac{f(d_h)}{1 + AG + iA\eta} dd_h . \tag{13.13}$$

The simulated susceptibility spectra for MNPs—ignoring a distribution $f(d_h)$—in a Maxwell-type medium are depicted in Figure 13.3. As for a Newtonian medium, an increase of the viscosity causes a shift of the susceptibility spectrum to lower frequencies (Figure 13.3(a)). An increase of the shear modulus G causes an increase of the real part at high frequencies, while its zero-field value is not affected, and a shift of the position of the maximum in the imaginary part to higher frequencies, while its amplitude increases (Figure 13.3(b)). The limit of a Newtonian fluid is reached for $G \to \infty$.

Figure 13.4 depicts the simulated results for a Voigt-Kelvin model. As before, an increase of the viscosity causes a shift of the spectrum to lower frequencies. However, an increase of the shear modulus G results in a decrease of the real part at low frequency and of the amplitude of the maximum of the imaginary part. In contrast to the Maxwell model, the real part of the susceptibility values at high frequencies decreases to zero, independent of shear modulus. An increase of G causes a shift of the frequency at which the maximum in the imaginary part occurs to higher frequencies and a decrease of its amplitude. Note that this behavior is qualitatively similar to the case when a static magnetic field is superimposed [40].

The same general behavior for the complex susceptibility of a Maxwell and a Voigt-Kelvin system was also derived by Tschöpe et al. [22] by solving the regarding equations of motion. Importantly, information on what type of viscoelastic model one deals with

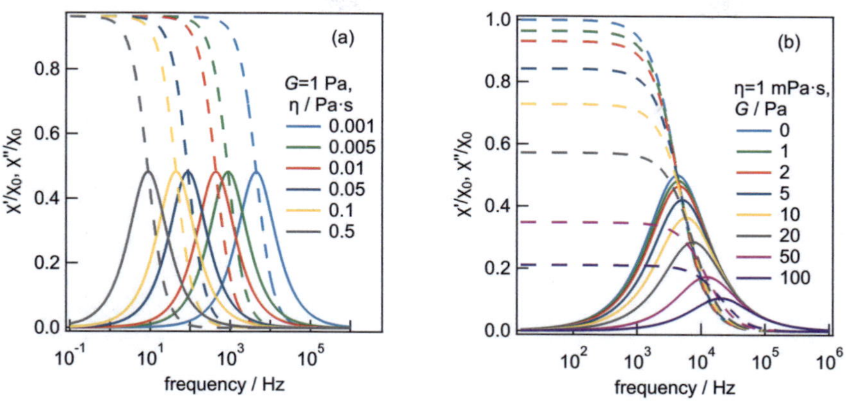

Figure 13.4: Simulated spectra for MNPs in a Voigt-Kelvin-type medium. (a) Variation of viscosity and (b) variation of shear modulus (d_h = 46 nm, T = 296 K).

can be seen from the general characteristics of the complex spectrum, which significantly differs for a Maxwell and a Voigt-Kelvin system.

Theoretical models for the complex susceptibility of thermally blocked MNPs in Maxwell- and Voigt-Kelvin-type matrices were proposed by Raikher and colleagues [41–43]. Here the authors solved the regarding equations of motion including thermal agitation. In the following, we will focus on the Voigt-Kelvin model [41].

The equation of motion is given by

$$I\ddot{\vartheta} + \zeta\dot{\vartheta} + K\vartheta = y(t) \tag{13.14}$$

with the angle between magnetic moment and applied magnetic field ϑ, the moment of inertia of the particle I, the rotational friction coefficient $\zeta = 6\eta V_h$, the linear elastic restoring parameter $K = 6GV_h$ and the stochastic driving torque due to thermal energy $y(t)$. The magnetic torque is assumed to be small. For the dynamic susceptibility, the following expressions were derived:

$$\chi_\alpha(\omega) = \chi_{0,\alpha}\left(1 + i\omega\int_0^\infty d\tau e^{i\omega\tau}G_\alpha(\tau)\right). \tag{13.15}$$

with

$$\chi_{0,\parallel} = \frac{nm^2}{k_BT}\exp\left(-\frac{k_BT}{K}\right)\left[\cosh\left(\frac{k_BT}{K}\right) - 1\right] \text{ and } \chi_{0,\perp} = \frac{nm^2}{k_BT}\exp\left(-\frac{k_BT}{K}\right)\sinh\left(\frac{k_BT}{K}\right)$$

as well as

$$G_\parallel(\tau) = \frac{\left(\cosh\left(\frac{k_BT}{K}\exp\left(-\frac{t}{\tau_K}\right)\right) - 1\right)\exp(i\omega t)}{\cosh\left(\frac{k_BT}{K}\right) - 1} \text{ and } G_\perp(\tau) = \frac{\sinh\left(\frac{k_BT}{K}\exp\left(-\frac{t}{\tau_K}\right)\right)\exp(i\omega t)}{\sinh\left(\frac{k_BT}{K}\right)}.$$

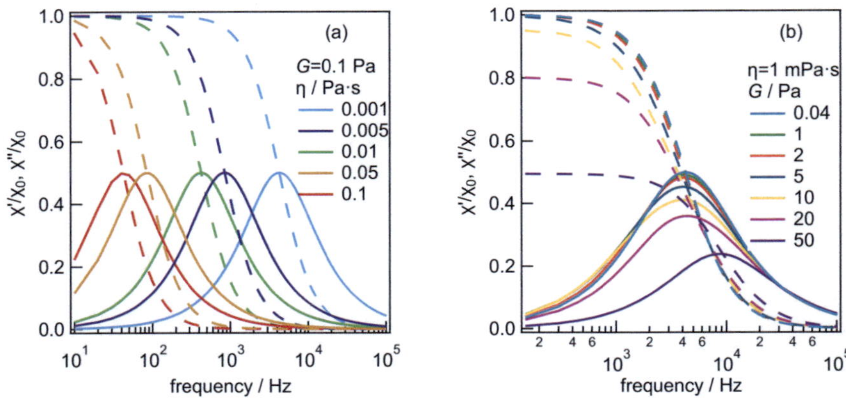

Figure 13.5: Spectra for MNPs in a Voigt-Kelvin-type medium calculated with model by Raikher et al. (a) Variation of viscosity and (b) variation of shear modulus (d_h = 46 nm, T = 296 K).

The total susceptibility is then given by

$$\chi_{tot}(\omega) = \frac{1}{3}\left(\chi_{\parallel} + 2\chi_{\perp}\right). \tag{13.16}$$

The symbol α in Equations (13.15) denotes the orientations parallel (||) or perpendicular ($\perp$) to the excitation field. The time constant $\tau_K = \zeta/K$. Note that the equation for the Brownian relaxation time in the study by Raikher et al. [41] differs from the expression in Equation (13.1) by a factor of 2 (this also results in a shift of the frequency of the maximum in the imaginary part when solving Equations (13.15) and (13.16) in the limit of vanishing elasticity), so that we tentatively replaced k_BT in Equation (13.15) by $2k_BT$ [25, 44].

Lateron, Rusakov et al. [45] extended their theoretical work, which for the case described above is limited to the case of planar (1D) rotation, to complete angular space (2D). The equations for the complex susceptibility they obtained qualitatively differ, however, from the ones derived for the 1D case (Equations (13.15) and (13.16)) and from the experimental data, which will be described in Section 13.4.2. Therefore, spectra measured on blocked MNPs in a Voigt-Kelvin-type matrix were analyzed with Equations (13.15) and (13.16).

Figure 13.5 displays the ac susceptibility spectra numerically calculated with Equations (13.15) and (13.16) for the same set of parameters as for Figures 13.3 and 13.4. Qualitatively, the same behavior as in Figure 13.4 is obtained. In comparison with the modified Debye model (Equation (13.13)), the numerical model by Raikher et al. requires considerably higher computational effort, especially when extending it by parameter distributions and implementing it in fitting routines.

13.3.3 Magnetorelaxometry of MNPs in Newtonian fluids

For a sample consisting of identical MNPs, an exponential decay of the relaxation signal is expected:

$$B(t) \sim \exp\left(-\frac{t}{\tau_{eff}}\right) \tag{13.17}$$

If there are distributions of core and hydrodynamic size, generally the cluster magnetic moment superposition model (CMSM)—as originally proposed by Eberbeck et al. [6]—is applied. Here the decay of the net magnetic moment of the sample is given by [6, 11]

$$B(t) = \Phi M_s \int_0^\infty f(V_h) \int_0^\infty f(V_c) V_c L(V_c) \left[1 - \exp\left(-\frac{t_{mag}}{\tau_{eff,H}}\right)\right] \exp\left(-\frac{t}{\tau_{eff}}\right) dV_c dV_h \tag{13.18}$$

with a system factor Φ and the Langevin function $L(V_c)$. The Langevin function is given by $L(\xi) = \coth(\xi) - 1/\xi$ with the Langevin parameter $\xi = mB/(k_B T)$. Note that the effective relaxation time $t_{eff,\,H}$ comprises the Brownian and Néel relaxation times in a static magnetic field. If the MNPs are thermally blocked, t_{eff} and $t_{eff,\,H}$ can be replaced by t_B and $t_{B,\,H}$, respectively, and integration has to be carried out over V_h only.

13.3.4 Magnetorelaxometry of MNPs in viscoelastic matrices

Currently, there are no models, which describe the complete magnetorelaxation signal for a given type of viscoelasticity. In the studies by Raikher and coworkers [41, 45], an equation for the effective relaxation time of thermally blocked MNPs in a Voigt-Kelvin-type matrix is provided:

$$\tau_{eff,VK} = \tau_B\left(1 + \frac{K}{2k_B T}\right) \tag{13.19}$$

Here t_B is the Brownian relaxation time in a purely viscous medium (Equation (13.1)), and $K = 6GV_h$ is the elastic restoring parameter. Note that again a difference of a factor of 2 is found comparing the studies by Raikher et al. [41] and Rusakov and Raikher [45]. Equation (13.19) already includes the correction.

In order to analyze experimental MRX curves measured on MNPs in a Voigt-Kelvin-type matrix, Equation (13.18) with expression (19) for the effective relaxation time can be applied.

13.3.5 Magnetic field dependence of Brownian relaxation time

Equations (13.1) and (13.2) provide the zero-field expressions for the Brownian and Néel relaxation times. In standard ACS measurements, where ac magnetic fields with amplitudes of well below 1 mT are applied, i.e., the Langevin parameter $\xi = mB/(k_BT) << 1$, they can reliably be used in the analysis. However, as soon as ξ becomes larger, significant changes of the relaxation time constants occur.

The dependence of the Brownian relaxation time of thermally blocked MNPs in large ac magnetic fields was theoretically studied by Yoshida and Enpuku [46] by solving the Fokker-Planck equation. Based on the simulated ACS spectra, they derived a set of empirical equations, which can be used to fit experimental data. Here the Brownian relaxation time in dependence of Langevin parameter ξ is given by [11, 40]

$$\tau_{B,H} = \frac{\tau_B}{\sqrt{1 + 0.126\xi^{1.72}}}, \tag{13.20}$$

i.e., the Brownian relaxation time decreases with increasing ac field amplitude. Similar expressions were also derived by Gratz and Tschöpe [47] as well as by Fock et al. [48].

The Brownian relaxation time for the case of a superimposed static magnetic field—either parallel or perpendicular to the ac probing field—were theoretically derived by Martsenyuk et al. [8] and are given by

$$\tau_{\text{par}} = \frac{d \ln L(\xi)}{d \ln \xi}\tau_{\text{B}} \text{ and } \tau_{\text{perp}} = \frac{2L(\xi)}{\xi - L(\xi)}\tau_{\text{B}}. \tag{13.21}$$

The situation is more difficult for the Néel relaxation since one has to account for the generally random distribution of easy axes. The Néel relaxation of MNPs with their easy axes parallel to an applied static magnetic field is given by [2]

$$\tau_{N,H} = \frac{\sqrt{\pi}\tau_{N0}}{\sigma^{3/2}(1-h^2)}\left[(1+h)\exp\left(-\sigma(1+h)^2\right) + (1-h)\exp\left(-\sigma(1-h)^2\right)\right]^{-1} \tag{13.22}$$

with the normalized magnetic field $h = B/B_{\text{k}}$, $B_{\text{k}} = 2K/M_{\text{s}}$ and $\sigma = KV_{\text{c}}/(k_{\text{B}}T)$. There is no expression for the Néel relaxation time of an ensemble of randomly oriented MNPs in a large ac magnetic field.

13.4 Experimental results

13.4.1 ACS and MRX measurements on MNPs in Newtonian media

ACS and MRX measurements were performed on a series of samples with different viscosities. Here water-glycerol mixtures were prepared. First, SHP-25 from Ocean Nanotechnology was studied. These MNPs are single core, thermally blocked nanoparticles consisting of magnetite and having a nominal (geometric) core diameter of 25 nm with a narrow size distribution [40]. Figure 13.6 depicts the spectra of the imaginary part of the complex susceptibility for samples with glycerol contents ranging from 0 to 86.8 wt%. One discerns a pronounced maximum, which shifts with increasing viscosity to lower frequencies, as expected for Brownian-dominated MNPs. In contrast to

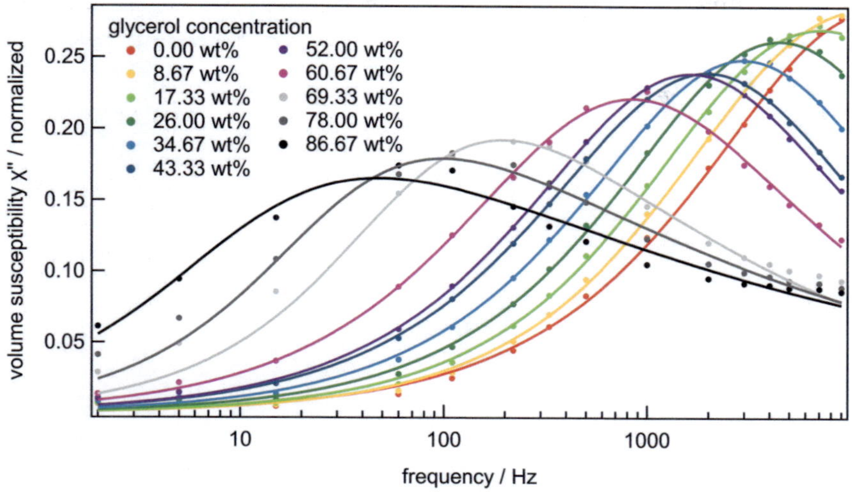

Figure 13.6: Imaginary part versus frequency for SHP-25 MNPs suspended in different water-glycerol mixtures. Numbers in inset give glycerol concentration. Symbols are measured data points, lines are fits with phenomenological Havriliak-Negami model (Equation (13.23)).

the theoretical expectations (cf. Figures 13.4(a) and 13.5(a)), its amplitude decreases, which is compensated by an increasing width of the imaginary part peak, which is related to an increase of the distribution of Brownian relaxation times.

In order to precisely determine the frequency f_{ch} at which the imaginary part peaks, data points were fitted with the phenomenological Havriliak-Negami model. Here the complex susceptibility is given by

$$\chi = \frac{\chi_{amp}}{\left(1 + (i\omega\tau)^{1-\alpha}\right)^{\beta}} + \chi_{\infty} \tag{13.23}$$

where χ_{∞} is the real-valued susceptibility at high frequencies, and χ_{amp} is the amplitude of the frequency-dependent part. The parameter α accounts for the width of the spectrum, while the parameter β reflects its asymmetry. For $\alpha = 0$ and $\beta = 1$, the Debye model is reconstituted.

The values for the dynamic viscosity were estimated from the measured imaginary part applying

$$\eta = \frac{k_B T}{6\pi f_{ch} V_h}, \tag{13.24}$$

which is obtained from Equation (13.1) and the condition $2\pi f_{\text{ch}}\tau_B = 1$.

To determine the viscosity values from the MRX curves depicted in Figure 13.7, Equation (13.18) was applied. The parameters of core diameter distribution and of the anisotropy energy density were estimated from measurements on an immobilized (by freeze-drying in a mannitol matrix) reference sample. Knowing these parameters, the

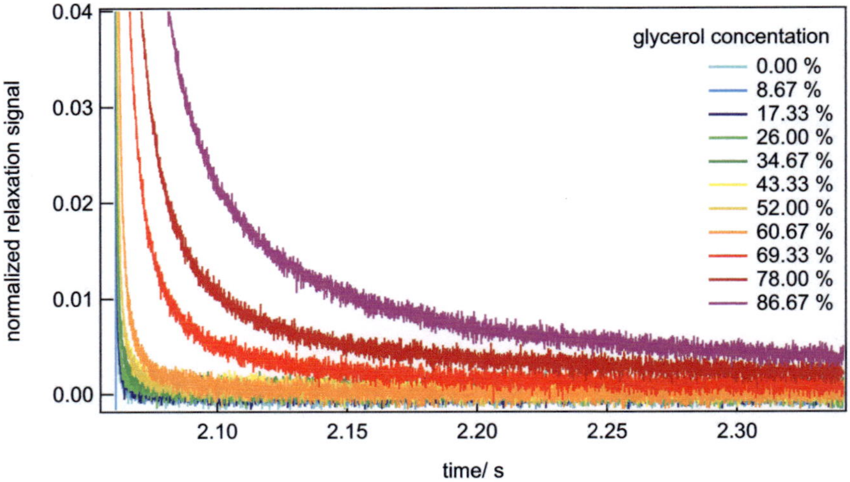

Figure 13.7: Normalized relaxation curves measured on SHP-25 MNPs suspended in different water-glycerol mixtures.

parameters of the distribution of hydrodynamic diameter were obtained from the MRX curve measured on an aqueous suspension, i.e., with known viscosity. Then keeping these parameters constant, the viscosity is the only free parameter for fitting the relaxation curves for SHP-25 MNPs in water-glycerol mixtures.

The viscosity values, estimated from ACS and MRX data, are displayed in Figure 13.8 versus the glycerol concentration. Good agreement with the theoretical values calculated with the equation by Cheng [49] is found.

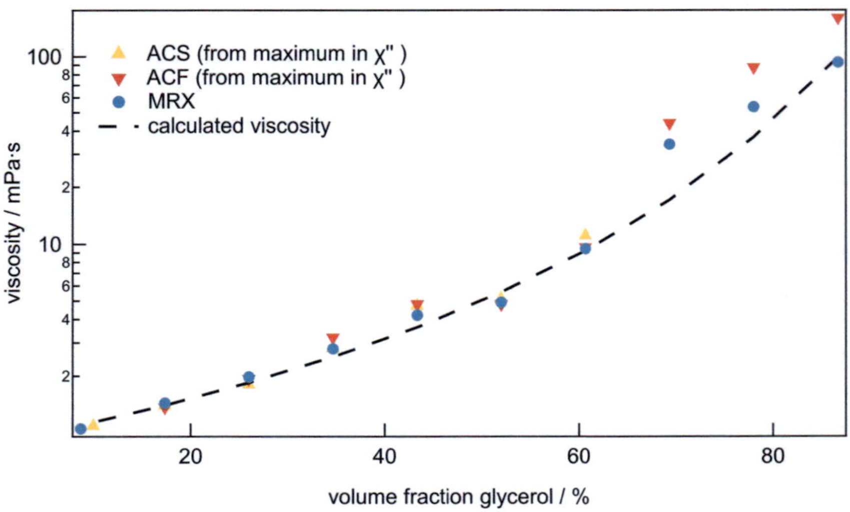

Figure 13.8: Viscosity values determined from ACS, ACF (refers to fluxgate-based ACS setup) and MRX measurements on SHP-25 versus concentration of glycerol. Dashed line displays theoretical viscosity calculated using the equation by Cheng [49].

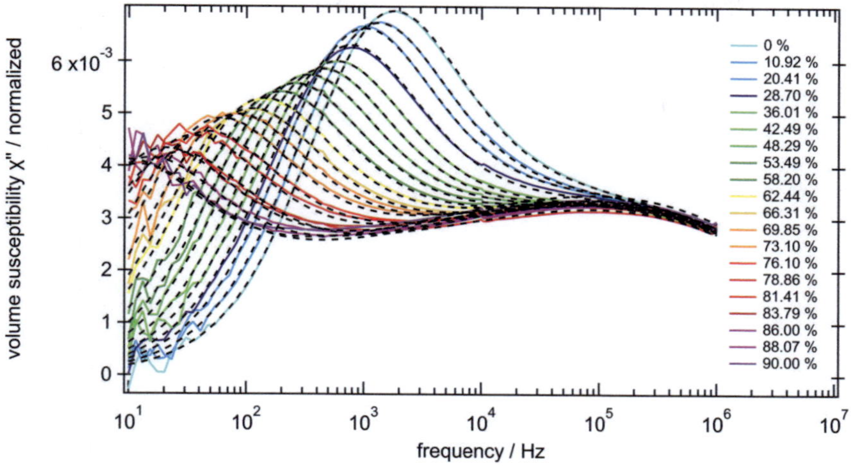

Figure 13.9: Imaginary part versus frequency for FeraSpin XL MNPs suspended in different water-glycerol mixtures. Numbers in inset give glycerol concentration. Symbols are measured data points, lines are fits with Equation (13.11).

In order to test, how reliably dynamic viscosities can be determined from ACS measurements on MNP systems, which exhibit Brownian and Néel contributions, measurements on a viscosity series of FeraSpin XL from nanoPET Pharma GmbH were performed. FeraSpin XL is a size fraction of FeraSpin R with a mean hydrodynamic diameter of about 60 nm. FeraSpin R is a multi-core particle system comprising of densely packed iron oxide cores with sizes of 5–7 nm [50]. The imaginary parts of the measured ACS spectra are shown in Figure 13.9. The aqueous sample measurement displays a maximum at about 1.8 kHz and a shoulder around 100 kHz. With increasing viscosity, the frequency of the low-frequency maximum shifts to lower frequencies, indicating that it is caused by Brownian-dominated MNPs while the wide high-frequency maximum remains unchanged, i.e., it is related to Néel relaxation.

To fit the measured spectra, Equation (13.11) is applied. For simplicity, the correlation coefficient ρ is set to zero (i.e., we ignore a correlation between core and hydrodynamic size). As for the analysis of the MRX curves in Figure 13.7, core parameters are independently estimated from relaxation curves of the immobilized reference samples, while the parameters of the hydrodynamic size distribution were obtained from the analysis of the ACS spectrum measured on the aqueous suspension of FeraSpin XL. Dashed lines in Figure 13.9 display the fitted spectra. Note that—for all samples except the aqueous one—viscosity is the only free parameter. In contrast to the viscosity series measured on SHP-25, the standard deviation of the lognormal distribution σ_h of hydrodynamic diameters (and Brownian relaxation times) remained nearly constant ($\sigma_h \approx 0.28$). Consequently, the observed decay in the amplitude of the maximum is caused by a gradual decrease of the Brownian-dominated MNP portion with increasing

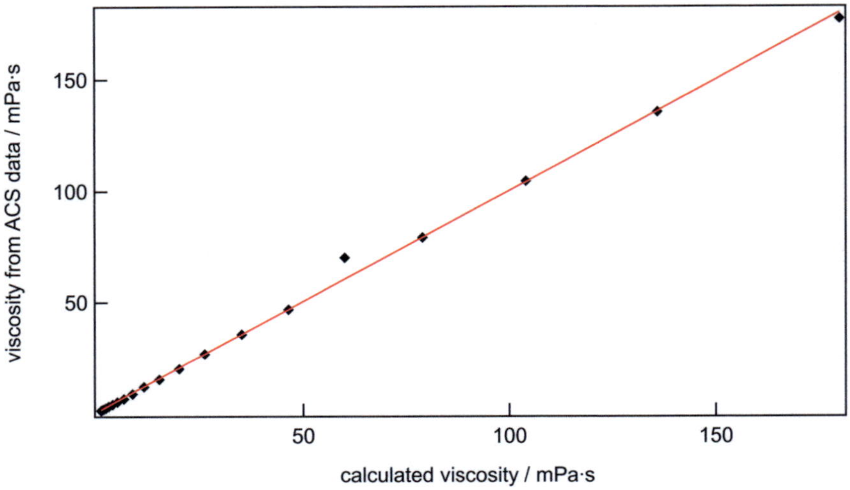

Figure 13.10: Dynamic viscosities determined from ACS spectra on FeraSpin XL viscosity series versus calculated values.

viscosity (transition from Brownian to Néel relaxation). As Figure 13.10 shows, excellent agreement between experimental and theoretical values is found again.

13.4.2 Investigation of gelation dynamic of aqueous gelatin suspensions

As Voigt-Kelvin-type matrix system, aqueous gelatin solutions are chosen. As demonstrated by Tschöpe et al. [22] using Ni nanorods as magnetic markers, rheological parameters of aqueous gelatin solutions from dynamic opto-magnetic measurements can well be estimated on the basis of the Voigt-Kelvin model. Here single-core $CoFe_2O_4$ synthesized at the University of Cologne [21] are applied as markers. They are surrounded by a PAA shell, have mean core diameters of 15 nm and a hydrodynamic diameter—when suspended in DI water—of 18 nm. Due to the comparably high anisotropy energy density, $CoFe_2O_4$ nanoparticles of this size are well thermally blocked.

Similarly to the procedure executed by Tschöpe et al. [22], $CoFe_2O_4$ MNPs were suspended in aqueous gelatin solutions with different gelatin contents (2.5, 5, 7.5 and 10 wt%). These suspensions were first heated up to 40 °C, thus being in the sol state, and the hydrodynamic diameter was determined from the position of the maximum in the ACS imaginary part and the known viscosity at 40 °C. The viscosity values at 40 °C were independently determined using an Anton Paar SVM 3000 Stabinger viscosimeter. Knowing the dynamic viscosity at this temperature, the ACS spectra measured at 40 °C were fitted with the Debye model extended by inserting a lognormal distribution of hydrodynamic diameters. For the 2.5 wt% sample, a mean hydrodynamic diameter of

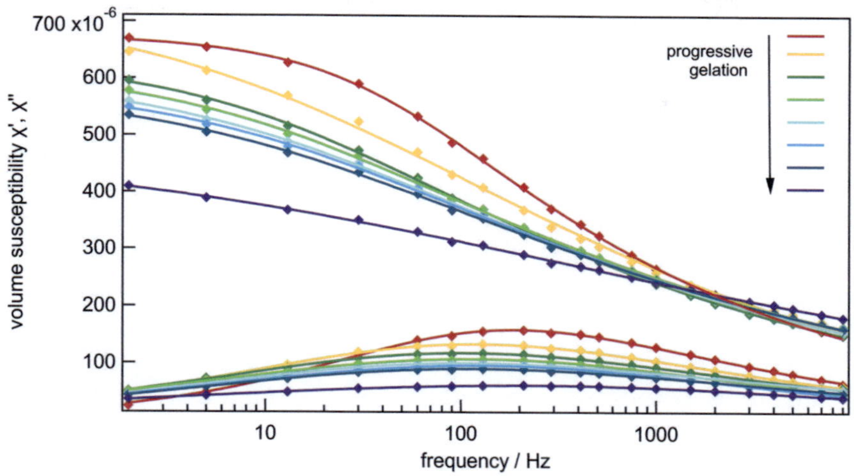

Figure 13.11: Real and imaginary parts of ACS spectra recorded at different stages of the gelation process.

37 nm was determined; for the 5 wt% sample, a value of 46 nm was found. The increase in hydrodynamic size compared to the value of the aqueous sample at room temperature is attributed to an adsorbent gelatin layer. Repeated heating of the samples to 40 °C after gelation always resulted in the same hydrodynamic size, so that we assume that the thickness of the absorbent gelatin layer does not change with time.

Figure 13.11 depicts real and imaginary part of the sample with 5 wt% gelatin, measured at different stages of the gelation process. Apparently, the magnitude of the real part at low frequencies continuously decreases with increasing gelation time while it approaches zero at high frequencies, as expected for a Voigt-Kelvin-type system.

Figure 13.12(a) and (b) shows the ACS imaginary parts measured on the (a) 2.5 and (b) 5 wt% samples. For the 2.5 wt% gelatin sample, a gradual shift of the peak frequency

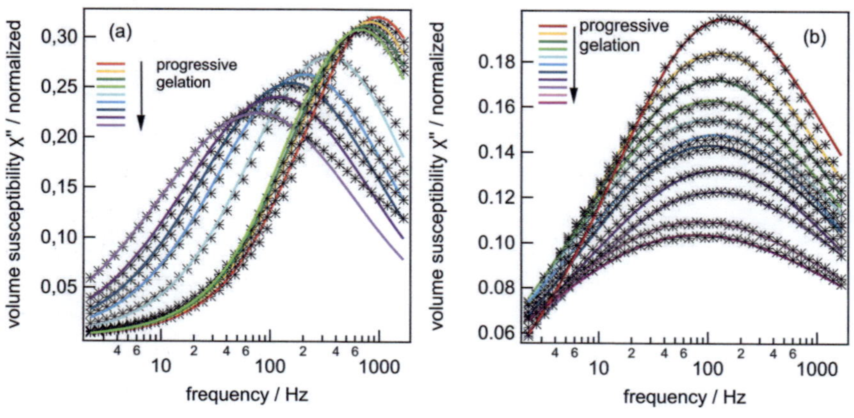

Figure 13.12: Evolution of ACS imaginary part of (a) sample with 2.5 wt% gelation and (b) with 5 wt% gelatin. Symbols show data points, lines are fits with Equations (13.15) and (13.16).

to lower frequencies and of its magnitude is observed. At the same time, the width of the spectrum increases with increasing gelation time. In contrast, the spectrum of the imaginary part measured on the 5 wt% gelatin sample first shows a shift of the peak frequency to lower frequencies but after a certain time, it reverses and increases again. But similarly to the 2.5 wt% sample case, the magnitude of the maximum decreases and the width increases with increasing gelation time.

The behavior observed for the 5 wt% sample is qualitatively similar to that observed by Tschöpe et al. [22] applying an oscillating magnetic field to Ni nanorods and detecting their response by a magneto-optic technique.

Reminding the fundamentals of a Voigt-Kelvin model (Section 13.3.2), the observed behavior is very clear: An increase of the (local) dynamic viscosity of the medium causes a shift of the imaginary part to lower frequencies. An increase of the shear modulus results in a decrease of the ACS magnitude and a shift of the position of the χ" maximum to higher frequencies. The latter effect is very pronounced for the sample with 5 wt% gelatin. But at the same time, a significant increase of the width of the spectrum is discernable. In order to extract values for viscosity η and shear modulus G, the measured spectra of the imaginary part were fitted with Equations (13.15) and (13.16). Since they do not consider distributions of parameters, they were generalized by inserting lognormal distributions of hydrodynamic diameter $f(d_h)$ and (local) dynamic viscosity $f(\eta)$. Obtained parameters for η and G are shown in Figure 13.13 as a function of gelation time.

Obviously, the same general trend is found for both samples. The viscosity continuously increases with rising gelation time, while the shear modulus first increases, followed by a plateau and then starts rising again. The main difference between both samples lies in the different magnitudes of the (nano)rheological parameters. While viscosity and shear modulus of the 2.5 wt% sample at 1000 min amount to about 10 mPa s and 8 Pa, respectively, they are 30 mPa s and 9 Pa for the 5 wt

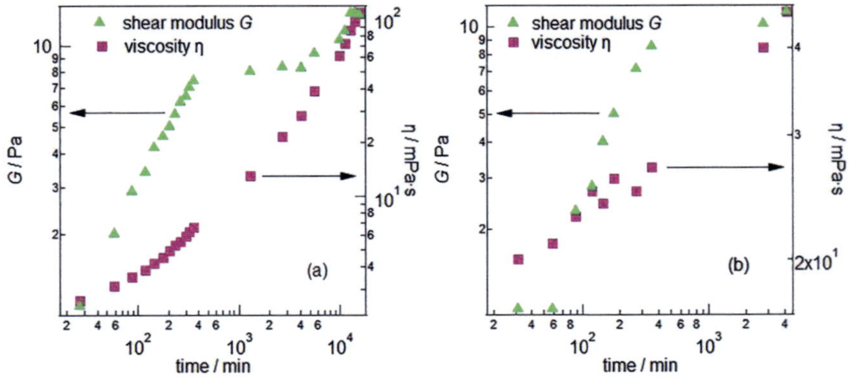

Figure 13.13: Viscosity η and shear modulus G as a function of gelation time for the samples with (a) 2.5 and (b) 5 wt% gelatin.

% sample. The different ratio of viscosity and shear modulus causes that the elastic term starts dominating for the 5 wt% gelatin sample at high gelation times.

Similar findings were obtained when analyzing the measured ACS spectra with the modified Debye model (Equation (13.13) extended by a distribution of local viscosities). Figure 13.14 displays the fitting results obtained with the numerical model by Raikher et al. [41] and the modified Debye model. The viscosities estimated by both approaches agree very well. Regarding shear modulus, its evolution with gelation time is qualitatively the same but the absolute values differ by a factor of about two. At this stage, it remains open what the reason is. As pointed out in Section 13.3.2, k_BT in the original equations given in the paper by Raikher et al. [41] was replaced by $2k_BT$ so that—in the limiting case of vanishing elasticity—the well-known expressions for the Brownian relaxation time and the standard Debye model are re-established.

According to Normand et al. [51], there are four phases in the gelation kinetics of gelatin. Comparing Figure 13.13 with their findings, the time interval up to 300 min can be attributed to phase 2, which is characterized by the gel formation and a rapid increase of G and η. The time up to about 4000 min may be related to phase 3, which—according to Normand et al.—is characterized by the extension of existing cross-links in the network rather than the formation of new ones. The rise of G at longer times—as discernable in Figures 13.13 and 13.14—was also observed by the authors applying a stress-controlled rheometer technique.

The ACS spectra measured on samples with 7.5 and 10 wt% gelatin could not be analyzed since the characteristic features were outside the accessible frequency window.

The same samples were also investigated by magnetorelaxometry. The aqueous gelatin solution samples were heated to 40 °C for about 60 min, and after quickly

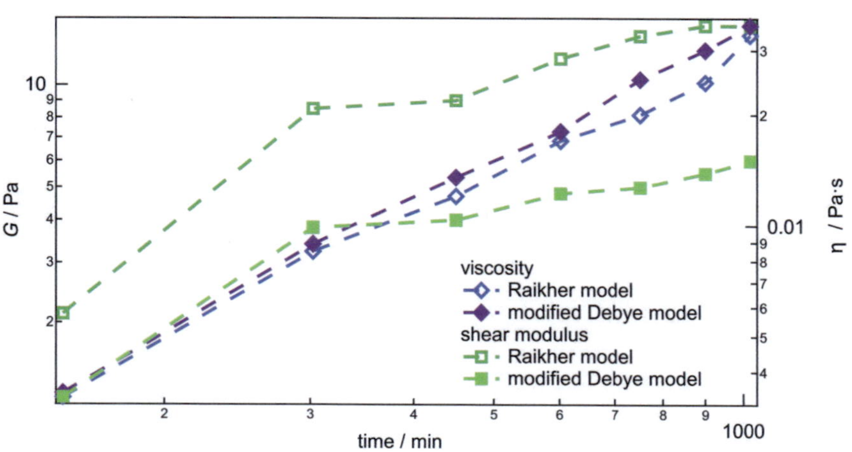

Figure 13.14: Comparison of temporal evolution of viscosity and shear modulus values determined from measured ACS spectra with numerical model by Raikher et al. [41] and modified Debye model.

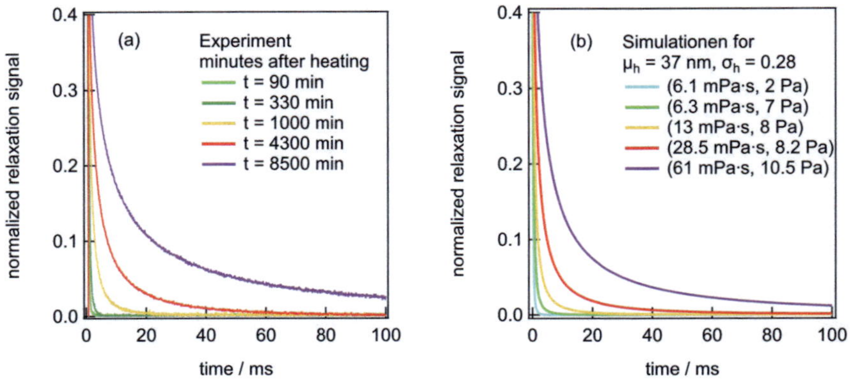

Figure 13.15: Normalized (a) measured and (b) simulated MRX curves of the samples with 2.5 wt% gelatin. The numbers in brackets in the legend of (b) denote the applied values for viscosity and shear modulus.

cooling them down to 23 °C, successive MRX measurements were performed. The measured MRX curves, measured on the 2.5 wt% gelatin sample and normalized to the signal before the magnetic field pulse is switched off, are depicted in Figure 13.15(a). With increasing gelation time, the decay of the relaxation signal monotonously slows down. An independent estimation of viscosity η and shear modulus G by fitting the measured MRX curves with Equations (13.18) and (13.19) is not possible. Therefore, Figure 13.15(b) shows the normalized MRX curves simulated with Equations (13.18) and (13.19) using the parameters obtained from the analysis of the corresponding ACS spectra. A reasonable agreement is found. A qualitatively similar behavior was observed for the 5 wt% gelatin sample. The behavior measured for the 7.5 and 10 wt% gelatin samples is more complex. The not normalized MRX curves for the 7.5 wt% gelatin sample are depicted in Figure 13.16(a). The change of the curves with increasing gelation is no longer monotonous and even a crossing can be observed. As for the ACS case, this may be caused by the different effect of viscosity and shear modulus changes on the MNP dynamics. Figure 13.16(b) shows the MRX curves measured on the sample with 10 wt% gelatins at different times after reaching 23 °C. Apparently, no significant differences between curves and the reference curve measured on a freeze-dried sample of $CoFe_2O_4$ nanoparticles are discernable, meaning that the dynamics of this sample is dominated by the Néel mechanism.

13.4.3 Investigation of magnetic field dependence of Brownian relaxation time

All ACS measurements described in Section 13.4.1 and 13.4.2 were carried out at small ac field amplitudes (200 µT, i.e., $\xi << 1$). To extract nanorheological parameters from

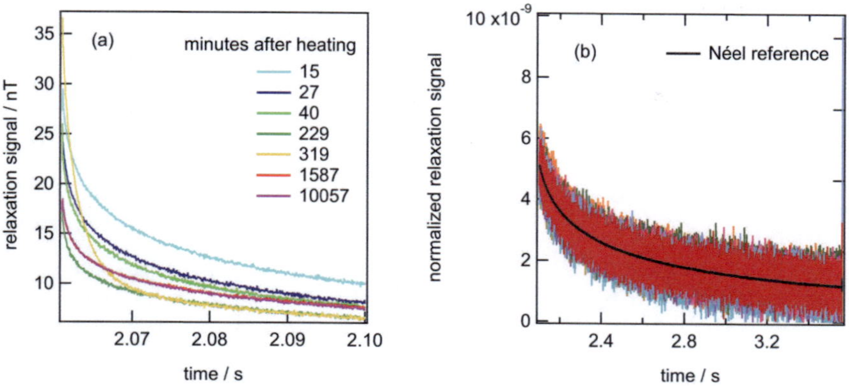

Figure 13.16: MRX curves measured on (a) 7.5 and (b) 10 wt% gelatin samples at different times of the gelation process.

measurements at higher ac field amplitudes or large-magnitude superimposed dc fields, the effect of magnetic field amplitude on the Brownian relaxation time has to be separated. The magnetic field-dependence of the Brownian relaxation time was investigated by measuring the ACS spectra with the setup originally developed for measuring the dynamics in a rotating magnetic field. Experimental details were briefly summarized in Section 13.2.

Experimental studies of the dependence of Brownian and Néel relaxation times of iron-oxide single-core MNPs (SHP-20 and SHP-25 from Ocean Nanotech) on ac field amplitude and on a superimposed dc magnetic field—either parallel or perpendicular to the ac probing field—were presented in the study by Dieckhoff et al. [40]. It was demonstrated that Equations (13.20) and (13.21) can well be applied. In order to extend the range of the Langevin parameters ξ, ACS measurements were also performed at Ni nanorods [52]. Combining their magnetic moment of the order of 10^{-17} Am^2 and the maximum field amplitude of 9 mT, which can be applied with the given setup, Langevin parameters up to 80 are covered. For comparison, the spherical $CoFe_2O_4$ nanoparticles used for the nanorheological studies have magnetic moments of the order of $7{\cdot}10^{-19}$ Am^2, providing a Langevin parameter at room temperature and 9 mT of approximately 1.5. Caused by shape anisotropy, these Ni nanorods with lengths of about 270 nm are strongly blocked so that dynamics are solely determined by the Brownian mechanism.

Figure 13.17(a) depicts the spectra of the imaginary part measured on an aqueous suspensions of Ni nanorods for different amplitudes of the ac magnetic field. To reduce dipolar interactions between nanorods, the nanorods volume fraction amounted to $5{\cdot}10^{-5}$. With increasing ac field amplitude, the position of the maximum shifts to higher frequencies and at the same its magnitude decreases. With the assumption that the maximum occurs at $\omega\tau_B = 1$, Equation (13.20) can be applied to fit the dependence of the

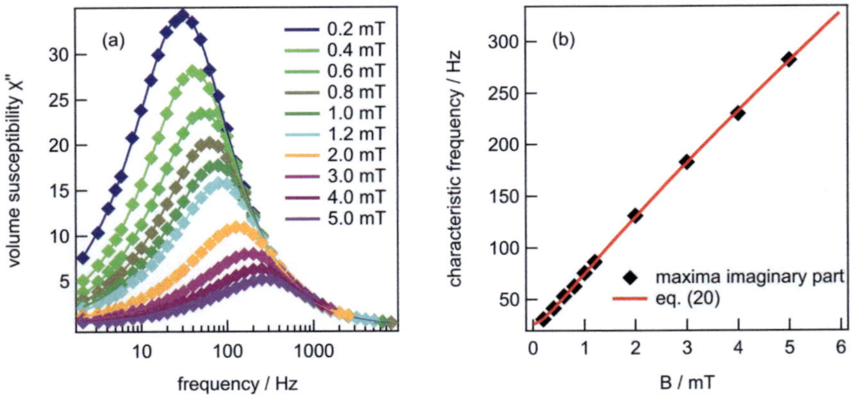

Figure 13.17: (a) Imaginary part versus frequency measured on Ni nanorod sample for different amplitudes of the ac magnetic field. (b) Extracted characteristic frequency versus field amplitude. Solid line shows fit with Equation (13.20) and assuming $\omega_{ch}\tau_B = 1$.

characteristic frequency f_{ch} on ac magnetic field amplitude. The best fit to the experimental data is shown in Figure 13.17(b). As can be seen, Equation (13.20) can be applied for the ac magnetic field amplitude dependence of the Brownian relaxation time over a large range of Langevin parameters.

In order to determine the effect of an oscillating magnetic field—as applied by Tschöpe et al. [22, 39] for their nanorheological studies using magneto-optical detection—on the Brownian relaxation time, we also performed ACS measurements in a small ac magnetic field and a perpendicular static magnetic field with magnitudes up to 9 mT on the aqueous Ni nanorod suspension. The characteristic frequency versus static magnetic field magnitude is shown in Figure 13.18. Note that fitting the characteristic frequency (or Brownian relaxation time) with Equation (13.21) with the two parameters ξ and t_B is only unique at low ξ values. Therefore, the dashed line in Figure 13.18 was calculated with the parameters from the fit of the data in Figure 13.17(b). The good agreement between experimental results and model indicates that Equation (13.21) can be applied to determine the effect of a perpendicularly applied static magnetic field on the Brownian relaxation time. This is important for the determination of rheological parameters from ACS measurements since they are included only in the zero-field Brownian relaxation time t_B.

13.5 Discussion

The use of thermally blocked MNPs as local probes for nanorheological investigations is straight forward for viscous media. The Brownian relaxation time τ_B, which is proportional to the dynamic viscosity, can be determined by any dynamic magnetic

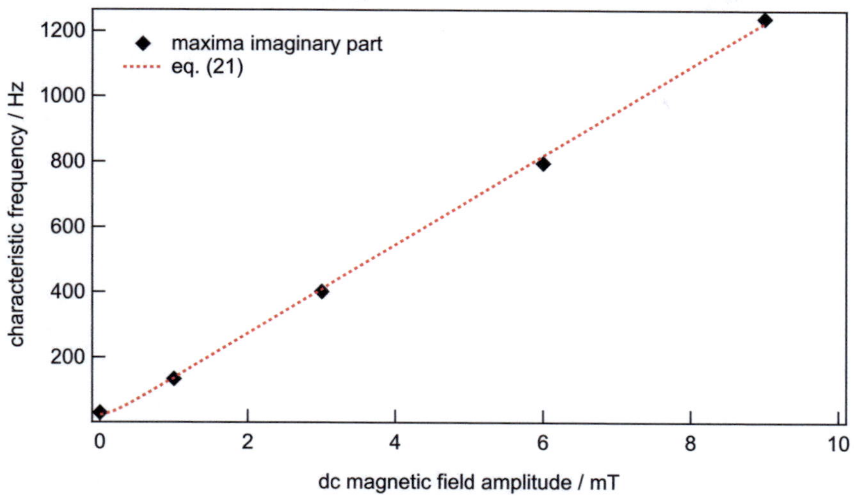

Figure 13.18: Dependence of the characteristic frequency on the magnitude of the perpendicular static magnetic field magnitude. Dashed line shows dependence calculated with Equation (13.21) using the parameters for ξ and τ_B from Figure 13.17(b).

measurement method. Here we focused on measurements of the complex (ac) susceptibility and the magnetorelaxation signal. The advantage of ACS is that a rather large range of relaxation times can be covered, while MRX measurements are generally restricted to relaxation times above a few 100 μs.

The situation is more complex for viscoelastic media. As we have demonstrated for the cases of a Maxwell and a Voigt-Kelvin model as the two limiting cases, viscosity and shear modulus have a very different effect on the ac susceptibility spectrum. A qualitative analysis of the ACS spectrum already gives some indication on the dominating viscoelastic model. The ACS spectrum of blocked MNPs in any viscoelastic matrix can basically be described by Equation (13.12). In contrast to numerical models for a Maxwell [42] or Voigt-Kelvin system [41], the computational effort is much smaller. The analysis procedure is straight forward for systems without parameter distributions: Using measured values of real and imaginary parts, the complex shear modulus or complex viscosity can be calculated as a function of frequency. If distributions (e.g., of the hydrodynamic size) have to be considered, the equation system is still solvable, proposed that the distribution—as generally determined from a fit of the ACS spectrum measured on a sample with MNPs in a Newtonian fluid with known viscosity—does not change. For the temporal evolution of the gelation process, as studied in this contribution, this is certainly a coarse assumption.

Comparing ACS and MRX for nanorheological studies applying MNPs as local probes, ACS has the advantage that there are theoretical models for Maxwell, Voigt-Kelvin and Jeffrey systems [41–43] describing the complex susceptibility spectrum.

As we have described in Section 13.3.2, the Debye model can easily be modified for non-Newtonian media by inserting a complex viscosity/shear modulus. In contrast, for the analysis of MRX data, there are currently only expressions for the effective relaxation time, which depend on both viscosity and shear modulus, i.e., their independent determination from a single MRX curve is not possible. For comparison, in an ACS measurement, real and imaginary parts of the complex susceptibility as a function of frequency can be used for the independent extraction of rheological parameters.

In conventional rheology, the relationship between stresses and strains is measured in order to determine the complex shear modulus. As a result, frequency-dependent values for storage G' and loss modulus G'' are obtained. In Section 13.4.2, we have demonstrated (similar results were obtained by Tschöpe et al. [22]) that experimental spectra could only be fitted by implementing a distribution of (local) dynamic viscosity. The main difference between conventional rheology and nanorheology is that MNPs are used in the latter one as probes for the local (nano)environment. Ignoring interactions between MNPs, the measured ACS spectrum is the superposition of the contributions from the individual nanoparticles inside the sample. Since—in a complex matrix as an aqueous gelatin solution—each MNP senses its own local environment, measured spectra cannot be described by a single value of complex viscosity (or shear modulus). Injecting MNPs as local probes into a matrix, the MNPs' response crucially depends on the way they are embedded in the matrix and on their size compared to the characteristic length scales of the matrix material [24, 53]. For example, if the nanoprobes are larger than the mesh size of a gel network, their response is expected to reflect the macrorheological properties of the matrix [22]. If they are "swimming" inside a mesh, they will sense the viscosity of the liquid. Thus, dynamic magnetic measurement techniques, such as ACS, provide more information than just the rheological parameters of the matrix on the nanoscale. Also, their dynamics strongly depends on whether they are functionalized so that they can be directly linked to the surrounding matrix [54, 55].

While magnetoviscous effects—caused by dipolar interactions and thus structure formation—were observed in conventional rheology [28], we did not observe any indications on magnetoviscous effects in dynamic magnetic measurements. Importantly, one should keep in mind that the magnetoviscous effect as theoretically proposed by Shliomis [27] does not occur in a dynamic magnetic measurement with stationary sample and that the Brownian relaxation time itself depends on magnetic field strength. For all samples investigated so far, the change of the ac susceptibility spectrum with magnetic field amplitude could well be described by the field-dependent Brownian relaxation time. Performing ac susceptibility measurements on samples with significantly higher MNP concentrations—as, e.g., done by Fischer et al. [56], the additional problem arises that dipolar interactions directly causes a shift of the ACS spectrum toward lower frequency [57, 58], thus pretending a magnetoviscous effect.

13.6 Conclusions

Dynamic magnetic measurement techniques such as ACS and MRX in combination with thermally blocked MNPs are a powerful tool to study the interaction between MNPs and the surrounding matrix. Applying appropriate models, such measurements can be used to determine rheological parameters on the nanoscale. However, one should keep in mind that these parameters, extracted from the MNP dynamics, may not be the same as the ones obtained from conventional rheology. But since MNPs sense their local environment on the nanoscale, dynamic magnetic measurements provide important information on the MNPs' embedding in and interaction with the matrix.

This contribution focused on ACS and MRX measurements. However, similarly, the dynamic magnetic response of MNPs can also be studied in oscillating—as performed by Tschöpe et al. [22, 39]—or rotating magnetic fields.

The methodology described in this contribution can be applied to different matrix systems, as, e.g., cells in biomedicine, other gel-like materials such as xanthan [23] and liquid crystals.

Acknowledgment: Financial support by the Deutsche Forschungsgemeinschaft, DFG Priority Program 1681 (LU800/4-1, LU800/4-2, LU800/4-3) is gratefully acknowledged.
Author contribution: All the authors have accepted responsibility for the entire content of this submitted manuscript and approved submission.
Research funding: None declared.
Conflict of interest statement: The authors declare no conflicts of interest regarding this article.

References

1. Brown WF Jr. Thermal fluctuations of a single-domain particle. Phys Rev 1963;130:1677–86.
2. Coffey WT, Cregg PJ, Kalmykov YP. On the theory of Debye and Néel relaxation of single domain ferromagnetic particles. Adv Chem Phys 1993;83:263–464.
3. Néel L. Some theoretical aspects of rock-magnetism. Adv Phys 1955;4:191–243.
4. Raikher YL, Shliomis MI. The effective field method in the orientational kinetics of magnetic fluids and liquid crystals. Adv Chem Phys 1994;87:595–751. 1994.
5. Fannin PC, Charles SW. On the calculation of the Néel relaxation time in uniaxial single-domain ferromagnetic particles. J Phys D Appl Phys 1994;27:185–8.
6. Eberbeck D, Wiekhorst F, Steinhoff U, Trahms L. Aggregation behaviour of magnetic nanoparticle suspensions investigated by magnetorelaxometry. J Phys Condens Matter 2006;8:S2829–46.
7. Ludwig F, Heim E, Schilling M. Characterization of magnetic nanoparticles by analyzing the magnetization and relaxation dynamics using fluxgate magnetometers. J Appl Phys 2007;101: 113909.
8. Martsenyuk MA, Raikher YL, Shliomis MI. On the kinetics of suspensions of ferromagnetic particles. Sov Phys JETP 1974;38:413–6.

9. Fannin PC, Scaife BKF, Charles SW. The measurement of the frequency dependent susceptibility of magnetic colloids. J Magn Magn Mater 1988;72:95–108.
10. Ahrentorp F, Astalan AP, Jonasson C, Blomgren J, Qi B, Thompson Mefford O, et al. Sensitive high frequency ac susceptometers in magnetic nanoparticle applications. AIP Conf Proc 2010;1311: 213–23.
11. Ludwig F, Eberbeck D, Löwa N, Steinhoff U, Wawrzik T, Schilling M, et al. Characterization of magnetic nanoparticle systems with respect to their magnetic particle imaging performance. Biomed Eng 2013;58:535–45.
12. Eberbeck D, Wiekhorst F, Steinhoff U, Trahms L. Quantification of biomolecule agglutination by magnetorelaxometry. Appl Phys Lett 2009;95:213701.
13. Heim EFL, Schilling M. Binding assays with streptavidin-functionalized superparamagnetic nanoparticles and biotinylated analytes using fluxgate magnetorelaxometry. J Magn Magn Mater 2009;321:1628–31.
14. Dieckhoff JMS, Ludwig F. Protein detection with magnetic nanoparticles in a rotating magnetic field. J Appl Phys 2014;115:024701.
15. Remmer H, Dieckhoff J, Schilling M, Ludwig F. Suitability of magnetic single- and multi-core nanoparticles to detect protein binding with dynamic magnetic measurement techniques. J Magn Magn Mater 2015;380:236–40.
16. Astalan AP, Ahrentorp F, Johansson C, Larsson K, Krozer A. Biomolecular reactions studied using changes in brownian rotation dynamics of magnetic particles. Biosens Bioelectron 2004;19: 945–51.
17. Chung SH, Hoffmann A, Bader SD, Liu C, Makowski L, Chen L. Biological sensors based on brownian relaxation of magnetic nanoparticles. J Appl Phys 2004;85:2971–3.
18. Schrittwieser S, Pelaz B, Parak WJ, Lentijo-Mozo S, Soulantica K, Dieckhoff J, et al. Homogeneous biosensing based on magnetic particle labels. Sensors 2016;16:828.
19. Enpuku K, Shibakura M, Arao Y, Mizoguchi T, Kandori A, Hara M, et al. Wash-free detection of C-reactive protein based on third-harmonic signal measurement of magnetic markers. Jpn J Appl Phys 2018;57:090309.
20. Bacri J-C, Dumas J, Gorse D, Perzynski R, Salin D. Ferrofluid viscometer. J Phys Lattr 1985;46: L1199–1205. https://doi.org/10.1051/jphyslet:0198500460240119900.
21. Roeben E, Roeder L, Teusch S, Effertz M, Deiters UK, Schmidt AM. Magnetic particle nanorheology. Colloid Polym Sci 2014;292:2013–23.
22. Tschöpe A, Birster K, Trapp B, Bender P, Birringer R. Nanoscale rheometry of viscoelastic soft matter by oscillating field magneto-optical transmission using ferromagnetic nanorod colloidal probes. J Appl Phys 2014;116:184305.
23. Sriviriyakula T, Bogrena S, Schaller V, Jonasson C, Blomgren J, Ahrentorp F, et al. Nanorheological studies of xanthan/water solutions using magnetic nanoparticles. J Magn Magn Mater 2019;473: 268–71.
24. Hess M, Rochels P, Roeben E, Zylla M, Webers S, Wende H, et al. Size effects on rotational particle diffusion in complex fluids as probed by magnetic particle nanorheology. Phys Chem Chem Phys 2019;21:26525–39.
25. Remmer H, Dieckhoff J, Tschöpe A, Roeben E, Schmidt AM, Frank L. Dynamics of $CoFe_2O_4$ single-core nanoparticles in viscoelastic media. Phys Procedia 2017;75:1150–7.
26. Odenbach S. Magnetoviscous effects in ferrofluids. Berlin: Springer; 2002.
27. Shliomis MI. Effective viscosity of magnetic suspensions. Sov Phys JETP 1977;34:1291–4.
28. Nowak J, Wolf D, Odenbach S. A rheological and microscopical characterization of biocompatible ferrofluids. J Magn Magn Mater 2014;354:98–104.
29. Kötitz R, Matz H, Trahms L, Koch H, Weitschies W, Rheinländer T, et al. SQUID based remanence measurements for immunoassays. IEEE Trans Supercon 1997;7:3678–81.

30. Matz H, Drung D, Hartwig S, Groß H, Kötitz R, Müller W, et al. A SQUID measurement system for immunoassays. Appl Supercond 1998;6:577–83.
31. https://qd-europe.com/de/en/product/materials-science/magnetometers/ac-susceptometer.
32. https://qd-europe.com/de/en/product/materials-science/magnetometers/mpms3-squid-magnetometer.
33. Dieckhoff J, Schilling M, Ludwig F. Fluxgate based detection of magnetic nanoparticle dynamics in a rotating magnetic field. Appl Phys Lett 2011;99:112501.
34. Shliomis M, Stepanov V. Frequency dependence and long time relaxation of the susceptibility of the magnetic fluids. J Magn Magn Mater 1993;122:176–81.
35. Svedlindh P, Jonsson T, Garcia- Palacios JL. Intra-potential-well contribution to the AC susceptibility of a noninteracting nano-sized magnetic particle system. J Magn Magn Mater 1997; 169:323–34.
36. Ludwig F, Balceris C, Johansson C. The anisotropy of the AC susceptibility of immobilized magnetic nanoparticles—the influence of intra-potential well contribution on the AC susceptibility spectrum. IEEE Trans Magn 2017;53:6101004.
37. Ludwig F, Balceris C, Jonasson C, Johansson C. Analysis of AC susceptibility spectra for the characterization of magnetic nanoparticles. IEEE Trans Magn 2017;53:6100904.
38. DiMarzio EA, Bishop M. Connection between the macroscopic electric and mechanical susceptibilities. J Chem Phys 1974;60:3802–11.
39. Gratz M, Tschöpe A. Size effects in the oscillatory rotation dynamics of Ni nanorods in poly(ethylene oxide) solutions. Macromolecules 2019;52:6600–12.
40. Dieckhoff J, Eberbeck D, Schilling M, Ludwig F. Magnetic-field dependence of Brownian and Néel relaxation times. J Appl Phys 2016;119:043903.
41. Raikher YL, Rusakov VV, Coffey WT, Kalmikov YP. Dynamic susceptibilities of an assembly of dipolar particles in an elastic environment. Phys Rev E 2001;63:031402.
42. Raikher YL, Rusakov VV. Orientational kinetics of dipolar particles in a Maxwell fluid matrix: inertialess limit for the rotary microrheology. Phys Rev B 2005;72:061406.
43. Rusakov VV, Raikher YL. Magnetic response of a viscoelastic ferrodispersion: from a nearly Newtonian ferrofluid to a Jeffreys ferrogel. J Chem Phys 2017;147:124903.
44. Remmer H, Roeben E, Schmidt AM, Schilling M, Ludwig F. Dynamics of magnetic nanoparticles in viscoelastic media. J Magn Magn Mater 2017;427:331–5.
45. Rusakov VV, Raikher YL. Orientational brownian motion in a viscoelastic medium. Colloid J 2017; 79:264–9.
46. Yoshida T, Enpuku K. Simulation and quantitative clarification of AC susceptibility of magnetic fluid in nonlinear brownian relaxation region. Jpn J Appl Phys 2009;48:127002.
47. Gratz M, Tschöpe A. Optical transmission versus ac magnetization measurements for monitoring colloidal Ni nanorod rotational dynamics. J Phys D Appl Phys 2017;50:015001.
48. Fock J, Balceris C, Costo R, Zeng L, Ludwig F, Hansen MF. Field-dependent dynamic responses from dilute magnetic nanoparticle dispersions. Nanoscale 2018;10:2052–66.
49. Cheng N-S. Formula for the viscosity of a glycerol–water mixture. Ind Eng Chem Res 2008;47: 3285–8.
50. Ludwig F, Wawrzik T, Yoshida T, Gehrke N, Briel A, Eberbeck D, et al. Optimization of magnetic nanoparticles for magnetic particle imaging. IEEE Trans Magn 2012;48:3780–3.
51. Normand V, Muller S, Ravey J-C, Parker A. Gelation kinetics of gelatin: a master curve and network modeling. Macromolecules 2000;33:1063–71.
52. Remmer H, Gratz M, Tschöpe A, Ludwig F. Magnetic field dependence of Ni nanorod brownian relaxation. IEEE Trans Magn 2017;53:6101204.
53. Cai L-H, Panyukov S, Rubinstein M. Mobility of nonsticky nanoparticles in polymer liquids. Macromolecules 2011;44:7853–63.

54. Roeder L, Bender P, Kundt M, Tschöpe A, Schmidt AM. Magnetic and geometric anisotropy in particle-crosslinked ferrohydrogels. Phys Chem Chem Phys 2015;17:1290–8.
55. Landers J, Roeder L, Salamon S, Schmidt AM, Wende H. Particle-matrix interaction in cross-linked PAAm-hydrogels analyzed by mössbauer spectroscopy. J Phys Chem C 2015;119:20642–8.
56. Fischer B, Wagner J, Schmitt M, Trieu V, Hempelmann R. Dependence of brownian relaxation on the volume fraction and an external magnetic field. Z Phys Chem 2006;220:69–77.
57. Ivanov AO, Zverev VS, Kantorovich SS. Revealing the signature of dipolar interactions in dynamic spectra of polydisperse magnetic nanoparticles. Soft Matter 2016;12:3507–13.
58. Ivanov AO, Camp PJ. Theory of the dynamic magnetic susceptibility of ferrofluids. Phys Rev E 2018; 98:050602(R).

Gareth J. Monkman*, Dirk Sindersberger, Nina Prem,
Andreas Diermeier and Tamara Szecsey

14 Dielectric behaviour of magnetic hybrid materials

Abstract: The objectives of this work include the analysis of electrical and magnetic properties of magneto-elastic hybrid materials with the intention of developing new techniques for sensor and actuator applications. This includes the investigation of dielectric properties at both low and high frequencies. The behaviour of capacitors whose dielectrics comprise magnetic hybrid materials is well known. Such interfacial magnetocapacitance can be varied according to magnetic content, magnetic flux density and the relative permittivity of the polymer matrix together with other dielectric content. The basic function of trapping electrical charges in polymers (electrets) is also established technology. However, the combination of magnetoactive polymers and electrets has led to the first electromagnetic device capable of adhering to almost any material, whether magnetically susceptible or not. During the course of this research, in addition to dielectrics, electrically conductive polymers based on (PDMS) matrices were developed in order to vary the electrical properties of the material in a targeted manner. In order to ensure repeatable results, this demanded new fabrication techniques hitherto unavailable. The 3D printing of silicones is far from being a mature technology and much pioneering work was necessary before extending the usual 3 d.o.f. to include orientation about and diffusion of particles in these three axes, thus leading to the concept of 6D printing. In 6D printing, the application of a magnetic field can be used during the curing process to control the particulate distribution and thus the spatial filler particle density as desired. Most of the devices (sensors and actuators) produced by such methods contain levels of carbonyl iron powder (CIP) embedded magnetic filler of up to 70 wt%. Contrary to this, a hitherto neglected research area, namely magnetoactive polymers (MAPs) having significantly lower magnetic particle concentrations (1 to 3 wt% CIP) were also investigated. With filler concentrations lower than 3 wt%, structures are formed which are completely absent at higher filler levels. CIP concentrations in the range of 1wt% demonstrate the formation of toroidal structures. Further development of coherent rings with a compact order results as filler concentrations increase towards 2 wt%. Above 3 wt% the structure eventually disintegrates to the usual random order found in traditional MAP with higher CIP content. Structured samples containing 1%–3 wt% CIP were investigated with the aid of X-ray tomography where solitary ring structures can be observed and eventually the

***Corresponding author: Gareth J. Monkman,** Mechatronics Research Unit, OTH-Regensburg, Regensburg, Germany, E-mail: gareth.monkman@oth-regensburg.de
Dirk Sindersberger, Nina Prem, Andreas Diermeier and Tamara Szecsey, Mechatronics Research Unit, OTH-Regensburg, Regensburg, Germany

This article has previously been published in the journal Physical Sciences Reviews. Please cite as G. J. Monkman, D. Sindersberger, N. Prem, A. Diermeier, and T. Szecsey "Dielectric behaviour of magnetic hybrid materials" *Physical Sciences Reviews* [Online] 2020, 5. DOI: 10.1515/psr-2019-0121 | https://doi.org/10.1515/9783110569636-014

formation of capillary doubles. Over wavelengths ranging from 1 to 25 μm, spectroscopic analysis of thin film MAP samples containing 2 wt% CIP revealed measurable magnetic-field-dependent changes in IR absorption at a wavenumber 2350 (λ = 4.255 μm). This was found to be due to the diamagnetic susceptibility of atmospheric carbon dioxide (CO_2). Consequently, the first potential application for sparse matrix MAPs was found.

Keywords: cappillary doublet, elastomer dielectric, diamagnetic, FTIR, interfacial magnetocapacitance, magnetoactive polymer, PDMS, toroid

14.1 Introduction

Magnetoactive polymers (MAPs) are elastomeric composites comprising micrometre-sized ferromagnetic or paramagnetic particles distributed within a mechanically compliant, nonmagnetic matrix [1]. Their mechanical [2–4] and electrical [5, 6] properties have been extensively investigated, though research into the behaviour of small quantities or even single filler particles is still in its infancy [7].

The project commenced with investigations into the effects of an applied magnetic field on the dielectric properties of MAPs. It rapidly became clear that although the electrical capacitance of an MAP could be changed by the influence of an external magnetic field, this was clearly attributable to interfacial capacitive effects caused by changes in the inter-particle distance between discrete magnetic content rather than any changes in dielectric permittivity.

Much of the basic theory and measurements concerning interfacial magnetocapacitance have been thoroughly investigated [5], although it is often (incorrectly) called magneto-dielectric effect [8–11].

Whilst embedded within an elastomer, movement of the magnetic particles is highly restricted. Because magnetic saturation is likely to be achieved long before the elastic limit of the polymer matrix is reached, magnetically influenced changes in capacitance are relatively small. Nevertheless, a decrease in distance between the particles along the magnetic flux lines, with increasing magnetic field strength, can be measured. However, it should be noted that only a change in capacitance takes place, not a change in the dielectric permittivity of the polymer matrix. This is not a true magnetodielectric effect as the relative permittivity is a function of the polymer material alone, which remains uninfluenced by the magnetic field.

Although magnetocapacitance can exist without multiferroic coupling, the term magnetodielectric refers specifically to single-phase materials having ferroelectric and ferromagnetic or antiferromagnetic order. Alternatives comprise composite materials combining conventional ferroelectrics and ferromagnetics segregated on a nanoscale level [12]. Lawes states categorically: “great care should be taken to separate intrinsic magnetodielectric coupling from interfacial magnetocapacitance” [13]. Consequently, the magnetodielectric effect implies electromagnetic coupling at the atomic, or at least molecular, level where the relative permittivity of the dielectric may change subject to an externally applied magnetic field. At the other end of the mesoscopic scale, where micrometre-sized particles are distributed within a polymer dielectric matrix, changes in capacitance are due purely to rearrangements in the magnetic particle structure or a shift of the magnetic domains within a single particle of sufficient size. The relative permittivity of the dielectric matrix remains constant [14]. Consequently, neither two-phase MAPs [15] nor ferromagnetic–ferroelectric heterostructures employing permanently magnetized hard magnetic materials [16] and ferroelectric content such as magnetoactive electrets [17] can be considered to be magnetodielectrics. Using pulsed magnetic fields, a slight magnetoelectret effect [18] may be observed, but in a MAP, due to the electrical conductivity of the particles, the induced electrical charge rapidly diminishes.

The often used term "dielectric constant" is also incorrect in that ε_r is a function of frequency (i.e. not a constant). In fact, at very high frequencies (i.e. light) $\varepsilon_r = n^2$ where n is the refractive index. For almost all polymeric materials, this value is considerably lower than that measured at lower frequencies. The correct term for ε_r is relative permittivity and the absolute permittivity ε is the product of free space permittivity and relative permittivity ε_0 ε_r [19]. For clarity, this (correct) nomenclature is used throughout this work.

The 3D printing of magnetoactive, electroactive or pure silicones and the newly developed 6D printing of MAPs [20] offer many possibilities for the production of suitable prototypes with complex geometries. For the 3D and 6D printing of the aforementioned polymer combinations, a special miniature extruder was developed. This allows the combination of the respective components whilst simultaneously performing degassing of the mixture in order to achieve cavity-free compounds [21].

Printing of a 3D silicone structure is augmented by the addition of magnetic content, which must be achieved prior to curing. Control of the additional 3 degrees of freedom (orientation and diffusion gradient of the particles) is realized by means of an externally applied magnetic field. This results in 6D printing. In addition to magnetic content, MAPs may also be made partly or wholly electrically conducting by the addition of carbon black, graphite [21] or aluminium graphite [22].

As most researchers have hitherto concentrated on maximizing magnetic content, MAPs with lower magnetic concentrations have been an area of neglected interest. During the course of this work, sparsely populated polymer matrices were found to have some very interesting, and often surprising, characteristics. Because of the lack of a true magnetoelectric effect, spectral investigations from infrared to ultraviolet reveal little, if any, magnetic-field-induced changes in transmission. One exception was discovered with MAP containing less than 3 wt% CIP content where the interaction with diamagnetic CO_2 was found to result in transmission changes at a wavelength of 4.255 μm [23]. These new and interesting results will be presented and discussed later in this chapter.

14.2 Sample fabrication and preparation

Addition-curing RTV-2 silicone rubbers [24] may be cross-linked at room temperature and can be used immediately after demoulding. The electrical and mechanical properties of silicone compounds and particularly poly(dimethylsiloxane) (PDMS) are well known [25].

Initially two components A (base material) and B (catalyst with platinum complex) of a RTV-2-silicone, typically SF00 or SF13 (Silikonfabrik), must be combined equally, as illustrated in Figure 14.1. Cross-linking commences at temperatures above 10 °C. Throughout this work curing was achieved at room temperature (22 °C), though it may be accelerated using higher temperatures (up to 200 °C).

In order to produce cavity-free silicone structures, a miniature extruder was developed, as depicted in Figure 14.2. This provides cavity-free mixtures suitable for 3D and 6D printing [20, 22].

Following cavity-free mixing, the silicone may be combined with the desired particulate. For MAPs, carbonyl iron particles can be used (CIP SQ from BASF). The CIP used are both mechanically and magnetically soft with diameters between 3.9 and 5 µm. CIP SQ has a pure iron (Fe) content of up to 99.8%. This differs considerably from MAP containing nanoparticles or micro-particles in solution where ionic strength must be considered [27]. In traditional MAPs, the CIP are embedded in a PDMS matrix and make up between 10 and 90 wt% of the mass of the entire mixture [2]. In this work, a much lower CIP concentration is employed, and the PDMS matrix is relatively soft (Shore A 00 or A 13 hardness). Finally, the uncured mixture may be poured into a mould or printed onto a substrate as desired and cavities deliberately added by injection.

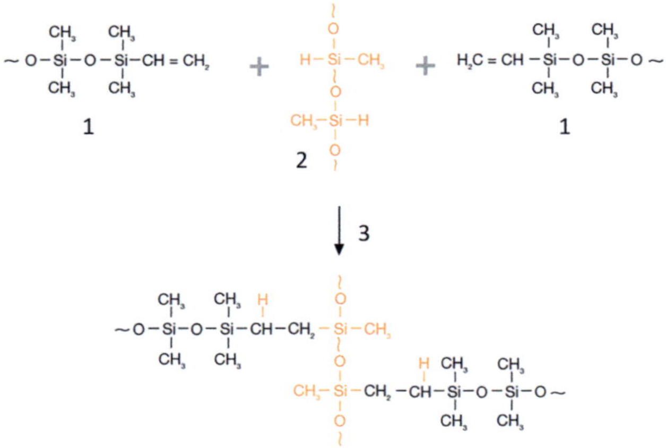

Figure 14.1: Cross-linking of silicone base and silicone catalyst [26]. 1.) End of polymer chain, 2.) cross-linker, 3.) Pt catalyst.

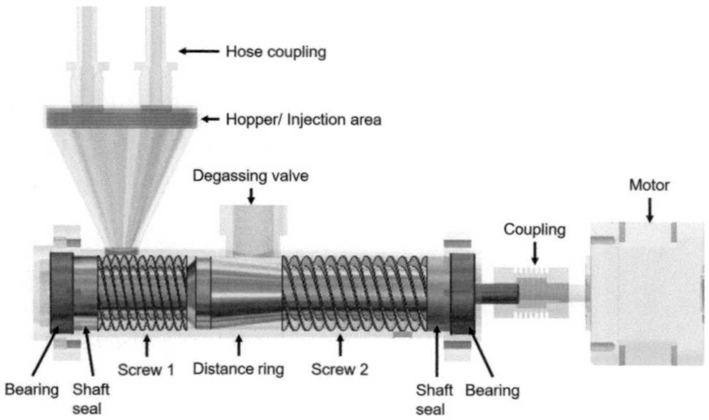

Figure 14.2: Miniature silicone extruder.

A large range of samples were prepared with carbonyl iron (CIP) contents ranging from 1 to 12% by weight. In order to eliminate the possibility of inter-particle magnetism being solely responsible for the structure formation, nickel and silver particle fillers were also used. As similar structures emerged, it could be concluded that this was neither a purely magnetic function nor something exclusively related to iron. Nevertheless, with magnetically susceptible particles, the structure can be influenced by an externally applied magnetic field.

14.2.1 Magnetization measurements

The magnetization behaviour of similar samples from the same production batch was measured using a SQUID magnetometer (Quantum Design MPMS XL) including MPMS RSO Controller and digital R/G Bridge. Measurements were carried out to the working limits of the measuring system; whereby only portions of the samples produced were analysed in order to remain within the limits of magnetic moments.

The resulting range of permeabilities measured are comparable to those found with magnetorheological fluids where values lie typically between 3 and 7 for magnetizable content in the range of 10–30 wt% [28]. Ferrofluids have much lower permeability, around 1.2 [29], and it is known that magnetoactive elastomers containing nanoparticles exhibit only very limited magnetoactive effects [17].

The relative permeabilities measured are not the permeabilities of individual iron particles. The values shown in Figure 14.3 represent the bulk permeabilities of the material samples. This shows that even at only 1 wt% CIP content, the permeability is not proportionally lower than that with 12 wt%.

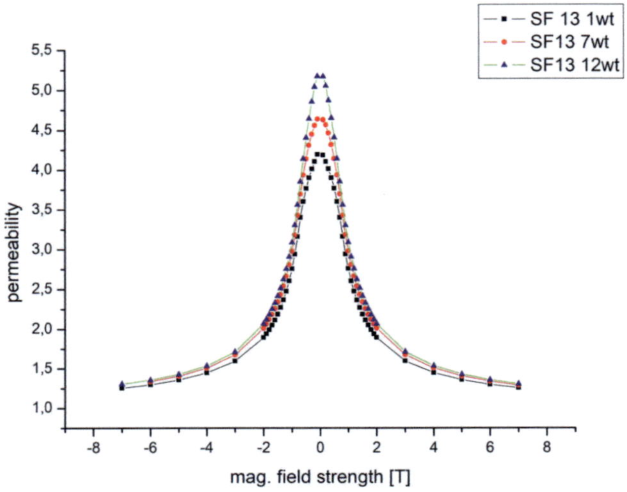

Figure 14.3: Magnetic permeability of SF13 samples containing 1, 7 and 12 wt% CIP.

14.2.2 Interfacial magnetocapacitance

In an electrorheological or magnetorheological fluid, the particles may freely rotate and translate [30]. This is not the case in an MAP where the elastic matrix limits such movement.

Figure 14.4 shows the influence of an externally applied magnetic field on the effective electrical capacitance of an MAP. Here C_n represents the increase in capacitance for the MAP subjected to an external magnetic field compared to that without a

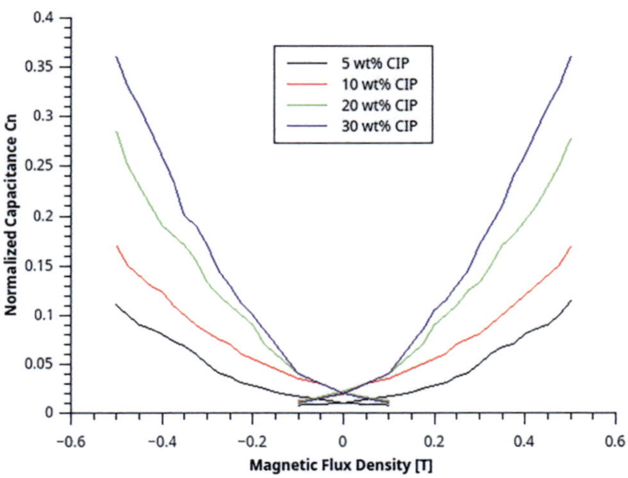

Figure 14.4: Capacitance as a function of applied magnetic flux density.

magnetic field and that measured while the MAP is subjected to an external magnetic field.

As may be observed from Figure 14.4, the electrical capacitance increases with rising magnetic flux density. The slight decrease about the zero point where the two plots cross is where the magnetic field changes from assisting fields (N-S) to opposing fields (N-N or S-S). For the purposes of interfacial magnetocapacitance, the use of PDMS as a matrix has now been largely superseded by boron-organo-silicon oxide polymers [31].

14.2.3 Internal toroidal structural formation

During vacuum evacuation, bubbles rise to the surface before being released. With very sparse filler particle concentrations, particles tend to gather on the boundary layers of ascending cavities [32]. A similar effect, previously intended for the attachment of lower-density talcum particles, has been documented [33] as has the self-assembly of particles into ring structures in two-phase fluids [34].

The hydrodynamic interaction of the higher-viscosity polymer tends to hinder the ascent of the heavier particles under the influence of gravitational acceleration causing them to lag behind the basic flow field [35]. The addition of the filler following mixing of the silicone components suggests that incipient cross-linking aids in stripping the ring from the bubble. Due to the velocity of cavity ascent, rather than a homogeneous distribution of particles over the spherical surface, a circumferential collection and subsequent ring (Torus) formation are created, as shown in Figure 14.5a.

The point at which the filler particle ring surrounding the rising cavity breaks free depends on the rate of ascent. The diameter of the toroid being a function of bubble dimension depends on the surface tension of the polymer, the mixing strategy and the

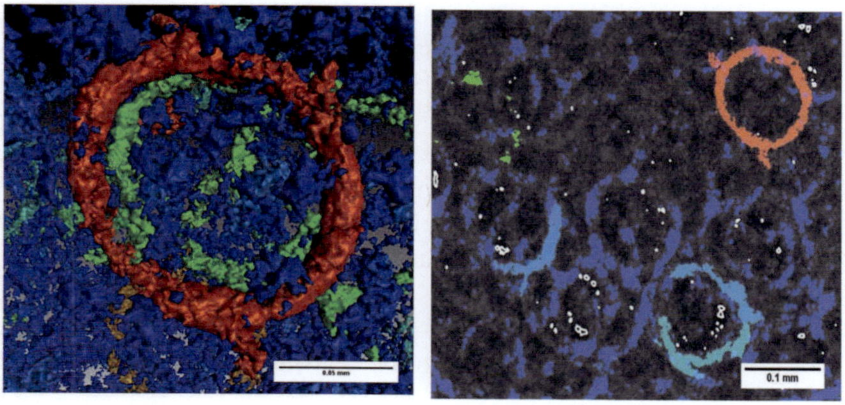

Figure 14.5: X-ray tomography of toroidal formations in SF13 with a) 1 wt% and b) 2 wt% CIP.

degree of cross-linking at the point of release. In Figure 14.5a, for 1 wt% CIP samples, average toroidal diameters of 80–90 μm resulted.

Cavity ascent is subject to purely laminar flow as the Reynolds factor {1} is far too low for turbulence to develop, thus maintaining a stable geometry.

$$\mathrm{Re} = \frac{V\rho L}{\mu} \tag{14.1}$$

Reynolds factor is a function of flow rate v, medium density ρ, the characteristic length L and dynamic viscosity μ. For an 80 μm diameter bubble rising at a velocity of 0.1 m/s through a medium with a dynamic viscosity of $\mu = 1500$ mPa/s at 23° (the polymer prior to curing), then expression {1} yields a maximum Reynolds number much lower than 1, thus guaranteeing laminar flow [36].

Expression {2} gives the volume of the torus formed,

$$V_r = 2\pi^2 R\alpha^2 \tag{14.2}$$

where R is the radius of the torus and α the radius of its tubular cross section. However, cavities do not always rise vertically but can also rise at an angle to the vertical, thus forming regular ring layers [32].

As long as the bubbles are so spaced that they do not make contact with one another, then depending on the flow regime, the rising bubble will collect particles in a volume represented by an oblique cylinder with radius R and height h, the volume for which is given in {3}.

$$V_C = \int_0^h \pi R^2 dx = \pi R^2 h \tag{14.3}$$

Dividing {2} by {3} gives the volume fraction ϕ in expression {4}

$$\phi = \frac{2\pi\alpha^2}{R^2} \tag{14.4}$$

Clearly, the particulate material has an influence on the resulting ring dimensions. When for 1 wt% nickel or silver filler particles are used, ring diameters with an average value of about 24 μm are formed. For 1 wt% CIP, the median diameter is about 80 μm. Inserting a mean toroidal radius of $R = 40$ μm and a cylinder height $h = 100$ μm, for particles of average radius $r = 2.5$ μm in expression {4} reveals a volume fraction for a ring of single particles of 2.45%. However, the density of iron is 7.86 kg/m^3 and that of silicone around 0.96 kg/m^3 [37]. Consequently, the mass fraction is 8.2 times lower or 0.3%. This represents the absolute minimum particle mass faction required before torus production can commence and assumes that all free particles within V_C (3) are gathered by the rising cavity. At the other end of the scale, it is logical to assume that there must be a maximum mass concentration for which toroid production is possible.

14.2.4 Capillary doublet formation

Increasing the CIP concentration towards 2 wt% results in a growth in toroid diameter until contiguation takes place as shown in Figure 14.5b. At this point the ring diameters were found to be around 130 µm. The enlargement of the outer diameter of the toroid caused by a higher weight percentage of filler material, in this case 2 and 3 wt%, can also be observed for nickel. This can be easily explained by the fact that the bubbles expand slightly as the weight of the collected particles causes them to be shed towards the interface between two connecting cavities.

This leads to axially symmetrical CIP containing volumes, known as capillary doublets [32, 38], which are formed at the interface between two rings in contact. This effect can be seen in the X-ray tomograph of Figure 14.6.

The nucleation and growth of capillary bridges, formed in colloidal crystals during the liquid phase separation process, are described step by step by Cheng and Wang [34]. The formation of capillary doublet structures is schematically illustrated in Figure 14.7. This is a familiar geometry in electronic engineering [39].

There are many mathematical derivations for the volume of a capillary bridge. Between two spheres they are catenoidal, but the upper and lower surfaces can be concave or convex, as shown in Figure 14.8. Expression {5} appears frequently as the basis of the relationship between capillary width and volume [40, 41]. Applying a somewhat simplified analysis to the model illustrated in Figure 14.7 reveals some approximate but interesting results.

$$d = \frac{s}{2}\left(\sqrt{1 + \frac{2V}{\pi R s^2}} - 1\right) \tag{14.5}$$

Transposing expression {5} for volume V gives {6}:

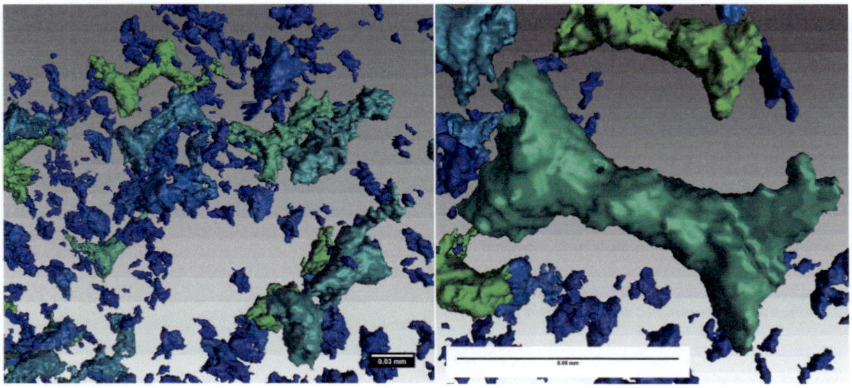

Figure 14.6: X-ray tomography showing capillary doublets at slightly above 2 wt%.

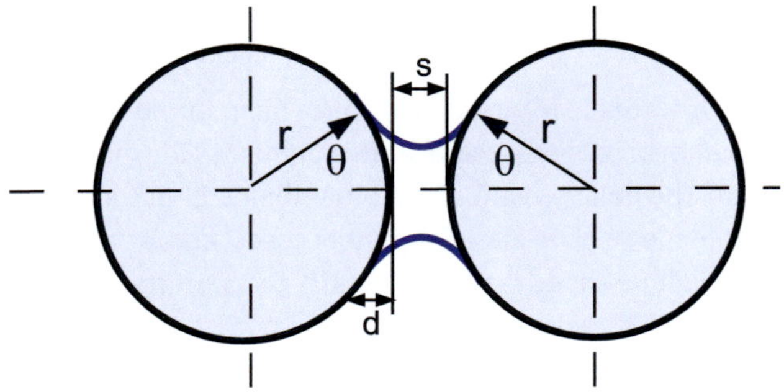

Figure 14.7: Formation of concave capillary doublets through bubble contiguation.

$$V = \frac{\pi R s^2}{2}\left(\left(\frac{2d}{s}+1\right)^2 - 1\right) \tag{14.6}$$

Taking approximate measurements from Figures 14.5 and 14.8 and inserting values for $d \approx s \approx 15$ µm and $R \approx 35$ µm in {6} yields a volume V_b of 9.896 × 10–14 m^3.

Assuming the formation of capillary bridges between all neighbouring cavities, the entire volume from which the particles are collected can be roughly modelled as the cylindrical volume given by {3}. Using the values extracted from Figure 14.8, the resulting cylindrical volume V_c of 5.03 × 10–13 m^3 can be calculated. The quotient between V_b and V_c represents the volume fraction in the capillary bridge, in this case ca. 19.67%.

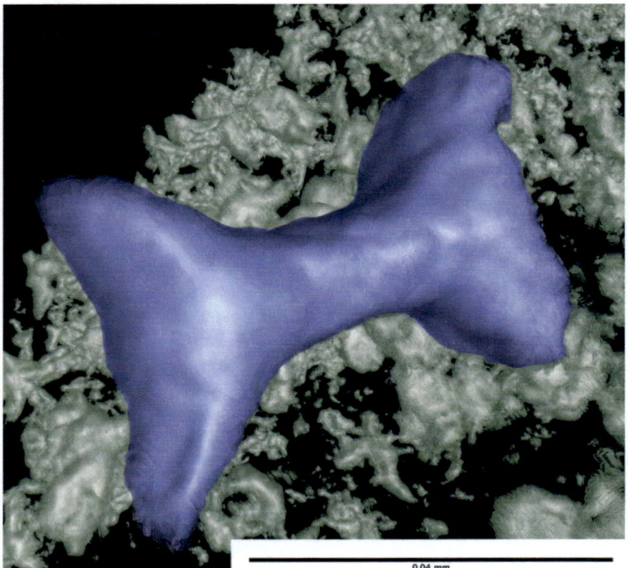

Figure 14.8: A single isolated carbonyl iron powder (CIP) capillary doublet after curing.

Figure 14.9: X-ray tomography showing a random distribution of CIP in SF00 3 wt%.

However, as previously mentioned, the density of carbonyl iron is 7.86 kg/m^3 and that of silicone around 0.96 kg/m^3. Consequently, the weight percentage is only 2.4%.

Given the observed onset of capillary doublet formation at 1.5 wt% which reaches a fully developed condition at around 2 and 3 wt% being the limit, the earlier calculated value is not unrealistic.

As higher CIP (>3 wt%) concentrations are reached, these structures then disappear with the usual random distribution of agglomerated particles within the polymer matrix shown in Figure 14.9 remaining. A similar situation also exists where the percentage of nickel or silver exceeds a threshold of about 3 wt%. The absence of any visible structure is typical for the usual high CIP concentration MAP, in which an arbitrary distribution of filler particles in spatially separated aggregates prevails.

14.2.5 Spectral analysis

The dielectric properties of MAP at high frequencies are important to the design of modern sensory elements associated with applications such as instrumentation, telecommunications and soft robotics. However, most current research has concentrated on lower frequency ranges [42] with a few investigations at microwave levels [43].

Optical properties concerning MAP surface structures have been investigated [44], [45] as has X-ray scattering [46]. Consequently, spectroscopy at wavelengths shorter than 1 µm has not been considered during this work.

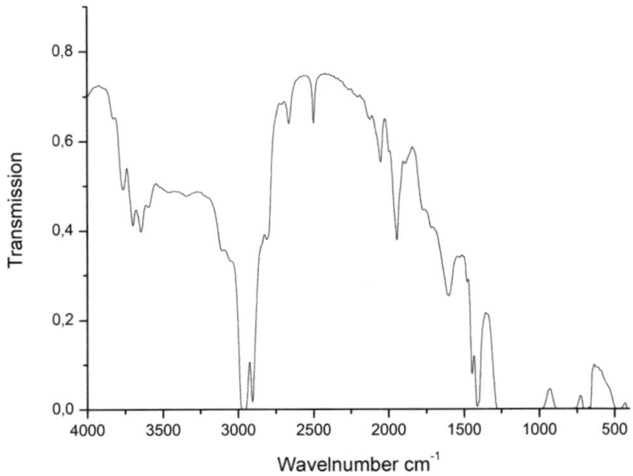

Figure 14.10: FTIR spectrum, PDMS alone.

The transmission/absorption behaviour of PDMS alone (without magnetic filler) in the infrared spectrum is well known [47]. MAPs have occasionally been the subject of spectroscopic analysis, usually on a secondary basis in order to verify other findings [48, 49]. The more usual MAP compounds containing higher ferromagnetic or paramagnetic content (40–80 wt%) have previously been subjected to IR spectral investigation but without decisive results with regard to the influence of an applied magnetic field [3, 50].

In transmission spectroscopy, the effective path length of the infrared radiation passing through the sample is determined by both the thickness and orientation of the sample to the directional plane of the IR beam [51]. Figure 14.10 shows the FTIR spectrum of PDMS sans CIP which is in agreement with other findings [52]. Spectra of samples containing 7, 12, 17, 23 and 35 wt% CIP reveal no changes in individual wavelengths but merely a shift of the entire spectrum due to opacity variations. Other findings concerning FTIR spectra of polyurethane-based magnetorheological polymers with considerably higher CIP content reveal similar results [3].

Nevertheless, Figure 14.11 reveals a small, but easily observable, difference at a wavelength of 4.255 µm (wavenumber 2350 cm^{-1}) when the sample is subjected to a radial magnetic field. The black curve is the initial condition without magnetic field. The red curve denotes the condition following the application of a magnetic field, and blue represents the relaxation of the MAP immediately following removal of the magnetic field. This difference occurs exclusively for samples containing 2 wt% CIP.

The experiments were repeated for the same samples but under the influence of a transverse magnetic field, where the effects were slightly lower but appeared at the same wavelength. In this case the effect commences with samples containing more

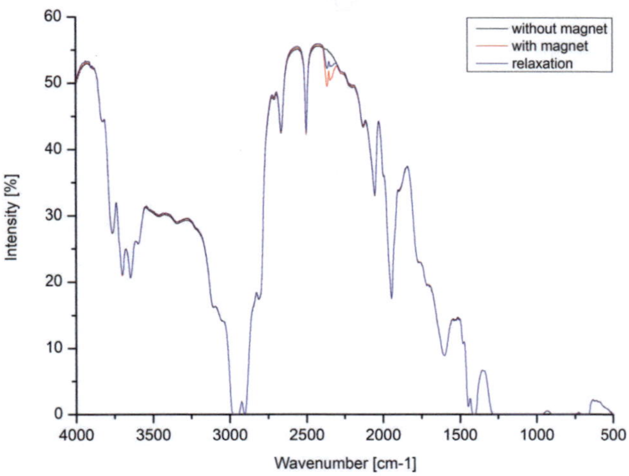

Figure 14.11: FTIR spectrum, MAP with 2 wt% of CIP in radial magnetic field.

than 1.5 wt% CIP and reaches a maximum at 2 wt% CIP. However, the effects are apparent at 1.5 wt% only under the influence of the transverse magnetic field.

The difference between IR transmission in the 2 wt% CIP samples, with and without magnetic field, is shown more clearly in the expanded graph of Figure 14.12 for PDMS with 2 wt% carbonyl iron powder (CIP).

The intensity changes shown in Figure 14.11 (expanded in Figure 14.12) are not large compared to changes generally seen throughout the entire IR spectrum. However, they were found to be observable and repeatable for over 40 samples each containing 2 wt% CIP. As illustrated in Figure 14.12, these intensity differences are observable only at a wavelength of 4.255 µm. Everywhere else in the spectrum there are only changes in average transmission level over the entire spectrum due to general material opacity.

However, samples containing 1 and 3 wt% CIP are largely unaffected by the influence of either a radial or a transverse magnetic field and merely resemble the

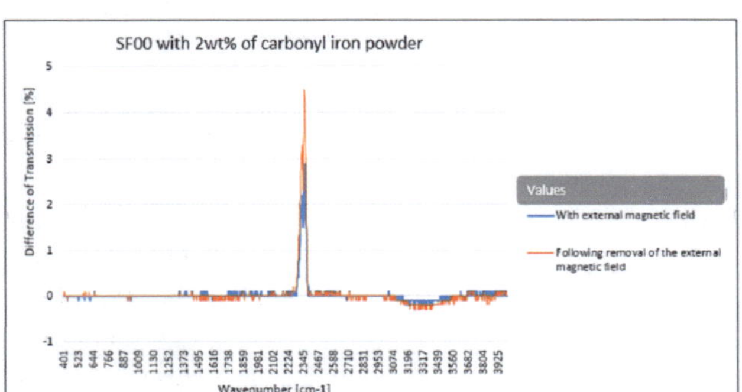

Figure 14.12: Transmission difference following application and removal of magnetic field.

characteristics of higher CIP concentration MAP. Clearly, the IR absorption in MAP samples having 2 wt% CIP content shows the greatest magnetic field dependency, while those with 1.5 wt% exhibit changes only in a transverse magnetic field. Although the effects of agglomeration and clustering cannot be entirely ruled out, samples containing 1.0, 2.5 and 3 wt% reveal no measurable influence.

The wave number band 2300–2400 cm^{-1} is indicative of carbon dioxide absorption, and this effect disappears when the same measurements are made in a pure dry nitrogen atmosphere. However, the amount of CO_2 in air is merely 0.04% volume. Nevertheless, CO_2 is known to be diamagnetic with a molar magnetic susceptibility of $-20.88 \cdot 10^{-6}$ cm^3/mol [53] which is greater than that of copper $-6.4 \cdot 10^{-6}$ cm^3/mol [54]. Nitrogen is also diamagnetic but has a magnetic susceptibility of only $-12 \cdot 10^{-6}$ cm^3/mol [54], and oxygen is paramagnetic with a magnetic susceptibility of $3335 \cdot 10^{-6}$ cm^3/mol [53]. Consequently, it is almost certain that the diamagnetic effects of CO_2 are responsible for the change in spectral absorption at this wavelength. Further experiments with 2 wt% MAP containing other diamagnetic materials such as copper instead of CIP revealed no such magnetic-field-induced changes in the IR spectra. Consequently, it can be concluded that the effect results from the interaction between diamagnetic CO_2 and the 2 wt% CIP magnetoactive elastomer. From Figures 14.5 and 14.6, it should also be noted that 2 wt% is the CIP concentration at which capillary doublets start to occur.

14.3 Conclusions

During the course of this research project, many developments have been made, which were necessary for the fabrication of test samples in order that repeatable investigations could be performed. The need for 3D printing of silicones led to the enhancement to 6D printing in order to achieve the required particle distribution within MAPs. For small batch production, the development and integration of a miniature extruder with automatic decavitation were necessary. This enabled the consistent production of a large number of test samples on which the remainder of this research was based.

Extensive spectral investigations concerning MAP with low CIP concentrations have been carried out. X-ray tomography has revealed how, with relatively low CIP concentrations, the formation of cavities in MAP causes the development of ring structures during gas evacuation. With increasing CIP concentration inter-cavity capillary doublets are formed which result in clearly measurable magnetic-field-dependent changes in IR absorption at a wavelength of 4.255 μm. This is almost certainly due to interactions between diamagnetic atmospheric CO_2 and the capillary doublet structures formed exclusively in MAP with CIP mass fraction between 1.5 and 3 wt%. Although the full effects (also on other diamagnetic gases) remain to be investigated, this inevitably has potential implications for future gas sensor devices.

The ring structures resulting from this research also represent microinductivities which can be fabricated in a targeted manner, thus enabling new applications in the high-frequency radio field. Furthermore, these anisotropic but well-organized structures have many magnetic-field-dependent implications for optical, thermal, acoustic and medical applications. Capillary doublet geometries offer many possibilities for microantennae designs at sub-millimetre wavelengths.

Acknowledgments: The authors would like to express their thanks to the German Research Federation (DFG) for financial support within the SPP1681 (Grant MO 2196/2-1) research programme and for the micro-computertomograph (Grant INST 102/11-1 FUGG). A special word of thanks to Birgit Striegl for the X-Ray analysis, Helmut Körner (Universität Regensburg) for the SQUID measurements and Manfred Röhrl for assistance with the Raman and FTIR spectroscopy.
Author contribution: All the authors have accepted responsibility for the entire content of this submitted manuscript and approved submission.
Research funding: The authors would like to express their thanks to the German Research Federation (DFG) for financial support within the SPP1681 (Grant MO 2196/2-1) research programme and for the micro-computertomograph (Grant INST 102/11-1 FUGG).
Conflict of interest statement: The authors declare no conflicts of interest regarding this article.

References

1. Odenbach S. Microstructure and rheology of magnetic hybrid materials. Arch Appl Mech 2016;86:269.
2. Li WH, Zhou Y, Tian TF. Viscoelastic properties of MR elastomers under harmonic loading. Rheol Acta 2010;49:733–40.
3. Xu Y, Gong X, Xuan S, Zhang W, Fan Y. A high-performance magnetorheological material: preparation, characterization and magnetic-mechanic coupling properties. Soft Matter 2011;7: 5246.
4. Stoll A, Mayer M, Monkman G, Shamonin M. Evaluation of highly compliant magneto-active elastomers with colossal magnetorheological response. J Appl Polym Sci 2014;131. https://doi.org/10.1002/app.39793.
5. Bica I. The influence of the magnetic field on the elastic properties of anisotropic magnetorheological elastomers. J Ind Eng Chem 2012;18:1666–1669.
6. Petcharoen K, Sirivat A. Magneto-electro-responsive material based on magnetite nanoparticles/ polyurethane composites. Mater Sci Eng 2016;C61:312–23.
7. Metsch P, Schmidt H, Sindersberger D, Kalina K-A, Brummund J, Auernhammer G, et al. Field-induced interactions in magneto-active elastomers: a comparison of experiments and simulations. Smart Mater Struct 2020.
8. Semisalova AS, Nikolai S, Perov G, Stepanov GV, Kramarenko EY, Khokhlova AR. Strong magnetodielectric effects in magnetorheological elastomers. Soft Matter 2013;9:11318.
9. Belyaeva IA, Kramarenko EY, Shamonin M. Magnetodielectric effect in magnetoactive elastomers: transient response and hysteresis. Polymer 2017;127:119–28.

10. Kostrov SA, Shamonin M, Stepanov GV, Kramarenko EY. Magnetodielectric response of soft magnetoactive elastomers: effects of filler concentration and measurement frequency. Int J Mol Sci 2019. MDPI.
11. Bica I, Anitas EM. Magnetodielectric effects in hybrid magnetorheological suspensions based on beekeeping products. J Ind Eng Chem 2019;77:385–92.
12. Catalan G. Magnetodielectric effect without multiferroic coupling. Appl Phys Lett 2006;88:102902.
13. Lawes G, Kimura T, Varma CM, Subramanian MA, Rogado N, Cava RJ, et al. Magnetodielectric effects at magnetic ordering transitions. Prog Solid State Chem 2009;37:40–54.
14. Guo F, Du C-b, Li R-p. Viscoelastic parameter model of magnetorheological elastomers based on Abel Dashpot. Adv Mech Eng 2014.
15. Varga Z, Filipcsei G, Zrínyi M. Smart composites with controlled anisotropy. Polymer 2005;46: 7779–87.
16. Böse H, Hesler A, Monkman G. Magnetorheologische Kompositmaterialien mit hartmagnetischen Partikeln, Verfahren zu deren Herstellung sowie deren Verwendung. Deutsche Patent DE 10 2007 028 663 A1. Priority: 21.06.2007. European Patent EP 2 160 741 B1 Granted: 17.08.2011.
17. Monkman GJ, Sindersberger D, Diermeier A, Prem N. The magnetoactive electret. Smart Mater Struct 2017.
18. Bhatnagar CS. The magnetoelectret. Indian J Pure Appl Phys 1964;2:331–2.
19. IEEE Std 211. IEEE standard definitions of terms for radio wave propagation; 1997. Reaffirmed 2003 – INSPEC Accession Number: 6010439.
20. Sindersberger D, Diermeier A, Prem N, Monkman GJ. Printing of hybrid magneto active polymers with 6 degrees of freedom. Mater Today Commun 2018.
21. Prem N, Chavez Vega J, Böhm V, Sindersberger D, Monkman GJ, Zimmermann K. Properties of polydimethylsiloxane and magnetoactive polymers with electro conductive particles. Macromol Chem Phys 2018.
22. Prem N, Sindersberger D, Monkman GJ. Mini-extruder for 3D magnetoactive polymer printing. Adv Mater Sci Eng 2019a. Hindawi.
23. Prem N, Sindersberger D, Monkman GJ. Infrared spectral analysis of low concentration magnetoactive polymers. J Appl Polym Sci 2019b.
24. Wacker gmbh. Processing Rt V-2 silicone rubbers. München: Hg. v. Wacker GmbH. Wacker Chemie; 2009.
25. Kuo ACM. Poly(dimethylsiloxane) – polymer data handbook. Oxford: Oxford University Press; 1999.
26. Domininghaus H, Elsner P, Eyerer P, Hirth T. Kunststoffe: Eigenschaften und Anwendungen. Berlin, Heidelberg: Springer; 2012. VDI-Buch.
27. Nguyen N-V, Wu J-S, Jen C-P. Effects of ionic strength in the medium on sample preconcentration utilizing nano-interstices between self-assembled monolayers of gold nanoparticles. BioChip J 2018;12:317–25.
28. Simon TM, Reitich F, Jolly MR, Ito K, Banks HT. The effective magnetic properties of magnetorheological fluids. Math Comput Model 2001;33:273–84.
29. Mayer D, Polcar P. A novel approach to measurement of permeability of magnetic fluids. Przeglad Elektrotechniczny 2012. 0033-2097. R. 88 NR 7b.
30. Block H, Kelly JP. Electrorheology. J Phys D Appl Phys 2000;21:1661. 11.
31. Monkman GJ, Striegl B, Prem N, Sindersberger D. Electrical properties of magnetoactive Boron-organo-silicon oxide polymers. Macromol Chem Phys 2020.
32. Sindersberger D, Prem N, Monkman GJ. Structure formation in low concentration magnetoactive polymers. AIP Adv 2019;9:035322.
33. Beaussart A, Parkinson L, Mierczynska-Vasilev A, Ralston J, Beattie DA. Effect of adsorbed polymers on bubble-particle attachment. Langmuir 2009;25:13290–4. American Chemical Society.

34. Cheng T-L, Wang YU. Shape-anisotropic particles at curved fluid interfaces and role of Laplace pressure: a computational study. J Colloid Interface Sci 2013;402:267–78.
35. Mileva E. Solid particle in the boundary layer of a rising bubble. Colloid Polym Sci 1990;268: 375–83.
36. Millett PC, Wang YU. Diffuse interface field approach to modeling and simulation of self-assembly of charged colloidal particles of various shapes and sizes. Acta Mater 2009;57:3101–9.
37. Roberts C, Graham A, Nemer M, Phinney L, Garcia R, Stirrup E. Physical properties of low-molecular weight polydimethylsiloxane fluids. Sandia Report SAND2017-1242. Sandia National Laboratories; 2017.
38. Soulié F, El Youssoufi MS, Cherblanc F, Saix C. Capillary cohesive local force: modelling and experiment. In: Peigney M, editor Towards optimal bounds on the recoverable strains in polycrystalline shape memory alloys; 2012.
39. Gromit. Electronics for dogs. Sparkford, Somerset: Haynes; 2010.
40. Gladkyy A, Schwarze R. Comparison of different capillary bridge models for application in the discrete element method – soft condensed matter. Cornell University; 2014.
41. Gabrieli F, Lambert P, Cola S, Calvetti F. Micromechanical modelling of erosion due to evaporation in a partially wet granular slope. Int J Numer Anal Meth Geomech 2011;36:918–43.
42. Balasoiu M., Bica I. Composite magnetorheological elastomers as dielectrics for plane capacitors: effects of magnetic field intensity. Results Phys 2016;6:199–202. Elsevier.
43. Kuznetsova IE, Kolesov VV, Zaitsev BD, Fionov AS, Shihabudinov AM, Stepanov GV, et al. Electrophysical and acoustic properties of magnetic elastomers structured by an external magnetic field. Bull Russ Acad Sci Phys 2017.
44. Forster E, M Mayer, R Rabindranath, H Böse, G Schlunck, GJ Monkman, und M Shamonin. Patterning of ultrasoft, agglutinative magnetorheological elastomers. J Appl Polym Sci 2013;128: 2508–15.
45. Nadasi H, May K, Eremin A, Stannarius R. Magnetoresponsive dispersions of anisometric pigment particles and gels; 2016. DFG SPP1681 Tagung, Benediktbeuern.
46. Schatte S, Seliger J, Prevost S, Gradzielski M. Magnetic nanocubes in worm-like micellar gels; 2016. DFG SPP1681 Tagung, Benediktbeuern.
47. Hamciuc C, Hamciuc E, Okrasa L. Silica/polyimide-polydimet hylsiloxane hybrid films. Thermal and electrical properties. Macromol Res 2011;19:250–60.
48. Fuchs A, S Joko, G Faramarz, C Mert Bahadir, Y Liu. Surface polymerization of iron particles for magnetorheological elastomers. J Appl Polym Sci 2010;117:934–42.
49. A Garcia-Márquez, D Arnaud, H Benoît, G Daniel, B-C Sylvie, D Bertrand. Iron oxide nanoparticle-containing main-chain liquid crystalline elastomer: towards soft magnetoactive networks. J Mater Chem 2011;21:8994.
50. Wei B, Gong X, Jiang W. Influence of polyurethane properties on mechanical performances of magnetorheological elastomers. J Appl Polym Sci 2009.
51. Griffiths P, De Haseth JA. Fourier transform infrared spectrometry. Hoboken, NJ: Wiley-Interscience; 2007. 0471194042.
52. Lambert J, Biele C, Marsmann HC. Strukturaufklärung in der organischen Chemie, vol 2. München: Pearson; 2012. 9783868941463. Auflage.
53. Havens GG. The magnetic susceptibilities of some common gases. Phys Rev 1933;43:992.
54. Martienssen W. Numerical data and functional relationships in science and technology – diamagnetic susceptibility – new Landolt-Börnstein series II/16. Heidelberg: Springer; 1986.

Dmitry Borin* and Gennady Stepanov

15 Magneto-mechanical properties of elastic hybrid composites

Abstract: The paper gives an overview of tunable elastic magnetic composites based on silicon rubber matrix highly filled with a magnetic soft and hard filler. The magnetic soft phase, which is represented by iron microparticles, allows active control of the physical properties of the composites, while the magnetically hard phase (e.g. neodymium–iron–boron alloy microparticles) is mainly responsible for passive adjustment of the composite. The control is performed by the application of an external magnetic field *in situ*, and passive adjustment is performed by means of pre-magnetization in order to change material remanent magnetization, i.e. the initial state. The potential and limits of active control and passive tuning of these composites in terms of their magneto-mechanical behavior are presented and discussed.

Keywords: anisotropy, magnetorheology, smart materials, soft magnetic composite

15.1 Introduction

The combination of magnetic microparticles and an elastic soft polymeric matrix makes it possible to obtain an intelligent material with a wide range of tunable properties. Similar to concentrated suspensions of magnetic microparticles, known as magnetorheological fluids, composites in which the liquid carrier medium is replaced by an elastic matrix are known as magnetorheological elastomers [1]. Other frequently used designations for them are magnetoactive or magnetosensitive elastomers [2]. The principal effect of a magnetic field on these composites is a change in their viscoelastic properties. To quantify this change, which is also referred to as the magnetorheological effect (MR effect), it is appropriate to use the change in the corresponding material parameter under the influence of an external magnetic field $\vec{H}$. The absolute MR effect is the difference between the material parameter in the magnetic field and that without it and the MR relative effect is the ratio of this difference to the parameter in the zero field. Other important magnetically controlled effects are the magneto-deformational (strictional) effect [3–10] and the rather poorly investigated shape memory effect [2, 11, 12]. Furthermore, the attention of researchers is attracted by the possibility of controlled modification of surface roughness of such composites [13–15], as well as

***Corresponding author: Dmitry Borin,** Institute of Mechatronic Engineering, Technische Universität Dresden, Dresden, 01062 Germany, E-mail: dmitry.borin@tu-dresden.de. https://orcid.org/0000-0003-3842-1487

Gennady Stepanov, State Research Center GNIICHTEOS, 105118, Moscow, Russia

This article has previously been published in the journal Physical Sciences Reviews. Please cite as: D. Borin and G. Stepanov "Magneto-mechanical properties of elastic hybrid composites" *Physical Sciences Reviews* [Online] 2020, 5. DOI: 10.1515/psr-2019-0126 | https://doi.org/10.1515/9783110569636-015

their dielectric properties [16–18]. This set of unique features makes these composites attractive for various technical implementations, such as damping devices and suspensions, activators, sensors and grippers, including medical applications (see e.g. [19–24]).

As a matrix of elastic magnetic composites, it is proposed to use various materials: from biocompatible hydrogels to silicon compounds and vulcanized rubbers (see e.g. [25–28]). The choice of matrix and magnetic filler depends on the intended application of the material and it must be ensured that the material properties are sufficiently controllable, e.g. that the MR effect is significant. Thus, at high modulus of elasticity (from ~1 MPa), even a high degree of filling of the composite with magnetic particles is not a guarantee of variability of controlled properties, i.e. the relative effect is very small [26]. Ultra-soft, gel-like matrix, having the modulus of elasticity of several kPa or lower, guarantees a high relative effect with a sufficient filling of the material with magnetic powder, but the absolute value of the increase of modules in the field may be insignificant [27]. The use of silicone compounds as a matrix highly filled with magnetic powder of microparticles has proven to be a promising option for magneto-controlled composites [2, 12, 29, 30]. Magnetic-soft iron powders are commonly used as a composite filler. Alternatively, approaches using magnetically hard powders were proposed [31–34]. First of all, this allows one to change the initial state of the composite by changing its remanence. Despite the remanence, the application of an external magnetic field allows for reversible changes in the properties of these materials [35–37]. The use of complex powders, in particular mixtures of soft magnetic and hard magnetic particles, expands the possibilities of active control, while maintaining the ability for passive tuning of the material [9, 38, 39].

In this contribution, we present an overview of recent results obtained within experimental characterization of magneto-controlled composites based on silicone rubber highly filled with different types of magnetic filler. The work is focused on magnetic properties and MR effect of the composites. Also, essential information about the components and the process of manufacturing composites is briefly presented. This article is correspondingly divided into three parts: chapter 15.2 deals with components and fabrication steps, chapter 15.3 presents magnetic properties and in chapter 15.4 magneto-mechanical response of the composites is considered.

15.2 Material composition and fabrication

The basic components of controllable magnetic elastic composites are a polydimethylsiloxane (PDMS) matrix and a magnetic filler. In Figure 15.1, chemical components of the matrix, polymerization reaction, schematic representation of the filled cross-linked matrix as well as optical microscopy images of various particle powders are provided. Below a brief description of the manufacturing process is given, while more details can be found in the study by Stepanov et al. [40].

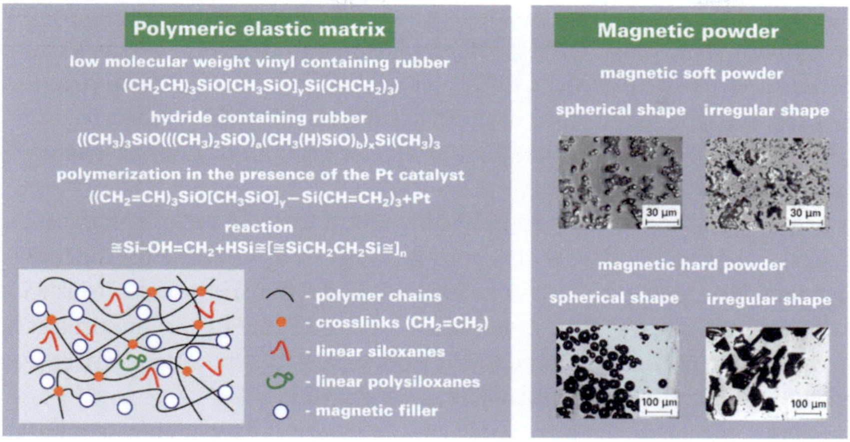

Figure 15.1: Typical components used for hybrid magnetic elastic composites: chemicals, polymerization reaction and schematic representation of the filled PDMS matrix (left) and exemplary optical microscopy images of magnetically soft and hard microparticles of the powders of various shape (right).

15.2.1 Composite manufacturing

The main technological stages of composite manufacturing include (I) preparation of magnetic filler (powder modification) (II) mixing of filler and liquid components of matrix (III) vacuumization of composition and its casting into molds, and, finally (IV) polymerization (cross-linking) process.

The aim of stage I is to ensure the compatibility of the powder microparticles with the silicone matrix and to prevent its possible impact on the cross-linking process. This is accomplished by coating the particle surface with a layer of siloxane liquid. Detailed information on particle treatment is given, e.g. in the studies [34, 38]. The determining criterion when choosing a soft magnetic filler is its high saturation magnetization and minimum coercivity. According to this criterion, iron and iron oxide (e.g. Fe_2O_3, Fe_3O_4) microparticle powder is commonly used as a magnetic soft filler, i.e. a ferro-/ferri-magnetic materials. Iron powders with microparticles of various sizes and shapes are commercially widely available, e.g. spherical carbonyl iron powders from BASF. The choice of the hard magnetic phase is based on the high remanence of the powder material as well as its high coercivity. Powder metallurgical technologies are usually used in the manufacture of permanent magnets, due to this, hard magnetic materials are also commercially available in powder form. Samarium cobalt (SmCo) and neodymium–iron–boron (NdFeB) alloys belong to materials with high coercivity, also known as the so called high-energy ones. Moreover, there is a technology for obtaining spherical microparticles for NdFeB alloys [41]. The use of spherical powders is preferable in terms of comparing experimental results with theoretical studies. On the other hand, the use of particles with a shape other than spherical is interesting from a practical point of view. When microscopic mechanisms of magnetic interactions for spherical and non-spherical particles differ, this will affect the macroscopic properties of composites. Thus, preference may be given to magnetically hard powders based on NdFeB, which are available with particles of different shapes (e.g. Magnequench™ powders).

In stage II, the treated filler is mechanically mixed with the initially liquid components of the polymer matrix. For this purpose, one can use a stirrer for the coarse dispersion and, if necessary 3-roll dispersing machine for fine dispersion. The matrix can be diluted with silicon oil in order to tune the elasticity of the composite. In addition, in order to prevent the composition from separation, the suspension can be enriched with structural additives that increase its viscosity.

Degassing of the mixture in stage III is necessary to remove air bubbles in order to obtain a defect free sample. The resulting degassed mixture is poured into a mold with a particular geometry, which is dictated by the further purpose of the sample. The mold walls can be coated with a surfactant layer that blocks the polymerization of the silicone polymer. Figure 15.2 shows examples of composite samples of various shapes.

The curing process (stage IV) can take place both at room and at elevated temperature. Heating increases the reaction rate and therefore can have a positive effect on the homogeneity of particle distribution in the matrix volume. As an alternative to heating in a convection furnace, microwave heating can be used, which accelerate the set of temperature inside the sample [34]. Also effective at preventing particle sedimentation is the rotation of the sample at low speeds (e.g. ~10 rpm) during the cross-linking process.

Optionally, during the stage IV, the sample can be placed in a homogeneous external magnetic field (without rotation). In this case, an anisotropic, i.e. structured in the direction of the applied field, composite will be obtained. The use of various filler concentrations and conditions of magnetic field application at this stage is used to obtain different types of internal structure morphologies [42–45]. Strategies of the targeted structuring of magnetic microparticles in a polymer composite are recently discussed in the study by Borin [46]. Composites pre-structured during the process of curing are not the focus of this work.

15.2.2 Composite pre-magnetization

The pre-magnetization of the specimens containing magnetically hard phase is realized using a uniform field provided by the electromagnet B-E25 (Bruker, Germany).

Figure 15.2: Composite samples of various shapes: disk, rod, cylinder and plate.

Typically external magnetizing fields are used up to $B_M = 1500$ mT are used. Cylindrical and rod-shaped samples are magnetized along the vertical axis and disk-as well as plate-like samples are magnetized in the direction perpendicular to the disk face. The pre-magnetization results in an induced microstructural anisotropy due to the particles rearrangements when the matrix is soft enough as shown in [37]. It's worth noting that effective magnetic fields inside the samples with different geometries and various pre-magnetizations are not equal for identically applied external fields, due to the samples own demagnetizing fields. However, it is not always possible to take this issue into account, since demagnetization factors for the composites with an induced micro-structural anisotropy remain unknown. For that reason, we only consider external fields in this work.

15.2.3 Additional processing

In addition to the above specified manufacturing steps, it is necessary to mention the additional processing of samples, namely their adaptation for subsequent fixation in experimental setup. Depending on what kind of experiment the sample of certain geometry is intended for, it is additionally supplied with structural elements made of solid polyethylene. For example, for strong fixation of rod specimens in the clamps of testing machines, annular plugs are glued using cyanoacrylate adhesive to their ends (see e.g. rod specimen in Figure 15.2). Solid polyethylene disks with diameters equal to or greater than the diameter of the specimen diameter shall be glued to the ends of the cylindrical specimens. Thin disks for magnetic measurements are completely fixed in a hollow thin-walled cylindrical holder with a height equal to the thickness of the sample (see Figure 15.3). Shear test specimens (e.g. disks and plates) are usually glued directly to the measurement geometry of the testing equipment.

15.3 Magnetic properties

A common method for determining the magnetic properties of magnetic hybrid composites is to obtain their magnetization curves by means of vibrating sample magnetometry. This method measures the magnetic moment of a sample oscillating with a given frequency and amplitude between a set of pickup coils in an external uniform magnetic field. We utilize a Lake Shore 7407 magnetometer and a nickel sphere with 3 mm diameter and a magnetic moment of $m = 6.92$ mA m^2 at external field $H = 398$ kA/m and $T = 298$ K is used as a calibration sample. The method can provide an accuracy of the magnetic moment measurements better than 1% of reading ±0.2% of full scale. For composites magnetic characterization, disk-shaped specimens were used (diameter 4.7 mm, height 1–1.5 mm). The sample plane in the experiment is perpendicular to the external magnetic field, as shown in Figure 15.3.

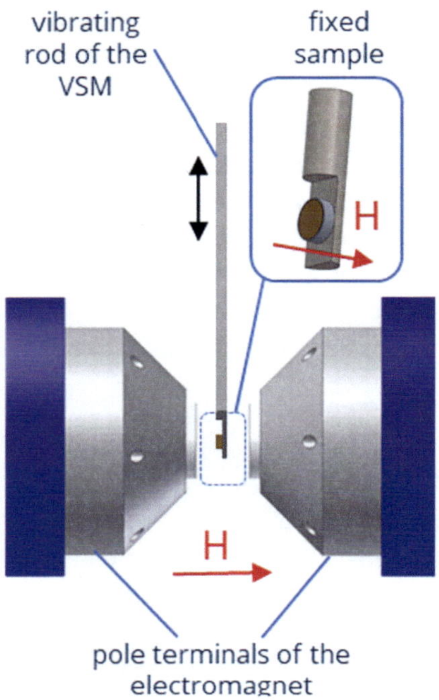

Figure 15.3: Sample orientation in the measurements of its magnetic properties on a vibrating sample magnetometer.

The magnetic properties of the elastic composites are primarily determined by the material and concentration of the filler's powder, but the matrix in which the powder particles are embedded can also influence the magnetization processes. Depending on the rigidity of the matrix, the filler particles will have different mobility levels. If the matrix is soft enough, the microparticles will move, rotate and form structures by the effect of the applied magnetic field. The microstructural rearrangement of the particles results in the correspondingly forced changes of the macroscopic behavior of the composite material. These statements are confirmed both from an experimental and theoretical point of view [2, 8, 35–37, 47–50]. In Figure 15.4, measured magnetization curves of two samples with different matrices filled with 20 vol% of spherical iron microparticles are shown. Despite the fact that the filler is a soft magnetic material, the magnetization curve of the composite based on a soft matrix is a hysteresis. In particular, the magnetization of the soft sample in a decreasing field is higher, than the magnetization in the increasing field. As mentioned above, such behavior is related to the hysteresis movement of particles inside the matrix. Curves run is reproduced by repeated measurements if negative magnetic fields are not used. In the case of a negative field, irreversible particle movements relative to the matrix (in particular, particle rotation) are possible. In this case, the so-called training of the sample (several repeated runs) is required to achieve reproducibility of the magnetization curves. Such behavior is associated not only with the mobility of particles (particle-matrix interaction), but also

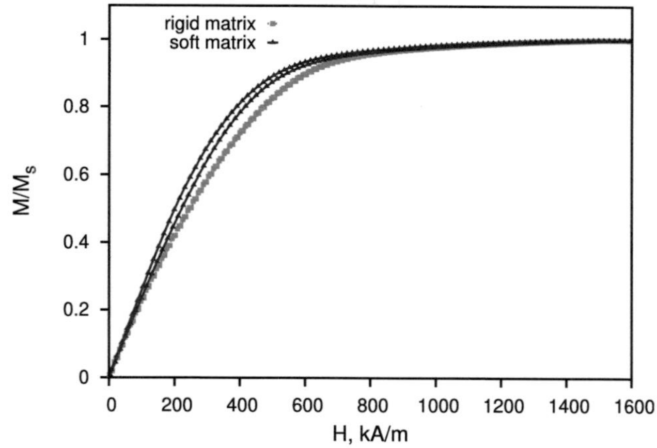

Figure 15.4: Magnetization curves of the elastic composites with ~20 vol% of iron microparticles (BASF CC powder) embedded into a rigid ($E \sim 1$ MPa) and soft ($E \sim 0.3$ MPa) PDMS matrix. Magnetization of the samples M is normalized to their saturation magnetization M_s.

with possible matrix defects. As an analogy of such behavior, we may mention the Mullins effect known from the mechanical response of filled rubbers [51].

Another important characteristic of composites based on a soft magnetic filler is the initial magnetic susceptibility χ_{ini}. The susceptibility χ_{ini} is defined as the slope of the initial linear part of the magnetization curve. For magnetic composites χ_{ini} is a function of magnetic powder concentration. Previously, it was proposed to use the so-termed Maxwell–Garnett approximation [52] to estimate the magnetic susceptibility of the composite depending on the volume concentration of the magnetic phase [4, 53, 54]:

$$\chi_{ini} = \frac{3\phi}{1-\phi}, \tag{15.1}$$

where ϕ is the filler volume concentration. Equation (15.1) provides χ/ϕ close to three for diluted composites ($\phi << 1$). Experimental measurements (Figure 15.5) show that Eq. (15.1) satisfactorily predicts values of χ_{ini} only for solid matrix composites, in which structuring and movement of particles is restricted, at filler concentrations not exceeding ~15 vol%.

It is obvious that structuring significantly increases initial magnetic susceptibility. Comparing the results obtained for elastic composites with the results obtained for samples based on the liquid matrix, we assume that the factor determining the susceptibility is the size of structures formed in the magnetic field, i.e. mobility of the particles. In the case of elastomers, the effect of structuring is restricted due to the elastic forces immobilizing the particle movements.

Composite samples containing a magnetically hard phase are characterized by a magnetic hysteresis loop with significantly different from zero remanence M_r and

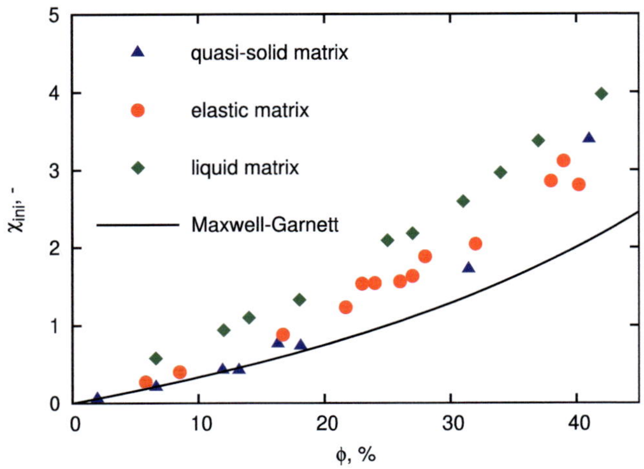

Figure 15.5: Measured initial magnetic susceptibility χ_{ini} of the composites based on different matrices as a function of the BASF CC carbonyl iron powder volume concentration ϕ. Results are given for specimens based on an epoxy resin (quasi-solid, $E \sim 2$ GPa), PDMS (elastic, $E \sim 0.3$ MPa), and silicon oil (liquid, $\eta = 0.35$ mPa s). For comparison, the Maxwell–Garnett approximation (Eq. 15.1) is as well provided.

coercive force H_c. Figure 15.6 demonstrates the initial magnetization curves for two types of hard magnetic powder embedded into epoxy resin comparing to magnetization of the carbonyl iron powder. Remarkable is the fact that spherical magnetically hard powder has a much higher saturation magnetization and a wider hysteresis loop

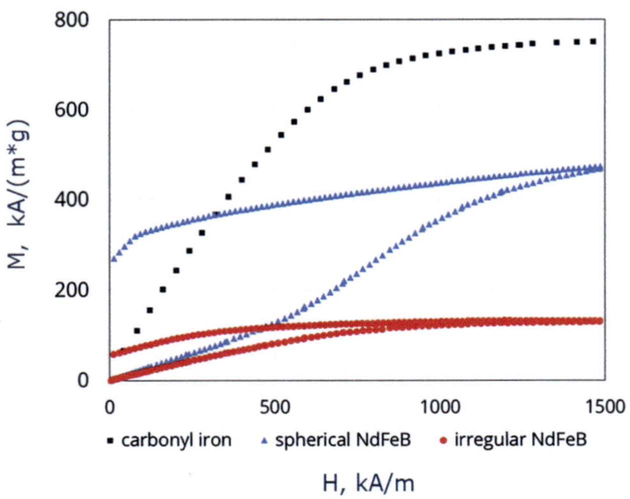

Figure 15.6: Initial magnetization curves of the carbonyl iron powder and two types of NdFeB-alloy powder embedded into epoxy resin.

than powder with irregular particles. Powders are commercial products and such difference in their properties is probably related to the technology of their production and exact composition. In Figure 15.7, dependencies of hard magnetic powders remanence on the applied magnetization field are given. In the composites with an elastic matrix structuring and rotation of particles can be forced by an external magnetic field. This can in turn significantly influence the magnetization process, as shown in theoretical and experimental studies [35, 36, 49, 50, 55–58]. Figure 15.8 presents full magnetic hysteresis curves for composites based on spherical magnetically hard powder embedded into the matrices of various elasticities. The processes of magnetization, rotation and structuring of particles are reflected in the material behavior. Similar to the behavior of soft magnetic filler, hard magnetic particles are also capable of structuring in a sufficiently soft matrix.

The hysteresis loop of soft composite is much narrow and asymmetrical than the loop of a composite with a hard matrix. This is determined by the direction of primary magnetization of the magnetic elastomer. Moreover, at a certain relationship of the remanence of the magnetic filler, the magnetic field strength applied at re-magnetization and the elasticity of the polymeric matrix, the nominal coercive force may have a negative value, i.e. the magnetization of the material is positive in the negative magnetic field as discussed in the study by Stepanov et al. [49]. The effect is caused by rotation of the magnetized particle to its initial state by elastic forces of the polymeric matrix. Another effect that can be seen in the process of magnetizing soft samples with hard magnetic phase is the appearance of significant difference between first consecutive magnetization loops (training effect) [55]. This effect is demonstrated in Figure 15.9. Furthermore, recently, an influence of the under-magnetization of hard magnetic

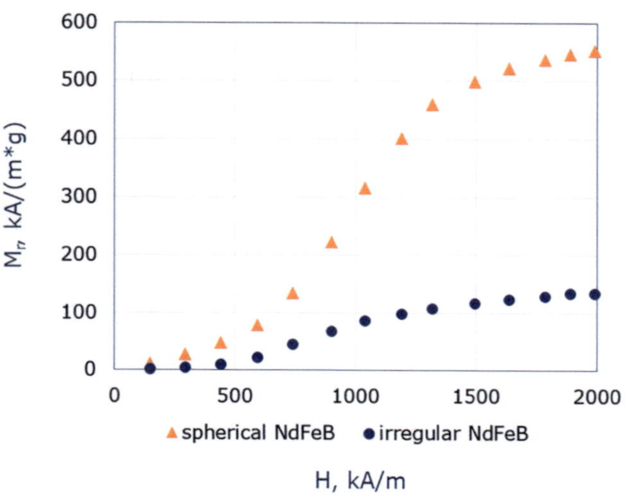

Figure 15.7: Dependency of remanence M_r (by weight) on the external field B_M which was applied to magnetize two different types of magnetically hard powder.

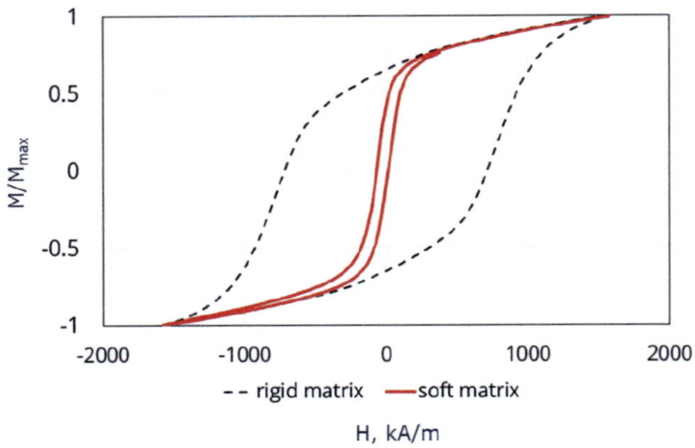

Figure 15.8: Magnetization loops of the composite filled with ~40 vol% of spherical NdFeB-alloy embedded into a rigid matrix ($E > 1000$ kPa) and soft matrix ($E \sim 60$ kPa). Initial curves are not shown. M_{max} is smaller than saturation magnetization, which is not reachable for the powder utilizing a Lake Shore 7407 magnetometer.

powder on the training effect in elastic composites was analyzed experimentally and theoretically [58]. In the case of composites with a complex filler, i.e. powder based on a mixture of hard and soft magnetic microparticles, the first-order reversal curve (FORC) measuring method is an effective tool for investigating magnetic interactions within the

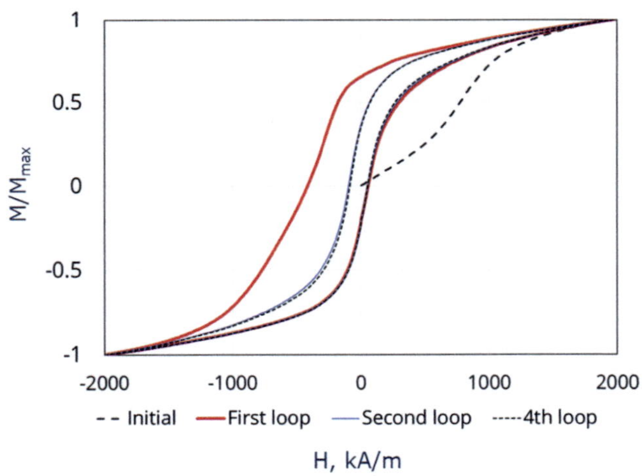

Figure 15.9: Magnetization loops of the composite filled with spherical NdFeB-alloy powder at concentration of ~11 vol%. The tensile modulus of the filled composite is $E \sim 130$ kPa. The presence of training effect is clearly demonstrated: Gradual diminution of discrepancy between consecutive magnetization loops. The training effect manifests itself most strongly between the first and second loops, while it quickly fades out after that.

material. In the context of magnetic elastomers, the FORC method was firstly used in the study by Linke et al. [59]. It has been demonstrated that FORC distributions are also subject to significant influence of structuring and rotation processes of hard magnetic particles dominating over rather energetically unfavorable domain processes within the particles. In addition, the FORC diagrams reflect both irreversible and reversible contributions of the magnetic soft phase to the magnetization of the major loop.

From an applied point of view, the remanence of the composite is an important aspect of irreversible magnetization processes. This parameter determines the initial state of the material, including its mechanical behavior, and it plays a key role in the passive tuning of the magneto-mechanical properties of a composite [33]. To prevent a change of the initial state of a material with the complex filler, the magnetic field used for active control should not change the remanence. The initial magnetization curves of the hard magnetic phase provide a possibility to estimate the range of the magnetic field, which should be used both for passive adjustment of the composite and for its active control. Beside, taking into account the fact that such undesirable process as the "training effect" which is the peculiarity of complete magnetization cycles, it is expedient not to use the change of magnetic field polarity when controlling under-magnetized composites with hard magnetic phase. Thus, for a practical estimation of possibility of adjustment of a material containing a various ratio of hard magnetic and soft magnetic phases, it is at least necessary to estimate remanence of a magnetically hard phase and initial curves of magnetization of the pre-magnetized composite, i.e. minor half-loops of the hysteresis. The issue of the mutual influence of the soft and hard magnetic components on the macroscopic magnetization of the material is considered in details experimentally and theoretically in very recent work [60].

15.4 Magneto-mechanical response

There are several approaches to measure the magneto-mechanical properties of hybrid elastic composites in conventional setups with the addition of an external homogeneous magnetic field (see e.g. Figures 15.10 and 15.11).

In the past rheometric devices were adapted to the measurement of magneto-rheological fluids with commercial magnetic cells [61–63]. Therefore, such devices are very often proposed to use for magnetic elastomers. It uses a plate-plate configuration with a disk-shaped specimen (Figure 15.10a). Critical issues of the method are discussed in the study by Borin et al. [64]. This method should be avoided without special modifications made in a commercial cell that is not originally designed for elastic materials. As shown in the study by Borin et al. [64], when using a rheometer, measuring cell configuration for rod-shaped samples can be used effectively for quasi-static elongation and torsional tests as well as for oscillating torsional tests (Figure 15.10b). From the field of solid mechanics, a universal test machine can be adopted to conduct a quasi-static and dynamic axial and shear loading (Figure 15.11).

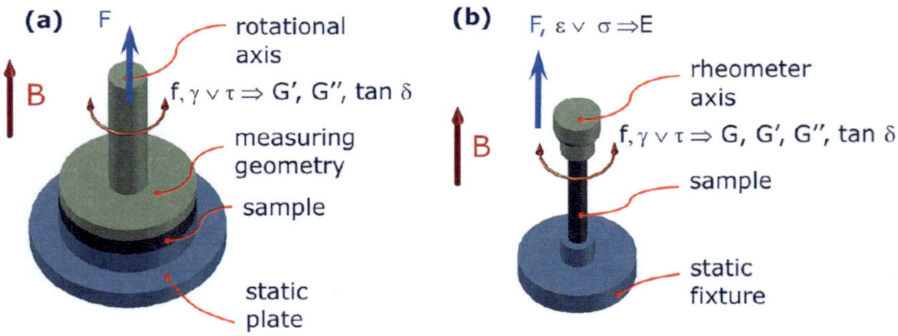

Figure 15.10: Configurations of the experimental setup using rotational rheometer: (a) Plate-plate configuration for the disk-shaped sample; (b) testing of the rod-shaped specimen; B – external magnetic field applied during the test, F – axial mechanical force, f – oscillating frequency; γ – shear deformation, τ – shear stress, G' – storage shear modulus, G'' – loss shear modulus, tanδ – loss factor, ϵ – tensile deformation, σ – tensile stress, E – tensile modulus.

Results of the axial and shear loading provide an information on the tensile modulus E and shear modulus G correspondingly. To the best of our knowledge, a direct systematic comparison of the tensile and shear moduli of magnetic elastic composites obtained using different experimental methods is missing from scientific publications. Certain aspects on this topic were addressed e.g. in the study by Han et al. [65] as well as recently by Borin et al. [39]. On the other hand, the fact, that there are no commercially available magnetic cells for tensile testing machines as well as for rod-shaped specimen rheometry is essential. The easiest solution for creating a homogeneous magnetic field is to use a cylindrical magnetic coil of a given size. The coil provides a magnetic field oriented parallel to the direction of the axial mechanical force F. However, in order to realize the possibility of changing the direction of the applied field relative to

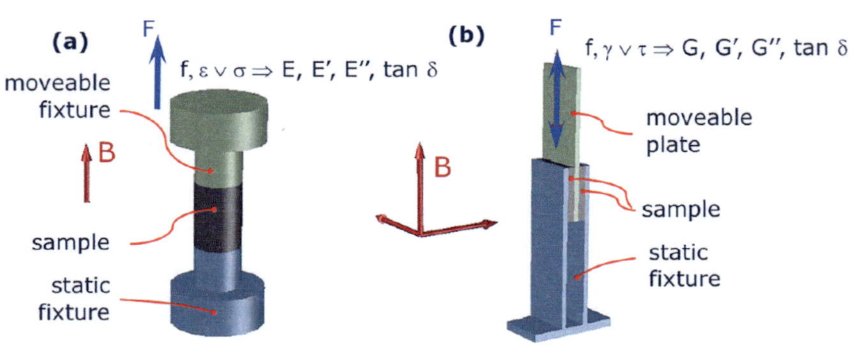

Figure 15.11: Configurations of the experimental setup using universal test machine: (a) Axial loading of the cylindrical specimen; (b) shear test of the plate-shaped composite using a layout with two coupled shear gaps; E' and E'' – storage tensile and loss tensile modulus correspondingly, other designations are as given in the caption of Figure 15.10.

mechanical stress, it is necessary to use a magnetic yoke with appropriate geometry. In the case of a tensile testing machine, this feature is most easily implemented for the shear test using a layout with two coupled shear gaps (Figure 15.11b). However, this configuration requires that the two samples be fully identical, including the conditions of their fixation in the measuring cell. This is not always feasible, especially in the context of soft composites. When using a combination of the rotational rheometer and conventional electromagnetic coil, the direction of the magnetic field with respect to mechanical stress is also different in the case of tensile and torsional loading (Figure 15.10b). The critical factor determining the correctness of the results obtained using this arrangement is the lack of magnetic elements in the construction of the rheometer, e.g. couplings, rotational axis etc. However, standard commercial device configurations (e.g. various modifications of Anton Paar and Thermo Scientific rheometers) contain magnetizing structural elements that distort the results of torque and axial force measurements. Another important methodological factor is the need to critically assess the raw data received from the instrument. Unfortunately, this is not always possible with commercial devices with proprietary software.

Taking the above into account, the most suitable experimental method to evaluate the magneto-mechanical response of elastic hybrid elastomers is the tensile and compression testing (see e.g. Figure 15.12). The method can provide an easier fixation of the sample and custom made magnetic cells can be as well more easily integrated into the setup. Within the method either quasi-static or dynamic loading is possible.The maximum displacement s used for basic characterization of the composites should be limited to the area of linear deformation and to avoid significant influence of radial deformation of the specimen on the measurement results. In non-quasi-static

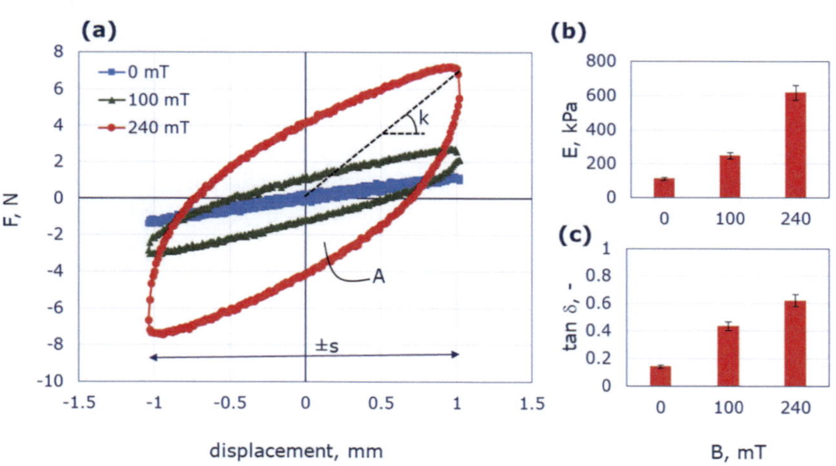

Figure 15.12: Force-displacement loops (a), storage modulus E' (b) and loss factor η (c) measured at $f = 0.1$ Hz, $\epsilon \sim 5\%$ for a magnetic hybrid elastic composite based on carbonyl iron microparticles with concentration ~40 vol% at various magnetic fields.

measurement, a dependence of axial force F on displacement s is experimentally obtained. The ratio F/s gives the stiffness of the material. To determine the tensile modulus E, the measured forces F and displacements s are transformed into stress σ and strain ϵ values, respectively, taking into account the geometry of the specimen. The modulus E is when defined as the slope of the stress-strain curve. The modulus E is when defined as the slope of the stress-strain curve. As a result of the dynamic loading of a viscoelastic material one obtains hysteretic $F(s)$ loops which allow to access such material parameters as the storage tensile modulus E' and the loss factor η ($\tan\delta$). The loss factor in turn represents a ratio between the loss and storage TENSILE MODULUS $\eta = \tan\delta = E''/E'$. The slope k of the major axis of the force–displacement loop provides the dynamic stiffness of the specimen, while the area A within the loop provides the dissipated energy per cycle of loading $A = \int F ds = \pi\eta k\Delta s^2$ (see as well Figure 15.12a). The storage modulus E' is obtained from the slope of the main axis of the stress–strain loop by obtaining values of σ and ϵ from the force F and displacement s, respectively. Figure 15.12 shows an example of a carbonyl iron-based composite response (particle concentration ~40 vol%) resulting from dynamic loading in an external homogeneous magnetic field of various induction.

For quantitative estimation of influence of a magnetic field on viscoelastic properties of a composite material, it is convenient to use the MR effect calculated from the experimentally obtained values of a certain physical parameter of a material. For example, for the elastic storage modulus E' and the loss factor η, the relative active MR effect R_a under the externally applied during the measurements magnetic field B is calculated correspondingly as follows:

$$R_{a\_E'} = \frac{E'(B) - E'_0}{E'_0}, \tag{15.2}$$

$$R_{a\_\eta} = \frac{\eta(B) - \eta_0}{\eta_0}, \tag{15.3}$$

where E'_0 and η_0 are storage and loss factor at zero field ($B = 0$). For samples containing a magnetically hard component, the concept of passive MR effect (R_p) is introduced, which takes into account the effect of pre-magnetization on the viscoelastic properties of the composite. Similar to Eqs. (15.2) and (15.3), this relative effect is quantified as

$$R_{p\_E'} = \frac{E'(B_M) - E'_0}{E'_0}, \tag{15.4}$$

$$R_{p\_\eta} = \frac{\eta(B_M) - \eta_0}{\eta_0}, \tag{15.5}$$

where B_M denotes a flux density of the external field applied to magnetize a composite sample, prior to it's characterization, i.e. no field is applied during the measurements.

Below is an overview of the experimental results obtained for samples of hybrid elastic composite with different magnetic fillers in the context of magnetorheological effect. All given results are obtained in the axial cyclic loading of cylindrical samples at a frequency of 1 Hz and maximal deformation of 5%. Concentration of the powder in all considered below samples is ~40 vol%. The matrix composition is selected so that the initial module of the all samples is E_0 ~ 120–140 kPa. Figures 15.13 and 15.14 show an influence of the externally applied field B and the field B_M, used for the pre-

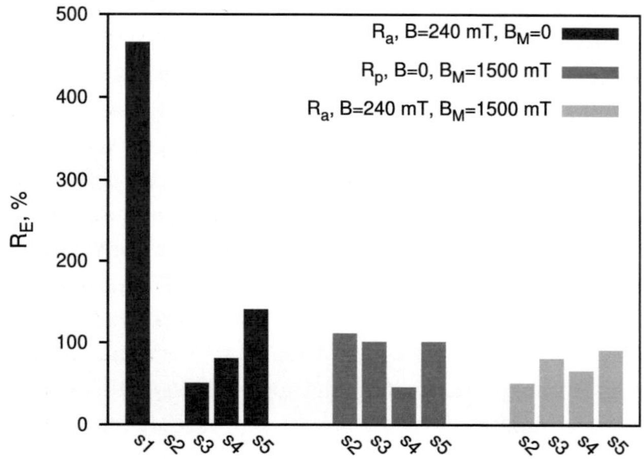

Figure 15.13: Active and passive MR effects (R_a and R_p) respective elastic storage modulus E' for composites with various filler.

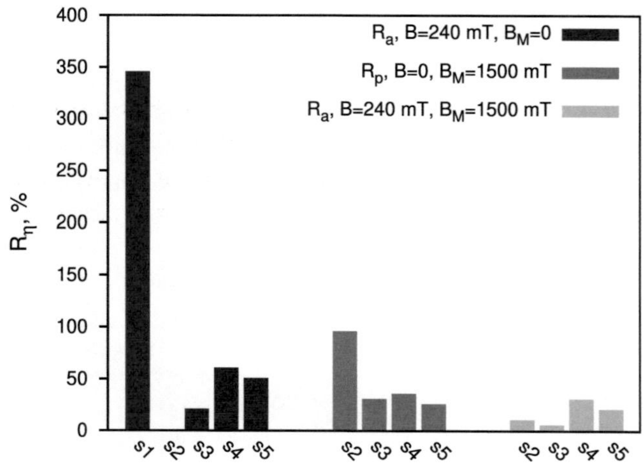

Figure 15.14: Active and passive MR effects (R_a and R_p) respective loss factor η for composites with various filler.

magnetization, on the elastic modulus E and loss factor η, respectively, for hybrid composites with a various filler (Table 15.1).

The influence of different ratios of magnetically hard and soft phases on the magneto-mechanical response of hybrid composites was considered in particular in the study by Borin et al. [38]. The current overview provides representative results obtained for samples of mixed composition with 25% carbonyl iron. Sample s1 serves more as a reference one, as its passive magnetic tuning is not possible.

First consider MR effect of the composites without pre-magnetization. The field B is not high enough to change initial state of the samples s2 and s2 containing magnetically hard phase. In this case, the sample s3 containing magnetically hard particles of irregular shape reacts to an external field, while the sample s2 with spherical magnetically hard particles remains uncontrollable. It is noteworthy that the magnetic susceptibility of particles in sample s3 is lower than that of particles in the sample s2. This mismatch of the reaction of the composite to the external magnetic field should be explained in terms of the variable microstructure. As demonstrated in performed microCT studies, see e.g. the study by Borin et al. [37], particles of irregular shape are rotating reversibly in already moderate fields, while their STRUCTURING is not occurred as well as a structuring of the spherical magnetically hard particles. The rotation of spherical particles can not be tracked using microCT method, due to obvious reasons, however, it can be assumed according to the results of the magnetic measurements [49]. Thus, the rotation of irregularly shaped particles in the applied field is sufficient to make the originally non-magnetized composite magnetically controlled. In the sample s1, the filler particle material has the highest magnetic susceptibility (see Figure 15.6). Due to the high dipole–dipole interaction induced by the magnetic field, the particles are aggregated in the matrix. It is obvious that as a result the MR effect of this composite significantly exceeds the MR effect of the sample s3. When to use complex filling, that is to add to composites with magnetically hard particles carbonyl iron, it is possible to increase the MR effect considerably. Thus, sample s4, in which part of spherical NdFeB-alloy particles is replaced by carbonyl iron (25%) becomes actively controlled. In sample s5 the same fraction of irregular NdFeB-alloy particles is replaced by carbonyl iron and its MR effect also becomes higher than in the sample s3. It can be assumed that the MR effect in the sample s4 is caused by formation of structures of

Table 15.1: Composition of specimens filler. Overall powder concentration in each specimen is ~40 vol%.

Specimen	Carbonyl iron	Spherical NdFeB	Irregular NdFeB
s1	100%	0%	0%
s2	0%	100%	0%
s3	0%	0%	100%
s4	25%	75%	0%
s5	25%	0%	75%

particles, while in the sample s5, in addition to the structuring, the rotation of irregularly shaped particles additionally affects the magneto-mechanics of the composite.

Application of much higher external fields to a composite containing a magnetically hard phase caused appearance of the remanence M_r (see Figure 15.7). Besides the appearance of remanence, the microstructure of such a composite is irreversibly altered [37]. Figures 15.13 and 15.14 as well demonstrate how the magnetizing field B_M changes an initial state of the composite specimens filled with various powders. Changes of up to approximately 100% in material elasticity are possible. Replacing the magnetically hard fraction with a soft one reduces the possibility of the passive tuning. This is most noticeable for a spherical NdFeB-alloy particle-based composite (specimen s4). On the other hand, pre-magnetization provides active control of sample s2, but reduces the range of active control of composites with mixed filler s4 and s5. Comparing an influence of the pre-magnetization on the modulus E' with the influence on the loss factor η one observes that the results for the samples s2 and s3 are inconsistent. While pre-magnetization increases the control range of the elasticity, their loss factor becomes almost uncontrollable.

Obviously, the explanations for the observed behavior of composites with different fillers are related to microstuctural changes induced by the applied magnetic field. It should be distinguished between the processes of the structuring and rotation of magnetic microparticles inside the matrix. Both processes are confirmed within microstructural observations, e.g. in the studies by Borin and coworkers [37, 66]. Use of the complex mixed powder provides broad passive tuning and remote active control of the composite which can be chosen for the demand of certain application. The NdFeB-alloy powder with an irregular morphology provides enhanced tuning of the hybrid elastic composite viscoelastic response despite weaker magnetization and lower magnetic susceptibility of the powder particles than compared to spherical NdFeB-alloy particles.

15.5 Summary and outlook

An overview of the basic magneto-mechanical properties of the hybrid elastic composite highly filled with various types of magnetic microparticles is given. Composite based on carbonyl iron particles is compared with specimens containing magnetically hard NdFeB-alloy powders of two different morphology and with specimens containing mixed powder of carbonyl iron and NdFeB-alloy particles. Material based on spherical NdFeB powder only is not controllable in a non-magnetized state, while when using NdFeB particles of irregular shape the composite responds to an external field without to be pre-magnetized. It is related to a rotation of particles within the matrix as follows from performed elsewhere microstrucural observations, see e.g. [37, 66]. The addition of magnetically soft carbonyl iron microparticles into the composite with NdFeB powder enhances the active MR effect of the non-magnetized samples significantly, while the presence of magnetically hard particles still provoke the composite to have a remanence and, consequently, adjustable initial state. Premagnetization of the composite based on a mixed powder slightly restricts the possibilities of its active control.

The presented results on the evaluation of active control and passive tuning of magnetic hybrid elastic composites show wide prospects of their potential applications. However, despite great efforts in the field of experimental characterization of these composites, theoretical approaches are still based on simplified assumptions, e.g. [7, 8, 47, 67]. On the other hand, the studies on modeling of magnetic hybrid composites are currently limited to elastic polymers with one type of magnetic particles and does not consider the complex mixed filling of matrices with magnetically soft and hard powders [68, 69]. This is due not only to the complexity of understanding the physical processes of interaction between magnetic particles of different types and the elastic polymer matrix. It is also problematic that the use of different experimental techniques leads to controversial results, the correct interpretation of which is difficult. Thus, in addition to proposition of further theoretical approaches, first and foremost ways must be found to develop and implement standardized mechanical testing methods to experimental characterization of magnetic hybrid elastic materials. Furthermore, since a high cyclic stability of the material will obviously required for a technological application of the magnetic hybrid materials, procedures will also have to be developed to investigate and quantify the change of the materials by repeated magnetic stimulation. The corresponding data obtained have to be in turn reflected back into the synthesis process in order to enable higher cyclic stability with constant MR effect. In this context, the morphology of magnetic microparticles and their chemical-physical coupling with the polymer matrix will play an important role. The first steps in this respect are e.g. studies of the magneto-mechanical coupling of single domain particles into viscoelastic polymeric matrices [70].

Acknowledgment: The authors would like to thank Prof. Stefan Odenbach for providing laboratory facilities of the Chair of Magnetofluiddynamics, Measuring and Automation Technology at Tecnische Universität Dresden.
Author contribution: All the authors have accepted responsibility for the entire content of this submitted manuscript and approved submission.
Research funding: Financial support by Deutsche Forschungsgemeinschaft (DFG) under Grant BO 3343/2-1 within SPP1681 and DFG-RFBR PAK907 providing the basis for our investigations is gratefully acknowledged. G.S. would like to acknowledge the support of RFBR under Grant 16-53-12009.
Conflict of interest statement: The authors declare no conflicts of interest regarding this article.

References

1. Wereley N. Magnetorheology: advances and applications. Cambridge: RSC Publishing; 2014.
2. Stepanov GV, Borin DY, Raikher YL, Melenev PV, Perov NS. Motion of ferroparticles inside the polymeric matrix in magnetoactive elastomers. J Phys Condens Matter 2008;20:204121.
3. Lanotte L, Ausanio G, Iannotti V, Pepe G, Carotenuto G, Netti P, et al. Magnetic and magnetoelastic effects in a composite material of Ni microparticles in a silicone matrix. Phys Rev B 2001;63: 054438.
4. Martin JE, Anderson RA, Read D, Gulley GL. Magnetostriction of field-structured magnetoelastomers. Phys Rev E 2006;69:051507.
5. Danas K, Kankanala SV, Triantafyllidis N. Experiments and modeling of iron-particle-filled magnetorheological elastomers. J Mech Phys Solid 2012;60:120.
6. Zubarev AY, Borin DY. Effect of particle concentration on ferrogel magnetodeformation. J Magn Magn Mater 2015;377:373–7.
7. Zubarev A, Chirikov D, Stepanov G, Borin D. Hysteresis of ferrogels magnetostriction. J Magn Magn Mater 2017;431:120–2.
8. Zubarev A, Chirikov D, Stepanov GV, Borin D. On the theory of hysteretic magnetostriction of soft ferrogels. Phys A 2018;498:86–95.
9. Borin DY, Odenbach S, Stepanov GV. Stress induced by the striction of magnetoactive elastic composites. J Magn Magn Mater 2019;470:85–8.
10. Stolbov O, Raikher Y. Magnetostriction effect in soft magnetic elastomers. Arch Appl Mech 2019; 89:63–76.
11. Nikitin LV, Mironova LS, Stepanov GV, Samus AN. The influence of magnetic field on the elastic and viscous properties of magnetoelastics. Polym Sci 2001;43:443.
12. Böse H. Viscoelastic properties of silicone-based elastomers. Int J Mod Phys B 2007;21:4790.
13. Glavan G, Salamon P, Belyaeva IA, Shamonin M, Drevensek-Olenik I. Tunable surface roughness and wettability of a soft magnetoactive elastomer. J Appl Polym Sci 2018;135:46221.
14. Chen S, Dong S, Wang X, Li W. Magneto-induced surface morphologies in magnetorheological elastomer films: an analytical study. Smart Mater Struct 2019;28.045016.
15. Glavan G, Kettl W, Brunhuber A, Shamonin M, Drevensek-Olenik I. Effect of material composition on tunable surface roughness of magnetoactive elastomers, Polymers 2019;11:594.
16. Kchit N, Bossis G. Electrical resistivity mechanism in magnetorheological elastomer. J Phys D Appl Phys 2009;42:105505.
17. Stepanov GV, Semerenko DA, Bakhtiiarov AV, Storozhenko PA. Magnetoresistive effect in magnetoactive elastomers. J Supercond Nov Magnetism 2013;26:1055–9.
18. Belyaeva IA, Kramarenko EY, Shamonin M. Magnetodielectric effect in magnetoactive elastomers: transient response and hysteresis. Polymer 2017;127:119–28.
19. Lanotte L, Ausanio G, Hison C, Iannotti V, Luponio C. The potentiality of composite elastic magnets as novel materials for sensors and actuators. Sens Actuators A Phys 2003;106:56.
20. Borin D, Stepanov G, Mikhailov V, Gorbunov A. The damping device based on magnetoactive elastomer. Magnetohydrodynamics 2007;43:437.
21. Li Y, Li J, Li W, Du H. A state-of-the-art review on magnetorheological elastomer devices. Smart Mater Struct 2014;23:123001.
22. Volkova T, Böhm V, Kaufhold T, Popp J, Becker F, Borin D, et al. Motion behaviour of magneto-sensitive elastomers controlled by an external magnetic field for sensor applications. J Magn Magn Mater 2017;431:262–5.
23. Becker TI, Zimmermann K, Borin DY, Stepanov GV, Storozhenko PA. Dynamic response of a sensor element made of magnetic hybrid elastomer with controllable properties. J Magn Magn Mater 2018;449:77–82.

24. Alekhina YA, Makarova LA, Kostrov SA, Stepanov GV, Kazimirova EG, Perov NS, et al. Development of magnetoactive elastomers for sealing eye retina detachments. J Appl Polym Sci 2019;136:47425.
25. Lokander M, Stenberg B. Performance of isotropic magnetorheological rubber materials. Polym Test 2003;22:245–51.
26. Kalio M. The elastic and damping properties of magnetorheological elastomers. Espoo: VTT Publications; 2005.
27. Filipcsei G, Csetneki I, Szilgyi A, Zrinyi M. Magnetic field-responsive smart polymer composites. Adv Polym Sci 2007;206:137–89.
28. Abrougui Mariem M, Lopez-Lopez Modesto T, Duran Juan DG. Mechanical properties of magnetic gels containing rod-like composite particles. Phil Trans Royal Soc A 2018;377. https://doi.org/10.1098/rsta.2018.0218.
29. Stepanov GV, Abramchuk SS, Grishin DA, Nikitin LV, Kramarenko EY, Khokhlov AR. Effect of a homogeneous magnetic field on the viscoelastic behavior of magnetic elastomers. Polymer 2007;48:488–95.
30. Chertovich AV, Stepanov GV, Kramarenko EY, Khokhlov AR. New composite elastomers with giant magnetic response. Macromol Mater Eng 2010;295:336–41.
31. Ausanio G, Hison C, Iannotti V, Luponio C, Lanotte L. Elastomagnetic effect in novel elastic magnets. J Magn Magn Mater 2004;272:2069.
32. Borin DY, Stepanov GV. Oscillation measurements on magnetoactive elastomers with complex composition. J Optoelectron Adv Mater 2013;3-4:249–53.
33. Borin DY, Stepanov GV, Odenbach S. Tuning the tensile modulus of the magnetorheological elastomer with magnetically hard powder. J Phys Conf Ser 2013;412:012040.
34. Stepanov GV, Borin DY, Kramarenko EY, Bogdanov VV, Semerenko DA, Storozhenko PA. Magnetoactive elastomer based on magnetically hard filler: synthesis and study of viscoelastic and damping properties. Polym Sci 2014;56:603–13.
35. Stepanov GV, Borin DY, Storozhenko PA. Rotation of magnetic particles inside the polymer matrix of magnetoactive elastomers with a hard magnetic filler. J Magn Magn Mater 2017;431:138–40.
36. Stepanov GV, Borin DY, Bakhtiiarov AV, Storozhenko PA. Magnetic properties of hybrid elastomers with magnetically hard fillers: rotation of particles. Smart Mater Struct 2017;26:035060.
37. Schümann M, Borin DY, Huang S, Auernhammer GK, Müller R, Odenbach S. A characterisation of the magnetically induced movement of NdFeB-particles in magnetorheological elastomers. Smart Mater Struct 2017;26:095018.
38. Borin DY, Stepanov GV, Dohmen E. Hybrid magnetoactive elastomer with a soft matrix and mixed powder. Arch Appl Mech 2019;89:105–17.
39. Borin DY, Stepanov GV, Dohmen E. On anisotropic mechanical properties of heterogeneous magnetic polymeric composites. Phil Trans Royal Soc A 2019;377:20180212.
40. Stepanov GV, Borin DY, Bakhtiiarov AV, Storozhenko PA. Hybrid magnetic elastomers prepared on the basis of a SIEL-grade resin and their magnetic and rheological properties. Phys Sci Reviews this issue 2020.
41. Magnequench. MQP-S-11-9-20001-070 isotropic powder. Maerial description; 2009.
42. Borin D, Günther D, Hintze C, Heinrich G, Odenbach S. The level of cross-linking and the structure of anisotropic magnetorheological elastomers. J Magn Magn Mater 2012;324:3452–345.
43. Günther D, Borin D, Günther S, Odenbach S. X-ray micro-tomographic characterization of field-structured magnetorheological elastomers. Smart Mater Struct 2012;21:015005.
44. Borbáth T, Günther S, Borin D, Gundermann T, Odenbach S. X-ray microCT analysis of magnetic field-induced phase transitions in magnetorheological elastomers. Smart Mater Struct 2012;21:105018.

45. Borin D, Odenbach S, Iskakova L, Zubarev A. Non-ergodic tube structures in magnetic gels and suspensions. Soft Matter 2018;14:8537–44.
46. Borin D Targeted patterning of magnetic microparticles in a polymer composite. Phil Trans Royal Soc A 2020;378:20190256.
47. Zubarev AY, Chirikov DN, Borin DY, Stepanov GV. Hysteresis of the magnetic properties of soft magnetic gels. Soft Matter 2016;12:6473.
48. Odenbach S. Microstructure and rheology of magnetic hybrid materials. Arch Appl Mech 2016;86: 269–79.
49. Stepanov GV, Borin DY, Bakhtiiarov AV, Storozhenko PA. Negative coercivity of magnetic elastomers filled with magnetically hard particles. J Magn Magn Mater 2020;498:166125.
50. Stepanov GV, Borin DY, Bakhtiiarov AV, Storozhenko PA. Influence of the size of magnetic filler particles on the properties of hybrid magnetic elastomer with magnetically hard filler. J Magn Magn Mater 2020;498:166071.
51. Mullins L. Softening of rubber by deformation. Rubber Chem Technol 1969;42:339–62.
52. Garnett JCM. Colours in metal phases and in metallic films. Phil Trans Royal Soc Lond 1904;203: 385–91.
53. Martin JE, Venturini E, Gulley GL, Williamson J. Using triaxial magnetic fields to create high susceptibility particle composites. Phys Rev E 2004;69:021508.
54. Bossis G, Volkova O, Lacis S, Meunier A. Magnetorheology: fluids, structures and rheology. In: Odenbach S, editor. Ferrofluids. Magnetically controllable fluids and their applications, Lecture notes in physics. Berlin Heidelberg: Springer; 2002, vol. 594.
55. Vaganov M, Borin D, Odenbach S, Raikher Y. Effect of local elasticity of the matrix on magnetization loops of hybrid magnetic elastomers. J Magn Magn Mater 2018;459:411866.
56. Vaganov M, Borin D, Odenbach S, Raikher YL. Modeling the magnetomechanical behavior of a multigrain magnetic particle in an elastic environment. Soft Matter 2019;15:4947.
57. Vaganov M, Borin D, Odenbach S, Raikher YL. Mesomagnetomechanics of hybrid elastomer composites: Magnetization of elastically trapped particles. J Magn Magn Mater 2020;499:166249.
58. Vaganov M, Borin D, Odenbach S, Raikher YL. Training effect in magnetoactive elastomers due to undermagnetization of magnetically hard filler. Physica B 2020;578:411866.
59. Linke JM, Borin DY, Odenbach S. First-order reversal curve analysis of magnetoactive elastomers. RSC Adv 2016;6:100407–16.
60. Becker TI, Stolbov OV, Borin DY, Zimmermann K, Raikher YL. Basic magnetic properties of magnetoactive elastomers of mixed content. Smart Mater Struct 2020;29:075034.
61. Laeuger J, Wollny K, Stettin H, Huck S. A new device for the full rheological characterization of magneto-rheological fluids. Int J Mod Phys B 2005;19:1353–9.
62. Laun HM, Schmidt G, Gabriel C, Kieburg C. Reliable plate–plate MRF magnetorheometry based on validated radial magnetic flux density profile simulations. Rheol Acta 2008;47:1049–59.
63. Gabriel C, Kieburg C, Laun HM. Clutch and brake related testing of magnetorheological fluids using the BASF twin gap magnetocell. Appl Rheol 2010;20:41778.
64. Borin D, Kolsch N, Stepanov G, Odenbach S. On the oscillating shear rheometry of magnetorheological elastomer. Rheol Acta 2018;57:217–27.
65. Han Y, Hong W, Faidley LAE. Field-stiffening effect of magneto-rheological elastomers. Int J Solid Struct 2013;50:2281–8.
66. Schümann M, Borin D, Morich J, Odenbach S. Reversible and non-reversible motion of NdFeB-particles in magnetorheological elastomers. J Intell Mater Syst Struct 2020. https://doi.org/10.1177/1045389X20949703.
67. Zubarev A, Musikhin A, Chirikov D, Borin D. Elastic stress in ferrogels with chain aggregates. J Magn Magn Mater 2020;498:166126.

68. Kalina K, et al. submitted to this issue.
69. Menzel A, et al. submitted to this issue.
70. Hess M, Gratz M, Remmer H, Webers S, Landers J, Borin D, et al. Scale-dependent particle diffusivity and apparent viscosity in polymer solutions as probed by dynamic magnetic nanorheology. Soft Matter 2020. https://doi.org/10.1039/c9sm00747d.

Kerstin Birster, Rouven Schweitzer, Christoph Schopphoven and Andreas Tschöpe*

16 Magnetic torque-driven deformation of Ni-nanorod/hydrogel nanocomposites

Abstract: Nickel (Ni) nanorods were prepared by the anodized aluminum oxide (AAO) template method and dispersed in poly(acrylamide) (PAM) hydrogels. The deformation of the magnetoresponsive composites was studied with particular attention to the consequences of finite magnetic shape anisotropy as compared to rigid dipoles on the field-dependent torque. For comparison with experiments, the composite was described as an elastic continuum with a local magnetic torque density, applied by discrete particles and determined by the local orientation of their magnetic anisotropy axis with respect to the magnetic field. The mean magnetic moment of the single domain particles m and their volume density in the composite φ_{vol} were derived from the static field-dependent optical transmission (SFOT) of linear polarized light. The mechanical coupling between the particles and their viscoelastic environment was retrieved from the rotational dynamics of the nanorods using oscillating field-dependent optical transmission (OFOT) measurements. Field- and orientation-dependent magnetization measurements were analyzed using the Stoner–Wohlfarth (SW) model and a valid parameter range was identified by introducing an effective anisotropy constant K_A as a new empirical model parameter. This adapted SW-model for quantitative description of the field- and orientation dependence of the magnetic torque was validated by measuring the local rotation of nanorods in a soft elastic hydrogel. Finally, torsional and bending deformation of thin magnetically textured composite filaments were computed and compared with experiments.

Keywords: actuator, anisotropy, bending, ferromagnetic, magnetic soft matter, magnetic torque, nanocomposite, nanorod, shape-programmable, torsion

16.1 Introduction

Dispersing magnetic particles in a nonmagnetic matrix enables the transmission of forces and torques to the material without contact. In a mechanically soft environment, the propulsion of the magnetic inclusions entails elastic deformations and such composites constitute a particular class of shape-programmable matter. Compared to

***Corresponding author: Andreas Tschöpe,** Experimentalphysik, Universität des Saarlandes, Campus E2 6, D-66123 Saarbrücken, Germany, E-mail: a.tschoepe@nano.uni-saarland.de

Kerstin Birster, Rouven Schweitzer and Christoph Schopphoven, Experimentalphysik, Universität des Saarlandes, Campus D2 2, D-66123 Saarbrücken, Germany

This article has previously been published in the journal Physical Sciences Reviews. Please cite as: K. Birster, R. Schweitzer, C. Schopphoven and A. Tschöpe "Magnetic torque-driven deformation of Ni-nanorod/hydrogel nanocomposites" *Physical Sciences Reviews* [Online] 2020, 5. DOI: 10.1515/psr-2019-0089 | https://doi.org/10.1515/9783110569636-016

other possible stimuli, magnetic actuation stands out because—in principle—complex deformation patterns can be achieved by a suitable combination of composite structure and time-variable magnetic field [1–6]. The versatility of possible motion patterns depends first of all on the magnetic properties of the particles. Soft magnetic ferrite microparticles, commonly used in magnetorheological elastomers (MRE) [7], adopt a multi-domain structure with vanishing remanence. The magnetization of such particles increases and finally saturates with increasing applied field. Deformations are controlled by magnetic dipolar interactions between the inclusions [8] and limited to axial elongation and compression of the composite. Linear stretching and contraction was also observed in ferrogels, in which the mechanical compliance of the hydrogel matrix is adapted to the lower magnetic dipolar forces between the ferrite nanoparticles [9]. The magnetoelastic coupling on the microscopic scale and its consequences on the macroscopic response of such composite materials is the topic of several chapters in the present volume.

In order to fully exploit the potential of magnetic actuation, however, it is necessary to apply magnetic forces as well as torques. The latter requires a significant magnetic anisotropy. Using soft magnetic multidomain or superparamagnetic nanoparticles, anisotropy was achieved by field-induced self-organization of the particles into linear aggregates in the liquid precursor state of the composite, which was then frozen during solidification [4, 6, 10]. An alternative and direct approach is the use of hard ferromagnetic particles with high remanence and coercivity [11, 12]. Programming the desired shape variation then involves the computation of the density and orientation distribution of magnetic particles. A corresponding methodology was derived for the case of non-interacting magnetic dipoles in an elastic continuum [2]. Magnetically responsive soft materials with programmed texture were fabricated using a two-step molding process [2] or 3D-printing of ferromagnetic domains [13].

A direct coupling between rigid magnetic dipoles and the elastic continuum is an adequate approximation for hard ferromagnetic particles and driving fields far below their coercivity [14]. In the general case of finite magnetic anisotropy, however, the magnetic moments rotate out of the anisotropy axes, which reduce the magnetic torque per particle as compared to a rigid dipole. The present chapter reviews recent studies on the effect of finite magnetic anisotropy on the torque-based deformation of Ni-nanorod/hydrogel composites. Due to their size and shape, the Ni nanorods are uniaxial ferromagnetic single domain particles with a total magnetic moment preferably oriented along the cylinder axis. By alignment of nanorods in the liquid precursor solution and during the polymerization process of the hydrogel network, textured nanocomposites can be obtained. These materials allow detailed studies on the magnetic anisotropy of the nanorods and the torque-driven local rotation in a soft elastic environment. The central part is devoted to a quantitative semi-empirical model for the field- and angular dependence of the magnetic torque acting on the Ni nanorods. Model calculations are compared with experimental results of the field-dependent torsion and bending of textured Ni-nanorod/hydrogel filaments.

16.2 Synthesis and physical properties of Ni nanorods

16.2.1 Synthesis

The synthesis of Ni nanorods makes use of anodized aluminum oxide (AAO) templates with a regular cylindrical pore structure [15–17], Figure 16.1 (left). Anodization of electropolished aluminum foils in 1 M H_2SO_4 at $U = 15$ V, following the two-step procedure by Masuda and Satoh [18], and further processing in 0.1 M H_3PO_4 [19] provided an oxide layer with ordered cylindrical channels and a mean pore diameter $D_p \approx 20$ nm, Figure 16.1 (left/top). The nanopores were filled with Ni by pulsed electrodeposition [20] which allowed the variation of the rod length by controlling the number of current pulses [17], Figure 16.1 (left/middle). The alumina templates were slowly dissolved in dilute NaOH solution at pH 11.5 to which poly(vinylpyrrolidone) (PVP) was added to prevent strong aggregation of the nanoparticles, Figure 16.1 (left/bottom). The nanorod dispersions were purified by repeated centrifugation and redispersion in bi-distilled water. For dispersion in PAM precursor solution, the nanorods were further functionalized with poly(acrylic acid) and stabilized at pH 8. Details on the synthesis protocol can be found in references [16, 17, 21].

Poly(acrylamide) (PAM) hydrogels were used as soft elastic matrix for the composite materials. The stiffness could be adjusted by varying the amount of a monomer stock solution (27 wt% acrylamide and 0.364 wt% *N,N′*-methylenebisacrylamide) in the precursor mixture. Copolymerization was started by adding the catalyst 0.2 wt% *N,N,N′,N′*-tetramethylethylenediamine (TEMED) and the initiator 0.18 wt% ammonium persulfate (APS). For fabrication of magnetically textured composites, the nanorod-precursor

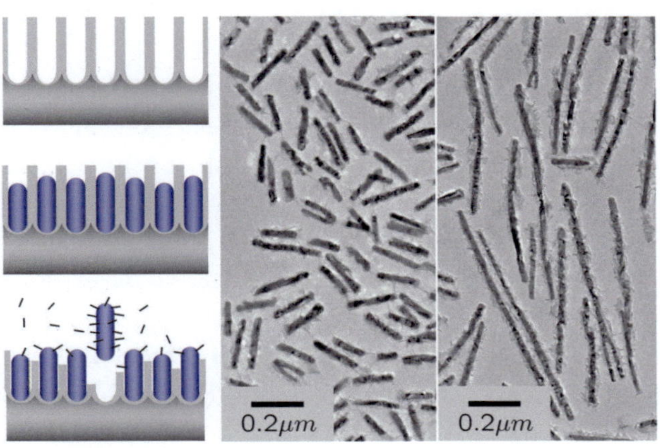

Figure 16.1: The synthesis of Ni nanorods involves the preparation of nanoporous aluminum oxide templates by anodic oxidation (left/top), pulsed electrodeposition of Ni (left/middle), and release of the nanorods by dissolution of the oxide template in the presence of PVP surfactant (left/bottom). TEM images of short (middle) and long (right) Ni nanorods reveal their cylindrical shape but also irregular structures (dendrites) as well as residues of the alumina template at the particle surface.

solution was exposed to a static magnetic field and the aligned nanorods were fixed during polymerization.

16.2.2 Structure

Transmission electron microscopy (TEM) images provide detailed information about the structure of the metallic Ni core. The majority of particles exhibited a linear cylindrical shape with core length L_c and diameter D_c, Figure 16.1 (middle and right). Occasionally there were also irregular structures such as kinks and branches into dendrites. Besides the Ni core with its nanocrystalline substructure, TEM images also revealed a faint contrast at the particle surface which was attributed to oxide residues from the alumina templates [17].

Colloids with different mean lengths of the nanorods in the range of $\langle L_c \rangle \approx 100\,\mathrm{nm} - 600\,\mathrm{nm}$ were synthesized by variation of the electrodeposition pulse numbers [17]. The polydispersity is characterized by the standard deviation of the length distribution and exhibited a linear correlation with the mean value, $\sigma_L \approx 0.23\langle L_c\rangle$. The diameter of the nanorods is determined by the template pore diameter. The preparation protocol and selected anodization conditions resulted in $\langle D_c \rangle \approx 18\,\mathrm{nm} - 26\,\mathrm{nm}$ and $\sigma_L \approx 0.16\langle D_c\rangle$.

16.2.3 Magnetic anisotropy

According to micromagnetic simulations, cylindrical Ni nanorods adopt a single domain state below a critical diameter $D_{cr} \approx 3.5\,(4\pi A/\mu_0 M_s^2)^{1/2} \approx 40\,\mathrm{nm}$ [16, 22], where $M_s = 488 \cdot 10^3\,\mathrm{A/m}$ is the saturation magnetization and $A = 3.4 \cdot 10^{-12}\,\mathrm{J/m}$ the exchange constant of metallic Ni [23]. For a typical length L_c = 350 nm, diameter D_c = 20 nm and corresponding volume $V \approx 1 \cdot 10^{-22}\,\mathrm{m}^3$, the total magnetic moment has the magnitude $m = MsV \approx 5 \cdot 10^{-17}\,\mathrm{Am}^2$ and preferably points along the cylinder axis. The magnetic shape anisotropy responsible for this axial alignment can be described by the Stoner–Wohlfarth (SW) model [24] with the shape anisotropy constant $K_s = \mu_0 M_s^2 (1 - 3N_{||}(n))/4$, where $N_{||}(n)$ is the demagnetization factor along the major principle axis of a homogeneously magnetized prolate spheroid with aspect ratio $n = L/D$ [25]. Within this approximation, the uniaxial shape anisotropy dominates over the weak cubic magnetocrystalline anisotropy $|K_1| = 5\,\mathrm{kJ/m}^3$ of Ni [23] for $N_{||}$<0.31, i.e., aspect ratio $n > 1.1$, and the overall anisotropy constant $K_A \approx K_S$. The energy barrier between the two equivalent magnetization states in opposite directions along the major axis of a particle with volume V, $\Delta E_S = K_S V \approx 4 \cdot 10^{-18}\,\mathrm{J}$ by far exceeds thermal energy so that Ni nanorods of the given size are ferromagnetic.

Due to their large magnetic moment Ni nanorods can be readily aligned by an external magnetic field when suspended in a liquid [26]. This aligned state can be preserved if the precursor of a hydrogel is dissolved in the same liquid volume and polymerized in the presence of the alignment field. The obtained textured nanorod/hydrogel composite with particles fixed in a rigid matrix allowed the investigation of their anisotropic magnetic properties. For ferrogels with uniaxial anisotropy, the experimental results agreed qualitatively with the predictions of the SW-model, e.g., with regard to the single domain state and the angular dependence of the relative remanence m_r/m_S [15]. However, reliable prediction of magnetic actuation requires a model that is also quantitatively consistent. With this specific objective, the field- and angular dependence of reversible magnetization changes (magnetic moment remains in the same local energy minimum) on the one hand and the angular dependence of irreversible switching (magnetic moment changes to the energy minimum in the opposite direction) on the other hand are the most important issues [16, 27].

According to the SW-model, the hysteresis cycle of uniaxial ferromagnets involves both irreversible (switching) and reversible magnetization changes. In particular, the reversible magnetization properties are revealed by the upper branches of the hysteresis curves in the first quadrant measured at different angles Θ between the texture axis and the direction of the applied field, Figure 16.2 (left). The reversible magnetization could be consistently described by the SW-model for all angles $\Theta \leq 70^{\circ}$ when the theoretical anisotropy constant $K_S = 73\ \mathrm{kJ/m^3}$ for an idealized spheroid was replaced by a slightly smaller empirical effective anisotropy constant $K_A = 63\ \mathrm{kJ/m^3}$ [27]. The result from a simultaneous regression analysis is shown in Figure 16.2 (left). The consistency of the SW-model, however, did not hold for $\Theta > 70^{\circ}$. In particular, the prediction of the SW-model for $\Theta = 90^{\circ}$ is a linear increase in magnetization up to the coercive field with constant normalized susceptibility $\tilde{\chi} = \chi/M_S = M_S/(2K_A)$, Figure 16.2 (left, insert). The non-linear gradual approach to saturation, observed in experiment, is a significant qualitative difference from the SW-model which obviously could not be resolved by modifying the value of K_A but could be well described by introducing a distribution function for the anisotropy constant [27].

Irreversible switching of the magnetization results in a reversal of the magnetic torque direction. In most cases, this effect is counterproductive for torque-driven magnetic actuation. To prevent such magnetization reversal the applied field should remain below a critical threshold, determined by the switching field distribution of the magnetic particles. The SW-model does not offer useful guidance in this aspect. The prediction for the switching field at $\Theta = 180^{\circ}$, $H_C = 2K_A/(\mu_0 M_S) = 2580\ \mathrm{Oe}$ was much larger than the mean of the switching field distribution derived from experiments $\bar{H}_C = 960\ \mathrm{Oe}$ [27]. The SW-model typically overestimates the switching field because only magnetization reversal of a homogeneously magnetized spheroid by coherent rotation is considered. However, micromagnetic simulations suggested reversal of the cylindrical nanowires by localized nucleation of a transverse domain wall at fields significantly lower than H_C for coherent rotation [28]. Moreover, for practical purposes, the mean switching field is not a useful threshold value because 50% of the particles already contribute a reversed

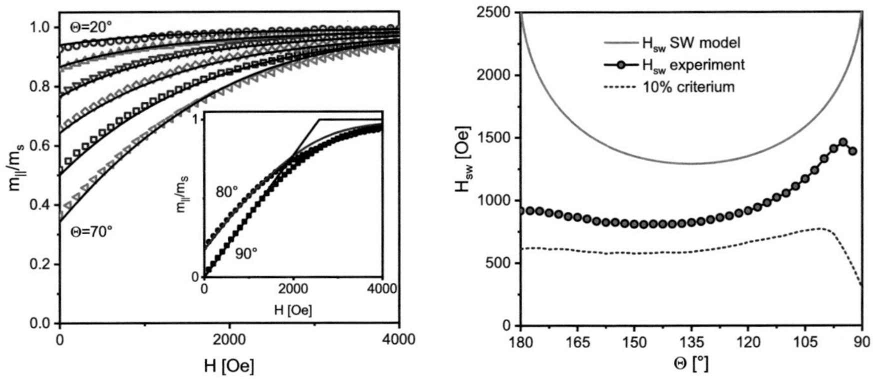

Figure 16.2: Magnetization of Ni-nanorods aligned and fixed in a rigid gelatin matrix: (left) upper branches of hysteresis curves measured at different texture angles $\Theta \leq 70^\circ$ could be consistently described by the SW-model (solid lines) using an effective anisotropy constant $K_A = 63$ kJ/m³. The measurement at $\Theta = 90^\circ$ (black dots, insert) did not show the expected linear increase but could be reproduced by introducing a distribution of magnetic anisotropy constants. (right) The critical switching field H_{sw} was much lower than the prediction by the SW-model, particularly at $\Theta = 180^\circ$. A critical threshold field, at which 10% of the nanorods were irreversibly remagnetized, was obtained at ≈ 600 Oe (dashed line). Reprinted with permission from [27].

magnetic torque. Instead, an empirical threshold was defined corresponding to a low acceptable fraction of switched particles with $H_{c,10\%} = 600$ Oe, Figure 16.2 (right). As a guideline, orientation of Ni nanorod axes below 90° are safe because only reversible magnetization changes occur. By design of an actuator component, the orientation may exceed 90° at zero field, but the particles should rotate to $\Theta < 90^\circ$ either by the deformation of the actuator or local rotation in the elastic matrix [29, 30] before reaching the threshold in order to prevent significant magnetization reversal.

16.2.4 Optical anisotropy

The field-induced alignment of Ni nanorods in a liquid dispersion medium was revealed in the Langevin-type normalized magnetization $m(H)/m_S = L(\zeta) = \coth \zeta - 1/\zeta$ with $\zeta = m\mu_0 H/k_B T$ [31]. An alternative and very useful experimental approach for the measurement of particle orientation uses the anisotropy of the optical extinction cross sections of the metallic nanorods $C_{\text{ext},L}$ and $C_{\text{ext},T_{1,2}}$ for longitudinal and the two transversal polarization directions, respectively. The transmittance of a dilute colloid can be described by the Beer–Lambert law, $\tau = I/I_0 = \exp(-Ns\langle C_{\text{ext}}\rangle)$, where I_0 and I are the light intensities before and after passing an optical path s through a dispersion of N particles per unit volume in a transparent medium. At zero magnetic field, isotropic orientation distribution of the nanorods is expected with the ensemble average extinction cross section $\langle C_{\text{ext}}\rangle_\times = (C_{\text{ext},L} + C_{\text{ext},T_1} + C_{\text{ext},T_2})/3$. With increasing external field H, alignment of the nanorods against thermal energy results in a characteristic field-dependent transmittance. The ensemble average $\langle C_{ext}\rangle(H)$ depends on the direction of polarization with respect to the applied field (e.g., perpendicular $\perp$ or parallel $||$) and is determined by the second moment of the distribution function, given by $\langle\cos^2\beta\rangle = 1 + 2/\zeta^2 - 2\coth(\zeta)/\zeta$. The transmitted intensity $I_\perp$, normalized to the zero field intensity I_x, increased with magnetic field whereas $I_{||}/I_x$ decreased, as expected for the lower electrical polarizability of the nanorods along the short as compared to the long rod axis, Figure 16.3 (left) [26]. By analyzing such static field-dependent optical transmission (SFOT) measurements, the mean magnetic moment per particle m and the particle density N in the colloid could be obtained. A reliable quantitative analysis is only possible on the basis of correct values for the extinction cross section. Parameters were calculated for spheroidal particles in the electrostatic approximation (EA) or using the separation of variables method (SVM) and were also obtained by FEM simulations for spheroids and capped cylinders [32]. The mean magnetic moment per particle was shown to be robust and independent of the chosen model whereas the particle densities were significantly different. For future use of the convenient EA-model analysis, a correction function for the particle density was derived.

Quantitative analysis of absolute transmittance data indicated the presence of an additional contribution to optical extinction, which was independent of the magnetic field [33]. This contribution was attributed to magnetic-optically inactive aggregates of Ni nanorods, e.g., nanorod dimers with compensating anti-parallel magnetic moments. An increasing contribution of this extinction background and correlated decrease of field-dependent optical extinction by individual nanorods was observed

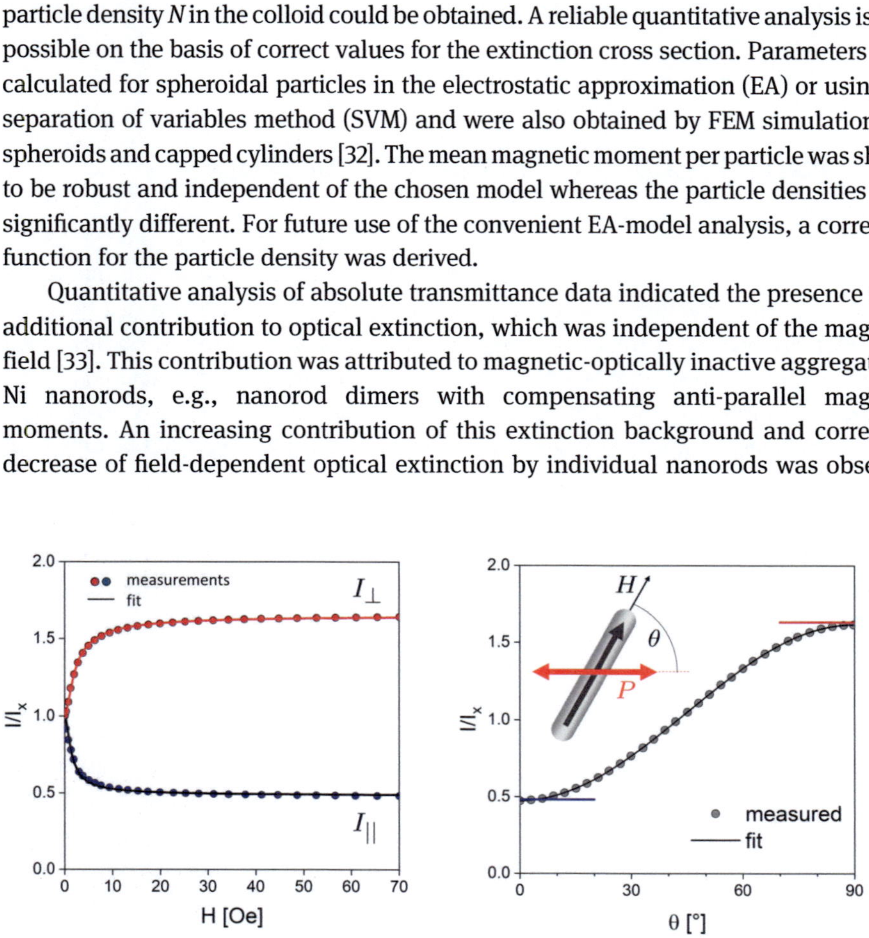

Figure 16.3: (left) Static field-dependent optical transmission (SFOT), normalized to the transmission at zero field, for polarization perpendicular (upper branch) and parallel (lower branch) to the magnetic field. Regression analysis provided the mean magnetic moment per particle $m = 5.2(2) \cdot 10^{-17}\ \mathrm{Am}^2$. (right) If all nanorods are aligned parallel, e.g., at saturation field, the transmittance of the nanorod dispersion depends on the angle θ between the direction of the applied field (parallel to the rod axis) and the polarization direction. This relation was used for optical measurements of the mean nanorod orientation.

after destabilization of a nanorod colloid by the addition of salt. The time-dependence of the extinction components could be described with the Smoluchowski coagulation model [33].

With increasing magnetic field, the transmission saturates as all nanorods in the colloid align along the field. At a sufficiently large value of the Langevin parameter, e.g., $\zeta > 30$, the transmitted intensity depends on the orientation angle of the rod axis with respect to polarization direction as $I(\theta) = I_{||} + (I_{\perp} - I_{||})\sin^2(\theta)$, Figure 16.3 (right). The inverse relation was used to determine the mean orientation of nanorods in a transparent matrix from optical transmission measurements in a variety of experimental methods [17, 21, 27, 31, 34–36].

16.3 Particle-matrix mechanical coupling

The magnetic hysteresis observed with Ni nanorod hydrogels as compared to the superparamagnetic behavior in a liquid medium points to the influence of the matrix properties on the magnetization behavior. Lowering the gelatine content in a hydrogel resulted in an increase in the initial susceptibility [15]. This effect was attributed to the additional torque-driven rotation of the nanorod axis into field direction enabled by the mechanical compliance of the matrix. Systematic changes in the hysteresis of an isotropic nanorod/hydrogel composite, in particular a decrease in coercivity, could be described by an extended SW-model, which included an additional term associated with the deformation energy of the elastic matrix, $E = E_Z + E_S + K_v G\vartheta^2/2$. Here, E_Z and E_S are the Zeeman and shape anisotropy energy, considered in the regular SW-model, G denotes the shear modulus of the matrix and K_v is the hydrodynamic shape factor for rotation of the magnetic particle [29, 37]. Energy minimization can be translated to torque balances,

$$m\mu_0 H \sin\phi = 2K_A(m/M_s)\sin\psi\cos\psi = K_v G\vartheta, \tag{16.1}$$

with the condition $\vartheta + \psi + \phi = \Theta$, Figure 16.4. Three quantities, m, K_A and K_v constitute the minimum set of physical parameters characteristic for a given batch of nanorods which needs to be characterized by experiments and specified in the model. This extended SW-model as well as other protocols were applied for quantitative analysis of magnetization data to deduce the shear modulus of gelatin matrices [29, 38].

Significant field-induced deformations of actuators require adjustment of the matrix stiffness to the magnetic forces and torques acting on the magnetic inclusions. With regard to Ni nanoparticles at moderate volume fraction, this could be achieved by using soft hydrogels. However, the concomitant mesh size of the polymeric network, $\xi \approx (k_B T/G)^{1/3}$ [39] (e.g., $\xi \approx 16$ nm for $G = 1$ kPa), must not be too large in order to trap the particles firmly and prevent slippage. Hence, particle-matrix mechanical coupling is a critical issue in nanoparticle/soft hydrogel composites which needs to be carefully characterized. As a particular benefit offered by magnetic particles, the coupling can be actively probed by applying magnetic fields and analyzing the response using concepts developed in microrheology. Several different approaches regarding measurement techniques, types of magnetic tracer particles and models for data analysis were evaluated in a round Robin test [40] and are described in separate contributions within this volume.

Particle/matrix mechanical interaction of Ni nanorods in Newtonian fluids was investigated using AC magnetization measurements, dynamical light scattering and optical transmission in a rotating magnetic field [31]. The rotational diffusion coefficients D_r, obtained from nanorod colloids with average length $\langle L_c \rangle = 100–240$ nm showed good agreement. Yet, the absolute values were smaller than expected from their geometric size by a factor ~2 independent of the method. The AC susceptibility of blocked magnetic dipoles in a Newtonian fluid monitors Brownian relaxation with the characteristic frequency $\omega_B = 2D_r = 2k_BT/\xi_r$ in the low field (Debye) limit. The rotational friction coefficient $\xi_r = K_V\eta$ depends on a shape- and size-dependent particle factor K_V and the viscosity η of the dispersion medium. Due to the large magnetic moments of Ni nanorods and corresponding Langevin parameter $\zeta > 10$ for a typical AC field amplitude of $H_0 = 10$ Oe, the relaxation spectra revealed strong non-linear effects. A field-dependent increase of the characteristic relaxation frequency was observed and could be well described by an empirical relation derived from numerical solutions of the Fokker–Planck equation [31, 36, 41, 42]. The same theoretical approach was used to model the frequency shift in the Ni nanorod relaxation spectra obtained from AC magnetic field-dependent optical transmission (ACOT) measurements [36]. The derived analysis of ACOT in the non-linear regime was employed for measuring the adsorption of gelatin on Ni nanorods. The increase in the rotational friction coefficient $\xi_r \sim K_V$ caused by adsorption of BSA on the nanorod surface could also be detected in the

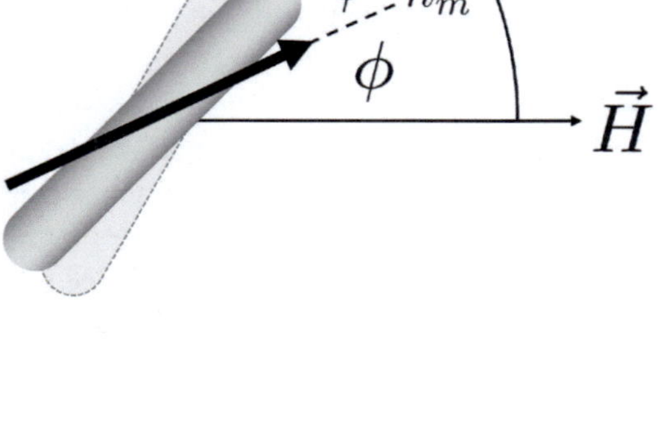

Figure 16.4: Definitions of angle variables: ϕ, between the direction of the magnetic moment ($\vec{n}_m$) and applied field $\vec{H}$, ψ, between magnetic moment ($\vec{n}_m$) and the direction of the anisotropy axis (equal to the nanorod axis $\vec{n}_r$), and ϑ, between nanorod axis at applied field ($\vec{n}_r$) with respect to the initial orientation at zero field ($\vec{n}_r^0$). In the experiment, the sum of all angles is determined by the orientation of the texture axis Θ.

phase lag between the mean particle orientation and the direction of a rotating magnetic field [35].

The transfer of both types of driving fields, alternating or rotating, to investigate rotational relaxation of nanorod dispersions in general viscoelastic media is not straightforward. Theoretical models for the dynamics of magnetic dipoles in canonical Maxwell-[43], Voigt-Kelvin or Jeffrey [44, 45] materials were developed and used for analysis of AC susceptibility of $CoFe_2O_4$ nanoparticles in the low field limit [46, 47]. A model-independent analysis was derived from the Di Marzio–Bishop model for the dielectric susceptibility of viscoelastic glasses and applied in the analysis of low field magnetic AC susceptibility of $CoFe_2O_4$ particles in PEG solutions [48]. An alternative, favorable in particular for probe particles with large magnetic moment, is the use of an oscillating driving field, i.e., a field of constant magnitude H_0 close to saturation at Langevin parameter $\zeta > 30$. The direction of the field vector, described by the time-dependent angle $\beta(t) = \beta_0 \exp(i\omega t)$, oscillates periodically in a fixed plane within a narrow angular range $\pm\beta_0$ [34, 49]. Driven by the magnetic field, the nanorods oscillate at the same frequency with mean nanorod orientation angle $\theta(t) = \theta_0 \exp(i\omega t - \delta) = \theta_0^* \exp(i\omega t)$ and phase shift δ. The rotational motion was monitored by optical transmission and the complex OFOT response function $X^*(\omega) = \theta_0^*/\beta_0$ derived. Early measurements were analyzed in terms of elementary mechanical models such as the Maxwell model (micellar solutions) or the Voigt-Kelvin model (hydrogels) [34]. Recently, the simple relationship $X^*(\omega) = (1 + KG^*(\omega))^{-1}$ between the OFOT response function and the complex dynamic modulus G^* was derived for a general linear viscoelastic continuum as matrix. The parameter $K = K_V/(m\mu_0 H_0)$ depends on the particle factor K_V, magnetic moment m and the magnetic field $\mu_0 H_0$. Because of the polydispersity of the Ni nanorods geometry (varying K_V) as well as magnetic moment (varying m) the parameter K is not a single-valued constant. A distribution function $P(K)$ was introduced,

$$X^*(\omega) = \int_0^\infty P(K)(1 + KG^*(\omega))^{-1} dK, \tag{16.2}$$

that was assumed to be characteristic for a particular batch of nanorods. For the analysis of OFOT measurements, $P(K)$ was calibrated by a reference measurement in a Newtonian fluid with $G^*(\omega) = i\eta_0\omega$ for constant viscosity η_0, Figure 16.5 (left). In order to avoid any model-dependent bias in the later analysis, $P(K)$ (inset) was obtained by numerical inversion with Tikhonov regularization [17]. Because the OFOT relaxation spectrum was only slightly broadened as compared to the Debye function, the distribution $P(K)$ was fairly narrow. The mean $\overline{K}$ was found to be systematically larger than expected from their size according to TEM by a factor 2–4. Nickel surface oxide, polymer surfactants and in particular the irregular residues from the alumina templates were supposed as the origin of the significant increase in hydrodynamic friction of the nanorods [17]. An effective hydrodynamic size (L_h, D_h) of the nanorods was estimated by matching $\overline{K}$ assuming a core–shell geometry with constant shell thickness. However, analysis of optical transmission experiments were all based on the calibrated $P(K)$ or $\overline{K}$ directly and independent of this geometric model.

Using eq. (16.2), the OFOT response function of a nanorod dispersion in a 1.6 wt% poly(ethylene oxide) (PEO, $M_w = 1000$ kg/mol) solution was translated into the dynamic modulus, Figure 16.5 (right). The OFOT spectra were independent of the oscillation amplitude β_0 which confirmed the linear response regime. This method was applied in a study on the crossover between the macroscopic continuum limit and probe size-dependent local viscoelastic properties in PEO solutions. By varying the polymer radius of gyration $R_g \approx 11$–62 nm and Ni nanorod hydrodynamic length $L_h \approx 170$–740 nm an empirical scaling relation for the zero shear rate viscosity, $\eta_0^{\mathrm{OFOT}}/\eta_0^{\mathrm{macro}} = \exp(-5.6R_g/L_h)$ was derived [17]. Furthermore, systematic changes in the local as compared to the macroscopic dynamic modulus were observed in viscoelastic semi-dilute entangled solutions. Both effects could be explained by a reduction in the effective polymer entanglement density (larger transient mesh size) in the vicinity of the nanorods. In this particular matrix system, size-dependent coupling was evident even for Ni nanorods with a hydrodynamic length as large as $L_h = 740$ nm. Rather strong reduction of the local viscosity by two orders of magnitude was observed for spherical $CoFe_2O_4$ tracer particles with hydrodynamic diameter $d_h \sim 23$ nm dispersed in the same polymer solutions within the round Robin test [40].

OFOT measurements and their analysis based on eq. (16.2) make it possible to observe the capture of Ni nanorods in the hydrogel's polymeric network during copolymerization of the acrylamide/bis-acrylamide precursor solution, Figure 16.6 (left). Only a few seconds after addition of the initiator APS, polymerization was evident from the rapid increase in the dynamic modulus. The crossover between

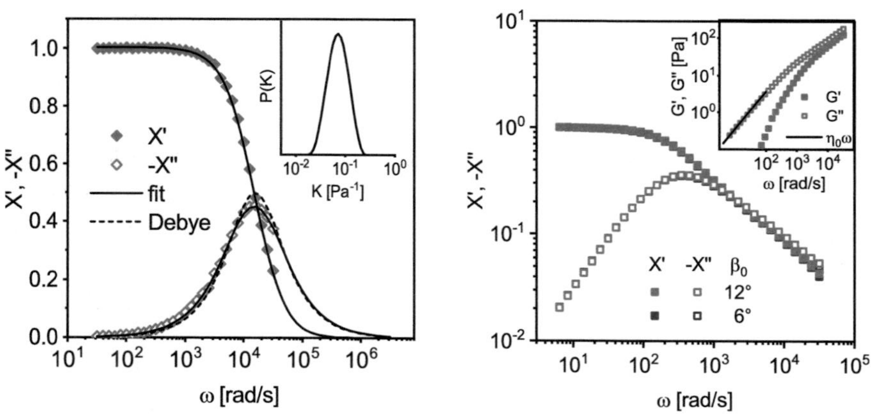

Figure 16.5: (left) OFOT response function of Ni nanorods in water for calibration. The distribution function $P(K)$ (insert) was derived by numerical inversion and used to translate the OFOT spectrum (right) of nanorods in a PEG solution (loglog-plot) into the dynamic modulus G^* (insert). Reproduced, with permission, from [17].

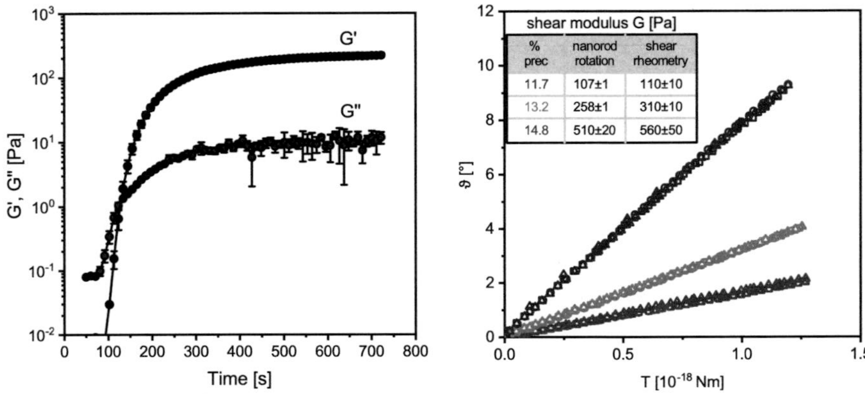

Figure 16.6: (left) Dynamic modulus at f = 10 Hz during chemical gelation of PAM at a composition of 16 wt% stock solution after addition of initiator APS. (right) Rotation angle of nanorods in PAM hydrogels with different amount of stock solution as function of magnetic torque (see Section 16.4) and derived shear modulus in comparison with results from macroscopic shear rheometry (Thermo Fisher Scientific HAAKE MARS II, CP60/2°).

storage and loss modulus indicated the formation of an elastic crosslinked PAM network. After the hydrogel matrix had reached a stable elasticity, the mechanical particle/matrix coupling could also be investigated by quasistatic field-dependent optical transmission measurements of linear polarized light as described in Section 16.2.4. The rotation angle ϑ increased proportional to the magnetic torque T, as expected for a linear elastic matrix. The shear modulus could be calculated from the slope, $G = K_V^{-1}(d\vartheta/dT)^{-1}$, and the values obtained for PAM hydrogels with different composition showed good agreement with results, determined by macroscopic shear rheometry, Figure 16.6 (right). Please notice that the magnetic torque used for the abscissa were calculated using the SW-model, though the successful quantitative description of the magnetization, shown in Section 16.2.3, does not necessarily prove that this model is also correct for the field- and angle dependence of the magnetic torque. This inference is valid only under the assumption of coherent rotation of the magnetization. Because the applied torque is of key importance for the envisaged quantitative modeling of field-induced deformation, this topic will be addressed in more detail in the following section.

16.4 Magnetic torque

While magnetization measurements as shown in Section 16.2.3 characterized the magnetization of the magnetic particles in field direction, $m_{||}$, the magnetic torque on a single nanorod is determined by the component of its magnetic moment perpendicular to the applied field, $T = m_{\perp}\mu_0 H$. The magnetization of a homogeneous spheroidal particle, as assumed in the SW-model, rotates coherently in an homogeneous external field and $m_{tot}^2 = m_{\perp}^2 + m_{\|}^2$ is constant. The experimentally observed continuous approach to saturation for nanorods aligned perpendicular to the field contradicted the prediction by the SW-model. However, the magnetization properties were consistent with the SW-model in the angular range of $\Theta \leq 70°$. In order to evaluate the SW-model for the quantitative description of the magnetic torque, the rotation of nanorods in a soft PAM hydrogel was measured optically as function of the magnetic field for different texture angles Θ, Figure 16.7 (left). The rotation was completely reversible as shown by the full symbols at 70°, which represent the data for the reverse curve recorded with decreasing field. This confirmed the purely elastic interaction between the nanorods and the hydrogel without indication of creep on the time scale of these experiments. The rotation increased with field and with larger angle between texture axis and field direction. Because the orientation of the nanorods was known for each given field, the magnetic torque for each data point could be calculated based on the SW-model using eq. (16.1). The same data, plotted not as function of the magnetic field but as function of the calculated torque, fell on a common master curve, Figure 16.7 (right). Irrespective of the axis orientation, the nanorods rotated proportional to the applied torque as expected for a linear elastic matrix. This consistent linear behavior and the agreement between derived shear modulus with macroscopic results (Figure 16.6 (right)) confirmed the SW-model as a reliable description of the field- and orientation-dependent magnetic torque on the Ni nanorods in the restricted range of $\Theta = 20–70°$. On the macroscopic scale, the ensemble of nanorods induce a local torque density, which depends on the local initial magnetic texture, the macroscopic deformation at the local volume element and particle density. The latter is an important aspect not only because it determines the total torque on the composite but particularly with regard to interparticle interactions.

16.5 Particle density

The field-induced change in mechanical stiffness (magnetorheological effect) and deformation (magnetostrictive effect) of magnetic gels and elastomers depend sensitively on the local particle arrangement and the resulting balance of magnetic and elastic forces on the microscopic scale [8, 50–53]. Alignment of particles and formation of closely packed particle chains along the field direction has an important impact on the mechanical properties [54] but—in comparison to a liquid environment—is hampered by the elastic restoring forces. Particular features of the magnetoelastic balance are the occurrence of two metastable states [55] with a resulting hysteretic behavior [56, 57] and magnetic field-dependent plasticity in soft elastomers [58]. Further interparticle effects arise from the inhomogeneous local magnetic field in the vicinity of a magnetic particle. The feedback in dipolar interactions and local structure results in a strong and complex coupling of the elastic strain and magnetic fields, which were computed by FEM simulations and used to calibrate macroscopic constitutive equations [59]. Finally, elastic interaction, mediated by the overlapping deformation fields of neighboring particles also affect their relative motion [60–62]. In order to minimize interference by the very complex magnetomechanical interparticle interactions, the experimental studies on Ni nanorod/hydrogel composites were performed at particle concentrations of $\varphi_{vol} \leq 10^{-4}$. The potential contributions of magnetic dipolar, elastic and electrostatic interparticle interactions at such low concentrations were estimated.

The rotation of Ni nanorods in a series of soft PAM-composites with identical gel composition was virtually independent of particle volume fraction in the range $\varphi_{vol} = 8 \cdot 10^{-7} - 6.4 \cdot 10^{-5}$ [21] which suggested that interparticle interactions were either negligible or compensating each other. Assuming the depolarization factor $N = -1$, the maximum mean demagnetizing field was estimated to be $H_s^{\max} = -\varphi_{\text{vol}} M_{s,\text{Ni}} \approx -0.45\,\text{Oe}$ at the largest volume fraction. With regard to the field applied during deformation measurements of > 1000 Oe, magnetic dipolar interaction are negligible up to a Ni volume fraction of $\varphi_{\text{vol}} = 10^{-3}$. Henkel plots, derived from demagnetization remanence measurements of textured hydrogels, did not show the signatures of dipolar fields, in contrast to filled AAO templates in which Ni nanorods are densely packed in a 2D layer [16]. Electrostatic interaction is essential to increase the colloidal stability of the magnetic nanorods in liquid dispersion [33]. In PAM hydrogels, the electrostatic screening length was estimated to $\lambda \approx 6$ nm. At typical particle separation ≥100 nm in the composites, Coulomb interaction between the charged nanorods are effectively screened [21]. To evaluate elastic interactions, the torque-driven rotation of two cylindrical inclusions as function of distance and orientation of the particle- and rotation axis was investigated by FEM simulations [21]. Compared with an isolated particle, a small increase or reduction of particle rotation was found depending on their spatial configuration. In a homogeneously distributed uniaxial composite, the different contributions combined and partially compensated each other to an overall effect on the order of 0.1% at a volume fraction of $\varphi_{\text{vol}} = 4 \cdot 10^{-4}$. Direct observation and measurement of Ni nanorod rotation by Laser scanning confocal microscopy also did not provide any evidence for significant elastic interparticle interaction [21]. Based on the estimates and experimental results, it was proposed that interparticle interactions were not relevant in Ni nanorod composites at particle volume fraction $\varphi_{\text{vol}} \leq 10^{-4}$ and could be neglected in the quantitative modeling of field-induced deformations.

16.6 Macroscopic deformation experiments

With regard to the weak total torque expected for a Ni nanorod volume fraction of $\varphi_{\text{vol}} \approx 10^{-4}$, experiments were performed on thin composite filaments, taking advantage of the high sensitivity of this geometry [21]. The magnetically textured

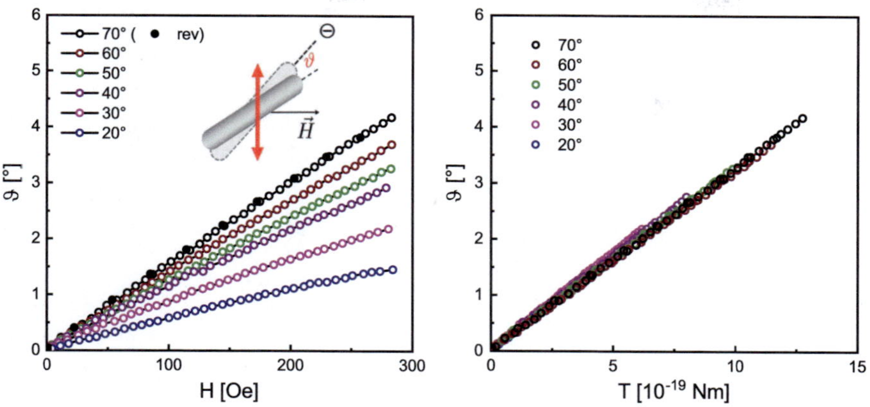

Figure 16.7: (left) Nanorod rotation angle ϑ as function of applied magnetic field H for different angles Θ between the texture axis and field direction. For maximum sensitivity, polarization of the laser was at 45° with respect to Θ. The rotation was reversible during the experiment indicated by the reverse curve (70°, black markers). (right) The same data, plotted as function of the magnetic torque, fell on a common master curve. Adapted, with permission, from Schopphoven et al. [21].

PAM-nanorod filaments were fabricated by mixing the nanorods in a precursor solution and sucking the solution into a thin PTFE tube immediately after starting the polymerization process by addition of APS. Guided by the information on the polymerization kinetics, Figure 16.6, the specimens were mounted in the magnetic assembly for alignment of the nanorods within a few seconds and left for 15 min. For deformation measurements, the composite filaments were pushed out of the tube, suspended at the top end either in a sealed chamber or immersed in water to minimize gravitational effects. The deformation in a horizontal homogeneous magnetic field, i.e., perpendicular to the filament axis, were quantified by analysis of recorded video images.

Experimental results were compared with calculations for an elastic thin cylinder with volume distributed magnetic torque density $\tau_m = M_s \varphi_{\mathrm{vol}} \mu_0 H \sin(\phi)$. The essential kernel provided by the SW-model is the relation between ϕ and the characteristic parameters for the nanorods (m, K_A, K_V) for a given field H, orientation angle Θ and matrix shear modulus G as described in Section 16.3. The texture was assumed to be constant within the cross section of area A so that $\Theta = \Theta(s)$ and $\tau_m(s)$ is a function of the linear position s along the deformed cylinder axis. The 1D-problem was further simplified by only considering magnetic textures that were either perpendicular to the cylinder axis (pure torsion) or in the plane, defined by the cylinder axis and the field direction (pure bending). The new variable $\omega(s)$ was introduced for the macroscopic torsion or bending angle, respectively, Figure 16.8 (right). The required additional relationship

$$d^2\omega/ds^2 = K \sin(\phi(\Theta(s), \omega)) \qquad (16.3)$$

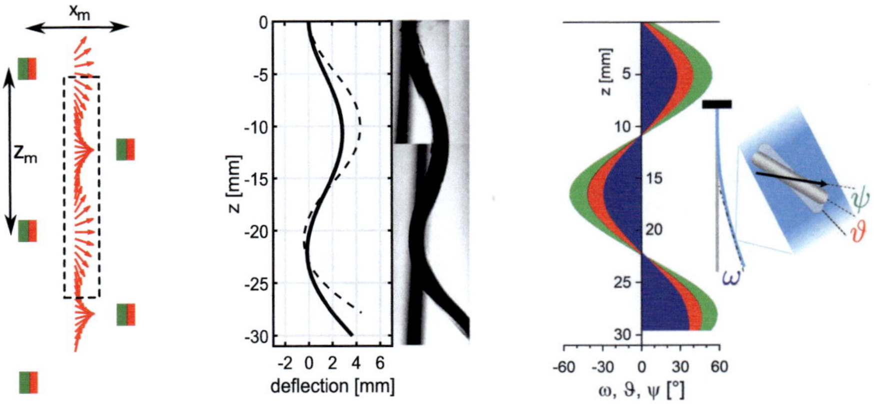

Figure 16.8: (left) Configuration of two rows of permanent magnets used for fabrication of a textured bending composite inside the dashed area. Texture angle Θ is indicated by red arrows. (middle) Calculated deflection of a Ni-nanorod/PAM-composite ($m = 1.7 \cdot 10^{-16}$ Am^2, $K_A = 63$ kJ/m^3, $K_V = 3.5 \cdot 10^{-19}$ m^3, $\varphi_{\text{vol}} = 5.4 \cdot 10^{-5}$, length $L = 32$ mm, diameter $D = 0.84$ mm, $E_{\text{mod}} = 350$ Pa, $B = 146$ mT) and calculated deflection under assumption of a rigid dipole model (dashed line). (right) Macroscopic bending angle $\omega(s)$, local rotation angle of the nanorods $\vartheta(s)$ and deflection of the magnetic moment from the anisotropy axis $\psi(s)$ add up and determine the orientation of the magnetic moment with respect to the applied field.

is an ordinary differential equation, derived from the St. Venant torsion or Euler–Bernoulli bending of a thin cylinder with volume distributed torque. The prefactor depends on the problem, $K = M_s\varphi_{\text{vol}}\mu_0 HA/GI_p$ with shear modulus G and polar moment of inertia I_p for torsion and $K = M_s\varphi_{\text{vol}}\mu_0 HA/EI_A$ with elastic modulus E and area moment of inertia I_A for bending, respectively. The set of eqs. (16.1) and (16.3) with $\omega + \vartheta + \psi + \phi = \Theta$ was solved by fixpoint iteration.

Various profiles $\Theta(s)$ for the orientation of the anisotropy axis were imprinted by polymerization in static magnetic fields of specific geometry. Ni-nanorod/PAM-composites with homogeneous texture perpendicular to the filament axis were prepared and the field-dependent torsion measured. The nonlinear increase of the total torsion angle, observed in experiment, was reproduced by the model calculations. The results were also consistent for different orientations of the texture axis [21].

In a second example, a profile with sinusoidal modulation of the texture axis was prepared using two rows of NdFeB permanent magnets, magnetized in the same direction and mutually shifted by half the spacing, Figure 16.8 (left). The magnetic field was characterized by motorized scanning with transversal and longitudinal Hall probes. By adjusting the spacing z_m between the permanent magnets and the separation x_m of the two rows, a sinusoidal modulation of the field direction was achieved, with maximum values of about 80° and −80° with respect to the horizontal, shown in Figure 16.8 (left) by red arrows. In a horizontal homogeneous magnetic field of 146 mT, the composite showed a sinusoidal bending, which was well reproduced by the model calculation, Figure 16.8

(middle). The extended SW-model allowed a more detailed analysis of the underlying physical effects. In particular, the various contributions to the rotation of the magnetic moment toward the field direction, which determines the magnetic torque, could be broken down, Figure 16.8 (right). The combined local rotation of the nanorods by ϑ and deflection of the magnetic moment from the anisotropy axis by ψ were of similar magnitude as the macroscopic bending angle ω. The computed deflection for rigid dipole are also plotted as dashed line in Figure 16.8 (middle) and obviously overestimated the magnetic torque. As pointed out by these results, deflection of the magnetic moment and local rotation of the active particles is not negligible in magnetoelastic soft composites of particles with finite magnetic anisotropy in a matix with low elastic modulus.

16.7 Conclusions

Ni nanorods, synthesized by the AAO-template method, were used as magnetic phase in magnetoresponsive hydrogels. The magnetization was analyzed using the Stoner–Wohlfarth model and an empirical anisotropy constant derived. Taking advantage of the nanorods' optical anisotropy, transmission measurements in static and time-dependent magnetic fields enabled the characterization of the nanorods' mean magnetic moment, concentration and the hydrodynamic particle–matrix interaction. Optical measurements of local nanorod rotation in a soft elastic matrix verified the SW-model for prediction of the magnetic torque in a restricted parameter range. Comparison of model calculations with macroscopic deformation experiments on thin composite filaments revealed a significant effect of the finite magnetic anisotropy and local particle rotation on the magnetic torque.

Acknowledgment: We thank C. Wagner, Physics Department, Saarland University, Germany, for access to the rheometer equipment. This work was financially supported by the German National Science Foundation DFG (Priority program SPP 1681, grant TS62/4-3).
Author contribution: All the authors have accepted responsibility for the entire content of this submitted manuscript and approved submission.
Research funding: This article was supported by German National Science Foundation DFG (Priority program SPP 1681, grant TS62/4-3).
Conflict of interest statement: The authors declare no conflicts of interest regarding this article.

References

1. Szabó D, Szeghy G, Zrínyi M. Shape transition of magnetic field sensitive polymer gels. Macromolecules 1998;31:6541–8.

2. Lum GZ, Ye Z, Dong X, Marvi H, Erin O, Hu W, et al. Shape-programmable magnetic soft matter. Proc Natl Acad Sci 2016;113:E6007–15.
3. Tokarev I, Minko S. Stimuli-responsive hydrogel thin films. Soft Matter 2009;5:511–24.
4. Kim J, Chung SE, Choi S-E, Lee H, Kim J, Kwon S. Programming magnetic anisotropy in polymeric microactuators. Nat Mater 2011;10:747–52.
5. Diller E, Zhuang J, Zhan Lum G, Edwards MR, Sitti M. Continuously distributed magnetization profile for millimeter-scale elastomeric undulatory swimming. Appl Phys Lett 2014;104:174101.
6. Huang H-W, Sakar MS, Petruska AJ, Pané S, Nelson BJ. Soft micromachines with programmable motility and morphology. Nat Commun 2016;7:12263.
7. Jolly MR, Carlson JD, Munoz BC. A model of the behaviour of magnetorheological materials. Smart Mater Struct 1996;5:607.
8. Zubarev A. Magnetodeformation of ferrogels and ferroelastomers: effect of microstructure of the particles' spatial disposition. Physica A 2013;392:4824–36.
9. Zrinyi M, Barsi L, Buki A. Deformation of ferrogels induced by nonuniform magnetic fields. J Chem Phys 1996;104:8750–6.
10. Collin D, Auernhammer GK, Gavat O, Martinoty P, Brand HR. Frozen-in magnetic order in uniaxial magnetic gels: preparation and physical properties. Macromol Rapid Commun 2003;24:737–41.
11. Raikher YL, Rusakov VV, Coffey WT, Kalmykov YP. Dynamic susceptibilities of an assembly of dipolar particles in an elastic environment. Phys Rev E 2001;63:031402.
12. Monz S, Tschöpe A, Birringer R. Magnetic properties of isotropic and anisotropic $CoFe_2O_4$-based ferrogels and their application as torsional and rotational actuators. Phys Rev E 2008;78:021404.
13. Kim Y, Yuk H, Zhao R, Chester SA, Zhao X. Printing ferromagnetic domains for untethered fast-transforming soft materials. Nature 2018;558:274–9.
14. Zhao R, Kim Y, Chester SA, Sharmama P, Zhao X. Mechanics of hard-magnetic soft materials. J Mech Phys Solid 2019;124:244–63.
15. Bender P, Günther A, Tschöpe A, Birringer R. Synthesis and characterization of uniaxial ferrogels with Ni nanorods as magnetic phase. J Magn Magn Mater 2011;323:2055–63.
16. Bender P, Krämer F, Tschöpe A, Birringer R. Influence of dipolar interactions on the angular-dependent coercivity of nickel nanocylinders. J Phys D Appl Phys 2015;48:145003.
17. Gratz M, Tschöpe A. Size effects in the oscillatory rotation dynamics of Ni nanorods in poly(ethylene oxide) solutions. Macromolecules 2019;52:6600–12.
18. Masuda H, Satoh M. Fabrication of gold nanodot array using anodic porous alumina as an evaporation mask. Jpn J Appl Phys 1996;35:L126–9.
19. Li AP, Müller F, Birner A, Nielsch K, Gösele U. Hexagonal pore arrays with a 50–420 nm interpore distance formed by self-organization in anodic alumina. J Appl Phys 1998;84:6023–6.
20. Nielsch K, Müller F, Li A-P, Gösele U. Uniform nickel deposition into ordered alumina pores by pulsed electrodeposition. Adv Mater 2000;12:582–6.
21. Schopphoven C, Birster K, Schweitzer R, Lux C, Huang S, Kästner M, et al. Elastic deformations in semi-dilute ni nanorod/hydrogel composites. Arch Appl Mech 2019;89:119–32.
22. Ross CA, Hwang M, Shima M, Cheng JY, Farhoud M, Savas TA, et al. Micromagnetic behavior of electrodeposited cylinder arrays. Phys Rev B 2002;65:144417.
23. Skomski R, Coey JMD. Permanent magnetism. Bristol, UK; Philadelphia, PA: Institute of Physics Publishing; 1999.
24. Stoner EC, Wohlfarth EP. A mechanism of magnetic hysteresis in heterogeneous alloys. IEEE Trans Magn 1991;27:3475.
25. Osborn JA. Demagnetizing factors of the general ellipsoid. Phys Rev 1945;67:351–7.
26. Klein T, Laptev A, Günther A, Bender P, Tschöpe A, Birringer R. Magnetic-field-dependent optical transmission of nickel nanorod colloidal dispersions. J Appl Phys 2009;106:114301.

27. Schopphoven C, Tschöpe A. Magnetic anisotropy of nickel nanorods and the mechanical torque in an elastic environment. J Phys D Appl Phys 2018;51:115005.
28. Ivanov YP, Vázquez M, Chubykalo-Fesenko O. Magnetic reversal modes in cylindrical nanowires. J Phys D Appl Phys 2013;46:485001.
29. Bender P, Tschöpe A, Birringer R. Determination of the shear modulus of gelatine hydrogels by magnetization measurements using dispersed nickel nanorods as mechanical probes. J Magn Magn Mater 2013;346:152–60.
30. Stepanov GV, Borin DY, Storozhenko PA. Rotation of magnetic particles inside the polymer matrix of magnetoactive elastomers with a hard magnetic filler. J Magn Magn Mater 2017;431:138–40.
31. Günther A, Bender P, Tschöpe A, Birringer R. Rotational diffusion of magnetic nickel nanorods in colloidal dispersions. J Phys Condens Matter 2011;23:325103.
32. Krämer F, Gratz M, Tschöpe A. Analysis of the static magnetic field-dependent optical transmission of Ni nanorod colloidal suspensions. J Appl Phys 2016;120:044301.
33. Tschöpe A, Krämer F, Birster K, Gratz M, Birringer R. Quantification of magneto-optically active nanorods and inactive aggregates in nickel nanorod colloids. Colloids Interface Sci Commun 2016; 10–11:11–4.
34. Tschöpe A, Birster K, Trapp B, Bender P, Birringer R. Nanoscale rheometry of viscoelastic soft matter by oscillating field magneto-optical transmission using ferromagnetic nanorod colloidal probes. J Appl Phys 2014;116:184305.
35. Schrittwieser S, Ludwig F, Dieckhoff J, Tschöpe A, Günther A, Richter M, et al. Direct protein detection in the sample solution by monitoring rotational dynamics of nickel nanorods. Small 2014;10:407–11.
36. Gratz M, Tschöpe A. Optical transmission versus ac magnetization measurements for monitoring colloidal Ni nanorod rotational dynamics. J Phys D Appl Phys 2017;50:015001.
37. Schulz L, Schirmacher W, Omran A, Shah VR, Böni P, Petry W, et al. Elastic torsion effects in magnetic nanoparticle diblock-copolymer structures. J Condes Matter Phys 2010;22:346008.
38. Bender P, Tschöpe A, Birringer R. Magnetization measurements reveal the local shear stiffness of hydrogels probed by ferromagnetic nanorods. J Magn Magn Mater 2014;372:187–94.
39. Rubinstein M, Colby R. Polymer physics, 1st ed. Oxford: Oxford University Press; 2003.
40. Hess M, Gratz M, Remmer H, Webers S, Landers J, Borin D, et al. Scale-dependent particle diffusivity and apparent viscosity in polymer solutions as probed by dynamic magnetic nanorheology. Soft Matter 2020;16:7562–75.
41. Yoshida T, Enpuku K. Simulation and quantitative clarification of ac susceptibility of magnetic fluid in nonlinear brownian relaxation region. Jpn J Appl Phys 2009;48:127002.
42. Remmer H, Gratz M, Tschöpe A, Ludwig F. Magnetic field dependence of Ni nanorod Brownian relaxation. IEEE Trans Magn 2017;53:1–4.
43. Raikher Y, Rusakov V. Magnetic rotary microrheology in a Maxwell fluid. J Magn Magn Mater 2006; 300:e229–33.
44. Raikher YL, Rusakov VV. Theory of Brownian motion in a Jeffreys fluid. J Exp Theor Phys 2010;111: 883–9.
45. Rusakov VV, Raikher YL. Magnetic response of a viscoelastic ferrodispersion: from a nearly Newtonian ferrofluid to a Jeffreys ferrogel. J Chem Phys 2017;147:124903.
46. Remmer H, Dieckhoff J, Tschöpe A, Roeben E, Schmidt AM, Ludwig F. Dynamics of CoFe2O4 single-core nanoparticles in viscoelastic media. Phys Procedia 2015;75:1150–7.
47. Remmer H, Roeben E, Schmidt AM, Schilling M, Ludwig F. Dynamics of magnetic nanoparticles in viscoelastic media. J Magn Magn Mater 2017;427(Suppl C):331–5.
48. Roeben E, Roeder L, Teusch S, Effertz M, Ulrich KD, Schmidt AM. Magnetic particle nanorheology. Colloid Polym Sci 2014;292:2013–23.

49. Wilhelm C, Browaeys J, Ponton A, Bacri J-C. Rotational magnetic particles microrheology: the maxwellian case. Phys Rev E 2003;67:011504.
50. Stolbov OV, Raikher YL, Balasoiu M. Modelling of magnetodipolar striction in soft magnetic elastomers. Soft Matter 2011;7:8484–7.
51. Ivaneyko D, Toshchevikov V, Saphiannikova M, Heinrich G. Mechanical properties of magneto-sensitive elastomers: unification of the continuum-mechanics and microscopic theoretical approaches. Soft Matter 2014;10:2213–25.
52. Metsch P, Kalina KA, Spieler C, Kästner M. A numerical study on magnetostrictive phenomena in magnetorheological elastomers. Comput Mater Sci 2016;124:364–74.
53. Stolbov OV, Raikher YL. Magnetostriction effect in soft magnetic elastomers. Arch Appl Mech 2019;89:63–76.
54. Schümann M, Gundermann T, Odenbach S. Microscopic investigation of the reasons for field-dependent changes in the properties of magnetic hybrid materials using x-ray microtomography. Arch Appl Mech 2019;89:77–89.
55. Biller AM, Stolbov OV, Raikher YL. Modeling of particle interactions in magnetorheological elastomers. J Appl Phys 2014;116:114904.
56. Biller AM, Stolbov OV, Raikher YL. Mesoscopic magnetomechanical hysteresis in a magnetorheological elastomer. Phys Rev E 2015;92:023202.
57. Puljiz M, Huang S, Kalina KA, Nowak J, Odenbach S, Kästner M, et al. Reversible magnetomechanical collapse: virtual touching and detachment of rigid inclusions in a soft elastic matrix. Soft Matter 2018;14:6809–21.
58. Becker TI, Böhm V, Chavez Vega J, Odenbach S, Raikher YL, Zimmermann K. Magnetic-field-controlled mechanical behavior of magneto-sensitive elastomers in applications for actuator and sensor systems. Arch Appl Mech 2019;89:133–52.
59. Kalina KA, Metsch P, Kästner M. Microscale modeling and simulation of magnetorheological elastomers at finite strains: a study on the influence of mechanical preloads. Int J Solid Struct 2016;102-103:286–96.
60. Puljiz M, Huang S, Auernhammer GK, Menzel AM. Forces on rigid inclusions in elastic media and resulting matrix-mediated interactions. Phys Rev Lett 2016;117:238003.
61. Puljiz M, Menzel AM. Forces and torques on rigid inclusions in an elastic environment: resulting matrix-mediated interactions, displacements, and rotations. Phys Rev E 2017;95:053002.
62. Puljiz M, Menzel AM. Displacement field around a rigid sphere in a compressible elastic environment, corresponding higher-order Faxén relations, as well as higher-order displaceability and rotateability matrices. Phys Rev E 2019;99:053002.

Malte Schümann and Stefan Odenbach*

17 The microstructure of magnetorheological materials characterized by means of computed X-ray microtomography

Abstract: Magnetorheological materials are a class of "smart materials", where mechanical material properties can be tuned by the application of externally applied fields. To accomplish the magneto-sensitive quality, magnetic particlesare distributed in a host matrix. In the last year's interest gained in materials based on solid matrices. In contrast to fluid systems, within a solid matrix, the particles are fixed within the material. This enables an evaluation of the structures formed by the particles by means of computed X-ray microtomography. As known from past investigations, the arrangement and movement of the magnetic particles within the matrix play a major role in determining the overall material properties. Computed X-ray microtomography proved to be a convenient tool, providing important new knowledge about those materials. This paper gives an overview of the application of the method of computed X-ray microtomography on several kinds of solid magnetorheological materials, the broad possibilities of data evaluation, and fundamental results obtained with this method and the described materials.

Keywords: magnetorheological effect; particle structure; smart materials; X-ray tomography.

17.1 Introduction

17.1.1 Magnetorheological materials

The core feature of magnetorheological materials is the tunability of mechanical properties, for instance, Young's moduli, by the external application of magnetic fields. The embedment of magnetic microparticles within a surrounding matrix realizes the magnetic properties of the material.

The first smart materials reacting on magnetic fields were realized as suspensions of magnetic particles in appropriate carrier liquids. Rabinov [1, 2] accomplished the embedment of magnetic micron-sized particles, obtaining the first magnetorheological suspension. This fluid material featured strong yield stress in the presence of magnetic

***Corresponding author: Stefan Odenbach,** Faculty of Mechnical Engineering, TU Dresden, Dresden, Germany, E-mail: Stefan.Odenbach@tu-dresden.de
Malte Schümann, Faculty of Mechanical Engineering, TU Dresden, Dresden, Germany, E-mail: Malte.Schuemann@tu-dresden.de

This article has previously been published in the journal Physical Sciences Reviews. Please cite as: M. Schümann and S. Odenbach "The microstructure of magnetorheological materials characterized by means of computed X-ray microtomography" *Physical Sciences Reviews* [Online] 2021, 6. DOI: 10.1515/psr-2019-0105 | https://doi.org/10.1515/9783110569636-017

fields, thus exhibiting magnetorheological properties. Aiming for long-term stability without sedimentation led to the transition to magnetic nanoparticles. Papell [3] earned a patent on the realization of so-called ferrofluids: A stable suspension of magnetic nanoparticles in a carrier liquid featuring magnetically controllable flow and fluid properties.

Regardless of their numerous applications and usages [4, 5], fluid matrix materials exhibit a range of disadvantages and technical difficulties as leakage, sedimentation, and the overall challenge of realizing a long-term stable material. This issue was addressed by switching from liquid to solid matrix materials. The idea was to provide a more stable and easily feasible magnetorheological material offering comparable properties and features.

The most obvious solution is the usage of a soft elastomer as a host matrix. This approach leads to an elastic and mechanically stable material, where mechanical properties such as Young's modulus can be tuned by magnetic fields. These magnetorheological elastomers are an important subject of research for about 15 years [6–13]. This class of materials brings its own difficulties and most of its behavior characteristics and properties are not fully understood yet. Nevertheless, several technical and scientific applications of magnetorheological elastomers were successfully realized [14–22]. More information on technical applications of magnetorheological elastomers can be found in the articles "Field-controlled vibrational behavior of magnetoactive elastomers used as adaptable sensor elements" and "Actuator systems based on a controlled particle-matrix interaction in magnetic hybrid materials with the application for locomotion and manipulation", both found within this issue.

Besides the choice of the matrix material, the properties of the embedded magnetic particles play a major role in the macroscopic material behavior. In addition to the most commonly used iron particles, magnetically hard particles are of special interest in the research of the past few years [23–27]. Most recent investigations even apply mixtures of magnetically soft and hard particles to combine the features and advantages of both particle types [25, 28]. Information on the impact of the composition of magnetorheological elastomers on their properties can be found in the article "Synthesis of MGE with a wide range of magnetically controlled properties", found in this issue as well. Continuing the pass of searching for suitable matrix material led to the development of magnetorheological foams [29–33], and magnetorheological gels [6, 34–36], adding further aspects and functional properties to the obtained materials.

17.1.2 Evolution of evaluation methods

Already at the beginning of research regarding magnetorheological materials, it became apparent, that the magnetically induced movement and rearrangement of the magnetic particles within the matrix form the key influence determining the macroscopic material behavior. Thus, besides the evaluation of global material properties in dependence of several material parameters, research focused on the evaluation of the internal particle structure.

Changes in the internal structure have been observed by microscopy [11, 37–39] or small-angle neutron scattering (SANS) [40, 41]. These methods provide either a two-dimensional projection of the internal structure or a statistical average over the particle ensemble.

With technical advances in tomography and digital image evaluation, computed X-ray microtomography turned out to be one of the most convenient methods for this task [12, 23, 42–48]. It enables a nondestructive evaluation of three-dimensional information and thus allows the evaluation of geometrical information of several thousands of particles at once. Furthermore, it is possible to combine the measurement with additional triggers such as mechanical strain, and most importantly: magnetic fields.

The high resolution of X-ray microtomography combined with sophisticated algorithms in digital image processing provides the possibility to obtain single-particle resolution. These recent advances allow the tracking of particles after the application of magnetic fields or mechanical loads. The obtained detailed geometrical information of the particle structures can serve as input data for theoretical descriptions of the material. Furthermore, the experimental data can provide a benchmark for proving the results of numerical simulations of the material behavior [49–57].

The scope of this contribution will be to outline the technical possibilities of computed X-ray microtomography for evaluating particle structures inside magneto-rheological materials and to describe the used methods of digital image processing. This description of the evaluation of microscopic particle distributions will be accompanied by investigations of the mechanical properties of those materials to provide the link between microscopic structural changes and macroscopic material behavior.

17.2 Computed X-ray microtomography

The method of X-ray absorption tomography is based on a mathematical theorem by Johann Radon [58] stating, that the two-dimensional distribution of a quantity A can be evaluated if the line integrals of this quantity are known for all angles of observation.

Lambert-Beer's attenuation law determines the attenuation of X-rays when passing through the material.

$$I(y) = I_0 \cdot e^{-\mu y} \tag{17.1}$$

I denote the transmitted intensity, I_0 the initial X-ray intensity, μ the linear absorption coefficient for X-rays, and y the thickness of the passed material. The term in the exponential function changes into a line integral over the absorption coefficients along the pass of the X-rays, if the sample is composed of different materials with different absorption coefficients.

$$I(Y) = I_0 \cdot e^{-\int_0^Y \mu(y)dy} \tag{17.2}$$

The negative logarithm of the ratio between the transmitted and the incident X-ray intensity is called the projection:

$$p(Y) = -\ln\left(\frac{I(Y)}{I_0}\right) = \int_0^Y \mu(y)dy \tag{17.3}$$

It provides the line integral of the absorption coefficient along the pass, the X-ray's pass took through the sample. As a result, if X-ray absorption images are recorded under different observation angles, Radon's theorem can be discretely realized.

To get a three-dimensional distribution of the absorption coefficients, a reconstruction of the projection images is conducted. Commonly, algorithms based on the Fourier slice theorem are used. With this approach, the Fourier transform of a parallel projection under a certain observation angle is equal to a line within the two-dimensional Fourier transform. Thus, if a Fourier transformation of the projection data is taken for many observation angles, a discrete representation of the two-dimensional Fourier transform of the distribution of the linear absorption coefficients in the sample is obtained.

If conventional cone-beam X-ray tubes are used to capture the absorption images, the Fourier Slice Theorem can be implemented with the Feldkamp–Davis–Kress (FDK) algorithm. Within the resulting 3D images, the gray value corresponds to the spatial density distribution of the material, as the linear absorption coefficient depends on the density. Those images can be processed to extract the distribution of particles inside a magnetorheological elastomer. A rendered 3D model of reconstructed tomography data is shown in Figure 17.1, depicting particle structures in a magnetorheological elastomer.

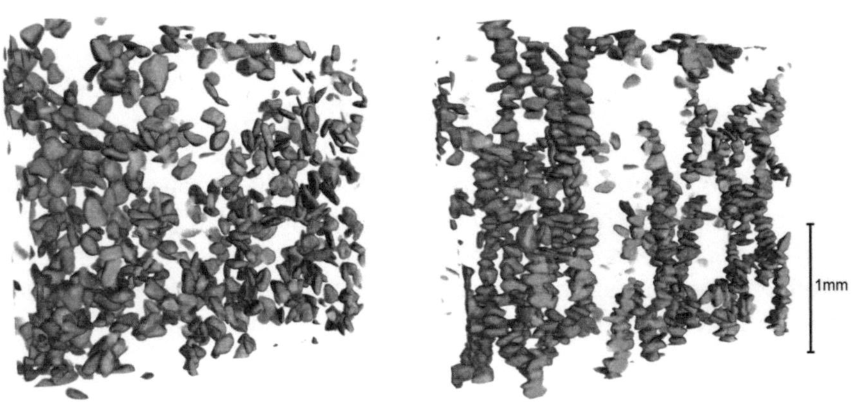

Figure 17.1: The images show magnetic particles obtained from reconstructed tomography data. The pictures are excerpts of a dataset counting approx. 12.500 particles. The left side shows the initial isotropic particle structure, the right side shows the chain-like structure, as a result of an application of a 2T magnetic field before tomography. The images show the necessity of a particle separation process after binarization to evaluate the individual particles [12, 23].

Many complex aspects like the energy dependence of the absorption coefficient and the general treatment of artefacts in tomographic data sets are not discussed within this introduction of computed X-ray microtomography. More detailed information regarding these points are given in [59, 60].

The tomographic data sets presented in this contribution were recorded using the home-built tomography setup TomoTU [42]. This setup uses a conventional nano-focus cone-beam X-ray tube with a maximum acceleration voltage of 160 kV and a focus spot size of 1.5 μm. A sample manipulation stage, consisting of various positioning stages, a rotational stage to rotate the sample enables the capture of absorption images from different observation angles. Here, a directly converting X-ray detector (Shad-O-Box 6K HS) with 2304 × 2940 pixels and a pixel size of 49.5 μm is used.

By varying relative positions of the X-ray tube, sample, and detector, a magnification of up to 20 can be realized. This enables a spatial resolution of about 2 μm in the final tomogram. The setup of TomoTU is pictured in Figure 17.2. The insert shows a sample stage, capable to apply different stimuli on magnetorheological elastomer samples during tomography.

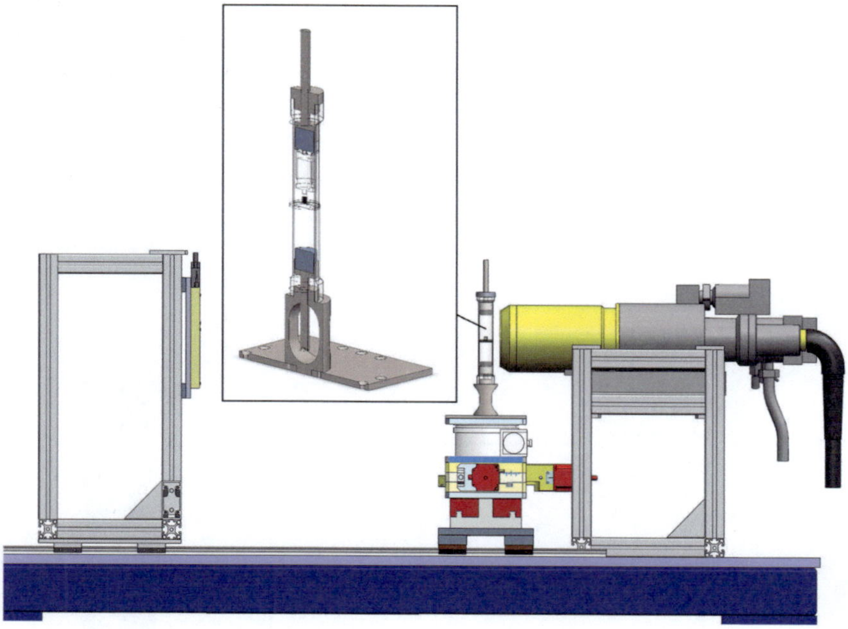

Figure 17.2: This CAD model visualizes the main components of the used tomography setup TomoTU [42]: the detector unit (left), the rotating sample stage (middle), and the X-ray source (right). The insert shows the sample setup which provides the stimuli to the sample. Two permanent magnets provide magnetic fields of up to 270 mT during tomography and stress can be applied to the sample by a plunger [12, 47].

17.2.1 Application of magnetic fields and mechanical stress

The sample stage shown in Figure 17.2 contains two permanent magnets positioned below and above the sample holder, their distance can be varied. This setup can provide homogeneous magnetic fields of up to 270 mT with homogeneity of 94% within the sample dimensions [23]. Furthermore, this setup is equipped to apply mechanical loads to the sample. With this feature experiments with coupled external magnetic and mechanical stimuli can be conducted [47]. To ensure comparable magnetic fields within accompanying mechanical measurements, the used stress-strain-testing device is equipped with two permanent magnets in the same manner. This setup was successfully used in several previous works, providing valid information about the behavior of the investigated materials under influence of magnetic fields [23, 45, 47, 61].

17.2.2 Digital image processing and particle separation

As visible in Figure 17.1, a direct evaluation of the particle geometry and location from reconstructed tomography datasets is not possible. The raw data as a result of the reconstruction consists of gray value images. A simple thresholding does not lead to a satisfying extraction of the particles, as those particles appear very often connected to each other. To accomplish a precise separation of the particles without altering their actual shape, sophisticated image processing is necessary. Furthermore, the usage of highly absorbent neodymium-iron-boron (NdFeB) particles lead to pronounced reconstruction artefacts and decrease the overall image quality of the raw data. The applied particle separation procedure, as described in this chapter, was conducted with Matlab and DipImage [62]. The process with its intermediate steps is visualized in Figure 17.3.

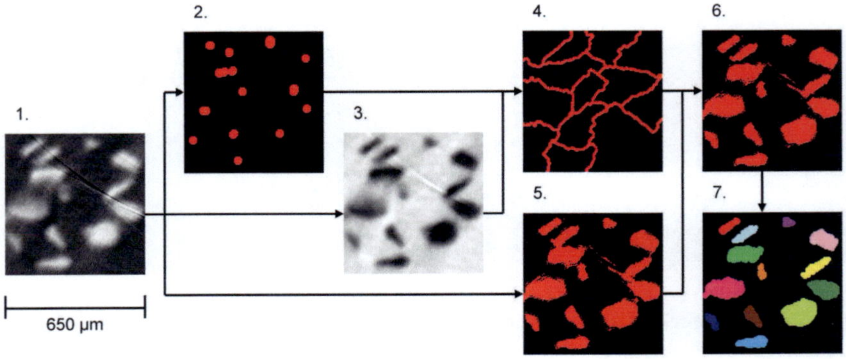

Figure 17.3: Separation process of the particles, converting raw gray value images, as given by the reconstructed tomography data, to precisely separated and labeled particles. An explanation regarding the individual steps is given in the text.

- Image 1: Gray value image as a result of tomography data reconstructions. The particles are visible, along with artifacts, noise, and blur.
- Image 2: After an application of a moderate median filter to reduce noise, local maxima indicating the particles are detected. Several particles will exhibit two or more maxima, which would result in over separation. As those maxima are very close to each other, this issue can be solved by a binary dilatation, which joins those maxima. After these steps exactly one maximum for every particle is present.
- Image 3: The negation of the initial image forms the base for the following watershed algorithm.
- Image 4: With images 2 and 3 a seeded watershed algorithm is applied, providing the borders which will separate the particles [63].
- Image 5: A threshold is applied to accomplish the transition from a gray value image to a binary image. With the tomography results presented here, a local threshold was calculated. As seen in image 1, the space between particles very close to each other is often brighter than the rest of the background. This issue leads to unsatisfactory results when applying a fixed threshold to the whole image. For the calculation of the local threshold, an additional strongly smoothed version of the original image is involved. The local threshold is calculated with

$$D_5 = D_1 > (A \cdot D_{1\,\mathrm{smoothed}} + B) \tag{17.4}$$

where A is a factor, B is a global offset and D denotes the images according to Figure 17.3. A and B have to be adapted to the type and quality of the provided raw data to accomplish optimal results. The offset B is chosen to match the resulting proportion of detected particle area to the actual volume concentration of particles, which is determined during synthesis. The resulting binary image shows the particles, still touching, with rough boundaries and artifacts.

- Image 6: Image 4, showing the borders, is subtracted from the binary image 5. This finally provides clearly separated particles.
- Image 7: A combination of binary erosion and dilatation eliminates most remaining reconstruction artifacts, smoothes the particle surfaces, and restores the particle shape to a quality which corresponds to the visual appearance of the particles – found in the initial gray value image – As the last step, labeling is conducted to distinguish the individual particles.

The final data obtained by the separation process enables an evaluation of all desired geometrical information for every individual particle. This procedure is applicable to samples containing several 10 thousands of particles, evaluating all at once. Furthermore, the obtained geometrical information is accurate enough to distinguish the particle from its neighbors. This enables the recovery of the same particle in another data set and thus enables particle tracking. Figure 17.4 visualizes the results of particle separation process by comparing a 3D particle structure before and after separation.

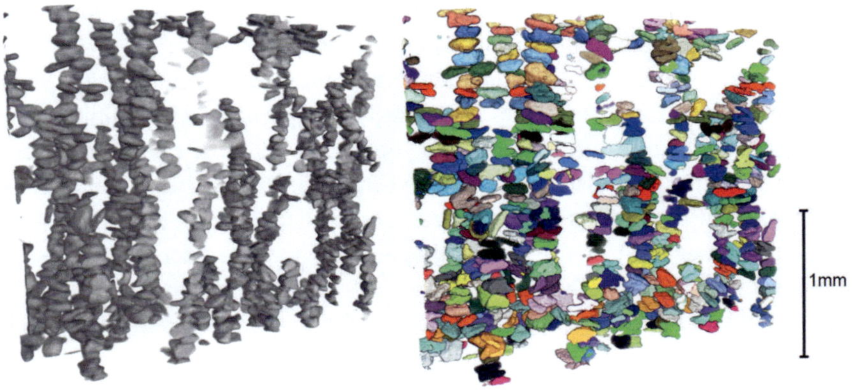

Figure 17.4: These images visualize the impact of the particle separation process on the dataset already shown in Figure 17.1. The sample contains 40 wt% of NdFeB-particles after a magnetization at 2 T. The left side shows the results of a simple thresholding, the right side shows the result of the particle separation process. The pictures are excerpts of a dataset counting approx.12.500 particles [23].

17.3 Data evaluation

17.3.1 Global structure evaluation

As a first step tomographic data sets enable an evaluation of the overall spatial distribution of magnetic particles inside a magnetorheological elastomer. This is of special interest for anisotropic magnetorheological elastomers, as those which were synthesized in presence of a magnetic field. The magnetic particle interaction induced by the applied magnetic field leads to a structure formation of the particles, as the particles are able to move freely within the liquid matrix as long as the crosslinking process is not finished. This leads to anisotropic magnetorheological elastomers with fixed chain-like or rod-like particle structures

With such samples, the dependence of the obtained particle structure on characteristic parameters is of central interest. In previous investigations, the impact of the particle concentration and the magnetic field applied during crosslinking on the resulting anisotropic particle chain structures were conducted [42, 43]. It was shown that varying particle concentration can lead to various columnar, rod-like, and tubular structures [42]. The variation of the magnetic field applied during crosslinking leads to increasing chain diameters and chain-to-chain distances with stronger fields [43].

With recent investigations, a visual insight into the obtained reconstructed tomography data sets remains an important benchmark to enable a critical check regarding the found results.

17.3.2 Particle positions, homogeneity, and pair correlation function

Concerning image processing, the analysis of structures inside anisotropic magneto-rheological elastomers, as outlined in the previous chapter, has been quite simple. An analysis on a single particle basis requires the separation of the particles as described in Section 17.2.2. After the successful separation of the particles, the position of every individual particle can be evaluated. With this information, the spatial homogeneity can be examined. As sedimentation of the particles is a known problem within sample synthesis, the validation of spatial homogeneity is a crucial necessity for almost every study presented here [23, 47, 48, 61, 64].

The pair correlation function (PCF) can be used to evaluate particle structures based on the particle positions. For theoretical descriptions of the material behavior, the spatial distribution of the particles is of central interest [65, 66]. The PCF is a statistical method which describes the relation between macroscopically measurable density ρ and the arrangement of microscopic objects as the probability to find an object at the radius r around a reference object [67, 68]. The method evaluates near- and long-range order or crystal-like structures of object ensembles. Thus, the PCF is widely used to characterize atomic structures in magnetic materials [69–71] and magnetically altered particle structures in ferrofluids [72, 73]. Up to this point, no direction depending on information can be obtained. Thus, the benefit of this method directly applied to anisotropic structures is limited. Two-dimensional approaches (radial distribution function, RDF) can be applied to evaluate macroscopic deformation of particle structures [47, 66] as well, but is limited to the evaluation of distances between particle chains, whereas the PCF enables an evaluation of three-dimensional data. To obtain information regarding different spatial directions, the method was modified to satisfy this requirement [61].

To realize a direction-dependent quality, a new control volume was introduced. Only particles with a limited angle between their directional vector and the considered spatial direction were taken into account for evaluation. The new control volume was defined as an intersection of the original spherical shell with a cone, opening in the considered direction. A more detailed explanation of the direction depending approach is given in [61]. Figure 17.5 visualizes the control volumes for the three spatial directions used here. The PCF is calculated separately for every one of the three spatial directions, giving information on the distance distribution of particles regarding the three directions. A similar approach has been chosen in ref. [74].

17.3.3 Particle size distribution, shape, and orientation

In addition to the particle positions, geometrical information of the individual particles can be evaluated from the tomography data. This procedure is applicable for larger particles, as those are most often anisotropically shaped. Furthermore, larger particles

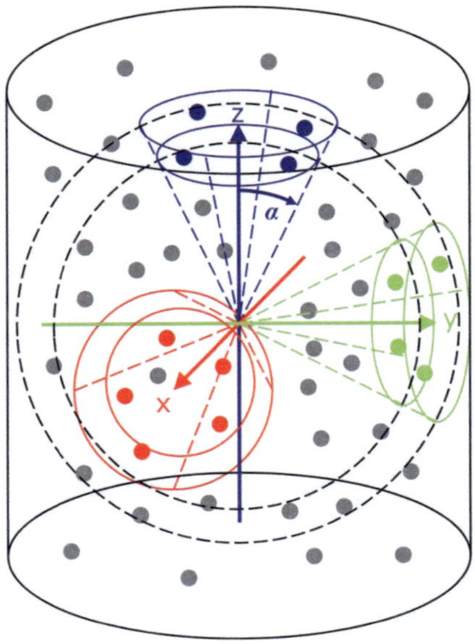

Figure 17.5: To realize a direction-dependent approach of the pair correlation function (PCF), a new control volume (full colored lines) was introduced as an intersection of the spherical shell (dashed gray lines) and a cone (dashed colored lines). Only objects located within this new control volume are taken into account for evaluation. Only one cone for each direction is shown here to enhance visibility. For calculation, a pair of cones for positive and negative direction is used. [61]

are necessary to obtain a sufficient resolution to precisely evaluate the geometrical information of the particles. Up to this point of technical advances with the used setup, this accounts for particles larger than approx. 40 µm. The geometrical properties which can be evaluated include i.e., particle volume or surface area. The shape of most particles can be approximated with an ellipsoid, enabling the evaluation of dimensions and directions of the three ellipsoid diameters of every individual particle.

Zingg's classification [75, 76] provides a convenient approach in evaluating particle shapes. The values of the three ellipsoid diameters as obtained by tomography were used to calculate the aspect ratios a/b and c/b. The longest ellipsoid diameter refers to a, the intermediate to b, and the short ellipsoid diameter to c. The graph depicted in Figure 17.6 shows the aspect ratios of approx. 11.000 particles, giving an impression of the distribution of particle shapes found in a magnetorheological elastomer containing NdFeB-particles with particle sizes ranging from 100–200 µm.

Furthermore, the gathered geometrical information of the particles can be used to make global statements about changes in the internal arrangement of the particles induced by magnetic fields. With anisotropically shaped particles their rotation to align to magnetic fields can be observed. This alignment of particles is visible in Figure 17.4. To evaluate the orientation of the particles, the angle of the ellipsoid axes can be measured. To calculate a characteristic angle between the particle and the magnetic field, the angle of the longest ellipsoid axis, called β, is evaluated. The here used NdFeB-particles feature a strong magnetic anisotropy leading to their orientation perpendicular to the applied magnetic fields. A phenomenon which is discussed in detail in ref. [23].

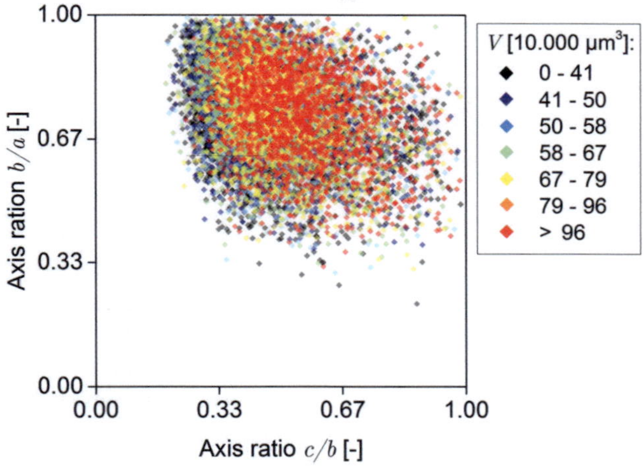

Figure 17.6: Zingg's classification of particle shapes depending on their volume applied to approx. 11.000 NdFeB particles, present in a magnetorheological elastomer.

17.3.4 Particle tracking

Aside global statements about the particle positions and geometrical information of the particles, an identification of particles allow tracking of particle motion between two situations. This procedure demands a very precise separation of the particles, as changes in measured geometrical properties of one particle from one to the other data set prohibits a correct assignment of the same particle in two data sets. Furthermore, it is necessary to equip the sample with a fixed spatial reference. For the specific experiments on magnetorheological elastomers described here, two small copper wires attached to the samples were used. The fixed position of this reference enables registration of several tomographic data sets, captured at different external stimuli.

For the first step, all particles in a near neighborhood of the original particle position as present in the first data set are taken into account in the second data set. This neighborhood has to be larger than the maximal possible translation of the particle from one situation to the other. In the next step, all those found particles are taken into account for an assignment to the original particle. To check the quality of the assignment of the particles between the two different data sets, a quality function which compares certain characteristics like the particle volume V and surface A, it's asphericity $P2A$ [29] and position x, y, z, was introduced [12]. This enables a validation of whether a correct assignment has been achieved or not. The following two Eqs. (17.5) and (17.6) were used to quantify the quality of a particle assignment. The particles considered for an assignment are named i and j. The smaller the values of G_{position} and G_{shape}, the higher the possibility of a correct assignment becomes. Thus the best fitting candidate for an assignment is chosen for every particle.

$$G_{\text{postion}} = \sqrt{(x_j - x_i)^2 + \left(y_j - y_i\right)^2 + (z_j - z_i)^2} \tag{17.5}$$

$$G_{\text{shape}} = \sqrt{\left(\frac{V_i - V_j}{V_i}\right)^2 + \left(\frac{A_i - A_j}{A_i}\right)^2 + \left(\frac{P2A_i - P2A_j}{P2A_i}\right)^2} \tag{17.6}$$

Using this procedure of particle assignment, it is possible to allocate and identify several thousand particles in one sample and to quantify changes in their spatial location and orientation induced by the external stimuli. Now it is possible not just to evaluate the present distribution of particle angles, but to calculate the exact rotational angle for every individual particle from one data set to the other. Furthermore, translation of the particles can be evaluated and particle trajectories can be observed [23]. Figure 17.7 shows correctly assigned particles from two datasets, one captured with and one without an application of a 240 mT field.

Initially, this procedure was limited to very small particle concentrations [45]. Recent advances enabled particle tracking at a particle concentration of $\Phi = 40$ wt%. Now, this method can be applied to materials which exhibit magnetorheological

Figure 17.7: Here, the assigned particles without (green) and with a 240 mT field (red) are shown. The black lines represent the trajectory of the assigned particles. A rotation of the particles is visible as well. [23]

effects [23], which allows a direct linking of results concerning mechanical behavior and particle movement.

17.3.5 Tomography data as input for calculation and simulation

As stated before, the experimental data can be used as data input for theoretical investigations [46, 49–51] and can act as a benchmark for theoretical predictions and simulations [52–55].

The combination of experimentally obtained results providing a comprehensive description of macroscopic and microscopic material properties and internal particle structures with theoretical modeling and simulation gives the scale bridging approach required for a deeper understanding of the magnetic field-driven complex behavior of magnetorheological materials. Only by combining theoretical and experimental investigations of identical materials gives access to key parameters, for instance, the internal magnetic fields, which will provide fundamental new results regarding magnetorheological materials in the future. More information on using tomographically captured particle data as an input for theoretical approaches can be found in the articles "Multiscale modeling and finite element simulation of magneto-rheological elastomers based on experimental data" and "Modeling and theoretical description of magnetic hybrid materials - bridging from meso-to macro-scales" within this issue.

17.4 Magnetorheological foams

Magnetoelastic foams are a relatively new kind of magnetic field-responsive smart materials developed in the last few years [30–33]. Owing to their high elasticity, soft foams loaded with magnetic particles show a great potential for use as magneto-rheological material. Because of their porous structure, the arrangement of particles in the polymer leads to highly complicated structures and complicated magnetic and mechanic properties. The mechanical properties of foams are strongly influenced by their structure. The pore size distribution can be interpreted as a universal description of the foam structure in general. Detailed knowledge of the distribution and its dependence on given process parameters are the basis for a better understanding of the properties of the foam and the foaming process. The foam structure can be evaluated by Computed X-ray microtomography. Figure 17.8 shows an excerpt of evaluated foam pores, captured by X-ray microtomography.

It was found, that the Weibull cumulative distribution allows a well-fitting description of the pore volume distribution of the polyurethane foam discussed in refs. [29] and [77]. Furthermore, it was shown, that the pore volume distribution function is changed in a specific way if carbonyl iron particles are incorporated. Figure 17.9 shows

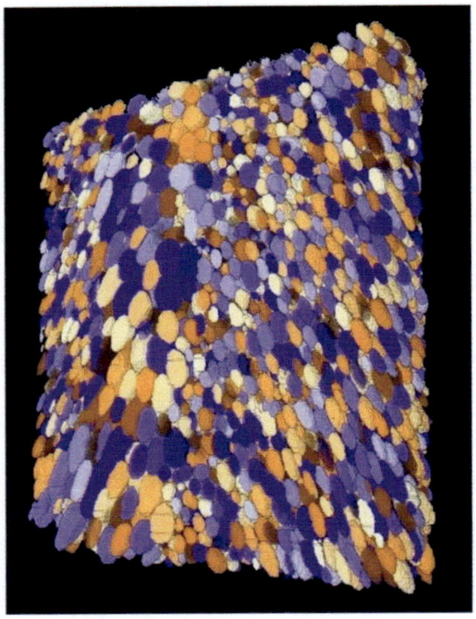

Figure 17.8: Foam pores of a polyurethane foam sample as obtained by tomography after the separation process [29].

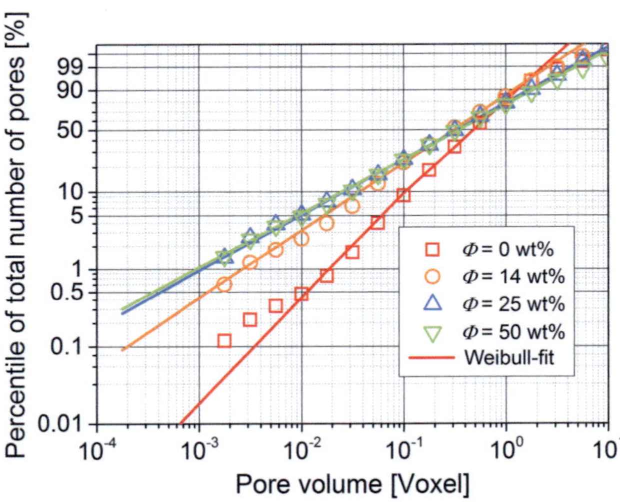

Figure 17.9: A Weibull plot showing pore volume distribution functions of polyurethane foams with different amounts of carbonyl iron particles. A strong increase in the number of small pores with the addition of particles is visible [29].

the pore volume distribution functions of polyurethane foams with varying concentrations of carbonyl iron particles.

The observed effects can be explained with known foam-stabilization mechanisms. The drainage of liquid foam is retarded by the particles. As a result, the process of merging smaller pores to larger ones is decelerated [29]. Furthermore, the impact of a magnetic field applied during the foaming process of polyurethane foams loaded with carbonyl iron particles was investigated [77]. An impact of the field on the pore volume distribution function is apparent and elongation of the pores as a result of field application was verified [77].

17.5 Magnetorheological elastomers containing carbonyl iron particles

Carbonyl iron powder is the most widely used choice for magnetic particles when synthesizing magnetorheological elastomers. Those materials provide strong and fully reversible magnetorheological effects, meaning an increase of Young's modulus induced by an applied magnetic field [47, 78, 79]. The magnetorheological effects of MRE is calculated from Young's moduli measured with an applied magnetic field E_B and without E_O, using

$$\mathrm{MRE} = \frac{E_B - E_0}{E_B}. \tag{17.7}$$

Computed X-ray microtomography offers an insight into the changes in microscopic particle structure which is highly responsible for the macroscopic mechanical behavior. A reversible formation of particle chains can be observed if the magnetic field is applied during tomography. Figure 17.10 shows the particle structures present in a magnetorheological elastomer with 40 wt% of carbonyl iron particles. The application of a 250 mT field to this material leads to a strong rearrangement of the particles to particle chains. This behavior is accompanied by a measured magnetorheological

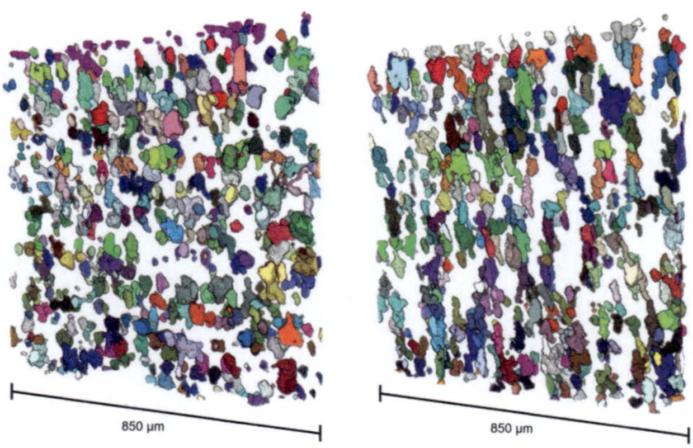

Figure 17.10: The images show a 3D excerpt of the reconstructed tomography data after the separation process. Here 40 wt% of carbonyl iron particles were used. On the left, the initial particle microstructure without a magnetic field is shown, on the right the same sample at a field of 250 mT, applied vertically. Here, particle chains are present [47].

effect of MRE = 633% ± 55%. The samples in these investigations feature Young's modulus of $E_O \approx 10$ kPa.

The evaluation of particle angles from these tomographic data sets shows a significant rotation of the particles and an impact on the radial distribution function if a magnetic field is applied [47]. The application of mechanical stress during tomography was incorporated to give a more realistic reflection of the material behavior at the measurement of Young's moduli, where mechanical stress is present as well. An evaluation of the radial distribution function of the particle structure in presence of mechanical stress of $\varepsilon = 12\%$ and the magnetic field shows a significant shift of the first maximum indicating the distance of the particles within the chains. This shift represents a reduction of particle distance and thus compression of the particle chains [47]. Figure 17.11 shows these results.

The evaluation of the volume of every individual particle enables an analysis of the particle size distribution, as conducted in [51, 64]. Thus it was shown, that the particle size has a severe impact on the magnetorheological properties of magnetorheological elastomers composed from particles with different particle size distributions [64]. Figure 17.12 shows 3D models of reconstructed tomography data sets from samples composed of four different sieve fractions of carbonyl iron particles. As seen in Figure 17.13 their size distribution can be evaluated by computed X-ray microtomography With these samples it was shown, that increasing particle size leads to increasing magnetorheological effects [64], from MRE = 30 ± 5.5% for the smallest particles to MRE = 56 ± 7% for the largest particles. The particle concentration of $\Phi = 40$ wt% was kept constant. With the evaluation of the particle shape distribution by means

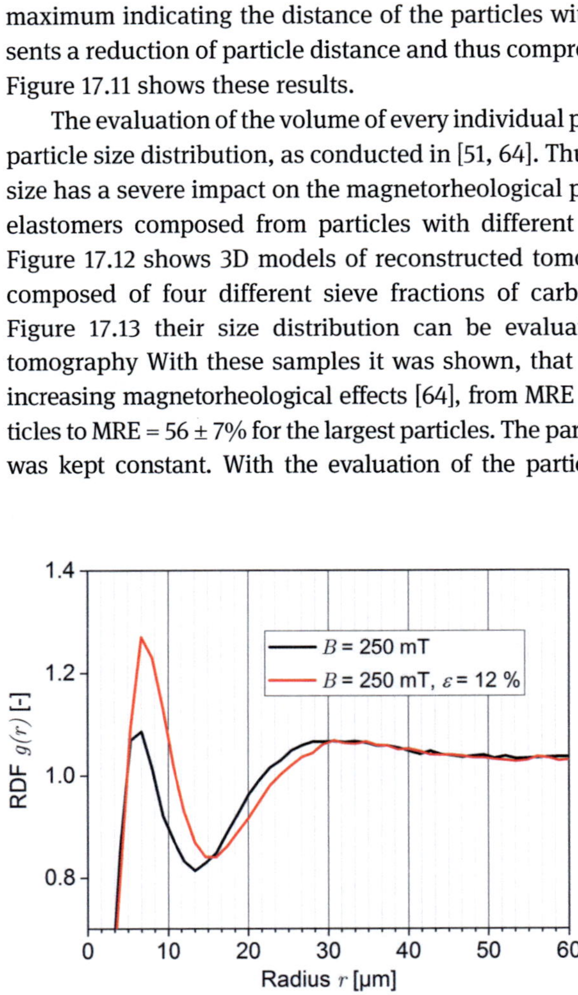

Figure 17.11: The graph shows the radial distribution function of the particle ensemble in presence of a 250 mT field applied in an axial direction, with and without $\varepsilon = 12\%$ mechanical strain. The distribution function $g(r)$ is calculated perpendicular to the field direction. The shift of the maximum to smaller distances indicates compression of the particle chains. [47]

Figure 17.12: From left to right, excerpts of the 3D models of reconstructed tomography data sets visualizing particles with of size fractions of 80–100 µm, 63–80 µm, 40–63 µm, and 20–40 µm [64].

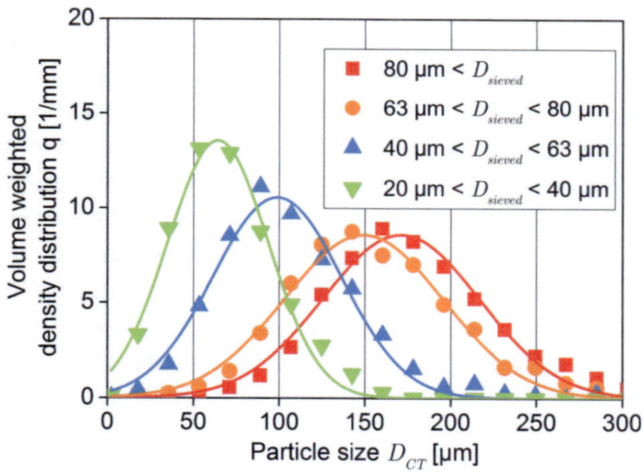

Figure 17.13: Volume weighted size distributions of the particle size classes obtained by tomography. The results correspond to the visualized particles in Figure 17.12. The particle sizes are given as the largest ellipsoid diameter. Plotted are the measurement results and a Gauss fit. [64]

of Zingg's classification, it was shown, that the particle shape remained nearly constant through the particle size classes, indicating that the found effects can be clearly addressed to the particle size and not their shape.

17.6 Magnetorheological elastomers containing neodymium-iron-boron particles

The usage of magnetically hard NdFeB-particles provides additional functionality to magnetorheological elastomers compared to those containing magnetically soft iron particles. The possibility of remanent magnetization enables nonreversible structure formation and change in mechanical properties by the application of strong magnetic fields. As shown in ref. [23] the known features like particle rotation and magneto-rheological effect remain mostly reversible when using small magnetic fields. The tomography data corresponding to the results presented in the following paragraphs was already shown in Figures 17.1 and 17.4. The used samples feature a Young' modulus of $E_0 \approx 5$ kPa and exhibiting a magnetorheological effect of $MRE = 60\%$ after magnetization at 2 T. The NdFeB-particles sizes range from 100–200 μm. As stated before, the particles feature a strong magnetic anisotropy leading to their orientation perpendicular to the applied magnetic fields. This behavior was analyzed by means of X-ray diffraction measurements [23].

Figure 17.14 shows the distribution of the angle between the longest ellipsoid axis of the particles and the direction of the magnetic field applied to the elastomer called β, for six different magnetic field situations. First, the angle has been evaluated for all particles, for a situation without any field applied. Afterward, a magnetic field of 240 mT has been applied and again, the particle orientation has been evaluated. Finally, the magnetic field has been removed to check, whether hysteresis in particle orientation appears. As one can see, the application of the magnetic field forces an alignment of the long major axis of the particles with the magnetic field direction, increasing the percentile of the total number of particles with small angles. After switching of the magnetic field, most of the particles rotate back to their initial position and only a small portion remains a bit more aligned with the magnetic field direction, a matter of fact which shows that the particles are well connected with the polymeric network, which exhibits a strong restoring force, nearly disabling any kind of hysteretic behavior. This mostly fully reversible rotation was further evaluated by scanning-

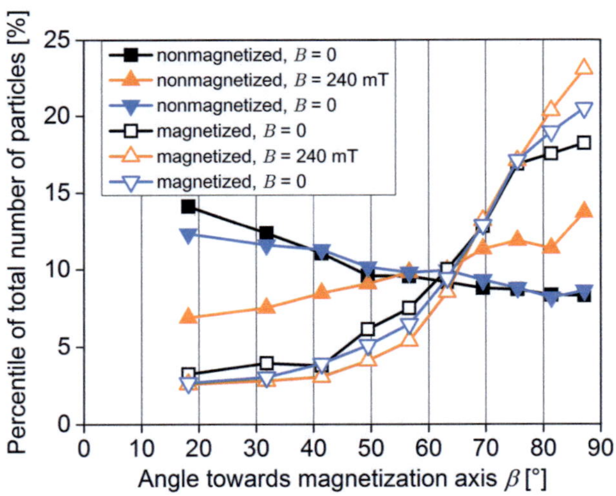

Figure 17.14: The frequency distribution of the particle angle β of all evaluated particles indicating a rotation of the particles induced by the magnetic fields [23].

confocal microscopy [23, 80, 81], proving a fully elastic deformation of the surrounding matrix, leading to a reversible rotation. This knowledge can be used to assign the observed nonreversible effects to the nonreversible magnetic behavior of the magnetically hard particles.

After the sample is magnetized at a 2 T field, the particles feature a pronounced remanence, leading to a nonreversible particle chain formation. Furthermore, the magnetization leads to a strong increase of the amount of particles with a large angle, indicating a severe orientation of the particles perpendicular to the applied field. Even at that state, an application of the 240 mT field leads to further rotation, which is partly reversible if the field is removed.

With the conduct of particle tracking, the rotation of the particles as seen in Figures 17.1 and 17.4 can be evaluated in detail. To do so, the difference of the particle angles between before and during application of the 240 mT field is calculated as $\Delta\beta = \beta_B - \beta_O$. The results are visualized in Figure 17.15. With these results a mean rotational angle of the particles can be evaluated, a result which is not within reach without particle tracking. Furthermore, a clear dependence of the initial particle angle β_O on the resulting rotational angle $\Delta\beta$ was found, and a significant impact of the particle shape [23].

As stated before, because of the remanent magnetization of the neodymium-iron-boron particles, the chain formation induced by a 2 T field is nonreversible. The direction-dependent PCF enables a convenient evaluation of the chain formation process. Figure 17.16 shows results for the PCF in chain direction of a NdFeB-particle loaded elastomer after application of magnetic field up to 2 T.

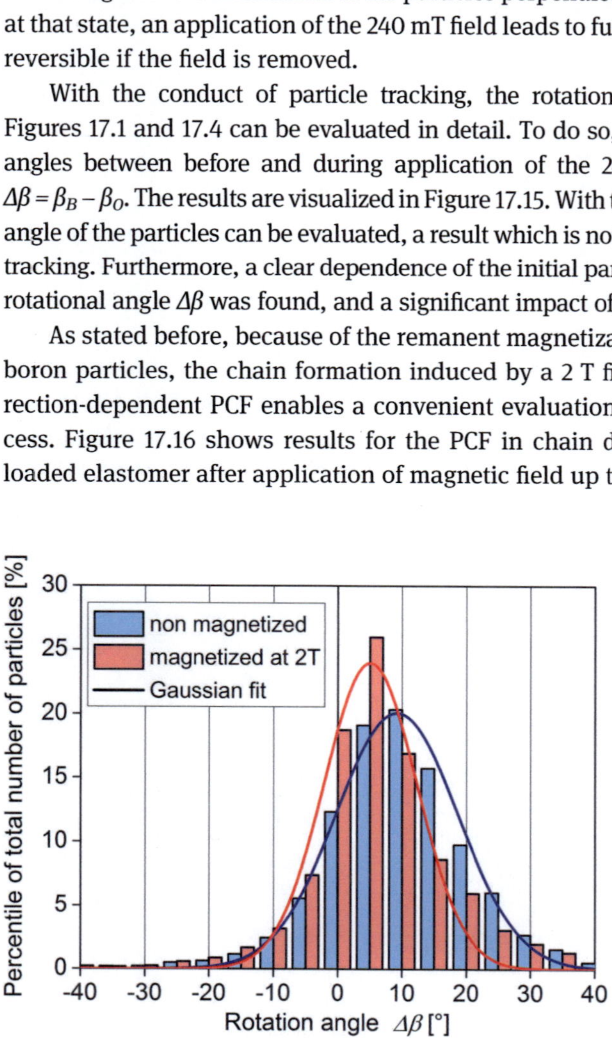

Figure 17.15: Particle tracking enables the calculation of the change of the angle for every particle individually, providing the rotational angle $\Delta\beta$. Regarding the Gaussian fit, the results show a significant rotation of every particle of approx. 10° for the unmagnetized material, and approx. 5° after magnetization at 2T [23].

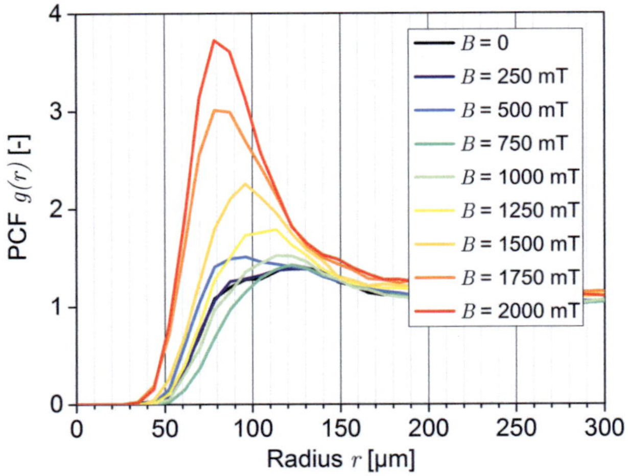

Figure 17.16: PCF regarding the z-direction for the particle structure obtained by tomography after application of magnetic fields up to 2 T oriented along the z-axis [61].

The results shown in Figure 17.16 correspond to the particle structures shown in Figure 17.17. As visible, the particle chains are not as prominent as those found with carbonyl iron particles as seen in Figure 17.10. Nevertheless, the evaluation of the direction-dependent PCF offers a high sensitivity detecting anisotropy, even for less pronounced particle chains.

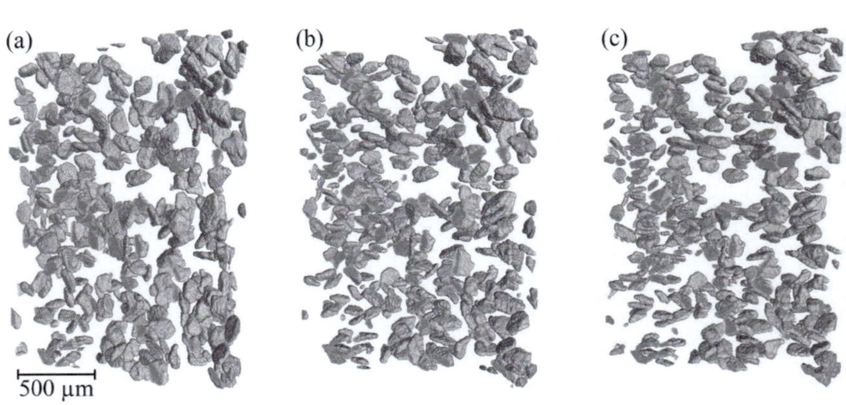

Figure 17.17: Excerpts of three reconstructed 3D models obtained by tomography showing the particle structure. (a) Shows the initial structure before application of any fields, (b) was captured after application of $B = 1250$ mT, (c) after $B = 2000$ mT. At this point particle chains in the field direction, in this picture: vertically, are clearly visible [61].

17.7 Summary and outlook

The microscopic arrangement and movement of the embedded magnetic particles is the key parameter governing the macroscopic mechanical behavior of magnetorheological materials. Computed X-ray microtomography provides a sophisticated method to evaluate the particle structure, giving information on particle distribution, individual particle positions, and geometrical properties of the particles. With advanced image processing methods, the obtained 3D information can be processed to enable evaluation methods ranging from overall structure analysis to evaluations based on single-particle information and even particle tracking. With this toolbox, the results obtained by accompanying measurements regarding the macroscopic mechanical behavior can be linked to the observed changes in particle structure. This combination provides a comprehensive and coherent description of the material properties and enables a deeper understanding of the underlying physical mechanisms found with magnetorheological materials.

Besides outlining the method of X-ray microtomography applied to magnetorheological materials, the presented work gave an overview of findings, which were not within reach without the accomplished recent advances with computed X-ray microtomography and the accompanying digital image processing methods.

The presented results show the potential of this method and enable future investigations to systematically analyze the important characteristics of magnetorheological materials as choice and embedment of particles, varying matrices, and synthesis methods. A comprehensive comparison of these results will provide important knowledge to approach one of the main goals of research regarding those materials: A targeted tailoring of magnetorheological materials to fit their desired properties.

Acknowledgments: We gratefully thank Günter Auernhammer and Robert Müller for conducting auxiliary measurements with our materials and by providing important insights to our investigations. Furthermore, we like to thank Klaus Zimmermann and his group for supplying novel sample materials and the group of Andreas Menzel and Hartmut Löwen for collaboration regarding using our experimental data for theoretical modeling.
Author contributions: All the authors have accepted responsibility for the entire content of this submitted manuscript and approved submission.
Research funding: This work was supported by Deutsche Forschungsgesellschaft (DFG) within the project OD18/21 in the frame of the priority program SPP1681 'Field controlled particle-matrix interactions: synthesis multiscale modeling and application of magnetic hybrid materials'.
Conflict of interest statement: The authors declare no conflicts of interest regarding this article.

References

1. Rabinow J. The magnetic fluid clutch. Electr Eng 1948;67:1167.
2. Rabinow J. U.S. Patent No. 2,575,360. Washington, DC: U.S. Patent and Trademark Office; 1951.
3. Papell SS. U.S. Patent No. 3,215,572. Washington, DC: U.S. Patent and Trademark Office; 1965.
4. Berkovski B, Bashtovoy V. Magnetic fluids and applications handbook. New York: Begell House; 1996, vol 36.
5. Raj K, Moskowitz R. Commercial applications of ferrofluids. J Magn Magn Mater 1990;85:233–45.
6. Filipcsei G, Csetneki I, Szilágyi A, Zrínyi M. Magnetic field-responsive smart polymer composites. In: Oligomers-polymer composites-molecular imprinting. Berlin, Heidelberg: Springer; 2007: 137–89 pp.
7. Jolly MR, Carlson JD, Muñoz BC, Bullions TA. The magnetoviscoelastic response of elastomer composites consisting of ferrous particles embedded in a polymer matrix. J Intell Mater Syst Struct 1996;7:613–22.
8. Böse H. Viscoelastic properties of silicone-based magnetorheological elastomers. Int J Mod Phys B 2007;21:4790–7.
9. Stepanov GV, Abramchuk SS, Grishin DA, Nikitin LV, Kramarenko EY, Khokhlov AR. Effect of a homogeneous magnetic field on the viscoelastic behavior of magnetic elastomers. Polymer 2007; 48:488–95.
10. Kallio M. The elastic and damping properties of magnetorheological elastomers. Espoo: VTT; 2005. ISBN 951-38-6447-2.
11. Stepanov GV, Borin DY, Raikher YL, Melenev PV, Perov NS. Motion of ferroparticles inside the polymeric matrix in magnetoactive elastomers. J Phys Condens Matter 2008;20:204121.
12. Schümann M, Gundermann T, Odenbach S. Microscopic investigation of the reasons for field-dependent changes in the properties of magnetic hybrid materials using X-ray microtomography. Arch Appl Mech 2019;89:77–89.
13. Bica I, Anitas EM, Averis LME, Kwon SH, Choi HJ. Magnetostrictive and viscoelastic characteristics of polyurethane-based magnetorheological elastomer. J Ind Eng Chem 2019;73:128–33.
14. Li Y, Li J, Li W, Du H. A state-of-the-art review on magnetorheological elastomer devices. Smart materials and structures 2014;23:123001.
15. Choi SB, Li W, Yu M, Du H, Fu J, Do PX. State of the art of control schemes for smart systems featuring magneto-rheological materials. Smart Mater Struct 2016;25:043001.
16. Deng HX, Gong XL, Wang LH. Development of an adaptive tuned vibration absorber with magnetorheological elastomer. Smart Mater and structures 2006;15:N111.
17. Böse H, Rabindranath R, Ehrlich J. Soft magnetorheological elastomers as new actuators for valves. J Intell Mater Syst Struct 2012;23:989–94.
18. Yoo JH, Wereley NM. Design of a high-efficiency magnetorheological valve. J Intell Mater Syst Struct 2002;13:679–85.
19. Zimmermann K, Naletova VA, Zeidis I, Turkov VA, Kolev E, Lukashevich MV, et al. A deformable magnetizable worm in a magnetic field—a prototype of a mobile crawling robot. J Magn Magn Mater 2007;311:450–3.
20. Kumbhar SB, Chavan SP, Gawade SS. Adaptive tuned vibration absorber based on magnetorheological elastomer-shape memory alloy composite. Mech Syst Signal Process 2018; 100:208–23.
21. Yang J, Sun SS, Zhang SW, Li WH. Review of structural control technologies using magnetorheological elastomers. Current Smart Mater 2019;4:22–8.
22. Ahamed R, Choi SB, Ferdaus MM. A state of art on magneto-rheological materials and their potential applications. J Intell Mater Syst Struct 2018;29:2051–95.

23. Schümann M, Borin DY, Huang S, Auernhammer GK, Müller R, Odenbach S. A characterisation of the magnetically induced movement of NdFeB-particles in magnetorheological elastomers. Smart Mater Struct 2017;26:095018.
24. Koo JH, Dawson A, Jung HJ. Characterization of actuation properties of magnetorheological elastomers with embedded hard magnetic particles. J Intell Mater Syst Struct 2012;23:1049–54.
25. Stepanov GV, Borin DY, Bakhtiiarov AV, Storozhenko PA. Magnetic properties of hybrid elastomers with magnetically hard fillers: rotation of particles. Smart Mater Struct 2017;26:035060.
26. Borin, DY, Stepanov, GV, Odenbach, S. Tuning the tensile modulus of magnetorheological elastomers with magnetically hard powder. Int J Phys Conf 2013;412:012040.
27. Stepanov GV, Borin DY, Bakhtiiarov AV, Storozhenko PA. Influence of the size of magnetic filler particles on the properties of hybrid magnetic elastomer with magnetically hard filler. J Magn Magn Mater 2020;498:166071.
28. Borin DY, Stepanov GV. Oscillation measurements on magnetoactive elastomers with complex composition. J Optoelectron Adv Mater 2013;15:249–53.
29. Schümann M, Günther S, Odenbach S. The effect of magnetic particles on pore size distribution in soft polyurethane foams. Smart Mater and structures 2014;23:075011.
30. Sorrentino L, Aurilia M, Forte G, Iannace S. Anisotropic mechanical behavior of magnetically oriented iron particle reinforced foams. J Appl Polym Sci 2011;119:1239–47.
31. Gong Q, Wu J, Gong X, Fan Y, Xia H. Smart polyurethane foam with magnetic field controlled modulus and anisotropic compression property. RSC Adv 2013;3:3241–8.
32. Volpe V, D'Auria M, Sorrentino L, Davino D, Pantani R. Injection molding of magneto-sensitive polymer composites. Mater Today Commun 2018;15:280–7.
33. D'Auria M, Davino D, Pantani R, Sorrentino L. Polymeric foam-ferromagnet composites as smart lightweight materials. Smart Mater Struct 2016;25:055014.
34. Wei B, Gong X, Jiang W, Qin L, Fan Y. Study on the properties of magnetorheological gel based on polyurethane. J Appl Polym Sci 2010;118:2765–71.
35. Reinicke S, Döhler S, Tea S, Krekhova M, Messing R, Schmidt AM, et al. Magneto-responsive hydrogels based on maghemite/triblock terpolymer hybrid micelles. Soft Matter 2010;6:2760–73.
36. Selzer L, Odenbach S. Effects of carbonyl iron particles on the rheological behavior of nanocomposite hydrogels. J Magn Magn Mater 2020:166394. https://doi.org/10.1016/j.jmmm.2020.166394.
37. Stepanov, G, Borin, D, Odenbach, S. Magnetorheological effect of magneto-active elastomers containing large particles. Int J Phys Conf 2009;149:012098.
38. Boczkowska A, Awietjan SF, Wroblewski R. Microstructure–property relationships of urethane magnetorheological elastomers. Smart Mater Struct 2007;16:1924.
39. Kallio M, Lindroos T, Aalto S, Järvinen E, Kärnä T, Meinander T. Dynamic compression testing of a tunable spring element consisting of a magnetorheological elastomer. Smart Mater Struct 2007;16:506.
40. Pop LM, Odenbach S. Investigation of the microscopic reason for the magnetoviscous effect in ferrofluids studied by small angle neutron scattering. J Phys Condens Matter 2006;18:S2785.
41. Balasoiu M, Lebedev VT, Orlova DN, Bica I, Raikher YL. SANS investigation of a ferrofluid based silicone elastomer microstructure. Int J Phys Conf 2012;351:012014.
42. Günther D, Borin DY, Günther S, Odenbach S. X-ray micro-tomographic characterization of field-structured magnetorheological elastomers. Smart Mater and structures 2011;21:015005.
43. Borbáth T, Günther S, Borin DY, Gundermann T, Odenbach S. XμCT analysis of magnetic field-induced phase transitions in magnetorheological elastomers. Smart Mater and structures 2012;21:105018.
44. Borin D, Günther D, Hintze C, Heinrich G, Odenbach S. The level of cross-linking and the structure of anisotropic magnetorheological elastomers. J Magn Magn Mater 2012;324:3452–4.

45. Gundermann T, Odenbach S. Investigation of the motion of particles in magnetorheological elastomers by X-μCT. Smart materials and structures 2014;23:105013.
46. Gundermann T, Cremer P, Löwen H, Menzel AM, Odenbach S. Statistical analysis of magnetically soft particles in magnetorheological elastomers. Smart Mater Struct 2017;26:045012.
47. Schümann M, Odenbach S. In-situ observation of the particle microstructure of magnetorheological elastomers in presence of mechanical strain and magnetic fields. J Magn Magn Mater 2017;441:88–92.
48. Schümann M, Morich J, Kaufhold T, Böhm V, Zimmermann K, Odenbach S. A mechanical characterisation on multiple timescales of electroconductive magnetorheological elastomers. J Magn Magn Mater 2018;453:198–205.
49. Tarama M, Cremer P, Borin DY, Odenbach S, Löwen H, Menzel AM. Tunable dynamic response of magnetic gels: impact of structural properties and magnetic fields. Phys Rev 2014;90:042311.
50. Pessot G, Cremer P, Borin DY, Odenbach S, Löwen H, Menzel AM. Structural control of elastic moduli in ferrogels and the importance of non-affine deformations. J Chem Phys 2014;141:015005.
51. Pessot G, Schümann M, Gundermann T, Odenbach S, Löwen H, Menzel AM. Tunable dynamic moduli of magnetic elastomers: from characterization by x-ray micro-computed tomography to mesoscopic modeling. J Phys Condens Matter 2018;30:125101.
52. Spieler C, Kästner M, Goldmann J, Brummund J, Ulbricht V. XFEM modeling and homogenization of magnetoactive composites. Acta Mech 2013;224:2453–69.
53. Metsch P, Kalina KA, Spieler C, Kästner M. A numerical study on magnetostrictive phenomena in magnetorheological elastomers. Comput Mater Sci 2016;124:364–74.
54. Kalina KA, Metsch P, Kästner M. Microscale modeling and simulation of magnetorheological elastomers at finite strains: a study on the influence of mechanical preloads. Int J Solid Struct 2016;102:286–96.
55. Kalina KA, Brummund J, Metsch P, Kästner M, Borin DY, Linke JM, et al. Modeling of magnetic hystereses in soft MREs filled with NdFeB particles. Smart Mater Struct 2017;26:105019.
56. Metsch P, Kalina KA, Brummund J, Kästner M. Two-and three-dimensional modeling approaches in magneto-mechanics: a quantitative comparison. Arch Appl Mech 2019;89:47–62.
57. Kalina KA, Metsch P, Brummund J, Kästner M. Development of a macro-model for magnetorheological elastomers based on microscopic simulations. Proc Appl Math Mech 2019;19: e201900288.
58. Radon J. Über die Bestimmung von Funktionen durch ihre Integralwerte längs gewisser Mannigfaltigheiten. *Ber. Verh. Sachs. Akad. Wiss.* Leipzig, Math Phys Klass 1917;69.
59. Kalender WA. Computed tomography: fundamentals, system technology, image quality, applications. New York: John Wiley & Sons; 2011.
60. Buzug TM. Computed tomography: from photon statistics to modern cone-beam CT. Berlin: Springer Science & Business Media; 2008.
61. Schümann M, Morich J, Günther S, Odenbach S. The evaluation of anisotropic particle structures of magnetorheological elastomers by means of pair correlation function. J Magn Magn Mater 2020. https://doi.org/10.1016/j.jmmm.2020.166537.
62. DIPImage. Available from: http://www.diplib.org [Accessed 22 Jan 2020].
63. Beucher S. Use of watersheds in contour detection. In: Proceedings of the International Workshop on Image Processing. CCETT; 1979.
64. Winger J, Schümann M, Kupka A, Odenbach S. Influence of the particle size on the magnetorheological effect of magnetorheological elastomers. J Magn Magn Mater 2019;481: 176–82.
65. Zubarev AY. Effect of chain-like aggregates on ferrogel magnetodeformation. Soft Matter 2013;9: 4985–92.

66. Weeber R, Kantorovich S, Holm C. Deformation mechanisms in 2d magnetic gels studied by computer simulations. Soft Matter 2012;8:9923–32.
67. Gray CG, Gubbins KE. Fundamentals. In: Theory of molecular fluids. Oxford: Oxford University Press; 1984, vol 1.
68. Allen MP, Tildesley DJ. Computer simulation of liquids. Oxford: Clarendon; 1987.
69. Villain J. Theory of one- and two-dimensional magnets with an easy magnetization plane. II. The planar, classical, two-dimensional magnet. J Phys 1975;36:581–90.
70. Abe R. Critical behaviour of pair correlation function in ising ferromagnets. Prog Theor Phys 1967; 38:568–75.
71. O´Leary WP. Partial structure factors and pair correlation functions for an amorphous magnetic alloy. J Phys F: Met Phys 1975;5:L175.
72. Elfimova EA, Ivanov AO. Pair correlations in magnetic nanodispersed fluids. J Exp Theor Phys 2010; 111:146–56.
73. Martin G, Bradbury A, Chantrell R. Particle pair correlation in a magnetically saturated ferrofluid. IEEE Trans Magn 1986;22:1137–9.
74. Ryzhkov AV, Melenev PV, Balasoiu M, Raikher YL. Structure organization and magnetic properties of microscale ferrogels: the effect of particle magnetic anisotropy. J Chem Phys 2016;145:074905.
75. Zingg T. Beitrag zur Schotteranalyse [Doctoral dissertation]. Zurich: ETH Zurich; 1935.
76. Krumbein WC. Measurement and geological significance of shape and roundness of sedimentary particles. J Sediment Res 1941;11:64–72.
77. Schümann M, Seelig N, Odenbach S. The effect of external magnetic fields on the pore structure of polyurethane foams loaded with magnetic microparticles. Smart Mater Struct 2015;24:105028.
78. Kwon SH, Lee JH, Choi HJ. Magnetic particle filled elastomeric hybrid composites and their magnetorheological response. Materials 2018;11:1040.
79. Stoll A, Mayer M, Monkman GJ, Shamonin M. Evaluation of highly compliant magneto-active elastomers with colossal magnetorheological response. J Appl Polym Sci 2014;131. https://doi.org/10.1002/app.39793.
80. Roth M, Schilde C, Lellig P, Kwade A, Auernhammer GK. Colloidal aggregates tested via nanoindentation and quasi-simultaneous 3D imaging. Eur Phys J E 2012;35:124.
81. Roth M, Franzmann M, D'Acunzi M, Kreiter M, Auernhammer GK. Experimental analysis of single particle deformations and rotations in colloidal and granular systems; 2011. arXiv preprint arXiv: 1106.3623.

Supplementary Material: The online version of this article offers supplementary material (https://doi.org/10.1515/j_psr-2019-0105).

Rudolf Weeber*, Patrick Kreissl and Christian Holm

18 Magnetic field controlled behavior of magnetic gels studied using particle-based simulations

Abstract: This contribution provides an overview of the study of soft magnetic materials using particle-based simulation models. We focus in particular on systems where thermal fluctuations are important. As a basis for further discussion, we first describe two-dimensional models which demonstrate two deformation mechanisms of magnetic gels in a homogeneous field. One is based on the change of magnetic interactions between magnetic particles as a response to an external field; the other is the result of magnetically blocked particles acting as cross-linkers. Based on the qualitative behavior directly observable in the two-dimensional models, we extend our description to three-dimensions. We begin with particle-cross-linked gels, as for those, our three-dimensional model also includes explicitly resolved polymer chains. Here, the polymer chains are represented by entropic springs, and the deformation of the gel is the result of the interaction between magnetic particles. We use this model to examine the influence of the magnetic spatial configuration of magnetic particles (uniaxial or isotropic) on the gel's magnetomechanical behavior. A further part of the article will be dedicated to scale-bridging approaches such as systematic coarse-graining and models located at the boundary between particle-based and continuum modeling. We will conclude our article with a discussion of recent results for modeling time-dependent phenomena in magnetic-polymer composites. The discussion will be focused on a simulation model suitable for obtaining AC-susceptibility spectra for dilute ferrofluids including hydrodynamic interactions. This model will be the basis for studying the signature of particle–polymer coupling in magnetic hybrid materials. In the long run, we aim to compare material properties probed locally via the AC-susceptibility spectra to elastic moduli obtained for the system at a global level.

Keywords: AC-susceptibility; computer simulations; gel deformation; magnetic gels; multiscale modeling; simulation models.

18.1 Introduction

In recent years, there has been an increased trend towards miniaturization in diverse fields of application. This includes technical applications, like sensing and actuation

---

***Corresponding author: Rudolf Weeber,** Institute for Computational Physics, University of Stuttgart, Stuttgart, Germany, E-mail: rudolf.weeber@icp.uni-stuttgart.de
Patrick Kreissl and Christian Holm, Institute for Computational Physics, University of Stuttgart, Stuttgart, Germany

This article has previously been published in the journal Physical Sciences Reviews. Please cite as: R. Weeber, P. Kreissl and C. Holm "Magnetic field controlled behavior of magnetic gels studied using particle-based simulations" *Physical Sciences Reviews* [Online] 2021, 6. DOI: 10.1515/psr-2019-0106 | https://doi.org/10.1515/9783110569636-018

[1, 2], so-called 'lab on a chip' devices, or microrobotics [3, 4] and biomedical treatment. Realizing such systems often relies on the availability of suitable 'smart materials', the properties of which can change under external stimuli. Magnetic composite materials are a fascinating example for such smart materials. They typically consist of magnetic nano- or micro-particles which are embedded in a polymer suspension or gel matrix [5–8]. They combine the magnetic properties of a ferrofluid [9, 10] with the elastic and viscoelastic properties of polymers [11]. Application and modification of an external magnetic field allows for the contactless dynamic control of some mechanical properties of the magnetic composites, such as their motion, shape deformation, or (anisotropic) stiffness. These special properties as well as the fact that the human body is transparent to magnetic fields makes magnetic composite materials particularly promising candidates for bio-medical applications [12]. The possible applications range from targeted drug delivery with externally controlled release to magnetic resonance-guided placement and surveillance of artery stents. Moreover, there are cases where treatment is based on the direct injection of magnetic particles or composites into the human body [13] – for example for imaging purposes, dissolving blood clots, or even for cancer treatment via directed chemotherapy [14–16] or externally induced local hyperthermia [17]. All of these applications are affected by or actually rely on the interaction of magnetic particles with their non Newtonian surrounding. Therefore, a deep understanding of the underlying coupling mechanisms is key when it comes to tailoring magnetic particles or composite materials to specific use cases. Examples of typical experimental methods for investigating magnetic composite materials include magnetic particle spectroscopy (MPS) [18], small-angle neutron scattering (SANS) [19–21], small-angle X-ray scattering (SAXS) [22, 23], cryo-transmission electron microscopy (cryo-TEM) [24, 25], and X-ray microtomography [26] which give valuable insight into their structure. Unfortunately however, in many systems, the details of this coupling are not fully known and difficult to determine experimentally.

In such cases, computer simulations can be a valuable tool, as they allow studying well-defined systems, the properties of which can be easily modified. Depending on the length-scales one is interested in, computational modeling requires different levels of detail. On the larger scales – mostly relevant for technical applications – continuum descriptions of magnetic composites [27, 28] can be solved using, e.g., finite element methods (FEM) [29, 30]. On the mesoscale, models include single magnetic particles. A theoretical approach for this scale is density functional theory (DFT) [31]. In this manuscript, we focus on particle-based molecular dynamics approach for simulating the magnetic, deformational and (visco)-elastic response of magnetic gels and magnetic particles in polymer suspensions. We will start with a study of the magnetic field-induced deformation of two-dimensional gel models in Sec. 18.2. The deformation mechanisms in three dimensions will be discussed in Sec. 18.3. Scale-bridging approaches to reduce the computational effort and to transfer knowledge between different levels of descriptions will be given in Sec. 18.4. Finally, in Sec. 18.5 we present a basic computational strategy for modeling frequency dependent susceptibility measurements on magnetic nanoparticles in polymer solution.

18.2 Direct observation of magnetic-field induced deformation in two-dimensional models

Three distinct mechanisms can lead to the deformation of a magnetic gel in a magnetic field. First, an inhomogeneous magnetic field exerts a force on magnetic moments. This force points along the field gradient. As a consequence, a magnetic gel can be deformed by fixing it at one end and applying a field gradient along the direction in which a deformation is desired [32]. More complex deformations can be achieved, by making the magnetic gel stiffer in certain regions, e.g., through more intense polymerization.

The second and third mechanisms for field-controlled deformation of magnetic gels pertain to homogeneous magnetic fields. This has the advantage of not causing a migration of the gel, even if it is not fixed to a substrate. In a homogeneous field, a deformation can be mediated by the interaction between the magnetic particles. These interactions are, on average, small and isotropic in the field-free case, as the magnetic particles are randomly aligned. In an external field, the magnetic moments in the gel get aligned parallel to the field. This results in an anisotropic interaction between the particles throughout the system. These interactions in turn lead to a repositioning of the magnetic particles and a strain of the surrounding polymer matrix. Macroscopically, this is described by minimizing the sum of two energetic contributions: (I) the energy needed for the elastic deformation and (II) the demagnetization energy [33]. The behavior has also been observed experimentally [34, 35]. This macroscopic description neglects the details of the spatial distribution of the magnetic particles in the system and their local mobility. It has been shown that both factors can influence the deformation [29, 36–38]. A further mechanism for field-induced deformation in homogeneous magnetic fields exists in a special class of magnetic gels. In these gels, magnetically blocked particles act as the cross-linkers of the gel [39–41]. In these so-called particle cross-linked gels, the polymer chains are covalently bound to specific points on the surface of the magnetic particles. Once an external field rotates the magnetic particles, the attached polymer chains have to follow. Their ends wrap around the magnetic particles, effectively shortening the length of the polymers. This is leads to shrinkage of the gel.

Although experimental systems are, with very few exceptions [42, 43], three-dimensional, we will begin by introducing two-dimensional simulation models. The deformation mechanisms described above can be observed directly – and visually –, in two-dimensional models. For three-dimensional models, visual inspection is typically no longer useful and the analysis is carried out purely quantitatively. The simulations discussed below have been carried out with the ESPResSo molecular dynamics software [44–46].

18.2.1 Deformation through the change of magnetic interactions

Let us turn to a model demonstrating the deformation of magnetic gels through changing the magnetic interactions between the particles. We call this model I in Ref. [37], where a more technical description can be found. The system has we boundary conditions. The model gel is constructed by placing particles on a regular square lattice. These particles will be the nodes of the network. They are then connected by polymers represented via bead–spring models. Four polymer chains are attached to each node. The network is made magnetic by assigning a randomly aligned magnetic moment to every second particle in the chains. Once the simulation starts, the particles can move freely, but the connections along the chains and at the nodes are maintained. Also the magnetic moments can rotate freely. This combination of a 2D model but 3D rotation is called quasi-two dimensional. The magnetic particles interact via the dipole–dipole interaction. This is suitable for single-domain particles, for magnetizable particles we refer the reader to Ref. [47]. There is no direct coupling between the rotation of the magnetic moments and the polymer chains. We hence describe magnetic particles which are either only loosely connected to the polymer backbone, or which realign their moment internally via the Néel mechanism. Entropy is of importance both, for polymers and for nanometer-size magnetic particles. Therefore, thermal fluctuations are taken into account for both, translation and rotation using the Langevin equation [48]. Due to the fluctuations, for quantitative observations, one has to average over many snapshots. When no external field is applied, the gel keeps a roughly square-like shape originating from its initial construction, see Figure 18.1. When a field is applied, the sample deforms strongly. It elongates parallel to the field direction and shrinks in the perpendicular one. While all chain orientations are equally favorable in the field-free case, this is no longer true in an external magnetic field, where the magnetic moments in the chains are aligned parallel to the field. In all chains aligned roughly parallel to the external field, the magnetic moments are in a head-to-tail configuration which is energetically favorable. In chains oriented perpendicular to the external field, the magnetic moments would be in a parallel side-by-side configuration which is highly unfavorable. Hence, those chains bend towards the field direction. Lastly, the gel is thinner at the ends than in the middle. Macroscopically, this can be related to a lowering of the demagnetization energy which is low for an ellipse aligned parallel to the local magnetization within the gel.

As can be seen in Figure 18.2, the overall area of the gel decreases, when it is exposed to a magnetic field. In this figure, we compare the area of the gel with and without applied fields for different strengths of the dipolar interaction. We use the dipolar interaction parameter

$$\lambda = \frac{\mu_0 m^2}{4\pi\sigma^3} \frac{1}{k_B T}, \tag{18.1}$$

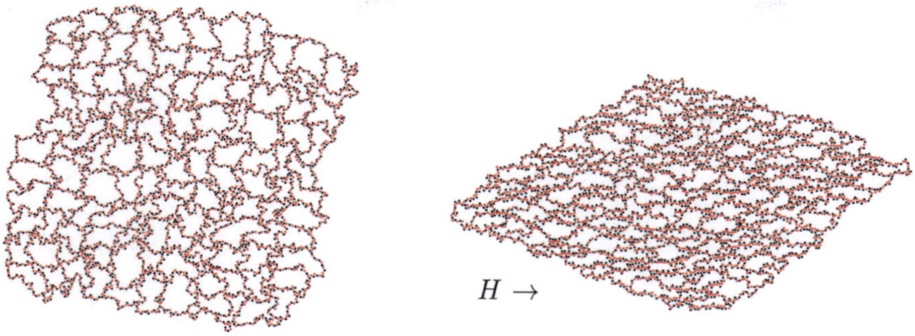

Figure 18.1: Snapshots of gel Model I from Ref. [37] based on polydisperse polymers, i.e., those with randomized chain length. Particles carrying a magnetic moment are highlighted with red color. The model describes a situation in which the interaction between the magnetic particles in the gel is strong compared to the thermal energy. In this case, aligning the dipoles by applying an external magnetic field will result in an interaction between them which is direction dependent throughout the entire system. In combination with the elasticity provided by the network of chains, this results in a deformation of the gel. The field-free case is shown on the left. A field is applied along the x-direction, on the right-hand side. The gel elongates parallel to the external field and shrinks in the perpendicular direction. Chains aligned perpendicular to the field would lead to unfavorable dipolar interactions between the magnetic particles in those chains. Hence, the chains bend towards the field direction. This figure is based on data from Ref. [37].

which compares the interaction energy of two touching magnetic particles of diameter σ and magnetic moment m with the thermal energy $k_B T$. As can be seen the shrinkage of the gel saturates for increasing dipole strengths. This is due to excluded volume and entropic forces limiting a further compression of the gel. Once the nonmagnetic

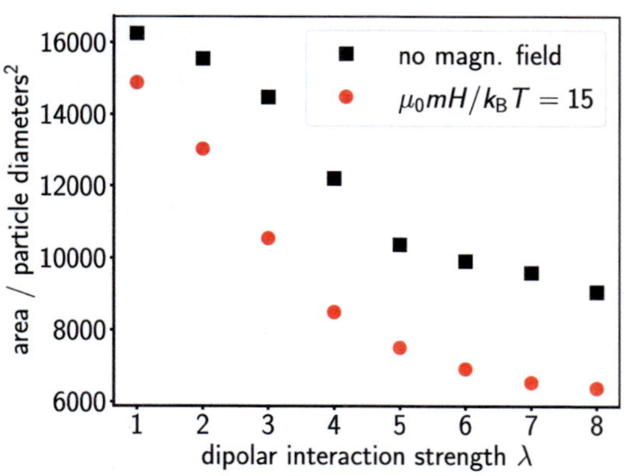

Figure 18.2: Area of a gel with and without an external magnetic field applied, for different strength of the dipole–dipole interaction λ. It can be seen that for increasing dipole moments, the contraction of the gel saturates quickly, because the internal pressure, caused by entropy and excluded volume, prevents further shrinking. Even without an external magnetic field, the gel area decreases with an increase of λ. Adapted from Ref. [37] with permission from the Royal Society of Chemistry.

particles positioned in between the magnetic ones are pushed aside, a further compression could only be achieved by overcoming the steep excluded volume interactions. This 'hardening' of the gel has also been observed in other models [49]. For further details and a more quantitative discussion, we refer to Ref. [37].

18.2.2 Field-controlled deformation of particle cross-linked gels

Let us now turn to gels in which magnetic particles act as cross-linkers. Snapshots of a corresponding model (model II of Ref. [37]) are shown in Figure 18.3. The model is again constructed by initially placing the nodes of the network on a square lattice. Here, however, the nodes are the magnetic nanoparticles. Their magnetic moments are chosen randomly in the model plane, making the model fully two-dimensional. As with the previous model, the nodes are connected by bead–spring polymer chains. The beads, however, have a radius of one-tenth of that of the magnetic particles. This approaches experimentally realistic ratios of the magnetic particle diameter and the polymer persistence length, e.g., 10 versus 1 nm. Furthermore, in the model for particle cross-linked gels, the polymer chains are attached to specific spots on the surface of the magnetic particles. This mimics the covalent bonds which are found in corresponding experimental systems [40, 41]. Numerically, this is realized by a concept called 'virtual sites', to which the polymers are connected. These are placed on the surface of the

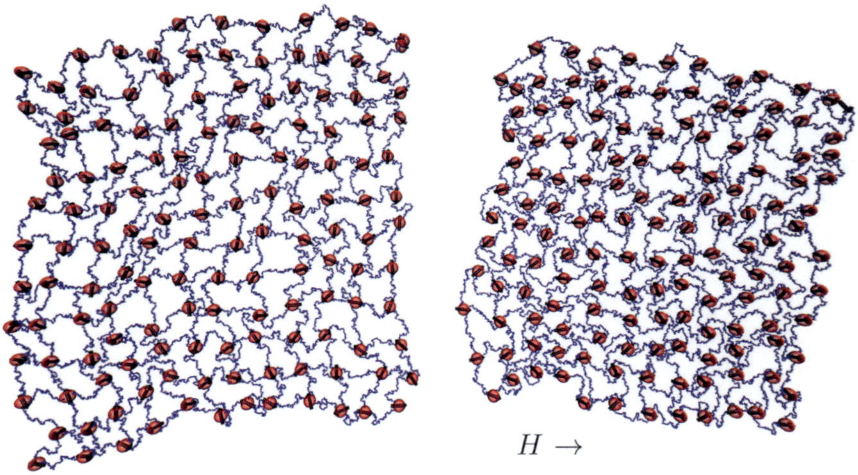

Figure 18.3: A two-dimensional simulation model of a gel is depicted, in which the magnetically blocked particles (red) act as cross-linkers. The polymer chains are bound to specific spots on the magnetic particle's surface. The field-free case is shown on the left-hand side. On the right-hand side, an external field is applied. Initially, the model is constructed on a square lattice, but the particles are free to move as the simulation starts. The magnetic moments initially are randomly aligned in the plane. In a homogeneous magnetic field (right-hand side), the gel shrinks isotropically. This is caused by the polymers wrapping around the magnetic particles, effectively shortening the polymer chains. This can be seen especially at the boundary. Reprinted from Ref. [54] with permission from Elsevier.

magnetic particles and rigidly follow their linear and rotational motion. Likewise, forces acting on the virtual sites, e.g., through a stress on the polymer chains, lead to a torque and force on the magnetic particles [45]. As in the previous model, thermal fluctuations are taken into account, and averaging is needed for quantitative observations. Also, as previously, open boundary conditions are applied.

For the field-free case (left image in Figure 18.3), the attachment points of the polymers are roughly oriented along the direction of the polymer chain. As the particles can move and rotate freely in the model plane, some thermal fluctuations compared to the initial, regular construction are observed. When a magnetic field is turned on (right-hand side of Figure 18.3), the gel shrinks. The mechanism is as follows: (I) the magnetic particles coalign to the external field, (II) the attachment points of the polymer chains rigidly follow that rotation, (III) this causes a stress on the polymers. To reduce this stress, the polymer chains wrap around the surface of the magnetic particles. (IV) This wrapping effectively reduces the length of the polymer chains causing the gel's shrinkage. The shrinkage is isotropic in the two-dimensional model, because all magnetic particles rotate around the same rotation axis, namely the one perpendicular to the model plane. In three dimensions this is no longer the case. It is important to note that this deformation mechanism does not depend on the magnetic interactions between the magnetic particles in the system. Rather, it depends on the interaction of the magnetically blocked particles with the external magnetic field, and on a direct coupling between the orientation of the magnetic moments and the polymer chains.

The shrinkage of the gel is limited by two factors, namely the excluded volume forces between the constituents of the system, and the entropic elasticity of the polymers. The latter in particular creates a torque on the magnetic particles which opposes their alignment to the external magnetic field. As a consequence, the magnetization of the gel is hindered. This can be seen from a plot of the magnetization versus the strength of the applied field (Figure 18.4). Here, the field strength is expressed in terms of the dimensionless Langevin parameter

$$\alpha = \frac{\mu_0 m H}{k_B T}, \tag{18.2}$$

which relates the magnetic interaction energy between a particle's dipole moment m and the field H to the thermal energy ($k_B T$). Here, μ_0 is the magnetic constant. We observe a magnetization which lags behind the one predicted by the Langevin law for non-interacting dipoles. This behavior is opposite to what is observed for ferrofluids [50]. Other causes of a hindered magnetization can either be constraints in the rearrangement of magnetic particles in a gel [51], or highly anisotropic magnetic particles [52, 53].

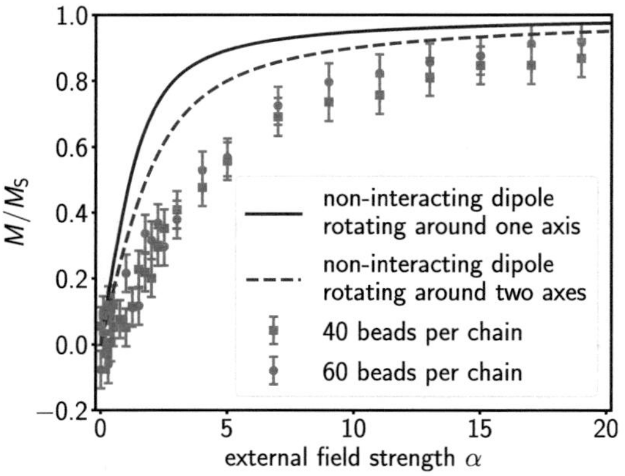

Figure 18.4: Magnetization curve for a gel with 40 and 60 beads per chain, respectively. The magnetization for a single dipole in quasi-2D and 2D are also plotted, for comparison. We observe that the magnetic response of the gel is strongly hindered compared to a single free dipole due to the elasticity of the polymer matrix. When the chains are longer, the gel can be magnetized more easily. At the same time, the amount of shrinkage is lower. It is worth mentioning that the actual error bars for large values of the Langevin parameters α (see Eq. (18.2)) are smaller than those plotted here. Adapted from Ref. [37] with permission from the Royal Society of Chemistry.

18.3 The deformation mechanisms in three dimensions

Now that the basic mechanisms for a ferrogel's deformation in a magnetic field have been illustrated and visualized in two dimensions, let us turn to three-dimensional models, more relevant for experimental systems.

18.3.1 Particle cross-linked gels

A three-dimensional model for particle cross-linked gels has been introduced in Refs. [54, 55]. Like its two-dimensional counterpart it is constructed by initially placing magnetically blocked particles on a regular lattice, then connecting those particles by bead-spring polymer chains. Also, the mechanism by which the polymer chains are bound to specific spots on the surface of the magnetic particles is the same as in the two-dimensional case. Thus, we have the same direct coupling between the magnetic particles rotation and the polymer chains. There are two important differences to the two-dimensional model discussed above. First, instead of open boundary conditions, we apply periodic ones. This is necessary to limit boundary effects which would be more pronounced than in two dimensions for a comparable number of particles in the simulation. Second, two network structures are studied. In the one, based on a diamond cubic lattice, four chains are connected to each magnetic particle. In the other, based on a simple cubic structure, six chains are connected to a magnetic particle.

Corresponding sketches can be seen in Figure 18.5. The molecular dynamics technique based on the Langevin equation is comparable to the two-dimensional model described in the previous section. However, due to the periodic boundaries, the system

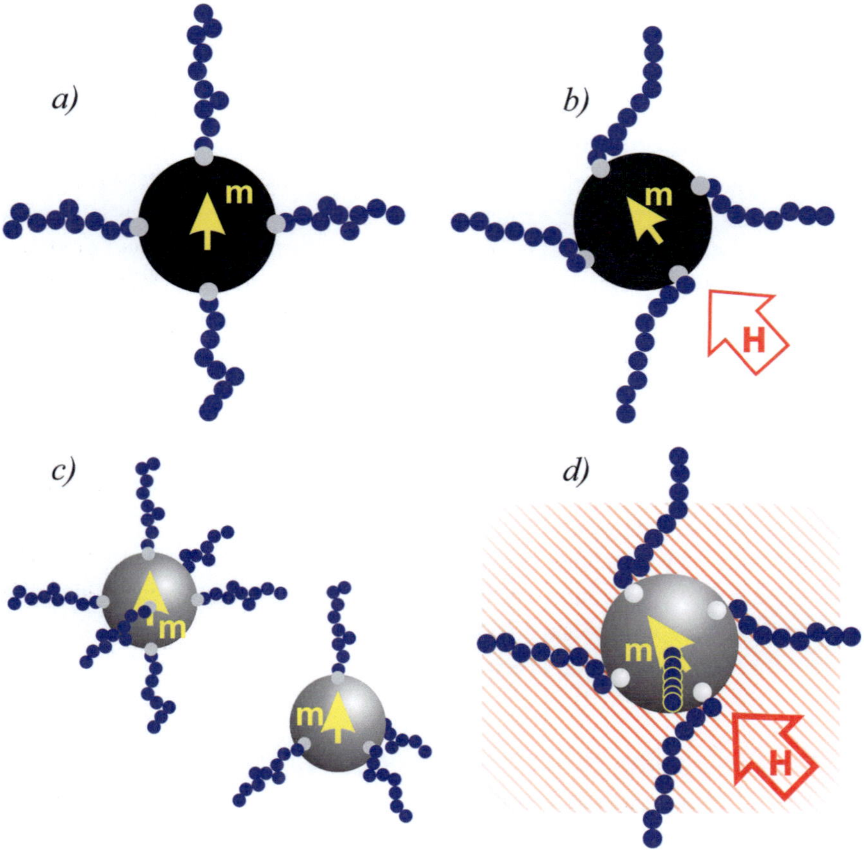

Figure 18.5: Sketch of magnetic nanoparticles in a gel. Upper row (a, b): 2D model, left without an applied external field, right with an applied external field, as marked with the red arrow. One can see four polymer chains rigidly bound to specific spots (light blue, virtual sites) on the surface of a magnetic nanoparticle. When a field is applied and the magnetic particle aligns its dipole to it, the polymer chains wrap around the particle. This exerts a stress on the chain, which leads to a shrinking of the ferrogel (isotropic in 2D). Lower row (c, d) 3D model, left without an external field, right with an external field applied. In sketch c) a 4-fold (diamond) and 6-fold (cubic) crosslinker is shown. In sketch d), an external magnetic field H is applied within the plane, as shown with the red arrow. Notice that only chains within the plane are subjected to the deformation, the polymer chains perpendicular to the plane are not affected. This leads to an anisotropic shrinking of the gel in 3D. Reprinted from Ref. [55], with the permission of AIP Publishing.

cannot deform freely. Instead, averages of the stress tensor were taken and new simulations were done iteratively with an adapted shape of the simulation box. In this way, the equilibrium shape for a given magnetic field was determined. Due to thermal fluctuations and a small system size, the stress tensor fluctuates significantly. Thus, many individual stress calculations are performed at each iteration.

Like in the two-dimensional case, the gel shrinks when it is exposed to an external field. In three dimensions, however, the particles do not all rotate around the same axis. Of the many axes possible, the one parallel to the applied field is special, since a particle's rotation around this axis does not change its interaction energy with the external field. As can be seen in Figure 18.5, this leads to an anisotropic shrinkage of the gel. When comparing the two network structures, we observe significantly different elasticities. This pertains both to the bulk modulus and the Poisson ratio. These differences are not fully understood, since there are counteracting trends at play. On the one hand, adding more polymer chains per magnetic particle enhances the gel's deformation mechanism, which relies on a coupling between the magnetic particles' orientation and the polymer chains. On the other hand, more polymer chains lead to a stiffer material, which counteracts the gel's shrinkage in a magnetic field. For more details, we refer to Ref. [55]. It is worth mentioning that the elasticity of magnetic composites can be strongly nonlinear [56] and frequency-dependent [11, 57]. In addition, some investigations on magnetoactive elastomers show hysteresis effects, suggesting that large rotations of the magnetic particles can lead to detachment of polymer chains from the particle's surface [58]. Both the two-dimensional and the three-dimensional models presented so far could be extended to incorporate this effect, by dynamically removing bonds where a prescribed force limit on the respective bond is exceeded. Using a sufficiently large gel sample in periodic boundary conditions these modeling approaches could be used to study how polymer detachment on a microscopic scale affects the elastomer on a macroscopic scale.

18.3.2 Gels deforming through changing magnetic interactions – role of the magnetic particle configuration

In Figure 18.6, a three-dimensional model gel is depicted which deforms through changes in the interaction between magnetic particles as they align to an external magnetic field (see Figure 18.1 for the two-dimensional model). For this deformation mechanism, the sample shape is of importance, as it affects the demagnetization field [59, 60]. Hence, we cannot use periodic boundary conditions. To limit surface effects, one has to use thousands of magnetic particles in such a model. As a consequence, it is no longer feasible to explicitly model the polymer chains in the bead–spring framework. Instead, the elasticity is provided by connecting the magnetic particles by entropic springs. These can either be viewed as a representation for individual polymers or as a coarse representation of an elastic background, in which the springs describe the ability for the particles to move locally. The model [38] was developed to study the interplay between the spatial configuration of the magnetic particles in the gel and the structure of the network. To obtain realistic configurations of magnetic particles, we use well-established simulations of ferrofluids similar to those in Refs. [52, 61–63]. Next, the desired sample shape is cut. Here, we will limit the discussion to spherical gels. The model is, however, also suitable to study the interplay between the particle microstructure and the sample shape. To provide the gel's elasticity, all pairs of particles are considered. Whether a polymer is added between the particles is decided based on a distance-dependent criterion. Likewise, the entropic elasticity of the polymer is chosen as a function of its end-to-end distance. Following polymer theory, longer polymers are represented by a weaker elastic spring. A snapshot of a small sample gel is shown in Figure 18.6. The systems used for magnetization and shape measurements are much larger, containing around 10,000 magnetic particles. The main challenge is to calculate the long-ranged dipole-dipole interactions. For the original work, a direct summation over all pairs of particles has been carried out on a graphics processor. By now, an alternative method scaling as $N \log N$ in the number of particles N is available [64]. To further accelerate the simulations, the molecular dynamics technique was supplemented by Monte Carlo moves rotating the entire sample. In particular, rotations around the axis of a magnetic field are likely, thermodynamically, but only happen at very long simulation time scales in a standard molecular dynamics simulation.

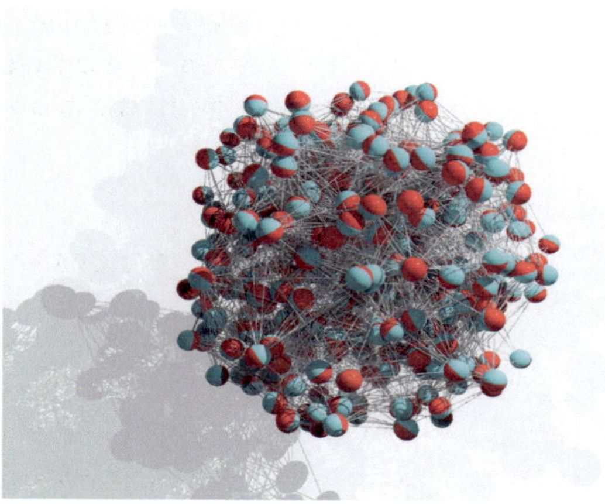

Figure 18.6: Model gel in which the polymers connecting the magnetic particles are represented as entropic springs. To visualize the orientations of the magnetic moments, their (magnetically) northern and southern hemispheres are drawn blue and red, respectively. The spatial configuration of the magnetic particles is obtained from a simulation of a standard ferrofluid. This way, correlations between adjacent magnetic particles observed in magnetic fluids are accounted for. A smaller system than that used for production simulations is shown, to improve visibility. Figure adapted from Ref. [38].

Let us turn to the influence of the spatial configuration of the magnetic particles before cross-linking. In experiments, it is possible to cure the gel around a previously prepared ferrofluid. This can be done in the absence of a magnetic field to produce a gel with an isotropic particle configuration or in a field to produce a uniaxial gel [65]. It is worth mentioning that, unless magnetically blocked particles are used and there exists a direct coupling between the particle orientation and the polymers, the gel still has no net magnetization, as long as no external field is applied. In Figure 18.7, we compare the deformation of isotropic and uniaxial gels placed in a magnetic field. As can be seen, in both cases, the gel elongates parallel to the applied field. This is in line with experimental findings obtained for a physical gel [34]. For the perpendicular direction, we see shrinkage for isotropic gels whereas no shrinkage is found for uniaxial gels. A possible explanation would be that in the uniaxial gel, chain-like clusters of magnetic particles are all aligned in a similar direction. Once the field is turned on, the gel rotates to align this direction parallel to the field. In the isotropic case, chain-like clusters are aligned in various directions. Once the field is applied, these chains will all bend towards the field direction. The polymers attached to those clusters will have to follow, which may lead to shrinkage in the direction perpendicular to the field. The rotation of chain-like clusters of magnetic particles into the field direction has also been described in a finite element model [29]. For further details on the model discussed here, we refer to Ref. [38].

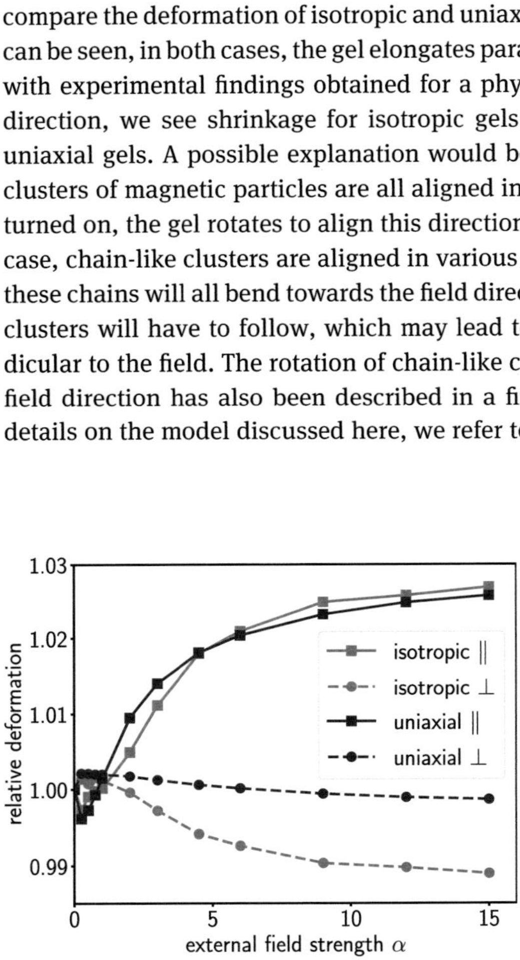

Figure 18.7: Relative deformation parallel and perpendicular to an applied field of spherical gel samples with isotropic and uniaxial micro-structures versus the strength of the magnetic field, which is expressed in terms of the Langevin parameter α (Eq. (18.2)). An elongation of the gel is observed parallel to the external field independent of the particle structure. A shrinkage in the perpendicular direction is observed for gels with an isotropic structure, whereas the gels with an uniaxial particle configuration do not deform significantly in the perpendicular direction. Adapted from Ref. [38].

18.4 Scale-bridging approaches

Simulations of magnetic gels tend to be computationally quite demanding. There are two main causes: first, the dipolar interactions are long-ranged. The force and torque on a dipolar particle in the system depends meaningfully on the position and orientation of all other dipole moments in the system. This problem is addressed by a number of specialized algorithms [64, 66–68]. Second, the polymeric nature of the gel results in a rather large number of degrees of freedom. This is the case in particular for bead-spring polymers. Here, each monomer bead results in two or three additional degrees of freedom, depending on the dimensionality. In models such as in Refs. [54, 55], 60 and more beads are used for each of more than a hundred polymer chains. One strategy to reduce the computational effort is to replace the bead–spring chains by entropic springs, modeled via a Gaussian potential. Then, each magnetic particle is connected to its neighbors by a bond with a distance-dependent potential. Harmonic potentials are often used as an approximation [38, 69]. For particle cross-linked gels, this is not sufficient, because the direct coupling between the magnetic particle's orientation and the polymers is not included. To resolve this, a scale-bridging approach was used in Ref. [70]. A pair potential as a function of distance and the angles of the magnetic moments was determined for two magnetically blocked particles connected by a polymer chain grafted to their surface. The polymer was modeled as a bead–spring chain and attached to specific spots on the two particles' surface (see Sec. 18.3.1). The probability to find the particles at a certain distance and at certain relative orientations was then evaluated using a histogram. In total, 10^{11} samples were collected. The resulting pair potential was then obtained numerically by Boltzmann-inversion of the probability distribution. Such techniques are sometimes referred to as 'systematic coarse-graining'. Finally, semiempirical closed-form expressions were fitted to the numeric data. These included the leading correlations. Most importantly, when one of the particles is rotated so that the polymer chain wraps around it, the equilibrium distance between the magnetic particles decreases. This correlation is the signature of the shrinking mechanism found in particle cross-linked gels (see Figure 18.3).

An alternative approach to model the elastic background is to construct a lattice of tracer points. The elasticity is provided by pairwise potentials between neighboring tracers and by potentials placed on the angles between connections ending on a site. This technique is particularly suitable for gels with a low particle concentration, where entropic springs between the particles would not capture the elastic background in sufficient detail. Using such a model, the buckling of a chain of magnetizable particles in a soft quasi two-dimensional gel has been studied [42]. A snapshot can be seen in Figure 18.8. A field is applied perpendicular to the chain of magnetic particles. As a result, the particles get magnetized perpendicular to the chain direction. To avoid the unfavorable dipole–dipole interactions, a free chain

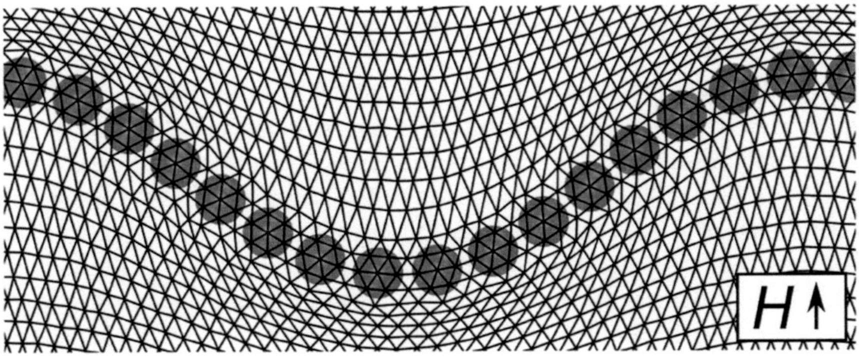

Figure 18.8: Simulation model for a chain of magnetizable particles embedded into a soft polymer gel. The system is quasi two-dimensional. The magnetic moments of the particles in the chain are aligned to an external magnetic field that points to the *y*-direction. As a consequence, the dipolar interactions between the particles in the chain are unfavorable. To avoid this, the magnetic particles tend to re-arrange. This is limited by the gel's elasticity. The minimization of elastic and magnetic energies makes the chain buckle with a wave-like shape. This behavior can be observed, both, in experiments and in a further theoretical model [42]. Figure adapted from Ref. [42] – published by the Royal Society of Chemistry.

would simply rotate parallel to the field direction. This is not possible due to the surrounding elastic matrix. Consequently, the chain buckles into a wave-like shape to reach a compromise between lowering the dipole–dipole interactions and not expending too much elastic energy. The simulation results are consistent with a second theoretical model and with experimental observations [42]. Modeling the elastic matrix as a network of tracer points or – somewhat related –, using tetrahedral volume elements [56], conceptually lie in between particle-based simulations and the continuum scale as studied using finite element methods [30, 71].

18.5 Obtaining AC-susceptibility spectra in computer simulations

Experimentally, AC-susceptibility measurements are an important tool to assess the properties of magnetic composite materials. The frequency-dependent susceptibility can be obtained by applying an external magnetic field with sinusoidal amplitude modulation to a probe of the material and measuring its magnetization response. The obtained AC-susceptibility spectra give insight into the magnetic properties of the investigated materials. However, using suitable theoretical models these spectra can also be used to assess frequency-dependent mechanical properties such as the local viscosity or mechanical storage and loss moduli [72–75]. Examples for such theories are extended Debye models [75–77], the Gemant–DiMarzio–Bishop model which has been adapted from dielectric spectroscopy [72, 78, 79], or the theoretical description developed by Raikher et al. [80].

The overall significance of AC-susceptometry for the experimental investigation of magnetic composites makes it an interesting target for studies using computer simulations. The main advantage of simulations is the degree of control that one gets over all aspects of the modeled system. Single aspects of the simulation model such as interaction potentials, length, or stiffness of surrounding polymers etc. can be easily changed, even where it might be quite difficult to do so experimentally. This gives the opportunity to study the signatures that these changes leave in the AC-susceptibility spectra, and to gain further insight into the details of the coupling mechanisms between magnetic particles and the surrounding matrix. For example, the impact of dipolar interactions on the AC-susceptibility has been studied in Ref. [81].

Here, we discuss a simple simulation model suitable for obtaining AC-susceptibility spectra for dilute ferrofluids consisting of magnetic nanoparticles immersed in a polymer suspension. Based on this model, we recently performed a simulation study [82] and compared its results to the experimental system of Ref. [72], a typical example for dilute composite magnetic systems. Such a system includes three main ingredients – the magnetic particles, the polymers, and the solvent. Resolving the system on a coarse-grained molecular dynamics scale, polymers are modeled using as bead–spring chains, where one MD bead corresponds to several monomer units in the experimental system. The beads are connected with harmonic bonds and interact via the purely repulsive Weeks–Chandler–Anderson (WCA) potential [83] to account for excluded volume. In addition, the beads are coupled to a thermalized lattice-Boltzmann (LB) hydrodynamics solver that is used to model the hydrodynamic interactions mediated by the solvent [84, 85]. The magnetic particles are included using the so-called 'raspberry' model [86, 87], i.e. the magnetic particles are homogeneously filled with fluid-coupling points that are rigidly fixed with respect to a central MD bead. This bead carries the mass, the moment of inertia tensor, and the magnetic moment of the particle and it is also used to model the monomer–particle interaction. If properly prepared, such raspberry-particles provide not only proper hydrodynamic interactions, but also correct coupling between translational and rotational mobility, which is key when it comes to studying the magnetization response of the system. Since we only consider magnetically blocked particles, the system's magnetization can solely decay due to Brownian relaxation. Since we consider a dilute ferrofluid, no dipolar interactions are included in the model. Note that this simplification also requires the dipolar interaction parameter λ (see Eq. (18.1)) to be sufficiently small. For the experimental system considered in the next paragraph, both conditions are fulfilled (for more details see Refs. [72, 82]). The MD simulation uses periodic boundary conditions.

Our basic setup is based on the experimental system of Ref. [72]. Therefore, we model spherical magnetic cobalt ferrite particles of radius $R = 7$ nm in polyethylene glycol (PEG) solutions of different concentration. For the solvent, water parameters are used except for the viscosity which we set to $\eta = 0.1\,\eta_{\text{water}}$ to speed up the relaxation of

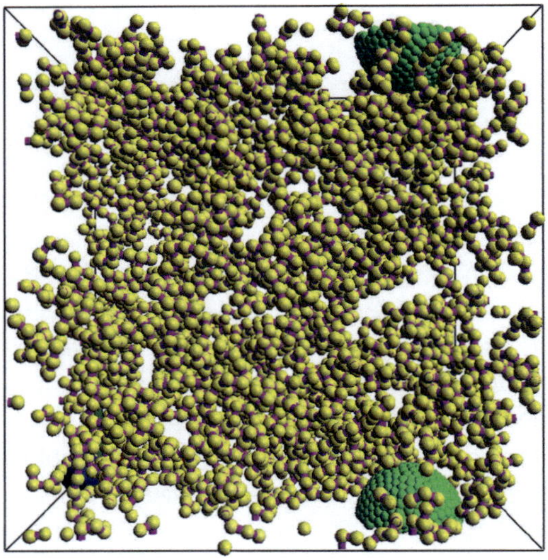

Figure 18.9: Representative snapshot of the simulation model used to reproduce AC-susceptometry. The magnetic particle is modeled as raspberry-particle (green) and surrounded by a polymer solution (yellow: Polymer beads, pink: Bonds). The volume fraction of polymers in this snapshot is $\phi = 0.05$. For clarity, LB fluid is not visualized. The simulation uses periodic boundary conditions.

the system's magnetization. The size of the polymer beads is chosen such that one MD bead corresponds to approximately six ethylene glycol units. A representative snapshot of the model is shown in Figure 18.9.

The straightforward way to measure the susceptibility of such a system using computer simulations would be to add an external magnetic field that interacts with the magnetic moment of the raspberry-particle. Modeling the field amplitude as a sinusoidal function and tracking the instantaneous magnetization response of the system via the vector component of the magnetic moment in field direction, one basically reproduces the experimental setup. While this approach to measuring AC-susceptibility works, there are some drawbacks to this method that render it virtually impracticable. Every simulation is performed at a certain frequency of the external field. Consequently, one obtains only one measurement point for the susceptibility per simulation. Therefore, to resolve an entire spectrum which typically spans several orders of magnitude, a huge amount of such simulations are necessary. Also, in every simulation the magnetization response has to be obtained with sufficient statistics. In order to have a good signal-to-noise ratio it would be advantageous to have a strong field–particle interaction. However, to stay in the linear regime this interaction has to be kept as small as possible, leading to significantly larger simulation times.

We therefore take a different approach to measuring AC-susceptibility. Using linear response theory, we can relate the time autocorrelation function of the magnetic moment of the raspberry particle to the system's susceptibility [88]. As a consequence, we can obtain the whole AC-spectrum from a single steady state simulation (without an external field). While these simulation runs require long sampling times to get sufficient statistics, they still offer a feasible way of measuring AC-susceptibility spectra. Figure 18.10 shows example results for AC-susceptibility spectra obtained in the described

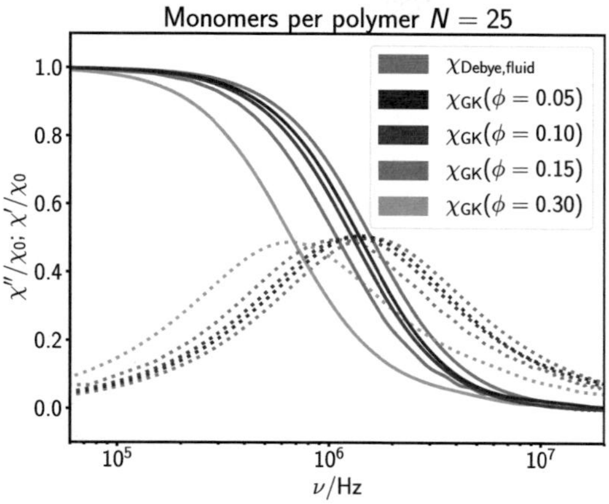

Figure 18.10: AC-susceptibility spectra obtained from simulations using linear response theory. Real parts of the complex susceptibilities (χ') are drawn with solid lines, imaginary parts (χ'') with dashed lines. For these simulations polymers with a length of $N = 25$ monomer beads were used and the polymer concentration varied. With increasing volume fraction ϕ of the polymers in solution, a shift towards lower frequencies occurs in the spectra. For reference, the simple Debye model is also plotted (green curve), corresponding to a purely Newtonian fluid without any polymers. The observed trend is in line with the experimental findings of Ref. [72].

manner, from simulations using polymers of a fixed length and varying the volume fraction of polymers in solution. With increasing polymer volume fraction we observe a shift of the spectra towards lower frequencies, which matches the experimental trends observed in Ref. [72]. Similarly, such shifts towards lower frequencies are experimentally observed with increasing polymer length, for a constant volume fraction [72]. In a recent simulation study [82], we showed that these shifts are qualitatively reproduced as well using the presented modeling approach. In the simulation model, polymers couple to the rotational behavior of the magnetic raspberry particle only hydrodynamically. Increasing the polymer volume fraction and chain length respectively strengthens the coupling within the polymer matrix, increasing the effective viscosity around the raspberry particle, which in turn leads to the frequency shifts in the spectra. This emphasizes the importance of hydrodynamic coupling in dilute composite magnetic systems [82].

The presented model will be the basis for further investigating the impact and importance that specific changes to the magnetic hybrid system have on the measured AC-susceptibility spectra. In the long run, we aim to directly measure elastic moduli for the system on a global level. Note that these measurements are restricted to small deformations when using the linear-response approach. Employing the already mentioned theoretical models will let us obtain the corresponding local observables from the AC-susceptibility – which will provide an opportunity for comparison.

18.6 Summary

In this article we provided an overview of particle-based simulations of magnetic gels and their behavior in an external magnetic field. Particular attention was paid to materials based on magnetic nanoparticles, such that thermal fluctuations are of importance. Two deformation mechanisms for gels in homogeneous magnetic fields were described and underpinned by simulation models. It was also shown that the spatial configuration of magnetic particles can influence the magneto-deformation behavior. Similarly, the structure of the polymer network can influence the material's elastic properties. We discussed scale-bridging techniques which may help to deal with the high computational cost of simulating soft magnetic materials. Here, both, systematic coarse-graining as well as models located somewhere in between particle-based and continuum methods can be of use. We concluded our discussion with an outlook on evaluating the AC-susceptibility spectra in soft magnetic materials. These spectra are experimentally accessible and are useful both for manipulating the systems' properties in a time-dependent fashion, and to probe the local interactions between the magnetic particles and the surrounding polymer. In conclusion, simulations are a valuable tool to study the microscopic mechanisms which are the basis for experimental observations. They may provide guidance for tailoring the materials towards desirable properties.

Author contributions: All the authors have accepted responsibility for the entire content of this submitted manuscript and approved submission.
Research funding: This research was financially supported by the German Research Foundation through the priority program SPP 1681 under grant HO 1108/23.
Conflict of interest statement: The authors declare no conflicts of interest regarding this article.

References

1. Becker TI, Böhm V, Vega JC, Odenbach S, Raikher YL, Zimmermann K. Magnetic-field-controlled mechanical behavior of magneto-sensitive elastomers in applications for actuator and sensor systems. Arch Appl Mech 2019;89:133–52.
2. Monz S, Tschöpe A, Birringer R. Magnetic properties of isotropic and anisotropic $CoFe_2O_4$-based ferrogels and their application as torsional and rotational actuators. Phys Rev E 2008;78:021404.
3. Schmauch MM, Mishra SR, Evans BA, Velev OD, Tracy JB. Chained iron microparticles for directionally controlled actuation of soft robots. ACS Appl Mater Interfaces 2017;9:11895–901.
4. Zimmermann K, Böhm V, Zeidis I. Vibration-driven mobile robots based on magneto-sensitive elastomers. In: Advanced intelligent mechatronics (AIM), 2011 IEEE/ASME international conference. IEEE, Budapest, Hungary; 2011:730–5 pp. https://doi.org/10.1109/AIM.2011.6027106.
5. Odenbach S. Microstructure and rheology of magnetic hybrid materials. Arch Appl Mech 2016;86: 269–79.

6. Weeber R, Hermes M, Schmidt AM, Holm C. Polymer architecture of magnetic gels: a review. J Phys Condens Matter 2018;30:063002.
7. Weeber R, Kreissl P, Holm C. Studying the field-controlled change of shape and elasticity of magnetic gels using particle-based simulations. Arch Appl Mech 2019;89:3–16.
8. Menzel AM. Mesoscopic characterization of magnetoelastic hybrid materials: magnetic gels and elastomers, their particle-scale description, and scale-bridging links. Arch Appl Mech 2018;89: 1–29.
9. Holm C, Weis J-J. The structure of ferrofluids: a status report. Curr Opin Colloid Interface Sci 2005; 10:133–40.
10. Odenbach S, editor. Colloidal magnetic fluids: basic, development and application of ferrofluids. Springer; 2009, vol 763.
11. Auernhammer GK, Collin D, Martinoty P. Viscoelasticity of suspensions of magnetic particles in a polymer: effect of confinement and external field. J Chem Phys 2006;124:204907.
12. Lyer S, Schreiber E, Tietze R, Mann J, Struffert T, Engelhorn T, et al. Nanotechnology and cancer treatment: magnetic nanoparticles for a new and innovative drug delivery system. Anticancer Res 2011;31:1991–2.
13. Qin J, Asempah I, Laurent S, Fornara A, Muller RN, Muhammed M. Injectable superparamagnetic ferrogels for controlled release of hydrophobic drugs. Adv Mater 2009;21:1354–7.
14. Tietze R, Zaloga J, Unterweger H, Lyer S, Friedrich RP, Janko C, et al. Magnetic nanoparticle-based drug delivery for cancer therapy. Biochem Biophys Res Commun 2015;468:463–70.
15. Janko C, Ratschker T, Nguyen K, Zschiesche L, Tietze R, Lyer S, et al. Functionalized superparamagnetic iron oxide nanoparticles (spions) as platform for the targeted multimodal tumor therapy. Front Oncol 2019;9:59.
16. George B, Fink M, Ermert H, Rupitsch SJ, Lyer S, Alexiou C. Guiding and accumulation of magnetic nanoparticles employing high intensity focused ultrasound for drug targeting applications. Curr Dir Biomed Eng 2019;5:589–92.
17. Engelmann UM, Roeth AA, Eberbeck D, Buhl EM, Neumann UP, Schmitz-Rode T, et al. Combining bulk temperature and nanoheating enables advanced magnetic fluid hyperthermia efficacy on pancreatic tumor cells. Sci Rep 2018;8:13210.
18. Draack S, Lucht N, Remmer H, Martens M, Fischer B, Schilling M, et al. Multiparametric magnetic particle spectroscopy of $CoFe_2O_4$ nanoparticles in viscous media. J Phys Chem C 2019;123: 6787–801.
19. Wiedenmann A, Keiderling U, Meissner M, Wallacher D, Gahler R, May RP, et al. Low-temperature dynamics of magnetic colloids studied by time-resolved small-angle neutron scattering. Phys Rev B 2008;77:184417.
20. Klokkenburg M, Erné BH, Wiedenmann A, Petukhov AV, Philipse AP. Dipolar structures in magnetite ferrofluids studied with small-angle neutron scattering with and without applied magnetic field. Phys Rev E 2007;75:051408.
21. Rosman R, Janssen JJM, Rekveldt MT. Interparticle correlation in Fe_3O_4 ferrofluids, studied by the small-angle neutron-scattering technique. J Appl Phys 1990;67:3072.
22. Nack A, Seifert J, Passow C, Wagner J. Hindered nematic alignment of hematite spindles in poly (n-isopropylacrylamide) hydrogels: a small-angle x-ray scattering and rheology study. J Appl Crystallogr 2018;51:87–96.
23. Kruse T, Krauthäuser H-G, Spanoudaki A, Pelster R. Agglomeration and chain formation in ferrofluids: two-dimensional x-ray scattering. Phys Rev B 2003;67:094206.
24. Butter K, Bomans PHH, Frederik PM, Vroege GJ, Philipse AP. Direct observation of dipolar chains in iron ferrofluids by cryogenic electron microscopy. Nat Mater 2003;2:88–91.
25. Tang J, Tong Z, Xia Y, Liu M, Lv Z, Gao Y, et al. Super tough magnetic hydrogels for remotely triggered shape morphing. J Mater Chem B 2018;6:2713–22.

26. Schümann M, Odenbach S. *In-situ* observation of the particle microstructure of magnetorheological elastomers in presence of mechanical strain and magnetic fields. J Magn Magn Mater 2017;441:88–92.
27. Potisk T, Svenšek D, Pleiner H, Brand HR. Continuum model of magnetic field induced viscoelasticity in magnetorheological fluids. J Chem Phys 2019;150:174901.
28. Puljiz M, Huang S, Kalina KA, Nowak J, Odenbach S, Kästner M, et al. Reversible magnetomechanical collapse: virtual touching and detachment of rigid inclusions in a soft elastic matrix. Soft Matter 2018;14:6809–21.
29. Stolbov OV, Raikher YL, Balasoiu M. Modelling of magnetodipolar striction in soft magnetic elastomers. Soft Matter 2011;7:8484–7.
30. Metsch P, Kalina KA, Spieler C, Kästner M. A numerical study on magnetostrictive phenomena in magnetorheological elastomers. Comput Mater Sci 2016;124:364–74.
31. Cremer P, Heinen M, Menzel A, Löwen H. A density functional approach to ferrogels. J Phys Condens Matter 2017;29:275102.
32. Zrínyi M. Intelligent polymer gels controlled by magnetic fields. Colloid Polym Sci 2000;278: 98–103.
33. Raikher YL, Stolbov OV. Magnetodeformational effect in ferrogel samples. J Magn Magn Mater 2003;258/259:477.
34. Gollwitzer C, Turanov A, Krekhova M, Lattermann G, Rehberg I, Richter R. Measuring the deformation of a ferrogel in a homogeneous magnetic field. J Chem Phys 2008;128:164709.
35. Backes S, Witt MU, Roeben E, Kuhrts L, Aleed S, Schmidt AM, et al. Loading of pnipam based microgels with cofe2o4 nanoparticles and their magnetic response in bulk and at surfaces. J Phys Chem B 2015;119:12129–37.
36. Ivaneyko D, Toshchevikov V, Saphiannikova M, Heinrich G. Effects of particle distribution on mechanical properties of magneto-sensitive elastomers in a homogeneous magnetic field. Condens Matter Phys 2012;15:33601.
37. Weeber R, Kantorovich S, Holm C. Deformation mechanisms in 2D magnetic gels studied by computer simulations. Soft Matter 2012;8:9923–32.
38. Weeber R, Holm C. Interplay between particle microstructure, network topology and sample shape in magnetic gels – a molecular dynamics simulation study. *Arxiv preprint 1704.06578*, 2017.
39. Ilg P. Stimuli-responsive hydrogels cross-linked by magnetic nanoparticles. Soft Matter 2013;9: 3465–8.
40. Messing R, Frickel N, Belkoura L, Strey R, Rahn H, Odenbach S, et al. Cobalt ferrite nanoparticles as multifunctional cross-linkers in PAAm ferrohydrogels. Macromolecules 2011;44:2990–9.
41. Roeder L, Reckenthaler M, Belkoura L, Roitsch S, Strey R, Schmidt A. Covalent ferrohydrogels based on elongated particulate cross-linkers. Macromolecules 2014;47:7200–7.
42. Huang S, Pessot G, Cremer P, Weeber R, Holm C, Nowak J, et al. Buckling of paramagnetic chains in soft gels. Soft Matter 2016;12:228–37.
43. Puljiz M, Huang S, Auernhammer GK, Menzel AM. Forces on rigid inclusions in elastic media and resulting matrix-mediated interactions. Phys Rev Lett 2016;117:238003.
44. Limbach HJ, Arnold A, Mann BA, Holm C. ESPResSo – an extensible simulation package for research on soft matter systems. Comput Phys Commun 2006;174:704–27.
45. Arnold A, Lenz O, Kesselheim S, Weeber R, Fahrenberger F, Röhm D, et al. ESPResSo 3.1 – molecular dynamics software for coarse-grained models. In: Griebel M, Schweitzer MA, editors. Meshfree methods for partial differential equations VI, volume 89 of lecture notes in computational science and engineering. Berlin, Heidelberg: Springer; 2013:1–23 pp.
46. Weik F, Weeber R, Szuttor K, Breitsprecher K, de Graaf J, Kuron M, et al. ESPResSo 4.0 – an extensible software package for simulating soft matter systems. Eur Phys J Spec Top 2019;227: 1789–816.

47. Biller A, Stolbov O, Raikher YL. Modeling of particle interactions in magnetorheological elastomers. J Appl Phys 2014;116:114904.
48. Kubo R. The fluctuation-dissipation theorem. Rep Prog Phys 1966;29:255–84.
49. Annunziata MA, Menzel AM, Löwen H. Hardening transition in a one-dimensional model for ferrogels. J Chem Phys 2013;138:204906.
50. Ivanov AO, Kuznetsova OB. Magnetogranulometric analysis of ferrocolloids: second-order modified mean field theory. Colloid J 2006;68:430–40.
51. Minina E, Sánchez P, Likos C, Kantorovich S. The influence of the magnetic filler concentration on the properties of a microgel particle: zero-field case. J Magn Magn Mater 2017;459:226.
52. Weeber R, Klinkigt M, Kantorovich S, Holm C. Microstructure and magnetic properties of magnetic fluids consisting of shifted dipole particles under the influence of an external magnetic field. J Chem Phys 2013;139:214901.
53. Kantorovich S, Pyanzina E, Sciortino F. The influence of shape anisotropy on the microstructure of magnetic dipolar particles. Soft Matter 2013;9:6594–603.
54. Weeber R, Kantorovich S, Holm C. Ferrogels cross-linked by magnetic nanoparticles – deformation mechanisms in two and three dimensions studied by means of computer simulations. J Magn Magn Mater 2015;383:262–6.
55. Weeber R, Kantorovich S, Holm C. Ferrogels cross-linked by magnetic particles: field-driven deformation and elasticity studied using computer simulations. J Chem Phys 2015;143:154901.
56. Cremer P, Löwen H, Menzel AM. Superelastic stress–strain behavior in ferrogels with different types of magneto-elastic coupling. Phys Chem Chem Phys 2016;18:26670–90.
57. Pessot G, Schümann M, Gundermann T, Odenbach S, Lowen H, Menzel AM. Tunable dynamic moduli of magnetic elastomers: from characterization by x-ray micro-computed tomography to mesoscopic modeling. J Phys Condens Matter 2018;30:125101.
58. Linke JM, Borin DY, Odenbach S. First-order reversal curve analysis of magnetoactive elastomers. RSC Adv 2016;6:100407–16.
59. Morozov K, Shliomis M, Yamaguchi H. Magnetic deformation of ferrogel bodies: procrustes effect. Phys Rev E 2009;79:040801.
60. Zubarev AY. On the theory of the magnetic deformation of ferrogels. Soft Matter 2012;8:3174–9.
61. Ilg P, Kröger M. Anisotropic self-diffusion in ferrofluids studied via brownian dynamics simulations. Phys Rev E – Stat Nonlinear Soft Matter Phys 2005;72:031504.
62. Cerdà JJ, Kantorovich S, Holm C. Aggregate formation in ferrofluid monolayers: simulations and theory. J Phys Condens Matter 2008;20:204125.
63. Jager S, Klapp SHL. Pattern formation of dipolar colloids in rotating fields: layering and synchronization. Soft Matter 2011;7:6606–16.
64. Weeber R, Nestler F, Weik F, Pippig M, Potts D, Holm C. Accelerating the calculation of dipolar interactions in particle based simulations with open boundary conditions by means of the P2NFFT method. J Comput Phys 2019;391:243–58.
65. Collin D, Auernhammer G, Gavat O, Martinoty P, Brand H. Frozen-in magnetic order in uniaxial magnetic gels: preparation and physical properties. Macromol Rapid Commun 2003;24:737.
66. Bródka A. Ewald summation method with electrostatic layer correction for interactions of point dipoles in slab geometry. Chem Phys Lett 2004;400:62–7.
67. Cerdà JJ, Ballenegger V, Lenz O, Holm C. P3M algorithm for dipolar interactions. J Chem Phys 2008; 129:234104.
68. Nestler F. An NFFT based approach to the efficient computation of dipole–dipole interactions under various periodic boundary conditions. Appl Numer Math 2016;105:25–46.
69. Pessot G, Löwen H, Menzel AM. Dynamic elastic moduli in magnetic gels: normal modes and linear response. J Chem Phys 2016;145:104904.

70. Pessot G, Weeber R, Holm C, Löwen H, Menzel AM. Towards a scale-bridging description of ferrogels and magnetic elastomers. J Phys Condens Matter 2015;27:325105.
71. Attaran A, Brummund J, Wallmersperger T. Development of a continuum model for ferrogels. J Intell Mater Syst Struct 2016;28:1358–75.
72. Roeben E, Roeder L, Teusch S, Effertz M, Deiters UK, Schmidt AM. Magnetic particle nanorheology. Colloid Polym Sci 2014;292:2013–23.
73. Gratz M, Tschöpe A. Optical transmission versus ac magnetization measurements for monitoring colloidal Ni nanorod rotational dynamics. J Phys Appl Phys 2017;50:015001.
74. Remmer H, Dieckhoff J, Tschöpe A, Roeben E, Schmidt AM, Ludwig F. Dynamics of $CoFe_2O_4$ single-core nanoparticles in viscoelastic media. Phys Procedia 2015;75:1150–7.
75. Remmer H, Roeben E, Schmidt AM, Schilling M, Ludwig F. Dynamics of magnetic nanoparticles in viscoelastic media. J Magn Magn Mater 2017;427:331–5.
76. Ludwig F, Guillaume A, Schilling M, Frickel N, Schmidt A. Determination of core and hydrodynamic size distributions of $CoFe_2O_4$ nanoparticle suspensions using ac susceptibility measurements. J Appl Phys 2010;108:033918.
77. Feyen M, Heim E, Ludwig F, Schmidt AM. Magnetic nanorotors with tailored field-induced dynamics. Chem Mater 2008;20:2942–8.
78. DiMarzio EA, Bishop M. Connection between the macroscopic electric and mechanical susceptibilities. J Chem Phys 1974;60:3802–11.
79. Niss K, Jakobsen B, Olsen NB. Dielectric and shear mechanical relaxations in glass-forming liquids: a test of the gemant-dimarzio-bishop model. J Chem Phys 2005;123:234510.
80. Raikher YL, Rusakov V, Coffey W, Kalmykov YP. Dynamic susceptibilities of an assembly of dipolar particles in an elastic environment. Phys Rev E 2001;63:031402.
81. Ivanov AO, Zverev VS, Kantorovich SS. Revealing the signature of dipolar interactions in dynamic spectra of polydisperse magnetic nanoparticles. Soft Matter 2016;12:3507–13.
82. Kreissl P, Holm C, Weeber R. Frequency-dependent magnetic susceptibility of magnetic nanoparticles in a polymer solution: a simulation study. Soft Matter 2021;17:174–83.
83. Weeks JD, Chandler D, Andersen HC. Role of repulsive forces in determining the equilibrium structure of simple liquids. J Chem Phys 1971;54:5237.
84. McNamara GR, Zanetti G. Use of the Boltzmann-equation to simulate lattice-gas automata. Phys Rev Lett 1988;61:2332–5.
85. Krüger T, Kusumaatmaja H, Kuzmin A, Shardt O, Silva G, Viggen EM. The lattice Boltzmann method: principles and practice. Cham: Springer; 2017.
86. Lobaskin V, Dünweg B. A new model for simulating colloidal dynamics. New J Phys 2004;6:54.
87. Fischer LP, Peter T, Holm C, de Graaf J. The raspberry model for hydrodynamic interactions revisited. I. Periodic arrays of spheres and dumbbells. J Chem Phys 2015;143:084107.
88. Kubo R. Statistical-mechanical theory of irreversible processes. I. general theory and simple applications to magnetic and conduction problems. J Phys Soc Jpn 1957;12:570–86.

Oleg V. Stolbov, Pedro A. Sánchez, Sofia S. Kantorovich and Yuriy L. Raikher*

19 Magnetostriction in elastomers with mixtures of magnetically hard and soft microparticles: effects of nonlinear magnetization and matrix rigidity

Abstract: In this contribution, a magnetoactive elastomer (MAE) of mixed content, i.e., a polymer matrix filled with a mixture of magnetically soft and magnetically hard spherical particles, is considered. The object we focus on is an elementary unit of this composite, for which we take a set consisting of a permanent spherical micromagnet surrounded by an elastomer layer filled with magnetically soft microparticles. We present a comparative treatment of this unit from two essentially different viewpoints. The first one is a coarse-grained molecular dynamics simulation model, which presents the composite as a bead-spring assembly and is able to deliver information of all the microstructural changes of the assembly. The second approach is entirely based on the continuum magnetomechanical description of the system, whose direct yield is the macroscopic field-induced response of the MAE to external field, as this model ignores all the microstructural details of the magnetization process. We find that, differing in certain details, both frameworks are coherent in predicting that a unit comprising magnetically soft and hard particles may display a nontrivial reentrant (prolate/oblate/prolate) axial deformation under variation of the applied field strength. The flexibility of the proposed combination of the two complementary frameworks enables us to look deeper into the manifestation of the magnetic response: with respect to the magnetically soft particles, we compare the linear regime of magnetization to that with saturation, which we describe by the Fröhlich–Kennelly approximation; with respect to the

***Corresponding author: Yuriy L. Raikher,** Laboratory of Physics and Mechanics of Soft Matter, Institute of Continuous Media Mechanics, Russian Academy of Sciences (Ural Branch), Perm, Russia, E-mail: raikher@icmm.ru. https://orcid.org/0000-0002-6167-6528
Oleg V. Stolbov, Laboratory of Physics and Mechanics of Soft Matter, Institute of Continuous Media Mechanics, Russian Academy of Sciences (Ural Branch), Perm, Russia. https://orcid.org/0000-0001-9088-7909
Pedro A. Sánchez, Institute of Natural Sciences and Mathematics, Ural Federal University, Ekaterinburg, Russia; and Wolfgang Pauli Institute, University of Vienna, Vienna, Austria. https://orcid.org/0000-0003-0841-6820
Sofia S. Kantorovich, Institute of Natural Sciences and Mathematics, Ural Federal University, Ekaterinburg, Russia; Faculty of Physics, Computational and Soft Matter Physics, University of Vienna, Vienna, Austria; and Research Platform MMM, University of Vienna, Vienna, Austria. https://orcid.org/0000-0001-5700-7009

This article has previously been published in the journal Physical Sciences Reviews. Please cite as: O. V. Stolbov, P. A. Sánchez, S. S. Kantorovich and Y. L. Raikher "Magnetostriction in elastomers with mixtures of magnetically hard and soft microparticles: effects of nonlinear magnetization and matrix rigidity" *Physical Sciences Reviews* [Online] 2020, 5. DOI: 10.1515/psr-2020-0009 | https://doi.org/10.1515/9783110569636-019

polymer matrix, we analyze the dependence of the reentrant deformation on its rigidity.

Keywords: magnetoactive elastomer, magnetically hard microparticles, magnetostriction effect

19.1 Introduction

The embedding of solid micro- and nanoparticles with magnetic properties into elastic polymer matrices became in recent years one of the most successful approaches for the design of "smart" materials, i.e., materials with a predefined response to external stimuli [1, 2]. The addition of the magnetic component allows to control on the fly the rheological properties of the viscoelastic polymer medium by means of applied external fields. Closely related to ferrofluids and magnetorheological fluids [3–5], these polymer-based materials include gels, whose structure is swollen by a liquid background [6–10], and elastomers, which are dry rubber-like materials [5, 11–16]. The latter, known generically as magnetically active elastomers (MAEs), are attracting large research efforts in recent years due to the broad range of applications that their magnetically controlled physical properties are inspiring [17–23]. A main part of such applications is related to their strong response to external fields that leads to large variations of their shape and mechanical properties. For example, MAEs are used to design controllable vibrational absorbers and mounts with tunable stiffness [24–26], soft actuators and micromanipulators [27], force sensors and artificial muscles for soft robotics [28, 29], coatings with tunable wettability [30, 31] or optical properties [32], tunable radiation absorbers [33], or biomedical implants [34].

The overall response of MAEs to external fields is determined by a mechanically constrained but substantial rearrangement of their embedded magnetic particles [5, 35]: for instance, under uniform external fields, magnetically hard particles possessing a permanent magnetic moment tend to align with the direction of the field, whereas magnetically soft particles tend to acquire induced magnetic moments in the same direction; in both cases, if the particle density is large enough to let interparticle interactions to be significant, particles will tend to assemble into straight chains parallel to the field. However, such rearrangements necessarily involve some degree of local deformation of the polymer matrix, either elastic [35, 36] or inelastic [37]. Therefore, macroscopic changes in the properties of MAEs as a response to external fields are the result of the interplay between the field-induced assembly of their magnetic particles and the mechanical constraints of the polymer matrix. Such changes include giant magnetorheological effects entailing large field-induced increases of the elastic modulus [13, 16, 38–40] and large magnetostriction effects, corresponding to variations in the shape of the sample [20, 38, 41, 42]. Under elastic regimes, magnetostriction in MAEs is often fully reversible, leading to magnetic shape memory effects [43, 44].

Even though the measurement of macroscopic magnetorheological and magnetostriction effects with current techniques is straightforward, the accurate fundamental characterization of their underlying microscopic mechanisms, and thus, the rational design of materials with taylored sophisticated properties, still is a serious challenge. Direct observations of the field-induced microstructural changes within MAEs became available only in recent years, either by means of optical microscopy [25, 44–47] or by cutting-edge techniques such as X-ray computerized microtomography [36, 37, 48, 49]. However, so far these techniques provide only static information on the internal microstructure. Application of particle tracking methods [35, 50, 51] is a promising experimental approach to achieve dynamic characterization that is still under development.

Following the growing interest in these systems, large theoretical research efforts have been devoted to MAEs in recent years. Classical approaches to the modeling of rubber-like materials are based on numerical solving of constitutive equations describing their elastic properties. In the case of MAEs, such continuum description can be applied not only to the polymer matrix, but also to the distribution of embedded magnetic particles. This implies to define, on the basis of microscopic motivations or phenomenological approaches, constitutive equations for both the elastic and the magnetic properties [52, 53]. The simplest approximations within this framework assume a linear elastic behavior, along with linear or nonlinear magnetic properties, and a weak magnetoelastic coupling. The latter implies treating the magnetic forces as mechanical loads, which allows to solve the mechanical and magnetic equations separately. More accurate approaches involve taking into account the nonlinearity of elastic response at finite strains [54–60]. Further important refinements are the consideration of geometry of the boundaries of finite samples [58, 61–63], anisotropies in the distribution of magnetic particles [64, 65], or strong magnetoelastic couplings that impose simultaneous solving of the elastic and magnetic equations [63, 66–68]. Continuum approaches have the main advantage of enabling direct comparison to macroscopic properties. However, they generally lack detailed descriptions of the material microstructure.

A widely used alternative to full continuum theoretical models is the explicit representation of the magnetic particles based on the dipole approximation [11, 69–73]. This allows both to naturally incorporate microstructural details by means of the discrete distribution of particles and to treat interparticle magnetic interactions as pair potentials, with the main drawback being significantly higher calculation costs. This approach can be combined with different approximations to treat microstructure and interactions, such as bead-spring network representations of the material [51, 74–76] or hybrid mean field models [77–79]. The simplest dipolar approach, which assigns a point dipole moment to each magnetic particle, is a reasonable approximation when the density of magnetic particles inside the elastomer is not high and, thus, mutual magnetization between particles is weak. However, different corrections might be needed when such effect is significant [70, 80, 81].

While the plethora of existing theoretical approaches for the study of MAEs keeps growing [60, 73, 82], important experimental aspects such as microscopic inelastic responses [37] or polydispersity of the magnetic component [83] remain poorly studied. In addition, the experimental search for MAEs with enhanced or more sophisticated magnetoelastic behaviors brings in complex characteristics that pose additional theoretical challenges. An example of a MAE material of increased complexity is the one obtained by mixing inside the polymer matrix two types of magnetic microparticles, with different sizes and magnetic properties, in order to achieve a combined active and passive magnetic control of the response [84–86]. The magnetic mixture consists of a relatively low fraction of large spherical microparticles of NdFeB alloy, which are magnetically hard (MH), and a high fraction of smaller carbonyl iron microparticles, which are magnetically soft (MS). In such a mixture, both MH and MS particles respond to external fields (active control), whereas MH ones can be permanently magnetized and affect the surrounding MS particles even in the absence of applied field (passive control). Very recently, we introduced the first theoretical study on the behavior of such magnetically hard + soft elastomers (HSMEs) [87]. Using a twofold modeling strategy, which combines a minimal continuum analytical description and a bead-spring computer simulation model, we analyzed the magnetostriction of a representative elementary cell of such material, consisting of a central MH particle surrounded by a cloud of MS ones, all being mechanically interconnected by the elastic matrix. As a first approximation, we assumed linear elasticity and magnetization under weak magnetoelastic coupling conditions for the continuum model, whereas for the bead-spring representation we adopted simple dipolar particles, also with linear magnetization of the MS ones. Both approaches provided the same qualitative behavior for the two cases we analyzed: nonmagnetized and magnetized central particle. In the first case, an axial elongation of the elementary volume in the direction of the field that grows parabolically with the strength of the latter was established. In the second case, we found an unusual behavior: due to the fact that the field of the central particle breaks the symmetry of the system, a reentrant axial deformation arises, in which the cell adopts prolate–oblate–prolate shapes as the strength of an applied field antiparallel to the central dipole increases.

On the basis of our preliminary characterization of the magnetostriction of a HSME elementary cell [87], in this work we analyze several parameters affecting its behavior. Here, we perform such analysis mainly by means of our continuum magnetoelastic description, whereas simulations with the bead-spring model are used only for consistency checks on a single set of parameters. Both models are conveniently modified to study the effects of a nonlinear magnetization of the MS particles. In addition, we also study the effects of a moderate variation in their initial susceptibility and the impact of different rigidities of the elastic matrix. We show that the saturation magnetization of the MS particles only has significant qualitative effects at high rigidities. Moreover, we find that the reentrant axial deformation tends to be hindered as the elastic modulus is increased, with only the axial elongation remaining at high rigidities. Finally, we observe that a moderate decrease of the initial susceptibility tends to favor the reentrant behavior, broadening the region of deformation into oblate profiles.

19.2 System and modeling approaches

19.2.1 System parameters

Typical HSME samples are synthesized using NdFeB MH particles of diameter $d_h \approx 50$ μm and saturation magnetization $M_h \approx 800$ emu. They are combined with MS particles of carbonyl iron with diameter $d_s \approx 5$ μm and high initial magnetic susceptibility, χ_0. Here, we will sample three different values of $\chi_0 = \{0.15, 0.2, 0.24\}$, where the highest one corresponds to the limiting value $\chi_0^\star \sim 3/4\pi$. The volume fraction of MS particles is around $\rho_s \approx 0.3$. In order to study the effects of rigidity of the polymer matrix, here we also sample several values for its shear modulus, G, comprised between 10^5 and 10^7 dyn/cm^3. For the external field, we sample field strengths up to $1.9 \cdot 10^4$ Oe that is the same order of magnitude of typical saturation fields used for these materials [36]. Finally, for our elementary HSME cell, we take an ideally spherical MH particle and a homogeneous spherical elastic shell around it of 25 μm. The latter contains the aforementioned volume fraction of embedded MS particles.

19.2.2 Qualitative description

Figure 19.1 shows the schematic representations of the elementary cell of a HSME defined for our analytical magnetoelastic and bead-spring simulation models. Here, we use Figure 19.1a, corresponding to the continuum description, to qualitatively describe the behavior observed in both models. In such scheme, the central dark disc represents the MH particle that, when magnetized, carries a point dipole $\vec{\mu}_h$ in its center. The orientation of this dipole, corresponding to an arbitrary direction along the magnetic easy axis of the MH particle, defines the symmetry axis of the cell.

The shadowed region around the central disc represents an incompressible elastic shell in which an assembly of implicit MS particles is embedded, whereas the thick solid circle indicates the boundary of this shell when it is unperturbed. Independently of the central particle being magnetized or not, as we showed in Reference [87], when an external field $\vec{H}_0$ is applied parallel to the symmetry axis of the system, the elastic shell tends to deform in such a way that its initially circular boundary adopts a prolate shape (dashed line). The prolateness grows with the strength of the field monotonically. When the MH particle is magnetized, a weak external field antiparallel to $\vec{\mu}_h$

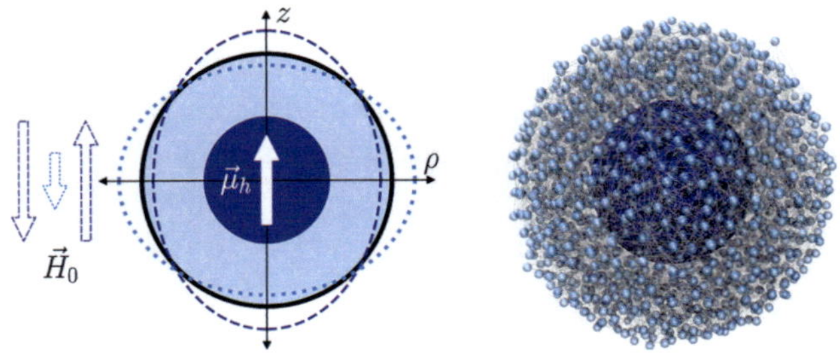

(a) (b)

Figure 19.1: (a) Scheme of the HSME elementary cell as represented in the continuum magnetoelastic model. Central dark disc represents the MH particle with dipole moment $\vec{\mu}_h$ pointing in z direction, light annulus corresponds to the elastic shell with MS particles, and its unperturbed boundary is indicated by the solid circle. Dashed and dotted ellipsoids show the shell deformation according to the strength and direction of the external field, $\vec{H}_0$, indicated by corresponding dashed and dotted arrows. (b) Snapshot of an unperturbed HSME elementary cell in the bead-spring simulation model, with a large dark central sphere representing the MH particle and a cloud of small light spheres representing the MS ones, connected by a network of elastic springs depicted as semitransparent lines. Radius of the MS particles has been halved to ease the visualization.

tends to deform the shell into an oblate shape (dotted line). This effect increases with the strength of the field up to a point of maximum oblateness, then decreases until the circular shape is recovered, and finally, an increasing prolateness can be observed again until the critical field required to force the inversion of the central dipole is attained. This latter event is out of the scope of the present study. The scheme in Figure 19.1b illustrates how the MH and MS particles are explicitly represented by soft spheres of different size in the bead-spring model, with the same type of arrangement established for the continuum description: a central big sphere representing the MH particle and a spherical shell of small MS particles, all highly interconnected by elastic springs representing the mechanical constraints imposed by the polymer matrix.

In the following sections, we briefly describe each model, underlining the modifications introduced to study the effects of the aforementioned parameters.

19.2.3 Continuum analytical approach

In order to define the analytical equations of our continuum model, here we assume a weak magnetoelastic coupling. Therefore, we can split the problem into components, which are described separately in the next two sections. In addition, the axial symmetry allows us to adopt, without loss of generality, a two-dimensional representation of the system. For both magnetostatic and elastic contributions, we obtain variational equations that are solved numerically using finite elements. Such calculation is performed with the *FEniCS* computing platform [88].

19.2.3.1 Magnetostatic problem

We consider the elementary cell outlined in Figure 19.1a, placed in an external homogeneous magnetic field $\vec{H}_0$. As pointed out earlier, the cell consists of a magnetically hard core region, $\Omega_1\,(r < r_1)$, and a magnetically soft shell region, $\Omega_2\,(r_1 < r < r_2)$, whereas the empty region external to the cell is denoted as Ω_3. Without loss of generality, $\vec{H}_0$ points along the Oz axis. The magnetically hard core has magnetization, M_h, also coaligned with Oz. In this geometry, the magnetically soft shell is a continuum medium that is reversibly magnetizable. Here, we introduce a nonlinear magnetization for this shell by taking the empirical Fröhlich–Kennelly law [89–91],

$$\chi(H) = \frac{\chi_0 M^{(\text{sat})} H}{M^{(\text{sat})} + \chi_0 H}, \tag{19.1}$$

where χ_0 is the initial susceptibility of the shell and $M^{(\text{sat})}$ is its saturation magnetization.

The magnetostatic problem in the absence of charges or currents is described by two Maxwell equations with their respective boundary conditions:

$$\nabla \times \vec{H} = 0, \quad \nabla \cdot \vec{B} = 0, \quad [\vec{H}_\tau] = 0, \quad [B_n] = 0. \tag{19.2}$$

Here, $\vec{B}$ is the magnetic flux density, and subscripts n and τ denote the components of a vector normal and tangential to the surface of the discontinuity boundary, respectively. Square brackets denote the difference between the values of a quantity on the two sides of the boundary.

The first equation of set (2) shows that $\vec{H}$ is a potential field, which can be expressed as a superposition of an external uniform field $\vec{H}_0$ and the gradient of a scalar potential ψ:

$$\vec{H} = \vec{H}_0 - \nabla\psi. \tag{19.3}$$

With allowance for the rotational symmetry around Oz, we may use cylindrical coordinates, so that the potential ψ depends only on the radial distance ρ and axis coordinate z. We require that potential $\psi(\rho, z)$, first, vanishes at the external boundary of the cell: $\psi|_{\rho=r_2} = 0$; and second, is periodic along the Oz axis: $\psi(\rho, z) = \psi(\rho, z + h)$, where h is the cell period.

The solution of Equation (19.2) is equivalent to finding an extremum of the energy functional for the entire volume V [92]:

$$\int_V \vec{B} \cdot \delta\vec{H}\, dV = 0, \tag{19.4}$$

where here and hereinafter δ denotes the variation of a quantity.

The vector of magnetic flux density is defined in each region of the model cell as follows:

$$\vec{B} = \begin{cases} \vec{H}_0 - \nabla\psi + 4\pi\vec{M}_h & \text{for } \Omega_1, \\ (1 + 4\pi\chi(H))\left(\vec{H}_0 - \nabla\psi\right) & \text{for } \Omega_2, \\ \vec{H}_0 - \nabla\psi & \text{for } \Omega_3. \end{cases} \tag{19.5}$$

Taking into account the relations:

$$\delta\vec{H} = -\nabla\,\delta\psi \quad \text{and} \quad \int_V \vec{H}_0 \cdot \nabla\delta\psi dV = 0,$$

Equation (19.4) can be transformed into

$$\frac{1}{4\pi}\int_V \nabla\psi \cdot \nabla\delta\psi dV = \int_{V_{\Omega_2}} \vec{M}_h \cdot \nabla\delta\psi dV + \int_{V_{\Omega_3}} \chi(H)\left(\vec{H}_0 - \nabla\psi\right) \cdot \nabla\delta\psi dV. \tag{19.6}$$

Finally, this variational equation is solved numerically for the potential $\psi(\rho, z)$, finding solutions for each sampled value of the initial susceptibility, χ_0.

19.2.3.2 Elastic problem

Having once obtained the solution of the magnetostatic problem, i.e., having found the distribution of the magnetic field inside the magnetically soft shell, one can calculate how the shell would deform under the resulting magnetic forces. In order to do that, we need to formulate the equations for a magnetoelastic medium, assuming balance between magnetic and elastic forces:

$$\nabla \cdot \tilde{\sigma} + \nabla \cdot \tilde{\sigma}_m = 0, \tag{19.7}$$

where $\tilde{\sigma}$ denotes the elastic stress tensor and $\tilde{\sigma}_m = \frac{1}{4\pi}\vec{B}\vec{H} - \frac{1}{8\pi}H^2\tilde{g}$ is the Maxwell stress tensor. In case of equilibrium, the pressure on both sides of the outer boundary Γ should be the same. Thus, one obtains:

$$\vec{n} \cdot \tilde{\sigma}\big|_\Gamma = \vec{n} \cdot (\tilde{\sigma}_m^{(e)} - \tilde{\sigma}_m^{(i)})\big|_\Gamma = 2\pi M_n^2 \vec{n}\big|_\Gamma, \tag{19.8}$$

where $\vec{n}$ is the vector normal to the outer boundary and $(.)^{(i)}$ and $(.)^{(e)}$ denote internal (inside the shell) and external (outside the shell) values. Then we write Hooke's law and the relation between strain tensor $\tilde{e}$ and displacement vector $\vec{u}$ as

$$\tilde{\sigma} = \lambda\,\mathrm{tr}\,(\tilde{e})\tilde{g} + 2G\tilde{e}, \quad \tilde{e} = \frac{1}{2}\left(\nabla\vec{u} + \nabla\vec{u}^T\right), \tag{19.9}$$

where $\tilde{g}$ is unity tensor, G stands for the shear modulus, and the Lamé coefficient λ characterizes the compressibility of the material and is related to its volume elastic modulus as $K = \lambda + 2G/3$.

In order to obtain a variational form of the magnetoelastic problem using the principle of virtual work, we have to multiply Equations (19.7) and (19.8) by $\delta\vec{u}$ and integrate:

$$\int_V (\nabla \cdot \tilde{\sigma} + \nabla \cdot \tilde{\sigma}_m^{(i)}) \cdot \delta\vec{u} dV - \int_S \vec{n} \cdot (\tilde{\sigma} + \tilde{\sigma}_m^{(i)} - \tilde{\sigma}_m^{(e)}) \cdot \delta\vec{u} dS = 0. \tag{19.10}$$

Employing Gauss–Ostrogradsky theorem, after simplifications, we come to a so-called weak variational form:

$$\int_V (\lambda \operatorname{tr}(\tilde{e}) \operatorname{tr}(\delta\tilde{e}) + 2G\tilde{e} \cdot \cdot \delta\tilde{e} + \tilde{\sigma}_m^{(i)} \cdot \cdot \delta\tilde{e}) dV = \int_S \left(2\pi M_n^2 \vec{n} + \vec{n} \cdot \tilde{\sigma}_m^{(i)} \right) \cdot \delta\vec{u} dS. \tag{19.11}$$

As pointed earlier, the presence of a magnetic field transforms our initially spherically symmetric problem into an axisymmetrical one. Therefore, here we also use cylindrical coordinates, (ρ, z), and solve the problem numerically with finite elements, obtaining $\vec{u}(\rho, z)$ for the quarter of the main cross section of the cell. In order to do this, we need to apply the Dirichlet boundary conditions

$$u_\rho \big|_{\rho=0} = 0, \quad u_z \big|_{z=0} = 0, \quad \vec{u} \big|_{r=r_1} = 0, \tag{19.12}$$

which mean that the shell is immobile at the shell–core boundary, and the symmetry requirement applies at the boundaries $\rho = 0$ and $z = 0$. For all calculations, the ratio $\lambda/G = 10^3$ was fixed.

19.2.4 Bead-spring model

Our bead-spring model is designed for molecular dynamics (MD) simulations. The solid magnetic particles are represented as soft spheres, assuming that they are always surrounded by an elastic layer of polymer material that prevents them to get into close contact. This assumption is consistent with the weak magnetoelastic coupling established for the continuum model. The soft core pair interaction is defined by a truncated and shifted Lennard-Jones potential, also known as Weeks–Chandler–Andersen (WCA) interaction [93]:

$$U_{\mathrm{WCA}}(r) = \begin{cases} U_{\mathrm{LJ}}(r) - U_{\mathrm{LJ}}(r = r_{\mathrm{cut}}), & r < r_{\mathrm{cut}} \\ 0, & r \geq r_{\mathrm{cut}} \end{cases}, \tag{19.13}$$

where $r = \vec{r}_i - \vec{r}_j$ is the center-to-center distance between the pair of particles i and j, $U(r) = 4\epsilon_{\mathrm{LJ}}[(d/r)^{12} - (d/r)^6]$ is the conventional Lennard-Jones potential, r_c is the truncation length, set to $r_c = 2^{1/6}d$ in order to make the interaction purely repulsive, and d is the center-to-center excluded distance, defined by the characteristic diameter of each particle, d_i and d_j, as $d = (d_i + d_j)/2$.

As mentioned earlier, the mechanical constraints imposed by the polymer matrix are represented by a network of elastic springs, with a simple harmonic potential

$$U_{\mathrm{S},i}(r) = \frac{1}{2} k_i (r - L_i)^2, \tag{19.14}$$

where r is the distance between the connected points, k_i is the elastic constant of the spring, and L_i its equilibrium length. The connections points are the centers of the MS particles and a set of fixed anchoring points randomly distributed on the surface of the MH one that remains permanently immobile. In order to ease the fitting of the elastic properties of this spring network, we take the elastic constants to be proportional to their corresponding equilibrium lengths, with the average equilibrium length of all the springs, $\langle L\rangle$, as scaling factor:

$$k_i = \bar{k}\frac{L_i}{\langle L\rangle} . \tag{19.15}$$

In this way, the only fitting parameter for the whole network is the constant factor $\bar{k}$. The direct comparison of the deformations obtained in the continuum and the bead-spring model [87] showed that the spring network of the latter fits rather well the elastic properties defined for simple mass–spring networks [94]. The bulk modulus of such networks is

$$K_{MS} = \frac{n\langle S\rangle\langle kL^2\rangle}{18} = \frac{n\bar{k}\langle S\rangle\langle L^3\rangle}{18\langle L\rangle} , \tag{19.16}$$

where n is the number density of connection points and $\langle S\rangle$ is the average number of springs connected to each of them. Assuming spatial isotropy and a Poisson ratio for the simple mass–spring network of $\nu = 1/4$, then the shear modulus can be defined as

$$G_{\mathrm{MS}} = \frac{3K_{\mathrm{SB}}(1-2\nu)}{2(1+\nu)} = \frac{n\langle S\rangle\langle kL^2\rangle}{30} = \frac{n\bar{k}\langle S\rangle\langle L^3\rangle}{30\langle L\rangle} . \tag{19.17}$$

The magnetic properties of the particles are represented as point dipoles located at their centers. The moment of the dipole corresponding to the MH particle, $\vec{\mu}_h = \mu_h\hat{k}$, is fixed to $\mu_h = M_h V_h$, where M_h is its magnetization, which we take as constant, and V_h its volume. In the same way, the dipole moment of the ith MS particle is given by

$$\vec{\mu}_i = \vec{M}_i V_s , \tag{19.18}$$

where V_s is its volume and $\vec{M}_i$ its magnetization. In this case, according to its magnetically soft nature, $\vec{M}_i$ is defined as

$$\vec{M}_i = \chi_i \vec{H}_{\mathrm{int}} , \tag{19.19}$$

where χ_i is the field-dependent susceptibility and $\vec{H}_{\mathrm{int}}$ the internal field inside the particle, which is parallel to the net external field at its position, $\vec{H}_{\mathrm{ext}}$. Following the Fröhlich–Kennelly nonlinear magnetization introduced earlier for the continuum model, χ_i is given by Equation (19.1) and the modulus of the internal field, $H_{\mathrm{int}} = ||\vec{H}_{\mathrm{int}}||$, is given by $H_{\mathrm{ext}} = ||\vec{H}_{\mathrm{ext}}||$ as

$$H_{\mathrm{int}} = H_{\mathrm{ext}} - \frac{4\pi}{3}\frac{\chi_0 H_{\mathrm{int}}}{1+\frac{\chi_0}{M_{\mathrm{sat}}}H_{\mathrm{int}}} , \tag{19.20}$$

where χ_0 is the initial susceptibility of the MS material and M_{sat} its saturation magnetization. From this expression, we obtain:

$$H_{int} = \frac{1}{6\chi_0}\Big[3\chi_0 H_{ext} - 4\pi\chi_0 M_{sat} - 3M_{sat} + \left(9\chi_0^2 H_{ext}^2 - 24\pi\chi_0^2 H_{ext} M_{sat} + 18\chi_0 H_{ext} M_{sat} + 16\pi^2\chi_0^2 M_{sat}^2 + 24\pi\chi_0 M_{sat}^2 + 9M_{sat}^2\right)^{1/2}\Big]. \quad (19.21)$$

Here we consider only two contributions to $\vec{H}_{ext}$: the externally applied field, $\vec{H}_0$, and the field generated by the dipole of the MH particle, when it is magnetized, at the center of the MS one, $\vec{H}_h^{(i)}$,

$$\vec{H}_{ext} = \vec{H}_0 + \vec{H}_h^{(i)}. \quad (19.22)$$

The latter is defined as

$$\vec{H}_h^{(i)} = \frac{3\vec{r}_i(\vec{\mu}_h \cdot \vec{r}_i)}{r_i^5} - \frac{\vec{\mu}_h}{r_i^3}, \quad (19.23)$$

where $\vec{r}_i$ is the vector connecting the center of the MH particle to the center of the polarized one and $r_i = ||\vec{r}_i||$. In this way, we disregard mutual magnetization between MS particles when calculating their induced dipoles. However, we fully take into account the dipole–dipole interaction between any pair of magnetized particles,

$$U_{dd}(ij) = -3\frac{(\vec{\mu}_i \cdot \vec{r}_{ij})(\vec{\mu}_j \cdot \vec{r}_{ij})}{r^5} + \frac{(\vec{\mu}_i \cdot \vec{\mu}_j)}{r^3}, \quad (19.24)$$

where μ_i, μ_j are their respective dipole moments, $\vec{r}_{ij} = \vec{r}_i - \vec{r}_j$ is the vector connecting their centers and $r = \|\vec{r}_{ij}\|$. Finally, MS particles also experience the Zeeman interaction with the external applied field. However, since their dipoles are induced, the effective interaction corresponds to one half of the conventional Zeeman potential [92]:

$$U_H = -\frac{1}{2}\vec{\mu}_i \cdot \vec{H}_0. \quad (19.25)$$

We perform our simulations with the package ESPResSo 4.1 [95], using MD with a Langevin thermostat [96]. Therefore, we perform Langevin dynamics (LD) simulations, integrating the Langevin translational and rotational equations of motion with the velocity Verlet algorithm [96, 97]. In difference with most usual LD simulations, we work under a quasi-athermal regime by setting a very small thermal energy in the system – around 10^2 times smaller than the average elastic energy of each spring under deformations produced by moderate applied fields. Therefore, our simulations correspond to an energy minimization with slight thermal fluctuations. The latter help the system to relax without getting trapped into high-energy local minima.

As is usual in coarse-grained simulations, we use a system of dimensionless units. Hereinafter, we denote dimensionless parameters with a tilde symbol, $\tilde{X}$. We take the diameter of the MS particles as reference length scale, so that $\tilde{d}_s = 1$ and $\tilde{d}_h = 10$, and the shear modulus of the matrix, G, as the reference scale for magnetic parameters, so that dimensionless field is defined as $\tilde{H} = H/\sqrt{G}$, magnetization as $\tilde{M} = M/\sqrt{G}$, and dipole moment as $\tilde{\mu} = \mu/\sqrt{G}d_s^3$. The latter definitions also apply to the results of the continuum model. Since here we are not interested in dynamics, for simplicity we take the Langevin translational and rotational friction coefficients as unity.

Each simulation run starts by placing and fixing the position and orientation of the MH particle in a simulation box with open boundaries. $N_a = 99$ connection points for the springs are randomly assigned to its surface. Around it, $N_s = 2{\cdot}10^3$ MS particles are randomly placed inside a spherical shell of dimensionless thickness 5. The latter are let to relax inside the shell by simply following their steric interactions, so any soft core overlap is removed. Then the spring network is built up by randomly choosing pairs of connecting points according to the following rules: first, the distance between them is not larger than $d_{\text{cut}} = 6$; second, none of them has more than $s_{\max} = 6$ springs attached. These arbitrary rules provide a good compromise between locality and isotropy of the elastic constraints acting on each MS particle, on the one hand, and the computational load, on the other. The result of such buildup procedure is a highly connected network with $\langle S\rangle \approx 6$, $\langle \tilde{L}\rangle \approx 4$ and $\langle \tilde{L}^3\rangle \approx 93$ [87]. Taking into account that the dimensionless number density of connecting points is $\tilde{n} = 6(N_s + N_a)/7\pi\tilde{d}_h^3$, we can use Equation (19.17) in dimensionless units to fit the elastic prefactor $\bar{k}$:

$$\bar{k} = \frac{30\tilde{G}\langle\tilde{L}\rangle}{\tilde{n}\langle S\rangle\langle\tilde{L}^3\rangle} = \frac{35\pi\tilde{G}\tilde{d}_h^3\langle\tilde{L}\rangle}{(N_s + N_a)\langle S\rangle\langle\tilde{L}^3\rangle} \approx 0.4. \tag{19.26}$$

Finally, with the setup described earlier, we set the central dipole $\tilde{\mu}_h$ and external field $\tilde{H}_0$ to their selected values, performing a final relaxation run of $5{\cdot}10^5$ integration steps, using a time step $\delta\tilde{t} = 0.01$. Only the final configuration obtained from each run is analyzed. For each set of sampled parameters, statistics are collected from 60 runs with independent initial configurations. In this case, we only sample different fields for two cases: a nonmagnetized central particle, $\tilde{\mu}_h = 0$, and a magnetized central particle with the lowest sampled matrix rigidity, $G = 10^5$ dyn/cm^3, corresponding to $\tilde{\mu}_h = 1324.6$.

19.3 Results and discussion

We start the discussion by considering the simplest case, that is when the central particle in the HSME elementary unit is nonmagnetic: $\tilde{\mu}_h = 0$. The first task is to find a correct common basis to compare the magnetically induced deformations predicted by the continuum and the bead-spring models. Whereas in the former the outer edge of the matrix is perfectly defined and the deformations are easy to visualize, in the bead-spring model no explicit outer boundary exists (see Figure 19.1b) since it is rendered by the discrete positions of MS particles. To find commensurate terms for that comparison, we define a virtual boundary of the bead-spring system as follows. First, the convex hull of all particles in the system is calculated. Then, by assuming that under any moderate deformation the elastic shell keeps an ellipsoidal profile, we perform a least-squares fit of an ellipsoid to that convex hull.

Taking advantage of the afore-introduced "ellipsoid terms," we characterize the deformations of the shell boundary by means of a single parameter, defined as $\Delta c^* = \langle (c - c_0)/c_0 \rangle$, where c is the distance (along the line parallel to the external field) from the center of the MH particle to the point where this line intersects the outer shell boundary , c_0 is the value of that distance when $\tilde{\mu}_h = 0$ and $\tilde{H}_0 = 0$; angle brackets denote the average over independent runs. Thus, Δc^* is positive for stretching of the cell along the field and negative in opposite case.

Figure 19.2 shows the dependence of Δc^* on $\tilde{H}_0$ for both models at $\tilde{\mu}_h = 0$. Note that the sign of $\tilde{H}_0$ indicates its orientation with respect to the reference axis. The curves are perfectly symmetric with respect to the ordinate axis ($\tilde{H}_0 = 0$), i.e., the unperturbed state of the system.

The results of the continuum model presented in Figure 19.2 correspond to several values of the elasticity modulus G. Note that the curves are plotted *vs.* nondimensional field magnitude $\tilde{H}_0 = H_0/\sqrt{G}$, so that one and the same abscissa point at different G's renders different dimensional values of the field. Had the calculation been done with

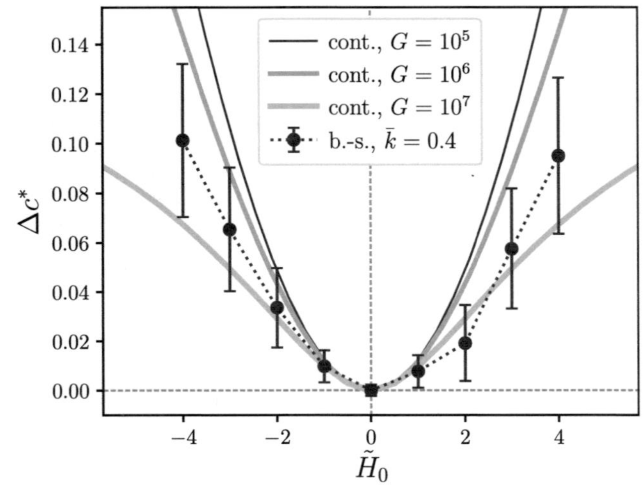

Figure 19.2: Longitudinal deformation parameter, Δc^*, as a function of the applied field, $\tilde{H}_0$, for the case of a nonmagnetized central particle, $\tilde{\mu}_h = 0$. Results of the continuum model for different values of the elastic modulus G are rendered by solid lines, data provided by the bead-spring model for $\bar{k} = 0.4$ are shown by symbols with error bars. Dotted curve connecting the symbols is a guide for the eye.

the linear magnetization law ($M^{(\text{sat})} = \infty$ in Eq. (19.20)), all the curves would have coincided. However, the nonlinear magnetization dependence (that we allow for here) removes this degeneration since the saturation magnetization is scaled with G as well. Due to that, the nondimensional saturation magnetization is lower for stiffer matrices and, as a result, the nonlinearities become more distinct. This is easily visible in Figure 19.2, where the curves are presented, which have been calculated for $\tilde{M}^{(\text{sat})} = \frac{1500\,\text{emu/cm}^3}{\sqrt{G}}$, see the curve for $G = 10^7$ dyn/cm^3. The curve rendered by the bead-spring model has been calculated for parameter $\tilde{k} = 0.4$, and the comparison implies that in the considered case of $\tilde{\mu}_h = 0$, the effective modulus that one may attribute to this system should lie inside the interval 10^6–10^7 dyn/cm^3.

Figure 19.3 shows the results on $\Delta c^\star(\tilde{H}_0)$ obtained when the central particle in the system bears permanent magnetic moment $\tilde{\mu}_h = \frac{4\pi 800}{3\sqrt{G}}$, where $4\pi/3$ is the nondimensional volume of the MH particle and 800 emu/cm^3 its magnetization. In this case the results of both models, although qualitatively similar, are quantitatively rather different. In both approaches, the essential effect of the magnetic field of the MH particle is to shift the minimum of the parabolic profile to negative values, thus producing oblate shape of the cell under inverted field. As already explained, with the employed scaling scheme, the increase of elastic modulus entails reduction of all the magnetic contribution, and it is no surprise that this shift becomes smaller. This tendency is clearly visible when comparing the curves rendered by the continuum model with one another in Figure 19.2.

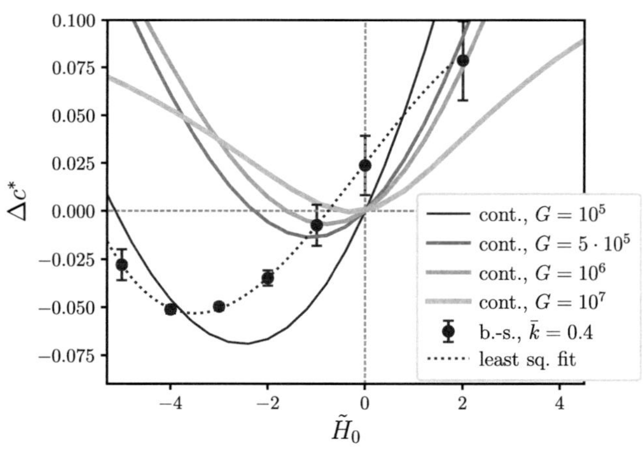

Figure 19.3: Longitudinal deformation parameter, Δc^*, as a function of the applied field, $\tilde{H}_0$, when the central particle has a magnetization of 800 emu/cm^3.
Solid lines correspond to the results of the continuum model for different values of the elastic modulus, G, symbols with error bars to the results of the bead-spring model for $\bar{k} = 0.4$. Dotted curve is a cubic spline fit to the latter.

It is instructive, however, to compare the continuum model with the bead-spring one. The latter demonstrates an "ambivalent" behavior, as follows from Figure 19.3. Indeed, under negative field the bead-spring model curve resembles that of the continuum one with elastic modulus about 10^5 dyn/cm^3, whereas under positive magnetization it rather displays similarity with the continuum curves corresponding to much higher elasticities: $G \sim 10^6$–10^7 dyn/cm^3.

To understand such field-tuned softening/stiffening, we recall that at $H_0 < 0$ the external field around the poles of the MH particle substantially compensates the field of the core and, thus, the field in the shell is on the average reduced. Under those conditions, the MS particles are less magnetized that entails lower aggregation and, consequently, makes the bead-spring shell to become effectively softer. When the external field is in the $H > 0$ range, in the "polar" zones the core field adds to the external one. This makes the MS particle aggregation in those zones stronger that, in turn, induces higher stresses inside the inter-bead-spring mesh and, by that, reduces its ability to deformations. As a consequence, the overall stiffness of the shell increases, as it is seen from comparison of the bead-spring (dashed) and continuum (solid) curves in the right-hand part of Figure 19.3. The field-modulated elastic modulus is an essential feature of the bead-spring model; note that in the continuum consideration such an effect is entirely absent. Meanwhile, as Figure 19.3 shows, the contribution of field-tuning – and this effect is most probably present in real magnetic elastomers – turns out to be sufficiently strong and because of that appeals for further investigation.

Figure 19.4 presents the dependence of the field-induced anisometry of the considered cell on the, this time dimensional, values of elasticity modulus; here only the results of continuum model are presented. In this diagram, the shaded curvilinear triangle corresponds to the combination of parameters under which the cell is oblate; in the points that make the borders of the triangle, the cell is spherical; outside the shaded area the cell responds to the applied field by elongation. As expected, with the increase of elastic modulus, the region of oblateness becomes more narrow and virtually disappears at about $G = 10^7$ dyn/cm^3. Note also that the dominating part of the triangle lies to the right of the dashed line that corresponds to the magnetic switching of the MH core of the shell.

When relating the afore-presented results to a real situation, one essential issue is to be clarified concerning the response of the MH core to the inverted (negative, in our notation) field. Indeed, under such a field a particle with permanent magnetic moment $\vec{\mu}_h$ residing in a compliant matrix is well able to rotate mechanically in order to turn $\vec{\mu}_h$ to the actual direction of the field. To get an estimation for the negative field strength when it happens, we consider a spherical single-domain particle of uniform magnetization M_h sitting in an elastic environment of shear modulus G. For the orientational-dependent energy of the system in the inverse field (antiparallel to $\vec{\mu}_h$) we have

$$U = M_h H V_p \cos\vartheta + 3GV_p\vartheta^2, \tag{19.27}$$

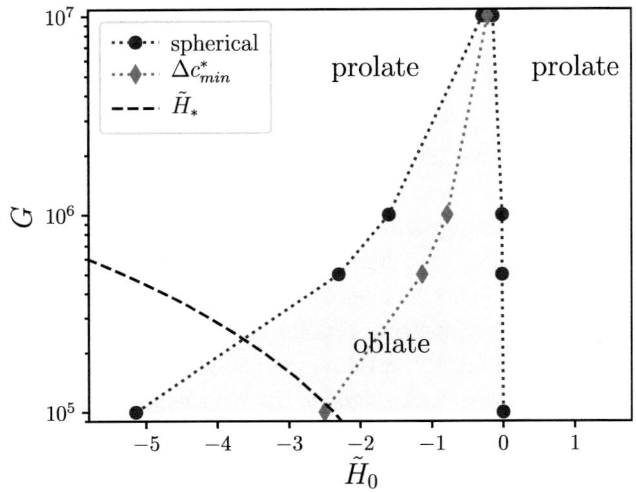

Figure 19.4: Deformation diagram of the HSME elementary unit as a function of the applied field, $\tilde{H}_0$, and the elastic modulus of the matrix, G, as predicted by the continuum model. Shadowed region corresponds to oblate deformations, being delimited by curves of no effective deformation (filled circles). Curve of filled diamonds indicates the maximum oblate deformation. Outside the oblate region, the system deforms into prolate shapes. Dashed curve is the critical field $\tilde{H}_*$ that would invert the orientation of the central dipole moment: results on the left of this curve are unphysical.

where V_p is the particle volume. As the particle is single-domain and magnetically hard, the magnetic moment is "frozen" into its body, so that the angle ϑ describes simultaneously: the deviation of the magnetic moment from the direction of the field and the angular displacement of the particle from its initial position.

Expanding (27) for small angular perturbations ($\vartheta = 0$ is the initial state), one gets

$$U \approx M_h H V_p \left(1 - \frac{1}{2}\vartheta^2 + \frac{1}{24}\vartheta^4\right) + 3GV_p\vartheta^2 .$$

Differentiation with respect to ϑ yields

$$\partial U/\partial\vartheta \approx -M_h H\vartheta\left[1 - \frac{1}{6}\vartheta^2\right] + 6G\vartheta ,$$

so that the condition of minimum is

$$\vartheta\left[6G - M_h H\left(1 - \frac{1}{6}\vartheta^2\right)\right] = 0 . \tag{19.28}$$

From (19.28) it is easy to find out that the particle dwells in the initial state $\vartheta = 0$ at $H < H_*$ and acquires a nonzero angle deviation (commences mechanical rotation) at $H > H_*$ with the critical field $H_* = 6G/M_h$; in nondimensional form it is $\tilde{H}_* = 6\sqrt{G}/\tilde{M}_h$. The latter dependence is plotted in Figure 19.4 by dashed line. According to the definition of $H > H_*$, the region $|\tilde{H}| > \tilde{H}_*$ (to the left) is unphysical since there the particle mechanical rotation should occur. At $M_h = 800$ emu, at the lowest value used here $G = 10^5$ dyn/cm^2, the critical field is $\tilde{H}_\star(G) = -2.37$, see the point on the abscissa axis of Figure 19.4. As seen, the minima of the presented curves and the full reentrant shrinking effect are justified only for the cells with elasticity $G > 6{\cdot}10^5$ dyn/cm^2. Although the existence of the minima (see Figure 19.3) of the bead-spring model and the continuum one for $G = 10^5$ dyn/cm^2 is doubtful, the occurrence of negative cell shrinking effect falls well in the physical region.

It is worth of noting, however, that afore-obtained expression, in fact, underestimates the strength of the inverse field when referring to the external applied one. Indeed, according to its derivation, $\tilde{H}_\star$ is the field experienced by the MH core of the cell. Due to that, the absolute value of minimal external field strength capable of initiating the particle rotation, exceeds $\tilde{H}_\star$ by the strength of the demagnetizing field generated in the MS shell. This means that when transforming the scaling of abscissa in Figure 19.4 to the units of external magnetic field, the dashed line would shift yet further to the left, thus yet widening the range of applicability of our model.

The afore-presented consideration describes an isolated elementary cell and its magnetomechanical response, so that it refers to the microscopic scale, i.e., that with reference distances of micron order. The problem of correct transferring of those results on the behavior of macroscopic HSME samples is nontrivial and requires special studies. Indeed, bridging the micro- and macroscales might be performed in various ways, which set spans from a simple Voigt approach [98] to sophisticated self-consistent schemes [68, 72]. However, if not to qualify for quantitative predictions, we infer that the revealed reentrant behavior would self-average to zero in the HSMEs, which were not subjected to the initializing external magnetization, whereas the reentrant response should be present in the premagnetized HSMEs provided the testing field is applied along the direction of the already existing magnetization.

We base this conclusion on the fact that a large number of small MS particles surrounding a single much greater MH grain screens out the permanent dipole field of the latter, so that with respect to magnetic interactions the MH grains become to a large extent isolated from one another. This effectively splits a macroscopic sample into an assembly of almost independent entities, each of which in the same way interacts with the two acting macroscopic fields: the external and the demagnetizing (determined by the overall shape of the sample) ones. To the contrary, in mechanical aspect the cells are never isolated and are always in contact. Evidently, on the one hand, a spherical cell is not appropriate as a precise building block for any macroscopic sample. However, on the other hand, in qualitative aspect, the field-induced mechanical responses of spherical and cuboid cells (of which any shape could be constructed with no voids) should be similar. Given that and taking into account that typical MAEs are virtually incompressible, we arrive at the conclusion that a magnetically initiated HSME sample, where all the MH magnetic moments are aligned, should in general reproduce the mechanical response of the elementary cell.

Acknowledgment: P.A.S. and S.S.K. acknowledge support by the DFG grant OD 18/24-1 and by the Act 211 of the Government of the Russian Federation, contract No. 02.A03.21.0006. Computer simulations were carried out at the Vienna Scientific Cluster.
Author contribution: All the authors have accepted responsibility for the entire content of this submitted manuscript and approved submission.
Research funding: This research was funded by DFG grant OD 18/24-1 and by the Act 211 of the Government of the Russian Federation, contract No. 02.A03.21.0006.
Conflict of interest statement: The authors declare no conflicts of interest regarding this article.

References

1. Filipcsei G, Csetneki I, Szilágyi A, Zrínyi M. Magnetic field-responsive smart polymer composites. In: Advances in polymer science. Oligomers - polymer composites - molecular imprinting. Berlin, Heidelberg: Springer; 2007, vol 206:137–89 pp.
2. Stuart MAC, Huck WTS, Genzer J, Muller M, Ober C, Stamm M, et al. Emerging applications of stimuli-responsive polymer materials. Nat Mater 2010;9:101–13.
3. Odenbach S. Ferrofluids. In: Buschow KHJ, editor. Handbook of magnetic materials. Berlin, Heidelberg: Elsevier; 2006, vol 16:127–208 pp.
4. de Vicente J, Klingenberg DJ, Hidalgo-Alvarez R. Magnetorheological fluids: a review. Soft Matter 2011;7:3701–10.
5. Odenbach S. Microstructure and rheology of magnetic hybrid materials. Arch Appl Mech 2016;86: 269–79.
6. Shiga T, Okada A, Kurauchi T. Magnetroviscoelastic behavior of composite gels. J Appl Polym Sci 1995;58:787–92.
7. Zrínyi M, Barsi L, Büki A. Ferrogel: a new magneto-controlled elastic medium. Polym Gels Netw 1997;5:415–27.
8. Zrínyi M. Intelligent polymer gels controlled by magnetic fields. Colloid Polym Sci 2000;278: 98–103.
9. Thévenot J, Oliveira H, Sandre O, Lecommandoux S. Magnetic responsive polymer composite materials. Chem Soc Rev 2013;42:7099–116.
10. Roeder L, Bender P, Kundt M, Tschope A, Schmidt AM. Magnetic and geometric anisotropy in particle-crosslinked ferrohydrogels. Phys Chem Chem Phys 2015;17:1290–8.
11. Jolly MR, Carlson JD, Muñoz BC, Bullions TA. The magnetoviscoelastic response of elastomer composites consisting of ferrous particles embedded in a polymer matrix. J Intell Mater Syst Struct 1996;7:613–22.
12. Ginder JM, Nichols ME, Elie LD, JL Tardiff. Magnetorheological elastomers: properties and applications. In: Wuttig MR, editor. Smart structures and materials: smart materials technologies, Proc. SPIE. Bellinham WA, USA: SPIE; 1999, vol 3675:131–8 pp.
13. Gong XL, Zhang XZ, Zhang PQ. Fabrication and characterization of isotropic magnetorheological elastomers. Polym Test 2005;24:669–76.
14. Varga Z, Filipcsei G, Zrínyi M. Magnetic field sensitive functional elastomers with tuneable elastic modulus. Polymer 2006;47:227–33.
15. Fuchs A, Zhang Q, Elkins J, Gordaninejad F, Evrensel C. Development and characterization of magnetorheological elastomers. J Appl Polym Sci 2007;105:2497–508.

16. Chertovich AV, Stepanov GV, Kramarenko EY, Khokhlov AR. New composite elastomers with giant magnetic response. Macromol Mater Eng 2010;295:336–41.
17. Carlson JD, Jolly MR. MR fluid, foam and elastomer devices. Mechatronics 2000;10:555–69.
18. Li W, Zhang X. Research and applications of MR elastomers. Recent Pat Mech Eng 2008;1:161–6.
19. Li WH, Zhang XZ, Du H. Magnetorheological elastomers and their applications. In: Visakh PM, Thomas S, Chandra AK, Mathew AP, editors. Advanced structured materials, Advances in Elastomers I. Blends and Interpenetrating Networks. Heidelberg, New York: Springer; 2013, vol 11: 357–74 pp.
20. Stepanov GV, Kramarenko EY, Semerenko DA. Magnetodeformational effect of the magnetoactive elastomer and its possible applications. J Phys: Conf Ser 2013;412:012031.
21. Li Y, Li J, Li W, Du H. A state-of-the-art review on magnetorheological elastomer devices. Smart Mater Struct 2014;23:123001.
22. Ubaidillah JS, Purwanto A, Mazlan SA. Recent progress on magnetorheological solids: materials, fabrication, testing, and applications. Adv Eng Mater 2015;17:563–97.
23. Shamonin M, Kramarenko EY. Highly responsive magnetoactive elastomers. In: Domracheva N, Caporali M, Rentschler E, editors. Novel magnetic nanostructures. unique properties and applications. micro & nanotechnologies series. Amsterdam: Elsevier; 2018:221–45 p.
24. Deng H-X, Gong X-L, Wang L-H. Development of an adaptive tuned vibration absorber with magnetorheological elastomer. Smart Mater Struct 2006;15:111–6.
25. Abramchuk S, Kramarenko EY, Stepanov GV, Nikitin LV, Filipcsei G, Khokhlov AR, et al. Novel highly elastic magnetic materials for dampers and seals: Part I: Preparation and characterization of the elastic materials. Polym Adv Technol 2007;18:883–90.
26. Sun TL, Gong XL, Jiang WQ, Li JF, Xu ZB, Li WH. Study on the damping properties of magnetorheological elastomers based on cis-polybutadiene rubber. Polym Test 2008;27:520–6.
27. Böse H, Rabindranath R, Ehrlich J. Soft magnetorheological elastomers as new actuators for valves. J Intell Mater Syst Struct 2012;23:989–94.
28. Volkova TI, Böhm V, Kaufhold T, Popp J, Becker F, Borin DY, et al. Motion behaviour of magneto-sensitive elastomers controlled by an external magnetic field for sensor applications. J Magn Magn Mater 2017;431:262–5.
29. Becker TI, Böhm V, Chavez Vega J, Odenbach S, Raikher YL, Zimmermann K. Magnetic-field-controlled mechanical behavior of magneto-sensitive elastomers in applications for actuator and sensor systems. Arch Appl Mech 2019;89:133–52.
30. Huang X, Sun Y, Soh S. Stimuli-responsive surfaces for tunable and reversible control of wettability. Adv Mater 2015;27:4062–8.
31. Sorokin VV, Sokolov BO, Stepanov GV, Kramarenko EY. Controllable hydrophobicity of magnetoactive elastomer coatings. J Magn Magn Mater 2018;459:268–71.
32. Koch I, Granath T, Hess S, Ueltzhöffer T, Deumel S, Jauregui Caballero CI, et al. Smart surfaces: magnetically switchable light diffraction through actuation of superparamagnetic plate-like microrods by dynamic magnetic stray field landscapes. Adv Opt Mater 2018;6:1800133.
33. Kuznetsova IE, Kolesov VV, Fionov AS, Kramarenko EY, Stepanov GV, Mikheev MG, et al. Magnetoactive elastomers with controllable radio-absorbing properties. Mater Today Commun 2019;21:100610.
34. Alekhina YA, Makarova LA, Kostrov SA, Stepanov GV, Kazimirova EG, Perov NS, et al. Development of magnetoactive elastomers for sealing eye retina detachments. J Appl Polym Sci 2019;136:47425.
35. Gundermann T, Odenbach S. Investigation of the motion of particles in magnetorheological elastomers by X-μCT. Smart Mater Struct 2014;23:105013.
36. Schümann M, Borin DY, Huang S, Auernhammer GK, Müller R, Odenbach S. A characterisation of the magnetically induced movement of NdFeB-particles in magnetorheological elastomers. Smart Mater Struct 2017;26:095018.

37. Sánchez PA, Gundermann T, Dobroserdova A, Kantorovich SS, Odenbach S. Importance of matrix inelastic deformations in the initial response of magnetic elastomers. Soft Matter 2018;14: 2170–83.
38. Ginder JM, Clark SM, Schlotter WF, Nichols ME. Magnetostrictive phenomena in magnetorheological elastomers. Int J Mod Phys B 2002;16:2412–8.
39. Abramchuk SS, Grishin DA, Kramarenko EY, Stepanov GV, Khokhlov AR. Effect of a homogeneous magnetic field on the mechanical behavior of soft magnetic elastomers under compression. Polym Sci Ser A 2006;48:138–45.
40. Stoll A, Mayer M, Monkman GJ, Shamonin M. Evaluation of highly compliant magneto-active elastomers with colossal magnetorheological response. J Appl Polym Sci 2014;131:39793.
41. Bednarek S. The giant linear magnetostriction in elastic ferromagnetic composites within a porous matrix. J Magn Magn Mater 2006;301:200–7.
42. Guan X, Dong X, Ou J. Magnetostrictive effect of magnetorheological elastomer. J Magn Magn Mater 2008;320:158–63.
43. Böse H. Viscolelastic properties of silicone based magnetorheological elastomers. Int J Mod Phys B 2007;21:4790–7.
44. Stepanov GV, Abramchuk SS, Grishin DA, Nikitin LV, Kramarenko EY, Khokhlov AR. Effect of a homogeneous magnetic field on the viscoelastic behavior of magnetic elastomers. Polymer 2007; 48:488–95.
45. Bellan C, Bossis G. Field dependence of viscoelastic properties of MR elastomers. Int J Mod Phys B 2002;16:2447–53.
46. Stepanov GV, Borin Dyu, Raikher YL, Melenev PV, Perov NS. Motion of ferroparticles inside the polymeric matrix in magnetoactive elastomers. J Phys Condens Matter 2008;20:204121.
47. An H-N, Picken SJ, Mendes E. Direct observation of particle rearrangement during cyclic stress hardening of magnetorheological gels. Soft Matter 2012;8:11995–2001.
48. Günther D, Borin DY, Günther S, Odenbach S. X-ray micro-tomographic characterization of field-structured magnetorheological elastomers. Smart Mater Struct 2012;21:015005.
49. Watanabe M, Takeda Y, Maruyama T, Ikeda J, Kawai M, Mitsumata T. Chain structure in a cross-linked polyurethane magnetic elastomer under a magnetic field. Int J Mol Sci 2019;20:2879.
50. Borbáth T, Günther S, Borin DY, Gundermann T, Odenbach S. X-μCT analysis of magnetic field-induced phase transitions in magnetorheological elastomers. Smart Mater Struct 2012;21:105018.
51. Pessot G, Schümann M, Gundermann T, Odenbach S, Löwen H, Menzel AM. Tunable dynamic moduli of magnetic elastomers: from characterization by X-ray micro-computed tomography to mesoscopic modeling. J Phys Condens Matter 2018;30:125101.
52. Brigadnov IA, Dorfmann A. Mathematical modeling of magneto-sensitive elastomers. Int J Solid Struct 2003;40:4659–74.
53. Dorfmann A, Brigadnov IA. Constitutive modelling of magneto-sensitive Cauchy-elastic solids. Comput Mater Sci 2004;29:270–82.
54. Dorfmann A, Ogden RW. Magnetoelastic modelling of elastomers. Eur J Mech A Solids 2003;22: 497–507.
55. Dorfmann A, Ogden RW. Nonlinear magnetoelastic deformations of elastomers. Acta Mech 2004; 167:13–28.
56. Dorfmann A, Ogden RW, Saccomandi G. Universal relations for non-linear magnetoelastic solids. Int J Non Lin Mech 2004;39:1699–708.
57. Bustamante R, Dorfmann A, Ogden RW. Universal relations in isotropic nonlinear magnetoelasticity. Q J Mech Appl Math 2006;59:435–50.
58. Raikher YL, Stolbov OV. Numerical modeling of large field-induced strains in ferroelastic bodies: a continuum approach. J Phys Condens Matter 2008;20:204126.

59. Ponte-Castañeda P, Galipeau E. Homogenization-based constitutive models for magnetorheological elastomers at finite strain. J Mech Phys Solid 2011;59:194–215.
60. Mukherjee D, Bodelot L, Danas K. Microstructurally-guided explicit continuum models for isotropic magnetorheological elastomers with iron particles. Int J Non Lin Mech 2020;120:103380.
61. Kankanala SV, Triantafyllidis N. On finitely strained magnetorheological elastomers. J Mech Phys Solid 2004;52:2869–908.
62. Bustamante R, Dorfmann A, Ogden RW. A nonlinear magnetoelastic tube under extension and inflation in an axial magnetic field: numerical solution. J Eng Math 2007;59:139–53.
63. Hasebe N, Wang XF, Nakanishi H. On magnetoelastic problems of a plane with an arbitrarily shaped hole under stress and displacement boundary conditions. Q J Mech Appl Math 2007;60: 423–42.
64. Zubarev A. Magnetodeformation of ferrogels and ferroelastomers; effect of microstructure of the particles spatial disposition. Physica A 2013;392:4824–36.
65. Kalina KA, Metsch P, Kästner M. Microscale modeling and simulation of magnetorheological elastomers at finite strains: a study on the influence of mechanical preloads. Int J Solid Struct 2016;102–103:286–96.
66. Metsch P, Kalina KA, Spieler C, Kästner M. A numerical study on magnetostrictive phenomena in magnetorheological elastomers. Comput Mater Sci 2016;124:364–74.
67. Kalina KA, Brummund J, Metsch P, Kästner M, Borin DY, Linke JM, et al. Modeling of magnetic hystereses in soft MREs filled with NdFeB particles. Smart Mater Struct 2017;26:105019.
68. Metsch P, Kalina KA, Brummund J, Kästner M. Two- and three-dimensional modeling approaches in magneto-mechanics: a quantitative comparison. Arch Appl Mech 2019;89:47–62.
69. Han Y, Hong W, Faidley LE. Field-stiffening effect of magneto-rheological elastomers. Int J Solid Struct 2013;50:2281–8.
70. Biller AM, Stolbov OV, Raikher YL. Dipolar models of ferromagnet particles interaction in magnetorheological composites. J Optoelectron Adv Mater 2015;17:1106–13.
71. Romeis D, Metsch P, Kästner M, Saphiannikova M. Theoretical models for magneto-sensitive elastomers: a comparison between continuum and dipole approaches. Phys Rev E 2017;95: 042501.
72. Menzel AM. Mesoscopic characterization of magnetoelastic hybrid materials: magnetic gels and elastomers, their particle-scale description, and scale-bridging links. Arch Appl Mech 2018;89: 17–45.
73. Khanouki MA, Sedaghati R, Hemmatian M. Experimental characterization and microscale modeling of isotropic and anisotropic magnetorheological elastomers. Composites B: Eng Times 2019;176:107311.
74. Ivaneyko D, Toshchevikov VP, Saphiannikova M, Heinrich G. Magneto-sensitive elastomers in a homogeneous magnetic field: a regular rectangular lattice model. Macromol Theory Simul 2011; 20:411–24.
75. Ivaneyko D, Toshchevikov V, Saphiannikova M. Dynamic moduli of magneto-sensitive elastomers: a coarse-grained network model. Soft Matter 2015;11:7627–38.
76. Sánchez PA, Minina ES, Kantorovich SS, Kramarenko EY. Surface relief of magnetoactive elastomeric films in a homogeneous magnetic field: molecular dynamics simulations. Soft Matter 2019;15:175–89.
77. Ivaneyko D, Toshchevikov V, Saphiannikova M, Heinrich G. Mechanical properties of magneto-sensitive elastomers: unification of the continuum-mechanics and microscopic theoretical approaches. Soft Matter 2014;10:2213–25.
78. Romeis D, Toshchevikov V, Saphiannikova M. Elongated micro-structures in magneto-sensitive elastomers: a dipolar mean field model. Soft Matter 2016;12:9364–76.

79. Romeis D, Toshchevikov V, Saphiannikova M. Effects of local rearrangement of magnetic particles on deformation in magneto-sensitive elastomers. Soft Matter 2019;15:3552–64.
80. Biller AM, Stolbov OV, Raikher YL. Modeling of particle interactions in magnetorheological elastomers. J Appl Phys 2014;116:114904.
81. Biller AM, Stolbov OV, Raikher YL. Mesoscopic magnetomechanical hysteresis in a magnetorheological elastomer. Phys Rev E 2015;92:023202.
82. Nam TH, Petríková I, Marvalová B. Experimental characterization and viscoelastic modeling of isotropic and anisotropic magnetorheological elastomers. Polym Test 2020;81:106272.
83. Winger J, Schümann M, Kupka A, Odenbach S. Influence of the particle size on the magnetorheological effect of magnetorheological elastomers. J Magn Magn Mater 2019;481: 176–82.
84. Borin DY, Stepanov GV. Oscillation measurements on magnetoactive elastomers with complex composition. J Optoelectron Adv Mater 2013;15:249–53.
85. Linke JM, Borin DY, Odenbach S. First-order reversal curve analysis of magnetoactive elastomers. RSC Adv 2016;6:100407–16.
86. Borin DY, Stepanov GV, Dohmen E. Hybrid magnetoactive elastomer with a soft matrix and mixed powder. Arch Appl Mech 2019;89:105–17.
87. Sánchez PA, Stolbov OV, Kantorovich SS, Raikher YL. Modeling the magnetostriction effect in elastomers with magnetically soft and hard particles. Soft Matter 2019;15:7145–58.
88. Alnæs MS, Blechta J, Hake J, Johansson A, Kehlet B, Logg A, et al. The FEniCS project version 1.5. Arch. Numerical Software 2015;3:9–23.
89. Kennelly AE. Magnetic reluctance. Trans Am Inst Electr Eng 1891;8:485–517.
90. Bozorth RM. Ferromagnetism. Hoboken NJ, USA: Wiley-IEEE Press; 1993.
91. Lee CH, Reitich F, Jolly MR, Banks HT, Ito K. Piecewise linear model for field-responsive fluids. IEEE Trans Magn 2001;37:558–60.
92. Landau LD, Lifshitz EM, Pitaevskii LP. Electrodynamics of continuous media, 2nd ed. New York: Pergamon Press; 1984.
93. Weeks JD, Chandler D, Andersen HC. Role of repulsive forces in determining the equilibrium structure of simple liquids. J Chem Phys 1971;54:5237–47.
94. Kot M, Nagahashi H, Szymczak P. Elastic moduli of simple mass spring models. Vis Comput 2015; 31:1339–50.
95. Weik F, Weeber R, Szuttor K, Breitsprecher K, de Graaf J, Kuron M, et al. ESPResSo 4.0 – an extensible software package for simulating soft matter systems. Eur Phys J Spec Top 2019;227: 1789–816.
96. Berendsen HJC. Simulating the physical world: hierarchical modeling from quantum mechanics to fluid dynamics. Cambridge: Cambridge University Press; 2007.
97. Rapaport DC. The art of molecular dynamics simulation, 2nd ed. Cambridge: Cambridge University Press; 2004.
98. Biller AM, Stolbov OV, Raikher YuL. Elastic properties of magnetorheological elastomer: description with the two-particle mesoscopic model. IOP Conf Ser Mater Sci Eng 2017;208:012007.

Andrey Zubarev*, Anton Musikhin and Dmitry Chirikov

20 Internal structures and mechanical properties of magnetic gels and suspensions

Abstract: We present results of theoretical and computer study of linear chain-like and complicated labyrinth structures in magnetic gels and suspensions as well as effect of these internal structures on macroscopic elastic properties of the composites. Our results show that at a certain threshold deformation, the structures experience a rupture which provokes a fall down of the macroscopic elastic stress, induced by the deformation. This effect is detected for both shear and tensile deformations. The results of calculations are compared with experimental data's.

Keywords: elastic properties; internal structures; magnetic gels and suspensions.

20.1 Introduction

Magnetic polymers and suspensions attract considerable interest of researches and engineers due to rich set of their physical properties promising for many high technologies [1–17].

One of the most interesting, from the scientific point of view, and valuable from the viewpoint of practical applications, features of ferrogels, and magnetic suspensions is their ability to change, in very wide ranges, shape, size, rheological, and other physical properties under the action of an external magnetic field. Macroscopic magnetomechanical effects in these composites are determined by morphology of internal spatial disposition of the particles in the host polymer either liquid medium [14–24].

The basic problems of the theory of composite materials are analysis of cooperative effects produced by many interacting particles of the filler. For this purpose, various intuitive and heuristic models with different fitting parameters have been proposed in literature (see, for example, [21, 25]).

Here we present results of theoretical and computer studies of two types of these structures – linear chains, percolating a magnetopolymer sample, as well as topologically complicated labyrinth structures.

***Corresponding author: Andrey Zubarev,** Ural Federal University, Lenina Ave 51, 620083, Ekaterinburg, Russia; and M. N. Mikheev Institute of Metal Physics of the Ural Branch of the Russian Academy of Sciences, Ekaterinburg, Russia, E-mail: A.J.Zubarev@urfu.ru. https://orcid.org/0000-0001-5826-9852
Anton Musikhin and Dmitry Chirikov, Ural Federal University, Lenina Ave 51, 620083, Ekaterinburg, Russia

This article has previously been published in the journal Physical Sciences Reviews. Please cite as: A. Zubarev, A. Musikhin and D. Chirikov "Internal structures and mechanical properties of magnetic gels and suspensions" *Physical Sciences Reviews* [Online] 2021, 6. DOI: 10.1515/psr-2020-0010 | https://doi.org/10.1515/9783110569636-020

The first type of the structures is typical for the composites with low volume concentration (about several per cent) of the embedded particles; the second type of the structures appear in the systems with relatively high (more than 20–30%) concentrations. Appearance of these internal structures significantly changes the sensitivity of the gels and suspensions to mechanic, electrical, magnetic, and other external impacts; changes the kinetics of the internal transport phenomena and chemical reactions, the rate of cell proliferation, other physical, chemical, and biological phenomena in these systems. This opens perspectives of the tunable synthesis of the magnetically controlled sensors; scaffolds for the growth of cell tissues with willing structure and properties; artificial muscles; dumpers, and other materials for biological and industrial applications.

20.2 Ferrogels with chain structures

In this part we consider a ferrogel with embedded identical non Brownian magnetizable spheres. Diameter of the particle is supposed much more than the characteristic size of the gel net cells; therefore, with respect to the particles, the gel can be considered as a continuous medium with some certain elastic properties. Usually these chains are being formed during the ferrogel synthesis under the action of an external magnetic field, applied while the gel polymerization; they are typical for the systems with a low either moderate volume concentrations of magnetic particles (usually when the concentration is not more than several per cent). Some photos of these chains can be found, for example in Refs. [26, 27]. In the systems with higher concentrations, the particles can form topologically complicated branched and labyrinth structures, considered in the third part of this paper.

Depending on condition of the material synthesis, the chains can be either shorter than the sample or percolate it (see, for example, Ref. [19]). The composites with the relatively short chains were studied theoretically in Refs. [18, 22]. Here we consider the situation when the chains percolate the sample, bridging its opposite boundaries.

Let us consider linear chains, consisting of magnetizable particles embedded in an elastic incompressible medium with the rigidity modulus G_0. We suppose that the chains bridge opposite boundaries of the composite. The magnetic field H_0 is applied along the chains. The composite is either is sheared in the direction, parallel to the boundaries (i.e. perpendicular to the chains) or stretched parallel to the field (see Figure 20.1).

The total (measured) stress in the composite can be presented as [29]

$$\sigma = \gamma G_0 + \sigma_m. \tag{20.1}$$

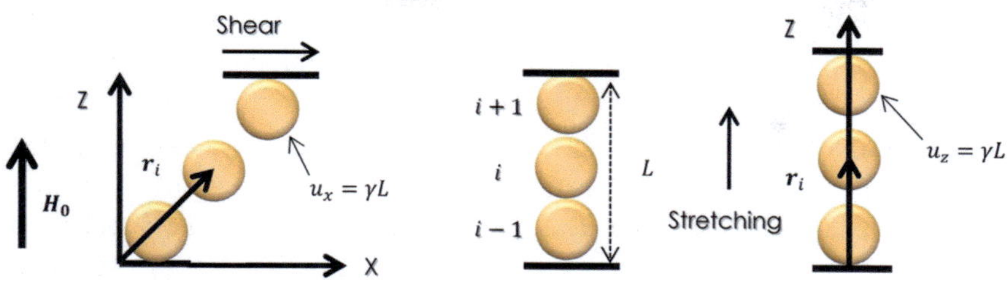

Figure 20.1: Sketch of the chain in the non-sheared (middle), sheared (left), and stretched (right) sample. Cited from Ref. [28].

Here γ is either shear or tensile strain, σ_m is magnetic part of the stress, produced by the aggregates. By using the results [29, 30] for the stress in a system of chain-like polymer macromolecules, we get the following estimate:

$$\sigma_m = \sigma_{xzm} = \frac{2\varphi}{Vn}\left[\sum_{i=1}^{n-1} F_{x\,i,i+1} \cdot R_{zi,i+1}\right], \text{ or}$$

$$\sigma_m = \sigma_{zzm} = \frac{2\varphi}{Vn}\left[\sum_{i=1}^{n-1} (F_{zi,i+1} + f_{i,i+1}) \cdot R_{zi,i+1}\right]. \tag{20.2}$$

The first relation of (2) is magnetic shear stress, second one is magnetic tensile stress; φ is volume concentration of the particles, n is number of the particles in the chain, $V = \pi d^3/6$ is the particle volume; $F_{x,zi,i+1}$ are x,z-components of force of magnetic attraction between the i-th particle to the $(i+1)$-th one; $R_{zi,i+1}$ is the distance between centers of the corresponding particles along the axis Oz; $f_{i,i+1}$ are the elastic forces that arise in the thin polymer layers between the neighboring particles, defined in Refs. [22, 31]:

$$f_{i,i+1} = G_1 d(u_{zi+1} - u_{zi}). \tag{20.3}$$

where u_{zi} is displacement of the i-th particle along the axis Oz. Note, when the composite experiences the tensile deformation, the thickness of the layer between two neighbor particles changes and the forces of resistance of the medium to the particles displacement along the axis Oz significant influences on the macroscopic tensile stress of the composite. On the contrary, under the shear deformation, the distance between the particles changes insignificantly. That is why in the first formula of Eq. (20.2) we do not take into account these elastic forces in the interparticle gaps.

Parameter $G_1 \sim G_0$ is the elastic modulus of the polymer in the gap between the particles. Generally speaking, due to change of the polymer conformation in the gap, the modulus G_1 can significantly differ from the modulus G_0 of the pure matrix [31].

We will ignore effects of magnetic interaction between the chains. This approximation is based on the results of Ref. [32], which demonstrate that the effects inside the chains play a dominant role in the formation of macroscopic properties of the composites, compared with the effects of the interchain interaction.

Let us introduce the Cartesian coordinate system, shown in Figure 20.1, with the origin at the lower pole of the first particle, axes Oz and Ox aligned along the applied field $\boldsymbol{H}_0$ and direction of the shear respectively.

We suppose that the system experiences either a macroscopic shear along the axis Ox either uniaxial deformation along the axis Oz with the upper plate displacement equals γL (see Figure 20.1), where L is the sample thickness. It is supposed that the first and the last n-th particles are fixed on the "low" and "upper" plates of the measuring gap, respectively.

For simplicity we estimate energy $U_{i+1,i}$ of the magnetic interaction between the particles by using the simplest dipole–dipole approximation:

$$U_{i+1,i}(\mathbf{r}) = -\frac{\mu_0 V^2}{4\pi}\left[3\frac{(\mathbf{M}\cdot(\mathbf{r}_{i+1}-\mathbf{r}_i))^2}{R^5_{i,i+1}} - \frac{M^2}{R^3_{i,i+1}}\right]. \tag{20.4}$$

Here μ_0 is the vacuum magnetic permeability, $\mathbf{M}$ is the particle magnetization. Strictly speaking, $\mathbf{M}$ depends on the particle position (its number i) in the chain, $\mathbf{r}_{i+1}$ and $\mathbf{r}_i$ are radius-vectors of the corresponding particles, $R_{i,i+1} = |\mathbf{r}_{i+1} - \mathbf{r}_i|$. The vector difference $\mathbf{u}_k = \mathbf{r}_k - \mathbf{r}_{k0}$ is displacement of the k-th particle, $\mathbf{r}_{k0}$ is its radius–vector before the sample deformation. We suppose that in the nondeformed composites the particles are densely situated in the chains, which are parallel to the applied field $\mathbf{H}_0$ and bridge the opposite boundaries of the sample.

Generally speaking, magnetization $\mathbf{M}$ of the particle depends on its position in the chain. This fact makes the calculations too cumbersome. Here we use the simplest approximation with $\mathbf{M}$ identical for all particles in the chain. It was shown in Ref. [33] that this approximation leads to not significant deviations from the strict approach with different magnetizations of the particles.

To determine the magnetic force $\mathbf{F}_i$, which acts on the i-th particle from the other particles in the chain, we will take into account the interaction only between the nearest neighbors in the chain. In the frame of this approximation:

$$\mathbf{F}_i = \mathbf{F}_{i,i+1} + \mathbf{F}_{i,i-1}, \mathbf{F}_{i,i+1} = -\nabla U_{i+1,i}(\mathbf{r}), \mathbf{F}_{i,i-1} = -\nabla U_{i-1,i}(\mathbf{r}). \tag{20.5}$$

In order to estimate the magnetization $\mathbf{M}$ of the particle, we consider each particle as situated in a uniform magnetic field $\mathbf{H}_\Sigma$, consisting of the external field $\mathbf{H}_0$ and sum of the fields $\mathbf{H}_{\Sigma i}$ created by the other particles of the chain in the center of the given one, averaged over all these particles:

$$\mathbf{H}_\Sigma = \mathbf{H}_0 + \frac{1}{n}\sum_{i=1}^{n}\mathbf{H}_{\Sigma i}. \tag{20.6}$$

Magnetic field $\mathbf{H}^{(ins)}$ inside the particle can be found from the classical relation [34]:

$$\mathbf{H}^{(ins)} + \frac{\mathbf{M}}{3} = \mathbf{H}_\Sigma. \tag{20.7}$$

Here 1/3 is demagnetizing factor of the spherical particle. In its turn, the magnetization $\mathbf{M}$ and the field $\mathbf{H}^{(ins)}$ can be estimated by using the empirical Frohlich–Kennelly relation [35]:

$$M = \chi \mathbf{H}^{(ins)}, \chi = \frac{\chi_0 M_s}{M_s + \chi_0 H^{(ins)}}. \tag{20.8}$$

Here χ_0 and M_s are initial susceptibility of the particle material and its saturated magnetization respectively; χ is the particle susceptibility in the internal field $\mathbf{H}^{(ins)}$.

Combining Eqs. (20.6)–(20.8), we obtained a system of nonlinear equations for the magnetization $\mathbf{M}$ components, which was solved numerically.

The equations of the stationary displacement of the particles in the chain can be presented in the following form [22, 31]:

$$\beta(\mathbf{u} - \mathbf{u}_i) + \mathbf{F}_i = 0, \quad i = 2, \ldots, n-1. \tag{20.9}$$

Here $\beta \sim G_0 d$ is a coefficient of the host polymer elastic resistance to the particle displacement, $\mathbf{u}$ is the displacement vector of the matrix. The first part of the equation reflects the elastic force acting on the i-th particle. The force arises from the fact that the particle displacement $\mathbf{u}_i$ is not equal to the displacement of the composite $\mathbf{u}$ far from the chain at the level of this particle. The second term in Eq. (20.9) corresponds to the magnetic forces acting on each particle. At the solution of system (9) we must take into account the condition of the particles noninterpenetration $R_{i,i+1} \geq d$. For maximal simplification of mathematics and transparence of the physical results, we estimate the elastic resistance β of the particles displacement as for the single particle. The similar approach is often used in hydromechanics of magnetic suspensions with the chain-like aggregates (see, for example, [34])

The explicit forms for the forces $\mathbf{F}_i$ can be obtained combining Eqs. (20.4) and (20.5) and substituting therein magnetization $\mathbf{M}$, found from (6–8). The details are omitted for brevity.

Note that the force $\mathbf{F}_i$, acting on the i-th particle, depends on the positions $\mathbf{r}_{i-1}$, $\mathbf{r}_{i+1}$ of the neighbor particles, therefore, on their displacements $\mathbf{u}_{i-1}$, $\mathbf{u}_{i+1}$. This means that the relation (9) presents a system of equations with respect to displacement of each particle in the chain.

We have solved system (9) numerically and obtained displacements $\mathbf{u}_i$ of each particle in the sheared sample as well as in the stretched one. After that for both types of the deformation the macroscopic stress σ has been determined by using Eqs. (20.1) and (20.2). Some results are presented in Figure 20.2.

When the deformation γ achieves a certain threshold magnitude ($\gamma = 0.2$ for the shear and $\gamma = 0.047$ for the tensile deformations; see Figure 20.2), the distance between the central particles in the chain becomes large compared to the gaps between the other particles. The gap in the center of the chain is so large that the stresses, at this

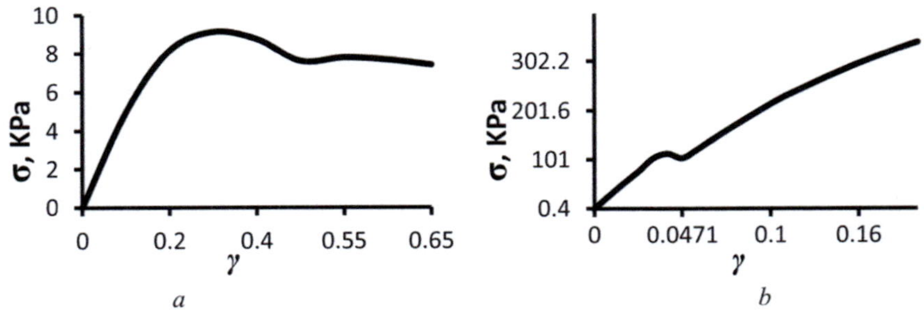

Figure 20.2: The results of stress calculations depending on the magnitude of the deformation. a: Shear stress. System parameters: $n = 10$, $G_0 = 3.2$ kPa, $H_0 = 100$ kA/m, $\varphi = 25\%$. b: Tensile stress. Parameters: $\varphi = 15\%$, $G_1 = G_0 = 210$ kPa, $H_0 = 350$ kA/m. Cited from Ref. [28].

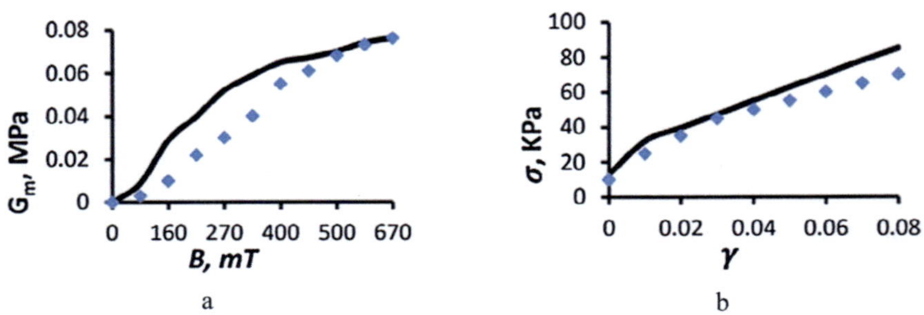

Figure 20.3: a: Shear modulus G_m versus the applied magnetic field induction B. System parameters: $\varphi = 10\%$, $\gamma = 0.1$, $n = 10$, $G_0 = 60$ kPa. Line – theory, dots – experiments [18]. b: total elastic stress σ versus the sample elongation γ. Parameters: $\varphi = 15\%$, $G_0 = 200$ kPa, $H_0 = 123$ kA/m. Dots – experiments [23]. Reproduced from Ref. [28].

deformation γ, drops down; it can be interpreted as a chain breaking. After this threshold value of γ, the gap in the chain "vanishes", since the particles evenly regroup, that leads to increase of the stresses (Figure 20.2). Note that the similar behavior of the stress also has been detected in computer simulations [24].

Comparison of our calculations of the magnetic part $G_m = \sigma_m/\gamma$ of the magnetic elastic modulus with experimental data of Ref. [18] and the total tensile stress σ with experiments of Ref. [23] are presented in Figure 20.3. The agreement between theoretical and experimental results is quite reasonable.

20.3 Internal structures in dense magnetic suspensions

The strong dependence of the rheological properties of magnetorheological suspension (MRSs) on a magnetic field is explained by the aggregation of the particles into heterogeneous anisotropic structures oriented along applied magnetic field (see, for example, overviews in Ref. [14, 16]). When these structures overlap the gap, filled with magnetorheological suspension, and form “bridges” between the walls of the gap, the rheological state of the suspension changes from the liquid to the quasi elastic one. Majority of theoretical models of MRSs are based on the concept that the particles form either the linear chains or separate dense ellipsoidal columns (see Refs. [15–17]). It should be noted that the separate discrete aggregates were observed in computer simulations with low concentration of magnetic particles [36, 37]. However, experiments [38–40] demonstrate that more topologically complicated structures (labyrinths, tubes, etc.) can appear is these systems when the particles concentration is high enough. With high concentration of particles, these complicated structures were also detected in computer simulations [41].

There is no a way to describe theoretically aggregation in MRSs with nonsmall concentration of the particles without some a′ priori suggested simplifications. On the other hand, the systems with volume concentration of particles in the range 20–40% are the most interesting from the viewpoint of their practical applications. In this work we suggest a computer model of structural transformations in magnetorheological fluids with 30% of the particles volume concentration.

The computer simulations are based on numerical solution of a system of differential equations of the particles motion. Magnetic interaction of each particle with the all other ones is taken into account. Based on the results of the suspension structuring, the shear stress as a function of the sample shear deformation is calculated.

We consider a system of 1600 identical non Brownian magnetizable spheres with diameter d_p each. The particles are situated in a cylinder with boundaries impenetrable for the particles. The diameter of a cylinder D is much more than its length L. At the onset of the simulations, we suppose chaotic, the “hard spheres gas” spatial distribution of the particles. To eliminate the particles overlapping, we “warm up” the system (details of this numerical procedure can be found in Refs. [42, 43]).

We take into account the Stokes hydrodynamic forces acting on the particles in carrier fluid and ignore hydrodynamic interaction between the particles. The last simplification cannot lead to any physical mistakes, since only mechanically equilibrium structures are in focus of our attention here. In contrast, magnetic interaction between all particles in the system is taken into account. Neglecting inertia, the equation of motion of the ith particle is:

$$\mathbf{F}_h^i + \sum_{j\neq i}\left(\mathbf{F}_m^{i,j} + \mathbf{F}_{st}^{i,j}\right) = 0, \quad i,j = 1,\ldots,N. \tag{20.10}$$

Here N is total number particles in the system; $\mathbf{F}_h^i$ is hydrodynamic force acting on the i-th particle; $\mathbf{F}_m^{i,j}$ and $\mathbf{F}_{st}^{i,j}$ are respectively forces of magnetic and sterical interaction between the i-th and -jth particles.

Ignoring the hydrodynamic interaction between the particles, the viscous force acting on the i-th particle can be calculated by using the Stokes formula:

$$\mathbf{F}_h^i = -3\pi d_p \eta \frac{d\mathbf{r}_i}{dt}. \tag{20.11}$$

Here η is viscosity of the carrier liquid; t is time. Approximation (11) means that the hydrodynamical mobility of the particle in the suspension is the same as that of a single particle.

Substituting relation (11) into Eq. (20.10), we get the system of ordinary differential equations of the first order:

$$\frac{d\mathbf{r}_i}{dt} = \frac{1}{3\pi d_p \eta} \sum_{j \neq i} \left(\mathbf{F}_m^{i,j} + \mathbf{F}_{st}^{i,j} \right), \quad i,j = 1, \ldots, N. \tag{20.12}$$

We estimate the magnetic force $\mathbf{F}_m^{i,j}$ based on the following considerations.

First, we take into account the mutual magnetization of the particles. Due to this effect, the particle magnetization $\mathbf{M}$ deviates from the applied field $\mathbf{H}$. Secondly, for mathematical simplicity, we propose the uniform magnetization $\mathbf{M}$ inside the particle. Note that this approach has been used successfully in Ref. [44] to estimate interaction between magnetizable particles. Then, for maximal simplification of calculations, we will consider the limiting cases of the weak field, corresponding to the linear law of the particle magnetization as well as the opposite situation of the strong field and saturated magnetization of the particles.

20.3.1 Weak magnetic field

In the case of the linear magnetization, the relation

$$\mathbf{M} = \chi \mathbf{H}^{(ins)} \tag{20.13}$$

is held. Here χ is initial magnetic susceptibility of the particle material; $\mathbf{H}^{(ins)}$ is magnetic field inside the particle.

By using the well-known relation [45] between magnetic field $\mathbf{H}_i^{(ins)}$ inside the magnetizable i-th particle, its magnetization $\mathbf{M}_i$ and external magnetic field $\mathbf{H}$, taking into account that the demagnetizing factor of a spherical particle is 1/3, after simple transformations one gets:

$$\mathbf{H}_i^{(ins)} + \frac{\mathbf{M}_i}{3} = \mathbf{H} + \sum_{j \neq i} \mathbf{H}_{i,j}^{(d)}. \tag{20.14}$$

In the simplest dipole approximation:

$$\mathbf{H}_{i,j}^{(d)} = \frac{3(\mathbf{r}_{i,j} \cdot \mathbf{m}_j) r_{i,j} - r_{i,j}^2 \mathbf{m}_j}{4\pi r_{i,j}^5}. \tag{20.15}$$

Here $V_p = \pi d_p^3/6$ is the particle volume, $\mathbf{m}_i = \mathbf{M}_i V_p$ is magnetic moment of the i-th particle, to be determined.

Substituting the dipole magnetic field $\mathbf{H}_{i,j}^{(d)}$ into Eq. (20.14) and taking into account Eq. (20.13), we come to the system of linear algebraic equations for the magnetic moments $\mathbf{m}_i$:

$$\frac{8(3+\chi)}{\chi d_p^3}\mathbf{m}_i + \sum_{j\neq i}\left[\frac{\mathbf{m}_j}{r_{i,j}^3} - \frac{3(\mathbf{r}_{i,j}\cdot\mathbf{m}_j)\mathbf{r}_{i,j}}{r_{i,j}^5}\right] = 4\pi\mathbf{H}, \quad i,j = 1,\ldots,N. \tag{20.16}$$

For a large number N of the particles, the most effective method of this system solution is the Gauss–Seidel method.

Having system (16) solved, we can calculate the magnetic forces $\mathbf{F}_m^{i,j}$ of interaction between the particles. In the framework of the model of the dipole–dipole interaction, the force is:

$$\mathbf{F}_m^{i,j} = \frac{3\mu_0}{4\pi r_{i,j}^5}\left[\mathbf{r}_{i,j}(\mathbf{m}_i\cdot\mathbf{m}_j) + \mathbf{m}_i(\mathbf{m}_j\cdot\mathbf{r}_{i,j}) + \mathbf{m}_j(\mathbf{m}_i\cdot\mathbf{r}_{i,j}) - 5\mathbf{r}_{i,j}\frac{(\mathbf{m}_i\cdot\mathbf{r}_{i,j})(\mathbf{m}_j\cdot\mathbf{r}_{i,j})}{r_{i,j}^2}\right]. \tag{20.17}$$

To estimate the sterical force $\mathbf{F}_{st}^{i,j}$ we use the Weeks–Chandler–Andersen potential [46] (see, also, details in Ref. [43]).

System (12) can be solved numerically, for example, by using the Euler method. At the calculations below the time step has been chosen so that the particle displacement for a given step would not exceed $d_p/20$.

The principal difference between structures in ferrofluids and magnetorheological fluids is in the fact that in ferrofluids the particles, due to the intensive Brownian motion, can achieve thermodynamically equilibrium disposition quite fast. In magnetorheological fluids the non Brownian particles can be "frozen" in some local energetic minimums. One of these typical "frozen" states is illustrated in Figure 20.4. In our computer simulations we observe not chains or individual bulk drops, but branched labyrinth-shaped structures. Qualitatively the similar structures were also observed in laboratory experiments [38]. As can be seen in Figure 20.4, the magnetized particles are concentrated near the cylinder periphery. The reason of that is the repulsive magnetic force acting between the particles at their lateral disposition. Because the cylinder walls are impenetrable, the particles accumulate at the periphery.

Now we are in a position to study the elastic properties of this system at shear deformations, in the direction perpendicular to the applied magnetic field. To this end, we fix the particles that touch the upper and lower bases of the cylinder. All other particles are free; hydrodynamic, magnetic, and steric forces act on them. We suppose that magnetic field is applied along the axis of the cylindrical container with the suspension; the diameter D of the cylinder is supposed much more than its length L.

Let us introduce the Cartesian coordinate system with the axis z aligned along the sample axis and axes x, y perpendicular to that. After this shear, the x and y coordinates of the particles fixed on the upper base of the cylinder, will be changed as:

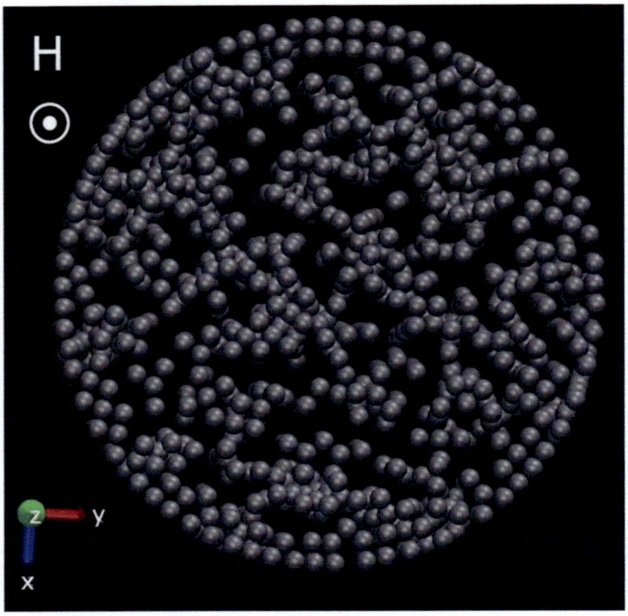

Figure 20.4: Typical screenshots made using VMD (visual molecular dynamics) of "frozen" particles under magnetic field after integration of equations of the particles motion. The magnetic field is directed along the axis of cylinder. Parameters of the system: volume concentration of the particles $\varphi = 0.3$; number of the particles in computer simulations $N = 1600$; dimensionless diameter of the cylinder $D/d_p = 26.1$; dimensionless height of the cylinder $L/d_p = 5.22$; magnetic susceptibility of the particles $x = 1000$. Cited from Ref. [43].

$$\Delta x_i = -\frac{y_i \gamma L}{D}, \quad \Delta y_i = \frac{x_i \gamma L}{D}. \tag{20.18}$$

It should be stressed that the relation (18) is valid provided that the strong inequality $D >> L$ is true.

We integrate Eq. (20.12) of the particles motion so that all particles that don't touch the upper base of the cylinder come to a new "frozen" state, corresponding to the macroscopical deformation of the "sample".

Let us select a ring on the upper base of the cylinder with the outer radius $D/2$, and ring thickness δ which is much less than its radius. In this way stress σ on the disk rim can be determined as [43]:

$$\sigma = \frac{2}{\pi D\delta} \sum_{i=1}^{k} \frac{\left[\mathbf{r}_i \times \mathbf{F}^{(up),i}\right]_y}{r_i} = \frac{2}{\pi D\delta} \sum_{i=1}^{k} \frac{y_i F_x^{(up),i} - x_i F_z^{(up),i}}{\sqrt{x_i^2 + y_i^2}} \tag{20.19}$$

Some results of calculations of the stress are shown in Figure 20.5.

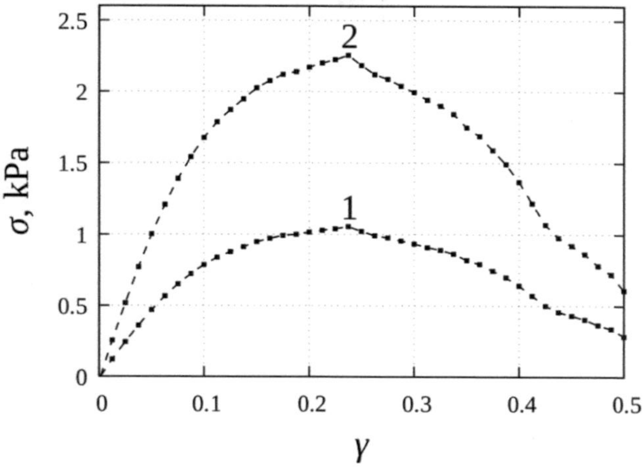

Figure 20.5: The stress σ on the disk rim in the MR fluid versus shear γ on the disk rim at the linear magnetization of the particles. Vector of the magnetic field H is perpendicular to shear. Parameters of the system the same as in Figure 20.4. Line 1 – magnetic field $H = 65$ kA/m; line 2 – $H = 95$ kA/m. Reproduced from [43].

20.3.2 Strong magnetic field

We will consider now the opposite limiting case of very strong magnetic field, when magnetization of the particles is fully saturated. To this end we will solve again Eq. (20.12) with the saturated magnetic moment $\mathbf{m}$ of each particle, aligned along the field. The dependence of the saturated stress σ_s versus the shear strain γ is shown in Figure 20.6.

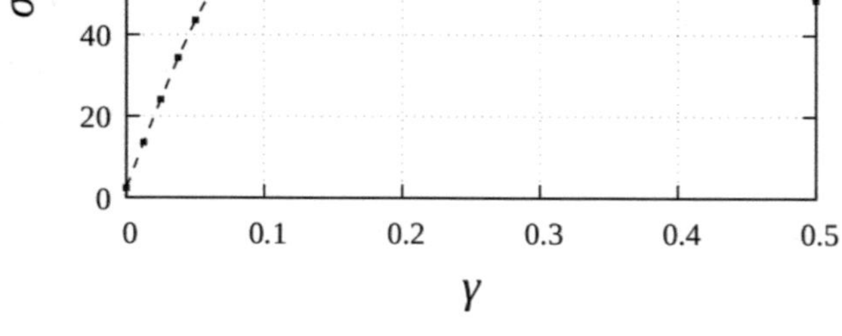

Figure 20.6: Limiting case of very strong magnetic field. The saturated stress σ_s on the disk rim versus shear γ on the disk rim. Parameters of the system are the same as in Figures 20.4 and 20.5. Cited from Ref. [43].

Let us consider a plate-shaped sample with the walls perpendicular the axis z and impenetrable for the particles. Periodic boundary conditions are specified along the x and y axis. Magnetic field is applied along the axis z. In the computer simulations, the upper wall of the cell is shifted along the x axis in small steps, the lower wall remains motionless. The stress σ, acting on the upper base of the plate, can be calculated as:

$$\sigma = \frac{1}{L^2} \sum_{i=1}^{k} F_x^{(up),i}. \tag{20.20}$$

The results in Figure 20.7 also demonstrate the non-monotonic dependence of the stress σ versus global shear γ with a maximum at a certain deformation.

It is interesting to consider a finite-sized object, namely the sphere. The wall of this sphere is impenetrable for the particles. Unlike the infinite gap, in this case, we can perform the calculations in single precision on a CUDA-capable graphics card. The implementation is optimized for large systems of several thousand particles. It makes use of one thread per particle. This method, implemented in the software package ESPResSo, allows us to perform calculations on GPU using NVIDIA CUDA Toolkit. In the computer simulation, we have used the GeForce RTX 2070 video card. Calculations on our hardware show that GPU performance is an order of magnitude greater than CPU performance. Therefore, we can perform the computer experiment for the sphere with a much larger number of particles. The results of these simulations can be seen in Figure 20.8. The "frozen" branched structures are well observed.

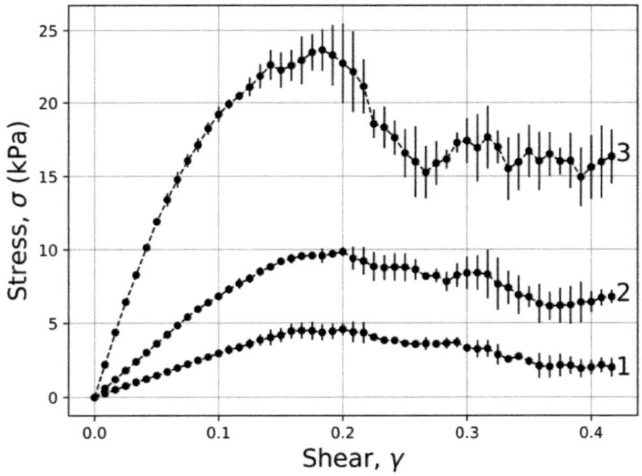

Figure 20.7: Limiting case of very strong magnetic field. The stress σ on the upper wall of the plate in MRS versus shear γ. Vertical bars illustrate standard deviation. Number of the particles in computer simulations $N = 2760$; dimensionless thickness of the plate $H/d_p = 13$; dimensionless length along x and y axis $L/d_p = 19.25$; saturated magnetization of the material of the particles $M_s = 1655$ kA/m 1 – volume concentration of the particles $\varphi = 5\%$; 2 – $\varphi = 10\%$; 3 – $\varphi = 30\%$.

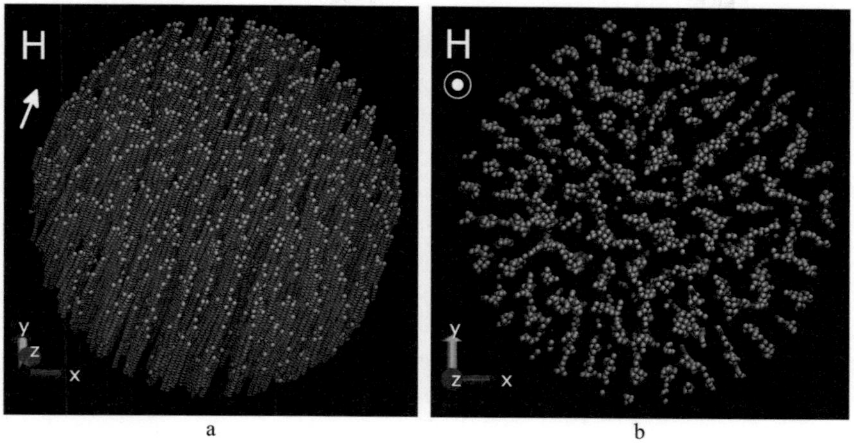

Figure 20.8: Typical screenshots made using VMD of the particles situated in the sphere. Parameters of the system: volume concentration of the particles $\varphi = 10\%$; number of the particles $N = 27462$; dimensionless diameter of the sphere $D/d_p = 65$.
a: General view. b: View along the direction of the magnetic field.

For the both limiting cases the computer results demonstrate the non-monotonic dependence of the stress σ on the shear γ with a maximum. The reason of this non-monotonous behavior is destruction of the internal structures at large enough γ, which causes decrease of the stress σ. The maximal value of the stress can be associated with the static yield stress, corresponding to transition of the system from the elastic to the flow regime behavior. We have not calculated the stress at higher values of γ, since the decreasing dependence of the stress on the shear corresponds to the mechanical instability of the systems which requires a special study.

Note that the nonmonotonic dependence of the stress versus stationary shear in the MRSs was theoretically described in the work [47] on the basis of the model of dense discrete ellipsoidal aggregates in the MRSs. The similar results for ferrogels is discussed in the part 2 of the paper.

Our results demonstrate that in the concentrated suspensions the labyrinth structures are expected rather than the ellipsoidal ones; for these structures the non-monotonic dependence of σ on γ is also expected.

20.4 Conclusions

Effect of internal chain-like and branched heterogeneous structures on elastic properties of soft and liquid composites of magnetizable particles is considered. Our results demonstrate that in the both types of composites the rupture of the internal structures provokes decrease of the macroscopic stress as function on the global shear. The results of computer simulations of dense magnetic suspensions demonstrate that appearance, under external field action, of branched labyrinth structures in these systems is more probable than appearance of discrete column-like aggregates considered in traditional theoretical models of magnetic suspensions.

Author contributions: All the authors have accepted responsibility for the entire content of this submitted manuscript and approved submission.
Research funding: This work was funded by Ministry of Education and Science of the Russian Federation, projects FEUZ-2020-0051, 02.A03.21.0006 as well as to the, Russian Foundation of Basic Researches, projects 19-31-90003, 20-02-00022.
Conflict of interest statement: The authors declare no conflicts of interest regarding this article.

References

1. Filipcsei G, Csetneki I, Szilagyi A, Zrınyi M. Magnetic field-responsive smart polymer composites. Adv Polym Sci 2007;206:137.
2. Boczkowska A, Awietjan SF. Tuning active magnetorheological elastomers for damping applications. Mater Sci Forum 2010;636–637:766.
3. van Bruggen M, van Zon J. Theoretical description of a responsive magneto-hydrogel transduction principle. Sens Actuators, A 2010;158:240–8.
4. van Berkum S, Dee JT, Philipse Albert P, Erne BH. Frequency-dependent magnetic susceptibility of magnetite and cobalt ferrite nanoparticles embedded in PAA hydrogel. Int J Mol Sci 2013;14: 10162–77.
5. Bañobre-López M, Piñeiro-Redondo Y, de Santis R, Gloria A, Ambrosio L, Tampieri A, et al. Poly (caprolactone) based magnetic scaffolds for bone tissue engineering. J Appl Phys 2013;109: 07B313.
6. Das B, Mandal M, Upadhyay A, Chattopadhyay P, Karak N. Bio-based hyperbranched polyurethane/Fe_3O_4 nanocomposites: smart antibacterial biomaterials for biomedical devices and implants. Biomed Mater 2013;8:035003.
7. Gloria A, Russo R, d'Amora U, Zeppetelli S, d'Alessandro T, Sandri M, et al. Magnetic poly(ε-caprolactone)/iron-doped hydroxyapatite nanocomposite substrates for advanced bone tissue engineering. J R Soc Interface 2013;10:20120833.
8. Li Y, Huang G, Zhang X, Li B, Chen Y, Lu T, et al. Magnetic hydrogels and their potential biomedical applications. Adv Funct Mater 2013;23:660–72.
9. Panseri S, Cunha C, D'Alessandro T, Sandri M, Giavaresi G, Marcacci M, et al. Intrinsically superparamagnetic Fe-hydroxyapatite nanoparticles positively influence osteoblast-like cell behaviour. J Nanobiotechnol 2012;10:32.
10. Zeng XB, Hu H, Xie LQ, Lan F, Jiang W, Wu Y, et al. Magnetic responsive hydroxyapatite composite scaffolds construction for bone defect reparation. Int J Nanomed 2012;7:3365.
11. Singh RK, Patel KD, Lee JH, Lee EJ, Kim JH, Kim TH, et al. Potential of magnetic nanofiber scaffolds with mechanical and biological properties applicable for bone regeneration. PloS One 2014;9: e91584.
12. Lopez-Lopez MT, Durán JDG, Iskakova LY, Zubarev AY. Mechanics of magnetopolymer composites: a review. J Nanofluids 2016;5:479–95.

13. Weeber R, Hermes M, Schmidt AM, Holm C. Polymer architecture of magnetic gels: a review. J Phys Condens Matter 2018;30:063002.
14. Bossis G, Volkova O, Lacis S, Meunier A. Magnetorheology: fluids, structure and rheology. In: Odenbach S, editor. Ferrofluids, magnetically controllable fluids and their applications. Berlin: Springer; 2002.
15. Muhammad A, Yao X, Deng Z. Review of magnetorheological (MR) fluids and its applications in vibration control. J Mar Sci Appl 2006;5:17.
16. de Vicente J, Klingenberg DJ, Hidalgo-Alvarez R. Magnetorheological fluids: a review. Soft Matter 2011;7:3701.
17. Hiemenz GJ, Choi YT, Wereley NM. Semi-active control of vertical stroking helicopter crew seat for enhanced crashworthiness. AIAA J Aircraft 2007;44:1031.
18. Jolly MR, Carlson JD, Muñoz BC, Bullions TA. The magnetoviscoelastic response of elastomer composites consisting of ferrous particles embedded in a polymer matrix. J Intell Mater Syst Struct 1996;6:613.
19. Danas K, Kankanala SV, Triantafyllidis N. Experiments and modeling of iron-particle-filled magnetorheological elastomers. J Mech Phys Solid 2012;60:120.
20. Han Y, Hong W, Faidley LE. Field-stiffening effect of magneto-rheological elastomers. Int J Solid Struct 2013;50:2281.
21. Agirre-Olabide I, Kuzhir P, Elejabarrieta MJ. Linear magneto-viscoelastic model based on magnetic permeability components for anisotropic magnetorheological elastomers. J Magn Magn Mater 2018;446:155–61.
22. Borin D, Chirikov D, Zubarev A. Shear elasticity of magnetic gels with internal structures. Sensors 2018;18:2054.
23. Bellan C, Bossis G. Field dependence of viscoelastic properties of MR elastomers. Int J Mod Phys B 2002;16:2447.
24. Cremer P, Lowen H, Menzel A. Tailoring superelasticity of soft magnetic materials. Appl Phys Lett 2015;107:171903.
25. Cristensen R. Mechanics of composite materials. New York: Wiley; 1979.
26. Coquelle E, Bossis G. Mullins effect in elastomers filled with particles aligned by a magnetic field. Int J Solid Struct 2006;43:7659–72.
27. Saxena P, Pelteret JP, Steinmann P. Modelling of iron-filled magneto-active polymers with a dispersed chain-like microstructure. Eur J Mech Solid 2015;50:132–51.
28. Zubarev A, Musikhin A, Chirikov D, Borin D. Elastic stress in ferrogels with chain aggregates. J Magn Magn Mater 2020;498:166126.
29. Doi M, Edwards SF. The theory of polymer dynamics. New York: Oxford University Press; 1986.
30. Larson RG. The structure and rheology of complex fluids. New York, USA: Oxford University Press; 1999.
31. Zubarev AY, Iskakova LY, Lopez-Lopez MT. Towards a theory of mechanical properties of ferrogels. Effect of chain-like aggregates. Physica A 2016;455:98–103.
32. Coquelle E, Bossis G, Szabo D, Giulieri F. Micromechanical analysis of an elastomer filled with particles organized in chain-like structure. J Mater Sci 2006;41:5941–53.
33. Zubarev AY, Chirikov DN, Borin DY, Stepanov GV. Hysteresis of the magnetic properties of soft magnetic gels. Soft Matter 2016;12:6473–80.
34. Martin JE, Anderson RA. Chain model of electrorheology. J Chem Phys 1996;104:4814–27.
35. Bozorth RM. Ferromagnetism. New York: Wiley; 1993.
36. Bossis G, Lancon P, Meunier A, Iskakova L, Kostenko V, Zubarev A. Kinetics of internal structures growth in magnetic suspensions. Physica A 2013;392:1567–76.
37. Wang Z, Holm C, Muller HW. Molecular dynamics study on the equilibrium magnetization properties and structure of ferrofluids. Phys Rev E 2002;66:021405.

38. Borin D, Gunther D, Hintze C, Heinrich G, Odenbach S. The level of cross-linking and the structure of anisotropic magnetorheological elastomers. J Magn Magn Mater 2012;324:3452.
39. Gunther D, Borin DY, Guenther S, Odenbach S. X-ray micro-tomographic characterization of field-structured magnetorheological elastomers. Smart Mater Struct 2012;21:015005.
40. Borin D, Odenbach S, Iskakova L, Zubarev A. Non-ergodic tube structures in magnetic gels and suspensions. Soft Matter 2018;14:8537.
41. Wang Z. Creep and recovery of magnetorheological fluids: experiments and simulations. J Rheol 2014;58:1725.
42. http://espressomd.org/wordpress/.
43. Zubarev A, Chirikov D, Borin D. Internal structures and elastic properties of dense magnetic fluids. J Magn Magn Mater 2020;498:166129.
44. Biller AM, Stolbov OV, Raikher YL. Mesoscopic magnetomechanical hysteresis in a magnetorheological elastomer. Phys Rev E 2015;92:023202.
45. Landau LD, Lifshitz EM. Electrodynamics of continuum media. London: Pergamon Press; 1960.
46. Heyes DM, Okumura H. Some physical properties of the Weeks-Chandler-Andersen fluid. Mol Simulat 2006;32:45.
47. Bossis G, Lemaire E, Volkova O. Yield stress in magnetorheological and electrorheological fluids: a comparison between microscopic and macroscopic structural models. J Rheol 1997;41:687–704.

Harald Pleiner and Helmut R. Brand*

21 Symmetry aspects in the macroscopic dynamics of magnetorheological gels and general liquid crystalline magnetic elastomers

Abstract: We investigate theoretically the macroscopic dynamics of various types of ordered magnetic fluid, gel, and elastomeric phases. We take a symmetry point of view and emphasize its importance for a macroscopic description. The interactions and couplings among the relevant variables are based on their individual symmetry behavior, irrespective of the detailed nature of the microscopic interactions involved. Concerning the variables we discriminate between conserved variables related to a local conservation law, symmetry variables describing a (spontaneously) broken continuous symmetry (e.g., due to a preferred direction) and slowly relaxing ones that arise from special conditions of the system are considered. Among the relevant symmetries, we consider the behavior under spatial rotations (e.g., discriminating scalars, vectors or tensors), under spatial inversion (discriminating e.g., polar and axial vectors), and under time reversal symmetry (discriminating e.g., velocities from polarizations, or electric fields from magnetic ones). Those symmetries are crucial not only to find the possible cross-couplings correctly but also to get a description of the macroscopic dynamics that is compatible with thermodynamics. In particular, time reversal symmetry is decisive to get the second law of thermodynamics right. We discuss (conventional quadrupolar) nematic order, polar order, active polar order, as well as ferromagnetic order and tetrahedral (octupolar) order. In a second step, we show some of the consequences of the symmetry properties for the various systems that we have worked on within the SPP1681, including magnetic nematic (and cholesteric) elastomers, ferromagnetic nematics (also with tetrahedral order), ferromagnetic elastomers with tetrahedral order, gels and elastomers with polar or active polar order, and finally magnetorheological fluids and gels in a one- and two-fluid description.

Keywords: dissipation function, ferromagnetic nematics, hydrodynamics, linear irreversible thermodynamics, magnetorheological fluids, reversible dynamics

---

***Corresponding author: Helmut R. Brand,** Department of Physics, University of Bayreuth, Bayreuth, Germany, e-mail: brand@uni-bayreuth.de

Harald Pleiner, Max Planck Institute for Polymer Research, Mainz, Germany, e-mail: pleiner@mpip-mainz.mpg.de

This article has previously been published in the journal Physical Sciences Reviews. Please cite as: H. Pleiner and H. R. Brand "Symmetry aspects in the macroscopic dynamics of magnetorheological gels and general liquid crystalline magnetic elastomers" *Physical Sciences Reviews* [Online] 2020, 5. DOI: 10.1515/psr-2019-0109 | https://doi.org/10.1515/9783110569636-021

21.1 General aspects of macroscopic dynamics

21.1.1 Macroscopic variables

In a microscopic dynamic description, all (or a very large number of) microscopic degrees of freedom are taken into account. Macroscopic dynamics makes use of only those rather few dynamic variables that are not relaxed to their equilibrium values on long, macroscopic time scales. They come in three classes: (A) *conserved quantities* that can only be transported and (B) *symmetry variables*, describing a spontaneously broken continuous symmetry, e.g., a preferred direction that breaks orientational symmetry, but whose orientation is arbitrary, cannot relax at all in the homogeneous limit. Class A and B are the true hydrodynamic variables since for any excitation with frequency ω and wave vector k, there is $\omega \to 0$ for $k \to 0$. Class (C) *macroscopic variables* comprise all quantities that only slowly relax and therefore interact with the hydrodynamic variables on macroscopic time scales (where the many microscopic degrees of freedom have already relaxed to their equilibrium values). There are no general criteria to identify such macroscopic variables, and their existence depends on the nature of the individual system.

Examples for conserved quantities are the mass density ρ (or individual mass densities in a mixture without chemical reactions), the momentum density g_i, and the energy density ε. Among the symmetry variables, we consider the nematic director n_i, the vector p_i describing the orientation of a polar preferred direction, and m_i denoting the direction of a spontaneous magnetization. In active systems with polar order, e.g., moving bird flocks or fish schools, the preferred direction is given by the orientation f_i of an active nonzero velocity. In solid (elastic) media, translational symmetry is broken and characterized by the displacement vector u_i. Macroscopic variables are the magnetization M_i in a magnetizable medium, relative velocities w_i in a two-fluid system (different constituents moving with two different velocities), and relative rotations Ω_i among different preferred directions or between a preferred direction and rotations of an elastic body. In the framework of SPP1681 we have worked on many of these systems, which will be discussed in detail, in the following [1–15].

Throughout this chapter, we use the notation m_i, g_i, n_i etc., to characterize vectorial quantities, with indices i running from 1 to 3 in three dimensions. This is a standard notation used in physical hydrodynamics [16].

The variables described above are space-time fields, e.g., $\varepsilon(\mathbf{r},t)$, that live on macroscopic time and length scales. In the Eulerian description, they are volume densities (of the total quantities of the whole system). The basic assumption to derive general equations for those variables is the applicability of thermodynamics. We start with the local formulation of the *first law of thermodynamics*, which can be interpreted as energy balance of the system. It relates changes of all the variables discussed above to changes of the entropy density $d\sigma$

$$d\varepsilon = Td\sigma + \mu_A dA + \psi_k^B d\nabla_k B + \mu_C dC \tag{21.1}$$

with T the temperature. The prefactors of the differentials are called (thermodynamic) conjugates.

The energy density is assumed to be a scalar, meaning it does not change under rotations. Thus, all the terms added up in Eq. (21.1) must be scalars. This is trivially fulfilled for $Td\sigma$ and for $\mu d\rho$, where $\mu_A dA \to \mu d\rho$ with μ the chemical potential. If A is a vector, like the momentum density, also μ_A has to be a vector connected by the scalar product, $\mu_A dA \to v_i dg_i$ with v_i the velocity. The B variables must not change the energy, so their contribution in Eq. (21.1) already contains a gradient. As a consequence also their conjugates must be vectors. Generally the B variables are vector-like quantities by themselves, e.g., $B \to n_i$, with $\psi_k^B d\nabla_k B \to \psi_{ik}^n d\nabla_k n_i$. In the case of the nematic director, external (electric or magnetic) fields are important and an additional contribution is added, $h_i^{nI} dn_i$, where h_i^{nI} is due to the dielectric or diamagnetic anisotropy of the nematic phase. Often the two contributions are combined into $h_i^n dn_i$, with $h_i^n = h_i^{nI} - \nabla_k \psi_{ik}^n$ and n_i belonging to class C.

Similarly, when the gradient of the displacement vector $\nabla_k u_i$ is replaced by a symmetric second-rank strain tensor U_{ij}, eliminating solid body rotations that must not contribute to the energy density, U_{ij} belongs to class C with $\mu_C dC \to \Phi_{ij} dU_{ij}$, where Φ_{ij} is the elastic stress tensor. For a thorough discussion of the relation of u_i with U_{ij} in the nonlinear case, we refer to the study by Pleiner et al. [17].

For the magnetization M_i in a magnetizable system, we have $\mu_C dC \to h_i^M dM_i$ with h_i^M the internal magnetic field. In a ferromagnetic system, $M_i = M m_i$ can be split into the unit vector m_i, denoting the direction of the spontaneous magnetization, which is a class B variable with $\mu_B dB \to h_i^m dm_i$, and into the strength of magnetic order M, a class C variable with $\mu_C dC \to h^M dM$. For active polar order, $F_i = F f_i$, the direction f_i is a class B variable with $\mu_B dB \to h_i^f df_i$, while F relaxes to its constant stationary value provided by the active entities of the system and is a class C variable. [18]. Relative rotations are rotations of, e.g., the ferromagnetic direction δm_i relative to rotations of the elastic media, $2\Omega_{ij} \equiv \nabla_i u_j - \nabla_j u_i$. They are linearly defined by $\Omega_i = m_j \Omega_{ij} - \delta m_i$ and are class C variables with $\mu_C dC \to L_i^\Omega d\Omega_i$. For a nonlinear definition of relative rotations, refer the study by Menzel et al. [19].

The *statics of a system* is then given by the phenomenological relation between the thermodynamic quantities and the variables. These relations involve static susceptibilities, e.g., compressibility, specific heat, thermal expansion for the scalar variables, Frank-type rank-4 tensors with Frank coefficients for B variables, and the elastic tensor containing elastic moduli or a rank-2 rotational tensor for the C variables. An efficient way of setting up these static relations is to use a phenomenological *energy functional* of all variables, which is given as follows:

$$E = \int dV \varepsilon(\{\sigma, A, \nabla_k B, C\}) \tag{21.2}$$

from which the conjugates follow by variational derivation according to Eq. (21.1). The crucial point is that the functional is restricted by symmetries, in particular space inversion and time reversal symmetry.

Before we discuss these symmetries in the following, we give here the general form of the *dynamic equations* for examples of the different classes of variables:

$$\mathbf{A} \qquad (\partial/\partial t)\rho + \nabla_k j_k^{\rho} = 0 \tag{21.3}$$

$$\mathbf{A} \qquad (\partial/\partial t)g_i + \nabla_j \sigma_{ij} = 0 \tag{21.4}$$

$$\mathbf{B} \qquad (\partial/\partial t)n_i + Y_i^n = 0 \tag{21.5}$$

$$\mathbf{C} \qquad (\partial/\partial t)U_{ij} + X_{ij}^{(el)} = 0 \tag{21.6}$$

$$\mathbf{C} \qquad (\partial/\partial t)\Omega_i + X_i^{\Omega} = 0 \tag{21.7}$$

$$\mathbf{C} \qquad (\partial/\partial t)M_i + Y_i^M = 0 \tag{21.8}$$

defining the mass current density $j_k^{(\rho)}$ and the stress tensor σ_{ij} for *A* variables and the (quasi-) currents Y_i^n and $X_{ij}^{(el)}$, $X_i^{(\Omega)}$, and Y_i^M for the others. The *A* variables show a divergence in the dynamics, while the *B* variables come with a gradient in the statics. As an effect, both types of variables give rise to the truly hydrodynamic behavior, $\omega(k \to 0) \to 0$, while the *C* variables do not.

21.1.2 Spatial inversion symmetry: statics

In three-dimensional space, not only rotations but also spatial inversion, S_I where $\boldsymbol{r}\to-\boldsymbol{r}$, is an important symmetry operation. Physical quantities behave differently under S_I, either they are invariant ("symmetric", or "even") with the signature $\epsilon_S = +1$ or they change sign (they are "antisymmetric" or "odd") with $\epsilon_S = -1$. The signatures $\epsilon_S = \pm$ are the only possibilities since when inversion is applied twice, the original state is regained, $S_I^2 = +1$.

(True) scalar quantities have $\epsilon_S = +1$, e.g., the variables ρ, σ, and ε, and also the conjugates T and μ. (True) vectors are odd with $\epsilon_S = -1$, in particular, the polarization and the polar preferred direction, electric fields, and the gradient ∇_i, as well as the velocity-type quantities like g_i, v_i, but also a relative velocity w_i. There is a different kind of vectors, called *axial vectors*, that are even under S_I with $\epsilon_S = +1$. Among them, there is the vorticity $2\omega_i = \epsilon_{ijk}\nabla_j v_k$, the magnetization M_i, and any magnetic field H_i. Relative rotations Ω_i are also axial vectors. Slightly more complicated is the case of nematic order. Usually the director n_i is treated as a vector with the additional constraint that all equations involved are invariant under the replacement $n_i\to-n_i$. Obviously, the director cannot be a polar vector and has to be treated as an axial one, as a necessary condition. We note that the $n_i\to-n_i$ invariance is a stronger constraint.

Since the energy density in Eq. (21.1) has $\epsilon_S = +1$, the conjugates of A and C class variables must have equal signatures, while for B class variables, the S_I behavior of the conjugates is opposite to that of the variables due to the gradient involved. For the director, it means ψ^n_{ij} has $\epsilon_S = -1$ but h^n_i has $\epsilon_S = +1$. Regarding the dynamics, Eqs. (21.3)–(21.8), the time t is S_I symmetric, and therefore, the quasi-currents of B and C class variables have the same spatial signature as the appropriate variables. For A class variables, this is opposite due to the gradient involved.

Depending on the symmetry of a given system, it can be S_I symmetric ("centrosymmetric") or not. Prominent examples for the latter are *chiral phases*, like cholesteric liquid crystals. Such phases show optical activity (rotation of the plane of polarization of light) that is described by a pseudoscalar quantity q_0 with $\epsilon_S = -1$. The use of q_0 in the macroscopic dynamic equations exactly characterizes those contributions and couplings that are specific for the chiral system and that would be forbidden in a achiral phase without a q_0. Well-known examples are the linear twist term in the Frank energy in cholesterics that give rise to the helical ground state and the static couplings between director rotations and, e.g., the thermal degree of freedom leading to the static Lehmann effect [20], absent in the achiral nematic phase. The origin of the existence of a q_0 can be microscopic, when chiral molecules are present, or it is a result of a complicated internal structure like in the smectic C_{B2} phase [21]. In the latter case, q_0 and $-q_0$ lead to two chiral structures with opposite handedness but with the same energy for the ground state: ambidextrous chirality [22].

The existence of a polar vector p_0 also leads to a noncentrosymmetric phase, like polar nematics [23, 24]. Again, the occurrence of p_0 denotes those contributions that would not be allowed in a centrosymmetric phase. Such a phase is not chiral.

A third possibility to break S_I symmetry in a phase is due to the existence of *tetrahedral (or octupolar) order* described by a third-rank tensor order parameter $T_{ij} = \sum_{\alpha=1}^4 n_i^{(\alpha)} n_j^{(\alpha)} n_k^{(\alpha)}$, where the $n_i^{(\alpha)}$ are the position vectors of the corners of a tetrahedron [25]. T_{ijk} has $\epsilon_S = -1$ since the $n_i^{(\alpha)}$ are true vectors. The hydrodynamic variables related to tetrahedral order are the 3-D rigid rotations $d\Gamma_i \sim \epsilon_{ipq} T_{pkl} dT_{qkl}$ of the T_{ijk} structure. This is a class B variable with $\psi^B_k d\nabla_k B \to \psi^\Gamma_{ik} d\nabla_k \Gamma_i$ and

$$(\partial/\partial t)\Gamma_i + Y_i^\Gamma = 0 \tag{21.9}$$

The variable Γ_i is S_I symmetric as is the quasi-current Y_i^Γ , while ψ^Γ_{ik} is odd. Similar to the nematic case, one can introduce a h_i^Γ that contains $\nabla_j \psi^\Gamma_{ij}$ and is even under S_I. There are subtleties in the nonlinear regime due to the non-Abelian nature of three-dimensional rotations, for details refer the study by Brand and Pleiner [26]. A phase that only contains tetrahedral order is neither chiral nor polar. If there is additionally nematic order (along a 2-fold symmetry axis of the tetrahedron), a D_{2d} phase occurs that is still not chiral (and not polar) but shows *ambidextrous helicity* [27]. Only when the tetrahedral structure is suitably combined with a biaxial nematic order, a chiral (but still nonpolar) phase D_2 can occur, where the q_0 is due to the different preferred directions involved [28]. Again, this is an example for *ambidextrous chirality*.

21.1.3 Time reversal symmetry

Under time reversal T_R, where $t \to -t$, all macroscopic quantities are either even (T_R symmetric) with $\epsilon_T = +1$ or they are odd (T_R antisymmetric) with $\epsilon_T = -1$ since $T_R^2 = +1$. There is no need to discriminate among the A,B,C case variables because the gradient is T_R symmetric, $\epsilon_T(\nabla_i) = +1$. The energy density is T_R symmetric, and therefore, the conjugates have the same T_R signature as the variables, $\epsilon_T(e) = \epsilon_T(C_e)$ for any variable e with conjugate C_e. Examples for T_R symmetric variables are ρ, σ, n_i, p_i, U_{ij}, and Ω_i, as well as their conjugates μ, T, h_i^n, h_i^p, Φ_{ij}, and L_i^Ω. Odd quantities with $\epsilon_T = -1$ are g_i and all types of velocities, including relative velocities and vorticity, M_i, and h_i^M, as well as magnetic fields.

T_R is closely related to the *second law of thermodynamics*. The latter states that reversible, time reversal symmetric processes with $\epsilon_T = +1$ must not change the entropy – they are nondissipative, while irreversible processes with $\epsilon_T = -1$ must increase the entropy – they are dissipative. Therefore, one can uniquely split up any current or quasi-current into a reversible (superscript R) and an irreversible part (superscript D), e.g., for a variable e, the following equation can be written:

$$(\partial e/\partial t) + Y_e^R + Y_e^D = 0. \tag{21.10}$$

If time is reversed, Eq. (21.10) reads $(\partial e/\partial t) + Y_e^R - Y_e^D = 0$. independent of the signature $\epsilon_T(e)$.

Thus, $\epsilon_T(Y_e^R) = \epsilon_T(e)$ and $\epsilon_T(Y_e^D) = -\epsilon_T(e)$. In particular,

$$\epsilon_T\left(Y_e^R\right) = \epsilon_T(\partial e/\partial t) = \begin{cases} -1 & \text{for } T_R - \text{even } e \\ +1 & \text{for } T_R - \text{odd } e \end{cases} \tag{21.11}$$

$$\epsilon_T\left(Y_e^D\right) = -\epsilon_T(\partial e/\partial t) = \begin{cases} +1 & \text{for } T_R - \text{even } e \\ -1 & \text{for } T_R - \text{odd } e \end{cases} \tag{21.12}$$

Among the reversible currents, σ_{ij}^R ,Y_i^{MR} are T_R even with $\epsilon_T = +1$, while $j_i^{\rho R}$, Y_i^{nR}, and $X_{ij}^{(el)R}$ are odd, with $\epsilon_T = -1$. On the other hand, the dissipative currents σ_{ij}^D, Y_i^{MD} have $\epsilon_T = -1$, while $j_i^{\rho D}$, Y_i^{nD}, and $X_{ij}^{(el)D}$ are even, $\epsilon_T = +1$.

We add the entropy balance equation (which is a conservation law in the reversible case) and the energy density conservation law to obtain the following equations:

$$\partial\sigma/\partial t + j_i^\sigma = 2R/T \tag{21.13}$$

$$\partial\varepsilon/\partial t + j_i^\varepsilon = 0 \tag{21.14}$$

with the dissipation function R as the entropy source term. Putting all macroscopic dynamic equations, including Eqs. (21.3)–(21.8), into the Gibbs relation, Eq. (21.1), one gets a representation of the dissipation function in terms of the dissipative currents and thermodynamic forces

$$\begin{aligned} 2R = -j_i^{\sigma D}\nabla_i T - \sigma_{ij}^D A_{ij} + Y_i^{nD} h_i^n + Y_i^{MD} h_i^M \\ + X_{ij}^{(el)D}\Phi_{ij} + X_i^{\Omega D} L_i^{\Omega} + Y_i^{\Gamma D} h_i^{\Gamma} > 0 \end{aligned} \tag{21.15}$$

Note that for case C variables, the thermodynamic forces are just the conjugates, while for case A variables (the conserved variables), the forces are the gradients of the conjugates $2A_{ij} = \nabla_j v_i + \nabla_i v_j$. Case B variables (symmetry variables) have been written in the form of class C variables, e.g., with h_i^n , the conjugate to dn_i, as force, rather than using ψ_{ij}^n, the conjugate to $d\nabla_j n_i$. There is no dissipative contribution $\sim j_i^{\rho D}$ since the mass density current is the momentum density $\rho v_i = g_i$, which is reversible.

In Eq. (21.15), pure divergence terms have been omitted since they do not contribute to the total entropy production $\sim \int R dV$ in the bulk. R is even under T_R since in the opposite case, one would get $R < 0$ for time reversed dissipative processes in contradiction to thermodynamics.

The dissipation function R can be used as a potential for dissipative currents. Within *linear irreversible thermodynamics*, with a linear relationship between currents and forces, one can set up phenomenologically R as a harmonic function of the thermodynamic forces. The dissipative currents then follow by partial derivation of R with respect to the forces. Examples are $j_i^{\sigma D} = -\partial R/\partial \nabla_i T$, $Y_i^{nD} = \partial R/\partial h_i^n$, and $X_i^{(el)D} = \partial R/\partial \Phi_{ij}$. If R contains nonvanishing Φ_{ij} contributions, e.g., $\sim \Phi_{ij}\Phi_{kl}$, the resulting $X_{ij}^{(el)D}$ describes relaxation of the strain tensor. This is the hydrodynamic description of *viscoelasticity* of non-Newtonian fluids or relaxing gels. In solids or chemically bonded elastomers, only gradients of the elastic stresses act as forces, e.g., $R \sim (\nabla_j \Phi_{ij})(\nabla_l \Phi_{kl})$, and the strains behave as conserved quantities ("permanent elasticity") showing diffusion rather than relaxation.

Generally, the dissipative currents are related to the forces by linear relations

$$Y_a^D = \zeta_{ab}^D F_b \tag{21.16}$$

where a and b denote the variables involved. By the very existence of the potential R, ζ_{ab}^D is symmetric, $\zeta_{ab}^D = \zeta_{ba}^D$. Of course, such relations have to fulfill the T_R symmetry properties discussed above, in particular, $\epsilon_T(\zeta_{ab}^D) = \epsilon_T(ab)$. If the (dissipative) transport parameter ζ_{ab}^D is constant or only depends on structural properties that are T_R symmetric, like the nematic director, only variables of equal T_R signature can couple, $\epsilon_T(a) = \epsilon_T(b)$. On the other hand, if there is a T_R-odd quantity present, like the magnetization or a magnetic field, an odd power of them in ζ_{ab}^D leads to $\epsilon_T(\zeta_{ab}^D) = -1$, and the variables a and b must behave oppositely $\epsilon_T(a) = -\epsilon_T(b)$.

For the *reversible dynamics*, there are somewhat different rules. Since the reversible currents must not contribute to R, they have to fulfill the following equation:

$$0 = -j_i^{\rho R}\nabla_i\mu - j_i^{\sigma R}\nabla_i T - \sigma_{ij}^R A_{ij} + Y_i^{nR}h_i^n + Y_i^{MR}h_i^M + X_{ij}^{(el)R}\Phi_{ij} + X_i^{\Omega R}L_i^{\Omega} \tag{21.17}$$

replacing Eq. (21.15) for the dissipative currents. Since the bilinear products of thermodynamic forces and reversible currents add up to zero, there is no potential, from which the latter can be derived.

The reversible currents come in two parts, $Y_a^R = Y_a^0 + \tilde{Y}_a^R$, where Y_a^0 are the non-phenomenological, material-independent, symmetry-given contributions, and $\tilde{Y}_a^R$ are the phenomenological parts characterized by (reversible) transport coefficients. The former comprise advection, convection, and rotational covariant derivatives, in addition to, e.g., the isotropic pressure p and the Ericksen stress in the stress tensor, and read as follows [2, 16, 29]:

$$j_i^{\rho 0} = \rho v_i \tag{21.18}$$

$$j_i^{\sigma 0} = \sigma v_i \tag{21.19}$$

$$j_i^{\varepsilon 0} = (\varepsilon + p)v_i \tag{21.20}$$

$$\sigma_{ij}^0 = g_i v_j + \delta_{ij}p - \Phi_{ij} + \Phi_{jk}U_{ik} + \Phi_{ik}U_{jk} + \frac{1}{2}\left(\psi_{ki}^n\nabla_j n_k + \psi_{kj}^n\nabla_i n_k\right) \tag{21.21}$$

$$Y_i^{n0} = v_k\nabla_k n_i + \epsilon_{ijk}n_j\omega_k \tag{21.22}$$

$$Y_i^{M0} = v_k\nabla_k M_i + \epsilon_{ijk}M_j\omega_k \tag{21.23}$$

$$X_{ij}^{(el)0} = v_k\nabla_k U_{ij} - A_{ij} + U_{kj}\nabla_i v_k + U_{ki}\nabla_j v_k. \tag{21.24}$$

with $p = -\varepsilon + \mu\rho + T\sigma + v_i g_i + h_i^M M_i$ containing all extensive variables and their conjugates. The stress tensor has been symmetrized in order to guarantee angular momentum conservation using the condition that Eq. (21.1) is invariant under rotations and that a divergence of an antisymmetric term is irrelevant [16]. All non-phenomenological contributions together fulfill Eq. (21.17).

Concerning the phenomenological parts, we first note that $\tilde{j}_i^{\rho R} = 0$ because $(\partial\rho)/(\partial t) + \nabla_i g_i = 0$. For the other currents, one has to set up the linear current-force relations explicitly

$$\tilde{Y}_a^R = \zeta_{ab}^R F_b \tag{21.25}$$

In accordance with Eq. (21.17), there must be a counter term $\tilde{Y}_a^R = \zeta_{ab}^R F_a$ such that $\zeta_{ab}^R + \zeta_{ba}^R = 0$ indicating antisymmetry with respect to a and b. Of course, the relation Eq. (21.25) has to be invariant under T_R with the result $\epsilon_T(\zeta_{ab}^R) = \epsilon_T(\tilde{Y}_a^R F_b)$ or $\epsilon_T(\zeta_{ab}^R) = -\epsilon_T(ab)$ due to $\epsilon_T(F_b) = \epsilon_T(C_b) = \epsilon_T(b)$. If ζ_{ab}^R is constant or only depends on structural properties that are T_R symmetric, like the nematic director, only variables of opposite T_R signature can couple, $\epsilon_T(a) = -\epsilon_T(b)$. On the other hand, if there is a T_R-odd quantity present, like the magnetization or a magnetic field, an odd power in ζ_{ab}^R leads to $\epsilon_T(\zeta_{ab}^R) = -1$, and variables a and b of the same T_R signature can couple. In such systems, it is possible to have dissipative as well as reversible self-couplings (or cross-couplings between two given variables), when the dissipative and reversible transport parameters have different T_R signatures. An example is (isotropic) heat conduction in ferromagnetic systems, where $j_i^{\sigma R} = -\kappa \nabla_i T$ and $\tilde{j}_i^{\sigma R} = -\kappa^T \epsilon_{ijkM_k} \nabla_j T$ [12]; we note that the latter contribution also exists when linearized since in a ferromagnetic system $\langle M_i \rangle \neq 0$.

Time reversal symmetry is often overlooked and sometimes mistreated as shown in the study by Brand et al. [1, 11].

21.1.4 Spatial inversion symmetry: dynamics

Of course, spatial inversion S_I is relevant for the dynamics as well. From Eqs. (21.3)–(21.8), we deduce that the physical currents of class A variables (e.g., j_i^ρ and σ_{ij}) have an S_I signature opposite to that of the variables, $\epsilon_S(j_a) = -\epsilon_S(a)$, while for all other variables, the quasi-currents have $\epsilon_S(Y_a) = +\epsilon_S(a)$. This applies to the reversible parts, as well as well as to the irreversible ones. Writing the second law of thermodynamics in the form of Eqs. (21.15) and (21.17), the forces for class A variables are gradients of the conjugates (defined in Eq. (21.1)) with the result $\epsilon_S(F_a) = -\epsilon_S(C_a) = -\epsilon_S(a)$, while for the other variables, the forces are just the conjugates resulting in $\epsilon_S(F_a) = \epsilon_S(C_a) = \epsilon_S(a)$.

For the dissipative and reversible transport parameters $\zeta_{ab}^{(D,R)}$ defined in Eqs. (21.16) and (21.25), this implies $\epsilon_S(\zeta_{ab}) = +\epsilon_S(ab)$ for all types of variables. Thus, if ζ_{ab} is a constant or only depends on structural properties that are S_I symmetric, like the nematic director, only variables of equal S_I signature can couple, $\epsilon_S(a) = +\epsilon_S(b)$. On the other hand, if there is a S_I- odd quantity present, like the pseudoscalar q_0 or the tetrahedral order T_{ijk}, an odd power of those in ζ_{ab} leads to $\epsilon_S(\zeta_{ab}) = -1$, and the variables a and b must behave oppositely $\epsilon_S(a) = -\epsilon_S(b)$. Examples are the dissipative dynamic Lehmann effect in chiral systems, where, e.g., temperature gradients (polar vectors) can couple to director rotations (axial vectors) [11, 30] and flow-induced reversible (heat) fluxes in tetrahedral systems, where the nonpolar A_{ij} couples to the polar vector $j_i^{\sigma R}$ [31].

21.2 Nematic order and magnetism

21.2.1 Isotropic and uniaxial magnetic elastomers and gels

Isotropic ferrogels are systems without a preferred direction combining the properties of a gel and of a conventional magnetic liquid. The hydrodynamic description of isotropic ferrogels has been presented in the study by Jarkova et al. [32]. Quite recently, a continuum model for ferrogels from an engineering perspective has been developed [33]. In parallel microscopic studies of the field-controlled change of shape and elasticity of magnetic gels using particle-based simulations have been advanced; these developments have been elucidated in the study by Weeber et al. [34]. A large effort has been dedicated to the mesoscopic characterization of magnetic hybrid materials such as magnetic gels and elastomers over the last few years, compare, for example, the study by Menzel [35] for a recent review. Important progress along these lines has been achieved for the tunable dynamic moduli of magnetic elastomers on the mesoscale by combining experimental data from X-ray tomography with coarse-grained dipole – spring modeling [36]. From an applied point of view, there has been a special emphasis on the magnetic field–controlled mechanical behavior of magnetosensitive elastomers in applications for actuator and sensor systems [37].

Uniaxial magnetic gels with a permanent magnetic moment have been synthesized by performing the cross-linking process in an external magnetic field [38] and afterward characterized with respect to their optical, magnetic, and mechanical properties [38]. For the hydrodynamic description of uniaxial magnetic gels and elastomers, we refer to the study by Bohlius et al. [39]. In the study by Menzel [40], it has been demonstrated how one can bridge the scales from particles to macroscopic length scales in a uniaxial magnetic gel. Very recently, the structure and the magnetooptical response of anisotropic fibrillous organoferrogels with mobile magnetic nanoparticles have been investigated [41]. While in this study no magnetic hysteresis has been found but an optical hysteresis is detected, revealing a complex interplay between the gel and the mobile magnetic particles [41].

21.2.2 Magnetic nematic elastomers and gels

Nematic elastomers with a magnetic degree of freedom are described by the director, the strain field, relative rotations between director and elastic rotations, and by a magnetization variable that is zero in equilibrium [2]. We concentrate here on the magnetic effects. There are reversible dynamic couplings between director rotations and the magnetization dynamics, $(\partial n_i/\partial t) \sim h_j^M$ and $(\partial M_i/\partial t) \sim h_j^n$, and similarly between relative rotations and the magnetization, $(\partial \Omega_i/\partial t) \sim h_j^M$ and $(\partial M_i/\partial t) \sim L_j^\Omega$, all with reversible material tensors of the form $\sim\epsilon_{ijk}n_k$. The symbol $\sim$ in the relations above indicates that the vectorial quantities on the left and on the right hand side are connected by second-rank tensors. It is the different T_R signature of the variables involved and the antisymmetry of the material tensors that allow these couplings. Applying an external magnetic field, rotations of the director and relative rotations are induced (through h_i^M). Through similar reversible couplings, external shear flow induces relative rotations.

In ordinary nematics flow alignment of the director is a prominent feature: shear flow rotates the director at a finite angle w.r.t. the flow direction. This angle is independent of the shear rate Σ and depends on a reversible flow parameter. In the present case, shear flow also induces relative rotations, which induce a magnetization, and finally, all three variables, n_i, Ω_i, and M_i are oriented, independent of the shear rate. Only the magnitude of the magnetization and the relative rotations are proportional to Σ. Since the system is elastic, the shear rate has to be oscillatory, $\Sigma\cos(\omega_s t)$. Since there is static coupling between relative rotations and strains, the relative rotations show a phase shift compared to the shear rate indicating that in this complicated system, flow alignment is no longer a pure reversible feature.

If a chiral agent is added or a chiral nematogenic molecule is used, a chiral magnetic nematic elastomer or gel is obtained [3]. The only new structural element compared to the achiral case is the pseudoscalar q_0 that is odd under S_I. A well-known consequence is the occurrence of a linear (nematic) twist term in the energy density functional, giving rise to a helical ground state, as well as bilinear coupling terms between twist and temperature and density changes and strains. The latter describe the temperature, density, and strain dependence of the helical pitch. They also give rise to the static part of the Lehmann effect. Note that there are no static couplings to changes of the magnetization due to the odd T_R signature of the latter.

In the dynamics, the presence of q_0 allows couplings between polar currents and axial forces (and vice versa) like, e.g., $j_i^{\sigma R} \sim q_0 h_j^M$ and $Y_i^{MR} \sim q_0 \nabla_j T$. Applying a temperature gradient along the helical axis, $\nabla_z T$, the growth of M_z is induced that saturates into a finite stationary magnetization, $M_z^{(stat)} \sim q_0 \nabla_z T$. Instead of the temperature gradient, also an external electrical field can be used, thus creating an electric field–induced magnetization [3]. The elastic degree of freedom is irrelevant for these effects.

21.2.3 Ferromagnetic nematics

Ferromagnetic nematic liquid crystals are fluids with two different kinds of internal order, a nematic one due to the ordering of standard nematogenic molecules and a ferromagnetic one due to the ordering of magnetic nanoplatelets made of complicated iron oxide compounds. Such suspensions have been realized by Mertelj et al. [42, 43]. In equilibrium, the two preferred directions n_i and m_i are parallel. Thus, there is an energetic penalty, $A_2[(n{\cdot}m)^2{-}1]$, for deviations from the ground state. In addition, there are various (reversible and irreversible) dynamic couplings of these two variables. The full hydrodynamic description for ferromagnetic nematics has been given in the study by Potisk et al. [9], which is a generalization of an earlier work [44] on "ferronematics", i.e., magnetic nematic liquid crystals without permanent magnetization. From a microscopic point of view, the study of the structure and rheology of hybrid mixtures of magnetic nanoparticles in liquid crystals using particle resolved simulations has started a few years ago. Quite recently, the focus in this area has been on the transport properties in liquid crystal-magnetic colloid mixtures [45].

A particularly interesting cross-coupling is the (simplified) dissipative dynamics [8, 44]

$$Y_i^{nD} = \frac{1}{\gamma_1} h_i^n + \chi_{ij}^D h_j^m \tag{21.26}$$

$$Y_i^{mD} = b_\perp^D h_i^m + \chi_{ji}^D h_j^n \tag{21.27}$$

with the symmetric material tensor $\chi_{ij}^D = \chi_2^D m_k n_k (\delta_{ij} - n_i n_j)$. The crucial point for its existence is that χ_{ij}^D is odd under T_R, as well as under the replacement $n_i \to -n_i$. For the experiment considered below, the force h_i^n is due to the Frank rotational elastic energy and a surface anchoring energy, while h_i^m describes the orienting force of an external field on a ferromagnet; of course, both forces contain contributions from the A_2 energy given above.

In the studies by Potisk et al. [8, 9], magnetooptic response experiments are described. On a sample of homogeneous structure $n \| m$, a field is applied perpendicularly that rotates the magnetization toward the field direction. Due to the coupling between m and n, also the director starts to rotate, which is hindered by the surface anchoring and the subsequent deformation of the director field. Finally, a stationary state is reached that depends on the field strength. Sending light through the sample (along the field direction), using a polarizer and analyzer, the measured phase shift between the ordinary and extraordinary beams allows to monitor the final director orientation, as well as the switching dynamics. It turns out that the director relaxation (γ_1) and the magnetization relaxation ($b_\perp{}^D$) alone are not sufficient to describe the results, but a nonzero χ_2^D is mandatory. Indeed, χ_2^D turns out to be rather large of the order of γ_1. This is a clear indication that, in such complicated systems, dynamic cross-couplings are important.

There is a reversible counterpart, $\chi_{ij}^R = \chi^R \epsilon_{ijk} n_k$, which is even under T_R and leads to a vanishing entropy production due to the antisymmetry of ϵ_{ijk} . The effect of adding those terms to Eqs. (21.26) and (21.27), not yet detected experimentally, would be an out-of plane rotation of the director [9].

Another area to detect the complexity of ferromagnetic nematics is flow. In ordinary nematics, there are basically two flow effects, viscosity and flow alignment due to a reversible coupling between (symmetric) shear flow and director rotations. In ferromagnetic nematics, there are additional cross-couplings possible due to the odd T_R signature of the magnetization. In particular, there are dissipative couplings between flow and the nematic and magnetic forces, h_i^n and h_i^m, and a reversible one involving h_i^m and a reversible version of the viscous coupling [10]. As a result, the velocity profile in a (simple) shear experiment generally deviates from linearity, when a field is present, and effective viscosities become field dependent increasing by a factor up to two in accordance with experiments [46]. In ordinary nematics, there are 3 Miesowicz viscosities according to the three possibilities to fix the director relative to the shear geometry (along the velocity field, along the velocity gradient, and perpendicular to both). In ferromagnetic nematics, there are nine different ways to fix independently (by different external fields) the orientation of n and m relative to shear flow. Due to the various couplings among the variables, these effective viscosities are rather complicated functions of the reversible and irreversible transport parameters involved [10].

21.2.4 Ferromagnetic nematics with tetrahedral order

When tetrahedral order is added to ferromagnetic nematics, three preferred structures exist – the nematic director n_i, the magnetization m_i, and the tetrahedral structure T_{ijk} (for the latter, refer the study by Fel [25] and the discussion around Eq. (21.9)). The interesting point in this situation is that m_i and T_{ijk} have opposite symmetry properties,$\epsilon_S = +1$ $\epsilon_T = -1$, and $\epsilon_S = -1$, and $\epsilon_T = +1$, respectively. We will only consider the case where the director is fixed to be parallel to one of the 2-fold symmetry axes of T_{ijk}. This is the structure of a D_{2d} phase [26]. The hydrodynamic variables are the rotations of the director and the rotation of the tetrahedral structure about the director. Rotations of the magnetization are independent degrees of freedom, but due to an energetic coupling between n_i and m_i, the two directions are parallel in equilibrium, and the phase is uniaxial. For the full hydrodynamics of such a phase, refer the study by Potisk et al. [12].

In the statics, the most prominent feature in the D_{2d} phase is the linear gradient term in the energy density $\sim T_{ijk} n_i \nabla_j n_k$. It favors helical structures of n_i and T_{ijk} (about a second, perpendicular 2-fold tetrahedral symmetry axis). For the magnetization, there is an analogous linear gradient term, $\sim T_{ijk} m_i \nabla_j m_k$. The sum of the two linear terms favors a combined helical structure of n_i and m_i (and T_{ijk}), thereby preserving n_i and m_i remaining parallel. Since the phase is achiral (there is no pseudoscalar quantity present), this is ambidextrous helicity.

In the reversible dynamics, there are very special couplings that only exist since all three ordered structures are involved. An example is the coupling between the thermal degree of freedom and director rotations, $j_i^{\sigma R} = \xi_{ij}^{Tn} h_j^n$ and $Y_i^{nR} = \xi_{ji}^{Tn} \nabla_j T$. Here, ξ_{ji}^{Tn} contains linearly n_k, m_l, and T_{pqr}, thereby accommodating the $n_i \to -n_i$ invariance, ensuring the reversibility of the currents (due to m_i), and compensating for the polarity of the heat current and the temperature gradient (due to T_{pqr}). Since ξ_{ji}^{Tn} is symmetric, the two contributions to Eq. (21.17) cancel as required.

In the dissipative dynamics, the magnetization couples, e.g., to the thermal degree of freedom, $j_i^{\sigma D} = \psi_{ji}^{TD} h_j^m$ and $Y_i^{mD} = -\psi_{ji}^{TD} \nabla_j T$. Here, ψ_{ji}^{TD} contains linearly m_l (ensuring the irreversibility of the currents) and T_{pqr} (compensating for the polarity of the heat current and the temperature gradient). Since ψ_{ji}^{TD} is symmetric, the two contributions to Eq. (21.15) are equal and add up as required. They describe that rotations of the magnetization create heat currents, while temperature gradients drive the dynamics of the magnetization.

21.3 Elasticity with nonnematic order

In this section, we discuss various elastic systems that show either a polar preferred direction or an ordered tetrahedral structure – finally together with a permanent magnetization. Differences can be traced back to the different symmetry signatures of the order parameters involved.

21.3.1 Polar and active polar gels

If in a nematic liquid crystal the director n_i is replaced by a polar vector p_i, the direction of a permanent polarization, a polar (nematic) phase results. Due to the different symmetry properties, $\epsilon_S(n_i) = +1$ and $\epsilon_S(p_i) = -1$, the polar phase behaves differently compared to the nematic one. In particular, polar phases tend to form splay textures [23] due to a linear splay term $\nabla_i p_i$ in the energy functional. Neither in nematics nor in ferromagnetic systems linear splay ($\nabla_i n_i$ or $\nabla_i m_i$) is possible.

In the study by Brand et al. [5], the hydrodynamics of polar gels is given. It turns out that the reversible dynamics is isomorphic to that of nematic gels [47] and need not to be repeated here. In the statics, there is the standard piezoelectric coupling between polarization and strain. Here, in the uniaxial case, it contains three static susceptibilities. There is another static coupling between polar textures and elasticity, $\psi_{ij}^p \sim U_{kl}$ and $\Phi_{ij}^{(el)} \sim \nabla_k p_l$, with $\psi_{ij}^p \equiv (\partial\varepsilon)/(\partial\nabla_i p_j)$. Relative rotations between p_i and the network, $\Omega_i \equiv \delta p_j \Omega_{ij} - \delta p_i$, couple to bend distortions of the polarization, $L_i^\Omega \sim \delta_{ik}^\perp\, p_j \nabla_j p_k$ and $\psi_{ij}^p \sim \delta_{ik}^\perp\, p_j \Omega_k$. This effect is not possible in nematic gels (because of the $n_i \to -n_i$ invariance) and neither in ferrogels (due to $\epsilon_T(m_i) = -1$).

Variants of polar ordered systems are *active polar* ones, where biological entities (e.g., fish schools, bird flocks, locust swarms, bacteria, etc.) move coherently in a specific direction, without a head to tail symmetry. The preferred direction is given by the velocity, $F_i = F f_i$, of the active entities. If the movement stops, also the order vanishes. To maintain the active velocity, the entities have to provide energy. This energetic intake is dissipated in the system and defines a nonequilibrium state with $F = F_0$. Nevertheless, it can be described in the usual hydrodynamic way just adding the driving force [18].

In the study by Pleiner et al. [4], we investigate an active polar, viscoelastic system with a relaxing elasticity, as, e.g., occurs in movements of bacteria colonies in a gel background. The additional variables are the relaxing strain tensor U_{ij}, rotations of the preferred direction, δf_i with $f_i \delta f_i = 0$, relaxation of F (toward F_0), and relative rotations between f_i and the network, $\Omega_i = f_j \Omega_{ij} - \delta f_i$. Note that Ω_i has the same symmetry signatures as f_i. Apart from the active velocity, there is also a passive one, $v_i^{(1)}$, describing movements of the passive background. In such a two-fluid system, the question arises, with which velocity a given variable is convected or transported (Eqs. 21.18–21.24). It turned out that there is no general principle to answer this question, and generally, those transport or convection velocities are material dependent, containing passive and active parts [48, 49]. In a linearized passive dynamics, the transport terms drop out, but in the active case, with a constant active velocity in the stationary state, even the linearized theory contains transport etc. due to the active velocity, and this active transport is an important part of the dynamics. We mention that sometimes a polarization vector is used to describe active transport [50], but due to the different T_R signature of the polarization compared to a velocity, the transport becomes "irreversible" violating thermodynamics and opening up an unphysical dissipation channel [4].

Results involving the active polar order and the elasticity are coupled relaxations of compressional strains, U_{zz}, with F and shear strains U_{xz} with relative rotations Ω_x, where F relaxes to F_0 and the other quantities to zero. The sound spectrum of a dynamical system is another important aspect of the physics involved. In simple fluids, one has (ordinary) sound with $\omega^2 = c_1^2 k^2$, with ω and k being the frequency and wave number of the linear excitation, respectively. It is purely reversible, and dissipation only enters at higher k-orders. If in addition the active velocity is taken into account, a second sound excitation arises that is coupled to the first one. However, the full sound spectrum is generally no longer invariant under $\omega \to -\omega$, indicating a nonequilibrium situation [18]: In the stationary state, wave propagation parallel or antiparallel to the active velocity is different. The elastic degree of freedom gives, by coupling to flow, another sound-like excitation (shear elastic wave) that couples to the other sound-like excitations. However, in the present case, elasticity is relaxing, and dissipation does not come at higher k powers but is already present to order k^0, which makes the final sound spectrum for this material very complicated [4].

21.3.2 Tetrahedral and ferromagnetic tetrahedral gels

In the study by Brand and Pleiner [6], we consider tetrahedral gels and elastomers, i.e. elastic media with a tetrahedral ordered structure described by the rank-3 tensor T_{ijk}

(already discussed above). A liquid phase that only contains tetrahedral order, T_d, is optically isotropic since the rank-2 dielectric tensor cannot couple to T_{ijk}. When combined with (isotropic) elasticity, a εT_d phase arises. Although the time reversal and space inversion properties of polarization and tetrahedral order are the same, the properties of εT_d are quite different from the polar (nematic) phase. One reason is that $T_{iik} = 0$ and εT_d is apolar. On the other hand, the rank-3 properties of T_{ijk} allow for specific couplings to the elastic tensors, in particular, $j_i^{\sigma D} \sim T_{ijk}\Phi_{jk}^{(el)}$ and $X_{ij}^{(el)D} \sim T_{ijk}\nabla_k T$ describe a dissipative coupling between elasticity and the thermal degree of freedom. The specific aspect of this coupling lies in the geometry – elastic shear stresses induce a heat current perpendicular to the shear plane, and a temperature gradient induces growth of perpendicular elastic shear strains. For a relaxing elasticity, this results finally in a stationary induced strain. Out-of-plane phenomena are characteristic for tetrahedral order.

If chirality is present (due to a pseudoscalar q_0), a T phase occurs in the liquid case [28] and a εT in the elastic case. The combination $q_0 T_{ijk}$ gives a S_I-positive rank-3 tensor that allows, particularly, couplings of the elastic degree of freedom with rotations of the tetrahedral structure, $\delta\Gamma_i$, and with relative rotations, Ω_i, (in the statics and in the dissipative dynamics) and with flow (in the reversible dynamics). Here, $\Omega_i = \epsilon_{ijk}\Omega_{jk} - \delta\Gamma_i$ means rotations of the tetrahedral orientation with respect to the elastic medium.

If a tetrahedral gel or elastomer is in addition ferromagnetic, the different time reversal and space inversion signatures of the tetrahedral structure and the permanent magnetization allow for a very rich hydrodynamic theory [13]. We assume the magnetization to be rigidly connected to one of the 2-fold symmetry axes of the tetrahedral structure. This preferred direction is taken as the z-direction. Relative rotations here mean combined rotations of m_i and T_{ijk} relative to the elastic network.

We concentrate on the interplay of elasticity and magnetization in the presence of tetrahedral order. First, shear strains create (statically) magnetization patterns $\nabla_z m_x \sim U_{yz}$ and $\nabla_z m_y \sim U_{xz}$ with the magnetization perpendicular to the shear plane. Compressional strains lead to magnetization patterns in the perpendicular plane $(\nabla_x m_y + \nabla_y m_x) \sim U_{zz}$. This should not be mixed up with the standard magnetostriction, where compressional strains U_{zz} or $U_{xx} + U_{yy}$ result in changes of the magnitude of the magnetization, $\delta M/M_0$. In addition, there is an indirect coupling of temperature gradients to elastic shear stresses, mediated by relative rotations, such that $\nabla_x T$ and $\nabla_y T$ create $\Phi_{yz}^{(el)}$ and $\Phi_{xz}^{(el)}$, respectively.

In the case of a transient network, a temperature gradient along the preferred axis triggers temporal changes of the strains, which finally result in a stationary elastic shear stress, $\Phi_{xy}^{(el)}$, in the perpendicular plane. This is due to a reversible as well as an irreversible coupling. In addition, there are also planar compressional stationary strains ($\Phi_{yy}^{(el)} = -\Phi_{xx}^{(el)}$) due to reversible couplings.

The transverse directions x and y introduced in the previous paragraphs are not determined without a suitable external force. The system is uniaxial and transversely isotropic, optically. Of course, the tetrahedral structure has a fourfold (improper) symmetry, only in optical measurements this cannot be detected. However, if one applies the external temperature gradient perpendicularly (e.g., $\nabla_x T$), the transverse isotropy is broken, which can be detected optically. Measuring the ratio $\Phi_{yz}^{(el)}/\Phi_{xz}^{(el)}$ of the induced stationary elastic strains, one gets the orientation of the transverse tetrahedral structure relative to the externally defined x-direction [13].

21.4 Magnetorheological fluids

21.4.1 A minimal hydrodynamic model

In the study by Potisk et al. [14], we discuss a simple, minimal model to describe the static and dynamic properties of magnetorheological fluids (MRFs). The main ingredient is transient elasticity (relaxing strain variable) that is induced by an external magnetic field. We model this by assuming the elastic moduli and the relaxation times to be proportional to M^2, where M_i is the magnetization, which is zero in equilibrium and in the absence of an external field. Obviously, this description can only be applied to small and intermediate fields, since for high fields there are saturation effects and the elastic moduli will not grow indefinitely. Important is also magnetostriction that provides a static coupling between elasticity and magnetization. In order to make this magnetorheological model as simple as possible, thermal and concentration effects have been neglected, e.g., disregarding temperature gradients and sedimentation. Isotropy of the material properties is assumed, thereby also eliminating relative rotations as variables.

From the symmetry point of view, this model is like a magnetic, viscoelastic suspension. The most important aspects are not due to symmetry reasons but come from internal material properties, like the tendency of the magnetic particles to form chains (and induce elasticity) in the presence of external fields. Another example for a special internal material property is the field dependence of the viscosity in some magnetic fluids [51]. The influence of this property on the threshold behavior in thermal instabilities has been investigated theoretically in the study by Pérez et al. [7].

The material is assumed to be within two parallel plates with the magnetic field $H_i = \delta_{iz} H_0$ perpendicular to them. First, we have applied this magnetorheological model to external *static shear deformations* (strains) $S_{ij} = S\delta_{iz}\delta_{jx}$. The two forces applied lead to three coupled nonlinear equations for the elastic shear stress, Ψ_{xz}, the magnetization parallel to the field, M_z, and the component M_x perpendicular to the external field (compare Eqs. 21.29–21.31 in ref.[14]). Analysis of these equations shows that all three quantities are nonvanishing as a consequence of the magnetostrictive coupling (in this geometry and with two external forces) of M_x, M_z, and Ψ_{xz} in the energy, as well as in the resulting minimizing equations. The resulting elastic shear stress, Ψ_{xz}, is proportional to H_0^2 and initially grows linearly as a function of S, then reaches a maximum (the "static yield stress") at a field-independent yield strain, and diminishes beyond. This is the unstable region, where elasticity breaks down and gives way to a liquid-like response. The tilt of the magnetization, characterized by a nonzero M_x, increases with S. If a normal pressure is applied additionally, the system becomes more rigid, and the static yield stress (as well as the appropriate yield strain) increases considerably. These effects are qualitatively in accordance with experiments.

Under *external shear flow*,$v_x = \dot{\gamma} z$ with shear rate $\dot{\gamma}$, one has to consider the full dynamic equations for the magnetization, the elasticity, and flow. The most important cross-coupling is between the magnetization and flow. It comes in two parts, the first is the convection term $\epsilon_{ijk} M_j \omega_k$ for $\dot{M}_i$ in Eq. (21.23) and the second is a phenomenological contribution to the magnetization current $Y_i^{MR} \sim c_{ijk}^R A_{jk}$. Both contributions are reversible since c_{ijk}^R contains an odd number of M_i factors. The shear part of the latter, c_2^R, is the analog to the flow alignment parameter in nematic liquid crystals and has been given in the context of ferrofluids in the studies by Müller and Liu [52] and Müller et al. [53]. The counter terms, necessary to guarantee zero entropy production, then enter the full stress tensor, σ_{ij}, in addition to the viscous stress due to the external flow and the elastic stress.

For stationary shear flow, where $\dot{\gamma} = \dot{\gamma}_0$ is a constant, the full stress tensor shows, as a function of $\dot{\gamma}_0$, a steep increase up to, what is called the dynamic yield shear stress, beyond which the increase is much slower. The dynamic yield shear stress increases quadratically with the external field H_0. The behavior at higher $\dot{\gamma}_0$ shows shear thinning and therefore fits better to a Casson or Herschel-Bulkley model [54], rather than to the standard Bingham model.

For oscillating shear flow, $\dot{\gamma} = \dot{\gamma}_0 \cos\omega t$, the linear response of the system is described by a complex shear modulus, $G = G' + iG''$, whose real and imaginary parts describe the reactive (reversible) and dissipative response, respectively. As a function of the frequency, the storage modulus G' increases quadratically until it reaches a constant plateau value. The loss modulus G'' increases linearly up to maximum and decreases until a minimum is reached and finally grows linearly again according to a simple viscous behavior. These features correspond fairly well to experimental findings [55, 56]. The shear modulus $|G|$ can exhibit two plateaus, at low and at intermediate frequencies, related to the two relaxation processes involved, strain relaxation and magnetic relaxation. The Cox-Merz rule, an empirical law relating nonlinear stationary to linear high frequency properties [57, 58], is often fulfilled in polymer dynamics but not in our model of MRFs. The difference is probably due to the fact that the columnar structure is destroyed under large steady shear but not when small amplitude oscillatory shear is applied.

21.4.2 A two-fluid description

In the minimal model of MRFs, we have used only one velocity field to describe the dynamics of all variables. Similar to the case of suspensions of large particles or in the dynamics of phase separation, we allow in a two-fluid description [15] macroscopic movements of the magnetic particles relative to the background fluid. Therefore, we introduce two density variables, ρ_1 and ρ_2 (as in a binary mixture), and two momenta $g_i^{(1)} = \rho_1 v_i^{(1)}$ and $g_i^{(2)} = \rho_2 v_i^{(2)}$. Compared to the minimal model, there are now two new variables, the concentration $\phi = \rho_2/\rho$ and the relative velocity $w_i = v_i^{(1)} - v_i^{(2)}$ that come together with the conserved variables, total density $\rho = \rho_1 + \rho_2$ and total momentum $g_i = g_i^{(1)} + g_i^{(2)}$. The concentration is even under T_R and S_I, while the relative velocity is odd under both.

This two-fluid description is different from the case of active polar gels in Section 21.3.1, where the second velocity is related to a broken symmetry and hence a class B variable. Here, the relative velocity w_i is simply a slowly relaxing variable of class C with the conjugate h_i^w ($\mu_C dC \to h_i^w dw_i$) and the balance equation $(\partial/\partial t)w_i + Y_i^w = 0$. The conjugate quantity to ϕ is the osmotic pressure Π and the (linearized) dynamic equation reads $\rho_0 (\partial/\partial t)\phi + \nabla_i j_i^\phi = 0$ (for the general case, refer the study by Pleiner et al. [15]). The conjugates h_i^w and have the same symmetry signatures as w_i and ϕ, respectively. The currents j_i^ϕ and Y_i^w are odd under S_I, while the second law of thermodynamics requires $j_i^{\phi R}$ and Y_i^{wD} to be odd under T_R and $j_i^{\phi D}$ and Y_i^{wR} to be even.

The general problem of any two-fluid theory regarding the form of the transport velocities, e.g., how to generalize Eqs. 21.18–21.24, cannot be discussed here but refer the studies by Pleiner and Harden [48, 49]. However, as part of our two-fluid model of MRFs, we choose velocity $v_i^{(1)}$ to transport or convect the variables ρ_1 and $g_i^{(1)}$, while for the magnetization M_i and the strain U_{ij}, we take $v_i^{(2)}$ (and heat is transported by the mean velocity $v_i = g_i/\rho$). In Pleiner et al. [15], we show how this choice fits well into the general scheme without violating any thermodynamic rule, if one requires $v_i^{(c)} = (\rho_1/\rho)v_i^{(2)} + (\rho_2/\rho)v_i^{(1)}$ to be the transport velocity for ϕ and w_i.

In the statics, $w_i \sim h_i^w$ and does not couple to other degrees of freedom (in a linearized description), while the concentration couples to the magnetization and the (trace of the) strain, $\delta\Pi \sim (2\chi_m M_i^0 \delta M_i + \tilde{\chi}_u \delta U_{kk})$, and the counter terms (to make sure ε is a potential) are $h_i^M = 2\chi_m M_i^0 \delta\phi$ and $\Phi_{ij} = 2\tilde{\chi}_u \delta_{ij} \delta\phi$, with $\tilde{\chi}_u = M_0^2 \chi_u$ in accordance with the similar form of the elastic moduli.

In the dynamics, the relative velocity reversibly couples to the magnetization, $Y_i^{wR} = \xi_{ijk} \nabla_j h_k^M$ and $X_i^{MR} = \xi_{kji} \nabla_j h_k^w$, with ξ_{ijk} containing only odd powers of M_i, as well as irreversibly, $Y_i^{wD} = -\nabla_j (\gamma_{jik} h_k^M)$ and $X_i^{MD} = \gamma_{kji} \nabla_j h_k^w$, where γ_{ijk} is even in M_i. The relative velocity also couples to flow $Y_i^{wD} = -\nabla_j (\nu_{jikl}^{(c)} A_{kl})$ and $\sigma_{ij}^D = -\nu_{ijkl}^{(c)} \nabla_l h_k^w$, where $\nu_{ijkl}^{(c)}$ is even in M_i. The concentration couples to magnetization, $j_i^{\phi D} = -\alpha_{kij} \nabla_j h_k^M$ and $Y_i^{MD} = -\nabla_k (\alpha_{ijk} \nabla_j \Pi)$, where α_{ijk} is odd in M_i in order to be dissipative.

A direct comparison with the one-fluid model can be made for a stationary shear flow in the parallel plate geometry. Assuming as boundary conditions at the moving plate for the background flow, $v_x^{(1)} = \Gamma_x$, for the particle phase, $v_i^{(2)} = 0$, and for the strain, $\nabla_z U_{xz} = 0$, one finds a nonzero and asymmetric relative velocity field w_x, whose maximum depends on the external shear strain rate Γ_x in a complicated way. As a result, the flow profile (of $v_x^{(1)}$) in the flow channel is not linear (in contrast to the one-fluid case). In addition, the stress-strain relation, although quite similar to the one-fluid case for small and large Γ_x, differs for intermediate Γ_x values, in particular, the somewhat unphysical overshoot present in the one-fluid case is no longer there. The introduction of a two-fluid description considerably improves the quality of the macroscopic model.

Bridging the gap between MRFs and soft magnetorheological gels, we would like to refer to the recent review of the understanding of the interplay between single particle motion, internal deformation, and matrix properties, in particular, concerning the buckling of chains and the matrix deformations around inclusions [59].

21.5 Summary and perspective

In this compact review, we have first outlined the approach of macroscopic dynamics based on the use of linear irreversible thermodynamics and the behavior under symmetry operations including inversion, time reversal, and rotations. The variables of interest come in three groups: conservation laws, variables associated with spontaneously broken continuous symmetries, and macroscopic variables, which relax on a sufficiently long time scale to be of hydrodynamic interest.

One field covered is the interaction of nematic order with magnetism. It includes the description of isotropic and uniaxial magnetic elastomers and gels, as well as ferromagnetic nematics, as they have become available experimentally a few years ago. Stimulated by this progress, we also investigated the effect of additional tetrahedral/octupolar order. Another related topic is the study of elasticity with nonnematic order including polar and active polar gels, as well as tetrahedral and ferromagnetic tetrahedral gels.

To examine the effect of larger particles in a carrier fluid, we have presented a minimal one-fluid model of MRFs, the results of which are in accord with a large body of experimental literature. To account for segregation effects between the carrier fluid and the particles, we generalized our approach very recently and gave a two-fluid description of MRFs.

As a perspective, we just mention two classes of systems for which the presented approach will be useful. One is the field of active magnetic gels and elastomers. Quite recently, it has become clear that magnetotactic bacteria fall into this class of systems. Another direction for the near future is the study of fluid- or gel-like systems, which show simultaneously magnetic and electric order. We are thinking, for example, of a system, which is ferroelectric and ferromagnetic simultaneously. Candidates are the ferromagnetic nematic liquid crystals already existing with a solvent, which contains ferroelectric particles. One might thus obtain a liquid multiferroic system with three types of order: ferromagnetic, ferroelectric, and nematic.

Acknowledgment: The authors thank the Deutsche Forschungsgemeinschaft for partial support over the last few years via the SPP 1681. The authors take pleasure to thank Tilen Potisk and Daniel Svenšek for the fruitful collaboration within the SPP1681 and Stefan Odenbach for initiating and coordinating this stimulating program.
Author contribution: All the authors have accepted responsibility for the entire content of this submitted manuscript and approved submission.
Research funding: This research was supported by the Deutsche Forschungsgemeinschaft through SPP1681.
Conflict of interest statement: The authors declare no conflicts of interest regarding this article.

References

1. Brand HR, Pleiner H, Svenšek D. Reversible and dissipative macroscopic contributions to the stress tensor: active or passive?. Eur Phys J E 2014;37:83.
2. Brand HR, Pleiner H. Macroscopic behavior of ferronematic gels and elastomers. Eur Phys J E 2014; 37:122.
3. Brand HR, Fink A, Pleiner H. Macroscopic behavior of ferrocholesteric liquid crystals and ferrocholesteric gels and elastomers. Eur Phys J E 2015;38:65.
4. Pleiner H, Svenšek D, Brand HR. Hydrodynamics of active polar systems in a (Visco)elastic background. Rheol Acta 2016;55:857.
5. Brand HR, Pleiner H, Svenšek D. Macroscopic behavior of polar nematic gels and elastomers. Eur Phys J E 2016;39:105.
6. Brand HR, Pleiner H. On the influence of a network on optically isotropic fluid phases with tetrahedral/octupolar order. Eur Phys J E 2017;40:34.
7. Pérez LM, Bragard J, Díaz P, Mancini HL, Laroze D, Pleiner H. Magneto-viscous effect on thermal convection thresholds in an Oldroyd magnetic fluid. J Magn Magn Mater 2017;444:432.
8. Potisk T, Svenšek D, Brand HR, Pleiner H, Lisjak D, Ostermann N, et al. Dynamic magneto-optic coupling in a ferromagnetic nematic liquid crystal. Phys Rev Lett. 2017;119:097802.
9. Potisk T, Mertelj A, Sebastián N, Ostermann N, Lisjak D, Brand HR, et al. Magneto-optic dynamics in a ferromagnetic nematic liquid crystal. Phys Rev E 2018;97:012701.
10. Potisk T, Svenšek D, Pleiner H, Brand HR. Effects of flow on the dynamics of a ferromagnetic nematic liquid crystal. Phys Rev E 2018;97:042705.
11. Brand HR, Pleiner H, Svenšek D. Dissipative versus reversible contributions to macroscopic dynamics: the role of time-reversal symmetry and entropy production. Rheol Acta 2018;57:773.
12. Potisk T, Pleiner H, Brand HR. Dynamic interplay of nematic, magnetic, and tetrahedral order in ferromagnetic nematic phases. Phys Rev E 2018;98:042703.
13. Potisk T, Pleiner H, Brand HR. Influence of tetrahedral order on ferromagnetic gel phases. Eur Phys J E 2019;42:35.
14. Potisk T, Svenšek D, Pleiner H, Brand HR. Continuum model of magnetic field induced viscoelasticity in magnetorheological fluids. J Chem Phys 2019;150:174901.

15. Pleiner H, Svenšek D, Potisk T, Brand HR. Macroscopic two-fluid effects in magnetorheological fluids. Phys Rev E 2020;101:032601.
16. Martin PC, Parodi O, Pershan PS. Unified hydrodynamic theory for crystals, liquid crystals, and normal fluids. Phys Rev A 1972;6:2401.
17. Pleiner H, Liu M, Brand HR. The structure of convective nonlinearities in polymer rheology. Rheol Acta 2000;39:560.
18. Pleiner H, Svenšek D, Brand HR. Active polar two-fluid macroscopic dynamics. Eur Phys J E 2013;36: 135.
19. Menzel A, Pleiner H, Brand HR. Nonlinear relative rotations in liquid crystalline elastomers. J Chem Phys 2007;126:234901.
20. Brand HR, Pleiner H. New theoretical results for the Lehmann effect in cholesteric liquid crystals. Phys Rev A 1988;37:2736.
21. Cladis PE, Brand HR, Pleiner H . Macroscopic properties of smectic C_G liquid crystals. Eur Phys J B 1998;6:347.
22. Cladis PE, Brand HR, Pleiner H. Fluid biaxial banana smectics: symmetry at work. Liq Cryst Today 1999;9:1.
23. Pleiner H, Brand HR. Spontaneous splay phases in polar nematic liquid crystals. Europhys Lett 1989;9:243.
24. Brand HR, Pleiner H, Ziebert F. Macroscopic dynamics of polar nematic liquid crystals. Phys Rev E 2006;74:021713.
25. Fel LG. Tetrahedral symmetry in nematic liquid crystals. Phys Rev E 1995;52:702.
26. Brand HR, Pleiner H. Macroscopic behavior of non-polar tetrahedratic nematic liquid crystals. Eur Phys J E 2010;31:37.
27. Pleiner H, Brand HR. Low symmetry tetrahedral nematic liquid crystal phases: ambidextrous chirality and ambidextrous helicity. Eur Phys J E 2014;37:11.
28. Pleiner H, Brand HR. Tetrahedral order in liquid crystals. Braz J Phys 2016;46:565.
29. Pleiner H, Brand HR. Hydrodynamics and electrohydrodynmaics of liquid crystals. In: Buka A, Kramer L, editors. Pattern formation in liquid crystals. New York: Springer; 1996;15–67 pp.
30. Leslie FM. Some thermal effects in cholesteric liquid crystals. Proc Roy Soc A 1968;307:359.
31. Brand HR, Pleiner H, Cladis PE. Flow properties of the optically isotropic tetrahedratic phase. Eur Phys J E 2002;7:163.
32. Jarkova E, Pleiner H, Müller H-W, Brand HR. Hydrodynamics of isotropic ferrogels. Phys Rev E 2003; 68:041706.
33. Attaran A, Brummund J, Wallmersperger T. Development of a continuum model for ferrogels. J Intell Mater Syst Struct 2017;28:1358.
34. Weeber R, Kreissl P, Holm C. Studying the field-controlled change of shape and elasticity of magnetic gels using particle-based simulations. Arch Appl Mech 2019;89:3.
35. Menzel AM. Mesoscopic characterization of magnetoelastic hybrid materials: magnetic gels and elastomers, their particle-scale description, and scale-bridging links. Arch Appl Mech 2019;89:17.
36. Pessot G, Schümann M, Gundermann T, Odenbach S, Löwen H, Menzel AM. Tunable dynamic moduli of magnetic elastomers: from characterization by X-ray micro-computed tomography to mesoscopic modeling. J Phys Condens Matter 2018;30:125101.
37. Becker TI, Böhm V, Chaves Vega J, Odenbach S, Raikher YL, Zimmermann K. Magnetic-field-controlled mechanical behavior of magneto-sensitive elastomers in applications for actuator and sensor systems. Arch Appl Mech 2019;89:133.
38. Collin D, Auernhammer GK, Gavat O, Martinoty P, Brand HR. Frozen-in magnetic order in uniaxial magnetic gels: preparation and physical properties. Macromol Rapid Commun 2003;24:737.
39. Bohlius S, Brand HR, Pleiner H. Macroscopic dynamics of uniaxial magnetic gels. Phys Rev E 2004; 70:061411.

40. Menzel AM. Bridging from particle to macroscopic scales in uniaxial magnetic gels. J Chem Phys 2014;141:194907.
41. Nadasi H, Corradi A, Stannarius R, Koch K, Schmidt AM, Aya S, et al. The role of structural anisotropy in the magnetooptical response of an organoferrogel with mobile magnetic nanoparticles. Soft Matter 2019;15:3788.
42. Mertelj A, Lisjak D, Drofenik M, Cŏpič M. Ferromagnetism in suspensions of magnetic platelets in liquid crystal. Nature 2013;504:237.
43. Mertelj A, Osterman N, Lisjak D, Cŏpič M. Magneto-optic and converse magnetoelectric effects in a ferromagnetic liquid crystal. Soft Matter 2014;10:9065.
44. Jarkova E, Pleiner H, Müller H-W, Brand HR. Macroscopic dynamics of ferronematics. J Chem Phys 2003;118:2422.
45. Shrivastav GP, Klapp SHL. Anomalous transport of magnetic colloids in a liquid crystal-magnetic colloid mixture. Soft Matter 2019;15:973.
46. Sahoo R, Rasna MV, Lisjak D, Mertelj A, Dahra S. Magnetodielectric and magnetoviscosity response of a ferromagnetic liquid crystal at low magnetic fields. Appl Phys Lett 2015; 106: 161905.
47. Brand HR, Pleiner H. Electrohydrodynamics of nematic liquid crystalline elastomers. Physica A 1994;208:359.
48. Pleiner H, Harden JL. In Nonlinear problems of continuum mechanics, special issue of notices of universities, South of Russia. Nat Sci 2003;46. arXiv:cond-mat/0404134.
49. Pleiner H, Harden JL. General nonlinear 2-fluid hydrodynamics of complex fluids and soft matter. AIP Proc 2004;708:46.
50. Marchetti MC, Joanny JF, Ramaswamy S, Liverpool TB, Prost J, Rao M, et al. Hydrodynamics of soft active matter. Rev Mod Phys 2013;85:1143.
51. Odenbach S. Magnetoviscous effects in ferrofluids Lecture Notes in Physics Monographs. Berlin: Springer; 2002.
52. Müller H-W, Liu M. Structure of ferrofluid dynamics. Phys Rev E 2001;64:061405.
53. Müller O, Hahn D, Liu M. Non-newtonian behaviour in ferrofluids and magnetization relaxation. J Phys Condens Matter 2006;18:S2623.
54. Susan-Resiga D. A rheological model for magnetorheological fluids. J Intell Mater Syst Struct 2009; 20:1001.
55. Lee JH, Lu Q, Lee JY, Choi HJ. Polymer-magnetic composite particles of Fe_3O_4/Poly(o-anisidine) and their suspension characteristics under applied magnetic fields. Polymers 2019;11:219.
56. Arief I, Mukhopadhyay PK. Magnetorheological Payne effect in bidisperse MR fluids containing Fe nanorods and Fe_3O_4 nanospheres: A dynamic rheological study. J Alloys Compd 2017;696:1053.
57. Cox WP, Merz EH. Correlation of dynamic and steady flow viscosities. J Polym Sci 1958;28:619.
58. Müller O, Liu M, Pleiner H, Brand HR. Transient elasticity and polymeric fluids: Small-amplitude deformations. Phys Rev E 2016;93:023114.
59. Auernhammer GK. Magnetorheological gels in two and three dimensions: understanding the interplay between single particle motion, internal deformations, and matrix properties. Arch Appl Mech 2019;89:153.

Andreas M. Menzel* and Hartmut Löwen

22 Modeling and theoretical description of magnetic hybrid materials—bridging from meso- to macro-scales

Abstract: Magnetic gels and elastomers consist of magnetic or magnetizable colloidal particles embedded in an elastic polymeric matrix. Outstanding properties of these materials comprise reversible changes in their mechanical stiffness or magnetostrictive distortions under the influence of external magnetic fields. To understand such types of overall material behavior from a theoretical point of view, it is essential to characterize the substances starting from the discrete colloidal particle level. It turns out that the macroscopic material response depends sensitively on the mesoscopic particle arrangement. We have utilized and developed several theoretical approaches to this end, allowing us both to reproduce experimental observations and to make theoretical predictions. Our hope is that both these paths help to further stimulate the interest in these fascinating materials.

Keywords: ferrogels, magnetorheological elastomers, magnetostriction, mediated interactions, nonlinear stress–strain behavior, scale-bridging

22.1 Introduction

The magnetic hybrid materials that we concentrate on in the following are referred to by various different terms, among them ferrogels, magnetosensitive elastomers, and magnetorheological elastomers. They consist of magnetic or magnetizable particles, typically of colloidal size, that are embedded in a permanently crosslinked, elastic, polymeric medium, possibly swollen by a solvent [1–8]. The particles are too large to migrate through their surrounding elastic environment so that their positions remain permanently fixed. For simplicity, we refer to these materials as magnetic gels and elastomers.

***Corresponding author: Andreas M. Menzel,** Institut für Physik, Otto-von-Guericke-Universität Magdeburg, Universitätsplatz 2, 39106 Magdeburg, Germany; Theoretische Physik II: Weiche Materie, Heinrich-Heine-Universität Düsseldorf, Universitätsstr. 1, 40225 Düsseldorf, Germany, E-mail: a.menzel@ovgu.de

Hartmut Löwen, Theoretische Physik II: Weiche Materie, Heinrich-Heine-Universität Düsseldorf, Universitätsstr. 1, 40225 Düsseldorf, Germany

This article has previously been published in the journal Physical Sciences Reviews. Please cite as: A. M. Menzel and H. Löwen "Modeling and theoretical description of magnetic hybrid materials—bridging from meso- to macro-scales" *Physical Sciences Reviews* [Online] 2020, 5. DOI: 10.1515/psr-2019-0088 | https://doi.org/10.1515/9783110569636-022

Mainly two different types of practical realizations of magnetic gels and elastomers can be distinguished. In the first ones, the spatial particle configurations are unstructured and basically isotropic [9–14]. The second ones contain anisotropic, frequently chain-like arrangements of the discrete particles [10, 14–17]. These aggregates can be generated by applying strong, homogeneous, external magnetic fields during the production process, while the particles are still free to move. In response to the directed magnetization, the anisotropic aggregates form. When in this state the elastic medium is built around the particles, their positions are locked in and the anisotropic structures and aggregates get stored permanently in the materials.

It is the magnetoelastic coupling arising from the combination of the two components, namely the magnetic or magnetizable particles and the elastic polymeric surrounding medium that leads to the fascinating characteristics of these materials. Two particularly outstanding features are the changes in global shape and in the overall mechanical stiffness that can be induced reversibly, from outside, on demand, and while the materials are used by applying homogeneous external magnetic fields. Magnetically inducing changes in shape is called magnetostriction, while magnetically achieving modifications of the mechanical properties is referred to as magneto-rheological effect.

Various theoretical studies were performed on systems containing regular lattice-type and more random spatial arrangements of the particles, as well as anisotropic and chain-like aggregates. These investigations have demonstrated that the positioning of the embedded particles relative to each other qualitatively influences the overall behavior, that is, for instance, whether a sample elongates or contracts along the axis of an applied magnetic field, or whether its mechanical stiffness under a certain type of deformation increases or decreases [18–25]. Accordingly, to understand the macroscopic behavior of these materials, it is of central importance to study their overall properties based on the particle interactions occurring on the mesoscopic scale. This is the major objective of our work.

Following this scope, we first need to develop appropriate tools to describe the interactions between the particles on the mesoscopic level. Especially, this refers to those interactions mediated by the elastic environment, which will be the main subject of Section 22.2. Afterward, we can link the mesoscopic characteristics to the overall material behavior. In this context, we address magnetostrictive and magneto-rheological effects in Section 22.3. Along similar lines, a statistical approach that we developed is overviewed in Section 22.4. Still one step further, we identified relations between the particle configuration and the overall nonlinear stress–strain behavior, which is outlined in Section 22.5. Finally, we include a list of several possible additional extensions of our considerations in Section 22.6.

22.2 Behavior of particulate inclusions in elastic environments on the mesoscopic scale

As a starting point, we need to understand how discrete individual particles behave in magnetic gels and elastomers when they are enclosed by the surrounding elastic medium. External magnetic field gradients impose forces on magnetic or magnetizable inclusions, while magnetic fields may impose torques or induce pairwise magnetic interactions between the particles. In all these cases, the particles push against the surrounding elastic medium and distort it. Elastic distortions are usually long-ranged. Therefore, the configurations of the other particles are affected by the changes in the surrounding elastic medium. More precisely, the other inclusions are displaced and rotated. Since the particles are typically much stiffer than the elastic surroundings, they counteract deformations of their own body shape. This additionally influences the state of distortion of the elastic environment, which in turn through the long-ranged character affects the configuration of all other particles. Therefore, the corresponding interactions between the particles mediated by the elastic surroundings significantly and mutually couple the positions and orientations of all particles to each other.

We considered explicitly the situation of a homogeneous, isotropic, infinitely extended and linearly elastic medium in which magnetizable, rigid and spherical particles are embedded under no-slip surface conditions. Within this framework, we derived explicit analytical expressions for the coupled displacements and rotations of all particles when they are subject to imposed forces and torques [26, 27]. The scheme of solution corresponds to an expansion in the inverse distance between the particles, which we have completed up to including the sixth order [28]. Our characterization was further extended to the overdamped dynamics of well-separated discrete particles in viscoelastic environments [29].

Figure 22.1 shows a set of three magnetizable particles embedded within a soft elastic gel in an experimental system. Rotating a nearly saturating external magnetic field within the plane of the particle configuration and magnetizing the particles accordingly, the magnetic interactions between them are altered. Consequently, the distances between the spherical inclusions change. As demonstrated, corresponding experimentally measured data points agree very well with our theoretical predictions [26]. There was one simultaneous overall fit parameter, the elastic shear modulus of the elastic environment namely. In this way, the suitability of the approach for micromechanical measurements was demonstrated. Together with dynamic extensions [29], our theoretical framework can further facilitate the interpretation of micro- and nanorheological measurements [30].

In a related context, we showed that two nearby particles under strong mutual magnetic attraction can overcome the counteraction of the elastic medium between them and jump into virtual contact [31]. This magnetomechanical collapse is reversible upon decreasing the mutual magnetic attraction. Under oblique or perpendicular

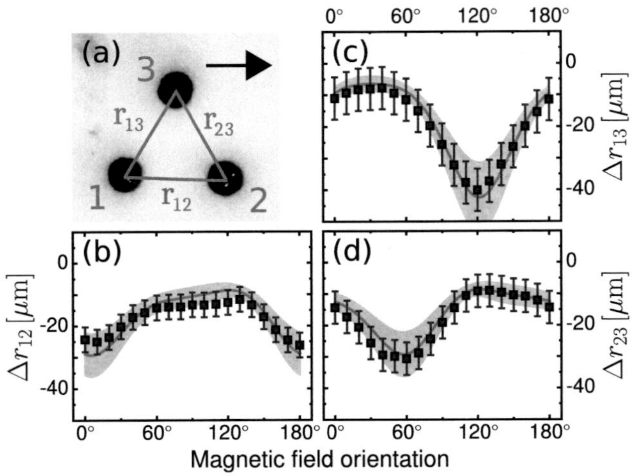

Figure 22.1: Interplay on the mesoscopic scale between the magnetic interactions of discrete particles and the resulting deformations of a surrounding elastic environment. (a) Far away from any boundary, three magnetizable nickel particles of diameters 208.5 ± 2.3 μm were embedded in a very soft elastic gel. An external magnetic field that magnetizes the particles was rotated clockwise in the particle plane, starting from the orientation of the black arrow. (b)–(d) Changes in the distances between the particles result from the modified magnetic interactions because of the field rotation. Data points and error bars represent the experimental measurements, the (red) lines follow from our theory with the shaded areas marking the associated uncertainty regime. From fitting the theory to the experimental data points, the shear modulus of the elastic gel was determined to be 76.3 ± 11.7 Pa. Reprinted figure with permission from Ref. [26]. Copyright 2016 by the American Physical Society (DOI:10.1103/PhysRevLett.117.238003).

magnetic fields, longer linear chains of magnetic particles in very soft magnetic gels showed a characteristic buckling type of distortion with multiple buckles along their contour [32]. Apart from that, in extension of a previous work on the stochastic motion of a particle through an extremely shear-thinning environment [33], we characterized the stochastic motion through shear-thickening surroundings while additionally driven, for instance, by a constant magnetic field gradient [34] [1]. Beyond the passive motion of individual spherical particles, we analyzed the diffusive behavior of more irregularly-shaped passive objects in fluid environments [35], and the consequences of additional self-propulsion in clusters of magnetic dipolar particles [36]. These considerations will be important when in the future the driven or active transport near or through biological membranes [37–39] will be investigated more intensely, for instance motivated by the use of magnetic particles as drug carriers. The distortions arising from thermophoretic effects while heating a particle in a surrounding elastic gel were analyzed as well [40], an aspect that might be relevant during hyperthermic cancer

1 In both studies [33, 34] the stochastic driving forces are rather thought to be imposed from outside with constant absolute value of their averaged squared amplitude than of thermal origin.

treatment. Finally, we also studied to lowest order the interactions mediated between discrete particles by a surrounding elastic medium nearby a rigid either free-slip or no-slip surface bounding the enclosing elastic environment [41].

22.3 From the mesoscopic particle configuration to overall magnetostrictive and magnetorheological effects

Next, to bridge from the mesoscopic positioning of the particles relatively to each other to the overall, possibly macroscopic behavior of the whole system, we concentrated on two different effects characteristic for magnetic gels and elastomers. First, these are magnetostrictive effects, that is, overall distortions when the systems are magnetized by a homogeneous external magnetic field. Second, we refer to magnetorheological effects, implying as a function of the applied magnetic field reversible changes in the global mechanical stiffness.

Concerning our investigations on magnetostriction, we followed the same theoretical path as outlined in Section 22.2. However, instead of considering an infinitely extended system, we now assumed the particles to be embedded in a linearly elastic finite-sized sphere. When the particles are magnetized, leading to each particle carrying for simplicity an identical dipole moment, they distort the surrounding elastic material through the resulting pairwise magnetic attraction or repulsion between them. We additionally included the interactions between the particles mediated by the distortions of their elastic environment to lowest order. Superimposing the contributions of all magnetized inclusions, we calculated the overall deformation of the system [42]. The underlying mathematical expressions are analytical and therefore contain the basically infinite amount of degrees of freedom involved in the distortion of the elastic sphere.

It turned out that the nature of the global deformation is strongly connected to the internal particle arrangement. For example, we evaluated simple cubic, body- and face-centered cubic, and rectangular regular lattice configurations of about one thousand particles with different homogeneous orientations of the magnetization. Whether the sphere elongates or contracts along the magnetization direction depends significantly on the mutual particle positioning, on the orientation of the magnetization axis, and on the value of the Poisson ratio quantifying the compressibility of the elastic material [42]. For randomized particle configurations, we found a tendency of elongation along the magnetization, in agreement with corresponding experimental observations [11, 43].

In order to address the magnetorheological effect, that is, resulting changes in the overall mechanical stiffness upon changes in the magnetization, we resorted to reduced dipole-spring models [44–49]. Each particle is represented by a sphere that again carries the same magnetic dipole moment at its center. To generate an elastic network, the spheres are linked by elastic springs. In this way, the complex elastic composite material is mapped to a discretized magnetoelastic network. The procedure allows to effectively numerically determine equilibrium ground states for different magnitudes of magnetization. For each ground state, the different mechanical moduli quantifying the stress necessary to achieve a certain type of deformation can be calculated in the framework of linear response.

It turned out that whether under weak induced magnetic interactions the mechanical moduli increase or decrease strongly depends on the particle configuration and on the orientation of the magnetization direction. We evaluated various different regular lattices and also randomized particle configurations [45, 47]. Also particle arrangements extracted from real experimental samples by X-ray tomographic means were evaluated [45, 46, 48]. Upon strong magnetization, we found an internal restructuring to set in. During this process, against the elastic restoring forces of the springs, particles collapse toward each other into virtual contact and form chain-like aggregates. We found this effect to be accompanied by a significant increase in the mechanical stiffness, see Figure 22.2, in qualitative agreement with corresponding experimental observations [13]. The dynamic moduli, quantifying the storage and loss parts of the dynamic response of the systems, were evaluated as a function of the magnetization and for different particle arrangements as well [47, 48].

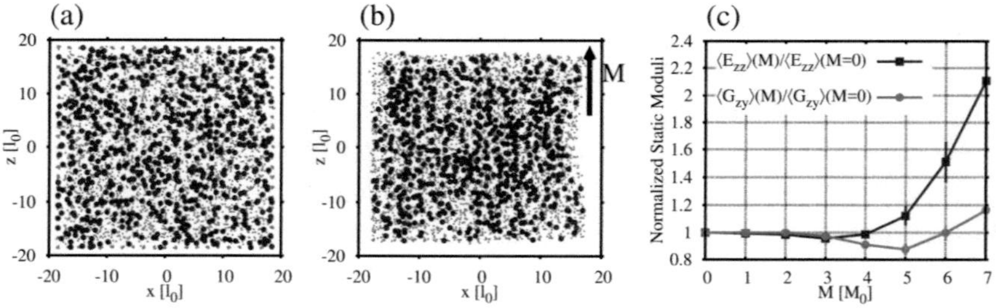

Figure 22.2: Dipole-spring approach to magnetic gels or elastomers describing the associated magnetorheological effect. (a) Corresponding side view, with the darker (blue) dots indicating magnetizable particles, the positions and volumes of which were extracted from a real experimental sample. Lighter (gray) dots mark additional interstitial network points of the elastic mesh of springs. The springs are not indicated for clarity. Here l_0 is a typical length scale set by the particle size. (b) When the system is strongly magnetized, by eye one can identify anisotropic structures arising along the axis of magnetization M. (c) We averaged the moduli E_{zz}, associated with elongations or contractions along the magnetization direction, and G_{zy}, connected to shears with M in the shear plane but displacements perpendicular to M, over several different numerical realizations of the elastic mesh of springs between the particles. For weak (homogeneous) magnetizations, the moduli tend to slightly decrease. They strongly increase when the magnetically-induced internal restructuring becomes significant. M_0 sets the unit of magnetization. The system in (b) was magnetized to $|M| = 7M_0$. Reproduced from Ref. [48]. (DOI:10.1088/1361-648X/aaaeaa).

As an extension, we evaluated the overdamped particle dynamics using dipole-spring systems [46, 49]. First, in qualitative agreement with corresponding experimental observations [50, 51], an increase of the eigenfrequencies of the systems was found for increasing magnetization [46]. However, second, it came as a bit of a surprise that, in spite of the overdamped nature, we could identify a shock-wave type of dynamics when a straight chain-like system contracts or elongates after switching on or off strong magnetic attractions between the particles [49]. The wave fronts separate high- and low-density regions.

Apart from that, it turned out that the investigated dipole-spring models prove similarly useful in the characterization of other types of complex elastic composite materials, particularly electrorheological elastomers [52, 53]. In the future, they may also play a certain role in explicitly linking mesoscopic theoretical characterizations to macroscopic theories [54–58]. To provide such a link in a first step [59], we have calculated from a simple homogeneous mesoscopic model picture the material coefficients of the static part of a macroscopic hydrodynamic theory [60] for anisotropic magnetic gels. In this way, we confirmed and illustrated the nature of the terms in the macroscopic approach and provided explicit expressions of the associated material parameters as functions of the mesoscopic properties.

22.4 Statistical particle-scale characterization and overall material properties

Another way to link the properties of the system from the level of the individual particles to the overall material behavior is given by statistical theories. To this end, we developed a classical density functional theory to describe the magnetorheological effect of a system of elastically linked particles. The statistical approach naturally includes thermal fluctuations in the individual particle positions.

In principle, density functional theory is exact in characterizing equilibrium systems [61, 62]. When solving for the equilibrium state, a particle-scale profile for the noise-averaged density distribution is obtained, related to the probability of finding a particle at a certain position in space. Nevertheless, approximations are often needed in reality to handle the mutual interactions between the particles.

To be able to apply the framework of classical density functional theory to elastic composite materials, we first had to clarify a central conceptual issue [63, 64]. Classical density functional theory was developed to describe liquid states. In the statistical characterizations of monodisperse liquids, the particles can effectively be treated as identical or indistinguishable, the latter at least from an averaging perspective. The situation is genuinely different for a given sample of an elastic composite material like a magnetic gel or elastomer. There, the colloidal particles are trapped by the surrounding elastic matrix and their positions relative to each other are fixed permanently.

This renders the particles distinguishable. From another perspective, when mapping the situation to simple dipole-spring networks, see Section 22.3, the elastic springs between the particles are permanent and neither cut nor reconnected over time. Therefore, each particle is permanently linked to a given set of specific other particles and solely elastically interacts with this subset. Consequently, again we can distinguish the particles by their relative positioning within this elastic interaction network. The solution was to map these discrete elastic spring interactions that distinguish between individual particles to effective potential interactions that do not distinguish between individual particles but lead to the same appearance of the system [63, 64].

So far, we have considered one-dimensional [63] and two-dimensional [64] particle arrangements using effective potential interactions. Changes in the system behavior upon magnetization were evaluated for attractive and repulsive magnetic particle interactions, respectively. We evaluated the associated density profiles and the magnetorheological effect, that is, the changes in the elastic moduli of the overall systems upon magnetization. For moderate amplitudes of magnetization, we found a decrease in the compressive elastic modulus in the one-dimensional configuration under mutual magnetic attraction of the particles [63]. Vice versa, the compressive and shear elastic moduli typically increased under mutual magnetic repulsion in the two-dimensional arrangements [64]. An exception are dense two-dimensional configurations, in which the magnetic repulsion can add to evade hard steric interactions between the particles. Moreover, as expected, we observed a positive magnetostrictive effect for the overall planar system under internal magnetic repulsion, that is, an increase in the system size [64]. For the one-dimensional configurations, we accounted for an additional embedding in a surrounding elastic environment by introducing an additional external anchoring potential [63], leading to the observations depicted in Figure 22.3.

All of our results were tested against Monte-Carlo simulations based on the effective potentials mentioned above and on genuine elastic spring interactions [63, 64]. We found good agreement in the parameter regimes that we concentrated on, see Figure 22.3. The extension from one- to two-dimensional configurations significantly reduced peculiar fluctuation effects that appeared in the one-dimensional arrangement. Further extensions to three-dimensional configurations will certainly allow for a more explicit comparison with actual experimental observations, possibly by using experimentally measured pair distribution functions for the particle arrangement as an input [65].

22.5 Linking the particle scale to the overall nonlinear stress–strain behavior

To establish a bridge from the particulate structures in magnetic gels and elastomers to their overall nonlinear stress–strain behavior, we turned to corresponding simulation

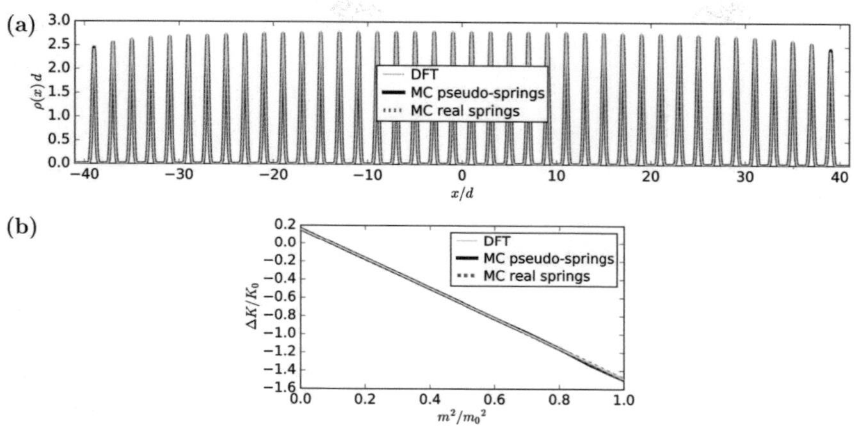

Figure 22.3: Statistical approach to magnetic gels and elastomers and to the magnetorheological effect via density functional theory, here for a one-dimensional model system of particles embedded in an elastic environment. (a) The density profile $\rho(x)$ of the magnetic particles identifies their location, with d the particle diameter. Since thermal fluctuations are included, the individual peaks have finite width. We first mapped the situation to a dipole-spring system ("real springs"), while the springs were then replaced by effective potential interactions ("pseudo-springs"). Monte-Carlo simulations ("MC") confirm the results of the density functional theory ("DFT"). (b) Rescaled changes $\Delta K/K_0$ of the compressive elastic modulus of the one-dimensional particle configuration as a function of increasing rescaled quadratic magnitude of the magnetic moments $(m/m_0)^2$ of the individual particles. The attractive magnetic particle interactions here lead to a decrease in the elastic modulus. Reproduced from Ref. [63]. (DOI:10.1088/1361-648X/aa73bd).

approaches [66, 67]. For this purpose, we considered rigid, finite-sized, spherical magnetic particles that carry permanent magnetic dipoles at their centers. These particles form chain-like aggregates with interparticle gaps of finite distance and are embedded under no-slip surface conditions into an elastic environment. The latter was discretized into a tetrahedral mesh, the elastic distortions of each tetrahedron being quantified by a nearly-incompressible nonlinear elastic deformation energy of the Neo-Hookean type[2]. Resulting systems contained several mutually shifted chain-like aggregates; see Figure 22.4(a) for a typical example system. We numerically clamped the systems at their faces where the chains start and end. Then we stretched the systems by small successive increments, allowing for an equilibration in each step of additional elongation. During this procedure, we recorded the overall forces necessary to achieve a certain amount of stretching. Figure 22.4(b) shows a resulting nonlinear force–strain curve averaged over 20 realizations of the systems.

---

2 There is a misplaced bracket in the expression for the elastic deformation energy density corresponding to the nearly-incompressible Neo-Hookean hyperelastic model as listed in Eq. (22.S1) of the supplementary material of Ref. [66] and in Eq. (22.1) of Ref. [67]. The first term should read $(\mu/2\,\mathrm{Tr}\,\{F^t{\cdot}F\}-3)$ [instead of $\mu/2(\mathrm{Tr}\,\{F^t{\cdot}F\}-3)$]. This is a pure typo and has been treated correctly in the simulations.

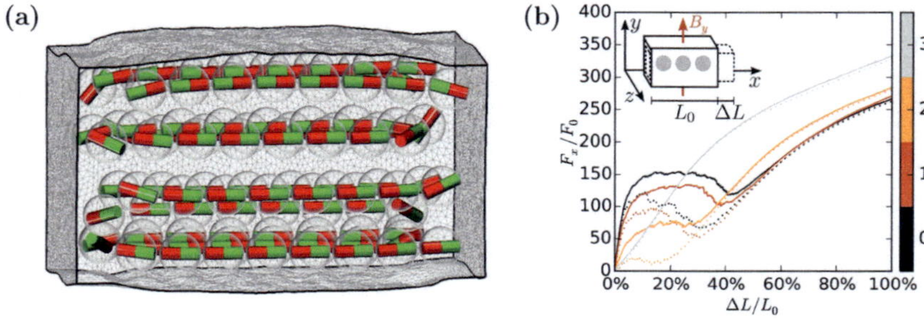

Figure 22.4: Nonlinear stress–strain behavior of numerical systems representing anisotropic magnetic gels or elastomers that contain chain-like particle aggregates. (a) Here, the particles carry permanent dipole moments indicated by the bar magnets that can freely rotate relatively to the particle frames. The surrounding elastic medium embeds the particles under no-slip surface conditions and is discretized into an elastic tetrahedral mesh. (b) Numerically clamping the systems at their faces, stretching them along the chain axes, and plotting the necessary rescaled stretching force F_x/F_0 averaged over several numerical realizations as a function of elongation $\Delta L/L_0$, pronounced nonlinear behavior emerges. This behavior results from the intrinsic coupling between magnetic and elastic properties. Applying a perpendicular magnetic field of rescaled magnitude B_y/B_0, the pronounced intermediate nonlinearity can be altered and even be removed approximately. Reprinted from Ref. [66], with the permission of AIP Publishing (https://doi.org/10.1063/1.4934698).

Our results reveal a pronounced nonlinear stress–strain behavior [66, 67], see the black curve in Figure 22.4(b). This behavior emerges from the interplay between the nonlinear magnetic particle interactions and the elastic properties of the surrounding medium. At low magnitudes of imposed strain, the magnetic particles are close to each other along the chains and any stretching needs to work against the resulting strong magnetic attraction. This leads to the initial steep increase of the curve. At a certain threshold and at the weakest point of all chains, the stretching force is strong enough to increase the gap between two particles and to match the magnetic attraction in magnitude. Once this barrier is mastered, the gap between the particles suddenly increases by a larger amount, because the magnetic attraction decreases with the further increasing distance between the particles. This leads to a drop of the force–strain curve. In fact the plateau emerging at intermediate strains in Figure 22.4(b) represents many successive such drops resulting from corresponding events throughout the system, until all chains are disintegrated.

Moreover, in the situation considered in Figure 22.4, the magnetic dipole of each particle is free to reorient relatively to the particle frame. Thus, when particles are separated from each other along the stretching axis, but lateral particles approach each other, the magnetic moments will flip to minimize the overall magnetic interaction energy. This effect additionally contributes to the appearance of the curves and to a big extent causes the trough that follows the plateau in Figure 22.4(b). Finally, when all aggregates are disintegrated at large strains, mainly the elastic medium is stretched further, which leads to a comparatively regular behavior.

From these considerations, it also becomes plausible that the nonlinear stress–strain behavior can be tuned by external magnetic fields, as for the colored curves in Figure 22.4(b). Strong perpendicular magnetic fields can even switch off the nonlinearity by inducing a perpendicular orientation of the magnetic moments from the start, see the light curve in Figure 22.4(b).

In addition to that, we also analyzed the situation for permanent magnetic dipole moments being anchored to the particle frames [67]. All magnetic moments initially point into the same direction along the chain axes, or they point into opposite directions for different chains. Overall, the effects resulting from the stretching and disintegration of the chains against their internal magnetic cohesion are still at work, in contrast to their unhindered flipping of magnetic moments.

From a more macroscopic point of view, the overall nonlinear response is reminiscent of the nonlinear stress–strain behavior of nematic liquid-crystalline monodomain elastomers [68, 69]. There, likewise, reorientation affects couple to elastic deformations and lead to the overall nonlinear stress–strain response of the systems.

22.6 Conclusions

In summary, we utilized and developed various theoretical tools and approaches to characterize the behavior of magnetic gels and elastomers on the mesoscopic scale of the individual magnetic or magnetizable particles embedded in the permanent elastic polymeric environment, and to connect the corresponding properties to the overall system behavior. Among our approaches were analytical calculations within the framework of linear elasticity theory, discretized dipole-spring models, classical density functional theory, and simulations of the nonlinear system behavior. We concentrated on the evaluation of interactions between the particles mediated by the elastic environment, resulting overall magnetostrictive and magnetorheological effects, that is, magnetically induced overall distortions and changes in the overall mechanical features of the systems, respectively, as well as overall nonlinear stress–strain properties. Several new types of behavior were revealed or quantified in this way.

Nevertheless, our work opens the way for several extensions in the future. On the particle scale, our analytical descriptions can be widened to characterize the interactions mediated by the elastic environment between elongated particles [70]. Furthermore, quantifying particle approaches from finite separation to very close locations still represent a major challenge [31, 71]. Similarly, extensions to maintain the volume of the overall system in the dipole-spring approaches under significant internal restructuring [47, 48] are desirable. Concerning our density functional theories [63, 64], the step towards three-dimensional particle configurations represents a natural next level. Eventually, so far our investigations on the nonlinear stress–strain behavior [66, 67] form a theoretical prediction, and we would be fascinated by corresponding experimental realizations.

Overall, we concentrated on characterizations on the mesoscopic scale and connections to the macroscopic material behavior. However, substantiating the mesoscopic models by scale connections [72] from still more microscopic analyses [73, 74] represents another important direction. Some of the methods developed in the theoretical characterization of magnetic gels and elastomers may also be helpful to investigate related problems in other types of elastic materials, for instance transport through possibly charged polymeric pores [75] or changes in adhesive properties through modified elastic moduli [76]. Finally, we point to another field of growing current interest concerning colloidal composite systems, namely active microswimmers in suspension [3, 77]. Many aspects of the descriptions developed in the present context will also be useful to study details within this related field, particularly when magnetically steerable elastic composite microswimmers are addressed [78].

Acknowledgment: We acknowledge the contributions of many students and postdocs with whom we had the pleasure to collaborate within this topical context and who contributed to our works in this field over the past few years. Among them are Elshad Allahyarov, Sonja Babel, Peet Cremer, Abdallah Daddi-Moussa-Ider, Lukas Fischer, Segun Goh, Borge ten Hagen, Andreas Kaiser, Giorgio Pessot, Mate Puljiz, and Mitsusuke Tarama. Moreover, we thank the German Research Foundation (DFG) for continuous support over the past few years via the SPP 1681, Günter K. Auernhammer, Dmitry Y. Borin, Thomas Gundermann, Christian Holm, Shilin Huang, Karl A. Kalina, Markus Kästner, Johannes Nowak, Stefan Odenbach, Christopher Passow, Malte Schümann, Joachim Wagner, and Rudolf Weeber for the fruitful collaborations within the SPP 1681, as well as Stefan Odenbach for the substantial and ongoing effort in establishing and coordinating this stimulating program.
Author contribution: All the authors have accepted responsibility for the entire content of this submitted manuscript and approved submission.
Research funding: German Research Foundation (DFG) through the SPP 1681, grant nos. ME 3571/3 and LO 418/16.
Conflict of interest statement: The authors declare no conflicts of interest regarding this article.

References

1. Filipcsei G, Csetneki I, Szilágyi A, Zrínyi M. Magnetic field-responsive smart polymer composites. Adv Polym Sci 2007;206:137–89.
2. Ilg P. Stimuli-responsive hydrogels cross-linked by magnetic nanoparticles. Soft Matter 2013;9: 3465–8.
3. Menzel AM. Tuned, driven, and active soft matter. Phys Rep 2015;554:1–45.
4. Odenbach S. Microstructure and rheology of magnetic hybrid materials. Arch Appl Mech 2016;86: 269–79.

5. Lopez-Lopez MT, Durán JDG, Iskakova LY, Zubarev AY. Mechanics of magnetopolymer composites: a review. J Nanofluids 2016;5:479–95.
6. Weeber R, Hermes M, Schmidt AM, Holm C. Polymer architecture of magnetic gels: a review. J Phys Condens Matter 2018;30:063002.
7. Menzel AM. Mesoscopic characterization of magnetoelastic hybrid materials: magnetic gels and elastomers, their particle-scale description, and scale-bridging links. Arch Appl Mech 2019;89:17–45.
8. Weeber R, Kreissl P, Holm C. Studying the field-controlled change of shape and elasticity of magnetic gels using particle-based simulations. Arch Appl Mech 2019;89:3–16.
9. Zrínyi M, Barsi L, Büki A. Deformation of ferrogels induced by nonuniform magnetic fields. J Chem Phys 1996;104:8750–6.
10. Jolly MR, Carlson JD, Muñoz BC. A model of the behaviour of magnetorheological materials. Smart Mater Struct 1996;5:607–14.
11. Gollwitzer C, Turanov A, Krekhova M, Lattermann G, Rehberg I, Richter R. Measuring the deformation of a ferrogel sphere in a homogeneous magnetic field. J Chem Phys 2008;128: 164709.
12. Evans BA, Fiser BL, Prins WJ, Rapp DJ, Shields AR, Glass DR, et al. A highly tunable silicone-based magnetic elastomer with nanoscale homogeneity. J Magn Magn Mater 2012;324:501–7.
13. Schümann M, Odenbach S. *In-situ* observation of the particle microstructure of magnetorheological elastomers in presence of mechanical strain and magnetic fields. J Magn Magn Mater 2017;441:88–92.
14. Schümann M, Gundermann T, Odenbach S. Microscopic investigation of the reasons for field-dependent changes in the properties of magnetic hybrid materials using X-ray microtomography. Arch Appl Mech 2019;89:77–89.
15. Collin D, Auernhammer GK, Gavat O, Martinoty P, Brand HR. Frozen-in magnetic order in uniaxial magnetic gels: preparation and physical properties. Macromol Rapid Commun 2003;24:737–41.
16. Günther D, Borin DY, Günther S, Odenbach S. X-ray micro-tomographic characterization of field-structured magnetorheological elastomers. Smart Mater Struct 2011;21:015005.
17. An H-N, Groenewold J, Picken SJ, Mendes E. Conformational changes of a single magnetic particle string within gels. Soft Matter 2014;10:997–1005.
18. Wood DS, Camp PJ. Modeling the properties of ferrogels in uniform magnetic fields. Phys Rev E 2011;83:011402.
19. Ivaneyko D, Toshchevikov VP, Saphiannikova M, Heinrich G. Magneto-sensitive elastomers in a homogeneous magnetic field: a regular rectangular lattice model. Macromol Theory Simul 2011; 20:411–24.
20. Stolbov OV, Raikher YL, Balasoiu M. Modelling of magnetodipolar striction in soft magnetic elastomers. Soft Matter 2011;7:8484–7.
21. Gong X, Liao G, Xuan S. Full-field deformation of magnetorheological elastomer under uniform magnetic field. Appl Phys Lett 2012;100:211909.
22. Han Y, Hong W, Faidley LE. Field-stiffening effect of magneto-rheological elastomers. Int J Solid Struct 2013;50:2281–8.
23. Zubarev AY. Effect of chain-like aggregates on ferrogel magnetodeformation. Soft Matter 2013;9: 4985–92.
24. Metsch P, Kalina KA, Spieler C, Kästner M. A numerical study on magnetostrictive phenomena in magnetorheological elastomers. Comput Mater Sci 2016;124:364–74.
25. Romeis D, Toshchevikov V, Saphiannikova M. Effects of local rearrangement of magnetic particles on deformation in magneto-sensitive elastomers. Soft Matter 2019;15:3552–64.
26. Puljiz M, Huang S, Auernhammer GK, Menzel AM. Forces on rigid inclusions in elastic media and resulting matrix-mediated interactions. Phys Rev Lett 2016;117:238003.

27. Puljiz M, Menzel AM. Forces and torques on rigid inclusions in an elastic environment: resulting matrix-mediated interactions, displacements, and rotations. Phys Rev E 2017;95: 053002.
28. Puljiz M, Menzel AM. Displacement field around a rigid sphere in a compressible elastic environment, corresponding higher-order Faxén relations, as well as higher-order displaceability and rotateability matrices. Phys Rev E 2019;99:053002.
29. Puljiz M, Menzel AM. Memory-based mediated interactions between rigid particulate inclusions in viscoelastic environments. Phys Rev E 2019;99:012601.
30. Remmer H, Roeben E, Schmidt AM, Schilling M, Ludwig F. Dynamics of magnetic nanoparticles in viscoelastic media. J Magn Magn Mater 2017;427:331–5.
31. Puljiz M, Huang S, Kalina KA, Nowak J, Odenbach S, Kästner M, et al. Reversible magnetomechanical collapse: virtual touching and detachment of rigid inclusions in a soft elastic matrix. Soft Matter 2018;14:6809–21.
32. Huang S, Pessot G, Cremer P, Weeber R, Holm C, Nowak J, et al. Buckling of paramagnetic chains in soft gels. Soft Matter 2016;12:228–37.
33. Menzel AM, Goldenfeld N. Effect of Coulombic friction on spatial displacement statistics. Phys Rev E 2011;84:011122.
34. Menzel AM. Velocity and displacement statistics in a stochastic model of nonlinear friction showing bounded particle speed. Phys Rev E 2015;92:052302.
35. Passow C, Ten Hagen B, Löwen H, Wagner J. Depolarized light scattering from prolate anisotropic particles: the influence of the particle shape on the field autocorrelation function. J Chem Phys 2015;143:044903.
36. Kaiser A, Popowa K, Löwen H. Active dipole clusters: from helical motion to fission. Phys Rev E 2015;92:012301.
37. Daddi-Moussa-Ider A, Goh S, Liebchen B, Hoell C, Mathijssen AJTM, Guzmán-Lastra F, et al. Membrane penetration and trapping of an active particle. J Chem Phys 2019;150:064906.
38. Daddi-Moussa-Ider A, Kaoui B, Löwen H. Axisymmetric flow due to a stokeslet near a finite-sized elastic membrane. J Phys Soc Jpn 2019;88:054401.
39. Daddi-Moussa-Ider A, Liebchen B, Menzel AM, Löwen H. Theory of active particle penetration through a planar elastic membrane. New J Phys 2019;21:083014.
40. Puljiz M, Orlishausen M, Köhler W, Menzel AM. Thermophoretically induced large-scale deformations around microscopic heat centers. J Chem Phys 2016;144:184903.
41. Menzel AM. Force-induced elastic matrix-mediated interactions in the presence of a rigid wall. Soft Matter 2017;13:3373–84.
42. Fischer L, Menzel AM. Magnetostriction in magnetic gels and elastomers as a function of the internal structure and particle distribution. J Chem Phys 2019;151:114906.
43. Filipcsei G, Zrínyi M. Magnetodeformation effects and the swelling of ferrogels in a uniform magnetic field. J Phys Condens Matter 2010;22:276001.
44. Annunziata MA, Menzel AM, Löwen H. Hardening transition in a one-dimensional model for ferrogels. J Chem Phys 2013;138:204906.
45. Pessot G, Cremer P, Borin DY, Odenbach S, Löwen H, Menzel AM. Structural control of elastic moduli in ferrogels and the importance of non-affine deformations. J Chem Phys 2014;141: 015005.
46. Tarama M, Cremer P, Borin DY, Odenbach S, Löwen H, Menzel AM. Tunable dynamic response of magnetic gels: impact of structural properties and magnetic fields. Phys Rev E 2014;90:042311.
47. Pessot G, Löwen H, Menzel AM. Dynamic elastic moduli in magnetic gels: normal modes and linear response. J Chem Phys 2016;145:104904.

48. Pessot G, Schümann M, Gundermann T, Odenbach S, Löwen H, Menzel AM. Tunable dynamic moduli of magnetic elastomers: from characterization by X-ray micro-computed tomography to mesoscopic modeling. J Phys Condens Matter 2018;30:125101.
49. Goh S, Menzel AM, Löwen H. Dynamics in a one-dimensional ferrogel model: relaxation, pairing, shock-wave propagation. Phys Chem Chem Phys 2018;20:15037–51.
50. Becker TI, Raikher YL, Stolbov OV, Böhm V, Zimmermann K. Dynamic properties of magneto-sensitive elastomer cantilevers as adaptive sensor elements. Smart Mater Struct 2017;26:095035.
51. Becker TI, Zimmermann K, Borin DY, Stepanov GV, Storozhenko PA. Dynamic response of a sensor element made of magnetic hybrid elastomer with controllable properties. J Magn Magn Mater 2018;449:77–82.
52. Allahyarov E, Löwen H, Zhu L. A simulation study of the electrostriction effects in dielectric elastomer composites containing polarizable inclusions with different spatial distributions. Phys Chem Chem Phys 2015;17:32479–97.
53. Allahyarov E, Löwen H, Zhu L. Dipole correlation effects on the local field and the effective dielectric constant in composite dielectrics containing high-k inclusions. Phys Chem Chem Phys 2016;18: 19103–17.
54. Allahyarov E, Menzel AM, Zhu L, Löwen H. Magnetomechanical response of bilayered magnetic elastomers. Smart Mater Struct 2014;23:115004.
55. Menzel AM. Hydrodynamic description of elastic or viscoelastic composite materials: relative strains as macroscopic variables. Phys Rev E 2016;94:023003.
56. Attaran A, Brummund J, Wallmersperger T. Modeling and finite element simulation of the magneto-mechanical behavior of ferrogels. J Magn Magn Mater 2017;431:188–91.
57. Attaran A, Brummund J, Wallmersperger T. Development of a continuum model for ferrogels. J Intell Mater Syst Struct 2017;28:1358–75.
58. Gebhart P, Wallmersperger T. A general framework for the modeling of porous ferrogels at finite strains. J Mech Phys Solid 2019;122:69–83.
59. Menzel AM. Bridging from particle to macroscopic scales in uniaxial magnetic gels. J Chem Phys 2014;141:194907.
60. Bohlius S, Brand HR, Pleiner H. Macroscopic dynamics of uniaxial magnetic gels. Phys Rev E 2004; 70:061411.
61. Evans R. Density functional theory for inhomogeneous fluids I: simple fluids in equilibrium. In: Cichocki B, Napiórkowski M, Piasecki J, editors. Lecture notes 3rd Warsaw School of Statistical Physics. Warsaw: Warsaw University Press; 2010:43–85.
62. Löwen H. Density functional theory for inhomogeneous fluids II: statics, dynamics, and applications. In: Cichocki B, Napiórkowski M, Piasecki J, editors. Lecture notes 3rd Warsaw School of Statistical Physics. Warsaw: Warsaw University Press; 2010:87–121.
63. Cremer P, Heinen M, Menzel AM, Löwen H. A density functional approach to ferrogels. J Phys Condens Matter 2017;29:275102.
64. Goh S, Wittmann R, Menzel AM, Löwen H. Classical density functional theory for a two-dimensional isotropic ferrogel model with labeled particles. Phys Rev E 2019;100:012605.
65. Gundermann T, Cremer P, Löwen H, Menzel AM, Odenbach S. Statistical analysis of magnetically soft particles in magnetorheological elastomers. Smart Mater Struct 2017;26:045012.
66. Cremer P, Löwen H, Menzel AM. Tailoring superelasticity of soft magnetic materials. Appl Phys Lett 2015;107:171903.
67. Cremer P, Löwen H, Menzel AM. Superelastic stress–strain behavior in ferrogels with different types of magneto-elastic coupling. Phys Chem Chem Phys 2016;18:26670–90.
68. Menzel AM, Pleiner H, Brand HR. Response of prestretched nematic elastomers to external fields. Eur Phys J E 2009;30:371.

69. Menzel AM, Pleiner H, Brand HR. On the nonlinear stress–strain behavior of nematic elastomers – materials of two coupled preferred directions. J Appl Phys 2009;105:013503.
70. Schopphoven C, Birster K, Schweitzer R, Lux C, Huang S, Kästner M, et al. Elastic deformations in semi-dilute Ni nanorod/hydrogel composites. Arch Appl Mech 2019;89:119–32.
71. Biller AM, Stolbov OV, Raikher YL. Modeling of particle interactions in magnetorheological elastomers. J Appl Phys 2014;116:114904.
72. Pessot G, Weeber R, Holm C, Löwen H, Menzel AM. Towards a scale-bridging description of ferrogels and magnetic elastomers. J Phys Condens Matter 2015;27:325105.
73. Weeber R, Kantorovich S, Holm C. Ferrogels cross-linked by magnetic nanoparticles – deformation mechanisms in two and three dimensions studied by means of computer simulations. J Magn Magn Mater 2015;383:262–6.
74. Weeber R, Kantorovich S, Holm C. Ferrogels cross-linked by magnetic particles: field-driven deformation and elasticity studied using computer simulations. J Chem Phys 2015;143:154901.
75. Allahyarov E, Löwen H, Taylor PL. Simulation study of ion diffusion in charged nanopores with anchored terminal groups. Electrochim Acta 2017;242:73–85.
76. Wang H, Jacobi F, Waschke J, Hartmann L, Löwen H, Schmidt S. Elastic modulus dependence on the specific adhesion of hydrogels. Adv Funct Mater 2017;27:1702040.
77. Bechinger C, Di Leonardo R, Löwen H, Reichhardt C, Volpe G, Volpe G. Active particles in complex and crowded environments. Rev Mod Phys 2016;88:045006.
78. Babel S, Löwen H, Menzel AM. Dynamics of a linear magnetic "microswimmer molecule". EPL (Europhys Lett) 2016;113:58003.

Supplementary Material: The online version of this article offers supplementary material (https://doi.org/10.1515/psr-2019-0088).

Nima H. Siboni*, Gaurav P. Shrivastav, Stavros D. Peroukidis and Sabine H. L. Klapp

23 Structure and rheology of soft hybrid systems of magnetic nanoparticles in liquid-crystalline matrices: results from particle-resolved computer simulations

Abstract: Hybrid mixtures composed of magnetic nanoparticles (MNP) in liquid crystalline (LC) matrices are a fascinating class of soft materials with intriguing physical properties and a wide range of potential applications, e.g., as stimuli-responsive and adaptive materials. Already in the absence of an external stimulus, these systems can display various types of orientationally disordered and ordered phases, which are enriched by self-assembled structures formed by the MNPs. In the presence of external fields, one typically observes highly nonlinear macroscopic behavior. However, an understanding of the structure and dynamics of such systems on the particle level has, so far, remained elusive. In the present paper we review recent computer simulation studies targeting the structure, equilibrium dynamics and rheology of LC-MNP systems, in which the particle sizes of the two components are comparable. As a numerically tractable model system we consider mixtures of soft spherical or elongated particles with a permanent magnetic dipole moment and ellipsoidal non-magnetic particles interacting via a Gay-Berne potential. We address, first, equilibrium aspects such as structural organization and self-assembly (cluster formation) of the MNPs in dependence of the orientational state of the matrix, the role of the size ratio, the impact of an external magnetic field, and the translational and orientational diffusion of the two components. Second, we discuss the non-equilibrium dynamics of LC-MNP mixtures under planar shear flow, considering both, spherical and non-spherical MNPs. Our results contribute to a detailed understanding of these intriguing hybrid materials, and they may serve as a guide for future experiments.

***Corresponding author: Nima H. Siboni,** Institut für Theoretische Physik, Technische Universität Berlin, Hardenbergstraße 36, 10623 Berlin, Germany, e-mail: hamidisiboni@tu-berlin.de. https://orcid.org/0000-0002-0305-5870

Gaurav P. Shrivastav, Institute für Theoretical Physics, Technische Universität Wien, Wiedner Hauptstr. 8-10, 1040 Vienna, Austria

Stavros D. Peroukidis, Department of Chemical Engineering, University of Patras, 26504 Patras, Greece; and Hellenic Open University, 26222, Patras, Greece

Sabine H. L. Klapp, Institut für Theoretische Physik, Technische Universität Berlin, Hardenbergstraße 36, 10623 Berlin, Germany

This article has previously been published in the journal Physical Sciences Reviews. Please cite as: N. H. Siboni, G. P. Shrivastav, S. D. Peroukidis and S. H. L. Klapp "Structure and rheology of soft hybrid systems of magnetic nanoparticles in liquid-crystalline matrices: results from particle-resolved computer simulations" *Physical Sciences Reviews* [Online] 2020, 5. DOI: 10.1515/psr-2019-0108 | https://doi.org/10.1515/9783110569636-023

Keywords: magnetic liquid crystal mixtures, molecular dynamics simulations, Monte Carlo simulations, rheology, structure

23.1 Introduction

Hybrid mixtures composed of magnetic nanoparticles (MNP) in liquid crystalline (LC) matrices have attracted attention about five decades: Already in 1970, Brochard and de Gennes predicted theoretically intriguing features of such "ferronematic" systems [1], including spontaneous magnetization and giant field induced effects. In fact, LC-MNP mixtures combine, in a synergistic way, various features of their constituents which are highly attractive already on their own: LCs, that is, molecular or colloidal systems of anisometric particles, show various ordered (liquid-crystalline) phases as well as phase separation, electric-optical switching behavior in external fields, and a nonlinear rheological behavior under shear, to name just a few examples. Suspensions of MNPs (or larger magnetic particles) show superparamagnetic behavior (see Refs. [2, 3] for reviews), giving rise to a wide range of applications. More fundamentally, magnetic particles are prototypes of self-assembling systems forming clusters, chains and other structures due to strong dipole-dipole interactions. This short list already suggests highly complex structural and dynamical behavior of LC-MNP mixtures.

The marked field sensitivity predicted by Brochard and de Gennes arises even for small concentrations of (spherical) MNPs and has been studied in many experiments involving different LC-MNP mixtures [4–10], including systems on the colloidal scale [11, 12]. Recently, new interest in LC-MNP mixtures was stimulated by the discovery [13, 14] of spontaneous magnetic ordering in a hybrid system of large, micron-sized magnetic plates embedded in a thermotropic LC. A crucial ingredient for the observed behavior are the strong local distortions of the LC director induced by the magnetic plates (whose size is much larger than that of the LCs). These distortions yield, in turn, pronounced anisotropic, effective (i.e., LC-mediated, elastic) interactions. Today, magnetic particles can be fabricated in a range of different sizes and in different shapes [5, 8, 9, 13, 15–18], and one would expect that these geometrical properties play a key role for the complex behavior of the mixture as a whole. On the one hand, the particle shape will influence the distortions and thus, the LC-mediated elastic interactions [19–24] between the MNPs. On the other hand, the size of the magnetic particles (affecting their magnetic moment) and their shape determine the self-assembly behavior, which crucially depends on the (shape-dependent) lowest-energy configuration of a pair of particles. For example, while two spherical MNPs arrange in a head-to-tail configuration, this is not the case for e.g., ellipsoidal MNPs [25, 26].

Despite the strong experimental interest in LC-MNP mixtures and the significant developments in the design of different composites from a chemical point of view, the microscopic (i.e., particle-resolved) structure of such systems is, in many cases, essentially unexplored. From the theoretical perspective, ferronematics involving large MNPs have mainly been studied on the basis of Landau–de Gennes free energy approaches [7, 27, 28], focusing on the impact of elastic distortions. While this continuum-type of approach is quite successful to explain, e.g., the spontaneous magnetization of the platelet system in Refs. [13, 14], it does not reveal particle-resolved information. This seems particularly important when it comes to mixtures in which the sizes of magnetic particles and LCs are comparable, such as lyotropic suspensions of colloid pigment rods and magnetite particles [11, 12]. The most appropriate theoretical method to study such systems on a particle level is many-particle computers simulations. However, until recently this type of method was not applied to LC-MNP mixtures, presumably due to the large computational effort required to handle the involved interactions which act on many length scales.

Driven by this lack, we have recently started a number of computer simulation studies targeting the structure, equilibrium dynamics and rheology of LC-MNP systems, in which the particle sizes of the two components are of the same order of magnitude [12, 29–34]. The present review summarizes main results and points out open questions. As a numerically tractable model system we consider mixtures of soft, spherical or elongated particles with an embedded permanent magnetic dipole moment and ellipsoidal non-magnetic particles interacting via a Gay-Berne (GB) potential. To analyze the resulting collective behavior, we have employed Monte Carlo (MC) and Molecular Dynamics (MD) simulations in equilibrium and under shear flow.

The article is organized as follows Section 23.2 is devoted to the equilibrium structure of mixtures involving spherical MNPs with different sizes relative to that of the LC particles. We review the phase behavior, cluster formation, and the impact of a constant magnetic field. In Section 23.3 we discuss the equilibrium dynamics in LC-MNP mixtures, particularly the translational motion in the zero-field case. In Section 23.4 we then turn to the non-equilibrium dynamic of mixtures driven by external shear flow. This is a particularly interesting issue in the context of rheology (for reviews of the shear-induced behavior of the pure components, such as the shear-induced ordering in pure LC systems or the magnetoviscous effects in systems solely composed of the MNPs, see [35–37].) Finally, we provide a brief outlook in Section 23.5.

23.2 Structure of LC-MNP mixtures

In this section we discuss the structure of mixtures of spherical MNPs and LCs in thermal equilibrium. We start by introducing our model systems and then summarize recent computer simulation studies based on the MC and equilibrium MD simulations [12, 29–31, 38].

23.2.1 Model and simulations details

Within our particle-resolved simulations the nonmagnetic LC component is represented by a GB rod and the MNP is modeled as a dipolar soft sphere (DSS). The latter has an embedded, permanent point dipolar moment of strength μ in the center. The GB rods are characterized by a length (l) to width (σ) ratio set to $l/\sigma = 3$. Pure GB systems [39] with this aspect ratio are very well studied and can serve as a reference. The DSS have a diameter σ_S. The components of the mixtures have sizes of the same order of magnitude, with $\sigma_s^\star = \sigma_S/\sigma$ ranging from one fourth of the width of the GB rods, i.e., $\sigma_s^\star = 1/4$ (in this case we refer to "small" MNPs), to two times, i.e., $\sigma_s^\star = 2$. We note that with much smaller or larger values of $\sigma_s^\star = \sigma_S/\sigma$, the simulations tend to become unrealistic. The spheres interact via a soft repulsive potential and a long range dipole-dipole potential. Long-range interactions are treated with the three-dimensional Ewald sum [40]. To measure the strength of the interactions we define the reduced dipole moment $\mu^\star = \mu/\sqrt{\epsilon_0 \sigma_s^3}$ and the reduced temperatures $T^\star = k_B T/\epsilon_0$ (where k_B is the Boltzmann constant and ϵ_0 is the soft-sphere energy parameter). Further we use the dipolar coupling parameter $\lambda = \mu^2/k_B T \sigma_s^3$ that takes values between $\approx$3.5 and $\approx$15, consistent with values characterizing real, strongly coupled MNPs [41–43]. Finally, a modified GB potential [30] has been used for the interaction between rods and spheres that depends only on the orientation of the rods and the inter-particle distance vector of the center of mass (CoM) of rods and spheres.

23.2.2 Mixtures with small MNPs

23.2.2.1 Structure and phase diagram

We start by considering LC-MNP systems with "small" MNPs, that is, $\sigma_s^\star = \sigma_S/\sigma = 1/4$ [31]. This size ratio is realistic, e.g., for colloidal systems of pigment nanorods (typical width 40 nm) and magnetic spheres (diameter approx. 10 nm) [12]. In Ref. [31], we have examined the self-assembly and equilibrium dynamics of the MNPs within the LC matrix which can be either isotropic or orientationally ordered.

The topology of the phase diagram that has been obtained is similar to the corresponding of the pure GB system. It contains three fully miscible phases, namely an isotropic (I), a nematic (N) and a smectic-B (SmB) phases. This already indicates that the MNPs do not exert a strong perturbation on the LC matrix at $\sigma_s^\star = 1/4$. The volume fraction ratio of the GB rods over MNP species is $\phi_r/\phi_S \cong 780$. This implies that we essentially examine the influence of the LC matrix on the MNPs rather than the opposite.

Under the conditions considered, the MNPs form clusters within the LC matrix, see Figure 23.1. The types of the clusters can be classified into four categories: i) chain-like clusters that consist of at least three particles (here, the end-chain particles have only one bond), ii) ring-like clusters (that form loops) in which all the particles have at least two bonds, iii) branched clusters (for which at least one particle of the cluster has three or more bonds with the other particles), and iv) "free", non-bonded particles or pairs of MNPs. We have found that the orientational state of the LC matrix (either isotropic or

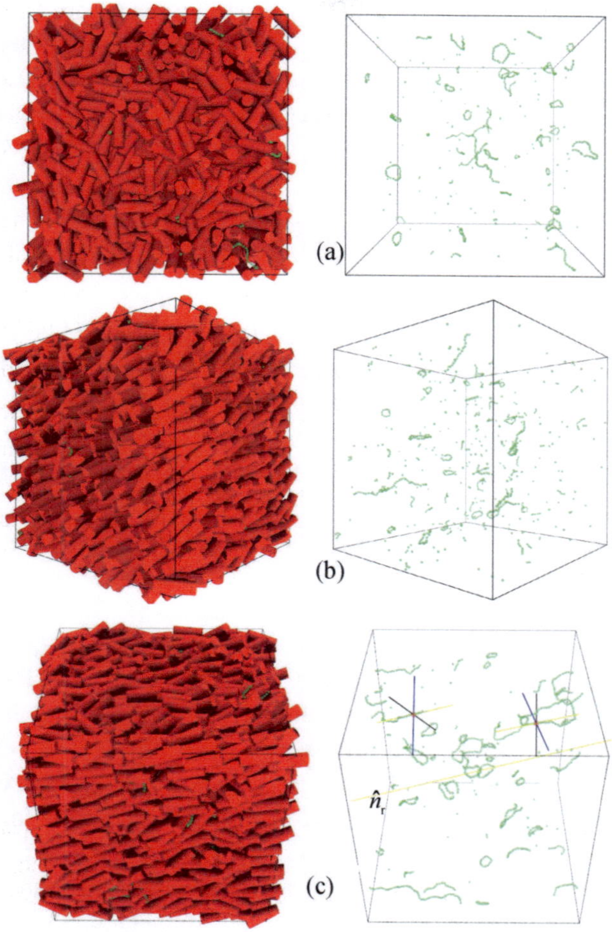

Figure 23.1: Representative snapshots of a GB-MNP mixture with $\sigma_s^* = 0.25$ and (concentration of rods) $x_r = 0.8$ in various states [(reduced temperature, number density)]: (a) isotropic state at $[(T^*, \rho^*) = (1.2, 0.34)]$ with $\lambda = 2.5$, (b) uniaxial nematic (N_u) state at $[(T^*, \rho^*) = (1.4, 0.44)]$ with $\lambda = 2.14$ and (c) uniaxial nematic (N_u) at $[(T^*, \rho^*) = (1.2, 0.44)]$ with $\lambda = 2.5$. The direction of $\hat{n}_r$ and the principal axis frames of a ring and a chain are also shown. For clarity, the rod species have been removed from the simulation box (right column). Reprinted from Ref. [31].

liquid crystalline) does not affect the type or the size of these clusters. Their size is mainly determined by the dipolar coupling parameter λ that for these systems takes values greater than 6. Similar types of clusters have been observed in monodispersed MNPs systems [44].

Still, an important finding for the mixed system is that the matrix in its nematic state promotes some degree of orientational order of the clusters (e.g., of the chains or rings). We have quantified these features by appropriate local order parameters [31].

Further, the observed alignment of the clusters along the director of the LC phase is enhanced at lower temperatures (i.e., larger values of λ), where more "tight" clusters are formed. Another interesting finding is that even though the matrix is uniaxial, the clusters (due to their non-uniaxial shape) do not rotate freely around the director of the phase.

23.2.2.2 Impact of an external field, relation to experiment

Starting from the LC-MNP model system with $\sigma_s^* = 1/4$, we have also investigated the impact of a uniform, external magnetic field on the structural behavior of the mixture. This study was motivated by a collaboration with experiments [12]. The latter had revealed that the magnetic field can induce orientational order to the matrix of non-magnetic rodlike particles (in this case, pigment rod) even when this matrix is isotropic in the field-free case. To understand this behavior on a particle level, we have employed MD simulations.

The MD simulation results not only describe qualitatively the behavior of the experimental system, but also provide a "bottom up" connection between the macroscopic properties (e.g., global orientational order parameters and birefringence) and microscopic inter-particle correlations. In particular, we have found that the MNPs self-assemble into 'rodlike' entities (i.e., chain-like clusters) along the direction of the magnetic field. These chain-like clusters hinder the rotational motion of the non-magnetic rods into perpendicular conformations with respect to the direction of the magnetic field. As a result, indirectly, nematic ordering is induced to the non-magnetic rods even though their zero-field state is an isotropic one. The main ingredient for the alignment of the rodlike species are, on the one hand, the size of the clusters (i.e., the length of the rodlike entities, which depends on λ), and on the other hand, the number per volume of the chain-like clusters [12]. In conclusion, our MD simulations clearly demonstrated the (hitherto only suspected) microscopic mechanisms that drive the macroscopic orientational behavior of such real systems. We note, however, that our model is not appropriate to describe suspensions of large, micron-sized magnetic particles in low molar mass liquid crystals (see for example Ref. [13]), where the MNPs induce strong distortions of the LC director field. For such systems theoretical approaches have been developed in which the LC is usually considered as continuum medium [45].

23.2.3 Mixtures with similar size components

Using MC simulations, we have also explored the structure and self-assembly of mixtures in which the diameter of the magnetic spheres is equal or greater than the width of the GB rods [29, 30]. The main motivation here was to understand the role of the rod-sphere size ratio on the stability of formed structures (e.g., rings and branched clusters), as well as the collective ordering of the MNPs and their influence on the LC matrix.

In the present section we focus on the case $\sigma_s^* = 1.0$ (MNP diameter equal to the width of the GB rods). The coupling parameter λ takes values between 3.5 and 15, corresponding to strongly coupled MNPs. As a reference system we also studied a mixture of GB rods with non-magnetic spheres. This reference system displays fully miscible, isotropic, nematic and smectic-B phases without any demixing or phase separation. Our MC simulations for the full system (LC matrix mixed with magnetic spheres) reveal a state diagram similar to the reference system. The volume fraction ratio of the GB rods over MNPs is $\phi_r/\phi_s \cong 12$. Interestingly, the tentative phase boundaries of the orientational ordered states (e.g., the nematic) are generally shifted towards higher temperatures (as compared to the reference system) indicating the enhancement of orientational order, despite the absence of an external magnetic field. An illustration of the microscopic structure of MNPs in various LC states is given in Figure 23.2. In the isotropic state, the MNPs self-assemble into (branched) worm-like chains that extend all over the simulation cell without orientational order. Note that due to the large value of λ, we do not observe the formation of finite-sized clusters. As the LC matrix undergoes a transition from the isotropic to the nematic state, the chains spontaneously "unwrap" transforming to essentially straight chains oriented along the nematic director. The dipole moments of the MNPs adopt a "head-to-tail" configuration forming ferromagnetic chains. Nevertheless, the net magnetization is zero since the chains adopt random "up-down" configurations.

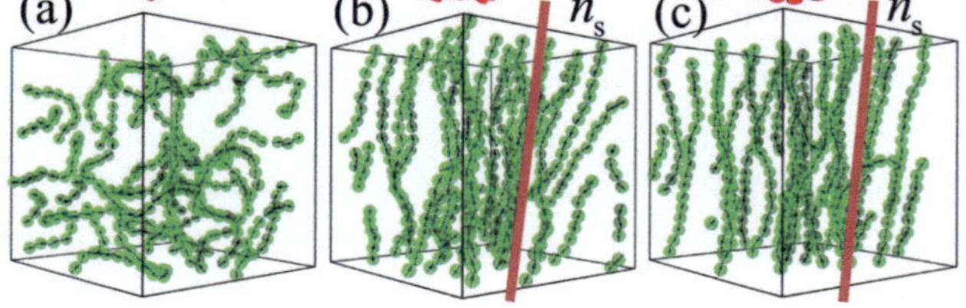

Figure 23.2: Representative simulation snapshots for mixtures with $\sigma_s^* = 1$ and (concentration) $x_r = 0.8$ in various states [(reduced temperature, number density)]: a) the isotropic (I) state $[(T^*, \rho^*) = (1.1, 0.34)]$ with $\lambda = 2.73$, b) uniaxial nematic (N_u) state $[(T^*, \rho^*) = (1.2, 0.40)]$ with $\lambda = 2.5$, c) uniaxial smectic-B (SmB) state $[(T^*, \rho^*) = (0.9, 0.40)]$ with $\lambda = 3.33$. In the bottom parts only the MNPs are shown for clarity. The directors $\hat{n}_r$ and $\hat{n}_s$ are indicated by thick lines. Reprinted from Ref. [29].

Taken altogether, the ferromagnetic chains stabilize the LC ordering (as demonstrated, for example, by the shift of the isotropic-to-nematic (I-N) phase transition towards higher temperatures) [29]. Vice versa, the LC matrix stabilizes the alignment of the chains along a common direction (i.e., the director of the nematic state), which is absent in pure MNPs fluids at the densities considered here. Thus, the LC matrix has a similar effect as an external field regarding the orientational ordering of the chains. The main difference is that an external magnetic field not only aligns the chains, but can also cause them to point along the same direction, yielding a globally ferromagnetic state.

The results discussed so far already indicate that the self-assembly of the MNPs, as well as their interplay with the LC matrix, are sensitive to the diameter ratio of the spheres relative to the GB rods' width. In what follows we focus on this effect for size ratios $\sigma_s^\star > 1.0$.

23.2.4 Mixtures with larger MNPs

For a mixture of GB rods and MNPs with diameter $\sigma_s^\star = 1.5$ and volume fraction ratio of GB rods over MNPs $\phi_r/\phi_s \cong 3.54$, the LC matrix exhibits isotropic, nematic and highly interdigitated smectic states. The nematic state is significantly promoted as compared to the system with $\sigma_s^\star = 1.0$. Nevertheless, the degree of alignment of the ferromagnetic chains is smaller. Furthermore, at lower temperatures, a layered state with interdigitated configuration of the rods is formed. This is in contrast to the system with $\sigma_s^\star = 1.0$, in which well-defined periodic layers are formed. This indicates that the MNPs exert an appreciable perturbation to the (partially) positionally ordered liquid crystalline state of the matrix.

By further increasing the diameter of MNPs to $\sigma_s^\star$ greater than 1.7, significant changes occurs in the morphology of the system, see Refs. [29, 30]. In particular, we refer to systems with $\sigma_s^\star = 1.7$ and $\sigma_s^\star = 2.0$ with volume fractions $\phi_r/\phi_s \cong 2.44$ and $\phi_r/\phi_s \cong 3.4$, respectively.

In these systems the perturbation of the MNPs to the LC matrix is so strong, that the LC matrix does not induce any significant alignment of the ferromagnetic chains in the nematic state. In addition, the MNPs destabilize the conventional smectic order (e.g., the smectic B state). For these systems, at lower temperatures, a lamellar structure occurs with alternating rod-rich and MNP-rich layers (microphase separation). The MNPs self-assemble into snake-like chains in an antiparallel manner, therefore, no net magnetization occurs. The most striking finding is that the mean alignment of the chains (i.e., the MNPs director) is perpendicular to the director of the rods giving rise to a biaxial lamellar state (L_b) (see Figure 23.3). This biaxial LC-MNP system differs from conventional biaxial LCs, that are generated by the shape of the particles [46] or from biaxial mixtures, of rods and disks [47]. In the case of LC-MNP systems, the biaxiality stems from the spontaneous organization of the ferromagnetic chains (of non-LC

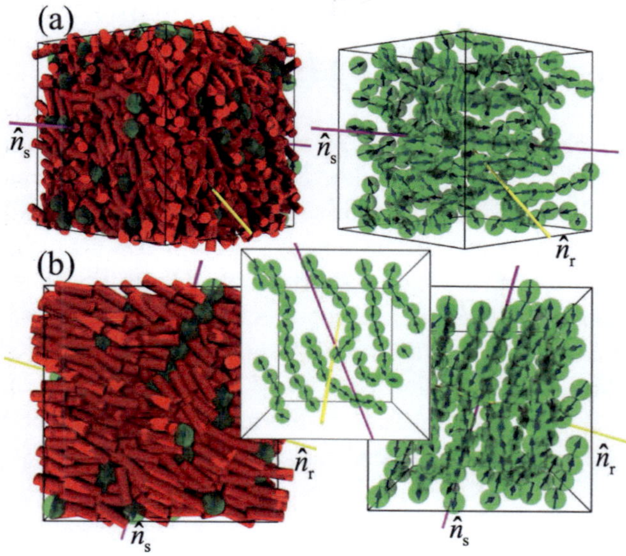

Figure 23.3: Representative simulation snapshots for mixtures with $\sigma_s^* = 2$ and (concentration) $x_r = 0.9$ in various states [(reduced temperature, number density)] a) the biaxial nematic (N_b) state $[(T^*, \rho^*) = (1.8, 0.338)]$ with $\lambda = 1.67$ and b) the biaxial lamellar state $[(T^*, \rho^*) = (1.2, 0.338)]$ with $\lambda = 2.5$. On the right sides only the MNPs are shown for clarity. Part b) additionally includes a snapshot of the structure in one dipolar layer. The directors $\hat{n}_r$ and $\hat{n}_s$ are indicated by thick lines. Reprinted from Ref. [29].

particles) within the LC matrix. The biaxial ordered hybrid also exhibits strong *anisotropic* sensitivity to external homogeneous magnetic field [30]. Upon applying a magnetic field to the LC-MNP matrix with $\sigma_s^* = 2.0$ in an otherwise isotropic state, orientational ordering is induced to the rodlike species. Interestingly, the director of the nematic state is perpendicular to the external field. On the other hand, the ferromagnetic chains align parallel to the direction of the magnetic field. Hence, a biaxial state can be induced in an initially isotropic state upon applying a homogeneous magnetic field. This is in contrast to systems with σ_s^* less than 1.5, where the field induces a nematic state with the LC director being parallel to that of the ferromagnetic chains.

From a physical point of view, one would expect that the complex morphologies also affect material properties such as light propagation and thermal transport in corresponding real soft matter magnetic systems. Recently, we have examined [38] the light propagation through MNPs in LC matrices. We have found that LC-MNP hybrids exhibit a strong magnetochiral dischroism (MCD) in a wide range of light wavelengths (in the visible regime from 500 to 750 nm). This behavior is generated by the significant changes in the magnetic MNP self-organization as the LC matrix undergoes a transition from an isotropic to a nematic state upon applying an external homogeneous magnetic field.

23.3 Equilibrium dynamics of mixtures of liquid crystals and magnetic nanoparticles

So far, we have focused on purely structural effects. We now turn to the equilibrium dynamics in the absence of a magnetic field or shear flow, focusing on the translational diffusion. It is well established by computer simulation studies (see, e.g., [43, 48, 49]) that pure magnetic dipolar fluids, even at low densities, can show interesting diffusional dynamics which strongly correlates with the structures they form. Therefore, and in view of the novel structures formed in our composite systems, we have performed various MD simulation studies [31, 32] to examine the equilibrium dynamics of LC-MNP systems with size ratios $\sigma_s^* = 1/4$ (Section 23.3.1) and $\sigma_s^* = 1.0$ (Section 23.3.2). Our results show that both, the orientational state of the matrix and the dipolar coupling strength, λ play important role on the translational dynamics.

23.3.1 Mixtures with similar size components

We start by discussing the case $\sigma_s^* = 1.0$, which we have investigated in most detail [32]. To elucidate the translational dynamics, we have analyzed the mean square displacement (MSD) of different species at different temperatures and densities. With changing temperatures, the strength of the dipolar coupling, λ, also changes which affects the size of dipolar chains [43]. Therefore, it is expected that the dynamics of the dipolar particles will becomes slower as the temperature is decreased. In fact, for pure dipolar fluids, a subdiffusive behavior is observed at intermediate times which then crosses over to the diffusive behavior at late times [48]. The situation in present case is, however, different: here, apart from the different chain sizes at different temperature, the LC matrix provides an anisotropic dense medium, which significantly modifies the dynamics of the MNPs [32].

We first consider the MSDs of the MNPs in the isotropic phase, i.e., at the temperatures and densities where the LC host matrix remains in the isotropic phase. The MSDs of the MNPs in the isotropic phase are plotted in Figure 23.4 for $\rho^* = 0.3$ and various values of λ. For all $\lambda < 11.25$, the MSDs show three different regimes, ballistic at short times, subdiffusive at intermediate times and diffusive at long times. At low λ, the subdiffusive regime is rather short, however, as λ is increased the subdiffusive regime grows, and for $\lambda > 9.0$ at $\rho^* = 0.3$, it spans the entire simulation time window. The subdiffusive behavior in this case is attributed to the combined effect of the complex environment by the LC matrix, on one hand, and the chain forming tendency of the dipolar fluids, on the other hand. The exponent of the MSD in the subdiffusive regime is obtained as ~0.6 [32].

In the nematic phase of the LC host matrix, the MSD of the MNPs can be decomposed into components parallel and perpendicular to the nematic director, pf the MNP

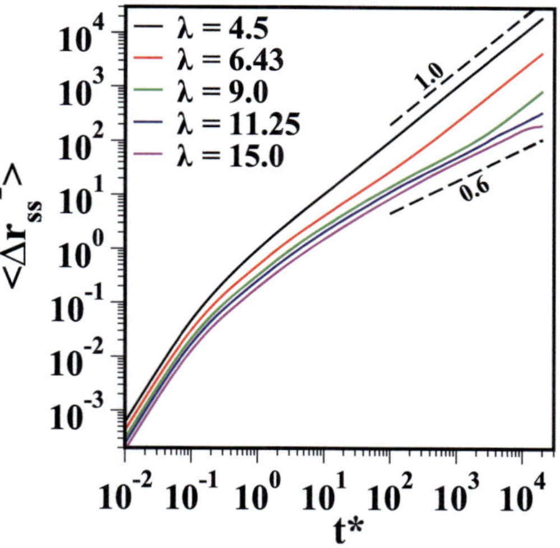

Figure 23.4: MSD of MNP in the isotropic regime for $\rho^* = 0.3$ at $\lambda = 4.5$, 6.43, 11.25, 15.0. At low λ, the MSD has a slope of 1.0 at large t (shown by black dashed line). At large λ, subdiffusive behavior is observed with an exponent 0.6. Reprinted from Ref. [32].

chains, $\hat{n}_S$. We fix $\lambda = 11.25$ (i.e., $T^* = 0.8$) and investigate the MSD for different ρ^*. The reason behind such a choice of λ is that at this value, the MSDs of the MNPs remain subdiffusive during the entire simulation time window. Figure 23.5(a) shows the component of the MSD of the MNPs parallel to $\hat{n}_S$. One observes an initial subdiffusive increase with time (reminiscent of the isotropic phase), which then crosses over to a normal diffusive behavior at long times (for all considered densities). The component of the MSD perpendicular to $\hat{n}_S$, plotted in Figure 23.5(b), remains subdiffusive even at long times. The exponent of subdiffusion is ~0.5, which is smaller than the observed exponent in the isotropic regime for the total MSD (at the same ρ^* and λ). The inset in Figure 23.5(b) illustrates the trapping of MNP chains in the transient cylindrical cavities formed by the LC matrix in the nematic phase. This leads to a slowing down of the

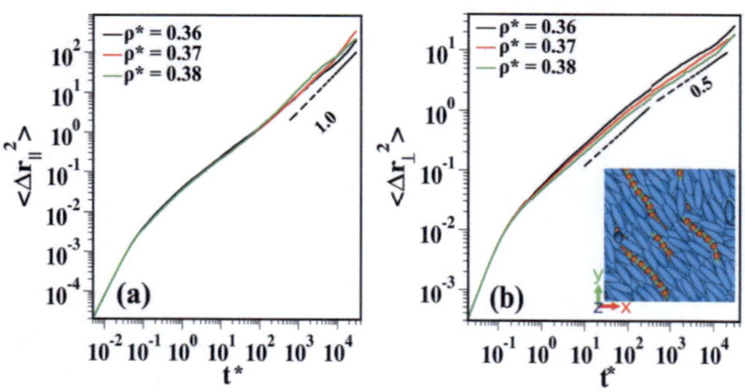

Figure 23.5: (a) The component of MSD parallel to the $\hat{n}_S$ for MNPs at the same values of ρ^* and λ as in (a). The black dashed line shows a slope 1.0. (b) The component of the MSD of the MNPs perpendicular to the $\hat{n}_S$ at $\rho^* = 0.36$, 0.37, 0.38 and $\lambda = 11.25$. The inset shows a top view of the snapshot for $\rho^* = 0.37$ and $\lambda = 11.25$. The MSD shows a subdiffusive behavior with exponent 0.5. Reprinted from Ref. [32].

translational dynamics of the MNPs in the direction perpendicular to the nematic director [32].

The trapping scenario is further clarified by investigating the velocity autocorrelation functions (VACF) for the MNPs. In the nematic phase, the component of the VACF for the MNPs parallel to the nematic directior, plotted in Figure 23.6(c) (shown by red dashed line) decays faster than the component in the perpendicular direction (represented by green dot-dashed line). The oscillations after the smooth initial decay are present in both the components. The oscillations in the parallel component arise due to the strong dipolar interactions which force the particles to remain in the chain. In the present case, the oscillations in the perpendicular component are the result of the confinement of the MNP chains in a narrow cylindrical channel formed by the LC matrix.

We also note that the dynamics of the LC particles remains unaffected by the MNPs. The MSD of the LC particles shows a diffusive behavior at long times in the isotropic phase [32, 50]. In the nematic phase, the components of the MSD of the LC particles remains diffusive in both, the direction parallel and perpendicular to the nematic director. Additionally, in the nematic direction the diffusion coefficient shows a non-monotonic behavior as a function of the density. Such a non-monotonic behavior occurs due to the faster movement of the LC particles along the director in the low density range of the nematic phase [50, 51].

23.3.2 Mixtures with small MNPs

Finally, we briefly consider the case $\sigma_s^* = 1/4$. Here, we find that the LC matrix in its nematic state not only induces orientational ordering of the clusters (see also

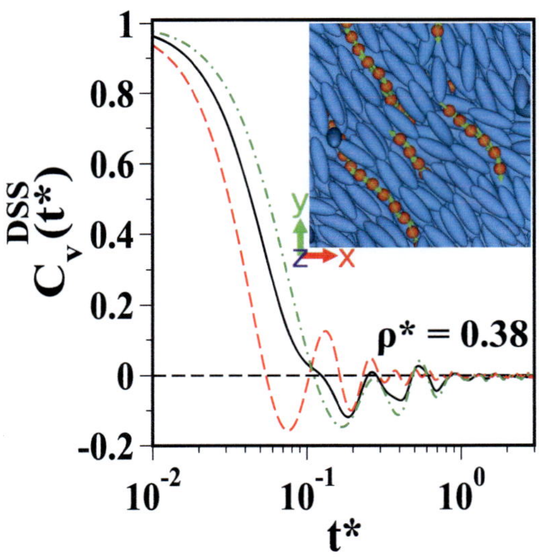

Figure 23.6: The VACF for MNP in the nematic regime at $\rho^* = 0.38$ and $\lambda = 11.25$. The red dashed and green dot-dashed lines represent the function $C_{v_\parallel}(t)$ and $C_{v_\perp}(t)$. The inset shows a top view of the snapshot for $\rho^* = 0.37$ and $\lambda = 11.25$. Reprinted from Ref. [32].

Section 23.2.2.1), but also promotes their motion along the director [31]. This is indicated by the fact that the long-time diffusion coefficient is largest in this direction. This is different at short times, where the MNPs travel faster perpendicularly to the nematic director. Regarding the impact of coupling, we find that the self-diffusion coefficient of the MNPs decreases by increasing λ, as the percentage of the "free" magnetic particles decreases. This gives place to the formation of larger clusters that slow down the mobility of the MNPs. Further, the so-called "non-gaussian" parameter that is zero in the perfect ballistic or diffusive regime, shows deviation from zero indicating non-trivial behavior of the MNPs. In summary, our MD simulations indicate the presence of complex translational dynamics of the MNPs that depend on the orientational state of the matrix, the corresponding MNP cluster formation, and on the size ratio.

23.4 Rheology of LC-MNP mixtures

While the equilibrium structure and dynamics of LC-MNP mixtures is important and quite intriguing, many applications of such soft hybrid systems, as well as some experiments, actually involve nonequilibrium conditions, particularly shear flow. Taking this as a motivation, we here present results from nonequilibrium molecular dynamics (NEMD) simulations for model mixtures composed of GB rods and spherical MNPs with $\sigma_s^* = 1.0$ (Section 23.4.1), as well as elongated MNPs (Section 23.4.2), in planar Couette shear flow.

23.4.1 Spherical MNPs

23.4.1.1 The flow curve

The rheological behavior of the LC-MNP mixtures is investigated by applying a steady shear (via Lee-Edwards boundary conditions) with a constant shear rate $\dot{\gamma}$. We calculate the evolution of the stress, σ_{xz}, as a function of strain. After reaching a steady state, the corresponding stress (σ_{xz}^{ss}) can be plotted as a function of the shear rate, yielding the so-called flow curve. Figure 23.7(a) shows the flow curves for the LC-MNP mixture for different λ. For all values of λ except $\lambda = 11.25$, we observe a crossover from "Newtonian" behavior, where σ_{xz}^{ss} is linearly proportional to $\dot{\gamma}^*$, to "non-Newtonian" behavior, where σ_{xz}^{ss} varies in a power-law manner as a function of $\dot{\gamma}^*$. This transition occurs at a "critical" shear rate, which is approximately given by $\dot{\gamma}_c^* = 0.02$ [33].

At shear rates $\dot{\gamma}^* \gtrsim \dot{\gamma}_c^*$ the stress varies as $\sigma_{xz}^{ss} \propto \dot{\gamma}^{*n}$ for all values of λ, where the power-law exponent is $n \approx 0.8$. This exponent, also called flow index, characterizes the non-Newtonian behavior of the LC-MNP mixture. Specifically, the fact that for $\dot{\gamma}^* \gtrsim \dot{\gamma}_c^*$, n is smaller than one, indicates a shear thinning. This behavior is also reflected by the average viscosity, defined as $\eta = \sigma_{xz}^{ss}/\dot{\gamma}^*$, which is plotted in Figure 23.7(b) as a function

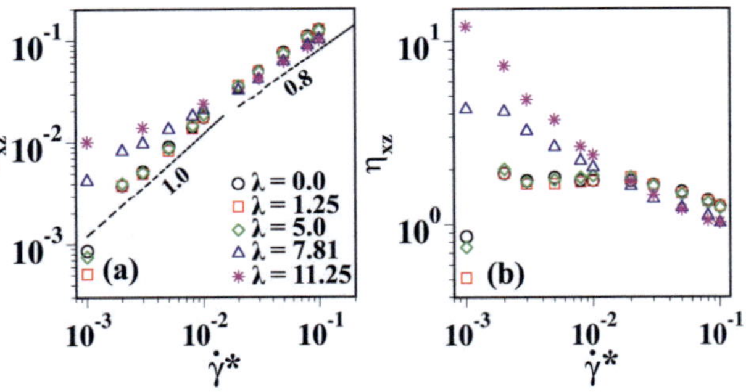

Figure 23.7: (a) The flow curve for the LC-MNP mixture at $\rho^* = 0.34$, $T^* = 0.8$ and $\lambda = 0.0, 1.25, 5.0, 7.81, 11.25$. Dashed lines represent slopes 1.0 and 0.8. (b) Average viscosities η_{xz}, obtained from σ_{xz}^{ss}, as a function of $\dot{\gamma}^*$ for the same λ values as in (a). Here, different symbols represent the same λ value as in (a). Reprinted from Ref. [33].

of $\dot{\gamma}^*$. For a wide range of complex fluids such as microgels, foams, and emulsions, the exponent n is considered to be a "material parameter" that weakly depends on the temperature and density [52].

23.4.1.2 Microstructure under shear

The nonlinear features observed in the flow curves discussed in the previous subsection already suggest profound structural changes. In the present section we focus, in particular, on the chain formation of magnetic particles. To this end we consider mixtures at $\lambda = 5.0$ and $\lambda = 11.25$, where the non-Newtonian behavior is particularly pronounced. Before we move ahead with the analysis, it should be noted that here in our LC-MNP mixture, the number density of the MNP is rather small ($\rho^*_{MNP} = 0.068$). In pure MNP fluids, at such low densities, the MNPs form chains with head-to-tail ordering of neighboring particles and the size of the chains depending on λ. These MNP chains are isotropically distributed, that is, there is no long-range orientational order [40, 43]. In our previous study of a LC-MNP mixture, we have shown that in the isotropic phase (such as one considered here at $\rho^* = 0.34$), the MNPs form isotropically distributed chains of significant length if $\lambda \geq 6.0$ [32]. Therefore, we expect that in equilibrium, sizes of the MNP chains should be rather small for $\lambda = 5.0$ and relatively larger for $\lambda = 11.25$.

We find that at the low shear rate, $\dot{\gamma}^* = 5 \times 10^{-3}$, LCs do not show the shear-induced alignment, see Figure 23.8(a). On the contrary, at the high shear rate, $\dot{\gamma}^* = 10^{-1}$, the LCs display a shear-induced alignment in the steady-state, see Figure 23.8(b). The MNP, plotted in Figure 23.8(c), (d), show no alignment in the direction of the shear at any of the shear rates considered [33].

In contrast to $\lambda = 5.0$, at large ($\lambda = 11.25$) the equilibrium configuration is characterized by large MNP chains. In the presence of shear, these MNP chains align with the

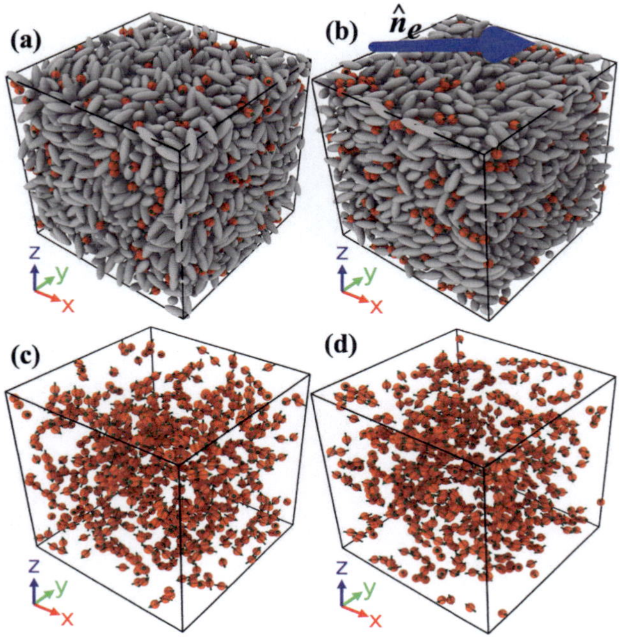

Figure 23.8: Snapshots of the LC-MNP mixture for $\lambda = 5.0$, $\rho^* = 0.34$ and $T^* = 0.8$ under shear. (a) and (b) illustrate the structure in the steady state ($\dot{\gamma}^* t^* = 15.0$) for $\dot{\gamma}^* = 5 \times 10^{-3}$ and 10^{-1}, respectively. The snapshots (c) and (d) show the MNPs alone at the parameters corresponding to (a), and (b) respectively. Reprinted from Ref. [33].

shear direction at both shear rates considered (see Figure 23.9(c), (d)). The response of the LCs depends on the shear rate: while the LC system at $\dot{\gamma}^* = 5 \times 10^{-3}$ (Figure 23.9(a)) displays weak alignment, the ordering at $\dot{\gamma}^* = 10^{-1}$ is more significant, see Figure 23.9(a). Here we already note that the observed shear-induced ordering of the two species is consistent with previous simulation studies on pure LCs [53–55] and pure dipolar fluids [56]. The difference in the present case is that the two components strongly influence each other.

These shear-induced changes in the microstructure are better quantified by investigating the nematic order parameters at different shear rates in the steady state for both the species of the mixture. The nematic order parameters are defined as the largest eigen value of the ordering matrices for the LCs and the MNPs. In Figure 23.10(a), we plot the steady-state nematic order parameters S^{ss} for the LC (S_e black circles) and the MNP (S_s red squares) for $\lambda = 5.0$ at different shear rates. We recall that at $\lambda = 5.0$, the flow curve reveals both, a Newtonian regime ($\dot{\gamma}^* < \dot{\gamma}_c^*$) and a non-Newtonian regime at high shear rates. As seen from Figure 23.10(a), the MNP do not develop any pronounced nematic ordering throughout the range of the shear rates considered. This is different for the LCs. In the Newtonian regime, the LCs do not show shear-induced ordering while at high shear rates, significant (*para-*)nematic ordering is observed. To this end

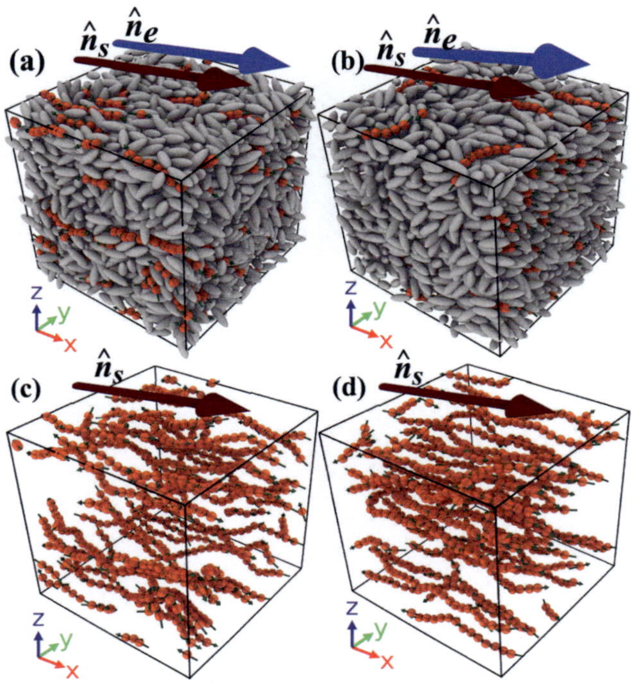

Figure 23.9: Snapshots of the LC-MNP mixture for $\lambda = 11.25$, $\rho^* = 0.34$ and $T^* = 0.8$ under shear. (a) and (b) illustrate the structure in the steady state ($\dot{\gamma}^* t^* = 15.0$) for $\dot{\gamma}^* = 5 \times 10^{-3}$ and 10^{-1}, respectively. The snapshots (c) and (d) show the MNPs alone at the parameters corresponding to (a), and (b) respectively. Reprinted from Ref. [33].

we note that the value of S_e related to the equilibrium I-N transition is 0.43 according to the Marier-Saupe theory [45]. This value of ($S_e = 0.43$) is represented by the horizontal light-blue dashed line in Figure 23.10. One sees that this (equilibrium) value is approached and finally exceeded when the shear rate becomes larger than $\dot{\gamma}_c^*$. We conclude that the non-Newtonian regime of the flow curve at $\lambda = 5.0$ is accompanied by orientational ordering of the LCs, but not the MNPs.

The situation at large λ, shown in Figure 23.10(b), is different. Here, the LC matrix exhibits a significant degree of ordering already at the lowest shear rate, $\dot{\gamma}^* = 10^{-3}$. Moreover, the nematic order parameter of the MNP is even larger. We understand these properties, which are in marked contrast to those observed at $\lambda = 5.0$, as a consequence of the pronounced chain formation of the MNP at $\lambda = 11.25$, see Figure 23.9(c), (d). Due to the strong correlation within the chains, these form rather stiff and long objects which align in the shear flow. The alignment of the chains, in turn, enhances the alignment of the non-magnetic LC particles, leading to relatively large values of S_e^{ss}.

Upon increase of the shear rate, the nematic order parameter of the MNP remains essentially constant in the range of the shear rates considered. For pure dipolar fluids

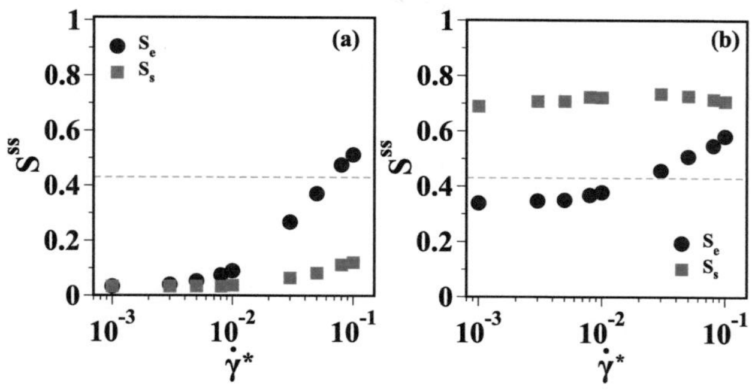

Figure 23.10: (a) Variation of the nematic order parameter in the steady state for $\lambda = 5.0$ for the LC (S_e) and the MNPs (S_s) as a function of $\dot{\gamma}^*$. The horizontal blue dashed line represents the critical value of the nematic order parameter at which I-N transition is observed. (b) Variation of the nematic order parameters for the LC (S_e) and the MNPs (S_s) in the steady state for $\lambda = 11.25$ as a function of $\dot{\gamma}^*$. Reprinted from Ref. [33].

under shear, it has been observed that the nematic order parameter decreases at high shear rates due to breaking of the chains [56]. In the present study, we do not observe such a behavior as the maximum shear rate is restricted to $\dot{\gamma}^* = 0.1$.

In summary, these results for mixtures with spherical MNPs indicate that the flow behavior is strongly affected by the strength of dipolar coupling among the MNPs. This property may find appealing technical applications in the situations where tunable flow properties are desired.

23.4.2 Elongated MNPs

In many real examples of LC-MNP mixtures, the MNPs are spherical (and hence different from the LCs). However, they can also be shape-anisotropic (which often goes along with a strong size difference relative to the LC particles) [5, 8, 15, 16]. The impact of non-spherical shape is indeed a timely issue in view of recent experimental realizations [9, 13, 17, 18], which have also motivated a number of analytical and numerical studies [57–60].

In the present section we discuss mixtures where the shape and size of the MNPs are *identical* to those of the LC particles. Clearly, this is a somewhat idealized situation compared to real systems. On the other hand, this choice simplifies a computational study because it eliminates effects induced by shape and size mismatch (see Refs. [61–66] for some examples).

The aforementioned recent studies involving nonspheriacl MNPs [57–60] are concerned with the effect of a magnetic field on the structural and optical properties. Here we rather focus on the response of the aforementioned mixtures to the mechanical perturbation (shear) in the presence of an external magnetic field. One should note that there is a key difference between optical and mechanical responses; unlike in the optical case, where light is transmitted through the sample without altering the structure, shearing itself can lead to structural changes [54, 67–70]. Therefore, the structure and rheology of such mixtures are affected by the interplay between the ordering induced by shear (see Ref. [71] and references therein) and the ordering induced by the magnetic field.

23.4.2.1 Model and simulation set-up

To isolate the role of the magnetic field-induced orientation of the MNPs on the structure and rheology, we consider a mixture where not only the shape and size of the MNPs and LCs are identical (both are described by GB particles with aspect ratio 3:1), but also they interact via the same pair potential. In this case, only the directions of the MNPs are coupled to the external magnetic field, and this coupling distinguishes the MNPs from the LCs. It is worth mentioning that the similarity of the pair potential has several consequences: (i) there is no specific anchoring between the LCs and the MNPs beyond anchoring among the LCs and the MNPs, (ii) polydispersity-induced phenomena are absent here, and (iii) unlike Refs. [12, 26, 31, 32, 43, 72], there is no structure formation due to the direct magnetic dipole-dipole interaction between the MNPs. These properties of the model enable us to isolate the role of the selectively controlled direction of particles and study its effect on the structure and rheology of the whole system.

We have implemented such a system in Ref. [34], where we focus on a temperature and density close to the isotropic state (in absence of the magnetic field and shear). The MNPs constitute 5% of the system's particles with magnetic dipoles which are embedded along their longest axes. The simulations are performed using NEMD in NVT-ensemble (for the simulation details see Ref. [34]).

Similar to a rheometer, the liquid is confined between two walls (along the z-axis, see Figure 23.11) and it is sheared by a relative motion of these walls (along the x-direction). These rigid and impenetrable walls are formed at relatively high temperatures (deep in the isotropic state) by a sudden freezing of two slabs of particles. The thickness of the walls is chosen to be larger than the GB potential cutoff in our simulation setup. These frozen slabs serve as solid, impenetrable, and rough walls, with which the system is sheared until it reaches the steady-state.

After reaching the steady state, the magnetic field is increased gradually in time (in a linear fashion) from zero to a saturation value where all the MNPs fully align with the field. This induces, strictly speaking, a time-dependence into the protocol which is, however, computationally advantageous (as compared to extensive steady-state simulations). In Ref. [34], we examined the effect of the field on the system for three different field directions: (i) along $\mathbf{n}^{(0)}$, (ii) perpendicular to $\mathbf{n}^{(0)}$ in the yz-plane (vorticity direction), and (iii) perpendicular to $\mathbf{n}^{(0)}$ in the xz-plane (shear plane), where

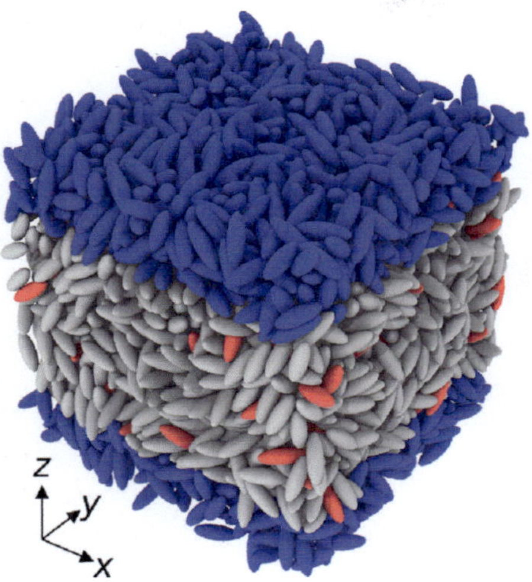

Figure 23.11: A snapshot of the simulation setup in absence of shear and magnetic field. The magnetic and non-magnetic particles are depicted by gray and red ellipsoids, respectively. The wall particles are distinguished by their blue color. The system, which is in an isotropic state, is confined by the walls along the *z*-axis and has periodic boundary conditions along *x*- and *y*-directions. Reprinted from Ref. [34].

$\mathbf{n}^{(0)}$ represents the nematic director in absence of the field. The shear stress, σ_{xz}, is calculated via summation of the x-component of the forces exerted by the liquid particles on the upper or lower wall particles.

23.4.2.2 Shear stress vs. magnetic field

Investigating the shear stress (or equivalently the viscosity) as a function of the applied external field, one observes that the present system shows an unusual and intriguing behavior [34], see Figure 23.12. The most pronounced effect of the magnetic field occurs in case (iii), where we observe, remarkably, a non-monotonic behavior. In contrast, there are no marked field-induced changes of σ_{xz} in the parallel case. We note that, due to our simulation protocol, the observed non-monotonicity is actually a time-dependent effect; yet we have checked that it occurs for a wide range of rates by which the field is changed.

In the subsequent discussion, we restrict ourselves to case (iii) and investigate the origins of the observed non-monotonicity as a function the magnetic field strength $H := |\mathbf{H}|$. More specifically, we find the signatures of the aforementioned non-monotonic behavior in the structure of the system. As shown by the snapshots in Figure 23.12, the magnetic field affects both the orientational and the positional order of the MNPs.

23.4.2.3 Orientational and positional ordering

A quantitative analysis of the nematic director of the MNPs ($\widehat{n}_{\mathrm{MNP}}$) reveals that there is a strong correlation between the magnetic field-induced deviation of $\widehat{n}_{\mathrm{MNP}}$ from $\mathbf{n}^{(0)}$ and the increase in the shear stress (see Figure 23.13). Interestingly by increasing H from

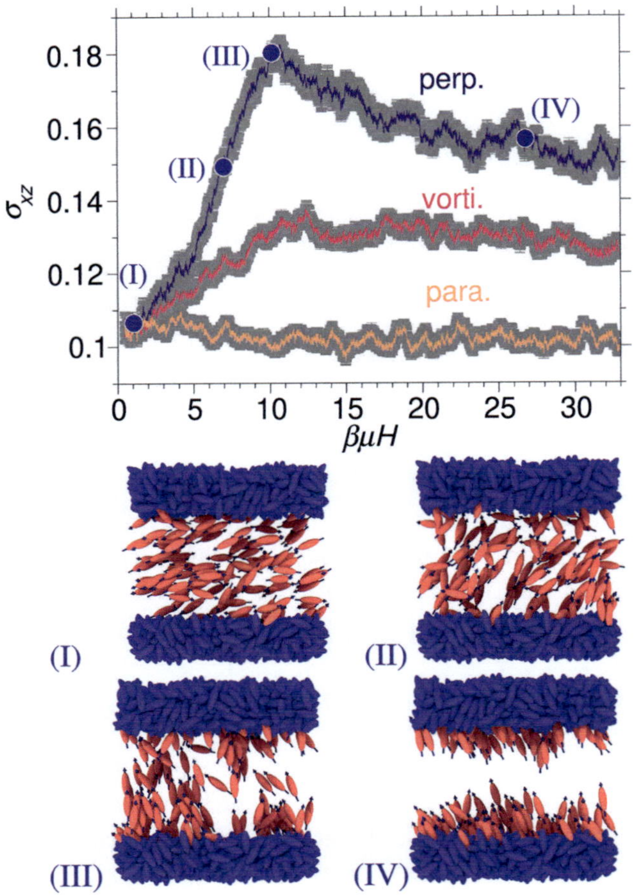

Figure 23.12: Upper panel: Shear stress as a function of the strength of the scaled magnetic field at three different directions: (i) the direction parallel to $\mathbf{n}^{(0)}$, (ii) the vorticity direction, which is parallel to the *y*-axis, and (iii) the direction perpendicular to $\mathbf{n}^{(0)}$ in the *xz*-plane. Here *H* is the magnetic field strength, μ is the magnetic dipole moment, and $\beta = 1/(k_B T)$ is the inverse thermal energy. Lower panel: Representative configurations associated with the four values of the field, which are indicated by bullets and the Roman numbers on the curve. Figures in both panels are adapted from Ref. [34].

zero, for small values of *H*, the orientations of the MNPs show a Fréederickz-like transition. For magnetic field strength larger than the threshold value for this transition, the MNPs start deviating from the shear-induced orientation, which leads to an increase in shear stress (see Figure 23.13). This correlation can be understood by considering that the misalignement of the MNPs restricts the shear-induced motion of the LCs, and therefore, increases the shear stress.

One should note that not only the orientational order of the MNPs, but also that of the LCs is affected by the magnetic field, although the LCs themselves are not susceptible to the magnetic field (see Figure 23.13). This is an indirect effect: The magnetic field re-orients the MNPs, which in turn, affects the orientation of the neighboring LCs due to the anisotropic steric interactions between the MNPs and LCs.

So far, we have focused on the correlation between the field-induced orientational ordering of the magnetic particles and the shear stress at small field strengths. Upon further increase of the magnetic field, a significant qualitative change in the spatial distribution of the MNPs occurs which correlates with the decrease of the shear stress. This qualitative change is illustrated by the four representative configurations at different values of μH in Figure 23.12: Starting from a relatively homogeneous distribution of MNPs at small H, further increase of H leads to a *spatial demixing* between MNPs and LCs, where the MNPs assemble at the walls. One can quantify these structural transformations by measuring the averaged number density profile of MNPs as a function of the external field. The results are plotted in Figure 23.14. At small field strengths (state I and II) a uniform spatial distribution of the MNPs (and thus, also the LCs) is observed. In qualitative contrast to that, at large field strengths (state points III and IV) one observes a pronounced double-peak structure of the MNPs density profile, reflecting the assembly of the MNPs at the walls. Using a quantitative scalar measure for the degree of inhomogeneity of MNPs, one can show that the demixing occurs in the same range of field strength as the drop in the shear stress [34]. This correlation can be understood as follows: as soon as the system demixes, that is, at large values of μH, the MNPs are concentrated close to the walls, and this provides a channel for the LCs in

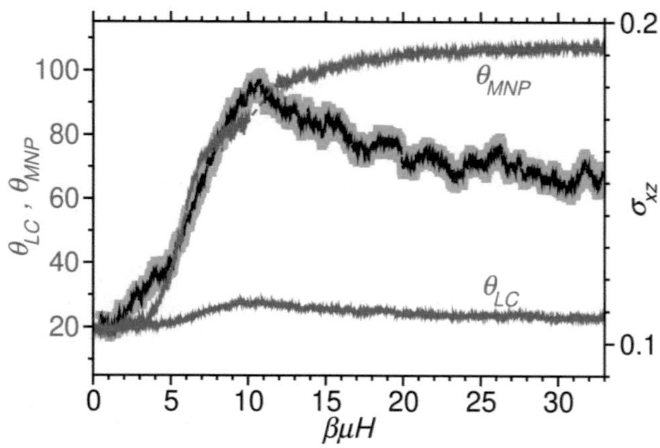

Figure 23.13: The angles between the nematic directors, i.e., $\hat{n}_{LC}$ and $\hat{n}_{MNP}$) with the shear direction (x-direction) are shown on the left vertical axis. In absence of the magnetic field (i.e., $H = 0$) the LCs and the MNPs are indistinguishable, and therefore the value of these angles coincide. The nematic director remains undistorted up to a threshold value of the magnetic field (here $\beta\mu H_{th} \simeq 3.3$), and for $H > H_{th}$ it starts deviating from the $H = 0$ direction. This deviation increases as the magnetic field increases and eventually all the MNPs are fully aligned with the direction of the magnetic field. This is Reminiscent of Fréderickz transition. The shear stress is shown on the right vertical axis as a function of H. Reprinted from Ref. [34].

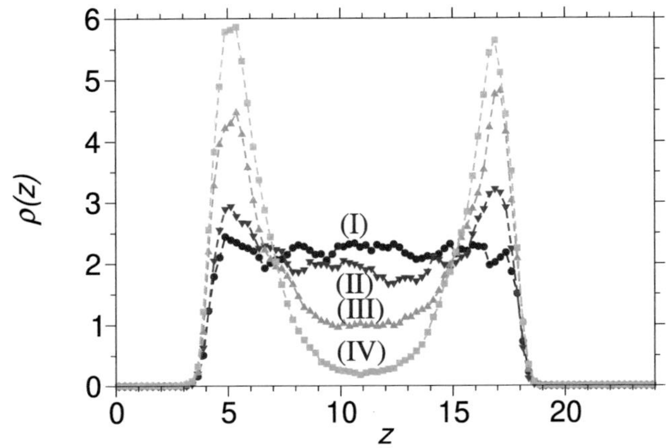

Figure 23.14: The density profiles of the MNPs at different magnetic field values, where $\rho(z_0) := n(z_0 - \delta z/2, z_0 + \delta z/2)/(L^2 \delta z)$, where n is the instantaneous number of MNPs between planes $z = z_0 - \delta z/2$ and $z = z_0 + \delta z/2$, and $\delta z = 0.25$ is the discretization resolution along the z-axis. The roman numbers refer to the same numbers in Figure 23.12. For small fields, a relatively uniform spatial distribution is obtained, in contrast to strong fields where a double-peaked profile emerges. The figure is adapted from Ref. [34].

which they can flow without (orientational and positional) disturbances from the MNPs.

There remains the question as to why the system demixes at all. In Ref. [34] we have proposed to explain the demixing within a quasi-equilibrium picture, namely as a competition between mixing and the packing entropies. A large packing entropy is achieved when the particles of similar orientations are neighbors, whereas a higher mixing entropy is obtained via a uniform distribution of particles over the whole system. As long as the misalignment between the nematic director of the two species is small, the mixing entropy dominates and all the particles (including the MNPs) are distributed uniformly in the system. In contrast, at large misalignments (i.e. large fields), the system gains entropy by packing the particles of similar orientations together. Hence, demixing occurs.

23.5 Outlook

In this article, we have summarized recent computer simulation studies of soft composites of LCs and MNPs in equilibrium and under shear flow. The results indicate a surprisingly rich and complex behavior of such systems already under the simplified conditions assumed in our models. This indicates, together with many recent observations from the experimental side (see, e.g., [11, 13, 14, 58]), that LC-MNP are ideal candidates for soft materials which are tunable by external fields, but also show self-assembly (by the involved interactions), and dynamical self-organization under driving forces such as shear.

There are several future directions of research which we would like to propose. First, the study of mixtures involving anisotropic MNPs with significant dipolar interactions (which we have neglected in Section 23.4.2) is only in its infancy. It is indeed well established that the shape of magnetic particles strongly affects the lowest-energy configurations of a pair of particles. For example, magnetic rods with longitudinal dipoles tend to form anti-parallel side-by-side configurations [25, 26] rather than the head-tail configurations familiar from magnetic spheres. This, in turn, determines the type of self-assembled clusters as well as the structures formed in a (constant) magnetic field. It seems therefore very promising to perform computational studies of interacting non-spherical MNPs of different size in LC matrices. One could also consider particles where the dipole is displaced from the center (similar to the studies for spherical MNPs [73–78]) or along the shortest axis, as done in recent experiments [79]. Depending on different dipole embeddings, we speculate to find intriguing structures which can be stabilized or destabilizes by an external magnetic field. The latter can also be time-dependent, an example being a planar rotating field which is known to lead to new structure formation (such as sheets) already in pure dipolar systems [80, 81]. Further, given the observations discussed in the present article it is clear that all these structures will also have a profound impact (and vice versa) on the rheology. Finally, for a better comparison with real systems, it will be important to overcome the computational difficulties when studying mixtures which much smaller or larger MNPs (relative to the size of the LCs).

Acknowledgments: Financial support from the German Science Foundation (DFG) via priority programme SPP 1681 is gratefully acknowledge.
Author contribution: All the authors have accepted responsibility for the entire content of this submitted manuscript and approved submission.
Research funding: The study was financially supported from the German Science Foundation (DFG) via priority programme SPP 1681.
Conflict of interest statement: The authors declare no conflicts of interest regarding this article.

References

1. Brochard F, de Gennes P. Theory of magnetic suspensions in liquid crystals. J Phys 1970;31:691.
2. Odenbach S. Magnetic fluids – suspensions of magnetic dipoles and their magnetic control. J Phys Condens Matter 2003;15:S1497.
3. Wells J, Kazakova O, Posth O, Steinhoff U, Petronis S, Bogart LK, et al. Standardisation of magnetic nanoparticles in liquid suspension. J Phys Appl Phys 2017;50:383003.
4. Liebert L, Martinet A. Coupling between nematic lyomesophases and ferrofluids. J Phys Lett 1979; 40:363.

5. Chen S-H, Amer NM. Observation of macroscopic collective behavior and new texture in magnetically doped liquid crystals. Phys Rev Lett 1983;51:2298.
6. Neto AF, Saba M. Determination of the minimum concentration of ferrofluid required to orient nematic liquid crystals. Phys Rev A 1986;34:3483.
7. Podoliak N, Buchnev O, Buluy O, D'Alessandro G, Kaczmarek M, Reznikov Y, et al. Macroscopic optical effects in low concentration ferronematics. Soft Matter 2011;7:4742.
8. Kopčanský P, Tomašovičová N, Koneracká M, Závišová V, Timko M, Džarová A, et al. Structural changes in the 6CHBT liquid crystal doped with spherical, rodlike, and chainlike magnetic particles. Phys Rev E 2008;78:011702.
9. Buluy O, Nepijko S, Reshetnyak V, Ouskova E, Zadorozhnii V, Leonhardt A, et al. Magnetic sensitivity of a dispersion of aggregated ferromagnetic carbon nanotubes in liquid crystals. Soft Matter 2011;7:644.
10. Kredentser S, Buluy O, Davidson P, Dozov I, Malynych S, Reshetnyak V, et al. Strong orientational coupling in two-component suspensions of rod-like nanoparticles. Soft Matter 2013;9:5061.
11. May K, Stannarius R, Klein S, Eremin A. Electric-field-induced phase separation and homogenization dynamics in colloidal suspensions of dichroic rod-shaped pigment particles. Langmuir 2014;30:7070.
12. May K, Eremin A, Stannarius R, Peroukidis SD, Klapp SH, Klein S. Colloidal suspensions of rodlike nanocrystals and magnetic spheres under an external magnetic stimulus: experiment and molecular dynamics simulation. Langmuir 2016;32:5085.
13. Mertelj A, Lisjak D, Drofenik M, Čopič M. Ferromagnetism in suspensions of magnetic platelets in liquid crystal. Nature 2013;504:237.
14. Mertelj A, Osterman N, Lisjak D, Čopič M. Magneto-optic and converse magnetoelectric effects in a ferromagnetic liquid crystal. Soft Matter 2014;10:9065.
15. Kyrylyuk AV, Hermant MC, Schilling T, Klumperman B, Koning CE, Van der Schoot P. Controlling electrical percolation in multicomponent carbon nanotube dispersions. Nat Nanotechnol 2011;6: 364.
16. Podoliak N, Buchnev O, Bavykin DV, Kulak AN, Kaczmarek M, Sluckin TJ. Magnetite nanorod thermotropic liquid crystal colloids: synthesis, optics and theory. J Colloid Interf Sci 2012;386:158.
17. Martinez-Pedrero F, Cebers A, Tierno P. Dipolar rings of microscopic ellipsoids: magnetic manipulation and cell entrapment. Phys Rev Appl 2016;6:034002.
18. Kredentser S, Kulyk M, Kalita V, Slyusarenko K, Reshetnyak VY, Reznikov YA. Magneto-induced anisotropy in magnetic colloids of superparamagnetic magnetite nanoparticles in an external magnetic field. Soft Matter 2017;13:4080.
19. Poulin P, Stark H, Lubensky T, Weitz D. Novel colloidal interactions in anisotropic fluids. Science 1997;275:1770.
20. Poulin P, Weitz D. Inverted and multiple nematic emulsions. Phys Rev E 1998;57:626.
21. Meeker S, Poon W, Crain J, Terentjev E. Colloid-liquid-crystal composites: an unusual soft solid. Phys Rev E 2000;61:R6083.
22. Nazarenko V, Nych A, Lev B. Crystal structure in nematic emulsion. Phys Rev Lett 2001;87:075504.
23. Lapointe CP, Mason TG, Smalyukh II. Shape-controlled colloidal interactions in nematic liquid crystals. Science 2009;326:1083.
24. Evans JS, Beier CN, Smalyukh II. Alignment of high-aspect ratio colloidal gold nanoplatelets in nematic liquid crystals. J Appl Phys 2011;110:033535.
25. McGrother SC, Gil-Villegas A, Jackson G. The effect of dipolar interactions on the liquid crystalline phase transitions of hard spherocylinders with central longitudinal dipoles. Mol Phys 1998;95: 657.
26. Alvarez CE, Klapp SH. Percolation and orientational ordering in systems of magnetic nanorods. Soft Matter 2012;8:7480.

27. Zadorozhnii V, Vasilev A, Reshetnyak VY, Thomas K, Sluckin T. Nematic director response in ferronematic cells. Europhys Lett 2005;73:408.
28. Raikher YL, Stepanov VI, Zakhlevnykh AN. Mean-field description of the order-disorder phase transition in ferronematics. Soft Matter 2013;9:177.
29. Peroukidis SD, Klapp SH. Spontaneous ordering of magnetic particles in liquid crystals: from chains to biaxial lamellae. Phys Rev E 2015;92:010501.
30. Peroukidis SD, Lichtner K, Klapp SH. Tunable structures of mixtures of magnetic particles in liquid-crystalline matrices. Soft Matter 2015;11:5999.
31. Peroukidis SD, Klapp SH. Orientational order and translational dynamics of magnetic particle assemblies in liquid crystals. Soft Matter 2016;12:6841.
32. Shrivastav GP, Klapp SH. Anomalous transport of magnetic colloids in a liquid crystal-magnetic colloid mixture. Soft Matter 2019;15:973.
33. Shrivastav GP, Siboni NH, Klapp SHL. Steady-state rheology and structure of soft hybrid mixtures of liquid crystals and magnetic nanoparticles. Soft Matter 2020;16:2516.
34. Siboni NH, Shrivastav GP, Klapp SH. J Chem Phys 2020;152:024505.
35. Odenbach S. Recent progress in magnetic fluid research. J Phys Condens Matter 2004;16:R1135.
36. Torres-Díaz I, Rinaldi C. Recent progress in ferrofluids research: novel applications of magnetically controllable and tunable fluids. Soft Matter 2014;10:8584.
37. Felicia LJ, Vinod S, Philip J. Recent advances in magnetorheology of ferrofluids (magnetic nanofluids)-a critical review. J Nanofluids 2016;5:1.
38. Yannopapas V, Klapp SH, Peroukidis SD. Magneto-optical properties of liquid-crystalline ferrofluids. Opt Mater Express 2016;6:2681.
39. De Miguel E, Rull LF, Chalam MK, Gubbins KE. Liquid crystal phase diagram of the Gay-Berne fluid. Mol Phys 1991;74:405.
40. Schoen M, Klapp S. Rev Comp Chem 2007;24:1.
41. Klokkenburg M, Erné BH, Meeldijk JD, Wiedenmann A, Petukhov AV, Dullens RP, et al. In SituImaging of field-induced hexagonal columns in magnetite ferrofluids. Phys Rev Lett 2006;97: 185702.
42. Gerth-Noritzsch M, Borin DY, Odenbach S. Anisotropy of the magnetoviscous effect in ferrofluids containing nanoparticles exhibiting magnetic dipole interaction. J Phys Condens Matter 2011;23: 346002.
43. Sreekumari A, Ilg P. Slow relaxation in structure-forming ferrofluids. Phys Rev E 2013;88:042315.
44. Rovigatti L, Russo J, Sciortino F. Structural properties of the dipolar hard-sphere fluid at low temperatures and densities. Soft Matter 2012;8:6310.
45. De Gennes P-G, Prost J. The physics of liquid crystals, 83 Oxford: Clarendon Press; 1993.
46. Peroukidis SD, Vanakaras AG. Phase diagram of hard board-like colloids from computer simulations. Soft Matter 2013;9:7419.
47. Cuetos A, Galindo A, Jackson G. Thermotropic biaxial liquid crystalline phases in a mixture of attractive uniaxial rod and disk particles. Phys Rev Lett 2008;101:237802.
48. Jordanovic J, Jäger S, Klapp SH. Crossover from normal to anomalous diffusion in systems of field-aligned dipolar particles. Phys Rev Lett 2011;106:038301.
49. Jaeger S, Stark H, Klapp SH. Dynamics of cluster formation in driven magnetic colloids dispersed on a monolayer. J Phys Condens Matter 2013;25:195104.
50. de Miguel E, Rull LF, Gubbins KE. Dynamics of the Gay-Berne fluid. Phys Rev A 1992;45:3813.
51. Allen MP. Diffusion coefficient increases with density in hard ellipsoid liquid crystals. Phys Rev Lett 1990;65:2881.
52. Bonn D, Denn MM, Berthier L, Divoux T, Manneville S. Yield stress materials in soft condensed matter. Rev Mod Phys 2017;89:035005.

53. Germano G, Schmid F. Nematic-isotropic interfaces under shear: A molecular-dynamics simulation. J Chem Phys 2005;123:214703.
54. Ripoll M, Holmqvist P, Winkler R, Gompper G, Dhont J, Lettinga M. Attractive colloidal rods in shear flow. Phys Rev Lett 2008;101:168302.
55. Ripoll M, Winkler R, Mussawisade K, Gompper G. Mesoscale hydrodynamics simulations of attractive rod-like colloids in shear flow. J Phys Condens Matter 2008;20:404209.
56. Sreekumari A, Ilg P. Anisotropy of magnetoviscous effect in structure-forming ferrofluids. Phys Rev E 2015;92:012306.
57. Sebastián N, Osterman N, Lisjak D, Čopič M, Mertelj A. Director reorientation dynamics of ferromagnetic nematic liquid crystals. Soft Matter 2018;14:7180.
58. Potisk T, Mertelj A, Sebastián N, Osterman N, Lisjak D, Brand HR, et al. Magneto-optic dynamics in a ferromagnetic nematic liquid crystal. Phys Rev E 2018;97:012701.
59. Zarubin G, Bier M, Dietrich S. A ferronematic slab in external magnetic fields. Soft Matter 2018;14: 9806.
60. Zarubin G, Bier M, Dietrich S. Effective landau theory of ferronematics. J Chem Phys 2018;149: 054505.
61. Kikuchi N, Horbach J. Mobile particles in an immobile environment: Molecular dynamics simulation of a binary Yukawa mixture. Europhys Lett 2007;77:26001.
62. Zaccarelli E, Liddle SM, Poon WC. On polydispersity and the hard sphere glass transition. Soft Matter 2015;11:324.
63. Heckendorf D, Mutch K, Egelhaaf S, Laurati M. Size-dependent localization in polydisperse colloidal glasses. Phys Rev Lett 2017;119:048003.
64. Ferreiro-Córdova C, Wensink H. Spinodal instabilities in polydisperse lyotropic nematics. J Chem Phys 2016;145:244904.
65. Verhoeff A, Wensink H, Vis M, Jackson G, Lekkerkerker H. Liquid crystal phase transitions in systems of colloidal platelets with bimodal shape distribution. J Phys Chem B 2009;113:13476.
66. Speranza A, Sollich P. Simplified onsager theory for isotropic-nematic phase equilibria of length polydisperse hard rods. J Chem Phys 2002;117:5421.
67. Blaak R, Auer S, Frenkel D, Löwen H. Homogeneous nucleation of colloidal melts under the influence of shearing fields. J Phys Condens Matter 2004;16:S3873.
68. Mokshin AV, Barrat J-L. Shear-induced crystallization of an amorphous system. Phys Rev E 2008; 77:021505.
69. Mandal S, Gross M, Raabe D, Varnik F. Heterogeneous shear in hard sphere glasses. Phys Rev Lett 2012;108:098301.
70. Shrivastav GP, Chaudhuri P, Horbach J. Heterogeneous dynamics during yielding of glasses: effect of aging. J Rheol 2016;60:835.
71. Hess S, Kröger M. Regular and chaotic orientational and rheological behaviour of liquid crystals. J Phys Condens Matter 2004;16:S3835.
72. Schmidle H, Hall CK, Velev OD, Klapp SH. Phase diagram of two-dimensional systems of dipole-like colloids. Soft Matter 2012;8:1521.
73. Kantorovich S, Weeber R, Cerda JJ, Holm C. Ferrofluids with shifted dipoles: ground state structures. Soft Matter 2011;7:5217.
74. Klinkigt M, Weeber R, Kantorgovich S, Holm C. Cluster formation in systems of shifted-dipole particles. Soft Matter 2013;9:3535.
75. Morphew D, Chakrabarti D. Supracolloidal reconfigurable polyhedra via hierarchical self-assembly. Soft Matter 2016;12:9633.
76. Steinbach G, Gemming S, Erbe A. Non-equilibrium dynamics of magnetically anisotropic particles under oscillating fields. Eur Phys J E 2016;39:69.

77. Yener AB, Klapp SH. Self-assembly of three-dimensional ensembles of magnetic particles with laterally shifted dipoles. Soft Matter 2016;12:2066.
78. Rutkowski DM, Velev OD, Klapp SH, Hall CK. Simulation study on the structural properties of colloidal particles with offset dipoles. Soft Matter 2017;13:3134.
79. Martinez-Pedrero F, Cebers A, Tierno P. Orientational dynamics of colloidal ribbons self-assembled from microscopic magnetic ellipsoids. Soft Matter 2016;12:3688.
80. Jäger S, Klapp SH. Pattern formation of dipolar colloids in rotating fields: layering and synchronization. Soft Matter 2011;7:6606.
81. Yan J, Bae SC, Granick S. Rotating crystals of magnetic Janus colloids. Soft Matter 2015;11:147.

Karl A. Kalina, Alexander Raßloff, Maximilian Wollner, Philipp Metsch, Jörg Brummund and Markus Kästner*

24 Multiscale modeling and simulation of magneto-active elastomers based on experimental data

Abstract: In this contribution, we present a framework for the multiscale modeling and simulation of magneto-active elastomers (MAEs). It enables us to consider these materials on the microscopic scale, where the heterogeneous microstructure consisting of magnetizable particles and elastomer matrix is explicitly resolved, as well as the macroscopic scale, where the MAE is considered to be a homogeneous magneto-active body. On both scales, a general continuum formulation of the coupled magneto-mechanical boundary value problem is applied and the finite element method is used to solve the governing equations. Starting with an experimental characterization of the individual constituents, i.e. particles and matrix, microscopic constitutive models for both are formulated and adjusted to the experimental data separately. With that, properties of MAEs resulting from the microscopic constitutive behavior can be captured within the presented modeling approach. Secondly, to discuss general macroscopic properties of magnetically soft and hard MAEs, a computational homogenization scheme is used to calculate the composites' effective behavior for different geometrical arrangements of the particles on the microscale. Finally, the calculated effective response of a magnetically soft composite system is used to identify the parameters of a macroscopic magneto-elastic model. Using the calibrated model, the behavior of macroscopic MAEs is simulated for different sample geometries.

Keywords: finite element simulation, magneto-active elastomers, magneto-mechanics, multiscale

Mathematics Subject Classification 2010: 65C05, 62M20, 93E11, 62F15, 86A22

***Corresponding author: Markus Kästner**, Chair of Computational and Experimental Solid Mechanics, Institute of Soild Mechanics, TU Dresden, 01062 Dresden, Germany, E-mail: markus.kaestner@tu-dresden.de

Karl A. Kalina, Alexander Raßloff, Maximilian Wollner, Philipp Metsch and Jörg Brummund, Chair of Computational and Experimental Solid Mechanics, Institute of Soild Mechanics, TU Dresden, 01062 Dresden, Germany

This article has previously been published in the journal Physical Sciences Reviews. Please cite as: K. A. Kalina, A. Raßloff, M. Wollner, P. Metsch, J. Brummund and M. Kästner, "Multiscale modeling and simulation of magneto-active elastomers based on experimental data" *Physical Sciences Reviews* [Online] 2020, 5. DOI: 10.1515/psr-2020-0012 | https://doi.org/10.1515/9783110569636-024

24.1 Introduction

Magneto-active elastomers (MAEs) are a special class of composite materials that alter their effective macroscopic behavior if an external magnetic field is applied. Due to the strong coupling behavior, they are attractive for several engineering applications such as actuators and sensors [1–4], valves [5], tunable vibration absorbers [6], medical robots [7] or prosthetic and orthotic devices with controllable stiffness [8].

Basically, MAEs consist of a soft polymer matrix filled with micron-sized magnetizable particles that can be arranged in an unstructured distribution or, if external fields are applied during the manufacturing process, in a chain-like [9–11] or even more complex structure [12]. Regarding the resulting macroscopic behavior, this leads to isotropic and anisotropic properties, respectively. Besides the underlying particle distribution, the effective magneto-mechanical behavior of the MAE will depend on the properties of the individual constituents. If the elastomer is combined with magnetically soft fillers as carbonyl iron, a reversible behavior will be observed. In contrast to that, a strongly irreversible behavior results if magnetically hard fillers, e.g. NdFeB or $CoFe_2O_4$, with strong magnetic hysteresis are used [13–17].

24.1.1 Modeling approaches

Strategies regarding the modeling of MAEs can be divided into microscopic and macroscopic approaches. In the former, the heterogeneous microstructure consisting of particles and matrix is explicitly resolved whereas the composite is regarded as macroscopically homogeneous in the latter, see Figure 24.1(a).

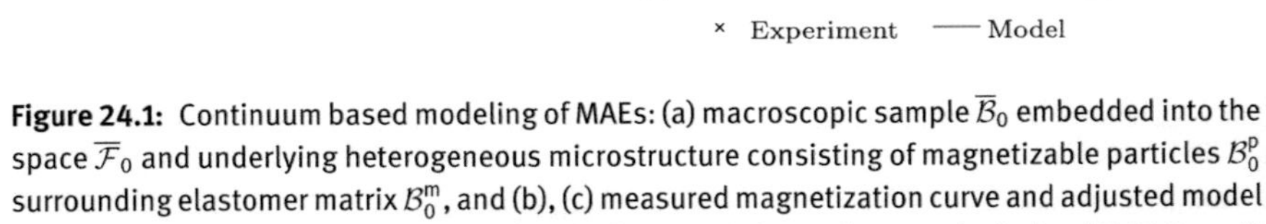

Figure 24.1: Continuum based modeling of MAEs: (a) macroscopic sample $\overline{\mathcal{B}}_0$ embedded into the free space $\overline{\mathcal{F}}_0$ and underlying heterogeneous microstructure consisting of magnetizable particles $\mathcal{B}_0^{\mathrm{p}}$ and surrounding elastomer matrix $\mathcal{B}_0^{\mathrm{m}}$, and (b), (c) measured magnetization curve and adjusted model (25) for magnetically soft carbonyl iron and major hysteresis loop of magnetically hard NdFeB particles and parameterized model (26)–(29). Experimental data given in (b) and (c) are taken from Spieler et al. [18] and Linke et al. [14], respectively.

Microscopic approaches

The first class of microscopic approaches is represented by particle-interaction models which are based on an energy minimization and the assumption, that the particles are described by magnetic dipoles [19–21]. Within these models, the polymer matrix is considered to be a continuous elastic medium or it is represented by springs between the particles, see Menzel [22] or Menzel and Löwen [23]. Particle-interaction models are numerically quite efficient but, due to the dipole assumption, their usage is limited to dilute systems. An extension of particle-interaction models to tightly filled systems is shown by Biller et al. [24], where a multipole expansion is used to calculate the particles' mutual magnetic interactions.

The second class of microscopically motivated approaches are continuum based models which resolve the local magnetic as well as mechanical fields within the composite [13, 25–30]. Thus, these approaches are not limited to the modeling of systems with a low amount of magnetic filler particles. Furthermore, they can be adapted to several kinds of dissipative materials, easily [13]. Microscopic continuum approaches also offer the possibility to analyze the stability of magneto-active materials under different loading conditions [31, 32]. However, the computationally expensive numerical solution of the underlying magneto-mechanical boundary value problem (BVP) which is usually done with the finite element (FE) method is a clear disadvantage of these models: in order to predict the effective material behavior of realistic magneto-active composites, an appropriate homogenization scheme as introduced by Chatzigeorgiou et al. [33] has to be used. A tool to capture not only the microstructure but also the shape of the macroscopic samples is the FE^2 method as performed by Keip and Rambausek [34, 35].

Macroscopic approaches

Macroscopic models represent an entirely different approach for the description of MAEs. Since the microstructure is not explicitly resolved and the composite is considered as a homogeneous continuum, they enable the representation of real structures with a reasonable computational effort. To capture the effects of the underlying microstructure implicitly, magneto-mechanical coupling terms have to be incorporated in these models. A variety of macroscopic approaches for the isotropic and transversely isotropic magneto-elastic case [9, 36–38] or the rate-dependent magneto-viscoelastic case [39–41] can be found in the literature.

Naturally, all of these models are purely phenomenological, i.e. fitting parameters have to be determined. This is done based on experimental results [9, 41] or by using data generated from a microscopic model combined with analytical [42, 43] or computational homogenization approaches [44]. If the fitting is done with experimental data, it is important to note, that at least inhomogeneous mechanical fields occur independent of the macroscopic sample geometry [35, 44, 45]. Moreover, if no ellipsoidal samples [46, 47] are used within the experiments, also the magnetization is inhomogeneous, e.g. in the case of widely used cylindrical shaped specimens. Due to this, the parameterized macroscopic model contains the influence of the sample geometry, i.e. the model describes the behavior of the MAE-sample and not the constitutive behavior of the pure MAE composite material. The fitting based on data generated from microscopic models avoids the described difficulty since it allows the identification of the parameters independent of any macroscopic sample geometry.

24.1.2 Content

In this contribution, a multiscale modeling approach based on Metsch et al. [29, 30] and Kalina et al. [13, 28, 44] is presented. Results of these works are summarized and partially supplemented in the present article. Within the developed multiscale modeling framework, MAEs are described based on a general continuum formulation, where the governing equations are solved with the FE method.

In order to achieve a deep understanding of the MAEs' effective behavior, the constitutive properties of the individual phases on the microscale as well as their interactions have to be investigated. Starting with an experimental characterization of the individual constituents, microscopic constitutive models are formulated and fitted to the measured data. This includes models for magnetically hard and soft filler particles and an adjusted hyperelastic model for a specific silicone elastomer. To discuss general macroscopic properties of magnetically soft and hard MAEs in a second step, computational homogenizations are preformed to calculate the composites' effective behavior for different microscopic particle arrangements. The approach is usable for the 2D and the 3D case, where both cases are compared. Finally, a macroscopic isotropic model for magnetically soft MAEs is parameterized by using homogenized data. It is used to analyze the behavior of MAEs for several sample geometries.

The organization of the paper is as follows: In Section 24.2, a general magneto-mechanical continuum formulation and the applied numerical framework are given. The microscopic constitutive models including their parametrization are described in Section 24.3. In Section 24.4, the multiscale approach is applied to study the behavior of several microstructures and in Section 24.5, a macroscopic model and its parametrization is shown. After a discussion of the results, the paper is closed by concluding remarks and an outlook to necessary future work.

24.2 Theoretical framework

In this section, the basic equations of the coupled magneto-mechanical BVP are summarized. This includes the field equations, general constitutive relations and a scale transition scheme which connects the micro- and the macroscale. Furthermore, an FE formulation, which is necessary for the numerical solution, is introduced. The theoretical framework is given in a condensed form. For more details, the reader is referred to the works [13, 28–30, 44].

24.2.1 Continuum formulation

According to the basic concept of continuum mechanics, a material body occupies the current and the reference configurations $\mathcal{B} \subset \mathcal{R}^3$ and $\mathcal{B}_0 \subset \mathcal{R}^3$ with boundaries $\partial\mathcal{B}$ and $\partial\mathcal{B}_0$, respectively. In the following, associated Eulerian and Lagrangian fields are denoted by lower- and uppercase letters. Vectors and higher-order tensors are indicated by boldface italic characters. The norms of vectors and second order tensors are written as $|\boldsymbol{v}| := \sqrt{v_l v_l}$ and $\|\boldsymbol{T}\| := \sqrt{T_{kl}T_{kl}}$, respectively. Therein, the tensor coordinates are given with respect to a Cartesian coordinate system and the Einstein summation convention is applied. Furthermore, the symbols ∇ and $\nabla_{\boldsymbol{X}}$ denote partial derivatives of a quantity $(\bullet)$ with respect to the spatial and reference configurations. Derivatives with respect to a tensor quantity $\boldsymbol{t}$ are given by $\partial_{\boldsymbol{t}}(\bullet)$ within the running text. Finally, the jump across material surfaces of discontinuity $\mathcal{S}$ and $\mathcal{S}_0$, with unit normal vectors $\boldsymbol{n}$ and $\boldsymbol{N}$ pointing from subdomains $\mathcal{B}^-$ to $\mathcal{B}^+$ and $\mathcal{B}_0^-$ to $\mathcal{B}_0^+$, is denoted by $[\![(\bullet)]\!] = (\bullet)^+ - (\bullet)^-$.

24.2.1.1 Field equations

Regarding the MAEs' microstructure, the embedded magnetizable particles are multidomain particles with a mean diameter of several micro meters, which is considerably larger than the polymer chain length. Thus, the assumption of a continuum and the following field equations are valid on the both considered scales of the material.

Kinematics

The first set of equations is given by kinematic relations. Each material point of the considered body is identified by its position vector $\boldsymbol{X} \in \mathcal{B}_0$ at time $t = t_0$. Due to the displacement $\boldsymbol{u}(\boldsymbol{X}, t) := \boldsymbol{\varphi}(\boldsymbol{X}, t) - \boldsymbol{X}$ of these points, the actual position changes to $\boldsymbol{x} \in \mathcal{B}$ at $t > t_0$. Therein, $\boldsymbol{\varphi}$ is a mapping function that is continuous in space and time, i.e. $[\![\boldsymbol{\varphi}]\!] = \boldsymbol{0}$ holds on $\mathcal{S}$. The deformation gradient $\boldsymbol{F}$ and its determinant J are defined by the relations

$$\boldsymbol{F} := (\nabla_{\boldsymbol{X}}\boldsymbol{\varphi})^{\mathrm{T}} \quad \text{and} \quad J := \det(\boldsymbol{F}) > 0. \tag{24.1}$$

Finally, the symmetric right Cauchy–Green deformation tensor $\boldsymbol{C} := \boldsymbol{F}^{\mathrm{T}}\cdot\boldsymbol{F}$ is introduced as a further kinematic quantity which is free of rigid body motions.

Maxwell equations and mechanical balance laws

The second set of relations are the two stationary magnetic Maxwell equations and the mechanical balance laws. For vanishing current densities $\boldsymbol{j}$ as well as neglected mechanical body forces $\varrho\boldsymbol{f}$ and inertia terms $\varrho\ddot{\boldsymbol{x}}$, these equations and corresponding jump conditions are given by

$$\nabla\cdot\boldsymbol{b} = 0 \quad \text{with} \quad \boldsymbol{n}\cdot[\![\boldsymbol{b}]\!] = 0 \quad \text{on } \mathcal{S}, \tag{24.2}$$

$$\nabla\times\boldsymbol{h} = \boldsymbol{0} \quad \text{with} \quad \boldsymbol{n}\times[\![\boldsymbol{h}]\!] = \boldsymbol{0} \quad \text{on } \mathcal{S}, \tag{24.3}$$

$$J\varrho - \varrho_0 = 0, \tag{24.4}$$

$$\nabla\cdot\boldsymbol{\sigma}^{\text{tot}} = \boldsymbol{0} \quad \text{with} \quad \boldsymbol{n}\cdot[\![\boldsymbol{\sigma}^{\text{tot}}]\!] = -\hat{\boldsymbol{t}} \quad \text{on } \mathcal{S} \quad \text{and} \tag{24.5}$$

$$\text{skw}\boldsymbol{\sigma}^{\text{tot}} = \boldsymbol{0} \tag{24.6}$$

with respect to the current configuration $\mathcal{B}$ [48–50]. In the equations above, $\boldsymbol{b}$ and $\boldsymbol{h}$ denote the magnetic induction and field which are linked by the permeability of free space μ_0 and the magnetization $\boldsymbol{m}$ via the equation

$$\boldsymbol{b} = \mu_0(\boldsymbol{h} + \boldsymbol{m}). \tag{24.7}$$

The quantities ϱ and ϱ_0 denote the mass density with respect to $\mathcal{B}$ and $\mathcal{B}_0$, respectively. Furthermore, the symbols $\boldsymbol{\sigma}^{\text{tot}}$ and $\hat{\boldsymbol{t}}$ indicate the total Cauchy stress tensor and a mechanical traction vector on $\mathcal{S}$, where $\boldsymbol{\sigma}^{\text{tot}}$ is given as the sum of the mechanical and magnetic stress tensors [28–30, 49].

Besides the introduced Eulerian quantities, Lagrangian fields related to $\mathcal{B}_0$ are defined. With them, an objective formulation of constitutive models is ensured. Furthermore, they are necessary for a total-Lagrangian FE formulation [28], which is applied here. The Lagrangian magnetic quantities $\boldsymbol{B}$, $\boldsymbol{H}$ and $\boldsymbol{M}$ as well as the total first Piola–Kirchhoff stress tensor $\boldsymbol{P}^{\text{tot}}$ and the corresponding mechanical traction vector $\hat{\boldsymbol{p}}$ are obtained from the pullback operations [28]

$$\boldsymbol{B} := J\boldsymbol{F}^{-1}\cdot\boldsymbol{b}, \quad \boldsymbol{H} := \boldsymbol{F}^{\text{T}}\cdot\boldsymbol{h} \quad \text{and} \quad \boldsymbol{M} := \boldsymbol{F}^{\text{T}}\cdot\boldsymbol{m} \quad \text{as well as} \tag{24.8}$$

$$\boldsymbol{P}^{\text{tot}} := J\boldsymbol{F}^{-1}\cdot\boldsymbol{\sigma}^{\text{tot}} \quad \text{and} \quad \hat{\boldsymbol{p}} := J|\boldsymbol{F}^{-\text{T}}\cdot\boldsymbol{N}|\hat{\boldsymbol{t}}. \tag{24.9}$$

It should be noted, that the introduced operations are not unique, but the choice according to Eqs. (24.8) and (24.9) leads to a preserved structure of the Eqs. (24.2)–(24.5) in their Lagrangian form [36, 38], e.g. $\nabla_{\boldsymbol{X}}\cdot\boldsymbol{B} = 0$.

In order to reduce the set of equations that has to be solved, it is favorable to introduce a magnetic potential. If, for example, the magnetic scalar potential φ which is defined by $\boldsymbol{H} =: -\nabla_{\boldsymbol{X}}\varphi$ is used, Eq. (24.3) is satisfied automatically and only Eq. (24.2) remains to solve the magnetic part of the BVP. Likewise, the magnetic vector potential $\boldsymbol{A}$ which is defined by the relation $\boldsymbol{B} =: \nabla_{\boldsymbol{X}}\times\boldsymbol{A}$ is usable. In contradiction to φ, now Eq. (24.2) is satisfied automatically and Eq. (24.3) has to be solved. A common way to ensure uniqueness of $\boldsymbol{A}$, is to demand the satisfaction of the Coulomb gauge $\nabla_{\boldsymbol{X}} \cdot \boldsymbol{A} = 0$. Note that this equation is fulfilled automatically in the 2D case but needs to be considered as a constraint in the 3D case [51].

Constitutive equations

Finally, the set of field equations is completed by the constitutive relations. In order to ensure thermodynamic consistency, the second law of thermodynamics has to be fulfilled for all values of the independent constitutive variables and their rates at any time t. Thus, the Clausius–Duhem inequality (CDI), which is derived from the balance of internal energy, the entropy balance, the second law of thermodynamics as well as the previously introduced balance equations has to be evaluated [49]. Assuming temperature fields that are homogeneous in space and time, the CDI in its Lagrangian form reduces to

$$-\dot{\Omega} + \boldsymbol{H} \cdot \dot{\boldsymbol{B}} + \boldsymbol{P}^{\mathrm{tot}} : \dot{\boldsymbol{F}} \geq 0 \quad \text{with} \quad \Omega := \varrho_0 \Psi\left(\boldsymbol{C}, \boldsymbol{B}, \boldsymbol{Q}^{\alpha}, \boldsymbol{Z}^{\beta}\right) + \frac{1}{2\mu_0 J} \boldsymbol{C} : (\boldsymbol{B} \otimes \boldsymbol{B}) \tag{24.10}$$

in the magneto-mechanical case. Therein, Ω denotes the so called amended free energy density function [36] which is decomposed into the specific Helmholtz free energy Ψ, that depends on $\boldsymbol{C}$ and $\boldsymbol{B}$ as well as possible vector or tensor-valued internal variables $\boldsymbol{Q}^{\alpha}$, $\boldsymbol{Z}^{\alpha}$ and a material independent free space part which follows from the pull back of the magnetic free field energy $\mu_0^{-1} \boldsymbol{b} \cdot \boldsymbol{b}$. Following the procedure of Coleman and Noll [52], the constitutive relations

$$\boldsymbol{P}^{\mathrm{tot}} = 2 \frac{\partial \Omega}{\partial \boldsymbol{C}} \cdot \boldsymbol{F}^{\mathrm{T}}, \quad \boldsymbol{H} = \frac{\partial \Omega}{\partial \boldsymbol{B}} \quad \text{and} \quad -\frac{\partial \Omega}{\partial \boldsymbol{Q}^{\alpha}} \cdot \dot{\boldsymbol{Q}}^{\alpha} - \frac{\partial \Omega}{\partial \boldsymbol{Z}^{\alpha}} : \dot{\boldsymbol{Z}}^{\alpha} \geq 0 \tag{24.11}$$

result from the evaluation of (24.10). The described formulation is suitable if the magnetic vector potential $\boldsymbol{A}$ is used, since $\boldsymbol{B}$ – which serves as independent constitutive variable in $\boldsymbol{\Omega}$ – could be derived directly from $\boldsymbol{A}$.

If instead a magnetic scalar potential formulation is chosen, it is favorable to use $\boldsymbol{H}$ instead of $\boldsymbol{B}$ as an independent constitutive variable. Therefore, a further amended free energy density function Ω^{*} is introduced by the Legendre–Fenchel transformation

$$\begin{aligned} \Omega^{*}\left(\boldsymbol{C}, \boldsymbol{H}, \boldsymbol{Q}^{\alpha}, \boldsymbol{Z}^{\beta}\right) &:= \inf_{\boldsymbol{B}} \left\{ \Omega\left(\boldsymbol{C}, \boldsymbol{B}, \boldsymbol{Q}^{\alpha}, \boldsymbol{Z}^{\beta}\right) - \boldsymbol{H} \cdot \boldsymbol{B} \right\} \\ &= \varrho_0 \Psi^{*}\left(\boldsymbol{C}, \boldsymbol{H}, \boldsymbol{Q}^{\alpha}, \boldsymbol{Z}^{\beta}\right) - \frac{\mu_0}{2} J \boldsymbol{C}^{-1} : (\boldsymbol{H} \otimes \boldsymbol{H}). \end{aligned} \tag{24.12}$$

Regarding Eq. (24.11), the second relation is now replaced by $\boldsymbol{B} = -\partial_{\boldsymbol{H}} \Omega^{*}$, where the other relations stay unchanged. As already stated, the presented set of equations holds on the micro- and the macroscale. The representation of the individual properties of the materials is inserted via a suitable choice of the free energy functions Ψ or Ψ^{*}.

24.2.1.2 Scale transition scheme

To connect microscopic and macroscopic quantities, a magneto-mechanical homogenization scheme is used. Depending on the choice of the independent variables, i.e. $\{\boldsymbol{u}, \boldsymbol{A}\}$ or $\{\boldsymbol{u}, \varphi\}$, the scale transition scheme has to be adjusted. In this work, the set of equations is exemplarily given for the scalar potential formulation. The reader is referred to Metsch et al. [29] or Kalina et al. [13] for the case of a vector potential formulation.

An effective macroscopic quantity $(\bar{\bullet})$, that is related to the macroscale, is identified from the microscopic field distribution within a representative volume element (RVE) by the volume average $\langle(\bullet)\rangle$. To ensure a physically meaningful scale transition, the equivalence of the macroscopic and the averaged microscopic energies which is also known as the Hill–Mandel condition [53] has to be fulfilled. In the magneto-mechanical case this relation is given by

$$\langle \boldsymbol{P}^{\text{tot}} : \dot{\boldsymbol{F}} \rangle - \langle \boldsymbol{B} \cdot \dot{\boldsymbol{H}} \rangle = \overline{\boldsymbol{P}}^{\text{tot}} : \dot{\overline{\boldsymbol{F}}} - \overline{\boldsymbol{B}} \cdot \dot{\overline{\boldsymbol{H}}}. \tag{24.13}$$

This condition is fulfilled if, for instance, periodic displacement and potential spaces in combination with antiperiodic fluxes are chosen:

$$\mathcal{U}(\overline{\boldsymbol{F}}) := \left\{ \boldsymbol{u} \in \mathcal{R}^d \middle| \boldsymbol{u} = \left(\overline{\boldsymbol{F}} - \boldsymbol{I}\right) \cdot \boldsymbol{X} + \tilde{\boldsymbol{u}} \text{ with } \tilde{\boldsymbol{u}}^+ = \tilde{\boldsymbol{u}}^- \right\}, \tag{24.14}$$

$$\mathcal{P}(\overline{\boldsymbol{H}}) := \left\{ \varphi \in \mathcal{R} \middle| \varphi = -\overline{\boldsymbol{H}} \cdot \boldsymbol{X} + \tilde{\varphi} \text{ with } \tilde{\varphi}^+ = \tilde{\varphi}^- \right\} \text{ and} \tag{24.15}$$

$$(\boldsymbol{N} \cdot \boldsymbol{P}^{\text{tot}})^+ = -(\boldsymbol{N} \cdot \boldsymbol{P}^{\text{tot}})^- \quad \text{and} \quad (\boldsymbol{N} \cdot \boldsymbol{B})^+ = -(\boldsymbol{N} \cdot \boldsymbol{B})^-. \tag{24.16}$$

In Eqs. (24.14)–(24.16), $(\bullet)^+$ and $(\bullet)^-$ are values on associated opposing boundaries $\partial\mathcal{B}_0^{\alpha+}$ and $\partial\mathcal{B}_0^{\alpha-}$, $\alpha \in \{1, \ldots, d\}$ of the RVE with the dimension d. The symbol $(\tilde{\bullet})$ marks a fluctuation quantity.

24.2.2 Numerical solution

The introduced nonlinear magneto-mechanical BVP is solved by using a total Lagrangian FE formulation. As in the previous subsection, the general equations are exemplarily given for the case of a scalar potential formulation. It is based on the weak forms

$$0 = \int_{\mathcal{B}_0} \boldsymbol{P}^{\text{tot}} : \delta\boldsymbol{F} \, \mathrm{d}V_0 - \oint_{\partial\mathcal{B}_0} \hat{\boldsymbol{p}}^{\text{tot}} \cdot \delta\boldsymbol{u} \, \mathrm{d}S_0 \quad \text{and} \tag{24.17}$$

$$0 = \int_{\mathcal{B}_0} \boldsymbol{B} \cdot \delta\boldsymbol{H} \, \mathrm{d}V_0 + \oint_{\partial\mathcal{B}_0} \hat{\eta} \delta\varphi \, \mathrm{d}S_0 \tag{24.18}$$

of the Lagrangian versions of Eqs. (24.2) and (24.5), where $\delta\boldsymbol{u}$ and $\delta\varphi$ denote the virtual displacement and scalar potential which have to vanish on the essential boundaries $\partial\mathcal{B}_0^u$ and $\partial\mathcal{B}_0^\varphi$, respectively [30]. The virtual deformation gradient $\delta\boldsymbol{F}$ and the virtual Lagrangian magnetic field $\delta\boldsymbol{H}$ are defined by $\delta\boldsymbol{F} := (\nabla_X \delta\boldsymbol{u})^{\mathrm{T}}$ and $\delta\boldsymbol{H} := -\nabla_X \delta\varphi$, respectively. Furthermore, $\widehat{\boldsymbol{p}}^{\mathrm{tot}}$ and $\widehat{\eta}$ denote prescribed nominal stress vectors including magnetic and mechanical loads on $\partial\mathcal{B}_0^p$ and fluxes $\boldsymbol{B}{\cdot}\boldsymbol{N}$ on $\partial\mathcal{B}_0^\eta$, respectively.

The weak forms according to Eqs. (24.17) and (24.18) are solved numerically by approximating the configuration $\mathcal{B}_0$ with n^{el} finite elements capturing the domains $\mathcal{B}_0^e$, in which the independent primary fields $\{\boldsymbol{u}, \varphi\}$ and their virtual counterparts $\{\delta\boldsymbol{u}, \delta\varphi\}$ are approximated with the shape functions N^α. Here, α denotes the global node number. Due to the arbitrariness of $\delta\boldsymbol{u}$ and $\delta\varphi$, the mechanical and magnetic residues

$$\mathbf{R}^\alpha := \bigcup_{e=1}^{n^{\mathrm{el}}} \left[\int_{\mathcal{B}_0^e} \nabla_X N^\alpha \cdot \boldsymbol{P}^{\mathrm{tot}} \mathrm{d}V_0 - \int_{\partial\mathcal{B}_0^{p,e}} N^\alpha \widehat{\boldsymbol{p}}^{\mathrm{tot}} \mathrm{d}S_0 \right] = 0 \quad \text{and} \tag{24.19}$$

$$\mathrm{R}^\alpha := \bigcup_{e=1}^{n^{\mathrm{el}}} \left[\int_{\mathcal{B}_0^e} \nabla_X N^\alpha \cdot \boldsymbol{B} \mathrm{d}V_0 - \int_{\partial\mathcal{B}_0^{\eta,e}} N^\alpha \widehat{\eta} \mathrm{d}S_0 \right] = 0 \tag{24.20}$$

follow from the weak forms given in Eqs. (24.17) and (24.18). Within the FE code, the set of coupled nonlinear Eqs. (24.19) and (24.20) is solved by applying a Newton–Raphson scheme at each time step t_n. To this end, the linear algebraic system

$$\begin{bmatrix} \mathbf{K}_{uu}^{\alpha\beta} & \mathbf{K}_{u\varphi}^{\alpha\beta} \\ \mathbf{K}_{\varphi u}^{\alpha\beta} & \mathrm{K}_{\varphi\varphi}^{\alpha\beta} \end{bmatrix} \begin{bmatrix} \Delta\boldsymbol{u}^\beta \\ \Delta\varphi^\beta \end{bmatrix} = - \begin{bmatrix} \mathbf{R}^\alpha \\ \mathrm{R}^\alpha \end{bmatrix} \quad \text{with } \alpha, \beta \in \left\{1, 2, \ldots, n^{\mathrm{nd}}\right\} \tag{24.21}$$

has to be solved for the total number of global nodes nnd in each iteration. Therein, $\mathbf{K}_{xy}^{\alpha\beta}$ with x, y being u or φ, denote the tangent terms which follow from Gâteaux-derivations of $\mathbf{R}^\alpha$ and R^α with respect to $\Delta\boldsymbol{u}$ and $\Delta\varphi$, respectively [28, 30]. If a homogenization problem according to the subsection 24.2.1.2 is solved, the primary fields $\boldsymbol{u}$ and φ and thus the nodal values $\boldsymbol{u}^\alpha$ and φ^α have to be elements of the spaces $\mathcal{U}(\overline{\boldsymbol{F}})$ and $\mathcal{P}(\overline{\boldsymbol{H}})$ according to (24.14) and (24.15).

All simulations were performed within a MATLAB-based in-house FE framework using the PARDISO solver project [54, 55] or, if large 3D problems are considered, the software tool FEniCS [56, 57].

24.3 Characterization of the constituents

As shown in Figure 24.1(a), the MAEs' heterogeneous microstructure consists of subdomains $\mathcal{B}_0^{\mathrm{p}}$ and $\mathcal{B}_0^{\mathrm{m}}$ denoting the particle and the matrix material phases. In the following section, the behavior of these individual constituents is characterized by using experimental results. Based on this, constitutive models which describe the behavior of an MAE at the microscopic level are given for both phases.

24.3.1 Particles

The magnetizable particles are very stiff compared to the soft elastomer matrix, so that $\boldsymbol{C} \approx \boldsymbol{I}$ holds in $\mathcal{B}_0^{\mathrm{p}}$. Therefore, the magneto-mechanical coupling effects in the particles are negligible and the free energy functions introduced in Eqs. (24.10) and (24.12) could be divided into a purely mechanical part and a purely magnetic part [13, 27, 28, 58, 59], i.e.

$$\Psi(\boldsymbol{C}, \boldsymbol{B}, \boldsymbol{Q}^\alpha) := \Psi^{\mathrm{mech}}(\boldsymbol{C}) + \Psi^{\mathrm{mag}}(\boldsymbol{B}, \boldsymbol{Q}^\alpha) \quad \text{or} \tag{24.22}$$

$$\Psi^*(\boldsymbol{C}, \boldsymbol{H}, \boldsymbol{Q}^\alpha) := \Psi^{*\mathrm{mech}}(\boldsymbol{C}) + \Psi^{*\mathrm{mag}}(\boldsymbol{H}, \boldsymbol{Q}^\alpha). \tag{24.23}$$

Independent of the magnetic properties of the particles, the mechanical part is described by a compressible neo-Hookean potential

$$\Psi^{\mathrm{mech}}(I_1, J) := \frac{1}{2\rho_0}\left[\mu(I_1 - \ln J^2 - 3) + \frac{\lambda}{2}(J^2 - \ln J^2 - 1)\right] \tag{24.24}$$

with $I_1 := \operatorname{tr} \boldsymbol{C}$ and λ and μ denoting the Lamé parameters. The particles stiffness should be chosen such that it is large compared to the one of the matrix. With that, they are represented as nearly rigid in the FE calculation. The description of the magnetic part depends on the properties of the particles, where a distinction between magnetically soft and hard behavior is done in the following.

24.3.1.1 Magnetically soft particles

Since magnetically soft materials do not exhibit a load dependent behavior, the actual state of the material only depends on the chosen primary magnetic quantity, i.e. $\boldsymbol{B}$ or $\boldsymbol{H}$. Exemplarily, a suitable description is shown for a scalar potential formulation in the following.

Regarding a polycrystalline material as carbonyl iron or nickel with a uniform distribution of the anisotropic crystals inside the particle, an isotropic behavior is observed at the microscopic level where the particles are represented as a continuum. Thus, $\Psi^{*\mathrm{mag}}$ only depends on the norm of the Lagrangian magnetic field $|\boldsymbol{H}|$. In order to capture the characteristic saturation effect of ferromagnetic materials, the Langevin function

$$\Psi^{*\mathrm{mag}}(\boldsymbol{H}) := -\frac{\mu_0}{2\varrho_0}\int_0^{|\boldsymbol{H}|} M^{\mathrm{s}}\left[\coth(\alpha H) - (\alpha H)^{-1}\right]\mathrm{d}H, \tag{24.25}$$

is used [30, 44]. Therein, M^{s} and α denote the saturation magnetization and a scaling factor, respectively. Experimental data for carbonyl iron and the fitted model with $M^{s} = 841\ \mathrm{kA\ m^{-1}}$ and $\alpha = 2.18 \times 10^{-5}\ \mathrm{mA^{-1}}$ are given in Figure 24.1(a).

24.3.1.2 Magnetically hard particles

In contrast to magnetically soft materials, the behavior of magnetically hard materials as NdFeB strongly depends on the load history. Thus, internal variables $\boldsymbol{Q}^{\alpha}$ are used to describe the actual state of the material.

A phenomenological constitutive model which is based on the work of Bergqvist [60] and has been published by Kalina et al. [13] is shown in the following. Again, polycrystalline materials with isotropic properties are regarded. Within the model, the pinning of domain wall motion is considered as the cause of hystereses in ferromagnets [61]. Since it is not sufficient to represent the pinning within a real polycrystalline material by only one pinning strength k, it is modeled by N independent so called pseudoparticles [60]. Each of them is characterized by the individual strength k^{α} and a related weight ω^{α} which can be interpreted as the probability that a magnetic moment in the material belongs to the respective pseudoparticle. In this case, the free energy function is represented by

$$\varrho_0 \Psi^{\mathrm{mag}}(\boldsymbol{B}, \boldsymbol{M}^{\alpha}) := -\sum_{\alpha=1}^{N} \omega^{\alpha} \boldsymbol{B} \cdot \boldsymbol{M}^{\alpha} + \sum_{\alpha=1}^{N} \omega^{\alpha} \int_{0}^{|\boldsymbol{M}^{\alpha}|} 1/y \tanh^{-1}(M^{\alpha}/M^{\mathrm{s}}) \mathrm{d}M^{\alpha}. \tag{24.26}$$

In the equation above, $\boldsymbol{M}^{\alpha}$ denotes a part of the magnetization belonging to the weight ω^{α}. According to the general constitutive relations given in Eq. (24.11), the equations

$$\boldsymbol{H} = \frac{\partial \Omega}{\partial \boldsymbol{B}} = \frac{1}{\mu_0 J} \boldsymbol{C} \cdot \boldsymbol{B} - \sum_{\alpha=1}^{N} \omega^{\alpha} \boldsymbol{M}^{\alpha} \quad \text{and} \tag{24.27}$$

$$\mathcal{D} = -\sum_{\alpha=1}^{N} \frac{\partial \Omega}{\partial \boldsymbol{M}^{\alpha}} \cdot \dot{\boldsymbol{M}}^{\alpha} = \sum_{\alpha=1}^{N} \omega^{\alpha} \boldsymbol{B}^{\mathrm{i}\alpha} \cdot \dot{\boldsymbol{M}}^{\alpha} \geq 0 \tag{24.28}$$

follow in which $\boldsymbol{B}^{\mathrm{i}\alpha} = \boldsymbol{B} - y^{-1} \tanh^{-1}(|\boldsymbol{M}^{\alpha}|/M^{\mathrm{s}}) |\boldsymbol{M}^{\alpha}|^{-1} \boldsymbol{M}^{\alpha}$ is defined as the irreversible induction field and $\mathcal{D}$ denotes the rate of dissipation. Applying the principle of maximum dissipation [62], the rate independent behavior of the material is given by the evolution equations for $\boldsymbol{M}^{\alpha}$, the switching conditions ϕ^{α} and the related Karush–Kuhn–Tucker conditions:

$$\dot{\boldsymbol{M}}^{\alpha} = \lambda^{\alpha} \frac{\boldsymbol{B}^{\mathrm{i}\alpha}}{|\boldsymbol{B}^{\mathrm{i}\alpha}|}, \phi^{\alpha} := |\boldsymbol{B}^{\mathrm{i}\alpha}| - k^{\alpha} \quad \text{and} \quad \lambda^{\alpha} \phi^{\alpha} = 0 \wedge \lambda^{\alpha} \geq 0 \wedge \phi^{\alpha} \leq 0. \tag{24.29}$$

Therein, λ^{α} and k^{α} denote a Lagrange parameter and the pinning strength, respectively. Experimental data for NdFeB and the adopted model with $N = 10$ are shown in Figure 24.1(c). The fitted set of parameters ω^{α}, k^{α}, y and M^{s} are given in Kalina et al. [13].

In order to incorporate the introduced constitutive model into a FE procedure according to subsection 24.2.2, the set of equations given in (24.29) has to be solved and the algorithmically consistent material tangent $\partial_{\Delta B}\Delta \boldsymbol{H}$ has to be calculated at each integration point. This is done by an implicit Euler scheme which is adapted from the theory of plasticity.

24.3.2 Polymer matrix

Besides the particles, the MAEs' effective behavior essentially depends on the properties of the polymer matrix. In order to achieve large coupling effects in the magnetoactive composite material, the stiffness of the matrix has to be preferably low but suitable regarding technical applications. A common choice for the production of MAEs are e.g. silicone elastomers [9–11, 47]. In order to consider a realistic behavior of such a material within the presented modeling approach, uniaxial tension tests for Zhermack silicone ZA 8 LT shore 8 from Troll Factory have been carried out. The geometry of the sample with circular cross section and the jig for the load application are taken from Dohmen [63].

24.3.2.1 Sample production and testing

The samples were manufactured by mixing a base liquid, a catalyst liquid, and varying amounts of silicone oil together. Furthermore, a slight amount of silicone-based white dye is added to colorize the samples for local strain measurements. To study the influence of silicone oil on the mechanical properties, five sets a six samples, with the oil volume fractions $\phi_{\text{oil}} = \{0, 10, 20, 30, 40\}\%$ were produced. The remaining volume fractions are distributed equally between base and catalyst. After assembling and stirring all chemicals into a beaker at room temperature, the compound was depressurized in a vacuum chamber for 10 min. In each manufacturing cycle, the liquid suspension was filled into two separate polytetrafluoroethylene (PTFE)-moulds, where metal caps with holes—the connectors for later mechanical testing—are placed inside. After waiting at least 2 h, the two samples are demoulded.

All manufactured samples were tested with a tensile test machine, where the local strains are measured via digital image correlation with the tool Aramis from GOM. To this end, patterns are sprayed on the sample surfaces, cf. the photograph in Figure 24.2(b). The maximum applied local stretch in the loading direction was $\lambda_{\parallel} \approx 2$. Before the recording of the stress-strain curves, all samples were preconditioned to avoid non-reproducibility due to the Mullins effect [64]. The samples were pulled with a displacement rate of 20 mm min^{-1}. An increase of the rate up to 200 mm min^{-1} had almost no influence on the measured stress–strain curves. Thus, the material could be regarded as ideally elastic. Altogether, the measured stress–strain curves shown in Figure 24.2(a) reveal a highly nonlinear response for all amounts of silicone oil. Furthermore, the recorded perpendicular stretches $\lambda_{\perp}$ indicate a nearly ideal incompressible behavior within the examined strain range, cf. Figure 24.2(b).

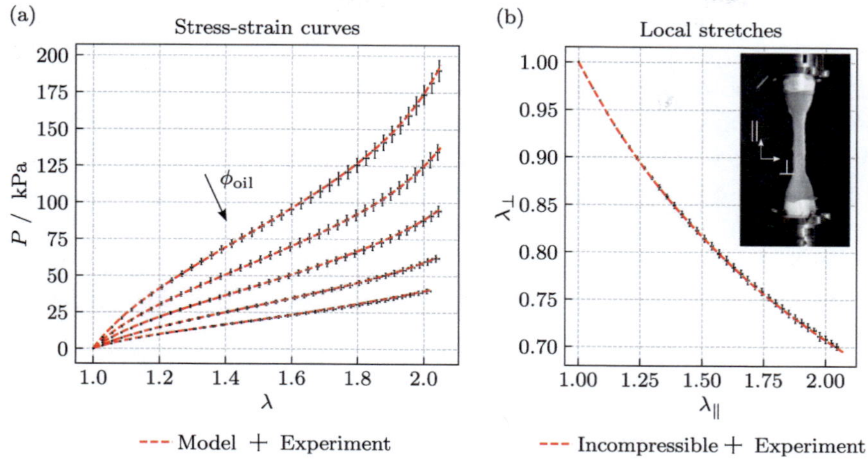

Figure 24.2: Characterization of a silicone elastomer by uniaxial tension tests: (a) experimentally determined stress strain curves for the amounts of silicone oil $\phi_{oil} = \{0, 10, 20, 30, 40\}\%$ and adjusted Ogden model (24.31), and (b) measured local stretches $\lambda_\|$ and $\lambda_\perp$ for $\phi_{oil} = 0\%$ in the load direction and perpendicular as well as ideal incompressible stretch curve. The directions $\|$ and $\perp$ are shown in the photograph of a tested specimen with sprayed patterns for the local strain measurement.

Finally it turns out, that the resulting shear modulus of the polymer, given in Table 24.1, could be adjusted in a specific way by the amount of added silicone oil within the production process. The dependence of μ_{init} on ϕ_{oil} is however nonlinear and could be described by

$$\mu_{init}(\phi_{oil}) = a \exp(-b\phi_{oil}) + c. \tag{24.30}$$

The parameters in Eq. (24.30) are determined in such a way, that $\mu_{init}(100\%) = 0\text{kPa}$ and yield $a = 103.91$ kPa, $b = 0.032$ and $c = -4.23$ kPa.

Table 24.1: Identified parameters of the Ogden model (24.31) for different amounts of silicone oil ϕ_{oil} and related initial shear moduli μ_{init}. The compression moduli κ are given for a Poisson's ratio of $\nu = 0.49$.

ϕ_{oil}	μ_{init}/kPa	μ_1/kPa	μ_2/kPa	μ_3/kPa	α_1	α_2	α_3	κ/kPa
0 %	99.17	−22.67	24.17	9×10^{-5}	−6.61	2	18.78	4.93×10^3
10 %	71.79	−15.88	19.33	8.16×10^{-5}	−6.58	2.03	18.29	3.56×10^3
20 %	52.09	−11.78	12.31	4.07×10^{-5}	−6.67	2.08	18.18	2.58×10^3
30 %	33.55	−13.63	2.81	–	−3.99	4.53	–	1.67×10^3
40 %	25.19	−6.85	1.72	–	−6.85	2.00	–	1.25×10^3

24.3.2.2 Model and parameter identification

Within the microscopic modeling approach, the polymer matrix is regarded as non-magnetizable. Furthermore, due to the negligible rate dependency of the characterized silicone elastomer, possible internal variables $\boldsymbol{Z}^{\alpha}$ do not have to be considered, so that Ψ within $\mathcal{B}_0^{\mathrm{m}}$ only depends on $\boldsymbol{C}$. If the polymer is regarded as isotropic, it follows $\Psi(\boldsymbol{C}) = \Psi(I_1, I_2, I_3) = \Psi(\lambda_1, \lambda_2, \lambda_3)$, where I_i and λ_i denote the invariants of the deformation tensor $\boldsymbol{C}$ and the principal stretches, respectively.

To describe the observed strongly nonlinear response, the ideal incompressible model

$$\begin{aligned} \varrho_0 \Psi &= \varrho_0 \Psi^{\mathrm{dev}}\left(\lambda_1^{\mathrm{iso}}, \lambda_2^{\mathrm{iso}}, \lambda_3^{\mathrm{iso}}\right) - p(J-1) \\ &:= \sum_{p=1}^{N} \frac{\mu_p}{\alpha_p}\left(\lambda_1^{\mathrm{iso}\alpha_p} + \lambda_2^{\mathrm{iso}\alpha_p} + \lambda_3^{\mathrm{iso}\alpha_p} - 3\right) - p(J-1) \end{aligned} \tag{24.31}$$

of the Ogden type is used, where $\lambda_i^{\mathrm{iso}} := \lambda_i J^{-1/3}$ are the isochoric principal stretches according to the Flory-split [65]. In the equation above, μ_p, α_p and p denote parameters of the Ogden model and the hydrostatic pressure, respectively. The adjusted model (31) and the measured uniaxial tension tests for all analyzed amounts of silicone oil are shown in Figure 24.2(a). Within the parameter identification, a model with $N = 3$ sets has been chosen for $\phi_{\mathrm{oil}} = \{0, 10, 20\}\%$. In case of $\phi_{\mathrm{oil}} = \{30, 40\}\%$, the adjustment is already accurate with $N = 2$ sets. Furthermore, to describe the initial shear modulus μ_{init} exactly and to ensure a physically meaningful behavior, the conditions

$$0 = \frac{1}{2}\sum_{p=1}^{N} \alpha_p \mu_p - \mu_{\mathrm{init}} \quad \text{and} \quad \alpha_p \mu_{(p)} \geq 0 \; \forall \, p \in 1 \ldots N \tag{24.32}$$

are used in the least square fitting algorithm as constraints. The determined parameter sets are given in Table 24.1. Altogether, the model is able to describe the highly nonlinear behavior up to the measured strains of approximately 100% quiet well.

To incorporate the parameterized model into a standard FE simulation, according to section 24.2.2, a compressible model is used instead of the ideal incompressible one. Thus, the latter term in Eq. (24.31) is replaced by the volumetric energy function $\varrho_0 \Psi^{\mathrm{vol}}(J) := \frac{\kappa}{4}(J^2 - 2\ln J - 1)$, with κ denoting the initial compression modulus. In order to ensure a stable FE calculation with a nearly incompressible stress response, a Poisson's ratio of $\nu = 0.49$ and, consequently, a compression modulus of $\kappa = 2/3\mu_{\mathrm{init}}(1+\nu)/(1-2\nu)$ is chosen. With that, the deviation to the parameterized incompressible model (24.31) is below 1.5% for stretches up to $\lambda = 1.7$ and below 4.5 % for the largest measured strains, respectively.

If only processes within the range of small or moderate strains are regarded, the stress–strain curve could be described by a comparatively simple neo-Hookean model which results from (24.31) for $N = 1$ and $\alpha_p = 2$. The corresponding Young's modulus follows to $E = 2\mu_{\mathrm{init}}(1 + \nu)$.

24.4 Study on the effective behavior of MAEs

Besides the previously characterized properties of the particles and matrix, the effective behavior of MAEs results from the arrangement and the interplay of the components on the microscale. In order to analyze the microscopically-driven behavior, the homogenization scheme given in subsection 24.2.1.2 is used to discuss basic macroscopic effects in the following. This is done within three studies for MAEs filled with magnetically soft and hard particles, where the polymer matrix is described by a simplified neo-Hookean potential in a first step.

24.4.1 Homogenization of magnetically soft MAEs

At first, the resulting properties of magnetically soft MAEs are discussed. Thereby, the magnetic behavior of carbonyl iron is regarded and the polymer matrix is described by a reduced neo-Hookean model. Thus, the resulting effective behavior is fully reversible but highly nonlinear.

24.4.1.1 Analyses of the MR effect

One of the most popular properties of MAEs is their ability to change the effective stiffness if an external field is applied. This phenomenon, also known as the magnetorheological (MR) effect, is examined for 3D cubic and random microstructures with varying particle volume fractions within this subsection. The parameters of the neo-Hookean elastic model describing the polymer matrix are chosen to $E = 100$ kPa and $\nu = 0.49$.

In order to determine the magnetically induced MR-effect, several simulations have to be performed for each analyzed microstructure. If the behavior under shearing is investigated, this can be done by increasing an effective field $\overline{\boldsymbol{H}}$ to a specific value $\overline{H}^{\max}$ in a first step and applying a shear deformation while holding $\overline{\boldsymbol{H}}$ constant, afterwards. Finally, the resulting effective mechanical shear stress which is related to the initial shear modulus $\overline{G}$ is calculated from the numerical homogenization. In order to ensure a virtual linear stress-strain-curve, a maximum shear deformation of $\overline{\gamma} = 1\%$ is applied. The described procedure is repeated for maximum magnetic fields $\overline{H}^{\max}$ in the range of 0–2000 kA m^{-1}, so that $\overline{G}$ could be determined as a function of $\overline{\boldsymbol{H}}$. According to the sketch given in Figure 24.3(b), the prescribed effective magnetic field and deformation are chosen as

$$[\overline{H}_K] = \begin{bmatrix} 0 \\ \overline{H}^{\max} \\ 0 \end{bmatrix}, \quad [\overline{F}_{lK}] = \begin{bmatrix} 1 & \overline{\gamma} & 0 \\ 0 & 1 & 0 \\ 0 & 0 & 1 \end{bmatrix}, \qquad (24.33)$$

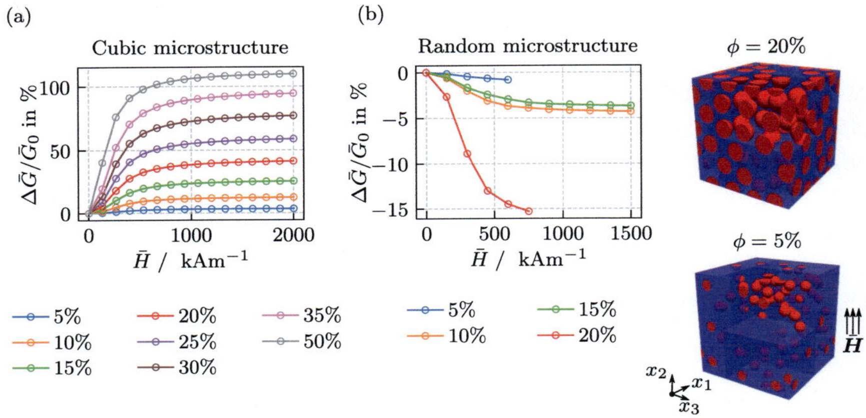

Figure 24.3: Predicted magnetorheological effect considering a simple shear deformation with $\bar{\gamma} = 1\%$: (a) simplified cubic microstructures with different particle volume fractions ϕ, and (b) random microstructure and considered 3D microstructures with 100 spherical particles.

where $[(\bullet)]$ denotes the specification of tensor coordinates with respect to a Cartesian frame. Consequently, the shear stress $\overline{P}_{21}$ is proportional to $\overline{G}$ which is changed by the magnetic loading. The FE meshes of the investigated cubic and random RVEs were generated with an in-house script for the tool HyperMesh. In case of the random RVE geometries with 100 included spherical particles, the minimum particle distance was set to $d/5$ within the positioning algorithm, cf. Figure 24.3(b).

Cubic microstructure

Initially, the behavior of cubic microstructures with varying particle volume fractions $\phi = \{10, 15, 20, 25, 30, 35, 50\}\%$ are analyzed. For all volume fractions, a magnetically induced stiffening could be observed, where the magnitude of the MR-effect increases with the particle volume fraction, see Figure 24.3(a). Regarding the curve $\Delta\overline{G}/\overline{G}_0$ over $\overline{H}$, the typical quadratic behavior could be observed for small magnetic fields. For large values of $\overline{H}$, a clear saturation effect is visible which leads to an overall S-shaped curve for each particle volume fraction. The predicted curves are in qualitative accordance with experimental results [66, 67].

Random microstructure

Furthermore, the behavior of more realistic MAEs with underlying random particle distributions is investigated. Within the analyses, the particle volume fractions $\phi = \{5, 10, 15, 20\}\%$ are investigated. Regarding the results given in Figure 24.3(b), several findings could be made. Equal to the cubic microstructure, the typical S-shaped curve is predicted. However, it stands out that the predicted MR-effect is negative for all considered microstructures which is in contradiction to experimental observations.

Furthermore, the predicted effect seems not to increase in a systematic way, since e.g. the decrease in stiffness of $\phi = 15\%$ is less compared to the one of $\phi = 10\%$ whereas it again increases for $\phi = 20\%$. Results which demonstrate a similar unsystematic response of the composites' MR-effect are shown by Gebhart et al. [45] for 2D random microstructures.

Since the overall MR-effect depends on the microscopic magnetic interactions—which increases for closer particle distances—and the properties of the polymer, a reason for the observed decrease in stiffness cannot be stated easily. Probably, a positive effect only occurs if the initial particle distance is reduced to lower values, because the magnetic attraction force increases nonlinear with respect to the particle distance [24, 68]. However, this supposition cannot be confirmed at the moment, since the performed simulations are getting already unstable for the analyzed RVEs with a minimum particle distance of $d/5$. Due to that, the simulated curves for $\phi = \{5, 20\}\%$ are not completed up to $\overline{H} = 1500\,\mathrm{kA\,m^{-1}}$.

24.4.1.2 Comparison of 2D and 3D simulations

Due to the enormous computational cost of realistic 3D simulations which explicitly resolve the magnetizable particles, many computational analyses based on microscopic approaches are restricted to the 2D case. Therefore, in order to assess whether those predictions are still sensible and can provide an understanding of principle microscopic mechanisms inside MAEs, a comparison of 2D and 3D simulations according to Metsch et al. [30] is shown here. A simplified cubic microstructure containing only one spherical or even circular particle is investigated, whereby RVEs with the same geometry (Cubic^{2D} and Cubic$^{3D}_{d}$) as well as the same particle-volume fraction (Cubic^{2D} and Cubic$^{3D}_{\phi}$) of $\phi \approx 20\%$ are compared.

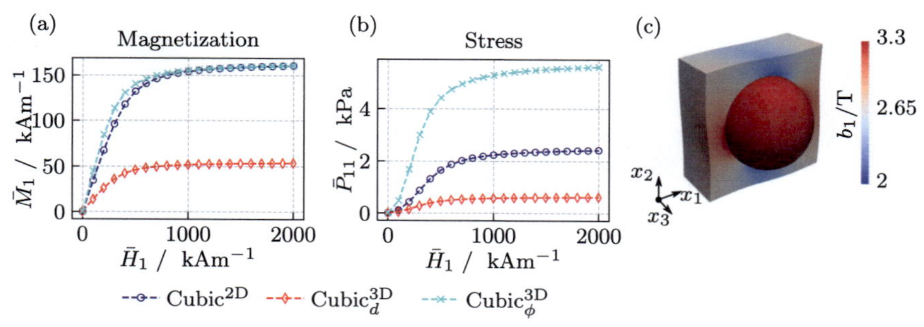

Figure 24.4: Comparison of cubic RVEs: (a) computed effective magnetization $\overline{\boldsymbol{M}}$ and (b) mechanical stress $\overline{\boldsymbol{P}}^{\mathrm{tot}}$ for $\overline{\boldsymbol{F}} = \boldsymbol{I}$, and (c) local induction field within the 3D geometry Cubic$^{3D}_{\phi}$ for shear loading.

While the latter two RVEs predict a virtually coinciding $\overline{M}$–$\overline{H}$ curve, its maximum value is reduced by a factor of three for the Cubic$_d^{3D}$ microstructure, cf. Figure 24.4(a). This effect is directly related to the differences in ϕ, whereas the variations in Cubic^{2D} and Cubic$_\phi^{3D}$ are a result of differences in the related cylindrical and spherical inclusions.

Comparing the effective mechanical stresses for $\overline{\boldsymbol{F}} = \boldsymbol{I}$ in Figure 24.4(b), a qualitative agreement can be found for all RVEs, i.e. a positive, S-shaped component $\overline{P}_{11}$ in dependence of $\overline{H}_1$. In contrast to that, considerable differences become apparent in the quantitative results. Due to the differing demagnetization factors and the mechanical behavior of cylinders and spheres, the magnetically induced particle interactions vary considerably even if the particle volume fraction is held constant.

24.4.2 Homogenization of magnetically hard MAEs

In contrast to magnetically soft MAEs, the usage of magnetically hard fillers as NdFeB in the production process leads to composites with irreversible and strongly path-dependent behavior. For instance, measurements presented by Linke et al. [14] show, that the overall magnetization loops of MAEs based on a soft silicone elastomer matrix filled with NdFeB are significantly smaller compared to the loops of NdFeB particles fixed in a stiff epoxy resin matrix, cf. Figure 24.5(a). In the following, the results presented in Kalina et al. [13, 69] are summarized. Therein, the elastomer matrix is represented by a neo-Hookean model with $E = 100$ kPa and $\nu = 0.49$.

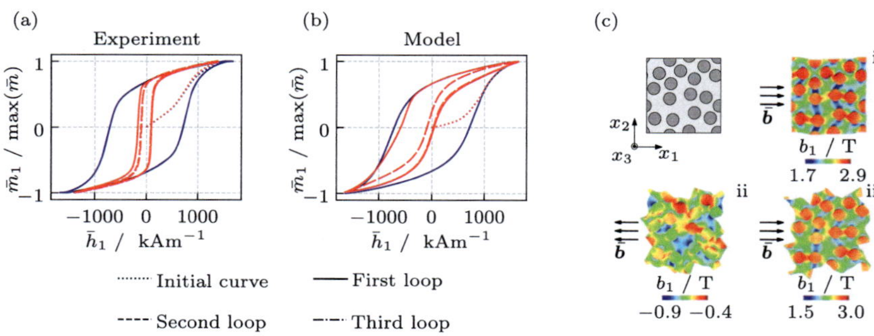

Figure 24.5: Magnetic hystereses of composites filled with NdFeB particles: (a) measured effective behavior of epoxy sample (blue lines) and MAE sample (red lines) with a low modulus elastomer of $E^m \approx 64$kPa, (b) calculated hysteresis loops for epoxy and MAE with $E^m = 100$ kPa, and (c) evolution of the MAE microstructure during the first hysteresis loop and local b_1-field. Experimental data are taken from Linke et al. [14] and Kalina et al. [13].

24.4.2.1 Magnetization loops

In order to investigate the effective behavior, numerical homogenizations with RVEs containing randomly distributed circular inclusions are examined. During the simulation, the macroscopic deformation is fixed, i.e. $\overline{\boldsymbol{F}} = \boldsymbol{I}$. The predicted initial magnetization curve as well as the first and second hysteresis loops of the MAE are compared to the hysetereses of a composite with NdFeB particles embedded into an epoxy resin matrix, see Figure 24.5(b). In accordance with the measurement, the second major hysteresis loop of the MAE is significantly smaller.

As demonstrated in Kalina et al. [13] from a theoretical point of view, this effect results from magnetically induced particle rotations within the polymer. Due to these rotations, the inversion of the MAE's overall magnetization is expedited compared to the epoxy sample, in which the inversion completely results from local domain switching processes within the mechanically fixed particles. Furthermore, the particle rotations insert stored strain energy into the polymer matrix. This energy storage appears after the first inversion of the external magnetic field and leads to an inverse rotation of the particles if the external field is equal to zero. Thus, the switching of $\overline{\boldsymbol{m}}$ within the following ascending branch and the second hysteresis loop is faster than the initial process and consequently, an asymmetrically shaped first hysteresis loop occurs. Finally, the simulations indicate that the reversion of $\overline{\boldsymbol{m}}$ is a result of particle rotations and internal domain conversion processes. The evolution of the MAE microstructure during the first hysteresis loop and the related local b_1-field are given in Figure 24.5(c).

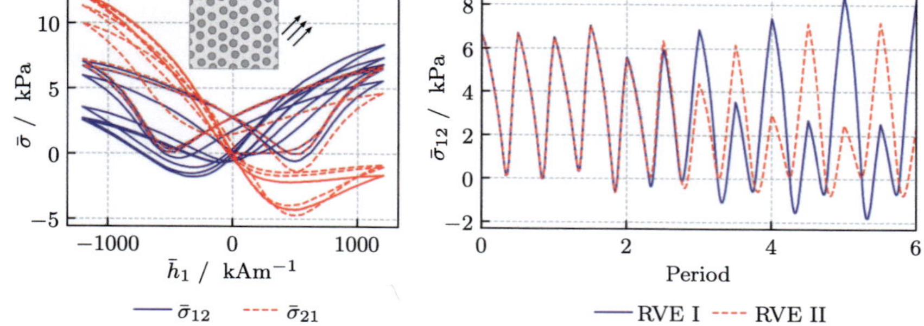

Figure 24.6: Simulation of a magnetically hard MAEs: (a) mechanical butterfly hystereses for an alternating magnetic load in the 45°-direction, and (b) comparison of the stress response of two RVEs with almost identical microstructure which only differs in the initial position of one single particle.

24.4.2.2 Mechanical hysteresis loops

To analyze the mechanical response of magnetically hard MAEs, imperfect hexagonal microstructures, i.e. perfectly hexagonal distributions modified by a small geometric perturbation of the particles reference position, cf. the microstructure shown in Figure 24.6(a). They are loaded with an alternating effective induction $\overline{\boldsymbol{B}}$ in the 45°-direction where the effective deformation $\overline{\boldsymbol{F}}$ is fixed.

Regarding the resulting mechanical stress response of the system, butterfly hystereses appear. The loops reveal a stress $\overline{\boldsymbol{\sigma}}$ which is initially symmetric and gets asymmetric after several load cycles, see $\overline{\sigma}_{12}$ and $\overline{\sigma}_{21}$ in Figure 24.6(a). As in the case of magnetic hysteresis, this is an effect of the local particle rotations in the mathematically positive or negative direction, respectively. Since these rotations are not in equilibrium, a resulting magnetically-induced torque, which has to be compensated by a non-symmetric mechanical stress tensor is observed. Furthermore, the effect of smaller hystereses, appearing in the system's magnetic response, also appears in the butterfly loops. The described effects become dominant with an increase of the local particle rotations. However, after five load cycles a constant hysteresis is reached in the system's following loading periods.

Finally, the magneto-mechanical response is calculated for two nearly identical imperfect hexagonal RVEs which differ in the reference position of only one particle. While the magnetic answer of both systems almost coincides and is thus not depicted, a strong variation of the mechanical response is detected when the microstructural particle rotations become dominant, cf. Figure 24.6(b). According to that, the presented study indicates a non-systematic behavior with respect to the stress response.

24.5 Macroscopic modeling approach

In the previous section, it is shown that the effective behavior of MAEs can be calculated by computational homogenizations. However, in order to simulate the behavior of real MAE samples under complex loading conditions in a computationally efficient way, it is necessary to use a macroscopic approach.

24.5.1 Macroscopic models

In this subsection, a macroscopic model similar to Kalina et al. [44] is discussed and extended for the magnetically nonlinear case. In contrast to the most previously published approaches in this field, it is constructed and parameterized by using results from computational homogenizations. As a consequence, the model contains the properties of the microstructure implicitly and is nonetheless computationally efficient compared to costly FE^2-schemes [34, 35]. The described strategy is often denoted as decoupled multiscale scheme [70].

24.5.1.1 Microscopic simulations—Generation of the data basis

The macroscopic modeling approach is realized for the description of 2D isotropic MAEs with an underlying random microstructure, where the minimum particle-particle distance is set to 1.1 times the diameter. For statistical purpose, 10 monodisperse RVEs, each containing $N = 400$ particles are analyzed. Exemplarily, the volume fraction is chosen to $\phi = 30\%$ in the following. The reader is referred to Kalina et al. [44] for a study which evaluates several particle volume fractions.

A suitable database for the parameter identification of the macroscopic model is generated by three purely mechanical load cases (M1–M3) and six coupled magneto-mechanical load cases (MM1–MM6), cf. the schematic depiction in Figure 24.7(b). Within

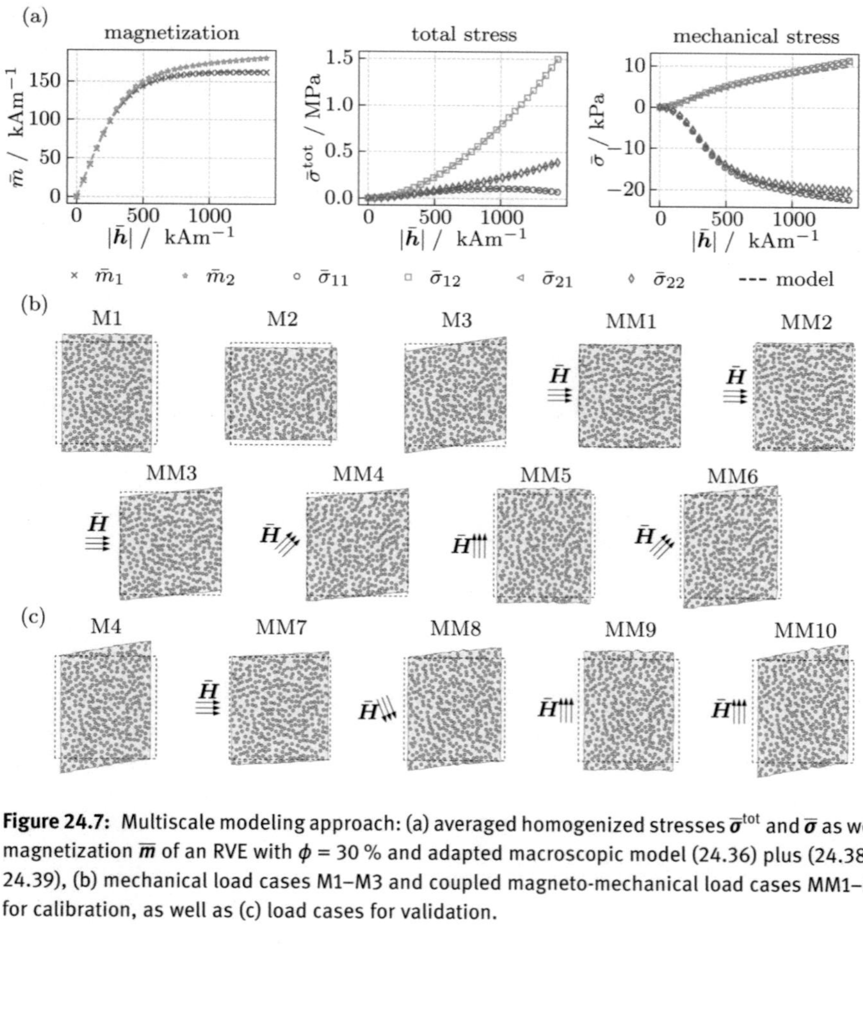

Figure 24.7: Multiscale modeling approach: (a) averaged homogenized stresses $\bar{\sigma}^{tot}$ and $\bar{\sigma}$ as well as magnetization $\bar{m}$ of an RVE with $\phi = 30$ % and adapted macroscopic model (24.36) plus (24.38, 24.39), (b) mechanical load cases M1–M3 and coupled magneto-mechanical load cases MM1–MM6 for calibration, as well as (c) load cases for validation.

the coupled load cases, the effective magnetic field $\overline{\boldsymbol{H}}$ and deformation $\overline{\boldsymbol{F}}$ are applied simultaneously, where the maximum effective stretches and shears are prescribed to different values: $\overline{\lambda}^{\max} = \{1.05, 1.1, 1.15\}$ and $\overline{\gamma}^{\max} = \{0.05, 0.1, 0.15\}$. With that, an accurate description of the macroscopic model in a wide range of deformations is guaranteed. The maximum effective field is applied to $|\overline{\boldsymbol{H}}^{\max}| = 1500\mathrm{kAm}^{-1}$. Load cases for the validation of the model, M4 and MM7–MM10, are shown in Figure 24.7(c).

Within the microscopic simulations, the properties of the constituents are described by the behavior of magnetically soft carbonyl iron particles according to Figure 24.1(b) and a silicone elastomer with $\phi_{\mathrm{oil}} = 40\%$ which is described by the Ogden model (24.31), see the stress–strain curve given in Figure 24.2(a).

24.5.1.2 Model description

In the macroscopic constitutive model, the general split of the amended free energy $\overline{\Omega}^*(\overline{\boldsymbol{C}}, \overline{\boldsymbol{H}})$ according to Eq. (24.12) into material dependent part $\overline{\varrho}_0 \overline{\Psi}^*(\overline{\boldsymbol{C}}, \overline{\boldsymbol{H}})$ and free space part $\overline{\Omega}^{*\mathrm{free}}(\overline{\boldsymbol{C}}, \overline{\boldsymbol{H}}) = -\mu_0/2\overline{J}\overline{\boldsymbol{C}}^{-1} : (\overline{\boldsymbol{H}} \otimes \overline{\boldsymbol{H}})$, that does not depend on any material properties, is applied. To capture the effects of the underlying microstructure which is not resolved within the macroscopic approach, magneto-mechanical coupling terms have to be incorporated in $\overline{\Psi}^*(\overline{\boldsymbol{C}}, \overline{\boldsymbol{H}})$. Regarding an isotropic, purely elastic and magnetically soft MAE, it can be described in terms of the following six magneto-mechanical invariants:

$$\overline{I}_1^{\mathrm{iso}} := \overline{J}^{-2/3}\,\mathrm{tr}\overline{\boldsymbol{C}}, \overline{I}_2^{\mathrm{iso}} := \frac{\overline{J}^{-4/3}}{2}\left[\mathrm{tr}^2\overline{\boldsymbol{C}} - \mathrm{tr}\left(\overline{\boldsymbol{C}}^2\right)\right], \overline{I}_3 := \overline{J}^2,$$
$$\overline{I}_4 := |\overline{\boldsymbol{H}}|^2, \overline{I}_5 := \overline{\boldsymbol{C}}^{-1} : \left(\overline{\boldsymbol{H}} \otimes \overline{\boldsymbol{H}}\right), \overline{I}_6 := \overline{\boldsymbol{C}}^{-2} : \left(\overline{\boldsymbol{H}} \otimes \overline{\boldsymbol{H}}\right). \tag{24.34}$$

To facilitate a systematic discussion of the macroscopic constitutive model, the specific Helmholtz free energy function is decomposed into purely mechanic, coupling and purely magnetic parts:

$$\overline{\Psi}^*\left(\overline{\boldsymbol{C}}, \overline{\boldsymbol{H}}\right) := \Psi^{*\mathrm{mech}}\left(\overline{I}_1^{\mathrm{iso}}, \overline{I}_2^{\mathrm{iso}}, \overline{I}_3\right) + \Psi^{*\mathrm{coup}}\left(\overline{I}_1^{\mathrm{iso}}, \overline{I}_2^{\mathrm{iso}}, \ldots, \overline{I}_6\right) + \Psi^{*\mathrm{mag}}(\overline{I}_4). \tag{24.35}$$

The chosen ansatzes are discussed in the following, where a magnetically linear model according to [44] and a magnetically nonlinear one are presented.

In the absence of magnetic fields, the MAE's behavior is characterized by the purely mechanical part of the model. To capture the nonlinear stress–strain behavior for effective stretches in the range of $\overline{\lambda} = 0.9$ up to 1.15. This part is described by an Ogden type model, so that the mechanical invariants have to be replaced by the principal stretches $\overline{\lambda}_i^{\mathrm{iso}}$ and $\overline{J}$. With that, the mechanical part is given by

$$\overline{\varrho}_0 \overline{\Psi}^{*\text{mech}} := \sum_{p=1}^{N} \frac{\mu_p}{\alpha_p} \left(\sum_{i=1}^{3} \overline{\lambda}_i^{\text{iso}^{\alpha_p}} - 3 \right) + \frac{K}{4} \left(\overline{J}^2 - \ln \overline{J}^2 - 1 \right). \tag{24.36}$$

Note that the effective nonlinear stress–strain behavior results from the Ogden model, which is used within the microscopic simulations, cf. Figure 24.2. To reduce the number of parameters in (24.36), only $N = 1$ term is used.

Using Eq. (24.34) and keeping the properties of the predicted macroscopic magneto-mechanical MAE response in mind, i.e. a quadratic dependence of $\overline{\Psi}^{\text{coup}}$ and $\overline{\Psi}^{\text{mag}}$ on the magnetic field for $|\overline{\boldsymbol{H}}| \to 0$, the linear model with

$$\mu_0^{-1} \overline{\varrho}_0 \overline{\Psi}^{*\text{coup}} := -C_1 \overline{I}_5 + C_2 \overline{I}_6 + C_3 \left(\overline{J} - 1 \right) \overline{I}_6 \quad \text{and} \quad \overline{\varrho}_0 \overline{\Psi}^{*\text{mag}} := \mu_0 C_4 \overline{I}_4 \tag{24.37}$$

is chosen. It has shown to be suitable for the description of the deformation dependent effective behavior within the examined range of magneto-mechanical load cases but is only valid for small magnetic fields up to $|\overline{\boldsymbol{H}}| \approx 250\text{kAm}^{-1}$. A possible extension of the linear model (24.37) for large values of $|\overline{\boldsymbol{H}}|$ is given by the nonlinear model with

$$\begin{aligned} \mu_0 \overline{\Psi}^{*\text{coup}} := & -\frac{\gamma_1}{\delta_1} \ln \left[\cosh \left(\mu_0 \delta_1 \sqrt{\overline{I}_5} \right) \right] + \gamma_2 \ln \left(1 + \mu_0^2 \delta_2 \overline{I}_6 \right) \ldots \\ & + \frac{1}{2} \gamma_3 \left(\overline{J} - 1 \right) \tanh^2 \left(\mu_0 \delta_3 \sqrt{\overline{I}_5} \right) \quad \text{and} \end{aligned} \tag{24.38}$$

$$\mu_0 \overline{\Psi}^{*\text{mag}} := \gamma_4 \ \ln \left(1 + \mu_0^2 \delta_4 \overline{I}_4 \right). \tag{24.39}$$

Therein, the well-known saturation behavior that results from the magnetization of the ferromagnetic particles, cf. Figure 24.1(b) is incorporated. The chosen functions enable the accurate description of the magnetization as well as the stress tensors, respectively. Altogether, a total number of 11 parameters have to be determined in the nonlinear case. In contrast to other macroscopic constitutive MAE models, e.g. Bustamante et al. [71] or Haldar et al. [41], a volumetric sensitivity has been added to $\overline{\Psi}^{\text{coup}}$.

24.5.1.3 Parameter identification

Besides the development of the macroscopic model, the crucial task in the presented approach is the identification of the mechanical, coupling and magnetic model parameter sets—in the following denoted as $\underline{\boldsymbol{\kappa}}^{\text{mech}}$, $\underline{\boldsymbol{\kappa}}^{\text{coup}}$ and $\underline{\boldsymbol{\kappa}}^{\text{mag}}$, respectively. A high accuracy of the models (24.36) plus (24.37) and (24.36) plus (24.38, 24.39) is ensured if they describe the effective homogenized magnetization $\overline{\boldsymbol{m}}$ as well as the total and mechanical stress tensors $\overline{\boldsymbol{\sigma}}^{\text{tot}}$ and $\overline{\boldsymbol{\sigma}}$ in the linear and nonlinear magnetic regime, respectively.

The proposed parameter identification according to Kalina et al. [44] features a stepwise algorithm based on four least square optimizations. At first, the parameter sets $\underline{\boldsymbol{\kappa}}^{\text{mech}}$, $\underline{\boldsymbol{\kappa}}^{\text{coup}}$ and $\underline{\boldsymbol{\kappa}}^{\text{mag}}$ are identified independently from $i = 1 \ldots N$ homogenized data sets defined by the correlated tensor fields $\{\overline{\boldsymbol{F}}_i, \overline{\boldsymbol{H}}_i\} \rightarrow \{\overline{\boldsymbol{b}}_i, \overline{\boldsymbol{\sigma}}_i{}^{\text{tot}}\}$. Thus, the following three separate optimizations have to be performed:

(i.i) Determine $\underline{\boldsymbol{\kappa}}^{\text{mech}}$ from the purely mechanical load cases,

(i.ii) find $\underline{\boldsymbol{\kappa}}^{\text{coup}}$ from the magneto-mechanical load cases by setting $\underline{\boldsymbol{\kappa}}^{\text{mech}}$ from (i.i) constant, and

(i.iii) identify $\underline{\boldsymbol{\kappa}}^{\text{mag}}$ from the magneto-mechanical load cases by fixing $\underline{\boldsymbol{\kappa}}^{\text{coup}}$ from the step (i.ii).

With that, both, the total stress tensor $\overline{\boldsymbol{\sigma}}^{\text{tot}}$ and the magnetization $\overline{\boldsymbol{m}}$ are described with high accuracy. However, since $\mathcal{O}(\overline{\boldsymbol{\sigma}}) \approx 10^{-2}\mathcal{O}(\overline{\boldsymbol{\sigma}}^{\text{tot}})$, see Figure 24.7(a), the mechanical stress tensor $\overline{\boldsymbol{\sigma}} = \overline{\boldsymbol{\sigma}}^{\text{tot}} - \overline{\boldsymbol{\sigma}}^{\text{m}}$ is not captured with sufficient precision using the identified set of parameters $\underline{\boldsymbol{\kappa}}^{\text{coup}}$ and $\underline{\boldsymbol{\kappa}}^{\text{mag}}$. Therefore, the quantities $\overline{\boldsymbol{\sigma}}^{\text{act}} := \overline{\boldsymbol{\sigma}} - \overline{\boldsymbol{\sigma}}^{\text{mech}}$ and $\overline{\boldsymbol{m}}$ are fitted simultaneously according to the final optimization

$$\{\underline{\boldsymbol{\kappa}}^{\text{coup}}, \underline{\boldsymbol{\kappa}}^{\text{mag}}\} = \arg\left\{ \min_{\underline{\boldsymbol{\kappa}}^{\text{coup}}\underline{\boldsymbol{\kappa}}^{\text{mag}}} \sum_{i=1}^{N} \left(W_i^{\sigma} \left\| \overline{\boldsymbol{\sigma}}^{\text{act}}\left(\overline{\boldsymbol{F}}_i, \overline{\boldsymbol{H}}_i, \underline{\boldsymbol{\kappa}}^{\text{mech}}, \underline{\boldsymbol{\kappa}}^{\text{coup}}, \underline{\boldsymbol{\kappa}}^{\text{mag}}\right) - \overline{\boldsymbol{\sigma}}_i^{\text{act}} \right\|^2 \ldots \right.\right.$$
$$\left.\left. + W_i^{m} \left| \overline{\boldsymbol{m}}\left(\overline{\boldsymbol{F}}_i, \overline{\boldsymbol{H}}_i, \underline{\boldsymbol{\kappa}}^{\text{coup}}, \underline{\boldsymbol{\kappa}}^{\text{mag}}\right) - \overline{\boldsymbol{m}}_i \right|^2 \right) \right\} \tag{24.40}$$

which is denoted as step (ii) in the following. Therein, W_i^m and W_i^{σ} are weighting factors for the single error sums which are determined as the inverse of the maximum value of the components $\overline{\sigma}_{kl}^{\text{act}}$ and $\overline{m}_k$ for each load case, respectively. Step (ii) naturally causes an optimization of the model with respect to both fields, $\overline{\boldsymbol{\sigma}}$ and $\overline{\boldsymbol{m}}$, but also guarantees an accurate description of $\overline{\boldsymbol{\sigma}}^{\text{tot}}$. The determined values from the steps (i.i)–(i.iii) are used as starting values in the highly nonlinear optimization.

The procedure is exemplarily applied for the nonlinear model (36) plus (38, 39). In order to determine the parameter sets $\underline{\boldsymbol{\kappa}}^{\text{mech}}$, $\underline{\boldsymbol{\kappa}}^{\text{coup}}$ and $\underline{\boldsymbol{\kappa}}^{\text{mag}}$, the load cases M1–M3 as well as MM1–MM6 are used. With that, the parameters follow to the values given in Table 24.2. Figure 24.7(a) exemplarily depicts the calibrated model and the homogenized data for the load case MM4. The remaining load cases M4 and MM7–MM10 are used to validate

Table 24.2: Identified parameters of the macroscopic mechanical model (36) as well as the nonlinear magnetic model (38 and 39). The isotropic model is adapted to homogenized data of an RVE with random particle distribution for $\phi = 30\,\%$.

K/MPa	μ_1/kPa	α_1/–
1.80	−8.33	−13.01

γ_1/T	δ_1/T^{-1}	γ_2/T^2	δ_2/T^{-2}	γ_3/T^2	δ_3/T^{-1}	ζ_1/T^2	ξ_1/T^{-2}
33	2.65	7.32×10^{-03}	6.76	2.97×10^{-02}	2.40	1.30×10^{-02}	5.77

the fitted model which achieves errors below 15 % for all considered load cases and field strengths. Altogether, the developed macroscopic approach is able to reproduce the homogenized response with a good accuracy up to the highly nonlinear magnetic regime.

24.5.2 Simulation of the magnetostrictive effect

With the microscopically based, adapted macro-model, the magnetostrictive (MS) effect of MAEs, i.e. the free deformation of a sample due to an external magnetic field, is analyzed. According to Figure 24.8(a), it is defined as

$$\overline{\varepsilon} = \frac{l - l_0}{l_0}, \tag{24.41}$$

with l and l_0 denoting the sample length in the reference and deformed configurations. Within the macroscopic simulation, the sample is embedded into a free space domain $\overline{\mathcal{F}}_0$ of adequate size, which enables the application of a homogeneous, external far-field $\overline{\boldsymbol{h}}^{\infty}$ [35, 44, 45, 72], see Figure 24.1(a). To this end, $\overline{\varphi} = \widehat{\overline{\varphi}}$ is prescribed on the exterior boundary $\partial\overline{\mathcal{F}}_0$ of the free space. Furthermore, displacements on $\partial\overline{\mathcal{F}}_0$ are prevented for simplicity, which however does not lead to a loss of generality [73]. Since the Lagrangian description of the BVP requires the calculation of a deformation field within the whole domain, $\overline{\mathcal{F}}$ is described as a neo-Hookean-elastic and non-magnetizable medium with $\overline{E}^{\text{free}} = 1\text{kPa}$ and $\overline{\nu}^{\text{free}} = 0.4$: This guarantees a negligible mechanical influence on the sample. In order to analyze the influence of the MAEs' geometry, elliptical samples with aspect ratios shifted from values between $l_0/h_0 = 1/4$ and 4 are considered, where h_0 is the sample's height. For the loading of the samples, a maximum far-field of $|\overline{\boldsymbol{h}}|^{\infty} = 1500\,\text{kA}\,\text{m}^{-1}$ is applied in the horizontal direction.

Evaluating the FE simulations, a positive MS effect is predicted for all geometries, where, according to Figure 24.8(b), the maximum elongation appears for a circular sample

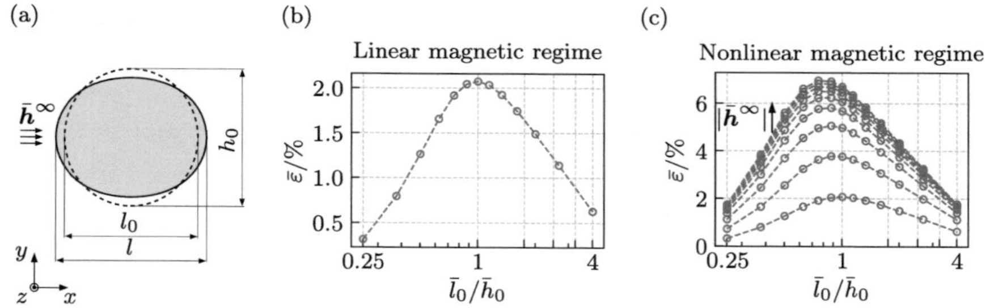

Figure 24.8: FE simulation of MS effects in elliptical MAE samples: (a) undeformed configuration (dashed lines) and deformed geometry (solid lines) for a circular sample, (b) and (c) overall MS strain $\overline{\varepsilon} = \frac{l-l_0}{l_0}$ depending on the sample shape. In (b), the linear magnetic regime for $|\overline{\boldsymbol{h}}^{\infty}| = 300\,\text{kA}\,\text{m}^{-1}$ and in (c), the nonlinear magnetic regime for $|\overline{\boldsymbol{h}}^{\infty}| = \{300, 450, \ldots 1500\}\,\text{kA}\,\text{m}^{-1}$ are depicted, respectively.

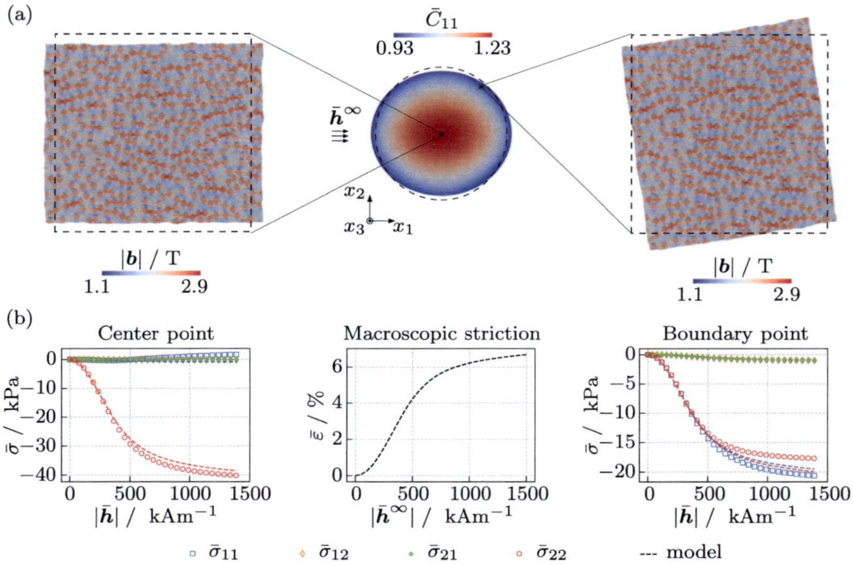

Figure 24.9: Decoupled multiscale scheme for a circular MAE: (a) inhomogeneous deformation of the sample due to an applied external field $\overline{\boldsymbol{h}}^\infty$ and local $\boldsymbol{b}$-field in the corresponding RVE at the center and a boundary point. In (b): comparison of the RVE's effective response to the macro-model according to Eqs. (24.36), (24.38) and (24.39) for localization BCs at these two points and macroscopic elongation $\overline{\varepsilon} = \Delta l / l_0$ of the sample.

within the linear magnetic regime at $|\overline{\boldsymbol{h}}^\infty| = 300\,\mathrm{kA\,m^{-1}}$ [44, 45]. Increasing the external field to values where saturation effects are relevant, the situation changes. Now, the maximum effect is predicted for the sample with a ratio $l_0/h_0 = 3/4$, see Figure 24.8(c). This shift of the maximum MS effect from the ratio 1/1 to a value below is in good agreement with results of Keip and Rammbausek [35], who perform FE^2-simulations for periodic microstructures, at the saturation point. According to the presented findings, a significant correlation of the sample geometry and the resulting magneto-mechanical response is apparent. Consequently, the overall behavior results from microstructural as well as macroscopic properties and a separation of both effects is hardly achievable in experiments. Furthermore, strongly inhomogeneous deformations $\overline{\boldsymbol{F}}$ occur inside the sample even for elliptical or, in the 3D case, ellipsoidal geometries, cf. Figure 24.9(a).

Finally, in order to validate the macroscopic approach, the decoupled multiscale scheme is applied exemplarily for the circular sample. To this end, the effective deformation gradient $\overline{\boldsymbol{F}}$ and magnetic field $\overline{\boldsymbol{H}}$ in quadrature points close to the sample's center and a boundary point at an angle of 45° starting from the x_1-axis are stored within the macroscopic FE simulation, respectively. The temporal evolutions of these tensor fields are applied in a computational homogenization to the RVE – this process is called localization. With that, the prediction of the macroscopic model can be compared to the response of the RVE. The norm of the local $\boldsymbol{b}$-field in the deformed RVEs and the calculated effective mechanical stress $\overline{\boldsymbol{\sigma}}$ of the macro-model and the homogenization are shown in Figure 24.9(a) and (b), respectively. Therein, only small deviations can be seen. This confirms the reliability of the macroscopic approach.

24.6 Conclusions

In this contribution, a continuum-based multiscale modeling approach for MAEs is presented. Applying the FE method for the solution of the nonlinear magneto–mechanical BVP, these materials are described on the micro- as well as the macroscale, where the two scales are connected via a computational homogenization scheme. The present work summarizes results given in Metsch et al. [29, 30] and Kalina et al. [13, 28, 44] and complements these findings at several points.

In order to incorporate realistic properties of the constituents on the microscale, experimental data are used to parameterize models for magnetically soft and hard filler particles as well as an elastic silicone elastomer matrix. For all of them, the chosen models are able to describe the measured response with high accuracy. By performing computational homogenizations, the effective behavior of magnetically soft as well as hard MAEs is predicted and qualitatively compared to experimental results. In the case of magnetically soft MAEs, the well-known magnetorheological effect is analyzed for 3D cubic and random microstructures. Furthermore, realistic 3D and simplified 2D simulations are compared, where a qualitative agreement but a significant quantitative deviation of both approaches is observed. In the case of magnetically hard MAEs, the experimentally determined effect of smaller hysteresis [14] is analyzed and can be explained by particle rotations on the microscale. In addition, a non-systematic behavior with respect to the MAEs' mechanical response is detected. Finally, a decoupled multiscale approach [70] which implicitly includes the microstructures' influence is presented for isotropic magnetically soft MAEs. This includes the summary of the magnetically liner model given in Kalina et al. [44] and an expansion to the magnetically nonlinear case, wherein saturation effects are considered. Exemplarily, the nonlinear model is calibrated by using homogenized data of 2D random microstructures for several load cases. The validity of the parameterized macroscopic model could be verified by further load cases. With the calibrated macroscopic model, the MS effect of several macroscopic samples are analyzed and compared with experiments.

Altogether, the presented modeling strategy has shown to be an adequate tool to predict the effective behavior of magnetically soft and hard MAEs, where the material is described without simplifications as the dipole assumption. Furthermore, the developed decoupled multiscale approach enables the implicit embedding of the microstructures' influence on the macroscale. A disadvantage of the strategy are however the necessary microscopic simulations for the calibration of the macroscopic model which are still quite costly, especially in the 3D case. The presented modeling strategy is adoptable for other particle-matrix-based magneto-active composites as e.g. ferrogels, see Birster et al. [74], Gebhart and Wallmersperger [72], Gebhart et al. [45] or Weeber et al. [75]. An essential task for the future is the expansion of the modeling approach to MAEs with chain-like microstructures. Furthermore, geometric data of real microstructures [16] should be used as input instead of artificially constructed distributions.

Acknowledgments: The computations were performed on a PC-Cluster at the Center for Information Services and High Performance Computing (ZIH) at TU Dresden. The allocated computing time is greatfully acknowleged. Finally, the authors want to thank Swen Blobel, Dr. Eike Dohmen and Dr. Sven Werdin for the support in the production and testing of the polymer samples.
Author contribution: All the authors have accepted responsibility for the entire content of this submitted manuscript and approved submission.
Research funding: The present study is funded by the German Research Foundation, Priority Program (SPP) 1681 grants KA 3309/2-1/3. This support is gratefully acknowledged.
Conflict of interest statement: The authors declare no conflicts of interest regarding this article.

References

1. Becker TI, Böhm V, Vega JC, Odenbach S, Raikher YL, Zimmermann K. Magnetic-field-controlled mechanical behavior of magneto-sensitive elastomers in applications for actuator and sensor systems. Arch Appl Mech 2019;89:133–52.
2. Schubert G. Manufacture, characterisation and modelling of magneto-rheological elastomers. PhD thesis. University of Glasgow; 2014.
3. Tian TF, Li WH, Deng YM. Sensing capabilities of graphite based MR elastomers. Smart Mater Struct 2011;20:025022.
4. Volkova TI, Böhm V, Kaufhold T, Popp J, Becker F, Borin DY, et al. Motion behaviour of magneto-sensitive elastomers controlled by an external magnetic field for sensor applications. J Magn Mater 2016;431:262–5.
5. Böse H, Rabindranath R, Ehrlich J. Soft magnetorheological elastomers as new actuators for valves. J Intell Mater Syst Struct 2012;23:989–94.
6. Carlson JD, Jolly MR. MR fluid, foam and elastomer devices. Mechatronics 2000;10:555–69.
7. Kim Y, Parada GA, Liu S, Zhao X. Ferromagnetic soft continuum robots. Sci Robot 2019;4:eaax7329.
8. Guðmundsson Í. A feasibility study of magnetorheological elastomers for a potential application in prosthetic devices. Master's thesis. University of Iceland; 2011.
9. Danas K, Kankanala SV, Triantafyllidis N.. Experiments and modeling of iron-particle-filled magnetorheological elastomers. J Mech Phys Solid 2012;60:120–38.
10. Ginder JM, Nichols ME, Elie LD, Tardiff JL. Magnetorheological elastomers: properties and applications. In MR Wuttig, editor, Smart structures and materials 1999: smart materials technologies. International Society for Optics and Photonics, SPIE; 1999, vol 3675, 131–138.

11. Hiptmair F, Major Z, Haßlacher R, Hild S. Design and application of permanent magnet flux sources for mechanical testing of magnetoactive elastomers at variable field directions. Rev Sci Instrum 2015;86:085107.
12. Martin JE, Anderson RA, Williamson RL. Generating strange magnetic and dielectric interactions: classical molecules and particle foams. J Chem Phys 2003;118:1557–70.
13. Kalina KA, Brummund J, Metsch P, Kästner M, Borin DY, Linke JM, et al. Modeling of magnetic hystereses in soft mres filled with NdFeB particles. Smart Mater Struct 2017;26:105019–31.
14. Linke JM, Borin DY, Odenbach S. First-order reversal curve analysis of magnetoactive elastomers. RSC Adv 2016;6:100407–16.
15. Sánchez PA, Gundermann T, Dobroserdova A, Kantorovich SS, Odenbach S. Importance of matrix inelastic deformations in the initial response of magnetic elastomers. Soft Matter 2018;14: 2170–83.
16. Schümann M, Odenbach S. The microstructure of magnetorheological materials characterized by means of computed X-ray microtomography. Phys Sci Rev 2019. Submitted.
17. Vaganov MV, Borin DY, Odenbach S, Raikher YL. Modeling the magnetomechanical behavior of a multigrain magnetic particle in an elastic environment. Soft Matter 2019;15:4947–60.
18. Spieler C, Kästner M, Goldmann J, Brummund J, Ulbricht V. XFEM modeling and homogenization of magnetoactive composites. Acta Mech 2013;224:2453–69.
19. Cremer P, Löwen H, Menzel AM. Tailoring superelasticity of soft magnetic materials. Appl Phys Lett 2015;107:171903.
20. Romeis D, Toshchevikov V, Saphiannikova M. Elongated micro-structures in magneto-sensitive elastomers: a dipolar mean field model. Soft Matter 2016;12:9364–76.
21. Romeis D, Metsch P, Kästner M, Saphiannikova M. Theoretical models for magneto-sensitive elastomers: a comparison between continuum and dipole approaches. Phys Rev 2017;95:042501.
22. Menzel AM. Mesoscopic characterization of magnetoelastic hybrid materials: magnetic gels and elastomers, their particle-scale description, and scale-bridging links. Arch Appl Mech 2019;89: 17–45.
23. Menzel AM, Löwen H. Modeling and theoretical description of magnetic hybrid materials – bridging from meso- to macro-scales. Phys Sci Rev 2019. Submitted.
24. Biller AM, Stolbov OV, Raikher YL. Modeling of particle interactions in magnetorheological elastomers. J Appl Phys 2014;116:114904.
25. Galipeau E, Ponte Castañeda P. The effect of particle shape and distribution on the macroscopic behavior of magnetoelastic composites. Int J Solid Struct 2012;49:1–17.
26. Galipeau E, Ponte Castañeda P. A finite-strain constitutive model for magnetorheological elastomers: magnetic torques and fiber rotations. J Mech Phys Solid 2013;61:1065–90.
27. Javili A, Chatzigeorgiou G, Steinmann P. Computational homogenization in magneto-mechanics. Int J Solid Struct 2013;50:4197–216.
28. Kalina KA, Metsch P, Kästner M. Microscale modeling and simulation of magnetorheological elastomers at finite strains: a study on the influence of mechanical preloads. Int J Solid Struct 2016;102–103:286–96.
29. Metsch P, Kalina KA, Spieler C, Kästner M. A numerical study on magnetostrictive phenomena in magnetorheological elastomers. Comput Mater Sci 2016;124:364–74.
30. Metsch P, Kalina KA, Brummund J, Kästner M. Two- and three-dimensional modeling approaches in magneto-mechanics: a quantitative comparison. Arch Appl Mech 2018;89:47–62.
31. Rudykh S, Bertoldi K. Stability of anisotropic magnetorheological elastomers in finite deformations: a micromechanical approach. J Mech Phys Solid 2013;61:949–67.
32. Goshkoderia A, Rudykh S. Stability of magnetoactive composites with periodic microstructures undergoing finite strains in the presence of a magnetic field. Compos B Eng 2017;128:19–29.

33. Chatzigeorgiou G, Javili A, Steinmann P. Unified magnetomechanical homogenization framework with application to magnetorheological elastomers. Math Mech Solid 2012;19:193–211.
34. Keip M-A, Rambausek M. A multiscale approach to the computational characterization of magnetorheological elastomers. Int J Numer Methods Eng 2016;107:338–60.
35. Keip M-A, Rambausek M. Computational and analytical investigations of shape effects in the experimental characterization of magnetorheological elastomers. Int J Solid Struct 2017;121:1–20.
36. Dorfmann A, Ogden RW. Nonlinear magnetoelastic deformations of elastomers. Acta Mech 2004; 167:13–28.
37. Bustamante R. Transversely isotropic nonlinear magneto-active elastomers. Acta Mech 2010;210: 183–214.
38. Kankanala SV, Triantafyllidis N. On finitely strained magnetorheological elastomers. J Mech Phys Solid 2004;52:2869–908.
39. Saxena P, Hossain M, Paul S. A theory of finite deformation magneto-viscoelasticity. Int J Solid Struct 2013;50:3886–97.
40. Saxena P, Hossain M, Paul S. Nonlinear magneto-viscoelasticity of transversally isotropic magneto-active polymers. Proc R Soc Lond Math Phys Eng Sci 2014;470:20140082.
41. Haldar K, Kiefer B, Menzel A. Finite element simulation of rate-dependent magneto-active polymer response. Smart Mater Struct 2016;25:104003.
42. Lefèvre V, Danas K, Lopez-Pamies O. Two families of explicit models constructed from a homogenization solution for the magnetoelastic response of mres containing iron and ferrofluid particles. Int J Non Linear Mech 2020;119:103362.
43. Mukherjee D, Bodelot L, Danas K. Microstructurally-guided explicit continuum models for isotropic magnetorheological elastomers with iron particles. Int J Non Lin Mech 2019;120:103380.
44. Kalina KA, Metsch P, Brummund J, Kästner M. Development of a macro-model for magnetorheological elastomers based on microscopic simulations. Int J Solids Struct 2020; 193–194:200–12.
45. Gebhart P, Attaran A, Wallmersperger T. Multiphysics modeling of fluid-saturated porous ferrogels at finite strains. Phys Sci Rev 2019. Submitted.
46. Pössinger T. Experimental characterization, modeling and simulation of magneto-rheological elastomers. PhD thesis. Ecole Polytechnique; 2015.
47. Bodelot L, Voropaieff J-P, Pössinger T. Experimental investigation of the coupled magneto-mechanical response in magnetorheological elastomers. Exp Mech 2018;58:207–21.
48. Jackson JD. Klassische elektrodynamik, 4 ed. Berlin: Walter de Gruyter; 2006.
49. Eringen AC, Maugin GA. Electrodynamics of continua I: foundations and solid media. New York: Springer; 1990.
50. de Groot SR, Suttorp LG. Foundations of electrodynamics. Amsterdam: North-Holland; 1972.
51. Stark S, Semenov AS, Herbert B. On the boundary conditions for the vector potential formulation in electrostatics. Int J Numer Methods Eng 2015;102:1704–32.
52. Coleman BD, Noll W. The thermodynamics of elastic materials with heat conduction and viscosity. Arch Ration Mech Anal 1963;13:167–78.
53. Hill R. Elastic properties of reinforced solids: some theoretical principles. J Mech Phys Solid 1963; 11:357–72.
54. Petra CG, Schenk O, Anitescu M. Real-time stochastic optimization of complex energy systems on high-performance computers. IEEE Comput Sci Eng 2014a;16:32–42.
55. Petra CG, Schenk O, Lubin M, Gärtner K. An augmented incomplete factorization approach for computing the schur complement in stochastic optimization. SIAM J Sci Comput 2014b;36: C139–62.
56. Alnæs M, Blechta J, Hake J, Johansson A, Kehlet B, Logg A, et al. The Fenics project version 1.5. Arch Numer Soft, 3, 2015.

57. Logg A, Mardal K-A, Wells G. Automated solution of differential equations by the finite element method: the FEniCS book. Springer Science & Business Media, 2012, vol 84.
58. Danas K. Effective response of classical, auxetic and chiral magnetoelastic materials by use of a new variational principle. J Mech Phys Solid 2017;105:25–53.
59. Ponte Castañeda P, Galipeau E. Homogenization-based constitutive models for magnetorheological elastomers at finite strain. J Mech Phys Solid 2011;59:194–215.
60. Anders B. Magnetic vector hysteresis model with dry friction-like pinning. Phys B Condens Matter 1997;233:342–7.
61. Jiles DC, Atherton DL. Theory of ferromagnetic hysteresis. J Magn Mater 1986;61:48–60.
62. Lubliner J. Plasticity theory. Courier Corporation; 2008.
63. Dohmen E. Zum Einsatz magnetischer Hybridmaterialien in adaptiven Leichtbaustrukturen. PhD thesis. TU Dresden; 2016.
64. Mullins L. Effect of stretching on the properties of rubber. Rubber Chem Technol 1948;21:281–300.
65. Flory PJ. Thermodynamic relations for high elastic materials. Trans Faraday Soc 1961;57:829–38.
66. Jolly MR, Carlson JD, Muñoz BC. A model of the behaviour of magnetorheological materials. Smart Mater Struct 1996a;5:607.
67. Jolly MR, Carlson JD, Muñoz BC, Bullions TA. The magnetoviscoelastic response of elastomer composites consisting of ferrous particles embedded in a polymer matrix. J Intell Mater Syst Struct 1996b;7:613–22.
68. Puljiz M, Huang S, Kalina KA, Nowak J, Odenbach S, Kästner M, et al. Reversible magnetomechanical collapse: virtual touching and detachment of rigid inclusions in a soft elastic matrix. Soft Matter 2018;14:6809–21.
69. Kalina KA, Brummund J, Metsch P, Kästner M. Modeling and simulation of hysteresis effects in magnetorheological elastomers. Proc Appl Math Mech 2018;18:e201800319.
70. Terada K, Kato J, Hirayama N, Inugai T, Yamamoto K. A method of two-scale analysis with micro-macro decoupling scheme: application to hyperelastic composite materials. Comput Mech 2013; 52:1199–219.
71. Bustamante R, Dorfmann A, Ogden RW. Numerical solution of finite geometry boundary-value problems in nonlinear magnetoelasticity. Int J Solid Struct 2011;48:874–83.
72. Gebhart P, Wallmersperger T. A general framework for the modeling of porous ferrogels at finite strains. J Mech Phys Solid 2019;122:69–83.
73. Vogel F, Bustamante R, Paul S. On some mixed variational principles in magneto-elastostatics. Int J Non Linear Mech 2013;51:157–69.
74. Birster K, Schweitzer R, Tschöpe A. Magnetic field-induced deformations of ni-nanorod/hydrogel composites. Phys Sci Rev 2019. Submitted.
75. Weeber R, Kreissl P, Holm C. Studying the field-controlled change of shape and elasticity of magnetic gels using particle-based simulations. Arch Appl Mech 2018;89:3–16.

Philipp Gebhart, Abdolhamid Attaran and
Thomas Wallmersperger*

25 Multiphysics modeling of porous ferrogels at finite strains

Abstract: Porous ferrogels are a new class of magnetoactive composite materials that consist of a polymeric hydrogel matrix with embedded magnetizable particles. The mutual particle interaction within the soft elastic matrix enables ferrogels to deform and alter their material characteristics upon magnetic stimulation. Due to these unique properties, ferrogels have attracted significant attention for potential uses in a variety of engineering applications, especially in biomedical engineering and microfluidics. Therefore, it is crucial to develop precise mathematical models capturing the complex material behavior of ferrogels, which spans over multiple length scales. The aim of this work is to present suitable modeling approaches for porous ferrogels. Following the hierarchical structure of scales, we present modeling frameworks for two different scenarios: (i) the modeling of ferrogels at the macroscale level and (ii) the modeling of ferrogels at the microscale level. Regarding the constitutive modeling of ferrogels, we limit our attention to locally nondissipative isotropic material response. For both modeling approaches, we provide comprehensive variational principles and briefly discuss relevant ingredients of a stable finite element implementation. In each section, numerical simulations are outlined in order to demonstrate the capabilities and relevant features of each modeling approach. Main emphasis of the numerical studies lies on the investigation of the macroscopic shape effect as well as on the characterization of the magnetomechanical material response of ferrogels with random monodisperse microstructures.

Keywords: computational homogenization, finite element analysis, large deformations, magnetoactive polymers, magnetomechanics, magneto-poro-mechanics, porous ferrogels

25.1 Introduction

In recent years, there has been a growing interest in the study of the behavior of field-responsive or so-called smart polymers. Porous ferrogels are a special class of these

***Corresponding author: Thomas Wallmersperger,** Institute of Solid Mechanics, TU Dresden, Dresden, Germany, E-mail: thomas.wallmersperger@tu-dresden.de
Philipp Gebhart and Abdolhamid Attaran, Institute of Solid Mechanics, TU Dresden, Dresden, Germany, E-mail: philipp.gebhart@tu-dresden.de (P. Gebhart), hamid.attaran@gmail.com (A. Attaran)

This article has previously been published in the journal Physical Sciences Reviews. Please cite as: P. Gebhart, A. Attaran, and T. Wallmersperger "Multiphysics modeling of porous ferrogels at finite strains" *Physical Sciences Reviews* [Online] 2020, 5. DOI: 10.1515/psr-2019-0091 | https://doi.org/10.1515/9783110569636-025

materials consisting of a porous polymeric hydrogel matrix with dispersed microsized or nanosized magnetizable particles. The magneto-poro-mechanical behavior of these materials is a complex phenomenon that spans over multiple length scales and essentially depends on (i) the constitutive behavior of the individual components, (ii) their morphology and microstructural arrangement as well as (iii) the macroscopic geometry of the specimen. Typically, multidomain ferromagnetic particles with soft magnetic material characteristics are used as filler particles resulting in reversible field-induced deformation mechanisms. Based on the arrangement of the magnetic particles within the polymer matrix, ferrogels can exhibit isotropic or anisotropic material behavior on the macroscale. In case of a random distribution of the particles within the polymer matrix, which can be realized by the synthesis of the ferrogel in the absence of a magnetic field, the macroscopic material response will be isotropic. If the ferrogel is instead synthesized in the presence of a magnetic field, the magnetizable particles tend to arrange themselves in chain-like structures leading to a directional anistropy of the material on the macroscale. Due to their broad range of synthetically fabricable properties as well as their ability to exhibit large deformations and alter their effective material behavior upon external magnetic stimulation, ferrogels have received considerable attention in different branches of engineering in recent times. Possible applications in biomedical engineering comprise on-demand drug and cell delivery, see e.g. the studies by Hu et al. [1], Zhao et al. [2] and Cezar et al. [3], or regeneration of damaged biological tissues, see e.g. the studies by Li et al. [4] and Culver et al. [5]. An overview of further potential applications in biomedical engineering is given in the studies by Li et al. [4] and Datta [6]. Moreover, ferrogels have potential to be applied in microfluidics as actuators for active flow control, see the study by Hong et al. [7]. Due to their increasingly gaining importance in engineering, mathematical modeling of ferrogels has been an area of active research in the last few years.

In general, one can distinguish between microscopic modeling approaches, where the heterogeneous microstructure of the composite material is explicitly resolved, and macroscopic modeling approaches, where – under the assumption of scale separation – the composite is considered as a macroscopical homogenous continuum.

Particle interaction models at the microscale level have been proposed by Weeber et al. [8], Tarama et al. [9], Cremer et al. [10] and Fischer and Menzel [11], among others. These modeling approaches are based on the minimization of potentials, where the magnetic interaction between the particles is described via mutually interacting magnetic dipoles. Note that these dipole models are only suitable for the modeling of dilute systems with relatively low particle volume fractions, while at higher particle volume fractions, the assumption of magnetic dipoles is not valid anymore. In order to extend the applicability of these models to systems with higher volumetric particle concentrations, a multipole expansion can be used, see e.g. the study by Biller et al. [12, 13]. Furthermore, appropriate scale transition techniques have been presented by Menzel [14] and Pessot et al. [15] in order to bridge between the microscale and macroscale.

Continuum-based modeling approaches for magnetoactive polymers at the microscale level were proposed by Galipeau and Ponte Castañeda [16, 17], Javili et al. [18], Kalina et al. [19, 20], Danas [21] and Metsch et al. [22, 23], among others. The fundamentals in the mathematical modeling of field-matter interaction in deformable continua were developed by Landau et al [24], Truesdell and Toupin [25], Brown [26], Tiersten [27], Coleman and Dill [28], de Groot and Suttorp [29], Maugin and Eringen [30], Eringen and Maugin [31], Pao [32], Hutter et al. [33] and Kovetz [34], among others. In contrast to the particle interaction models, the magnetic and mechanical fields are fully resolved within the continuum models. Therefore, continuum-based modeling approaches are not a priori restricted to the modeling of dilute systems with low particle volume fractions. These microscale models are usually embedded into an appropriate scale transition scheme as developed by Chatzigeorgiou et al. [35] in order to predict the effective material behavior of the composite at the macroscale. However, please note that these frameworks are limited to analyze the material behavior of heterogeneous microstructure at a single material point of the macrostructure. In order to account for macrostructural effects, FE^2-techniques were developed by Keip and Rambausek [36, 37] and Rambausek et al. [38], where at each quadrature point of the macroscopic body, a microscopic boundary value problem with a microscopically representative volume element is solved.

The continuum modeling of ferrogels at the macroscale level has been addressed in the recent works of Nedjar [39], Attaran et al. [40, 41] and Gebhart and Wallmersperger [42]. These models have in common that the ferrogel is considered as a magnetoactive solid-fluid mixture on the macroscale. However, the parametrization of phenomenological constitutive models for magnetoactive polymers based on experiments poses some major challenges due to the crucial macrostructural geometry influence of the specimen as extensively discussed in the study by Keip and Rambausek [36].

The present work focuses on the multiscale continuum-based modeling of porous ferrogels. We present theoretical and computational frameworks for the modeling of ferrogels at the macroscale and microscale level. The field-matter interaction in both models is described based on the statistical model by de Groot and Suttorp [29] in the nonrelativistic approximation. For a comprehensive overview on the field-matter interaction in deformable continua, we refer to the studies by Maugin and Eringen [30], Eringen and Maugin [31], Pao [32], Hutter et al. [33], Kovetz [34], Kankanala and Triantafyllidis [43], Ogden and Steigmann [44], among others.

The work is organized as follows: Section 25.2 covers the macroscale modeling of porous ferrogels. In sections 25.2.1 and 25.2.2, kinematics, field equations and modeling assumptions are outlined within a consistent thermodynamical framework. A phenomenological constitutive model for isotropic porous ferrogels in an enthalpy-based setting is developed in section 25.2.3. Section 25.2.4 introduces an incremental variational principle that governs the solution of the saddlepoint problem. In section 25.2.5, the modeling framework is applied to investigate the macroscopic shape effect for initially spheroidal shaped specimens.

Section 25.3 covers the microscale modeling of ferrogels. To this end, Section 25.3.1 and 25.3.2 first summarize kinematics and field equations for the quasi-static case of finite magnetomechanics in a Lagrangian geometric setting. Constitutive models for the polymeric gel matrix and the soft magnetizable particles in an energetic constitutive arrangement are presented in section 25.3.3. A suitable energy-based scale transition scheme is introduced in section 25.3.4 in order to bridge between the microscale and the macroscale. A minimization principle for the solution of the boundary value problem and details of the finite element implementation are discussed in section 25.3.5. The developed framework is applied in section 25.3.6 to investigate the effective material response of ferrogels with random monodisperse particle distributions.

Section 25.4 closes the paper with some concluding remarks as well as an outlook to possible future work.

25.2 Macroscale modeling of ferrogels

In this chapter, we propose a framework for the macroscopic modeling of porous ferrogels at finite strains. We assume separation of the length scales of macroscale and microscale such that the ferrogel is treated as a multiphasic mixture on the macroscale consisting of magnetoactive solid phase (s), a fluid phase (f) and additional mobile particles (m). Within this theory each spatial point is simultaneously occupied by all constituents. For an extensive review of mechanics of multiphase continua, we refer to Biot [45, 46], Lewis and Schrefler [47] and Coussy [48, 49], among others.

Notation: At this point, we want to introduce some basic symbolic notations in order to facilitate the readability of the text in the following chapters. Throughout the text scalar quantities are denoted by nonbold symbols, while vectors and higher-order tensors are indicated by bold symbols. The spatial gradient of a field is defined by $\nabla_{\boldsymbol{x}}(\cdot)$, its spatial divergence as $\operatorname{div}(\cdot)$ and its spatial rotation as $\operatorname{curl}(\cdot)$. In analogy $\nabla_{\boldsymbol{X}^\alpha}(\cdot)$, $\operatorname{Div}_\alpha(\cdot)$ and $\operatorname{Curl}_\alpha(\cdot)$ denote these operations with respect to the reference coordinates of component α.

25.2.1 Kinematics

Consider a multiphase mixture B consisting of $\alpha = \{s, f, m\}$ components, where each body B^α consists of a fixed set of material particles $P^\alpha \in B^\alpha$ embedded in the Euclidean space at time $t \in \mathbb{R}^+$. Let their reference configurations be denoted by $\mathcal{B}^\alpha \subset \mathbb{R}^3$ and the current configuration by $\mathcal{S} \subset \mathbb{R}^3$. The corresponding referential and current positions of a material point P^α in the Euclidian space are given by $\boldsymbol{X}^\alpha \in \mathcal{B}^\alpha$ and $\boldsymbol{x} \in \mathcal{S}$, respectively. The motion of the material body B^α is a time-dependent family of configurations described by the bijective deformation map as follows:`

$$\boldsymbol{\varphi}^\alpha := \begin{cases} \mathcal{B}^\alpha \times T \to \mathcal{S} \subset \mathbb{R}^3 \\ (\boldsymbol{X}^\alpha, t) \mapsto \boldsymbol{x} = \boldsymbol{\varphi}^\alpha(\boldsymbol{X}^\alpha, t) \end{cases} \tag{25.1}$$

which maps the referential position $\boldsymbol{X}^\alpha \in \mathcal{B}^\alpha$ of a particle P^α of component α to its current position $\boldsymbol{x} \in \mathcal{S}$ at time $t \in T \subset \mathbb{R}^+$. The material velocity and acceleration fields are given as follows:

$$\boldsymbol{v}^\alpha(\boldsymbol{X}^\alpha, t) = \frac{\partial}{\partial t}\boldsymbol{\varphi}^\alpha(\boldsymbol{X}^\alpha, t) \quad \text{and} \quad \boldsymbol{a}^\alpha(\boldsymbol{X}^\alpha, t) = \frac{\partial}{\partial t}\boldsymbol{v}^\alpha(\boldsymbol{X}^\alpha, t). \tag{25.2}$$

A fundamental quantity in finite strain kinematics is the deformation gradient $\boldsymbol{F}^\alpha$ which is defined as the material gradient of the nonlinear deformation map:

$$\boldsymbol{F}^\alpha = \nabla_{X^\alpha}\boldsymbol{\varphi}^\alpha(\boldsymbol{X}^\alpha, t). \tag{25.3}$$

Hence, the mechanical compatibility condition $\mathrm{Curl}_\alpha \boldsymbol{F}^\alpha = \boldsymbol{0}$ is a priori fulfilled. In order to ensure a bijective relation between referential and current position and to avoid self-penetration of the body, the Jacobian $J^\alpha = \det \boldsymbol{F}^\alpha > 0$ of the deformation gradient has to be strictly positive. In order to describe deformations without being influenced by rigid body motions, the right Cauchy-Green tensor $\boldsymbol{C}^\alpha$ and left Cauchy-Green tensor $\boldsymbol{b}^\alpha$ can be introduced as follows:

$$\boldsymbol{C}^\alpha = \boldsymbol{F}^{\alpha T} \cdot \boldsymbol{F}^\alpha \quad \text{and} \quad \boldsymbol{b}^\alpha = \boldsymbol{F}^\alpha \cdot \boldsymbol{F}^{\alpha T}. \tag{25.4}$$

For the sake of simplicity and a more compact notation, we will drop the superscript s for quantities referring to the solid component, whereas quantities with the superscript β refer to component $\beta = \{f, m\}$.

25.2.2 Field equations

In this section, we derive all required field equations to describe fluid and particle transport in a magnetoactive solid skeleton material. Within the developed macroscopic theory, the solid skeleton is chosen as a reference component. The field equations are formulated in the Lagrangian description of the solid skeleton.

25.2.2.1 Maxwell equations

The reduced form of Maxwell's equations for the magnetostatic case in absence of free electric currents is given by Gauss's law for magnetism and Ampère's law;

$$\operatorname{Div}\mathbb{B} = 0 \quad \text{and} \quad \operatorname{Curl}\mathbb{H} = \mathbf{0}, \tag{25.5}$$

where $\mathbb{B}$ denotes the Lagrangian magnetic induction and $\mathbb{H}$ the Lagrangian magnetic field strength[1]. At a surface of material discontinuity $\mathcal{S}_d$, the corresponding jump conditions read as follows:

$$[\![\mathbb{B}]\!] \cdot \boldsymbol{N} = 0 \quad \text{and} \quad [\![\mathbb{H}]\!] \times \boldsymbol{N} = \mathbf{0}. \tag{25.9}$$

Here, $[\![(\cdot)]\!] = (\cdot)^{+} - (\cdot)^{-}$ denotes the jump of a quantity with regard to direction of the normal vector $\boldsymbol{N}$ to the interface $\mathcal{S}_d$. In a material body, the magnetic quantities are linked via the following general relation:

$$\mathbb{B} = \mu_0 J \boldsymbol{C}^{-1} \cdot (\mathbb{H} + \mathbb{M}), \tag{25.10}$$

where $\mathbb{M}$ denotes the Lagrangian magnetization and μ_0 the magnetic permeability of free space. By the introduction of a magnetic scalar potential ϕ such that

$$\mathbb{H} = \nabla_{\boldsymbol{X}} \phi, \tag{25.11}$$

Ampère's law is a priori satisfied. The continuity condition across a material interface associated with the potential reads

$$[\![\phi]\!] = 0. \tag{25.12}$$

25.2.2.2 Balance of mass

The conservation of mass is formulated separately for each component of the multi-phase mixture. For the solid phase as a closed thermodynamical system, the local Lagrangian form of the balance of mass is given as follows:

$$\rho_0^s = J \rho^s, \tag{25.13}$$

---

1 **Geometric transformations between Lagrangian and Eulerian tensor fields**
The geometrical mappings of infinitesimal line, area and volume elements between the reference and the current configuration are given as follows:

$$\mathrm{d}\boldsymbol{x} = \mathbf{F} \cdot \mathrm{d}\boldsymbol{X}, \quad \mathrm{d}\boldsymbol{a} = \operatorname{cof}\mathbf{F} \cdot \mathrm{d}\boldsymbol{A} \text{ and } \mathrm{d}v = J\mathrm{d}V. \tag{25.6}$$

The push-forward operations for the magnetic field quantities read as follows:

$$\mathbb{b} = \mathbb{B} \cdot \operatorname{cof}\boldsymbol{F}^{-1}, \quad \mathbb{h} = \mathbb{H} \cdot \boldsymbol{F}^{-1} \quad \text{and} \quad \mathbb{m} = \mathbb{M} \cdot \boldsymbol{F}^{-1}. \tag{25.7}$$

A Piola transformation of the first total Piola-Kirchhoff stress tensor leads to the symmetrical total Cauchy stress tensor

$$\boldsymbol{\sigma}^{\text{tot}} = \boldsymbol{P}^{\text{tot}} \cdot \operatorname{cof}\boldsymbol{F}^{-1}. \tag{25.8}$$

where ρ_0^s and ρ^s denote the initial and current partial mass density of the solid, respectively. The partial mass density is linked to effective mass density ρ_e^s via $\rho^s = \rho_e^s n^s$, where $n^s = \frac{\mathrm{d}v^s}{\mathrm{d}v}$ is the volume fraction of the solid component. Next, consider the conservation of mass of component β. By choosing a Lagrangian description of the solid skeleton, the considered representative volume element is thermodynamically open with respect to component β. Therefore, the local form of the conservation of mass of component β reads as follows:

$$\dot{m}^\beta = -\mathrm{Div}\,\boldsymbol{J}^\beta, \tag{25.14}$$

in terms of the partial mass density of component β per unit reference volume of the solid, which is related to the current partial mass density ρ^β via $m^\beta = J\rho^\beta$. The operator $\dot{(\cdot)} = \frac{\partial}{\partial t}(\cdot)|_X$ defines the material time derivate with respect to the solid phase. Furthermore, $\boldsymbol{J}^\beta = m^\beta \boldsymbol{F}^{-1} \cdot (\boldsymbol{v}^\beta - \boldsymbol{v})$ denotes the material mass flux vector of component β.

25.2.2.3 Balance of linear and angular momentum

The local form of the balance of linear and angular momentum for the multicomponent body reads as follows:

$$\rho_0^s \boldsymbol{a} + \sum_\beta m^\beta \boldsymbol{a}^\beta = \mathrm{Div}\boldsymbol{P}^{\mathrm{tot}} + \left(\rho_0^s + \sum_\beta m^\beta\right)\boldsymbol{g} \quad \text{and} \quad \mathrm{skew}\left[\boldsymbol{P}^{\mathrm{tot}} \cdot \boldsymbol{F}^T\right] = \boldsymbol{0}, \tag{25.15}$$

where $\boldsymbol{g}$ is the gravitational acceleration and $\boldsymbol{P}^{\mathrm{tot}}$ is the total first Piola-Kirchhoff stress tensor, which contains the contributions of the additional ponderomotive body force and body couple arising in magnetomechanics. The total first Piola-Kirchhoff stress tensor can be additively decomposed

$$\boldsymbol{P}^{\mathrm{tot}} = \boldsymbol{P} + \boldsymbol{P}^{\mathrm{pon}} \tag{25.16}$$

into a mechanical contribution $\boldsymbol{P}$ and a pondermotive contribution $\boldsymbol{P}^{\mathrm{pon}}$, respectively. The specific form of the ponderomotive stress tensor is nonunique and depends on the chosen theory of field-matter interaction, see the studies by Pao [32] and Hutter et al. [33]. For the statistical model by de Groot and Suttorp [29] in the nonrelativistic approximation, the ponderomotive first Piola-Kirchhoff stress tensor takes the following form:

$$\boldsymbol{P}^{\mathrm{pon}} = \frac{1}{\mu_0 J}\left[\boldsymbol{F}\cdot\mathbb{B}\otimes\mathbb{B} - \frac{1}{2}[\boldsymbol{C}:(\mathbb{B}\otimes\mathbb{B})]\,\boldsymbol{F}^{-T}\right] + (\mathbb{M}\cdot\mathbb{B})\,\boldsymbol{F}^{-T} - \boldsymbol{F}^{-T}\cdot\mathbb{M}\otimes\mathbb{B}. \tag{25.17}$$

Furthermore, the ponderomotive first Piola-Kirchhoff stress tensor $\boldsymbol{P}^{\mathrm{pon}} = \boldsymbol{P}^{\mathrm{mag}} + \boldsymbol{P}^{\mathrm{max}}$ can be separated into a magnetization stress

$$\boldsymbol{P}^{\mathrm{mag}} = (\mathbb{M}\cdot\mathbb{B})\,\boldsymbol{F}^{-T} - \boldsymbol{F}^{-T}\cdot\mathbb{M}\otimes\mathbb{B} \tag{25.18}$$

and a Maxwell stress tensor

$$\boldsymbol{P}^{\max} = \frac{1}{\mu_0 J}\left[\boldsymbol{F} \cdot \mathbb{B} \otimes \mathbb{B} - \frac{1}{2}[\boldsymbol{C} : (\mathbb{B} \otimes \mathbb{B})]\,\boldsymbol{F}^{-T}\right] \tag{25.19}$$

resulting in a symmetrical Cauchy-like Maxwell stress tensor when pushed forward to the current configuration via the Piola transformation (25.8). Note that the Maxwell stress also exists in free space, whereas all other stress contributions vanish in vacuum. The jump conditions at a surface of material discontinuity are given as follows:

$$[\![\boldsymbol{P}^{\text{tot}}]\!] \cdot \boldsymbol{N} = -\boldsymbol{T}, \tag{25.20}$$

where $\boldsymbol{T}$ denotes the nominal mechanical traction vector.

25.2.2.4 Dissipation inequality

The dissipation inequality postulates that the temporal change in the free energy of the multicomponent system must be less than the power of the external actions. The local form of the Clausius-Duhem inequality under isothermal conditions reads

$$\mathcal{D} = -\dot{\Psi} + \boldsymbol{P}^{\text{tot}} : \dot{\boldsymbol{F}} + \mathbb{H} \cdot \dot{\mathbb{B}} + \sum_\beta \mu^\beta \dot{m}^\beta + \sum_\beta \left[-\nabla_{\boldsymbol{X}} \mu^\beta + \boldsymbol{F}^T \cdot \left(\boldsymbol{g} - \boldsymbol{a}^\beta\right)\right] \cdot \boldsymbol{J}^\beta \geq 0, \tag{25.21}$$

in terms of a total energy density function Ψ per unit reference volume. Here, μ^β denotes the chemical potential of component β, which is the conjugate variable to the partial mass density m^β and driving force for the evolution of m^β. For a more detailed derivation of (25.21), we refer to Gebhart and Wallmersperger [42]. The dissipation inequality can be split into a local and a convective part as follows:

$$\mathcal{D}_{\text{loc}} = -\dot{\Psi} + \boldsymbol{P}^{\text{tot}} : \dot{\boldsymbol{F}} + \mathbb{H} \cdot \dot{\mathbb{B}} + \sum_\beta \mu^\beta \dot{m}^\beta \geq 0 \tag{25.22a}$$

$$\mathcal{D}_{\text{con}} = \sum_\beta \left[-\nabla_{\boldsymbol{X}} \mu^\beta + \boldsymbol{F}^T \cdot \left(\boldsymbol{g} - \boldsymbol{a}^\beta\right)\right] \cdot \boldsymbol{J}^\beta \geq 0 \tag{25.22b}$$

leading to sharper restrictions to the thermodynamical consistency of the constitutive equations.

25.2.3 Constitutive modeling

In the following, we will restrict ourselves to the constitutive modeling of porous magnetoactive gels with local nondissipative material behavior and without mobile particles. The constitutive relations that connect the total stress tensor, magnetic field strength and chemical potential of the fluid with its dual variables in the material configuration are provided by the evaluation of the local dissipation inequality according to Coleman and Noll [50]. We consider an enthalpic constitutive arrangement such that the total first Piola-Kirchhoff stress tensor $\boldsymbol{P}^{\text{tot}}$, the Lagrangian magnetic induction $\mathbb{B}$ and the chemical potential of the fluid μ^f are given as follows:

$$\boldsymbol{P}^{\text{tot}} = \frac{\partial\tilde{\Psi}(\boldsymbol{C},\mathbb{H},m^f)}{\partial\boldsymbol{F}}, \quad \mathbb{B} = -\frac{\partial\tilde{\Psi}(\boldsymbol{C},\mathbb{H},m^f)}{\partial\mathbb{H}} \quad \text{and} \quad \mu^f = \frac{\partial\tilde{\Psi}(\boldsymbol{C},\mathbb{H},m^f)}{\partial m^f} \tag{25.23}$$

in terms of an objective total energy-enthalpy density function $\tilde{\Psi}(\boldsymbol{C},\mathbb{H},m^f)$ per unit reference volume. We restrict ourselves to an isotropic material response. For isotropic materials, the material symmetry condition is given as follows:

$$\tilde{\Psi}\left(\boldsymbol{Q}\cdot\boldsymbol{C}\cdot\boldsymbol{Q}^T,\boldsymbol{Q}\cdot\mathbb{H},m^f\right) = \tilde{\Psi}\left(\boldsymbol{C},\mathbb{H},m^f\right) \quad \forall\,\boldsymbol{Q}\in\mathcal{SO}(3). \tag{25.24}$$

Furthermore, we assume an additive decomposition of the total energy-enthalpy density function into a purely elastic part, a poroelastic part, a magnetoelastic part and a free space contribution as follows:

$$\tilde{\Psi}\left(\boldsymbol{C},\mathbb{H},m^f\right) = \tilde{\Psi}_{\text{elas}}(\boldsymbol{C}) + \tilde{\Psi}_{\text{poro}}\left(\boldsymbol{C},m^f\right) + \tilde{\Psi}_{\text{mag}}(\boldsymbol{C},\mathbb{H}) + \tilde{\Psi}_{\text{vac}}(\boldsymbol{C},\mathbb{H}). \tag{25.25}$$

25.2.3.1 Elastic contribution

The purely elastic behavior of the solid skeleton is characterized by a three parametric polyconvex neo-Hookean material model:

$$\tilde{\Psi}_{\text{elas}}(\boldsymbol{C}) = \frac{G}{2}\left[\left(\|\boldsymbol{F}\|^2-3\right)-2\ln J\right] + \frac{\kappa}{\beta^2}\left(\beta\ln J + J^{-\beta}-1\right), \tag{25.26}$$

where $G > 0$ and $\kappa > 0$ denote the shear and bulk modulus in the reference configuration, respectively, while $\beta > 0$ is an empirical coefficient, see the study by Ogden and Hill [51].

25.2.3.2 Poroelastic contribution

The energy contribution of the pore fluid is given by a model adapted from the classical small-strain theory of Biot [45] as follows:

$$\tilde{\Psi}_{\text{poro}}\left(\boldsymbol{C},m^f\right) = \frac{M}{2}\left[b(J-1) - \frac{m^f}{\rho_e^f}\right]^2 \tag{25.27}$$

where M and b denote the Biot modulus and the Biot coefficient, respectively. Approximations for the Biot modulus and Biot coefficient are given in the works of Rice and Cleary [52] and Coussy [53]. More complex constitutive models to describe the poromechanical behavior can be found in e.g. the studies by Gajo and Denzer [54], Nedjar [55] and Voung et al. [56]. It should be mentioned that a Biot modulus $M \to \infty$ and a Biot coefficient $b = 1$ correspond to an incompressible solid phase in case of an also incompressible fluid phase.

25.2.3.3 Magnetoelastic contribution

For the magnetoelastic contribution, we choose an isotropic tensor function based on one purely magnetic and two coupling invariants, which reads as follows:

$$\tilde{\Psi}_{\mathrm{mag}}(\boldsymbol{C}, \mathbb{H}) = -\frac{1}{2}\mu_0 \left[\alpha_1 \boldsymbol{C}^{-1} : (\mathbb{H} \otimes \mathbb{H}) + \alpha_2 \boldsymbol{C}^{-2} : (\mathbb{H} \otimes \mathbb{H}) + \alpha_3 |\mathbb{H}|^2\right]. \tag{25.28}$$

Here, α_1, α_2 and α_3 denote dimensionless material parameters. This phenomenological constitutive model is only able to accurately capture the material behavior at moderate magnetic fields. If saturation effects at higher magnetic fields should be captured sufficiently, more complex energy-enthalpy density functions should be used, see e.g. the constitutive models for magnetorheological elastomers presented in the studies by Lefèvre et al. [57] and Mukherjee et al. [58].

25.2.3.4 Free space contribution

The enthalpy stored in the underlying free space – where the multicomponent mixture is embedded in – is given as follows:

$$\tilde{\Psi}_{\mathrm{vac}}(\boldsymbol{C}, \mathbb{H}) = -\frac{1}{2}\mu_0 J \boldsymbol{C}^{-1} : (\mathbb{H} \otimes \mathbb{H}). \tag{25.29}$$

25.2.3.5 Material fluid mass flux vector

The material fluid mass flux vector is defined by the constitutive law

$$\boldsymbol{J}^f = \partial_{\mathfrak{B}} \tilde{\Phi}(\boldsymbol{C}, \mathfrak{B}) \tag{25.30}$$

in terms of an objective dissipation density function $\tilde{\Phi}$ that has to be convex and smooth in $\mathfrak{B} = [-\nabla_X \mu^f + \boldsymbol{F}^T \cdot (\boldsymbol{g} - \boldsymbol{a}^f)]$, which can be interpreted as the fluid driving force. Assuming Stokes flow on the microscale, we describe the fluid flow through the porous solid skeleton by a spatial isotropic Darcy-type flow governed by the following equation:

$$\tilde{\Phi}(\boldsymbol{C}, \mathfrak{B}) = \frac{1}{2}\left[\rho_e^f\right]^2 k J \boldsymbol{C}^{-1} : (\mathfrak{B} \otimes \mathfrak{B}), \tag{25.31}$$

where $k > 0$ denotes the spatial hydraulic permeability.

25.2.4 Variational formulation

The solution of the initial boundary value problem in the quasi-static case, where mechanical inertia effects and time-dependent electromagnetic couplings are neglected, is governed by a rate-type variational principle whose time-discrete formulation is given as follows:

$$\{\boldsymbol{\varphi}, \phi, \mu^f, m\} = \arg\left\{ \inf_{\boldsymbol{\varphi} \in \mathcal{V}_{\boldsymbol{\varphi}}} \sup_{\phi \in \mathcal{V}_{\phi}} \sup_{\mu^f \in \mathcal{V}_{\mu^f}} \inf_{m^f \in \mathcal{V}_{m^f}} \tilde{\Pi}^{\Delta}\left(\boldsymbol{C}, \mathbb{H}, m^f\right) \right\}. \tag{25.32}$$

It determines the deformation map of the solid skeleton $\boldsymbol{\varphi}$, the magnetic scalar potential ϕ, the partial mass density of the fluid m^f and the chemical potential of the fluid μ^f at current time t as a saddlepoint of the time-discrete potential $\tilde{\Pi}^{\Delta}$. In the absence of external loading contributions, the time-discrete potential $\tilde{\Pi}^{\Delta}$ is given as follows:

$$\tilde{\Pi}^{\Delta}\left(\boldsymbol{C}, \mathbb{H}, \mu^f, m^f\right) = \int_{\mathcal{B}} \tilde{\Psi}\left(\boldsymbol{C}, \mathbb{H}, m^f\right) - \Delta t\, \tilde{\Phi}\left(\boldsymbol{C}_n, \mathfrak{B}\right) - \mu^f\left(m^f - m^f_n\right) dV, \tag{25.33}$$

within the time interval $[t_n, t]$ with length $\Delta t = t - t_n$. Note that the dissipation density function is evaluated at a frozen deformation at time t_n. The admissable trial solution spaces for the primary variables are given as follows:

$$\begin{aligned} \mathcal{V}_{\boldsymbol{\varphi}} &:= \{\boldsymbol{\varphi} \in H^1(\mathcal{B}) | \boldsymbol{\varphi} = \overline{\boldsymbol{\varphi}} \quad \text{on} \quad \partial\mathcal{B}_{\boldsymbol{\varphi}}\}, \\ \mathcal{V}_{\phi} &:= \{\phi \in H^1(\mathcal{B}) | \phi = \overline{\phi} \quad \text{on} \quad \partial\mathcal{B}_{\phi}\}, \\ \mathcal{V}_{\mu^f} &:= \{\mu^f \in H^1(\mathcal{B}) | \mu^f = \overline{\mu}^f \quad \text{on} \quad \partial\mathcal{B}_{\mu^f}\}. \\ \mathcal{V}_{m^f} &:= \{m^f \in L^2\} \end{aligned} \tag{25.34}$$

The saddlepoint problem (32) can be solved in a sequential manner. First, we introduce a reduced time-discrete potential

$$\tilde{\Pi}_{\text{red}}\left(\boldsymbol{C}, \mathbb{H}, \mu^f\right) = \int_{\mathcal{B}} \tilde{\pi}_{\text{red}}\left(\boldsymbol{C}, \mathbb{H}, \mu^f\right) dV \tag{25.35}$$

through the local condensation of the partial mass density of the fluid m^f given as follows:

$$\tilde{\pi}_{\text{red}}\left(\boldsymbol{C}, \mathbb{H}, \mu^f\right) = \inf_{m^f \in \mathcal{V}_{m^f}} \left[\tilde{\Psi}\left(\boldsymbol{C}, \mathbb{H}, m^f\right) - \Delta t\, \tilde{\Phi}\left(\boldsymbol{C}_n, \mathfrak{B}\right) - \mu^f\left(m^f - m^f_n\right)\right], \tag{25.36}$$

see also the studies by Böger et al. [59] and Teichtmeister et al. [60]. With the definition of the reduced time-discrete potential (25.35) at hand, the global solution of the initial boundary value problem is then given as follows:

$$\{\boldsymbol{\varphi}, \phi, \mu^f\} = \arg\left\{ \inf_{\boldsymbol{\varphi} \in \mathcal{V}_{\boldsymbol{\varphi}}} \sup_{\phi \in \mathcal{V}_{\phi}} \sup_{\mu^f \in \mathcal{V}_{\mu^f}} \tilde{\Pi}_{\text{red}}\left(\boldsymbol{C}, \mathbb{H}, \mu^f\right) \right\}. \tag{25.37}$$

25.2.5 Representative boundary value problems

In this section, the presented macroscopic modeling framework is applied to investigate the magnetomechanical response of soft isotropic ferrogels at steady state. This numerical study is motivated by the experiments conducted in the studies by Diguet

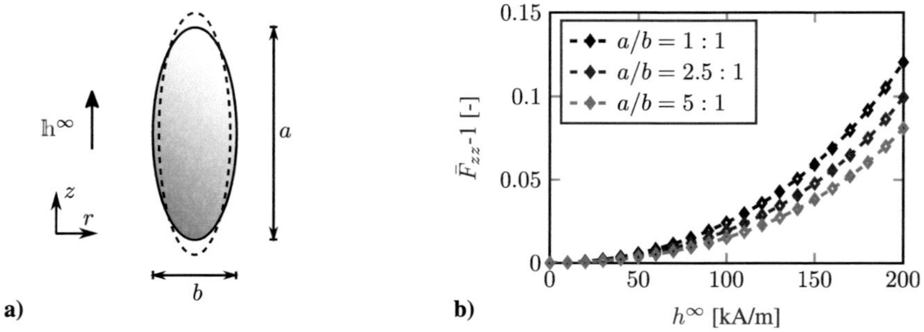

Figure 25.1: Macroscale modeling of ferrogels. (a) Boundary value problem and (b) macroscopic deformation response of initially spheroidal ferrogel specimens with aspect ratios of a/b = {1:1, 2.5:1, 5:1} in direction of the applied magnetic field.

et al. [61] and Bodelot et al. [62]. In detail, we consider spheroidal shaped specimens with aspect ratios of a/b = {1:1, 2.5:1, 5:1}. This specific shape is chosen because only for ellipsoidal shaped bodies a homogeneous distribution of the magnetic fields within the specimen can be realized. The free space surrounding the material body is modeled as a hyperelastic pseudosolid with negligible elastic properties and magnetic properties identical to free space. Throughout the domain, a homogeneous external magnetic field is applied which is aligned in parallel with the major axis of the spheroid, see Figure 25.1a. The free space box surrounding the body is of sufficient size such that the magnetic self-field of the body vanishes on the boundary of the free space. Furthermore, we prescribe homogeneous Dirichlet boundary conditions for the chemical potential of the fluid on the surface of the ferrogel domain. The material parameters used for the numerical study are listed in Table 25.1.

In Figure 25.1b, the macroscopic deformation of the specimens in direction of the applied magnetic field is plotted over the external applied magnetic field strength.

Table 25.1: Macroscale modeling of ferrogels. Material parameters of the ferrogel.

Parameter	Symbol	Value	Unit
Shear modulus	G	25	[kPa]
Bulk modulus	κ	50	[kPa]
Elastic coefficient	β	1	[-]
Biot modulus	M	100	[kPa]
Biot coefficient	b	1	[-]
Effective mass density of the fluid	ρ_e^f	1000	[kg/m³]
Hydraulic permeability coefficient	k	900	[mm^2/MPas]
Permeability of free space	μ_0	$4\pi \times 10^{-7}$	[N/A^2]
Magnetic material parameter	α_1	0.35	[-]
Magnetic material parameter	α_2	0.1	[-]
Magnetic material parameter	α_3	0.15	[-]

For all investigated specimen aspect ratios, a positive magnetostriction is observable which is quadratically dependent on the applied magnetic field. The magnetostrictive effect increases with decreasing aspect ratio and reaches its maximum for an aspect ratio of a/b = 1:1. This is in line with the analytical model based on the concept of demagnetizing energy proposed by Raikher and Stolbov [63, 64].

The spatial distribution of the deformation gradient component F_{zz} in the rotational symmetry plane for all three specimens at an external magnetic field strength of h^{∞} = 200 kA/m is depicted in Figure 25.2a. In the initially spherical shaped specimen, we observe the development of a boundary layer where the deformation is significantly lower than in the center of the specimen. For higher aspect ratios, the region of maximal deformation shifts to the tips of the specimens at both ends of the major axis, while the field distribution of the deformation gradient component F_{zz} is nearly homogeneous along the minor axis of the bodies. These field distributions of F_{zz} – caused by the jump of the magnetic field quantities at the boundary of the material body – seem to be characteristic for this specific load case as these deformation patterns were also reported in the numerical studies of Keip and Rambausek [36] and Kalina et al. [65]. Due to the induced deformation, the shape of the specimen differs from an exact spheroid to some extend. This consequently results in an inhomogeneous magnetization field m_z within the bodies, see Figure 25.2b. For all three specimens, the region where the magnetization component m_z reaches its minimum is located at both ends of the major axis. Furthermore, it can be seen that aspect ratios far from a/b = 1:1 lead to significantly smaller deviations of the magnetization component m_z in the center of the body and along its minor axis. The spatial distribution of the mechanical Cauchy stress tensor component σ_{zz} in all three specimens is depicted in Figure 25.2c. It was shown by Kalina et al. [65] that the mechanical stress tensor is a crucial quantity when it comes to the fitting of macroscopic constitutive models for magnetoactive polymers. For all three specimens, the maximum of the mechanical cauchy stress σ_{zz} is located at the ends of the major axis. Furthermore, we observe that larger aspect ratios lead to a significantly higher field homogeneity of σ_{zz} in the center of the specimen and along its minor axis. This is due to the shape of the ponderomotive traction vector $\boldsymbol{t}^{\mathrm{pon}} = [\![\boldsymbol{\sigma}^{\mathrm{pon}}]\!] \cdot \boldsymbol{n}$ at the interface between the free space and the material body which strongly depends upon the geometry of the body. The mechanical stress tensor field intrinsically depends upon the pondermotive traction vector as $\boldsymbol{\sigma} \cdot \boldsymbol{n} = \boldsymbol{t}^{\mathrm{pon}}$ holds on the boundary of the body in absence of mechanical tractions.

These results show that for spheroidal shaped specimens, aspect ratios far from a/b = 1:1 lead to significantly more homogeneous field distributions within the body. Thus, prolate spheroids are preferable in an experimental setup. Furthermore, a relaxation of the assumption of field homogeneity in the postprocessing of experimental data would lead to a more accurate reconstruction of the internal magnetization and stress state of the material body, as proposed by Keip and Rambausek [36].

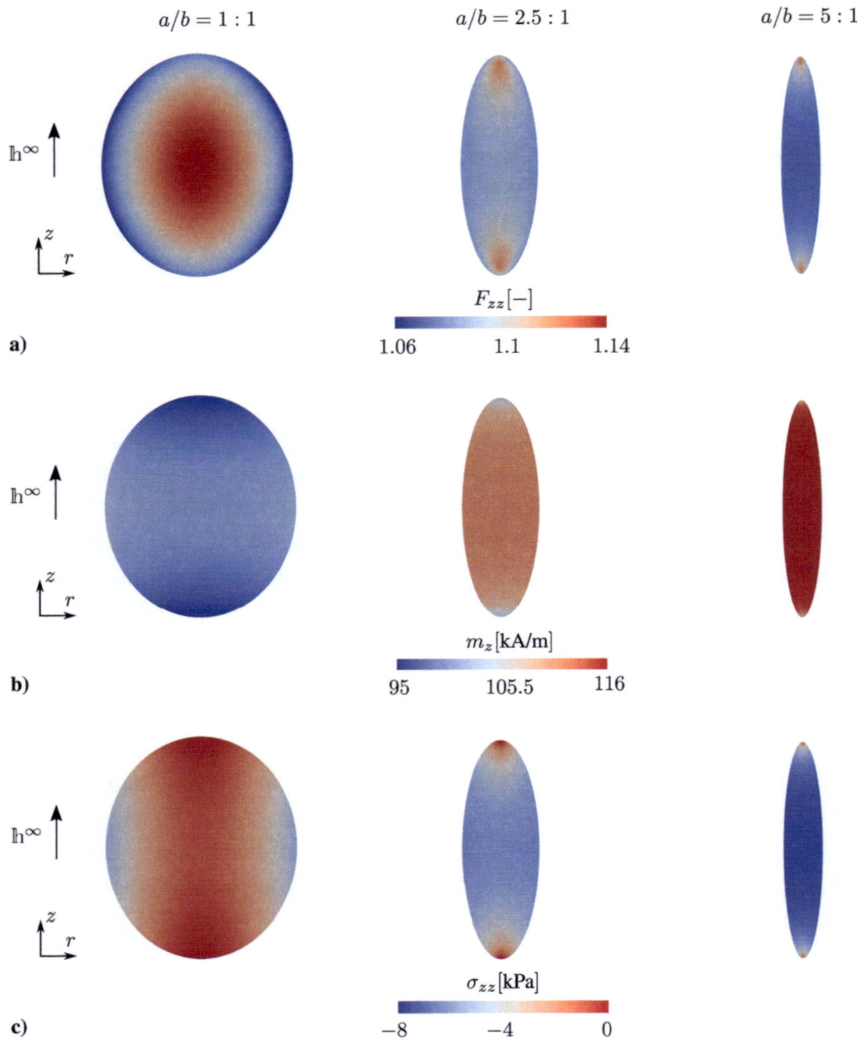

Figure 25.2: Macroscale modeling of ferrogels. Spatial distribution of (a) the deformation gradient component F_{zz}, (b) the magnetization component, m_z (c) the mechanical cauchy stress component σ_{zz} in the initially spheroidal ferrogel specimens with aspect ratios of $a/b = \{1:1,\ 2.5:1,\ 5:1\}$ at an external magnetic field strength $\mathbb{h}^\infty = h^\infty \boldsymbol{e}_z$ with $h^\infty = 200$ kA/m.

25.3 Microscale modeling of ferrogels

In this section, we propose a microscopic continuum-based framework for the modeling of ferrogels with nondissipative material behavior. Assuming separation of length scales, this modeling approach is embedded into a suitable energy-based computational homogenization scheme following Chatzigeorgiou et al. [35] in order to bridge between microscale and macroscale. This modeling technique allows us to analyze the effective material behavior of the microstructure at a single material point of the macrostructure without the influence of any macrostructural effects.

25.3.1 Kinematics

Consider a representative volume element $\mathcal{RVE}$ consisting of a fixed set of material particles $P \in \mathcal{RVE}$ embedded in the Euclidean space at time $t \in \mathbb{R}^+$. Let its reference configuration be denoted by $\mathcal{B} \subset \mathbb{R}^3$ and its current configuration by $\mathcal{S} \subset \mathbb{R}^3$. If $\boldsymbol{X} \in \mathcal{B}$ and $\boldsymbol{x} \in \mathcal{S}$ denote the referential and current position of a material point P, respectively, in the Euclidian space, the deformation $\boldsymbol{\varphi}$ is the smooth bijective mapping

$$\boldsymbol{\varphi} := \begin{cases} \mathcal{B} \times T \to \mathcal{S} \subset \mathbb{R}^3 \\ (\boldsymbol{X}, t) \mapsto \boldsymbol{x} = \boldsymbol{\varphi}\,(\boldsymbol{X}, t) \end{cases} \tag{25.38}$$

of the material body onto a time sequence of configurations in space. The tensor field

$$\boldsymbol{F} = \nabla_X \boldsymbol{\varphi}(\boldsymbol{X}, t). \tag{25.39}$$

denotes the deformation gradient, which linearly maps material tangent vectors to associated deformed spatial tangent vectors.

25.3.2 Field equations

In the following, we summarize the local field equations for the quasi-static case of magnetomechanics. In the subsequent formulation, free current densities as well as mechanical body forces are neglected. With respect to a Lagrangian description in the reference configuration, the boundary value problem is governed by the following set of equations:

$$\begin{array}{lll} 1. & \text{Gauss's law for magnetism} & \operatorname{Div}\mathbb{B} = 0 \\ 2. & \text{Ampère's law} & \operatorname{Curl}\mathbb{H} = \mathbf{0} \\ 3. & \text{Balance of mass} & \rho_0 = J\rho \\ 4. & \text{Balance of linear momentum} & \operatorname{Div}\boldsymbol{P}^{\text{tot}} = \mathbf{0} \\ 5. & \text{Balance of angular momentum} & \operatorname{skew}\left[\boldsymbol{P}^{\text{tot}} \cdot \boldsymbol{F}^T\right] = \mathbf{0} \\ 6. & \text{Dissipation inequality} & -\dot{\Psi} + \boldsymbol{P}^{\text{tot}} : \dot{\boldsymbol{F}} + \mathbb{H} \cdot \dot{\mathbb{B}} \geq 0 \end{array} \tag{25.40}$$

By introducing a magnetic vector potential $\boldsymbol{A}$ such that

$$\mathbb{B} = \operatorname{Curl}\boldsymbol{A}, \tag{25.41}$$

Ampère's law is a priori satisfied. This definition – in combination with the continuity condition $(9)_1$ on the normal component of the magnetic induction – leads to a tangential continuity of the magnetic vector potential

$$[\![\boldsymbol{A}]\!] \times \boldsymbol{N} = \boldsymbol{0} \tag{25.42}$$

across a material interface.

25.3.3 Constitutive modeling

We consider an energetic constitutive arrangement such that the microscopic total first Piola-Kirchhoff stress tensor $\boldsymbol{P}^{\text{tot}}$ and the Lagrangian magnetic field strength $\mathbb{H}$ are given as follows:

$$\boldsymbol{P}^{\text{tot}} = \frac{\partial \Psi(\boldsymbol{C}, \mathbb{B})}{\partial \boldsymbol{F}} \quad \text{and} \quad \mathbb{H} = \frac{\partial \Psi(\boldsymbol{C}, \mathbb{B})}{\partial \mathbb{B}} \tag{25.43}$$

in terms of an objective total energy density function $\Psi(\boldsymbol{C}, \mathbb{B})$. Following the concept of Miehe et al. [66] and Ethiraj and Miehe [67], we assume that the total energy density function Ψ is polyconvex with respect to the deformation gradient $\boldsymbol{F}$ for $\mathbb{B} = \boldsymbol{0}$ and convex with respect to the Lagrangian magnetic induction $\mathbb{B}$.

We restrict ourselves to an isotropic material response of the constituents. For isotropic materials, the material symmetry condition is given as follows:

$$\Psi\big(\boldsymbol{Q} \cdot \boldsymbol{C} \cdot \boldsymbol{Q}^T, \boldsymbol{Q} \cdot \mathbb{B}\big) = \Psi(\boldsymbol{C}, \mathbb{B}) \quad \forall\, \boldsymbol{Q} \in \mathcal{SO}(3). \tag{25.44}$$

Furthermore, we assume an additive decomposition of the total energy density function into a purely elastic part, a magnetic particle part and a free space contribution as follows:

$$\Psi(\boldsymbol{C}, \mathbb{B}) = \Psi_{\text{elas}}(\boldsymbol{C}) + \Psi_{\text{mag}}(\mathbb{B}) + \Psi_{\text{vac}}(\boldsymbol{C}, \mathbb{B}). \tag{25.45}$$

25.3.3.1 Elastic contribution

The purely elastic behavior of the constituents is modeled by a two parametric neo-Hookean material model as follows:

$$\Psi_{\text{elas}}(\boldsymbol{C}) = \frac{1}{2}\left[G\big(||\boldsymbol{F}||^2 - \ln J^2 - 3\big) + \frac{\lambda}{2}(J^2 - \ln J^2 - 1)\right], \tag{25.46}$$

where $\lambda > 0$ and $G > 0$ denote the first and second Lamé parameter, respectively, see the study by Ciarlet [68]. In the small strain regime, these are linked to the Young's modulus E and Poisson's ratio ν via $\lambda = \frac{\nu E}{(1+\nu)(1-2\nu)}$ and $G = \frac{E}{2(1+\nu)}$.

25.3.3.2 Magnetic contribution

The soft magnetic behavior of the ferromagnetic particles is described by a Langevin-type model, which reads as follows:

$$\Psi_{\text{mag}}(\mathbb{B}) = \frac{\mu_0 m_s^2}{3\chi}\left\{\ln\left[\frac{3\chi}{\mu_0 m_s}|\mathbb{B}|\right] - \ln\left[\sinh\left(\frac{3\chi}{\mu_0 m_s}|\mathbb{B}|\right)\right]\right\}. \tag{25.47}$$

Here, m_s denotes the magnetic saturation and χ the magnetic susceptibility which is linked to the relative magnetic permeability via $\mu_r = 1/(1-\chi)$. Note that there are more than one definitions of magnetic susceptibility in the literature.

25.3.3.3 Free space contribution

The energy stored in the underlying free space – where the $\mathcal{RVE}$ is embedded in – is given as follows:

$$\Psi_{\text{vac}}(\boldsymbol{C}, \mathbb{B}) = \sup_{\mathbb{H}}\left[\tilde{\Psi}_{\text{vac}}(\boldsymbol{C}, \mathbb{H}) + \mathbb{B}\cdot\mathbb{H}\right] = \frac{1}{2}\mu_0^{-1}J^{-1}\boldsymbol{C} : (\mathbb{B}\otimes\mathbb{B}). \tag{25.48}$$

25.3.4 Homogenization framework

Computational homogenization schemes enable to bridge scales between microscale and macroscale by the application of suitable micro-to-macro transition concepts, linking the microscopic response of the heterogeneous microstructure directly with the macroscopic overall response in a homogenized sense. Within this work, we make use of an energy-based formulation based on the unified magnetomechanical homogenization framework outlined in the study by Chatzigeorgiou et al. [35].

25.3.4.1 Definition of macrovariables

The macroscopic Lagrangian magnetic induction $\overline{\mathbb{B}}$ and magnetic field strength $\overline{\mathbb{H}}$ are related to their microscopic counterparts via averages over the volume or the boundary of the $\mathcal{RVE}$ as follows:

$$\overline{\mathbb{B}} = \frac{1}{V_0}\int_{\mathcal{B}} \mathbb{B}\, dV = \frac{1}{V_0}\int_{\partial\mathcal{B}} \boldsymbol{N}\times\boldsymbol{A}\, dA \quad \text{and} \tag{25.49}$$

$$\overline{\mathbb{H}} = \frac{1}{V_0}\int_{\mathcal{B}} \mathbb{H}\, dV = \frac{1}{2}\frac{1}{V_0}\int_{\partial\mathcal{B}} \boldsymbol{X}\times\mathbb{H}\times\boldsymbol{N}\, dA. \tag{25.50}$$

In analogy, the macroscopic deformation gradient $\overline{\boldsymbol{F}}$ and the total first Piola-Kirchhoff stress tensor $\overline{\boldsymbol{P}}^{\text{tot}}$ are defined as follows:

$$\overline{\boldsymbol{F}} = \frac{1}{V_0} \int_{\mathcal{B}} \boldsymbol{F} \, dV = \frac{1}{V_0} \int_{\partial\mathcal{B}} \boldsymbol{\varphi} \otimes \boldsymbol{N} \, dA \quad \text{and} \tag{25.51}$$

$$\overline{\boldsymbol{P}}^{\text{tot}} = \frac{1}{V_0} \int_{\mathcal{B}} \boldsymbol{P}^{\text{tot}} \, dV = \frac{1}{V_0} \int_{\partial\mathcal{B}} \boldsymbol{P}^{\text{tot}} \cdot \boldsymbol{N} \otimes \boldsymbol{X} \, dA . \tag{25.52}$$

25.3.4.2 Hill-Mandel macrohomogeneity condition

Following conceptually the study by Hill [69], the macrohomogenity condition (Hill-Mandel lemma) in the magnetomechanical context is given as follows:

$$\frac{1}{V_0} \int_{\mathcal{B}} \boldsymbol{P}^{\text{tot}} : \dot{\boldsymbol{F}} + \mathbb{H} \cdot \dot{\mathbb{B}} \, dV = \overline{\boldsymbol{P}}^{\text{tot}} : \dot{\overline{\boldsymbol{F}}} + \overline{\mathbb{H}} \cdot \dot{\overline{\mathbb{B}}} \tag{25.53}$$

Note that we assume an additive decomposition of the primary microscopic variables into linear macroscopic contributions and superimposed fine-scale fluctuation fields

$$\boldsymbol{\varphi} = \overline{\boldsymbol{F}} \cdot \boldsymbol{X} + \tilde{\boldsymbol{\varphi}} \quad \text{and} \quad \boldsymbol{A} = \frac{1}{2} \overline{\mathbb{B}} \times \boldsymbol{X} + \tilde{\boldsymbol{A}}. \tag{25.54}$$

The micro-macro consistency of power densities is satisfied by a suitable set of boundary conditions, i.e. periodic ones for the deformation map and the vector potential and antiperiodic ones for the total traction vector and the tangential magnetic field strength

$$[\![\boldsymbol{\varphi}]\!] = \overline{\boldsymbol{F}} \cdot [\![\boldsymbol{X}]\!] \quad \text{and} \quad \boldsymbol{T}^{\text{tot}}(\boldsymbol{X}^-) = -\boldsymbol{T}^{\text{tot}}(\boldsymbol{X}^+) \tag{25.55}$$

$$[\![\boldsymbol{A}]\!] = \frac{1}{2} \overline{\mathbb{B}} \times [\![\boldsymbol{X}]\!] \quad \text{and} \quad \mathbb{H} \times \boldsymbol{N}(\boldsymbol{X}^-) = -\mathbb{H} \times \boldsymbol{N}(\boldsymbol{X}^+) \tag{25.56}$$

at corresponding points on opposing boundaries of the microstructure. Here, $[\![(\cdot)]\!] = (\cdot)^+ - (\cdot)^-$ denotes the jump of a quantity with respect to opposite boundaries of the $\mathcal{RVE}$.

25.3.5 Variational formulation

The numerical implementation of the magnetomechanical boundary value problem is based on an energy formulation with periodic Dirichlet boundary conditions conceptually in line with the studies by Kalina et al. [19] and Miehe et al. [70]. The deformation map and the Lagrangian magnetic vector potential are determined by the variational principle as follows:

$$\{\boldsymbol{\varphi}, \boldsymbol{A}\} = \arg \left\{ \inf_{\boldsymbol{\varphi} \in \mathcal{V}_{\boldsymbol{\varphi}}} \inf_{\boldsymbol{A} \in \mathcal{V}_{\boldsymbol{A}}} \int_{\mathcal{B}} \Psi(\boldsymbol{C}, \mathbb{B}) \, dV \right\}. \tag{25.57}$$

In this minimization principle, the admissible function space for the deformation field and the magnetic vector potential satisfy the Dirichlet boundary conditions

$$\begin{aligned} \mathcal{V}_{\boldsymbol{\varphi}} &:= \left\{ \boldsymbol{\varphi} \in H^1(\mathcal{B}) \middle| [\![\boldsymbol{\varphi}]\!] = \overline{\boldsymbol{F}} \cdot [\![\boldsymbol{X}]\!] \quad \text{on} \quad \partial\mathcal{B} \right\} \\ \mathcal{V}_{\boldsymbol{A}} &:= \left\{ \boldsymbol{A} \in H(\text{Curl}, \mathcal{B}) \middle| [\![\boldsymbol{A}]\!] = \frac{1}{2} \overline{\mathbb{B}} \times [\![\boldsymbol{X}]\!] \quad \text{on} \quad \partial\mathcal{B} \right\} \end{aligned} \tag{25.58}$$

on the surface ∂B of the microstructure. In three dimensions, the solution of the boundary value problem (25.57) is nonunique in terms of the gauge symmetry transformation $\boldsymbol{A} \to \boldsymbol{A} + \nabla\varphi$. From a numerical point of view, this will lead to an ill-conditioned stiffness matrix. For a well-posed mathematical formulation, additional gauge conditions can be imposed on the vector potential, such as the Coulomb gauge enforcing $\text{Div}\boldsymbol{A} = 0$, see the study by Semenov et al. [71]. For a more comprehensive treatment of variational principles in magnetomechanics, we refer to the works of Vogel et al. [72, 73], Miehe et al. [70, 74], Bustamante et al. [75] and Šilhavý [76].

25.3.6 Representative boundary value problems

In this section, the microscale modeling framework is applied to investigate the effective material behavior of ferrogels with a random monodisperse particle distribution. In this study, we consider microporous ferrogels, where the characteristic length of the pores within the polymer matrix is much smaller than the size of the magnetizable particles. In detail, we analyze the effective material response of two-dimensional unit cells with three different particle volume fractions $\phi = \{10\%, 20\%, 30\%\}$ for two representative combined magnetomechanical load states. The unit cells comprise approximately 500 magnetizable particles in order to obtain a representative volume element with an isotropic material response, see the studies on representativity of unit cells of Danas [21]. This determines the geometrical dimensions of the different unit cells. The chosen set of material parameters for the constituents used in this study is given in Table 25.2.

Table 25.2: Microscale modeling of ferrogels. Material parameters of the constituents.

Parameter	Symbol	Particle	Matrix	Unit
Young's modulus	E	200×10^6	100	[kPa]
Poisson's ratio	ν	0.3	0.3	[-]
Permeability of free space	μ_0	$4\pi \times 10^{-7}$	$4\pi \times 10^{-7}$	[N/A^2]
Magnetic susceptibility	χ	0.9	0	[-]
Magnetic saturation	m_s	875	-	[kA/m]

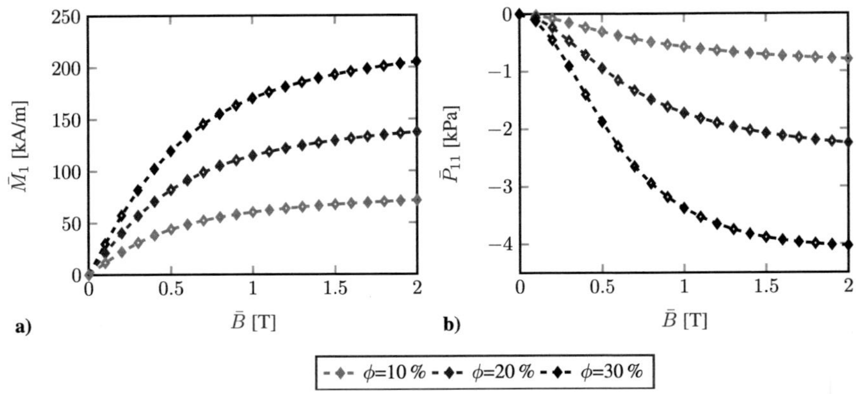

Figure 25.3: Microscale modeling of ferrogels. Effective material behavior of random monodisperse microstructures for load case I: (a) Magnetization $\overline{M}_1$ versus magnetic induction $\overline{B}$ and (b) mechanical stress $\overline{P}_{11}$ versus magnetic induction $\overline{B}$.

25.3.6.1 Load Case I

As a first load case, we consider a magnetically driven load state with a prevented macroscopic deformation of the microstructures characterized by

$$[\overline{\boldsymbol{F}}] = \begin{bmatrix} 1 & 0 \\ 0 & 1 \end{bmatrix} \quad \text{and} \quad [\overline{\mathbb{B}}] = \begin{bmatrix} \overline{B} \\ 0 \end{bmatrix}, \tag{25.59}$$

where the macroscopic magnetic induction $\overline{B}$ is linearly increased from 0 T to a maximum of 2 T within 20 increments. The effective magnetization behavior of the microstructures is depicted in Figure 25.3a. Due to the material isotropy of the unit cells – caused by the unstructured particle distribution within the polymer matrix – the magnetization $\overline{M}_2$ vanishes. Furthermore, a monotonic increase of the magnetization with increasing particle volume fraction can be observed. The magnetization responses of the different microstructures also indicate that the saturation magnetization of the composite is linearly dependent on the particle volume fraction and can be calculated via a simple rule of mixtures $\overline{m}_s = \phi\, m_s$. In Figure 25.3b, the macroscopic mechanical stress $\overline{P}_{11}$ is plotted over the applied magnetic induction. Initially, a quadratic dependency of the mechanical stress on the magnetic induction is observable, while at higher magnetic fields, saturation of the magnetization leads to a saturation of the mechanical stress as well.

Negative values of the mechanical stress $\overline{P}_{11}$ along with a prevented macroscopic deformation indicate the tendency of the unit cells to expand in the direction of the applied magnetic induction. This is in agreement with experimental observations in the study by Gollwitzer et al. [77] and Filipcsei and Zrínyi [78]. The spatial distribution of the local magnetic induction b_1 within the unit cells in Figure 25.4 exhibits considerably

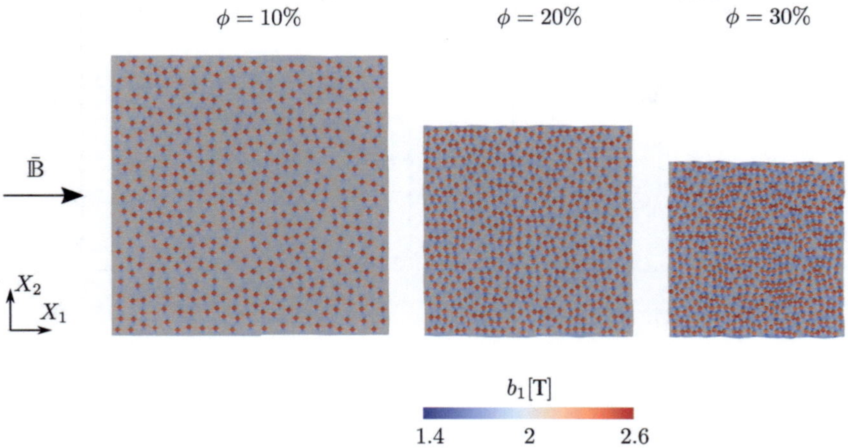

Figure 25.4: Microscale modeling of ferrogels. Spatial distribution of the local magnetic induction b_1 in the unit cells with particle volume fraction $\phi = \{10\%, 20\%, 30\%\}$ at a macroscopic magnetic induction of $\overline{\mathbb{B}} = \overline{B}\boldsymbol{e}_1$ with $\overline{B} = 2\,\mathrm{T}$.

stronger mutual particle interactions at higher particle volume fractions. This effect arises due to the decreasing particle distances with increasing particle volume fractions.

25.3.6.2 Load Case II

In the second load case, the microstructures are subjected to a combined magneto-mechanical load state characterized by

$$[\overline{\boldsymbol{F}}] = \begin{bmatrix} 1 & 0.1 \\ 0 & 1 \end{bmatrix} \quad \text{and} \quad [\overline{\mathbb{B}}] = \begin{bmatrix} 0 \\ \overline{B} \end{bmatrix} \tag{25.60}$$

where – in analogy to the previous load case – the macroscopic magnetic induction $\overline{B}$ is linearly increased from 0 T to a maximum of 2 T within 20 increments. The macroscopic magnetization behavior of the microstructures is shown in Figure 25.5a. A comparison with the results of the previous load case given in Figure 25.3a indicates that the magnetization response is nearly independent of the macroscopic deformation state, see also the studies of Kalina et al. [19]. Due to the macroscopic simple shear deformation of the microstructure, the mechanical shear stress $\overline{P}_{12}$ is positive for all three unit cells at vanishing magnetic fields, see Figure 25.6b.

We observe an initial quadratic dependency of the mechanical shear stress $\overline{P}_{12}$ on the magnetic induction with saturation effects occurring at higher magnetic fields. The monotonic increase of the mechanical shear stress $\overline{P}_{12}$ with the magnetic induction for the unit cells with particle volume fractions of $\phi = \{10\%, 20\%\}$ at a fixed macroscopic simple shear deformation indicates a positive magnetorheological effect. This observation coincides with the experimental results in the study by Jolly et al. [79]. In

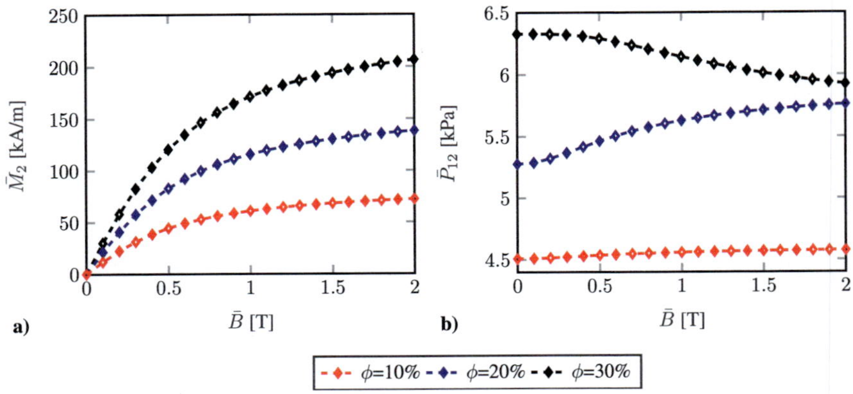

Figure 25.5: Microscale modeling of ferrogels. Effective material behavior of random monodisperse microstructures for load case II: (a) Magnetization $\overline{M}_2$ versus magnetic induction $\overline{B}$ and (b) mechanical stress $\overline{P}_{12}$ versus magnetic induction $\overline{B}$.

contrast, the decrease of the mechanical shear stress $\overline{P}_{12}$ with increasing magnetic induction indicates a negative magnetorheological effect for the unit cell with 30% particle volume fraction. Similar unsystematic behavior was found in the numerical studies of Kalina et al. [80] for three-dimensional unit cells with random monodisperse microstructures. In general, the magnetorheological effect is highly sensitive to morphological changes of the microstructure and strongly depends upon the material

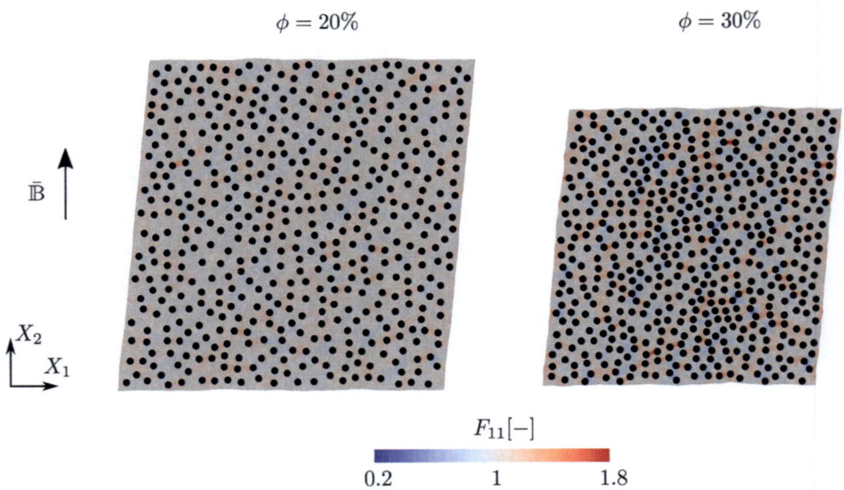

Figure 25.6: Microscale modeling of ferrogels. Spatial distribution of the local deformation gradient component F_{11} in the unit cells with particle volume fraction $\phi = \{20\%, 30\%\}$ at a macroscopic magnetic induction of $\overline{\mathbb{B}} = \overline{B}\boldsymbol{e}_2$ with $\overline{B} = 2\,\mathrm{T}$.

behavior of the polymer matrix. A detailed analysis of this effect and its influencing factors is beyond the scope of the paper and a task for future investigations. Figure 25.6 shows the spatial distribution of the local deformation gradient component F_{11} in the unit cells with a particle volume fraction of $\phi = \{20\%, 30\%\}$. Due to the fact that the particles try to rearrange themselves in energetically advantageous configurations, we observe particle movements in direction perpendicular to the applied magnetic field in order to form chain-like substructures. This leads to locally large deformations within the microstructure, especially at high particle volume fractions. In reality, this could cause debonding of the particles from the matrix material. From a numerical point of view, this can result in convergence issues triggered by mesh distortions.

25.4 Conclusions

The present work covers the multiphysics modeling of porous ferrogels at finite strains. Following the hierarchical structure of scales, we present theoretical and computational frameworks for two different modeling strategies: (i) the modeling of ferrogels at the macroscale level and (ii) the modeling of ferrogels at the microscale level. Prototypical constitutive models for locally nondissipative material behavior are derived in an enthalpy-based constitutive arrangement on the macroscale and an energy-based constitutive arrangement on the microscale. For both modeling approaches, a rigorous variational treatment is given, providing a canonically compact model structure. This consequently leads to a symmetric and compact formulation of the strongly coupled nonlinear multiphysics problem, which is ideally suited for an efficient finite element implementation. Representative boundary value problems outline the relevant features and capabilities of each modeling approach. The study on the macroscopic shape effect reveals that the field homogeneity is significantly higher in spheroidal shaped specimens with larger aspect ratios. Therefore, prolate spheroids are preferable in the experimental characterization of magnetoactive polymers. The proposed microscale model is embedded into a suitable energy-based scale transition scheme that allows to analyze the effective material response of the microstructure without the influence of any macrostructural effects. The study on the effective material behavior of random monodisperse microstructures for two load cases, which characterize the magnetostrictive and magnetorheological effect of the composite, reveals a qualitatively good agreement with experimental findings and theoretical studies in literature.

Future work will be devoted to the extension and parametrization of the constitutive model at the macroscale level based on a comprehensive data set generated from the microscale model. This approach allows the identification of an intrinsic material parameter set independent of macrostructural shape effects. From a numerical point of view, the development and implementation of a mesh update algorithm should be a future task since the mesh quality at large deformations has crucial influence on the convergence and stability of the finite element algorithm.

Acknowledgments: The financial support of the German Research Foundation (DFG) through the Priority Programme (SPP) 1681, grant WA 2323/8, is gratefully acknowledged. The computations were performed on a High Performance Computing (HPC) system at the Center for Information Services and High Performance Computing (ZIH) at TU Dresden. The authors thank the ZIH for generous allocations of compute resources.
Author contribution: All the authors have accepted responsibility for the entire content of this submitted manuscript and approved submission.
Research funding: None declared.
Conflict of interest statement: The authors declare no conflicts of interest regarding this article.

References

1. Hu S-H, Liu T-Y, Liu D-M, Chen S-Y. Nano-ferrosponges for controlled drug release. J Contr Release 2007;121:181–9.
2. Zhao X, Kim J, Cezar CA, Huebsch N, Lee K, Bouhadir K, et al. *Active scaffolds for on-demand drug and cell delivery. Proc Natl Acad Sci USA 2011;108:67–72.*
3. Cezar CA, Kennedy SM, Mehta M, Weaver JC, Gu L, Vandenburgh H, et al. *Biphasic ferrogels for triggered drug and cell delivery. Adv Healthcare Mater 2014;3:1869–76.*
4. Li Y, Huang G, Zhang X, Li B, Chen Y, Lu T, et al. *Magnetic hydrogels and their potential biomedical applications. Adv Funct Mater 2013;23:660–72.*
5. Culver HR, Clegg JR, Peppas NA. Analyte-responsive hydrogels: intelligent materials for biosensing and drug delivery. Acc Chem Res 2017;50:170–8. PMID: 28170227.
6. Datta P. 17 – magnetic gels. In: Pal K, Banerjee I, editors Polymeric Gels, Woodhead Publishing Series in Biomaterials. Woodhead Publishing; 2018, pp. 441–65.
7. Hong S, Jung Y, Yen R, Chan HF, Leong KW, Truskey GA, et al. *Magnetoactive sponges for dynamic control of microfluidic flow patterns in microphysiological systems. Lab Chip 2014;14:514–21.*
8. Weeber R, Kantorovich S, Holm C. Deformation mechanisms in 2D magnetic gels studied by computer simulations. Soft Matter 2012;8:9923–32.
9. Tarama M, Cremer P, Borin DY, Odenbach S, Löwen H, Menzel AM. Tunable dynamic response of magnetic gels: impact of structural properties and magnetic fields. Phys Rev E 2014;90:042311.
10. Cremer P, Löwen H, Menzel AM. Tailoring superelasticity of soft magnetic materials. Appl Phys Lett 2015;107:171903.
11. Fischer L, Menzel AM. Magnetostriction in magnetic gels and elastomers as a function of the internal structure and particle distribution. J Chem Phys 2019;151:114906.
12. Biller AM, Stolbov OV, Raikher YL. Modeling of particle interactions in magnetorheological elastomers. J Appl Phys 2014;116:114904.
13. Biller AM, Stolbov OV, Raikher YL. Mesoscopic magnetomechanical hysteresis in a magnetorheological elastomer. Phys Rev E 2015;92:023202.
14. Menzel AM. Bridging from particle to macroscopic scales in uniaxial magnetic gels. J Chem Phys 2014;141:194907.
15. Pessot G, Weeber R, Holm C, Löwen H, Menzel AM. Towards a scale-bridging description of ferrogels and magnetic elastomers. J Phys Condens Matter 2015;27:325105.

16. Galipeau E, Ponte Castañeda P. The effect of particle shape and distribution on the macroscopic behavior of magnetoelastic composites. Int J Solid Struct 2012;49:1–17.
17. Galipeau E, Ponte Castañeda P. A finite-strain constitutive model for magnetorheological elastomers: magnetic torques and fiber rotations. J Mech Phys Solid 2013;61:1065–90.
18. Javili A, Chatzigeorgiou G, Steinmann P. Computational homogenization in magneto-mechanics. Int J Solid Struct 2013;50:4197–216.
19. Kalina KA, Metsch P, Kästner M. Microscale modeling and simulation of magnetorheological elastomers at finite strains: a study on the influence of mechanical preloads. Int J Solid Struct 2016;102:286–96.
20. Kalina KA, Brummund J, Metsch P, Kästner M, Borin DY, Linke JM, et al. *Modeling of magnetic hystereses in soft MRES filled with NDFEB particles. Smart Mater Struct 2017;26:105019–31.*
21. Danas K. Effective response of classical, auxetic and chiral magnetoelastic materials by use of a new variational principle. J Mech Phys Solid 2017;105:25–53.
22. Metsch P, Kalina KA, Spieler C, Kästner M. A numerical study on magnetostrictive phenomena in magnetorheological elastomers. Comput Mater Sci 2016;124:364–74.
23. Metsch P, Kalina KA, Brummund J, Kästner M. Two- and three-dimensional modeling approaches in magneto-mechanics: a quantitative comparison. Arch Appl Mech 2018;89:47–62.
24. Landau LD, Bell J, Kearsley M, Pitaevskii L, Lifshitz E, Sykes J. Electrodynamics of continuous media. Elsevier, 2013, vol 8.
25. Truesdell C, Toupin R. The classical field theories. Berlin, Heidelberg: Springer Berlin Heidelberg; 1960. pp. 226–858.
26. Brown WF. Magnetoelastic interactions. Berlin: Springer, 1966, vol 9.
27. Tiersten HF. Coupled magnetomechanical equations for magnetically saturated insulators. J Math Phys 1964;5:1298–318.
28. Coleman BD, Dill EH. Thermodynamic restrictions on the constitutive equations of electromagnetic theory. Z Angew Math Phys 1971;22:691–702.
29. De Groot SR, Suttorp LG. Foundations of electrodynamics. Amsterdam: North-Holland Publishing Company; 1972.
30. Maugin GA, Eringen AC. On the equations of the electrodynamics of deformable bodies of finite extent. J Mécanique 1977;16:101–47.
31. Eringen AC, Maugin GA. Electrodynamics of continua I: foundations and solid media. New York: Springer-Verlag; 1990.
32. Pao Y-H. Electromagnetic forces in deformable continua. In: Nemat-Nasser S, editor, Mechanics today. Pergamon Press, 1978, vol 4, pp. 209–305.
33. Hutter K, Ven AA, Ursescu A. Electromagnetic field matter interactions in thermoelastic solids and viscous fluids. Springer-Verlag Berlin Heidelberg; 2006, vol 710.
34. Kovetz A. Electromagnetic theory. Oxford University Press; 2000.
35. Chatzigeorgiou G, Javili A, Steinmann P. Unified magnetomechanical homogenization framework with application to magnetorheological elastomers. Math Mech Solid 2014;19:193–211.
36. Keip M-A, Rambausek M. Computational and analytical investigations of shape effects in the experimental characterization of magnetorheological elastomers. Int J Solid Struct 2017;121: 1–20.
37. Keip M-A, Rambausek M. A multiscale approach to the computational characterization of magnetorheological elastomers. Int J Numer Methods Eng;107:338–60.
38. Rambausek M, Göküzüm FS, Nguyen LTK, Keip M-A. A two-scale FE-FFT approach to nonlinear magneto-elasticity. Int J Numer Methods Eng; 117:1117–42.
39. Nedjar B. A theory of finite strain magneto-poromechanics. J Mech Phys Solid 2015;84:293–312.
40. Attaran A, Brummund J, Wallmersperger T. Development of a continuum model for ferrogels. J Intell Mater Syst Struct 2017;28:1358–75.

41. Attaran A, Brummund J, Wallmersperger T. Modeling and finite element simulation of the magneto-mechanical behavior of ferrogels. J Magn Magn Mater 2017;431:188–91.
42. Gebhart P, Wallmersperger T. A general framework for the modeling of porous ferrogels at finite strains. J Mech Phys Solid 2019;122:69–83.
43. Kankanala S, Triantafyllidis N. On finitely strained magnetorheological elastomers. J Mech Phys Solid 2004;52:2869–908.
44. Ogden R, Steigmann D. Mechanics and electrodynamics of magneto- and electro-elastic materials. CISM International Centre for Mechanical Sciences. Wien: Springer-Verlag; 2011, 527.
45. Biot MA. General theory of three-dimensional consolidation. J Appl Phys 1941;12:155–64.
46. Biot MA. Theory of finite deformations of porous solids. Indiana Univ Math J 1972;21:597–620.
47. Lewis RW, Schrefler BA. The finite element method in the static and dynamic deformation and consolidation of porous media. John Wiley & Sons; 1998.
48. Coussy O. Poromechanics. John Wiley & Sons; 2004.
49. Coussy O. Mechanics and physics of porous solids. John Wiley & Sons; 2010.
50. Coleman BD, Noll W. The thermodynamics of elastic materials with heat conduction and viscosity. Arch Ration Mech Anal 1963;13:167–78.
51. Ogden RW, Hill R. Large deformation isotropic elasticity: on the correlation of theory and experiment for compressible rubberlike solids. Proc R Soc Lond A Math Phys Sci 1972;328: 567–83.
52. Rice JR, Cleary MP. Some basic stress diffusion solutions for fluid-saturated elastic porous media with compressible constituents. Rev Geophys 1976;14:227–41.
53. Coussy O. Mechanics of porous continua. John Wiley & Sons; 1995.
54. Gajo A, Denzer R. Finite element modelling of saturated porous media at finite strains under dynamic conditions with compressible constituents. Int J Numer Methods Eng 2011;85:1705–36.
55. Nedjar B. Formulation of a nonlinear porosity law for fully saturated porous media at finite strains. J Mech Phys Solid 2013;61:537–56.
56. Vuong A-T, Yoshihara L, Wall W. A general approach for modeling interacting flow through porous media under finite deformations. Comput Methods Appl Mech Eng 2015;283:1240–59.
57. Lefèvre V, Danas K, Lopez-Pamies O. Two families of explicit models constructed from a homogenization solution for the magnetoelastic response of MRES containing iron and ferrofluid particles. Int J Non Linear Mech 2020;119:103362.
58. Mukherjee D, Bodelot L, Danas K. Microstructurally-guided explicit continuum models for isotropic magnetorheological elastomers with iron particles. Int J Non Linear Mech 2020;120:103380.
59. Böger L, Nateghi A, Miehe C. A minimization principle for deformation–diffusion processes in polymeric hydrogels: constitutive modelling and Fe implementation. Int J Solid Struct 2017;121: 257–74.
60. Teichtmeister S, Mauthe S, Miehe C. Aspects of finite element formulations for the coupled problem of poroelasticity based on a canonical minimization principle. Comput Mech 2019;64: 685–716.
61. Diguet G, Beaugnon E, Cavaillé J. Shape effect in the magnetostriction of ferromagnetic composite. J Magn Magn Mater 2010;322:3337–41.
62. Bodelot L, Voropaieff J-P, Pössinger T. Experimental investigation of the coupled magneto-mechanical response in magnetorheological elastomers. Exp Mech 2018;58:207–21.
63. Raikher Y, Stolbov O. Magnetodeformational effect in ferrogel samples. J Magn Magn Mater 2003; 258–259:477–9.
64. Raikher YL, Stolbov OV. Deformation of an ellipsoidal ferrogel sample in a uniform magnetic field. J Appl Mech Tech Phys 2005;46:434–443.
65. Kalina KA, Metsch P, Brummund J, Kästner M. A macroscopic model for magnetorheological elastomers based on microscopic simulations. Int J Solid Struct 2020;193–194:200–12.

66. Miehe C, Vallicotti D, Zäh D. Computational structural and material stability analysis in finite electro-elasto-statics of electro-active materials. Int J Numer Methods Eng;102:1605–37.
67. Ethiraj G, Miehe C. Multiplicative magneto-elasticity of magnetosensitive polymers incorporating micromechanically-based network kernels. Int J Eng Sci 2016;102:93–119.
68. Ciarlet PG. Mathematical elasticity: Volume I: three-dimensional elasticity. North-Holland; 1988.
69. Hill R. On constitutive macro-variables for heterogeneous solids at finite strain. Proc R Soc Lond A Math Phys Sci 1972;326:131–47.
70. Miehe C, Rosato D, Kiefer B. Variational principles in dissipative electro-magneto-mechanics: a framework for the macro-modeling of functional materials. Int J Numer Methods Eng;86: 1225–76.
71. Semenov AS, Kessler H, Liskowsky A, Balke H. On a vector potential formulation for 3D electromechanical finite element analysis. Commun Numer Methods Eng 2006;22:357–75.
72. Vogel F, Bustamante R, Steinmann P. On some mixed variational principles in electro-elastostatics. Int J Non Linear Mech 2012;47:341–54. Nonlinear Continuum Theories.
73. Vogel F, Bustamante R, Steinmann P. On some mixed variational principles in magneto-elastostatics. Int J Non Linear Mech 2013;51:157–69.
74. Miehe C, Vallicotti D, Teichtmeister S. Homogenization and multiscale stability analysis in finite magneto-electro-elasticity. application to soft matter ee, me and mee composites. Comput Methods Appl Mech Eng 2016;300:294–346.
75. Bustamante R, Dorfmann A, Ogden R. On variational formulations in nonlinear magnetoelastostatics. Math Mech Solid 2008;13:725–45.
76. Šilhavý M. A variational approach to nonlinear electro-magneto-elasticity: convexity conditions and existence theorems. Math Mech Solid 2018;23:907–28.
77. Gollwitzer C, Turanov A, Krekhova M, Lattermann G, Rehberg I, Richter R. Measuring the deformation of a ferrogel sphere in a homogeneous magnetic field. J Chem Phys 2008;128:164709.
78. Filipcsei G, Zrínyi M. Magnetodeformation effects and the swelling of ferrogels in a uniform magnetic field. J Phys Condens Matter 2010;22:276001.
79. Jolly MR, Carlson JD, Muñoz BC, Bullions TA. The magnetoviscoelastic response of elastomer composites consisting of ferrous particles embedded in a polymer matrix. J Intell Mater Syst Struct 1996;7:613–22.
80. Kalina KA, Raßloff A, Wollner M, Metsch P, Brummund J, Kästner M. Multiscale modeling and simulation of magneto-active elastomers based on experimental data. Phys Sci Rev 2020.

Tatiana I. Becker*, Yuriy L. Raikher, Oleg V. Stolbov, Valter Böhm and Klaus Zimmermann

26 Magnetoactive elastomers for magnetically tunable vibrating sensor systems

Abstract: Magnetoactive elastomers (MAEs) are a special type of smart materials consisting of an elastic matrix with embedded microsized particles that are made of ferromagnetic materials with high or low coercivity. Due to their composition, such elastomers possess unique magnetic field-dependent material properties. The present paper compiles the results of investigations on MAEs towards an approach of their potential application as vibrating sensor elements with adaptable sensitivity. Starting with the model-based and experimental studies of the free vibrational behavior displayed by cantilevers made of MAEs, it is shown that the first bending eigenfrequency of the cantilevers depends strongly on the strength of an applied uniform magnetic field. The investigations of the forced vibration response of MAE beams subjected to in-plane kinematic excitation confirm the possibility of active magnetic control of the amplitude-frequency characteristics. With change of the uniform field strength, the MAE beam reveals different steady-state responses for the same excitation, and the resonance may occur at various ranges of the excitation frequency. Nonlinear dependencies of the amplification ratio on the excitation frequency are obtained for different magnitudes of the applied field. Furthermore, it is shown that the steady-state vibrations of MAE beams can be detected based on the magnetic field distortion. The field difference, which is measured simultaneously on the sides of a vibrating MAE beam, provides a signal with the same frequency as the excitation and an amplitude proportional to the amplitude of resulting vibrations. The presented prototype of the MAE-based vibrating unit with the field-controlled "configuration" can be implemented for realization of acceleration sensor systems with adaptable sensitivity. The ongoing research on MAEs is oriented to the use of other geometrical forms along with beams, e.g. two-dimensional structures such as membranes.

***Corresponding author: Tatiana I. Becker**, Technical Mechanics Group, Faculty of Mechanical Engineering, Technische Universität Ilmenau, Ilmenau, Germany, E-mail: tatiana.becker@tu-ilmenau.de

Yuriy L. Raikher and Oleg V. Stolbov, Laboratory of Physics and Mechanics of Soft Matter, Institute of Continuous Media Mechanics, Ural Branch of Russian Academy of Sciences, Perm, Russia, E-mail: yuriy.raikher@gmail.com (Y.L. Raikher), oleg100@gmail.com (O.V. Stolbov)

Valter Böhm, Faculty of Mechanical Engineering, Ostbayerische Technische Hochschule Regensburg, Regensburg, Germany, E-mail: valter.boehm@oth-regensburg.de

Klaus Zimmermann, Technical Mechanics Group, Faculty of Mechanical Engineering, Technische Universität Ilmenau, Ilmenau, Germany, E-mail: klaus.zimmermann@tu-ilmenau.de

This article has previously been published in the journal Physical Sciences Reviews. Please cite as: T. I. Becker, Y. L. Raikher, O. V. Stolbov, V. Böhm and K. Zimmermann "Magnetoactive elastomers for magnetically tunable vibrating sensor systems" *Physical Sciences Reviews* [Online] 2020, 5. DOI: 10.1515/psr-2019-0125 | https://doi.org/10.1515/9783110569636-026

Keywords: adaptable sensor unit, amplification ratio, bending vibration, eigenfrequency, magnetoactive elastomer, magnetic field control

26.1 Introduction

Magnetoactive elastomers (MAEs) consist of magnetically soft and/or magnetically hard particles of microsize that are embedded into a nonmagnetic elastic matrix (Figure 26.1). This class of smart materials has potentially an extending field of application in diverse technical areas. Just like magnetorheological fluids and ferrofluids [1, 2], MAEs strongly respond to magnetic fields. Experimental investigations on MAEs with magnetically soft particles show that their mechanical properties can be changed sufficiently, when they are subjected to an external magnetic field, e.g. [3, 4]. It is found that the elastic modulus of these materials is "tunable", which means its effective increase in the presence of an applied magnetic field [5–7]. The reasons for this active effect of "tuning" are interactions between the different types of magnetic filler and the matrix in a microscopic scale and the subsequent structure formation of the particles.

Recent studies on MAEs with embedded magnetically hard particles show that they possess striking properties, particularly remanent magnetization, even in the absence of an external field [8–10]. Once such freshly synthesized elastomer when subjected to a strong magnetic field acquires a remanent magnetization and becomes actually an elastic magnet with modified properties as compared to the original elastomer. This effect of passive magnetic control of properties depends on ferromagnetic properties and concentration of magnetically hard particles as well as the field magnitude applied during the initial magnetization [11]. A special type of MAEs with a mixed magnetic filler made of magnetically hard and soft particles is of particular research interest at present. Such elastomers with a mixed magnetic filler show not just pure passive and active effects that elastomers containing only one type of particles would have [12–14]. The interplay of two fillers different in their ferromagnetic properties results in a complex distribution of the magnetic field, which in turn influences the field reaction of the material itself.

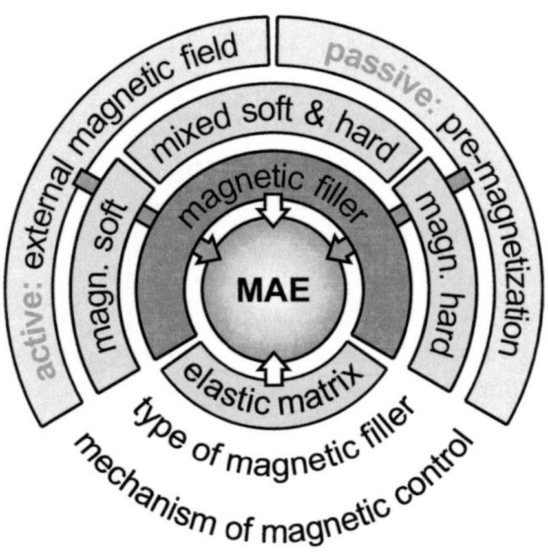

Figure 26.1: MAE material systems with different magnetic fillers.

Magnetic and viscoelastic properties of these materials, depending on magnetic field, are studied using various experimental approaches in [6, 14–16]. Specific features of MAEs can be also identified by the magnetization analysis based on first-order reversal curve measurements [17, 18]. To shed light on the particle characterization and structuring, the imaging technique such as X-ray microtomography is used [19]. The extracted data of the position, rotation and size of individual particles gives an input for theoretical modeling and simulation studies of MAEs. From a methodological point of view, these investigations follow two main modeling approaches: dipole-interaction models [20–23] and finite element methods based on a continuum formulation [24, 25]. Mesoscopic scale models, that allow to predict shape deformations of MAEs with magnetically hard and soft particles in a magnetic field, are developed in [26, 27].

Due to their exceptional field-dependent properties, smart materials like MAEs offer great potential for designing intelligent technical applications. The fact that a magnetic field might be exerted contactlessly makes them unique for a number of applications in the area of vibration isolation and damping [28, 29]. A variable differential mount apparatus using an MAE for improving driving comfort of a vehicle is presented in [30]. In the field of biomechanics, prosthetic and orthotic devices with shock absorption springs made of magnetic elastomers can be adjusted to a level corresponding to an activity of the user [31]. Magnetorheological fluids in terms of their application in medical and rehabilitation devices are discussed in [32, 33]. By using the influence of a magnetic field on the mechanical deformation of MAEs, various actuation and locomotion systems can be developed [34, 35]. A method to program the magnetic actuation in order to achieve complex time-varying shapes of MAEs is proposed in [36]. This may lead to a vast number of miniature soft devices mimicking biological beating patterns, e.g. an artificial cilium. A nature-inspired flexible gripper made of an MAE with the remote actuation is shown in [37]. Another example is a soft robotic form-fit gripper based on the field-induced plasticity effect of MAEs [38, 39]. This gripper is able to adapt its shape to the geometry of gripped objects, and the shape remains preserved after a mechanical load when a large magnetic field is present. As soon as the magnetic field is removed, the original shape of the MAE gripper is restored. Various film structures of anisotropic MAEs with a directional field response can be used for applications in soft robotics [40].

An object of this study is to assess the potential of MAEs to design a sensor system with an adjustable sensitivity controlled by a magnetic field. Following our previous work [41], we study an oscillatory motion of MAE macro beams that might serve as functional elements for fabricating sensors of several kinds, e.g., acceleration, force and pressure sensors. The concept to design a system incorporating a sensor element made of MAE to detect mechanical stimuli of the environment, is presented in [7]. According to it, an MAE beam structure with a mixture of magnetically soft and hard particles shows bending vibrations induced by an external excitation of its base. This response can be identified by measurements of the magnetic field distortion using Hall sensors that are spatially distributed around the vibrating element. The resulting signals then could be converted in a facile way into electromagnetic ones for measuring and analyzing. The key point is that an adaptivity of the MAE sensor element can be realized by means of an externally applied magnetic field, which leads to the change of the material properties, above all the mechanical compliance, due to the complex particle–matrix interactions. This means that the sensor element displays different behavior without and with an external field, and its "sensitivity" can be adjusted depending on the given field magnitude. Note that the proposed sensor differs from other oscillatory schemes employing MAEs, see [42, 43], for example, by using a uniform, no gradient magnetic field whose magnitude, therefore, does not depend on the position of the MAE. In other words, the field works as a tunable bias factor defining the conditions under which the sensor operates [41].

In essence, the most comprehensible working process of the scheme [7] is the camertone-type oscillation of the MAE cantilever, which in general has a wide set of eigenfrequencies. However, if the beam is sufficiently massive, its basic (the lowest) eigenfrequency is quite far even from the first overtone. This can facilitate the filtering of the signal and the analysis of the system under monitoring. Hence, the proposed concept requires detailed investigations of free bending vibrations of MAEs. Section 26.2 shows the experimental results on the vibration characteristics of MAE beams of different geometry and magnetic filling dependent on an applied, almost homogeneous magnetic field. It also includes the theoretical analysis based on the developed magneto-mechanical model for determination of eigenfrequencies of MAE beams. Section 26.3 presents the prototype of an MAE-based vibrating unit and results on its forced bending vibrations without and with an external uniform magnetic field. The detection of the MAE vibrations using methods of the magnetic field sensing are discussed in Section 26.4. Finally, conclusions are outlined in Section 26.5.

26.2 Magnetic field-dependent vibration characteristics of MAEs

Considering the dynamic behavior of functional sensor elements made of MAEs [38], one of the major issues is the dependence of the material properties on an externally applied magnetic field that primarily influences vibration characteristics. The design of MAE vibration elements involves the determination of both geometry and material properties. Dimensions of an MAE beam and its material composition (i.e., elastic matrix material, concentration and type of the magnetic filler) are to be chosen as to make the first eigenfrequency be in a specific range assigned in advance corresponding to the controlling magnetic field. In what follows, we present experimental and theoretical methods of determination of the first eigenfrequency of MAE beams, taking into account magneto-mechanical interactions [38].

26.2.1 Experimental investigation of free bending vibrations of MAE cantilevers in a uniform magnetic field

To get a qualitative notion of the problem, consider an MAE beam that hangs vertically being fixed by its upper end and is excited by imposing a small horizontal displacement on the lower free end [41]. During the bending vibrations of the beam on account of its continuity, the displacement amplitude is maximal at the free end and falls gradually down to zero at the fixed end. The same relates to the rotation angle between the tangent to the neutral axis of the beam and the vertical. Knowing the eigenfrequencies and corresponding modes of vibrations of the MAE beam, one can predict the emergence of resonance, and therefore permit or avoid it by relevant variation of an applied magnetic field.

26.2.1.1 Materials and methods

The oligomer base for the MAEs used in tests is silicone Alpa-Sil Classic (Alpina Co.) with Shore-A hardness 6–8, in which the magnetically soft particles of carbonyl iron powder (CIP)-CC (BASF Co.) with an averaged diameter of 6 μm are mechanically stirred up. Due to the high viscosity of the paste-like mixture and the short polymerization time, isotropic beam samples with a homogeneous distribution of the magnetic filler are produced. Four different volume concentrations of the CIP are used: $\phi = 0, 0.1, 0.2$ and 0.3. The material parameters are listed in Table 26.1.

The geometry of an MAE beam is shown in Figure 26.2. It has rounded edges, so as to smooth the magnetic field distortions at the corners. The cross-section of the beam is

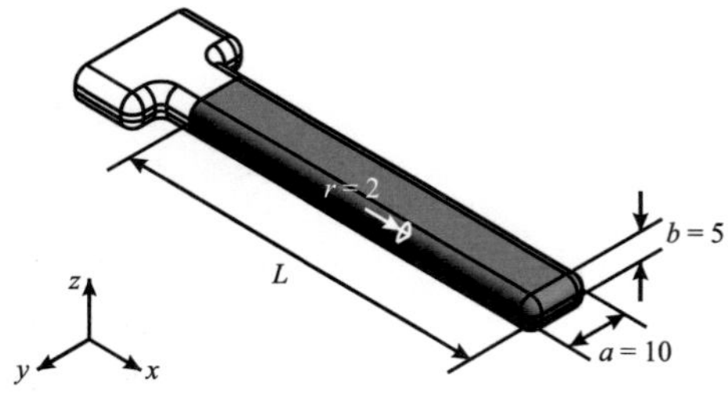

Figure 26.2: Geometry of the MAE beam, dimensions are given in mm.

Table 26.1: Material parameters of MAE beam samples.

Composition No.	**1**	**2**	**3**	**4**
CIP volume concentration, ϕ	0	0.1	0.2	0.3
Mass density, ρ (10^3 kg m^{-3})	1.06	1.74	2.42	3.10

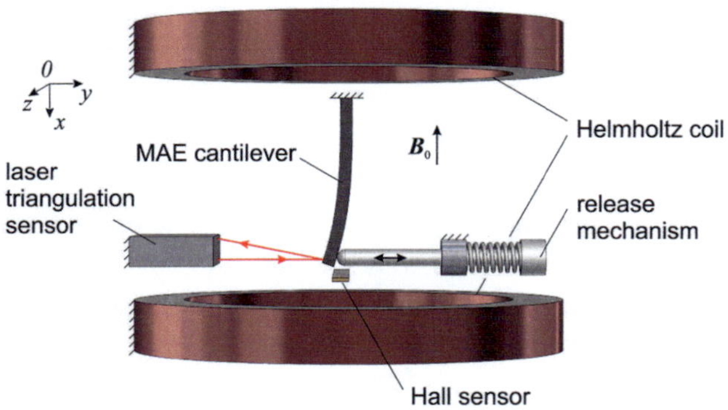

Figure 26.3: Experimental setup used for the investigation of free bending vibrations of an MAE cantilever in the presence of a uniform magnetic field. Reprinted by permission from [41].

nearly a rectangle of $a \times b$ = 10 mm × 5 mm with a corner radius of r = 2 mm. The cross-sectional area is A = 46.57 mm^2. Its area moment of inertia equals to I_z = 89.21 mm^4. Samples of three different beam lengths are prepared: L = 40, 60 and 80 mm. The "transverse" part of the T-form is designed for fixing the sample inside the support. Altogether, 12 MAE beam samples of three different lengths and four different compositions are used in the experimental studies.

The setup shown in Figure 26.3 is used to investigate the vibration behavior of an MAE beam in the presence of an external magnetic field. It consists of a pair of Helmholtz coils placed symmetrically along the vertical x axis. The magnetic flux density B_0 induced by the coils in the relevant region between them is up to 60 mT and points straight vertically. This is proven experimentally by measurements in air with a Hall sensor and also simulated using finite element method. Both tests render the magnetic field homogeneity not worse than 96.5 % [41].

An MAE beam is held by its top end in a fixed nonmagnetic support and positioned in the middle between the Helmholtz coils (Figure 26.3). Such construction makes, therefore, a kind of cantilever. In equilibrium, the central axis of the beam is co-aligned with the symmetry axis of the coil. The initial deflection of the beam in the x-y plane is created using a mechanical release mechanism. It consists of a threaded rod which is screwed against the pressure spring to adjust an initial deflection of the cantilever's free end of 5 mm along the y axis. When the trigger is pulled, the tensioned spring snaps the rod back, and the cantilever is released.

Time-dependent displacement of the cantilever's free end is measured using a laser triangulation sensor. The preliminary investigations show that due to the symmetry of the beam, the bending vibrations take place principally in the x-y plane of the initial deflection. Additionally, the distortion of the applied magnetic field close to the free end of the sample can be measured by a Hall sensor [41].

26.2.1.2 Results of free bending vibrations of MAE cantilevers

Experimental examination of the free bending vibrations of the MAE cantilevers in the vertical x-y plane is performed for various magnitudes of an applied uniform magnetic field from zero up to 60 mT in steps of 10 mT. For each beam sample and each field magnitude, the vibration behavior of the beam is analyzed based on 12 repetitions for both orientations in the fixed support.

The results indicate that, in contrast to the cantilevers made of pure silicone, the oscillatory dynamics of each sample comprising CIP particles can be changed essentially by means of the uniform magnetic field. Figure 26.4 illustrates the displacement measurements for the MAE cantilever with a CIP volume concentration of $\phi = 0.3$ and length of $L = 60$ mm for the two extreme values of B_0. As it could be seen, the vibrations are lightly damped, and their period is with a good accuracy isochronous, i.e., it is independent of the amplitude. This enables us to treat the vibrating system under study as a linear one.

From the data underlying Figure 26.4 (b), it follows that in the absence of the field the period of vibrations is 0.245 s, and the first eigenfrequency is $f_1(0) = 4.08$ Hz. For $B_0 = 60$ mT, we find $f_1(B_0) = 6.74$ Hz. Thus, for this MAE beam, the maximum relative change of the first eigenfrequency is $R_{f_1} = f_1(B_0)/f_1(0) - 1 = 65.21\%$. This relative change of the eigenfrequency can be related to the field-induced effective increase of the Young modulus E of the MAE material from $E(0) = 0.69$ MPa to $E_{\text{eff}}(B_0) = 2.36$ MPa. Note that the observed tunability of the vibration characteristics is reversible when a moderate magnetic field of 60 mT is removed. For strong fields, irreversible effects of the magnetic particle movement in the elastic matrix may take place.

Using the logarithmic decrement, which is the ratio of any two successive amplitudes, the dimensionless damping ratio D is obtained. For all cantilevers used in tests, $D < 0.05$ that corresponds to the underdamped case, and it decreases depending on the field magnitude. Just as for a system with finite number of degrees of freedom, it may be

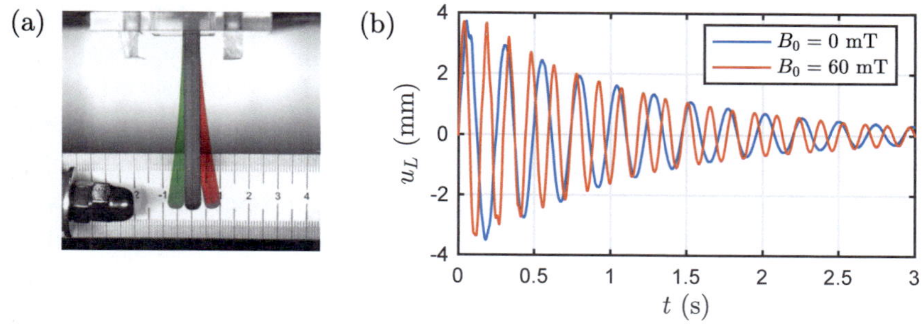

Figure 26.4: Free bending vibrations of the MAE cantilever of $L = 60$ mm and $\phi = 0.3$: (a) Overlapped sequential frames of the video recorded at 750 fps without magnetic field; (b) time-dependent displacement $u_L(t)$ of the free end for $B_0 = 0$ and 60 mT. Reprinted by permission from [41].

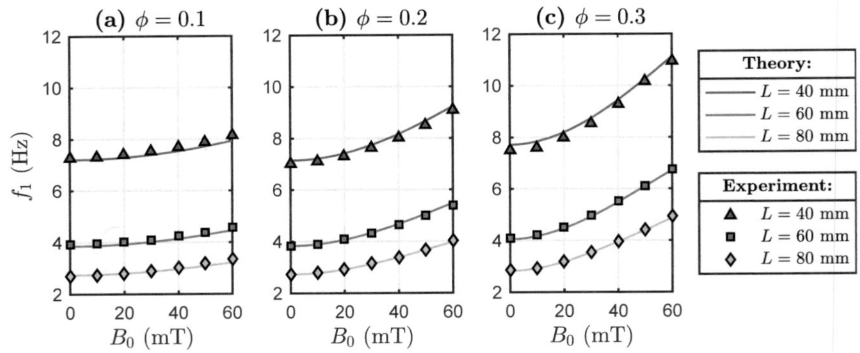

Figure 26.5: The first eigenfrequency f_1 of the free bending vibrations depending on the magnetic field magnitude B_0 for MAE cantilevers of different lengths L with CIP volume concentrations: (a) $\phi = 0.1$; (b) $\phi = 0.2$; (c) $\phi = 0.3$. Experimental values (markers) represent the mean values of 12 measurements. Theoretical dependencies (solid lines) are obtained for parameters $\chi_e = 79$ and $k_m = 0.72$. Reprinted by permission from [38].

assumed that the damping ratio $D = \pi\beta_1 f_1$ is proportional to the internal damping coefficient β_1 during harmonic vibrations at the eigenfrequency f_1.

The first eigenfrequency $f_1(B_0)$ of the free bending vibrations obtained from the displacement measurements for all MAE cantilevers is shown in Figure 26.5. These dependencies increase monotonically with the magnitude of the uniform magnetic field. The maximum relative change $R_{f1} = 73.56\%$ is observed for the longest sample of $L = 80$ mm and with a maximum CIP concentration of $\phi = 0.3$. Thus, the vibration characteristics of the MAE beams and, above all, their mechanical compliance can be controlled and reversibly tuned by means of an external uniform magnetic field.

26.2.2 Magneto-mechanical modeling for determination of MAE eigenfrequencies

The elastic forces and torques acting on every beam element are induced due to a combination of elastic, inertial and gravity forces. This case is well known, and the pertinent equations of motion could be taken from the literature [44, 45]. For a cantilever made of an MAE, one may now modify the behavior of the system by subjecting it to a uniform magnetic field along the vertical (up or down) direction. This transforms the mechanical problem into a magneto-mechanical one, where it is needed to consider, among usual forces and torques, the one that is caused by the ponderomotive (magnetic) force [41].

26.2.2.1 Model expression for the magnetic torque

Due to magnetostatic reasons, the MAE cantilever when positioned straight and parallel to the applied field has the configuration of the lowest magnetic energy. Any

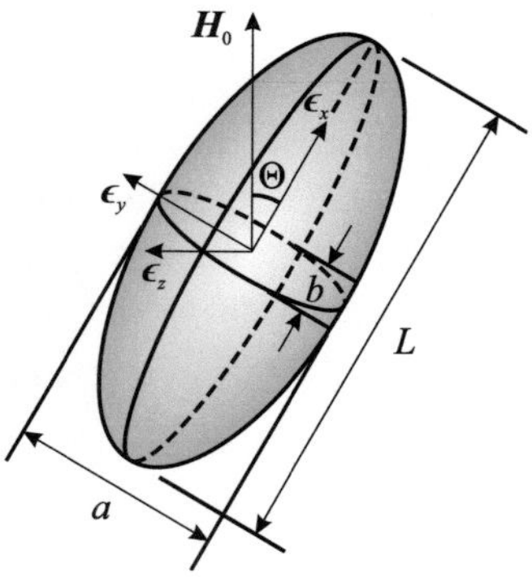

Figure 26.6: MAE ellipsoid in a uniform magnetic field used to derive the model expression of the magnetic torque. Reprinted by permission from [38].

distortions of this shape increase the magnetic energy, and thus induces ponderomotive forces striving to restore the configuration. Evidently, these forces affect the cantilever's eigenfrequency, the greater the stronger the magnetic field. In other words, by measuring the frequency change of the bending vibrations, one is able to extract information on the field strength, and vice versa. By knowing the magnitude of the applied uniform field, it is possible, based on the extended magneto-mechanical model presented below, to obtain theoretically the eigenfrequencies of MAE vibrating units [41].

In such a sample, the magnetization depends not only on the magnetic susceptibility of an MAE material, but also on the shape of the beam. The magnetic poles appearing on its ends produce a demagnetizing field, which is proportional and directed opposite to the magnetization. For the prismatic MAE beam used in the experiment (Figure 26.2), the magnetic field is not uniform and varies from place to place in the beam, being maximum in the middle. In order to calculate analytically the field inside a prismatic MAE beam, the approximation of its shape by a magnetizable general ellipsoid with the same dimensions $L > a > b$ of the principle axes is used (Figure 26.6). The magnetic susceptibility χ of the MAE is assumed to be constant and independent of the magnetic field and strain. Those simplifications are justified by the fact that a low-to-moderate magnetic field range (<100 mT) and small deformations of the MAE beam are considered [38].

Assume a uniform external magnetic field $\boldsymbol{H}_0 = H_0(\cos\Theta\,\epsilon_x + \sin\Theta\,\epsilon_y)$ applied along the vertical axis, where $\boldsymbol{\epsilon}_{x,y,z}$ are the unit vectors directed along the ellipsoid principle axes, and $\boldsymbol{\Theta}$ is the angle between $\boldsymbol{\epsilon}_x$ and the vertical (Figure 26.6). Then, the magnetic field $\boldsymbol{H}$ inside the ellipsoid is shown to be uniform and defined by [41, 46]:

$$\boldsymbol{H} = H_0\left(\frac{\cos\Theta}{1+\chi N_{xx}}\epsilon_x + \frac{\sin\Theta}{1+\chi N_{yy}}\epsilon_y\right), \tag{26.1}$$

where N_{xx} and N_{yy} are components of the diagonalized demagnetization tensor that depend only on the aspect ratios a/L and b/L of the ellipsoid [47].

The magnetic susceptibility χ in (26.1) should be taken as an effective value χ_e rendering that of the MAE with the given volume concentration ϕ of the CIP filler. This dependence can be approximated by the Lichtenecker formula [48], which is quite reliable for the composites with ϕ up to 0.3. Starting from its conventional expression for the magnetic permeability $\mu(\phi) = \mu_e^{\phi}$, one can obtain an expression in terms of susceptibility:

$$\chi(\phi) = (\chi_e + 1)^{\phi} - 1. \tag{26.2}$$

For the linear magnetization law, the magnetic moment of the ellipsoid is defined as $\boldsymbol{m} = \chi(\phi)V_{el}\boldsymbol{H}$ with the ellipsoid volume $V_{el} = \pi abL/6$.

The magnetic energy of the ellipsoid tilted under the angle Θ to the direction of the field is $U_{mag} = -\frac{1}{2}\mu_0 \boldsymbol{m}\boldsymbol{H}_0$, where μ_0 is vacuum magnetic permeability. As seen from (26.1), the direction of the magnetic moment m does not coincide with that of the applied field H_0, due to which the ellipsoid experiences a field-induced torque $T = T\epsilon_z$ along the z axis. Its magnitude T is determined by the first derivative of the magnetic energy with respect to the rotation angle. Assuming small angular deviations $\Theta \ll 1$, the following expression for the ponderomotive torque can be obtained [38]:

$$T = -\frac{\partial U_{mag}}{\partial \Theta} = \mu_0 H_0^2 V_{el}\, \Theta\, W\left(\chi_e, \frac{a}{L}, \frac{b}{L}\right), \tag{26.3}$$

where the notation W called the form-factor is introduced as:

$$W\left(\chi_e, \frac{a}{L}, \frac{b}{L}\right) = \frac{\chi^2(\phi)(N_{xx} - N_{yy})}{(1+\chi(\phi)N_{xx})(1+\chi(\phi)N_{yy})}. \tag{26.4}$$

This means that under weak perturbations the ponderomotive torque T acting on the ellipsoid is quadratic in the applied field strength and linear in the rotation angle, while the form-factor W takes into account the magnetic properties and the anisotropy of the beam. Note that the smallest component of the demagnetization tensor always corresponds to the longest dimension of the body. Therefore, in (26.4) the difference N_{xx}–N_{yy} is negative that ensures that the torque T is restoring.

26.2.2.2 Equations of motion of vibrating MAE cantilever

To evaluate the eigenfrequencies of an elongated MAE cantilever, the thin-rod approximation is considered [44]. A uniform magnetic field H_0 is directed vertically, i.e., is coaligned with the gravity field g. Under these conditions, it is evident that in equilibrium the cantilever dwells in a straight vertical configuration, and its field-induced magnetic moment m is parallel to the applied field.

During the bending vibrations of the MAE cantilever in the vertical x-y plane, there are three factors that generate restoring torques striving to recover the initial equilibrium configuration: the elasticity of the material, gravity and ponderomotive (magnetostatic) torque. All of them are included in the dynamic equations of motion, in which the rotary inertia is accounted for, while the viscous forces (dissipation) are neglected. In other words, keeping in mind the experimental evidence shown in Figure 26.4, we assume that the beam restores its equilibrium configuration by way of lightly damped small vibrations [41].

The set of equations of motion for the in-plane small free bending vibrations is as follows [38]:

$$\rho A \frac{\partial^2 u(x,t)}{\partial t^2} = \frac{\partial Q_y(x,t)}{\partial x},$$

$$\rho I_z \frac{\partial^2 \vartheta(x,t)}{\partial t^2} = \frac{\partial M_z(x,t)}{\partial x} + Q_y(x,t) + \rho g A (x-L)\vartheta(x,t)$$
$$+ k_m \mu_0 A H_0^2 W\left(\chi_e, \frac{a}{L}, \frac{b}{L}\right)\vartheta(x,t), \tag{26.5}$$

$$\frac{\partial \vartheta(x,t)}{\partial x} = \frac{M_z(x,t)}{EI_z}, \qquad \frac{\partial u(x,t)}{\partial x} = \vartheta(x,t).$$

Here, the coordinate x varies from 0 to L, t is the time, $u(x, t)$ is the horizontal displacement of the beam points in the Oy direction, $\vartheta(x, t)$ is the rotation angle around the z axis, $Q_y(x, t)$ is the component of the shear force, $M_z(x, t)$ is the bending moment in the Oz direction. The other parameters entering (2.2.2) are: ρ the mass density of the MAE, A the cross-sectional area of the beam, I_z the area moment of inertia, and E the Young modulus of the MAE.

To compensate the approximation discrepancy between the geometry of an MAE prismatic beam and ellipsoid, an adjustable parameter k_{m}, appearing as a multiplier to the term of the ponderomotive torque, is introduced in the second equation of (26.5). The other parameter χ_{e}, namely the effective magnetic susceptibility of CIP, enters the set of equations in a nonlinear way, as is seen from (26.2) and (26.4).

This model is applied for the determination of eigenfrequencies of the MAE cantilevers by considering the fixed-free boundary conditions: no deflection and rotation at $x = 0$, and no bending moment and shear force at $x = L$.

The numerical method of initial parameters is used [44, 41]. The calculations are performed for the material and geometrical parameters corresponding to the MAE cantilevers used in the experiment, see Section 26.2.1.1. For the parameters χ_{e} and k_{m} entering the theoretical model, different sets of values in the ranges $k_{\mathrm{m}} < 1$ and

$40<\chi_e<140$ are examined. The latter is the range of magnetic susceptibility measured for compacted CIP [49].

26.2.3 Comparison of experimental and theoretical results of the first eigenfrequency

According to [41], the proposed model is applied for interpretation of the experimental data. As a first step, by fitting the experimental eigenfrequencies $f_1(0)$ of the MAE cantilevers measured in zero magnetic field (Figure 26.5), the Young moduli E of the MAE compositions with different CIP volume concentrations ϕ are evaluated (Table 26.2).

Then the first eigenfrequencies of the MAE cantilevers are calculated numerically for different magnitudes of the applied magnetic field using the corresponding averaged values of the Young modulus $E(\phi)$. For each particular sample of length L and concentration ϕ, the parameters χ_e and k_m entering the theoretical model are obtained from fitting the experimental frequency data. It turned out from calculations that values $\chi_e = 79$ and $k_m = 0.72$ are "universally" valid, i.e., they come out virtually whatever sample is chosen for fitting. Comparison between experimental and numerical data on the first eigenfrequency as a function of the cantilever length is given in Figure 26.5. It can be seen that the numeric solution with the above-given particular values of χ_e and k_m fairly well (percent error below 5.6%) describe the measurement results [41].

The essence of the undertaken modification is the inclusion of an expression for the ponderomotive torque, which is exerted by the applied uniform magnetic field on the magnetized MAE cantilever, to the dynamical equations of thin rods. The fitting procedure yields the value of $k_m = 0.72$ below unity. At the first sight, this implies that the effective magnetic torque density is somewhat smaller than that generated by the model ellipsoid. Note, however, that the magnetic term in (26.5) is the product $k_m W$, where the form-factor W is given by (26.4). Although W is a nonlinear function of the magnetic

Table 26.2: Young moduli of the MAE cantilevers of different length L [41].

Composition No.	1	2	3	4
CIP volume concentration, ϕ	0	0.1	0.2	0.3
Young modulus, E (10^5 Pa):				
$L = 40$ mm	2.09	3.24	4.12	6.17
$L = 60$ mm	2.08	3.37	4.34	6.84
$L = 80$ mm	2.22	2.83	4.41	6.88
Averaged young modulus, $E(\phi)$ (10^5 Pa)	2.13	3.14	4.29	6.63

susceptibility, the leading contribution of the product is $k_m\chi^2(\phi)$. Therefore, it is mostly the value of this product that matters for the adjustment, so that greater k_m means smaller the effective susceptibility χ_e. As to the latter, the used value $\chi_e = 79$ is rendered by a more precise fitting procedure that takes into account the full nonlinear dependence $W(\chi_e)$. This value looks quite reasonable if it is to be compared with the data from [49], which puts for measured permeability values of compacted CIP in the range from 40 to 140.

From symmetry considerations, it follows that the dependence of $f_1(B_0)$ for a weak applied field should be quadratic. Qualitatively, this can be seen from the initial parts of the curves in Figure 26.5. A fairly good agreement obtained demonstrates the reliability of the developed magneto-mechanical model for determination of the eigenfrequencies of an MAE cantilever.

26.3 Forced vibration response of MAE in a uniform magnetic field

The possibility to change in a great extend the vibration characteristics of MAE beams by means of an active magnetic control can be potentially implemented for realization of sensor systems with adjustable sensitivity. A prototype of an MAE-based vibrating unit for detection of external mechanical stimuli is presented in Figure 26.7. Its main functional element is a double-fixed MAE beam placed in the inner region of the Helmholtz coil. In this section, the focus is on the experimental investigation of forced bending vibrations displayed by the MAE under the influence of a uniform magnetic field. This will pave the way to adjust the vibration response of such unit, and thus to realize its adaptable sensitivity. In addition, a theoretical approach using an oscillator model with a cubic stiffness term is considered in order to describe the resulting nonlinear behavior of the MAE beam.

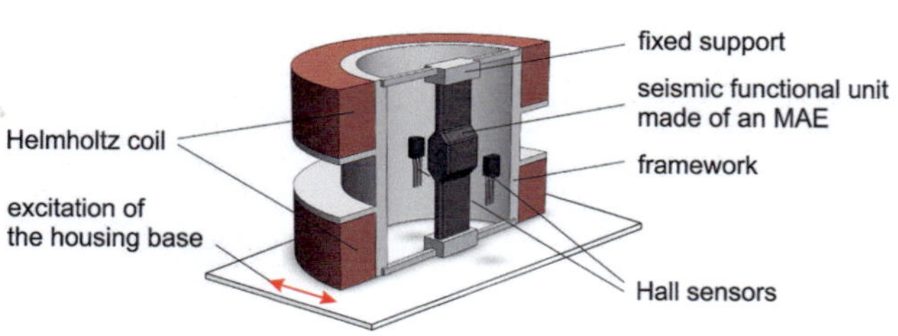

Figure 26.7: Prototype of an MAE-based vibrating sensor element with magnetic field-adjustable vibration characteristics for detection of external mechanical stimuli. Reprinted by permission from [38].

26.3.1 Experimental investigation of forced bending vibrations of the MAE unit

26.3.1.1 Experimental setup and methods

The setup used to investigate the forced bending vibrations of a double-fixed MAE beam in the presence of an external magnetic field is shown schematically in Figure 26.8. The MAE beam is fastened vertically at both ends to a housing. Its kinematic excitation is realized in the following way. The housing is attached by a rod to an electrodynamic vibration test system. It is controlled by a waveform generator, whose sinusoidal signal is strengthened using a power amplifier. The core of the vibration test system moves in the manner described by the harmonic function $u_0(t) = u_0 \cos(2\pi\Omega t)$ in the horizontal direction along the rod's axis, i.e., perpendicular to the beam's axis in equilibrium. The excitation amplitude u_0 is specified to be in a narrow range of 0.4 ± 0.02 mm. The excitation frequency Ω is set to a value from the range of 5 Hz up to 22 Hz. For each measurement, the frequency is held constant, and then it is changed gradually only in one direction (either upwards or downwards) for subsequent measurements. To record large responses near resonance, the minimum step size for the frequency setting is 0.01 Hz. To detect resulting in-plane bending vibrations of the MAE beam, the laser triangulation sensor is used. Its output signal is transmitted to a digital oscilloscope and analyzed using a PC software [50].

The magnetic field is produced by a Helmholtz coil, consisting of two parallel-mounted circular coils that have a radius of 56.5 mm and a winding number of 855. The cross-section area of the copper winding is of 30×21.5 mm^2. The coils are placed coaxially around the MAE housing with a separation of 35 mm. Note that this Helmholtz coil is about twice smaller than the one used in test of free bending vibrations (Figure 26.3) and, in contrast to it, can be regulated using a standard power supply. This follows our

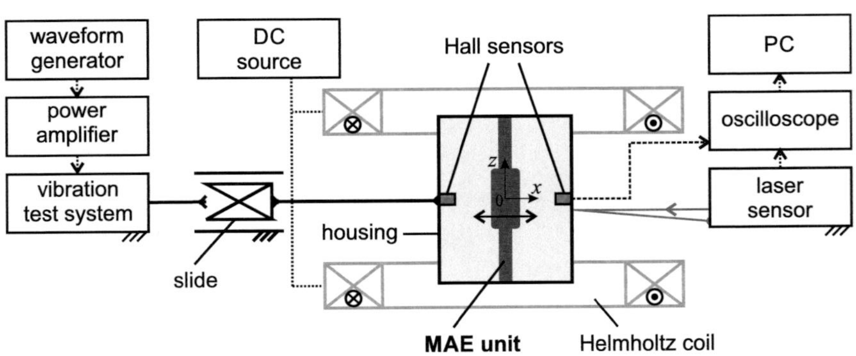

Figure 26.8: Schematic diagram of the setup used to study forced in-plane bending vibrations of a double-fixed MAE beam in a field of the Helmholtz coil. Reprinted by permission from [50].

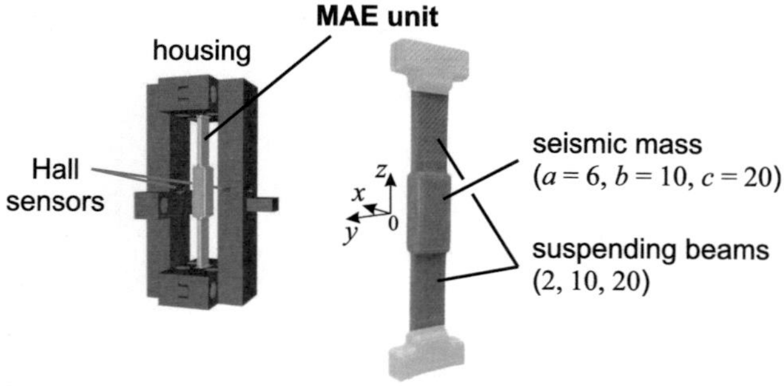

Figure 26.9: The geometry of the MAE vibrating unit and its double-sided fixation inside the housing. Dimensions are given in mm with respect to *x*, *y* and *z* axes [50].

attempt to realize a practical design of the intended prototype without laboratory limitations. In the absence of the MAE, the current of $I = 2.4$ A in each coil creates an almost uniform magnetic field inside the inner region between the coils. The field is measured using a Hall effect sensor attached to a positioning system and also analyzed using finite element method (ANSYS Maxwell software). The results show that the magnetic induction at the center point is $B_0 = 31.62$ mT. In the inner centric spherical volume with a diameter of 30 mm, the Helmholtz coil provides a homogeneity of 95%.

To detect changes of the magnetic field caused by the presence of an MAE, two Hall effect sensors (DRV5053, Texas Instruments Inc.) are fixed in the middle of the housing's sides (Figures 26.7 and 26.8). In equilibrium, the distance between each sensor and the center point of the MAE beam is $d = 12$ mm. Both sensors measure the vertical component of magnetic field, when the external field of the same direction is applied. Their sensitivity is 23 mV/mT, that allows measuring fields with magnitudes up to 34.78 mT. Such a sensitivity range is sufficient, as in this configuration field distortions on the sides of the MAE are negative, and so the measured field values are smaller than the ones applied [50].

The geometry of the MAE beam is shown in Figure 26.9. The sample is utilized in the form of a double-fixed elastic beam with a large central part as a seismic mass suspended on two sides. The overall length of the beam without including the fixing parts is 60 mm. The cross-sectional dimensions 2 mm × 10 mm of two suspending beams are specified by numerical simulations using the finite element method (ANSYS Modal analysis) on condition that the basic eigenfrequency of the unit, being far below the second one, corresponds to the transverse bending. Simulation results show that for the selected sample geometry, its first and second eigenfrequencies corresponding to the free bending and torsion vibrations are 8.24 and 18.65 Hz, respectively. Also, all edges are rounded with a radius of 1 mm, so as to smooth magnetic field distortions at the corners. The MAE beam is made with the use of the same materials and synthesis methods as described in Section 26.2.1.1. The mass content of the CIP magnetic filler is

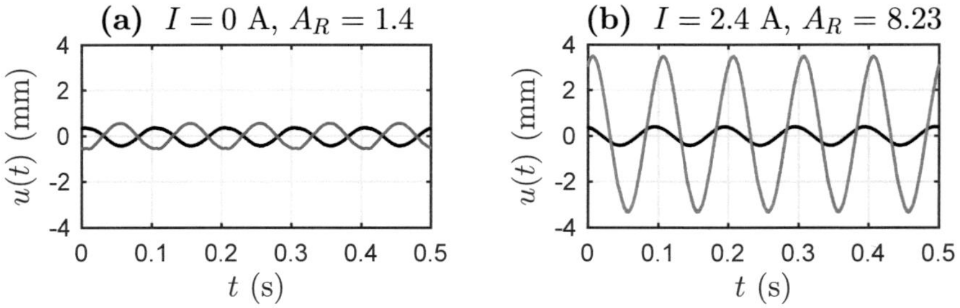

Figure 26.10: Steady-state displacement $u(t)$ as a time function measured at the middle of the MAE unit for the currents: (a) $I = 0$ A (blue line); (b) $I = 2.4$ A (orange line). Black line shows the input excitation with $\Omega = 10$ Hz and $u_0 = 0.4$ mm. Reprinted by permission from [50].

of 76.2 wt%, which is nearly $\phi = 30$ vol%. This is chosen due to the fact that considerable changes of material properties in a moderate magnetic field occur with increasing CIP concentration, as shown in Figure 26.5.

26.3.1.2 Results of forced bending vibrations of the MAE unit

The steady-state response of the forced, lightly damped in-plane vibrations of the MAE unit is studied experimentally for various excitation frequencies of the housing [50]. Figure 26.10 shows the displacement recorded by the laser sensor at the middle of the MAE unit for the same applied excitation with $\Omega = 10$ Hz in the following two cases: (a) without an external magnetic field; (b) with the field generated by a current of $I = 2.4$ A in the Helmholtz coil.

It can be seen that the response vibrations have the same frequency as that of the excitation with a phase shift. Their amplitudes are strongly dependent on the applied field strength. Consider the amplification ratio A_R defined as the ratio of the response amplitude to the forcing amplitude, which is of $u_0 = 0.4 \pm 0.02$ mm. For the case without magnetic field, the ratio A_R is 1.4 and the phase shift is about 170° corresponding to the high frequency response after the resonance, see Figure 26.10 (a). In the magnetic field, we have $A_R = 8.23$ and a phase shift of about 39°, which refers to the frequency response before the resonance, see Figure 26.10 (b).

Amplification ratio curves as a function of the excitation frequency are presented in Figure 26.11. It can be seen that these curves have the same qualitative behavior. The application of the magnetic field generated by a current of $I = 2.4$ A leads to a shift of the curve's peak to the right. This "tuning" effect is related to the magnetic field-induced change of the vibration characteristics of the MAE unit, primarily its first eigenfrequency of free bending vibrations. The higher the strength of a magnetic field, the stronger shift along the Ω axis it would cause. As a result, the forced vibration response of the MAE unit can be actively controlled by the strength of an external magnetic field,

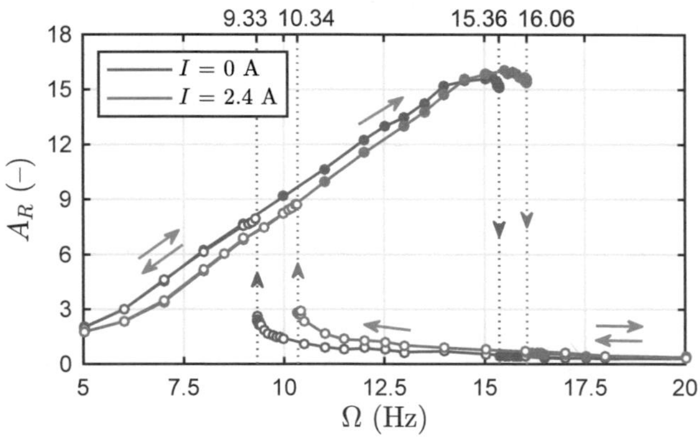

Figure 26.11: Amplification ratio A_R against the excitation frequency Ω for two currents I. Thick and white markers correspond to a step increase and decrease of Ω, respectively. Frequency values of quasi-discontinuous jumps in the response amplitude are ticked on the top axis. Reprinted by permission from [50].

in this case by changing the current supplied to the coil. The "configuration" of the MAE unit can be switched reversibly between the low/high and resonance responses.

A certain asymmetry of the curves of the amplification ratio near the resonance peak shows that there is an element of nonlinearity. Both curves are bent to the right side that is related to the hardening type of behavior. At specific ranges of the excitation, two stable stationary oscillations may exist depending on whether the frequency Ω increases or decreases stepwise. This bifurcation results in the appearance of quasi-discontinuous jumps in the response amplitude. By increasing the current in the coil, the apparent stiffness of the MAE unit increases, consequently, the hardening effect of nonlinearity decreases. As seen in Figure 26.11, the frequency width of the bifurcation range changes from 6.03 to 5.72 Hz. As will be seen in the next section, in stronger fields the nonlinear system would tend to approach the linear one [50].

For the case of Ω = 10 Hz without applied magnetic field, two stable steady-state responses exist, and Figure 26.10 (a) corresponds to the one of the lower branch, as the frequency decreases. When I = 2.4 A, there is only one stable response shown in Figure 26.10 (b).

26.3.2 Modeling of nonlinear system behavior under forced vibrations

A similar nonlinear phenomenon of the resonance hysteresis takes place in mechanics for an oscillating spring–mass system with a cubic stiffness term, which motion is described by Duffing equation [51]. This equation can provide the simplest possible model for the observed forced vibrations of the MAE beam with reasonable qualitative

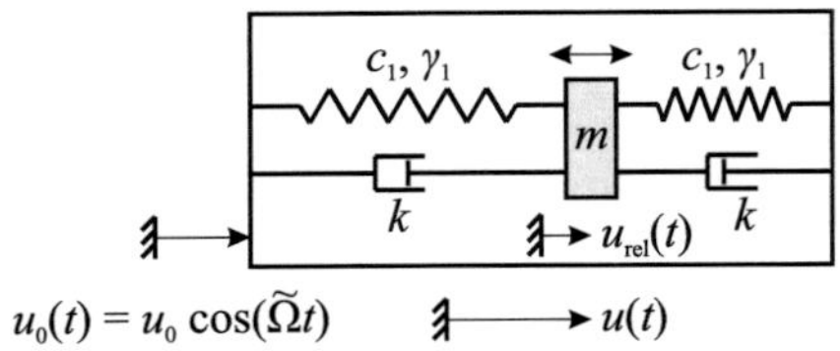

Figure 26.12: Simplified mechanical model with one degree of freedom for the forced vibrations of an MAE beam.

agreement. It is also applicable to study frequency response of a steel cantilever in the nonuniform magnetic field of two permanent magnets [52].

Assume that due to the symmetry, the central seismic mass m of the MAE beam moves translationally as a solid body in the horizontal direction. As it is connected by the same two suspending beams to a rigid housing, these beams can be modeled as two springs placed on the left and right sides of the mass (Figure 26.12). The bending stiffness of the suspending beams depends on the Young modulus E of the material, the area moment of inertia I_z of the cross-section and length of the beams. Undergoing large deflections, the behavior of the bending stiffness becomes nonlinear. This can be seen as a nonlinear progressive spring force with a cubic stiffness expressed by

$$F_c = c_1 u_{\mathrm{rel}} + \gamma_1 u_{\mathrm{rel}}^3, \tag{26.6}$$

where c_1 and γ_1 are positive spring constants of its linear and cubic terms.

The relative displacement $u_{\mathrm{rel}}(t)$ is the displacement of the mass with respect to the housing, which is in turn subjected to the kinematic excitation $u_0(t) = u_0 \cos(\tilde{\Omega} t)$ in the horizontal direction (Figure 26.12). Here, $\tilde{\Omega}$ is the angular frequency of excitation. The absolute vibrational motion of the mass is described by the function $u(t) = u_{\mathrm{rel}}(t) + u_0(t)$ and can be measured with a laser triangulation sensor.

Considering that vibrations of the suspending beams are lightly damped as shown in Figure 26.4 (b), the damping force is modeled as being proportional to the relative velocity of the mass with a low viscous damping coefficient k. Equation of motion of the mass relative to an inertial reference frame is given by

$$m(\ddot{u}_{\mathrm{rel}}(t) + \ddot{u}_0(t)) = -2c_1 u_{\mathrm{rel}}(t) - 2\gamma_1 u_{\mathrm{rel}}^3(t) - 2k\dot{u}_{\mathrm{rel}}(t).$$

This equation represents only a single mode of vibration. The angular eigenfrequency of the translating motion is $\omega_0 = \sqrt{2c_1/m}$. The damping ratio, cubic stiffness coefficient and the frequency ratio are introduced as dimensionless quantities in the following way:

$$D = \frac{k}{m\omega_0}, \quad \mathrm{Y} = \frac{2\gamma_1 u_0^2}{m\omega_0^2}, \quad \eta = \frac{\tilde{\Omega}}{\omega_0}. \tag{26.7}$$

Equation of motion for the relative displacement can be written in form of the Duffing equation

$$\ddot{u}_{\mathrm{rel}}(t) + 2D\omega_0 \dot{u}_{\mathrm{rel}}(t) + \omega_0^2 u_{\mathrm{rel}}(t) + \mathrm{Y}\frac{\omega_0^2}{u_0^2} u_{\mathrm{rel}}^3(t) = u_0 \eta^2 \omega_0^2 \cos\left(\tilde{\Omega} t\right). \tag{26.8}$$

To obtain approximately a steady-state solution of (26.8), we suppose that it is proportional to the excitation amplitude with an amplification ratio $A_{R,\text{rel}}$ and has an induced phase shift φ_{rel}:

$$u_{\text{rel}}(t) = u_0 A_{R,\text{rel}} \cos\left(\tilde{\Omega}t - \varphi_{\text{rel}}\right).$$

Substituting this solution form into (26.8) and using an expression $\cos^3(\alpha) = 3/4\cos(\alpha) + 1/4\cos(3\alpha)$, we equate the sine and cosine terms to zero separately:

$$\begin{aligned} u_0\omega_0^2\cos\left(\tilde{\Omega}t - \varphi_{\text{rel}}\right)&: \quad -\eta^2 A_{R,\text{rel}} + A_{R,\text{rel}} + \frac{3\gamma}{4}A_{R,\text{rel}}^3 = \eta^2\cos(\varphi_{\text{rel}}), \\ u_0\omega_0^2\sin\left(\tilde{\Omega}t - \varphi_{\text{rel}}\right)&: \quad 2D\eta A_{R,\text{rel}} = \eta^2\sin(\varphi_{\text{rel}}). \end{aligned} \tag{26.9}$$

Note that for an approximate solution, the cosine term of triple angle remains unbalanced.

Denoted by η_*^2, the squared frequency of free translating vibration, dependent on the dimensionless amplitude, is as follows:

$$\eta_*^2 = 1 + \frac{3\gamma}{4}A_{R,\text{rel}}^2. \tag{26.10}$$

By excluding the phase shift in (26.9), the dependence of the relative amplification ratio regarding the housing on the frequency ratio is

$$A_{R,\text{rel}} = \frac{\eta^2}{\sqrt{\left(\eta_*^2 - \eta^2\right)^2 + (2D\eta)^2}}.$$

For the absolute motion of the mass, we have

$$u(t) = u_{\text{rel}}(t) + u_0(t) = u_0 A_{R,\text{rel}}\cos\left(\tilde{\Omega}t - \varphi_{\text{rel}}\right) + u_0\cos\left(\tilde{\Omega}t\right) = u_0 A_R\cos\left(\tilde{\Omega}t - \varphi\right),$$

where the absolute amplification ratio A_R is defined by

$$A_R = \sqrt{\frac{\eta_*^4 + (2D\eta)^2}{\left(\eta_*^2 - \eta^2\right)^2 + (2D\eta)^2}}. \tag{26.11}$$

The phase shift between the excitation and absolute motion equals to

$$\tan(\varphi) = \frac{2D\eta^3}{\eta_*^4 - \eta_*^2\eta^2 + (2D\eta)^2}.$$

As is seen from (26.11), the obtained solution of the amplification ratio A_R with respect to an inertial frame can be conveniently represented on a graph as a function of the squared frequency ratio η^2. Figure 26.13 shows the comparison of these theoretical dependencies for the spring–mass model with the experimental results of forced vibrations of the MAE beam (taken from Figure 26.11). The calculation parameters of the model for both cases without and with magnetic field are the following:

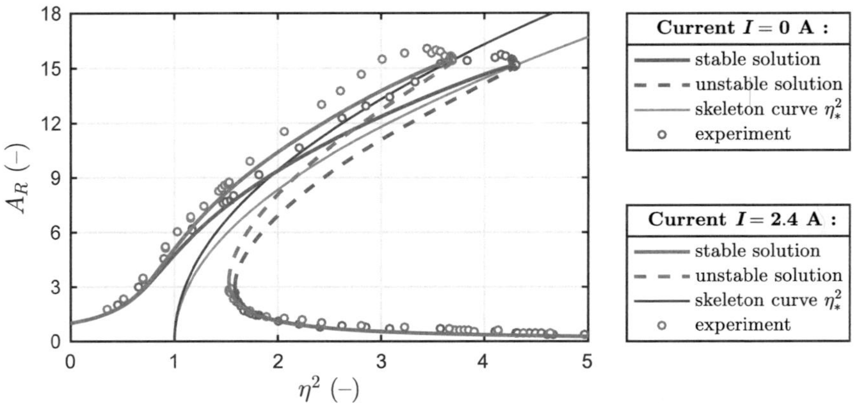

Figure 26.13: Comparison of the theoretical (lines) and experimental (markers) dependencies of the amplification ratio A_R as a function of η^2 for two currents I.

$$
\begin{aligned}
I = 0\,\mathrm{A}: \quad & f_0 = \frac{\omega_0}{2\pi} = 7.41\,\mathrm{Hz}, \quad \gamma = \frac{41.5\mathrm{s}^{-2}}{\omega_0^2} = 0.0191, \quad D = 0.0685; \\
I = 2.4\,\mathrm{A}: \quad & f_0 = \frac{\omega_0}{2\pi} = 8.36\,\mathrm{Hz}, \quad \gamma = \frac{41.5\mathrm{s}^{-2}}{\omega_0^2} = 0.015, \quad D = 0.0623.
\end{aligned}
\tag{26.12}
$$

These parameters are chosen so that the theoretical frequency values of the quasi-discontinuous jumps in the response amplitude coincide with the experimental ones. In the range between these frequencies, as Figure 26.13 shows, Eq. (26.11) gives three solutions: with a large amplitude (upper branch), with a small amplitude (lower branch) and with an intermediate amplitude. The latter is unstable and can not be implemented practically. One of two stable solutions from the lower or upper branch can be observed depending on whether the excitation frequency increases or decreases monotonously as done in our experiment.

The considered model explains clearly the influence of a magnetic field on the nonlinear effect of MAE vibrations. The range of bifurcation, where both stable solutions exist, depends on the slope of the so-called skeleton curve defined by (26.10). For $\gamma > 0$, it bends to the right characterizing hardening type of behavior. Since the eigenfrequency of free vibration increases in the presence of a magnetic field, the value of γ decreases as the square of the frequency, and the damping ratio D is inversely proportional to it, see (26.7). Therefore, the slope of the skeleton curve relative to the vertical gets smaller, as the current in the Helmholtz coil, i.e. the applied field strength, increases. In the limiting case of strong magnetic field, the nonlinear forced response would tend to approach the behavior of a linear undamped system with a resonance frequency $\eta \simeq 1$ independent of the amplitude.

26.4 Detection of MAE vibrations by magnetic field sensing

With respect to the intended application of MAE beams as functional sensing elements of an acceleration sensor (Figure 26.7), the focus of this section is on the unambiguous determination of MAE forced vibrations. Such a vibrating unit placed in an external magnetic field distorts the field distribution over the space around it [50]. By using suitable strategies and ranges of the magnetic field sensing, this information can be used to determine its vibrational motion.

26.4.1 Theoretical analysis of MAE magnetostatics

Consider a coordinate system with the origin at the center point of the MAE vibrating unit in equilibrium, so that the x-z plane is the plane of its vibrations (Figure 26.8). As in [50], let us simplify the geometry of the unit and take into account only the central seismic mass. It is a rectangular parallelepiped with dimensions of a = 6 mm, b = 10 mm, c = 20 mm in x, y and z directions, respectively (Figure 26.9).

When a uniform magnetic field B_0 is applied vertically along the z axis, the magnetization M at the center of the parallelepiped points in the same direction. The demagnetizing field is, however, directed opposite to the magnetization. Due to the symmetry, the magnetic field along the x axis has only a vertical component $B_z(x)$. Its value inside the parallelepiped is bigger than B_0. For the region $|x| > a/2$ outside the body, $B_z(x)$ is smaller that B_0 and tends asymptotically to it with increasing distance.

The strength $H_z(x)$ of the static magnetic field around the parallelepiped along the x axis can be calculated using the approximation of a uniform magnetization as follows [53]:

$$H_z(x) = H_0 + \frac{M}{\pi} \sum_{i=1,2} \arctan\left(\frac{(-1)^{i+1} b x_i}{c\sqrt{x_i^2 + (b^2 + c^2)/4}} \right), \tag{26.13}$$

where $x_i = x + (-1)^i a/2$, $i = 1,2$.

This analytical solution shows that $H_z(0) = H_0 - 0.26M/\pi$. For the linear magnetization law, we have $M = \chi H_z(0)$, where the magnetic susceptibility $\chi(\phi) = 2.72$ of the MAE with a concentration of $\phi = 0.3$ is approximated by (26.2) taking, as before, $\chi_e = 79$. Thus, the magnetization of the material at the center is $M = 2.23\,H_0$. It can also be shown that the solution (26.13) satisfies the Maxwell equations in both regions inside and outside the parallelepiped and the boundary conditions. In particular, on the left and right sides $x = \pm a/2$, the tangential component H_z of the field strength is continuous, while B_z undergoes a jump.

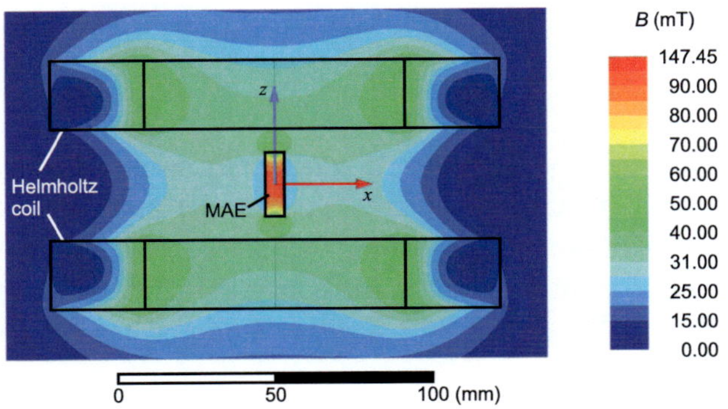

Figure 26.14: Magnetic induction of an MAE rectangular parallelepiped placed in the middle of the Helmholtz coil as a result of the numerical solution using magnetostatic analysis (ANSYS Maxwell software). Simulation parameters are: excitation current in the coil $I = 2.4$ A; dimensions of the parallelepiped $a = 6$ mm, $b = 10$ mm, $c = 20$ mm; magnetic susceptibility of MAE $\chi = 2.72$.

The numerical solution for the distribution of the magnetic field magnitude around the MAE parallelepiped is obtained using magnetostatic analysis (ANSYS Maxwell software) and shown in Figure 26.14. The simulation parameters of the Helmholtz coil are the same as for the coil used in experiments, see Section 26.3.1.1.

When the parallelepiped is displaced from the center point by a distance u_{rel} along the x axis, it is closer to the right Hall sensor with a distance of $d - u_{\text{rel}}$ between them (Figure 26.8). So, the distance to the left Hall sensor is $d + u_{\text{rel}}$. In equilibrium, the distance between each sensor and the center point of the MAE beam is $d = 12$ mm. Both sensors, right and left, are fixed inside the housing so that they measure vertical field components only $B_{z1}(d - u_{\text{rel}})$ and $B_{z2}(d + u_{\text{rel}})$, respectively. As closer the magnetized body is to the sensor, as more field distortion it produces. Therefore, $B_{z1} < B_{z2}$ for $u_{\text{rel}} > 0$, and vice versa.

The difference of these field values is calculated using (26.13) as

$$\Delta B(u_{\text{rel}}) = B_{z1}(d - u_{\text{rel}}) - B_{z2}(d + u_{\text{rel}}) = \\ = \frac{2.23}{\pi} B_0 \sum_{i=1,2} \left(\begin{array}{l} \arctan\left(\dfrac{(-1)^{i+1} b (d_i - u_{\text{rel}})}{c\sqrt{(d_i - u_{\text{rel}})^2 + (b^2 + c^2)/4}} \right) \\ -\arctan\left(\dfrac{(-1)^{i+1} b (d_i + u_{\text{rel}})}{c\sqrt{(d_i + u_{\text{rel}})^2 + (b^2 + c^2)/4}} \right) \end{array} \right), \qquad (26.14)$$

where $d_i = d + (-1)^i a/2$, $i = 1,2$, and $|u_{\text{rel}}| \le d - a/2$.

The series expansion of this dependence at zero contains odd powers only, and the coefficients of the third and higher powers are negligibly small. Hence, the difference of the measured field values can be well approximated as a linear function of the displacement u_{rel}:

$$\Delta B(u_{rel}) \simeq 0.0156\text{mm}^{-1} \cdot B_0 u_{rel}. \tag{26.15}$$

Thus, the considered strategy of the magnetic field measurement at two specific points around the MAE vibrating unit provides a linear relationship between the kinematic variable u_{rel} and magnetic field magnitude ΔB measured by the electrical signals of the Hall sensors.

26.4.2 Experimental results of magnetic field sensing

Figure 26.15 shows the experimental results for the steady-state response of the MAE unit for an excitation frequency of $\Omega = 16$ Hz and a current in the coil of $I = 2.4$ A. In the absence of the magnetic unit, this current generates an almost uniform magnetic field with a magnitude of $B_0 = 31.62$ mT at the center point. The horizontal displacement $u(t)$ at the middle point of the MAE unit is recorded by a laser triangulation sensor. The amplitude of this response is $\tilde{u} = 6.19$ mm. So, the amplification ratio is in this case $A_R = 15.48$, which means that the resonance response takes place. For the relative motion of the MAE unit about the housing, the amplitude is $\tilde{u}_{rel} = 6.29$ mm.

The magnetic field distortion caused by the MAE vibrations is simultaneously measured by two Hall sensors that are fixed on the sides of the housing. As can be seen in Figure 26.15, the difference $\Delta B(t)$ of the field values of both sensors has the same frequency as that of the excitation. Moreover, its amplitude is of $\Delta\tilde{B} = 4.03$ mT, which is 12.74% of the applied field $B_0 = 31.62$ mT. Note that this distortion is slightly bigger than the change of 9.81% obtained using the theoretical approximation by Eq. (26.15).

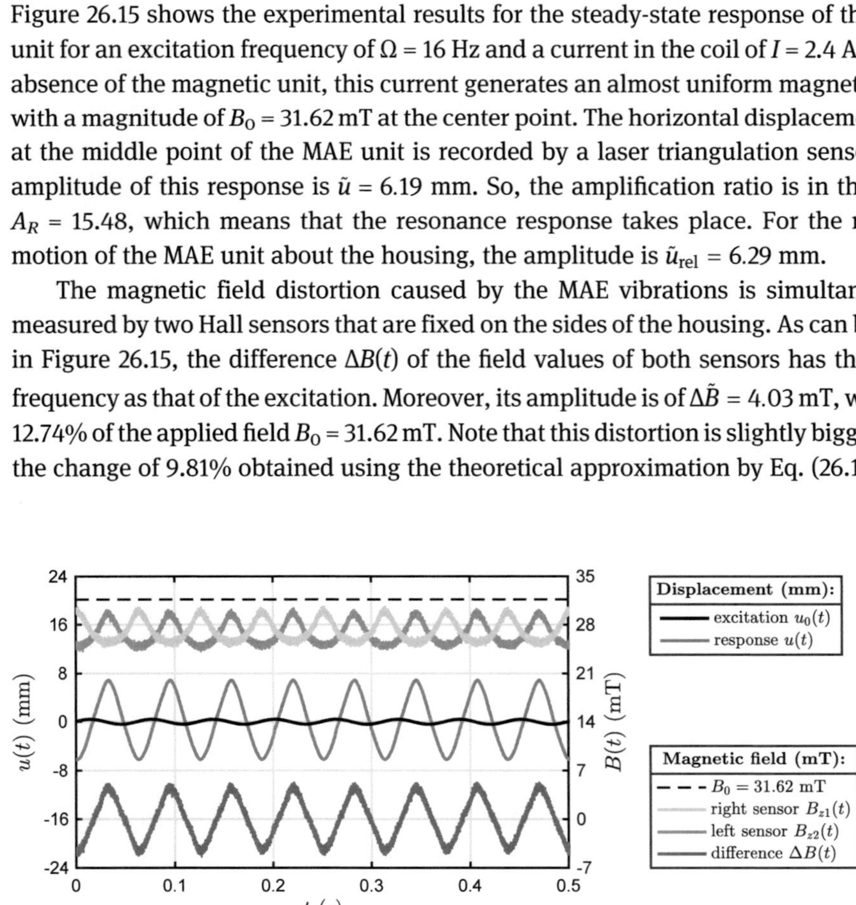

Figure 26.15: Steady-state response of the MAE vibrating unit recorded by a laser triangulation sensor (orange line) and two Hall sensors (blue line) for an excitation frequency $\Omega = 16$ Hz, excitation amplitude $u_0 = 0.4$c mm and current of $I = 2.4$ A.

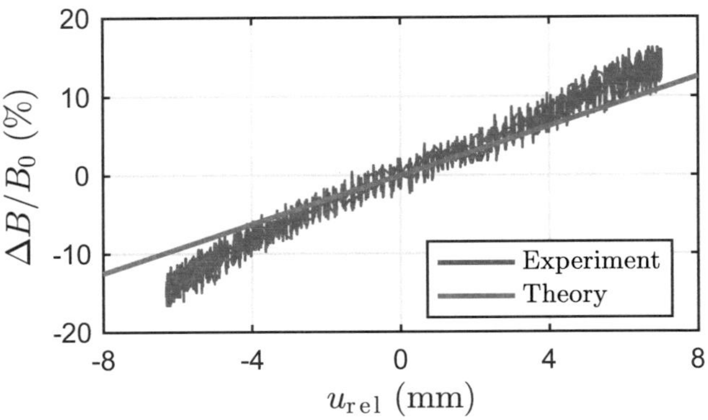

Figure 26.16: Linear relation of the magnetic field distortion ΔB on the displacement u_{rel} of the MAE vibrating unit relative to its housing. Theoretical line is defined by (26.15).

The experimental results confirm that during the forced bending vibrations of the MAE unit, there is a linear dependence between the relative displacement and the occurring magnetic field distortion (Figure 26.16). Some disparity of the slopes of the experimental and theoretical lines is caused by the simplifications assumed during the analytical calculations. In particular, Eq. (26.15) describes only the field change produced by the central seismic part of the MAE unit, and is derived using approximation of its uniform magnetization.

26.5 Conclusions

This work presents the results of experimental and theoretical investigations on MAEs towards an approach of their potential application as vibrating sensor elements with adaptable sensitivity. Starting with the experimental and model-based studies of the free vibrational behavior of cantilevers made of MAEs, it is shown that the first bending eigenfrequency of the cantilevers depends strongly on the strength of an applied uniform magnetic field. The developed theoretical model predicts reliably the dependencies of the vibration characteristics on the geometric configuration and the material parameters of the chosen MAE.

The experimental investigation of the forced vibration response of a double-fixed MAE beam under kinematic excitation confirms the possibility of active magnetic control of the amplitude-frequency characteristics. With change of the uniform field strength, the MAE beam reveals different steady-state responses for the same excitation. The resonance may occur at various ranges of the excitation frequency. Nonlinear dependencies of the amplification ratio on the excitation frequency are obtained for two different applied fields. The model of an oscillating springâ€œmass system with a cubic stiffness term is employed to analyze the influence of the magnetic field on the nonlinear phenomenon of the resonance hysteresis. It is shown that the steady-state vibrations of MAE beams can be detected based on the magnetic field distortion. The field difference, which is measured simultaneously on the sides of a vibrating MAE beam, provides a signal with the same frequency as the excitation. The field amplitude is proportional to the amplitude of resulting vibrations.

The presented prototype of the MAE-based vibrating unit with the field-controlled "configuration" can be implemented for realization of acceleration sensor systems with adaptable sensitivity. Such an MAE unit transforms mechanical stimuli of its housing into bending vibrations that, in turn, could be converted in a facile way into electromagnetic signals of Hall sensors for measuring and analyzing. Due to the exceptional field-dependent properties, the MAE vibrating unit can be used both in a resonance mode or in a regime of low/high responses. The ongoing research on MAEs is oriented to the use of other geometrical forms along with beams, e.g. two-dimensional structures such as membranes.

Acknowledgment: This research is funded by the Deutsche Forschungsgemeinschaft (DFG) in the framework of the priority program SPP 1681 under the projects BE 6553/1-1 and ZI 540-17/3. The support of the research association PAK 907 between the DFG and the Russian Foundation for Basic Research (RFBR) is also gratefully acknowledged (project 19-52-12045). We thank our scientific colleagues and their teams in the research association PAK 907: D. Borin (Dresden), G. Stepanov (Moscow), S. Kantorovich and A. Zubarev (Ekaterinburg). We thank all working groups of SPP 1681 for the great scientific work they are performing in the respective subprojects. We thank specially S. Odenbach (Dresden) for his efforts in the coordination of both research programs.
Author contribution: All the authors have accepted responsibility for the entire content of this submitted manuscript and approved submission.
Research funding: None declared.
Conflict of interest statement: The authors declare no conflicts of interest regarding this article.

References

1. Carlson JD, Jolly MR. MR fluid, foam and elastomer devices. Mechatronics 2000;10:555–69.
2. Bossis G, Volkova O, Lacis S, Meunier A. Magnetorheology: fluids, structures and rheology. In: Odenbach S, editor. Ferrofluids: Magnetically Controllable Fluids and Their Applications, Springer-Verlag Berlin Heidelberg: Lect Notes Phys, LNP; 2002, vol 594, pp. 202–30.
3. Jolly MR, Carlson JD, Munoz BC, Bullions TA. The magnetoviscoelastic effect of elastomer composites consisting of ferrous particles embedded in a polymer matrix. J Intell Mater Syst Struct 1996;7:613–22.
4. Kallio M. The elastic and damping properties of magnetorheological elastomers, Espoo: VVT Publications; 2005, vol 565:146 p.

5. Varga Z, Filipcsei G, Zrínyi M. Magnetic field sensitive functional elastomers with tuneable elastic modulus. Polymer 2006;47:227–33.
6. Lee D, Lee M, Jung N, Yun M, Lee J, Thundat T, et al. Modulus-tunable magnetorheological elastomer microcantilevers. Smart Mater Struct 2014;23: 055017.
7. Volkova TI, Böhm V, Kaufhold T, Popp J, Becker F, DYu B, et al. Motion behaviour of magneto-sensitive elastomers controlled by an external magnetic field for sensor applications. J Magn Magn Mater 2017;431:262–5.
8. Koo JH, Dawson A, Jung HJ. Characterization of actuation properties of magnetorheological elastomers with embedded hard magnetic particles. J Intell Mater Syst Struct 2010;23:1049–54.
9. Kramarenko EY, Chertovich AV, Stepanov GV, Semisalova AS, Makarova LA, Perov NS, et al. Magnetic and viscoelastic response of elastomers with hard magnetic filler. Smart Mater Struct 2015;24: 035002.
10. Böse H, Hesler A, Monkman G. Magnetorheological composite materials comprising hard magnetic particles, method for the production thereof and use thereof. United States: Google Patent DE Patent 102007028663 A1; 2007.
11. Stepanov GV, Borin DY, Storozhenko PA. Rotation of magnetic particles inside the polymer matrix of magnetoactive elastomers with a hard magnetic filler. J Magn Magn Mater 2017;431:138–40.
12. Borin DY, Stepanov GV. Oscillation measurements on magnetoactive elastomers with complex composition. J Optoelectron Adv Mater 2013;15:249–53.
13. Becker TI, Stolbov OV, Borin DY, Zimmermann K, Raikher YuL. Basic magnetic properties of magnetoactive elastomers of mixed content. Smart Mater Struct 2020;29: 075034.
14. Borin D, Stepanov G, Dohmen E. Hybrid magnetoactive elastomer with a soft matrix and mixed powder. Arch Appl Mech 2019;89:105–17.
15. Remmer H, Roeben E, Schmidt AM, Schilling M, Ludwig F. Dynamics of magnetic nanoparticles in viscoelastic media. J Magn Magn Mater 2017;427:331–5.
16. Schümann M, Borin D, Huang S, Auernhammer GK, Müller R, Odenbach S. A characterisation of the magnetically induced movement of NdFeB-particles in mag-netorheological elastomers. Smart Mater Struct 2017;26: 095018.
17. Linke JM, Borin DY, Odenbach S. First-order reversal curve analysis of magnetoactive elastomers. RSC Adv 2016;6:100407–16.
18. Vaganov MV, Linke J, Odenbach S, Raikher YuL. Model FORC diagrams for hybrid magnetic elastomers. J Magn Magn Mater 2017;431:130–3.
19. Odenbach S. Microstructure and rheology of magnetic hybrid materials. Arch Appl Mech 2016;86: 269–79.
20. Biller AM, Stolbov OV, Raikher YuL. Mesoscopic magnetomechanical hysteresis in a magnetorheological elastomer. Phys Rev E 2015;92: 023202.
21. Tarama M, Cremer P, Borin DY, Odenbach S, Löwen H, Menzel AM. Tunable dynamic response of magnetic gels: impact of structural properties and magnetic fields. Phys Rev E 2014;90: 042311.
22. Goh S, Menzel AM, Löwen H. Dynamics in a one-dimensional ferrogel model: relaxation, pairing, shock-wave propagation. Phys Chem Chem Phys 2018;20:15037.
23. Weeber R, Kreiss P, Holm C. Studying the field-controlled change of shape and elasticity of magnetic gels using particle-based simulations. Arch Appl Mech 2019;89:3–16.
24. Gebhart P, Wallmersperger T. A general framework for the modeling of porous ferrogels at finite strains. J Mech Phys Solid 2019;122:69–83.
25. Metsch P, Kalina KA, Brummund J, Kästner M. Two- and three-dimensional modeling approaches in magneto-mechanics: a quantitative comparison. Arch Appl Mech 2019;89:47–62.
26. Raikher YuL, Stolbov OV. Numerical modeling of large field-induced strains in ferroelastic bodies: continuum approach. J Phys Condens Matter 2008;20:204126.

27. Biller AM, Stolbov OV, Raikher YuL. Modeling of particle interactions in magnetorheological elastomers. J Appl Phys 2014;116. https://doi.org/10.1063/1.4895980. 114904.
28. Kozlowska J, Boczkowska A, Czulak A, Przybyszewski B, Holeczek K, Stanik R, et al. Novel MRE/CFRP sandwich structures for adaptive vibration control. Smart Mater Struct 2016; 25: 035025.
29. Yang J, Sun SS, Du H, Li WH, Alici G, Deng HX. A novel magneto-rheological elastomer isolator with negative changing stiffness for vibration reduction. Smart Mater Struct 2014;23. https://doi.org/10.1088/0964-1726/23/10/105023. 105023.
30. Kim MS, Yang KM, Lee SH, Yoon JH, Jeong UC, Yang IH, Oh JE. Variable differential mount apparatus using magnetorheological elastomer. U.S. Patent 8,844,914 B2; 2014.
31. Thorsteinsson F, Gudmundsson I, Lecomte C. Prosthetic and orthotic devices having magnetorheological elastomer spring with controllable stiffness. U.S. Patent 9,078,734 B2; 2015.
32. Oh J-S, Choi S-B. State of the art of medical devices featuring smart electro-rheological and magneto-rheological fluids. J King Saud Univ Sci 2017;29:390–400.
33. El Wahed AK, Balkhoyor LB. The performance of a smart ball-and-socket actuator applied to upper limb rehabilitation. J Intell Mater Syst Struct 2018;29:2811–22.
34. Zimmermann K, Naletova VA, Zeidis I, Turkov VA, Kolev E, Lukashevich MV, et al. A deformable magnetizable worm in a magnetic field – A prototype of a mobile crawling robot. J Magn Magn Mater 2007;311:450–3.
35. Zimmermann K, Zeidis I, Böhm V, Kaufhold T, Volkova T, Krautz M, et al. Mechanics of actuators based on magnetic hybrid materials with application for robotics, fluid control and sensor technology. IFToMM Probl Mech 2014;57:23–41.
36. Lum GZ, Ye Z, Dong X, Marvi X, Erin O, Hu W, et al. Shape-programmable magnetic soft matter. Proc Natl Acad Sci Unit States Am 2016;113:E6007–15.
37. Gao W, Wang L, Wang X, Liu H. Magnetic driving flowerlike soft platform: biomimetic fabrication and external regulation. ACS Appl Mater Interfaces 2016;8:14182–9.
38. Becker TI, Böhm V, Chavez Vega J, Odenbach S, Raikher YL, Zimmermann K. Magnetic-field-controlled mechanical behavior of magneto-sensitive elastomers in applications for actuator and sensor systems. Arch Appl Mech 2019;89:133–52.
39. Zimmermann K, Chavez Vega J, Becker TI, Witte H, Schilling C, Köhring S, et al. An approach to a form-adaptive compliant gripper element based on magneto-sensitive elastomers with a bioinspired sensorized surface. Int Sci J Probl Mech 2019;75:23–38.
40. Schmauch MM, Mishra SM, Evans BA, Velev OD, Tracy JB. Chained iron microparticles for directionally controlled actuation of soft robots. ACS Appl Mater Interfaces 2017;9:11895–901.
41. Becker TI, Raikher YuL, Stolbov OV, Böhm V, Zimmermann K. Dynamic properties of magneto-sensitive elastomer cantilevers as adaptive sensor elements. Smart Mater Struct 2017;26: 095035.
42. Ramanujan RV, Lao LL. The mechanical behavior of smart magnet–hydrogel composites. Smart Mater Struct 2006;15:952–6.
43. Abramchuk S, Kramarenko E, Grishin D, Stepanov G, Nikitin L, Filipcsei G, et al. Novel highly elastic magnetic materials for dampers and seals: Part II. Material behavior in a magnetic field. Polym Adv Technol 2007;18:513–8.
44. Svetlitsky VA. Dynamics of Rods. Berlin Heidelberg: Springer-Verlag; 2005:448 p.
45. Magnus K, Popp K, Sextro W. Schwingungen. Wiesbaden: Springer Vieweg; 2013:298 p.
46. Kneller E. Ferromagnetismus. Berlin: Springer; 1962:816 p.
47. Osborn JA. Demagnetizing factors of the general ellipsoid. Phys Rev 1945;67:351–7.
48. Zakri T, Laurent JP, Vauclin M. Theoretical evidence for 'Lichtenecker's mixture formulae' based on the effective medium theory. J Phys D Appl Phys 1998;31:1589–94.
49. Bozorth RM. Ferromagnetism. Wiley-IEEE Press; 1993:992 p.

50. Becker TI, Böhm V, Schale F, Zimmermann K. Vibrating sensor unit made of a magnetoactive elastomer with field-adjustable characteristics. J Magn Magn Mater 2020;498:166196. https://doi.org/10.1016/j.jmmm.2019.166196.
51. Guckenheimer J, Holmes P. Nonlinear oscillations, dynamical systems, and bifurcations of vector fields. Appl Math Sci. 1983;42:462 p.
52. Moon FC, Holmes PJ. A magnetoelastic strange attractor. J Sound Vib 1979;65:275–96.
53. Pshenichnikov AF. Magnetic field in the vicinity of a single magnet. Magnetohydrodynamics 1993; 29:37–40.

Jhohan Chavez*, Valter Böhm, Tatiana I. Becker, Simon Gast, Igor Zeidis and Klaus Zimmermann

27 Actuators based on a controlled particle-matrix interaction in magnetic hybrid materials for applications in locomotion and manipulation systems

Abstract: The paper deals with the investigation of magneto-sensitive elastomers (MSE) and their application in technical actuator systems. MSE consist of an elastic matrix containing suspended magnetically soft and/or hard particles. Additionally, they can also contain silicone oil, graphite particles, thermoplastic components, etc., in various concentrations in order to tune specific properties such as viscosity, conductivity and thermoelasticity, respectively. The focuses of investigations are the beneficial properties of MSE in prototypes for locomotion and manipulation purposes that possess an integrated sensor function. The research follows the principle of a model-based design, i.e. the working steps are ideation, mathematical modelling, material characterization as well as building first functional models (prototypes). The developed apedal (without legs) and non-wheeled locomotion systems use the interplay between material deformations and the mechanical motion in connection with the issues of control and stability. Non-linear friction phenomena lead to a monotonous forward motion of the systems. The aim of this study is the design of such mechanical structures, which reduce the control costs. The investigations deal with the movement and control of 'intelligent' mechanisms, for which the magnetically field-controlled particle-matrix interactions provide an appropriate approach. The presented grippers enclose partially gripped objects, which is an advantage for handling sensitive objects. Form-fit grippers with adaptable contour at the contact area enable a uniform pressure distribution on the surface of gripped objects. Furthermore, with the possibility of active shape adaptation, objects with significantly differing geometries can be gripped. To realise the desired active shape adaptation, the effect of field-induced plasticity of MSE is used. The first developed prototypes mainly confirm the functional principles as such without direct application. For this, besides the ability of locomotion and manipulation itself, further technological possibilities have to be added to the systems.

***Corresponding author: Jhohan Chavez**, Department of Mechanical Engineering, Technical Mechanics Group, TU Ilmenau, Ilmenau, Germany, e-mail: jhohan-harvey.chavez-vega@tu-ilmenau.de
Valter Böhm, Ostbayerische Technische Hochschule Regensburg, Fakultät Maschinenbau, Regensburg, Germany
Tatiana I. Becker, Simon Gast, Igor Zeidis and Klaus Zimmermann, Technische Universität Ilmenau, Technische Mechanik, Ilmenau, Germany

This article has previously been published in the journal Physical Sciences Reviews. Please cite as: J. Chavez, V. Böhm, T. I. Becker, S. Gast, I. Zeidis and K. Zimmermann "Actuators based on a controlled particle-matrix interaction in magnetic hybrid materials for applications in locomotion and manipulation systems" *Physical Sciences Reviews* [Online] 2020, 5. DOI: 10.1515/psr-2019-0087 | https://doi.org/10.1515/9783110569636-027

The first steps are therefore being taken towards integrated MSE based adaptive sensor systems.

Keywords: apedal locomotion systems, field-controlled particle-matrix interaction, field-induced plasticity, form-fit gripper, integrated sensor-actuator systems, magneto-sensitive elastomer

27.1 Introduction

Both in macrotechnology (mechanical engineering, automotive engineering, precision engineering) and in particular in microtechnology (minimally invasive medical technology, microbiology), technical solutions are sought which are essentially determined by multifunctional materials and structural elements with dynamically variable functions. The technical basis is no longer formed by structurally delimitable modules with separate basic functions for the processing of energy, material and information. Rather, there is a far-reaching functional integration that begins with the material and is defined by it. Due to the great progress made in the fields of numerical mathematics, control and information technology (hardware and software), nonlinear effects in material behaviour that have been "rather combated" nowadays hardly pose any problems in application. On the contrary, nonlinear behaviour, anisotropy with respect to elongation, viscosity, magnetisation and, generally speaking, controllable material properties are desired in order to make completely new technical solutions possible. Magneto-sensitive elastomers (MSEs) belong to that class of smart materials whose mechanical properties can be controlled by means of magnetic fields. The MSE consist of an elastic matrix containing suspended magnetically soft and/or hard particles [1, 2]. Additionally, they can also contain silicone oil, thermoplastic components, graphite particles, etc. in various concentrations in order to tune specific properties such as viscosity [3–9], thermoelasticity [10, 11] and conductivity ([12] and the work of Monkman et al., *Dielectric behavior of magnetoactive hybrid materials* published in this special issue) respectively. Many investigations on MSE have been based on experiments using hard magnetic particles [13, 14]. For the investigations in this paper, the use of soft magnetic particles with a small degree of remanent magnetisation is preferable. The application of an external magnetic field results in a number of physical changes to the polymer including field stiffening and magnetostriction [15].

In the present work, the magnetically generated and controlled deformations are the central aspect in the application of MSE for locomotion (Section 27.3) and manipulation (Section 27.4). At the end of the 20th century, when nobody used the modern term 'soft robotics' yet, basic research began on the use of magnetically controllable materials (ferrofluid, magneto-rheological fluid, MSE, etc.) for the two tasks mentioned above in robotics. The technical implementation of locomotion based on the earthworm peristaltic using ferrofluid was first published by Saga & Nakamura [16, 17]. At about the same time as the authors, Kimura [18] investigated the use of ferroelastomers for locomotion purposes. Also the work of Zrinyi [19, 20] and Kashima [21] provided the experimental basis for mobile robots from exclusively compliant structures.

In the just mentioned papers, the experimental part dominates the research. Naletova and Zeidis [22–25] developed mathematical-mechanical models for wormlike locomotion systems made of MSE. They used methods of continuum mechanics and mechanics of rigid-body systems (especially asymptotic methods of non-linear mechanics).

Sensors (e.g. accelerometer) based on materials with magnetically controllable parameters were introduced in 2008 by Phan [26]. Petterson in 2010 described and evaluated a robot gripper that utilises the effects of a magnetorheological fluid [27]. The property of the so-called field-induced plasticity [3, 28] makes possible to use MSE in gripper systems. Chavez [5, 29] developed a form-fit gripper, which is described in detail in Section 27.4.

27.2 Experimental results on material characterization

When a magnetic field is present in the MSE, the iron particles tend to align with the lines of the magnetic field flux [15, 30]. By removing the applied magnetic field, the particles return to their original position within the elastomeric matrix. The magnetic remanence, as well as the plastic deformation, after removing small or moderate magnetic field is negligible. In the experiments performed, the MSE samples contain silicone oil to provide a higher viscosity and allow the magnetic particles embedded in the MSE to move more freely, so greater effects of the influence of the magnetic field on the mechanical behaviour can be observed. In addition, the grade of compliance of the MSE with silicone oil in its original state before the application of the magnetic field is greater. This is an attractive property for applications in soft robotics (specifically in form-fit grippers). In order to study the mechanical behaviour of MSE materials, several test pieces are created for compression and tensile experiments. The experimental results enable to calculate an E-modulus. The E-modulus varies depending on the volume concentration of magnetic particles. In the manufacturing of MSE samples, Neukasil RTV 26 elastomer was used as a basis component mixed with the cross-linking agent A7. For a purely elastomeric specimen of this material, the hardness obtained is 7 in the scale Shore A. However, as mentioned above, the elastomeric matrix may contain a proportion of silicone oil to provide custom properties to the material. From previously presented studies [31–33], it is concluded that the recommended volume concentration of silicone oil in the specimen is 45%. In the present work, the volume fraction of the silicone oil is calculated for the polymeric part of the MSE and not for the global volume. The polymeric part comprises the base of pure elastomer and silicone oil. Therefore, as an example, for the supposed case in which an MSE specimen has 50 vol.% of iron, the other 50 vol.% of the volume corresponds to the polymeric part.

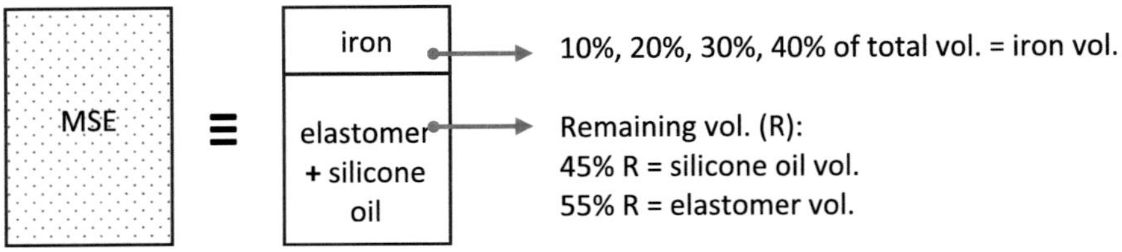

Figure 27.1: Volume distribution of the components of the MSE.

Since the silicone oil is 45% of the elastomeric part, the fraction of silicone oil is 22.5 vol.% of the total volume of the MSE. This is graphically explained in Figure 27.1.

The ferromagnetic material used is carbonyl iron powder (CIP) grade CEP CC. The purity according to manufacturer data (BASF) is 99.5%, and the average particle size lies between 6 and 10 μm. Cylindrical samples are manufactured for the compression (diameter: 16 mm, height: 12 mm) and tensile (diameter: 5 mm, height: 80 mm) tests. These geometries are shown in Figure 27.2. Additionally, samples for tensile tests have "heads" which are useful for clamping during the test, but do not have a large influence on the force measurement during the elongation. The iron concentration of the MSE samples varies from 10 to 40 vol.%.

Four samples are manufactured and tested for the same volume concentration of iron, and the results shown in this section correspond to the average value obtained for these four samples. Each specimen is tested four times. The highest coefficient of variation found for the experimental results is 2.5%, which confirms the repeatability of the experiments and the homogeneity of the samples.

For the compression tests, the samples are deformed up to 0.25 strain with 0.2 mm/s as recommended by ISO 7743. The deformation is performed under different magnetic field conditions. The magnetic field present in the specimens is not uniform. It has a gradient since a permanent magnet is used as a source (Figure 27.3). The permanent magnet is a Neodymium cylinder (diameter: 70 mm, height: 35 mm), magnetisation N45, Nickel-plated coated, residual magnetism (Br): 1.32–1.37 T, coercive field strength (bHc): 860–995 kA/m. The permanent magnet creates a magnetic field gradient where the intensity decreases at a greater distance from the magnet. For simplicity, a constant magnetic field for the whole cylinder volume of MSE is considered. This constant value is defined by the averaged magnetic field in vertical component. This simplification implies an idea of uniform magnetic field for the MSE materials. For tensile tests, the samples are only tested without the presence of magnetic fields, since their dimensions are more significant. On these samples, the gradient of magnetic field is greater when using the magnet as a source.

The results for tensile and compression test on the MSE samples are shown in Figure 27.4. From recommendations of standard norms, tensile tests comprehend a greater strain than compression tests. However for elastomers with particles located within the elastomer matrix, a large elongation concludes in breaking off the samples.

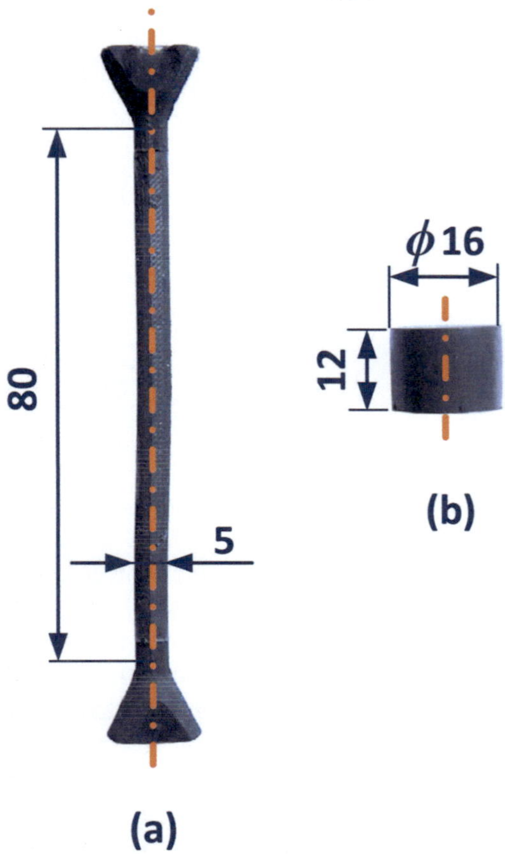

Figure 27.2: MSE cylindrical samples for tensile (a) and compression (b) tests.

Figure 27.3: Experimental setup for compression of MSE cylindrical samples in presence of a magnetic field produced by permanent magnet. Its magnitude varies with distance. 1: load cell, 2: upper plate for compression, 3: MSE sample, 4: bottom plate for compression, 5: permanent magnet.

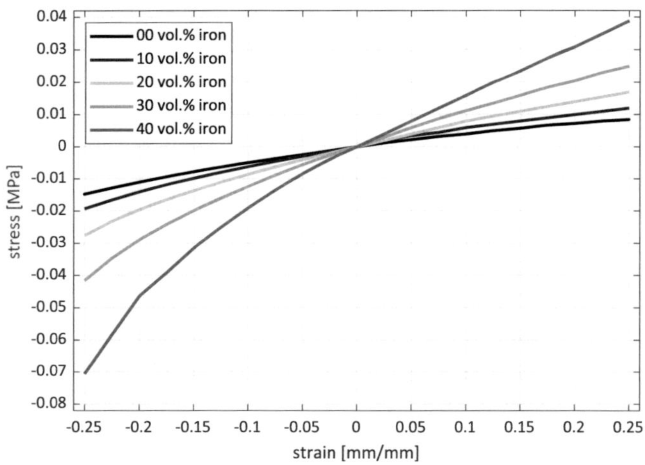

Figure 27.4: Results of tensile and compression test on the MSE samples, in absence of a magnetic field.

Thus, in this work results are presented for the same level of strain for both tensile and compression, to avoid the rupture of the samples. The positive and negative values of strain and stress correspond to tensile and compression, respectively. Due to the viscoelastic properties of the MSE materials, every compression or tensile experiment exhibits a hysteresis. However, for this work only the results for the increase of deformation are shown. Results for the relief of deformation are omitted since the experiments are performed with a quasi-static deformation.

As expected, samples with a bigger volume fraction of iron lead to a bigger stiffness. Additionally, pure polymeric samples (elastomer base and silicone oil) are also tested and their results are shown as a black curve in Figure 27.4. These samples do not experience a field-induced stiffness in presence of a magnetic field, since they do not contain ferromagnetic particles.

If the material is modelled with a linear behaviour, the E-modulus can be calculated as the ratio between the stress and strain. Nevertheless, as illustrated on the experimental results, a linear model suits only to the tensile part.

The results for 0.25 strain compression tests with a magnetic field present are summarized in Figure 27.5. The compression is realised in the presence of four different magnitudes of the magnetic field, which depends on the distance between the permanent magnet and the MSE sample (Figure 27.3). It is notable that the required force for the same deformation becomes greater as the magnetic field increases. This is very favourably used for form-fit grippers based on MSE and explained in Section 27.4. For a higher concentration of iron particles, the effect of field-stiffening is amplified. Samples with 30 and 40 vol.% iron are the most beneficial for further investigations. Nonetheless, the ease of manufacturing MSE materials decrease for smaller amount of pure elastomer part. Therefore, MSE materials with 30 vol.% are considered for the next paragraphs.

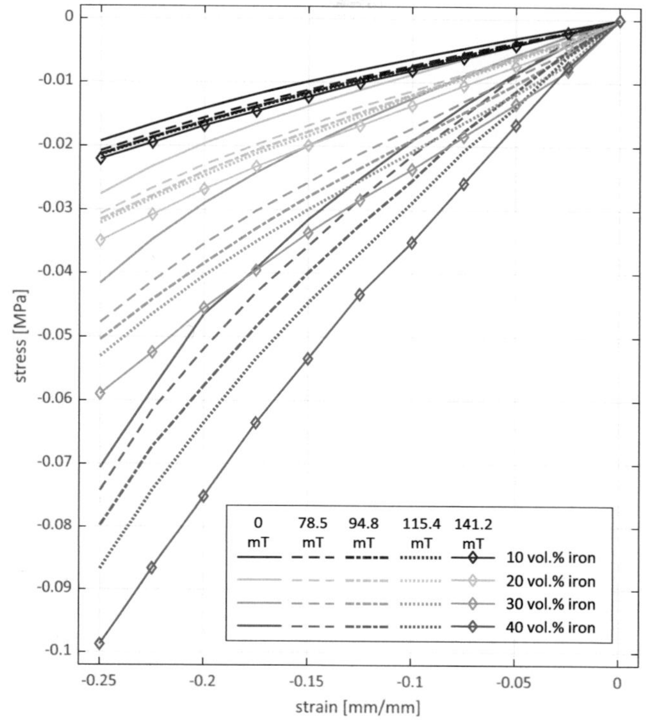

Figure 27.5: Results of compression test on the MSE samples, with four different magnitudes of the magnetic field.

A linear material behaviour is a poor model for the MSE. A hyperelastic model is more appropriate for the MSE. For simplicity the chosen model is Neo-Hookean since it comprehends a single parameter (initial shear modulus). To obtain the parameter for the material model, the compression experiments are simulated using finite element (FE) tools (Ansys Workbench ®). After a few iterations, a parameter for the material model which enables a proper fit with the experimental results is found. Figure 27.6 shows the experimental and simulated results of compression for the MSE with 30 vol.% iron for three magnitudes of magnetic field. However, this material model also has disadvantages. The Neo-Hookean model does not conclude as a stress-strain curve with inflections as in experimental measurements and these inflections are becoming more pronounced for bigger magnetic fields. Nevertheless, the difference between stress-train curves from simulations and experimental results is small compared to the measured forces. The most perceptible error is found in the results for compression tests developed in the presence of the greatest magnetic field (average of 141.2 mT) in MSE.

The parameters obtained for the material model are used in Section 27.4 for MSE based form-fit grippers.

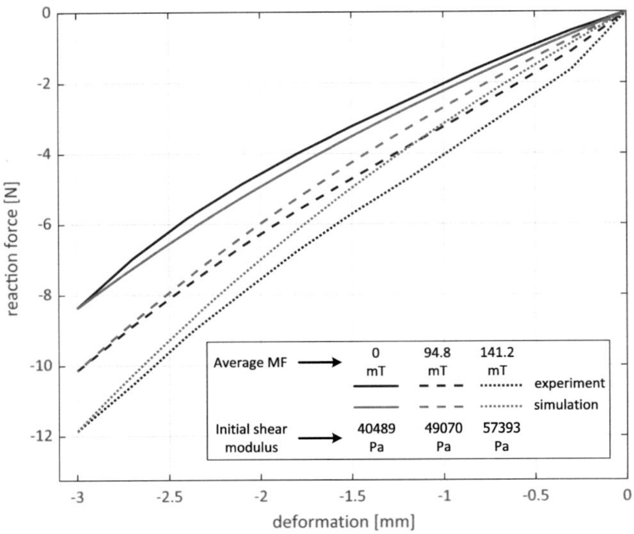

Figure 27.6: Results for compression on MSE with 30 vol.% iron for three values of magnetic field. Blue curves: experimental results. Red curves: simulations using Neo-Hookean material model.

Borin et al. present on their work *Magneto-mechanical properties of elastic hybrid composites* in the present special issue, another approach to the properties of magnetic composites with different polymeric matrixes.

27.3 Non-pedal locomotion systems based on MSE

Basic research on the deformation of membrane-enveloped wormlike ferrofluid structures and on the behaviour of MSE materials in an external magnetic field forms the basis for the development of novel apedal locomotion systems. These locomotion systems, often inspired by biological examples [16, 24, 34], can be divided into two classes. Locomotion on free ferrofluid surfaces by a locally and temporally changing magnetic field is called *passive*, when a non-magnetic object is moved due to the deformation of this surface. With *active* locomotion systems, the movement is generated in the system itself. MSE objects experience a change of position, see Figure 27.7.

Such active systems form the primary goal of the research, with the design focus on the realisation of an apedal locomotion. A model-based approach leads to the development of systems whose locomotion is based on high-frequency excitation of MSE. By the specific control of different oscillation modes of the locomotion systems, the definition of the motion direction succeeds beside the adjustment of the velocity. Uniaxial and planar locomotion systems with only one electromagnetic drive are developed.

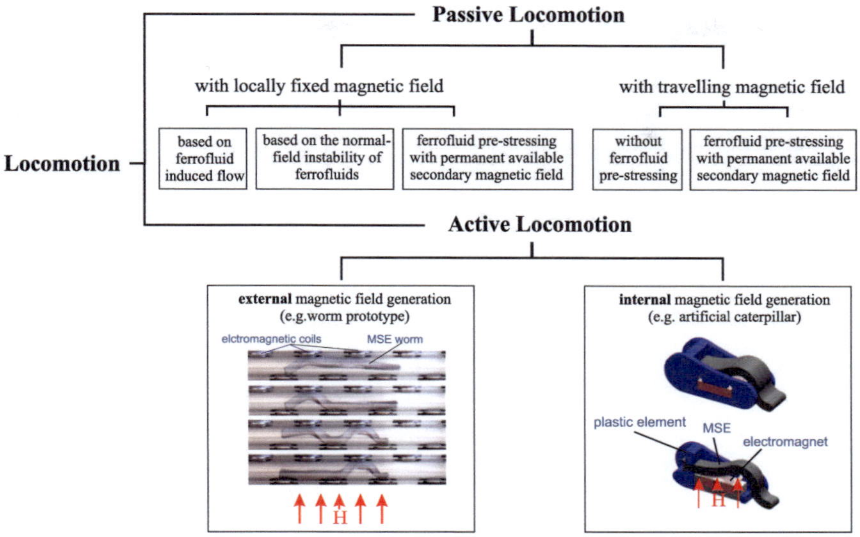

Figure 27.7: Active and passive locomotion systems realised with magneto-sensitive elastomer.

27.3.1 A basic experiment with locomotive MSE

In order to understand how magnetic field induced motion can be realised with MSE and whether this motion is stable in the mechanical sense, a basic experiment is considered. The system consists of two material points with masses *m* connected by an MSE spring of stiffness *k*. Under the influence of an external magnetic field, which changes periodically, a harmonic internal force *F*(*t*) is generated, see Figure 27.8.

The motion of the system is governed by the differential equations.

$$m\ddot{x}_1 + k(x_1 - x_2) = F(\dot{x}_1) + F(t),$$

$$m\ddot{x}_2 + k(x_2 - x_1) = F(\dot{x}_2) - F(t), \tag{27.1}$$

where $F(t) = F_a(1 + \cos\psi)$, with $\psi = \nu t$. The Coulomb friction force acting on the material points with number *i* yields

$$F(\dot{x}_i) = \begin{cases} mg\mu_-, & \text{if } \dot{x}_i < 0, \\ F_0, & \text{if } \dot{x}_i = 0, \\ -mg\mu_+, & \text{if } \dot{x}_i > 0, \end{cases} \tag{27.2}$$

where μ_- and μ_+ are anisotropic friction coefficients with $\mu_- \geq \mu_+ \geq 0$. The force F_0 satisfies the relation $-mg\mu_+ < F_0 < mg\mu_-$.

To reduce the number of parameters that characterise the system, the dimensionless variables are introduced

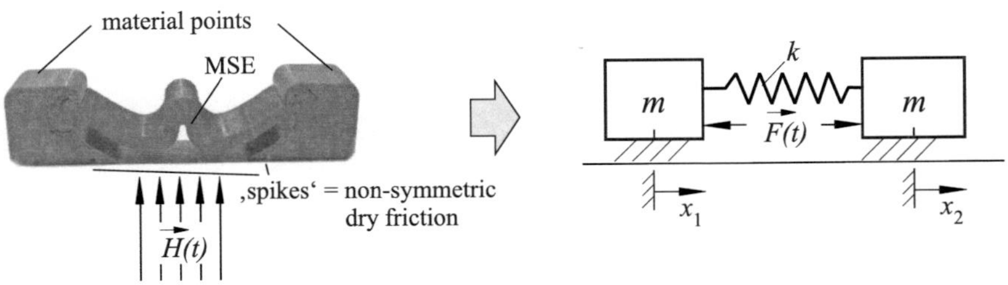

Figure 27.8: From the real system to the mechanical model.

$$x_i^\star = \frac{x_i}{L}, \quad t^\star = t\omega, \quad v^\star = \frac{v}{\omega}, \quad \dot{x}_i^\star = \frac{\dot{x}_i}{L\omega}, \quad F^\star = \frac{F}{mg\mu_-},$$

$$\mu_0^\star = \frac{F_0}{mg\mu_-}, \quad \varepsilon = \frac{mg\mu_-}{kL}, \quad \alpha = \frac{F_a}{mg\mu_-}, \quad \mu = \frac{\mu_+}{\mu_-}\ (0 \le \mu \le 1), \tag{27.3}$$

where $\omega^2 = k/m$ and L is the scale of length used for the procedure of non-dimensionalisation.

Proceeding with the dimensionless variables in Equations (27.1) and (27.2) and then omitting the asterisks, identifying the variables $x_i^\star$ and $t^\star$, it follows

$$\begin{aligned} \ddot{x}_1 + x_1 - x_2 &= \varepsilon[\alpha(1 + \cos\psi) + r_1], \\ \ddot{x}_2 + x_2 - x_1 &= \varepsilon[-\alpha(1 + \cos\psi) + r_2], \end{aligned} \tag{27.4}$$

where

$$r_i = \begin{cases} 1, & \text{if } \dot{x}_i < 0, \\ \mu_0, & \text{if } \dot{x}_i = 0, \\ -\mu, & \text{if } \dot{x}_i > 0. \end{cases} \tag{27.5}$$

Assuming $\varepsilon \ll 1$, $\alpha \sim 1$, $|x_2 - x_1| \sim 1$, the procedure of averaging [35] is applied to system (4). For this purpose new variables are introduced: the velocity of the centre of mass $V = (\dot{x}_1 + \dot{x}_2)/2$ and the deviation relatively to the centre of mass $z = (x_2 - x_1)/2$. Replacing $z = a\cos\varphi$, $\dot{z} = -a\sqrt{2}\sin\varphi$, $\varphi = t\sqrt{2} + \vartheta$, where V, a, ϑ – slow variables, system (4) takes the so-called standard form:

$$\begin{aligned} \dot{V} &= \frac{\varepsilon}{2}\left(F\left(V + a\sqrt{2}\sin\varphi\right) + F\left(V - a\sqrt{2}\sin\varphi\right)\right), \\ \dot{a} &= \frac{\varepsilon}{2\sqrt{2}}\sin\varphi\left(F\left(V + a\sqrt{2}\sin\varphi\right) - F\left(V - a\sqrt{2}\sin\varphi\right) + 2\alpha(1 + \cos\psi)\right), \\ \dot{\varphi} &= \sqrt{2} + \frac{\varepsilon}{2a\sqrt{2}}\cos\varphi\left(F\left(V + a\sqrt{2}\sin\varphi\right) - F\left(V - a\sqrt{2}\sin\varphi\right) + 2\alpha(1 + \cos\psi)\right), \\ \dot{\psi} &= \nu. \end{aligned} \tag{27.6}$$

The system (6) is investigated in a vicinity of the main resonance $\nu = \sqrt{2} + \varepsilon\Delta$ $(\Delta \ne 0)$. Introducing a new slow variable $\xi = \psi - \varphi$, after averaging on a fast variable φ it yields

$$\dot{V} = \begin{cases} \varepsilon, & \text{if } V < -a\sqrt{2}, \\ -\varepsilon\left(\dfrac{1+\mu}{\pi}\arcsin\dfrac{V}{a\sqrt{2}} - \dfrac{1-\mu}{2}\right), & \text{if } |V| \le a\sqrt{2}, \\ -\varepsilon\mu, & \text{if } V > a\sqrt{2}, \end{cases} \tag{27.7}$$

$$\dot{a} = \begin{cases} -\dfrac{\varepsilon}{\sqrt{2}}\left(\dfrac{1+\mu}{\pi}\sqrt{1-\dfrac{V^2}{2a^2}} + \dfrac{\alpha}{2}\sin\xi\right), & \text{if } |V| \le a\sqrt{2}, \\ -\varepsilon\dfrac{\alpha}{2\sqrt{2}}\sin\xi, & \text{if } |V| > a\sqrt{2}, \end{cases} \tag{27.8}$$

$$\dot{\xi} = \varepsilon\left(-\frac{\alpha}{2a\sqrt{2}}\cos\xi + \Delta\right). \tag{27.9}$$

By the definition of the steady-state motion, the variables V, a and ξ are constant according to the averaged equations. From Equations (27.7–27.9) it can be found

$$\begin{gathered} V = \frac{\sin\Phi}{|\Delta|}\sqrt{\frac{\alpha^2}{4} - E^2\cos^2\Phi}, \quad \Delta \ne 0, \\ a = \frac{1}{|\Delta|\sqrt{2}}\sqrt{\frac{\alpha^2}{4} - E^2\cos^2\Phi}, \quad \alpha \ge 2E\cos\Phi, \\ \sin\xi = -\frac{2E\cos\Phi}{\alpha}, \cos\xi = \frac{2\Delta}{\alpha|\Delta|}\sqrt{\frac{\alpha^2}{4} - E^2\cos^2\Phi}, \\ E = \frac{1+\mu}{\pi}, \quad \Phi = \frac{\pi(1-\mu)}{2(1+\mu)}. \end{gathered} \tag{27.10}$$

The motion of the system with the velocity V from Equation (27.10) is stable.

The result of the numerical integration of the exact system (4) is given in Figure 27.9. The following values of parameters for an MSE are taken: $m = 3{\cdot}10^{-3}$ kg, $k = 3$ N/m, $F_a = 3{\cdot}10^{-3}$ N, $\nu = 47\,\mathrm{s}^{-1}$, $L = 1$ cm, $\mu_- = 0.1$, $\mu_+ = 0.05$. So, the values of the dimensionless parameters are: $\varepsilon = 0.1$, $\mu = 0.5$, $\alpha = 1$, $\Delta = 0.8$.

The first equation of Equation (27.10) gives the value for the dimensionless velocity of the centre of mass $V = 0.17$. This value corresponds to the dimensional velocity $V = 5.4$ cm/s.

For a nonlinear spring with a small cubic nonlinearity, the expression for the spring force F_{el} in dimensionless variables is

$$F_{\mathrm{el}} = -x - \varepsilon\beta x^3, \quad \beta \sim 1, \tag{27.11}$$

here β is the nonlinearity measure. In this case the expression for the stationary velocity V is

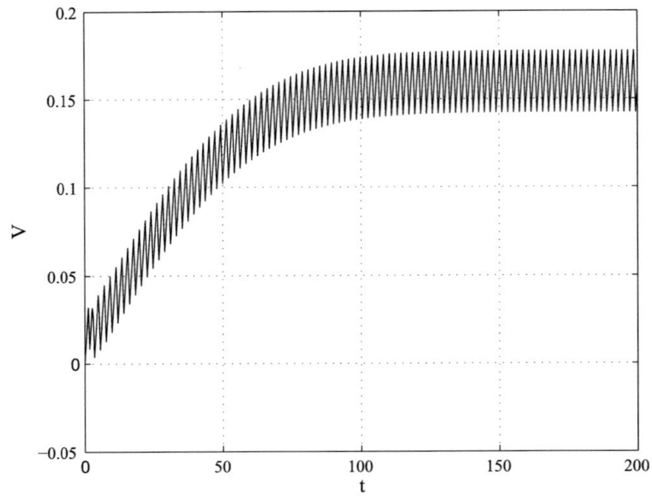

Figure 27.9: The velocity V of the centre of mass vs. time t, following from the numerical integration of the exact equations of motion. The stationary velocity is nearly equal to 0.17.

$$V = a\sqrt{2}\sin\Phi, \tag{27.12}$$

and the amplitude a is determined from the equation

$$a\left|3a^2\beta - \Delta\sqrt{2}\right| = \sqrt{\frac{\alpha^2}{4} - E^2\cos^2\Phi}. \tag{27.13}$$

Remark: In the investigations given above, the motion with a periodic excitation force, which has a zero mean value, is considered. In this case, the motion of the centre of mass requires asymmetry. Here, this asymmetry is achieved due to different friction coefficients, when changing the direction of motion (anisotropic surface). The internal force itself is assumed to be harmonic (i.e. symmetrical). The motion is only possible in the direction of less friction. In [36] a two-material point system with different masses is considered. There, the variable part of the periodic excitation force is considered asymmetrical, provided that it has zero average (e.g. increase and decrease of the magnetic field occur at different velocities). It is shown that in this case it is possible moving not only on an isotropic surface, but even on an anisotropic one to the side of more friction.

Thus, by changing the magnetic field, it is possible to control not only the velocity of the system, but also the direction of motion, see Section 27.3.2.

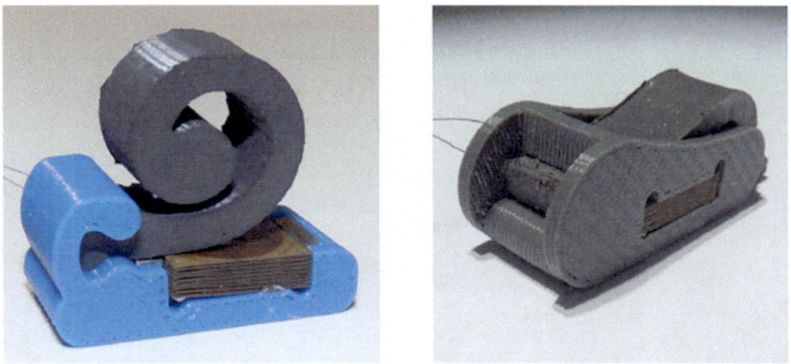

Figure 27.10: Vibration driven locomotion systems for rectilinear motion based on MSE – 'Type 1' (left) and 'Type 2' (right).

27.3.2 Vibration driven mobile systems for rectilinear motion

For *rectilinear* uni- and biaxial motion, two vibration-driven locomotion systems based on MSE are developed. To explore the mechanical properties of MSEs under magnetic fields, please refer to the work *Magnetoactive elastomers for magnetically tunable vibrating systems* written by Becker et al. and published in this special issue. The locomotion of these systems (Figure 27.10) is caused by a periodic deformation of a compliant MSE body. The deformation is generated by integrated micro-coils. A periodically alternating magnetic field causes an oscillation of the MSE. The friction force between the supporting surface (ground) and the system changes periodically due to a change of the normal force. The oscillation mode can be varied in dependence of the pulsing frequency in a wide range.

The qualitative description of the mechanical performance of the two systems is provided by transient dynamical analyses by means of the FE-Software ANSYS®, see Figures 27.11 and 27.12 for the simulation of 'Type 1'.

Figure 27.13 shows the velocity at various driving frequencies ranging from 1 to 30 Hz when the robot is actuated using forces $F = 0.006$ N (top, blue) and $F = 0.010$ N (bottom, green). The FE-simulations confirm the result published in [36] that the direction of movement of the robot and its velocity depend essentially on the relation between the eigenfrequency of the system and the driving frequency. There are multiple driving frequency ranges, which cause the robot locomotion in opposite directions from the initial position. At $F = 0.010$ N, the robot moves in the positive x-direction at the frequency ranges between 5 … 6 Hz and 10 … 13 Hz, and in the negative x-direction between 6 … 10 Hz and at over 13 Hz, Figure 27.13. At low frequencies, the internal force is nearly equal to the excitation force. At higher driving frequencies, particularly in the vicinity of the natural frequencies, a marked change in the characteristics of the internal forces is observed. The system 'Type 1' achieves speeds in the range of 2 … 3 mm/s.

In contrast to 'Type 1', the locomotion system 'Type 2' converts directly the forced vibrations of the MSE into a propulsive force due to its contact with the ground

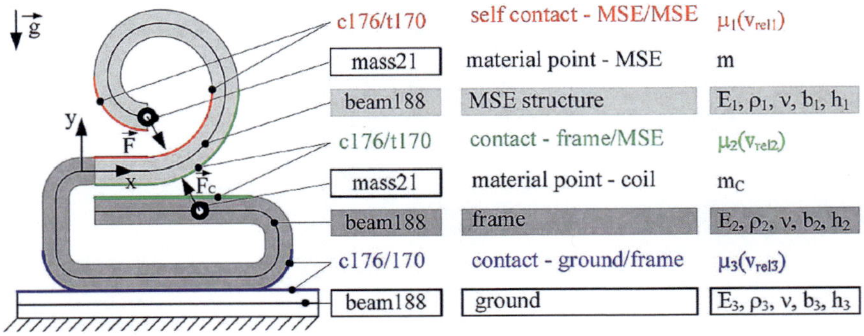

Figure 27.11: The mechanical model of the biaxial locomotion system 'Type 1' with ANSYS-elements used in simulation and the corresponding parameters.

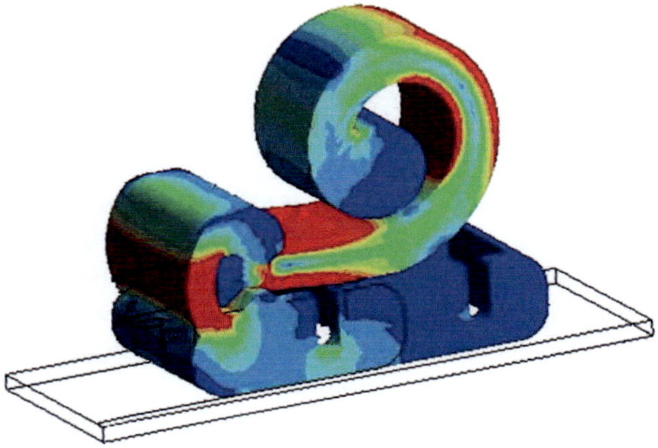

Figure 27.12: The equivalent stresses after von Mises for the locomotion system 'Type 1'.

(Figure 27.14, left). Only uniaxial locomotion is possible. The images from a high-speed camera show selected motion sequences with and without contact of the MSE to the ground (Figure 27.14, right). Depending on the friction pairing between the elastomer and the ground, the maximum velocity achieved with 'Type 2' is about 80 mm/s. Once again, there is great potential for optimisation both in the shape of the MSE and in the control *via* the periodically changing magnetic field of the miniature coil.

27.3.3 Amoeboid locomotion systems based on MSE

Conventional terrestrial locomotion systems with legs or wheels have a limited field of application, and they are difficult to miniaturize. Therefore, for *planar* motion biologically inspired locomotion systems following the paradigm of an amoeba are

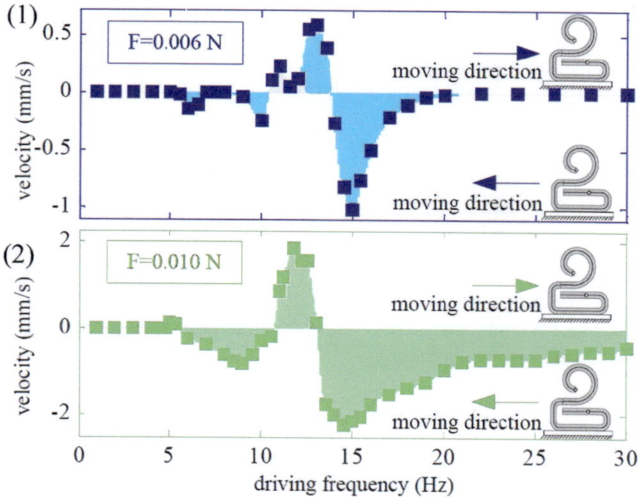

Figure 27.13: Velocity of the robot in dependent on the driving frequency for the actuating forces (1) $F = 0.006$ N (blue) and (2) $F = 0.010$ N (green).

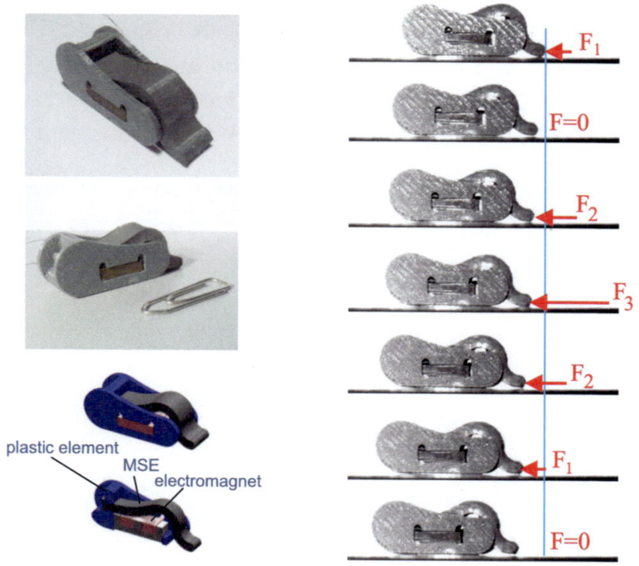

Figure 27.14: Apedal locomotion system 'Type 2' with parameters. Mass: 7.3 g, length: 37 mm, width: 12 mm, height: 13 mm (left); and motion sequences with driving forces (right).

developed, which use MSE and controllable by magnetic fields. The movement of a natural amoeba is characterised by cytoplasmic streaming and continuous changing of the body shape [34, 37]. The periodic change of the shape of the technical locomotion system is also realised by means of a vibration drive using permanent magnets and

Figure 27.15: Planar locomotion systems based on MSE: the biological inspired prototype 'Moving star' (left) and the mobile system 'Triangular amoeba'(right).

miniature coils. According to the classification in Figure 27.7, these are active locomotion systems, Figure 27.15.

Using the example of the 'Moving star', the development process from the biological example (amoeba proteus) *via* the mechanical model to the technical amoeba is shown in Figure 27.16. From the point of view of mechanics, the amoeba means the realisation of a fluid flow (caused by pressure gradients) and the change in the flexibility of a body with variable contours. In the technical implementation, the forces are not generated by pressure differences but magnetically. The electromagnetic driving force is generated between activated and non-activated coils and the envelope body made of MSE. This driving force generates a locally differently pronounced deformation of the flexible body. In this area, the static friction force between the locomotion

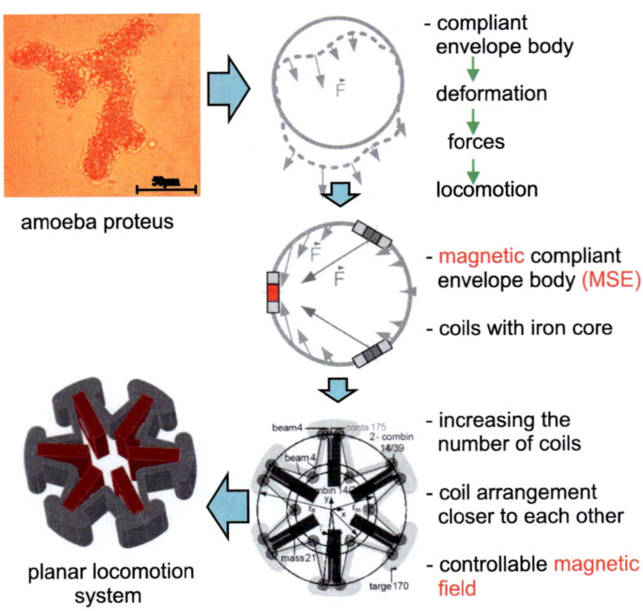

Figure 27.16: From the biological example to the prototype (photo of amoeba proteus by Danja Voges †).

system and the ground is locally exceeded. Thus, a periodic driving force leads to the locomotion of the system.

With the coils of size 11.5 × 11.5 × 3 mm integrated in the MSE and arranged at 60° to each other, a sufficiently close relative position to each other is achieved. This leads to an increase in the electromagnetic forces and ensures the energy efficiency during motion. The direction of motion of the system is determined by the choice of coils activation. The electromagnets each have an iron core. The flexibility of the MSE body is ensured on the one hand by the material itself, and on the other hand by the shape of the body. The connecting elements between the coil carriers are designed as joints with distributed compliance [38].

A model-based simulation of the system performance is provided by transient dynamical analyses by means of FE-Method. Within the model, the micro coils are treated as material points. Their connection to the compliant body is ensured by rigid beam elements. The MSE body, with a diameter of 40 mm is assembled with beam elements in the range of great curvature and with discrete spring-damper-elements at the intersection of different cross areas. The actuating force is represented by single forces acting on each mass point. The value and the direction of these forces depend on the relative position of the coils and, of course, the control algorithm. Forces acting on the compliant body are neglected due to the small permeability of it. Figure 27.17 (right) shows exemplarily the obtained results for the following case: two coils A and B in Figure 27.17 (left) are actuated simultaneously and alternate periodically. The system moves in y-direction. The focus is on the velocity of the centre point of the system in dependence on the actuating frequency. It can be seen, that the direction of motion can be controlled by means of the actuating frequency. The maximum average velocity of the 'Moving star' is 5 mm/s at the driving frequency of 22 Hz. The results of the numerical simulation using FE-Method and the measurements during the experiment are in good qualitative and quantitative agreement.

In the prototype 'Triangular amoeba', the idea of realisation of a simple actuation is based on the use of permanent magnets, which are attached to the shaft of a rotary drive inside the MSE structure. The dynamic excitation of the MSE takes place by means

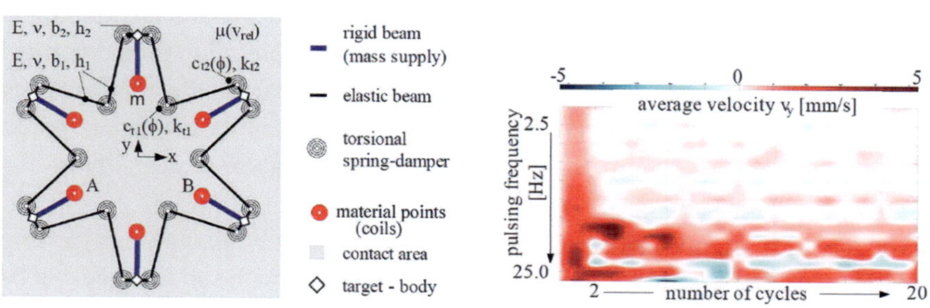

Figure 27.17: The model elements for the FE simulation with parameters (left) and the velocity in y-direction depending on the pulsing frequency (right).

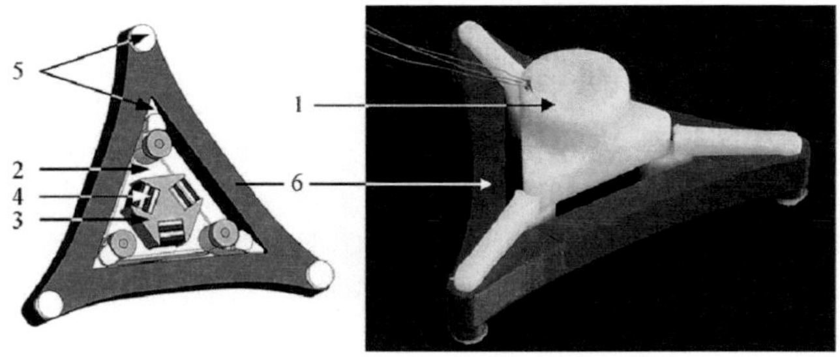

Figure 27.18: Structure of the locomotion system 'Triangular amoeba'. 1: motor, 2: housing, 3: rotor, 4: permanent magnet, 5: plastic body, 6: MSE.

of an alternating magnetic field. The MSE structure reacts only passively to a magnetic field and does not operate actively as an actuator. To be able to move in the plane, the system uses an asymmetric geometry of the compliant MSE. The structure consists of addition-curing silicone with a Shore hardness A of 8. The mass ratio of elastomer to CIP (same particles as presented in Section 27.2) is 3:1 for the MSE part. The realised prototype has an outer diameter of 67.5 mm and a height of 23 mm. The mass of the system is 21.6 g, Figure 27.18.

Modal analyses and transient motion analyses based on a multibody system model are performed to describe the locomotion system. Figure 27.19 (top) shows the excitation control forces at the points A, B and C of the MSE. The results of an exemplary selected transient simulation are given in Figure 27.19 (bottom, right). At an excitation frequency of 1.5 Hz, the system performs a rotary motion. The theoretical results agree with the experimental data and show that with the mentioned type of control and the selected asymmetrical basic MSE structure, a locomotion in the plane is possible.

27.4 Form-fit grippers

The use of MSE in form-fit grippers is of great interest in the field of soft robotics [12, 31, 32, 39]. Depending on the elastomeric base and the amount of silicone oil added, the material can be soft in its original state and allow a higher adaptability when deformed by objects with certain geometries. After applying a magnetic field, the MSE becomes harder due to the field-induced stiffening effect. The increase of its stiffness after the application of the field is favourable as it keeps the deformation (and therefore the geometry) from the object to grip.

A scheme for this manipulation system is shown in Figure 27.20. At the beginning (i) the MSE based grippers are found in their original state and are soft, so they can be deformed by flexible objects without the need to exert a great force (ii). After switching on a magnetic field in the vicinity of the MSE end-effectors, the rigidity of the MSE material increases and due to the field-induced plasticity the deformation can be

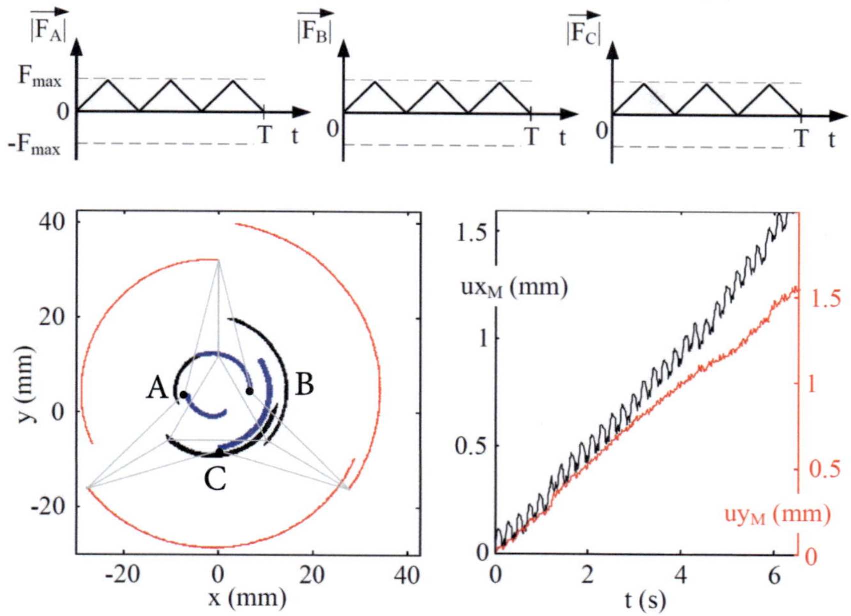

Figure 27.19: Applied control functions of the forces in the points A, B and C (top) and paths of the nodes after 100 actuation periods for an excitation frequency of $f = 1.5$ Hz with associated movement of the centre during a circular motion of the system (bottom).

recorded (iii). The shape adaptability depends on the composition of the MSE and its geometry. As long as the magnetic field is not removed, the deformation will remain intact in the MSE end-effectors, even when the object to grab is removed (iv). This can be used to transport elements with the same geometry (v ... n–1). Finally, by removing the magnetic field, the mechanical properties are almost entirely recovered, without a noticeable plastic deformation (n). To develop an MSE end-effector capable to grip various shapes, the investigations are focused on three complex end-effector shapes. These prototypes are fabricated and tested with two objects in order to examine the quality of their shape adaptability. The deformations are simulated by using the mechanical behaviour models introduced in Section 27.2. Previous studies show that promising geometries for the final effectors are hollow rotational-symmetric bodies. The hollow half spherical MSE end-effector has a thinner MSE layer in the area in contact with the object to grip in order to allow compliance on the deformation [32], see Figure 27.21a. The half spherical MSE is modified and two new geometries are presented, both with a cylinder-like shape but each with different curvatures in their designs. The most relevant geometric details on the design of the three mentioned MSE prototypes are shown in Figure 27.21.

The second studied geometry (Figure 27.21b) is named as a cylindrical-convex geometry, since the lateral layer (green dotted curve) corresponds to an elliptical

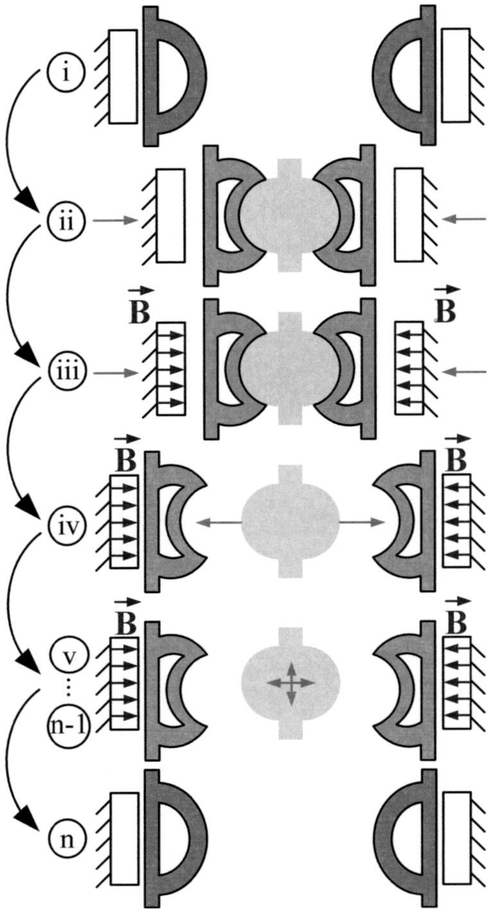

Figure 27.20: Gripping process by using a form-fit MSE based end-effector.

geometry, which centre lies outside of the MSE sample. The third geometry (Figure 27.21c) is a cylindrical-concave sample and differs from the second on the lateral face. In this case the centre of the elliptical geometry lies inside the MSE sample. For the fabrication of the samples, the composition studied in detail in Section 27.2 (30 vol.% iron and 45 vol.% silicone oil for the polymeric part) is preferable.

The steps (i–iii) from the gripping process presented in Figure 27.20 are of interest in the present work in order to compare the shape adaptability of the three MSE end-effectors. It is not possible to detect how good the shape adaptability of the MSE samples is, when the objects are still surrounded (see [ii], Figure 27.20). Nevertheless, the hyperelastic material model presented on Section 27.2 can be used to calculate an approximation of the deformations during the manipulation and compare qualitatively this simulation with the experimental results.

For the experiments, a sphere of 16 mm diameter and a cylinder of 20 mm diameter are used as exemplars for handled objects. An 8 mm indentation of both geometries

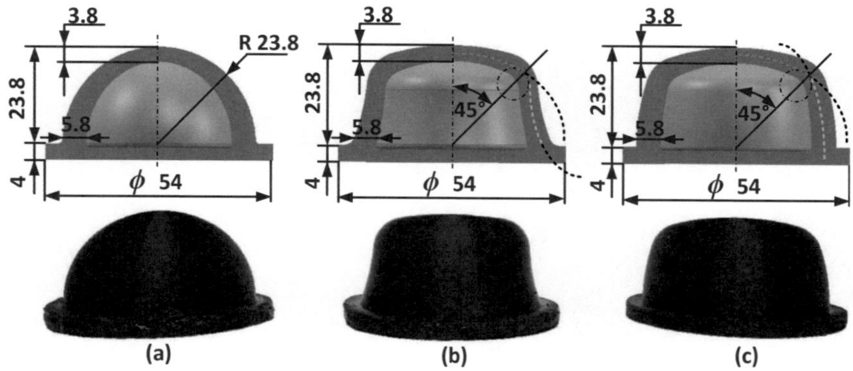

Figure 27.21: Above: cross section and geometric details of the three studied samples. Below: photo of the manufactured samples. Cross-sections: (a) semi-spherical, (b) cylindrical-convex, (c) cylindrical-concave.

into the MSE samples is accomplished in the experiments. The experimental setup is shown in Figure 27.22. The actuator (a) provides displacement in the vertical axis to the elements bonded to it. The handled object (b) is fixed to a load cell (c) to measure the reaction forces during deformation. The MSE geometries (d) are placed on a structure (e) to allow the deformation. After the deformation, the magnetic field is applied to the MSE material by moving the magnet (f) closer to the sample. Finally, the shape adaptation quality is measured in radial direction of the sample using a triangulation laser sensor (g).

In the tests, the MSE end-effector is firstly deformed by the gripped object, and the shape is recorded on the MSE material after the application of the magnetic field through the permanent magnet, see Figure 27.20. For the simulations, the surrounding geometry of the MSE end-effector just after the deformation without the application of the magnetic field is numerically calculated by using the hyperelastic material model. The results are shown in Figure 27.23. A reason for the difference between theoretical and experimental results lies on the simplification of the model, which does not consider the encapsulated air. For further work, a coupling between a hyperelastic material model and a magnetostatic analysis is pending in order to enable a change of the initial shear modulus on the hyperelastic model depending on the magnitude of magnetic field.

27.5 Future work and conclusion

Instead of a detailed summary, which usually follows at this point, this concluding section presents a specific aspect of future work with initial ideas. From the presented results it is exhibited that the concept for locomotion and manipulation systems promises a real application in future. Nevertheless, there are still missing features on

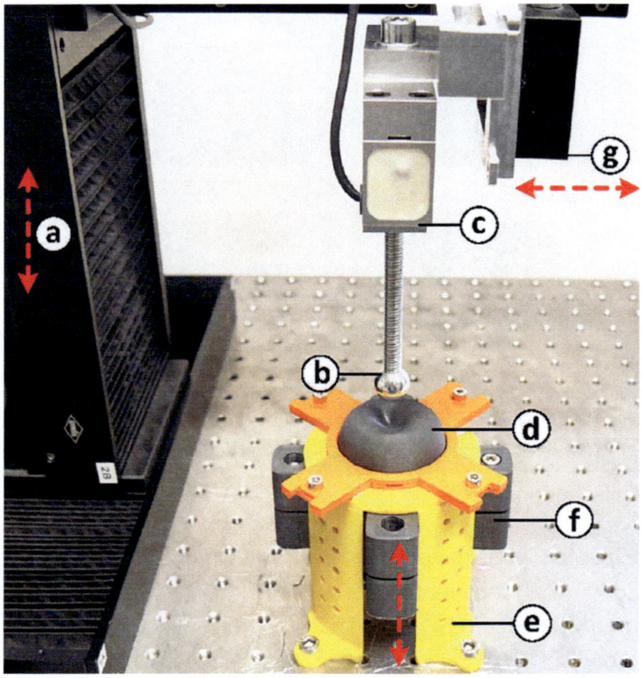

Figure 27.22: Experimental setup for shape adaptability and its components, a: displacement actuator, b: handled object, c: load cell, d: MSE sample, e: structure, f: encapsulated permanent magnet, g: triangulation laser sensor.

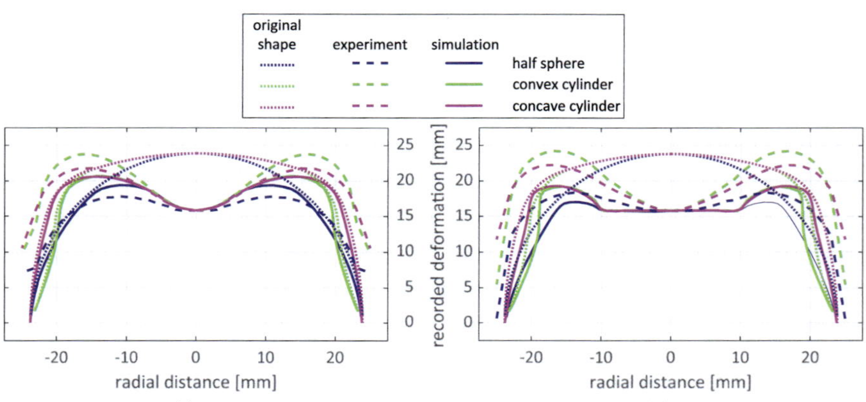

Figure 27.23: Experiments and simulations for shape adaptability of the three MSE end-effector prototypes for (a) sphere of 16 mm diameter and (b) cylinder of 20 mm diameter as gripped object.

these systems to achieve a technical implementation. One of the most needed attributes is a sensor integration on the introduced actuators, so integrated sensor-actuator systems can be achieved. Thus, this section describes a first concept for a sensory

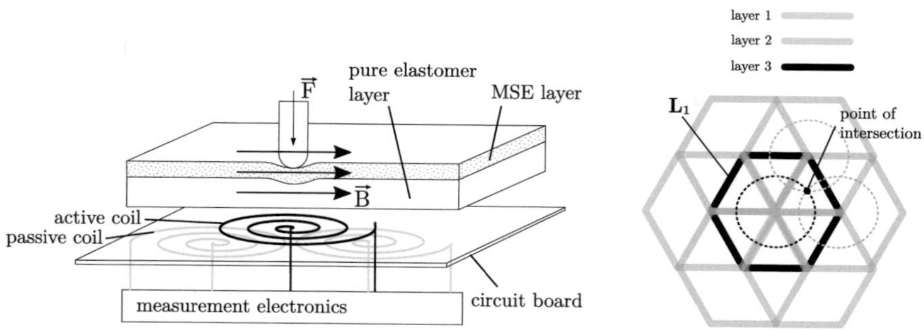

Figure 27.24: Tactile sensor concept (left), example of a planar coil array with seven hexagonal coils (right).

element which can be attached to the described prototypes. Additionally, a short summary of the results is given at the end of this chapter.

27.5.1 Concept of an MSE based tactile sensor

One aim of the future work is to combine sensitive effects and an active control of MSE by varying the magnetic field. The combination of MSE samples and inductors is investigated in [40, 41]. The following ideas and first design steps deal as well with an application incorporating both features.

The sensor concept shown in Figure 27.24 (left) is capable of sensing tactile information and controlling elasticity at once. In order to obtain information on a deformation, the sensor incorporates a circuit board with multiple planar spiral coils. The sensor consists of an elastic (MSE and pure elastomer) and a rigid (circuit board) part. The elastomer sample is made by a thin MSE layer and a pure elastomer base layer. The coils are connected to measurement electronics that is capable of a sequenced measurement of inductance. Therefore, every single inductance is time shifted part of an L-C-tank circuit driven by its eigenfrequency. Every deformation induces a change of reluctance near the spiral coil, thereby affecting the inductance. The measurements are performed with a high frequency alternating magnetic field (≈ 2 MHz) at low amplitudes. Thus, an external induced magnetic field is superposed with the low magnitude field of the coils, controlling the elasticity of the MSE layer. The features of the sensor are a planar mapping and force measurement of an indentation as well as a tunable elasticity.

The example shown in Figure 27.24 (right) is used to visualize the principle for a planar mapping of a shifted indentation of constant force. This setup consists of three circuit board layers holding seven hexagonal coils in total. The change of inductance for a planar shifted indentation has a radial dependency to the centre of a spiral coil [42]. It is assumed, that for one coil all values of equal change of inductance can be fitted with an ellipse. It is necessary to determine the ellipses of one coil per layer that are close to the indentation.

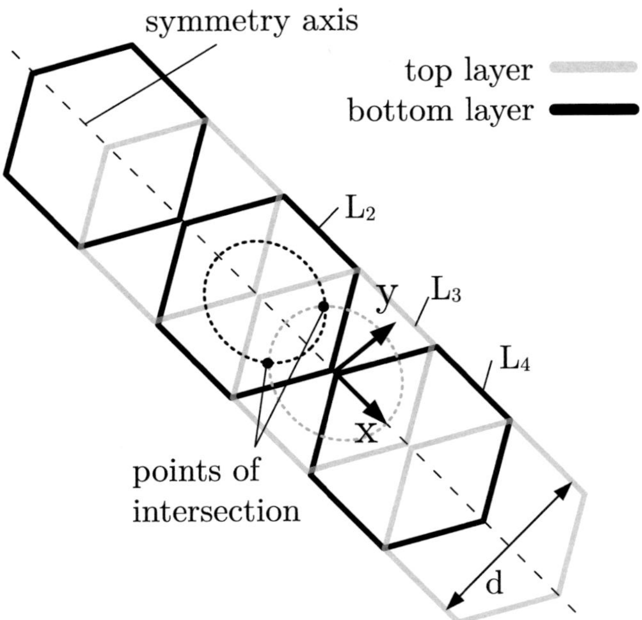

Figure 27.25: Realised array of coils for a linear tactile sensor.

Thus, one point of intersection can be found for every planar shifted indentation within the region of the coil L_1 (Figure 27.24, right). The coordinates of this point correspond to the planar *force-applied position*. For this sensor concept, it is required that there is still a positive change of inductance outside of the coil dimensions.

A first realised approach is the linear coil array shown in Figure 27.25. This circuit board holds multiple hexagonal coils arranged in a line. Due to its symmetry, the points of intersection have the same x-coordinate. The experimental setup consists of a three axis linear drive holding the indenter and a fixed tactile sensor with a layered composition as shown in Figure 27.24. For the given parameters listed in Table 27.1, there is a positive change of inductances measured outside of the coil dimensions, Figure 27.26. As a result, this setup is suitable for a linear positioning of an indentation.

Further information on the deformation can be obtained by measuring the inductance of a series connection of multiple coils. Next steps focus on using this data for evaluating the *magnitude of the force*.

27.5.2 Compact summary of the results

The use of MSE allows the realisation of actuator systems with a complex and adaptable mechanical behaviour, primarily resulting from the magnetically field-controlled particle-matrix interactions. The implementation of the beneficial properties of MSE in prototypes for locomotion and manipulation purposes is considered. The

Table 27.1: Experimental parameters.

Parameters	Values
Indenter shape	Ball
Indenter diameter	5 mm
Coil diameter d	23 mm
Coil turns	30
Indentation	1 mm
MSE thickness	2 mm
Pure elastomer thickness	6 mm
Distance between layers	0.5 mm
Capacitance	120 pF
Grid dimension	2 mm
Measurement range *y*	−26 to 26 mm
Measurement range *x*	−6 to 6 mm

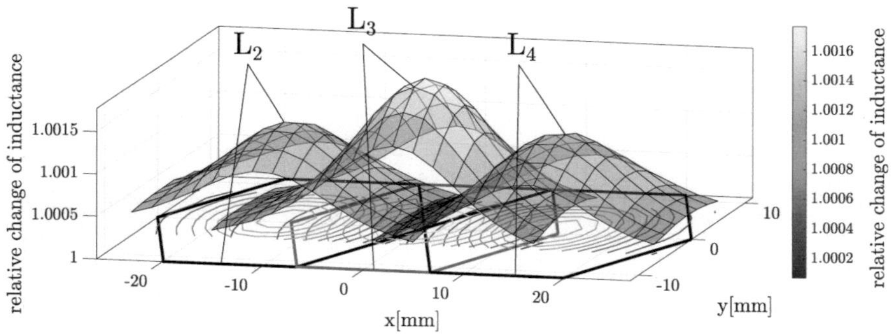

Figure 27.26: Experimental data of relative inductance change of a linear coil array.

investigations are formed by the steps: idea generation, model-based simulation, material synthesis, material characterization and building of prototypes. For technical solutions, two properties which are achievable through the systematic integration of the particles in the matrix and their interactions are important above all: the ability to create complex deformations by appropriate shaping of the objects due to the design of the magnetic fields and the realisation of a reversible variable, by magnetic fields controllable, mechanical compliance. The development of integrated sensor-actuator systems is defined as a new research subject with great potential.

Acknowledgement: This work was supported by Deutsche Forschungsgemeinschaft (DFG) within the project ZI540-17/1-3, in the framework of the priority program SPP 1681. We thank our partners in the SPP 1681 for the inspiring cooperation. Special thanks go to the teams of Prof. S. Odenbach (TU Dresden), Prof. M. Schilling (TU

Braunschweig), Dr. R. Müller (IPHT Jena), Prof. S. Dutz (TU Ilmenau) and Prof. G. Monkmann (OTH Regensburg). Financial support by DFG under the project BE6553/1-1 within PAK 907 is also gratefully acknowledged. This work was carried out with the helpful support of numerous German and foreign scientists, who always showed a great interest in our work. We especially thank Prof. J. Steigenberger (Ilmenau), Prof. N.N. Bolotnik (Moscow) and Prof. V.A. Naletova (Moscow).
Author contribution: All the authors have accepted responsibility for the entire content of this submitted manuscript and approved submission.
Research funding: This work was supported by Deutsche Forschungsgemeinschaft (DFG) within the project ZI540-17/1-3 (SPP 1681) and BE6553/1-1 (PAK 907).
Conflict of interest statement: The authors declare no conflicts of interest regarding this article.

References

1. Becker TI, Raikher YuL, Stolbov OV, Böhm V, Zimmermann K. Dynamic properties of magneto-sensitive elastomer cantilevers as adaptive sensor elements. Smart Mater Struct 2017;26:095035.
2. Böse H, Rabindranath R, Ehrlich J. Soft magnetorheological elastomers as new actuators for valves. J Intell Mater Syst Struct 2012;23:989–94.
3. Melenev PV, Raikher YL, Rusakov VV. Structure-mechanical model for plasto-elastic behavior of soft magnetic elastomers. Polym Sci 2010;52:430–5.
4. Odenbach S. Microstructure and rheology of magnetic hybrid materials. Arch Appl Mech 2016;86: 269–79.
5. Zimmermann K, Böhm V, Kaufhold T, Chavez Vega J, Becker T, Odenbach S, et al. Investigations and simulations on the mechanical behavior of magneto-sensitive elastomers in context with soft robotic gripper application. Int Sci J IFToMM Probl Mech 2016;65:13–26.
6. Volkova TI, Böhm V, Kaufhold T, Popp J, Becker F, Borin DYu, et al. Motion behaviour of magneto-sensitive elastomers controlled by an external magnetic field for sensor applications. J Magn Magn Mater 2017;431:262–5.
7. Nadzharyan TA, Sorokin VV, Stepanov GV, Bogolyubov AN, Kramarenko EYu. A fractional calculus approach to modeling rheological behavior of soft magnetic elastomers. Polymer 2016;92: 179–88.
8. Bica I. The influence of the magnetic field on the elastic properties of anisotropic magnetorheological elastomers. J Ind Eng Chem 2012;18:1666–9.
9. Tian TF, Li WH, Alici G. Study of magnetorheology and sensing capabilities of mr elastomers. J Phys Conf 2013;412:012037.
10. Zając P, Kaleta J, Lewandowski D, Gasperowicz A. Isotropic magnetorheological elastomers with thermoplastic matrices: structure, damping properties and testing. Smart Mater Struct Mar 2010; 19:045014.
11. Morgenstern AH, Calascione TM, Fischer NA, Lee TJ, Wentz JE, Nelson-Cheeseman BB. Thermoplastic magnetic elastomer for fused filament fabrication. AIMS Mater Sci 2019;6: matersci-06-03-363.
12. Prem N, Chavez Vega J, Böhm V, Sindersberger D, Monkman GJ, Zimmermann K. Properties of polydimethylsiloxane and magnetoactive polymers with electroconductive particles. Macromol Chem Phys 2018;219:1800222.

13. Becker TI, Raikher YL, Stolbov OV, Borin D, Stepanov G, Böhm V, et al. Magnetic state of pre-magnetised magnetic hybrid elastomers with a complex filler. In: Proceedings of the 6th Colloquium of SPP 1681, Benediktbeuern, September 26–28, 2018; 2018:10–11 pp.
14. Becker TI, Zimmermann K, Borin DY, Stepanov GV, Storozhenko PA. Dynamic response of a sensor element made of magnetic hybrid elastomer with controllable properties. J Magn Magn Mater 2018;449:77–82.
15. Han Y, Mohla A, Huang X, Hong W, Faidley LE. Magnetostriction and field stiffening of magneto-active elastomers. Int J Appl Mech 2015;7:1550001.
16. Saga N, Nakamura T. Elucidation of propulsive force of microrobot using magnetic fluid. J Appl Phys 2002;91:7003–5.
17. Saga N, Nakamura T. Development of a peristaltic crawling robot using magnetic fluid on the basis the locomotion mechanism of the earthworm. Smart Mater Struct 2004;13:566–9.
18. Kimura T, Umehara Y, Kimura F. Magnetic field responsive silicone elastomer loaded with short steel wires having orientation distribution. Soft Matter 2012;8:6206.
19. Zrinyi M, Barsi L, Büki A. Deformation of ferrogels induced by nonuniform magnetic fields. J Chem Phys 1996;104:566–9.
20. Zrinyi M, Barsi L, Szabo D, Kilian H-G. Direct observation of abrupt shape transition in ferrogels induced by nonuniform magnetic field. J Chem Phys 1997;106:5685–92.
21. Kashima S, Miyasaka F, Hirata K. Novel soft actuator using magnetorheological elastomer. IEEE Trans Magn 2012;48:1649–52.
22. Naletova VA, Zimmermann K, Zeidis I, Turkov VA, Kalmykov SA, Lukashevich MV. Dynamics of a prolate magnetizable elastic body in a cylindrical channel. In: 7th EUROMECH solid mechanics conference, Lisbon, Portugal, September 7–11, 2009: 2009:259–60 pp.
23. Zimmermann K, Zeidis I, Naletova VA, Kalmykov SA, Turkov VA. Modelling of a locomotion device based on deformable thin rod with magnetizable composite. In: 20th international conference on soft magnetic materials, Kos Island, Greece, September 18–22, 2011:1 p.
24. Zimmermann K, Zeidis I, Naletova VA, Turkov VA. Modelling of worm-like motion systems with magneto-elastic elements. Phys Status Solidi 2004;1:3706–9.
25. Zimmermann K, Zeidis I, Naletova VA, Turkov VA, Stepanov GV, Lukashevich MV. Undulation of magnetizable elastic body in magnetic field. In: Third Moscow international symposium on magnetism, Moscow, Russia, June 25–30; 2005; 2005:86–9 pp.
26. Phan KL, Mauritz A, Homburg FGA. A novel elastomer-based magnetoresistive accelerometer. Sensor Actuator Phys 2008;145–146:109–15.
27. Pettersson A, Davis S, Gray JO, Dodd TJ, Ohlsson T. Design of a magnetorheological robot gripper for handling of delicate food products with varying shapes. J Food Eng 2010;98:332–8.
28. Melenev P, Raikher Y, Stepanov G, Rusakov V, Polygalova L. Modeling of the field-induced plasticity of soft magnetic elastomers. J Intell Mater Syst Struct 2011;22:531–8.
29. Chavez Vega J, Kaufhold T, Schümann M, Böhm V, Zimmermann K, Odenbach S. Soft robotic gripping of sensible objects using field-induced plasticity of magneto-sensitive elastomers. In: 11th European magnetic sensors and actuators conference, Torino, 2016; 2016:259–60 pp.
30. Han Y, Hong W, Faidley L. Field-stiffening effect of magneto-rheological elastomers. Int J Solid Struct 2013;50:2281–8.
31. Chavez Vega J, Kaufhold T, Böhm V, Becker T, Zimmermann K, Martens M, et al. Field-induced plasticity of magneto-sensitive elastomers in context with soft robotic gripper applications. Proc Appl Math Mech 2017;17:23–6.
32. Chavez Vega J, Schorr P, Scharff M, Schale F, Böhm V, Zimmermann K. Towards magneto-sensitive elastomers based end-effectors for gripping application technologies. In: 2019 IEEE international conference on mechatronics (ICM), Ilmenau, March, 18–20, 2019; 2019:217–22 pp.

33. Becker TI, Böhm V, Chavez Vega J, Odenbach S, Raikher Y, Zimmermann K. Magnetic-field-controlled mechanical behavior of magneto-sensitive elastomers in applications for actuator and sensor systems. Arch Appl Mech 2019;89:133–52.
34. Condeelis J. Life at the leading edge: formation of cell protrusion. Annu Rev Cell Biol 1993;9: 414–40.
35. Bogolyubov NN, Mitropolski YA. Asymptotic methods in the theory of nonlinear oscillations. New York: Gordon and Breach Science; 1961.
36. Zimmermann K, Zeidis I, Bolotnik N, Pivovarov M. Dynamics of a two-module vibration-driven system moving along a rough horizontal plane. Multibody Syst Dyn 2009;22:199–219.
37. Stossel ThP. On the crawling of animal cells. Science 1993;260:1086–94.
38. Zentner L, Linß S. Compliant systems. New York: DeGruyter; 2019.
39. Zimmermann K, Chavez Vega J, Becker T, Witte H, Schilling C, Köhring S, et al. An approach to a form-adaptive compliant gripper element based on magneto-sensitive elastomers with a bioinspired sensorized surface. Int Sci J IFToMM Probl Mech 2019;75:23–38.
40. Alekhina Y, Makarova L, Rusakova T, Semisalova A, Perov N. Properties of magnetorheological elastomers in crossed ac and dc magnetic fields. J Sib Fed Univ Math Phys 2017;10:45–50.
41. Kawasetsu T, Horii T, Ishihara H, Asada M. Flexible tri-axis tactile sensor using spiral inductor and magnetorheological elastomer. IEEE Sensor J 2018;18:5834–41.
42. Kawasetsu T, Horii T, Ishihara H, Asada M. Size dependency in sensor response of a flexible tactile sensor based on inductance measurement. In: IEEE sensors; 2017:3 p. https://doi.org/10.1109/icsens.2017.8233908.

Christine Gräfe, Elena K. Müller, Lennart Gresing, Andreas Weidner, Patricia Radon, Ralf P. Friedrich, Christoph Alexiou, Frank Wiekhorst, Silvio Dutz and Joachim H. Clement*

28 Magnetic hybrid materials interact with biological matrices

Abstract: Magnetic hybrid materials are a promising group of substances. Their interaction with matrices is challenging with regard to the underlying physical and chemical mechanisms. But thinking matrices as biological membranes or even structured cell layers they become interesting with regard to potential biomedical applications. Therefore, we established *in vitro* blood-organ barrier models to study the interaction and processing of superparamagnetic iron oxide nanoparticles (SPIONs) with these cellular structures in the presence of a magnetic field gradient. A one-cell-type–based blood-brain barrier model was used to investigate the attachment and uptake mechanisms of differentially charged magnetic hybrid materials. Inhibition of clathrin-dependent endocytosis and F-actin depolymerization led to a dramatic reduction of cellular uptake. Furthermore, the subsequent transportation of SPIONs through the barrier and the ability to detect these particles was of interest. Negatively charged SPIONs could be detected behind the barrier as well as in a reporter cell line. These observations could be confirmed with a two-cell-type–based blood-placenta barrier model. While positively charged SPIONs heavily interact with the apical cell layer, neutrally charged SPIONs showed a retarded interaction behavior. Behind the

***Corresponding author: Joachim H. Clement,** Department of Internal Medicine II, Hematology and Medical Oncology, Jena University Hospital, Jena, Germany, E-mail: joachim.clement@med.uni-jena.de. https://orcid.org/0000-0002-6601-2456

Dr. Christine Gräfe, Elena K. Müller and Lennart Gresing, Department of Internal Medicine II, Hematology and Medical Oncology, Jena University Hospital, Jena, Germany, E-mail: christine.graefe@med.uni-jena.de (C. Gräfe), elena.mueller@med.uni-jena.de (E.K. Müller), lennart.gresing@med.uni-jena.de (L. Gresing)

Andreas Weidner and Prof. Silvio Dutz, Institute of Biomedical Engineering and Informatics (BMTI), Technische Universität Ilmenau, Ilmenau, Germany, E-mail: andreas.weidner@tu-ilmenau.de (A. Weidner), silvio.dutz@tu-ilmenau.de (S. Dutz)

Patricia Radon and Dr. Frank Wiekhorst, Physikalisch-Technische Bundesanstalt, Berlin, Germany, E-mail: patricia.radon@ptb.de (P. Radon), frank.wiekhorst@ptb.de (F. Wiekhorst)

Dr. Ralf P. Friedrich and Prof. Christoph Alexiou, Department of Otorhinolaryngology, Head and Neck Surgery, Section of Experimental Oncology and Nanomedicine (SEON), Else Kröner-Fresenius-Stiftung-Professorship, Universitätsklinikum Erlangen, Erlangen, Germany, E-mail: ralf.friedrich@uk-erlangen.de (R.P. Friedrich), christoph.alexiou@uk-erlangen.de (C. Alexiou)

This article has previously been published in the journal Physical Sciences Reviews. Please cite as: C. Gräfe, E. K. Müller, L. Gresing, A. Weidner, P. Radon, R. P. Friedrich, C. Alexiou, F. Wiekhorst, S. Dutz and J. H. Clement "Magnetic hybrid materials interact with biological matrices" *Physical Sciences Reviews* [Online] 2020, 5. DOI: 10.1515/psr-2019-0114 | https://doi.org/10.1515/9783110569636-028

blood-placenta barrier, negatively charged SPIONs could be clearly detected. Finally, the transfer of the *in vitro* blood-placenta model in a microfluidic biochip allows the integration of shear stress into the system. Even without particle accumulation in a magnetic field gradient, the negatively charged SPIONs were detectable behind the barrier. In conclusion, *in vitro* blood-organ barrier models allow the broad investigation of magnetic hybrid materials with regard to biocompatibility, cell interaction, and transfer through cell layers on their way to biomedical application.

Keywords: biocompatibility, magnetic nanoparticles, particle-cell interaction

28.1 Introduction

A broad and highly promising branch of nanotechnology is covered by nanomedicine—the field of health and medicine involving the development and application of nanoscale objects for diagnostic and therapeutic purposes as well as for monitoring and preventing diseases. Thus, it is expected that nanomedicine will improve early diagnosis and treatment of a wide range of diseases by development of better devices, highly specific and efficient drugs, and innovative therapies [1]. More precisely, the utilization of nanoparticle formulations and nanodrugs can help to overcome obstacles like the appearance of severe side effects, instability of drugs, drug delivery to difficult-to-reach sites, and bioavailability, while at the same time increasing therapeutic efficiencies [2]. Passive drug targeting usually is achieved by enhanced permeability and retention (EPR) effects often prevailing in fenestrated tissues and leaky vessels of tumors combined with low lymphatic drainage [3]. By specific conjugation of nanoparticles with ligands directed toward cellular surface or environmental markers, particles can be targeted actively to sites of interest such as tumors, metastases, or inflammation [4]. A horizontal scan published in 2015 by Noorlander et al [5] identified clearly assigned nanomedicinal products already approved for clinical use by the European Medicines Agency (EMA) and Food and Drug Administration [5]. The authors identified most of the products being designated to therapeutic applications, especially cancer treatment, whereas nanomedicinal objects directed toward diagnosis and vaccines constitute a small percentage only. As proposed by Pelaz et al. [1], a major scientific challenge is the lack of knowledge about the behavior of nanoparticles inside living organisms. In order to bridge the gap between laboratory and clinics, an intensive and multidisciplinary cooperation between physicists, chemists, biologists, pharmacists, and physicians is of vital importance.

A special class of nanoparticles is constituted by superparamagnetic iron oxide nanoparticles (SPIONs), whose magnetization appears to be zero in the absence of an external magnetic field, whereas they behave similarly to a paramagnet under the influence of an external magnetic field [6]. Cores of such particles are made of the two main forms, magnetite (Fe_3O_4) or its oxidized product maghemite (γ-Fe_2O_3), and usually comprise multiple domains with single magnetic domain sizes below 12–15 nm in order to achieve the superparamagnetic character [7, 8]. While its saturation magnetization is moderate only [9], in contrast to other solid nanoparticles including cobalt and nickel, the essential element iron is naturally occurring in the human body and therefore not associated with any intrinsic risk [10]. The daily uptake of iron via dietary products is estimated to be 1–2 mg, whereby upon its metabolization, it is used as an essential cofactor of several enzymes and proteins [11]. However, not only iron deficiency constitutes a general problem in human health but also iron excess and accumulation within cells and organelles can result in pathological disorders [12]. Usually, SPIONs are equipped with an additional coating surrounding the iron oxide core in order to maintain colloidal stability especially in aqueous solutions and to protect the magnetite core from further oxidation. Such coatings are adopted according to the application and are also important for particles' biocompatibility and biodegradability. Additionally, they open the possibility of ligand conjugation for further functionalization [13, 14]. For instance, polymers such as polysaccharides and polyethylene glycol (PEG) are frequently used materials showing good biocompatibility [15], while silica coatings allow low loss rates of transmitted light and therefore are widely used for bioimaging and biosensing purposes [16].

The superparamagnetic property of SPIONs opens up new and highly attractive applications especially in the biomedical field. Thus, they can be utilized for site-specific drug delivery during magnetic drug targeting, as contrast agents for magnetic-based imaging techniques, as heat inducers in hyperthermia anticancer therapy, in magnetic tissue engineering, and for cell labeling and *in vivo* cell tracking [1, 13, 14, 17]. As the first nanomedicinal product, ferumoxil (also known as Lumirem®—a siloxane-coated SPION of 300 nm) was approved by the EMA as an MRI contrast agent in 1994, and since then, four more SPION products have been introduced into clinics until today [5]. The shortening effect of SPIONs on especially transverse relaxation time T2 induces a signal loss during MRI, which results in a negative (i.e., dark) contrast image [18]. In addition, approaches utilizing aminosilane-coated SPIONs together with application cycles of alternating magnetic fields are approved for efficient hyperthermia treatment of tumor entities such as glioblastoma and other entities [19]. Thus, clinical trials revealed an improvement of median survival time from 6.2 to 13.4 months without any complications compared to conventional therapy of glioblastoma [20]. Further studies indicate that hyperthermia treatment prior to radiotherapy induces a radiosensitizing effect on glioma cells [18].

A currently arising new nanomedical sector is formed by the fusion of diagnostic and therapeutic applications, called "theranostics" [21]. Particularly, theranostics allows both monitoring of pharmacokinetics via methods such as imaging and therapeutic treatment of diseases including cancer. Therefore, multistep procedures are eliminated, which in turn accelerates disease treatment and pave the way for individualized medicine. Especially SPIONs are highly predestined candidates for implementations of theranostics as their multifunctional potential easily allows the combination of contrast-enhancing features in MRI with magnetic drug targeting and hyperthermia-based cancer therapy [13]. Impressively, preclinical investigations performed by Hayashi et al. [22] using SPIONs heterofunctionalized with PEG and folic acid demonstrated efficient particle accumulation in myeloma cancer tissues of mice, resulting in enhanced contrast during MRI. At the same time, the application of an external alternating magnetic field specifically induced local hyperthermia, promoted tumor shrinkage, and significantly prolonged overall survival of diseased animals.

Despite the various beneficial and promising functions of SPIONs in biomedicine, there are still several obstacles limiting their unconditional use. A major aspect comprises the fact that studies have revealed that only a small proportion of actively targeted SPIONs actually reach target tissue [4]. To overcome this problem by magnetic drug targeting, magnetic field configurations allowing elevated penetration depths are critical as the magnetic field strength shrinks cubically with growing distance. Furthermore, premature burst of drug cargoes, especially at instillation time points, impairs delivery efficiencies to target sites and may evoke cytotoxic side effects [23]. The so-called protein corona comprising an adsorptive layer of ambient proteins on the particles' surface immediately after contact with the environmental medium can critically influence biological effects of SPIONs and thus hamper intended purposes [24]. This phenomenon also impedes the comparability of data obtained from both *in vitro* and *in vivo* studies using similar particle formulations and might induce their premature immunologic clearance from the bloodstream caused by the reticuloendothelial system (RES). Finally, in several cases, the fate of internalized SPIONs into cells and their biodistribution are not clear, whereas for a safe and effective clinical application, such information is of vital importance. Hence, more standardized studies related to these issues are imperatively needed [13].

28.1.1 Cellular barriers

Cellular barriers define a compartment for organs and tissues and allow protection from harmful materials and influences as well as control of entry and release of substances and molecules. Most of the cellular barriers in the human body form an interface between peripheral blood and organs, like the brain, the placenta, or the eye. But also the skin, the air-lung barrier, or the intestinal barrier have to be mentioned.

28.1.1.1 Blood-brain barrier

As the center of the central nervous system (CNS), the brain constitutes a highly complex organ controlling all other body parts and functions. Despite its mass with only 2% relative to the total body mass, its demand for 20% of the body's nutrients is considerably high [25]. For its sufficient supply on the one hand and its protection from harmful stimuli on the other hand, the blood-brain barrier (BBB) forms a selective physiologic barrier between the peripheral circulatory system and the CNS. Thus, the maintenance of homeostatic concentrations of ions, such as Na^+, K^+, and Ca^{2+}, is considered an essential prerequisite for proper neuronal functions [26].

28.1.1.1.1 General structure

The BBB is maintained by a complex interplay between several components of the so-called "neurovascular unit." The central element of this unit is embodied by the endothelial cells of cerebral microvessels, also referred to as brain microvascular endothelial cells (BMECs). Compared to peripheral endothelial cells, BMECs are featured by tight cell-cell junctions lacking fenestration, increased contents in mitochondria, and minimal pinocytotic activity [26, 27]. In addition to ubiquitously abundant adherence junctions, the presence of tight junctions sealing intercellular clefts between the adjacent cells is a characteristic of BMECs. The formation of such tight junctions is accomplished on the one hand by transmembrane molecules such as junctional adhesion molecule 1, occludin, and claudin and on the other hand by membrane-associated guanylate kinase (MAGUK) proteins, which coordinate cytoplasmic protein-protein interactions via multiple protein-binding domains [26]. MAGUK proteins comprise *zonula occludens* (ZO) constructs with three subforms (ZO-1–3) joining transmembrane tight junction units with the actin cytoskeleton. As tight junctions are located in apical cellular segments, they not only restrict paracellular permeability to molecules of a few nanometers only [28] and induce strongly elevated transendothelial electrical resistance (TEER; i.e., measurement of electrical resistance across the cellular layer [29]) but also allow cell polarity and asymmetric distribution of membrane constituents [26]. Caused by this asymmetric and site-specific expression of receptors, transporters, and enzymes facing luminal (i.e., apical membrane) or abluminal (i.e., basolateral membrane) sides, the BBB additionally presents a metabolic barrier modulating the activity of toxic and neuroactive compounds. With regard to TEER, the dense abundance of tight junctions in BMECs mediates resistances of more than 1,000 $\Omega\ cm^2$ *in vivo* compared to 2–20 $\Omega\ cm^2$ in peripheral capillaries [30, 31]. Apart from endothelial cells (i.e., BMECs), further cell types are involved in the formation of the neurovascular unit. In this way, granular and filamentous pericytes irregularly cover abluminal membranes of BMECs forming the closest association with the endothelium. As pericytes possess contractile proteins, they are believed to be involved in the regulation of capillary blood flow [26]. Additionally, pericyte-derived angiopoietin has been shown to induce endothelial expression of occludin enhancing BBB tightness [32]. The 30- to 40-nm-thick basal lamina encloses both pericytes and BMECs [33]. The composition of this lamina is covered by collagen type IV, heparin sulfate, proteoglycans, laminin, fibronectin, and other components of the extracellular matrix and affects BMECs' intracellular signaling pathways and tight junction expressions via interactions with endothelial integrins [33]. Finally, astrocyte end feet connected to the basal lamina support the BBB and provide the cellular link to neurons. There is strong evidence that astrocytes influence BBB formation by regulating endothelial expression of tight junction proteins, distinct transporters, and specialized enzyme systems via secreted factors including transforming growth factor beta, glial cell–derived neurotrophic factor, and basic fibroblast growth factor [34, 35].

28.1.1.1.2 Ways across the BBB

For a sufficient supply of the brain with nutrients and the efficient efflux of metabolic waste products on the one hand and its protection from harmful compounds on the other hand, the BBB constitutes not only a physical barrier but also a metabolic and transport barrier too. Small gaseous molecules such as O_2 and CO_2 can freely pass lipid membranes of the BBB as well as small lipophilic agents including alcohol, cocaine base, and barbiturates. For the transport of small hydrophilic compounds across the BBB, specific transport systems on luminal and abluminal endothelial surfaces are available. The two major superfamilies of such transporters cover ATP-binding cassette (ABC) transporters and solute carriers. In contrast to the transporter-mediated passage, hydrophilic molecules larger than 400–500 Da, such as peptides and proteins, usually are excluded from transcellular trafficking, unless they specifically undergo receptor- or adsorption-mediated transcytosis [18, 36]. Adsorption-mediated transcytosis relies on nonspecific interactions of ligands with surface moieties expressed on luminal membrane sides of endothelial cells. It is predominantly mediated by caveolae- and clathrin-mediated endocytotic processes as well as endocytotic processes independent from clathrin and caveolin. The specific binding of ligands to BMECs' surface receptors, such as insulin receptors, transferrin receptors, and low-density lipoprotein (LDL) receptor–related proteins, is mandatory for receptor-mediated transcytosis. In this way, the brain is efficiently supplied with insulin, transferrin-bound iron, LDL, lactoferrin, and many more.

The tight regulation of transport systems across the BBB often constitutes a considerable obstacle for efficient delivery of therapeutic drugs into the brain. Therefore, specific strategies for overcoming this hurdle have been developed. The injection of hyperosmolar solutions of mannitol, lactamide, saline, or others into the brain-supplying carotid artery has been shown to transiently disrupt cerebral tight junctions by shrinking endothelial cells, which in turn improves the delivery of chemotherapeutic drugs into the CNS of patients with various types of brain tumors [37]. Similarly, chemical destabilization of the BBB by alkylglycerols efficiently limits BBB opening to 3–15 min [38]. Another strategy to temporarily access the brain comprises acoustic cavitation of polymer- or lipid-shelled microbubbles by focused ultrasound [39]. Resulting oscillations of microbubbles induce localized disintegration of tight junctions conveying brain entry for coapplied drugs. Despite indicated conventional approaches facilitating paracellular drug transport, strategies enhancing transcellular carrying have also been developed. These include drug modification toward a more lipophilic character as well as the use of prodrugs, such as L-3,4-dihydroxyphenylalanine, capable for crossing the BBB via transport systems and being converted into active forms on entry [40]. Recently emerged approaches use nanoparticle carriers for enhanced drug delivery into the CNS [13, 36, 41]. For instance, doxorubicin-loaded liposomes functionalized with cationic surface peptides have been shown to efficiently cross the BBB via adsorption-mediated transcytosis in brain tumor–bearing rats [42]. For metallic nanoparticles including SPIONs, passive diffusion, clathrin-mediated transcytosis, and trans-synaptic transport have been demonstrated as mechanisms for overcoming the BBB, whereby the additional application of an external magnetic field enhances SPION accumulation in cerebral perivascular zones within mice [38]. The conjugation of nanoparticle formulations with targeting moieties such as monoclonal antibodies, peptides, and specific plasma proteins encourages transcellular passage through BBB-forming endothelial cells too. Coupling of transferrin or transferrin-binding antibodies to particle surfaces are typical candidates for promoting receptor-mediated transcytosis [43]. Similarly, insulin, lactoferrin, folate, heparin-binding epidermal growth factor (EGF), and integrin $\alpha_v\beta_3$ have been utilized as ligands for CNS-directed delivery [13, 38, 41, 43]. Another common attempt includes particle coating with polyoxyethylene sorbitan monooleate (polysorbate 80). This nonionic surfactant is expected to adsorb apolipoproteins A-I, B-100, and E, mimicking LDL particles, which interact with LDL receptors leading to receptor-mediated transcytosis through BBB-forming endothelial cells [44]. With respect to brain tumor–specific targeting, tumor-directed ligands, e.g., lactoferrin, neutrophilin-1, EGF, antibodies directed toward vascular endothelial growth factor or EGF receptors including mutant EGF receptor variant III and many others have been suggested to be conjugated to nanoparticles' surface [18].

28.1.1.2 Blood-placenta barrier

Pregnant women are daily exposed not only to foreign substances like occupational or environmental materials but also to prescribed maternal medication [45] since prescribed drug use during pregnancy is common in many developed countries [46]. Owing to the placenta being indispensable for a pregnancy, any interference with its function can lead to severe adverse effects on the development of the unborn child [47, 48]. In order to prevent new scandals, like the severe thalidomide-induced birth defects in the 1960s [49], it is important to understand transport mechanisms of different substances through the blood-placenta barrier (BPB) to the fetus and furthermore potential harmful consequences of these materials inside the unborn child and the placenta. Understanding these mechanisms might also help to develop novel drugs that are able to selectively treat complications in the mother, the fetus, or the placenta without harming or affecting the others [50].

28.1.1.2.1 General structure and functions of the BPB

The BPB, which has the highest interspecies variability among mammals, is a fascinating multifunctional organ [51]. This highly effective structure is responsible for the bidirectional transfer of important substances, like carbon dioxide, oxygen, water, nutrients, vitamins, and hormones, between maternal and fetal blood circulations. Besides these vital materials, xenobiotics also can be transported across this barrier [45].

The placenta is a hemochorial organ, where the fetal tissue directly comes in contact with maternal blood circulation [47]. The actual cellular barrier, which separates the maternal blood in the intervillous space from the fetal blood in the placental vessels, consists of a continuous layer of the syncytiotrophoblast with some individual cytotrophoblasts, a thin layer of chorionic connective tissue and the endothelium of the fetal capillary system [47, 52]. The syncytiotrophoblast layer, which is formed out of cytotrophoblasts by syncytial fusion, was found to be the rate-limiting barrier component [53]. There is only one syncytiotrophoblast in a placenta, which is maintained by the incorporation of cytotrophoblasts throughout gestation [54]. The villous core is filled with mesenchymal cells during early pregnancy, which can later differentiate into various cell types, like endothelial cells and blood cells, macrophages, myofibroblasts, and fibroblasts, which can all be found in this space [47, 54]. Fibroblasts are responsible for the secretion of typical matrix proteins like collagen type I and III and proteoglycans. Furthermore, Hofbauer cells, the macrophages of the placenta, are also present in the villous stroma during pregnancy [54]. During the course of pregnancy, the barrier between maternal and fetal blood supply decreases in thickness from over 50 μm in the second month to less than 5 μm by week 37 of pregnancy owing to thinning of the syncytiotrophoblast layer and spreading of the cytotrophoblasts [52]. This in turn leads to enhanced transport of substances, especially nutrients, through the barrier, which are needed for the growth of the fetus especially in the later stages of gestation [47].

28.1.1.2.2 Transport mechanisms across the BPB

The most important function of the placenta is the exchange of nutrients, oxygen, and other substances between maternal and fetal blood circulation [45]. To date, several mechanisms are known for the exchange of endogenous as well as exogenous substances through the BPB: passive diffusion, facilitated diffusion, active carrier-mediated transport, and endocytic pathways, especially pinocytosis [47]. Many transporters, among these ABC efflux pumps, were found to be expressed in the fetal capillary endothelium as well as in the syncytiotrophoblast layer. Here, most of the carriers are located on the apical side facing maternal blood, which may help to protect the fetus from putatively cytotoxic or teratogenic substances. Furthermore, the basolateral side of the syncytiotrophoblasts expresses transporters different from the apical ones, which indicates a polarized transport system across the biological barrier [52]. Depending on the thinning of the BPB and the concurrent minimization of the maternal-fetal diffusion distance throughout the course of pregnancy [45], the exchange mechanisms across this barrier might also depend on the gestation time in which the tissue comes in contact with the substance of interest. Furthermore, the characteristics of the compound and its protein-binding capacity also influence its capability to cross the placental barrier [45].

28.1.2 *In vitro* barrier models

The BBB constitutes a complex network of several collaborating components, making a plain isolation for functional analyses complicated. In order to experimentally study different aspects with regard to BBB cell biology and screening for CNS drug permeability, diverse models have been developed. Especially, high-throughput drug screening draws on parallel artificial membrane permeability assays as non–cell-based surrogate models composed of filters with lipid membranes [27]. In context of functional studies, on the one hand, *in vivo* models include animal testing usually using mice and rats as well as guinea pigs, rabbits, dogs, and monkeys [54]. Owing to profound differences in anatomy, physiology, and genetics, the obtained results cannot be directly extrapolated to the human organism [20], though approaches using freshly isolated capillaries from brain tissues have been utilized for *ex vivo* studies for several decades [55, 56]. As the constant availability of required human tissues is limited and animal testing should be minimized, several cell culture–based *in vitro* models representing the BBB have been developed for multifunctional analyses. The deliberate isolation of primary BMECs from porcine, bovine, murine, or ratty sources and their implementation in transwell-based studies have contributed to valuable insights into physiologic and pathophysiologic processes regarding the BBB [55, 57, 58]. With regard to setups representing the human BBB, only few models are available. These are either based on stem cell–derived endothelial cells or immortalized human BMEC lines [55]. For the latter case, the lentiviral-immortalized cell line hCMEC/D3 is most widely spread and well characterized. While monocultured cell layers of hCMEC/D3 gain TEER values of 30–50 $\Omega\ cm^2$, co-culture with astrocytes raise TEER values to 60 $\Omega\ cm^2$ [55, 59]. Anyway, a comparative study by Eigenmann et al [59] comprising hCMEC/D3, human brain microvascular endothelial cells (HBMECs), and two other cell lines revealed that while all four endothelial cell lines specifically expressed the adherence junction protein VE-cadherin, tight junction protein ZO-1 was only confirmed in hCMEC/D3 and HBMECs. Focusing on TEER and molecular paracellular permeability, the monoculture of HBMECs attained cell layers of highest tightness in this study. Advanced approaches applying shear stress to endothelial cells by the integration of flow conditions to the apical and/or basolateral side superiorly mimic the *in vivo* situation of the BBB. As a consequence, endothelial cells respond by structural and functional remodeling and differentiation and result in higher TEER values as compared to static transwell models [27, 60]. However, no model exactly mimics the full expression pattern of enzymes, transporters, receptors, and other structural proteins of an *in vivo* BBB [34, 55], and no gold standard *in vitro* model exists.

Since novel medical products like nanoparticles are rapidly developed, there is a requirement for valid and predictive models to mimic the behavior of these materials in the blood-placenta barrier (BPB). Furthermore, the need for medications that are available for pregnant women increases, for which the preclinical study using non–*in vivo* models is also important [61].

In vivo models using rodents are applied in current research, especially to give information about the biodistribution of materials in whole organisms [50]. But owing to the high species-to-species differences in placental constitution [51], the data gained from these models cannot be readily extrapolated to the human organism [50].

The dual perfused *ex vivo* human placenta perfusion model, which was first described by Panigel [62] in the 1960s, is based on the usage of the term placentae to study placental function and translocation rates of substances and particles across the BPB. This experimental model allows studying the transfer of substances in the placenta in an organized placental tissue and the simultaneous investigation of physiochemical and pharmacokinetic factors that influence the transfer [63]. While this model maintains the complexity of an intact placenta and is therefore very close to the *in vivo* situation, the transport studies are technically challenging, the exposure time is limited to 4–8 h, and large amounts of substances or particles are required for the investigations [64]. All in all, this *ex vivo* model can provide information about the transplacental translocation rate, acute cytotoxicity and possible fetal exposure of substances, and also the potential role of transporters in the placental barrier [47, 50]. Despite all this, a huge drawback of the model is that owing to the usage of the term placentae, the behavior of investigated substances in the first trimester and during the course of pregnancy cannot be studied sufficiently using only this model [47].

Transwell systems, isolated plasma membrane vesicles, and placental tissue explants are classical ways to study the BPB *in vitro*, which are well-established alternatives to *ex vivo* and *in vivo* models and furthermore allow higher throughput [61]. For the transwell system, there are several possible human cell lines, like BeWo, Jar, or JEG-3 [64], but until now, most *in vitro* studies of transfer across the BPB were performed using the BeWo cell line, in particular the b30 clone [65–70], which was developed in the 1980s [71]. These human choriocarcinoma-derived cells with strong resemblance to cytotrophoblastic cells have the ability to form confluent monolayers on permeable transwell inserts and are therefore a suitable *in vitro* model to study the transfer and also special transport mechanisms of different substances, e.g., nanoparticles [48, 50, 61, 64, 65, 70]. Combined, *in vitro* models, especially transwell-based ones, are suitable for the prescreen of a plethora of compounds and substances in advance to other more complex models and furthermore have the ability for mechanistic transfer studies. Nevertheless, there is still need for improvement of this *in vitro* model, which lacks anatomic integrity and blood flow, in order to increase its predictive value [47, 50].

28.1.3 Nanoparticle-barrier interactions

28.1.3.1 Biocompatibility/toxicity

As the variety of biomedicinal nanoparticles is huge, so are their biological effects. According to the particle type, they can influence cell physiology in many different ways. Obviously, particles characterized by an intensive cellular interaction bring along a great capability for influencing both vitality and physiology of cells. The particle's ability to penetrate biological barriers and their resulting biodistribution are crucial factors determining nanoparticle-induced outcomes at the systemic level. Apart from the particle size, shape, dose, incubation duration, and type of exposed cells, especially surface coating and functionalization play a substantial role in determining biological effects [13, 72]. Thus, many studies have identified a cationic surface charge of nanoparticle formulations as a decisive factor mediating cytotoxicity via various mechanisms [10, 72–74], including the induction of nanoscale holes within plasma membranes, which promotes substantial membrane damage resulting in cell death [75]. Fischer et al. [73] demonstrated that besides the cationic surface net charge, the charge density and flexibility of the three-dimensional structure of the coating molecules also essentially determine biological effects. Endosomal escape of cationic nanoparticle formulations explicitly results in the presence of bare particles within the cytosol, which in turn can induce direct interaction of particles with cellular organelles and proteins including the actin cytoskeleton inducing destabilization and cell cycle arrest [10]. Apart from cationic surface charges, anionic nanoparticles are associated with cytotoxic effects, too, whereas neutral particles are largely nontoxic [74, 76, 77]. A reasonable cause for this is provided by the respective studies which show the low adherence of particles coated by dextran, starch, or other neutral materials at the polar plasma membrane, whereas anionic nanoparticles can still cluster at the sparse membrane areas constituting positive charges [76, 78]. Based on the fact that the hydrodynamic size of nanoparticles (and particle agglomerates) affects particles' reactivity—whereby higher surface-to-volume ratios render elevated surface energies—small particles tend to be more toxic than larger ones [74, 79]. Finally, the formation of a protein corona can also influence biological effects of distinct nanoparticles. After depletion of coating material, cells are exposed to the particle's often inorganic core, which in the case of SPIONs is *per se* considered biocompatible in contrast to silver, cadmium, or other metals [10, 80]. However, during further degradation, increasing contents of free iron ions can cross mitochondrial membranes reacting with hydrogen peroxide and oxygen during the Fenton reaction, giving rise to highly reactive hydroxyl radicals [10, 80]. The resulting oxidative stress can provoke radical damage of DNA and other components including the cytoskeleton, which brings on inflammatory processes and cytotoxic events [74].

28.1.3.2 Attachment and endocytosis

Uptake mechanisms for nanoparticles into the human body are versatile as they can enter via different routes such as the lungs after inhalation, the skin by dermal application, or by enteral resorption upon ingestion [81]. In terms of biomedical applications, direct intravenous administration is widely used—especially in case of cancer nanotherapeutics [82]. The nanoparticles' biodistribution upon this systemic administration modality is determined according to particle size, shape, surface charge, and other surface properties [83]. While for passive targeting, advantage is taken of the EPR effect, active targeting by grafting distinctive ligands, markers, or functional groups onto particles' surface can considerably affect their accumulation and elevate local concentrations at specific target sites. Similarly, magnetic drug targeting in case of SPIONs and other magnetic nanoparticles displays another relevant option for actively directing particle distribution. As a consequence of their size in a typical range of subcellular components and biological molecules [84], nanoparticles can show intensive cellular interactions and be taken up into cells via distinct endogenous uptake mechanisms. Usually, nanoparticles are taken up into cells by active endocytotic pathways, though the cellular entry via passive diffusion has been described as well, especially in case of cationic particle formulations [76, 80]. While large particles with exceeding sizes of 500 nm can only be engulfed by specialized cells including macrophages and neutrophils via phagocytosis, the ingestion of smaller particles via pinocytotic vesicles can be actioned by virtually all types of eukaryotic cells [85–87]. The binding affinity mediated by hydrophobic or electrostatic interactions and receptor-ligand binding to cellular surfaces plays a crucial role in the first step of particle internalization. In this context, diverse studies revealed that cationic surface charges of nanoparticles show intensive interactions with anionic phospholipids or protein domains originating from the glycocalyx (e.g., sialic acids) of cellular plasma membranes [88], whereas neutral and anionic nanoparticles show less pronounced membrane adsorption and subsequent particle internalization [76]. Additionally, the elevated abundance of negatively charged phosphatidylserine in the cytosolic leaflets of the plasma membrane, endosomes, and lysosomes is thought to strengthen the entry of cationic proteins to endocytotic pathways [89]. Subsequent endocytotic ingestion is distinguished into macropinocytosis, caveolin-dependent endocytosis, clathrin-mediated endocytosis, and pathways independent from both clathrin and caveolin with further subclassifications [90, 91]. Most studies focus on clathrin-mediated endocytosis when investigating uptake mechanisms for particles below 200 nm [92, 93], though other processes and even an interplay of several pathways might occur [94]. Regardless of whether particle internalization is induced via plasma membrane invaginations at clathrin-coated pits (i.e., clathrin-mediated endocytosis), cholesterol-rich domains (i.e., caveolin-dependent endocytosis), or other active processes, membrane-budded vesicles transfer the nanoparticle cargo into early endosomes first [87]. Within these compartments, which possess a mild acidified milieu, receptors can be recycled, and digestion off the received material by hydrolysis, as well as its sorting into delivery to specific intracellular compartments or transcytotic pathways, is initiated [95]. Next, the cargo is directed to acidic late endosomes with increasing proteolytic activity before it finally enters lysosomes. For nanoparticle cargoes directed to distinct subcellular targets, pathways avoiding the lysosomal destination of the nanoscaled vehicle are indispensable for successful implementations. Indeed, endosomal escape has been shown for some nanoparticle formulations

[13, 96]. Especially, a strongly cationic particle surface charge seems to promote the escape from endosomes to reach the cytoplasm via destabilization of endosomal membranes by mechanisms including ion-pair formation with anionic lipids and endosomal/lysosomal buffering, swelling, and rupture (i.e., the "proton sponge effect") [97]. Besides the resulting cytosolic delivery, targeting of distinct cellular compartments such as the nucleus, mitochondria, and Golgi has also been pursued [82]. As mentioned earlier, besides surface chemistry, the level of nanoparticle interaction with cells is significantly influenced by particles' properties including size and shape as well as the cellular type itself and the microenvironment as it can induce the protein corona formation, which in turn provides the particle with a new biological identity [13, 80]. In terms of particle size, gathered data emphasize that, in general, smaller particles are internalized more intensively, although a minimum size seems beneficial for efficient induction of particle uptake probably owing to sufficient cross-linking of membrane receptors [80, 86, 93, 98]. In this regard, Win and Feng [99] observed that polystyrene nanoparticles of 100 nm show most pronounced particle internalization into adenocarcinomal cells, while the uptake of larger (200–1000 nm) particles is gradually decreased and lowest for 50 nm.

The aim of this approach is to investigate the interaction of multifunctional hybrid materials with cellular barriers. The cell membrane can be described as a hybrid matrix, so does the cellular composition of a distinct barrier in the human body. The membranous lipid bilayer with its incorporated proteins, carbohydrates, and fatty acids is on the one hand perfectly structured and on the other hand highly dynamic. With regard to the potential of SPIONs as multifunctional hybrid nanomaterials applicable in biomedicine, their interaction with cellular barriers is of large impact. In order to study the cell-nanoparticle interaction, we established two *in vitro* blood-barrier models with regard to the medical need and the grade of complexity. On the one hand, a BBB model is based on one representative cell type, and on the other hand, a blood-placenta model consists of two cell types. Furthermore, we evolved the models from a static to a fluidic setup. The SPIONs that were applied to the *in vitro* barrier models were selected especially for their different coatings and surface charge. The models were run in the presence of a magnetic field gradient, with a special focus on the putative accumulation of nanoparticles behind the barriers.

28.2 Materials and methods

28.2.1 Characterization of SPIONs

To determine the hydrodynamic diameter of SPIONs, dynamic light scattering was applied using a Zetasizer nano series ZS (Malvern Instruments, Herrenberg, Germany). Scattered He-Ne laser light (633 nm) detected at 173°, during three independent measurements with 12 replicates each, was used to determine intensity-weighted hydrodynamic diameters. Additionally, zeta potentials were measured

	fluidMAG-D	fluidMAG-PEI	fluidMAG-CMX	fluidMAG-DEAE	SEON^{LA-BSA}
Supplier	chemicell	chemicell	chemicell	chemicell	AG Alexiou
Core material	magnetite	magnetite	maghemite	magnetite	magnetite
Shell material	starch	PEI, 750 kDa	carboxymethyl-dextran, sodium	starch / DEAE	lauric acid + bovine serum albumin
Hydrod. diameter	141 nm	157 nm	150 nm	132 nm	157 nm
ζ-potential	-11 ± 7 mV	+47 ± 7 mV	-24 ± 6 mV	+36 ± 4 mV	-19 ± 4 mV

	nano-screen MAG-G/D	nano-screen MAG-G/PEI	nano-screen MAG/G-CMX	Sodium Citrate	SEON^{LA-HSA}
Supplier	chemicell	chemicell	chemicell	AG Dutz	AG Alexiou
Core material	magnetite	magnetite	maghemite	magnetite	magnetite
Shell material	starch, fluorochrom(476/490 nm)	PEI, 750 kDa fluorochrom(476/490 nm)	carboxymethyl-dextran, sodium	sodium citrate	lauric acid + human serum albumin
Hydrod. diameter	139 nm	135 nm	132 nm	116 nm	97 nm
ζ-potential	-10 ± 4 mV	+54 ± 9 mV	+36 ± 4 mV	-35.4 mV	-21.3 mV

Figure 28.1: Characteristics of experimentally used SPIONs. SPION = superparamagnetic iron oxide nanoparticle; BSA = bovine serum albumin; DEAE = diethylamine ethyl; PEI = polyethylenimine.

using the same instrument but based on laser Doppler velocimetry. Particle concentrations were determined by measuring the saturation magnetization of the prepared samples by means of vibrating sample magnetometry (PMC MicroMag 3900 VSM – Lakeshore Cytotronics Inc., Westerville, USA). The properties of the nanoparticles used are presented in Figure 28.1. Nanoparticles were applied in concentrations as indicated.

28.2.2 Cell culture

Cell culture preparations and respective solutions were handled under laminar flow hoods. Equipment was autoclaved and sterilized by rapidly acting ethanol-based wipe disinfection or filtration by using 0.2-µm pore size membrane filters. If not stated otherwise, all cells were cultured at 37 °C in a water-saturated atmosphere supplemented with 5% CO_2.

28.2.2.1 Cell lines: An established cell line of HBMECs was used for building up the *in vitro* BBB model. This adherent cell line originally isolated from cortical capillaries has been immortalized by introduction of the SV40 large T antigen [100]. HBMECs were kindly provided by Werner Reichardt from Ernst-Abbe-Hochschule Jena. HBMECs were cultivated using RPMI 1640 cell culture medium supplemented with 10% (v/v) fetal bovine serum (FBS) and 1% penicillin/streptomycin (Pen/Strep: 10,000 U/ml/10,000 µg/ml).

Adherent MCF-7 (DSMZ GmbH, Braunschweig, Germany) cells constitute an epithelial cell line derived from the pleural effusion of a female patient with a mammary gland adenocarcinoma. MCF-7 cells were cultured in Dulbecco's Modified Eagle Medium (DMEM) provided with 10% (v/v) FBS.

The adherent, epithelial-like human choriocarcinoma cell line BeWo (DSMZ), which was first isolated by Pattillo and Gey [101] in 1968, was selected as an *in vitro* model for the BPB [65, 70]. Furthermore, primary human placental pericytes (hPC-PL, PromoCell GmbH, Heidelberg, Germany), which are multipotent mesenchymal-like cells often found in association with small blood vessels, were used for supporting BeWo cells and extending the *in vitro* model. These cells were first cultivated in Pericyte Growth Medium (PGM, PromoCell GmbH, Heidelberg, Germany) according to the purchaser's recommendations, while BeWo cells were grown in DMEM supplemented with GlutaMAX-I and 10% (v/v) FBS. Since pericytes and BeWo cells had to be cultivated in the same medium for the coculture model, both were later kept in DMEM. For some experiments, BeWo cells were cultivated in

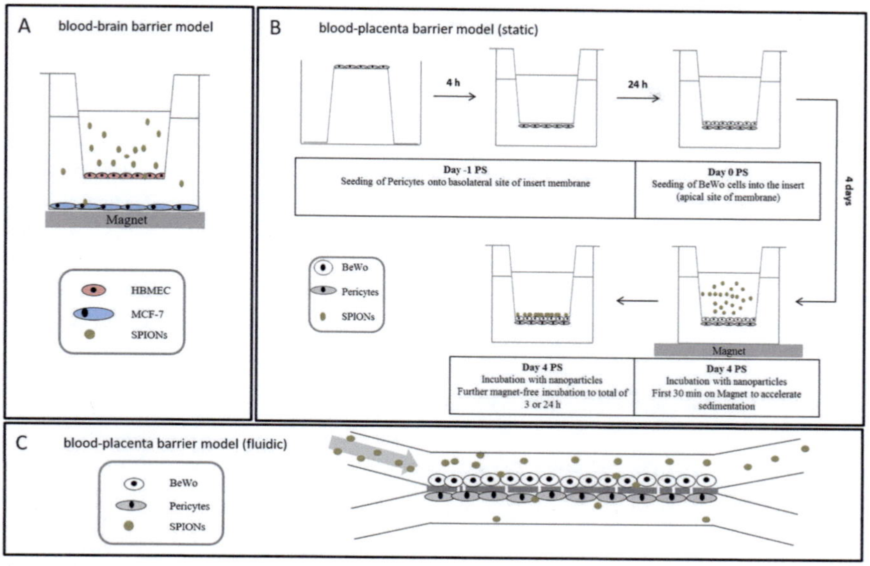

Figure 28.2: The established *in vitro* barrier models. (A) *In vitro* blood-brain barrier model; (B) *in vitro* blood-placenta barrier model—setup of the static system; (C) *in vitro* blood-placenta barrier model in a microfluidic biochip. PS = post seeding; HBMEC = human brain microvascular endothelial cell; SPION = superparamagnetic iron oxide nanoparticle.

pericyte-conditioned medium (PCM), which was collected after 3–4 days of pericyte cell cultivation. After centrifugation (10 min at 1000 rcf), the medium was sterile filtrated (0.2-µm pore size), supplemented with 1% Pen/Strep, and stored at 4 °C until further usage. At confluence, the cells were detached using trypsin-EDTA and subsequently subcultured in a ratio of 1:2 to 1:5 with fresh medium. BeWo cells were used for experiments between passages 8 and 30, while pericytes were used in passages 6–10.

In regular intervals, cell cultures were tested with regard to mycoplasmic contaminations using the commercial PCR-based Venor™ Gem mycoplasma detection kit (Sigma-Aldrich Chemie GmbH, Steinheim, Germany).

28.2.3 Transwell-based generation of an *in vitro* BBB model

For the creation of *in vitro* barrier models, transwell permeable supports (diameter: 6.5 mm) with a PET membrane and a pore size of 3.0 µm (Corning, Inc., Corning, USA) were used.

In order to generate the *in vitro* BBB model, subconfluent HBMECs were harvested and resuspended in a seeding medium being composed of RPMI 1640 supplied with 10% FBS and 1% Pen/Strep. After 30-min equilibration of 24-well transwell inserts within the seeding medium, the equilibration medium was removed, and per insert, 100 µl containing 240,000 cells was added into the insert (apical membrane site/donor compartment). Subsequently, the cell-equipped transwell inserts were carefully placed into 24-well companion plates filled with 600 µl of seeding medium per well (basolateral site/acceptor compartment) (Figure 28.2A). Cellular growth and cell integrity were verified regularly by means of TEER measurements.

28.2.4 Transwell-based generation of an *in vitro* BPB model

All steps were performed using DMEM supplemented with 1% Pen/Strep. For the coculture model, 350,000 pericytes in 80 µl were seeded onto the basolateral site of inserts that were placed upside down into 12-well plates and let to adhere for 4 h while adding the fresh medium every 30 min [102, 103]. Afterward, the inserts were transferred into 24-well plates and cultivated for additional 24 h. Subsequently, 50,000–400,000 BeWo cells were seeded into the inserts onto the apical site of membranes in 100 µl of medium. For monocultures, only BeWo cells were used. The cells were cultured for 3–5 days post seeding (PS), during which the medium in both the donor and the acceptor compartment was changed every other day. Cellular growth and cell integrity were verified regularly by means of TEER measurements. The timeline of seeding and incubation of the transwell model is shown in Figure 28.2B.

28.2.5 Microfluid-based *in vitro* blood-placenta model

For building up the *in vitro* BPB, 200,000 BeWo cells (1.0×10^5 cells/cm^2) were seeded into the donor compartment (upper channel) of a microfluidic biochip [104] (microfluidic chipshop GmbH, Jena, Germany) and incubated for 24 h at 37 °C, 5% CO_2 in a humidified atmosphere. Afterward, 350,000 human placental pericytes (2.3×10^5 cells/cm^2) (hPC-PL) were seeded into the acceptor compartment (lower channel) and incubated for another 48 h under the same conditions. For incubation under continuous flow, the biochip was connected to an Ismatec peristaltic pump (Cole-Parmer, Wertheim, Germany), and the cell barrier was incubated for additional 96 h at a flow rate of 44 µl/min corresponding to a shear stress rate of 0.5 dyn/cm^2 (0.05 Pa) (Figure 28.2C). Shear stress rate was calculated according to Raasch et al. [104]. For a static incubation, the biochip was not connected to the pump but incubated applying the same conditions. DMEM + 15% FBS + 1% Pen/Strep was used in all experiments. The cell barrier integrity was routinely verified by molecular permeability assay using sodium fluorescein (NaFl), histological analysis, as well as immunofluorescence staining of the tight junction-associated protein ZO-1 of BeWo cells. SPIONs were added to the upper compartment to a final concentration of 100 µg/ml, and biochips were incubated with a single circulation of the upper compartment at a constant flow rate of 44 µl/min. The medium of the upper compartment was immediately sampled after one flow through. The content of the lower compartment was also collected, as well as the membrane was excised for further investigations. All samples were analyzed for their absolute iron content at the PTB Berlin using a commercially available magnetic particle spectroscopy (MPS) device (see Section 28.2.12.1).

28.2.6 Cell viability assays

Different assays were used in order to verify cellular viability of HBMECs after exposure to diverse types of SPIONs. While acute cytotoxicity for SPION incubations of up to 24 h was investigated using biochemical and flow cytometry–based assays, SPION-associated long-term effects on cell viability were analyzed by means of real-time cell analysis (RTCA) using the xCELLigence system (see Section 28.2.7).

28.2.6.1 PrestoBlue™ assay: The PrestoBlue™ assay (Invitrogen, Karlsruhe, Germany) is based on the reduction of a nonfluorescent resazurin-based reagent to fluorescent resorufin by metabolically active cells and is used to analyze cell viability on nanoparticle incubation.

In order to test the SPION-specific cytotoxic effect on HBMECs, per well of a 96-well black-walled µ-Clear® plate, 15,000 cells resuspended within RPMI 1640 cell culture medium supplied with 10% FBS were seeded in triplicate. In case of BeWo cells and pericytes, 20,000 cells were applied per well in triplicate. If necessary, 1% Pen/Strep was added to the culture medium. On cultivation overnight, SPIONs dispersed within 18 µl aqua bidest. were added to a final volume of 90 µl, resulting in final SPION concentrations of 5–100 µg/cm^2 (corresponding to 19–378 µg/ml) and incubated for 3 h or 24 h. Positive and negative controls were always obtained by adding 18 µl aqua bidest. only or 0.1% Triton X-100 to cell-seeded wells, respectively. Additionally, cell-free wells containing the cell culture medium and 18 µl of the respective SPION formulations were carried along as background controls.

According to the manufacturer's protocol, PrestoBlue™ reagent supplied as a 10× solution was added into each well and incubated at 37 °C for 30–60 min. The emitted fluorescence at 600 nm (10 to 40 nm bandwidth) upon excitation with 545 nm (20-nm bandwidth) was detected using the CLARIOstar microplate reader (BMG LABTECH GmbH, Orthenberg, Germany). The measured values of nanoparticle-treated cells were background corrected and compared to diluent-treated controls.

28.2.6.2 SYTOX® Red dead cell staining: The principle of SYTOX® dead cell staining is based on the inability of a high-affinity, nucleic acid–intercalating fluorescent dye to pass intact cell membranes, whereas cells with compromised plasma membranes are easily penetrated. Thus, damaged and dead cells can be identified by bright fluorescence signals and can be distinguished from nonstained vital cells.

In scope of investigating the cytotoxic effect of different nanoparticles, 80,000–350,000 cells were seeded into 12-well plates and incubated overnight. After incubation with indicated concentrations of nsMAG/G particles for 3 or 24 h, the cells including the supernatant were harvested by treatment with accutase. After washing twice with PBS supplemented with 2 mM EDTA (PE), the cells were resuspended within 500 µl of 2.5 nM SYTOX® Red (diluted in PE buffer) and incubated at 4 °C for 15 min. Pure SPION solutions were stained analogously in order to verify both nonspecific interactions between particles and dye and the spectral overlap of fluorescently labeled SPIONs into the SYTOX® Red channel. Additionally, these only-SPION–containing samples were utilized for setting up gates excluding free particles from data acquisition during flow cytometric analysis. Without washing, at least 10,000 cellular events per sample were analyzed by the use of a FACS Calibur cytometer (BD Biosciences, San Jose, USA) with SYTOX® Red staining detected with a 661/16 nm bandpass filter upon excitation with 633 nm and fluorescently labeled SPIONs detected with a 585/42 nm bandpass filter upon excitation with 488 nm.

28.2.7 Real-time cell analysis

The RTCA via the xCELLigence DP (ACEA Biosciences Inc., San Diego, USA) presents a noninvasive approach for monitoring cellular proliferation, size, morphology, and attachment. This method was used to analyze the dynamic effects of SPIONs on cell proliferation and viability especially for long-term incubations.

Here, this method was used to detect the toxicity of the investigated nanoparticles to the cells of interest. For the experiments, after measuring the background of medium-containing wells, the cells were seeded into 16-well E plates (HBMEC and BeWo: 35,000 cells, pericytes: 30,000 cells/well). After sedimentation for 30 min at RT, the cells were monitored for the first 24 h; afterward, SPIONs dispersed within 10 µl aqua bidest. were added, resulting in final concentrations of 25–100 µg/cm^2 (corresponding to 25–100 µg/ml). Alternatively, negative control cells were incubated with 10 µl aqua bidest. only, whereas cell free wells containing the cell culture medium were treated with the respective SPION solutions as background controls. Cell index progression was monitored for further 72 h. The cell index was monitored each 30 s for the initial 30 min, afterward each 30 min for approximately 96 h. For analysis of the received data, RTCA software 1.2 (Roche Diagnostics GmbH, Penzberg, Germany) was used.

28.2.8 Analysis of the cell layer integrity

The cell layer integrity of HBMECs seeded on transwell membranes resembles a critical aspect in verifying the *in vitro* BBB model's condition and quality. While repetitive TEER measurements were used in order to monitor the progression of this tightness parameter in vital cell cultures, end point determinations such as molecular permeability assays or microscopic analysis of fixed cells with subsequent fluorescent staining or histological cross sections were used to get a more complete insight into the cell layer functionality and composition.

28.2.8.1 TEER measurements: For continuously verifying the tightness of a cellular barrier, measuring the TEER is a well-established method and vitally important for the evaluation of *in vitro* BBB models [57]. As permeability barriers restrict the movement of ions through the cell layer, an increased TEER indicates an *in vitro* barrier of elevated tightness.

By using chopstick electrodes connected to an epithelial voltohmmeter (EVOM Epithelial Voltohmmeter, World Precision Instruments, Berlin, Germany), the TEER values of HBMECs and BeWo/pericyte layers were determined. Thus, per insert, the mean of three TEER measurements determined at three different positions was aligned to the effective membrane area (i.e., 0.33 cm^2 membrane area including pores). Further, reference TEER measurements of cell-free membrane inserts with respective cell culture media served as background resistances and were subtracted from sample values.

28.2.8.2 Molecular permeability assay: The determination of paracellular permeability of cellular barriers was performed using NaFl (376 Da; Sigma-Aldrich Chemie GmbH, Steinheim, Germany). As NaFl is small, freely diffusible, and nontoxic, this fluorescent molecule is frequently utilized as a highly sensitive paracellular tracer for both *in vitro* and *in vivo* studies [57, 105].

After preparation of the transwell models as described, the cells were incubated with 100 µl of 2.5 µM NaFl diluted within phenol red–free medium and 10% FBS at the apical site, whereas companion plates at the basolateral site were filled with 600 µl of phenol red–free medium supplied with 10% FBS. The NaFl-exposed inserts were incubated at 37 °C under orbital shaking (90 rpm, 30-mm amplitude) for 10–60 min, whereby at defined time points, inserts including the incubation medium were transferred to a new well containing 600 µl of fresh phenol red–free medium with 10% FBS, and incubation was continued as stated above. Fluorescence intensities of samples obtained from the basolateral medium upon 10-, 30-, and 60-min incubation were transferred into black-walled 96-well µ-Clear® plates in triplicate and measured by using the CLARIOstar microplate reader (λ_{ex} = 460/9 nm, λ_{em} = 515/20 nm). NaFl permeability coefficients P_{NaFl} were calculated from applied and detected concentrations (c in nM) and volumes (V in cm^3) as described by Audus and Borchardt [106] according to the following equation (28.I):

$$P_{NaFl} = \frac{c_{acceptor} \cdot V_{acceptor}}{t \cdot A \cdot c_{donor}} \qquad 28.I$$

with t and A defining the incubation time (s) and the effective diffusion area (cm^2), respectively. Alternatively, NaFl restraints were calculated from NaFl permeability coefficients as the ratio of cell-free and cell-grown insert membranes.

28.2.9 Histological analysis of cross sections

Thin microtome sections of 5–15 µm were prepared from the membranes of the transwell inserts to get a broader image of the thickness and integrity of the cell layers as well as the interaction of nanoparticles with the barrier. After preparation of the transwell model and nanoparticle incubation as described earlier, the membranes were cut out of the inserts and embedded in 1% agarose dissolved in PBS in a truncated 1.5-ml Eppendorf tube after washing with PBS and fixation with 10% formalin solution. The membranes were incubated for at least 20 min at 4 °C for consolidation; afterward, the agarose blocks containing the membranes were transferred into embedding cassettes and incubated in 10% formalin for 1–2 days. After removing the formalin with distilled water for 4 × 15 min, the blocks were dehydrated using the automatic tissue processor Leica TP1020 (Leica Biosystems Nussloch GmbH, Nussloch, Germany). Directly afterward, the blocks were embedded into paraffin using a Leica EG1160 embedding center (Leica Biosystems Nussloch GmbH, Nussloch, Germany) and cooled at −20 °C for at least 2 h. Sections of 15-µm thickness of the membrane-containing blocks were prepared using the Leica RM 2165 automated rotary microtome (Leica Biosystems Nussloch GmbH, Nussloch, Germany). The sections were first transferred to a RT water bath to remove air bubbles and then transferred to a 40 °C water bath for stretching before they were applied to microscopy slides and dried at 37 °C overnight. For microscopic analysis, the sections were stained with Nuclear Fast Red and Prussian blue. The paraffin was removed from the dried samples by using xylene, followed by a descending ethanol series to rehydrate the sections. Subsequently, the sections were stained with a 2% potassium ferrocyanide solution for 10 min. After washing with distilled water, the cell nuclei were stained with Nuclear Fast Red for additional 10 min before washing for 1 min with running tap water. Before embedding the sections with Entellan®. New, they were dehydrated using an ascending ethanol series. The slides were analyzed using the microscope Axiovert 25, the camera AxioCam HRc and the software AxioVision SE64 4.9 (Carl Zeiss GmbH, Jena, Germany).

28.2.9.1 Prussian blue staining: Prussian blue resembles a dark blue pigment generated by the iron chelating reaction of ferrocyanides. Thus, the staining is commonly used to visualize ferric iron within biological samples, such as iron-based nanoparticles within cellular environments.

In order to stain SPIONs in cellular samples, a protocol adapted from the study by Schlorf et al. [107] was used. In detail, samples fixed by incubation with 10% formalin solution for 15 min were permeabilized by a 10-min incubation in 0.1% Triton X-100 (dissolved in D-PBS) and washed twice with D-PBS. Next, a freshly prepared solution of 2% (w/v) potassium ferrocyanide dissolved in 1 M hydrochloric acid was applied and incubated at 37 °C for 10 min before two washing steps using D-PBS were carried out. Unless otherwise stated, the cells were counterstained by sample incubation with the eosin Y-containing DiffQuick II solution for 30 s. As the resulting cytoplasmic red staining was not stable, microscopic images were acquired immediately. Alternatively, cells were counterstained using Nuclear Fast Red solution and subtracted to a dehydrating alcoholic series in order to achieve stable dyeing.

28.2.10 Fluorescent staining for confocal laser scanning microscopy

Confocal laser scanning microscopy (cLSM) upon (immuno)fluorescence staining offers the possibility to specifically view distinct cellular structures and their three-dimensional distribution. The spatial pinhole allows the acquisition of signals from the confocal plain only, thus eliminating out-of-focus light. On the one hand, cytoskeletal staining was used for the sake of studying both the cell integrity and the cellular uptake of fluorescently labeled SPIONs. On the other hand, the expression and distribution of the tight junction protein ZO-1 upon immunofluorescence staining was additionally used in order to assess the tightness of transwell-cultured cell layers. For all approaches, the confocal laser scanning microscope LSM 510 META and the appendant software ZEN 2009 6.0 SP2 (both Carl Zeiss Microscopy GmbH, Jena, Germany) were used.

28.2.10.1 Phalloidin staining: Phalloidin is a bicyclic heptapeptide strongly binding filamentous actin (F-actin) preventing its depolymerization. Labeled with a fluorescent tag, this molecular dye is prevalently used for visualization of F-actin *in vitro*.

For the investigation of the cellular uptake mechanism of SPIONs, 200,000 to 400,000 cells were resuspended in the appropriate cell culture medium and seeded on glass cover slips, which had been flamed and placed into 24-well plates. Upon cell cultivation overnight, SPIONs were applied and incubated for 3 h. If appropriate, endocytosis inhibitors were added as indicated and preincubated for 60 min before SPION addition. The samples were washed three times with D-PBS before they were fixed at RT by applying a 10% formalin solution for 15 min. Next, formalin was removed by another three washing steps with D-PBS, and the cells were permeabilized by a 10-min treatment with 0.1% Triton X-100 (diluted with D-PBS). The samples were incubated with the D-PBS–based staining solution containing 19 ng/ml Alexa Fluor® 633 Phalloidin (Invitrogen) and either 10 ng/ml DAPI I (Abbott Laboratories) or 12 µg/ml Hoechst 33258 (Invitrogen). Afterward, the cells were washed with D-PBS again and analyzed microscopically without embedding. The cells were scanned layer by layer with a constant slice distance of 1 µm, each resulting in z-stacked images of 10–18 slices. Acquired image stacks were quantitatively analyzed using MATLAB® R2013a software (MathWorks, Natick, USA), wherein the amount of internalized SPIONs was calculated from overlapping signals derived from the cytoskeletal F-actin and the SPION channels. Actin channels were binarized using the threshold algorithm as described by Otsu [108], multiplied with original SPION channels, and integrated to total SPION intensities per image slice. For comparability of multiple microscopic field views containing divergent numbers of cells, particle amounts were normalized to the cell-representing F-actin signal.

28.2.10.2 Immunofluorescence staining of ZO-1: In order to quantify the integrity of the barrier layer in the *in vitro* coculture transwell model, both tight junctions and adherens junctions were visualized using antibodies against ZO-1 and β-catenin, respectively, which were already shown to contribute to cell-cell contacts in the human placenta [109]. The peripherally located ZO-1 is associated with cytoplasmic parts of cell-cell contacts after formation of these junctions and is therefore applicable to visualize tight junctions in cell layers [110]. Since β-catenin is part of the cadherin-catenin complex, which is present at adherens junctions, it can be used as a reliable marker for cell-cell interactions [109].

After preparation of the monoculture or coculture transwell models accordingly, the cells were washed, fixed, and permeabilized. Before staining, unspecific binding sites were blocked by incubation of the inserts in 5% BSA solution for 1 h at RT using an orbital shaker. The membranes were cut out of the inserts and cut into half. Each half was incubated with either rabbit anti–ZO-1 or rabbit anti–β-catenin in a 1:100 dilution in 1% BSA and 0.1% Triton X-100 in PBS for 1 h in a wet chamber at 37 °C. After washing twice with PBS, the membrane halves were incubated with a staining solution containing the AlexaFluor® 488–labeled goat anti-rabbit secondary antibody (1:200 in 1% BSA and 0.1% Triton X-100 in PBS), 0.19 µg/ml AlexaFluor® 633 Phalloidin, and 12 µg/ml Hoechst 33258. After staining, the membranes were washed twice and embedded on microscopy slides before analysis. For cocultures, the orientation of embedding depended on the cells of interest.

Moreover, samples stained without primary antibodies and samples not stained with primary or secondary antibodies were carried along and served as controls for specific secondary antibody binding and autofluorescence, respectively. Upon staining, the samples were washed with D-PBS; membranes were cut out of retainers and embedded on glass slides using water-based mounting medium prior to microscopic analysis.

28.2.11 Flow cytometry

Flow cytometry is an analytic method based on the principles of light scattering and light excitation/emission of fluorochromes, which is an essential technique to analyze many different parameters of cells simultaneously, like cell viability using DNA-binding dyes or expression of different cell surface markers using targeting antibodies. Here, this method was used to characterize pericytes by investigation of surface markers and to analyze cytotoxic effects of SPIONs on cells. For the experiments, the FACS Calibur cytometer (BD Biosciences, San Jose, USA) was used, and the results were evaluated using FlowJo™ software (FlowJo, LLC, Ashland, USA).

28.2.11.1 Flow cytometry–based nanoparticle-cell interaction: For the purpose of investigating the interaction of fluorescently labeled SPIONs with cells and its correlation to cell viability, 80,000 cells/cm^2 were seeded into 12-well plates in duplicate. Following overnight culture, the cells were incubated with fluorescently labeled SPIONs dispersed within 100 µl aqua bidest. resulting in final concentrations of 25 µg/cm^2 or 50 µg/cm^2 (corresponding to 94.5 µg/ml or 189 µg/ml) for up to 24 h. Negative controls were treated analogously using 100 µl aqua bidest. only. After incubation time was completed, sample cell culture media were collected, and the cells were harvested using HyQTase™. Next, the cells were spun down by a 5-min centrifugation step at 300 rcf (4 °C), washed with ice-cold PE buffer, and centrifuged as mentioned before. For investigating the correlation between SPION interaction with cells and cytotoxicity, cell pellets were additionally treated with SYTOX® dead cell staining and directly analyzed as described earlier. Otherwise, cell pellets were fixed using 10% formalin for 15 min, washed using ice-cold PE buffer, and resuspended in 500 µl of PE buffer again. Finally, per sample, at least 10,000 cellular events were analyzed by flow cytometry, wherein cellular loading with fluorescent-labeled SPIONs was detected with a 585/42-nm bandpass filter upon excitation with 488 nm. Additionally, cell-free samples containing SPIONs only were utilized for setting up gates, excluding free particles from data acquisition during flow cytometric analysis.

28.2.12 Detection and quantification of SPIONs

For studying the passage of SPIONs through HBMEC layers, complete basolateral acceptor compartments were analyzed for the presence of SPIONs either by magnetic particle or atomic absorption spectroscopy (AAS). Both approaches resemble highly sensitive methods of quantifying superparamagnetic or elementary iron, respectively.

For the experimental investigations, HBMECs were plated out on transwell membranes as described earlier. After five days of cultivation, barrier tightness was confirmed by means of TEER measurements (see above). If indicated, one day prior to the particle incubation, MCF-7 cells with a density of 350,000 cells/well were seeded into 24-well plates and cultivated overnight. On the day of incubation experiment, HBMEC-grown inserts were transferred to implied MCF-7–grown or cell-free 24-well companion plates each containing 600 µl of RPMI 1640 medium supplemented with 10% FBS and 1% Pen/Strep. Unless stated otherwise, 100 µg/cm^2 of SPIONs dissolved within 165 µl (equal to 200 µg/ml) of medium supplemented with 10% FBS was added. For the first 30 min, particle incubation was carried out on top of a block magnet (210 mT at 3-mm distance, field gradient: 6.8 T/m) in order to bring the particles in close proximity to the cell layer localized on the transwell membrane. Thereafter, the incubation was continued without the magnet until the indicated incubation time was complete. In order to precisely quantify the amount of SPIONs in the respective compartments, samples were analyzed by MPS and AAS. Additionally, cell layer integrity upon SPION exposure was verified as well as by microscopic analysis upon fluorescence staining.

28.2.12.1 Magnetic particle spectroscopy: MPS represents a sensitive magnetic detection method that allows for the quantification of the superparamagnetic nanoparticle iron content without being affected by biological components such as cells or the suspension medium [111–113]. It is based on the nonlinear magnetic susceptibility response of magnetic nanoparticles exposed to an oscillating magnetic field. Thus, odd harmonics of magnetic moments A_i of the detected time-dependent signal are Fourier transformed, yielding the MPS spectrum. As the SPION-specific amplitude signal is proportional to the applied SPION amounts, MPS provides the opportunity to precisely determine SPION contents of biological samples [114].

For the quantification of SPIONs within the distinct compartments of the transwell system, samples were processed as described by Gräfe et al. [115]. Briefly, complete apical donor and basolateral acceptor compartment media were collected upon incubation time. By avoiding the application of any metallic materials during preparations, porous membranes including cellular layers were cut off retainers and homogenized within 800 µl aqua bidest. by using a ceramic scalpel and the gentleMACS™ dissociator. Sample volumes were reduced to 20–50 µl by centrifugal vacuum concentration using the Speed Vac™ SPD111 (Thermo Fisher Scientific Inc., Waltham, USA) at 100 × g and 40 °C. In case of the BPB, the two cell layers located on the apical and basolateral site of the transwell membrane were individually trypsinated using 100 and 600 µl in the donor and acceptor compartment, respectively, and the reaction was stopped by adding the same amounts of the medium supplemented with FBS. The cells were resuspended and lysed in 20 µl of 10% sodium dodecyl sulfate (Carl Roth GmbH, Karlsruhe, Germany) after centrifugation at 400 rcf for 5 min. MPS spectra were measured using a commercial MPS device (Bruker Biospin, Rheinstetten, Germany) operating at an oscillating magnetic field B_{drive} of 25 mT and a frequency f_0 of 25 kHz. The third harmonic A_3 of the MPS spectrum was used for iron quantification by normalization to the corresponding $A_{3,ref}$ of a reference sample of known iron amount. In turn, the reference sample's iron amount was cross-validated by photometry (510 nm) upon phenanthroline staining by dissolving in hydrochloric acid, reduction by hydroxylamine, and addition of 1,10-phenanthroline monohydrate. Furthermore, A_5/A_3 was recorded in order to verify the magnetic behavior including particle agglomeration. Analogous samples without the addition of SPIONs served as background controls from which the limit of detection (LOD) was calculated according to equation 28.II:

$$\mathrm{LOD} = \bar{X}_{\text{background}} + 3 \cdot \mathrm{STDV}_{\text{background}} \qquad \text{28.II}$$

where $\bar{X}$ and STDV represent mean and standard deviation of background measurements, respectively, according to McNaught and Wilkinson [116].

28.2.12.2 Atomic absorption spectroscopy: In addition to the quantitative estimation of SPIONs via MPS, AAS was applied in order to verify the SPIONs' interaction with and passage through HBMEC layers based on the spectroanalytical detection of elementary iron. Thus, iron concentrations of analyzed samples were determined on measuring the absorbance at the characteristic and highly sensitive wavelength of 248.3 nm.

To this end, complete media of both donor and acceptor compartments were collected, and SPIONs were pelleted by centrifugation at 20,000 × g for 45 min. Upon discarding the excessive supernatant, 25 µl of the sediment-containing solution was dissolved in 162.5 µl of 32% HCl. In contrast, MCF-7 cells seeded into bottom wells of 24-well companion plates as well as cut outs of cell-covered transwell membranes were directly dissolved in 187.5 µl of 32% HCl. The samples were supplemented with 62.5 µl of 10% trichloroacetic acid and incubated for 5 min at RT before precipitated proteins and debris were removed by a 5-min centrifugation at 3600 × *g*. Finally, the supernatants were transferred into conic AAS tubes, and iron contents were analyzed by using the AAS-5 FL supplied by Analytik Jena AG, Jena, Germany. For the quantification, a calibration curve with defined iron concentrations (0–50 µmol/l) was prepared. Samples exceeding these concentrations were diluted in aqua bidest. and measured again.

For testing the accuracy of measurement, precision controls containing 6.5 mg/ml $FeCl_3$ were also measured. The LOD was calculated as described above.

28.2.13 Statistical analyses

Data of repetitive independent experiments with multiple replicates each are presented as weighted mean ± (weighted) standard deviation. Statistical significance tests were performed using Prism 6 (GraphPad Software, La Jolla, USA) applying a one- or two-way analysis of variance (ANOVA) with 95% confidence intervals, followed by a multiple comparison test and correction according to Dunnett [117, 118], Tukey [119], or Sidak [120]. Differences are considered as statistically significant for $p < 0.05$ (*), $p < 0.01$ (**), $p < 0.001$ (***), or $p < 0.0001$ (****).

28.3 Results and discussion

28.3.1 Establishment of a suitable *in vitro* BBB model for SPION-cell interaction studies

28.3.1.1 SPION-associated effects on cell viability

The *in vitro* BBB model should be established with the cell line HBMEC. At first, the acute cytotoxic effects of SPION exposure on HBMECs for 3 h were tested using the PrestoBlue™ assay. For concentrations of up to 100 µg/cm^2, no noticeable effect on HBMECs' viability can be observed after 3 h or 24 h of incubation with neutral fluidMAG-D and anionic SEON$^{LA\text{-}BSA}$ particles resulting in a relative cell viability of 90.0 ± 2.5% and 79.7 ± 1.9% after 24 h of incubation with the respective SPIONs. The exposure of HBMECs to up to 100 µg/cm^2 of cationic fluidMAG-DEAE for 24 h induced a slight concentration-dependent reduction in cellular viability to 67.1 ± 4.2% compared to controls. Cationic PEI-coated SPIONs strongly affect cell viabilities as incubations of 100 µg/cm^2 for 3 h as well as for 24 h trigger a relative decrease to 49.4 ± 3.6% and 13.1 ± 1.7%, respectively [115].

28.3.1.2 Binding of SPIONs to cells and particle uptake

Cellular binding and uptake of SPIONs are closely connected with cytotoxicity. While biocompatible particles are often characterized by lower cellular affinities, highly interactive particle types are usually associated with cytotoxic effects. In order to study these aspects for selected SPIONs, both Prussian blue staining and flow cytometry were applied. Within a 3-h incubation, the cellular binding of neutral fluidMAG-D particles is low. In contrast, cells exposed to anionic SEON$^{LA\text{-}BSA}$, cationic DEAE-coated and PEI-coated SPIONs exhibit a strong Prussian blue stain with the highest intensity for the latter. After 24 h, extensive Prussian blue staining of fluidMAG-D–incubated

HBMECs was observed and indicates an enhanced cellular binding of these neutral particles compared to exposures for 3 h. Similarly, anionic SEON$^{LA\text{-}BSA}$ and cationic fluidMAG-DEAE and fluidMAG-PEI showed a pronounced particle adherence as indicated by a more intense staining pattern upon 24 h. In all cases, the particles formed clusters, whereby fluidMAG-PEI was again most notable.

In order to monitor the temporal progress of cellular SPION loading in context of quantity, flow cytometry of HBMECs exposed to fluorescently labeled SPIONs was performed. HBMECs incubated with neutral nano-screenMAG-G/D show low fluorescence upon incubation for 5 min (11 ± 1 relative fluorescence units [RFU]), which gradually increases during incubation for 30 and 180 min with cellular fluorescence of 37 ± 4 and 228 ± 22 RFU, respectively. Additional to cellular fluorescence labeling, side scatters (SSC) were recorded in order to monitor SPION loading via this granularity-representing parameter. However, for nano-screenMAG-G/D incubation for up to 180 min, the SSC shows only minor changes to 173 ± 10 AU compared to control cells with 162 ± 11 AU. After particle incubation for 24 h, both the cellular fluorescence loading and the SSC strongly expand to 877 ± 68 RFU and 329 ± 20 AU, respectively. In contrast, cationic fluidMAG-PEI shows the highest cellular labeling affinities of 2878 ± 156 RFU immediately after particle addition to the cells (5 min) with a gradual decrease during 30 min, 60 min, and 180 min to 2273 ± 291, 1834 ± 65, and 719 ± 504 RFU, respectively. Analogously, SSC rapidly increases from 162 ± 11 to 896 ± 34 AU upon incubation with nano-screenMAG-G/PEI for 5 min and subsequently decreases to 651 ± 2 AU upon the 180-min incubation.

Flow cytometry allows the analysis of cellular SPION loadings in a (semi)quantitative manner. However, it is not possible to gain information of the actual particle uptake into the cells. This is why in the next step, cLSM was used in order to identify the spatial distribution of fluorescently labeled SPIONs within fixed, fluorescently stained HBMECs. By specifically blocking endocytotic pathways with diverse inhibitors, the uptake mechanisms for both starch- and PEI-coated SPIONs were investigated in detail. Figure 28.3 summarizes this analysis, whereby the image acquisition layer by layer was used to study the colocalization of SPIONs with the intracellular F-actin cytoskeleton within the limits of optical resolution. Based on this premise, internal SPIONs were quantified by integrating the particles' fluorescence intensities of F-actin overlapping signals. During microscopic analysis, SPIONs have not been observed colocated with the cell's nucleus. The effects of specific inhibitors on the internalization of neutral fluidMAG-D into HBMECs are presented in Figure 28.3A. The SPION uptake into HBMECs is massively compromised by approximately 90% when particle incubation is carried out at 4 °C instead of 37 °C, indicating an energy-dependent uptake mechanism for this particle type. Through blocking of caveolin- and clathrin-dependent endocytotic pathways by polyene macrolide filipin (Sigma-Aldrich, Taufkirchen, Germany) and chlorpromazine hydrochloride (Sigma-Aldrich, Taufkirchen, Germany), the amount of internalized starch-coated particles is significantly reduced to 39 ± 10% and 53 ± 34%, respectively. While the single-drug treatment of HBMECs with the fungal toxin

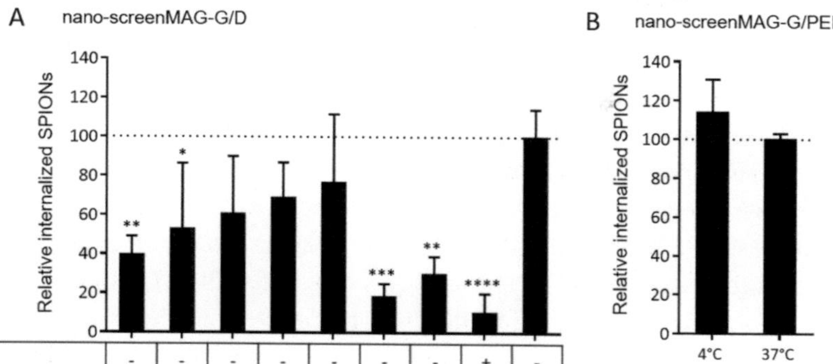

4°C	-	-	-	-	-	-	-	+	-
Filipin (2µg/ml)	+	-	-	+	+	-	+	-	-
Chlorpromazine (5µg/ml)	-	+	-	+	-	+	+	-	-
Cytochalasin D (0.5µg/ml)	-	-	+	-	+	+	+	-	-

Figure 28.3: Cellular uptake mechanisms of SPIONs analyzed by confocal laser scanning microscopy. HBMECs were seeded on glass cover slips placed within 24-well plates with a seeding density of 165,000 cells/cm^2. After preincubation of cells with indicated inhibitors for 60 min, nano-screenMAG-G/D or -G/PEI were added, resulting in a final concentration of 50 µg/cm^2 (corresponding to 60 µg/ml), and incubated for 3 h. Cells were fixed, permeabilized, and stained with DAPI and Alexa Fluor® 633 Phalloidin. Internalized SPIONs were identified by SPION-derived fluorescence overlapping with cells' actin-derived signals. (A) Quantitative analysis of internalized nano-screenMAG-G/D into HBMECs subjected to indicated incubation conditions. (B) Quantitative analysis of internalized nano-screenMAG-G/PEI into HBMECs subjected to indicated incubation conditions. Shown are means ± standard deviation of two independent experiments with three microscopic fields of view each. Statistical significance of indicated samples compared to controls was tested by one-way ANOVA followed by Dunnett's multiple comparison, wherein differences are considered statistically significant for $p < 0.05$ (*), $p < 0.01$ (**), $p < 0.001$ (***), and $p < 0.0001$ (****). Filipin, Chlorpromazine: endocytosis inhibitors; cytochalasin D: inhibitor of actin polymerization. SPION = superparamagnetic iron oxide nanoparticle; HBMEC = human brain microvascular endothelial cell; ANOVA = analysis of variance; PEI = polyethylenimine.

cytochalasin D (Sigma-Aldrich, Taufkirchen, Germany) does not impair particle uptake significantly (61 ± 29%), the combination of this F-actin depolymerizing substance with chlorpromazine hydrochloride efficiently decreases the SPION internalization by 82%. In contrast, the reduction of the incubation temperature from 37 to 4 °C does not impair the presence of internalized nano-screenMAG-G/PEI into HBMECs as relative SPION internalization is 100 ± 3 and 114 ± 17%, respectively (Figure 28.3B).

These data confirm that SPIONs are internalized by HBMECs through endocytosis and thus encourage using this cell line as the barrier model.

28.3.1.3 Establishment and optimization of the BBB model

In order to obtain a BBB-representing *in vitro* test model, the human cell line HBMEC was utilized. Therefore, transwell inserts comprising a porous membrane were used for generating HBMEC layers separating the upper donor compartment from the lower acceptor compartment. As the barrier integrity is a critical factor for all further investigations regarding particle interaction and passage through the barrier cells, initial experiments revealing optimal cell culture conditions were performed.

28.3.1.3.1 Testing of the transwell system and cell-seeding density

Different types of membranes and additional coatings for transwell inserts are available for the generation of a cellular transwell barrier system. Testing collagen-coated PTFE- and tissue culture–treated PET membranes, each seeded with cell numbers of 160,000–300,000 cells per insert, indicates distinct differences between the two transwell types. While cells seeded on PTFE membranes can easily be imaged by bright-field microscopy, the optically visible pores of PET membranes strongly impede the microscopic evaluation of cellular growth on the latter membrane type. However, investigating the cell barrier integrity regarding TEER revealed that HBMECs seeded on the PET membrane—irrespective of the cell seeding number—reach higher TEER values than HBMECs seeded on PTFE membrane inserts. For instance, seven days after seeding 240,000 cells per insert, HBMECs cultivated on PET membranes achieve TEER values of $53 \pm 2\ \Omega\ cm^2$, whereas PTFE-cultivated cells remain at $26 \pm 1\ \Omega\ cm^2$. Based on these findings, PET membrane inserts and a cell seeding number of 240,000 cells per insert were used for all further experiments in order to obtain cellular barriers with appropriate characteristics.

28.3.1.3.2 Influence of cell media supplements on HBMEC layer integrity

The influence of different cell media and media supplements on HBMEC layers was tested in context of their effect on diverse barrier integrity parameters. Figure 28.4 summarizes the results of these investigations. The Nuclear Fast Red staining of histological cross sections prepared from HBMEC layers cultivated under indicated media conditions (see Figure 28.4A) demonstrates that for nonsupplemented RPMI medium, a loose cell cluster is formed. By using astrocyte-conditioned medium (ACM), a more connected but tenuous cell layer is developed, whereas the addition of 10% FBS to ACM seems to strengthen the cellular layer to a more consolidated appearance. However, the cultivation of HBMEC transwell systems with RPMI medium containing 10% FBS induces the formation of a dense, continuous, and uniform cell barrier. Both the cytoskeletal and ZO-1 staining of HBMEC layers cultivated under the very same conditions shown in Figure 28.4B give similar results, whereby cells cultivated with nonsupplemented RPMI appear rounded and sparse in cell-cell contacts without ZO-1 proteins located in the peripheral cell areas. In comparison, ACM-cultivated HBMEC layers seem to be flat and connected intercellularly, which is highlighted in particular by the increased presence of ZO-1 at the cell margins. For both RPMI and ACM, the addition of 10% FBS seems to enhance the three-dimensional assembly of a tight cellular network, whereas the peripheral ZO-1 localization is still present but slightly less intense. By repetitive TEER measurements, the tightness of developing HBMEC layers was quantified over time (see Figure 28.4C). Notably, TEER values strongly increase during the first days after cell seeding and seem to reach a plateau between day

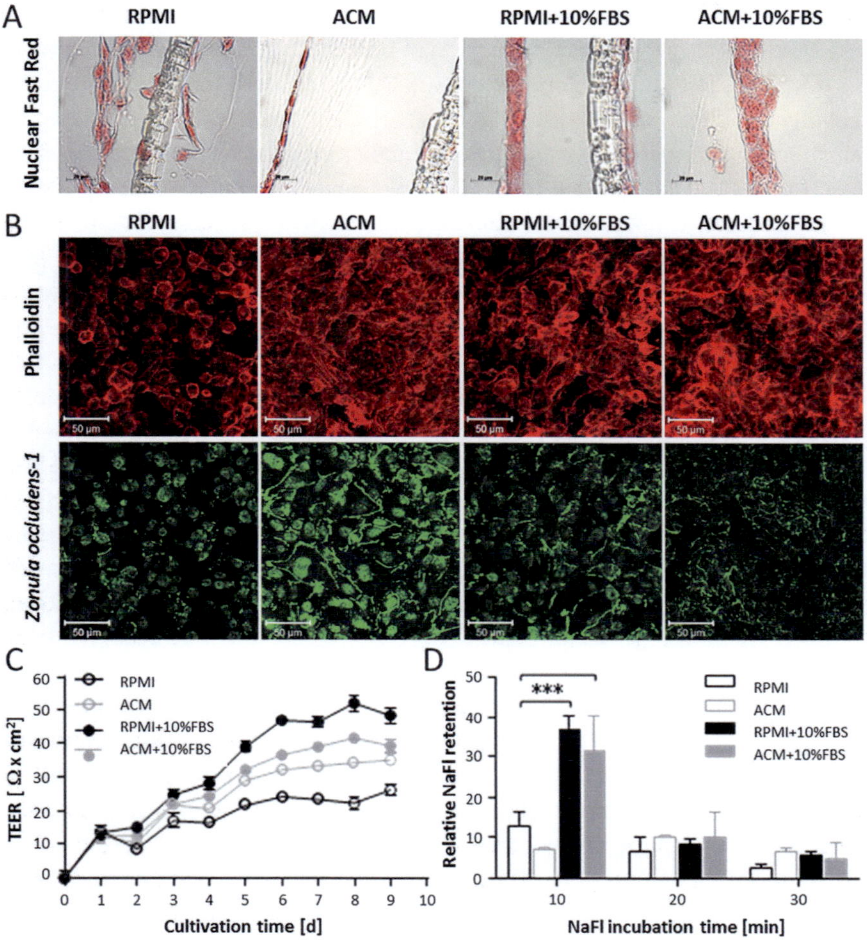

Figure 28.4: Influence of cell culture media and media supplements on barrier integrity parameters. HBMECs were seeded on transwell inserts and cultured for up to nine days. One day after cell seeding, media were replaced by indicated fresh media and cultivated further. (A) Nuclear Fast Red–stained histological cross sections of cell layers cultivated under indicated conditions for five days. (B) Fluorescent staining of filamentous actin (red) and *zonula occludens-1* (green) of cell layers cultivated under indicated conditions for 6 days. (C) Transendothelial electrical resistance (TEER) measurements of HBMEC layers. Shown are means of two replicate transwell inserts with three measurements each and background correction with cell-free membrane inserts. (D) Molecular retention of cell layers to sodium fluorescein (NaFl) upon dye incubation for 10, 30, or 60 min. Shown are means of two replicate transwell inserts cultivated for six days relative to cell-free inserts. Statistical significance among grouped samples was tested by two-way ANOVA followed by Dunnett's multiple comparison, where differences are considered statistically significant for $p < 0.001$ (***). [115]. ACM = astrocyte-conditioned medium; ANOVA = analysis of variance; HBMEC = human brain microvascular endothelial cell; FBS = fetal bovine serum.

five and seven, whereas on day six, the nonsupplemented RPMI medium results in the lowest TEER values of 24.0 ± 0.7 Ω cm^2, plain ACM-treated HBMECs achieve a TEER value of 31.9 ± 0.8 Ω cm^2. The FBS supplementation of ACM and RPMI elevates the prospective TEER values to 36.4 ± 0.6 and 46.9 ± 0.7 Ω cm^2, respectively. The HBMEC layer's retention to the small molecular dye NaFl was tested as another integrity parameter. The results shown in Figure 28.4D indicate that both nonsupplemented RPMI and ACM only allow a low barrier tightness with a 12.8 ± 3.4- and 7.1 ± 0.4-fold increase of the NaFl retention relative to cell-free transwell inserts after 10-min incubation with the molecular dye. Again, the addition of 10% FBS to ACM and RPMI medium showed the most pronounced effects with elevations of the cell layer's molecular retention of 37.0 ± 3.4- and 31.4 ± 9.2-fold, respectively.

Taking into account all the results regarding the analysis of the influence of cultivation conditions on barrier integrity, for all further transwell experiments, HBMEC layers were prepared by using the RPMI medium supplemented with 10% FBS and cultivation for 5–6 days.

Comparing the resulting TEER values of a maximum of 52 ± 2 Ω cm^2 to those in the literature reveals a good agreement with experimental setups using similar human cell models [57,121–123]. Nevertheless, it is to be noted that these human models using immortalized cell lines represent tightness parameters achieving only small fractions of the ones present *in vivo*, which are estimated to be 1000–2000 Ω cm^2 [31, 122]. Although comparable *in vitro* models utilizing primary cells of porcine or bovine origin generate up to 2500 Ω cm^2 [55], the transfer of experimental outcomes to human context is limited. Thus, instead of switching to nonhuman models, a further advancement of human models *in vitro* seems reasonable. As indicated by several studies, the shift from static to dynamic cell culture systems does not only mimic *in vivo* situations present at the BBB more closely, but also beneficially affects phenotypes of BMECs [27, 60, 124, 125]. Moreover, other sophisticated dynamic models utilize hollow fibers, which carry pulsatile flow and provide the framework for scaffolding BMECs in addition to astrocytes on luminal and abluminal surfaces [126].

28.3.1.4 Interaction of SPIONs with the BBB model

After having established an appropriate *in vitro* BBB model based on a HBMEC-seeded transwell system, the setup was used for SPION interaction studies. On the one hand, the particles' effects on barrier integrity were tested by various methods in a time-dependent manner in order to gain an insight into the consequences for the barrier itself. On the other hand, SPION passage through the barrier-forming cells was analyzed and quantified in a next step for evaluating the particles' barrier-penetrating ability.

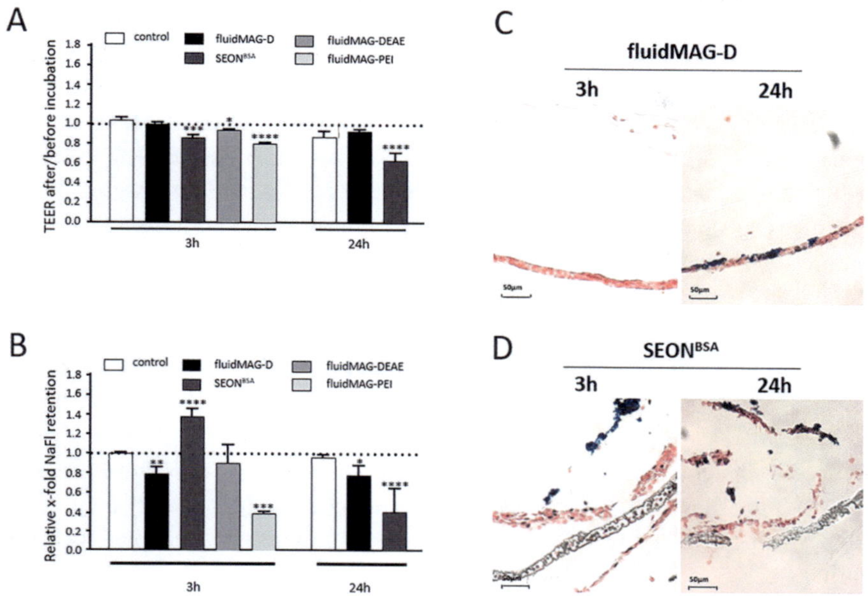

Figure 28.5: SPION-associated effects on barrier integrity parameters of HBMEC layers. HBMECs were seeded on transwell inserts and cultured for five days. Indicated SPIONs were added resulting in a final concentration of 100 μg/cm^2 (corresponding to 200 μg/ml) and incubated for 3 h or 24 h as indicated, whereby the first 30 min were carried out on top of a block magnet. (A) Ratios of transendothelial electrical resistance (TEER) measurements of cell layers before and after SPION incubation. Shown are means of two to four independent experiments with two replicate inserts each. (B) Molecular retention of SPION-incubated cell layers to sodium fluorescein (NaFl). Shown are means of two to four independent experiments with two replicate inserts each. Statistical significance of indicated samples compared to controls was tested by one-way ANOVA for fluidMAG-DEAE and fluidMAG-PEI and two-way ANOVA for fluidMAG-D and SEON^{LA-BSA}. Both cases were followed by Dunnett's multiple comparison, where differences are considered statistically significant for $p < 0.05$ (*), $p < 0.01$ (**), $p < 0.001$ (***), and $p < 0.0001$ (****). (C and D) Nuclear Fast Red– and Prussian blue–stained histological cross sections of cell layers incubated with fluidMAG-D (C) or SEON^{LA-BSA} (D) as stated above. HBMEC = human brain microvascular endothelial cell; ANOVA = analysis of variance; SPION = superparamagnetic iron oxide nanoparticle; DEAE = diethylamine ethyl; PEI = polyethylenimine.

28.3.1.4.1 SPION-associated effects on barrier integrity

Maintaining the physiological integrity of the BBB is an essential premise for keeping the brain's homeostasis. In order to investigate consequences of SPION exposure on the HBMEC-based *in vitro* model, TEER measurements before and after incubation with diverse types of SPIONs as well as molecular permeability assays and microscopic analysis of histological cross sections were performed. Data summarized in Figure 28.5 show that SPIONs differently affect the barrier integrity after particle exposure for 180 min. Both TEER measurements and molecular retentions to NaFl shown in Figure 28.5A and B reveal strong and significant alterations under the influence of cationic fluidMAG-PEI with a decrease in TEER values (relative to initial values) to 0.80 ± 0.03 and NaFl retentions to 0.40 ± 0.02 (relative to control cells). While a slight but significant decrease in TEER values to 0.94 ± 0.01 is observed for fluidMAG-DEAE particles, no influence of these cationic particles is detected by molecular permeability assays based on NaFl. While for exposure of HBMEC layers to fluidMAG-D for 3 and 24 h, TEER values do not show any significant changes, NaFl retentions are slightly but significantly reduced to 0.79 ± 0.07 and 0.78 ± 0.10, respectively. However, in case of incubation with $SEON^{LA\text{-}BSA}$, relative TEER values are significantly reduced to 0.86 ± 0.03, whereas the cell layer's retention toward NaFl elevates the NaFl retention to 1.38 ± 0.08 for 3-h incubations. An incubation of 24 h with these anionic particles is characterized by a strong and significant reduction in both relative TEER values (0.63 ± 0.08) and NaFl retentions (0.41 ± 0.24). Stained histological cross sections of particle-incubated transwell systems shown in Figure 28.5C and D provide a more comprehensive insight into the cell layers' conditions after particle exposure to neutral fluidMAG-D and anionic $SEON^{LA\text{-}BSA}$. Microscopic images imply that after incubation with fluidMAG-D, HBMEC layers keep an intact and continuous appearance for incubation times of up to 24 h. In contrast, cross sections of $SEON^{LA\text{-}BSA}$-incubated cells seem already slightly diminished after 3 h and strongly compromised after 24 h, where the layer's continuity is barely visible. Notably, HBMEC layers are not located directly on the transwell membrane, which might be a result of the multistep sample preparation for this method. Focusing on Prussian blue–stained SPIONs detected within the cross sections, it is striking that fluidMAG-D is hardly detectable in samples obtained after an incubation for 3 h but intensively abundant upon a 24-h incubation. In contrast, cellular layers exposed to $SEON^{LA\text{-}BSA}$ show extensive SPION staining as soon as 3 h after particle addition.

Using the established HBMEC-based model system, SPION-associated effects on barrier integrity are investigated first. Therefore, realistic SPION concentration prevailing at the biological barrier is difficult to estimate [127]. In general, particle dilution within the blood volume and clearance by the RES contribute to low global concentrations. However, targeting strategies—either by surface functionalization or magnetic forces—might bring on strongly elevated local particle amounts [10]. Owing to that, following investigations primarily address consequences of incubations with elevated SPION concentrations. In agreement with cytotoxicity data from two-dimensional cell culture experiments discussed above, neutral fluidMAG-D hardly influences the intactness of transwell HBMEC layers with respect to TEER values and NaFl retention, as well as microscopic analysis of histological cross sections for incubation of up to 24 h. A related study performed by Thomsen et al [121] using starch-coated SPIONs made similar observations. In contrast, the exposure to anionic $SEON^{LA\text{-}BSA}$ reduces TEER values in a time-dependent manner, while NaFl retention is significantly increased relative to controls before its pronounced reduction. Correlating these findings to microscopic investigations provides a reasonable explanation for the observed phenomenon: the rapid and comprehensive accumulation of $SEON^{LA\text{-}BSA}$ on

HBMEC layers during the first 3 h might obstruct the diffusion of NaFl molecules from the upper donor into the lower acceptor compartment, which gives rise to an apparently elevated retention capacity for this molecular dye. Following further SEON$^{LA\text{-}BSA}$ incubation, NaFl permeability is finally increased owing to the barrier-disrupting nature of these anionic SPIONs.

Taking all these factors together, data strongly indicate the possibility that intensively interacting SEON$^{LA\text{-}BSA}$ destabilizes the established cell layers by events such as disturbing cell-cell contacts, which eventually reduces barrier integrity and elevates endothelial permeability. Moreover, it has to be noted that the strongly damaged appearance of histological HBMEC cross sections upon SEON$^{LA\text{-}BSA}$ incubation is probably strengthened by the multistep sample preparation upon barrier destabilization. Do SEON$^{LA\text{-}BSA}$ particles specifically bind distinct cellular structures and surface components or are there general nonspecific interactions with cells that bring on the barrier disruption? Does immunogenicity in cells of human origin in response to the massive exposure to bovine albumin via the induction of permeability-increasing cytokines play a role? Based on the protein alignment of human and bovine albumin (UniProt identifiers: P02768 and P02769), an identity not exceeding 76.34% leaves scope for such a possibility [128]. Additionally, cytokine-inducing effects in human cell cultures have been described before [129]. In order to answer this question, further comprehensive analyses are still necessary.

With reference to cationic fluidMAG-PEI, more pronounced diminishing impacts on HBMEC layers are already detectable after a 3-h incubation, which manifest in the significant reduction of both TEER and NaFl retention and agrees to PEI particles' cytotoxic phenotype. Similarly, in accordance with biocompatibility observations, an exposure to cationic fluidMAG-DEAE entails no relevant detectable effects on barrier integrity. In spite of these interesting aspects, both cationic particle formulations are not suitable for the transwell test system assessing particle passage through the biological barrier. The reason for being that—probably based on excessive particle accumulation and agglutination within the cell culture medium with aggregates of up to 5 µm (data not shown)—neither fluidMAG-PEI nor fluidMAG-DEAE passes cell-free transwell membranes in an appropriate manner. Hence, for subsequent evaluations of SPION passage through the *in vitro* model representing the human BBB, neutral fluidMAG-D and anionic SEON$^{LA\text{-}BSA}$ are utilized.

28.3.1.4.2 Passage of SPIONs through the *in vitro* BBB

Because starch-coated fluidMAG-D minimally affects barrier integrities of HBMEC-based transwell models, the passage of these neutral particles was investigated in detail for incubation times of up to 3 h [115]. To this end, MPS was used for directly detecting and quantifying SPIONs within the distinct compartments of the transwell system with a high sensitivity. Figure 28.6 sums up the experimental data obtained. The particle standard curve presented in Figure 28.6A implies a close correlation of utilized fluidMAG-D and its detection and quantification via this magnetization

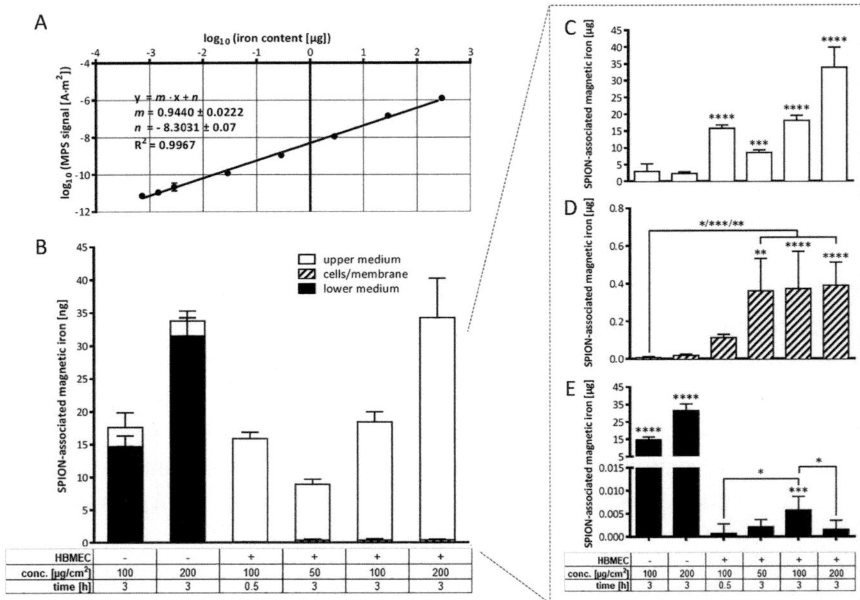

Figure 28.6: SPION distribution of distinct compartments of the *in vitro* blood-brain barrier model analyzed by magnetic particle spectroscopy (MPS). HBMECs were seeded on transwell inserts and cultured for 5 days. FluidMAG-D particles were added, resulting in a final concentration of 50–200 µg/cm² (corresponding to 100–400 µg/ml) and incubated for up to 3 h, whereby the first 30 min were carried out on top of a block magnet. (A) Standard correlation curve of fluidMAG-D diluted in the cell culture medium and measured by MPS. (B) Compartment-specific contents of magnetic iron determined by MPS. (C–E) SPION distribution in the upper compartment (C), cells/membrane fraction (D), and lower compartment (E) (all shown in more detail). Shown are means ± standard deviation of three independent experiments with three replicate inserts each. Statistical significance of samples compared to controls without SPIONs and among each other was tested by one-way ANOVA followed by Tukey's multiple comparison, wherein differences are considered statistically significant for $p < 0.05$ (*), $p < 0.01$ (**), $p < 0.001$ (***), and $p < 0.0001$ (****) [115]. SPION = superparamagnetic iron oxide nanoparticle; HBMEC = human brain microvascular endothelial cell; ANOVA = analysis of variance.

response–based method over multiple orders of magnitude ranging from few nanograms to several hundred micrograms. Thus, the lower detection limit is as small as 1.9 ng of iron for the biological sample. Analyses of compartment-specific contents of SPION-associated magnetic iron shown in Figure 28.6B demonstrate that with an average of 92.2 ± 1.5%, the great majority but not all utilized particles are recovered via this method including sample preparation. Furthermore, data imply that fluidMAG-D is predominantly found within the lower acceptor compartment after a 3-h incubation of cell-free transwell inserts with 100 and 200 µg/cm², as 83.3 ± 9.4% (14.7 ± 1.7 µg) and 93.1 ± 11.3% (31.5 ± 3.8 µg) of detected SPION-associated iron are found here, respectively. In the presence of HBMEC layers on the transwell membrane, most (95.9–99.3%)

of the fluidMAG-D are detected within the upper donor compartment (Figure 28.6C). During the incubation of blank transwell inserts with 100 μg/cm^2 of fluidMAG-D, particle amounts recovered within the cell-free membranes are 7.3 ± 5.4 ng, whereas for cell-studded inserts, the amount of magnetic iron within this compartment is significantly increased to 112.8 ± 17.4 and 371.9 ± 198.4 ng after 0.5 and 3 h, respectively (Figure 28.6D). A detailed analysis of lower acceptor medium of HBMEC-grown inserts incubated with 100 μg/cm^2 indicates a low (0.7 ± 2.1 ng) amount of magnetic iron after 0.5 h, though below the lower detection limit. However, extending the incubation time to 3 h results in the significant increase in magnetic iron to 5.8 ± 3.0 ng within this compartment. Strikingly, neither for the 3-h incubation with 50 μg/cm^2 nor for 200 μg/cm^2 fluidMAG-D, such pronounced increases are detectable (2.2 ± 1.6 and 1.6 ± 2.0 ng).

In order to verify the results of particle passage through the *in vitro* barrier model by MPS-based quantification of magnetic iron, AAS was applied, which allows the quantification of total iron. In addition to the detailed analysis of SPION presence within the barrier-forming HBMEC layer and lower acceptor medium, a second cell type (MCF-7) located on bottom wells mimicking underlying tissues was applied and analyzed for iron. Furthermore, the passage of fluidMAG-D was compared to that of SEON$^{LA\text{-}BSA}$ in a time-dependent manner. The results summarized in Figure 28.7 show the standard curve for fluidMAG-D derived from AAS quantification. While the linear correlation of utilized iron and AAS-detected signals are indicated for these particles in Figure 28.7A, the detection range is limited to a maximum of 3 μg, implicating additional dilution steps for quantifications of elevated iron amounts. The lower detection limit of this method is 2 ng. The compartment-specific iron contents shown in Figure 28.7B indicate again that the highest amounts of fluidMAG-D are found within the lower acceptor compartment and MCF-7 cells if cell-free transwell inserts are incubated for 3 h (20.5 ± 4.7 and 2.8 ± 1.2 μg) or 24 h (15.4 ± 5.8 and 8.0 ± 2.7 μg).

Compared to that, the amount of analogously applied anionic SEON$^{LA\text{-}BSA}$ into cell-free inserts is 5.2 ± 7.4 and 4.3 ± 2.7 μg for 3 h and increases to 9.5 ± 0.2 and 6.2 ± 6.0 μg for 24 h. However, if membrane inserts are covered with HBMEC layers, most detected SPIONs are present within the barrier-forming HBMEC compartment. Thus, an incubation with fluidMAG-D increases iron accumulation to 3.1 ± 0.3 μg after 24 h, whereas in case of SEON$^{LA\text{-}BSA}$, iron amounts within this compartment are significantly elevated to 10.0 ± 7.2 μg after 3 h and 10.9 ± 4.3 μg after 24 h. Focusing on iron contents detected within the lower acceptor medium, data imply that a 3-h incubation with fluidMAG-D results in a significant increase of iron (25.2 ± 12.7 ng) compared to control inserts without SPION treatments, while for SEON$^{LA\text{-}BSA}$, an elevated tendency is observed but without statistical evidence. Extending incubation times for both particle types to 24 h significantly elevates iron contents of acceptor medium compartments especially for anionic SEON$^{LA\text{-}BSA}$. Similarly, enhanced iron levels in well bottom-seeded lower MCF-7 cells are detectable with statistical evidence after 24 h for both fluidMAG-D (8.5 ± 3.6 ng) and SEON$^{LA\text{-}BSA}$ (11.2 ± 6.6 ng); for an incubation time of 3 h, a statistically significant increase is observable for SEON$^{LA\text{-}BSA}$-treated conditions only. In addition to the MPS-

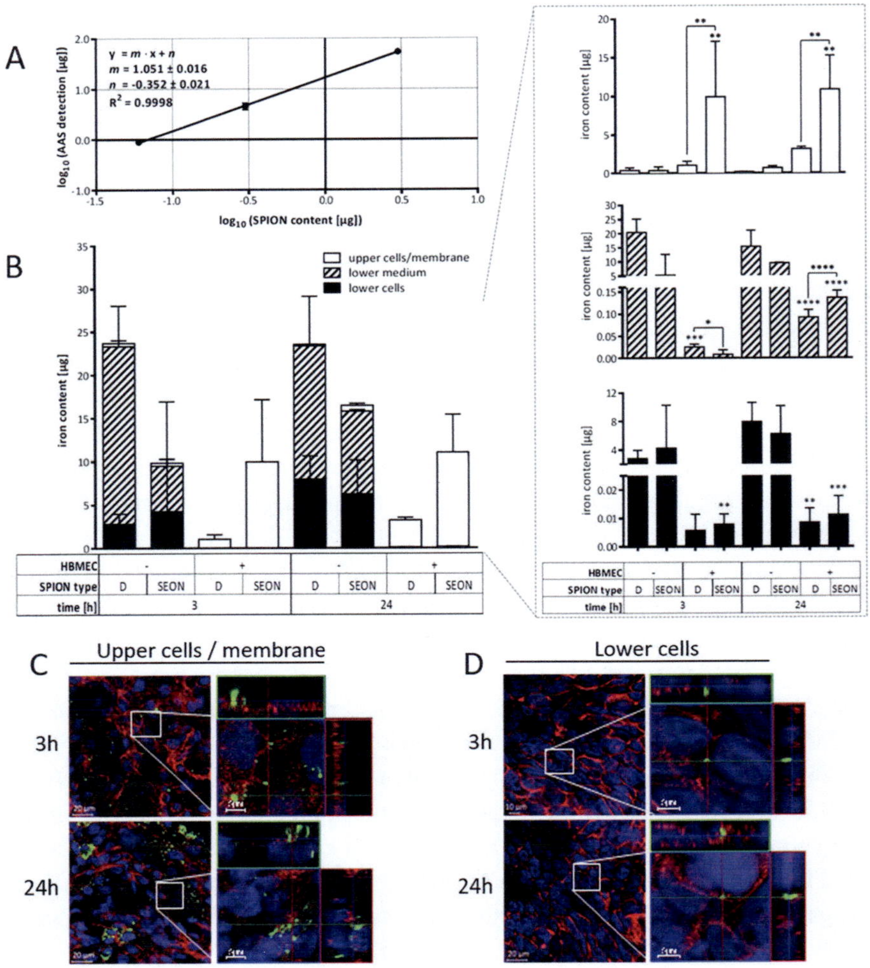

Figure 28.7: SPION distribution of distinct compartments of the *in vitro* blood-brain barrier model analyzed by atomic absorption spectroscopy (AAS). HBMECs were seeded on transwell inserts and cultured for five days. Indicated SPIONs were added, resulting in a final concentration of 100 μg/cm^2 (corresponding to 200 μg/ml) and incubated for 3 h or 24 h, whereby the first 30 min were carried out on top of a block magnet. (A) Standard correlation curve of fluidMAG-D diluted in the cell culture medium measured by AAS. (B) Compartment-specific contents of total iron determined by AAS. Shown are means ± standard deviation of three independent experiments with three replicate inserts each. (C and D) Fluorescent staining of barrier-forming HBMECs (upper cells/membrane, C) and MCF-7 cells seeded into well bottoms (lower cells, D) after incubation with nano-screenMAG-G/D (green). Nuclei and filamentous actin are stained with DAPI (blue) and Alexa Fluor® 633 Phalloidin (red), respectively. Samples were analyzed by confocal laser scanning microscopy. Statistical significance of grouped samples compared to controls without SPIONs and among each other was tested by two-way ANOVA followed by Dunnett's multiple comparison, where differences are considered statistically significant for $p < 0.05$ (*), $p < 0.01$ (**), $p < 0.001$ (***), and $p < 0.0001$ (****). ANOVA = analysis of variance; HBMEC = human brain microvascular endothelial cell; SPION = superparamagnetic iron oxide nanoparticle.

and AAS-based quantification of iron within the distinct compartments of the transwell system, microscopic analyses of both apical barrier–forming HBMECs and underlying lower MCF-7 cells after the incubation with fluorescently labeled starch particles were performed. On the one hand, images presented in Figure 28.7C demonstrate the presence of green particle signals in HBMEC layers after a 3-h incubation, which is further enhanced after 24 h. It is remarkable that nano-screenMAG-G/D colocalizes with intracellular F-actin staining for both incubation times. On the other hand, analogous SPION signals colocalizing with F-actin staining of lower MCF-7 cells beyond the barrier can microscopically be detected very rarely after both 3 and 24 h (Figure 28.7D).

Taken together, the passage of SPIONs through the BBB-representing *in vitro* model is demonstrated by diverse methods comprising MPS and AAS for highly sensitive quantification and cLSM for the optical detection and verification of particle internalization.

SPIONs' superparamagnetic properties are exploited in order to highly sensitively quantify particle amounts present within the distinct compartments of incubated HBMEC transwell systems, which in turn gives insights into particle passage at this biological barrier. By applying MPS, it is shown that this technique based on the nonlinear magnetic susceptibility response accurately and highly sensitively detects fluidMAG-D particles over multiple orders of magnitude. Utilizing the MPS-based direct detection of SPIONs within such barrier interaction studies provides profound insights into the passages' dynamics and underlying mechanisms [115]. Actually, the comparison of MPS, ultraviolet/visible spectroscopy, AAS, and atomic emission spectroscopy by Friedrich et al. [130] already identified MPS as the most sensitive technique.

Focusing on the here presented experimental outcome obtained from compartment-specific quantification of fluidMAG-D in the transwell model, data indicate that the presence of an HBMEC layer on the transwell membrane almost entirely prevents the translocation of fluidMAG-D from the upper into the lower compartment, which confirms the pronounced barrier tightness. Strikingly, the increase in particle concentrations from 50 to 100 μg/cm^2 and 200 μg/cm^2 (equal to 100–400 μg/ml) does not enhance SPION accumulation within the cellular fraction. This fact might suggest that HBMECs cannot bind and internalize more of the starch-coated particles during the first 3 h of incubation, potentially owing to a fully stretched cellular endocytosis apparatus. As discussed above, the slow cellular accumulation kinetics as well as the active clathrin-dependent uptake of fluidMAG-D into HBMECs corresponds well to this assumption. Moreover, another substantial factor affecting the bioavailability of SPIONs to cells must be seen in particle agglomeration. As high SPION concentrations diminish the nanoparticles' stability, decreased amounts of free particles contribute to less particle-cell interaction and the generation of false-negative or false-positive results, which have been discussed in the literature previously [131]. The fact that within the lower acceptor compartment a small but statistically significant amount of SPIONs is detectable after the 3-h incubation with 100 μg/cm^2 of fluidMAG-D only provides further proofs for the active particle transport across the BBB model via transcytotic processes. The absence of SPIONs within this compartment after 30 min implies the time-dependent character.

The overall comparison of data obtained from MPS and AAS shows elevated iron amounts for the latter. This is accounted for by the fact that MPS is based on the specific detection of intact superparamagnetic particles, while during AAS, degraded particles or particle fragments also contribute to the determined signal. Another relevant issue can be seen in potentially altered magnetic behavior of SPIONs during the incubation progress. Thus, pronounced particle agglomeration modifies the magnetic susceptibility response. As such, a shift is detected by ratios of the 5th and 3rd spectrum amplitude (i.e., A_5/A_3 ratio) recorded during MPS measurements, inaccuracies can be corrected by switching to references of convenient A_5/A_3 ratios [111]. An alternative method for the highly sensitive *in vitro* quantification of SPIONs, regardless of any particle labeling, might be seen in a technique introduced by Gunn et al [132]. There, authors utilized the standardized methodology of proton nuclear magnetic resonance to quantitatively detect SPIONs within biological samples for five orders of magnitude and iron concentrations of 10 ng/ml. However, the particles' disintegration as well as cell debris and cell digest hampers the unconditional application of this methodology [132]. Taken together, the data presented here demonstrate that the combination of the *in vitro* model with both SPION quantification methods (i.e., MPS and AAS) and integrity-evaluating assays allows the detailed investigation of the SPION passage through this biological barrier and the assessment of the resulting consequences on the barrier itself.

In order to develop the concept of a cellular barrier as a dynamic matrix further, we moved to a more complex mammalian cellular barrier, the BPB. The *in vitro* BPB consists of 2 cell types instead of one in the BBB model.

28.3.2 Establishment and optimization of an *in vitro* BPB model

As a first step toward the investigation of nanoparticle processing in the BPB, a transwell-based *in vitro* model of this barrier was established and optimized with regard to the cell culture medium, seeding density, and incubation time. Furthermore, the additional effect of pericytes on the barrier function was investigated comprehensively [133].

28.3.2.1 Medium adjustment for coculture of BeWo cells and pericytes

In order to create a transwell-based coculture BPB model, used cell types, BeWo, as well as pericytes needed to be cultivated in a uniform cell culture medium. For this purpose, both cells were cultured in DMEM + 10% FBS, instead of PGM in case of pericytes. After confirming that substitution of the cultivation medium did not lead to any morphologic alterations of the pericytes, alterations in the characteristic cell surface marker expression of pericytes evoked by the change in the cultivation medium were investigated by flow cytometry. According to the purchaser, pericytes should be CD31 and CD34 negative as well as CD105 and CD146 positive. Pericytes grown in PGM were also analyzed in passage four for comparison. Cells in all investigated passages and both media were negative for the markers CD31 and CD34. CD146 expression was stable over the observation time period, while CD105 expression depended on the cell passage. Its expression for cells in DMEM increased in early passages (<p6) and decreased again after p12. Furthermore, CD105 was only sparsely expressed in cells grown in PGM. Overall pericyte growth was dramatically reduced in higher passages. As a consequence, all further experiments were performed with pericytes cultured in DMEM and with a passage number between 6 and 10.

28.3.2.2 Effect of coculture on barrier formation and integrity

In the following section, the coculture model is compared to the monoculture with regard to the morphology of the cell layers, electrical resistances, and permeability. Different seeding densities of BeWo cells were used, and the model was investigated on day 3–5 post seeding (PS) in order to find optimal conditions for further studies using this *in vitro* model.

The seeding density directly influenced the thickness of the BeWo cell layers, especially for the monoculture. While the cell layer was found to be loosely packed for the monoculture, it was tighter and thinner for the coculture regarding all cell densities. The pericytes on the basolateral side of the membranes grew as a very thin layer. Afterward, the resulting barrier of the coculture in comparison to the monoculture was further investigated using TEER measurements and NaFl permeability measurements and by visualization of cell-cell contacts by cLSM. The transepithelial electrical resistances evoked by the cell barriers of monoculture and coculture models were measured for different BeWo densities during days 3–5 PS. The cocultivation of cell layers showed higher TEER values for all seeding densities than the corresponding monoculture samples. Pericytes cultivated alone did not produce any TEER values exceeding 15 Ω cm^2 but were able to increase values when cultivated together with BeWo cells. For the monoculture, the resistance values increased from 18 to 46 Ω cm^2 (day 4) with rising amounts of BeWo cells, while for the coculture, 200,000 cells showed the highest TEER measurements, with about 100 Ω cm^2 at day 4 PS. Measurement of the passage of the passive permeability marker NaFl across the cell barrier revealed a higher restraint of the marker for the coculture in comparison to the monoculture for days 3–5 PS. The barrier tightness for the permeability marker of the coculture model increased steadily throughout the observation time from 60- to 140-fold restraint, while the values for the monoculture did not exceed a 35-fold restraint.

In order to investigate the formation of cell-cell contacts in the *in vitro* barrier, both ZO-1 as a marker for tight junctions and β-catenin for adherens junctions were visualized by immunofluorescent staining and subsequent cLSM (Figure 28.8). Apically

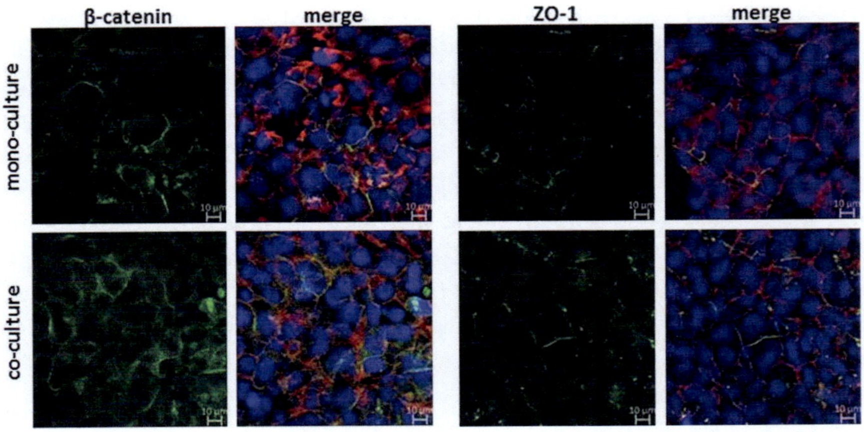

Figure 28.8: Comparison of expression of cell-cell contact markers β-catenin and ZO-1 for monoculture and coculture models using immunofluorescent staining. Transwell systems were prepared using 200,000 BeWo cells in monoculture and with 350,000 pericytes for coculture, and cultivated until day 4 PS; fixed and permeabilized cells were stained with rabbit anti–ZO-1 or β-catenin primary antibody followed by AlexaFluor® 488–labeled goat anti-rabbit secondary antibody (green), Hoechst 33258 (blue) and AlexaFluor® 633 Phalloidin (red) to visualize cell-cell contacts, cell nuclei, and cell cytoskeleton, respectively; fluorescence signals were acquired by cLSM; scale bar = 10 μm; PS = post seeding [133]. ZO = *zonula occludens*; cLSM = confocal laser scanning microscopy.

located BeWo cells were shown to form dense layers under both monoculture and coculture conditions, which was visualized using Hoechst (blue) and AlexaFluor® 633 Phalloidin (red) for counterstaining of cells. The cell-contact markers ZO-1 and β-catenin (green) were shown to be mostly peripherally localized in monoculture as well as in coculture, but were both expressed higher in the coculture. The adherens junction marker seemed to be expressed higher in the BeWo cells than the one for tight junctions.

Taken together, these data confirmed the development of an *in vitro* cell barrier using both monoculture and coculture. Since the coculture barrier was shown to produce higher TEER resistances, higher NaFl restraint and more cell-cell contacts than the monoculture, further experiments concerning the passage of SPIONs were conducted using the coculture transwell BPB model with 200,000 BeWo cells, which produced the highest electrical resistance values. Furthermore, based on the results for TEER and permeability measurements, day 4 PS and day 5 PS were chosen for investigating the passage of nanoparticles through the *in vitro* barrier.

28.3.2.3 Pericyte-associated effects on the BPB model

In order to further investigate the effect pericytes evoked on BeWo cells during cocultivation, different conditions were compared concerning TEER and cell layer

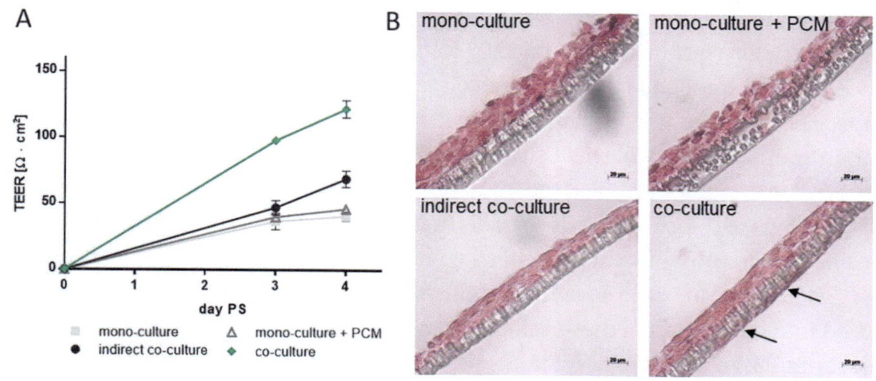

Figure 28.9: Investigation of the pericyte-associated effects in the BPB model using TEER and histologic slices. Transwell systems were prepared using 200,000 BeWo cells and (if applicable) 350,000 pericytes and cultivated for 3–4 days; conditions: coculture, monoculture, indirect coculture (pericytes on the bottom of the well), and monoculture + PCM. (A) TEER values were measured for day 3 and 4 PS; mean TEER values (Ω cm^2) ± SD ($n = 3$) are shown. (B) Histologic slices of transwell membranes were prepared for day 4 PS; transwell membranes were embedded in paraffin and sectioned, and cells were stained with Nuclear Fast Red; arrows mark pericytes grown on the basolateral side of the membrane; scale bars represent 20 µm; PS = post seeding [133]. BPB = blood-placenta barrier; TEER = transendothelial electrical resistance; PCM = pericyte-conditioned medium; SD = standard deviation.

morphology via histologic cross sections. Besides the monoculture and coculture conditions, an indirect coculture, where pericytes were seeded on the bottom of the wells in the same density as for the direct coculture, and a monoculture supplemented with PCM were used.

TEER values of all conditions increased during the observation period (day 3–4 PS) (Figure 28.9A). While resistances measured for the coculture (>100 Ω cm^2) were representatively higher than the ones measured for the monoculture, as already shown in the previous chapter, the values for the indirect coculture as well as for the monoculture + PCM were located at about 50 Ω cm^2, hence in the same range as the model with only BeWo cells. Transverse sections of the cell-bearing transwell membranes at day 4 PS also revealed looser cell layers for the monoculture + PCM, which is comparable to the monoculture. The cell layer of the indirect coculture shows a denser packing, similar to the findings for the coculture (Figure 28.9B).

These results emphasize the positive effect of pericytes on the barrier tightness in the *in vitro* model, which seems to be dependent on the direct cocultivation, where BeWo cells are localized on the apical membrane side and pericytes are localized on the basolateral one.

The commercial clone of BeWo was used to create a transwell *in vitro* BPB model with higher seeding densities of about 6×10^5 cells/cm^2. Investigations concerning TEER measurements and permeability revealed the formation of a barrier using BeWo cells (monoculture) after 4–5 days, where measured TEER values exceeded 50 Ω cm^2 and NaFl restraints were located at about 25-fold in comparison to cell-free inserts. The presence of tight junctions in the barrier cell layer was also verified by confocal microscopy. These values already indicate the formation of a tighter barrier using this clone than another study achieved using the b30 clone [65]. Histologic cross sections of the transwell membranes revealed the formation of a BeWo multilayer as soon as day 3 PS. Observations of the growth of BeWo cells in cell culture flasks showed that owing to the lack of growth inhibition after cell-cell contact, these cells tend to grow in multilayered patches before even forming confluent layers. Since these findings can be translated into the growth of these cells on transwell membranes, obtaining an intact monolayer of BeWo cells on membranes seemed impossible. Even researchers using the b30 clone suggested that the formation and maintenance of confluent monolayers is challenging owing to the lack of growth inhibition upon contact, and therefore, they used a BeWo multilayer for transport studies as a reproducible model since the integrity of the barrier is a critical point for *in vitro* transport studies across the BPB [70]. Furthermore, it is noted that the BPB in the human organism is composed of multiple layers in early stages of pregnancy, a syncytiotrophoblast layer, a confluent cytotrophoblast layer beneath, as well as the fetal blood vessels. Considering all the above mentioned aspects, in the present study, transport studies across the BPB were performed using a BeWo multilayer on transwell inserts rather than taking the change of subconfluent monolayers.

In order to reveal the effect of pericytes in the coculture model consisting of BeWo and pericytes, comparison between this setting and the monocultivation of BeWo cells in transwell inserts was performed, wherein the coculture model was shown to form a tighter barrier than the monoculture. Addition of pericytes to the BPB model led to higher TEER values and NaFl restraints as well as an increased expression of cell-cell contact markers. In comparison to other studies using the b30 clone, the coculture BPB model in this study created higher TEER values, which show the formation of a tighter cellular barrier compared to the b30 clone barrier [65, 66].

28.3.2.4 Effect of SPIONs on BPB cells

The effects of three differently charged SPIONs, neutral starch-coated particles, cationic PEI-coated particles, and anionic CMX-coated particles, onto the cells of the *in vitro* BPB model concerning cellular viability were evaluated in the following section using different methods. These experiments were performed in cell culture plates for each cell type independently prior to coculture transwell experiments. Concentration-dependent as well as incubation time-dependent effects of the particles were investigated in more detail.

28.3.2.4.1 Influence of SPIONs on cellular viability

Effects of SPIONs on the cellular viability of BeWo and pericytes were investigated using three different methods. The PrestoBlue™ cell viability assay and the SYTOX® red dead cell staining were used to reveal any cytotoxic effects for short-term nanoparticle exposure of 3 h, while the RTCA using the xCELLigence system showed changes in cellular behavior for up to 96 h. For both cell types, incubation with negatively charged CMX-coated particles did not lead to any inhibition of cellular viability for concentrations up to 100 µg/cm^2. Interestingly, SYTOX® staining even showed an increase in the viable population up to 110% for pericytes incubated with 50 µg/cm^2. For the treatment with neutral starch-coated particles, the results achieved from the PrestoBlue™ assay indicated an increasing cytotoxicity with increasing concentrations for both cell types since the viable cell population decreased to 60% for the highest concentration. Meanwhile, the viable population of analogously incubated cells analyzed by SYTOX staining did not decrease. Focusing on the PrestoBlue results, PEI-coated particles seemed to have no effect on the BeWo cells in any concentration used since the viability was always higher than 80%, while a significant decrease in the SYTOX-negative population to less than 50% could be seen for the flow cytometric analysis. The cellular viability of pericytes was shown to be decreased in a concentration-dependent manner, which could be seen for both experimental settings. Interestingly, the cellular viability upon incubation with PEI-coated particles obtained from flow cytometric experiments with values of less than 40% was significantly lower than the one from PrestoBlue™ assay (70–80%), which was already seen for the BeWo cells.

In order to investigate the long-term effects of the differently charged SPIONs on the cellular viability of BeWo cells and pericytes, cells incubated with fluidMAG particles were monitored for 96 h via impedance measurements using RTCA. In Figure 28.10, the relative cell indices (%) are depicted compared to the control sample (100%) for both cell types. Neither fluidMAG-D nor fluidMAG-CMX particles inhibited the cellular viability of BeWo cells for all concentrations investigated, wherein all values measured were above 85%. Interestingly, even increasing cell indices could be shown after about 24 h of incubation for the neutral starch-coated particles, indicating a stimulating effect on BeWo cells up to 120%. However, cellular viability of cells incubated with 100 µg/cm^2 of fluidMAG-PEI particles was strongly inhibited (50%), while lower concentrations of these particles even stimulated the cells up to more than 120% shortly after addition of nanoparticles in the same manner as starch-coated particles. In contrast, all particles showed a concentration-dependent cytotoxic effect on pericytes, where the viability of cells was decreased to less than 75% for all particles and all concentrations. Here again, fluidMAG-PEI particles influenced cellular viability to the strongest extent, with the highest concentration decreasing the viability by more than 80%.

In summary, it could be shown that cationic PEI-coated particles exhibited short- and long-term cytotoxic effects on both cell types, where pericytes showed the stronger reactions. Pericyte viability was furthermore influenced by all SPIONs in a

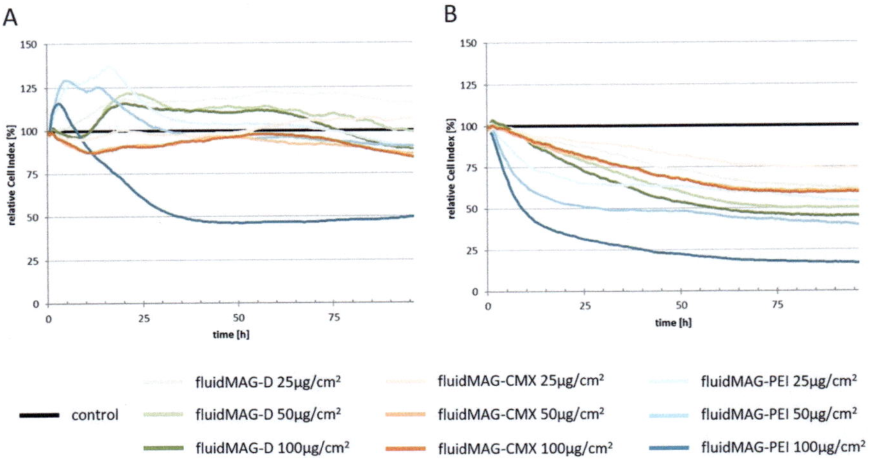

Figure 28.10: Viability of BeWo cells (A) and pericytes (B) after long-term incubation with SPIONs (96 h) measured by RTCA: 35,000/30,000 cells (BeWo/pericytes) were seeded in RTCA 16-well E plates; cell indices were recorded via impedance measurements at different time points over a period of 96 h; fluidMAG-D/PEI/CMX particles were added in different concentrations at time point 0; relative cell indices (%) obtained by normalization to time point 0 and dividing measured values by respective data from diluent-treated control cells; values represent means of two replicates. Shown are the results of one representative experiment of three independent experiments. RTCA = real-time cell analysis; SPION = superparamagnetic iron oxide nanoparticle; PEI = polyethylenimine.

concentration-dependent manner after long-term exposure, while BeWo cells were not affected.

The *in vitro* BPB model, with the direct cocultivation of BeWo cells and pericytes, was afterward used to study the behavior of three differently coated and thereby differently charged SPIONs. Neutral starch-coated particles consisted of a starch coating, cationic PEI-coated particles had polyethylenimine, and anionic CMX-coated particles had a carboxymethyldextran shell. For all particle types, intracellular localization could be shown by confocal microscopy owing to colocalization of particles with F-actin. As demonstrated with the BBB model, clathrin-dependent endocytosis is the most prominent mechanism by which nanomaterials enter cells [95]. However, cationic particles, especially PEI-coated ones, were also shown to create nanoscale holes in the negatively charged cell membrane, which leads to another entry port for these nanoparticles into the cells [134]. Whether these are the internalization mechanisms for the SPIONs used in this study needs to be investigated in further experiments.

In different experimental settings, the cytotoxicity of the particles for short- and long-term incubations was investigated. Cationic particles showed the strongest short- and long-term cytotoxic effects on both cell types. These results are in concordance with those of previous studies that investigated the impact of nanoparticle charge onto the cytotoxicity [77,135]. In general, pericyte viability was more influenced by incubations with all three particles than BeWo cells, which might be accounted for by the larger cellular surface of pericytes resulting in an increased interaction surface.

The finding of a higher interaction of PEI-coated particles in comparison to the other particles can be explained by the strong interaction of cationic particles with the negatively charged cell membrane [81]. This interaction may also lead to a perforation of the membrane and subsequently to the formation of nanoscale holes [95]. Besides a higher incorporation rate into the cells, these small holes in the cell membrane also explain the higher cytotoxic effects of PEI-coated particles. Another factor influencing the toxicity of these particles is the "proton sponge effect," wherein cationic particles induce organelle rupture via osmotic swelling after uptake into cells by endocytic pathways [86]. In addition to short-term cytotoxicity screens, the long-term cytotoxicity of the three SPIONs also was investigated by RTCA, which showed pronounced long-term toxic effects of PEI-coated particles on BeWo cells for high concentrations and on pericytes for all concentrations used. Furthermore, increased relative cell indices were measured for BeWo cells incubated with starch-coated particles in all concentrations as well as for lower concentrations of PEI-coated ones after addition of NPs compared to diluent-treated samples. With this impedance-based method, the net cell adhesion on the culture plates is measured, which is influenced by the cell morphology, the cell count, and the movement of cells [136]. The subsequent rise in the cell index after NP addition could therefore be caused by different factors influencing this value. Besides the induction of cell proliferation, an increase in the cell size or an increased adhesion of the cells also could be accounted for the increased values. The detailed mechanism by which incubation with nanoparticles stimulates BeWo cells and thereby increases the measured cell indices by RTCA could not be identified in the present study and should therefore be further investigated.

28.3.2.5 Interactions of the SPIONs with the BPB model

First, the behavior and distribution of the SPIONs in the BPB model as well as the barrier integrity and morphology were investigated using histologic cross sections from transwell membranes of the coculture model after 3 and 24 h of particle exposure (Figure 28.11A). The nanoparticles that interacted with the barrier cells were visualized by Prussian blue staining. For all conditions, no disruption of the barrier integrity and no changes in cellular morphology were visible in comparison to untreated models. Regarding the rate of nanoparticle interaction, PEI-coated particles were shown to interact most intensively with the apically located BeWo cell layer of the barrier, while fluidMAG-CMX interaction was the weakest. For starch-coated particles, the interaction strongly increased for an incubation time of 24 h in comparison to the shorter exposure time, being consistent with observations obtained from interaction studies of SPIONs with BeWo cells alone. With this method, no nanoparticles could be visualized in the pericyte cell layer on the basolateral side of the transwell membrane. The integrity of the *in vitro* transwell barrier after SPION incubation for 3 h or 24 h was further

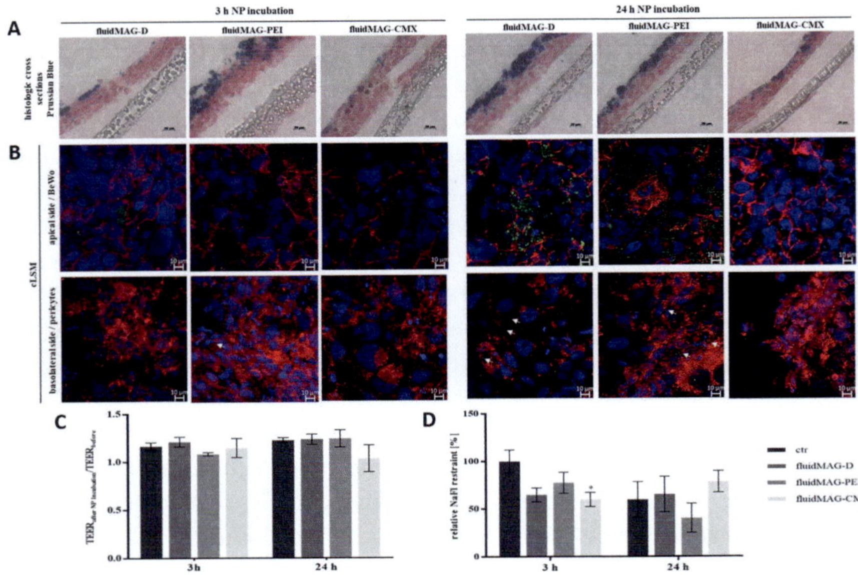

Figure 28.11: Analysis of the barrier integrity and morphology of the transwell coculture BPB model after exposure to different SPIONs for 3 h or 24 h. For the coculture model, 1.1×10^6 cells/cm² of pericytes were seeded onto the basolateral side of 24-well membrane inserts, and 6.1×10^5 cells/cm² of BeWo cells were seeded on the apical side of the insert membrane after 24 h. On day four post seeding (PS), barrier models were exposed to 100 µg/cm² (200 µg/ml) of D/PEI/CMX-coated SPIONs for 3 h or 24 h. (A) After SPION exposure, histologic cross sections of transwell models were prepared and stained with Nuclear Fast Red and Prussian blue. Scale bars represent 10 µm. (B) For the analysis by confocal laser scanning microscopy (cLSM), samples incubated with fluorescently labeled SPIONs (green) were fixed and stained with Hoechst 33258 (blue) and Alexa Fluor® 633 phalloidin (red). White arrows mark SPION aggregates in the pericyte cell layer. Scale bars represent 10 µm. (C) Transepithelial electrical resistance (TEER) values measured in triplicate per insert before and after SPION exposure were compared for each condition. Shown are the mean values of the quotient of measured TEER values before/after SPION incubation ± standard deviation of three independent experiments. (D) The passage of the permeability marker sodium fluorescein (NaFl) through the barrier after SPION incubation was measured in duplicate for each condition, and the calculated permeability coefficients were normalized to blank membranes. Shown are mean values of the *x*-fold NaFl retention ± standard deviation of three independent experiments. The significance of the results compared to control measurements without SPIONs was tested using two-way ANOVA followed by Tukey's multiple comparison test. Statistically significant differences are depicted as $^*p < 0.05$ [133]. SPION = superparamagnetic iron oxide nanoparticle; PEI = polyethylenimine; ANOVA = analysis of variance.

quantified using the transepithelial electrical resistance and the permeability measurement for the marker NaFl (Figure 28.11C). Comparison of TEER values after and before SPION incubation revealed no alteration of electrical resistance of the coculture model for all particles and both incubation times. Regarding molecular permeability, NaFl restraint was slightly lower for barriers incubated with particles for 3 h than for diluent-treated control barriers. After 24 h, only values for the incubation with cationic PEI-coated particles showed lower values than the control (Figure 28.11D). Since no nanoparticles could be visualized in the pericytes by histologic cross sections, for further investigation of the nanoparticle distribution in the two cell layers of the coculture BPB model, both sides of the transwell membranes incubated with fluorescently labeled particles (nsMAG/G-) for 3 and 24 h were investigated by cLSM (Figure 28.11B). Fluorescence signals of multiple z-layers (1-µm thick) were acquired, and one representative layer showing particles that penetrated the cell barrier is depicted for each condition for a 24-h incubation. As already described before, BeWo cells formed dense cell layers on the apical side of the membrane, while pericytes were found on the basolateral side. All three particle types could be visualized in the apical cell layer. For 24 h, only few NP spots (highlighted by white arrows) could be seen for the model incubated with CMX-coated particles, while D- and PEI-coated particles were visible all over the apical side of the cell barrier. For these two conditions, some nanoparticles also were visible in the basolateral pericyte cell layer, which are also highlighted by white arrows. For incubation with nsMAG/G-CMX, no particles could be seen on the basolateral side of the membrane. After 3 h of NP incubation, fewer particles were visible in the apical BeWo cell layer, while only PEI-coated particles were detected in the pericyte cell layer.

28.3.2.5.1 Passage of SPIONs through the *in vitro* BPB

In scope of studying the passage of SPIONs in the BPB model, the total amount of magnetic iron in each of the four compartments after incubation with nanoparticles for 3 and 24 h was quantified via MPS. The results of three independent experiments are depicted in Figure 28.12. PEI-coated particles could be shown to interact with the cells to the highest extent, indicated by detected SPION contents as high as 13.4 µg in the BeWo cell layer after 24 h. Consistent with these findings, less particles were detected in the upper donor compartment. Since CMX-coated particle interaction with BeWo cells was the lowest with 2.3 µg of detected SPIONs after 24 h, the highest amounts of iron (11.6 µg) could be found in the donor compartment for this condition. For starch-coated particles, an enhanced interaction could be shown by an increased detection of SPIONs in the BeWo layer from 0.5 µg after 3 h to 3.9 µg after incubation for 24 h. In the pericyte cell layer on the basolateral side of the transwell membrane, only small amounts of magnetic iron up to 7 ng could be detected. For an incubation time of 3 h, only CMX-coated particles could be detected in the pericyte cell layer with an amount of 6.6 ng since this was the only sample exceeding the lower detection limit of 1.8 ng. For 24 h of incubation, all particles could be reliably detected in the basolateral pericyte cell layer, where incubation with PEI-coated particles showed the lowest SPION content of 2.6 ng and the highest values were obtained for incubation with starch-coated particles with an iron amount of 4.5 ng. Nanoparticle amounts detected in the lower acceptor compartment were located in a range from 0 to 1 ng of iron. Since the lower detection limit for this compartment was calculated to be 0.9 ng, only starch-coated

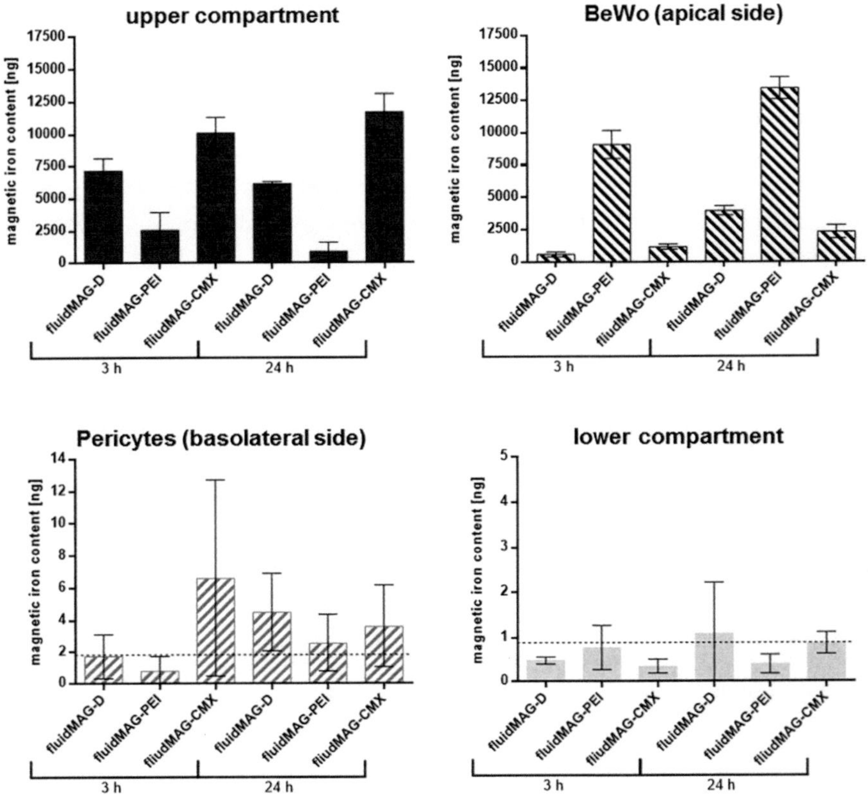

Figure 28.12: Investigation of the passage of SPIONs through the transwell BPB model after exposure for 3 h or 24 h via MPS. Transwell coculture systems were prepared using 200,000 BeWo cells/350,000 pericytes; on day 4 PS, barrier models were exposed to 100 µg/cm^2 of fluidMAG-D/PEI/CMX particles for 3 and 24 h; complete medium of lower and upper compartments was collected, and cells from the apical and basolateral side of the membrane were harvested by trypsination; samples were measured using a MP spectrometer; Nanoparticle spectra were normalized to corresponding nanoparticle solutions of known iron content. Shown are the iron contents of all four compartments (ng) $\pm$ SD from 2 to 3 independent experiments. For pericytes (basolateral side) and the lower compartment, the lower detection limit is depicted as a dotted line at 1.83 and 0.88 ng, respectively. PS = post seeding [133]. SPION = superparamagnetic iron oxide nanoparticle; BPB = blood-placenta barrier; MPS = magnetic particle spectroscopy; PS = post seeding; PEI = polyethylenimine; SD = standard deviation.

particles could be reliably measured in the acceptor compartment with an amount of 1.1 ng after 24 h.

In order to investigate the ability of the three SPIONs to cross cell-free transwell membranes, similar experiments as described with the *in vitro* model were also performed with blank inserts. While for D- and CMX-coated particles the majority of nanoparticles were able to cross the blank membrane, most of the PEI-coated particles were detected in the upper donor compartment, indicating a hindered passage across the transwell membrane.

Taken together, it could be shown that incubation with all three investigated SPIONs had no significant harmful effect on the barrier integrity of the *in vitro* BPB model. In different experiments, PEI-coated particle interaction with the apical BeWo cell layer could be shown to be the highest among the investigated particles, while CMX-coated ones showed the lowest interaction capacity. For starch-coated particles, a significant increase in the amount of interacting particles could be seen after 24 h. Nanoparticles could also be detected in the basolateral pericyte cell layer by cLSM and MPS. Furthermore, using MPS measurements, the distribution of nanoparticles in the four compartments of the transwell BPB model could be quantified.

In order to quantify the amount of nanoparticles in the upper and lower compartment of the transwell system as well as in both cell layers, analysis via MPS was performed. The detected amounts of particles in the upper compartment and the apical BeWo cell layer confirm the previous microscopic studies, where cationic particles showed the highest interaction capacity with the cells, while neutral particles interacted intermediately, and anionic particles showed the lowest interaction. Furthermore, in the basolateral pericyte cell layer, some particles were detected too. In contrast to the BBB model, almost all measured magnetic iron contents in the lower compartment were located below the detection limit, only neutral D-coated particles could be detected after a 24-h incubation with a magnetic iron amount of 1.1 ng. These results indicate that either no particles were able to cross the tight *in vitro* barrier or the magnetic properties might be altered owing to the transcellular passage of SPIONs and are therefore not detected by this method.

28.3.2.6 Passage of SPIONs through the *in vitro* BPB under flow conditions

In order to go a step forward, we transferred the *in vitro* blood-placenta model from the transwell setup to a microfluidic biochip as this allows adding shear stress to the cells, an important factor for placental function [137]. Shear stress rates below 2 dyn/cm^2 (0.2 Pa) are observed in the early stages of pregnancy, which is protective of maintenance of the trophoblast layer with its villous structures [138]. We established the biochip and investigated the interaction of different nanoparticles under fluidic conditions with a special focus on their passaging abilities.

After formation of the BeWo cell layer on the apical side of the membrane and the pericyte cell layer on the basolateral side of the membrane, the cell layers were examined by histologic slices. A thin barrier consisting of BeWo cells and pericytes was visible, which significantly reduced the permeability of the membrane for NaFl in comparison to the blank (46.2 vs. 345 nM NaFl). Incubation of the biochip with or without laminar flow affects neither the expression of the tight junction protein ZO-1 nor the overall cell morphology (Figure 28.13). The applied SPIONs do not disrupt the cell barrier on the biochip.

The putative passage of the SPIONs through the barrier was monitored by MPS after a single round of circulation without a magnetic field gradient and compared to

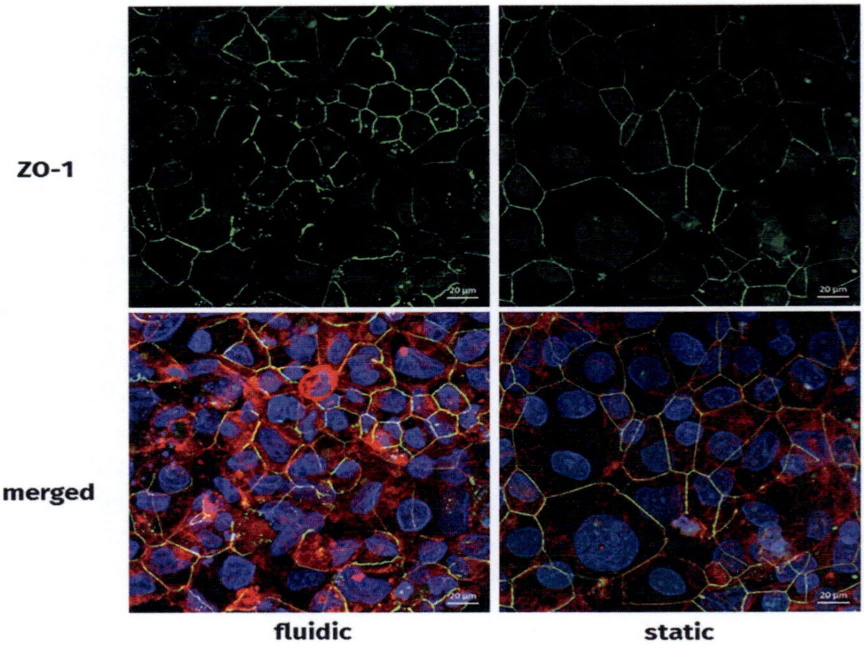

Figure 28.13: Effects of incubation with and without laminar flow on cell-cell contacts and cell morphology. The BeWo cells were stained with Alexa Fluor® 633 Phalloidin (red), Hoechst 33258 (blue), and *zonula occludens-1* (ZO-1) antibody (green), followed by analysis by confocal laser scanning microscopy (cLSM) for detection of F-actin, cell nuclei, and ZO-1. Both incubation regimes (with and without laminar flow) allowed similar cell morphology and localization of ZO-1.

the static situation. In that scenario, the nanoparticle-containing medium remains in the biochip for 3 h. The distribution within the upper donor channel, the lower acceptor channel, and the membrane was measured by MPS and was SPION specific (Figure 28.14). More than 90% of sodium citrate–coated nanoparticles remain in the upper donor channel independent of the presence of a cell layer or FBS. A low amount of sodium citrate nanoparticles was detectable in the barrier cells (2.5% ± 0.8% SD; without FBS 6.1% ± 1.8% SD). SEON$^{LA\text{-}HSA}$ nanoparticles displayed a different distribution. Almost all SEON$^{LA\text{-}HSA}$ nanoparticles remain in the upper donor channel. In the lower acceptor channel, less than 0.2% of SEON$^{LA\text{-}HSA}$ nanoparticles could be measured in the fluidic and the static setting each. This shows that the *in vitro* BPB under fluidic conditions is comparable to the static conditions with regard to the transportation behavior of the added nanoparticles.

BeWo cells cocultured with pericytes form a stable barrier inside the microfluidic biochip, confirming the observations from the static transwell system. Incubation of the biochip with or without laminar flow affects neither the expression of the tight junction protein ZO-1 nor the overall cell morphology. Histologic slices (not shown) reveal the formation of a thin barrier by the BeWo cells, which significantly reduced the

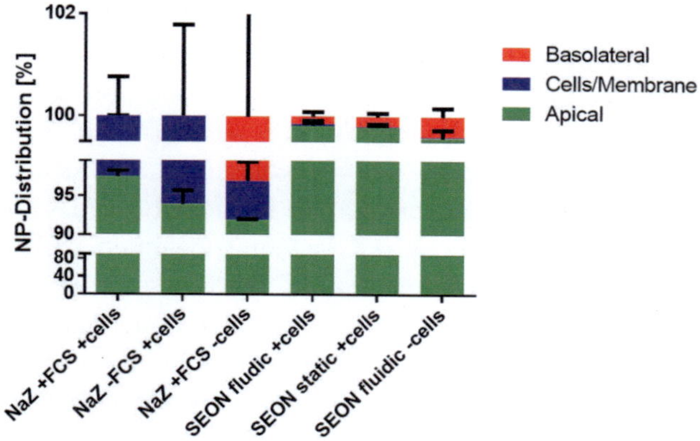

Figure 28.14: Quantification and passage of different nanoparticles in the BPB model by magnetic particle spectroscopy (MPS). After 7 days of incubation, the reservoir of the biochip was emptied. For the single round of circulation, the upper channel was filled with 800 µl of the nanoparticle solutions (100 µg/ml in the standard medium), respectively. The nanoparticle solution circulated one time through the upper channel of the biochip followed by 800 µl of fresh medium without nanoparticles in an incubator at 37 °C, 5% CO_2 in a humidified atmosphere. The flow through was collected in one sample (apical: upper channel). Subsequently, the medium of the lower channel (basolateral: lower channel) was collected, and the membrane carrying the BeWo cells and pericytes (cells/membrane) was excised. All samples were reduced to a final volume of ~50 µl and analyzed by MPS. All relative concentrations were calculated from the obtained absolute values. NaZ = sodium citrate nanoparticles; SEON = $SEON^{LA\text{-}HSA}$ nanoparticles. BPB = blood-placenta barrier.

permeability of the membrane for fluorescein sodium. Different nanoparticles neither reduce the vitality of BeWo cells in 2D cell culture (data not shown) nor disrupt the cell barrier on the biochip. MPS is a suitable technique to determine the nanoparticle concentrations for different compartments of the chip. The cellular accumulation of SPIONs as well as the nanoparticle passage depends on core-shell particle characteristics.

28.4 Conclusions

The here presented data establish experimental *in vitro* tools for the detailed evaluation of SPION interactions and passage through biological barriers including the assessment of SPION-associated consequences for barrier integrity. For the first time, MPS is adopted in this experimental biological setting analyzing SPION passage through biological barriers in order to achieve direct and highly sensitive particle detection within the distinct compartments. Hence, a profound insight into passage's dynamics is provided. Furthermore, the presented results impressively demonstrate that free fluidMAG-D traverse the barrier system in a time-dependent manner most likely based on clathrin-induced transcytotic processes. In addition, barrier integrity remains unaffected on particle passage, which is explained by the comparatively slow and gentle cellular accumulation and uptake. In contrast, data imply that nanoparticles with cationic surface charge as well as the BSA-coated nanoparticles intensively assemble within the barrier-forming cells and affect the BBB model in a barrier-disrupting manner. These observations are made for nonfunctionalized SPIONs and can be carried out for each individual SPION formulation in order to achieve a comprehensive risk assessment. For applications not targeting the brain or the placenta, particle penetration into and accumulation off-target upon repeated administration is strictly undesirable. Though, according to purposes when transition of SPIONs into, e.g., the brain is required, additional functionalization such as by ligand binding may be utilized for the enhanced particle delivery across the BBB and need to be evaluated as well. The here presented test systems provide an appropriate platform and allow an insight into the underlying mechanisms, thus contributing to a comprehensive understanding of particle interactions with the BBB and the BPB. In consequence, gained knowledge aids in the tailored design of SPIONs according to the particular application.

The presented investigations ignore the impact of the biomolecule corona and its pivotal role on biological effects. These aspects are investigated and discussed in detail in the study by Dutz et al. [139]. The role of the biomolecule corona should be considered during SPION functionalization as it may present opsonins or shield functional groups on the particle surface.

Utilizing and extending the here established systems, the inclusion of flow conditions by using a microfluidic biochip might mimic the *in vivo* situation even closer. Nevertheless, it has to be noted that the experimental setup has some limitations too. The results indicate that especially when testing the passage of free SPIONs across the barriers under inflammatory conditions, the extension from the monoculture system to a coculture or triculture system with astrocytes or/and pericytes is consequent. Even if such elevated models are more difficult in handling and more interference prone, they even closer represent the complexity of biological tissues and incorporate the eclectic interplay of different cell types during inflammatory states.

The detailed analysis and magnetic characterization of SPIONs after barrier penetration is of vital importance as this also influences tissue distribution, magnetic response during SPION applications such as imaging or therapeutic heating, and elimination processes. Hence, the presented *in vitro* barrier models in line with further verifications according to Barnes et al. [140] must be utilized for such clarifications in the future. Resulting knowledge can help to improve arrangement of magnets in order to achieve appropriate magnetic fields, which optimize SPION application efficacy. Pedram et al. [141] emphasized that the application of a magnetic field between 0.6 and 2.5 T and a minimal gradient and/or amplitude are crucial for not damaging the BBB. Furthermore, alternative approaches in magnet-assisted transport of SPIONs across barriers and into deeper tissues might solve problems in the delivery of therapeutic substances. A remarkable example can be seen in the novel method developed by

Shapiro et al [142], who use a two-magnet system in order to push magnetic nanomedicines into diseased tissue. Taken together, SPION-based therapies bear promising and versatile potentials to solve major biomedical problems.

Acknowledgments: We highly appreciate the continuous support of the DFG in the frame of the high priority program 1681. The success of this program is closely connected with Prof. Stefan Odenbach. Without his enthusiasm and outstanding engagement—at first in science but also in organizing affairs—the efforts within the SPP1681 would not be imaginable. We thank all colleagues for their support and the inspiring discussions during collaborations and meetings. Funded by the DFG, High Priority Program 1681: AL552/5-3; CL202/3-3; WI4230/1-3; DU1293/7-3. We further acknowledge the support of the Core Facility "Metrology of ultra-low magnetic fields" at Physikalisch-Technische Bundesanstalt, which receives funding from the Deutsche Forschungsgemeinschaft-DFG (funding code DFG KO 5321/3-1 and TR 408/11-1).
Author contribution: All the authors have accepted responsibility for the entire content of this submitted manuscript and approved submission.
Research funding: None declared.
Conflict of interest statement: The authors declare no conflicts of interest regarding this article.

References

1. Pelaz B, Alexiou C, Alvarez-Puebla RA, Alves F, Andrews AM, Ashraf S, et al. Diverse applications of nano-medicine. ACS Nano 2017; 11: 2313.
2. Onoue S, Yamada S, Chan HK. Nanodrugs: pharmacokinetics and safety. Int J Nanomed 2014; 9: 1025.
3. Matsumura Y, Maeda H. A new concept for macromolecular therapeutics in cancer chemotherapy: mechanism of tumoritropic accumulation of proteins and the antitumor agent smancs. Cancer Res 1986; 46: 6387.
4. Bazak R, Houri M, El Achy S, Kamel S, Refaat T. Cancer active targeting by nanoparticles: a comprehensive review of literature. J Cancer Res Clin Oncol 2015; 141: 769.
5. Noorlander CW, Kooi MW, Oomen AG, Park MV, Vandebriel RJ, Geertsma RE. Horizon scan of nanomedicinal products. Nanomedicine 2015; 10: 1599.
6. Bean CP, Livingston JD. Superparamagnetism. J Appl Phys 1959; 30: S120.
7. Caruntu D, Caruntu G, O'Connor CJ. Magnetic properties of variable-sized Fe_3O_4 nanoparticles synthesized from non-aqueous homogeneous solutions of polyols. J Phys D Appl Phys 2007; 40: 5801.
8. Qiang Y, Antony J, Sharma A, Nutting J, Sikes D, Meyer D. Iron/iron oxide core-shell nanoclusters for biomedical applications. J Nanoparticle Res 2006; 8: 489.
9. Blasiak B, van Veggel FCJM, Tomanek B. Applications of nanoparticles for MRI cancer diagnosis and therapy. J Nanomaterials 2013:148578. https://doi.org/10.1155/2013/148578.
10. Singh N, Jenkins GJ, Asadi R, Doak SH. Potential toxicity of superparamagnetic iron oxide nanoparticles (SPION). Nano Rev 2010; 1: 5358.

11. Kautz L, Nemeth E. Molecular liaisons between erythropoiesis and iron metabolism. Blood 2014; 124: 479.
12. Worwood M, Bain BJ, Bates I. Chapter 7 – Iron deficiency anaemia and iron overload A_2 – Lewis, S. Mitchell. Philadelphia: Churchill Livingstone; 2006, 131–60.
13. Santhosh PB, Ulrih NP. Multifunctional superparamagnetic iron oxide nanoparticles: promising tools in cancer theranostics. Cancer Lett 2013; 336: 8.
14. Wahajuddin AS. Superparamagnetic iron oxide nanoparticles: magnetic nanoplatforms as drug carriers. Int J Nano 2012; 7: 3445.
15. Yu M, Huang S, Yu KJ, Clyne AM. Dextran and polymer polyethylene glycol (PEG) coating reduce both 5 and 30 nm iron oxide nanoparticle cytotoxicity in 2D and 3D cell culture. Int J Mol Sci 2012; 13: 5554.
16. Alwi R, Telenkov S, Mandelis A, Leshuk T, Gu F, Oladepo S, et al. Silica-coated super paramagnetic iron oxide nanoparticles (SPION) as biocompatible contrast agent in biomedical photoacoustics. Biomed Optics Express 2012; 3: 2500.
17. Laurent S, Dutz S, Häfeli UO, Mahmoudi M. Magnetic fluid hyperthermia: focus on superparamagnetic iron oxide nanoparticles. Adv Colloid Interface Sci 2011; 166: 8.
18. Shevtov M, Multhoff G. Recent developments of magnetic nanoparticles for theranostics of brain tumor. Curr Drug Metab 2016; 17: 737.
19. MagForce Nanotechnologies AG. MagForce Nanotechnologies AG receives European regulatory approval for its NanoCancer therapy. http://www.magforce.de/en/presse-investoren/news-events/detail/article/magforce-nanotechnologies-ag-erhaelt-europaeische-zulassung-fuer-die-nano-krebsR-therapie.html [Accessed August 2017].
20. Su H, Wang Y, Gu Y, Bowman L, Zhao J, Ding M. Potential applications and human biosafety of nanomaterials used in nanomedicine. J Appl Toxicol 2018; 38: 3.
21. Angelakeris M. Magnetic nanoparticles: a multifunctional vehicle for modern theranostics. Biochim Biophys Acta (BBA) Gen Subj 2017; 1861: 1642.
22. Hayashi K, Nakamura M, Sakamoto W, Yogo T, Miki H, Ozaki S, et al. Superparamagnetic nanoparticle clusters for cancer theranostics combining magnetic resonance imaging and hyperthermia treatment. Theranostics 2013; 3: 366.
23. Brazel CS, Huang X. The cost of optimal drug delivery: reducing and preventing the burst effect in matrix systems. ACS Symp Ser 2014; 879: 267.
24. Mahmoudi M, Simchi A, Milani AS, Stroeve P. Cell toxicity of superparamagnetic iron oxide nanoparticles. J Colloid Interface Sci 2009; 336: 510.
25. Rolfe DF, Brown GC. Cellular energy utilization and molecular origin of standard metabolic rate in mammals. Physiol Rev 1997; 77: 731.
26. Hawkins BT, Davis TP. The blood–brain barrier/neurovascular unit in health and Disease. Pharmacol Rev 2005:57173. https://doi.org/10.1124/pr.57.2.4.
27. Bicker J, Alves G, Fortuna A, Falcão A. Blood–brain barrier models and their relevance for a successful development of CNS drug delivery systems: a review. Eur J Pharm Biopharm 2014; 87: 409.
28. Jain KK. Nanobiotechnology-based strategies for crossing the blood–brain barrier. Nanomedicine 2012; 7: 1225.
29. Srinivasan B, Kolli AR, Esch MB, Abaci HE, Shuler ML, Hickman JJ. TEER measurement techniques for in vitro barrier model systems. J Lab Autom 2015; 20: 107.
30. Butt AM, Jones HC, Abbott NJ. Electrical resistance across the blood–brain barrier in anaesthetized rats: a developmental study. J Physiol 1990; 429: 47.
31. Abbott NJ, Patabendige AAK, Dolman DEM, Yusof SR, Begley DJ. Structure and function of the blood–brain barrier. Neurobiol Dis 2010; 37: 13.

32. Hori S, Ohtsuki S, Hosoya KI, Nakashima E, Terasaki T. A pericyte-derived angiopoietin-1 multimeric complex induces occludin gene expression in brain capillary endothelial cells through Tie-2 activation in vitro. J Neurochem 2004; 89: 503.
33. Farkas E, Luiten PG. Cerebral microvascular pathology in aging and Alzheimer's disease. Prog Neurobiol 2001; 64: 575.
34. Takeshita Y, Ransohoff RM. Inflammatory cell trafficking across the blood–brain barrier (BBB): chemokine regulation and in vitro models. Immunol Rev 2012; 248: 228.
35. Abbott NJ, Ronnback L, Hansson E. Astrocyte-endothelial interactions at the blood–brain barrier. Nat Rev Neurosci 2006; 7: 41.
36. Fernandes C, Soni U, Patravale V. Nano-interventions for neurodegenerative disorders. Pharmacol Res 2010; 62: 166.
37. Doolittle ND, Miner ME, Hall WA, Siegal T, Hanson EJ, Osztie E, et al. Safety and efficacy of a multicenter study using intraarterial chemotherapy in conjunction with osmotic opening of the blood–brain barrier for the treatment of patients with malignant brain tumors. Cancer 2000; 88: 637.
38. Hersh DS, Wadajkar AS, Roberts N, Perez JG, Connolly NP, Frenkel V, et al. Evolving Drug Delivery strategies to overcome the blood brain barrier. Curr Pharm Des 2016; 22: 1177.
39. Hynynen K, McDannold N, Vykhodtseva N, Jolesz FA. Noninvasive MR imaging–guided focal opening of the blood–brain barrier in rabbits. Radiology 2001; 220: 640.
40. Simuni T, Hurtig H. Parkinson's disease: diagnosis and clinical management, chapter levadopa: a pharmacologic miracle four decades later. New York, NY, USA: Demos Medical Publishing; 2008.
41. Cheng Y, Morshed RA, Auffinger B, Tobias AL, Lesniak MS. Multifunctional nanoparticles for brain tumor imaging and therapy. Adv Drug Deliv Rev 2014; 66: 42.
42. Schnyder A, Huwyler J. Drug transport to brain with targeted liposomes. NeuroRX 2005; 2: 99.
43. Smith MW, Gumbleton M. Endocytosis at the blood–brain barrier: from basic understanding to drug delivery strategies. J Drug Target 2006; 14: 191.
44. Tian XH, Lin XN, Wie F, Feng W, Huang ZC, Wang P. Enhanced brain targeting of temozolomide in polysorbate-80 coated polybutylcyanoacrylate nanoparticles. Int J Nanomed 2011; 6: 445.
45. Myren M, Mose T, Mathiesen L, Knudsen LE. The human placenta—an alternative for studying foetal exposure. Toxicol In Vitro 2007; 21: 1332.
46. Daw JR, Hanley GE, Greyson DL, Morgan SG. Prescription drug use during pregnancy in developed countries: a systematic review. Pharmacoepidemiol Drug Saf 2011; 20: 895.
47. Buerki-Thurnherr T, von Mandach U, Wick P. Knocking at the door of the unborn child: engineered nanoparticles at the human placental barrier. Swiss Med Wkly 2012; 142: w13559.
48. Correia Carreira S, Walker L, Paul K, Saunders M. The toxicity, transport and uptake of nanoparticles in the in vitro BeWo b30 placental cell barrier model used within NanoTEST. Nanotoxicology 2015; 9: 66.
49. Kim JH, Scialli AR. Thalidomide: the tragedy of birth defects and the effective treatment of disease. Toxicol Sci 2011; 122: 1.
50. Muoth C, Aengenheister L, Kucki M, Wick P, Buerki-Thurnherr T. Nanoparticle transport across the placental barrier: pushing the field forward!. Nanomedicine 2016; 11: 941.
51. Enders AC, Blankenship TN. Comparative placental structure. Adv Drug Deliv Rev 1999; 38: 3.
52. Vähäkangas K, Myllynen P. Drug transporters in the human blood–placental barrier. Br J Pharmacol 2009; 158: 665.
53. Young AM, Allen CE, Audus KL. Efflux transporters of the human placenta. Adv Drug Deliv Rev 2003; 55: 125.
54. Huppertz B. The anatomy of the normal placenta. J Clin Pathol 2008; 61: 1296.

55. Helms HC, Abbott NJ, Burek M, Cecchelli R, Couraud PO, Deli MA, et al. In vitro models of the blood–brain barrier: an overview of commonly used brain endothelial cell culture models and guidelines for their use. J Cerebral Blood Flow Metab 2016; 36: 862.
56. Pardridge WM, Yang J, Eisenberg J, Tourtellotte WW. Isolation of intact capillaries and capillary plasma membranes from frozen human brain. J Neurosci Res 1987; 18: 352.
57. Deli MA, Ábrahám CS, Kataoka Y, Niwa M. Permeability studies on in vitro blood–brain barrier models: physiology, pathology, and pharmacology. Cell Mol Neurobiol 2005; 25: 59.
58. Wilhelm I, Krizbai IA. Vitro models of the blood–brain barrier for the study of drug delivery to the brain. Mol Pharm 2014; 11: 1949.
59. Eigenmann DE, Xue G, Kim KS, Moses AV, Hamburger M, Oufir M. Comparative study of four immortalized human brain capillary endothelial cell lines, hCMEC/D3, hBMEC, TY10, and BB19, and optimization of culture conditions, for an in vitro blood–brain barrier model for drug permeability studies. Fluids Barr CNS 2013; 10: 33.
60. Siddharthan V, Kim YV, Liu S, Kim KS. Human astrocytes/astrocyte-conditioned medium and shear stress enhance the barrier properties of human brain microvascular endothelial cells. Brain Res 2007; 1147: 39.
61. Wick P, Grafmueller S, Petri-Fink A, Rothen-Rutishauser B. Advanced human in vitro models to assess metal oxide nanoparticle–cell interactions. MRS Bull 2014; 39: 984.
62. Panigel M. Past, present, and future of placental perfusion experiments. Contrib Gynecol Obstet 1985; 13: 132.
63. Hutson JR, Garcia-Bournissen F, Davis A, Koren G. The human placental perfusion model: a systematic review and development of a model to predict in vivo transfer of therapeutic drugs. Clin Pharmacol Ther 2011; 90: 67.
64. Braakhuis HM, Kloet SK, Kezic S, Kuper F, Park MV, Bellmann S, van der Zande M, Le Gac S, Krystek P, Peters RJ, Rietjens IM, Bouwmeester H. Progress and future of in vitro models to study translocation of nanoparticles. Arch Toxicol 2015; 89: 1469.
65. Cartwright L, Poulsen MS, Nielsen HM, Pojana G, Knudsen LE, Saunders M, et al. In vitro placental model optimization for nanoparticle transport studies. Int J Nanomed 2012; 7: 497.
66. Ali H, Kalashnikova I, White MA, Sherman M, Rytting E. Preparation, characterization, and transport of dexamethasone-loaded polymeric nanoparticles across a human placental in vitro model. Int J Pharm 2013; 454: 149.
67. Poulsen MS, Mose T, Maroun LL, Mathiesen L, Knudsen LE, Rytting E. Kinetics of silica nanoparticles in the human placenta. Nanotoxicology 2015; 9: 79.
68. Kloet SK, Walczak AP, Louisse J, van den Berg HH, Bouwmeester H, Tromp P, et al. Translocation of positively and negatively charged polystyrene nanoparticles in an in vitro placental model. Toxicol In Vitro 2015; 29: 1701.
69. Li H, van Ravenswaay B, Rietjens IM, Louisse J. Assessment of an in vitro transport model using BeWo b30 cells to predict placental transfer of compounds. Arch Toxicol 2013; 87: 1661.
70. Heaton SJ, Eady JJ, Parker ML, Gotts KL, Dainty JR, Fairweather-Tait SJ, et al. The use of BeWo cells as an in vitro model for placental iron transport. Am J Physiol Cell Physiol 2008; 295: C1445.
71. van der Ende A, Du Maine A, Simmons CF, Schwartz AL, Strous GJ. Iron metabolism in BeWo chorion carcinoma cells. Transferrin-mediated uptake and release of iron. J Biol Chem 1987; 262: 8910.
72. Mahmoudi M, Laurent S, Shokrgozar MA, Hosseinkhani M. Toxicity evaluations of superparamagnetic iron oxide nanoparticles: cell "vision" versus physicochemical properties of nanoparticles. ACS Nano 2011; 5: 7263.
73. Fischer D, Li Y, Ahlemeyer B, Krieglstein J, Kissel T. In vitro cytotoxicity testing of polycations: influence of polymer structure on cell viability and hemolysis. Biomaterials 2003; 24: 1121.
74. Marano F, Hussain S, Rodrigues-Lima F, Baeza-Squiban A, Boland S. Nanoparticles: molecular targets and cell signaling. Arch Toxicol 2011; 85: 733.

75. Hong S, Leroueil PR, Janus EK, Peters JL, Kober MM, Islam MT, et al. Interaction of polycationic polymers with supported lipid bilayers and cells: nanoscale hole formation and enhanced membrane permeability. Bioconj Chem 2006; 17: 728.
76. Verma A, Stellacci F. Effect of surface properties on nanoparticle– cell interactions. Small 2010; 6: 12.
77. Villanueva A, Cañete M, Roca AG, Calero M, Veintemillas-Verdaguer S, Serna CJ. The influence of surface functionalization on the enhanced internalization of magnetic nanoparticles in cancer cells. Nanotechnology 2009; 20: 115103.
78. Ghinea N, Simionescu N. Anionized and cationized hemeundecapeptides as probes for cell surface charge and permeability studies: differentiated labeling of endothelial plasmalemmal vesicles. J Cell Biol 1985; 100: 606.
79. Sajid M, Ilyas M, Basheer C, Tariq M, Daud M, Baig N, Shehzad F. Impact of nanoparticles on human and environment: review of toxicity factors, exposures, control strategies, and future prospects. Env Sci Poll Res 2015; 22: 4122.
80. Treuel L, Jiang X, Nienhaus GU. New views on cellular uptake and trafficking of manufactured nanoparticles. J R Soc Interface 2013; 10: 20120939.
81. Hoet PH, Brüske-Hohlfeld I, Salata OV. Nanoparticles –known and unknown health risks. J Nanobiotech 2004; 2: 12.
82. Shi J, Kantoff PW, Wooster R, Farokhzad OC. Cancer nanomedicine: progress, challenges and opportunities. Nat Rev Cancer 2017; 17: 20.
83. Kiessling F, Mertens ME, Grimm J, Lammers T. Nanoparticles for imaging: top or flop?. Radiology 2014; 273: 10.
84. Canton I, Battaglia G. Endocytosis at the nanoscale. Chem Soc Rev 2012; 41: 2718.
85. Aderem A, Underhill DM. Mechanisms of phagocytosis in macrophages. Annu Rev Immunol 1999; 17: 593.
86. Kettiger H, Schipanski A, Wick P, Huwyler J. Engineered nanomaterial uptake and tissue distribution: from cell to organism. Int J Nanomed 2013; 8: 3255.
87. Alberts B, Johnson A, Lewis J, Raff M, Roberts K, Walter P. Molecular biology of the cell, 5th ed., Garland Sciences; 2007: 789–99.Chapter 13.
88. Schauer R. Sialic acids and their role as biological masks. Trend Biochem Sci 1985; 10: 357.
89. Yeung T, Gilbert GE, Shi J, Silvius J, Kapus A, Grinstein S. Membrane phosphatidylserine regulates surface charge and protein localization. Science 2008; 319: 210.
90. Conner SD, Schmid SL. Regulated portals of entry into the cell. Nature 2003; 422: 37.
91. Bohmer N, Jordan A. Caveolin-1 and CDC42 mediated endocytosis of silica-coated iron oxide nanoparticles in HeLa cells. Beilstein J Nanotechnol 2015; 6: 167.
92. Hillaireau H, Couvreur P. Nanocarriers' entry into the cell: relevance to drug delivery. Cell Mol Life Sci 2009; 66: 2873.
93. Iversen TG, Skotland T, Sandvig K. Endocytosis and intracellular transport of nanoparticles: present knowledge and need for future studies. Nano Today 2011; 6: 176.
94. Jiang X, Musyanovych A, Rocker C, Landfester K, Mailänder V, Nienhaus GU. Specific effects of surface carboxyl groups on anionic polystyrene particles in their interactions with mesenchymal stem cells. Nanoscale 2011; 3: 2028.
95. Sahay G, Alakhova DY, Kabanov AV. Endocytosis of nanomedicines. J Contr Rel 2010; 145: 182.
96. Sandin P, Fitzpatrick LW, Simpson JC, Dawson KA. High-speed imaging of rab family small GTPases reveals rare events in nanoparticle trafficking in living cells. ACS Nano 2012; 6: 1513.
97. Guo S, Huang L. Nanoparticles escaping RES and endosome: challenges for siRNA Delivery for cancer therapy. J Nanomater 2011; 11: 1.
98. Oh N, Park JH. Endocytosis and exocytosis of nanoparticles in mammalian cells. Int J Nanomed 2014; 9: 51.

99. Win KY, Feng SS. Effects of particle size and surface coating on cellular uptake of polymeric nanoparticles for oral delivery of anticancer drugs. Biomaterials 2005; 26: 2713.
100. Stins MF, Badger J, Kim KS. Bacterial invasion and transcytosis in transfected human brain microvascular endothelial cells. Microb Pathog 2001; 30: 19.
101. Pattillo RA, Gey GO. The establishment of a cell line of human hormone-synthesizing trophoblastic cells in vitro. Cancer Res 1968; 28: 1231.
102. Schiera G, Bono E, Raffa MP, Gallo A, Pitarresi GL, Di Liegro I, Savettieri G. Synergistic effects of neurons and astrocytes on the differentiation of brain capillary endothelial cells in culture. J Cell Mol Med 2003; 7: 165.
103. Xue Q, Liu Y, Qi H, Ma Q, Xu L, Chen W, et al. A novel brain neurovascular unit model with neurons, astrocytes and microvascular endothelial cells of rat. Int J Biol Sci 2013; 9: 174.
104. Raasch M, Rennert K, Jahn T, Gärtner C, Schönfelder G, Huber O. An integrative microfluidically supported in vitro model of an endothelial barrier combined with cortical spheroids simulates effects of neuroinflammation in neocortex development. Biomicrofluidics 2016; 10: 044102.
105. Kaya M, Ahishali B. *Assessment of permeability in barrier type of endothelium in brain using tracers: evans blue, sodium fluorescein, and horseradish peroxidase*. Totowa, NJ: Humana Press; 2011, 369–82.
106. Audus KL, Borchardt RT. Characterization of an in vitro blood–brain barrier model system for studying drug transport and metabolism. Pharm Res 1986; 3: 81.
107. Schlorf T, Meincke M, Kossel E, Glüer CC, Jansen O, Mentlein R. Biological properties of iron oxide nanoparticles for cellular and molecular magnetic resonance imaging. Int J Mol Sci 2011; 12: 12.
108. Otsu N. A threshold selection method from gray-level histograms. IEEE Trans Syst Man Cybern 1979; 9: 62.
109. Leach L, Lammiman MJ, Babawale MO, Hobson SA, Bromilou B, Lovat S, Simmonds MJ. Molecular organization of tight and adherens junctions in the human placental vascular tree. Placenta 2000; 21: 547.
110. Gottardi CJ, Arpin M, Fanning AS, Louvard D. The junction-associated protein, zonula occludens-1, localizes to the nucleus before the maturation and during the remodeling of cell–cell contacts. Proc Natl Acad Sci USA 1996; 93: 10779.
111. Poller WC, Löwa N, Wiekhorst F, Taupitz M, Wagner S, Möller K, et al. Magnetic particle spectroscopy reveals Dynamic changes in the magnetic behavior of very small superparamagnetic iron oxide nanoparticles During cellular uptake and enables Determination of cell-labeling efficacy. J Biomed Nanotechnol 2016; 12: 337.
112. Scharlach C, Kratz H, Wiekhorst F, Warmuth C, Schnorr J, Genter G, Ebert M, et al. Synthesis of acid-stabilized iron oxide nanoparticles and comparison for targeting atherosclerotic plaques: evaluation by MRI, quantitative MPS, and TEM alternative to ambiguous Prussian blue iron staining. Nanomedicine NBM 2015; 11: 1085.
113. Harms C, Datwyler AL, Wiekhorst F, Trahms L, Lindquist R, Schellenberger E, Mueller S, et al. Certain types of iron oxide nanoparticles are not suited to passively target inflammatory cells that infiltrate the brain in response to stroke. J Cereb Blood Flow Metab 2013; 33: 1.
114. Löwa N, Wiekhorst F, Metzkow S, Ludwig A, Trahms L. Magnetic particle spectroscopy for the quantification of magnetic nanoparticles in living cells. Biomed Tech (Berl) 2013;58. https://doi.org/10.1515/bmt-2013-4138.
115. Gräfe C, Slabu I, Wiekhorst F, Bergemann C, von Eggeling F, Hochhaus A. Magnetic particle spectroscopy allows precise quantification of nanoparticles after passage through human brain microvascular endothelial cells. Phys Med Biol 2016; 61: 3986.
116. McNaught AD, Wilkinson A. Compendium of chemical terminology, 2nd ed. Oxford, UK: Blackwell Science; 1997.

117. Dunnett CW. A multiple comparison procedure for comparing several treatments with a control. J Am Stat Ass 1955; 50: 1096.
118. Dunnett CW. New tables for multiple comparisons with a control. Biometrics 1964; 20: 482.
119. Tukey JW. Comparing individual means in the analysis of variance. Biometrics 1949; 5: 99.
120. Sidak Z. Rectangular confidence regions for the means of multivariate normal Distributions. J Am Stat Ass 1967; 62: 626.
121. Thomsen LB, Linemann T, Pondman KM, Lichota J, Kim KS, Pieters RJ. Uptake and transport of superparamagnetic iron oxide nanoparticles through human brain capillary endothelial cells. ACS Chem Neurosci 2013; 4: 1352.
122. Abbott NJ, Dolman DEM, Yusof SR, Reichel A. In vitro models of CNS barriers. New York, NY: Springer New York; 2014, 163–197.
123. Grover A, Hirani A, Pathak Y, Sutariya V. Brain-targeted delivery of docetaxel by glutathione-coated nanoparticles for brain cancer. AAPS Pharm Sci Technol 2014; 15: 1562.
124. Tarbell JM. Shear stress and the endothelial transport barrier. Cardiovasc Res 2010; 87: 320.
125. Cucullo L, Hossain M, Puvenna V, Marchi N, Janigro D. The role of shear stress in blood–brain barrier endothelial physiology. BMC Neurosci 2011; 12: 40.
126. Stanness K, Guatteo DJE. A dynamic model of the blood–brain barrier “in vitro”. Neurotoxicology 1996; 17: 481.
127. Kong SD, Lee J, Ramachandran S, Eliceiri BP, Shubayev VI, Lal R, Jin S. Magnetic targeting of nanoparticles across the intact blood–brain barrier. J Contr Rel 2012; 164: 49.
128. The UniProt Consortium. UniProt: the universal protein knowledge-base. http://www.uniprot.org/help/publications [Accessed April 2017].
129. Schlesinger JB, van Harmelen V, Alberti-Huber CE, Hauner H. Albumin inhibits adipogenesis and stimulates cytokine release from human adipocytes. Am J Physiol Cell Physiol 2006; 291: 27.
130. Friedrich RP, Janko C, Poettler M, Tripal P, Zaloga J, Cicha I, et al. Flow cytometry for intracellular SPION quantification: specificity and sensitivity in comparison with spectroscopic methods. Int J Nanomed 2015; 10: 4185.
131. Dusinska M, Boland S, Saunders M, Juillerat-Jeanneret L, Tran L, Pojana G, et al. Towards an alternative testing strategy for nanomaterials used in nanomedicine: lessons from NanoTEST. Nanotoxicology 2015; 9: 118.
132. Gunn J, Paranji RK, Zhang M. A simple and highly sensitive method for magnetic nanoparticle quantitation using ^{1}H-NMR spectroscopy. Biophys J 2009; 97: 2640.
133. Müller EK, Gräfe C, Wiekhorst F, Bergemann C, Weidner A, Dutz S, Clement JH. Magnetic nanoparticles interact and pass an in vitro co-culture blood-placenta barrier model. Nanomaterials 2018; 8: E108.
134. Hong S, Leroueil PR, Janus EK, Peters JL, Kober M-M, Is- lam MT, et al. Interaction of Polycationic Polymers with Supported Lipid Bilayers and Cells: Nanoscale Hole Formation and Enhanced Membrane Permeability. Bioconjugate Chemistry 2006; 17: 728.
135. Bähring F, Schlenk F, Wotschadlo J, Buske N, Liebert T, Bergemann C, et al. Suitability of viability assays for testing biological effects of coated superparamagnetic nanoparticles. IEEE Trans Magn 2013; 49: 383.
136. Kho D, MacDonald C, Johnson R, Unsworth CP, O’Carroll SJ, Du Mez E, et al. Application of xCELLigence RTCA biosensor technology for revealing the profile and window of drug responsiveness in real time. Biosensors 2015; 5: 199.
137. Morley LC, Beech DJ, Walker JJ, Simpson NAB. Emerging concepts of shear stress in placental development and function. Mol Hum Reprod 2019; 25: 329.
138. James JL, Cartwright JE, Whitley GS, Greenhill DR, Hoppe A. The regulation of trophoblast migration across endothelial cells by low shear stress: consequences for vascular remodelling in pregnancy. Cardiovasc Res 2012; 93: 152–61.

139. Dutz S, Weidner A, von der Lühe M, Gräfe C. Biehl P, Demut J, et al. Hybrid nanomaterials of biomolecule corona coated magnetic nanoparticles and their interaction with biological systems. Phys Sci Rev. accepted.
140. Barnes AL, Wassel RA, Mondalek F, Chen K, Dormer KJ, Kopke RD. Magnetic characterization of superparamagnetic nanoparticles pulled through model membranes. Biomagn Res Technol 2007; 5: 1.
141. Pedram MZ, Shamloo A, Alasty A, Ghafar-Zadeh E. Optimal magnetic field for crossing superparamagnetic nanoparticles through the brain blood barrier: a computational approach. Biosensors 2016; 6: 25.
142. Shapiro B, Dormer K, Rutel IB. A two-magnet system to push therapeutic nanoparticles. AIP Conf Proc 2010; 1311: 77–88.

Ralf P. Friedrich, Christina Janko, Harald Unterweger, Stefan Lyer and Christoph Alexiou*

29 SPIONs and magnetic hybrid materials: Synthesis, toxicology and biomedical applications

Abstract: In the past decades, a wide variety of different superparamagnetic iron oxide nanoparticles (SPIONs) have been synthesized. Due to their unique properties, such as big surface-to-volume ratio, superparamagnetism and comparatively low toxicity, they are principally well suited for many different technical and biomedical applications. Meanwhile, there are a numerous synthesis methods for SPIONs, but high requirements for biocompatibility have so far delayed a successful translation into the clinic. Moreover, depending on the planned application, such as for imaging, magnetic drug targeting, hyperthermia or for hybrid materials intended for regenerative medicine, specific physicochemical and biological properties are inevitable. Since a summary of all existing SPION systems, their properties and application is far too extensive, this review reports on selected methods for SPION synthesis, their biocompatibility and biomedical applications.

Keywords: nanomedicine; regenerative medicine; SPION production; T cell therapy; thrombolysis; vascular prostheses.

29.1 Introduction

Nanoparticles have been closely followed in science for quite some time, and the development of further particle systems with different properties is still in progress. Due to their high surface area to volume ratio, they possess unique physical and

***Corresponding author: Christoph Alexiou,** Department of Otorhinolaryngology, Head and Neck Surgery, Section of Experimental Oncology and Nanomedicine (SEON), Else Kröner-Fresenius-Stiftung-Professorship Universitätsklinikum, Erlangen, Germany, E-mail: C.Alexiou@web.de
Ralf P. Friedrich, Christina Janko, Harald Unterweger and Stefan Lyer, Department of Otorhinolaryngology, Head and Neck Surgery, Section of Experimental Oncology and Nanomedicine (SEON), Else Kröner-Fresenius-Stiftung-Professorship Universitätsklinikum, Erlangen, Germany, E-mail: Ralf.Friedrich@uk-erlangen.de (R.P. Friedrich), Christina.Janko@uk-erlangen.de (C. Janko), Harald.Unterweger@uk-erlangen.de (H. Unterweger), Stefan.Lyer@uk-erlangen.de (S. Lyer)

This article has previously been published in the journal Physical Sciences Reviews. Please cite as: R. P. Friedrich, C. Janko, H. Unterweger, S. Lyer, and C. Alexiou "SPIONs and magnetic hybrid materials: Synthesis, toxicology and biomedical applications" *Physical Sciences Reviews* [Online] 2021, 6. DOI: 10.1515/psr-2019-0093 | https://doi.org/10.1515/9783110569636-029

chemical properties. Especially superparamagnetic iron oxide nanoparticles (SPIONs), with their superparamagnetic properties and low toxicity, are suitable candidates for various applications, which include catalysis, imaging, environmental applications and medical applications [1]. The impact of nanoparticles is strongly influenced by the composition of the surface which is often modified through interactions with the direct environment [2]. In biomedical applications, nanoparticles that come into contact with biological fluids, immediately form a protein corona which can influence the colloidal stability and define the interactions between nanoparticles and cells or tissue [3, 4]. Moreover, by an external magnetic field, their magnetic properties enable a region specific accumulation, drug release or exerted force [5–10]. However, these behaviours are influenced by shape, size, crystal lattice, coating etc., and is thus highly dependent on the synthesis methods [11]. Hence, the planned application already suggests possible methods and coatings. While particles used for imaging should be highly stable, inert and circulate for an extended time in the vascular system, SPIONs intended for drug targeting or hyperthermia should possess a high magnetic saturation, be accumulated in the target area and interact with cells, tissue or structures. Investigation of these interactions, especially between nanoparticles and cells as well as the thorough characterization of the resulting nanoparticle-cell hybrids, is crucial for all particles intended for biomedical applications. Finally, suitable models for the intended application have to confirm and prove the suitability of the particles before further preclinical *in vivo* models are deployed.

In this review we describe the synthesis, physicochemical properties, functionalization and toxicological characterization of SPIONs and demonstrate possible particle-specific biomedical applications. In the recent past, we have produced several distinct SPIONs and extensively determined their biological impact and potential usage. The SPIONs were synthesized using iron salt precipitation techniques, stabilized by various substances and, if necessary, functionalized with varying therapeutics. The characterization of these SPION systems required an exact and reliable determination of the iron concentration, not only of the ferrofluid, but in particular of the cellular nanoparticle concentration. Thus, we compared different quantification methods to validate the measurements and evaluate the detection limits. Followed by an intense physicochemical characterization, the biological impact, such as cellular SPION uptake, cell proliferation and toxicity was intensively investigated in distinct cell lines with 2D and 3D cell culture methods by end point and real time analysis [12, 13]. Altogether, those methods lead to a better understanding of SPION-induced biological processes. Finally, the performance of functionalized SPIONs were tested in different disease models for cancer therapy and thrombolysis and as a tool for magnetically based cell guidance used for cell therapy and scaffold endothelialization.

29.2 Synthesis, concentration and storage of SPIONs

29.2.1 Lauric acid-coated SPIONs

In biomedical applications, SPIONs are generally well tolerated and used e.g. as contrast agents for magnetic resonance imaging (MRI), magnetic drug targeting and hyperthermia [9, 14, 15]. Nevertheless, there is a strong need for further improvement of the SPION surface to minimize toxicity and improve biocompatibility. Towards this goal, according to a modified alkaline precipitation protocol [16], we produced lauric acid-coated magnetite nanoparticles ($SPION^{LA}$), by dissolving Fe (II) chloride and Fe (III) chloride salts and subsequent coating with lauric acid [17]. The particles demonstrated a very high saturation magnetization (Ms) of 469 kA/m, almost the Ms value of 480 kA/m of magnetite [18], a hydrodynamic diameter of 44.6 ± 0.1 nm and a zeta potential of −50.6 ± 0.9 in water [19]. The surface coating with lauric acid successfully prevented an agglomeration of the particles in water. However, in more complex fluids like cell culture media without or with low concentrations of fetal calf serum (FCS), the particles tend to form agglomerates [19, 20]. Only after incubation in blood or in media with high amounts of FCS (10%), the colloidal stability of the particles increased, indicating the formation of a stabilizing protein corona. Nevertheless, in biomedical or technical application, where the SPIONs are supposed to be uniformly embedded in a water-based matrix, such as hydrogels and magnetically responsive composites, $SPION^{LA}$ particles are highly suitable [6].

29.2.2 Lauric acid- and bovine serum albumin (BSA)-coated SPIONs

The safe usage of SPIONs for direct vascular administration requires particles to be as stable and biocompatible as possible. Thus, we improved the existing $SPION^{LA}$ particles by an additional coating with bovine serum albumin ($SPION^{LA\text{-}BSA}$) [21]. We found, that above a BSA concentration of 5%, the particles became highly stable. TEM measurements revealed an individual core diameter of 7.64 ± 1.68 nm, DLS a hydrodynamic diameter in cell culture medium of 61.7 ± 1.15 nm and Fourier transform infrared spectroscopy (FTIR) measurements confirmed the protein core–shell structure. Moreover, magnetic measurements by superconducting interference device (SQUID) and vibrating sample magnetometer (VSM) proved that the addition of BSA had no significant influence on the magnetic behaviour of the nanoparticles. Notably, in contrast to $SPION^{LA}$, $SPION^{LA\text{-}BSA}$ demonstrated a significantly increased colloidal stability and did not lead to any macro- or microscopically observable aggregates in FCS-supplemented cell culture media or blood. Moreover, toxicity assays with Jurkat cells, an immortalized line of human T lymphocyte cells, revealed a strongly induced cell death in presence of $SPION^{LA}$ at concentrations of 100 µg/mL and above, whereas the corresponding amount of $SPION^{LA\text{-}BSA}$ did not induce excessive cell death. Finally, we investigated the therapeutic suitability of $SPION^{LA\text{-}BSA}$ as a carrier system for therapeutics. In *in vitro* assays using Jurkat cells, the binding of the cytostatic drug mitoxantrone (MTO), which is used in the treatment of e.g. acute leukemia, lymphoma, prostate and breast cancer [22, 23], exhibited excellent therapeutic potential, exceeding that of free MTO. In conclusion, with $SPION^{LA\text{-}BSA}$, we have synthesized nanoparticles that show great potential for medical applications.

29.2.3 Lauric acid- and human serum albumin (HSA)-coated SPIONs

For clinical applications, especially when particles are administered directly into the vascular system, the requirements for biological safety are extremely high [24]. Some formulations such as Resovist®, Sinerem® or Rienso® have been used in clinical applications in the past, but in the meantime, they have been withdrawn from the market and other SPIONs, e.g. for magnetic drug targeting or hyperthermia, have still not been approved. One prerequisite of such particles is that the effect on the human immune system is as low as possible. Hence, we developed a new formulation of SPIONs containing a corona of clinical approved human serum albumin ($SPION^{LA\text{-}HSA}$) [17]. The core–shell structure around the $SPION^{LA\text{-}HSA}$ particles was confirmed using freeze fracture transmission electron microscopy and showed a layer of organic material with a thickness of around 2 nm around the multicore particles (Figure 29.1(A)).

Dynamic light scattering (DLS) revealed a hydrodynamic diameter of 52.7 ± 0.8 nm with a very low polydispersity index (PDI) of 0.2 ± 0.03. The chemical structure of the surface, analysed by FTIR (FTIR), indicated the presence of albumin and iron oxides. Finally, we investigated the biocompatibility and cellular uptake of $SPION^{LA\text{-}HSA}$ by flow cytometry and microwave plasma atomic emission spectroscopy (MP-AES). The results clearly showed that $SPION^{LA\text{-}HSA}$ were taken up by the cells in a concentration-dependent manner and were well tolerated by the Jurkat cells. Cytotoxic effects were only observed at very high concentrations of 200 µg/mL and above and demonstrated that they were more biocompatible than the former $SPION^{LA\text{-}BSA}$ particles. To demonstrate the suitability of $SPION^{LA\text{-}HSA}$ as a drug carrier, we adsorbed mitoxantrone (MTO) on the particles ($SPION^{A\text{-}HAS*MTO}$) and found a high MTO binding efficiency of up to 98.9% at a loading dose of 200 $µg_{MTO}$/mL $SPION^{LA\text{-}HSA}$. The release kinetics of MTO from the particle system in cell culture media using different high-performance liquid chromatography-ultraviolet (HPLC-UV) assisted release assays displayed only a slow release from the particles with no observable burst release. Finally, we proved the feasibility of the final $SPION^{LA\text{-}HAS*MTO}$

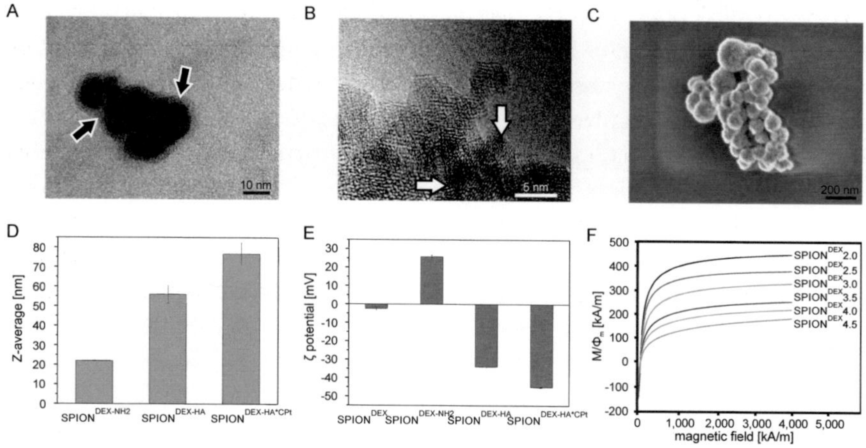

Figure 29.1: Physicochemical properties of SPIONs.
(A) Freeze fracture-TEM images of a single SPION$^{LA\text{-}HSA}$ cluster at a magnification of 150,000. The black arrows mark a clearly visible layer around the particle cluster. Figure modified from Zaloga et al. [17]. (B) High magnification TEM image of SPION^{Dex} particles in a polymer matrix with visible lattice planes (white arrors). Figure modified from Unterweger et al. [26]. (C) SEM picture of SPION^{Citrate}. The image of freeze-dried SPION^{Citrate} particles was taken with a Zeiss Auriga SEM (Carl Zeiss, Oberkochen, Germany) with an acceleration voltage of 1.5 kV. Figure modified from Mühlberger et al. [34]. (D–F) Size, zeta potential and magnetisation curves of SPION^{Dex} variants. (D, E) Change in size and zeta potential of the colloid after functionalization, esterification, and incorporation of the drug. (F) Magnetization curves after normalization to the volume fraction of the magnetic phase showing dependency on the dextran concentration during precipitation. (Bulk magnetite Fe_3O_4, Ms,bulk = 475 kA/m; bulk maghemite γ-Fe_2O_3, Ms,bulk ≈ 370 kA/m [35]) Figures modified from Unterweger et al. [26].

formulation for local drug delivery in a magnetically-based *in vitro* assay using fluorescence microscopy. Notably, a magnetic field gradient between 30 and 400 mT was sufficient to achieve a strong accumulation which led to a much faster apoptosis of cells which were closer to the magnet.

This study demonstrated the development of a hybrid lauric acid/human albumin coated ferrofluid and the feasibility of SPION$^{LA\text{-}HSA/MTO}$ as a drug targeting system for localized tumour therapy.

29.2.4 Dextran-coated SPIONs

The requirements for all nanoparticle-based applications include low toxicity and high colloidal stability, not only in water and cell culture media, but especially in more complex fluids such as blood. However, particles often have to be further functionalised with peptides, proteins, enzymes or therapeutics to accomplish specific features. Hence, we developed another SPION system coated with dextran, a biocompatible, biodegradable and water-dispersible branched polysaccharide [14, 25]. The $SPION^{Dex}$ particles were fabricated by co-precipitation of Fe(II) and Fe(III) salts in the presence of dextran T40 under argon atmosphere and subsequent crosslinking of the dextran layer. The extremely colloidal stable and biocompatible particles with an individual core size of 4.3 ± 0.9 nm, a hydrodynamic diameter of 79.6 ± 0.6 nm and a zeta potential of –3.2 mV demonstrated to be highly suitable for magnetic resonance imaging [25]. Because of the hydroxyl groups of dextran, various functionalities could be added to the particles. In a recent work, we evaluated the suitability of these particles as a directed multicarrier drug carrier system for hyaluronic acid (HA) and cis-diamminedichloridoplatinum(II) (CisPt) [26]. HA enables an active targeting strategy as it interacts with the CD44 receptor, which is expressed by many tumour-derived cells [27]. Moreover, it can form polymer–metal complexes with suitable drugs such as CisPt which is used as a chemotherapeutic to treat a wide variety of different cancers [28, 29]. Thus, we coated $SPION^{Dex}$ with hyaluronic acid (HA) after amination with NH_3 and then loaded it with cisplatin. TEM revealed an individual magnetic particle size which was dependent on the initial dextran concentration and had a mean value between 4.6 and 4.3 nm and a hydrodynamic diameter between 37.5 and 21.1 nm, respectively. After binding of HA and loading with cisplatin the hydrodynamic particle size increased to 77 nm with a zeta potential of –45 mV (Figure 29.1(D) and (E)). The magnetic properties, recorded by SQUID measurements, showed superparamagnetic behaviour, with a saturation magnetization dependent on the dextran concentration during precipitation (Figure 29.1(F)). Particles were highly stable with no sedimentation tendency or agglomeration over several weeks. Without cisplatin, toxicity measurements with Jurkat cells in flow cytometry and PC-3 cells in xCELLigence experiments showed high biocompatibility, whereas functionalization with cisplatin led to a dose-dependent induction of apoptosis, increase in membrane potential reduction and a high amount of DNA degradation due to induction of cell death. In conclusion, SPIONs with a dextran and a cisplatin-bearing HA coating represents a promising new approach in the treatment of cancer.

29.2.5 Citrate-coated SPIONs

Special applications require specific particle properties. While particles for imaging should circulate in the bloodstream for as long as possible, particles produced for magnetic drug targeting must be able to be magnetically enriched in the target area. Other applications, in which cells must be directed into a specific region, e.g. a tumour, by means of a magnet, require a sufficient quantity of intracellular magnetic particles, without having a too strong impact on cellular viability. Depending on the cell type, some cells are more prone than others to take up particles. Whereas phagocytes (i.e. monocytes, macrophages and neutrophils) are specialized in phagocytosis, a process to incorporate huge amounts of surrounding debris, other cells, like T lymphocytes, are considerably less efficient in the uptake of debris or, in this case, nanoparticles [30, 31]. Recently, we demonstrated the feasibility to load T cells with SPIONs, to remove excess uningested nanoparticles, and to attract loaded cells with an external magnetic field [32, 33]. However, the used SPIONs were either too cytotoxic (SPION^{LA}) or coated with non-human proteins (SPION$^{LA\text{-}BSA}$), which could lead to immune reactions when administered to humans [32]. Other experiments, using particles containing a human serum albumin coating (SPION$^{LA\text{-}HSA}$), resulted in an insufficient cellular particle uptake. Thus, for future particle-cell hybrid-based applications, SPIONs with a sufficiently high magnetic susceptibility, containing a cyto- and immunocompatible layer and an adequate cell loading capacity are needed. Therefore, we produced SPIONs by alkaline coprecipitation of iron(II) and iron(III) chloride and *in situ* coating with sodium citrate (SPION^{Citrate}) (Figure 29.1(C)) [34]. In water, the particles revealed a hydrodynamic diameter of 58 nm and a negative zeta potential of –48.5 mV. However, due to the colloidal instability in cell culture medium, the mean diameter increased to 430 nm. Interestingly, this instability was eliminated after incubation in blood, most likely due to the formation of a stabilizing protein corona.

29.2.6 Poly(acrylic acid-co-maleic acid) (PAM)-coated SPIONs

Finally, we synthesized SPION^{PAM} particles, another SPION particle system with a polymer coating of poly(acrylic acid-co-maleic acid) (PAM) [36]. PAM is a non-toxic polyelectrolyte, known to improve the colloidal stability of core shell nanoparticles to produce a very stable protective shell [37]. The particles, with a hydrodynamic diameter of 129 nm and a zeta potential of –41 mV at pH 7, were highly susceptible for cellular binding and uptake, even at low concentrations. Human umbilical vein endothelial cells (HUVECs) treated for 48 h with 30 µg/mL showed an iron load, as measured by atom emission spectroscopy (AES), of 156 pg/cell (Figure 29.7(A)–(D)). Despite a very high cellular particle load, the particles were highly cytocompatible, as shown by flow cytometry, and only caused very slight, non-significant cellular effects compared with non-treated control samples. Thus, SPION^{PAM} could be very beneficial for applications, which are dependent on a high cellular particle load such as magnetically-based steering of particle-loaded cells for endothelialization of vascular scaffolds. Moreover, due to their carboxyl groups, those particles could be easily functionalized with proteins or enzymes, such as tissue plasminogen activator (tPA), by amino-reactive ester reactions [36].

29.2.7 Concentration and purification of SPIONs

The successful utilization of iron oxide nanoparticles depend on a sufficient high iron concentration, colloidal stability, biocompatibility and purity of the ferrofluid. Hence, it is important to remove or reduce excess substances, such as ligands, buffer, coating etc. Additionally, the manufacturing process often prohibits a sufficient high final particle concentration; hence the concentration needs to be increased without inducing aggregation and instability to the system. Common methods, such as centrifugation, centrifugal ultrafiltration and magnetic purification for the concentration of particle suspensions are not very advantageous, as they often involve high shear stress and lead to particle agglomeration [38]. We have therefore evaluated the usage of tangential flow ultrafiltration for purification of the substances needed for SPION production and for concentration of the SPION fluid [39] (Figure 29.2). The main advantage of this method is that substances are tangentially passed across a filtration membrane with relatively low pressure and thereby avoids the formation of aggregates and local high concentrations by high centrifugal forces. Drying loss experiments proved that excess protein is rapidly removed from the suspension. Filtration, or up to four-fold concentration of $SPION^{LA\text{-}BSA}$, did not affect hydrodynamic cluster size or colloidal stability in cell culture media and blood. Additionally, the cellular uptake of particles, such as $SPION^{LA\text{-}BSA}$, was only slightly changed. At the same time, the saturation magnetization, if not normalized to the concentration, could be increased over 3.7-fold from 446.5 A/m to 1667.9 A/m and the achievable temperature T_{max} in magnetic hyperthermia was strongly enhanced from 44.4 °C up to 64.9 °C. In the past, we have confirmed the feasibility of this method using other SPIONs, like $SPION^{LA\text{-}HSA}$ and $SPION^{Dex}$. Hence, tangential flow ultrafiltration is advantageous for concentration and purification of complex SPION particle suspensions to enhance the potential for the usage in medical applications.

29.2.8 Storage of SPIONs

An underestimated problem regarding the use of nanoparticles is their alteration during storage which can lead to the formation of clusters and precipitates, colloidal instabilities and changes in magnetic properties and biocompatibility [40]. Hence, important physicochemical and biological parameters can change dramatically dependent on storage temperature and time resulting in non-reproducible experiments or even unsafe preclinical studies. The storage stability of the particles should therefore always be investigated in order to determine the time window of applicability. As an example, we investigated the changes during the storage of SPIONs containing multiple iron oxide

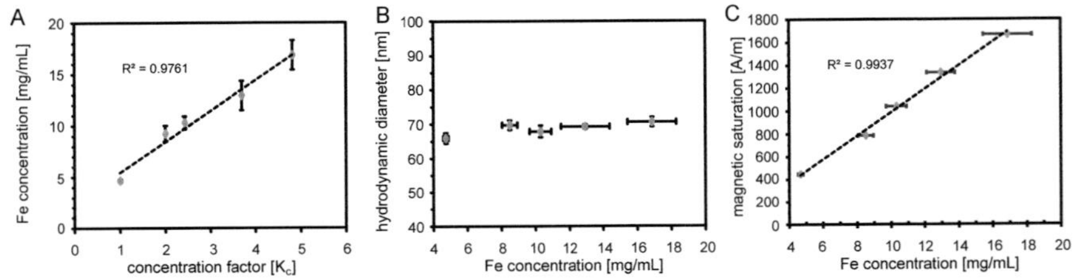

Figure 29.2: Effect of tangential flow ultrafiltration on $SPION^{LA\text{-}BSA}$ concentration.
(A) Iron concentration vs. concentration factor. (B) Hydrodynamic diameter of concentrated SPION samples measured by photon cross-correlation spectroscopy (PCCS). (C) Increase of the saturation magnetization after concentration by tangential flow ultrafiltration. Figures modified from Zaloga et al. [39].

cores and coated with lauric acid and albumin (SPION$^{LA\text{-}BSA}$) [41] (Figure 29.3). We regularly examined the physical properties of SPION$^{LA\text{-}BSA}$ stored for 12 weeks at different temperatures from 4 to 45 °C. Throughout the time, the particles were colloidally stable and photon cross-correlation spectroscopy (PCCS) did not reveal any clear and obvious change of the hydrodynamic size and surface charge. Moreover, transmission electron microscopy (TEM) did not show any changes in size and shape of iron oxide cores. However VSM (VSM) showed a time and temperature-dependent decrease in saturation magnetization by maximally 28.8% with the strongest effects at 45 °C and the slowest decrease at 4 °C. The suspension underwent a macroscopic color change over time, indicating an oxidation of the iron oxide cores. Despite an Ar atmosphere during preparation, X-ray diffraction (XRD) showed a peak shift with increasing storage time and temperature, corresponding to an oxidation from magnetite (Fe_3O_4) to maghemite (γ-Fe_2O_3). Biological effects on cellular uptake, phosphatidylserine exposure, plasma membrane integrity and mitochondrial membrane potential as well as DNA degradation due to storage conditions were investigated by flow cytometry. Analysis of the data showed that increasing storage temperature and time lead to a significant decrease in cellular uptake and toxicity. The example shows that particle properties can undergo significant changes depending on storage conditions. We therefore recommend including studies of storage effects in addition to regular SPION characterization.

29.3 Interactions between nanoparticles and cells

29.3.1 Cellular uptake of SPIONs

The use of particles, especially for biomedical applications, requires them to be as biocompatible as possible. However, to investigate the concentration-dependent nanoparticle-mediated cytotoxicity, an exact quantification not only of the SPION fluids, but

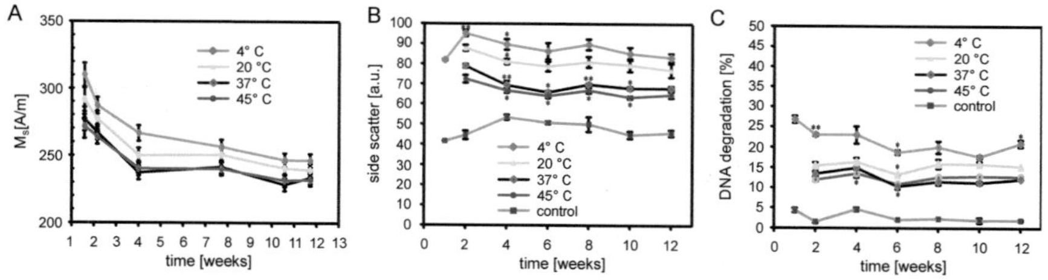

Figure 29.3: Effect of storage temperature and storage time on SPION$^{LA\text{-}BSA}$ particles. (A) Changes in saturation magnetisations (Ms) over time at different temperatures. The time points are given in weeks after synthesis. (B) Side scatter (SSC), as a marker of SPION uptake and (C) DNA degradation, after 48 h incubation of Jurkat cells with 300 µg/mL SPION aliquots with the indicated storage times and temperatures. The significance were determined using an unpaired *t*-test (*$p < 0.05$; **$p < 0.005$). Figures modified from Zaloga et al. [41].

especially of the intracellular SPION amounts, by reliable and reproducible methods is inevitable. For this purpose, different quantification methods, such as ultraviolet spectrophotometry (UVS), magnetic particle spectroscopy (MPS) atom adsorption/emission spectroscopy (AAS/AES) are available. Thus, a study, containing all three methods, allows a direct comparison and an evaluation of the shortcomings and advantages. For this purpose, we have incubated endothelial cells with three different nanoparticle systems, $SPION^{LA}$, $SPION^{LA\text{-}BSA}$ and the commercially available Rienso® and subsequently determined the cellular SPION amount [42]. The reliability of the applied UVS method, which uses the SPION absorption at 370 nm, and the magnetic particle spectroscopy (MPS) and AAS methods were successfully verified using different SPION dilutions and cell lysates. Whereas UVS and MPS quantifies iron-containing SPIONs, and MPS is additionally dependent on the magnetic moment of the nanoparticles, AAS determines the total iron content and can be used for all particles, even with very low magnetic moment. Generally, the detection threshold of AAS ($SPION^{LA\text{-}BSA}$ 0.6 μg_{Fe}/mL; $SPION^{LA}$ 0.6 μg_{Fe}/mL; Rienso® 0.6 μg_{Fe}/mL) and MPS ($SPION^{LA\text{-}BSA}$ 0.3 μg_{Fe}/mL; $SPION^{LA}$ 0.5 μg_{Fe}/mL; Rienso® 2.5 μg_{Fe}/mL) was far beneath the UVS thresholds ($SPION^{LA\text{-}BSA}$ 4.7 μg_{Fe}/mL; $SPION^{LA}$ 4.5 μg_{Fe}/mL; Rienso® 4.0 μg_{Fe}/mL). Thus, even the detection threshold of MPS for the detection of Rienso® particles, which have a very small core size, were lower with MPS than with UVS. However at concentrations of ~5 μg_{Fe}/mL and above, the UVS method is very reliable for all particles.

The performed *in vitro* experiments using AAS, MPS and UVS demonstrated that cellular uptake is dose-dependent and strongly depends on the SPION type. Interestingly, given a sufficiently high SPION concentration, the achieved cellular SPION amounts highly correlated with the side scatter data received from flow cytometry. Exemplarily, Figure 29.4(A)–(C) shows the correlation achieved with $SPION^{LA\text{-}BSA}$ particles. Thus, flow cytometry can not only be used to simultaneously measure multiple toxicity-related parameters such as cell number, cellular apoptosis and necrosis, membrane potential, DNA cycle and DNA damage and estimate possible effects on living organisms, but also provides a fast and reliable method to determine the absolute cellular amount of different SPION types.

29.3.2 Biocompatibility of SPIONs

The ability to determine the exact nanoparticle concentrations enable reliable dosage-dependent studies of the cytotoxic effects of nanoparticles. In order to estimate and classify the consequences of particle exposure, it is advisable to use various relevant cell lines and control particles, as the effects are not only dosage-dependent but also particle type and cell line dependent. In a recent study, we investigated the effects of three different superparamagnetic iron oxide particles ($SPION^{LA}$, $SPION^{LA\text{-}HSA}$, $SPION^{Dex}$) on HUVECs and on four human breast cancer cell lines (T-47D, BT-474, MCF7, MDA-MB-231) belonging to different subtypes [19]. We analyzed cellular SPION uptake, magnetic properties, cell proliferation and toxicity using atomic emission

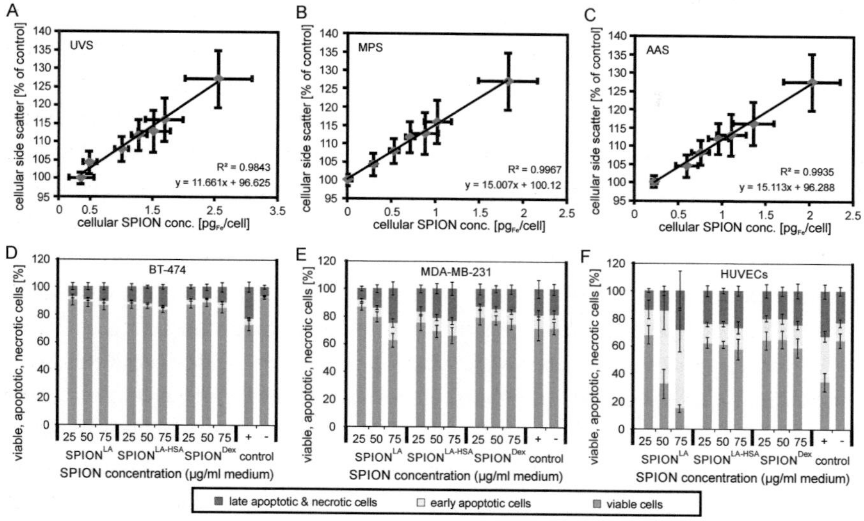

Figure 29.4: Cellular uptake and toxicity of SPIONs.
(A–C) Correlation between side scatter measurements and SPION-loads. HUVECs were cultured for 48 h in medium containing 0–100 μg_{Fe}/mL $SPION^{LA\text{-}BSA}$. (A–C) Relationship between cellular SPION-load as measured by (A) UVS, (B) MPS and (C) AAS and the normalized side scatter data delivered by flow cytometry. Data are expressed as the mean ± standard error ($n = 3$ with technical triplicates). R^2 represents the coefficient of determination. y describes the mathematic relationship between side scatter and cellular SPION content. Figures modified from Friedrich et al. [42]. (D–E) Sensitivity and viability of different cells after SPION treatment. Cells were incubated for 48 h with increasing amounts of $SPION^{LA}$, $SPION^{LA\text{-}HSA}$ and $SPION^{Dex}$. Cell viability was determined by annexin A5-FITC/PI staining and analyzed by flow cytometry. The amount of viable (Ax–PI–), apoptotic (Ax+PI–) and necrotic (PI+) cells are shown for (D) BT-474, (E) MDA-MB-231 and (F) HUVEC cells. Positive controls contain 2% DMSO, and negative controls represent the corresponding amount of H_2O instead of water-based ferrofluid. Data are expressed as the mean ± standard deviation ($n = 4$ with technical triplicates). Figures modified from Poller et al. [19].

spectroscopy, magnetic susceptometry, flow cytometry and microscopy. We also investigated the influence of different liquids on the stability of nanoparticles and found an increase in particle stability by increasing the protein concentration in the medium, suggesting the formation of a stabilizing protein corona. After the physico-chemical and biological characterization, we ascertained that cellular particle uptake depends significantly on the coating of the nanoparticles. Their effects on distinct cell types, revealed several particle- and cell type-as well as concentration- and time-dependent effects on particle uptake and toxicity (Figure 29.4(D)–(E)). These results, together with the respective magnetic properties of the SPIONs, allowed an assessment of the use of these particles for different medical applications. $SPION^{Dex}$ are highly colloidally stable, hardly absorbed by cells, have an extremely low *in vitro* and *in vivo* toxicity and are therefore a promising candidate for MRI imaging. In contrast, $SPION^{LA\text{-}HSA}$ have very good magnetic properties, are colloidally very stable, less toxic and are better absorbed by cells than $SPION^{Dex}$ and are therefore more suitable candidates for hyperthermia and/or magnetic drug targeting. In contrast, without the stabilizing albumin coating, $SPION^{LA}$ behaved entirely different, showing much stronger effects on cellular uptake and cellular toxicity. This study impressively demonstrated that particle synthesis and functionalization can strongly alter the biological effects of particles and that these effect can be highly cell type dependent, suggesting the use of different cell lines for particle characterization. Generally, multiparameter flow cytometry is a feasible method to investigate the biocompatibility not only of SPIONs but in principal of all kinds of metal or non-metal particles, e.g. cobalt ferrite nanoparticles, magnetic microgels and microspheres [43–45].

In addition to cell culture experiments performed under static conditions, it is advisable to analyze the effect of nanoparticles under dynamic *in vitro* and *ex vivo* conditions [46, 47]. In a systematic study with 10 nanoparticle systems intended for cardiovascular applications, we found that under dynamic conditions most particles were significantly better tolerated than under static conditions (~400 µg/ml vs. 100 µg/ml), indicating a sedimentation-based increase of toxicity by colloidally instable particles [46]. Experiments under flow conditions are also crucial to evaluate the suitability of particles for magnetically-based accumulation. We previously confirmed the applicability of serum albumin coated SPIONs for site directed accumulation and cellular uptake by an external magnet in an *in vitro* flow model [20]. In combination with a peristaltic pump, the arterial bifurcation model imitates the blood flow and can be used to analyze the endothelial accumulation of circulating SPIONs in presence or absence of an external magnet. In the absence of a magnetic field, no increase in accumulation of SPIONs was observed. In contrast, application of an external magnet led to an enhanced accumulation of SPIONs at the region in near proximity of the magnet. The accumulation led to an increased uptake of albumin-coated SPIONs. However, as expected, the SPIONs were well tolerated by endothelial cells and did not affect endothelial cell viability or attachment. In addition to histological staining, a complementary investigation by other techniques, such as micro-computed X-ray tomography (XµCT), enabled a more complete understanding of the distribution of magnetic nanoparticles after magnetic accumulation and a good possibility to classify and evaluate various magnetic particle systems [47].

29.3.3 Investigation of cytotoxic SPIONs

Unlike most SPIONs that have been stabilized by coating with proteins, lipids or polymers, SPIONs for biomedical applications, such as hyperthermia and/or drug targeting, require enhanced toxic properties. This can be achieved by covalent or adsorptive functionalization with suitable small molecules which not only alters the chemical and physical properties of the particles, but particularly increases the biological impact on cells. To analyze the interactions between SPIONs functionalized with toxic substances and cells, we have modified several SPIONs with known cytotoxic molecules.

Recently, we produced lauric acid and bovine serum albumin-coated SPIONs and loaded them with the cytotoxic drug MTO ($SPION^{LA\text{-}BSA*MTO}$) [48]. Since multicellular spheroids mimick a tumour environment, they might be more suitable to test the efficacy of therapeutic agents than two-dimensional cell culture models [49]. Therefore, we compared the cytotoxic effects of MTO-loaded and unloaded SPIONs and free MTO on the proliferation and viability on a previously established three-dimensional spheroid model formed by the human colon cancer cell line HT-29 [50]. The analysis of the amount of viable, apoptotic and necrotic cells by flow cytometry and imaging by fluorescence microscopy showed that both free MTO and $SPION^{LA\text{-}BSA*MTO}$ were able to effectively kill tumour cells. In contrast, unloaded SPIONs revealed a high biocompatibility and did not affect cellular viability and growth of the spheroids. The study demonstrated that 3D cell culture models are highly suitable for *in vitro* analysis of particle toxicity.

The SPION system was further optimized by exchange of the bovine serum albumin with human serum albumin [17]. The resulting $SPION^{LA\text{-}HSA}$ particles could be reproducibly synthesized using substances from cGMP (current good manufacturing practice) grade quality. Similar to $SPION^{LA\text{-}BSA*MTO}$, MTO was adsorbed to produce $SPION^{LA\text{-}HSA*MTO}$ particles. Using microwave-plasma assisted atomic emission spectroscopy (MP-AES) and flow cytometry, we investigated cellular uptake and biocompatibility of $SPION^{LA\text{-}HSA}$ and $SPION^{LA\text{-}HSA*MTO}$. Whereas $SPION^{LA\text{-}HSA}$ were highly cytocompatible, $SPION^{LA\text{-}HSA*MTO}$ demonstrated strong site-specific therapeutic effects. The use of nanoparticle-drug carrier systems requires an advanced analysis of the adsorption and desorption behaviour of the drug. Adsorption efficiency and desorption kinetics were therefore analyzed by dynamic light scattering (DLS), FTIR (FTIR), release experiments, surface titration and small-angle neutron and X-ray scattering (SANS and SAXS) [51]. The results showed a strong adsorption to the iron oxide surface, HSA and to $SPION^{LA\text{-}HSA}$ particles and the release of MTO from the particles were significantly slower than from the release from pure HSA. Moreover, the stability of the binding was indicated by the absence of a burst release.

The $SPION^{LA\text{-}HSA}$ particles are not only suitable as drug carrier for MTO, they can also be functionalized with other therapeutics such as paclitaxel (Ptx) which is an anti-proliferative and antitumourigenic agent and frequently used for the systemic treatment of ovarian, lung, breast and pancreatic cancer [52, 53]. In a recent study we synthesized $SPION^{LA\text{-}HSA*Ptx}$ particles and investigated the physicochemical properties by DLS, ζ potential measurements, infrared spectroscopy, isoelectric point titration, drug release quantification and magnetic susceptibility measurements. Cytotoxic effects were determined extensively by flow cytometry, IncuCyte® live-cell imaging, and growth experiments using two- and three-dimensional breast cancer cell cultures models (Figure 29.5). To address the heterogeneity of breast cancer, we used four different human breast cancer cell lines (T-47D, MCF-7, BT-474 and MDA-MB-231) corresponding to different subtypes [54, 55]. Binding of Ptx to $SPION^{LA\text{-}HSA}$ did not alter the high magnetization capability of $SPION^{LA\text{-}HSA}$. At 5 T, the samples showed a magnetization of ~460 kA/m, which is close to the bulk value of magnetite of 480 kA/m [56]. The hydrodynamic size of both samples was only changed very slightly (73.6 and 72.8 nm, respectively), indicating an incorporation of the hydrophobic Ptx into the HSA protein corona. The particles revealed a perfect colloidal stability in all kinds of aqueous solutions and in blood. Whereas $SPION^{LA\text{-}HSA}$ only showed negligible toxic effects, the cytotoxic efficiency, measured by flow cytometry and life cell imaging, of $SPION^{LA\text{-}HSA*Ptx}$ and free Ptx demonstrated to be highly dosage and cell line-dependent. In conclusion, the 2D cell culture experiments and 3D cell culture models using multicellular spheroids suggest that $SPION^{LA\text{-}HSA*Ptx}$ is a potential particle system for the treatment of different breast cancer tumours.

29.3.4 Haemocompatibility of SPIONs

The toxicological investigation of SPIONs with different primary cells and cell lines, e.g. by flow cytometry to determine different viability markers, is an important step in the assessment of the biocompatibility of nanoparticles [19]. However, to fulfill regulatory requirements needed for future translation into clinics, further particle optimization and in-depth characterizations is necessary [57]. In recent studies we have shown that sedimentation, agglomeration cellular uptake and viability is strongly dependent on the particle corona and that particles can be stabilized by serum albumin [19, 42]. Beside comprehensive toxicological investigations using impedance-based real-time cell analysis, flow cytometry and dynamic flow experiments, the determination of biocompatibility also requires particles to be examined in *ex vivo* experiments using human blood and isolated red blood cells (RBCs) (Figure 29.6(A) and (B)) [58].

In whole blood, both $SPION^{LA}$ and $SPION^{LA\text{-}BSA}$ did not induce haemolysis. However, in plasma-free RBCs, haemolysis occurred dose dependently after treatment with $SPION^{LA}$ but not with $SPION^{LA\text{-}BSA}$. These data, together with data from SDS polyacrylamide gel electrophoresis, suggests, that blood serum proteins quickly form a protein corona around $SPION^{LA}$ and $SPION^{LA\text{-}BSA}$ in whole blood, which protect RBCs from direct interaction with the nanoparticle surface, leading to a significant improvement of biocompatibility. Thus, the combination of intrinsic nanoparticle properties and the “biological identity”, as defined by the protein corona in the respective environment, determines the biological effects of the SPIONs. Consequently, only all physicochemical and biological assays in their entirety allow a detailed assessment of particle safety and, through a feedback loop, a systematic improvement of particle synthesis.

Beside SPION-mediated haemolytic effects, another important feature is the agglomeration behaviour in blood. Nanoparticles intended for *in vivo* applications, especially when applied into the vascular system, need to be stable and not form superstructures. It is therefore recommendable to perform a blood stability assay in which particles are incubated with ethylenediaminetetraacetic acid (EDTA)- or/and citrate-

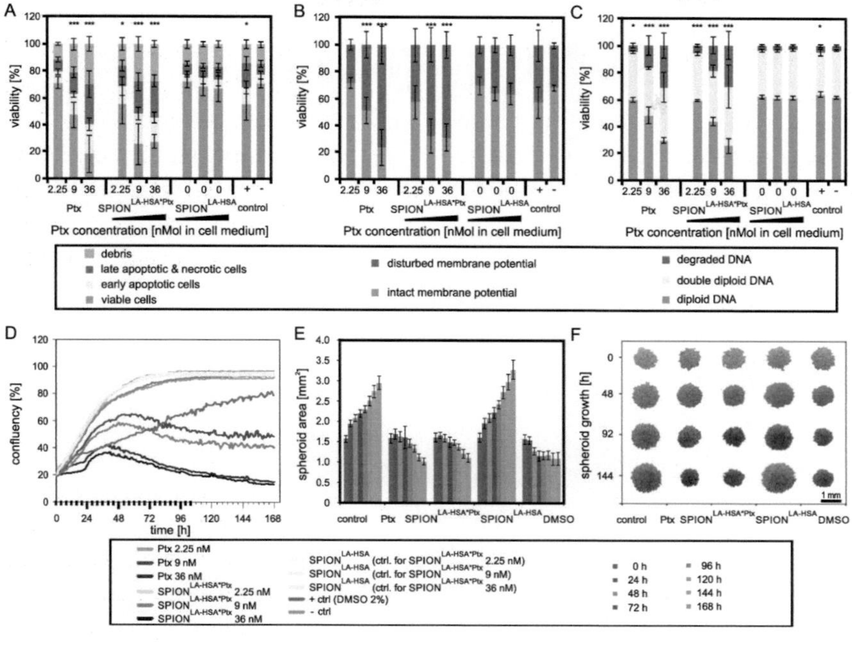

Figure 29.5: Effects of Ptx treatment on breast cancer cells.
(A–C) MDA-MB-231 cells were incubated for 48 h with increasing amounts of free Ptx, SPION$^{LA\text{-}HSA*Ptx}$ and SPION$^{LA\text{-}HSA}$, and analyzed by multiparameter flow cytometry. (A) Viability was determined by AxV–FITC and PI staining, yielding the percentage of viable (Ax−PI−), early apoptotic (Ax+PI−), and late apoptotic and necrotic (PI+) cells. (B) Mitochondrial membrane potential was analyzed by DiIC1(5) staining and distinguishes cells with intact (DiIC1(5) positive) and depolarized (DiIC1(5) negative) membranes. (C) DNA degradation and cell cycle were determined by PIT staining and showed the amount of degraded DNA, diploid DNA (G1 phase), and double-diploid DNA (synthesis/G2 phase). (D) Growth kinetics of MDA-MB-231 cells in the presence of increasing amounts of free Ptx, SPION$^{LA\text{-}HSA*Ptx}$ and SPION$^{LA\text{-}HSA}$. Cellular confluency was determined hourly using the IncuCyte® live-cell imaging system. (E, F) Effects of SPION$^{LA\text{-}HSA\text{-}Ptx}$ and Ptx on three-dimensional MDA-MB-231 spheroids treated with 36 nM free Ptx or particle-bound Ptx (SPION$^{LA\text{-}HSA*Ptx}$). (E) The diameter of the projected two-dimensional area of the spheroids during treatment. (F) Representative pictures of the spheroid during the first 144 h after treatment. Positive controls contain 2% DMSO, and negative controls represent the corresponding amount of solvent instead of drug or ferrofluid. Statistical significance are indicated with $*p < 0.01$, $**p < 0.001$ and $***p < 0.0001$, and were calculated via Student's *t*-test analysis. Figures modified from Lugert et al. [13].

stabilised blood (Figure 29.6(C)–(E)). The assay clearly demonstrated the absence of agglomeration with particles such as SPION$^{LA\text{-}HSA}$, SPION^{Dex} and SPION^{Citrate} [13, 19, 34].

Exogenous and endogenous stimuli, including microorganism, immune complexes, chemicals, dying cells etc., are known to lead to defense reactions by neutrophils [59–62]. They either remove non-endogenous substances by phagocytosis, inactivation by release of reactive oxygen species (ROS) or by immobilization. During immobilization, the eu- and heterochromatin of the neutrophils homogenize, the nuclear envelope and

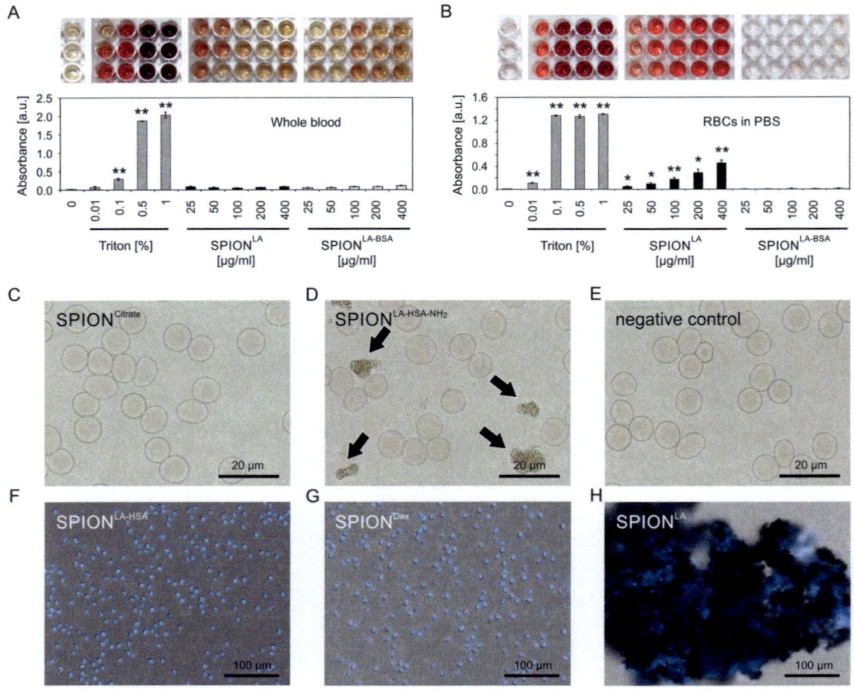

Figure 29.6: Haemocompatibility of nanoparticles.
(A–B) Protein corona of particles increases biocompatibility and reduces haemolysis. 450 µl of whole blood or red blood cells (RBCs) in peripheral blood smears (PBS) were incubated with 50 µl nanoparticle dilutions (in H_2O) for 3 h at 37 °C in Eppendorf tubes. The surfactant Triton X-100 dissolving cellular plasma membranes served as positive control for RBC lysis, PBS served as negative control. After centrifugation, haemolytic samples are characterized by appearance of free haemoglobin in the supernatant. Absorption of supernatants of whole blood or isolated RBCs was measured at 590 nm (*$p < 0.05$; **$p < 0.01$). Figures modified from Janko et al. [58]. (C–E) Blood stability of SPIONs. Freshly drawn human blood, stabilized against coagulation with citrate, were mixed with (C) $SPION^{Citrate}$, (D) $SPION^{LA\text{-}HAS\text{-}NH2}$ as positive control and (E) corresponding amount of H_2O as negative control. In contrary to the aminated particles, which are known to have a poor colloidal stability and form agglomerates [32] (marked by arrows), $SPION^{Citrate}$ did not show any sign of agglomeration. Figures modified from Mühlberger et al. [34]. (F–H) SPION-dependent NET formation. Non-stabilized SPIONs induce NET formation in protein-free buffer. Polymorphonuclear cells (PMN) were incubated with 200 µg/ml nanoparticles for 3 h in PBS and stained with Hoechst. Stable $SPION^{LA\text{-}HSA}$ and $SPION^{Dex}$ particles did not lead to NETosis, whereas non-stabilized $SPION^{LA}$ formed agglomerates and led to abundant NET formation. Figures modified from Bilyy et al. [64].

the granule membranes disintegrate and the decondensed chromatin decorated with cytoplasmic and granular proteins is released [59]. Inside this extracellular matrix called neutrophil extracellular traps (NETs), foreign substances can be embedded and degraded. A recent study investigated the NETosis formation after exposure to carbon and polystyrene nanoparticles, which are frequently present in the environment [63].

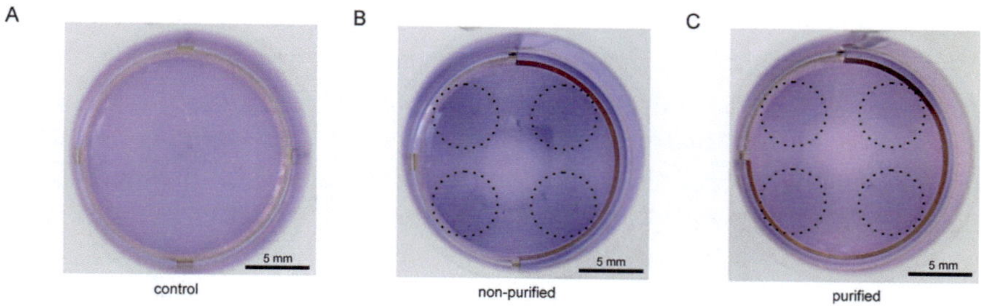

Figure 29.7: Visualization of magnetic attractability of SPION^{Citrate}-loaded cells.
EL4 cells were incubated for 24 h with SPION^{Citrate} at an iron concentration of 75 µg/mL. Subsequently, control cells (A) and cells not purified (B) and purified from excess particles (C) were stained with crystal violet and incubated for 48 hrs on a magnetic plate (approx. 0.5 T). Loaded cells, whether purified or not, are preferably sedimented in the area of the magnets. Magnets are depicted as interrupted circle lines. Figures modified from Muhlberger et al. [34].

Nanoparticles are size-dependently inactivated and neutralized by NET formation. Thus, particles in the size of 10 and 40 nm induced a very fast NETosis, whereas particles between 100 and 1000 nm led to a much slower reaction. Other experiments demonstrated that agglomerates of non-stable SPIONs (SPION^{LA}) induced NETosis [64]. However, stabilization with a biocompatible coting of HSA or dextran (SPION$^{LA\text{-}HSA}$, SPION^{Dex}) led not only to a highly reduced agglomeration tendency, but also to a significantly improved reduction of NET formation (Figure 29.6(F)–(H)). Interestingly, when using plasma or serum-containing medium, even SPION^{LA} were stabilized and showed reduced agglomeration and NETosis, suggesting that colloidal stable nanoparticles behave inert and are most likely cleared primarily by phagocytosis.

29.4 SPION applications

29.4.1 Navigation of particles-cell-hybrids for T cell-based therapy

There are some further auspicious applications, such as advanced cell therapies and controlled tissue assembly, which are not based on the targeted accumulation of particles but on the targeted navigation of cells loaded with particles [65, 66]. This approach is also suitable for advanced T cell therapy. It is known that tumour-infiltrating lymphocytes (TILs) have a large impact on the prognosis of patients, as T lymphocytes are responsible for the immune activation against tumours [67]. Magnetic targeting could be further developed to magnetic T cell targeting, a new application for cancer therapy, by triggering the immune response of T cells and direct the activated cells directly into the tumour by an externally applied magnetic field [68]. Aiming on such a T cell-based therapy, we have recently investigated the effects of SPIONs on T cells [32–34]. SPION^{Citrate} exhibited a high biocompatible and stability in blood.

Incubation of non-adherent T lymphocytes from mouse lymphoma (EL4 cells) with $SPION^{Citrate}$, with subsequent purification of excess particles by immunoaffinity chromatography, showed only a very slight effects on T cells viability which were clearly lower than with $SPION^{LA}$ and comparable with SPIONs with additional albumin coatings. Additionally, after SPION incubation, atomic emission spectroscopy revealed a cellular iron load of 1.02 ± 0.17 pg/cell. As demonstrated by live-cell imaging, this iron load was sufficient to effectively direct the cells by an external magnetic field (Figure 29.7). The data suggest that magnetically controlled T cells targeting might be further developed to a controlled immune therapy, such as a directed chimeric antigen receptor (CAR) T cell therapy for solid tumours [69].

29.4.2 SPION-based production of biocompatible vascular prostheses

In cardiovascular diseases, lack of adequate venous material for transplantation constitutes a common problem, resulting in a strong need for biocompatible vascular grafts. The market for vascular grafts is currently dominated by two synthetic materials: Polytetrafluoroethylene (PTFE) and polyester (PE) are routinely used to replace damaged vessels. However, the occlusion rate for small-lumen prostheses and vessels below 6–8 mm is very high [70]. The reason for this is primarily the interaction between the artificial materials and the immune system. The formation of an intima through an increased proliferation rate of smooth muscle cells at the interface between the prostheses and the body's own vessels often lead to a fast occlusion. A therapy of this so-called restenosis in which antiproliferative drugs as well as substances with an anti-thrombotic effect are administered systemically, are quite ineffective [71]. Based on the systemic drug administration, it is not possible to achieve a sufficiently high concentration at the necessary site without causing serious systemic side effects. The use of tissue-engineered hybrid materials containing a biocompatible scaffold and a cell-coated lumen may reduce the risk of thrombosis, intimal hyperplasia and calcification of the prostheses as the endothelialization of the vascular grafts counteracts a recurrence of vessel occlusion. However, the endogenous cell colonization is very inefficient, especially in the case of small-volume transplants. In the meantime, various methods have been established to colonize artificial transplants or transplants based on biological scaffolds with cells [72, 73]. The success of cell seeding, as a special form of tissue engineering, depends not only on the cell type, but essentially on the material, composition and shape of the scaffold. The colonization of a tubular scaffold requires special techniques, whereas a flat collagen membrane can be coated by simply applying a cell suspension. If a functional transplant is to be generated, the cells must not be damaged or otherwise impaired by high mechanical stress or temperature fluctuations during seeding. For example, a tubular scaffold can be generated from a planar scaffold previously coated with cells. Another possibility is coating using rotational forces. However, an impairment of the cells was observed, especially at high velocities [74]. Another approach is the *in vitro* endothelialization under flow conditions in a bioreactor [75]. The currently most promising technique is the magnetic cell seeding. The technique of magnetic vascular engineering, uses magnetically marked cells that are pulled by radial symmetric magnetic fields onto the inner wall of tubular vascular prostheses and held there until they are attached. This procedure is fast, gentle on the cells and leads to very homogeneous colonization of the surface [76].

We have evaluated the suitability of various nanoparticles for magnetic cell seeding. The requirements for such particles are very high as they need to be as non-toxic as possible and bind to or incorporate into cells in sufficient quantities to make them magnetically guidable. For instance, $SPION^{Dex}$ was shown to be perfectly stable and non-toxic, but the poor interaction with cells led to an insufficient cellular SPION amount [14, 19]. $SPION^{LA\text{-}BSA}$ particles were well absorbed by HUVECs [20, 42] and allowed a solid and homogeneous colonization of the scaffolds with cells even at low cellular iron concentrations. However, due to concerns about the bovine origin of BSA, we tested $SPION^{LA\text{-}HSA}$ containing human serum albumin [17] and $SPION^{PAM}$ nanoparticles coated with a polymer corona [36]. $SPION^{LA\text{-}HSA}$ were perfectly biocompatible but the cellular uptake were insufficient. In contrast, $SPION^{PAM}$ enabled a very high cellular SPION accumulation and were biocompatible with HUVECs, showing only very slight effects at high concentrations on the amount on viable, apoptotic and necrotic cells, mitochondrial membrane potential and DNA degradation in flow cytometry-based assays (Figure 29.8(A)–(D)). In consequence, $SPION^{PAM}$ were favored and used as a standard for further endothelialization experiments and methodological improvements.

Previous proof of concept experiments with transparent plastic tubes demonstrated the feasibility of the 3-dimensional magnetic cell seeding of vascular scaffolds (Figure 29.8(E)). Ongoing endothelialization experiments using ePTFE and polyester-based vascular prostheses showed that the method is applicable to prosthesis commonly used in clinics (Figure 29.8(F)). In order to provide a more cell compatible surface on the artificial scaffolds, we investigated the effect of different protein and peptide films such as p-lysine, fibrin, gelatin and collagen. In particular gelatin coating enhances the adhesion of cells to the scaffold. However, for a more durable coating and to facilitate the formation of an extracellular matrix (ECM) on the luminal surface, the biomolecules need to be better attached to the scaffold surface, either by covalent linkage or by applying a thicker ECM-mimicking matrix, such as gelatin hydrogels.

In contrast to gelatin hydrogels, produced in presence of the common chemical cross linker glutaraldehyde (GTA), we found that hydrogels fabricated by the reagent-free cross-linking technique using electron beam treatment, demonstrated strongly enhanced cell attachment and led to a confluent endothelium and a superior biocompatibility [77]. Further improvement of cell growth and cell adhesion was achieved by structured gelatin hydrogels produced by topographical patterning using a silicon template and stabilized by high-energy electron beam treatment [78]. Biocompatibility studies of the patterned hydrogels showed low toxicity to HUVECs

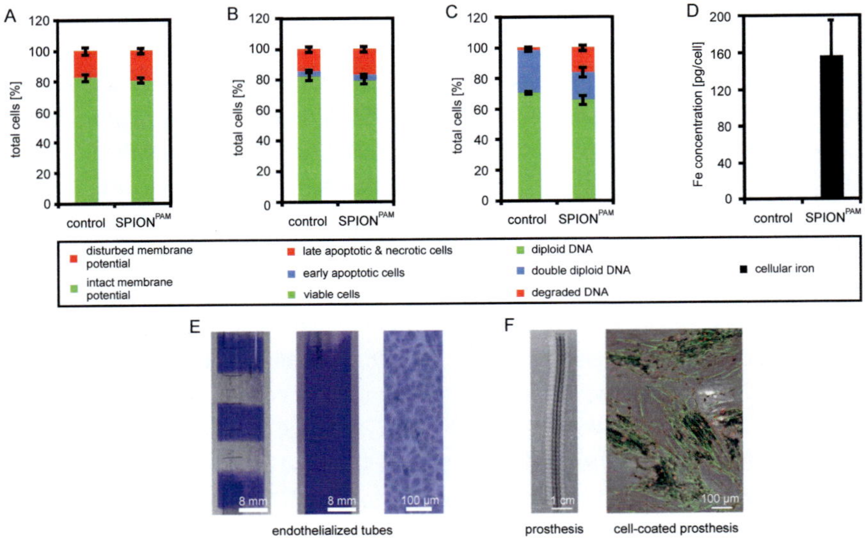

Figure 29.8: Viability and endothelial cell uptake of $SPION^{PAM}$ and magnetic-based scaffold endothelialization.
(A–C) Viability of HUVECs 48 h after incubation with poly(acrylic acid-co-maleic acid) (PAM)-coated SPIONs ($SPION^{PAM}$) measured by flow cytometry. (A) Integrity of mitochondrial membrane potential analyzed by DiIC1(5) staining. (B) Cell viability determined by Annexin V/propidium iodide staining. (PI+) necrotic/late apoptotic cells, (AxV+, PI–) early apoptotic cells, (AxV–, PI–) viable cells. (C) Cell cycle and DNA degradation determined by propidium iodide/tritonX analysis (PIT). The results were normalized to untreated control cells, set to 100%. (D) Cellular $SPION^{PAM}$ uptake after 48 h incubation with HUVECs measured by microwave plasma atomic emission spectroscopy (MP-AES). (E, F) HUVECs loaded with SPIONs colonize the wall of tubular scaffolds after magnetic cell seeding. (E) Endothelialized tubes stained with crystal violet. For better visualization, the settings of the instrument forced the cells to attach in a stripe-like pattern (left). Homogeneous endothelialization (middle). Micrographs of the confluent cell layer (right). (F) Vascutec Gelsoft prosthesis after endothelialization with HUVECs and 72 h incubation. Actin was stained with fluorescently labelled phalloidin (green) and nuclei with Hoechst (red).

and extremely high biocompatibility to fibroblasts (NHDFs). Moreover, the pattern led to an enhanced and directed growth for both cell types. Finally, the hydrogels also revealed a tunable biodegradation, suggesting the usage as future completely biodegradable scaffolds, where the material could be slowly replaced by endogenous extracellular matrix.

29.4.3 SPION-based thrombolysis

A common cardiovascular event is the formation of thrombi, which lead to the occlusion of blood vessels. Thrombi are biological matrices and consist of coagulated blood by the formation of a protein network of fibrin and thrombocytes as well as other blood components. The purpose of thrombus formation is to block damaged blood vessels in order to prevent excessive blood loss on the one hand and infections on the other. However, they can also be the cause of many life-threatening vascular diseases such as myocardial infarction, ischemic stroke, pulmonary embolism and leg vein thrombosis [79, 80]. Although the mechanisms leading to thrombosis have been well understood and better drugs have been developed, the therapeutic effect is still disappointing. The earlier the vascular occlusion is treated and lysis of the thrombus is initiated, the greater the chances of success of a therapy. This process known as thrombolysis is triggered by drugs that activate the body's own enzyme plasmin which is responsible for fibrinolysis. In particular, the so-called tissue plasminogen activator (tPA) is used for this purpose. Thrombolytics are administered either systemically via intravenous administration or, in special cases, locally via an intra-arterial catheter. The therapeutic time window after the onset of symptoms is very narrow and is only 4.5 h in patients treated with tPA [81]. After this time, the risk of multiple side effects, such as high bleeding risk due to coagulation disorder or anticoagulation, outweighs the therapeutic benefit [82]. TPA is currently the drug of choice for the treatment of thrombosis, but the development of novel drugs or methods of treatment or administration is urgently needed to improve treatment efficiency and reduce the side effects associated with current drugs. There are already some drugs that are used in certain indications. These include aspirin in combination with ADP $P2Y_{12}$ to prevent stent thrombosis [83], and thrombin and factor Xa inhibitors to prevent embolism [84]. However, despite its many side effects, tPA is still the only drug approved by the FDA for the treatment of acute ischemic stroke [85]. Alternative treatment methods, such as the targeted transport of therapeutics by nanosystems, are currently under development and there are also some reports on the manufacture of such systems. For example, there are nanoparticles containing L-arginine and L-aspartic acid to prevent platelet aggregation [86], nanoparticles coated with aspirin via the tetra peptide Arg–Gly–Asp–Val, which target activated platelets and reduce aspirin resistance [87], heparin-conjugated carbon nanocapsules [88], albumin nanoparticles loaded with a tPA plasmid and associated with microbubbles to contribute to the prevention of thrombosis [89], and SiO_2-coated magnetic nanoparticles functionalized with tPA and streptokinase [90]. In addition, the use of various thrombin inhibitors is being investigated. Hirulog, a hirudin analogue, was bound to lipid nanoparticles and directed to the thrombus using fibrin-binding peptides [91]. Some innovative strategies are aimed at targeting tPA to the thrombus: Magnetic nanoparticles coated with tPA or urokinase could be directed to the thrombus with an external magnet [92–94], a tPA transport system with gelatin and zinc acetate, could be activated at the thrombus by ultrasound [95] and dextran-coated nanoparticles coated with tPA could specifically reach the thrombus via additional functionalization with an activated FXIIIa-sensitive protein [96].

We are developing a treatment strategy using magnetic nanoparticles which would allow a targeted localization at the site of occlusion, creating a very high concentration of the thrombolytic agent on site, while the rest of the organism is only slightly exposed. We have produced different tPA-coated nanoparticles to investigate whether it is possible to draw magnetic hybrid materials carrying thrombolytics into a thrombus and dissolve it from the inside using static magnetic fields. Prior to that, we have carefully characterized their physicochemical properties, biocompatibility, stability, enzyme-activity and the migration behaviour and matrix penetration efficiency within dense fibrin matrices to evaluate their suitability for thrombolytic applications [36, 97].

While adsorptive binding bears the risk of premature drug release and poor reproducibility, covalent binding could possibly influence drug structure and efficacy [98]. We have therefore investigated possible covalent and non-covalent approaches for tPA binding to SPIONs and directly compared and evaluated their advantages and disadvantages [36]. A clinically available tPA (Actilyse) were purified by tangential flow filtration and coupled to polyacrylic acid-co-maleic acid coated SPIONs (SPIONPAM) by amino-reactive activated ester reaction or by adsorption (SPIONPAM*tPA.cov and SPIONPAM*tPA.ads, respectively). Compared to the adsorptive approach, binding efficiencies after covalent binding indicated a superior tPA attachment to the particles. Biocompatibility measurements using HUVECs did not show any differences between tPA-free and tPA-containing SPIONs, indicating that HUVECs only respond to the total SPION-load and were unable to differentiate between tPA-loaded or unloaded SPIONs. The enzymatic tPA activity, measured by an chromogenic S-2288 activity assay, revealed a much higher activity of covalently bound tPA compared to adsorptive bound tPA (Figure 29.9(A) and (B)). Remarkably, long-time storage of $\text{SPION}^{\text{PAM*tPA.cov}}$ at 4 °C led only to a weak reduction in activity, whereas $\text{SPION}^{\text{PAM*tPA.ads}}$ lost just about all enzymatic activity. Similar results were shown using the fibrin–agarose assay, where the fibrinolytic activity is investigated within a matrix. After 24 h, both particle systems still had the ability to dissolve the fibrin matrix, while after 40 d of storage, $\text{SPION}^{\text{PAM*tPA.ads}}$ had nearly completely lost their fibrinolytic activity (Figure 29.9(C)–(E)). Thus, covalent linkage significantly improved the long-term stability and reactivity compared to simple tPA adsorption. In conclusion, we have shown that by using an activated ester reaction, covalent binding of tPA to SPIONs have significant advantages over tPA which is simply bound by electrostatic adsorption.

These findings were confirmed with another SPION carrier system [97]. Different amounts of tPA were immobilized on carboxylated dextran nanoparticles ($\text{SPION}^{\text{Dex-COOH}}$) via adsorption or covalent binding due to a carbodiimide-mediated amide bond formation. TEM images showed a very similar core diameter of 3.7 nm for $\text{SPION}^{\text{Dex-COOH}}$ and 3.8 nm for $\text{SPION}^{\text{Dex-COOH*tPA}}$ and the hydrodynamic diameter increased from ~166 to ~170 nm, independent of the binding mechanism. They all were highly biocompatible, and showed high stability and no sign of agglomeration in aqueous medium, even after storage of the samples at 4 °C for five weeks. Superconducting quantum interference devise (SQUID) measurements showed similar magnetization curves with a saturation magnetization of about 391 kA/m, superparamagnetic behaviour, no remanence and no hysteresis. Similar to the $\text{SPION}^{\text{PAM*tPA}}$ system, the chromogenic S-2288 activity assay demonstrated that covalently bound tPA on $\text{SPION}^{\text{Dex-COOH*tPA.cov}}$ were more active than

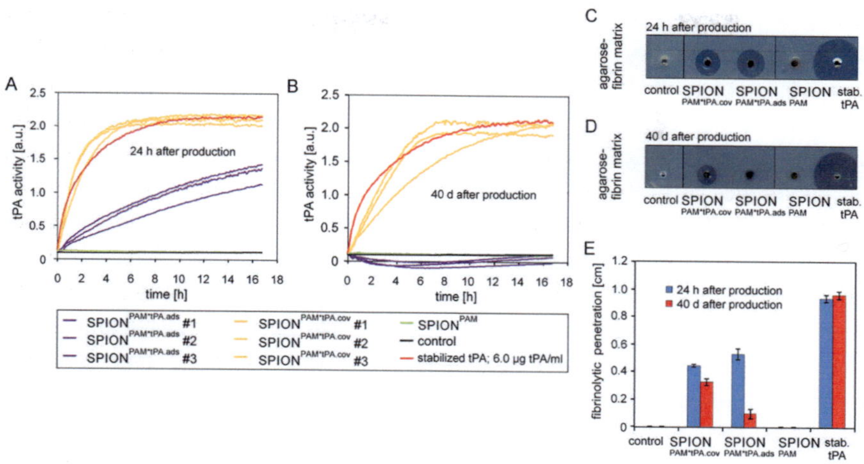

Figure 29.9: tPA activity of functionalized SPIONs with covalent ($SPION^{PAM*tPA.cov}$) and non-covalent ($SPION^{PAM*tPA.ads}$) tPA.
(A and B) tPA activity of functionalized SPIONs measured with the chromogenic S-2288 activity assay. (A) tPA activity measured after 24 h. (B) tPA activity measured after 40 days. (C, D) Activity measurement of tPA-functionalized SPIONs after 24 h and 40 days, measured with thrombus-mimicking fibrin-containing agarose plates. (E) tPA activity calculated by the distance between the wall of the sample holes and the edge of the fibrinolysis zones of functionalized SPIONs after 24 h and 40 days. Figure modified from Friedrich et al. [36].

adsorbed tPA on $SPION^{Dex\text{-}COOH*tPA.ads}$ particles. Furthermore, the efficiency of $SPION^{Dex\text{-}COOH*tPA.cov}$ to dissolve plasma clots was slightly higher than with $SPION^{Dex\text{-}COOH*tPA.ads}$, however, differences in fibrinolytic activity using agarose–fibrin matrices were not detectable. Notably, $SPION^{Dex\text{-}COOH*tPA.cov}$ could be attracted by an external magnet, navigate into thrombus-mimicking gels and effectively dissolve the fibrin matrix.

In summary, we have successfully produced biocompatible magnetic drug carriers functionalized with tPA for targeted thrombolysis under an external magnetic field. The enhanced drug targeted thrombolysis could be an enormous advantage in a clinical scenario in which the success of the treatment is highly dependent on a rapid effect. At the same time, the risks of a high-dose systemic administration of blood-thinning drugs might be reduced and the overall prognosis of the treatment greatly improved.

29.5 Conclusion and perspectives

Magnetic nanoparticles have a broad spectrum of properties that make them useful for many medical applications and could provide therapeutic possibilities to cure a wide range of different diseases or harmful conditions like cancer, atherosclerosis and thrombosis. However, the type of particle synthesis is responsible for the resulting chemical, physical and biological properties and not only determines the applicability in biomedical applications, but often also defines the type of application. The requirements for the particles are not always the same, but differ depending on whether they are to be used for imaging, diagnosis, magnetic drug targeting, hyperthermia or for magnetically based cell guidance. The actual synthesis of the raw particles is not the greatest challenge, as there are numerous physical, chemical and even biological methods available in the literature to achieve specific shapes, sizes and magnetic properties [1]. The greatest difficulty is to optimize and functionalize them for the respective applications. It is necessary to transfer the hydrophobic surface into a hydrophilic and above all into a biocompatible surface. The main goal here is to improve colloidal stability so that the particles do not agglomerate in the corresponding fluids, such as cell culture medium and especially blood. There are already numerous possibilities to achieve this stabilization and new strategies are constantly being added. In the past, fatty acids, peptides, gelatin and various polymers such as polyethylene glycol (PEG) and poly (N-isopropylacrylamide) (NIPAAM) were used for this purpose. The physicochemical characterization of these particles is usually straight forward. There are many established methods for this, including dynamic light scattering (DLS) and TEM for sizing, zeta potential for surface charge and VSM, SQUID, MPS, SANS, SAXS etc. for magnetic particle characterization. However, a reliable and comprehensive biological characterization to classify the biocompatibility of the particles is problematic. Until now, specific methods and clearly defined criteria and limits about the toxicology of nanoparticles are missing. Due to the strong self-absorption of the SPIONs, results obtained by absorption, luminescence or fluorescence measurements with plate photometric methods, e.g. MTT for the measurement of viability, have to be considered with utmost caution. Therefore, techniques based on the measurement of single cells, e.g. flow cytometry, are preferable. In addition, this method can be used to quickly determine a large number of relevant parameters, not only for viability and the number of apoptotic and necrotic cells, but also for further data on the cell size and, after appropriate standardization, the amount of internalized particles. However, even this method is not sufficient to obtain a complete toxicological classification to justify preclinical *in vivo* experiments. Therefore, further *in vitro* and *ex vivo* experiments, including 3D models and experiments in physiological fluids such as blood, must be performed to obtain a comprehensive classification of the biocompatibility.

In order to demonstrate the high demands and difficulties that can occur during synthesis, characterization and use of nanoparticles, we have exemplarily reviewed some of our recent approaches for the development of therapeutic and diagnostic particle systems. Beside a brief description of the syntheses of the different particle systems, we focused on the physicochemical and especially on a detailed biological characterization, such as cellular SPION uptake, cell proliferation, toxicity and biocompatibility and potential medical applications.

In conclusion, SPIONs can serve as a comprehensive nanotechnology platform for nanoparticle-based drugs and contrast agents. However there is a strong need for general, reliable, standardized and accepted protocols for the evaluation of nanoparticle safety toward human health to enable a fast translation of nanomedicine into clinics.

Author contributions: All the authors have accepted responsibility for the entire content of this submitted manuscript and approved submission.
Research funding: This work was supported by the Deutsche Forschungsgemeinschaft (SPP1681: Grant Number: AL 552/5).
Conflict of interest statement: The authors declare no conflicts of interest regarding this article.

References

1. Ali A, Zafar H, Zia M, Ul Haq I, Phull AR, Ali JS, et al. Synthesis, characterization, applications, and challenges of iron oxide nanoparticles. Nanotechnol Sci Appl 2016;9:49–67.
2. Gebauer JS, Malissek M, Simon S, Knauer SK, Maskos M, Stauber RH, et al. Impact of the nanoparticle-protein corona on colloidal stability and protein structure. Langmuir 2012;28: 9673–9.
3. Faunce TA, White J, Matthael KI. Integrated research into the nanoparticle-protein corona: a new focus for safe, sustainable and equitable development of nanomedicines. Nanomedicine 2008;3: 859–66.
4. Lesniak A, Fenaroli F, Monopoli MR, Aberg C, Dawson KA, Salvati A. Effects of the presence or absence of a protein corona on silica nanoparticle uptake and impact on cells. ACS Nano 2012;6: 5845–57.
5. Liu TY, Hu SH, Liu TY, Liu DM, Chen SY. Magnetic-sensitive behavior of intelligent ferrogels for controlled release of drug. Langmuir 2006;22:5974–8.
6. Deuflhard M, Eberbeck D, Hietschold P, Wilharm N, Muhlberger M, Friedrich RP, et al. Magnetically responsive composites: electron beam assisted magnetic nanoparticle arrest in gelatin hydrogels for bioactuation. Phys Chem Chem Phys 2019;21:14654–62.
7. Laurent S, Saei AA, Behzadi S, Panahifar A, Mahmoudi M. Superparamagnetic iron oxide nanoparticles for delivery of therapeutic agents: opportunities and challenges. Expet Opin Drug Deliv 2014;11:1449–70.
8. Tietze R, Zaloga J, Unterweger H, Lyer S, Friedrich RP, Janko C, et al. Magnetic nanoparticle-based drug delivery for cancer therapy. Biochem Biophys Res Commun 2015;468:463–70.
9. Tietze R, Lyer S, Durr S, Struffert T, Engelhorn T, Schwarz M, et al. Efficient drug-delivery using magnetic nanoparticles–biodistribution and therapeutic effects in tumour bearing rabbits. Nanomedicine 2013;9:961–71.
10. Dürr S, Janko C, Lyer S, Tripal P, Schwarz M, Zaloga J, et al. Magnetic nanoparticles for cancer therapy. Nanotechnol Rev 2013;2:395.
11. Hauser AK, Mathias R, Anderson KW, Hilt JZ. The effects of synthesis method on the physical and chemical properties of dextran coated iron oxide nanoparticles. Mater Chem Phys 2015;160: 177–86.
12. Durr S, Lyer S, Mann J, Janko C, Tietze R, Schreiber E, et al. Real-time cell analysis of human cancer cell lines after chemotherapy with functionalized magnetic nanoparticles. Anticancer Res 2012;32: 1983–9.

13. Lugert S, Unterweger H, Muhlberger M, Janko C, Draack S, Ludwig F, et al. Cellular effects of paclitaxel-loaded iron oxide nanoparticles on breast cancer using different 2d and 3d cell culture models. Int J Nanomed 2019;14:161–80.
14. Unterweger H, Dezsi L, Matuszak J, Janko C, Poettler M, Jordan J, et al. Dextran-coated superparamagnetic iron oxide nanoparticles for magnetic resonance imaging: evaluation of size-dependent imaging properties, storage stability and safety. Int J Nanomed 2018;13:1899–915.
15. Liu G, Gao J, Ai H, Chen X. Applications and potential toxicity of magnetic iron oxide nanoparticles. Small 2013;9:1533–45.
16. Khalafalla S, Reimers G. Preparation of dilution-stable aqueous magnetic fluids. IEEE Trans Magn 1980;16:178–83.
17. Zaloga J, Pöttler M, Leitinger G, Friedrich RP, Almer G, Lyer S, et al. Pharmaceutical formulation of hsa hybrid coated iron oxide nanoparticles for magnetic drug targeting. Eur J Pharm Biopharm 2016;101:152–62.
18. Hunt CP, Moskowitz BM, Banerjee SK. Magnetic properties of rocks and minerals. Rock physics & phase relations: a handbook of physical constants. Washington DC: American Geophysical Union; 1995, vol. 3:189–204 pp.
19. Poller JM, Zaloga J, Schreiber E, Unterweger H, Janko C, Radon P, et al. Selection of potential iron oxide nanoparticles for breast cancer treatment based on in vitro cytotoxicity and cellular uptake. Int J Nanomed 2017;12:3207–20.
20. Matuszak J, Zaloga J, Friedrich RP, Lyer S, Nowak J, Odenbach S, et al. Endothelial biocompatibility and accumulation of spion under flow conditions. J Magn Magn Mater 2014;380:20–6.
21. Zaloga J, Janko C, Nowak J, Matuszak J, Knaup S, Eberbeck D, et al. Development of a lauric acid/albumin hybrid iron oxide nanoparticle system with improved biocompatibility. Int J Nanomed 2014;9:4847–66.
22. Koeller J, Eble M. Mitoxantrone: a novel anthracycline derivative. Clin Pharm 1988;7:574–81.
23. LiverTox: clinical and research information on drug-induced liver injury. Mitoxantrone [Internet]. Bethesda, MD: National Institute of Diabetes and Digestive and Kidney Diseases; 2012. Available from: https://www.ncbi.nlm.nih.gov/books/NBK547931/ [Updated 19 Feb 2020].
24. Wolfram J, Zhu M, Yang Y, Shen J, Gentile E, Paolino D, et al. Safety of nanoparticles in medicine. Curr Drug Targets 2015;16:1671–81.
25. Unterweger H, Janko C, Schwarz M, Dezsi L, Urbanics R, Matuszak J, et al. Non-immunogenic dextran-coated superparamagnetic iron oxide nanoparticles: a biocompatible, size-tunable contrast agent for magnetic resonance imaging. Int J Nanomed 2017;12:5223–38.
26. Unterweger H, Tietze R, Janko C, Zaloga J, Lyer S, Durr S, et al. Development and characterization of magnetic iron oxide nanoparticles with a cisplatin-bearing polymer coating for targeted drug delivery. Int J Nanomed 2014;9:3659–76.
27. Mattheolabakis G, Milane L, Singh A, Amiji MM. Hyaluronic acid targeting of cd44 for cancer therapy: from receptor biology to nanomedicine. J Drug Target 2015;23:605–18.
28. Dasari S, Tchounwou PB. Cisplatin in cancer therapy: molecular mechanisms of action. Eur J Pharmacol 2014;740:364–78.
29. Zhang W, Tung CH. Cisplatin cross-linked multifunctional nanodrugplexes for combination therapy. ACS Appl Mater Interfaces 2017;9:8547–55.
30. Foroozandeh P, Aziz AA. Insight into cellular uptake and intracellular trafficking of nanoparticles. Nanoscale Res Lett 2018;13:339.
31. Zupke O, Distler E, Jurchott A, Paiphansiri U, Dass M, Thomas S, et al. Nanoparticles and antigen-specific t-cell therapeutics: a comprehensive study on uptake and release. Nanomedicine 2015;10: 1063–76.

32. Mühlberger M, Janko C, Unterweger H, Schreiber E, Band J, Lehmann C, et al. Functionalization of t lymphocytes for magnetically controlled immune therapy: selection of suitable superparamagnetic iron oxide nanoparticles. J Magn Magn Mater 2019;473:61–7.
33. Muhlberger M, Janko C, Unterweger H, Band J, Schreiber E, Lehmann C, et al. Non-magnetic chromatographic separation of colloidally metastable superparamagnetic iron oxide nanoparticles and suspension cells. J Chromatogr B Analyt Technol Biomed Life Sci 2019;1122–1123:83–9.
34. Mühlberger M, Janko C, Unterweger H, Friedrich R, Friedrich B, Band J, et al. Functionalization of t lymphocytes with citrate-coated superparamagnetic iron oxide nanoparticles for magnetically controlled immune therapy. Int J Nanomed 2019;14:8421–32.
35. Hunt CP, Singer MJ, Kletetschka G, TenPas J, Verosub KL. Effect of citrate-bicarbonate-dithionite treatment on fine-grained magnetite and maghemite. Earth Planet Sci Lett 1995;130:87–94.
36. Friedrich RP, Zaloga J, Schreiber E, Toth IY, Tombacz E, Lyer S, et al. Tissue plasminogen activator binding to superparamagnetic iron oxide nanoparticle-covalent versus adsorptive approach. Nanoscale Res Lett 2016;11:297.
37. Toth IY, Illes E, Bauer RA, Nesztor D, Szekeres M, Zupko I, et al. Designed polyelectrolyte shell on magnetite nanocore for dilution-resistant biocompatible magnetic fluids. Langmuir 2012;28: 16638–46.
38. Dalwadi G, Benson HA, Chen Y. Comparison of diafiltration and tangential flow filtration for purification of nanoparticle suspensions. Pharm Res (N Y) 2005;22:2152–62.
39. Zaloga J, Stapf M, Nowak J, Pottler M, Friedrich RP, Tietze R, et al. Tangential flow ultrafiltration allows purification and concentration of lauric acid-/albumin-coated particles for improved magnetic treatment. Int J Mol Sci 2015;16:19291–307.
40. Oberle A, Ludtke-Buzug K. Stability analysis of superparamagnetic iron oxide nanoparticles (spions) at 37 degrees c. Biomed Tech 2013;58(1 Suppl). https://doi.org/10.1515/bmt-2013-4099.
41. Zaloga J, Janko C, Agarwal R, Nowak J, Muller R, Boccaccini AR, et al. Different storage conditions influence biocompatibility and physicochemical properties of iron oxide nanoparticles. Int J Mol Sci 2015;16:9368–84.
42. Friedrich RP, Janko C, Poettler M, Tripal P, Zaloga J, Cicha I, et al. Flow cytometry for intracellular spion quantification: specificity and sensitivity in comparison with spectroscopic methods. Int J Nanomed 2015;10:4185–201.
43. Turcu R, Craciunescu I, Garamus VM, Janko C, Lyer S, Tietze R, et al. Magnetic microgels for drug targeting applications: physical–chemical properties and cytotoxicity evaluation. J Magn Magn Mater 2015;380:307–14.
44. Li W, Jan Z, Ding Y, Liu Y, Janko C, Pischetsrieder M, et al. Facile preparation of multifunctional superparamagnetic phbv microspheres containing spions for biomedical applications. Sci Rep 2016;6:23140.
45. Lucht N, Friedrich PR, Draack S, Alexiou C, Viereck T, Ludwig F, et al. Biophysical characterization of (silica-coated) cobalt ferrite nanoparticles for hyperthermia treatment. Nanomaterials 2019;9. https://doi.org/10.3390/nano9121713.
46. Matuszak J, Baumgartner J, Zaloga J, Juenet M, da Silva AE, Franke D, et al. Nanoparticles for intravascular applications: physicochemical characterization and cytotoxicity testing. Nanomedicine 2016;11:597–616.
47. Tietze R, Rahn H, Lyer S, Schreiber E, Mann J, Odenbach S, et al. Visualization of superparamagnetic nanoparticles in vascular tissue using xmuct and histology. Histochem Cell Biol 2011;135:153–8.
48. Hornung A, Poettler M, Friedrich RP, Zaloga J, Unterweger H, Lyer S, et al. Treatment efficiency of free and nanoparticle-loaded mitoxantrone for magnetic drug targeting in multicellular tumor spheroids. Molecules 2015;20:18016–30.

49. Thoma CR, Zimmermann M, Agarkova I, Kelm JM, Krek W. 3d cell culture systems modeling tumor growth determinants in cancer target discovery. Adv Drug Deliv Rev 2014;69-70:29–41.
50. Hornung A, Poettler M, Friedrich RP, Weigel B, Duerr S, Zaloga J, et al. Toxicity of mitoxantrone-loaded superparamagnetic iron oxide nanoparticles in a ht-29 tumour spheroid model. Anticancer Res 2016;36:3093–101.
51. Zaloga J, Feoktystov A, Garamus VM, Karawacka W, Ioffe A, Bruckel T, et al. Studies on the adsorption and desorption of mitoxantrone to lauric acid/albumin coated iron oxide nanoparticles. Colloids Surf B Biointerfaces 2018;161:18–26.
52. Stage TB, Bergmann TK, Kroetz DL. Clinical pharmacokinetics of paclitaxel monotherapy: an updated literature review. Clin Pharmacokinet 2018;57:7–19.
53. Kundranda MN, Niu J. Albumin-bound paclitaxel in solid tumors: clinical development and future directions. Drug Des Dev Ther 2015;9:3767–77.
54. Jiang G, Zhang S, Yazdanparast A, Li M, Pawar AV, Liu Y, et al. Comprehensive comparison of molecular portraits between cell lines and tumors in breast cancer. BMC Genom 2016;17(7 Suppl): 525.
55. Chekhun S, Bezdenezhnykh N, Shvets J, Lukianova N. Expression of biomarkers related to cell adhesion, metastasis and invasion of breast cancer cell lines of different molecular subtype. Exp Oncol 2013;35:174–9.
56. Brabers VAM. Progress in spinel ferrite research. Handbook of magnetic materials 8. Amsterdam – Lausanne – New York – Oxford – Tokyo: Elsevier; 1995, vol 8:199–344 pp.
57. Lyer S, Tietze R, Unterweger H, Zaloga J, Singh R, Matuszak J, et al. Nanomedical innovation: the seon-concept for an improved cancer therapy with magnetic nanoparticles. Nanomedicine 2015; 10:3287–304.
58. Janko C, Zaloga J, Pöttler M, Dürr S, Eberbeck D, Tietze R, et al. Strategies to optimize the biocompatibility of iron oxide nanoparticles – "spions safe by design". J Magn Magn Mater 2017; 431:281–4.
59. Fuchs TA, Abed U, Goosmann C, Hurwitz R, Schulze I, Wahn V, et al. Novel cell death program leads to neutrophil extracellular traps. J Cell Biol 2007;176:231–41.
60. Manfredi AA, Covino C, Rovere-Querini P, Maugeri N. Instructive influences of phagocytic clearance of dying cells on neutrophil extracellular trap generation. Clin Exp Immunol 2015;179:24–9.
61. Chen K, Nishi H, Travers R, Tsuboi N, Martinod K, Wagner DD, et al. Endocytosis of soluble immune complexes leads to their clearance by fcgammariiib but induces neutrophil extracellular traps via fcgammariia in vivo. Blood 2012;120:4421–31.
62. Maueroder C, Mahajan A, Paulus S, Gosswein S, Hahn J, Kienhofer D, et al. Menage-a-trois: the ratio of bicarbonate to co2 and the ph regulate the capacity of neutrophils to form nets. Front Immunol 2016;7:583.
63. Munoz LE, Bilyy R, Biermann MH, Kienhofer D, Maueroder C, Hahn J, et al. Nanoparticles size-dependently initiate self-limiting netosis-driven inflammation. Proc Natl Acad Sci U S A 2016;113: E5856–65.
64. Bilyy R, Unterweger H, Weigel B, Dumych T, Paryzhak S, Vovk V, et al. Inert coats of magnetic nanoparticles prevent formation of occlusive intravascular co-aggregates with neutrophil extracellular traps. Front Immunol 2018;9:2266.
65. Lee EA, Yim H, Heo J, Kim H, Jung G, Hwang NS. Application of magnetic nanoparticle for controlled tissue assembly and tissue engineering. Arch Pharm Res 2014;37:120–8.
66. Connell JJ, Patrick PS, Yu Y, Lythgoe MF, Kalber TL. Advanced cell therapies: targeting, tracking and actuation of cells with magnetic particles. Regen Med 2015;10:757–72.
67. Lanitis E, Dangaj D, Irving M, Coukos G. Mechanisms regulating t-cell infiltration and activity in solid tumors. Ann Oncol 2017;28:xii18–32.

68. Sanz-Ortega L, Rojas JM, Marcos A, Portilla Y, Stein JV, Barber DF. T cells loaded with magnetic nanoparticles are retained in peripheral lymph nodes by the application of a magnetic field. J Nanobiotechnol 2019;17. https://doi.org/10.1186/s12951-019-0440-z.
69. Newick K, O'Brien S, Moon E, Albelda SM. Car t cell therapy for solid tumors. Annu Rev Med 2017; 68:139–52.
70. Klinkert P, Post PN, Breslau PJ, van Bocke JH. Saphenous vein versus ptfe for above-knee femoropopliteal bypass. A review of the literature. Eur J Vasc Endovasc Surg 2004;27:357–62.
71. Hamon M, Lecluse E, Monassier JP, Grollier G, Potier JC. Pharmacological approaches to the prevention of restenosis after coronary angioplasty. Drugs Aging 1998;13:291–301.
72. Piterina AV, Cloonan AJ, Meaney CL, Davis LM, Callanan A, Walsh MT, et al. Ecm-based materials in cardiovascular applications: inherent healing potential and augmentation of native regenerative processes. Int J Mol Sci 2009;10:4375–417.
73. Khan OF, Sefton MV. Endothelialized biomaterials for tissue engineering applications in vivo. Trends Biotechnol 2011;29:379–87.
74. Hsu SH, Tsai I, Lin DJ, Chen DC. The effect of dynamic culture conditions on endothelial cell seeding and retention on small diameter polyurethane vascular grafts. Med Eng Phys 2005;27:267–72.
75. Melchiorri AJ, Bracaglia LG, Kimerer LK, Hibino N, Fisher JP. In vitro endothelialization of biodegradable vascular grafts via endothelial progenitor cell seeding and maturation in a tubular perfusion system bioreactor. Tissue Eng C Methods 2016;22:663–70.
76. Perea H, Aigner J, Hopfner U, Wintermantel E. Magnetically induced cell seeding of tubular scaffolds for vascular tissue engineering. Tissue Eng 2006;12:1085–6.
77. Wisotzki EI, Friedrich RP, Weidt A, Alexiou C, Mayr SG, Zink M. Cellular response to reagent-free electron-irradiated gelatin hydrogels. Macromol Biosci 2016;16:914–24.
78. Tadsen M, Friedrich RP, Riedel S, Alexiou C, Mayr SG. Contact guidance by microstructured gelatin hydrogels for prospective tissue engineering applications. ACS Appl Mater Interfaces 2019;11: 7450–8.
79. Jackson SP. Arterial thrombosis–insidious, unpredictable and deadly. Nat Med 2011;17:1423–36.
80. Dimarsico L, Cymet T. Pulmonary embolism–a state of the clot review. Compr Ther 2007;33: 184–91.
81. Tissue plasminogen activator for acute ischemic stroke. The national institute of neurological disorders and stroke rt-pa stroke study group. N Engl J Med 1995;333:1581–7.
82. Zhang J, Yang Y, Sun H, Xing Y. Hemorrhagic transformation after cerebral infarction: current concepts and challenges. Ann Transl Med 2014;2:81.
83. Depta JP, Bhatt DL. New approaches to inhibiting platelets and coagulation. Annu Rev Pharmacol Toxicol 2015;55:373–97.
84. Ageno W, Spyropoulos AC, Turpie AG. Role of new anticoagulants for the prevention of venous thromboembolism after major orthopaedic surgery and in hospitalised acutely ill medical patients. Thromb Haemostasis 2012;107:1027–34.
85. Chapman SN, Mehndiratta P, Johansen MC, McMurry TL, Johnston KC, Southerland AM. Current perspectives on the use of intravenous recombinant tissue plasminogen activator (tpa) for treatment of acute ischemic stroke. Vasc Health Risk Manag 2014;10:75–87.
86. Chen Y, Cui G, Zhao M, Wang C, Qian K, Morris-Natschke S, et al. Synthesis, nano-scale assembly, and in vivo anti-thrombotic activity of novel short peptides containing l-arg and l-asp or l-glu. Bioorg Med Chem 2008;16:5914–25.
87. Jin S, Wang Y, Zhu H, Wang Y, Zhao S, Zhao M, et al. Nanosized aspirin-arg-gly-asp-val: delivery of aspirin to thrombus by the target carrier arg-gly-asp-val tetrapeptide. ACS Nano 2013;7:7664–73.
88. Tang AC, Chang MY, Tang ZC, Li HJ, Hwang GL, Hsieh PC. Treatment of acute thromboembolism in mice using heparin-conjugated carbon nanocapsules. ACS Nano 2012;6:6099–107.

89. Ji J, Ji SY, Yang JA, He X, Yang XH, Ling WP, et al. Ultrasound-targeted transfection of tissue-type plasminogen activator gene carried by albumin nanoparticles to dog myocardium to prevent thrombosis after heart mechanical valve replacement. Int J Nanomed 2012;7:2911–9.
90. Tadayon A, Jamshidi R, Esmaeili A. Delivery of tissue plasminogen activator and streptokinase magnetic nanoparticles to target vascular diseases. Int J Pharm 2015;495:428–38.
91. Peters D, Kastantin M, Kotamraju VR, Karmali PP, Gujraty K, Tirrell M, et al. Targeting atherosclerosis by using modular, multifunctional micelles. Proc Natl Acad Sci U S A 2009;106: 9815–9.
92. Ma YH, Wu SY, Wu T, Chang YJ, Hua MY, Chen JP. Magnetically targeted thrombolysis with recombinant tissue plasminogen activator bound to polyacrylic acid-coated nanoparticles. Biomaterials 2009;30:3343–51.
93. Bi F, Zhang J, Su Y, Tang YC, Liu JN. Chemical conjugation of urokinase to magnetic nanoparticles for targeted thrombolysis. Biomaterials 2009;30:5125–30.
94. Zhou J, Guo D, Zhang Y, Wu W, Ran H, Wang Z. Construction and evaluation of fe(3)o(4)-based plga nanoparticles carrying rtpa used in the detection of thrombosis and in targeted thrombolysis. ACS Appl Mater Interfaces 2014;6:5566–76.
95. Kawata H, Uesugi Y, Soeda T, Takemoto Y, Sung JH, Umaki K, et al. A new drug delivery system for intravenous coronary thrombolysis with thrombus targeting and stealth activity recoverable by ultrasound. J Am Coll Cardiol 2012;60:2550–7.
96. McCarthy JR, Sazonova IY, Erdem SS, Hara T, Thompson BD, Patel P, et al. Multifunctional nanoagent for thrombus-targeted fibrinolytic therapy. Nanomedicine 2012;7:1017–28.
97. Heid S, Unterweger H, Tietze R, Friedrich RP, Weigel B, Cicha I, et al. Synthesis and characterization of tissue plasminogen activator-functionalized superparamagnetic iron oxide nanoparticles for targeted fibrin clot dissolution. Int J Mol Sci 2017;18. https://doi.org/10.3390/ijms18091837.
98. Veiseh O, Gunn JW, Zhang M. Design and fabrication of magnetic nanoparticles for targeted drug delivery and imaging. Adv Drug Deliv Rev 2010;62:284–304.

Index

https://doi.org/10.1515/9783110569636-030

Bei Fragen zur Produktsicherheit wenden Sie sich bitte an:
If you have any questions regarding product safety,
please contact:

Walter de Gruyter GmbH
Genthiner Straße 13
10785 Berlin
productsafety@degruyterbrill.com